THE CAMBRIDGE HANDBOOK OF EVOLUTIONARY PERSPECTIV[
ON HUMAN BEHAVIOR

The transformative wave of Darwinian insight continues to expand throughou
human sciences. While still centered on evolution-focused fields such as evolutio:
psychology, ethology, and human behavioral ecology, this insight has also influer
cognitive science, neuroscience, feminist discourse, sociocultural anthropol
media studies, and clinical psychology. This handbook's goal is to amplify the wave
by bringing together world-leading experts to provide a comprehensive and up-to-
date overview of evolution-oriented and influenced fields. While evolutionary psychology remains at the core of the collection, it also covers the history, current standing,
debates, and future directions of the panoply of fields entering the Darwinian fold. As
such, *The Cambridge Handbook of Evolutionary Perspectives on Human Behavior* is
a valuable reference not just for evolutionary psychologists, but also for scholars and
students from many fields who wish to see how the evolutionary perspective is
relevant to their own work.

LANCE WORKMAN is a visiting professor in psychology at the University of South
Wales, UK.

WILL READER is a senior lecturer in psychology at Sheffield Hallam University, UK.

JEROME H. BARKOW is a multidisciplinary researcher. He is Professor Emeritus of
Social Anthropology at Dalhousie University, Canada, where he has spent most of his
career.

THE CAMBRIDGE HANDBOOK OF EVOLUTIONARY PERSPECTIVES ON HUMAN BEHAVIOR

Edited by

Lance Workman
University of South Wales

Will Reader
Sheffield Hallam University

Jerome H. Barkow
Dalhousie University

CAMBRIDGE
UNIVERSITY PRESS

CAMBRIDGE
UNIVERSITY PRESS

University Printing House, Cambridge CB2 8BS, United Kingdom

One Liberty Plaza, 20th Floor, New York, NY 10006, USA

477 Williamstown Road, Port Melbourne, VIC 3207, Australia

314–321, 3rd Floor, Plot 3, Splendor Forum, Jasola District Centre,
New Delhi – 110025, India

79 Anson Road, #06–04/06, Singapore 079906

Cambridge University Press is part of the University of Cambridge.

It furthers the University's mission by disseminating knowledge in the pursuit of
education, learning, and research at the highest international levels of excellence.

www.cambridge.org
Information on this title: www.cambridge.org/9781316642825
DOI: 10.1017/9781108131797

© Cambridge University Press 2020

First published 2020

Printed in the United Kingdom by TJ International Ltd, Padstow Cornwall

A catalogue record for this publication is available from the British Library.

Library of Congress Cataloging-in-Publication Data
Names: Workman, Lance, editor. | Reader, Will, editor. | Barkow, Jerome H, editor.
Title: The Cambridge handbook of evolutionary perspectives on human behavior / edited
by Lance Workman, University of South Wales, Will Reader, Sheffield Hallam University,
Jerome H. Barkow, Dalhousie University, Halifax, Canada.
Description: Cambridge, United Kingdom ; New York, NY : Cambridge University Press,
[2020] | Series: Cambridge handbooks in psychology | Includes bibliographical references
and index.
Identifiers: LCCN 2019028602 (print) | LCCN 2019028603 (ebook) | ISBN 9781316642825
(hardback) | ISBN 9781108131797 (epub)
Subjects: LCSH: Evolutionary psychology. | Human behavior.
Classification: LCC BF698.95 .C36 2020 (print) | LCC BF698.95 (ebook) | DDC 155.7–dc23
LC record available at https://lccn.loc.gov/2019028602
LC ebook record available at https://lccn.loc.gov/2019028603

ISBN 978-1-316-64282-5 Hardback
ISBN 978-1-316-64281-8 Paperback

We would like to dedicate this handbook to Ethel and Bill Reader, Philip and Betty Barkow, and George and Margret Workman. And finally we would also like to pay special tribute to Anne Campbell – a true pioneer of evolutionary approaches.

Contents

PART X   EVOLUTION AND THE MEDIA

Figures

Tables

Contributors

Dwaipayan Adhya, Department of Psychiatry, University of Cambridge, UK

Jerome H. Barkow, Department of Sociology and Social Anthropology, Dalhousie University, Canada

Simon Baron-Cohen, Department of Psychology, University of Cambridge, UK

Deirdre Barrett, Harvard Medical School, Harvard University, USA

Louise Barrett, Department of Psychology, University of Lethbridge, Canada

Lucy Bates, School of Psychology, University of Sussex, UK

Alec T. Beall, Department of Psychology, University of British Columbia, Canada

Kent C. Berridge, Department of Psychology, University of Michigan, USA

Laura Betzig, anthropologist and historian, http://laurabetzig.org

Geoffrey Bird, Institute of Psychiatry, Psychology and Neuroscience, King's College London, UK

David F. Bjorklund, Department of Psychology, Florida Atlantic University, USA

Rebecca Brewer, Department of Psychology, Royal Holloway, University of London, UK

William M. Brown, School of Psychology, University of Bedfordshire, UK

Rebecca L. Burch, Human Development Department, OSWEGO State University of New York, USA

David M. Buss, Department of Psychology, University of Texas, USA

Abraham P. Buunk, Department of Psychology, University of Groningen, The Netherlands

Anne Campbell, Psychology Department, Durham University, UK

Jason Carroll, Cancer Research UK – Cambridge Research Institute, UK

Bernard Chapais, Department of Anthropology, University of Montreal, Canada

Mathias Clasen, Department of English, Aarhus University, Denmark

Kathryn Coe, Richard M. Fairbanks School of Public Health, Indiana University–Purdue University Indianapolis, USA

Luke Colquhoun, School of Psychology and Therapeutic Studies, University of South Wales, UK

Daniel Conroy-Beam, Department of Psychological and Brain Sciences, University of California, Santa Barbara, USA

Michael C. Corballis, Department of Psychology, University of Auckland, New Zealand

Martin Daly, Department of Psychology, Neuroscience and Behaviour, McMaster University, Canada

Corey L. Fincher, Department of Psychology, University of Warwick, UK

Maryanne L. Fisher, Department of Psychology, Saint Mary's University, Canada

Jo Fowler, School of Psychology and Therapeutic Studies, University of South Wales, UK

Adrian Furnham, Faculty of Brain Sciences, University College London, UK

Justin R. Garcia, Kinsey Institute and Department of Gender Studies, Indiana University, USA

Ellen C. Garland, School of Biology, University of St Andrews, UK

Paul Gilbert, Mental Health Research Unit, Derbyshire Healthcare NHS Foundation Trust, UK

Francisco R. Gómez, Department of Psychology, University of Lethbridge, Canada

Sebastian Grüneisen, Department of Developmental and Comparative Psychology, Max Planck Institute for Evolutionary Anthropology, Germany

Gregory Hanlon, Department of History, Dalhousie University, Canada

Sara Hughes, Department of Psychology, Sociology and Politics, Sheffield Hallam University, UK

Satoshi Kanazawa, Department of Management, London School of Economics and Political Science, UK

Jens Kjeldgaard-Christiansen, Department of English, Aarhus University, Denmark

Veronika Konok, Department of Ethology, Eötvös University, Hungary

Mark Kotter, Department of Clinical Neurosciences, University of Cambridge, UK

Dennis L. Krebs, Department of Psychology, Simon Fraser University, Canada

Morten L. Kringelbach, Department of Clinical Medicine, Aarhus University, Denmark

Peter J. LaFreniere, Department of Psychology, University of Maine, USA

Aicha Massrali, Department of Psychiatry, University of Cambridge, UK

Alex Mesoudi, Department of Biosciences, University of Exeter, UK

Ádám Miklósi, Department of Ethology, Eötvös University, Hungary

Alyson J. Myers, Department of Psychology, Florida Atlantic University, USA

Craig T. Palmer, Department of Anthropology, University of Missouri, USA

Arkoprovo Paul, Department of Psychiatry, University of Cambridge, UK

Irene M. Pepperberg, Department of Psychology, Harvard University, USA

Gretchen Perry, School of Social Work, University of Canterbury, New Zealand

Lanna J. Petterson, Department of Psychology, University of Lethbridge, Canada

Jack Price, Institute of Psychiatry, Psychology and Neuroscience, King's College London, UK

William J. Ray, Department of Psychology, Pennsylvania State University, USA

Will Reader, Department of Psychology, Sociology and Politics, Sheffield Hallam University, UK

Luke Rendell, School of Biology, University of St Andrews, UK

Catherine Salmon, Department of Psychology, University of Redlands, USA

Clemens Schwender, Filmuniversität Babelsberg Konrad Wolf, Germany

Scott W. Semenyna, Department of Psychology, University of Lethbridge, Canada

Peter K. Smith, Department of Psychology, Goldsmiths, University of London, UK

Rachelle M. Smith, College of Science and Humanities, Husson University, USA

Deepak Srivastava, Institute of Psychiatry, Psychology and Neuroscience, King's College London, UK

Eloise Stark, Department of Psychiatry, University of Oxford, UK

Steve Stewart-Williams, School of Psychology, University of Nottingham, Malaysia

Frank J. Sulloway, Department of Psychology, University of California, Berkeley, USA

Sandie Taylor, School of Psychology and Therapeutic Studies, University of South Wales, UK

Randy Thornhill, Department of Biology, University of New Mexico, USA

Frederick M. Toates, Department of Biological Sciences, The Open University, UK

Jessica L. Tracy, Department of Psychology, University of British Columbia, Canada

Alfonso Troisi, Department of of Systems Medicine, University of Rome Tor Vergata, Italy

Doug P. VanderLaan, Department of Psychology, University of Toronto, Canada

Paul L. Vasey, Department of Psychology, University of Lethbridge, Canada

Andreas Wilke, Psychology Department, Clarkson University, USA

Lance Workman, School of Psychology and Therapeutic Studies, University of South Wales, UK

Emily Wyman, School of Economics, University of Nottingham, UK

Preface

With the publication of the *Origin of Species*, Darwin realized that his theory of evolution by natural selection would have clear repercussions for the development of academic areas outside of biology. In particular, he predicted that our understanding of human behavior would be transformed as psychology dealt with the repercussions of evolutionary theory:

[I]n the distant future I see open fields for more important researches. Psychology will be based on a new foundation, that of the necessary acquirement of each mental power and capacity by gradation.
(Darwin, 1859, p. 458)

By "the necessary acquirement of each mental power and capacity by gradation," Darwin was suggesting that, in addition to physical features, human behavior and internal states can be seen as adaptations that were shaped by natural and sexual selection. In these two sentences Darwin introduces the concept of "evolutionary psychology"; that is, human psychological abilities arose to aid survival and reproduction in the deep ancestral past and, if we wish to understand human nature, then we need to consider how and why it evolved. Note, however, that he also used the phrase "in the distant future." Darwin was certainly prophetic in suggesting we would have to wait some time for this development. While there were a number of relatively unsuccessful attempts to integrate evolutionary thinking into psychology, the scientific development of evolutionary psychology only began in earnest during the 1990s (Barkow, Cosmides, & Tooby, 1992; Buss, 1995; Pinker, 1994). Such developments in the 1990s led, in turn, to a transformative wave of Darwinian insight during the early years of the twenty-first century. Today, this wave continues to expand throughout the human sciences. While still centered on evolution-focused fields such as evolutionary psychology, human behavioral ecology, ethology, and sociobiology, its influence is now being felt in cognitive, social, and developmental psychology, neuroscience, feminist theory, and sociocultural anthropology, as well as in psychiatry and media studies. The list continues to grow. The aim of this handbook is to aid the wave by providing comprehensive coverage of evolution-oriented and influenced fields through the writings of a mixture of world-leading experts and up-and-coming scholars.

Although the contributors to this handbook all share an interest in the relationship between evolution and the human condition, they come from a wide variety of backgrounds and certainly do not have a unitary view on all of the issues explored here. Some contributors are happy to call themselves evolutionary psychologists or human behavioral ecologists; others would shy away from, or even reject, such labels. But while they may not all speak with one voice, their voices all come from a Darwinian source, and it is fair to say that all have been swept up by this transformative wave. Maybe there will come a time, perhaps in the not too distant future, when to preface one's research area with the term "evolutionary" will be unnecessary: the new foundation will become the orthodoxy. We hope this handbook will provide one small step on the way to that state of affairs.

While the core of the handbook is the field of evolutionary psychology, our intention is to provide an up-to-date, comprehensive source of advanced literature covering evolutionary thought and the history, current standing, debates, and future directions of the panoply of fields within or entering the Darwinian fold. It is anticipated that *The Cambridge Handbook of Evolutionary Perspectives on Human Behavior* will become a reference tool not just for evolutionary psychologists and their graduate students, but also for scholars from many fields who wish to see how the evolutionary perspective is relevant to their own work.

The handbook is divided into 10 sections covering: the comparative approach; sociocultural anthropology; neuroscience; social psychology; cognitive psychology; developmental issues; sexual selection; psychopathology; applied issues; and, finally, the relationship between evolutionary theory and the media.

We have provided a brief introduction to each section. Each of these is designed to allow the reader to gain a broad grasp of the content of a given section and, for those unfamiliar with the subject matter, a brief glimpse into the work of the contributors. We have purposely kept these introductory sections brief as we did not want readers to wade through an essay before reading a series of essays.

Enjoy.

Lance Workman, Will Reader, and Jerome H. Barkow

References

Barkow, J. H., Cosmides, L., & Tooby, J., eds. (1992). *The Adapted Mind: Evolutionary Psychology and the Generation of Culture*. Oxford/New York: Oxford University Press.

Buss, D. M. (1995). *The Evolution of Desire*. New York: Basic Books.

Darwin, C. (1859). *On The Origin of Species by Means of Natural Selection, or the Preservation of Favoured Races in the Struggle for Life*. London: John Murray.

Pinker, S. (1994). *The Language Instinct: How the Mind Creates Language*. London: Penguin.

Acknowledgments

We would like to acknowledge the endless support we have received during the preparation of this handbook from Janka Romero, Emily Watton, Santosh Laxmi Kota, Manu Menon, Sudesh Kumar, John Marr and Stanly Emelson.

We would also like to acknowledge a number of past evolutionists who are no longer with us but who have been personally influential in the development of our thinking over many years. These include: Richard D. Alexander, Richard Andrew, Patrick Bateson, John Bowlby, Anne Campbell, Donald T. Campbell, Michael R. A. Chance, Daniel G. Freedman, William D. Hamilton, Judith Rich Harris, Robert Hinde, Aubrey Manning, John Maynard Smith, Elaine Morgan, Jaak Panksepp, George C. Williams, and Margo Wilson.

PART I

THE COMPARATIVE APPROACH

Prior to Darwin, humans lived in a different world from other species. While our machines were inhabited by ghosts, other creatures were simply machines devoid of internal states (Descartes, 1641). With the publication of the *Origin of Species* in 1859, however, people began to question this anthropocentric assumption of a discontinuity between "us" and "them." Thirteen years later in Darwin's final book, *The Expression of the Emotions in Man and Animals* (1872), he developed this argument of continuity between human and nonhuman species further by drawing on observations of parallels of expression and reaction in a wide range of species. It is fair to say that *The Expression of the Emotions* led directly to the development of "comparative psychology" and provided legitimacy to the study of animal behavior as a means to better understand ourselves (Workman, 2013). In 1894, Conway Lloyd Morgan formalized this approach in his book, *Introduction to Comparative Psychology*, setting out the ground rules for the comparative method.

During the twentieth century, comparative psychology subdivided into two main approaches. One approach focused on the internal states of animals and eventually developed into the new field of animal cognition, while the other attempted to exorcise mentalistic language from the field of animal behavior and eventually became known as *behaviorism*. Today, studies of animal cognition still draw on Darwin's conception of continuity between species, but, by integrating developments in ethology and neuroscience, they also relate specific cognitive abilities to the behavioral ecology of a population.

It is fair to say that Irene M. Pepperberg's research into avian cognition is a major contribution to the field of animal cognition. Pepperberg's work with African Grey parrots showed how a species that has not shared a common ancestor with our own lineage since the late Carboniferous period can nonetheless exhibit human-like vocal communication. We begin Part I with her chapter on human–Grey parrot comparisons in cognitive performance.

The old adage that "elephants never forget" is based on a large body of anecdotal evidence. In recent years, however, field and lab studies have begun to put some flesh on these anecdotal bones. Lucy Bates has spent a number of years observing and testing this social giant. Her chapter on the cognitive abilities in elephants reinforces and dispels the myths that have built up around the intellectual prowess of these largest of all land animals.

Another group of animals that, due to their apparent complex social behavior, has long fascinated us is the cetaceans. Like elephants, whales and dolphins are renowned for their apparent cognitive prowess. In the third and final chapter in Part I, Ellen C. Garland and Luke Rendell consider culture and communication among cetaceans. Is it possible that we can improve our understanding of the roots of human language and culture by studying creatures that evolved in the three-dimensional liquid world of the oceans? Garland and Rendell provide us with the current state of play with regard to these questions and suggest fertile areas for future research.

REFERENCES

Darwin, C. (1859). *On The Origin of Species by Means of Natural Selection, or the Preservation of Favoured Races in the Struggle for Life*. London: John Murray.

Darwin, C. (1872). *The Expression of the Emotions in Man and Animals*. London: HarperCollins.

Descartes, R. (1641). *Meditations on First Philosophy*, trans. by J. Cottingham (1996). Cambridge, UK: Cambridge University Press.

Morgan, C. L. (1894). *Introduction to Comparative Psychology*. London: Walter Scott.

Workman, L. (2013). *Charles Darwin: Mindshaper*. Basingstoke: Palgrave Macmillan.

1

Human–Grey Parrot Comparisons in Cognitive Performance

IRENE M. PEPPERBERG

1.1 INTRODUCTION

Animal cognition is, by itself, an incredibly broad field, encompassing a huge variety of taxa and involving many different topics in both the laboratory and nature. When asked to view the field through the lens of an evolutionary perspective relative to humans, most scientists focus on our nearest relatives, the great apes (e.g., note the preponderance of chapters devoted to nonhuman primates in Vonk & Shackelford, 2012). Convergent evolution, however, provides striking insights into how distantly related species have responded to similar social and ecological challenges, and comparisons between avian and primate species have demonstrated remarkable parallels in various capacities (Emery & Clayton, 2004; Pepperberg, 1999, 2013). In areas such as vocal learning, avian species, separated from humans by over 300 million years of evolution (e.g., Hedges et al., 1996), actually provide a better model for study than do apes (e.g., Bolhuis & Everaert, 2013; Chakraborty et al., 2015). Such vocal ability can also lead to the acquisition of a limited form of referential symbolic communication with humans, which in turn can further facilitate comparative studies of avian–human cognitive capacities (e.g., Pepperberg, 1999). Interestingly, researchers have recently demonstrated that avian neural systems are functionally comparable to those of the great apes (e.g., Güntürkün & Bugnyar, 2016; Olkowicz et al., 2016), thereby providing a clear basis for advanced avian cognitive capacities. Grey parrots (*Psittacus erithacus*), which are often at the center of studies on avian intelligence and communication (Pepperberg, 1999), will be the focus of this chapter.

First, however, a bit of background: I have studied the cognitive and communicative abilities of Grey parrots for almost 40 years. Via a modeling procedure – the model/rival or M/R technique, adapted from experiments by both Bandura (1971) and Todt (1975) – I have trained these birds to acquire some level of referential communication using the sounds of English speech, then used this communication code to examine their cognitive abilities. My oldest subject, Alex, learned to use vocal labels to identify a wide variety of objects, colors, and shapes, processed queries to judge category, relative size, quantity, and the presence or absence of similarity/difference in attributes, and showed label comprehension and a zero-like concept; he demonstrated some understanding of phonological awareness and numerical competence more comparable to that of young children than to other nonhumans (Pepperberg, 2012b). His requests (for specific objects or to be moved to a particular location) were intentional (Pepperberg, 1988). My younger birds (particularly Griffin) are acquiring similar vocal abilities, particularly with respect to labeling of objects, colors, and shapes (e.g., Pepperberg & Nakayama, 2016; Pepperberg & Shive, 2001; Pepperberg & Wilcox, 2000).

This chapter reviews their recent advances. For Alex, the focus will be on his final numerical studies: inferring the cardinality of new number labels from their order on the number line (Pepperberg & Carey, 2012) and the addition of small quantities (Pepperberg, 2012a). His data provide evidence for actual counting (see Section 1.2). For Griffin, the focus will be on tests involving delayed gratification (the ability to forgo an immediate reward to gain one of better quality; Koepke, Gray, & Pepperberg, 2015), the recognition of optical illusions involving amodal and modal completion (respectively, recognizing occluded shapes and Kanizsa figures; Pepperberg & Nakayama, 2016), and (along with Grey parrots belonging to a colleague) tasks requiring reasoning by exclusion (inferring where an item is hidden after being given information on where it is not; Pepperberg et al., 2013, 2019). In most of these studies, Grey parrots demonstrate capacities comparable to those of children aged three to five years.

1.2 NUMBER CONCEPTS: COUNTING

Some understanding of number is a widespread phenomenon (in nonhumans, from fish [Petrazzini et al., 2015] to bears [e.g., Vonk & Beran, 2012]; in humans, even in preverbal children [Wynn, 1990] and preliterate hunter–gatherer societies [e.g., Frank et al., 2008]). However, symbolic representation of number – the recognition that individual symbols represent exact, specific quantities – involves advanced capacities, once thought to be limited only to humans (reviewed in Pepperberg & Carey, 2012). Only a few nonhumans have demonstrated exact symbolic number representation: two apes, Matsuzawa's Ai (Matsuzawa, 1985) and Boysen's Sheba (Boysen & Berntson, 1989), and my subject, a Grey parrot, Alex (Pepperberg, 1987, 1994). Symbolic representation is important because it is a prerequisite for true counting, as defined by Gallistel and Gelman (1992) via several counting principles (CPs). CPs state that numerals must be applied in order to items in a set to be enumerated and in a one-to-one correspondence, that the last numeral in a count represents a set's cardinal value, and that the successor function (that each numeral is known to be exactly one more than the one before it and exactly one less than the one after it; e.g., Carey, 2009) must be understood. Determining whether nonhumans could acquire CPs has attracted widespread scientific interest.

Acquisition of CPs is not easy, and even children take several years to accomplish the task (Carey, 2009; Fuson, 1988). At about two years of age, most children can rattle off a series of number words (a "count list"), but they often confuse the order of the number labels and assign exact cardinal meaning only to "one"; other numerals mean "some" or "plural" (e.g., Barner & Bachrach, 2010). About nine months later, they learn "two"; other numerals are "more than two." A few months later, they master "three," then "four" (Wynn, 1992); they acquire a stable, accurate count list. Only then (generally somewhere between 3.5 and 4 years of age) do children *induce* the CPs and understand that each successive numeral in their count list is exactly one more than its predecessor. This induction separates them from those understanding only the exact meaning of a few small numbers ("one-," "two-," "three-," or "four-knowers"; Sarnecka & Carey, 2008); they can now encode cardinal value expressed by *any* numeral in their count list. Carey and I (Pepperberg & Carey, 2012) tested whether a Grey parrot could perform this induction; at the time, no nonhuman had been able to acquire this knowledge purely by induction.

We worked with Alex, who had previously been taught to use English count words ("one" through "sih" [six]) to label sets of one to six individual items exactly (production and comprehension; Pepperberg, 1987, 1994; Pepperberg & Gordon, 2005). He had also been taught to use the same count words to label Arabic numerals 1 through 6, but had had no training to associate any Arabic numeral with any specific quantity of items. Without training, he had subsequently inferred the relationship between the Arabic numerals and the sets of objects (Pepperberg, 2006b); that is, he had deduced the ordinality of the Arabic numerals by recognizing that an Arabic symbol had the same numerical value as its vocal label, comparing representations of quantity for which the labels stood and inferring their rank ordering based on these representations. So, for

example, given a green "5" and a yellow "2," he could answer both queries: "What color [is the] number [that is] bigger/smaller?" Notably, he had never been trained to recite number labels in order, nor had he even learned the number labels in order (see later in this section). Thus, even at this point, he appeared to exhibit numerical understanding far closer to that of children than other animals. However, he differed from humans and was like other nonhumans in that he had demonstrated no savings in his learning of larger numerals; that is, unlike four-year-old children, he did not acquire "five" and "six" via the successor function, without training. Why was Alex unlike children in this instance? Might the issue be Alex's difficulty not in learning the meaning of the numerical symbols, but rather in producing the English sounds? To generate any given English label, he had to learn to coordinate his syrinx, tracheal muscles, glottis, larynx, tongue height and protrusion, beak opening, and even esophagus (Patterson & Pepperberg, 1998). Might there be a way to dissociate vocal and conceptual learning to test this possibility?

Carey and I devised the following experiment (Pepperberg & Carey, 2012): I began by teaching Alex to identify vocally the Arabic numerals 7 and 8 in the absence of their respective quantities, divorcing the time needed to learn the speech patterns from any concept of number. After the labels were produced clearly, I trained him to understand that $6 < 7 < 8$; that is, where the new numerals fit on the number line. Without further training, he demonstrated that he understood the relationships among 7 and 8 and his other Arabic labels. Could he now, like children, spontaneously understand that "seven" represented one more physical object than "six," and that "eight" represented two more than "six" and one more than "seven," by labeling appropriate physical sets on first trials? Nothing in his training up to this point would provide specific information about the values of 7 and 8; they could refer to 10 and 20 items, respectively. Interestingly, all of his other numerals had been taught as either +1 or –1 than those he already knew (i.e., his first number labels were "three" and "four"; he was next taught "five" and "two," then "six" and "one"; Pepperberg, 1987, 1994); could he thus use past and present information to *induce* the cardinal meaning of the labels "seven" and "eight" – which he had learned to use for the Arabic numerals 7 and 8 – from their ordinal positions on an implicit count list? The answer was positive: Alex, like children, had created a representational structure that allowed him to encode the cardinal value expressed by any numeral in his count list (Carey, 2009); that is, to understand the successor function (NB: details of experimental design, including all controls for possible cuing, are in Pepperberg & Carey, 2012). Notably, the nonhuman primates tested so far (e.g., Boysen et al., 1993; Inoue & Matsuzawa, 2009) had to be trained to understand the ordinality of their numbers and have not yet inferred cardinality from ordinality.

This study provided the first demonstration that a nonhuman could engage in the bootstrapping process that underlies the construction of the integer list representation of number. Very much like young children, Alex could accomplish this task because he had true symbolic representation of his numerals, an ability fostered by his training in referential communication.

1.3 NUMBER CONCEPTS: ADDITION

Alex had also demonstrated a limited ability with respect to addition. By replicating part of a study on apes (Boysen & Berntson, 1989), I showed that he could watch his trainers hide two small sets of objects (quantities from zero to six) and, while they were still hidden, provide the *exact* vocal label for the total (i.e., requiring summation and symbolic representation of a hidden quantity; Pepperberg, 2006a). In contrast to most other addition studies with nonhumans, the task avoided use of only one token type of a standard size (e.g., whole marshmallows), which could allow evaluations based on contour and mass, not number (see discussion in Mix, Huttenlocher, & Levine, 2002). Overall, his data were comparable to those of young children (Mix et al., 2002) and, because he added to six, were more advanced than those of apes (Boysen & Hallberg, 2000).

Once Alex had acquired the numerals through 8, I returned to the addition task to learn if he could, like apes (Boysen & Berntson, 1989), sum three separate hidden sets or a set of two hidden Arabic numerals (Pepperberg, 2012a). These tasks would demonstrate further knowledge of the representational nature of the numerals. Addition of three sets would require two updates in memory rather than one; the study with Arabic numerals would determine whether he could spontaneously transfer to summing representations of quantities rather than physical quantities. As with the sets of items, he was sequentially shown two Arabic numerals initially hidden under cups and, in their consequent absence, was asked to vocally produce a label to indicate their sum. In a separate small set of trials, he was shown the same stimuli in the same manner, but was simultaneously presented with various Arabic numerals of different colors in randomized numerical order and asked for the color of the numeral representing the sum; the colors and positions of the numerals changed on each trial. The second set of trials ensured that Alex could not learn a particular pattern over time (e.g., "if I see X + Y, I say Z"). This procedure, with its additional step, would allow testing of the same sums many more times without training him to produce a specific response, unlike tasks given to other nonhuman subjects (Pepperberg, 2012a).

Because of Alex's death, the study did not contain enough trials to test all possible sums and combinations of addends or to repeat most queries. However, he received at least one trial for each sum from 1 to 8 for the Arabic numerals and at least one trial for each sum from one to six

for the three sets of objects (just by chance, the trials requiring summation to eight were scheduled for the end of the study and could not be completed). His results were statistically significantly correct (Pepperberg, 2012a). The lack of replication of the various sums over the trials emphasizes the first-trial nature of the results and shows that no training could have been involved. Notably, if Alex's numerals had only approximate meanings, his errors would likely have exhibited a range close to the correct response. In contrast, such was the case only once (Pepperberg, 2012a). Overall, his data surpassed what would be expected if he were using the kinds of systems employed by most nonhumans or preverbal infants – for example, analog magnitude systems or object files, which cannot represent any positive integer above 4 exactly (for a review, see Carey, 2009).

Alex had only three trials on queries requiring a color response. The small number of trials preclude real statistical power, but tended toward significance: he erred on the first trial, but was correct on the next two. The task was unlike any he had ever experienced before and he had received no prior training; thus, he might not initially have understood the point of the task. In contrast to the ape that had previously been studied (Boysen & Berntson, 1989), however, Alex had to indicate the label not for the sum, but rather for the color of the numeral that represented the correct numerical sum (an additional step), and the total summed quantity on which he was tested could reach 8. His results thus suggested an intriguing level of competence on yet another numerical task – again, one based on his capacity for symbolic representation (once more, all experimental details can be found in the published paper – Pepperberg, 2012a).

1.4 DELAYED GRATIFICATION: THE "MARSHMALLOW TEST"

Delay of gratification involves postponing immediately available rewards to gain more desirable future rewards, maintaining the choice during delay, and tolerating the frustration of this self-inflicted delay. For example, children (about four years of age) are told they can eat one marshmallow placed in front of them immediately, but that they could earn a second if they refrain from eating the first until the experimenter returns from running an errand (Mischel, Shoda, & Rodriguez, 1989). As such, the task purportedly tests cognitive capacities related to future planning and self-control, capacities often thought to be lacking in nonhumans (Emery & Clayton, 2004). Delayed gratification has thus become an important tool for studying comparative cognition, particularly from an evolutionary perspective (Hillemann et al., 2014; Koepke et al., 2015). Interestingly, even many children have trouble passing the test, suggesting that it indeed examines important cognitive processes.

The task might appear straightforward, but actually involves multiple competing strategies. In nature, subjects must evaluate whether the risks (e.g., losing out to competitors, experiencing energy depletion, succumbing to predation) outweigh the advantages of waiting to find a better or larger food source (Stephens, Kerr, & Fernádez-Juricic, 2004). In the laboratory, subjects might fail because, based on personal experience, they do not trust the experimenter to fulfill the promise of the better or larger reward, for example (Kidd, Palmeri, & Aslin, 2013).

Delayed gratification also consists of two independent but related components, originally tested together in children (Mischel, 1974). The first, *delay choice*, is the initial election to wait for the better future reward. The second, *delay maintenance*, is the ability to bridge the delay interval; that is, to inhibit continuously the impulse to take the immediate reward during the delay after the initial choice to wait (Mischel et al., 1989; Toner & Smith, 1977; Toner, Lewis, & Gribble, 1979). Many studies on nonhumans test these components independently, such that in some experiments subjects first choose between an immediate, lower-value food or delayed, higher-value food and cannot change their decision during the experimenter-specified delay. Results may be confounded because subjects are often unable to inhibit initial pointing to the better payoff (Boysen & Berntson, 1995; Hillemann et al., 2014), and such tasks do not test whether subjects can sustain their delay choice. Other experiments, using either an exchange or accumulation paradigm, do allow subjects to alter their initial choice at any time. In the former experiments, which involve waiting for *better*, subjects are given a less preferred food that they can keep intact throughout the delay and then exchange for something of greater value or end the trial by consumption. In the latter experiments, which involve waiting for *more*, a series of identical edibles is moved within the subject's reach at a fixed rate (usually seconds) per item. Food accumulates until the subject interrupts by taking the available rewards (for a review, see Koepke et al., 2015).

Interestingly, results may differ depending on whether the task involves waiting for *better* or *more*. Nonhuman primates succeed at both tasks, often waiting for up to 10 minutes; birds, however, generally succeed only for *better*, rarely waiting longer than a few seconds for *more* (for reviews, see Hillemann et al., 2013; Koepke et al., 2015). The reasons for these species differences are unclear, but they may involve differences in foraging behavior (Koepke et al., 2015). Grey parrots, like other previously tested avian species, had shown little aptitude for waiting for *more* (Vick, Bovet, & Anderson, 2010), but had never been tested on *better*. My students and I thus decided to examine this possibility.

We were particularly interested to see how Griffin would perform because, unlike previously tested nonhumans, he already had a rudimentary understanding of the vocal label "wait" (Koepke et al., 2015). Every day at noon he is given highly preferred cooked grains, initially too hot to eat, and is told to "wait," although hearing the label does not decrease his anticipatory actions. He evinces similar actions at my entry into the lab, expecting to be picked up and preened; he thus hears "wait" while I use hand sanitizer and remove my outdoor shoes. Again, the instruction has little effect on his behavior, but he may have associated the vocalization with a delay of something he desires. Thus, unlike most other nonhumans, avian or primate (e.g., Auersperg, Laumer, & Bugnyar, 2013; Dufour et al., 2007, 2012), he did not have to be trained extensively on the exchange paradigm usually used to test *better* before beginning the experiment; that is, learning to exchange a nonfood token or less preferred food item for something desirable – activities that may have affected the responses of other subjects (e.g., showing effects of training). Here, the question was whether Griffin would be able to *infer* that "wait" could be associated with an *alternative* choice.

Our study differed from others given to nonhumans in several additional ways (Koepke et al., 2015). First, unlike previously tested birds (Auersperg et al., 2013; Dufour et al., 2012; Hillemann et al., 2014), Griffin was not subjected to longer delays as the experiment progressed; he had no idea how long a given trial might take as trials *randomly* lasted 10, 40, 160, 320, 640, or 900 seconds. Thus, he was not trained to wait for progressively longer periods, but had to choose to wait as long as necessary. Second, given that reward visibility made waiting more difficult for children (Mischel & Ebbensen, 1970; Mischel et al., 1989), we tested Griffin in both visible and nonvisible conditions for each time delay. For both cockatoos (Auersperg et al., 2013) and corvids (Dufour et al., 2012; Hillemann et al., 2014), the more preferred reward was always visible unless cached; caching corvids were more successful (Dufour et al., 2012). Success with a nonvisible reward would suggest the need to maintain a mental representation of the preferred food during delays. Third, we introduced control trials for each time delay in which we presented the favored item first and asked Griffin to wait for a less favored one. If he understood that "wait" was an option and not a command, he should appropriately ignore our instruction and fail to wait on these trials. Success would show that he was not simply trained to wait until he received a cue from the experimenter. Finally, some experimenters (e.g., Bramlett et al., 2012; Drapier et al., 2005) have used much less desirable items as the immediate reward. The subjects in these studies might easily have lost interest in the items, the temptation to consume them might have been missing, and the point of delaying gratification would be less relevant. Here, we used seven items that Griffin strongly preferred over his basic diet, but for which he still had a range of preference; these items were carefully ranked and we never contrasted the most and least favorable items. Using a range of rewards also prevented Griffin from considering a particular treat as a signal to wait, forcing him to evaluate his choice on each and every trial, including control trials. He had to make a new decision for each set of items, further demonstrating that he understood the metrics of the task. As a separate issue, we also ensured that Griffin's ability to wait was not a consequence of satiation by using small rewards (e.g., half or less of a cashew) and performing only a few trials, separated by at least 15 minutes, on any given day. More than four trials per day occurred only when Griffin gave evidence of desiring more rewards (e.g., repeatedly vocally requesting a high-tier treat).

Griffin was successful, waiting for the preferred reward for up to 15 minutes (Koepke et al., 2015). For all delays on visible and nonvisible conditions, he waited on approximately 90 percent of the trials, and his success rate did not depend on delay length or reward pairing, and neither did it vary significantly with trial number (i.e., he was not learning the task). On all of the control trials – one in each of the different delay times and conditions – Griffin elected to eat the initial, preferred reward rather than wait for the less preferred item, therefore demonstrating his attention to reward type and his recognition of "wait" as a label for an action rather than a command. Furthermore, he did not wait because he had lost interest in the reward: on two 900-second (15-minute) delays, he failed once after waiting 740 seconds and once after waiting 815 seconds. He seemed to recognize the concept of delayed gratification, even at the onset of the experiment.

Of particular interest were his coping strategies for dealing with the delays, many of which were reminiscent of those used by children (Mischel & Ebbesen, 1970). He could not cover his eyes with his claws the way children would cover their eyes with their hands, but he might throw the cup containing the less favored treat across the room so as not to have it in view or, like children, move the cup just out of reach or talk to himself. He might preen extensively, and, like children, try to fall asleep while waiting. He also, like children (Steelandt et al., 2012), occasionally licked the treat, but did not eat it. When observed in children, these behavior patterns are often termed "self-distractive." Unlike corvids, which can cache (Dufour et al., 2012), hiding the immediate reward was not part of Griffin's repertoire. Thus, at least in a quality exchange paradigm, little difference exists among various avian species and primates (NB: experimental details are in Koepke et al., 2015). We still, however, must test whether Griffin can wait for *more* to claim total equivalence across species on delayed gratification overall.

Although other species, as noted above, have indeed succeeded on this task without experience in symbolic representation, our ability to test Griffin in exactly the same way as children – a procedure unlike that used for most other nonhumans – was dependent upon his comprehension of the verbal symbol "wait."

1.5 ROBUST RECOGNITION OF SHAPES: MODAL AND AMODAL COMPLETION

One of the few clear markers of higher-order cognition is the ability to transfer concepts across domains (Rozin, 1976). Alex demonstrated such behavior with respect to

absence and a zero-like concept (Pepperberg & Brezinksy, 1991; Pepperberg & Gordon, 2005), but only to a limited extent with concepts such as shape. Although he could transfer from uncolored three-dimensional (3D) wooden polygons of a single size to 3D polygons of various materials, sizes, and colors, he was limited to 3D regular polygons. Could Griffin, similarly trained, now demonstrate invariance over a much wider range of testing conditions, transferring, for example, from 3D objects to 2D drawings? Furthermore, could he transfer to the phenomenon of occluded objects (amodal completion) and imaginary, subjective contours (Kanizsa figures) as shown in Figure 1.1? Such testing images differ in more numerous and varied ways from training exemplars than those given to Alex, and for subjective contours, no such overlap occurs.

This study (Pepperberg & Nakayama, 2016) was particularly interesting because demonstrating such abilities in nonhumans has been difficult. Although neither a primate brain nor a visual system would seem a prerequisite for such capacities – results from many studies, from insects to nonhuman primates, are consistent with various animals responding appropriately to 2D objects that are visible to humans as partially occluded or as partly represented in their outline form by subjective contours – these studies are subject to a variety of alternative interpretations (reviewed in Pepperberg & Nakayama, 2016). However, with a parrot that understood symbolic representation – in other words, could vocally identify various shapes ("#-corner," where # = 1, 2, 3, 4, 6) – we could recreate a situation more like that given to humans, who in most cases are given a small number of trials involving several different stimuli and simply asked to label what they see. Furthermore, Griffin grew up in a very rich environment, more similar to that of preschool children. Thus, he saw and manipulated real-world 3D objects of all forms, materials, and colors, both in full view and occluded, during a period of over 16 years before the study commenced. Such experience may be a prerequisite for carrying out the tasks being studied (Stephan, Wilkinson, & Huber, 2013).

We asked Griffin to label paper depictions of stimuli, something he had never before been asked to do. To test for amodal completion (occlusion), we used variously

(a)   (b)

Figure 1.1 (a) Occluded shapes (amodal completion) and (b) subjective (Kanizsa, illusory) shapes (modal completion).

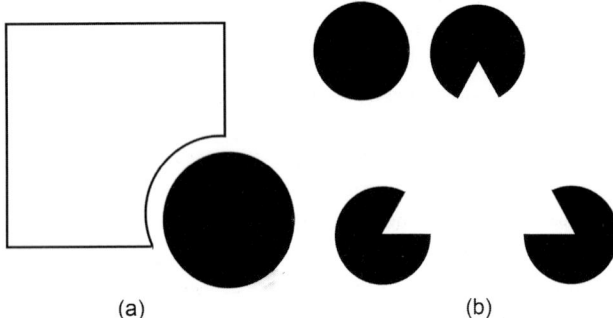

Figure 1.2 (a) Detached probe for amodal completion and (b) probe for modal completion.

colored regular polygons for each shape he could label, of different sizes, occluded by black circles (e.g., asking "What shape blue?"). Griffin had not been trained to label either circles or the color black, making it unlikely that he would attempt to label them. We also used other black shapes as occluders, to see if being able to label occluders would distract him from the task at hand. To ensure that Griffin was responding appropriately to occluded figures, we also asked him to identify irregular polygons ("detached probes") that were not occluded (e.g., regular polygons that looked as though a bite had been removed from them, with an appropriately sized adjacent circle; Figure 1.2a). For the subjective contours, Kanizsa figures were constructed using black "pac-men" to form regular polygons, again for each shape he could label, of different sizes, on colored paper. Controls ("probes," one or two for each of the #-cornered shapes) involved placing additional circles or "pac-men" near the Kanizsa figure so that Griffin could not simply quantify black objects (e.g., Figure 1.2b). Each trial was unique with respect to color, size of polygon, or size of occluder/ pac-men. He was given only 38 trials for each type of task.

We decided that the strongest way to test Griffin's abilities would be to track his responses to single presentations of each possible probe. Most experiments in the nonhuman literature use repeated presentation of identical probes for testing (e.g., Nagasaka & Wasserman, 2008; Nakamura et al., 2011) and either reward the subject for all probe trials (potentially encouraging guessing) or, after decreasing primary rewards to a set percentage similar to the proportion of probe trials, reward none of the probe/ test trials (potentially discouraging possible correct attempts). Our procedure would avoid any possible issues of familiarity, training, or encouraging either guessing or discouraging correct attempts.

Griffin identified the figures for amodal and modal completion at statistically significant levels and was correct on all probe trials. Importantly, he was correct on his very first trials, showing that no training was occurring and that transfer from 3D to 2D figures was immediate. Interestingly, for the detached probes (e.g., Figure 1.2a), where no occlusion occurred, he responded to figures never before seen (irregular polygons) with the number

of visible corners. For modal test stimuli, in contrast to amodally completed stimuli, there is *nothing* in common at an image level between the trained and test stimuli, yet his accuracy was identical to that for the amodal stimuli. He also was not quantifying pac-men or numbers of circles, because in no instance did an error correspond to the quantity of these objects (NB: experimental details can be found in Pepperberg & Nakayama, 2016).

Against Griffin's remarkable success, a substantial literature exists showing that, as noted above, with very few exceptions, animals either do not show these completion phenomena at all or show some degree of success only after having undergone considerable training with very closely related stimuli (reviewed in Pepperberg & Nakayama, 2016). In some instances where training was not an issue, success or failure may have involved mechanisms unrelated to the perceptual ones being examined here, and performance may have actually relied on mechanisms that do not match those of humans (e.g., luminance, aspect ratio: Minini & Jeffery, 2006; responses based on numbers of stimuli, stimulus generalization, reliance on local cues such as one angle in the stimuli, statistical averaging over thousands of trials: reviewed in Pepperberg & Nakayama, 2016). In other instances, actual tasks differed considerably among the laboratories (with respect to, e.g., motion, 2D vs. 3D stimuli, CRT vs. LCD monitors [i.e., flicker fusion effects] and pre-exposure to stimuli) with the consequence that results also varied considerably (again, reviewed in Pepperberg & Nakayama, 2016).

What could account for Griffin's success? Many other creatures must be able to solve problems involving at least some form of amodal completion in their daily lives (e.g., Lea, Slater, & Ryan, 1996; Regolin & Vallortigara, 1995). For example, processing partial clues about a potential predator and reacting is safer than not, even if some false alarms incur costs. Modal completion may rely on similar, early-level visual processing (e.g., imagine three black circles on a colored background that are occluded by a triangle of the same color; Nakayama, He, & Shimojo, 1995). As noted earlier, many species have demonstrated abilities consistent with, although not conclusively defined by, such processing. Griffin, in contrast, performed in a manner more sophisticated than other laboratory-based subjects. His results may be a consequence of two capacities, although others may also be involved. First, he understood symbolic representation: that a sound could stand for a physical object. Thus, his understanding that a 2D depiction could stand for a 3D representation of reality, including depth perception, would not be surprising. Note that baboons failed when tested for amodal completion in a task involving considerable training on 2D figures and forced, two-choice testing on 2D transfer stimuli (Deruelle et al., 2000), but did succeed (although only after several hundred training trials and 60 forced, two-choice testing trials) when given both training and testing stimuli that provided background depth cues indicating that the occluder was indeed in front of the targeted object (Fagot et al., 2006; see also Nagasaka, Brooks, & Wasserman, 2010, on bonobos). Second, as noted earlier, Griffin was raised in an extremely rich environment for a laboratory subject, providing him with the same kind of experience that seems to enable young children to succeed on such tasks (Kellman & Spelke, 1983).

1.6 STUDIES ON EXCLUSION

Inference by exclusion implies the ability to base a decision on the exclusion of potential alternatives. The task appears simple, but actually involves several levels of competence. The simplest level is based purely on avoidance. For example, a subject with experience in picking A when told "take A" and then shown A and B and told "take B" may avoid A merely because it simply knows not to choose A in the absence of hearing the appropriate cue, "A"; such a level of exclusion does not, however, necessarily lead to any understanding of the inferential relationship between "B" and B (Dixon, 1977). Likewise, a task may involve hiding two items in two containers, with one container subsequently being shown to be empty; to show strong inferential reasoning by exclusion, the subject's choice of the full container must not simply be a consequence of avoiding the container from which something has been extracted or the one most recently manipulated (reviewed in Mikolasch, Kotrschal, & Schloegl, 2012; Pepperberg et al., 2013). Even young children (approximately three years old) succeed on these relatively simple versions of the task, but their level of understanding is unclear (Hill, Collier-Baker, & Suddendorf, 2012). In more complicated tasks (e.g., Premack & Premack, 1994), the subject is shown that two *different* items are hidden, one is removed surreptitiously and shown to the subject, and the subject must then infer where the other must be. In the strongest sense, the subject also must be shown not to be working under the "maybe A, maybe B" assumption described by Mody and Carey (2016): in the two-position hiding, the subject might assume that either position is possible and eliminate one based on evidence, but not *conclusively* understand that B is the *only* logical choice, and rather assume that now it is just the more *likely* one.

My students and I tested four Grey parrots on tasks designed to examine at least some of their abilities regarding this form of reasoning. We note that previous studies with this species showed that the task is exquisitely sensitive to the conditions under which it is run. For example, very few of the birds in the Mikolasch, Kotrschal, and Schloegl (2011) study (which replicated that of Premack & Premack, 1994) succeeded, but their birds had to distinguish between identical cups and thus use spatial memory to infer where the remaining object was hidden – an additional cognitive process. In another experiment, auditory cues were used (either the empty or full cup was shaken), and Grey parrots succeeded only when the container was shaken horizontally, not vertically (Schloegl et al., 2012), probably because the vertical motion resembled that of the feeding behavior of a parent and thus provided a distracting confound.

In an attempt to clarify the abilities of Grey parrots on this task, my students and I performed two experiments on four birds (for additional details about the parrots, see Pepperberg et al., 2013). Two of the birds, Griffin and Arthur, had lived in a laboratory for almost all of their lives. Griffin was 16 years old at the time of the experiment and had demonstrated full object permanence (OP; a prerequisite for studies on exclusion) as a juvenile (Pepperberg, Willner, & Gravitz, 1997); he had been in the lab since he was 7.5 weeks old. Arthur was 13 years old and had been in the lab since he was 1 year old. Arthur had not been formally tested on OP, but, like Griffin, he had been the subject of a previous study that ensured he understood that the vocal phrase "Go pick up cup" was a request to make a choice and that items that had been hidden under the cups should still be available unless contrary evidence existed; he also had had training on referential speech. Two other Grey parrots, Pepper and Franco, lived in a suburban household (that of the Hartsfields) with two adult humans who had previously been trainers in the Pepperberg lab. Pepper, a female, was 15 years old and had lived with the Hartsfields since she was about 3 months old; Franco, a male, was 10 years old and had joined the household when he was 7 years old, having lived with another family previously. Pepper had received considerable training on referential communication, but had not been formally tested on production or comprehension; Franco had entered the Hartsfield household with the capacity to produce some human speech, but his referential knowledge was unknown. He subsequently had about three years of referential training, but no formal testing. Thus, we could examine whether extensive laboratory experience was a prerequisite for succeeding on the task. None of the birds were food nor water deprived.

The first experiment mostly replicated the Grey parrot study of Mikolasch, Kotrschal, and Schloegl (2011), which (as noted above) had replicated that of Premack and Premack (1994), to learn if our subjects could succeed on the basic task. Here, parrots watched an experimenter hide two equally desirable foods under two separate opaque cups (but unlike the previous study, using cups of two different colors), surreptitiously remove one of the foods, and then, in view of the birds, pocket/eat that food, leaving the birds to find the still baited cup. The experiment contained controls for various alternative explanations for the birds' behavior (e.g., olfactory cues, local enhancement, human cues, simple association rules based on what treat is handled; Mikolasch et al., 2011; Pepperberg et al., 2013). All birds succeeded at statistically valid levels (Pepperberg et al., 2013), but might they still have avoided a cup from which something had been removed rather than tracking the eaten/handled food?

Specifically, might the birds conceivably have focused on the lack of removal of *an* object, not on *the* specific object? Would the birds still be willing to return to a cup from which a favored reward was partially removed in contrast to a cup that held something to which they were indifferent? Thus, in a second experiment, some trials were run with one food slightly more preferred than the other, during which *two* items of each type were hidden in each cup and only one of the items was removed from one cup – randomly, either the more or less preferred food. If the birds understood what was happening, they should always go to the cup with the preferred food, as even one preferred food item would be a better result than two of the less preferred items. Sessions also included Experiment 1-type trials to see if the birds tracked when to use and when not to use exclusion. Thus, the birds would be rewarded for attending closely to all of the experimental aspects needed to infer how to receive their preferred treat.

All four birds succeeded on these trials, demonstrating that they did not simply avoid a cup from which something had been removed (Pepperberg et al., 2013). Three birds (Griffin, Arthur, and Pepper) also appeared to be able to switch between experimental conditions; that is, to understand something about when to use or when not to use exclusion. The fourth bird (Franco) also succeeded on these trials starting with two hidden items in each cup, but seemed to have difficulty switching between these trials and Experiment 1-type trials, actually making several errors on the latter. Possibly he was not closely attending and expected two treats to be in each cup on each trial, or possibly he lost interest in the task and ceased to attend closely to the procedure (for a discussion of "boredom" in Grey parrots, see Pepperberg & Gordon, 2005).

The results of these experiments demonstrate that Grey parrots can reason by exclusion to some degree. They do not simply avoid an empty container, but infer the most advantageous choice based on the specific context of the trial; they are at least at the level of three-year-old children. The data also showed that laboratory experience did not seem to be critical for success, although it is possible that at least some training in symbolic representation assisted all four birds. Hill, Collier-Baker, and Suddendorf (2012) have specifically argued for a relationship between capacities for symbolic representation and success on the cup task used here. The rationale is that subjects having undergone symbolic training could transfer their deductive/inferential reasoning across tasks, representing the hidden object(s) in some manner. (Again, all experimental details can be found in the published paper – Pepperberg et al., 2013.) Recently, my colleagues and I have shown that Grey parrots could pass the stronger test of exclusion, as defined and executed by Mody and Carey (2016), which involves four hiding places (two pairs) and probabilistic reasoning. Here, a reward is hidden in one cup of each pair (e.g., A, C), and one cup (e.g., B) is then shown to be empty. The subject should conclude that the reward is 100 percent likely in A and only 50 percent likely in either C or D, and so choose A. To ensure that Griffin was not simply choosing the cup next to the empty one, we performed an additional experiment in which for some trials he was given an incentive to gamble (to choose C or D) by being shown that a special, much preferred treat was being hidden in the 50-percent side. Griffin succeeded at all tasks (for details, see Pepperberg et al., 2019). His success places him beyond the level of five-year-old children.

1.7 CONCLUSION

All of these studies have demonstrated that Grey parrots can perform advanced cognitive tasks, comparable to those performed by nonhuman primates and young children. In Alex's number studies and Griffin's experiments on modal and amodal completion, success required their capacities for symbolic representation. Studies on delayed gratification and exclusion also could be seen to involve some aspects of such representation – for the former task, knowledge of the meaning of "wait"; for both tasks, mental representations of treats that were not visually present. Pepper and Franco, although not given the same formal training as Alex and Griffin, also had experience in labeling, as did Arthur (note, however, that Arthur's success in such labeling tasks was limited; Pepperberg & Wilkes, 2004). Symbolic training, which can be independent of vocal learning, has enabled several species to succeed on advanced cognitive tasks that would not likely otherwise have been possible (e.g., apes: Boysen, 2006; monkeys: Livingstone et al., 2014; Grey parrots: Pepperberg, 1999, 2013; Pepperberg & Carey, 2016). Premack (1983, 1984) specifically claimed that such training enhances nonhuman primates' abilities on specific tasks such as those requiring abstract judgment and analogic reasoning. Granted, many studies examining advanced cognitive capacities in nonhumans do not rely on training in symbolic representation (for reviews, see ten Cate & Healy, 2017; Vonk & Shackelford, 2012; Wasserman & Zentall, 2006), but such training does allow nonhumans to be tested in ways that are very similar to those used with humans, and thus provides a particularly strong method for examining comparative cognition via an evolutionary perspective.

REFERENCES

Auersperg, A. M. I., Laumer, I. B., & Bugnyar, T. (2013). Goffin cockatoos wait for qualitative and quantitative gains but prefer "better" to "more". *The Royal Society: Biology Letters*, **9**, 20121092.

Bandura, A. (1971). Analysis of modeling processes. In A. Bandura, ed., *Psychological Modeling*. Chicago, IL: Aldine-Atherton, pp. 1–62.

Barner, D., & Bachrach, A. (2010). Inference and exact numerical representation in early language development. *Cognitive Science*, **60**, 40–62.

Bolhuis, J. J., & Everaert, M., eds. (2013). *Birdsong, Speech, and Language*. Cambridge, MA: MIT Press.

Boysen, S. T. (2006). The impact of symbolic representations on chimpanzee cognition. In S. Hurley & M. Nudds, eds., *Rational Animals?* Oxford: Oxford University Press, pp. 506–511.

Boysen, S. T., & Berntson, G. G. (1989). Numerical competence in a chimpanzee (*Pan troglodytes*). *Journal of Comparative Psychology*, **103**, 23–31.

Boysen, S. T., & Berntson, G. G. (1995). Responses to quantity: Perceptual versus cognitive mechanisms in chimpanzees (*Pan troglodytes*). *Journal of Experimental Psychology: Animal Behavior Processes*, **21**, 82–86.

Boysen, S. T., & Hallberg, K. I. (2000). Primate numerical competence: Contributions toward understanding nonhuman cognition. *Cognitive Science*, **24**, 423–443.

Boysen, S. T., Berntson, G. G., Shreyer, T. A., & Quigley, K. S. (1993). Processing of ordinality and transitivity by chimpanzees (*Pan troglodytes*). *Journal of Comparative Psychology*, **107**, 208–215.

Bramlett, J. L., Perdue, B. M., Evans, T. A., & Beran, M. J. (2012). Capuchin monkeys (*Cebus apella*) let lesser rewards pass them by to get better rewards. *Animal Cognition*, **15**, 963–969.

Carey, S. (2009). *The Origin of Concepts*. New York: Oxford.

Chakraborty, M., Walløe, S., Nedergaard, S., et al. (2015). Core and shell song systems unique to the parrot brain. *PLoS ONE*, **10**, e0118496.

Deruelle, C., Barbet, I., Dépy, D., & Fagot, J. (2000). Perception of partly occluded figures by baboons (*Papio papio*). *Perception*, **291**, 1483–1497.

Dixon, L. S. (1977). The nature of control by spoken words over visual stimulus selection. *Journal of the Experimental Analysis of Behavior*, **27**, 433–442.

Drapier, M., Chauvin, C., Dufour, V., Uhlrich, P., & Thierry, B. (2005). Food exchange with humans in brown capuchin monkeys. *Primates*, **46**, 241–248.

Dufour, V., Pelé, M., Sterck, E. H. M., & Thierry, B. (2007). Chimpanzee (*Pan troglodytes*) anticipation of food return: Coping with waiting time in an exchange task. *Journal of Comparative Psychology*, **121**, 145–155.

Dufour, V., Wascher, C. A. F., Braun, A., Miller, R., & Bugnyar, T. (2012). Corvids can decide if a future exchange is worth waiting for. *Biology Letters*, **8**, 201–204.

Emery, N. J., & Clayton, N. S. (2004). The mentality of crows: Convergent evolution of intelligence in corvids and apes. *Science*, **306**, 1903–1907.

Fagot, J., Barbet, I., Parron, C., & Deruelle, C. (2006). Amodal completion by baboons (*Papio papio*): Contribution of background depth cues. *Primates*, **47**, 145–150.

Frank, M. C., Everett, D. L., Fedorenko, E., & Gibson, E. (2008). Number as a cognitive technology: Evidence from Pirahã language and cognition. *Cognition*, **108**, 819–824.

Fuson, K. (1988). *Children's Counting and Concepts of Number*. New York: Springer-Verlag.

Gallistel, C. R., & Gelman, R. (1992). Preverbal and verbal counting and computation. *Cognition*, **44**, 43–74.

Güntürkün, O., & Bugnyar, T. (2016). Cognition without cortex. *Trends in Cognitive Sciences*, **20**, 291–303.

Hedges, S. B., Parker, P. H., Sibley, C. G., & Kumar, S. (1996). Continental breakup and the ordinal diversification of birds and mammals. *Nature*, **381**, 226–229.

Hill, A., Collier-Baker, E., & Suddendorf, T. (2012). Inferential reasoning by exclusion in children. *Journal of Comparative Psychology*, **126**, 243–254.

Hillemann, F., Bugnyar, T., Kotrschal, K., & Wascher, C. A. F. (2014). Waiting for better, not for more: Corvids respond to quality in two delay maintenance tasks. *Animal Behaviour*, **90**, 1–10.

Inoue, S., & Matsuzawa, T. (2009). Acquisition and memory of sequence order in young and adult chimpanzees (*Pan troglodytes*). *Animal Cognition*, **12**(Suppl. 1), S59–S69.

Kellman, P. J., & Spelke, E. S. (1983). Perception of partially occluded objects in infancy. *Cognitive Psychology*, **15**, 483–524.

Kidd, C., Palmeri, H., & Aslin, R. N. (2013). Rational snacking: Young children's decision making on the marshmallow task is moderated by beliefs about environmental reliability. *Cognition*, **126**, 109–114.

Koepke, A., Gray, S. L., & Pepperberg, I. M. (2015). Delayed gratification: A Grey parrot (*Psittacus erithacus*) will wait for a better reward. *Journal of Comparative Psychology*, **129**, 339–346.

Lea, S. E. G., Slater, A. M., & Ryan, C. M. E. (1996). Perception of object unity in chicks: A comparison with the human infant. *Infant Behavior and Development*, **19**, 501–504.

Livingstone, M. S., Pettine, W. W., Srihasam, K., et al. (2014). Symbol addition by monkeys provides evidence for normalized quantity coding. *Proceedings of the National Academy of Sciences*, **111**, 6822–6827.

Matsuzawa, T. (1985). Use of numbers by a chimpanzee. *Nature*, **315**, 57–59.

Mikolasch, S., Kotrschal, K., & Schloegl, C. (2011). African Grey parrots (*Psittacus erithacus*) use inference by exclusion to find hidden food. *Biology Letters*, **7**, 875–877.

Mikolasch, S., Kotraschal, K., & Schloegl, C. (2012). The influence of local enhancement on choice performances in African Grey parrots (*Psittacus erithacus*) and jackdaws (*Corvus monedula*). *Journal of Comparative Psychology*, **126**, 399–406.

Minini, L., & Jeffery, K. J. (2006). Do rats use shape to solve "shape discriminations"? *Learning & Memory*, **13**, 287–297.

Mischel, W. (1974). Processes in delay of gratification. In L. Berkowitz, ed., *Advances in Experimental Social Psychology*, Vol. 7. New York: Academic Press, pp. 249–292.

Mischel, W., & Ebbesen, E. B. (1970). Attention in delay of gratification. *Journal of Personality and Social Psychology*, **16**, 329–337.

Mischel, W., Shoda, Y., & Rodriguez, M. I. (1989). Delay of gratification in children. *Science*, **244**, 933–938.

Mix, K., Huttenlocher, J., & Levine, S. C. (2002). *Quantitative Development in Infancy and Early Childhood*. New York: Oxford University Press.

Mody, S., & Carey, S. (2016). The emergence of reasoning by the disjunctive syllogism in early childhood. *Cognition*, **154**, 40–48.

Nagasaka, Y., & Wasserman, E. A. (2008). Amodal completion of moving objects by pigeons. *Perception*, **37**, 557–570.

Nagasaka, Y., Brooks, D. I., & Wasserman, E. A. (2010). Amodal completion in bonobos. *Learning & Motivation*, **41**, 174–186.

Nakamura, N., Watanabe, S., Betsuyaku, T., & Fujita, K. (2011). Do bantams (*Gallus gallus domesticus*) amodally complete partly occluded lines? An analysis of line classification performance. *Journal of Comparative Psychology*, **125**, 411–419.

Nakayama, K., He, Z. J., & Shimojo, S. (1995). Visual surface representation: A critical link between lower-level and higher level vision. In S. M. Kosslyn & D. N. Osherson, eds., *Invitation to Cognitive Science*. Cambridge, MA: MIT Press, pp. 1–70.

Olkowicz S., Kocourek, M., Lučan, R. K., et al. (2016). Birds have primate-like numbers of neurons in the forebrain. *Proceedings of the National Academy of Sciences*, **113**, 7255–7260.

Patterson, D. K., & Pepperberg, I. M. (1998). A comparative study of human and Grey parrot phonation: II. Acoustic and articulatory correlates of stop consonants. *Journal of the Acoustical Society of America*, **103**, 2197–2213.

Pepperberg, I. M. (1987). Evidence for conceptual quantitative abilities in the African Grey parrot: Labeling of cardinal sets. *Ethology*, **75**, 37–61.

Pepperberg, I. M. (1988). An interactive modeling technique for acquisition of communication skills: Separation of "labeling" and "requesting" in a psittacine subject. *Applied Psycholinguistics*, **9**, 59–76.

Pepperberg, I. M. (1994). Evidence for numerical competence in an African Grey parrot (*Psittacus erithacus*). *Journal of Comparative Psychology*, **108**, 36–44.

Pepperberg, I. M. (1999). *The Alex Studies*. Cambridge, MA: Harvard University Press.

Pepperberg, I. M. (2006a). Addition by a Grey parrot (*Psittacus erithacus*), including absence of quantity. *Journal of Comparative Psychology*, **120**, 1–11.

Pepperberg, I. M. (2006b). Ordinality and inferential abilities of a Grey parrot (*Psittacus erithacus*). *Journal of Comparative Psychology*, **120**, 205–216.

Pepperberg, I. M. (2012a). Further evidence for addition and numerical competence by a Grey parrot (*Psittacus erithacus*). *Animal Cognition*, **15**, 711–717.

Pepperberg, I. M. (2012b). Symbolic communication in the Grey parrot. In J. Vonk & T. K. Shackelford, eds., *The Oxford Handbook of Comparative Evolutionary Psychology*. New York: Oxford University Press, pp. 297–319.

Pepperberg, I. M. (2013). Interspecies communication with Grey parrots: A tool for examining cognitive processing. In G. Witzany, ed., *Biocommunication of Animals*, New York: Springer, pp. 213–232.

Pepperberg, I. M., & Brezinsky, M. V. (1991). Acquisition of a relative class concept by an African Grey parrot (*Psittacus erithacus*): Discriminations based on relative size. *Journal of Comparative Psychology*, **105**, 286–294.

Pepperberg, I. M., & Carey, S. (2012). Grey parrot number acquisition: The inference of cardinal value from ordinal position on the numeral list. *Cognition*, **125**, 219–232.

Pepperberg, I. M., & Gordon, J. D. (2005). Number comprehension by a Grey parrot (*Psittacus erithacus*), including a zero-like concept. *Journal of Comparative Psychology*, **119**, 197–209.

Pepperberg, I. M., & Nakayama, K. (2016). Robust representation of shape by a Grey parrot (*Psittacus erithacus*). *Cognition*, **153**, 146–160.

Pepperberg, I. M., & Shive, H. A. (2001). Simultaneous development of vocal and physical object combinations by a Grey parrot (*Psittacus erithacus*): Bottle caps, lids, and labels. *Journal of Comparative Psychology*, **115**, 376–384.

Pepperberg, I. M., & Wilcox, S. E. (2000). Evidence for a form of mutual exclusivity during label acquisition by Grey parrots (*Psittacus erithacus*)? *Journal of Comparative Psychology*, **114**, 219–231.

Pepperberg, I. M., & Wilkes, S. R. (2004). Lack of referential vocal learning from LCD video by Grey parrots (*Psittacus erithacus*). *Interaction Studies*, **5**, 75–97.

Pepperberg, I. M., Willner, M. R., & Gravitz, L. B. (1997). Development of Piagetian object permanence in a Grey parrot (*Psittacus erithacus*). *Journal of Comparative Psychology*, **111**, 63–75.

Pepperberg, I. M., Koepke, A., Livingston, P., Girard, M., & Hartsfield, LA. (2013). Reasoning by inference: Further studies on exclusion in Grey parrots (*Psittacus erithacus*). *Journal of Comparative Psychology*, **127**, 272–281.

Pepperberg, I. M., Gray, S. L., Cornero, F. M., Mody, S., & Carey, S. (2019). Logical reasoning by a Grey parrot (*Psittacus erithacus*)? A case study of the disjunctive syllogism. *Behaviour*, 156, 409–445.

Petrazzini, M. E. M., Agrillo, C., Izard, V., & Bisazza, A. (2015). Relative versus absolute numerical representation: Can guppies represent "fourness"? *Animal Cognition*, **18**, 1007–1017.

Premack, D. (1983). The codes of man and beasts. *Behavioral and Brain Sciences*, **6**, 125–167.

Premack, D. (1984). Possible general effects of language training on the chimpanzee. *Human Development*, **27**, 268–281.

Premack, D., & Premack, A. J. (1994). Levels of causal understanding in chimpanzees and children. *Cognition*, **50**, 347–362.

Regolin, L., & Vallortigara, G. (1995). Perception of partly occluded objects in young chicks. *Perception and Psychophysics*, **57**, 971–976.

Rozin, P. (1976). The evolution of intelligence and access to the cognitive unconscious. In J. M. Sprague & A. N. Epstein, eds., *Progress in Psychobiology and Physiological Psychology*, Vol. 6. New York: Academic Press, pp. 245–280.

Sarnecka, B. W., & Carey, S. (2008). How counting represents number: What children must learn and when they learn it. *Cognition*, **108**, 662–674.

Schloegl, C., Schmidt, J., Boeckle, M., Weiss, B. M., & Kotrschal, K. (2012). Grey parrots use inferential reasoning based on acoustic cues alone. *Proceedings of the Royal Society B: Biological Sciences*, **279**, 4135–4142.

Steelandt, S., Thierry, B., Broihanne, M.-H., & Dufour, V. (2012). The ability of children to delay gratification in an exchange task. *Cognition*, **122**, 416–425.

Stephan, C., Wilkinson, A., & Huber, L. (2013). Pigeons discriminate objects on the basis of abstract familiarity. *Animal Cognition*, **16**, 983–992.

Stephens, D. W., Kerr, B., & Fernández-Juricic, E. (2004). Impulsiveness without discounting: The ecological rationality hypothesis. *Proceedings of the Royal Society B: Biological Sciences*, **271**, 2459–2465.

ten Cate, C., & Healy, S., eds. (2017). *Avian Cognition*. Cambridge, UK: Cambridge University Press.

Todt, D. (1975). Social learning of vocal patterns and modes of their application in Grey Parrots. *Zeitschrift für Tierpsychologie*, **39**, 178–188.

Toner, I. J., & Smith, R. A. (1977). Age and overt verbalization in delay-maintenance behavior in children. *Journal of Experimental Child Psychology*, **24**, 123–128.

Toner, I. J., Lewis, B. C., & Gribble, C. M. (1979). Evaluative verbalization and delay maintenance behavior in children. *Journal of Experimental Child Psychology*, **28**, 205–210.

Vick, S. J., Bovet, D., & Anderson, J. R. (2010). How do African Grey parrots (*Psittacus erithacus*) perform on a delay of gratification task? *Animal Cognition*, **13**, 351–358.

Vonk, J., & Beran, M. J. (2012). Bears "count" too: Quantity estimation and comparison in black bears (*Ursus americanus*). *Animal Behaviour*, **84**, 231–238.

Vonk, J., & Shackelford, T. K., eds. (2012). *The Oxford Handbook of Comparative Evolutionary Psychology*. New York: Oxford University Press.

Wasserman, E. A., & Zentall, T. R., eds. (2006). *Comparative Cognition: Experimental Explorations of Animal Intelligence*. New York: Oxford University Press.

Wynn, K. (1990). Children's understanding of counting. *Cognition*, **36**, 155–193.

Wynn, K. (1992). Children's acquisition of the number words and the counting system. *Cognitive Psychology*, **24**, 220–251.

2

Cognitive Abilities in Elephants

LUCY BATES

2.1 INTRODUCTION

Everybody knows elephants are intelligent – everybody, that is, except for evolutionary psychologists. The popular characterization of these large, long-lived, very social mammals is that they have fantastic memories and are considered somehow "special," but the scientific evidence behind this reputation is somewhat lacking, with there having been relatively few attempts to study elephant cognition during the twentieth century. Fewer than 20 manuscripts detailing novel studies of elephant cognitive abilities had been published by the end of the first decade of this millennium (Byrne, Bates, & Moss, 2009), though a gradually increasing research effort is now resulting in progress.

So why, as psychologists, are we even interested in elephants, animals with which we have not shared a common ancestor for around 105 million years (Hedges, 2001; Murphy et al., 2001)? The answer rests at least in part on the fact that elephants share attributes with us that are thought to have been key in the evolution of our own intelligence, namely being long lived, highly social, and large brained. Evolutionary psychologists want to know if *humanlike* intellectual capacities have arisen in any other non-primate species, and so looking at those species that share similar "complex sociality" is a sensible place to start, given the key role of sociality in the evolution of our own intellect (Byrne & Bates, 2010; Byrne & Whiten, 1988; Dunbar, 1988; Tomasello, 2014).

Despite gaps in our knowledge, it is now possible to sketch out a view of the cognition of elephants and generate testable hypotheses of their abilities. This sketch, when considered alongside what we know about cetacean, great ape, and human cognition, should allow us to think more about what cognitive features may be necessary for and common among large-brained, social mammals. However, studying the cognitive capacities of elephants can be extremely challenging, as they do not make good laboratory animals.

It is therefore important to remember that most conclusions about elephant cognition are still tentative, and it is not always easy or productive to directly compare them to humans or other primates. Moreover, the design of a study is often crucial to gaining real insights into elephant cognition. Treating elephants like honorary primates with long noses and simply replicating laboratory studies from research in child development and primate cognition have not always proven fruitful. Instead, the studies that have paid the greatest dividends are those that consider problems that are actually relevant to elephants (Bates, 2018).

2.2 ELEPHANT SOCIETY

Elephants live in hierarchical societies, with the mother–calf unit at the foundation. Immature elephants remain dependent on their natal family groups for many years, with females maintaining close proximity for life and males becoming independent at around 10–15 years of age. Families are led by a matriarch (usually the oldest female, who may be over 60 years of age) and are composed of her close female kin (sisters, daughters, granddaughters, nieces) and their dependent offspring, typically numbering around 20 animals (varying from 3 to over 40) (Moss & Lee, 2011; Moss & Poole, 1983). Families form a fission–fusion society, breaking into smaller units for periods of hours or days. Dominance hierarchies are not generally considered to be strong in female savanna elephants, and we tend to consider females as egalitarian and cooperative (Moss & Lee, 2011). This egalitarian society involves a lot of play (Lee & Moss, 2014), allomothering of other females' calves, and a lot of touching, rubbing against, and/or caressing each other (Lee, 1987).

In African elephants (genus *Loxodonta*), two or more families may form bond groups, and bond groups can form clans or subpopulations within the population (Archie, Moss, & Alberts, 2006; Wittemyer, Douglas-Hamilton, & Getz, 2005; Wittemyer et al., 2009). Asian elephant (genus *Elephas*) social organization is somewhat looser, with smaller groups, less coherent core groups, and less social connectivity at the population level than savanna elephants (*Loxodonta africana*) (de Silva & Wittemyer, 2012).

Male society exhibits clearer dominance hierarchies based on sexual status and physical size and strength. Young, independent males seek out the company of other, preferably older, males and spend a lot of time sparring with peers, at least in part to ascertain their relative strength (Chiyo et al., 2011; Lee et al., 2011; O'Connell-Rodwell et al., 2011). Despite being sexually mature by their early 20s, males are not considered socially mature until much later. From their 30s onward, males enter annual "musth" phases – periods of greatly heightened testosterone that can last several months – that signal social and sexual maturity (Poole, 1987, 1989a; Poole & Moss, 1981). During musth, males become considerably more aggressive and wander far in search of sexually receptive females. Musth males seek to maintain exclusive access to any such females, chasing non-musth males away and even fighting with other males in musth, which can result in severe injury or even death for the losing male (Poole, 1987, 1989b).

2.3 ELEPHANT COGNITION

With my colleague, Richard Byrne, I have previously argued that many of the impressive cognitive abilities seen in primates are underpinned by efficient or specific perception, learning, and memory skills and that only relatively few species require or possess additional cognitive capacities such as insight or theory of mind (Bates & Byrne, 2015; Byrne & Bates, 2010). As we shall see in this chapter, there is a growing body of scientific data that supports the popular notion that elephants also have impressive perceptual, learning, and memory skills, but there are also some glimmers of evidence to suggest that elephants fall into the seemingly rarer category that have more going on than "merely" efficient perception, storage, and recall of information.

Elephants have the biggest *absolute* brain size of any land animal (Cozzi, Spagnoli, & Bruno, 2001; Shoshani, Kupsky, & Marchant, 2006), as well as fairly large brain sizes *relative* to their bodies (with an encephalization quotient of between 1.3 and 2.3, varying between sex and African and Asian genera) (Jerison, 1973). Within these big brains, the olfactory bulb is particularly well developed and the temporal lobe is disproportionately large, perhaps as would be expected in animals for which olfactory and auditory means of communication are particularly significant (Hakeem et al., 2005; Shoshani et al., 2006). But how do elephants use these large brains?

2.3.1 Perception, Learning, and Memory

Experimental studies have demonstrated that free-ranging savanna elephants possess sophisticated perceptual and categorization skills. They are able to categorize a single species (humans) into ethnic subclasses based on both olfactory (Bates et al., 2007) and, separately, auditory cues (McComb et al., 2014).

To test their olfactory abilities, my colleagues and I presented elephants with red-colored cloths that had been worn by either a Maasai warrior or a Kamba man, or a red-colored cloth that had not been worn by anyone. Maasai warriors occasionally spear elephants in the Amboseli area of Kenya, where the study was conducted, whereas the agricultural Kamba rarely pose any mortal threat to elephants in the area. Through our repeated measures design, it was apparent that elephants were able to determine the difference between cloths worn by Maasai and Kamba men based on their olfactory cues only. When detecting cloths worn by Maasai, the elephants rapidly fled the area, usually heading to an area of greater cover such as long grass. Cloths worn by Kamba men caused only mild alarm in comparison; the elephants did usually move away, but at a much reduced speed, over a significantly shorter distance, and taking much less time to subsequently relax (Bates et al., 2007). (Elephants' extreme olfactory sensitivity could have other implications for humans, with some captive savanna elephants now being trained as landmine "biosensors" in southern Africa because they have proven readily able to learn to detect and signal the smell of TNT [Miller et al., 2015].)

Karen McComb and colleagues, also working at Amboseli, showed that elephants from the same population also discriminate between Maasai and Kamba men by differences in their voices. They played back recordings of men saying, "Look, look, over there, a group of elephants is coming" in their own language, and again found the most heightened response of the elephants to Maasai men. They subsequently showed that the reaction was specific to the age and sex of the Maasai speaker, with less fear shown toward Maasai women and boys than adult men (McComb et al., 2014). Notably, however, the reactions observed to even adult male Maasai speakers were not as extreme as we noted in the olfactory presentations (Bates et al., 2007), and both McComb and I have suggested that this may be because warriors are unlikely to produce audible speech when hunting. Thus, hearing Maasai speaking means that they are probably not intending to threaten the elephants, whereas detecting them by olfaction in the absence of other cues could occur when the warriors are trying to conceal or hide themselves, which could end badly for the elephants.

It has also been shown, again through playback experiments, that wild savanna elephants discriminate between the long-distance rumble vocalizations made by other elephants. Females in Amboseli are familiar with the contact-call rumbles of around 100 other adult females (McComb et al., 2000) and can recognize the caller at distances of up to 2.5 km (McComb et al., 2003). Recently, Stoeger and Baotic (2017) showed that males can discriminate between the rumbles of familiar and unfamiliar females and that – unlike the females tested by McComb and colleagues – males prefer the rumbles of unfamiliar females. Both male and female elephants extract social information from rumble vocalizations, albeit with apparently different intentions.

There is also some suggestion that elephants use functionally referential vocal signals. Soltis and colleagues showed that savanna elephants in Samburu, Kenya, produce acoustically distinct alarm rumble vocalizations in response to detecting Samburu men or African bees, and when these two different alarm rumbles were played back to the elephants, they displayed different, and appropriate, behavioral reactions. They displayed vigilance and flight behaviors in response to both alarm rumble types, but headshaking was only observed in the bee alarm playback trials (Soltis et al., 2014).

While less is known about the behavior of free-ranging Asian elephants, Thuppil and Coss (2013) showed that they also have finely tuned discriminative abilities when it comes to detecting potential predators. Elephants retreated silently upon hearing playbacks of tiger growls, who occasionally prey on elephant calves, but they responded loudly and aggressively to leopard growls, who do not generally pose a threat.

Irie-Sugimoto and colleagues examined relative quantity judgments in five captive Asian elephants. The elephants had to choose from one of two baskets that contained different numbers of food items. All five elephants chose the larger quantity significantly more often than the smaller (Irie-Sugimoto et al., 2009). Elephants were as good at picking the larger quantity when it was only slightly bigger (e.g., 6:5) as when it was considerably bigger (e.g., 5:1), and performance did not vary with the total number of items presented (up to 12). In a second experiment, four additional Asian elephants watched and listened to the baskets being baited, but they could not see the final amounts in the baskets. Again, all elephants chose the basket containing the larger amount significantly more often than expected by chance. As in the first experiment, the elephants did not exhibit disparity or magnitude effects, in which performance declines with smaller differences between quantities or as the total quantity increases.

However, Perdue and colleagues argue that there were problems with the methodology adopted by Irie-Sugimoto and colleagues, including a low number of trials and repetitions overall, the possibility of cueing by the elephants' handlers, and the possibility that the elephants might have been able to see into the baskets in the second, apparently nonvisual, experiment. When Perdue and colleagues tested two captive savanna elephants in numerical tasks that eliminated these potential problems, disparity effects were apparent (Perdue et al., 2012). Whether this difference is based on the different methodology or reflects a difference between *Loxodonta* and *Elephas* is not yet known. Either way, these studies show that both species are able to keep track of a large number of items in immediate working memory (up to 15 items in the Perdue et al. study).

My colleagues and I tested working memory in savanna elephants with an "expectancy-violation" paradigm that involved moving urine deposits from known individuals to positions where they would be discovered by target individuals from the same family (Bates et al., 2008a). We varied the placement of deposits between feasible and impossible ("unexpected") locations and measured the elephants' responses to each situation. From the elephants' reactions, we concluded that they are able to continually track the locations of at least 17 family members in relation to themselves as either absent, present and moving in front of self, or present and moving behind self. This remarkable ability to hold in mind and regularly update information about the locations and movements of many others again must be supported by a large working memory capacity.

These discrimination and working memory skills are counterparts to the most famed elephant cognitive characteristic: long-term memory. McComb and colleagues (2000) were able to investigate elephants' long-term memory of other elephants' calls by playing back the contact rumbles of: (1) a female who had died, at 3 months and at 23 months after her death; and (2) a female who had left her natal family group to join another family. In both cases, playing back the call to the caller's original family resulted in contact calling by the test families, indicating that they still recognized the calls as familiar.

McComb and colleagues also showed that this knowledge is more pronounced in families with older matriarchs, indicating that social knowledge accrues with age (McComb et al., 2001). Families with older matriarchs focused their defensive bunching on genuine strangers and remained relaxed when hearing more familiar associates. This pattern is consistent with more recent results demonstrating the adaptive value of age. McComb and colleagues showed that auditory discriminations of lion threats improve with experience: older females demonstrated more appropriate decision-making in response to potential threats (in this case, in the form of hearing lion roars) (McComb et al., 2011). Younger matriarchs underreacted to hearing roars from male lions, elephants' most dangerous predators. Sensitivity to the roars of male lions increased with matriarch age, with the oldest, most experienced females showing the strongest responses to this danger.

Further demonstrations of long-term memory and increasing environmental knowledge come from studies of movement patterns and reactions to droughts (Blake et al., 2003; Leggett, 2006; Polansky, Kilian, & Wittemyer, 2015). Family groups with older matriarchs are better able to survive periods of drought because the older matriarchs lead their families over larger areas during droughts than those with younger matriarchs, again apparently drawing on their accrued knowledge (this time about the locations of permanent, drought-resistant sources of food and water) (Foley, Pettorelli, & Foley, 2008). Similarly, it has been shown that Asian elephant calves are more likely to survive if their grandmother is still present in the group. The calves of young mothers who resided in a group with their grandmother had an eight-times lower mortality risk than those whose grandmother was not present (Lahdenperä, Mar, & Lummaa, 2016).

Tests of learning and memory have also been conducted with captive Asian elephants. Rensch (1957) tested one zoo elephant in a visual discrimination, object-choice task. The female was slow to learn the task (as might be expected in such an abstract task), but once she mastered the object-choice paradigm, she learned each discrimination increasingly rapidly. She could remember the "correct" pattern from 20 pairs at the same time and retained this knowledge over the long term, still performing well a year later.

Arvidsson, Amundin, and Laska (2012) and Plotnik et al. (2014) separately showed that captive Asian elephants can also solve object-choice tests using olfactory cues, again with long-term retention (16 weeks) in the Arvidsson et al. study. While this ability is perhaps not surprising, it is interesting to note that the same elephants did not use auditory information to solve similar tests (Plotnik et al., 2014). There are probably few if any occasions when elephants need to rely on sound to locate food items in the wild. Hearing is obviously important in social communication and in detecting and locating dangers, but food is probably located entirely through scent and vision.

Taken together, these studies clearly demonstrate that perceiving, learning, and classifying ecologically relevant information appear to be highly refined skills in elephants, and this information can be retained and recalled over the long term.

2.3.2 Beyond Perception, Learning, and Memory: Problem-Solving, Causal Understanding, and Insight?

Problem-solving and causal understanding remain intriguing questions in elephants. Captive elephants in a Burmese logging camp were tested on a causal understanding task. Fifteen individuals were trained to remove food from a bucket only after they had touched the bucket lid, which was always placed on top of the bucket during training (Nissani, 2006). In the test phase, the lid was placed next to the bucket, eliminating the need to touch or remove it before accessing the food. Elephants almost always continued to touch the lid, however, which Nissani argued showed a lack of causal understanding.

Yet it is important to remember that logging elephants are trained to follow precise sequences of behavior and may be punished for deviating from these routines. The elephants may have understood that the lid was irrelevant and touching it was not a necessary part of the causal chain for retrieving the food in the test phase, but persisted because that is what they had previously been trained to do and experience had shown them not to alter from learned patterns. I would argue it is unwise to draw conclusions about problem-solving and causal understanding – or any other ability for that matter – from elephants that are not typically free to make their own choices.

Mizuno and colleagues report an example of spontaneous behavior to solve a problem, with captive Asian elephants accessing food by blowing it toward themselves (Mizuno et al., 2016). The study draws parallels between using the trunk in this novel way and tool use. Both Asian and savanna elephants have been seen to use different tool types, including using sticks to scratch parts of the body and throwing sticks, logs, and stones at other animals or human observers (Chevalier-Skolnikoff & Liska, 1993). Moreover, Asian elephants modify sticks before using them as fly switches (Hart & Hart, 1994; Hart et al., 2001), adding them to the small number of animals that *modify* tools. However, the cognitive demands of manufacturing fly switches are probably not great (Bates, Poole, & Byrne, 2008) and do not compare in complexity to the manufacture of tools by, say, chimpanzees (Boesch & Boesch, 1990; Goodall, 1986; Sanz, Call, & Morgan, 2009) or New Caledonian crows (Hunt, 1996, 2000).

Hart and colleagues report an unpublished study with captive Asian elephants that failed to show "insight" (Hart, Hart, & Pinter-Wollman, 2008). The elephants did not use sticks placed in their enclosure to obtain out-of-reach food rewards, despite the authors' belief that this should have been an easy task (since elephants frequently reach their trunks toward high food and separately pick up and hold sticks). Foerder and colleagues subsequently modified this task, providing blocks for the elephants to stand on as well as sticks (Foerder et al., 2011). Again none of the test elephants used the sticks, but one juvenile male did spontaneously stand on a block to reach the food. After solving this problem once, the elephant showed flexibility and generalization of the technique to other, similar problems by using the same box in different situations, or different objects in place of the box when it was not available.

Foerder and colleagues argued that expecting elephants to use a stick to reach food ignores the most important function of the trunk: it is primarily a sensory organ, and not a hand to reach for things. Elephants rarely use trunk-held tools to acquire food, as this would preclude the trunk's principal task of sniffing out the food. They stated: "when a stick is held in the trunk, the tip is curled backwards and may be closed, prohibiting olfactory and tactile feedback. These deficits might not deter the elephant from using a trunk-held tool for other tasks but they may inhibit the use of such tools to acquire food" (Foerder et al., 2011, p. 4). Indeed, elephants rarely use trunk-held tools to acquire food. It seems that disregarding natural behavior led to a false negative in the Hart et al. study.

2.3.3 Empathy and Theory of Mind?

In addition to their memory skills, elephants are famed for their sociality and care of family members. In an analysis of behavioral data taken from four decades of field observations of savanna elephants, my colleagues and I found that they clearly and frequently display empathy in the form of protection, comfort, and consolation, as well as by actively helping those who are in difficulty, such as assisting injured individuals to stand and walk or helping calves out of rivers or ditches with steep banks (Bates et al., 2008b; Lee, 1987). We argued that there was clear evidence of coalitions and alliances, cooperative problem-solving and helping, diagnosing animacy and goal-directedness in others, and understanding the physical competence, emotional state, and intended goals of others (Bates et al., 2008b). Observations of Asian elephants also demonstrate compassionate empathy and consolation, with more physical contact and communication occurring between elephants after a distressing event (Plotnik & de Waal, 2014).

Plotnik and colleagues showed in a neat experiment previously used to explore cooperation in primates that Asian elephants can learn to coordinate with a partner to simultaneously pull the two ends of a rope and bring a food reward within reach (Plotnik et al., 2011). The elephants would inhibit their own attempts to pull the rope for up to 45 seconds while waiting for the partner to arrive, and they understood that it was pointless to attempt the task if their partner could not access the rope. The elephants recognized that they needed to coordinate with a partner in order to obtain the shared goal of the food reward. Remarkably, one elephant showed considerable flexibility or perhaps even insight in her response: instead of pulling the rope with her trunk in conjunction with her partner, she simply placed her foot on her end of the rope, leaving her partner to pull the other end alone. This had the same effect of pulling the table closer, but meant that only one elephant exerted any effort.

Data on traditional theory of mind tasks among elephants are sketchy but tantalizing. Smet and Byrne present initial evidence that captive savanna elephants follow and understand human pointing gestures to locate hidden food in an object-choice task (Smet & Byrne, 2013, 2014b). The elephants seemingly understood that the human experimenter was pointing in order to communicate information to them about the location of a hidden object, choosing the baited bucket more often than would normally happen by chance when the experimenter pointed to it when using her arm on the same side as the bait, as well as when she used the opposite arm pointing across her body (Smet & Byrne, 2013). Five of the elephants tested also continued to succeed on the test even when the pointing was not sustained up to the moment it chose a bucket (Smet & Byrne, 2014b). Moreover, these elephants could discriminate the visual attention of the human experimenter, more often using visual gestures of their head and trunk to request food when the experimenter's face was oriented toward them (Smet & Byrne, 2014a). If the experimenter's face and body were turned away from them, the elephants engaged in fewer actions directed toward the experimenter: savanna elephants appear to take the status of the audience into account when signaling with gestures.

However, Plotnik and colleagues found no evidence for comprehension of human pointing in Asian elephants (Plotnik et al., 2013). However, this difference may be based on the slightly different methodology. Plotnik only gave the elephants the pointing cue for five seconds, the buckets were placed on a table that was moved toward the elephants after the pointing cue ceased, and both buckets had lids on them that had to be removed in order to search for and retrieve the food reward.

In contrast, Smet and Byrne placed buckets on the floor without lids and remained standing between them as the elephant approached to make its choice. Smet and Byrne argue that their design, with sustained or longer pointing times and repeatedly looking back from the elephant to the bucket, mimics typical or natural communicative behavior. Interestingly, Smet and Byrne argue that the use of the human pointing cue was novel and spontaneous in the captive savanna elephants they tested, stating that the handlers did not typically point to anything when in the company of the elephants (Smet & Byrne, 2013). In contrast, Plotnik states that the handlers of the captive Asian elephants he tested regularly point out objects and items for the elephants to attend to and pick up (J. Plotnik, 2018, pers. comm.), making the pattern of results even more surprising and interesting. Evidently, more research is required before we can draw conclusions about elephants' understanding of human pointing and any potential differences between the species in this ability, but the most recent research by Plotnik's team suggests that the elephants' experiences and ecological differences between African and Asian elephants – as well as the study methodology – may account for some of the differences recorded (Ketchaisri et al., 2019).

Social learning in its myriad forms is a key social behavior, and it is widely assumed that elephants learn from watching others, particularly older members of their social networks. Yet, curiously, there are virtually no experimental analyses of the role of social learning – its form, patterns, or relevance – in elephants, apart from one study that presented captive savanna elephants with a classic "two-action" feeding apparatus (Greco et al., 2013). The elephants did not copy the method used by the demonstrating animal, so there was no evidence of imitation, only local enhancement or observational conditioning. However, extractive foraging tasks may not be the best arena in which to investigate social learning in elephants, as very little of their feeding behavior in natural conditions requires such complex processing.

Other forms of information exchange are well documented between free-ranging savanna elephants (Lee & Moss, 1999), and it seems most unlikely that such a long-lived and highly social species would not learn from observing others, especially given what we know about the increased fitness of families with older matriarchs, which could suggest older elephants need to and do pass knowledge on to others (Foley et al., 2008; Lahdenperä et al., 2016; McComb et al., 2001, 2011). While there has not yet been any comprehensive consideration of culture in elephants, Fishlock and colleagues demonstrated the likelihood that travel paths to resources are traditional, with spatial knowledge transmitted from older relatives to younger individuals (Fishlock, Caldwell, & Lee, 2016).

Familial social network positions are also likely learned and traditional, with young females actively seeking to maintain the traditional place of their family in the wider social network when older individuals were lost to poaching (Goldenberg, Douglas-Hamilton, & Wittemyer, 2016). Moreover, elephants may possess social learning abilities that are absent in nonhuman primates, with evidence of vocal imitation in both a savanna elephant (Poole et al., 2005) and a captive Asian elephant (Stoeger et al., 2012). The savanna elephant was found to be imitating the engine sound of passing trucks, and the Asian elephant, housed alone for many years, mimicked the verbal commands of its human keepers. Furthermore, observational data suggest that older female savanna elephants *might* teach young, naive, nulliparous females how to behave when they come into estrous for the first time (Bates et al., 2010), although further exploration of this possibility is required.

A particularly enticing area of study concerns elephants' awareness of self, an apparently rare capacity in nonhuman animals. Two tests of self-awareness in Asian elephants have been published, relying on Gallup's "mark test" paradigm developed for apes (Gallup, 1970). Mirror self-recognition (MSR) is the ability to recognize a reflection in the mirror as oneself, and the mark test involves surreptitiously placing a colored mark on an individual's forehead that it could not see or be aware of without the aid of a mirror. If the individual uses the mirror to investigate the mark, the individual must recognize the reflection as itself, and so is argued to be self-aware. Most species tested only respond to the reflection as though it were a conspecific – another individual, not themselves: only great apes, magpies (*Pica pica*) and bottlenose dolphins (*Tursiops truncatus*) have been argued to pass the mark test (Anderson & Gallup, 2011; Prior, Schwarz, & Gunturkun, 2008; Reiss & Marino, 2001). Povinelli (1989) observed no signs of self-recognition in the two elephants he tested, but these elephants were only given a few days' prior exposure to the mirror before being tested. In contrast, apes that passed the test have typically had weeks or months of prior experience.

In the second experiment, Plotnik and colleagues report that one of the three adult females they tested did show MSR, suggesting that elephants do belong in the exclusive club of animals that have the capacity to recognize themselves in a mirror (Plotnik, de Waal, & Reiss, 2006; Plotnik et al., 2010). Elephants may have some concept of the self as an entity.

Of course, interpreting the MSR test remains controversial (Heyes, 1994; Heyes & Street, 1995), but a recent study has provided an intriguing counterpart. Adapting a design used with human children (Brownell, Zerwas, & Ramani, 2007; Moore et al., 2007), Dale and Plotnik (2017) explored Asian elephants' "body awareness," which is argued to fall on the same continuum of self-understanding as MSR (Dale & Plotnik, 2017; Moore et al., 2007). In this study, elephants had to walk onto a large rubber mat with a stick in front of it and pick up and pass the stick to their handlers. In the test conditions, the stick was tied onto the rubber mat with a short rope, so standing on the mat meant they could not pass the stick forwards. The elephants successfully passed the test, stepping off the mat in order to pass the stick significantly more in the test condition compared to controls, often even in the first block of trials. The elephants apparently understood that their bodies were obstructing the successful passing of the stick and removed themselves when necessary.

The extent of this capacity across the animal kingdom is currently unknown, given that such body-awareness tests have so far only been applied to two species (human children and elephants). It may be predicted that many animals must be aware of their bodies, but interestingly, the initial study with children showed a correlation between success on the body-awareness task and success in the mirror mark test, suggesting that success is not straightforward (Moore et al., 2007).

It would be impossible to discuss elephant cognition and behavior without mentioning the action for which they are perhaps best known: their apparent "mourning" of dead conspecifics. Observations of savanna elephants' reactions to dying or dead family members have been argued to support the notions that they have a concept of self and act empathically (Bates et al., 2008b; Douglas-Hamilton et al., 2006), and experimental data confirm that they recognize and prefer to interact with elephant bones than those of other animals (McComb, Baker, & Moss, 2006). Other than humans, chimpanzees are the only other species reported to react to dead family or group members as though they have an awareness of death (Anderson, Gillies, & Lock, 2010). However, the mechanisms and meanings of such behavior require further exploration. Finally, there is currently no evidence for elephants' capacity to understand others' knowledge or beliefs, because no such tests have yet been conducted with elephants. Explicit tests of their understanding of others' mental states are urgently required, but they will be challenging to design and conduct.

2.4 CONCLUSION

It is apparent that elephants of both extant genera possess very impressive memory, learning, and perceptual skills. Moreover, while research on elephant cognition is still in its infancy, the data that exist lend themselves to predicting that they, like great apes and perhaps some cetaceans and corvids, do possess some kind of qualitatively different cognitive skills. There are tantalizing indications that elephants might possess insight and theory of mind abilities.

However, much is still unknown, and firm conclusions are not yet feasible.

Almost nothing is known about elephants' understanding of physical causality. Their tool use may be rudimentary, but it is possible that they have a greater understanding of the physical world than we currently appreciate. Furthermore, we still know relatively little about how elephants communicate, particularly in the visual–gestural and chemosensory domains; we need to explore their insight and representational skills a great deal more. We also know virtually nothing about how they learn from others and what they know of others' minds. Do elephants imitate others and do they really teach naive individuals key skills? Do they have behavioral traditions or cultures? Do they understand the knowledge and beliefs of others? These questions, and many others, remain unanswered at present, but all point to exciting and necessary research studies.

One particularly important lesson that elephants can teach us is the importance of remembering other animals' ecologies. Not all species have forward-facing eyes and two hands, and our tests of their cognitive skills must take this into account. While our view of what is "smart" in animals may always be inherently biased by our own skills, we need to think beyond our own situation to really understand the evolution of cognition and intelligence. And in doing so, perhaps we may start to see more value in protecting species other than just our own.

REFERENCES

Anderson, J. R., & Gallup, G. G. (2011). Which primates recognize themselves in mirrors? PLoS Biology, 9(3), 2–4.

Anderson, J. R., Gillies, A., & Lock, L. C. (2010). Pan thanatology. Current Biology, 20(8), 349–351.

Archie, E. A., Moss, C. J., & Alberts, S. C. (2006). The ties that bind: Genetic relatedness predicts the fission and fusion of social groups in wild African elephants. Proceedings of the Royal Society, Series B, 273, 513–522.

Arvidsson, J., Amundin, M., & Laska, M. (2012). Successful acquisition of an olfactory discrimination test by Asian elephants, Elephas maximus. Physiology and Behavior, 105(3), 809–814.

Bates, L. A. (2018). Elephants – Studying cognition in the African savannah. In N. Bueno-Guerra & F. Amici, eds., Field and Laboratory Methods in Animal Cognition, A Comparative Guide. Cambridge, UK: Cambridge University Press, pp. 177–198.

Bates, L. A., & Byrne, R. W. (2015). Primate social cognition: What we have learned from nonhuman primates and other animals. In M. Mikulincer & P. R. Shaver, APA Handbook Personality and Social Psychology, Vol. 1. Washington, DC: American Psychological Association, pp. 47–78.

Bates, L. A., Sayialel, K. N., Njiraini, N. W., et al. (2007). Elephants classify human ethnic groups by odor and garment color. Current Biology, 17(22), 1–5.

Bates, L. A., Sayialel, K. N., Njiraini, N. W., et al. (2008a). African elephants have expectations about the locations of out-of-sight family members. Biology Letters, 4(1), 34–36.

Bates, L. A., Lee, P. C., Njiraini, N., et al. (2008b). Do elephants show empathy? Journal of Consciousness Studies, 15(10–11), 204–225.

Bates, L. A., Handford, R., Lee, P. C., et al. (2010). Why do African elephants (Loxodonta africana) simulate oestrus? An analysis of longitudinal data. PLoS ONE, 5(4), e10052.

Blake, S., Bouché, P., Rasmussen, H., Orlando, A., & Douglas-Hamilton, I. (2003). The Last Sahelian Elephants: Ranging Behavior, Population Status and Recent History of the Desert Elephants of Mali. Nairobi: Save the Elephants.

Boesch, C., & Boesch, H. (1990). Tool use and tool making in wild chimpanzees. Folia Primatologica, 54, 86–99.

Brownell, C. A., Zerwas, S., & Ramani, G. B. (2007). So big: The development of body self-awareness in toddlers. Child Development, 78(5), 1426–1440.

Byrne, R. W., & Bates, L. A. (2010). Primate social cognition: Uniquely primate, uniquely social, or just unique? Neuron, 65(6), 815–830.

Byrne, R. W., & Whiten, A. (1988). Machiavellian Intelligence: Social Expertise and the Evolution of Intellect in Monkeys, Apes and Humans. Oxford: Clarendon Press.

Byrne, R. W., Bates, L. A., & Moss, C. J. (2009). Elephant cognition in primate perspective. Comparative Cognition & Behavior Reviews, 4, 1–15.

Chevalier-Skolnikoff, S., & Liska, J. (1993). Tool use by wild and captive elephants. Animal Behaviour, 46, 209–219.

Chiyo, P. I., Archie, E. A., Hollister-Smith, J. A., et al. (2011). Association patterns of African elephants in all-male groups: The role of age and genetic relatedness. Animal Behaviour, 81(6), 1093–1099.

Cozzi, B., Spagnoli, S., & Bruno, L. (2001). An overview of the central nervous system of the elephant through a critical appraisal of the literature published in the XIX and XX centuries. Brain Research Bulletin, 54(2), 219–227.

Dale, R., & Plotnik, J. M. (2017). Elephants know when their bodies are obstacles to success in a novel transfer task. Scientific Reports, 7, 46309.

de Silva, S., & Wittemyer, G. (2012). A comparison of social organization in Asian elephants and African savannah elephants. International Journal of Primatology, 33(5), 1125–1141.

Douglas-Hamilton, I., Bhalla, S., Wittemyer, G., & Vollrath, F. (2006). Behavioural reactions of elephants towards a dying and deceased matriarch. Applied Animal Behaviour Science, 100(1–2), 87–102.

Dunbar, R. I. M. (1988). The social brain hypothesis. Evolutionary Anthropology, 6, 178–190.

Fishlock, V., Caldwell, C., & Lee, P. C. (2016). Elephant resource-use traditions. Animal Cognition, 19(2), 429–433.

Foerder, P., Galloway, M., Barthel, T., Moore, D. E., & Reiss, D. (2011). Insightful problem solving in an Asian elephant. PLoS ONE, 6(8), e23251.

Foley, C., Pettorelli, N., & Foley, L. (2008). Severe drought and calf survival in elephants. Biology Letters, 4(5), 541–544.

Gallup, G. G. (1970). Chimpanzees: Self-recognition. Science, 167, 86–87.

Goldenberg, S. Z., Douglas-Hamilton, I., & Wittemyer, G. (2016). Vertical transmission of social roles drives resilience to poaching in elephant networks. Current Biology, 26(1), 75–79.

Goodall, J. (1986). The Chimpanzees of Gombe. Cambridge, MA: Harvard University Press.

Greco, B. J., Brown, T. K., Andrews, J. R. M., Swaisgood, R. R., & Caine, N. G. (2013). Social learning in captive African elephants (Loxodonta africana africana). Animal Cognition, 16(3), 459–469.

Hakeem, A. Y., Hof, P. R., Sherwood, C. C., et al. (2005). Brain of the African elephant (Loxodonta africana): Neuroanatomy from magnetic resonance images. *The Anatomical Record Part A: Discoveries in Molecular, Cellular, and Evolutionary Biology,* **287**(1), 1117–1127.

Hart, B. L., & Hart, L. A. (1994). Fly switching by Asian elephants: Tool use to control parasites. *Animal Behaviour,* **48**, 35–45.

Hart, B., Hart, L. A., McCoy, M., & Sarath, C. (2001). Cognitive behaviour in Asian elephants: Use and modification of branches for fly switching. *Animal Behaviour,* **62**, 839–847.

Hart, B. L., Hart, L. A., & Pinter-Wollman, N. (2008). Large brains and cognition: Where do elephants fit in? *Neuroscience and Biobehavioral Reviews,* **32**(1), 86–98.

Hedges, S. B. (2001). Afrotheria: Plate tectonics meets genomics. *Proceedings of the National Academy of Sciences,* **98**(1), 1–2.

Heyes, C. M. (1994). Reflections on self-recognition in primates. *Animal Behaviour,* **47**, 909–919.

Heyes, C. M., & Street, G. (1995). Self-recognition in primates: Further reflections create a hall of mirrors. *Animal Behaviour,* **50**, 1533–1542.

Hunt, G. (1996). Manufacture and use of hook-tools by New Caledonian crows. *Nature,* **379**, 249–251.

Hunt, G. (2000). Human-like, population-level specialization in the manufacture of Pandanus tools by New Caledonian crows Corvus moneduloids. *Proceedings of the Royal Society, Series B,* **267**, 403–413.

Irie-Sugimoto, N., Kobayashi, T., Sato, T., & Hasegawa, T. (2009). Relative quantity judgment by Asian elephants (*Elephas maximus*). *Animal Cognition,* **12**(1), 193–199.

Jerison, H. (1973). *Evolution of the Brain and Intelligence.* New York: Academic Press.

Ketchaisri, O., Siripunkaw, C., & Plotnik, J. M. (2019). The use of a human's location and social cues by Asian elephants in an object-choice task. *Animal Cognition.* doi:10.1007/s10071-019-01283-0.

Lahdenperä, M., Mar, K. U., & Lummaa, V. (2016). Nearby grandmother enhances calf survival and reproduction in Asian elephants. *Scientific Reports,* **6**: 27213.

Lee, P. (1987). Allomothering among African elephants. *Animal Behaviour,* **35**(1), 278–291.

Lee, P. C., & Moss, C. J. (1999). The social context for learning and behavioural development among wild African elephants. In H. Box & K. Gibson, eds., *Mammalian Social Learning.* Cambridge, UK: Cambridge University Press, pp. 102–125.

Lee, P. C., & Moss, C. J. (2014). African elephant play, competence and social complexity. *Animal Behavior and Cognition,* **2**(2), 144.

Lee, P. C., Poole, J. H., Njiraini, N., Sayialel, K. N., & Moss, C. J. (2011). Male social dynamics: Independence and beyond. In C. J. Moss, H. Croze, & P. C. Lee, eds., *The Amboseli Elephants.* Chicago, IL: Chicago University Press, pp. 260–271.

Leggett, K. E. A. (2006). Home range and seasonal movement of elephants in the Kunene Region, northwestern Namibia. *African Zoology,* **41**(1), 17–36.

McComb, K., Moss, C., Sayialel, S., & Baker, L. (2000). Unusually extensive networks of vocal recognition in African elephants. *Animal Behaviour,* **59**(6), 1103–1109.

McComb, K., Moss, C., Durant, S. M., Baker, L., & Sayialel, S. (2001). Matriarchs act as repositories of social knowledge in African elephants. *Science,* **292**(5516), 491–494.

McComb, K., Reby, D., Baker, L., Moss, C., & Sayialel, S. (2003). Long-distance communication of acoustic cues to social identity in African elephants. *Animal Behaviour,* **65**(2), 317–329.

McComb, K., Baker, L., & Moss, C. (2006). African elephants show high levels of interest in the skulls and ivory of their own species. *Biology Letters,* **2**(1), 26–28.

McComb, K., Shannon, G., Durant, S. M., et al. (2011). Leadership in elephants: The adaptive value of age. *Proceedings of the Royal Society B: Biological Sciences,* **278**(1722), 3270–3276.

McComb, K., Shannon, G., Sayialel, K. N., & Moss, C. (2014). Elephants can determine ethnicity, gender, and age from acoustic cues in human voices. *Proceedings of the National Academy of Sciences,* **111**(14), 5433–5438.

Miller, A. K., Hensman, M. C., Hensman, S., et al. (2015). African elephants (*Loxodonta africana*) can detect TNT using olfaction: Implications for biosensor application. *Applied Animal Behaviour Science,* **171**, 177–183.

Mizuno, K., Irie, N., Hiraiwa-Hasegawa, M., & Kutsukake, N. (2016). Asian elephants acquire inaccessible food by blowing. *Animal Cognition,* **19**(1), 215–222.

Moore, C., Mealiea, J., Garon, N., & Povinelli, D. J. (2007). The development of body self-awareness. *Infancy,* **11**, 157–174.

Moss, C. J., & Lee, P. C. (2011). Female social dynamics: Fidelity and flexibility. In C. J. Moss, H. Croze, & P. C. Lee, eds., *The Amboseli Elephants.* Chicago, IL: Chicago University Press, pp. 205–223.

Moss, C. J., & Poole, J. H. (1983). Relationships and social structure of African elephants. In R. Hinde, ed., *Primate Social Relationships: An Integrated Approach.* Oxford: Blackwells, pp. 314–325.

Murphy, W. J., Eizirik, E., O'Brien, S., et al. (2001). Resolution of the early placental mammal radiation using Bayesian phylogenetics. *Science,* **294**, 2348–2351.

Nissani, M. (2006). Do Asian elephants (*Elephas maximus*) apply causal reasoning to tool-use tasks? *Journal of Experimental Psychology. Animal Behavior Processes,* **32**(1), 91–6.

O'Connell-Rodwell, C. E., Wood, J. D., Kinzley, C., et al. (2011). Male African elephants (*Loxodonta africana*) queue when the stakes are high. *Ethology Ecology & Evolution,* **23**, 388–397.

Perdue, B. M., Talbot, C. F., Stone, A. M., & Beran, M. J. (2012). Putting the elephant back in the herd: Elephant relative quantity judgments match those of other species. *Animal Cognition,* **15**(5), 955–961.

Plotnik, J. M., & de Waal, F. B. M. (2014). Asian elephants (*Elephas maximus*) reassure others in distress. *PeerJ,* **2**, e278.

Plotnik, J. M., de Waal, F. B. M., & Reiss, D. (2006). Self-recognition in an Asian elephant. *Proceedings of the National Academy of Sciences,* **103**(45), 17053–17057.

Plotnik, J. M., de Waal, F. B. M., Moore, D., & Reiss, D. (2010). Self-recognition in the Asian elephant and future directions for cognitive research with elephants in zoological settings. *Zoo Biology,* **29**(2), 179–191.

Plotnik, J. M., Lair, R., Suphachoksahakun, W., & de Waal, F. B. M. (2011). Elephants know when they need a helping trunk in a cooperative task. *Proceedings of the National Academy of Sciences,* **108**(12), 5116–5121.

Plotnik, J. M., Pokorny, J. J., Keratimanochaya, T., et al. (2013). Visual cues given by humans are not sufficient for Asian elephants (*Elephas maximus*) to find hidden food. *PLoS ONE,* **8**(4), e61174.

Plotnik, J. M., Shaw, R. C., Brubaker, D. L., Tiller, L. N., & Clayton, N. S. (2014). Thinking with their trunks: Elephants use smell but not sound to locate food and exclude nonrewarding alternatives. *Animal Behaviour,* **88**, 91–98.

Polansky, L., Kilian, W., & Wittemyer, G. (2015). Elucidating the significance of spatial memory on movement decisions by African savannah elephants using state-space models. *Proceedings of the Royal Society B: Biological Sciences,* **282**(1805), 20143042.

Poole, J. H. (1987). Rutting behaviour in African elephants: The phenomenon of musth. *Behaviour*, **102**, 283–316.

Poole, J. H. (1989a). Announcing intent: The aggressive state of musth in African elephants. *Animal Behaviour*, **37**, 140–152.

Poole, J. H. (1989b). Mate guarding, reproductive success and female choice in African elephants. *Animal Behaviour*, **37**, 842–849.

Poole, J. H., & Moss, C. J. (1981). Musth in the African elephant *Loxodonta africana*. *Nature*, **292**, 830–831.

Poole, J. H., Tyack, P. L., Stoeger-Horwath, A. S., & Watwood, S. (2005). Elephants are capable of vocal learning. *Nature*, **434**, 455–456.

Povinelli, D. J. (1989). Failure to find self-recognition in Asian elephants (*Elephas maximus*) in contrast to their use of mirror cues to discover hidden food. *Journal of Comparative Psychology*, **103**(2), 122–131.

Prior, H., Schwarz, A., & Gunturkun, O. (2008). Mirror-induced behavior in the magpie (*Pica pica*): Evidence of self-recognition. *PLoS Biology*, **6**(8), e202.

Reiss, D., & Marino, L. (2001). Mirror self-recognition in the bottlenose dolphin: A case of cognitive convergence. *Proceedings of the National Academy of Sciences*, **98**(10), 5937–5942.

Rensch, B. (1957). The intelligence of elephants. *Scientific American*, **196**(2), 44–49.

Sanz, C., Call, J., & Morgan, D. (2009). Design complexity in termite-fishing tools of chimpanzees (*Pan troglodytes*). *Biology Letters*, **5**(3), 293–296.

Shoshani, J., Kupsky, W. J., & Marchant, G. H. (2006). Elephant brain. Part I: Gross morphology, functions, comparative anatomy, and evolution. *Brain Research Bulletin*, **70**(2), 124–157.

Smet, A. F., & Byrne, R. W. (2013). African elephants can use human pointing cues to find hidden food. *Current Biology*, **23**(20), 2033–2037.

Smet, A. F., & Byrne, R. W. (2014a). African elephants (*Loxodonta africana*) recognize visual attention from face and body orientation. *Biology Letters*, **10**, 20140428.

Smet, A. F., & Byrne, R. W. (2014b). Interpretation of human pointing by African elephants: Generalisation and rationality. *Animal Cognition*, **17**(6), 1365–1374.

Soltis, J., King, L. E., Douglas-Hamilton, I., Vollrath, F., & Savage, A. (2014). African elephant alarm calls distinguish between threats from humans and bees. *PLoS ONE*, **9**(2), e89403.

Stoeger, A. S., & Baotic, A. (2017). Male African elephants discriminate and prefer vocalizations of unfamiliar females. *Scientific Reports*, **7**, 1–10.

Stoeger, A. S., Mietchen, D., Oh, S., et al. (2012). An Asian elephant imitates human speech. *Current Biology*, **22**(22), 2144–2148.

Thuppil, V., & Coss, R. G. (2013). Wild Asian elephants distinguish aggressive tiger and leopard growls according to perceived danger. *Biology Letters*, **9**(5), 20130518.

Tomasello, M. (2014). The ultra-social animal. *European Journal of Social Psychology*, **44**(3), 187–194.

Wittemyer, G., Douglas-Hamilton, I., & Getz, W. M. (2005). The socioecology of elephants: Analysis of the processes creating multitiered social structures. *Animal Behaviour*, **69**(6), 1357–1371.

Wittemyer, G., Okello, J. B. A., Rasmussen, H. B., et al. (2009). Where sociality and relatedness diverge: The genetic basis for hierarchical social organization in African elephants. *Proceedings of the Royal Society, Series B*, **276**(1672), 3513–3521.

3 Culture and Communication among Cetaceans

ELLEN C. GARLAND AND LUKE RENDELL

3.1 INTRODUCTION

Cetaceans live in an environment that is alien to us. They are adapted to live in a three-dimensional fluid world that has exerted evolutionary pressures vastly different from those we are shaped by as terrestrial beings, yet they share our mammalian heritage. For these reasons, they are profoundly relevant to any comparative analysis of human behavior, particularly aspects in which both they and we represent relative peaks across the mammalian order. Humans are visual creatures; in contrast, cetaceans rely heavily on their acoustic sensory system because a profound reliance on sound in the ocean for everything from communication to navigation is essential given the very limited light that penetrates more than 100 m or so. Sound travels much faster and further in water than air, allowing communication to occur over tens of kilometers and sometimes across an entire ocean basin. Marine niches vary in time and space in quite different ways to those on land. This all makes for a fascinating backdrop to the many parallels between the behaviors and cultures of cetaceans and humans. Cetaceans show some of the most sophisticated and complex vocal and cultural behavior we know outside humans, encompassing learning, shared traditions, and gene–culture coevolution (Whitehead & Rendell, 2014). Researching these phenomena remains a significant challenge, however. Most cetaceans are very large and not particularly amenable to captivity. Therefore, many studies rely on opportunistic observations and data collected from free-ranging wild animals in a vast and hostile three-dimensional environment. Nonetheless, scientists have focused on everything from learning in an individual animal through to the transfer of vocal displays among populations spread across an entire ocean basin, representing a huge range of spatial and temporal scales, from a local home range of, say, a few kilometers through to thousands of kilometers across an ocean basin. Exploration of the many parallels and contrasts between cetacean communication and culture on the one hand and human language and culture on the other has great potential for deepening our understanding of the evolutionary roots of language and

culture (Laland & Galef, 2009). Using a number of case studies, from cultural song revolutions and the spread of new feeding tactics in humpback whales, through vocal clans in sperm whales and individual signature whistles in bottlenose dolphins, to gene–culture coevolution in killer whale ecotypes, we highlight how new knowledge of cetacean lives is opening up our understanding of the evolution of culture and communication to a far broader comparative base. We conclude by suggesting future avenues of research to address fundamental, unanswered questions that may provide essential pieces to understanding the puzzle of the evolutionary roots of the human language and culture.

3.2 CULTURE IN ANIMALS

Controversy still exists over whether animals exhibit culture in any form (Laland & Galef, 2009). The study of human culture is deeply rooted in anthropology and the description of variation in human behavior. The complexity of human culture is undeniable and self-evident, particularly in our ability to continually increase the complexity and/or efficiency of technological advances, a process known as cumulative cultural evolution (Galef, 2009; Tomasello, 2009). Furthermore, the key characteristics of human culture – teaching, social imitation, and norming/conforming (Tomasello, 2009) – are suggested to be mostly absent or present only in reduced complexity in animals. While no one is arguing that we humans are not the pinnacles of cultural complexity, many researchers, including ourselves, are, however, suggesting that we humans are just one of a myriad of other species that exhibit something that could usefully be termed culture. Furthermore, from this perspective, the comparison of traits among vastly different lineages (e.g., primates, birds, and cetaceans) is essential to elucidating how culture can evolve – its drivers – in order to shed light on the roots of our own cultural complexity.

With that in mind, we view culture as a continuum from complex traditions (i.e., human) down to very simple ones (e.g., salmon homing behavior; see Sargeant & Mann, 2009). We define culture here as the acquisition of information or behavior through some form of social learning from conspecifics (Fragaszy & Perry, 2003; Rendell & Whitehead, 2001; Whiten, 2009). This may be a single trait culturally transmitted among individuals or represent a complex suite of traits that differs among populations and thus acts to identify and define each population. We acknowledge that some anthropologists may find our definition of culture less than satisfying, since it does not specify features such as language and shared meanings. But to include features that are only really observable in others through the medium of language renders culture as something that humans can have only by definition rather than observation.

In order to assert that culture is the underlying driver of an observed behavior, however, one must first evaluate the plausibility of other influences so as to understand whether an observed variation in behavior can be accounted for on a genetic or ecological basis such that cultural transmission is not a necessary condition for it to arise (Laland & Janik, 2006). This approach – termed the method of exclusion – has been highly influential, but also has some important weaknesses: causally relevant ecological or genetic variation can be missed; culture can be erroneously rejected in the presence of measurable but causally insignificant variations in genetics and/or ecology; and it pretends that behavior can develop from any one of these factors alone (Laland & Janik, 2006; Laland et al., 2009). Moving beyond exclusion-type approaches, it is important to evaluate the influence of different casual factors on the development of a particular behavior, as all behavior develops as an interaction between genetic blueprint and environmental experience (Laland & Janik, 2006). An individual may innovate a novel variation to a feeding technique in order to exploit a new prey item that is available because the environment has changed. Clearly, there are ecological influences in this situation; the diffusion of this novelty through the population, however, may still have significant social influence.

When we talk about transmission of information or behaviors, the direction of information flow is rather important. Different directions – vertical, oblique, or horizontal – may produce different cultural patterns or potentially confound the assignment of the trait to cultural processes. Vertical transmission is from a parent to their offspring, oblique transmission is from a nonparent model from the previous generation, and horizontal transmission is within-generation transmission. Vertical transmission may be difficult to distinguish from underlying genetic explanations (Krützen et al., 2005; Laland & Janik, 2006), while horizontal transmission is thought to be characterized by cultural conformity and the rapid spread of behavioral variants through a population (Garland et al., 2011; Sargeant & Mann, 2009; Whitehead, 2009). Horizontal transmission may only become apparent when examining geographic variation in a single behavior or suite of behaviors across multiple populations, unless researchers are fortunate enough to catch the spread of a new behavioral variant during their observations.

3.3 CETACEANS: AQUATIC MAMMALS

As terrestrial beings, the concept of living in the ocean is somewhat alien to us. The ocean is hostile; most of it is dark, all of it is saline, it presents serious thermal challenges to warm-blooded animals, oxygen levels are low, and there are no landmarks. For primates, it is difficult to move around with any efficiency. Nonetheless, this three-dimensional fluid world covering two-thirds of the planet has shaped the evolution of one mammalian order to such an extent that it is their home for their entire lives. Instead of a profound reliance on sight, as light does not penetrate deeper than 100 m or so, cetaceans rely heavily on their acoustic sensory system. Sound travels much faster and further in water than air, allowing for a reliance on sound for everything from communication to navigation, sometimes across thousands of kilometers.

Cetaceans are mammals – they breathe air. Their ability to hold their breath and dive to over 1 km in depth to capture prey is astonishing and requires some fairly specialized adaptations. Cetacea originated from a semiaquatic ancestor in the Eocene, approximately 50 million years ago (Thewissen et al., 2007). They are the basal node in the Cetacea/Artiodactyla clade, making cetaceans distantly related to the hippopotamus (Thewissen et al., 2007). Cetaceans are divided into two groups: the mysticetes or baleen whales and the odontocetes or toothed whales, which include dolphins and porpoises (see review in Berta & Sumich, 2006). This creates two very different groups. The baleen whales are large, highly migratory, and have plates of baleen to filter their food (krill and small shoaling fish) from the water column. In contrast, toothed whales (including dolphins) are mostly smaller in size, capture fast-moving fish or mammals, and have home ranges and are therefore less likely to undertake large migrations (although there are notable exceptions). These two diverse groups, mysticetes and odontocetes, produce fundamentally different types of vocalizations. The odontocetes produce mainly mid- to high-frequency tonal whistles and pulsed echolocation clicks (Herman & Tavolga, 1980). Mysticetes or baleen whales do not produce pulsed echolocation clicks – with the possible exception being low-frequency clicks produced by feeding humpback whales, *Megaptera novaeangliae* (Stimpert et al., 2007) – and their vocalizations are usually much lower in frequency, can be patterned, and are longer in duration than odontocete calls (Herman & Tavolga, 1980).

While the full diversity of the cetacean order is still not completely known, as new species are still being discovered on average every decade (Dalebout et al., 2002; Morin et al., 2017), more than 80 species are recognized. This diversity means they have the potential to provide additional insights in any comparative analysis; differences among cetacean species can be compared to the observed differences between humans and other apes, for example. However, studying cetaceans is challenging. Most species are not amenable to captivity, making controlled and repeatable experiments difficult. Researchers are often forced to rely on field observations and natural replicates found in the environment. This means it can take decades of painstaking data collection to finally reveal cultural phenomena. While this can slow the process of discovery, the ability to present a diverse array of cultural phenomena in cetaceans is an impressive testament to scientists' perseverance over the past four decades since studies of live free-ranging cetaceans began in earnest.

Given their reliance on acoustic communication in the ocean, each cetacean species has evolved a unique vocal communication system. Many cetaceans learn the sounds around them, and given our lonely status as a vocal learner in the primate lineage, this fact alone is significant. For example, a killer whale calf will learn the vocalizations unique to its pod (Ford, 1991). This shared set of vocalizations or vocal dialect can be learned only from another individual to maintain the consistently matched displays that characterize each pod (Deecke et al., 2000; Ford, 1991). These calls can change with time; individuals are able to modify their vocal displays over time to more closely match conspecifics (Crance et al., 2014). Examining vocal variation on a large geographic scale may reveal differences among groups, clans (groups that share similar dialects), or populations. Below, we explore a number of case studies to present the diversity of culturally transmitted vocal displays we currently know about in cetaceans. To supplement these accounts of vocal culture, we also present an example of the spread of novel foraging techniques throughout a population. The combination of foraging and vocal culture in some species presents a suite of cultural traditions with which populations can be identified. Finally, we describe how very recent evidence from killer whale genomic studies shows that gene–culture coevolution, where the dual inheritances of cultural and genetic information interact with each other, has been occurring in lineages other than our own. Each of the case studies presented below could form a book chapter in its own right (Whitehead & Rendell, 2014), so by necessity our account can provide only a summary and not all of the relevant details.

3.4 CASE STUDIES

3.4.1 Humpback Whales: Song Revolutions and Feeding Traditions

Cultural transmission in humpback whales has been documented in both song displays and foraging tactics (Allen et al., 2013; Garland et al., 2011; Noad et al., 2000). Both present clear examples of horizontal transmission within and, in the case of song, also between populations. Male humpback whales produce a long, complex, stereotyped, sexual display termed "song" (Glockner, 1983; Payne & McVay, 1971). Song functions in sexual selection through mate attraction and/or male–male social assortment (Darling et al., 2006; Payne & McVay, 1971; Smith et al., 2008). The song is arranged in a nested hierarchy (Herman

& Tavolga, 1980); single sounds are termed "units," and a few units are organized into a stereotyped "phrase" (Payne & McVay, 1971). Phrases are repeated to make a "theme," and a few different themes, each composed of a different set of stereotyped phrases, make up a song (Payne & McVay, 1971). At any one point in time, all males within a population will sing the same themes in the same order (Frumhoff, 1983; Payne & Payne, 1985; Payne et al., 1983). Thus, there is intense conformity to the current song within a population, but confusingly the song also continuously changes or evolves each year in a progressive, unidirectional way (Cato, 1991; Payne & Payne, 1985; Winn & Winn, 1978). All males must be making the same changes to the song (by adding or deleting the same units, phrases, or themes) to maintain the observed conformity. This is a paradox; humpback whale song is a display that is both the same (at a population level) but also constantly changing. Regardless, each year males must learn the changes to the song, and they must learn from those around them – this is an example of cultural evolution (Payne & Payne, 1985; Payne et al., 1983). This progressive cultural evolution allows the song to evolve over a number of seasons to decades (Payne & Payne, 1985).

However, humpback whale song also undergoes dramatic cultural revolutions, where the song display (termed "song type") from a neighboring population is rapidly adopted by all of the males in a population (Garland et al., 2011; Noad et al., 2000). Noad et al. (2000) documented the introduction of song from the western Australian population into the eastern Australian population in 1996–1997, most likely introduced by a small number of males. Males in the eastern Australian population rapidly switched their song to the new, novel song type. The uptake of this different song suggests a strong preference for novelty (Noad et al., 2000). This cultural phenomenon was viewed as a "one off" (Whitehead, 2009) until a striking pattern of song transmission emerged across the South Pacific Ocean. Garland et al. (2011, 2012, 2013a, 2013b, 2015) have since documented the horizontal cultural transmission of multiple song types that spread rapidly and repeatedly across the South Pacific. Multiple song types spread in a stepwise fashion from one population to next, from eastern Australia across to French Polynesia, over 11 years (Figure 3.1). Each song type took approximately two years to transit the region. Song exchange on shared migratory routes or feeding grounds and/or the movement of individuals between breeding grounds within or between seasons may be the underlying mechanisms allowing song exchange (Garland et al., 2011, 2012, 2013a, 2013b, 2015; Payne & Guinee, 1983), but the details remain unknown. The original cultural revolution discovered by Noad et al. (2000) continued to transit the Pacific, and it represents a cultural signal that spanned two ocean basins and seven years. This is the best – and possibly only – example of repeated population-wide

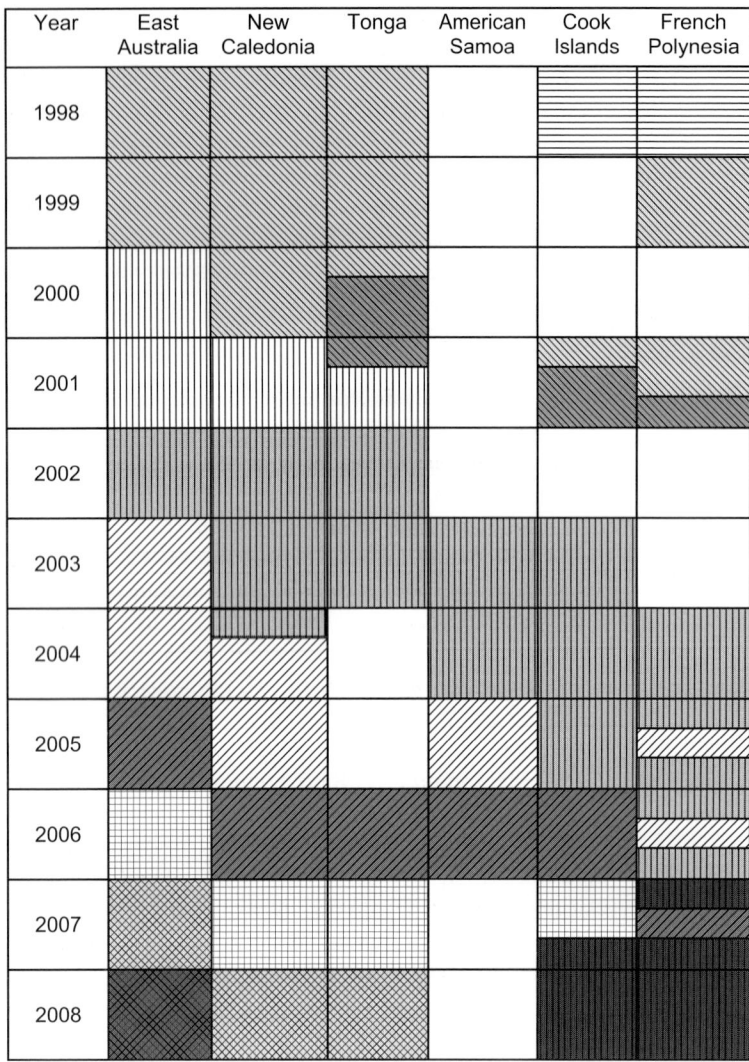

Figure 3.1 Song types identified in the South Pacific region from 1998 to 2008. Populations are listed from west to east across the region. Each hatching style represents a distinct song type. Unpatterned cells represent no data being available. Two patterns within a year/location indicate both song types were present (e.g., Tonga in 2000). In these cases, the seasons are broken into three periods (early, middle, or late) to indicate when a new song type was recorded. Different shades of gray underneath the same hatching style represent evolutionary song change – minor modifications to the same general song type. Reprinted from *Current Biology* **21**, Garland et al., "Dynamic horizontal cultural transmission of humpback whale song at the ocean basin scale," 687–691, Copyright (2011), with permission from Elsevier.

horizontal cultural transmission where behavioral variants were passed among populations in any nonhuman animals (Garland et al., 2011). Underlying genetic and ecological factors were discounted as the speed of song movement precluded genetic transmission and the same song type was used by populations located on breeding grounds with entirely different ecologies (e.g., shallow shelf waters vs. islands formed by seamounts). The rapid spread of song types (variants or phenotypes) was likened to the spread of human fashion trends, although there are a number of unanswered questions. The outcome of cultural transmission is clear, but the underlying mechanisms of individual learning are speculative at best. Understanding the drivers of this large-scale phenomenon and the individual learning strategies that may create such a process is a rich field for future research.

The second documented cultural tradition in humpback whales is in a different domain – foraging. Bubble-feeding is a common foraging technique that has been observed in a number of populations (Allen et al., 2013; Clapham, 2000; Hain et al., 1982). A novel and innovative modification to the technique, termed lobtail feeding, was observed

spreading throughout the Gulf of Maine population of humpback whales (Allen et al., 2013). From an initial individual whale observed utilizing this innovation on the summer feeding grounds, the technique diffused through the population over 27 years (1980–2007). As this technique is apparently specialized for feeding on a particular prey item (sand lance), network-based diffusion analysis (NBDA) was employed to tease apart the influences of social and ecological factors. The analysis indicated that support for models including a social transmission effect was orders of magnitude greater than support for models without it. Therefore, social transmission played an important role in the spread of this innovation within the Gulf of Maine population, without disregarding ecological factors, primarily prey abundance. The use of NBDA enabled the influences of both social and ecological factors to be quantified, highlighting that social learning allows individuals to adapt to changing ecological circumstances. Humpback whales are therefore capable of having multiple, independently evolving cultural traditions within a population (Allen et al., 2013), representing a peak in cultural complexity in nonhumans.

3.4.2 Sperm Whales: Vocal Clans and Cultural Hitchhiking

Sperm whales (*Physeter macrocephalus*) have a culturally transmitted communication system and a multilevel social structure. In tropical and subtropical waters, 8–12 females and their immature offspring form stable, matrilineal social "units," two or more of which may temporarily join together to form a "group" for hours to days (Christal et al., 1998; Whitehead, 2003; Whitehead & Weilgart, 2000). In contrast, after leaving their natal group at around six years of age, males disperse to high latitudes to grow sufficiently in order to attain social maturity, before making periodic trips back to the tropics to mate (Whitehead, 2003; Whitehead & Weilgart, 2000).

Despite their large size (10–17 m in length), sperm whales are toothed whales and thus are capable of producing echolocation clicks. Sperm whales often produce short series of clicks or "codas" during social interactions at or near the surface, or at the start of foraging dives (Whitehead & Weilgart, 1991). Codas are composed of a stereotyped sequence of 3–40 broadband clicks produced in particular temporal patterns (Watkins & Schevill, 1977). Multiple coda types exist, and they are characterized by the timing pattern between the clicks (the inter-click interval). Each unit has a distinct dialect made up of a number of different coda types (Gero et al., 2016a; Rendell & Whitehead, 2003), with the most common coda types shared among all adult members of the social unit (Gero et al., 2016b; Schulz et al., 2011). Calves learn the dialect of their social unit, although it may take several years for them to perfect this repertoire (Gero et al., 2016b; Schulz et al., 2011). At a broader scale, collections of units share similar dialects, and such dialect groups have been termed vocal clans (Rendell & Whitehead, 2003). Two different scenarios of vocal sharing apparently occur in different ocean basins. In the Pacific Ocean, multiple vocal clans live in sympatry. These clans are socially segregated based on their distinct dialects, and units only associate with other units that share the same dialect (Gero et al., 2016b; Rendell & Whitehead, 2003; Rendell et al., 2012). This system operates on an impressive scale: clans can range over thousands of kilometers of ocean, which they share with other (sympatric) clans; clans contain many thousands of whales; and clans result from the cultural transmission of vocal patterns (Rendell & Whitehead, 2003; Rendell et al., 2012).

In the Atlantic Ocean, in contrast, coda dialects initially appeared to vary geographically, so that different dialect groups were allopatric, with a single repertoire typically heard in any given area (Whitehead et al., 2012). However, recent work by Gero et al. (2016a) discovered a similar pattern of dialect sharing (clan structure) to the Pacific Ocean. Sympatric vocal clans do occur in the Atlantic, as two different vocal clans were identified in the eastern Caribbean. Units in each vocal clan were only observed associating with units of their own clan. Again, many years of data were required to discover this pattern – the study spanned the decade 2005–2015. Based on this evidence from both ocean basins, a higher-order level of social organization of cultural identity at the clan level is likely to be important to the species as a whole (Gero et al., 2016a).

At a lower level in the social organization, there also appears to be individual-specific variation in the timing of codas, which allows individual identification in a limited number of coda types (Antunes et al., 2011; Oliveira et al., 2016). Antunes et al. (2011) found that it was possible to discriminate among individuals for a single coda type only (the 5 R) in a study of three codas shared in the social unit (12 other coda types were present but had limited sample size). This led the authors to suggest there may be different functions among coda types and that individual information may be encoded at a finer scale within the wider stereotyped rhythm of a given coda type. Gero et al. (2016b) recently confirmed the presence of identity cues in sperm whale codas, as well as the stability of one socially learned coda type (the 1 + 1 + 3 coda) across an entire population over 30 years. This is a striking example of cultural transmission allowing high levels of conformity to be maintained across a large number of individuals who may not be continuously associated with each other (Gero et al., 2016b). As codas encode information about individuals, units, and clans, this allows discrimination among several levels of their social structure, providing support for the "social complexity hypothesis" (the idea that complex social structure is a driver of diversity in animal communication systems) among sperm whales (Gero et al., 2016b). Clearly, social structure and culture are interacting to produce and reinforce these vocal patterns.

The strong correlation between matrilineal social unit and vocal repertoire lends itself to investigations of gene–culture coevolution. Whitehead et al. (1998) investigated the population structure of sperm whales in the South Pacific. This revealed a striking pattern: South Pacific sperm whales had a non-geographically based population structure, with maternally related groups sharing similar coda repertoires. The strong correlation between mitochondrial DNA (mtDNA) haplotype and coda dialect led Whitehead et al. (1998) to suggest that parallel processes of matrilineal inheritance of a culturally transmitted coda repertoire and mtDNA haplotype occurred in sperm whales. To explain this phenomenon further, Rendell et al. (2012) investigated whether genetic differences explained vocal dialect variation. This is important: if mtDNA inheritance patterns correlate with the pattern of coda repertoires, this could indicate that there is a genetic basis for the (vocal) behavior and by inference that coda dialects are unlikely to be culturally based. Rendell et al. (2012) demonstrated that mtDNA haplotypes were widely shared among clans, indicating that the vocal variation was unlikely to have a largely genetic basis. Instead, this finding was consistent with females occasionally switching social units and, far more rarely, clans. Rendell et al. (2012) suggest that the hypothesis of generally vertical maternal cultural transmission, oblique transmission within matrilineal social units, and horizontal/oblique transmission when females occasionally switch clans was more consistent with the observed pattern. Under this scenario, statistical – but not causal – correlations with haplotype frequencies will build up due to the parallel vertical transmission of dialects and haplotypes (Rendell et al., 2012).

These are all indications that gene–culture coevolution is not, as once thought, restricted to humans (Whitehead, 1998). The low mtDNA diversity seen in the matrilineal whales, of which sperm whales are an example, could occur due to neutral or nearly neutral mtDNA loci hitchhiking on selected maternally inherited cultural traits (Whitehead, 1998). Given that gene–culture coevolution is more likely to occur when cultural features are transmitted with little error between generations (Laland, 1992), the apparently conformist vocal behavior of matrilineal whales provides an excellent environment for selection to act on variation between groups (Whitehead, 1998). Cultural differences exist at different levels within the social structure; this may allow multilevel cultural hitchhiking, where innovations have different speeds and so spread first through units and then the larger clans (Whitehead, 2005). This intriguing possibility still requires more evidence, but it presents a clear and exciting avenue for future studies.

3.4.3 Bottlenose Dolphins: Signature Whistles and "Sponging"

Bottlenose dolphins (*Tursiops truncatus* and *Tursiops aduncus*) are incredible vocal learners. Wild and captive studies have contributed to an understanding of the learning capabilities of these small, toothed whales. Bottlenose dolphins live in a fission–fusion society that is not matrilineally organized, but where group membership is instead fluid and changes quite rapidly on timescales of minutes to hours (see Connor et al., 2000). Females form close associations with other females – often, but not always, kin (Frère et al., 2010) – while males form multileveled alliances with other males to compete for reproductive access to females (Connor & Krützen, 2015; Connor et al., 1992). Unlike the large baleen whales, coastal bottlenose dolphins have smaller home ranges that allow researchers to conduct long-term, year-round studies into the behavior of free-ranging dolphins (there is a deep-water "offshore" form of this species, but very little is known about it).

Bottlenose dolphins produce signature whistles, an individually distinctive call based on a unique pattern of frequency modulations, which encode their individual identities independent of voice features (Caldwell & Caldwell, 1965; Janik et al., 2006). Signature whistles are developed early in life (Sayigh et al., 1990), and they are then stable for females for at least a decade (Sayigh et al., 1990), but may be modified by males to better match the signature whistles produced by their male alliance partners (Watwood et al., 2004). This convergence on a similar, shared signal is indicative of group conformity and conveys both alliance membership and individual identity. Signature whistles also act as a group cohesion call to maintain contact with social group members when separated (Janik & Slater, 1998). Female bottlenose dolphins use their signature whistles not only as individual identity cues, but also to facilitate mother–calf reunions (King et al., 2016). Calves develop their own signature whistles based on the signature whistles they hear in the environment, including their mothers' (Sayigh et al., 1995), but how closely they copy their mothers' whistles appears to be sex dependent (Sayigh et al., 1990).

A fascinating recent discovery was that bottlenose dolphins copy one another's signature whistles as a means of addressing specific social companions (Janik, 2000; King & Janik, 2013). This vocal matching acts as an affiliative signal allowing the dolphin to address specific social companions (King et al., 2014) who are typically close associates, such as mother–calf pairs or members of male alliances (King et al., 2013, 2018). This ability to use a learned signal as an individually specific label, in essence addressing a conspecific by using their "name," is exceptionally rare in nonhuman animals (King & Janik, 2013). Bottlenose dolphins have also been shown to produce referential signals by labeling objects with novel, learned whistle patterns (Richards et al., 1984). The cognitive ability to vocally label an object with a learned signal is complex in itself (Janik, 2009), but beyond that, these animals have the ability to recognize the whistles of conspecifics after 20 years of separation (Bruck, 2013). Clearly, this cetacean species is exceptional in its learning and memory capabilities.

A further and fundamental piece of evidence for culture in marine mammals is the existence of tool-using culture in a subset of wild bottlenose dolphins (Krützen et al., 2005; Sargeant & Mann, 2009). In Shark Bay, located on the west coast of Australia, a long-term study of bottlenose dolphins has been underway since 1984. Within this population, a small subset of related individuals place marine sponges over their beaks to probe the sediment for fish (Krützen et al., 2005; Mann & Sargeant, 2003). "Sponging" occurs in a particular habitat – the deep-water channels – but these habitats are also part of the foraging habitat of individuals that do not sponge (Krützen et al., 2005). This feeding technique has been passed from mother to calf (vertically) to allow an exploitation of this particular environment (Krützen et al., 2005, 2014). However, it is female calves only that acquire sponging, as male offspring of sponging mothers do not appear to show sponging in later life (Krützen et al., 2005; Mann & Sargeant, 2003). This could be due to the continued association between females and their mothers after weaning (Sargeant & Mann, 2009) and/or the lengthy time required to sponge; this time commitment may be too costly for males who must maintain alliances (Krützen et al., 2005). The vertical transmission of this foraging technique is restricted to a single matriline, making underlying genetic inheritance a potentially confounding factor (Laland & Janik, 2006). Krützen et al. (2005) investigated the likelihood of a single or multi-locus mode of inheritance. None of the 10 different modes of inheritance considered agreed with the data, particularly given the highly significant correlation between haplotype and sponging. The authors therefore concluded that the trait was unlikely to have a genetic basis. Further, a second sponging matriline has now been identified in the western section of Shark Bay, indicating sponging is a widespread foraging tactic in that region (Kopps et al., 2014). The interactions among social structure (fission–fusion), variable prey habitats in the environment, prolonged calf dependency, and high social tolerance result in correlations between calves' foraging tactics and their mothers' (Krützen et al., 2014; Sargeant & Mann, 2009). This tradition represents the first tool-use culture in marine mammals (Krützen et al., 2005, 2014), and is therefore a significant contribution to the animal culture debate.

3.4.4 Killer Whales: Dialects and Gene–Culture Coevolution

Killer whales (*Orcinus orca*) are a cosmopolitan species divided into different ecotypes based on feeding strategies (Ford et al., 1998). In British Columbia (Canada), northern resident killer whales are organized into pods that contain a number of related matrilines (Bigg et al., 1990; Ford, 1991). Each pod has a group-specific repertoire (dialect) of discrete calls (a pulsed vocalization) that function in group cohesion, communication, and identity (Ford, 1991). Within pods, subtle differences in call type structure exist among matrilines (Miller & Bain, 2000), but repertoires are generally stable through time (Ford, 1991). Call types are also shared among pods within an area, and these broader groupings are termed acoustic clans (Ford, 1991; Yurk et al., 2002). Deecke et al. (2000) found that call types could change in structure through time, and this change occurred in two different matrilines within the northern resident community. Of the two call types investigated, one decreased in similarity through time, while the other maintained similarity in structure over 13 years. This indicated that the call types underwent cultural drift and then horizontal transmission of call structure between the matrilines over the study period (Deecke et al., 2000). This ability to vocally learn and match call types in the wild has been demonstrated in controlled conditions (Abramson et al., 2018). Crance et al. (2014) found unidirectional cross-dialect learning in a captive setting over a period of six years by two juvenile males who increased their association with an unrelated male. Adult females did not change their dialects (acquire the male's calls) during the same time period. This indicates that there may be a sex or age bias in learning, as well as model bias – active selection of who to learn

from. The authors suggest that males may preferentially vocally converge if such associations improve prey capture or increase reproductive success, as has been demonstrated in bottlenose dolphins.

Young killer whales typically learn their repertoires from their mothers and matrilines (e.g., Filatova et al., 2015). However, oblique transmission of call types is at least possible (Crance et al., 2014), and horizontal transmission of call types between pods of free-ranging killer whales has been documented (Deecke et al., 2000). To understand how dialect groups may have evolved in the wild given the myriad of vocal learning strategies and potential vocal models (e.g., mother, matriline group, father), agent-based models have been used to investigate the cultural evolution of killer whale dialects (Filatova & Miller, 2015). A null model was created that simply included genetic transmission and an absence of learning; this did not produce the observed pattern of group-specific call repertoires. However, a complex mix of occasional innovation or random error, vocal divergence from related matrilines, and learning from the mother or matrilineal group produced a closer approximation of observed patterns (Filatova & Miller, 2015). This complex interaction of learning rules and the ability to produce similar outcomes (matching patterns observed in the wild) with different combinations of rules emphasizes the complex nature of such cultural evolution. To produce the discrete call types found in the wild, occasional innovation combined with random errors was required (Filatova & Miller, 2015). To produce the divergence of shared calls in related matrilines, a learning rule for a tendency to diverge from kin was required (Filatova & Miller, 2015). Innovation or higher levels of random error combined with divergence from kin produced punctuated steps in evolution toward discrete calls, as did rules to diverge from kin with random error assigned proportional to the matriline variance. There are both convergent and divergent mechanisms at work, just at different scales: within-group vocal convergence aids in group identity and cohesion, while between-group divergence may be accelerated due to sexual selection by a female preference for mates with dissimilar repertoires (Filatova & Miller, 2015). Clearly, cultural convergence, divergence, and transmission are complex, but agent-based models provide a means to produce testable hypothesizes that can be validated in controlled experiments on free-ranging whales.

Both horizontal and oblique transmission may disrupt gene–culture coevolution, as non-matrilineal transmission will dilute any correlation between genetic markers and a learned behavioral repertoire (Deecke et al., 2000). At a broader scale, however, gene–culture coevolution appears to have promoted rapid divergence among killer whale ecotypes (Foote et al., 2016). Ecotypes have different feeding specializations, allowing each to adapt to a narrow ecological niche after rapidly diversifying from a recent ancestor (Deecke et al., 2005; Foote et al., 2009; Ford et al., 1998; LeDuc et al., 2008; Riesch et al., 2012). For example, in the

North Pacific there are two sympatric ecotypes, the *resident*, fish-eating killer whales and the *transient*, mammal-eating killer whales (Ford et al., 1998). This separation is stable across the North Pacific in multiple subpopulations. Despite sympatry of the ecologically divergent ecotypes, their cultural differences in the form of learned feeding behaviors and repertoires result in sufficient reproductive isolation to create a situation of incipient speciation (Riesch et al., 2012). Given the matrilineal structure of pods, the cultural transmission of behavioral traits (feeding and vocal), and male-mediated gene flow, the interaction between ecological and behavioral variation and genome-level evolution can be investigated (Foote et al., 2016). Foote et al. (2016) found that allele frequencies have drifted apart among ecotypes, indicating that ecotypes have been mating assortatively and that reproductive isolation occurs quickly after the formation of a new ecotype. Genetic isolation of small, matrilineal founder groups (through bottleneck events) from an ancestral population and subsequent population and geographic expansion and strong genetic drift created these ecotypes (Foote et al., 2016). Even in sympatry, gene flow occurs almost exclusively within ecotypes; therefore, these ecological categories predict genetic structure better than does geography (Foote et al., 2016). To further investigate gene–culture coevolution, a number of genes involved in cold-water adaption processes (skin and adipose tissue development) were identified through comparing temperate North Pacific ecotypes to Antarctic ecotypes (Foote et al., 2016). Further to this, selection for genes that encode proteins associated with dietary variation (mammal vs. fish eating) and also several genes that encode proteins involved in reproduction with functional roles in fertilization may be candidate genes for postzygotic reproductive isolation (Foote et al., 2016). Both genetic drift and selection have promoted genome-wide shifts in the frequency of alternative alleles in each ecotype from initial founder effects, rapid reproductive isolation, and subsequent population expansion (Foote et al., 2009, 2016; LeDuc et al., 2008). The culturally inherited foraging tactics and ecological niches that flow from them have prompted significant evolutionary changes in killer whale ecotypes, as shown by the divergence in genes associated with environment, reproductive isolation, and diet. The ability of killer whales to adapt to novel habitats and ecological niches is a result of behavioral flexibility and the social learning of new niches through the matrilineal group (Riesch et al., 2012). These differences are then reinforced by shared vocal repertoires that provide group (vertical transmission) and clan (oblique and horizontal transmission) identity, which themselves evolve culturally together through time (Deecke et al., 2000).

3.5 FUTURE DIRECTIONS

We have presented evidence of a diverse array of culturally transmitted traits in a number of cetacean species. The aquatic environment has shaped their evolutionary trajectory in a very different direction to terrestrial mammals, but the existence of such traits in a mammalian group so distinct from our own primate lineage presents an important opportunity for comparative research. Despite a lack of hands for manipulating tools, cetaceans clearly display cultural traditions. The rich diversity of vocal culture is abundantly clear, with the occasional manipulation of objects (tools) also being present. Instead of using hands, other appendages (e.g., beak) are utilized to manipulate tools. As noted above, a subset of the Shark Bay bottlenose dolphin population forages using sponges. This contributes to the ever-increasing evidence of tool use in nonhuman primates (e.g., Whiten, 2015) and birds (e.g., Rutz et al., 2010).

If we think about the evolution of human culture and look back to important advances in our past, the manipulation of fire is somewhat prominent – some have even suggested it was instrumental to the advent of modern humans (Wrangham, 2009). Fire is not a marine phenomenon. If we are to truly have a comparative understanding of culture and the roots of human language, in our view we must let go of such terrestrial constructs and find analogous measures in the aquatic environment. Perhaps one measure that has been used extensively by humans to develop speech, vocal production learning, where the acoustic structure of a signal can be modified following experience with the signals of other individuals (Janik & Slater, 2000), may be more relevant to cetaceans. This rare ability to modify sound represents an important increase in evolutionary complexity in a communication system (Janik, 2014). Vocal production learning has been documented in bottlenose dolphins, Risso's dolphins (*Grampus griseus*; Favaro et al., 2016), beluga whales (*Delphinapterus leucas*; Ridgway et al., 2012), killer whales (Crance et al., 2014; Foote et al., 2006), and humpback whales (Garland et al., 2011; Noad et al., 2000), while a number of other species require more data to definitively assess this, but show such variation in their vocal repertoires as to suggest they are very likely vocal learners. Given the reliance on vocal communication in cetaceans and the rarity of this complex ability in other mammals (Janik & Slater, 1997), vocal production learning represents a robust domain for exploring comparative analyses without the constraints of a particular terrestrial or aquatic locale.

We have shown that vocal cultures along with novel foraging traditions provide strong evidence for cultural phenomena in the Cetacea. The study of cetaceans, however, is far from easy. These large aquatic mammals are not particularly amenable to captivity, and their oceanic lifestyle results in few observations, as they are rarely at the surface and are not always located close to land. The study of cetaceans is expensive, time-consuming, and a battle of perseverance, but the outcomes of such studies have provided unprecedented and unique examples of nonhuman cultural phenomena. Just because it is difficult does not mean it is not worth the effort. Investigations into gene–culture coevolution and cultural hitchhiking in sperm whales and killer whales are proving highly informative in understanding comparative mechanisms, and they present an exciting avenue for future studies that should include other matrilineal cetaceans (e.g., pilot whales [*Globicephala melas* and *Globicephala macrorhynchus*]). On the other hand, playback experiments are required in a number of species to understand the context in which certain behaviors are displayed. For example, in sperm whales, playback experiments are needed to understand the social and behavioral context in which particular coda types are exchanged (Gero et al., 2016b). Creativity (not least in identifying funding sources) is required when designing experiments to test patterns of cultural transmission. We advocate the use of computer models such as agent-based modeling of known phenomena to test various social learning scenarios (e.g., Filatova & Miller, 2015). Once a plausible learning scenario has been identified, this can be tested in targeted field experiments. The undertaking of large-scale field experiments on cetaceans is challenging in itself; having a model to test will hopefully reduce the uncertainty in such social learning experiments. For example, a novel song type could be seeded into a population of humpback whales through playback in the environment and the spread of the song throughout the population recorded. This does not address the important ethical question of whether we *should* manipulate nonhuman cultures in this way. Such a manipulation would not just affect the seeded population; the novel, seeded song would likely spread through multiple populations and may very well spread into another ocean basin. Careful consideration of what "controlled" experiments may actually do before letting them loose in the environment is required.

The diversity of cetaceans and their social structure, ecology, and phylogeny has resulted in a large diversity of vocal abilities, cognition, memory, foraging tactics, and culture. The examples of gene–culture coevolution provide rich grounds for comparative analyses with human language and similar punctuated steps in vocal evolution. While there is no single answer for how culture is displayed in this diverse order, this diversity in cultural traits is the key. These contrasts, from signature whistles to song revolutions and tool-assisted foraging, provide a myriad of peaks in the evolution and precursors of complex culture, which provide a rich foundation for comparative studies to understand the evolutionary roots of human language and culture.

3.6 ACKNOWLEDGMENTS

ECG was supported by a Royal Society Newton International Fellowship and LR was supported by the Marine Alliance for Science and Technology for Scotland (MASTS) pooling initiative, and their support is gratefully acknowledged. MASTS is funded by the Scottish Funding Council (grant reference HR09011) and contributing institutions.

REFERENCES

Abramson J. Z., Hernández-Lloreda M. V., García L., et al. (2018). Imitation of novel conspecific and human speech sounds in the killer whale (Orcinus orca). *Proceedings of the Royal Society B: Biological Sciences*, **285**, 20172171.

Allen, J., Weinrich, M., Hoppitt, W., & Rendell, L. (2013). Network-based diffusion analysis reveals cultural transmission of lobtail feeding in humpback whales. *Science*, **340**, 485–488.

Antunes, R., Schulz, T., Gero, S., et al. (2011). Individually distinctive acoustic features in sperm whale codas. *Animal Behaviour*, **81**, 723–730.

Berta, A., & Sumich, J. L. (2006). *Marine Mammals: Evolutionary Biology*, 2nd ed. Burlington, MA: Academic Press.

Bigg, M. A., Olesiuk, P. F., Ellis, G. M., Ford, J. K. B., & Balcomb III, K. C. B. (1990). Organization and genealogy of resident killer whales (*Orcinus orca*) in the coastal waters of British Columbia and Washington State. *Reports of the International Whaling Commission*, **12**, 383–405.

Bruck, J. N. (2013). Decades-long social memory in bottlenose dolphins. *Proceedings of the Royal Society B: Biological Sciences*, **280**, 20131726.

Caldwell, M. C., & Caldwell, D. K. (1965). Individualized whistle contours in bottlenosed dolphins (*Tursiops truncatus*). *Nature*, **207**, 434–435.

Cato, D. H. (1991). Songs of humpback whales: The Australian perspective. *Memoirs of the Queensland Museum*, **30**, 277–290.

Christal, J., Whitehead, H., & Lettevall, E. (1998). Sperm whale social units: Variation and change. *Canadian Journal of Zoology*, **76**, 1431–1440.

Clapham, P. (2000). The humpback whale: Seasonal feeding and breeding in a baleen whale. In J. Mann, R. C. Connor, P. L. Tyack, & H. Whitehead, eds., *Cetacean Societies: Field Studies of Dolphins and Whales*. Chicago, IL: University of Chicago Press, pp. 173–196.

Connor, R. C., & Krützen, M. (2015). Male dolphin alliances in Shark Bay. Changing perspectives in a 30-year study. *Animal Behaviour*, **103**, 223–235.

Connor, R. C., Smolker, R. A., & Richards, A. F. (1992). Two levels of alliance formation among male bottlenose dolphins (*Tursiops* sp.). *Proceedings of the National Academy of Sciences*, **89**, 987–990.

Connor, R. C., Wells, R. S., Mann, J., & Read, A. J. (2000). The bottlenose dolphin: Social relationships in a fission-fusion society. In J. Mann, R. C. Connor, P. L. Tyack, & H. Whitehead, eds., *Cetacean Societies: Field Studies of Dolphins and Whales*. Chicago, IL: University of Chicago Press, pp. 91–126.

Crance, J. L., Bowles, A. E., & Garver, A. (2014). Evidence for vocal learning in juvenile male killer whales, *Orcinus orca*, from an adventitious cross-socializing experiment. *Journal of Experimental Biology*, **217**, 1229–1237.

Dalebout, M. L., Mead, J. G., Baker, C. S., Baker, A. N., & Helden, A. L. (2002). A new species of beaked whale Mesoplodon perrini sp. n. (Cetacea: Ziphiidae) discovered through phylogenetic analyses of mitochondrial DNA sequences. *Marine Mammal Science*, **18**, 577–608.

Darling, J. D., Jones, M. E., & Nicklin, C. P. (2006). Humpback whale songs: Do they organize males during the breeding season? *Behaviour*, **143**, 1051–1101.

Deecke, V. B., Ford, J. K. B., & Spong, P. (2000). Dialect change in resident killer whales: Implications for vocal learning and cultural transmission. *Animal Behaviour*, **40**, 629–638.

Deecke, V. B., Ford, J. K. B., & Slater, P. J. B. (2005). The vocal behaviour of mammal-eating killer whales: Communicating with costly calls. *Animal Behaviour*, **69**, 395–405.

Favaro, L., Neves, S., Furlati, S., et al. (2016). Evidence suggests vocal production learning in a cross-fostered Risso's dolphin (Grampus griseus). *Animal Cognition*, **19**, 847–853.

Filatova, O. A., & Miller, P. J. O. (2015). An agent-based model of dialect evolution in killer whales. *Journal of Theoretical Biology*, **373**, 82–91.

Filatova, O. A., Samarra, F. I. P., Deecke, V. B., et al. (2015). Cultural evolution of killer whale calls: Background, mechanisms and consequences. *Behaviour*, **152**, 2001–2038.

Foote, A. D., Griffin, R. M., Howitt, D., et al. (2006). Killer whales are capable of vocal learning. *Biology Letters*, **2**, 509–512.

Foote, A. D., Newton, J., Piertney, S. B., Willerslev, E., & Gilbert, M. T. P. (2009). Ecological, morphological and genetic divergence of sympatric North Atlantic killer whale populations. *Molecular Ecology*, **18**, 5207–5217.

Foote, A. D., Vijay, N., Ávila-Arcos, M. C., et al. (2016). Genome–culture coevolution promotes rapid divergence of killer whale ecotypes. *Nature Communications*, **7**, 11693.

Ford, J. K. B. (1991). Vocal traditions among resident killer whales (*Orcinus orca*) in coastal waters of British Columbia. *Canadian Journal of Zoology*, **69**, 1454–1483.

Ford, J. K. B., Ellis, G. M., Barrett-Lennard, L. G., et al. (1998). Dietary specialization in two sympatric populations of killer whales (*Orcinus orca*) in coastal British Columbia and adjacent waters. *Canadian Journal of Zoology*, **76**, 1456–1471.

Fragaszy, D., & Perry, S. (2003). Towards a biology of traditions. In D. Fragaszy, & S. Perry, eds., *The Biology of Traditions: Models and Evidence*. Cambridge, UK: Cambridge University Press, pp. 1–32.

Frère, C. H., Krutzen, M., Mann, J., et al. (2010). Social and genetic interactions drive fitness variation in a free-living dolphin population. *Proceedings of the National Academy of Sciences*, **107**, 19949–19954.

Frumhoff, P. (1983). Aberrant songs of humpback whales (*Megaptera novaeangliae*): Clues to the structure of humpback songs. In R. Payne, ed., *Communication and Behavior of Whales*. Boulder, CO: Westview Press, pp. 81–127.

Galef, B. G. (2009). Culture in animals? In K. N. Laland & B. G. Galef, eds., *The Question of Animal Culture*. Cambridge, MA: Harvard University Press, pp. 222–246.

Garland, E. C., Goldizen, A. W. W., Rekdahl, M. L. L., et al. (2011). Dynamic horizontal cultural transmission of humpback whale song at the ocean basin scale. *Current Biology*, **21**, 687–691.

Garland, E. C., Lilley, M. S., Goldizen, A. W., et al. (2012). Improved versions of the Levenshtein distance method for comparing sequence information in animals' vocalisations: Tests using humpback whale song. *Behaviour*, **149**, 1413–1441.

Garland, E. C., Gedamke, J., Rekdahl, M. L., et al. (2013a). Humpback whale song on the southern ocean feeding grounds: Implications for cultural transmission. *PLoS ONE*, **8**, e79422.

Garland, E. C., Noad, M. J., Goldizen, A. W., et al. (2013b). Quantifying humpback whale song sequences to understand the dynamics of song exchange at the ocean basin scale. *Journal of the Acoustical Society of America*, **133**, 560–569.

Garland, E. C., Goldizen, A. W., Lilley, M. S., et al. (2015). Population structure of humpback whales in the western and central South Pacific Ocean as determined by vocal exchange among populations. *Conservation Biology*, **29**, 1198–1207.

Gero, S., Bøttcher, A., Whitehead, H., & Madsen, P. T. (2016a). Socially segregated, sympatric sperm whale clans in the Atlantic Ocean. *Royal Society Open Science*, **3**, 160061.

Gero, S., Whitehead, H., & Rendell, L. (2016b). Individual, unit and vocal clan level identity cues in sperm whale codas. *Royal Society Open Science*, **3**, 150372.

Glockner, D. A. (1983). Determining the sex of humpback whales (*Megaptera novaeangliae*) in their natural environment. In R. Payne, ed., *Communication and Behavior of Whales*. Boulder, CO: Westview Press, pp. 447–464.

Hain, J. H. W., Carter, G. R., Kraus, S. D., Mayo, C. A., & Winn, H. E. (1982). Feeding behavior of the humpback whale, *Megaptera novaeangliae*, in the Western North Atlantic. *Fishery Bulletin*, **80**, 259–268.

Herman, L. M., & Tavolga, W. N. (1980). The communication systems of cetaceans. In L. M. Herman, ed., *Cetacean Behaviour: Mechanisms and Functions*. New York: Wiley-Interscience, pp. 149–209.

Janik, V. M. (2000). Whistle matching in wild bottlenose dolphins (*Tursiops truncatus*). *Science*, **289**, 1355–1357.

Janik, V. M. (2009). Acoustic communication in delphinids. *Advances in the Study of Behavior*, **40**, 123–157.

Janik, V. M. (2014). Cetacean vocal learning and communication. *Current Opinion in Neurobiology*, **28**, 60–65.

Janik, V. M., & Slater, P. J. B. (1997). Vocal learning in mammals. *Advances in the Study of Behavior*, **26**, 59–99.

Janik, V. M., & Slater, P. J. B. (1998). Context-specific use suggests that bottlenose dolphin signature whistles are cohesion calls. *Animal Behaviour*, **56**, 829–838.

Janik, V. M., & Slater, P. J. B. (2000). The different roles of social learning in vocal communication. *Animal Behaviour*, **60**, 1–11.

Janik, V. M., Sayigh, L. S., & Wells, R. S. (2006). Signature whistle shape conveys identity information to bottlenose dolphins. *Proceedings of the National Academy of Sciences*, **103**, 8293–8297.

King, S. L., & Janik, V. M. (2013). Bottlenose dolphins can use learned vocal labels to address each other. *Proceedings of the National Academy of Sciences*, **110**, 13216–13221.

King, S. L., Sayigh, L. S., Wells, R. S., Fellner, W., & Janik, V. M. (2013). Vocal copying of individually distinctive signature whistles in bottlenose dolphins. *Proceedings of the Royal Society B: Biological Sciences*, **280**, 20130053.

King, S. L., Harley, H. E., & Janik, V. M. (2014). The role of signature whistle matching in bottlenose dolphins, *Tursiops truncatus*. *Animal Behaviour*, **96**, 79–86.

King, S. L., Guarino, E., Keaton, L., Erb, L., & Jaakkola, K. (2016). Maternal signature whistle use aids mother–calf reunions in a bottlenose dolphin, *Tursiops truncatus*. *Behavioural Processes*, **126**, 64–70.

King, S. L., Friedman, W. R., Allen, S. J., et al. (2018). Bottlenose dolphins retain individual vocal labels in multi-level alliances. *Current Biology*, **28**, 1993–1999.e3.

Kopps, A. M., Ackermann, C. Y., Sherwin, W. B., et al. (2014). Cultural transmission of tool use combined with habitat specializations leads to fine-scale genetic structure in bottlenose dolphins. *Proceedings of the Royal Society B: Biological Sciences*, **281**, 20133245.

Krützen, M., Mann, J., Heithaus, M. R., et al. (2005). Cultural transmission of tool use in bottlenose dolphins. *Proceedings of the National Academy of Sciences*, **102**, 8939–8943.

Krützen, M., Kreicker, S., Macleod, C. D., et al. (2014). Cultural transmission of tool use by Indo-Pacific bottlenose dolphins

(*Tursiops* sp.) provides access to a novel foraging niche. *Proceedings of the Royal Society B: Biological Sciences*, **281**, 20140374.

Laland, K. N. (1992). A theoretical investigation of the role of social transmission in evolution. *Ethology and Sociobiology*, **13**, 87–113.

Laland, K. N., & Galef, B. G. (2009). *The Question of Animal Culture*. Cambridge, MA: Harvard University Press.

Laland, K. N., & Janik, V. M. (2006). The animal cultures debate. *Trends in Ecology & Evolution*, **21**, 542–547.

Laland, K. N., Kendal, J. R., & Kendal, R. L. (2009). Animal culture: Problems and solutions. In K. N. Laland & B. G. Galef, eds., *The Question of Animal Culture*. Cambridge, MA: Harvard University Press, pp. 174–197.

LeDuc, R. G., Robertson, K. M., & Pitman, R. L. (2008). Mitochondrial sequence divergence among Antarctic killer whale ecotypes is consistent with multiple species. *Biology Letters*, **4**, 426–429.

Mann, J., & Sargeant, B. (2003). Like mother, like calf: The ontogeny of foraging traditions in wild Indian Ocean bottlenose dolphins (*Tursiops* sp.). In D. Fragaszy & S. Perry, eds., *The Biology of Traditions: Models and Evidence*. Cambridge, UK: Cambridge University Press, pp. 236–266.

Miller, P. J. O., & Bain, D. E. (2000). Within-pod variation in the sound production of a pod of killer whales, *Orcinus orca*. *Animal Behaviour*, **60**, 617–628.

Morin, P. A., Scott Baker, C., Brewer, R. S., et al. (2017). Genetic structure of the beaked whale genus Berardius in the North Pacific, with genetic evidence for a new species. *Marine Mammal Science*, **33**, 96–111.

Noad, M. J., Cato, D. H., Bryden, M. M., et al. (2000). Cultural revolution in whale songs. *Nature*, **408**, 537.

Oliveira, C., Wahlberg, M., Silva, M. A., et al. (2016). Sperm whale codas may encode individuality as well as clan identity. *Journal of the Acoustical Society of America*, **139**, 2860–2869.

Payne, K., & Payne, R. (1985). Large scale changes over 19 years in songs of humpback whales in Bermuda. *Zeitschrift für Tierpsychologie*, **68**, 89–114.

Payne, K., Tyack, P. L., & Payne, R. (1983). Progressive changes in the songs of humpback whales (*Megaptera novaeangliae*): A detailed analysis of two seasons in Hawaii. In R. Payne, ed., *Communication and Behavior of Whales*. Boulder, CO: Westview Press, pp. 9–57.

Payne, R., & Guinee, L. N. (1983). Humpback whale, *Megaptera novaeangliae*, songs as an indicator of "stocks". In R. Payne, ed., *Communication and Behavior of Whales*. Boulder, CO: Westview Press, pp. 333–358.

Payne, R. S., & McVay, S. (1971). Songs of humpback whales. *Science*, **173**, 587–597.

Rendell, L. E., & Whitehead, H. (2001). Culture in whales and dolphins. *Behavioral and Brain Sciences*, **24**, 309–382.

Rendell, L. E., & Whitehead, H. (2003). Vocal clans in sperm whales (*Physeter macrocephalus*). *Proceedings of the Royal Society B: Biological Science*, **270**, 225–231.

Rendell, L., Mesnick, S. L., Dalebout, M. L., Burtenshaw, J., & Whitehead, H. (2012). Can genetic differences explain vocal dialect variation in sperm whales, *Physeter macrocephalus*? *Behavior Genetics*, **42**, 332–343.

Richards, D. G., Wolz, J. P., & Herman, L. M. (1984). Vocal mimicry of computer-generated sounds and vocal labelling of objects by bottlenose dolphins (*Tursiops truncatus*): Evidence for vocal learning. *Journal of Comparative Psychology*, **98**, 10–28.

Ridgway, S., Carder, D., Jeffries, M., & Todd, M. (2012). Spontaneous human speech mimicry by a cetacean. *Current Biology*, **22**, R860–R861.

Riesch, R., Barrett-Lennard, L. G., Ellis, G. M., Ford, J. K. B., & Deecke, V. B. (2012). Cultural traditions and the evolution of reproductive isolation: Ecological speciation in killer whales? *Biological Journal of the Linnean Society*, **106**, 1–17.

Rutz, C., Bluff, L. A., Reed, N., et al. (2010). The ecological significance of tool use in New Caledonian crows. *Science*, **329**, 1523–1526.

Sargeant, B. L., & Mann, J. (2009). From social learning to culture: Intrapopulation variation in bottlenose dolphins. In K. N. Laland & B. G. Galef, eds., *The Question of Animal Culture*. Cambridge, MA: Harvard University Press, pp. 152–173.

Sayigh, L. S., Tyack, P. L., Wells, R. S., & Scott, M. D. (1990). Signature whistles of free-ranging bottlenose dolphins *Tursiops truncatus*: Stability and mother–offspring comparisons. *Behavioral Ecology and Sociobiology*, **26**, 247–260.

Sayigh, L. S., Tyack, P. L., Wells, R. S., Scott, M. D., & Irvine, A. B. (1995). Sex difference in signature whistle production of free-ranging bottlenose dolphins, *Tursiops truncatus*. *Behavioral Ecology and Sociobiology*, **36**, 171–177.

Schulz, T. M., Whitehead, H., Gero, S., & Rendell, L. (2011). Individual vocal production in a sperm whale (*Physeter macrocephalus*) social unit. *Marine Mammal Science*, **27**, 149–166.

Smith, J. N., Goldizen, A. W., Dunlop, R. A., & Noad, M. J. (2008). Songs of male humpback whales, *Megaptera novaeangliae*, are involved in intersexual interactions. *Animal Behaviour*, **76**, 467–477.

Stimpert, A. K., Wiley, D. N., Au, W. W. L., Johnson, M. P., & Arsenault, R. (2007). "Megapclicks": Acoustic click trains and buzzes produced during night-time foraging of humpback whales (*Megaptera novaeangliae*). *Biology Letters*, **3**, 467–470.

Thewissen, J. G. M., Cooper, L. N., Clementz, M. T., Bajpai, S., & Tiwari, B. N. (2007). Whales originated from aquatic artiodactyls in the Eocene epoch of India. *Nature*, **450**, 1190–1194.

Tomasello, M. (2009). The question of chimpanzee culture, plus postscript (chimpanzee culture, 2009). In K. N. Laland & B. G. Galef, eds., *The Question of Animal Culture*. Cambridge, MA: Harvard University Press, pp. 198–221.

Watkins, W. A., & Schevill, W. E. (1977). Sperm whale codas. *Journal of the Acoustical Society of America*, **62**, 1486–1490.

Watwood, S. L., Tyack, P. L., & Wells, R. S. (2004). Whistle sharing in paired male bottlenose dolphins, *Tursiops truncatus*. *Behavioral Ecology and Sociobiology*, **55**, 531–543.

Whitehead, H. (1998). Cultural selection and genetic diversity in matrilineal whales. *Science*, **282**, 1708–1711.

Whitehead, H. (2003). *Sperm Whales: Social Evolution in the Ocean*. Chicago, IL: University of Chicago Press.

Whitehead, H. (2005). Genetic diversity in the matrilineal whales: Models of cultural hitchhiking and group-specific non-heritable demographic variation. *Marine Mammal Science*, **21**, 58–79.

Whitehead, H. (2009). How might we study culture: A perspective from the ocean. In K. N. Laland & B. G. Galef, eds., *The Question of Animal Culture*. Cambridge, MA: Harvard University Press, pp. 125–151.

Whitehead, H., & Rendell, L. (2014). *The Cultural Lives of Whales and Dolphins*. Chicago, IL: University of Chicago Press.

Whitehead, H., & Weilgart, L. (1991). Patterns of visually observable behaviour and vocalizations in groups of female sperm whales. *Behaviour*, **118**, 275–296.

Whitehead, H., & Weilgart, L. (2000). The sperm whale: Social females and roving males. In J. Mann, R. C. Connor, P. L. Tyack, & H. Whitehead, eds., *Cetacean Societies: Field Studies of Dolphins and Whales*. Chicago, IL: University of Chicago Press, pp. 154–172.

Whitehead, H., Dillon, M., Dufault, S., Weilgart, L., & Wright, J. (1998). Non-geographically based population structure of South Pacific sperm whales: Dialects, fluke-markings and genetics. *Journal of Animal Ecology*, **67**, 253–262.

Whitehead, H., Antunes, R., Gero, S., et al. (2012). Multilevel societies of female sperm whales (*Physeter macrocephalus*) in the Atlantic and Pacific: Why are they so different? *International Journal of Primatology*, **33**, 1142–1164.

Whiten, A. (2009). The identification and differentiation of culture in chimpanzees and other animals: From natural history to diffusion experiments. In K. N. Laland & B. G. Galef, eds., *The Question of Animal Culture*. Cambridge, MA: Harvard University Press, pp. 99–124.

Whiten, A. (2015). Experimental studies illuminate the cultural transmission of percussive technologies in *Homo* and *Pan*. *Philosophical Transactions of the Royal Society of London B: Biological Science*, **370**, 169–177.

Winn, H. E., & Winn, L. K. (1978). The song of the humpback whale *Megaptera novaeangliae* in the West Indies. *Marine Biology*, **47**, 97–114.

Wrangham, R. W. (2009). *Catching Fire: How Cooking Made Us Human*. London: Profile Books Ltd.

Yurk, H., Barrett-Lennard, L., Ford, J. K. B., & Matkin, C. O. (2002). Cultural transmission within maternal lineages: Vocal clans in resident killer whales in southern Alaska. *Animal Behaviour*, **63**, 1103–1119.

PART II

Sociocultural anthropology is a heavily politicized field and has tended (since the 1970s) to either ignore or attack evolution-based efforts to understand human nature, cultural capacity, and culture itself. Efforts by evolutionists to reach out to these fields (e.g., Barkow, 2006) have met with little immediate success, though it is possible that opposition is waning as evolutionist anthropologists (e.g., Salazar, 2018) continue to write about issues that one would expect to be of interest to all anthropologists. However, as Buss and von Hippel (2018) have argued, aspects of our evolved psychology, such as the desire to defend prestige, ironically may be serving to prevent acceptance of evolutionary thought.

Evolutionist anthropologists, including the three contributors to Part II, have made major contributions to our understanding of human universals, the psychological infrastructure of our culturally patterned behavior, and the interaction between biological and cultural evolution. With the exception of the subfield of psychological anthropology, mainstream sociocultural anthropology in general still tends to ignore psychological processes. Perhaps this is why even anthropologists who do take an evolutionary perspective may pay relatively little attention to the evolved psychology underlying culturally patterned behavior (though there are exceptions, including Daniel Fessler and Jerome Barkow). Thus, it is not surprising that Betzig does not mention male evolved psychology in her chapter. Fortunately, this is hardly a problem: her data are entirely compatible with evolutionary psychology's findings about male sexuality (e.g., see Chapter 45 in this handbook; Barkow, 1981; Symons, 1979).

Betzig discusses eusociality and introduces us to a number of exceptionally cooperative species before focusing on the range of variation in cooperation in human societies. Much of her chapter reviews how, while our ancestral foraging societies were somewhat polygynous, with the most successful hunters having two or three wives, polygyny exploded with agriculture and early empires. She combs through the Bible, Herodotus, and numerous other ancient records to document how

powerful men have had sexual access to enormous numbers of women. Often these men were served by large numbers of eunuchs (she does not discuss whether the eunuchs were also sexually used, though her accounts frequently mention that they were often chosen from among the best-looking boys).

Chapais explicitly links evolved psychology to cultural practices. He does this in order to account for the numerous universals and near universals found by Donald Brown (1991) (and more recently discussed exhaustively by Christoph Antweiler, 2016). Perhaps the time is ending when it took courage for sociocultural anthropologists like Antweiler, Brown, and Chapais to write about universals; their highly politicized discipline has tended to see the subject as trivial (Geertz, 1965; Lévi-Strauss, 1983) or, worse, as the neocolonial imposition of Western concepts on the non-Western peoples of the world. (Alternatively, perhaps Antweiler, Brown, and Chapais are simply courageous scholars.) Certainly, the important subject of cross-cultural universals and near universals has been neglected by anthropologists (as Chapais points out). Chapais meets the various criticisms of the study of universals by focusing not on alphabetical lists, but on the underlying processes, including evolutionary psychology processes. Some universals interact with other universals, generating additional categories of universals. For example, his "fifth class comprises basic *adaptive processes regulating social relationships*, such as kin altruism, mutualistic cooperation, status competition, and the female bias in child care." These presumably are "genetically specified." His seventh class of social universals, "biparental families, lineage structures, status orders – are subsumed under the label of *consistent social arrangements* ..." "Accordingly, understanding the origins of social arrangements involves understanding how the aggregation of multiple individual social acts, social interactions, and social relationships translate into higher-level structural regularities, which exhibit their own properties ..." He gives us 10 classes of social interactions, the first six of which have psychological underpinnings. His system is somewhat complex, but leads to numerous insights. For example, he writes that: "The occurrence of absolute universals … is conditional upon features of the socioecological environment, but the latter have the particularity of having a panhuman distribution. In general, those features are found in all societies because they reflect basic properties of the physical world, of the human body, or of human groups."

Mesoudi begins with lucid discussions of the concept of culture and of the various schools of thought in the evolutionary human behavioral sciences and how these differ and overlap. There has been a long-standing tension between evolutionary psychology and its precursors on the one hand and those who view cultural behavior as adaptation without specifying the mediating evolved psychological mechanisms on the other (see Barkow, 1989, for a discussion). Somewhat related to the latter is the school of thought that is primarily interested in culture as an evolving system ("cultural evolutionists" or "gene–culture coevolution"). Mesoudi's discussion of "content biases" is a bridge between gene–culture coevolution and evolutionary psychology because a bias is in effect an evolved psychological mechanism. Mesoudi emphasizes the value of all approaches and is always positive, even when he discusses the politicization of mainstream sociocultural anthropology and sociology.

REFERENCES

Antweiler, C. (2016). *Our Common Denominator: Human Universals Revisited* (D. Kerns, Trans.). Oxford/New York: Berghahn Books.

Barkow, J. H. (1981). Evolution et sexualité humaine. In C. Crépault, J. Lévy, & H. Gratton, eds., *Sexologie Contemporaine*. Quebec: Les Presses de l'Université du Québec, pp. 103–118.

Barkow, J. H. (1989). The elastic between genes and culture. *Ethology and Sociobiology*, **10**(1–3), 111–129.

Barkow, J. H. (2006). Sometimes the bus does wait. In J. H. Barkow, ed., *Missing the Revolution: Darwinism for Social Scientists*. New York: Oxford University Press, pp. 3–59.

Boyd, R., & Richerson, P. J. (1985). *Culture and the Evolutionary Process*. Chicago, IL: University of Chicago Press.

Brown, D. E. (1991). *Human Universals*. New York: McGraw-Hill.

Buss, D. M, & von Hippel, W. (2018). Psychological barriers to evolutionary psychology: Ideological bias and coalitional adaptations. *Archives of Scientific Psychology*, **6**(1), 148–158.

Geertz, C. (1965). The impact of the concept of culture on the concept of man. In J. R. Platt, ed., *New Views of the Nature of Man*. Chicago, IL: University of Chicago Press, pp. 93–118.

Lévi-Strauss, C. (1983). *Le Regard Éloigné*. Paris: Plon.

Salazar, C. (2018). *Explaining Human Diversity: Cultures, Minds, Evolution*. London: Routledge.

Symons, D. (1979). *The Evolution of Human Sexuality*. New York: Oxford University Press.

4 Eusociality in Humans

LAURA BETZIG

It is difficult for many people to imagine that an individual's role in evolution is entirely contained in its contribution to vital statistics. It is difficult to imagine that an acceptable moral order could arise from vital statistics, and difficult to dispense with belief in a moral order in living nature. It is difficult to imagine that the blind play of the genes could produce man.

– George C. Williams (1966, p. 4)

At one end of the eusociality continuum lie members of the genus *Atta*, the leafcutter ants. Across the American tropics, surrounded by the forests, grasslands, and pastures they harvest, these ants live in enormous subterranean nests, where they grow fungus. Older, bigger, "major" workers come and go along trails outside, where they forage for plants used as fungus substrates or specialize as soldiers in nest defense; younger, smaller, "minor" workers stay on the nest, where they tend fungus gardens and feed brood. At the core of every *Atta* colony is a single queen, who lays an average of 20 eggs every minute, or around 28,800 eggs every day, or more than 10,000,000 eggs every year. Over the course of her approximately 15-year lifespan, she may produce as many as 150–200 million young. Workers occasionally lay trophic eggs that are fed to the queen; and rarely, their rudimentary ovaries produce male larvae and pupae after a queen is lost. But the vast majority of *Atta* workers never get a chance to reproduce: they live out their lives as parts of a sterile caste (Dijkstra & Boomsma, 2006; Hölldobler & Wilson, 2009, 2011).

Acorn woodpeckers sit at the other end of the eusociality continuum, as cooperative breeders. From northern South America to the North American west coast, *Melanerpes formicivorus* live in extended family groups made up of one or more breeding females, one or more breeding males, and as many as 10 nonbreeding adults who work as helpers at the nest. Both breeders and nonbreeders store acorns, maintain the hollowed-out branches and trees they use as acorn granaries, defend their territories, and feed the brood. But breeders often inflict costs. Dominant males father over three times as many young as their subordinates, and co-breeder females throw more than a third of each other's eggs out of the nest. Bigger groups fledge more offspring, but individuals on average do worse: the ratio of helpers to breeders goes up with the size of the group. Per capita, the number of fledglings produced goes down as group size goes up; per capita, pairs fledge more young than any other groups (Koenig & Mumme, 1987; Haydock & Koenig, 2002).

Homo sapiens societies span the eusociality range. For the more than 100,000 years of our prehistory, most of us lived in small groups and got help from our grandparents, fathers, mothers, and older sisters and brothers. Reproduction was often delayed or depressed in those who helped: in human children, the onset of direct reproduction was late; and in human grandmothers, the end of direct reproduction was early. But as a result, successful mothers were able to give birth to more children, at shorter intervals, with better rates of survival; and successful fathers had children by two or three wives. Reproductive variance was low. We were cooperative breeders (see Crespi, 2014; Foster & Ratneiks, 2005 on convergent evolution).

After we left Africa and settled into the Near East, reproductive variance went up. For the nearly 10,000 years of our history, many of us lived in large groups, and were worked for by often unrelated – and often nonreproductive – soldiers and slaves. Those workers were often facultatively or obligately sterile as celibates or eunuchs. Successful mothers continued to raise roughly a dozen children, but successful fathers raised hundreds of children, by hundreds of women. That made us eusocial, or "truly" social (on terms, see Sherman et al., 1995; Wilson, 1975).

4.1 COOPERATIVE BREEDERS

No woman is an island. All of us live in groups. And unlike the vast majority of mothers across taxa, women with children get help. Children in hunter–gatherer societies get help from their grandparents (Hawkes & Coxworth, 2013), from their fathers and fathers' relatives (Kaplan et al., 2000), and from older sisters and brothers (Kramer, 2005). But who helps, when they help, how they help, and how much they help can all vary a lot (Bentley & Mace, 2009; Hewlett &

Lamb, 2005; Hrdy, 1999, 2009; Sear & Mace, 2008; Strassmann & Kurapati, 2010; Turke, 1988; Voland, Chasiotis, & Schiefenhövel, 2005).

Some children get help from their sisters and brothers. Pumé foragers, who live on the savannas of the Orinoco Basin in Venezuela, hunt, fish, and collect mangos and other fruits and roots. Few, if any, children cover their own costs, but most contribute to family subsistence to some extent. Pumé girls aged 3–6 contribute 12.5 percent as much as an average adult, girls aged 7–10 contribute 31 percent, and girls aged 11–14 contribute around 60 percent, when both foraging work and domestic work (cooking and cleaning, water and firewood hauling, and so on) are counted. Those ratios are higher in other subsistence cultures, especially among farmers (Kramer & Greaves, 2011; compare Bird & Bird, 2002, on the Meriam of the Torres Strait in Australia).

Some children get help from their grandmothers and grandfathers. The Hadza of Tanzania, who live around Lake Eyasi in the Central Rift Valley, hunt and gather meat, fruit, honey, and *llekwa*, or *Vigna frutescens*, roots. Grandmothers are very much net producers. Every day, unmarried Hadza girls forage for an average of 2 hours and 53 minutes, childbearing Hadza women for an average 4 hours and 24 minutes, and post-reproductive Hadza women for an average of 7 hours and 7 minutes. Besides which, older women do harder work. Young women collect berries, grandmothers dig up roots (Hawkes, O'Connell, & Blurton Jones, 1989; compare Meehan, Quinlan, & Malcom, 2013, on the Aka of Central Africa).

Probably most importantly, children get help from their fathers and from their kin. In the tropical forests of eastern Paraguay, Aché foragers collect honey, palm starch, and fruit, but the bulk of their diets come from meat. Females are net food consumers across every age category, and married men, whether in monogamous or polygynous households, tend to be calorie consumers – the only exception being young, monogamously married men. The big providers – the big hunters – are single men. Single men under 30 produce a surplus of 19 kcal on an average day; for older single men aged from 30 to 54, the surplus is almost 55 kcal. In other words, young men and single men subsidize everybody else as they wait in the queue to reproduce (Hill & Hurtado, 2009; compare Weissner, 2002, on the Kalahari !Kung).

But most foragers eventually get a chance to become parents. For contemporary hunter–gatherers, both ranges and variances in reproductive success tend to hover in or around single digits. At the high end, Aché men report from 0 to 13 live births, with a variance of 15.05; and at the low end, the !Kung report completed fertilities of 0 to 12, with a variance of 8.6 for men. The figures are similar for forager women. At the low end, Aché women report a range of 0–12 live births, with a variance of just 3.57; and at the high end, Hadza women report 0–16 children born, with a variance of 7.7. Most contemporary foragers are fairly egalitarian – reproductively and otherwise.

Every one of these groups is at least a little bit polygynous. In these and other forager cultures, the most reproductively successful fathers have more children than the most reproductively successful mothers. And reproductive variances for women are consistently lower. The highest variances reported for forager women are lower than the lowest variances reported for forager men. But overall, sex differences among hunter–gatherers are consistently small (Betzig, 2012, 2016).

Genetic evidence backs this up. Elevated levels of diversity on the human X chromosome and extremely low diversity on the human Y chromosome suggest a consistent difference between the sexes in reproductive variance. Over evolutionary time, the effective breeding population of women is likely to have been consistently larger than the effective breeding population of men. Across contemporary populations including the Khoisan, a group in southern Africa that includes the !Kung, genetic estimates of the breeding sex ratio are consistently biased in favor of women. That is, fewer males than females have contributed genes to descendant generations. A larger fraction of the female population has reproduced; a larger fraction of the male population has not. *H. sapiens* has been polygynous (Hammer et al., 2008; Lippold et al., 2014).

After we first left Africa, genetic variance went up. Geographically diverse samples of Y chromosome sequences suggest a series of reproductive bottlenecks over the course of human evolution. One from around 40,000–60,000 years ago coincides with *H. sapiens'* move out of Africa into Eurasia by a small number of colonizers, after which the effective breeding population expanded. More females than males contributed to that expansion; again, the effective breeding population among women was more than twice the effective breeding population among men. And those differences took off with the Neolithic. A second bottleneck, from around 4,000 to 8,000 years ago, overlaps more or less with the origins of agriculture and with the origins of civilization. Genetically, this transition is dramatic. The effective breeding population among women became as high as 17 times the effective breeding population among men. Y chromosome evidence of extreme male lineage expansions across Europe and Asia coincides with those bottlenecks: subsistence intensification and social stratification seem to have produced much greater male reproductive variance. People became more polygynous (Karmin et al., 2015; Poznik et al., 2016).

4.2 EUSOCIAL BREEDERS

As soon as the first history was written, in the ancient Near East, some men had children by tens or hundreds of women, and those children were brought up, in part, by members of celibate or sterile castes. Across ancient civilizations – from the Near East to India to China to Greece to Rome and beyond – successful mothers continued to raise children in single digits, but successful fathers raised children by hundreds of women (Betzig, 1986, 2014).

4.2.1 In the Beginning

The first written accounts – the first histories – were put together in the Bible. Hebrew patriarchs, like Abraham, kept three or four women; judges, like Gideon, had tens; kings, like David and Solomon, accommodated as many as a thousand. Thousands of unmarried men worked as forced laborers or conscripted soldiers, and many of them were eunuchs (Betzig, 2005, 2014; Scheidel & Morris, 2009).

The Torah is full of polygynous patriarchs. Abraham's father, Terah, had children by at least two women. We know that because Abraham married his half-sister, Sarah, who was the daughter of his father but not the daughter of his mother. Abraham went on to have children by at least three women: Hagar, who was Sarah's Egyptian maid, became the mother of Ishmael; Sarah became the mother of Isaac; and Keturah, who married Abraham after Sarah died, became the mother of another six sons. Abraham's son, Isaac, took just one wife; and Isaac's son, Jacob, also known as Israel, had just one named daughter, Dinah – remembered as a rape victim of the Hivite, Shechem. But four named women gave him twelve named sons: six by his wife, Leah, two by his wife, Rachel, two by Leah's servant, Silpah, and two by Rachel's servant, Bilhah. They obeyed Abraham's mandate: they fathered multitudes (Genesis 17:5). "The descendants of Israel were fruitful and increased greatly; they multiplied and grew exceedingly strong; so that the land was filled with them" (Exodus 1:7).

After they stopped wandering and settled into Canaan, the judges rose up. Most patriarchs counted their sons in single digits; most judges counted an order of magnitude more. Gideon had 70 sons by his wives, along with at least one son by a concubine; Jair had 30 sons; Ibzan had 30 sons and 30 daughters; Abdon had 40 sons and 30 grandsons.

Then they set up a king. David made wives of Ahinoam, Abigail, Michal, Maacah, Haggith, Abital, Elgah, and Bathsheba, with more anonymous wives and concubines from Jerusalem. They gave him 19 named sons (Amnon, Chileab, Absalom, Adonijah, Shephatiah, Ithream, Shimea, Shobab, Nathan, Solomon, Ibhar, Elishua, Elpelet, Nogah, Nepheg, Japhia, Elishama, Eliada, and Eliphelet), with another named daughter, Tamar – remembered as a rape victim of her brother Amnon. David lived in an ivory palace, wrapped in cassia-scented robes and entertained by stringed instruments, with "ladies of honor" and "virgin companions" around him (Psalms 45:9, 14). His son, Solomon, built another house on a hewn stone foundation, with cedar pillars and beams capped with capitals of cast bronze, and he filled it with 700 נָשִׁים (nashim), or wives, and 300 פִּילַגְשִׁים (pilagshim), or concubines. Solomon's son, Rehoboam, who ruled over the kingdom of Judah, had 18 wives and 60 concubines, who gave him 60 daughters and 28 sons; and Rehoboam's son, Abijah, had 14 wives, who bore 16 daughters and 22 sons. Half a century later, Ahab, who ruled over the kingdom of Israel and made Jezebel his queen, had 70 sons in Samaria alone (Betzig, 2005).

Collecting women is what *made* men kings. That was made perfectly clear to David, who was told by a prophet: "You are the man. Thus says the Lord, the God of Israel, 'I anointed you king over Israel, and I delivered you out of the hand of Saul; and I gave you your master's house, and your master's wives into your bosom'" (2 Samuel 12:7–8).

Afterwards, this was made perfectly clear to David's son. Absalom stole the hearts of the men of Israel, and of their daughters and wives. He raised an insurrection against his father, and went in to 10 of his concubines. As Absalom was advised by David's traitor, Ahithophel: "Go in to your father's concubines, whom he has left to keep the house; and all Israel will hear that you have made yourself odious to your father, and the hands of all who are with you will be strengthened" (2 Samuel 16:21). So he ended up hung in an oak, with three darts in his chest.

Even the patriarchs had coveted their father's women. Reuben, who was Jacob's firstborn son by his first wife, Leah, is supposed to have lost his inheritance *because* he polluted his father's couch. One day while Jacob was away, Reuben went in and lay down with Bilhah, his stepmother Rachel's maid. And afterwards, Reuben's place in the genealogy was reassigned. "His birthright was given to the sons of Joseph the son of Israel," who was Rachel's firstborn son (1 Chronicles 5:1).

Before they crossed over the Jordan, Moses had warned that this would happen. If the Israelites had to set up a king, let him be a good one. He should not multiply horses for himself, and he should not multiply silver or gold. But most of all: "He shall not multiply wives for himself, lest his heart turn away" (Deuteronomy 17:16–17).

Generations later, the people of Israel received a warning from one of Gideon's sons. Jotham told the people a parable about trees wanting a king of their own. The good ones – honorable olives, sweet figs, and spirit-uplifting grapevines – declined, and only the thorny, prolific bramble was willing. As the bramble said to the trees: "If in good faith you are anointing me king over you, then come and take refuge in my shade; but if not, let fire come out of the bramble and devour the cedars of Lebanon" (Judges 9:15).

Generations after that, Samuel, who was the last judge of Israel, warned the people again. The king that they wanted would put their fields, flocks, sons, and daughters to his personal uses – and in the hands of his סָרִיס, or *sarisim*, or eunuchs. "And in that day you will cry out because of your king, whom you have chosen for yourselves; but the Lord will not answer you in that day" (1 Samuel 8:18).

Eunuchs show up 47 times, in 45 verses, in the Hebrew Bible – variously rendered as officers (12 times), chamberlains (13 times), eunuchs (17 times), or chief eunuchs, *Rab'saris* (3 times), under King James. They work as messengers and treasurers, administrators and military commanders, all over the Near East. They include a chief butler, chief baker, and captain of the guard under a pharaoh in Egypt; Jezebel's bed keepers and Ahab's summoners in Israel; in Judah, the chamberlains of Josiah and Zedekiah and the commander of Zedekiah's men of war; Sennacherib's chief eunuch, sent to fetch tribute to Assyria from Hezekiah in Judah; Nebuchadnezzar II's chief eunuchs, sent to bring members of Hezekiah's family to Babylon; and seven named chamberlains (Mehu'man, Biztha, Harbo'na, Bigtha and Abag'tha, Zethar and Carkas) under the Persian emperor Ahasuerus, probably Xerxes I, with another two eunuchs (Bigthan and Teresh) who guarded Ahasuerus' threshold, a eunuch (Hathach) appointed to wait on Esther, and two eunuchs (Sha-ash'gaz and Hegai) in charge of Ahasuerus' first and second harems. When David made his son, Solomon, his successor, he assembled the most powerful men in his kingdom – mighty men, seasoned warriors, princes, and *sarisim* (1 Chronicles 28:1). They were there to the end. When Sennacherib set siege to Jerusalem, Hezekiah was warned that "some of your own sons, who are born to you, shall be taken away; and they shall be eunuchs in the palace of the king" (2 Kings 20:18); and when Nebuchadnezzar set siege to Jerusalem again, and he carried away Jehoiachin, "the king's mother, the king's wives, his officials, and the chief men of the land, he took into captivity from Jerusalem to Babylon" (2 Kings 24:15; see Kadish, 1967; Retief, Cilliers, & Riekert, 2005).

By the biblical period, women were collected at Egyptian, Assyrian, Babylonian, and Persian courts. An early story in Genesis has Abraham give up his wife, Sarah, to the king of Egypt: "And when the princes of Pharaoh saw her, they praised her to Pharaoh. And the woman was taken into Pharaoh's house" (Genesis 12:15). In the same way, but later, women were requisitioned from all over the Near East in the Amarna letters – 382 clay tablets dug up out of the House of the Correspondence associated with the Eighteenth Dynasty pharaohs Amenhotep III and IV. Some Amarna correspondents were asked to send their daughters along ("prepare your daughter for the king, your lord, and prepare the contributions"); others offered their wives ("how, if the king wrote for my wife, how could I hold her back?"). When Amenhotep III married Gilukhepa, a daughter of the king of Mitanni, she brought along 317 marvelous harem women, with their hand-bracelets, foot-bracelets, earrings, and toggle-pins. And in another Amarna letter, Pharaoh ordered 40 extremely beautiful females from a Canaanite prince of Gezer, at a price of 40 shekels of silver apiece (*Amarna Letters*, nos. 25, 99, 254, 301, and 369). Archaeological evidence from the Nineteenth Dynasty pharaoh, Rameses II, suggests prolific results. On scarabs and ostraca, bas reliefs, and statues in and beyond his tomb in the Valley of the Kings, the names of at least 96 "king's sons" and "bodily king's sons" were written down; and there should have been roughly as many daughters (Fisher, 2001; Weeks, 2006).

Other women were captured in war. When the Assyrian emperor, Sennacherib, carried off 200,150 men, women, and children from 46 fortified cities in Judah in around 701 BCE, Hezekiah of Judah threw in his daughters, his harem, and his male and female musicians (Sennacherib, *Annals*). And when Nebuchadnezzar II of Babylon invaded Jerusalem in 597 BCE, he captured the king's mother, the king's wives, and the king's eunuchs from Judah's king, Jehoiachin (2 Kings 24:15). A decade later, when Nebuchadnezzar invaded again, he threatened to take of all of the wives and sons away from Judah's king, Zedekiah, and burn Jerusalem with fire (Jeremiah 38:23). Inventories in the Assyrian emperor Ashurbanipal's 20,000-plus clay tablet library list 13 governesses, 145 weavers, 52 maids, and 260 miscellaneous women (*Imperial Administrative Records*, nos. 21–26); and archaeological evidence of a House of the Palace Women in Nebuchadnezzar's Babylon mentions provisioners and overseers of the slave girls (*Court of Nebuchadnezzar II*). Another reference from the Bible has gold and silver vessels from Jerusalem's temple brought out for the wives and concubines of Belshazzar, his eventual successor (Daniel 5:2).

There were more harems in Persia. When Darius III went to war with Alexander the Great, he took along the queen mother, the queen, 365 woman companions (one for each day of the year), their children, a herd of eunuchs to guard them, 200 *propinqui* (close kin), and 15,000 *cognati* (more remote kin) (Curtius 3.3.14–24). Some may have descended from his mother, who had 80 brothers (Curtius 10.5.23), or from his ancestor, Artaxerxes II, who was remembered as the father of 118 sons (Justin 10.1.1). Others would have descended from Xerxes I. In the story of Esther, his Jewish princess, this Xerxes issued an order: "Let beautiful young virgins be sought out for the king." They were basted for six months with myrrh and another six months with spices, then herded into the palace. "In the evening she went, and in the morning she came back to the second harem in custody of Sha-ash'gaz the king's eunuch who was in charge of the concubines; she did not go in to the king again, unless the king delighted in her and she was summoned by name" (Esther 2:2, 14).

By then, hundreds of eunuchs had worked under emperors at Egyptian, Assyrian, Babylonian, and Persian courts. There are hints about eunuchs in Egyptian myths about the desert god, Seth – who shows up on a limestone macehead from the predynastic tomb of the Scorpion King. As early as the Sixth Dynasty pyramid texts, a story was told that Seth killed his older brother, Osiris, then fought against Osiris' son, Horus: Horus lost sight in one eye, and Seth's testicles became impotent (*Pyramid Texts*, no. 1463). That story had staying power. "Seth threw filth in the face of Horus, but Horus crushed the genitals of Seth," remembered chapter 17 of the *Book of the Dead*.

Eunuchs showed up in Assyrian palace decrees: they patrolled palace corridors or watched the front doors

(Tiglath-Pileser III, *Middle Assyrian Palace Decree*). By the eighth century BCE, they had taken over. Eunuchs worked in the imperial bodyguard, they commanded armies, they administered the "inner" and "outer" courts at the palace, and they governed provinces (Grayson, 1995). Toward the end of the sixth century BCE, Nebuchadnezzar sent another eunuch, Ash'penaz, to round up all of Judah's "youths without blemish, handsome and skillful in all wisdom, endowed with knowledge, understanding learning, and competent to serve in the palace," and to bring them to Babylon (Daniel 1:3–4).

Other castrates worked for emperors other than Xerxes in Persia. Xenophon, who was an admirer of Cyrus, thought his courtiers became "gentler when deprived of desire" (*Cyropaedia*, 7.5.63), but Diodorus, who admired Alexander, called Bagoas, who worked for Darius III, "a eunuch in physical fact but a militant rogue in disposition" (Diodorus 17.5.3). Boys were chosen for castration as war prisoners or assessed as tribute: Persian emperors ordered 500 eunuchs a year from Assyria and castrated good-looking Ionian boys (Herodotus 3.92, 6.32).

4.2.2 Eusocials Out West

Centuries before historians wrote about promiscuous Persian emperors, there were stories about promiscuous Greeks. Hesiod filled his *Theogony* with promiscuous deities, and Homer filled the *Iliad* and *Odyssey* with promiscuous men. Households in Athens and elsewhere held tens of slave women, who were looked after by eunuchs; households in Rome and beyond held hundreds of slave women, and the Roman civil service was filled up with thousands of celibate slave men and eunuchs (Betzig, 1992, 2014; Scheidel, 2009).

In the *Iliad*, Priam loses Paris and another 49 sons – 19 by his wife and the rest by the women of his house. Their children are smashed on the ground, "and their wives are dragged away by Achaean hands" (*Iliad* 22.64–65, 24.495–496). And in the *Odyssey*, Odysseus sails home to Penelope, his long-suffering wife – and to the 50 wool-carding maids captured on earlier raids. "You dogs!" he rants at the men lying about his house, "you courted my wife, and slept with my servants by force." Every one of those maids ends with a noose around her neck (*Odyssey* 14.203, 264, 22.421–473).

The usual arrangement was laid out in an Athenian speech. "We have *hetairai* (courtesans) for pleasure, *pallakai* (kept women) to care for our daily needs, with *gynaikes* (wives) to bear us legitimate children" (Demosthenes 59.122). And on the side there were *pornoi*, or whores, and "multitudes" of *douloi*, or slaves (Athenaeus, *Deipnosophistae* 6.13). There may have been 100,000 slaves in Classical Athens, close to half the population; most women worked indoors, and they were as always concentrated in rich men's homes. Aristotle took slavery for granted as part of the natural order: "The first and fewest possible parts of a family are master and slave"

(*Politics* 1253b), and Plato assumed well-to-do householders were hoarders: "Imagine one of these owners, a master of say some 50 slaves" (*Republic* 9.578d). Slave women in Athens, as elsewhere, gave *nothoi*, or bastards, to their masters. Aristonothos ("excellent bastard"), Notharchos ("leading bastard"), Kleinothos ("glorious bastard"), and Philonothos ("loving bastard") were some of their names (Ogden, 1996; Rawson, 2011).

εὐνοῦχοι (*eunuchi*), or eunuchs, looked after those bastards. As early as the eighth century BCE, Hesiod – who might have beaten Homer in a singing contest – had Kronos "harvest his father's genitals" with a long, jagged sickle (*Theogony* 178–182), and Homer – who might have gone blind after a visit to Ithaca – had Odysseus take care of his treacherous goatherd by "ripping away his genitals as raw meat" for his dogs (*Odyssey* 22.476–477). By the fifth century BCE, Aeschylus had Clytemnestra "brutally mangle" her husband's corpse; and Orestes was threatened with "mutilation" and the destruction of his seed (*Choephoroi* 439–40; *Eumenides* 185–190).

Herodotus (3.48, 8.105) remembered a tyrant from Corinth who sent 300 boys from the best families in Corfu to be "turned into eunuchs," and a trader from Chios who made a living "by the unholy trade of castrating any good looking boys" he could find. There were other eunuchs in Athens. In his *Politics* (1311b), Aristotle had a eunuch kill the tyrant of Cyprus in order to avenge an insult; and Plato's *Protagoras* (314c) had Callias, the most prosperous man in town, keep a eunuch guard at his door, who overheard the sophists inside.

There would be more polygynists, and more eunuchs, in Rome. His frenemy, Mark Antony, cozied up to the man who would be the first emperor, Augustus, with this letter: "Good luck to you if when you read this letter you have not been with Tertulla or Terentilla or Rufilla or Salvia Titisenia, or all of them. Does it matter where or in whom you get your erections?" (Suetonius, *Augustus* 69.2). Tacitus (*Annals* 6.1) lit into Tiberius: "He regained his secluded sea cliffs, for his criminal lusts shamed him. Their uncontrollable activity was worthy of an oriental tyrant." Seneca (*On Constancy* 18.2), who slept with the emperor's sister, disapproved of Gaius: "At a banquet, this is at a public gathering, using his loudest voice, Gaius taunted a senator with the way his wife behaved in sexual intercourse. Ye gods! What a tale for the ears of a husband! What a fact for an emperor to know!" Dio (61.31.1) remembered how Claudius was pimped by his mother and grandmother, who provided slaves, and may have been pimped by his wife: "In addition to her shameless behavior in general she at times sat as a prostitute in the palace herself and compelled other women of the highest rank to do the same." And Suetonius had fun with the last Caesar: Nero hosted soirees that lasted all night and all day; he threw parties in the Campus Martius with prostitutes from all over the city; he raped a vestal virgin; "he even desired illicit relations with his own mother" – or a spitting image of her (Suetonius, *Nero* 28.2).

There were an estimated 6 million slaves in a population of 60 million in the Roman Empire by Augustus' time, and most were owned by rich women and men. Great men saved places for hundreds of slaves in their family tombs, and many of those slaves were women. In papyri from Roman Egypt, the ages of female slaves at the time of sale hovered around a median of 19, and in Justinian's law books, slave women, who were legally unmarried, were rewarded for bearing healthy bastards (*Digest* 1.5.15, 21.1.14–15, 34.5.10(11)0.1). Those children were often freed, and they were lovingly commemorated on Roman tombs. Many grew up to become knights, or sat in the senate (Tacitus, *Annals* 13.27), and hundreds of *Augusti vernae* (*Aug Vern*) and *Augusti liberti* (*Aug Lib*) – slaves and freed slaves, members of the imperial family or *Familia Caesaris* – worked in the emperors' civil service. The numbers increased after the empire moved east. After Constantine settled in Constantinople, women were added to the imperial *gynaeceum*, or women's rooms (Eutropius 9.25; Zonaras 12.23). The fifth-century CE register of dignitaries, or *Notitia Dignitatum*, listed imperial *gynaecea* in Italy at Rome, Apulia, Aquileia, Ravenna, and Milan; at Carthage in North Africa; at Winchester in Britain; at Pannonia and Dalmatia in the Balkans; and at Arles, Lyons, Rheims, Tourney, Trier, Autun, and Vienne in Gaul; further east, there were *gynaecea* in Caesarea and Tyre. These women made cloth, and were sexually accessible to their masters (Lactantius, *On the Deaths of the Persecutors* 38.1–4).

Eunuchs were attached to every emperor after Julius Caesar. Maecenas, who worked as a regent for the first emperor, Augustus, was "attended in public" by a pair of eunuchs (Seneca, *Moral Letters* 114); and the eunuch, Lygdus, may have had Drusus, the second emperor's son, done in (Tacitus, *Annals* 4.10). Helicon, his Egyptian *cubicularius*, was especially close to the third emperor, Gaius – he played ball with him, had dinner with him, was with him when he was going to bed, and advised him on foreign affairs (Philo, *Embassy* 27.175); and the fourth emperor, Claudius, had a eunuch, Posides, who was honored at his British triumph (Suetonius, *Claudius* 28). Nero sent out his *cubicularii* to start the Great Fire in Rome, then put up "gangs of eunuchs" in the Golden House he built on the ruins (Tacitus, *Histories* 2.71). By the end of the first century CE, there were "troops of eunuchs" at court (Suetonius, *Titus* 7.1); and by the start of the second century CE, another eunuch, Favorinus, debated with Hadrian (Philostratus, *Lives of the Sophists* 1.8). By the third century CE, eunuchs were enlisted as treasurers and tax collectors (*Historia Augusta*, Severus Alexander, 23.6–7); and after 311 CE, when Diocletian was done in, eunuchs "who had chief authority at court and with the emperor" were slain (Lactantius, *On the Deaths of the Persecutors* 15).

Then Constantine moved his capital to Constantinople, and his bureaucracy was filled with eunuchs. There were 1,000 cooks, barbers, butlers, and waiters, "eunuchs more in number than flies around the flocks in spring," soon after the time of Constantine I (Libanius, *Orations* 18.130); by the tenth century CE, they "swarmed around the grand palace like flies around a cowshed in summer" under Constantine VII (Theophanes, *Continuatus* 318). All of 8 out of 18 administrative ranks were reserved for members of a sterile caste, and eunuchs consistently outranked the "bearded" civil service. A *praepositus sacri cubiculi*, or grand chamberlain, was set over the emperor's sacred bedchamber; a *comes sacrae vestis* kept the sacred wardrobe; a *sacellarius* was keeper of the purse; and a *castrensis sacri palatii* managed the sacred treasury, "for the property of the treasury is as it were the private property of the emperor," by law (*Digest* 43.8.2.4). Others ended up as governors or generals, counselors or consuls, and "eunuchs for the kingdom of heaven" were patriarchs and bishops in the Eastern Church (Claudian, *Against Eutropius* 1.417–462, 2.342–344, with reference to Matthew 19:12; see Tougher, 2008, 2017).

4.2.3 Eusocials Out East

An old rumor had India's first emperor, Chandragupta Maurya, who brought the Indus and Ganges together, get a look at Alexander the Great sometime around 326 BCE; and the first august emperor, Qin Shihuangdi, brought the warring states of China together and unified them roughly a century later, in 221 BCE. Emperors in both Asian empires counted their women by the thousands, and filled their bureaucracies with eunuchs (Betzig, 2014; Ebrey, 2002).

Kautilya, who was remembered as Chandragupta Maurya's advisor, is supposed to have put together an *Arthashastra*, or Text on Gain. "The king shall construct his harem consisting of many compartments, one within the other, enclosed by a parapet and a ditch, and provided with a door," he wrote; emperors were to outfit their women's quarters with maternity wards and residences for princes and princesses (Kautilya 1.20). And Chandragupta Maurya's grandson, Ashoka the "Sorrowless" emperor, is supposed to have won his throne over the bodies of 99 brothers by other mothers (*Dipavamsa* 6.22; *Mahavamsa* 5.20–21). Then he reassured subjects in rock edicts across his empire that he would look after their welfare, "whether I am eating or in the closed female apartments, in the inner chamber, in the royal rancho, on horseback or in pleasure orchards" (Ashoka, *Rock Edict* 6).

By roughly six centuries later, Vatsayana's *Kamasutra*, or Love Threads, advised Gupta emperors to rub ointments on themselves "to enable them to enjoy many women" in a night, and it instructed consorts to learn how to solve riddles, do arithmetic, act, sculpt, speak multilingually, and make parrots talk in order to keep an emperor interested, even though he had "thousands of other women" around him (*Kamasutra* 1.3, 5.6). The *Manusmriti*, or Laws of Manu, which would have been available to imperial Guptas, suggested that *vaishyas* (men of the next-to-lowest caste) should have access to both *vaishya* and *shudra* (lowest-caste) women; that *kshatryias* (men of the next-to-highest caste) should have access to *kshatriya*, *vaishya*, and *shudra* women; and that *brahmins* (highest-caste men) should have access to all women; "but for the *shudra*, the *shudra* alone has been ordained" (*Manusmriti* 3.13). Other harems fill the Hindu epics, probably completed at the Gupta courts. In the *Ramayana*, Ravana steals Rama's bride, Sita, away and hides her in "an array of palaces and mansions with thousands of women" of his own, but the monkey, Sugriva, helps Rama get Sita back, even though Sugriva himself is "addicted to sensual behavior and relegating duties to his ministers" (*Ramayana* 3.53.7–8, 4.28.1–8). So it goes in the *Mahabharata*, where there are harems in heaven (the moon has 27 wives), in the oceans (rivers "like rivaling concubines" run into them), and on earth – where pretty girls drink choice liquor and dance by the thousands when Arjuna marries the princess, Draupadi (*Mahabharata* 1(5) 19, 1(7)60, 1(13)191).

Many of those harems could have been guarded by eunuchs. The first part of Kautilya's *Arthashastra* had *varshadharas*, or "rain holders," regulate affairs to be conducive to the happiness of Mauryan emperors, their masters. And other passages asked *pandakas*, or "deviants," to spy on the houses of enemy kings (*Arthashastra* 1.20, 6.1).

In some of the same ways, Vatsayana's *Kamasutra* (2.9) advised Gupta emperors to tell *tritiya prakritis*, or "third genders," who often dressed as women, to keep their "desires secret" when they dressed as men; and the Laws of Manu called a Brahmin without knowledge "as fruitless as" a *shandha*, or effeminate man (*Manusmriti* 2.158, 4.205, 211, 9.79, 201). There are ambiguous genders in the *Rayamana* (6.49), where Rama, after wandering for 14 years, finds followers who are "neither man nor woman" waiting when he comes home to be crowned. And there are *klibas*, or "impotents," in the Great Bharat, where Arjuna, the hero of the Bhagavad-Gita, is cursed by the goddess Urvashi and ends up as Brihannala, or "Big Rod," the unmanned dancer (*Mahabharata* 3(41), 4(45)10; see Gannon, 2011; Kulke & Rothermund, 2004).

But eusociality was greater in China. Soon after Qin Shihuangdi unified that empire in 221 BCE, he connected 270 palaces together and filled them with 10,000 women – "beautiful women and bells and drums that he'd taken from the feudal rulers," remembered China's first historian, Sima Qian (*Shi ji*, 6). The Han Dynasty (206 BCE–220 CE) reformer, Wang Mang, sent out Grandees Without Specified Appointments and Government Agents, 45 of each, to select virtuous young ladies to be brought back to the palace: earlier Han emperors collected 3,000; later Han emperors collected 6,000 (*History of the Former Han Dynasty* 99C.13b). The Sui Dynasty (581–618 CE) emperor, Yangdi – who built the Grand Canal and rebuilt Qin Shinhuangdi's Great Wall – is supposed to have kept 100,000 women at Yangzhou alone; and the long-lived Tang Dynasty (618–907 CE) emperor, Xuanzong, had

palaces for 3,000, 8,000, and 40,000 women. The *Song Shi* names 65 children of the Song Dynasty (960–1279 CE) emperor, Huizong, but there were probably others: he was presented with a virgin every five to seven days (Ebrey, 2002).

Imperial agents had the art of selecting consorts down to a science. A handbook attributed to the Yellow Emperor, the *Plain Girl Classic*, recommends girls with silky hair and soft flesh, neither too tall nor too short, neither too fat nor too slight, who have never borne children; and *The Secrets of the Jade Chamber*, a Sui Dynasty handbook, is explicit: when it comes to women, one should be vexed only by their not being young, if possible between 14 or 15 and 18 or 19; in any event, they should not exceed 30. *Records Made with the Red Brush* kept track of the health and menstrual cycles of palace women: court ladies led consorts to the royal bedchamber on correct calendar days, and consorts who conceived were rewarded with gold rings. And Recorders of Imperial Intercourse were assigned civil service ranks – 6a under the Tang, 9a under the Song (Goldin, 2002; van Gulik, 1961).

Those women were provisioned and protected by a sterile caste. The first emperor set up a new eunuch agency, the Central Imperial Depot, or *Zhongchangshi*. Han emperors had at least 2,000 court eunuchs put to death. Tang emperors invested in a eunuch nobility: 1 prince, 27 dukes, 4 marquises, 2 earls, 4 viscounts, and 7 barons; and the *Song Shi* included 53 eunuch biographies – some, palace administrators; others, army commanders.

Those numbers continued to go up as the dynasties wore on. Taizu, the Ming Dynasty (1368–1644 CE) founder, knew eunuchs were too influential, so he put up an iron plaque outside his palace that said: "Eunuchs are forbidden to interfere in government affairs; and those who try will be killed" (*Ming shi* 304). But when the last Ming emperor gave way to the Qing, Beijing held an estimated 70,000 castrates, with another 30,000 scattered over the empire. Ming eunuchs filled four departments, eight bureaus, twelve directorates, and the *Dongchang*, Eastern Depot or secret police. Obligately sterile soldiers, like the admiral Zheng He, explored the Indian Ocean and the Persian Gulf; and obligately sterile administrators, like the architect Nguyen An, designed the Forbidden City in Beijing, which held more than 900 buildings and almost 9,000 rooms (Tsai, 1996, 2002).

Genetic evidence is consistent, again. Samples of Y chromosomes from contemporary men suggest an expansion of three East Asian male lineages within the last half-millennium. Those clades show up with high frequency across populations, and represent more than 40 percent of contemporary Han Chinese. Roughly 300,000,000 living men are descended from just three ancestor males who lived around 6,000 years ago, in the late Neolithic Age (Yan et al., 2014).

Even more striking evidence suggests a strong genetic legacy of the Mongol emperors who ruled China for roughly a century. A Y chromosome star cluster has been found in 8 percent of all men from the Pacific to the Caspian Sea, and in 1/200 of all men worldwide. It has been suggested that the star cluster originated in Mongolia approximately a millennium ago, and has been carried on by male-line descendants of the Yuan dynasty ancestor, Genghis Khan (Zerjal et al., 2003).

4.3 DISCUSSION

Across taxa, helpers and workers offer a number of reproductive benefits. They include building and maintaining living spaces, group and territory defense, food collection and preparation, and care and feeding of offspring.

Across taxa, helpers and workers pay a variety of reproductive costs. They include hormonal suppression of gamete (egg and sperm) production, gonad (testes) removal or gonad (ovary) reduction, egg destruction and egg consumption, and induced abortion and infanticide (Choe & Crespi, 1997; Koenig & Dickinson, 2004, 2016; Mann et al., 2000; Solomon & French, 1997; Stacey & Koenig, 1990; Wolfe & Sherman, 2007).

What makes it possible for breeders to collect those benefits? What makes it possible for breeders to inflict those costs? The simplest answer seems to be: a rich habitat. Wherever abundance is surrounded by dearth, it makes sense to stick around and help or work.

Most insect species are solitary, but most insects are social. Social insects flourish in the tropics, where plant and animal food abound year round. Insects who live alone – grasshoppers, cockroaches, beetles – tend to be foragers; they subsist on a scattered variety of dead vegetable matter, animal feces, and corpses. Insects who live in groups – termites, bees, wasps – tend to occupy closed spaces in or around steady supplies of food. *Atta* leafcutter colonies, the most eusocial of groups, live under the enormous canopies of American rainforests, along trunk routes that extend as far as 250 m, in nests that displace up to 40,000 kg of dirt, where they perennially grow fungus. Individuals wander across the landscape, groups gather in enclosed spaces; societies occupy the ecological center of a territory, solitaries occupy the periphery. As the great myrmecologists, Bert Hölldobler and Ed Wilson, sum up: "Solitary forms tend to prevail over social insects only in the more remote and transient of living spaces" (Hölldobler & Wilson, 2009, p. 6).

The same may be said of cooperatively breeding birds. Like eusocial insects, avian cooperative breeders tend to live nearer the equator, in the African and American tropics and Australian Pacific, where the temperature and rainfall are less variable and territories can be held all year. Migratory birds are bad candidates; permanent residents are good ones (Arnold & Owens, 1998; Cockburn, 2006). Acorn woodpeckers capture insects in early summer, and they eat sap, buds, and catkins in the spring, but acorns are always consumed. These woodpeckers occupy pine or oak tree granaries, with as many as 30,000 holes, drilled and defended by generations of woodpecker ancestors, and filled with food. Those granaries are in limited supply – so adults stay on the nest and help their parents reproduce as they wait for vacant sites and available mates. As the eminent ornithologists, Walt Koenig, Janis Dickinson, and Steve Emlen, conclude: "If nonbreeding helpers exist, it is at least in part because of some constraint on their ability to breed" (Koenig, Dickinson, & Emlen, 2016, p. 363).

H. sapiens societies play by the same rules. Around the world, hunter–gatherers settle down where resources are available year round; and as always, more sedentary societies are less fair. Foragers are attracted to productive environments, but avoid continental climates – where temperature and precipitation vary more from year to year and the seasons are more marked (Keeley, 1988). And foragers are often water tethered: on the Kalahari, camps are based around old river courses, and villages were built along salmon runs on the American northwest coast. Sedentary societies defend and inherit territories and differentiate lineage and rank (Kelly, 2013). And as always, group members with more assets have more reproductive success (Betzig, 2012, 2016). Peripatetic foragers – the Hadza or the Aché – might collect two or three wives at a time; sedentary foragers – Ainu chiefs in Japan or chiefs on the American northwest coast – often numbered their women and children in double digits.

As the intrepid anthropologist, Bob Carneiro, pointed out almost half a century ago, the first states rose up in river valleys, hemmed in by oceans and mountains: "They are all areas of circumscribed agricultural land" (Carneiro, 1970, p. 734). When, on the order of 10,000 years ago, the first empire rose up in the Near East, it was on the land between the Tigris and Euphrates rivers, bordered to the east by the Zagros Mountains and to the west by the Syrian Desert; and Egypt became an empire on the black land of the Nile, bounded by the red land of the Sahara that stretched across North Africa to the east and west. The Delian League Empire was a cluster of Aegean islands that sent grain and tribute to Athens' well-fortified port at Piraeus; and the Roman Empire sat in the Mediterranean Basin, with access to grain and tribute from Sicily and Egypt. Empires centered on the alluvial deposits of the Ganges and Indus were hemmed in by the Himalayas and the Hindu Kush; and set apart by the Gobi Desert to the north and the China Sea to the south, the Middle Kingdom flourished in Yellow River loess. Emperors collected hundreds or thousands of women, and empires were run by hundreds or thousands of eunuchs.

REFERENCES

Amarna Letters (2001). Translated by W. L. Moran. Baltimore, MD: Johns Hopkins University Press.

Arnold, K., & Owens, I. (1998). Cooperative breeding in birds: A comparative test of the life history hypothesis. *Proceedings of the Royal Society B*, **265**, 739–745.

Bentley, G., & Mace, R. (2009). *Substitute Parents: Biological and Social Perspectives on Alloparenting in Human Societies*. Oxford: Berghahn.

Betzig, L. L. (1986). *Despotism and Differential Reproduction: A Darwinian View of History*. New York: Aldine de Gruyter.

Betzig, L. L. (1992). Roman polygyny. *Ethology and Sociobiology*, **13**, 309–349.

Betzig, L. L. (2005). Politics as sex: The Old Testament case. *Evolutionary Psychology*, **3**, 326–346.

Betzig, L. L. (2009). But what is government itself but the greatest of all reflections on human nature? *Politics and the Life Sciences*, **28**, 102–105.

Betzig, L. L. (2012). Means, variances and ranges in reproductive success: Comparative evidence. *Human Behavior and Evolution*, **33**, 309–317.

Betzig, L. L. (2013). Darwin's question: How can sterility evolve? In K. Summers & B. Crespi, eds., *Human Social Evolution: The Foundational Works of Richard D. Alexander*. New York: Oxford University Press, pp. 365–374.

Betzig, L. L. (2014). Eusociality in history. *Human Nature*, **25**, 80–99.

Betzig, L. L. (2016). Mating systems. In T. Shackelford & V. Weekes-Shackelford, eds., *Encyclopedia of Evolutionary Psychological Science*. Berlin: Springer, pp. 1–11.

Bird, R. B., & Bird, D. (2002). Constraints on knowing or constraints on growing? Fishing and collecting by the children of Mer. *Human Nature*, **13**, 239–267.

Book of the Dead (1967). Translated by W. Budge. New York: Dover.

Carneiro, R. L. (1970). A theory of the origin of the state. *Science*, **169**, 733–738.

Choe, J., & Crespi, B. (1997). *Evolution of Social Behavior in Insects and Arachnids*. Chicago, IL: University of Chicago Press.

Cockburn, A. (2006). Prevalence of different modes of parental care in birds. *Proceedings of the Royal Society of London B*, **273**, 1375–1383.

Crespi, B. (2014). The insectan ape. *Human Nature*, **25**, 6–27.

Dijkstra, M. B., & Boomsma, J. J. (2006). Are workers of *Atta* leafcutter ants capable of reproduction? *Insectes Sociaux*, **53**, 136–140.

Ebrey, P. (2002). *Women and the Family in Chinese History*. London: Routledge.

Emlen, S. T. (1997). Predicting family dynamics in social vertebrates. In J. Krebs & N. Davies, eds., *Behavioural Ecology*, 4th ed. Oxford: Blackwell Scientific, pp. 228–253.

Fisher, M. (2001). *The sons of Ramesses II*. Wiesbaden: Harrassowitz.

Foster, K., & Ratneiks, F. (2005). A new eusocial vertebrate? *Trends in Ecology and Evolution*, **20**, 363–364.

Gannon, S. (2011). Exclusion as language and the language of exclusion: Tracing regimes of gender through linguistic representations of the "eunuch." *Journal of the History of Sexuality*, **20**, 1–27.

Goldin, P. (2002). *The Culture of Sex in Ancient China*. Honolulu, HI: University of Hawaii Press.

Grayson, A. K. (1995). Eunuchs in power. In M. Dietrich & O. Oretz, eds., *Festschrift für Wolfram Freiherrn*. Neukirchen: Alter Orient und Altes Testament, pp. 85–98.

Hammer, M. F., Mendez, F. L., Cox, M. P., Woerner, A. E., & Wall, J. D. (2008). Sex-biased evolutionary forces shape genomic patterns of human diversity. *PLoS Genetics*, **4**, e1000202.

Hawkes, K., & Coxworth, J. (2013). Grandmothers and the evolution of human longevity: A review of findings and future directions. *Evolutionary Anthropology*, **22**, 294–302.

Hawkes, K., O'Connell, J. F., & Blurton Jones, N. G. (1989). Hardworking Hadza grandmothers. In V. Standen & R. Foley, eds., *Comparative Socioecology: The Behavioural Ecology of Humans and Other Mammals*. London: Blackwell, pp. 341–366.

Haydock, J., & Koenig, W. D. (2002). Reproductive skew in the polygyandrous acorn woodpecker. *Proceedings of the National Academy of Sciences*, **99**, 7178–7183.

Hewlett, B. S., & Lamb, M. (2005). *Hunter–Gatherer Childhoods: Evolutionary, Developmental and Cultural Perspectives*. New York: Aldine de Gruyter.

Hill, K., & Hurtado, A. M. (2009). Cooperative breeding in South American hunter–gatherers. *Proceedings of the Royal Society B*, **276**, 3863–3870.

Hölldobler, B., & Wilson, E. O. (2009). *The Superorganism*. New York: Norton.

Hölldobler, B., & Wilson, E. O. (2011). *The Leafcutter Ants*. New York: Norton.

Hopkins, K. (1963). Eunuchs in politics in the later Roman Empire. *Proceedings of the Cambridge Philological Society*, **9**, 62–80.

Hrdy, S. B. (1999). *Mother Nature: A History of Mothers, Infants and Natural Selection*. New York: Pantheon Books.

Hrdy, S. B. (2009). *Mothers and Others: The Evolutionary Origins of Mutual Understanding*. Cambridge, MA: Harvard University Press.

Imperial Administrative Records: Palace and Temple Administration (1992). Edited by F. M. Fales & M. Postgate. Helsinki: University of Helsinki Press.

Kadish, G. (1967). Eunuchs in ancient Egypt? In *Studies in Honor of John A. Wilson*. Chicago, IL: Oriental Institute, pp. 55–72.

Kaplan, H., Hill, K., Lancaster, J. B., & Hurado, A. M. (2000). A theory of human life history. *Evolutionary Anthropology*, **9**, 156–185.

Karmin, M., Saaq, L., Vicente, M., et al. (2015). A recent bottleneck of Y chromosome diversity coincides with a global change in culture. *Genome Research*, **25**, 459–466.

Keeley, L. (1988). Hunter–gatherer economic complexity and population pressure: A cross-cultural analysis. *Journal of Anthropological Archaeology*, **7**, 373–411.

Kelly, R. (2013). *The Foraging Spectrum*, 2nd ed. New York: Cambridge University Press.

Koenig, W. D., & Mumme, R. (1987). *Population Ecology of the Cooperatively Breeding Acorn Woodpecker*. Princeton, NJ: Princeton University Press.

Koenig, W. D., & Dickinson, J. (2004). *Evolution of Cooperative Breeding in Birds*. New York: Cambridge University Press.

Koenig, W. D., & Dickinson, J. (2016). *Cooperative Breeding in Vertebrates*. New York: Cambridge University Press.

Koenig, W. D., Dickinson, J., & Emlen, S. T. (2016). Synthesis: Cooperative breeding in the twenty-first century. In W. D. Koenig & J. L. Dickinson, eds., *Cooperative Breeding in Vertebrates*. Cambridge, UK: Cambridge University Press, pp. 353–374.

Kramer, K. L. (2005). Children's help and the pace of reproduction: Cooperative breeding in humans. *Evolutionary Anthropology*, **14**, 224–237.

Kramer, K. L., & Greaves, R. D. (2011). Juvenile subsistence effort, activity levels, and growth patterns: Middle childhood among Pumé foragers. *Human Nature*, **22**, 303–326.

Kulke, H., & Rothermund, D. (2004). *A History of India*, 4th ed. London: Routledge.

Lippold, S., Xu, H., Ko, A., et al. (2014). Human paternal and maternal demographic histories: Insights from high-resolution Y chromosome and mtDNA sequences. *Investigative Genetics*, **5**, 13–30.

Mann, J., Connor, R., Tyack, P., & Whitehead, H. (2000). *Cetacean Societies*. Chicago, IL: University of Chicago Press.

Meehan, C. L., Quinlan, R., & Malcom, C. (2013). Cooperative breeding and maternal energy expenditure among Aka foragers. *American Journal of Human Biology*, **25**, 42–57.

Ogden, D. (1996). *Greek Bastardy*. London: Oxford University Press.

Poznik, G. D., Xue, Y., Mendez, F. L., et al. (2016). Punctuated bursts in human male demography inferred from 1,244 worldwide Y-chromosome sequences. *Nature Genetics*, **48**, 593–601.

Pyramid Texts (1995). Translated by J. P. Allen. Leiden: Brill.

Rawson, B. (1986). Children in the Roman *familia*. In B. Rawson, ed., *The Family in Ancient Rome*. Ithaca, NY: Cornell University Press, pp. 170–200.

Rawson, B. (2011). *A Companion to Families in the Greek and Roman Worlds*. London: Blackwell.

Retief, F., Cilliers, F., & Riekert, S. (2005). Eunuchs in the Bible. *Acta Theologica, Supplement*, **7**, 247–258.

Scheidel, W. (2009). A peculiar institution? Greco-Roman monogamy in global context. *History of the Family*, **14**, 280–291.

Scheidel, W., & Morris, I. (2009). *The Dynamics of Ancient Empires*. New York: Oxford University Press.

Sear, R., & Mace, R. (2008). Who keeps children alive? A review of the effects of kin on child survival. *Evolution and Human Behavior*, **29**, 1–18.

Sennacherib (1924). *Annals*, translated by D. D. Luckenbill. Chicago, IL: University of Chicago Press.

Sherman, P. W., Lacey, E., Reeve, K., & Keller, L. (1995). The eusociality continuum. *Behavioral Ecology*, **6**, 102–108.

Solomon, N., & French, J. (1997). *Cooperative Breeding in Mammals*. Cambridge, UK: Cambridge University Press.

Stacey, P., & Koenig, W. (1990). *Cooperative Breeding in Birds*. New York: Cambridge University Press.

Strassmann, B., & Kurapati, N. (2010). Are humans cooperative breeders? Most studies of natural fertility populations do not support the grandmother hypothesis. *Behavioral and Brain Sciences*, **33**, 35–39.

Tiglath-Pileser III (1995). *Middle Assyrian Palace Decree*, translated by M. T. Roth. Atlanta, GA: Scholars Press.

Tougher, S. (2008). *The Eunuch in Byzantine History and Society*. London: Routledge.

Tougher, S. (2017). *Roman Castrati: Eunuchs in the Roman Empire*. London: Bloomsbury Academic.

Tsai, S. H. (1996). *Eunuchs of the Ming Dynasty*. Albany, NY: SUNY Press.

Tsai, S. H. (2002). Eunuch power in imperial China. In S. Tougher, ed., *Eunuchs in Antiquity and Beyond*. Cardiff: Classical Press of Wales, pp. 221–233.

Turke, P. W. (1988). Helpers at the nest: Childcare networks on Ifaluk. In L. Betzig, M. Borgerhoff Mulder, & P. Turke, eds., *Human Reproductive Behavior*. Cambridge, UK: Cambridge University Press, pp. 173–188.

van Gulik, J. H. (1961). *Sexual Life in Ancient China*. Leiden: Brill.

Voland, E., Chasiotis, A., & Schiefenhövel, W. (2005). *Grandmotherhood: The Evolutionary Significance of the Second Half of Female Life*. New Brunswick: Rutgers University Press.

Weeks, K. (2006). *KV5: A Preliminary Report on the Excavation of the Tomb of the Sons of Rameses II in the Valley of the Kings*. New York: American University in Cairo Press.

Weissner, P. (2002). Hunting, healing, and Hxaro exchange: A long term perspective on !Kung (Ju/'hoansi) large-game hunting. *Evolution and Human Behavior*, **23**, 1–30.

Williams, G. C. (1966). *Adaptation and Natural Selection: A Critique of Some Current Evolutionary Thought*. Princeton, NJ: Princeton University Press.

Wilson, E. O. (1975). *Sociobiology: The New Synthesis*. Cambridge, MA: Belknap.

Wolff, J., & Sherman, P. (2007). *Rodent Societies*. Chicago, IL: University of Chicago Press.

Yan, S., Wang, C. C., Zheng, H. X., et al. (2014). Y chromosomes of 40% Chinese descend from three Neolithic super-grandfathers. *PLoS ONE*, **9**, e105691.

Zerjal, T., Xue, Y., Bertorelle, G., et al. (2003). The genetic legacy of the Mongols. *American Journal of Human Genetics*, **72**, 717–721.

5 The Nature and Psychological Foundation of Social Universals

BERNARD CHAPAIS

But are societies and cultures really so different at the level of forms and processes? Aren't they in some ways depressingly the same? Don't anthropologists, time after time in society after society, come against the same processes carried out under a variety of symbolic disguises? I think they do ... Once one gets behind the surface manifestations, the uniformity of human behavior and of human social arrangements is remarkable.

– Robin Fox (1989, pp. 18–19)

The extraordinary extent of human behavioral diversity, both in terms of forms and meanings, is matched only by its no less remarkable, but less salient degree of unity and redundancy in terms of motivations and functions. Paramount among the manifestations of humankind's behavioral unity are *absolute universals* – behavioral phenomena present in all societies. The most comprehensive repertoire of absolute universals was compiled by Brown (1991) based on a review of anthropological studies. Brown provided an enumerative description of "the traits that all people, all societies, all cultures have in common," or what he called the *Universal People* (Brown, 1991, p. 130) – a concept reminiscent of Fox's description of a hypothetical primeval group issued from Adam and Eve (Fox, 1989). Brown's description was later published in the form of an alphabetical list by Pinker (2002, p. 435). That list contained more than 400 elements and went as follows: "aesthetics, affection, age grades, age statuses, age terms, ambivalence, anthropomorphization, antonyms, baby talk, belief in supernatural/religion, ... black (color term), body adornment, childbirth customs, child care," and so on. A post hoc classification of those elements reveals that it comprises several broad categories: emotions and motivations (e.g., affection, sexual jealousy), cognitive abilities (classification, conjectural reasoning), phonetic components (morphemes, vowels), semantic linguistic components (numerals, synonyms), higher-order linguistic phenomena (baby talk, grammar), tools (weapons, toys), technical procedures (cooking, body adornment), domains of knowledge (time, disease interpretation), and various types of social universals, including social activities (rituals, trade), enduring social patterns (social stratifications, gender-based

divisions of labor), and concepts referring to social patterns (territoriality, ethnocentrism).

Such a rich repertoire of panhuman psychological and behavioral elements has the unquestionable merit of imposing a reflection on their causality and significance. Remarkably, however, the study of behavioral universals has been basically neglected by cultural anthropology (for notable exceptions, see Malinowski, 1944, Murdock, 1945; Kluckohn, 1953; Goodenough, 1970; Hockett, 1973; Fox, 1980, 1989; Brown 1991, 2001, 2004; Konner, 2010). The vast majority of cultural anthropologists have consistently been more interested in cultural differences than cross-cultural similarities and, if only for that reason, did not look for universals. Moreover, many anthropologists defined behavioral universals in terms of substance (content) – that is, in terms of patterns having the same form and meaning everywhere and for this reason failed to find such patterns. In Geertz's terms, "to say that 'religion,' 'marriage,' or 'property' are empirical universals is to say that they have the same content, and to say that they have the same content is to fly in the face of the undeniable fact that they do not" (1973, pp. 39–40). True, substantive universals are virtually nonexistent, but human universals exist at other levels of analysis. An examination of Brown's (1991, 2001, 2004) list shows that behavioral universals are in fact large, open-ended *categories*, each of which contains a vast number of distinct cultural forms that share motivational, functional, or structural commonalities – they are panhuman, culturally polymorphic categories (Chapais, 2014).

Even then, most cultural anthropologists dismissed the significance of such categories. Claude Lévi-Strauss, the father of French structuralist anthropology and one of the staunchest advocates of the role of universal "mental structures" in organizing cultural diversity, nonetheless discarded behavioral universals, and unequivocally so. After noting that "ethnologists, mostly American, have greatly enriched our inventory and proposed a list of universal traits: age gradings, sports, bodily adornment, calendar, toilet-cleanliness, community organization, cooking, cooperative labor, cosmology, courtship, dance, decorative art, and so on," he concluded that "in addition to the bizarre character of an alphabetical repertoire [he referred here to Murdock, 1945] those common denominators are only vague and meaningless categories" (Lévi-Strauss, 1983, p. 61, my translation). Those repertoires were indeed highly heterogeneous, and the absence of any conceptual framework integrating so many disparate elements into a coherent system rendered them extremely difficult to interpret. Brown himself noted the problem when he wrote that future analyses of universals carried out "at the deeper level of process or innate mechanism may presumably unearth universals that are at present wholly unknown and almost certainly would produce hierarchical orders among some set of universals, orders that distinguish the fundamental processes from their more superficial consequences" (Brown, 1991, p. 141).

Establishing such "hierarchical orders" linking mental and behavioral universals requires addressing three questions: (1) How do the various types of *behavioral* universals articulate with each other? This implies classifying them in such a way as to be able to differentiate between elementary ones, which are likely to have psychological underpinnings, and derived ones. (2) What are those psychological underpinnings; in other words, what is the nature of *psychological* universals? (3) How do psychological universals generate behavioral universals? In relation to the latter question, establishing lists and classifications of psychological universals and behavioral universals is one thing; uncovering the processes involved in how the former generate the latter is another. Taken together, those three questions set a rather formidable research program. In this chapter, I propose preliminary answers to the three questions by concentrating on *social universals* – a large subset of behavioral universals.

Four research fields and corresponding theoretical approaches are relevant to the present purpose: cross-cultural psychology, cognitive anthropology, the phylogenetic comparative perspective, and evolutionary psychology. In their review of psychological universals from a cross-cultural psychology perspective, Norenzayan and Heine (2005) define them as "core mental attributes shared at some conceptual level by all or nearly all non-brain-damaged adult human beings across cultures" (p. 763). They differentiate between three broad categories, listed here in decreasing degree of universality: *accessibility universals*, which involve the same mental processes (motivational, emotional, and cognitive), serving the same function, and being used to the same degree (i.e., by the same proportion of individuals) in all societies (e.g., the tendency to experience increased positive affect toward familiar objects); *functional universals*, which involve the same mental processes used for similar functions, but at different degrees across societies (e.g., attachment styles); and *existential universals*, which involve the same mental processes serving different functions in different societies (e.g., the criteria used to make similarity judgments about objects). Norenzayan and Heine (2005) are thus primarily concerned with the second question mentioned above: the nature of psychological universals.

A number of cognitive anthropologists have addressed the second and third questions. They aimed to identify the nature of the universal cognitive mechanisms and unitary organizing principles structuring cultural diversity in various domains, including color classifications (Kay, 2005), plant and animal classifications (e.g., Atran & Medin, 2008; Berlin, 1992), kinship terminologies (e.g., Hage, 1997; Jones, 2004, 2010; Kronenfeld, 2006), and religious beliefs (e.g., Atran, 2002; Boyer & Bergstrom, 2008). To my knowledge, however, cognitive anthropologists have not carried out similar analyses with regard to complex social patterns, nor explored the relations between cognitive universals and social universals.

The phylogenetic comparative perspective refers here to the comparison of human and nonhuman primate behavior. Traits that we share with other primates – whether psychological or behavioral – were likely part of our evolutionary heritage, in which case they have a panhuman distribution (e.g., Kappeler & Silk, 2010). The contribution of that method to the study of human universals is not limited to traits that we share with other species; it also reveals traits that are uniquely human. Interestingly, many such universals have been overlooked by researchers who concentrated their efforts on the anthropological or psychological literature. This is likely because recognizing the significance of the corresponding universals requires using nonhuman species as an out-group in relation to humans. Among those traits are the recognition of affinal kinship (in-laws), the existence of lifetime bonds between brothers and sisters (a seemingly trivial, but highly consequential trait), patrilineal kinship networks, the instructional control of others, and the parental influence on mate selection (Chapais, 2014, 2016a, 2016b). The comparative method thus contributes to the first two questions above.

As to evolutionary psychology, what is perhaps its most fundamental notion – that of evolved psychological mechanisms (hereafter, psychological adaptations) – implies the existence of complex psychological universals regulating behavior. That notion is thus directly relevant to Question 2. Psychological adaptations are defined as mechanisms designed by natural selection to solve specific problems of survival and reproduction faced by early humans on a recurrent basis, such as acquiring sexual partners, preventing sexual infidelity, caring for offspring, attaining status, avoiding incest, and trading resources and services (Buss, 2014; Tooby & Cosmides, 1992; for a review of psychological adaptations regulating social behavior, see Neuberg, Kenrick, & Schaller, 2010). By virtue of their universal character, psychological adaptations are expected to translate into some kind of behavioral universals, at least in some socioecological conditions, but little is known about the exact nature of those behavior patterns and conditions and the exact processes intervening in the production of social universals (Question 3).

In sum, most research on absolute universals focused on establishing repertoires of behavioral universals and defining broad types of psychological universals. Remarkably, there exists no conceptual framework integrating behavioral universals and no model of how psychological universals might generate behavioral universals. In what follows, I propose a classification of the subset of social universals in 10 categories, which differ in the level of analysis at which social phenomena are studied, and I address the three above questions: How do the 10 categories articulate with each other? What is the nature of the psychological universals underlying them? How do the latter produce social universals? I argue that the notion of psychological adaptation generates mental units that provide the interface between psychological universals and social universals – between human nature and the sociocultural realm.

5.1 SOCIAL UNIVERSALS

To achieve the present classification of social universals, I relied primarily on Brown's list, which I completed with examples drawn from other sources (Chapais, 2014, 2015, 2016a, 2016b). The resulting repertoire is certainly not exhaustive, but the assumption here is that the corresponding sample encompasses all possible types of social universals. The proposed framework was thus meant to accommodate all types of social universals and was built inductively. The rationale underlying it is that social phenomena may be studied at various levels of analysis, with the corresponding classes forming a hierarchical structure in which the elements at a given analytical level constitute the building blocks of elements at the next higher level. A particularly useful illustration of this principle is Hinde's (1991) *social complexity framework*, which comprises four levels of analysis, namely individual behaviors, the building blocks of social interactions, which are themselves the components of social relationships, which are the building blocks of social structures. The present framework includes these four classes plus six others.

The first class of social universals comprises *individual social acts* (the equivalent of Hinde's individual behaviors), which are themselves the building blocks of *social interactions*, the second class of social universals. Differentiating between these two classes is not immediately obvious. For instance, is "insulting" an individual social act or a social interaction? To answer that question, I use the following criterion: in social acts only the behavior of the actor is known, whereas in social interactions the behavior of both participants is specified. The label of "insulting," for instance, does not contain information about the target's reaction (e.g., ignoring, leaving, insulting in return, and so on). The accent is placed on the actor's behavior and motivation, without regard to the recipient's reaction, and for this reason the element is classified as an individual social

act. The same applies to "courting," which does not specify the target's response. In contrast, the expression "incest avoidance" refers to a social interaction because it implies both an incestuous attempt and a refusal by the target. Similarly, "toilet training" implies the act of teaching and the act of learning.

The class of individual social acts is particularly important because, as will be argued, it includes the most basic of social universals, the likely building blocks of all others, and hence those whose connections with psychosocial universals are expected to be among the most direct. It is noteworthy that Brown's (1991) list of absolute universals includes very few individual social acts. This is not because they are few in number, but because they are easier to identify when using a hypothetico-deductive approach, as discussed in Section 5.2.

Social interactions are the building blocks of the third class of social universals, *social relationships* (e.g., pair-bonds, friendships and task-specific cooperative partnerships). In that view, social relationships are emergent products of consistent patterns of social interactions between two individuals, those patterns being defined in terms of the nature, frequencies, and temporal patterning of their constituent social interactions (Hinde, 1991). Brown's (1991) list of universals includes little information on types of social relationships. These are nonetheless highly significant considering that even though nonhuman primates form enduring bonds (Mitani, 2009; Silk, 2010), several types of social relationships are uniquely human, notably leader–follower relationships, bonds between affines, and bonds between patrilineal kin (Chapais, 2014).

The fourth class of social universals, *social differentiation criteria*, includes the universal factors on the basis of which humans establish differences between themselves and others and among others and decide on the course of actions to be taken in a variety of contexts. They are the criteria involved in the selection of partners for social interactions and hence in the formation of social relationships. They comprise intrinsic attributes (e.g., sex, relative age, absolute age, type of kinship link, generation, group membership), physical characteristics (e.g., health status, physical formidability, beauty), physical skills (e.g., strength, speed, dexterity), intellectual skills (e.g., intelligence, oratory), and temperamental traits (e.g., fairness, reliability, braveness). Brown's (1991) list of absolute universals includes just a few social differentiation criteria (Table 5.1). A comprehensive list of those criteria adds significantly to the repertoire of social universals and raises questions about their origins.

The fifth class comprises basic *adaptive processes regulating social relationships*, such as kin altruism, mutualistic cooperation, status competition, and the female bias in childcare. The adaptive character of those processes, combined with the fact that they have deep phylogenetic roots – being present in other species, including nonhuman primates – and that, taken together, they govern all types of social relationships, strongly suggest that they are

Table 5.1 A classification of a large sample of social universals found in Pinker's (2002) alphabetical list of absolute universals drawn from Brown (1991). Primary social universals are likely to have psychological underpinnings. Derived social universals result from combinations of primary social universals.

Primary social universals

1 **Individual social acts.** Insulting, tickling, healing the sick, gossiping, gift-giving, mourning, courting, greeting, threatening,[a] joking, monitoring/imitating high-status individuals,[a] provisioning children,[a] stealing[a]

2 **Social interactions.** Misinforming/misleading others, toilet training, rape, weaning, food sharing, teaching, taking revenge, incest avoidance, mating, male conflicts over mate,[a] female conflicts over mate,[a] sibling conflicts,[a] mediation in conflicts, dominance interactions,[a] coalitions, social play

3 **Dyadic relationships.** Friendships,[a] pair-bonds,[a] liaisons,[a] mother–offspring bonds,[a] father–offspring bonds,[a] grandparent–grandchild bonds,[a] lifetime bonds between cross-sex siblings,[a] bonds between matrilineal kin and between patrilineal kin,[a] bonds between affines (e.g., brothers-in-law),[a] task-specific cooperative partnerships,[a] leader–follower relationships

4 **Social differentiation criteria.** Gender,[a] age difference with self,[a] others' absolute age,[a] others' relative ages,[a] paternal vs. maternal kin,[a] lineal vs. collateral kin,[a] lineal distance,[a] collateral distance,[a] ascending vs. descending generation,[a] affinal link,[a] same vs. distinct kin group,[a] physical formidability,[a] physical skills,[a] intellectual skills,[a] temperamental traits,[a] phenotypic similarity between others,[a] beauty,[a] health,[a] in-group vs. out-group membership

5 **Adaptive processes regulating social relationships.** Reciprocity (exchange), negative reciprocity (revenge), mutualistic cooperation, kin altruism,[a] nepotism, resource competition, status competition,[a] male sexual competition, male bias in lethal aggression and coalitionary violence, female sexual competition,[a] female mate choice,[a] female bias in childcare

6 **Moral dimensions of adaptive processes.** Fairness (equity), valuation of generosity, judging others, disapproval of stinginess, resentment of being exploited,[a] valuation of loyalty[a]

Derived social universals

7 **Consistent social arrangements.** Multimale–multifemale group,[a] kin groups, division of labor by age and sex, age grades, biparental families,

Continued

Table 5.1 Cont.

Derived social universals

household, status orders, prestige inequalities, kin-based structures (e.g., lineages, clans),[a] between-group alliances,[a] federated/nested structures of intergroup relations,[a] leadership/government

8 **Categories of social activities.** Games,[a] sports,[a] feasts, ceremonies, cooperative labor, trade, collective decision-making, capital punishment, intergroup conflicts,[a] religious rituals, rites of passage, intergroup visits,[a] collective performances (storytelling, dance, music, singing)

9 **Principles characterizing social life.** Political control by males, hospitality, collective identity, in-group favoritism (ethnocentrism, parochialism), ostracism, parental influence on mate selection (marital unions),[a] exogamy,[a] property, territoriality, economic inequalities, murder proscribed, classificatory kinship,[a] group-wide coordination,[a] multigroup coordination,[a] social inheritance (status, wealth)

10 **Abstract social notions.** Law, proscriptions, prescriptions, membership rules, corporate statutes (kin or age based), rights, etiquette, obligations, sanctions, humor

[a] From Chapais (2014, 2015, 2016a, 2016b).

genetically specified. The elements forming the next class of social universals – *moral dimensions of adaptive processes* – appear to be correlates of the previous class. The principles of fairness and equity, for example, regulate reciprocity, while the valuation of generosity pertains to altruism and reciprocity. This class thus includes the motivational and emotional aspects of adaptive processes regulating social relationships.

I refer to the first six classes of social universals as *primary* social universals to indicate not only that they are, structurally speaking, the most basic ones, but also that as such they are those most likely to have psychological underpinnings. The remaining four classes are grouped under the label of *derived social universals* to indicate that they result from various combinations of primary universals and therefore that they do not have direct psychological underpinnings.

The elements belonging to the seventh class of social universals (e.g., biparental families, lineage structures, status orders) are subsumed under the label of *consistent social arrangements*, with this expression replacing the less inclusive term of "social structure" in Hinde's (1991) framework. Those elements are derived in the sense that they describe emergent products of consistent patterns of social relationships. In that view, families, lineages, and a large number of such structures existed as mere social regularities – enduring networks of social relationships –

well before they were recognized as social entities per se and conceptualized. Accordingly, understanding the origins of social arrangements involves understanding how the aggregation of multiple individual social acts, social interactions, and social relationships translate into higher-level structural regularities that exhibit their own properties, the latter being the subject matter of sociological studies.

The same principle applies to *categories of social activities* (e.g., games, rituals, feasts, trade), which form the eighth class of social universals. These reflect the existence of recurrent, domain-specific patterns of social interactions, usually involving several individuals belonging to various social networks. Contrary to social arrangements, however, social activities refer to *events* rather than to enduring patterns of social relationships. The corresponding labels thus place the accent on the nature and content of social interactions rather than on social structure. They refer to a distinct level of analysis of the social realm.

The ninth class of social universals includes elements defined at a higher level of abstraction and a distinct level of analysis. *Principles characterizing social life* include things like male political control, hospitality, the parental influence on marital unions, property, and in-group favoritism. Such principles may characterize social relationships, social arrangements, or social activities. For example, hospitality pertains to various types of social relationships, the parental influence on marital unions characterizes parent–child relationships, and in-group favoritism describes social groups, regardless of how these are defined. All such principles stem from the properties of lower-level universals. Like social arrangements and social activities, they are the emergent products of the aggregation of multiple individual acts, which are themselves governed by the same universal set of social motivations.

Finally, social universals include a number of highly *abstract social notions* such as proscriptions, law, membership rules, and etiquette. Those constructs have much in common with the previous class in that they may be construed as extremely general principles characterizing social life. The reason they are placed in a distinct class is that they do not refer to any concrete social pattern, whereas the elements of the previous class, such as property and ostracism, do have a concrete dimension.

The present classification of social universals indicates that the 10 classes pertain to different aspects and levels of analysis of social life and closely articulate with each other within some sort of hierarchical structure, the most elementary components of which are represented by individual social acts. As noted earlier, all six classes of primary social universals are expected to be psychologically grounded and therefore any model of the psychological foundation of social universals must account for the six classes.

5.2 THE PSYCHOLOGICAL FOUNDATION OF SOCIAL UNIVERSALS

To address the question of the psychological underpinnings of social universals, I adopt a deductive, evolutionarily grounded approach. I posit that psychological adaptations pertaining to social behavior (hereafter, *psychosocial adaptations*) are the most fundamental and consequential mechanisms regulating social life. Based on this assumption, one expects psychosocial adaptations to play a central role in the production of primary social universals.

It should be noted right from the outset that the concept of psychological adaptation directly accounts for the existence of three of the six classes of primary social universals, namely social differentiation criteria (Class 4), adaptive processes regulating social relationships (Class 5), and the moral dimensions of the latter (Class 6). Adaptive processes regulating social relationships (e.g., reciprocity, status competition, kin altruism) meet the criteria of psychological adaptations and are construed as such by evolutionary psychologists (e.g., Neuberg et al., 2010). They are thus psychological universals, and rather complex ones. The reason they are included in Table 5.1 as social universals is that the corresponding terms simultaneously refer to the large functional categories of social patterns they generate, such as social exchange, revenge, and nepotism. The same applies to the moral dimensions of adaptive processes (e.g., the principle of fairness). Those dimensions imply the existence of motivational, emotional, and cognitive mechanisms regulating the adaptation (e.g., the capacity to detect cues of fairness or unfairness and to experience the associated emotions), and they simultaneously refer to the behavioral manifestations of those mechanisms in the sociocultural realm (e.g., the various concepts and social practices pertaining to fairness). Psychosocial adaptations also imply that social differentiation criteria – the factors involved in the selection of social partners – are genetically specified aspects of the adaptation. For example, gender, age, kinship, health, beauty, and status are integral parts of the adaptation regulating sexual attraction and mate selection (e.g., Buss, 2003). In sum, the hypothesis that psychological adaptations regulate social life predicts that social universals will include the resulting broad functional categories of social patterns, the moral principles governing those patterns, and the criteria on the basis of which social partners are selected.

The next question then concerns the relations between psychological adaptations and the most basic of the remaining three classes of primary social universals – individual social acts, the building blocks of social interactions and social relationships. Assuming that psychological adaptations evolved to address broad functional objectives (e.g., status attainment), it is not immediately obvious how such general objectives might translate, ultimately, into specific types of individual social acts. This question is fundamental because it concerns the extent to which human nature produces specific social contents, by opposition to general potentials and constraints (Tooby &

Cosmides, 1992). It is about the extent to which the evolved human architecture of the human mind determines: (1) specific types of social acts; (2) the socioecological contexts in which those acts take place; and (3) the identity (in terms of age, sex, kinship, status, and the like) of the social partners involved. In short, it is about circumscribing the behavioral deployment or behavioral scope of psychological adaptations. In relation to this issue, it has been suggested that psychological adaptations evoke distinct, genetically specified behavioral strategies in different socioecological contexts in the manner of a jukebox programmed to play different tunes in different environments (Tooby & Cosmides, 1992; see also Brown et al., 2011; Gangestad, Haselton, & Buss, 2006; Nettle, 2009). According to the model proposed here, psychological adaptations would produce, in addition to evoked behavioral outputs, a large number of universal categories of individual social acts, which are *not* genetically specified and, moreover, operate as templates for the production and adoption of cultural variants.

To illustrate the model, I focus in detail on a single psychosocial adaptation hypothesized to regulate status attainment. The following discussion summarizes and to some extent develops ideas presented elsewhere (Chapais, 2017).

5.2.1 Status Attainment

The notion of social status refers to an individual's position in various social orders in which one's rank correlates positively (but not necessarily linearly) with advantages such as access to desired resources and services, influence, power, and privileges. The evolutionary importance of status in humans is supported by its phylogenetic connection with dominance status in other primates (Chapais, 2015; Watts, 2010; Weisfeld & Dillon, 2012), by the universality of social hierarchies in human societies (Anderson, Hildreth, & Howland, 2015), and by several empirical studies reporting positive correlations between male status and fertility (e.g., Borgerhoff Mulder, 1987; Chagnon, 1988; Gurven & von Rueden, 2006; Irons, 1979; Kaplan & Hill, 1985; Smith, 2004; von Rueden, Gurven, & Kaplan, 2011). An influential evolutionary model of human status was proposed by Henrich and Gil-White (2001), who reported that in a wide array of domains, from hunting to shamanism, highly skilled individuals were the object of more admiration, benefited from more privileges, received more deference, were preferentially imitated by others, and were more influential even beyond their area of expertise; in short, they had more prestige. Henrich and Gil-White compared the manifestations of prestige-based status in humans and dominance status in both humans and nonhuman primates and concluded that the two phenomena rested on different psychologies and resulted from different selective pressures. On this basis, they argued for the existence of two alternative strategies of status attainment in humans: dominance, which

involves physical power, and prestige, which involves competence. Tracy, Shariff, and Cheng (2010), Cheng, Tracy, and Henrich (2010), and Cheng et al. (2013) later suggested that the two strategies rested on different kinds of pride: hubristic pride motivating physical dominance and authentic pride motivating prestige. In short, people would attain status either by intimidating others and being feared, by exhibiting competence and being attractive, or by using both strategies but separately.

In contrast with this dual model of status, I argued elsewhere (Chapais, 2015) that dominance is a domain of competence in itself and that as such it involves prestige. There would therefore be a single way to status: competence. The capacity to exercise dominance depends on the acquisition of competencies in a number of domains besides sheer physical strength (e.g., in making or handling weapons, recruiting allies based on ideational arguments, coordinating and leading them in various contexts, controlling information or resources affecting others' physical or psychological welfare, and using nonphysical entities to threaten or inflict costs to others). Individuals exhibiting the relevant skills are attractive and have prestige, like experts in any other domain, presumably because they are in a position to provide various types of services to their fellow group members, from efficient protection, to access to resources, to healing. In support of that view, empirical studies indicate that even dominance based on physical formidability (stature and strength) confers a high status because it reflects competence and social value, rather than because dominant individuals are feared by others. Taller men are perceived as having greater leadership abilities (Blaker et al., 2013; Murray & Schmitz, 2011; Re et al., 2013; von Rueden, 2014), and group members willingly grant a high status to physically formidable men owing to their leadership abilities and the services they may provide, such as punishing free-riders and negotiating with other groups (Lukaszewski et al., 2016). In nonhuman primates as well, high-ranking individuals, especially the alpha male and female, are uniquely positioned to provide specific types of benefits to subordinates, including efficient protection against aggressors, decisive support in conflicts, and access to monopolizable resources (mates or food). The high value of top-ranking individuals as potential partners may explain why subordinates are attracted to them, offering grooming and support in exchange for help and tolerance (reviewed in Chapais, 2015). From that perspective, dominants have *proto-prestige* and primate dominance hierarchies concurrently are prestige hierarchies.

According to this reasoning, competence is always involved in the attainment of status in humans, regardless of the domain of activity, agonistic or non-agonistic (see also Anderson & Kilduff, 2009a, 2009b). Consequently, individuals are expected to experience the emotion of pride both when feeling competent in the accomplishment of relevant tasks and when outperforming others. Correlatively, they are expected to experience the emotions of sadness and shame both when feeling incompetent and when being outperformed by others. Whatever the exact nature of the specific emotional system underlying status attainment, this brief discussion will suffice to illustrate the present model. Figure 5.1 illustrates its pyramidal structure, which comprises the four levels discussed below: (1) systemic social motives; (2) context-specific social motives; (3) universal social propensities; and (4) universal categories of individual social acts.

5.2.2 Systemic Social Motives as Design Features of Psychological Adaptations

The foundational principle of the proposed model is that achieving the general functional objective of a psychosocial adaptation (e.g., status attainment) requires achieving a specific number of sub-goals that form a biological system; in other words, all sub-goals must be achieved for the adaptation to work. I refer to those genetically specified design features of psychosocial adaptations as *systemic social motives*. In the case of status attainment, and based on how status was defined in Section 5.2.1, there would exist four such motives: (1) opportunistically acquiring and improving one's competence (in one or more domains of activity); (2) monitoring one's and others' relative position in the competence order; (3) seeking to outperform or outrank others whenever this is possible and advantageous – which depends on the context (e.g., Anderson et al., 2012) – and to maintain one's position in the face of challenges; and (4) signaling one's position in the status hierarchy in order to reap the benefits associated with it. The four motives are necessary and fully complementary aspects of status attainment: in the absence of either one of them, the system would not be operational. This implies that humans are motivationally and cognitively predisposed to acquire competence, pay attention to their relative level of competence, compete for status, and signal their status. The proximate mechanisms involved would be such that the corresponding behaviors are experienced as rewarding (increasing pride or reducing shame). A comparative phylogenetic perspective supports this proposition. Primate studies indicate that the four systemic motives have deep evolutionary roots. Status attainment in a primate dominance hierarchy involves the same components: acquiring fighting skills (notably, through play); knowing about one's relative rank and the rank of others; seeking to maintain one's rank in the face of challenges and to outrank others whenever opportunities arise (upon changes in relative physical or coalitionary power); and signaling one's rank (e.g., through displays).

Systemic social motives may be construed as a category of psychological universals. The assumption that they are design features of psychological adaptations means that they are absolute universals and characterize all humans. The systemic social motives of status attainment or their equivalents are not found in repertoires of behavioral universals (Brown, 1991) or discussed explicitly in syntheses written from the perspective of cross-cultural psychology (e.g., Berry et al., 2002; Smith & Bond, 1999).

5.2.3 The Socioecological Deployment of Systemic Social Motives

This principle is concerned with the conditions under which systemic social motives are expected to produce social universals. Except for extremely basic behavior patterns such as reflexes, human behavior is virtually always modulated by cultural learning and hence highly variable – social behavior is always sociocultural. A straightforward implication of this is that behavior patterns that are cross-culturally universal must be so because the socioecological factors promoting them are also universal. Social universals like status competition, courtship, and nepotism would exhibit a panhuman distribution not because they are generated by the human mind independently of the context, but because the contextual conditions under which they are observed are general enough to be met in all bisexual human groups (Chapais, 2014). In that view, and somewhat paradoxically, absolute social universals are no less context dependent than cross-culturally variable behavior patterns. It follows that to identify the social universals stemming from a psychosocial adaptation, one must first define the universal categories of contexts containing the cues activating the adaptation.

With respect to status attainment, the corresponding systemic motives are expected to be activated, by definition, when individuals are in a position to perform or witness *status-relevant activities*. These are defined as activities meeting the following three criteria: (1) they are complex enough to require various types and levels of competencies to be performed adequately; (2) they are socially valued and hence attract attention – in other words, performers attract various audiences; and (3) they allow assessments of relative competence to be made both by the participants themselves and by the observers – that is, the activities lend themselves to social comparisons and social rankings. Although they are extremely variable cross-culturally, status-relevant activities may be classified in three broad universal classes depending on how they lend themselves to competence assessments: dyadic competitive activities; cooperative activities; and individual activities. In *dyadic competitive activities* (e.g., real physical fights, insult contests, play-wrestling, oratory debates), contestants are in a position to directly assess their own relative competence based on their performance and the identification of a winner. Status-relevant *cooperative activities* include both noncompetitive activities (e.g., group hunting, house building, long-distance traveling, collective dancing) and competitive activities (e.g., intergroup conflicts, ball games). Participants in those activities can assess their relative competence by comparing their performances in the course of the activity and their respective contributions to the end result. In status-

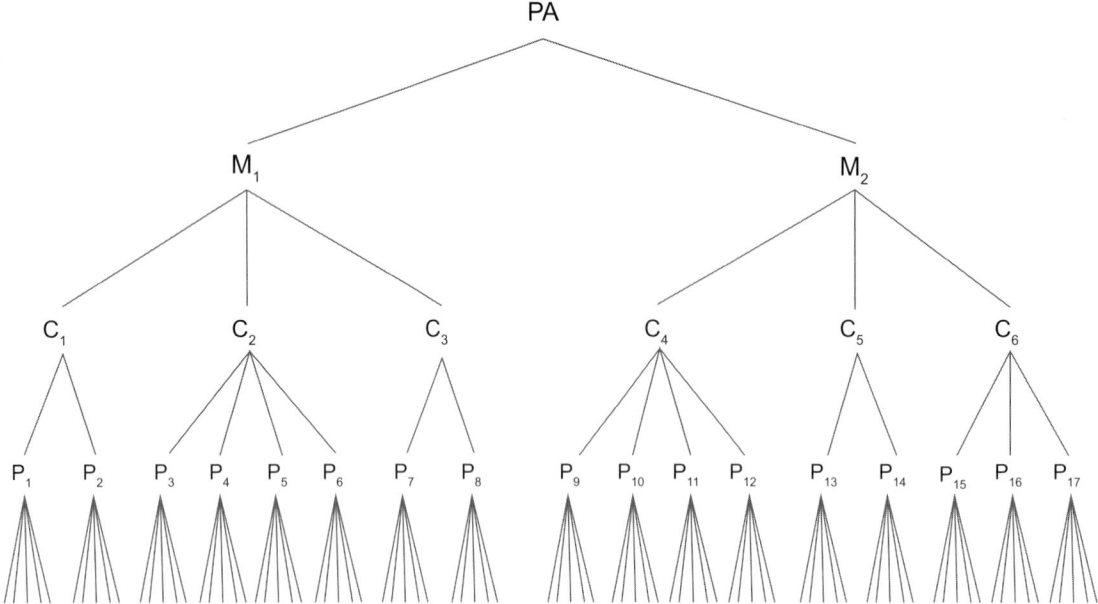

Propensity-congruent sets of sociocultural variants

Figure 5.1 A highly schematic representation of how psychological adaptations (PAs) generate universal categories of individual social acts. In this example, a PA comprises two systemic social motives (M), which are activated in three universal categories of socioecological contexts exhibiting the relevant cues, with this resulting in six context-specific social motives (C). The latter may be achieved through a number of cognitively different pathways, with this translating into as many distinct behavioral options, or universal social propensities (P). Each propensity in turn operates as a template for the production of a set of cultural variants congruent with the P – it generates an open-ended category of individual social acts.

relevant *individual activities* (e.g., solitary hunting, cooking, shamanistic activity, tool-making, storytelling), performers may assess their relative competence based on various correlates of the activity (e.g., type and quantity of game brought back, quality of tools made, frequency and extent of healing, reaction of audience) and their knowledge of the competence of others.

These three classes of status-relevant activities define the first three broad categories of universal contexts recruiting the adaptation. A fourth, more inclusive category of contexts includes all situations in which individuals witness or communicate a posteriori about status-relevant activities performed by others – whether individual, dyadic, or cooperative – and hence are in a position to evaluate and comment upon the performers' relative competence. Finally, a fifth universal category of contexts eliciting the adaptation comprises all situations in which individuals neither participate in nor witness or communicate about status-relevant activities, but are in a position to interact with others based on their own relative status or the relative status of others (e.g., by signaling their competence levels or by imitating higher-ranking individuals). In sum, it is assumed here that in all human societies individuals have opportunities to participate in individual, dyadic, and cooperative status-relevant activities, witness such activities performed by others, and interact with others based on their status, and that those five broad

contextual categories contain the cues activating the status-attainment adaptation.

Given that status attainment involves four systemic social motives, which may be actualized in five universal categories of contexts, this produces a four-by-five table of 20 different situations in which individuals may seek to achieve a specific status-relevant motive in a specific type of context, or 20 universal *context-specific social motives*. For instance, participants in a status-relevant individual activity such as hunting are in a position to achieve the four systemic motives: outperforming others (based on how performance is defined by their group); evaluating their own performance in relation to what they know about the performance of others; signaling their success (competence) on their return; and opportunistically improving their competence (e.g., by adopting more efficient behavioral procedures).

Context-specific social motives may be construed as another category of psychological universals that result from interactions between systemic motives and the universal socioecological contexts actualizing them; they are context-dependent psychological universals.

5.2.4 From Systemic Motives to Universal Social Propensities

It is assumed here that the whole set of human cognitive abilities is available for the actualization of psychological adaptations, from which it follows that context-specific motives may be achieved through a number of cognitively different pathways available to all humans. For example, the motive to signal one's competence in the context of individual activities may be achieved either by exhibiting the immediate markers of success (e.g., captured game), which is the equivalent of saying, "This reflects my competence level," or by attributing one's counter-performance to external factors, which is the equivalent of saying "This does not reflect my competence level." I refer to the cognitively distinct, universally available strategies stemming from systemic motives as *universal social propensities*. Table 5.2 presents a list of 40 such propensities pertaining to status attainment.

Table 5.2 A sample of universal social propensities (*n* = 40) relating to the status-attainment adaptation, classified according to five universal categories of contexts activating them (in bold) and the adaptations' four systemic social motives: acquiring competence (A); evaluating competence (E); competing for competence (C); and signaling competence (S). Reprinted from Chapais (2017).

Performing status-relevant individual activities

1 Evaluating one's performance relative to others' in retrospect (E)
2 Assessing costs and benefits of outperforming others (C)
3 Seeking to outperform others when benefits outweigh costs (C)
4 Exhibiting immediate markers of competence/success (showing off) (S)
5 Attributing one's counter-performance to external factors (S)
6 Opportunistically improving one's competence through individual learning (A)
7 Improving one's competence by copying experts (A)

Participating in status-relevant cooperative activities

8 Evaluating one's performance level in relation to participants (E)
9 Evaluating participants' relative performance levels (E)
10 Assessing costs and benefits of outperforming certain participants (C)
11 Seeking to outperform participants when benefits outweigh costs (C)
12 Seeking to conceal one's errors (incompetence) (S)
13 Seeking to associate with experts during activity (S)
14 Exhibiting immediate markers of competence/success (S)
15 Opportunistically improving one's competence through individual learning (A)
16 Improving one's competence by observing and copying experts (A)

Participating in dyadic competitive activities

17 Assessing rival's relative competence (E)
18 Assessing costs and benefits of outranking rival (C)

Continued

Table 5.2 Cont.

Participating in dyadic competitive activities

19 Seeking to outperform (outrank) rival when benefits outweigh costs (C)
20 Seeking to intimidate rival by exhibiting physical/behavioral markers of competence (C)
21 Seeking to conceal cues of fear/nervousness (C)
22 Mocking rival's errors (incompetence) (C)
23 Seeking to obtain audience's support against rival (C)
24 Exhibiting immediate markers of victory (S)
25 Opportunistically improving one's competence through individual learning (A)
26 Improving one's competence by copying experts (implies admiration of experts) (A)

Witnessing or communicating about others' status-relevant activities

27 Assessing own performance level in relation to others (E)
28 Assessing others' relative performance levels (E)
29 Pointing out errors of performers to third parties (C)
30 Mocking others' errors (incompetence) (C)
31 Attributing others' success to factors other than competence (e.g., chance, cheating) (C)
32 Improving one's competence by copying experts (implies admiration of experts) (A)

Interacting with others based on one's or others' relative competence

33 Envying others' status and its correlates (e.g., possessions) (C)
34 Resenting non-recognition of one's status (C)
35 Informing others about one's performance level and competence (S)
36 Exhibiting generosity (as correlate of competence) (S)
37 Exhibiting permanent (e.g., symbolic) markers of one's competence (S)
38 Adopting characteristic features of high-status individuals (S)
39 Deferring to higher-status individuals (S)
40 Seeking to associate/ally with high-status individuals (A, S)

As another example, when participating in dyadic competitive activities (physical, verbal, or intellectual) (Table 5.2, third context), individuals have cognitively different options: they may seek to outperform their rival, intimidate the latter by exhibiting various competence markers, mock the rival's errors (incompetence), or obtain the audience's support against the rival. Similarly, when witnessing status-relevant activities performed by others (Table 5.2, fourth context), observers may still compete for status with the performers, but indirectly. In a species in which theory of mind allows individuals to read and affect others' beliefs, thoughts, and opinions, it is possible to compete for status by devaluating others' reputation: one may point out the participants' errors to third parties, mock them, or attribute the participants' success to factors other than competence.

A basic assumption of the present model is that universal social propensities require neither social learning nor prior cultural innovations to emerge. They are construed as basic enough to be internally generated – individually learned in similar sets of circumstances – and hence available to all humans independently of their cultural background. Cognitively speaking, coming up with a social propensity would be comparable to coming up with the idea of taking cover when it is raining. Universal propensities would thus be the province of intuitive solutions to the actualization of systemic motives. Clearly, therefore, an important issue concerns the range and exact nature of the intuitive solutions pertaining to any systemic motive. Accordingly, the list of propensities in Table 5.2 is tentative and certainly not exhaustive.

If social propensities are individually learned, as posited here, they may be culturally countered. For example, even though it might be particularly tempting to outperform others, signal one's competence, or mock others' incompetence in certain circumstances, cultural norms might well prohibit such behaviors. Universal propensities are therefore best construed as *default propensities*: they reflect what people would do if they were not culturally encouraged, and strongly so, to do something else. In that view, human nature includes, in addition to a wide array of motivations, emotions, cognitive processes, and genetically specified behavioral proclivities (e.g., evoked behavioral outputs), a large number of context-specific, intuitive behavioral propensities. This particular domain of human nature remains to be fully explored and characterized.

Universal social propensities are not actual behavior patterns, they are not social universals. They are context-dependent psychological universals resulting from interactions between systemic social motives, universal socioecological contexts, and universally available cognitive processes. Psychological adaptations would thus produce various types of psychological universals differing in the level of analysis at which the psychological underpinnings of social behavior are studied. This view broadens substantially the notion of psychological universals. Aknin et al. (2013) reported having found "the first evidence for a possible psychological universal" (p. 646). Based on surveys and experiments conducted in different countries, they found that prosocial spending – experiencing emotional rewards when using one's financial resources to help others – was universal, even though the intensity of emotional rewards varied cross-culturally. They interpreted this pattern as a functional universal (Norenzayan & Heine, 2005) and concluded that the corresponding emotions were "deeply ingrained in human nature" and adaptive. From the perspective of the present model, prosocial spending would be a universal social propensity among many others stemming from the psychological adaptation regulating altruism or that regulating social exchange (e.g., Cosmides & Tooby, 2015).

Universal social propensities are not represented in Brown's (1991) list of absolute universals, presumably because their identification requires a deductive psychological framework such as the present one. Summed up over all psychological adaptations, universal social propensities are expected to add considerably to the list of absolute universals.

5.2.5 Social Propensities and the Production of Panhuman Categories of Social Acts

The model's last principle is concerned with how universal social propensities translate into actual behavior patterns. Social propensities would operate as templates for the production of behavioral variants compatible with the propensity, or *propensity-congruent social acts*. In a species endowed with cumulative culture, it is possible for individuals to actualize any social propensity in many different ways depending on the cultural materials they have at their disposal – behavior patterns, ideas, beliefs, and techniques. Considering that those materials are constantly evolving, universal social propensities would set permanent goals shared by all humans, while culture would generate constantly renewable means of achieving those goals in different contexts. This is expected to produce, for any propensity, an open-ended class of cultural expressions – behavioral or ideational – in harmony with it. Accordingly, the 40 propensities listed in Table 5.2 each sustain a specific but generic type of individual social act, which takes on different cultural expressions in different societies. For example, the proclivity to adopt the characteristic features of experts (#38 in Table 5.2) may express itself in the adoption of the attitudes, behavior patterns, language, bodily ornaments, clothing style, material possessions, or opinions of experts in any socially valued competence domain. Across societies, that motivation generates a highly polymorphic class of cultural expressions compatible with it. As another example, signaling one's competence in the context of individual, dyadic, or cooperative status-relevant activities (#4, #14, and #24 in Table 5.2) may be achieved by exhibiting the immediate markers of one's competence (e.g., captured game, war trophies). Again, the exact content of such markers varies considerably across activities and societies.

Principle #4 is germane to the question of how psychological adaptations articulate with culture. It offers a way of reconciling the unitary character of human nature with the extreme diversity of the sociocultural realm or, in Lévi-Strauss's terms, of "overcoming the apparent antinomy between the unity of humankind and the seemingly endless plurality of the forms under which we perceive it" (1983, p. 62). It also addresses Robin Fox's remark (epigraph) about the high level of behavioral redundancy characterizing human societies. In relation to this fundamental question, an important principle was formulated by Sperber (1985, 1996), who suggested that symbolic representations are evaluated positively or negatively depending on their degree of compatibility with psychological mechanisms. Positively evaluated variants would spread through the population, whereas negatively evaluated variants would have narrower distributions or be discarded. In that vein, Tooby and Cosmides (1992) further noted that those evaluation mechanisms should be domain-specific and differ across psychological adaptations. One may infer from this that the degree of appealingness of a cultural variant should be proportional to its degree of harmony with psychological adaptations. The exact steps and criteria involved, however, have not been specified.

In the present model, universal social propensities are the primary mental units involved. They are seen as setting strong constraints on both the mechanisms of creativity – the capacity to produce new associations out of previously unrelated elements (Laland & Reader, 2003) – and the mechanisms of social learning, with this resulting in the selective retention of innovations congruent with the adaptation's propensities. Although the literature on the psychology of creativity is relatively abundant (e.g., Fogarty, Creanza, & Feldman, 2015; Kaufman & Sternberg, 2010), the issue of how psychological adaptations bias creativity has received little attention. Recent evolutionarily oriented reviews (Carruthers & Picciuto, 2014; Fogarty et al., 2015; Gabora & Kaufmann, 2010) discuss the cognitive processes likely to account for the uniqueness of human creativity, but are silent about the constraints affecting the *content* of what is created. Similarly, the cognitive processes involved in social learning have received a fair amount of attention, notably in relation to their exact role in the cumulative dimension of culture (Dean et al., 2012; Legare & Nielsen, 2015; Tennie, Call, & Tomasello, 2009), but those studies do not address the question of how psychological adaptations might bias the adoption of cultural variants. As argued here, psychological adaptations would interface with cultural innovation and cultural transmission through the intermediary of a substantial number of motive-specific, context-specific, internally generated, and cognitively distinct social propensities.

*

I noted earlier that a satisfactory model of the psychological underpinnings of social universals must account for the six classes of primary social universals (Table 5.1). Heretofore, I have been concerned with four of the six classes, namely social differentiation criteria, adaptive processes regulating social relationships, the moral dimensions of those processes, and individual social acts. What about the remaining two classes of primary social universals: social interactions and social relationships? In the present model, they result from interactions between the first four classes. Indeed, individual social acts provide the very building blocks of social interactions and social relationships; adaptive processes and their moral correlates regulate the patterning of social interactions within social relationships; and social differentiation criteria govern the selection of social partners engaging in enduring relationships. The four major components of psychological adaptations thus define both the content and the patterning of social interactions and social relationships. It

would seem, therefore, that the present model goes a long way toward accounting for the existence of primary social universals. This conclusion, however, is based on the analysis of a single psychological adaptation and must await the investigation of the social propensities pertaining to other psychological adaptations (see Chapais, 2017, for an analysis of the social propensities stemming from a distinct psychological adaptation: the control of sexual fidelity).

5.3 CONCLUDING REMARKS

I end this chapter with two remarks, the first of which concerns the origins and causes of absolute universals, the second their significance. Brown (2001, 2004) identified three "general explanations for universals." First, some cultural universals, such as fire and cooking, would be "inventions that, due to their great antiquity and usefulness, have diffused to all societies." Second, other universals, like kinship terminologies and classifications of plants and animals, would "be reflections in culture of non-cultural features that are ubiquitous and important for one reason or another" (e.g., with regard to kinship terminologies, the fact that in all societies people reproduce and have various kin types). Third, other universals, such as romantic love and a preference for a smaller waist-to-hip ratio in women than in men, would "derive (more or less) directly from human nature or [be] features of human nature," the latter including the evolved psychological mechanisms studied by evolutionary psychologists (Brown, 2001, p. 163).

The present model is congruent with those explanations, but integrates them within a single framework in which all three are direct reflections of human nature. Indeed, the idea that non-cultural features exhibiting a panhuman distribution produce behavioral universals *implies* the existence of universal biopsychological mechanisms producing those patterns. The fact that humans classify their kin according to universal criteria (e.g., Chapais, 2016b; Goodenough, 1970; Jones, 2010) entails the existence of mental mechanisms underlying that proclivity. A unitary human nature is a necessary condition for contextual similarities to produce behavioral universals. Similarly, the hypothesis that particularly useful innovations have high diffusion rates implies the existence of unitary psychological criteria for appraising the usefulness of cultural innovations and, from there, their degree of appealingness. Thus, all three explanations attribute a central role to human nature in the production of social universals. So does the present model, which specifies that this results from psychological adaptations operating in panhuman contextual conditions and producing social propensities that define the relative degree of appealingness of cultural variants.

5.3.1 The Tip of the Iceberg

Much of the interest of anthropologists and psychologists in absolute universals lies in their implications regarding the role of human nature in regulating behavior. Absolute universals have always been construed as providing a particularly strong body of evidence in support of the biological underpinnings of human behavior. This is because the panhuman distribution of absolute universals suggests that they are *determined* by human nature. Determined not in the sense that human nature produces forms and meanings invariable across socioecological contexts – the highly variable expressions of absolute universals plainly contradicts that view; determined in the sense that the ubiquity of mental processes or behavior patterns in human societies suggests that their presence transcends the features of the socioecological environment – that they are *context-independent* categories.

This is not the case, however. The occurrence of absolute universals, as argued previously, is conditional upon features of the socioecological environment, but the latter have the particularity of having a panhuman distribution. In general, those features are found in all societies because they reflect basic properties of the physical world, of the human body, or of human groups. For instance, the occurrence of courtship, sexual competition, and stable breeding bonds – regardless of their cultural expressions – depends on the existence of bisexual groups, and nepotism requires the existence of kin groups, conditions that are satisfied in all human societies. The extremely basic nature of those prerequisites explains why absolute social universals have a general character (Table 5.1). Absolute universals would thus be *context-dependent* categories (Chapais, 2014).

If human nature generates universal behavioral regularities whenever the relevant socioecological conditions are themselves universal, it is expected to generate behavioral regularities whenever the relevant socioecological conditions are recurrent, regardless of those conditions having a wide of narrow distribution and of the proportion of individuals exhibiting the behavior pattern. To take a rather extreme and, for this reason, revealing example, a relatively small proportion of men are imprisoned in their lifetime all over the world and not all human societies have male prisons. Whenever they are found, however, male prisons create a specific and recurrent set of behavioral regularities, including strict dominance orders, strong ethnic alliances, intergang battles, high rates of violence, and rape (e.g., Sabo, Kupers, & London, 2001). That set of behavioral traits is associated with a number of socioecological features, including rejection from society, seclusion, unisexual population, high population density, room and board provided, limited repertoire of activities, and so on. In all likelihood, the behavioral correlates of that particular socioecological configuration reflect the operation of human nature in those circumstances; that is, they reflect what men tend to do naturally in this type of context and that, consequently, is extremely difficult to counter through imposed norms and sanctions (Chapais, 2017). In that sense, the behavioral characteristics of male prisons constitute *context-dependent universals*: behavior

patterns that are always present when the corresponding conditions are met (Chapais, 2014).

On this account, any cross-cultural behavioral regularity would express the operation of human nature in the corresponding contexts, and therefore absolute universals would constitute the tip of the iceberg as far as the manifestations of human nature are concerned. They would reflect human nature in highly specific and relatively rare types of socioecological contexts: those having a panhuman distribution. They would thus be no more and no less significant than any other context-dependent universal as far as their implications for the potency of human nature are concerned. Implied here, sociocultural anthropologists are dealing with human nature whenever they study cross-cultural regularities – a rather nonintuitive idea to many of them.

I conclude that if absolute universals are the most salient and spectacular manifestations of human nature, they nonetheless constitute a tiny fraction of those manifestations. Human nature is ubiquitous.

5.4 ACKNOWLEDGMENTS

I thank Donald Brown and the editors for helpful comments on the manuscript.

REFERENCES

Aknin, L. B., Dunn, E. W., Helliwell, J. F., et al. (2013). Prosocial spending and well-being: Cross-cultural evidence for a psychological universal. *Journal of Personality and Social Psychology*, **104**, 635–652.

Anderson, C., & Kilduff, G. J. (2009a). The pursuit of status in social groups. *Current Directions in Psychological Science*, **18**, 295–298.

Anderson, C., & Kilduff, G. J. (2009b). Why do dominant personalities attain influence in face-to-face groups? The competence-signaling effects of trait dominance. *Journal of Personality and Social Psychology*, **96**, 491–503.

Anderson, C., Willer, R., Kilduff, G. J., & Brown, C. E. (2012). The origins of deference: When do people prefer lower status? *Journal of Personality and Social Psychology*, **102**, 1077–1088.

Anderson, C., Hildreth, J. A. D., & Howland, L. (2015). Is the desire for status a fundamental human motive? A review of the empirical literature. *Psychological Bulletin*, **141**, 574–601.

Atran, S. (2002). *In Gods We Trust: The Evolutionary Landscape of Religion*. Oxford: Oxford University Press.

Atran, S., & Medin, D. (2008). *The Native Mind and the Cultural Construction of Nature*. Cambridge, MA: MIT Press.

Berlin, B. (1992). *Ethnobiological Classification: Principles of Categorization of Plants and Animals in Traditional Societies*. Princeton, NJ: Princeton University Press.

Berry, J. W., Poortinga, Y. H., Segall, M. H., & Dasen, P. R. (2002). *Cross-cultural Psychology: Research and Applications*, 2nd ed. Cambridge, UK: Cambridge University Press.

Blaker, N. M., Rompa, I., Dessing, I. H., et al. (2013). The height leadership advantage of men and women: Testing evolutionary psychology predictions about the perceptions of tall leaders. *Group Processes and Intergroup Relations*, **16**, 17–27.

Borgerhoff Mulder, M. (1987). On cultural and reproductive success: Kipsigis evidence. *American Anthropologist*, **89**, 617–634.

Boyer, P., & Bergstrom, B. (2008). Evolutionary perspectives on religion. *Annual Review of Anthropology*, **37**, 111–130.

Brown, D. S. (1991). *Human Universals*. Boston: McGraw Hill.

Brown, D. S. (2001). Human universals and their implications. In N. Roughley, ed., *Being Humans: Anthropological Universality and Particularity in Transdisciplinary Perspectives*. Berlin: Walter de Gruyter, pp. 156–174.

Brown, D. S. (2004). Human universals, human nature and human culture. *Daedalus*, 133, 47–54.

Brown, G. R., Dickins, T. E., Sear, R., & Laland, K. N. (2011). Evolutionary accounts of human behavioural diversity. *Philosophical Transactions of the Royal Society B*, **366**, 313–324.

Buss, D. (2003). *The Evolution of Desire: Strategies of Human Mating*. New York: Basic Books.

Buss, D. M. (2014). *Evolutionary Psychology: The New Science of the Mind*, 5th ed. New York: Routledge.

Carruthers, P., & Picciuto, E. (2014). The origins of creativity. In E. S. Paul & S. B. Kaufman, eds., *The Philosophy of Creativity: New Essays*. Oxford: Oxford University Press, pp. 199–223.

Chagnon, N. (1988). Life histories, blood revenge, and warfare in a tribal population. *Science*, **239**, 985–992.

Chapais, B. (2014). Complex kinship patterns as evolutionary constructions, and the origins of sociocultural universals. *Current Anthropology*, **55**, 751–783.

Chapais, B. (2015). Competence, and the evolutionary origins of status and power in humans. *Human Nature*, **26**, 161–183.

Chapais, B. (2016a). The evolutionary origins of kinship structures. *Structure and Dynamics*, **9**, 33–51.

Chapais, B. (2016b). Universal aspects of kinship. In T. K. Shackelford & V. A. Weekes-Shackelford, eds., *Encyclopedia of Evolutionary Psychological Sciences*. Berlin: Springer.

Chapais, B. (2017). Psychological adaptations and the production of culturally polymorphic social universals. *Evolutionary Behavioral Sciences*, **11**, 63–82.

Cheng, J. T., Tracy, J. L., & Henrich, J. (2010). Pride, personality, and the evolutionary foundations of human social status. *Evolution and Human Behavior*, **31**, 334–347.

Cheng, J. T., Tracy, J. L., Foulsham, T., Kingstone, A., & Henrich, J. (2013). Two ways to the top: Evidence that dominance and prestige are distinct yet viable avenues to social rank and influence. *Journal of Personality and Social Psychology*, **104**, 103–125.

Cosmides, L., & Tooby, J. (2015). Adaptations for reasoning about social exchange. In D. Buss, ed., *The Handbook of Evolutionary Psychology*. New York: John Wiley and Sons, pp. 625–668.

Dean, L. G., Kendal, R. L., Schapiro, S. J., Thierry, B., & Laland, K. N. (2012). Identification of the social and cognitive processes underlying human cumulative culture. *Science*, **335**, 1114–1118.

Fogarty, L., Creanza, N., & Feldman, M. W. (2015). Cultural evolutionary perspectives on creativity and human innovation. *Trends in Ecology and Evolution*, **30**, 736–754.

Fox, R. (1980). *The Red Lamp of Incest*. New York: Dutton.

Fox, R. (1989). The cultural animal. In R. Fox, ed., *The Search for Society*. New Brunswick, NJ: Rutgers University Press, pp. 1–34.

Gabora, L., & Kaufmann, S. B. (2010). In J. C. Kaufman & R. J. Sternberg, eds., *The Cambridge Handbook of Creativity*. Cambridge, UK: Cambridge University Press, pp. 279–300.

Gangestad, S. W., Haselton, M. G., & Buss, D. M. (2006). Evolutionary foundations of cultural variation: Evoked culture and mate preferences. *Psychological Inquiry*, **17**, 75–95.

Geertz, C. (1973). The impact of the concept of culture on the concept of man. In C. Geertz, ed., *The Interpretation of Cultures*. New York: Basic Books, pp. 33–54.

Goodenough, W. H. (1970). *Description and Comparison in Cultural Anthropology*. Chicago, IL: Aldine.

Gurven, M., & von Rueden, C. (2006). Hunting, social status, and biological fitness. *Social Biology*, **53**, 81–99.

Hage, P. (1997). Unthinkable categories and the fundamental laws of kinship. *American Ethnologist*, **24**, 652–667.

Henrich, J., & Gil-White, F. J. (2001). The evolution of prestige: Freely conferred deference as a mechanism for enhancing the benefits of cultural transmission. *Evolution and Human Behavior*, **22**, 165–196.

Hinde, R. A. (1991). A biologist looks at anthropology. *Man*, **26**, 583–608.

Hockett, C. F. (1973). *Man's Place in Nature*. New York: McGraw Hill.

Irons, W. (1979). Cultural and biological success. In N. A. Chagnon & W. Irons, eds., *Evolutionary Biology and Human Social Behavior: An Anthropological Perspective*. North Sciutate, MA: Duxbury Press, pp. 257–272.

Jones, D. (2004). The universal psychology of kinship: Evidence from language. *Trends in Cognitive Sciences*, **8**, 211–215.

Jones, D. (2010). Human kinship, from conceptual structure to grammar. *Behavioral and Brain Sciences*, **33**, 367–416.

Kaplan, H., & Hill, K. (1985). Hunting ability and reproductive success among the Ache foragers: Preliminary results. *Current Anthropology*, **26**, 131–133.

Kappeler, P. M., & Silk, J. B., eds. (2010). *Mind the Gap: Tracing the Origins of Human Universals*. Berlin: Springer.

Kaufman, J. C., & Sternberg, R. J., eds. (2010). *The Cambridge Handbook of Creativity*. Cambridge, UK: Cambridge University Press.

Kay, P. (2005). Color categories are not arbitrary. *Cross Cultural Research*, **39**, 72–78.

Kluckohn, C. (1953). Universal categories of culture. In *Anthropology Today: An Encyclopedic Inventory*. Chicago, IL: Chicago University Press, pp. 507–523.

Konner, M. (2010). Evolutionary foundations of cultural psychology. In S. Kitayama & D. Cohen, eds., *Handbook of Cultural Psychology*. New York: Guilford Press, pp. 77–105.

Kronenfeld, D. B. (2006). Issues in the classification of kinship terminologies: Toward a new typology. *Anthropos*, **101**, 203–219.

Laland, K. N., & Reader, S. M. (2003). Comparative perspective on human innovation. In M. J. O'Brien & S. J. Shennan, eds., *Innovation in Cultural Systems: Contributions from Evolutionary Anthropology*. Cambridge, MA: MIT press, pp. 40–51.

Legare, C. H., & Nielsen, M. (2015). Imitation and innovation: The dual engines of cultural learning. *Trends in Cognitive Sciences*, **19**, 688–699.

Lévi-Strauss, C. (1983). *Le Regard Éloigné*. Paris: Plon.

Lukaszewski, A. W., Anderson, C., Simmons, Z. L., & Roney, J. R. (2016). The role of physical formidability in human social status allocation. *Journal of Personality and Social Psychology*, **110**, 385–406.

Malinowski, B. (1944). *A Scientific Theory of Culture and Other Essays*. Chapel Hill, NC: University of North Carolina Press.

Mitani, J. C. (2009). Male chimpanzees form enduring and equitable social bonds. *Animal Behavior*, **77**, 633–640.

Murdock, G. P. (1945). The common denominators of culture. In R. Linton, ed., *The Science of Man in the World Crisis*. New York: Columbia University Press, pp. 123–142.

Murray, G. R., & Schmitz, J. D. (2011). Caveman politics: Evolutionary leadership preferences and physical stature. *Social Science Quarterly*, **92**, 1215–1235.

Nettle, D. (2009). Beyond nature versus culture: Cultural variation as an evolved characteristic. *Journal of the Royal Anthropological Institute*, **15**, 223–240.

Neuberg, S. L., Kenrick, D. T., & Schaller, M. (2010). Evolutionary social psychology. In R. Dunbar & L. Barrett, eds., *The Oxford Handbook of Evolutionary Psychology*. Oxford: Oxford University Press, pp. 761–796.

Norenzayan, A., & Heine, S. J. (2005). Psychological universals: What are they and how can we know? *Psychological Bulletin*, **131**, 763–784.

Pinker, S. (2002). *The Blank Slate*. New York: Penguin Books.

Re, D. E., DeBruine, L. B., Jones, B. C., & Perrett, D. I. (2013). Facial cues to perceived height influence leadership choices in simulated war and peace contexts. *Evolutionary Psychology*, **11**, 89–103.

Sabo, D., Kupers, T. A., & London, W., eds. (2001). *Prisons Masculinities*. Philadelphia, PA: Temple University Press.

Silk, J. B. (2010). Female chacma baboons form strong, equitable, and enduring social bonds. *Behavioral Ecology and Sociobiology*, **60**, 197–204

Smith, E. A. (2004). Why do good hunters have higher reproductive success? *Human Nature*, **15**, 343–364.

Smith, P. B., & Bond, M. H. (1999). *Social Psychology across Cultures*, 2nd ed. Boston, MA: Allyn and Bacon.

Sperber, D. (1985). Anthropology and psychology: Towards an epidemiology of representations. *Man*, **20**, 73–89.

Sperber, D. (1996). *Explaining Culture: A Naturalistic Approach*. Oxford: Blackwell.

Tennie, C., Call, J., & Tomasello, M. (2009). Ratcheting up the ratchet: On the evolution of cumulative culture. *Philosophical Transactions of the Royal Society B*, **364**, 2404–2415.

Tooby, J., & Cosmides, L. (1992). The psychological foundations of culture. In J. Barkow, L. Cosmides, & J. Tooby, eds., *The Adapted Mind: Evolutionary Psychology and the Generation of Culture*. New York: Oxford University Press, pp. 19–136.

Tracy, J. L., Shariff, A. F., & Cheng, J. T. (2010). A naturalist's view of pride. *Emotion Review*, **2**, 163–177.

von Rueden, C. (2014). The roots and fruits of social status in small-scale human societies. In J. T. Cheng, J. L. Tracy, & C. Anderson, eds., *The Psychology of Social Status*. Berlin: Springer, pp. 179–200.

von Rueden, C., Gurven, M., & Kaplan, H. (2011). Why do men seek status? Fitness payoffs to dominance and prestige. *Proceedings of the Royal Society*, **278**, 2223–2232.

Watts, D. P. (2010). Dominance, power, and politics in non-human and human primates. In P. M. Kappeler & J. B. Silk, eds., *Mind the Gap: Tracing the Origins of Human Universals*. Berlin: Springer, pp. 109–138.

Weisfeld, G. E., & Dillon, L. M. (2012). Applying the dominance hierarchy model to pride and shame, and related behaviors. *Journal of Evolutionary Psychology*, **10**, 15–41.

6

The Study of Culture and Evolution across Disciplines

ALEX MESOUDI

6.1 INTRODUCTION

Evolutionary scholars often appear to disagree strikingly about the value of "culture" in explaining human behavior. John Tooby, one of the founders and figureheads of evolutionary psychology, recently responded to the Edge.org question "What scientific idea is ready for retirement?" with the answer "Learning and culture" (Tooby, 2014). At around the same time, Joseph Henrich, Professor of Human Evolutionary Biology at Harvard, published a book titled *The Secret of Our Success: How Culture Is Driving Human Evolution, Domesticating Our Species, and Making Us Smarter* (Henrich, 2015). How can two highly reputable scholars, both committed to understanding human behavior within an evolutionary framework, come to such apparently opposing conclusions? Should the concept of "culture" be retired, or should it be heralded as the secret to humanity's evolutionary success?

"Culture" is also often partly why many scholars within the traditional social sciences and humanities are reluctant to accept or adopt evolutionary approaches to human behavior at all. Evolutionary theory is often assumed (incorrectly, as we will see below) to predict species-wide universals of human behavior and to be inconsistent with the extensive cultural diversity and historical contingency documented by anthropologists, archaeologists, historians, and other social scientists. Such debates are often played out in the context of the nature–nurture dichotomy, with "culture" and "learning" on the "nurture" side and "genes" or "evolution" on the "nature" side. For many social scientists, nurture/culture wins this contest every time, and evolution can be safely ignored.

In this chapter, I explore these issues. I begin by discussing different definitions of culture. I then review how culture has been understood and used within the main branches of the human evolutionary behavioral sciences (Laland & Brown, 2011): sociobiology, human behavioral ecology, evolutionary psychology, and cultural evolution/gene–culture coevolution. Finally, I discuss why, despite the increasing focus on culture within evolutionary approaches, the mainstream social sciences remain steadfast in their rejection of evolutionary theory as a useful tool for understanding human behavior.

6.2 WHAT IS CULTURE?

Scholars have proposed literally hundreds of definitions of culture (Baldwin et al., 2006; Kroeber & Kluckohn, 1952), and I will not review them all here. Instead, I will pick out four common uses that have relevance to the evolutionary approaches discussed later. The first two are examples, in my view at least, of unsuitable definitions of culture. The last two are the focus of later discussion.

6.2.1 Definition 1: "Culture Is Everything That Humans Do (and Other Species Don't)"

Many definitions of culture simply define it as everything that humans do, with the implicit or explicit additional assumption that it is what humans do and no other species does. For example, Cronk (1999) summarizes the typical definition of culture in anthropology textbooks as something similar to "Everything that people have, think, and do as members of a society" (p. 4). These usually derive from Tylor's (1871) influential definition of culture as "that complex whole which includes knowledge, beliefs, art, law, morals, custom, and any other capabilities and habits acquired by man [sic] as a member of society" (p. 1).

This kind of definition is problematic because, as Cronk (1999) points out, it is far too broad. Humans have, think, and do all kinds of things, from bipedal locomotion and menopause, to forming social hierarchies and obeying marriage rules, to producing late-night talk shows and sending animals into space. A concept that explains everything simultaneously explains nothing. And particularly within the evolutionary human sciences, there is vigorous debate – as we will see below – about the extent to which genetic evolution, individual learning, and other processes have shaped different forms of human activity. Labeling them all "culture" from the outset puts the cart before the horse.

Defining culture as "whatever is unique to humans and absent in other species" is also problematic. It makes a claim that precludes a priori any similarities between humans and other species in anything "cultural." In fact, there is a thriving field of study uncovering the cultural capacities of diverse species (Galef & Laland, 2005; Whiten et al., 2016). This comparative work provides valuable insights into the evolutionary origins and function of culture, insights that would be lost if we decided beforehand that culture is uniquely human. There are also many derived human traits, such as bipedalism, which most people would probably not class as "culture."

6.2.2 Definition 2: "Culture Involves Higher, More Civilized Forms of Activity"

This definition, often associated with literary critics such as Matthew Arnold (1869), is what we might colloquially call "high culture." This is the "culture" of Sunday newspaper supplements: opera, art galleries, fine dining, and high-brow literature written by novelists who wear unusual glasses.

This rather elitist definition (who decides what is high culture?) was understandably rejected by early social scientists who were interested in explaining *all* of human behavior, not just that which is deemed important by a certain section of society. Tylor's (1871) inclusive and broad definition of culture given in Section 6.2.1 is an example of a reaction against this elitist definition.

Evolutionary scholars would agree with this rejection of the elitist definition. But it is historically noteworthy here because, while Tylor eliminated the within-society elitist connotations of culture, he just moved the elitism up a level, to entire societies. Tylor advocated a theory of unilinear progressive social evolution, where entire societies can be classed as "higher" or "lower" along a ladder of increasing "evolutionary" complexity, from savagery to barbarism to civilization. Tylor and others (e.g., Morgan, 1877) placed contemporary hunter–gatherers and other small-scale societies at the bottom of this ladder and the Western British and American societies of which they (not uncoincidentally) were members at the top.

Many social scientists still associate evolution with this unilinear, progressive, society-level notion of social evolution espoused by these late nineteenth century scholars, which Boas (1940) and others later convincingly showed to be empirically untenable. It is important to note that this is very different from the idea of cultural evolution that we will come to later, which makes no claims about the inevitable progression of entire societies along ladders of complexity. This notion of ladder-like progress along stages is a very un-Darwinian idea that draws more from Herbert Spencer than Charles Darwin (Freeman, 1974).

6.2.3 Definition 3: "Culture as Between-Group Variation in Behavior"

This is a more useful definition, although still not without its problems. Culture in this sense describes systematic between-group variation in human (or other species') behavior. One might think of cultural variation in, say, marriage practices, means of subsistence, or religious customs, as documented by cultural anthropologists over the last century (Murdock, 1967). Cultural psychologists have similarly documented cultural variation in various psychological processes, along dimensions such as individualism–collectivism or analytic–holistic cognition (Heine, 2011). The term "culture" here may become synonymous with the "group" or "society" being described, be it a nation state (e.g., "Japanese culture") or an ethnic or linguistic group (e.g., "Hopi culture").

In the nonhuman literature, between-group differences in behavior that cannot be attributed to genes or environmental constraints are typically referred to as cultural "traditions" (Fragaszy & Perry, 2003). Tool-use traditions in chimpanzees are prominent examples of nonhuman cultural traditions, with some members of some chimpanzee groups habitually practicing, say, nut-cracking, and other groups failing to ever nut-crack despite the availability of nuts and rocks (Whiten et al., 1999).

While culture is frequently used in this sense, there are reasons to be cautious about this definition. Between-group variation in behavior may result from many different processes, only some of which certain scholars are happy calling "cultural." Culture in this sense is really only a description, not an explanation. We must also be careful not to essentialize these group differences, which tends to happen if "culture" is used as a group descriptor: there is much individual variation in behavior *within* Japan, and *within* Hopi communities, and much overlap with other groups (e.g., Koreans or Navajos). Really, it is a shorthand for the statistical signature of "relatively more between-group variation than within-group variation in the behavior of interest."

6.2.4 Definition 4: "Culture as Socially Learned Information"

This is perhaps the most common definition of culture used by evolutionary scholars, as well as being implicit within most social science definitions (e.g., the "acquired by man [sic]" part of Tylor's definition). Richerson and Boyd (2005) provide a good example:

Culture is information capable of affecting individuals' behavior that they acquire from other members of their species through teaching, imitation, and other forms of social transmission. (Richerson & Boyd, 2005, p. 5)

The terms "social transmission," "social learning," "cultural learning," and "cultural transmission" are used in this chapter synonymously to mean the nongenetic passing on of information from one individual to another.

"Information" here is a broad term referring to what we might colloquially call knowledge, beliefs, skills, attitudes, or norms.

This definition is not human-specific, leaving scope for the comparative study of nonhuman culture. It is also not elitist, and does not restrict culture to certain sections of societies, or some societies rather than others. Furthermore, it provides a causal explanation for (at least some of) the behavioral variation that we considered in the previous definition: those traits common to people in Japan, such as speaking Japanese or using chopsticks, are common because they are socially learned from other Japanese people. Thus, cultural variation is that part of behavioral variation that is generated by the social transmission of information from person to person within a particular group. The common distinction made by sociologists between "culture" and "society" is useful here: "culture" (our Definition 4) concerns a pool of socially transmitted information (beliefs, knowledge, attitudes, norms, etc.), whereas "society" describes a set of social relations and social structures that characterize a particular group (our Definition 3). These social relations and structures may be determined by culturally transmitted norms and values, as well as genetically inherited predispositions.

A particular kind of transmitted culture is *cumulative culture* (Dean et al., 2014). This term has arisen within comparative psychology to refer to the supposedly uniquely human ability to preserve and build up socially learned information over successive generations. So while other species may have transmitted culture, in the sense that they learn from one another (e.g., chimpanzees learn how to nut-crack from others) and may have cultural traditions (e.g., nut-cracking is common in some chimpanzee groups due to social learning), there is no sense in which nut-cracking technology accumulates in efficiency or complexity over time in the way that human technology does. The reasons for this are currently much debated (Dean et al., 2014; Kempe et al., 2014; Tennie et al., 2009), but probably require high-fidelity social learning of the kind only seen in humans. This notion of cumulative culture will become important when we consider cultural evolution below.

6.3 CULTURE IN THE EVOLUTIONARY HUMAN SCIENCES

6.3.1 Sociobiology

The modern resurgence of evolutionary thinking in the human sciences began with the final chapter of E. O. Wilson's *Sociobiology* (Wilson, 1975), where Wilson applied the theoretical tenets of social evolution theory (e.g., Hamilton, 1964) and ethology or behavioral ecology (e.g., Tinbergen, 1963) to our own species. Wilson argued that purportedly universal or near-universal patterns of human behavior such as aggression, territoriality, warfare, genocide, xenophobia, mating systems, homosexuality, the

sexual division of labor, and reciprocity all show homologues in other species, are partly genetically caused fitness-enhancing adaptations (or part of "human nature"), and can be explained in terms of sexual selection, kin selection, reciprocal altruism, and other principles devised and used to explain nonhuman behavior.

As has been well documented (Segerstråle, 2000), this led to fierce opposition from both social scientists and other evolutionary biologists (Kitcher, 1985; Sahlins, 1976; Sociobiology Study Group of Science for the People, 1976). The criticisms were many, including some unjustified political attacks on Wilson for the imagined consequences of his arguments (e.g., eugenics, patriarchy, ruthless capitalism) that he neither intended nor advocated. And Wilson should be given credit for at least *attempting* to bridge the gulf that then existed between the natural and social sciences, rather than simply assuming – as did (and do) many social scientists – that humans are so flexible as to make our evolutionary history irrelevant.

But the critics had a valid point when they charged Wilson of ignoring or downplaying culture in both the form of cultural variation and transmitted culture (Definitions 3 and 4). Wilson stated at the outset that "human qualities will be discussed insofar as they appear to be general traits of the species" (p. 548), seemingly dismissing cultural variation as irrelevant. Speculative claims were made about the genetic basis of phenomena such as socioeconomic class differences ("A key question ... is whether there exists a genetic predisposition to enter certain classes and to play certain roles. Circumstances can easily be conceived in which such genetic differentiation might occur," p. 554), which social scientists understandably balked at.

Wilson later lamented the omission of culture from his early writings, and a few years after *Sociobiology* he attempted to make amends. His book with Charles Lumsden, *Genes, Mind, and Culture*, presented mathematical models aiming to show how culture (in the form of socially transmitted "culturgens" controlled by "epigenetic rules") can be incorporated into a gene–culture coevolutionary framework (Lumsden & Wilson, 1981). Yet this was not positively received (Kitcher, 1985; Smith & Warren, 1982). Lumsden and Wilson's (1981) models showed that culture is kept on a tight leash by genetic evolution, and the suspicion was that the models were set up to ensure this sociobiology-like conclusion. As we will see later, other approaches to gene–culture coevolution (Boyd & Richerson, 1985; Cavalli-Sforza & Feldman, 1981) afforded more explanatory power and independence to transmitted culture, and were consequently received more positively. Lumsden and Wilson's (1981) work received much less attention.

6.3.2 Human Behavioral Ecology

As the label "human sociobiology" became increasingly taboo, it transformed into new fields, each of which attempted to address the criticisms of its forebear (Laland & Brown, 2011). One major problem was that sociobiological theory as applied to humans was vague and loose, offering broad generalizations that were seldom tested quantitatively against empirical data (Kitcher, 1985). In contrast, the field of (nonhuman) behavioral ecology applied many of the same evolutionary principles to non-human behavior, but using rigorous optimality models tested against carefully collected data (Krebs & Davies, 1984). Human behavioral ecologists sought to use similarly rigorous quantitative models and data to test evolutionary theories of human behavior (Borgerhoff Mulder & Schacht, 2012; Nettle et al., 2013; Winterhalder & Smith, 2000).

Of particular interest here is human behavioral ecologists' focus on cross-cultural comparisons. Human behavioral ecologists assume that human behavior is flexible enough to generate adaptive responses to diverse environmental conditions, in contrast to the sociobiological focus on behavioral universals or fixed genetically evolved predispositions. Consequently, human behavioral ecologists showed how marriage practices may vary in a biologically adaptive manner, with polyandry emerging in harsh environments where more than one man is needed to raise offspring (Crook & Crook, 1988; Smith, 1998) and polygyny emerging when men can monopolize resources and accumulate wealth, particularly via pastoralism (Borgerhoff Mulder, 1990). Another line of research has examined cross-cultural and historical variation in fertility, seeking to explain why societies differ in their average number of children and why this number has dropped over time as societies become richer (the "demographic transition": Borgerhoff Mulder, 1998; Lawson et al., 2012).

Human behavioral ecology addresses sociobiology's lack of cultural variation (Definition 3). Yet it does not address our Definition 4: culture in the sense of socially learned information. Human behavioral ecologists, like behavioral ecologists in general (Grafen, 1984), adopt the "phenotypic gambit." Behavior is assumed to be biologically adaptive, but the means by which this adaptiveness is achieved are tactically ignored. Adaptive behavior (e.g., polyandry) might arise as a result of genetically specified "if–then" mechanisms: if in harsh conditions, then adopt polyandry. Or it might arise due to individual learning: people use a genetically evolved but flexible intelligence to independently figure out that in their harsh environment, polyandry is the best option. Or it might arise due to social learning: people acquire polyandry norms from other members of society through observation or instruction, as the result of a historical process of cultural evolution. (Actually, these are not mutually exclusive alternatives: cultural evolution requires some kind of individual learning to provide innovation, while all forms of learning are at some level genetic adaptations.)

The point is that human behavioral ecologists intentionally choose to ignore these proximate mechanisms in favor of answering what is, to them, the more interesting

ultimate question: Is a particular behavior biologically adaptive? This is a perfectly reasonable – and often highly productive – scientific strategy. After all, we do not expect theories of human behavior to include interactions between atoms; there must be some limit to how far down our explanations go. Ultimate questions of evolutionary function and history can often be studied without considering proximate functions of development and mechanism (Tinbergen, 1963). Yet as we will see later, other evolutionary human scientists disagree that proximate factors such as learning and transmitted culture can be safely ignored when thinking about ultimate factors.

6.3.3 Evolutionary Psychology

Evolutionary psychology emerged in the 1990s as a fusion of sociobiological theory and cognitive science (Barkow et al., 1992; Buss, 2009; Pinker, 1997). Evolutionary psychologists typically focus on how natural selection has shaped psychological mechanisms during our species' evolutionary past to maximize inclusive fitness. It is assumed that the resulting cognitive adaptations evolved to deal with adaptive challenges in past environments (sometimes called the "environment of evolutionary adaptedness") and not necessarily current environments, in contrast to human behavioral ecologists. Key topics within evolutionary psychology have included cognitive adaptations for cheater detection that evolved for maintaining reciprocal interactions (Cosmides, 1989), sex differences in mating preferences that evolved due to differential parental investment (Buss et al., 1990), and patterns of homicide that can be predicted from kin selection (Daly & Wilson, 1988).

To some degree, early evolutionary psychologists shared human sociobiologists' preoccupation with behavioral universals rather than cultural variation (Tooby & Cosmides, 1992, pp. 88–93). Brown (1991) notably compiled a list of hundreds of "human universals," including division of labor, aggression, ethnocentrism, gossip, reciprocity, and play (interestingly, this reads much like Wilson's list above). In a prominent cross-cultural study, Buss et al. (1990) found consistent sex differences in mating preferences across 37 societies, predictable on the basis of differential parental investment. For example, men rated markers of fertility higher, such as youth, while women rated markers of offspring provisioning higher, such as wealth.

This focus on universals was an explicit reaction against what Tooby and Cosmides (1992) called the "Standard Social Science Model," which they characterized as a denial of human nature and an extreme cultural determinism where evolution played no role in shaping human behavior. While this was a laudable aim, some have argued that evolutionary psychologists swung too far the other way. As Laland and Brown (2011) point out, in Buss et al.'s (1990) cross-cultural study of mating preferences, much more variation (around 14 percent) in responses was explained by the cultural background of the participants than was explained by sex (only around 2 percent). Yet the study is typically cited as evidence for the universality of human mating strategies. The situation has improved since these early studies, however, with evolutionary psychologists now formulating specific hypotheses about particular patterns of cultural variation and testing them in targeted cross-cultural comparisons (Apicella & Barrett, 2016).

Tooby and Cosmides (1992) also proposed an important distinction between *transmitted culture*, which is our Definition 4 above, and *evoked culture*. The latter involves a universal, evolved human psychology that responds differently to different environmental conditions. In the context of our definitions above, Tooby and Cosmides are arguing that some "cultural" variation (Definition 3) may not in fact be generated by transmitted culture (Definition 4); it may be the result of genetically evolved, preexisting responses to predictable environmental cues. Note that this is different from Wilson's (1975) earlier speculations that genetic differences might explain behavioral differences between groups of people. Tooby and Cosmides explicitly disavowed this, instead arguing that people everywhere are genetically far too similar to explain any behavioral variation directly (which concurs with modern genetic data; Feldman, 2014). Genes instead generate a set of universal responses to predictable environmental variation.

An example of evoked culture that has received much recent attention involves pathogens (Nettle, 2009; Schaller, 2006). Thornhill, Fincher, and colleagues have argued that a country's current or past exposure to pathogens is a key environmental trigger that evokes different behavioral responses (Fincher & Thornhill, 2012; Fincher et al., 2008). For example, many East Asian countries exhibit high collectivism, which entails a wariness of contact with foreigners and other out-group members. Thornhill, Fincher, and colleagues view this as an adaptive, evoked response to high levels of pathogens, which are most often transmitted from out-groups who carry novel diseases to which the in-group has no immune defense. Positive correlations between collectivism and past and present pathogen prevalence are offered in support of this hypothesis (Fincher et al., 2008; although for critical analyses, see Currie & Mace, 2012; Hruschka & Henrich, 2013).

A similar notion to evoked culture has been proposed by cognitive anthropologists such as Dan Sperber (Claidière & Sperber, 2007; Claidière et al., 2014; Sperber, 1996). As with evolutionary psychologists, Sperber downplays the role of high-fidelity cultural transmission (Definition 4), instead arguing that cultural diversity and stability (Definition 3) emerge as people independently reconstruct representations based on preexisting cognitive biases, or "attractors." In a sense, this is like evoked culture, but where "the environment" is people's cognition, rather than something external like pathogens. An example is bloodletting (Miton et al., 2015): diverse societies have seemingly independently converged on the practice of cutting the skin near the location of an ailment to release "bad blood." Although this has no medical efficacy, it seems to be universally cognitively attractive and so has been "rediscovered" multiple times.

In sum, an early focus within evolutionary psychology on human universals (e.g., Brown, 1991) has given way to an appreciation that cultural variation (Definition 3) is not only perfectly consistent with an evolved human psychology (Barrett, 2015), but also that cultural variation can be utilized to test evolutionary predictions (Apicella & Barrett, 2016). In this, evolutionary psychology has converged with human behavioral ecology. What of Definition 4: transmitted culture? Evolutionary psychologists typically focus instead on evoked culture and either downplay the role of transmitted culture or take the human behavioral ecologists' approach that transmitted culture is simply a proximate mechanism by which human groups arrive at biologically adaptive (or ancestrally adaptive) behavioral equilibria. In this, both of these fields differ quite substantially from the field to which we turn next.

6.3.4 Cultural Evolution/Gene–Culture Coevolution

I noted above that E. O. Wilson's response to criticism that sociobiology failed to take culture seriously was to develop theoretical models of gene–culture coevolution (Lumsden & Wilson, 1981). These models explored how transmitted culture (Definition 4) and genetic inheritance jointly produce human behavior. While a step in the right direction, Lumsden and Wilson's models were heavily criticized for merely recapitulating Wilson's earlier claims that genes "hold culture on a leash" (Smith & Warren, 1982).

In parallel to Wilson's own attempts, a group of anthropologists and biologists were developing a body of theory that gave more explanatory power – and more independence – to transmitted culture. Cavalli-Sforza and Feldman (1981) and Boyd and Richerson (1985) used the mathematical techniques of population genetics to examine two questions. First, under what conditions does transmitted culture evolve? In other words, if we treat the capacity for transmitted culture as a trait that evolved just like any other, we can ask: What is its adaptive function relative to pure genetic adaptation and/or purely individual learning?

Boyd and Richerson (1985) developed models showing that transmitted culture is favored when environments change moderately quickly: too fast for genes to track, but not so fast that the culturally transmitted behavior is out of date (see also Aoki et al., 2005). Transmitted culture also evolves when individual learning is costly (Boyd & Richerson, 1985). Under such conditions, however, social learning evolves but does not increase the average fitness of the population. This phenomenon became known as

"Rogers' paradox" after Alan Rogers, the first person to clearly point it out (Rogers, 1988). The fact that social learning does not enhance average population fitness is not inherently paradoxical, but does contradict the common claim that humans are so ecologically and demographically successful because of transmitted culture.

Rogers' paradox occurs because the success of social learning is frequency dependent. When rare, social learners do well because they forego the costs borne by individual learners. But when common, and environments change, social learners will be copying other social learners' outdated information. At equilibrium, social and individual learners have equal fitness, which will be equal to the fitness of a population entirely composed of individual learners (which is fixed, because their learning is not dependent on others). Thus, social learning evolves, but does not enhance fitness in a way that could be described as the "secret to our success."

However, transmitted culture *can* enhance the average fitness of a population when: (1) individuals learn selectively, only copying others when individual learning is inaccurate or selectively copying successful individuals; and/or (2) socially learned traits can be accumulated over successive generations such that individuals can learn socially what they could never invent alone: this is the cumulative culture noted above (Boyd & Richerson, 1995; Enquist et al., 2007; Kendal et al., 2009). There is evidence that a range of species, including humans, show selective social learning (Hoppitt & Laland, 2013; Whiten et al., 2016), while as noted above only humans appear to possess cumulative culture (Dean et al., 2014; Tennie et al., 2009).

The second question is, assuming transmitted, selective, cumulative culture *has* evolved in humans, what consequences does this have for our understanding of human behavior? Do genes keep culture on a tight leash, such that it acts as a proximate mechanism that ultimately maximizes past or present inclusive fitness, as assumed by evolutionary psychologists and human behavioral ecologists? Or does it make genes irrelevant, as assumed by most mainstream social scientists?

Cultural evolutionists' answer to this is typically "neither." In Richerson and Boyd's words, "culture is on a leash, all right, but the dog on the end is big, smart and independent. On any given walk, it is hard to tell who is leading who" (2005, p. 194). The key argument here is that transmitted, cumulative culture (Definition 4) constitutes an evolutionary process in its own right (Campbell, 1965; Plotkin, 1995). In *On the Origin of Species*, Darwin (1859) specified three requirements for the evolutionary process: variation, inheritance, and selection. Applied to genetic evolution, these are genetic variation that arises through mutation and recombination, genetic inheritance via DNA replication, and natural selection due to competition between genes or individuals for survival and reproduction. Many scholars, beginning with Darwin himself, have noted that these three requirements are also met for

culture: there is cultural variation in beliefs, attitudes, skills, etc.; there is cultural inheritance (i.e., the transmitted culture of our Definition 4); and there are various selective forces that cause some cultural variants to be more likely to survive and reproduce than others (Mesoudi et al., 2004).

Cavalli-Sforza and Feldman (1981) and Boyd and Richerson (1985) modeled this process of cultural evolution given what is known about human cultural change and learning. Of particular interest were cases where the dynamics of cultural evolution appear different from those of genetic evolution. So Cavalli-Sforza and Feldman (1981) modeled horizontal cultural transmission (learning from an unrelated member of the same generation) and oblique cultural transmission (learning from an older nonparent) as well as gene-like vertical cultural transmission (learning from one's parents). Boyd and Richerson (1985) modeled processes such as conformity, where the most common cultural trait is preferentially adopted, and prestige bias, where one preferentially learns from people with high social status. "Guided variation" they defined as the intentional modification of acquired traits, a Lamarckian-like process that has no clear parallel in genetic evolution.

Boyd and Richerson's (1985) analyses suggested that while general learning heuristics such as conformity and prestige bias are broadly adaptive, they can sometimes lead to biologically maladaptive outcomes. For example, conformity can lead to the failure to switch to a superior product or tool when an inferior one is already established (Henrich, 2001). Prestige bias can lead to cultural hitchhiking, as people copy neutral or harmful traits from prestigious individuals because it is hard to figure out what causes social success (Henrich & Gil-White, 2001). This can be relatively trivial, such as copying the clothing style of prestigious celebrities, but also potentially drastic, such as copycat suicides that follow a highly publicized celebrity suicide (Mesoudi, 2009). Boyd and Richerson (1985) also suggest that prestige bias can account for the "demographic transition" – the systematic reduction in family size that has occurred at different times in different countries over the last several decades. If "family size" is a culturally transmitted trait and prestigious people have smaller families because they choose to invest more in attaining cultural prestige than having children, then family size will drop as people copy preferences for small families from prestigious others (Colleran, 2016). Finally, Bentley and colleagues (2004, 2007) have shown that various cultural traits change in a random fashion, akin to genetic drift. Examples include first names, dog breeds, and pottery decorations.

This explains cases of biologically maladaptive or adaptively neutral human behavior, answering the common criticism that evolutionary approaches are "adaptationist" and incorrectly assume that every human behavior is biologically adaptive (Gould & Lewontin, 1979). But the key message of cultural evolutionists is that the overall package of transmitted, cumulative culture *is* adaptive, and indeed is the key to our species' extraordinary ecological and demographic success over the last 10,000 years or so (Boyd et al., 2011; Henrich, 2015). While other species rely on the slow process of genetic evolution to adapt to novel environments, they argue that we have successfully colonized virtually every terrestrial environment due to cultural adaptations such as complex tool kits and social institutions. These cultural adaptations represent the accumulated wisdom of multiple generations of people, as beneficial modifications are selectively preserved and culturally transmitted to subsequent generations via relatively content-neutral social learning biases. This contrasts with some evolutionary psychologists' explanation of our species' success, which credits our content-rich, domain-specific cognition and the ability of single individuals to come up with solutions to adaptive problems "on the fly" (Pinker, 2010). For cultural evolutionists, cultural adaptations go beyond what any single individual could create alone, moving the explanation from individual cognition to population-level cultural evolution, and from evoked culture to cumulative, transmitted culture.

Cultural evolutionists also place more emphasis on cultural history, as opposed to genetic history. A group of scholars has borrowed phylogenetic methods – originally developed in biology to reconstruct the evolutionary history of species – and applied them to cultural traits based on the logic that these cultural traits evolve through a process of descent with modification just like species (Gray et al., 2007; Mace & Holden, 2005; O'Brien & Lyman, 2003; Pagel, 2009). Language is a good example of a socially learned trait that forms very long-lasting lineages: languages (English, French, etc.) are socially learned by children with relatively high fidelity, allowing such languages to persist for thousands of years. Bouckaert et al. (2012) reconstructed the cultural evolutionary history of the Indo-European language family, finding that it originally spread along with farming practices from present-day Turkey around 8,000 years ago. Similar questions about the historical spread of empires have been addressed by Turchin and colleagues (Turchin, 2003; Turchin et al., 2013). In these cases, we can see links emerging between our Definitions 3 and 4: contemporary cross-cultural variation (Definition 3) emerges as the result of the long-term transmission of cultural traits (Definition 4). Indo-European languages are transmitted from person to person, generation to generation, ultimately generating the cultural variation in languages that we see today.

Finally, note that this modern cultural evolution approach is very different from the progressive "social evolution" theories of the late 1800s (Morgan, 1877; Tylor, 1871). The modern focus is often on traits (languages, tools) that may diffuse across social boundaries rather than the transformation of entire monolithic societies, and the evolutionary process is one of branching diversification that may or may not result in increased (cultural) adaptation, rather than linear, inevitable progress along fixed stages.

6.4 COMPARISON OF APPROACHES

While all modern evolutionary approaches to human behavior seek to explain cultural variation (Definition 3), cultural evolution – much more than any other field – focuses on transmitted culture (Definition 4) as an explanation for this variation. Cultural evolutionists argue that complex behavioral traits, from technology to institutions, can be explained as cultural adaptations that evolve not genetically, but culturally. Genes provide the social learning apparatus that this rests on, as explored in models of the evolution of culture, but the real "explanatory action" is at the level of selectively preserved, culturally transmitted skills, beliefs, and knowledge. Evolutionary psychology, and similar approaches such as cultural attraction, focus on biologically evolved, content-rich, domain-specific cognitive adaptations for dealing with past ancestral challenges. Here, the explanatory action is very much with natural selection and genetic adaptation. Human behavioral ecologists share this latter assumption, although they differ in certain details from evolutionary psychology, such as focusing on behavior rather than cognition, and stressing current adaptiveness rather than ancestral adaptiveness.

These seemingly disparate approaches do not necessarily conflict if we consider them to be targeting different phenomena. Most cultural evolutionists would probably agree that learning is not entirely content free. Indeed, one prominent strand of cultural evolution research examines "content biases," presumed genetically evolved biases in human social learning that favor the acquisition and transmission of biologically adaptive information. Examples include biases for information about social interactions (Mesoudi et al., 2006), disgusting and potentially disease-carrying substances (Eriksson & Coultas, 2014), and dangerous animals (Barrett & Broesch, 2012). Content biases may also result from the individual transformation of information according to features of human cognition, such as the hierarchical structure of event knowledge, which favors increasingly "schematized" descriptions of events (Mesoudi & Whiten, 2004). This seems very similar to the idea of cultural attraction (Sperber, 1996).

Furthermore, cultural evolution research has demonstrated that content-neutral cultural transmission rules such as prestige bias can lead to biologically adaptive behavior in a way that is perfectly consistent with the other approaches. For example, Henrich and Henrich (2010) found that food taboos in Fiji prohibiting pregnant women from eating certain fish were: (1) adaptive, in the sense that these fish contained high levels of toxins that increased the chances of miscarriage; and (2) maintained via prestige-biased social learning, as women learned the taboos from prestigious "wise women" within the community. This seems perfectly consistent with, say, human behavioral ecologists' assumption that transmitted culture will, at a proximate level, result in what is ultimately biologically adaptive behavior.

Other research questions appear to require an answer explicitly in terms of transmitted culture, with genetic evolution relegated to the background. For example, the question "Why do people in England speak English, and people in France speak French?" surely must be answered in terms of the cultural evolutionary history of the Indo-European language family (Bouckaert et al., 2012), with these languages representing different tips of the branching language tree that diverged due to various cultural forces, including drift-like mutations, demographic shifts, borrowing from other languages, and ultimately the spread of ancestral languages from Anatolia with farming about 8,000 years ago. Evolutionary psychologists, in contrast, would be interested in the even more ultimate question of "How did the language faculty evolve biologically?" (Pinker, 1994). There is no conflict here, just different levels of explanation (Tinbergen, 1963).

Or another example might be, "How do we explain the rise and fall of specific empires throughout history?" that Turchin and colleagues (Turchin, 2003; Turchin et al., 2013) have been addressing using cultural evolution models tested against historical data. More relevant to contemporary societies might be, "Why do some modern countries function better than others?" that cultural evolution researchers are beginning to address in terms of the cultural transmission and evolution of inclusive institutions that reduce inequality (Henrich et al., 2010; Hruschka & Henrich, 2013; Matthews et al., 2016). Evolutionary psychologists, in contrast, would be more interested in the cognitive adaptations (e.g., reciprocity) that make such institutions possible in the first place.

Yet on closer inspection, this neat division becomes less clear-cut. Take language as an example. Recently, a group of cultural evolutionists have challenged Chomsky's (1965) notion of an innate universal grammar that structures all languages within specific constraints (Pinker, 1994). They argue that systematic grammatical regularities instead arise due to the repeated transmission of languages as they are learned by each new speaker (Chater et al., 2009; Christiansen & Chater, 2008; Kirby et al., 2007, 2008). In other words, rather than the brain having genetically evolved to constrain and structure languages, languages have instead culturally evolved to become more easily learnable. There is no innate "language acquisition device," just general cognitive constraints that are not specific to language learning. The "explanatory action" in this cultural evolution account shifts from genetic to cultural adaptation.

Another prominent example concerns cooperation. Another group of cultural evolutionists have argued that while standard sociobiological principles of kin selection and reciprocity can certainly explain *some* aspects of human cooperation, they cannot explain the large-scale, non-kin-based cooperation that characterizes human societies, from hunter–gatherer bands to historical empires to modern-day nation states, where large numbers of genetically unrelated people cooperate often with no expected return on their cooperation (Boyd & Richerson, 1985; Gintis et al., 2003; Henrich, 2004; Richerson et al., 2016; Turchin, 2015). They argue that cultural evolution is the key to this large-scale cooperation, because it allows altruism to be selected and favored at the level of the cultural group. Imagine that different cultural groups (e.g., tribes, empires, or nations) vary in the extent to which their members act altruistically to one another, such as sharing food or participating in collective group defense. If those groups are reasonably cohesive and compete with one another either directly (e.g., via warfare) or indirectly (e.g., for resources), then those groups that are more internally cooperative and feature self-sacrificial behavior for the good of the group will do better than less internally cooperative groups in which everyone is out for themselves. This kind of group selection does not work for genetic evolution because free-riders out-reproduce altruists and migration breaks down genetically homogenous groups (Williams, 1966). But it does work, at least in theory, for cultural groups, assuming those groups have culturally transmitted norms for dealing with free-riders (e.g., punishment) and biases for maintaining within-group cultural homogeneity in the face of migration (e.g., conformity).

These theories of the evolution of language and of cooperation are controversial. Many evolutionary psychologists and behavioral ecologists argue that kin selection and reciprocity can, in fact, explain large-scale human cooperation (Krasnow et al., 2013; West et al., 2011), and that language regularities *do* result from a genetically evolved, universal language instinct (Pinker, 1994). These are topics of ongoing empirical tests (Kirby et al., 2008; Lamba & Mace, 2011). And as noted earlier, a broader debate concerns the basis for our species' extraordinary ecological and demographic success compared to other primates: cultural evolutionists attribute this to our ability to rapidly culturally adapt to novel environments and accumulate complex technology and institutions via content-neutral transmission biases (Boyd et al., 2011; Henrich, 2015), while evolutionary psychologists credit our genetically evolved cognitive adaptations (Barrett et al., 2007; Pinker, 2010).

In sum, while there are large areas of agreement, evolutionary scholars disagree over how much explanatory power to assign to transmitted culture. Evolutionary psychologists and human behavioral ecologists typically treat transmitted culture as a proximate mechanism that helps to explain how we arrive at genetically adaptive behavior. Much cultural evolution work is consistent with this, as in the case of genetically evolved content biases. But many cultural evolutionists argue that transmitted culture also often changes the rules of the human evolutionary game, creating new equilibria (e.g., large-scale cooperation), spreading maladaptive behavior (e.g., copycat suicide, small family sizes), and replacing genetic evolution as a source of adaptation (e.g., resulting in complex technology and institutions or grammatical structure).

6.5 CULTURE, EVOLUTION, AND THE MAINSTREAM SOCIAL SCIENCES

What of the mainstream social sciences since their vehement reaction against sociobiology in the late 1970s (Sahlins, 1976; Sociobiology Study Group of Science for the People, 1976)? Has the increasing appreciation within evolutionary psychology of cultural variation (Apicella & Barrett, 2016) led to greater acceptance among social scientists? And what of the burgeoning field of cultural evolution (Mesoudi, 2011), which seems to speak directly to most social scientists' criticism that sociobiologists ignored transmitted culture? Has cultural evolution been welcomed as an advance over the sometimes crude genetic determinism of early sociobiology?

Sadly not. Evolutionary approaches to human behavior are still typically ignored or rejected by mainstream social scientists (Barkow, 2005; Horowitz et al., 2014). Evolutionary psychology has received many of the same criticisms that sociobiology received in the 1970s, including charges of genetic determinism, just-so storytelling, and hidden political agendas (Rose & Rose, 2000). Many of these criticisms are simply misguided (Kurzban & Haselton, 2006). For example, no evolutionary psychologist has ever claimed that genes entirely determine human behavior, with no environmental influence. The case of evoked culture is a good example of a clear and explicit gene × environment interaction: genes specify possible behavioral reactions to different environmental inputs. And evolutionary psychologists and anthropologists are often just as liberal as non-evolutionary scholars in their political leanings (Lyle & Smith, 2012) – not that this should matter for questions of science.

I suspect this rejection of anything "evolutionary" or "biological" is a continued legacy of the divide in the early twentieth century between biology and culture. In anthropology, Kroeber (1917) explicitly partitioned the "organic," or biological, from the "superorganic," or cultural. This split between biological and cultural anthropology persists today. Back in the 1910s, the desire on the part of the cultural side of the discipline to distance itself from the crude, inaccurate, and politically distasteful race theories of the time (Gould, 1996) was quite understandable. The progressive social evolution theories (Morgan, 1877; Tylor, 1871) were not much better, as discussed above. Yet modern evolutionary approaches to human behavior bear no resemblance to these early "evolutionary" approaches.

One might imagine that cultural evolution would be viewed more favorably by mainstream social scientists, given that it gives more explanatory power to transmitted culture. Yet this field is just as fiercely criticized (Fracchia & Lewontin, 1999; Ingold, 2007). Typical criticisms of cultural evolution, aside from it being confused with the aforementioned progress theories of social evolution, are that it denies individual agency, that it inappropriately reduces culture to a collection of unconnected "traits" that cannot be divorced from their proper social context,

and that it simplifies processes such as cultural transmission, which are too complex and ungeneralizable to fit into neat categories such as "horizontal" or "vertical." Fracchia and Lewontin (1999), for example, ask:

> Is culture "transmitted" at all? An alternative model, one that accords better with the actual experience of acculturation, is that culture is not "transmitted" but "acquired." Acculturation occurs through a process of constant immersion of each person in a sea of cultural phenomena, smells, tastes, postures, the appearance of buildings, the rise and fall of spoken utterances. (p. 73)

This kind of criticism has its roots in the wider rejection of the scientific method within sociocultural anthropology, sociology, and other more humanities-oriented social sciences. The "writing culture" movement of the 1980s (Clifford & Marcus, 1986) abandoned any pretense that ethnography should be a scientific methodology in favor of producing "thick descriptions" of other societies, more akin to literature than science. The "writing against culture" movement (Abu-Lughod, 1996) argued that the use of "culture" to describe a group of people (a version of our Definition 3) acts to essentialize between-group differences and ignore within-group diversity, much like the earlier concept of "race." Ironically, this has led mainstream social scientists to abandon the concept of "culture" just as it is becoming increasingly popular within the evolutionary sciences. The recent "ontological turn" within cultural anthropology (Viveiros de Castro, 2014) seemingly rejects any possibility of cross-cultural comparison or scientific investigation, seeking instead to understand each society entirely in its own ontological terms, and especially not with reference to, or use of, scientific concepts, which are viewed as products of Western power structures. Indeed, much of modern sociocultural anthropology has become a form of political activism, rather than a scientific endeavor (e.g., Allen & Jobson, 2016).

It is a real shame that these barriers persist, particularly in light of the emergence of the field of cultural evolution. As we have seen, the assumption of Kroeber and others that culture is separate to biology has been explored formally by cultural evolution modelers, as well as tested in a large body of empirical research. It is not enough to simply assert that culture is independent of biology; this is an empirical question. The models and findings reviewed above suggest that neither the genetic determinists nor the cultural determinists of the early twentieth century were correct. An acultural biology does not determine human behavior, but nor does an abiological culture. The cultural evolutionists would say that culture evolved and it is broadly adaptive, but it does not simply do our genes' bidding, because culture itself constitutes an evolutionary process whose very function is to be partially independent of our much slower genetic evolution.

In the absence of a scientific approach to culture within the social sciences, cultural evolutionists are proceeding to address questions and topics that have been of long-standing interest to social scientists, but using more rigorous quantitative methods: phylogenetic methods to examine the spread of language families (Bouckaert et al., 2012; Pagel, 2009); dynamical models and hypothesis testing to explore the historical rise and fall of empires (Turchin, 2003; Turchin et al., 2013); and multilevel selection theory to explain the origin and function of cooperative social institutions (Henrich, 2006, 2004; Hruschka & Henrich, 2013). The criticisms that these methods and concepts are too simplistic and reductionist (Fracchia & Lewontin, 1999; Ingold, 2007) represent a misunderstanding of the use of models, statistics, and hypothesis testing in science. Simple models do not imply that the modeler thinks that the real world really is simple. Rather, as in biology, simple models are the best way of understanding a complex reality (Servedio et al., 2014; Smaldino, 2016). One could adopt Fracchia and Lewontin's approach of embracing complexity, but then, how does one study the "immersion in a sea of the appearance of buildings?" Models of, say, vertical, horizontal, and oblique transmission may be gross simplifications, but they still provide useful insights that apply across societies (McElreath & Strimling, 2008). The same goes for lab and field experiments (Henrich et al., 2010; Mesoudi & Whiten, 2008) and cross-cultural comparisons (Mace & Holden, 2005). Of course, all societies are different, but they are not *so* different as to make comparisons invalid. This is not to say that such methods are perfect; witness the recent replication crisis within the behavioral sciences (Open Science Collaboration, 2015) and the growing realization that traditional statistical methods are flawed (McElreath, 2016). But this necessitates the adoption of better standards (e.g., preregistration) and methods (e.g., Bayesian statistics), rather than the wholesale abandonment of the scientific method.

6.6 CONCLUSIONS

"Culture" does not appear to be a concept ready to be retired. Within the evolutionary sciences, one might even say that it is a concept entering the prime of its life. Evolutionary psychologists are increasingly paying attention to cultural variation (Apicella & Barrett, 2016), as human behavioral ecologists have been doing for many years (Nettle et al., 2013). Cultural evolutionists are applying evolutionary methods and concepts to study cultural change in a way that fills a gap left by the shift in the mainstream social sciences and humanities away from scientific methodologies (Mesoudi, 2011). There are broad areas of agreement across these different evolutionary fields: all would agree, for example, that culture itself is a biological adaptation, and that learning is often guided to make the acquisition of biologically adaptive information more likely. There are also areas of disagreement, with cultural evolutionists willing to give transmitted culture more explanatory power and independence. Ultimately, these are empirical issues, as models, experiments, ethnographic observation, historical data, and cross-cultural comparisons are brought to bear on

key topics such as the evolution of technology, cooperation, language, and sociality.

REFERENCES

Abu-Lughod, L. (1996). Writing against culture. In R. G. Fox, ed., *Recapturing Anthropology: Working in the Present*. Santa Fe, NM: School of American Research Press, pp.137–162.

Allen, J. S., & Jobson, R. C. (2016). The decolonizing generation: (Race and) theory in anthropology since the Eighties. *Current Anthropology*, **57**, 129–148.

Aoki, K., Wakano, J. Y., & Feldman, M. W. (2005). The emergence of social learning in a temporally changing environment: A theoretical model. *Current Anthropology*, **46**, 334–340.

Apicella, C. L., & Barrett, H. C. (2016). Cross-cultural evolutionary psychology. *Current Opinion in Psychology*, **7**, 92–97.

Arnold, M. (1869). *Culture and Anarchy*. Oxford: Oxford University Press.

Baldwin, J. R., Faulkner, S. L., Hecht, M. L., & Lindsley, S. L. (2006). *Redefining Culture: Perspectives across Disciplines*. Mahwah, NJ: Lawrence Erlbaum.

Barkow, J. H. (2005). Introduction: Sometimes the bus does wait. In J. H. Barkow, ed., *Missing the Revolution: Darwinism for Social Scientists*. Oxford: Oxford University Press.

Barkow, J. H., Cosmides, L., & Tooby, J., eds. (1992). *The Adapted Mind: Evolutionary Psychology and the Generation of Culture*. Oxford: Oxford University Press.

Barrett, C., Cosmides, L., & Tooby, J. (2007). The hominid entry into the cognitive niche. In S. W. Gangestad & J. A. Simpson, eds., *Evolution of Mind*. New York: Guilford, pp. 241–248.

Barrett, H. C. (2015). *The Shape of Thought: How Mental Adaptations Evolve*. Oxford: Oxford University Press.

Barrett, H. C., & Broesch, J. (2012). Prepared social learning about dangerous animals in children. *Evolution and Human Behavior*, **33**, 499–508.

Bentley, R. A., Hahn, M. W., & Shennan, S. J. (2004). Random drift and culture change. *Proceedings. Biological Sciences*, **271**, 1443–1450.

Bentley, R., Lipo, C. P., Herzog, H. A., & Hahn, M. W. (2007). Regular rates of popular culture change reflect random copying. *Evolution and Human Behavior*, **28**, 151–158.

Boas, F. (1940). *Race, Language and Culture*. New York: Macmillan.

Borgerhoff Mulder, M. (1990). Kipsigis women's preferences for wealthy men: Evidence for female choice in mammals? *Behavioral Ecology and Sociobiology*, **27**, 255–264.

Borgerhoff Mulder, M. (1998). The demographic transition: Are we any closer to an evolutionary explanation? *Trends in Ecology & Evolution*, **13**, 266–270.

Borgerhoff Mulder, M., & Schacht, R. (2012). Human behavioral ecology. In *eLS*. Chichester: John Wiley & Sons.

Bouckaert, R., Lemey, P., Dunn, M., et al. (2012). Mapping the origins and expansion of the Indo-European language family. *Science*, **337**, 957–960.

Boyd, R., & Richerson, P. J. (1985). *Culture and the Evolutionary Process*. Chicago, IL: University of Chicago Press.

Boyd, R., & Richerson, P. J. (1995). Why does culture increase human adaptability? *Ethology and Sociobiology*, **16**, 125–143.

Boyd, R., Richerson, P. J., & Henrich, J. (2011). The cultural niche: Why social learning is essential for human adaptation.

Proceedings of the National Academy of Sciences, **108**, 10918–10925.

Brown, D. E. (1991). *Human Universals*. Philadelphia, PA: Temple University Press.

Buss, D. M. (2009). *Evolutionary Psychology: The New Science of the Mind*. Boston, MA: Pearson.

Buss, D. M., Abbott, M., Angleitner, A., et al. (1990). International preferences in selecting mates: A study of 37 cultures. *Journal of Cross-Cultural Psychology*, **21**, 5–47.

Campbell, D. T. (1965). Variation and selective retention in socio-cultural evolution. In H. R. Barringer, G. I. Blanksten, & R. W. Mack, eds., *Social Change in Developing Areas*. Cambridge, MA: Schenkman, pp. 19–49.

Cavalli-Sforza, L. L., & Feldman, M. W. (1981). *Cultural Transmission and Evolution*. Princeton, NJ: Princeton University Press.

Chater, N., Reali, F., & Christiansen, M. H. (2009). Restrictions on biological adaptation in language evolution. *Proceedings of the National Academy of Sciences*, **106**, 1015–1020.

Chomsky, N. (1965). *Aspects of the Theory of Syntax*. Cambridge, MA: MIT Press.

Christiansen, M. H., & Chater, N. (2008). Language as shaped by the brain. *Behavioral and Brain Sciences*, **31**, 489–509.

Claidière, N., & Sperber, D. (2007). The role of attraction in cultural evolution. *Journal of Cognition and Culture*, **7**, 89–111.

Claidière, N., Scott-Phillips, T. C., & Sperber, D. (2014). How Darwinian is cultural evolution? *Philosophical Transactions of the Royal Society B: Biological Sciences*, **369**, 20130368.

Clifford, J., & Marcus, G. E. (1986). *Writing Culture: The Poetics and Politics of Ethnography*. Berkeley, CA: University of California Press.

Colleran, H. (2016). The cultural evolution of fertility decline. *Philosophical Transactions of the Royal Society of London, Series B: Biological Sciences*, **371**, 20150152.

Cosmides, L. (1989). The logic of social exchange: Has natural selection shaped how humans reason? Studies with the Wason selection task. *Cognition*, **31**, 187–276.

Cronk, L. (1999). *That Complex Whole: Culture and the Evolution of Human Behavior*. Boulder, CO: Westview Press.

Crook, J., & Crook, S. J. (1988). Tibetan polyandry: Problems of adaptation and fitness. In L. Betzig, M. Borgerhoff Mulder, & P. Turke, eds., *Human Reproductive Behavior*. Cambridge, UK: Cambridge University Press, pp. 97–114.

Currie, T. E., & Mace, R. (2012). Analyses do not support the parasite-stress theory of human sociality. *Behavioral and Brain Sciences*, **35**, 83–85.

Daly, M., & Wilson, M. (1988). *Homicide, Foundations of Human Behavior*. Hawthorne, NY: Aldine de Gruyter.

Darwin, C. (1859). *On The Origin of Species by Means of Natural Selection, or the Preservation of Favoured Races in the Struggle for Life*. London: John Murray.

Dean, L. G., Vale, G. L., Laland, K. N., Flynn, E., & Kendal, R. L. (2014). Human cumulative culture: a comparative perspective. *Biological Reviews*, **89**, 284–301.

Enquist, M., Eriksson, K., & Ghirlanda, S. (2007). Critical social learning: A solution to Rogers' paradox of nonadaptive culture. *American Anthropologist*, **109**, 727–734.

Eriksson, K., & Coultas, J. C. (2014). Corpses, maggots, poodles and rats: Emotional selection operating in three phases of cultural transmission of urban legends. *Journal of Cognition and Culture*, **14**, 1–26.

Feldman, M. (2014). Echoes of the past: Hereditarianism and a troublesome inheritance. *PLoS Genetics*, **10**, e1004817.

Fincher, C. L., & Thornhill, R. (2012). Parasite-stress promotes in-group assortative sociality: The cases of strong family ties and heightened religiosity. *Behavioral and Brain Sciences*, **35**, 61–79.

Fincher, C. L., Thornhill, R., Murray, D. R., & Schaller, M. (2008). Pathogen prevalence predicts human cross-cultural variability in individualism/collectivism. *Proceedings. Biological Sciences*, **275**, 1279–1285.

Fracchia, J., & Lewontin, R. C. (1999). Does culture evolve? *History and Theory*, **38**, 52–78.

Fragaszy, D. M., & Perry, S., eds. (2003). *The Biology of Traditions: Models and Evidence*. Cambridge, UK: Cambridge University Press.

Freeman, D. (1974). The evolutionary theories of Charles Darwin and Herbert Spencer. *Current Anthropology*, **15**, 211–237.

Galef, B. G., & Laland, K. N. (2005). Social learning in animals: Empirical studies and theoretical models. *BioScience*, **55**, 489–499.

Gintis, H., Bowles, S., Boyd, R., & Fehr, E. (2003). Explaining altruistic behavior in humans. *Evolution and Human Behavior*, **24**, 153–172.

Gould, S. (1996). *The Mismeasure of Man*. London: W. W. Norton & Co.

Gould, S. J., & Lewontin, R. C. (1979). The Spandrels of San Marco and the Panglossian Paradigm: A critique of the adaptationist programme. *Proceedings. Biological Sciences*, **205**, 581–598.

Grafen, A. (1984). Natural selection, kin selection and group selection. In J. R. Krebs & N. B. Davies, eds., *Behavioral Ecology: An Evolutionary Approach*. Oxford: Blackwell Scientific, pp. 62–84.

Gray, R. D., Greenhill, S. J., & Ross, R. M. (2007). The pleasures and perils of Darwinizing culture (with phylogenies). *Biological Theory*, **2**, 360–375.

Hamilton, W. D. (1964). The genetical evolution of social behavior I and II. *Journal of Theoretical Biology*, 7, 1–52.

Heine, S. J. (2011). *Cultural Psychology*. New York: W. W. Norton & Co.

Henrich, J. (2001). Cultural transmission and the diffusion of innovations. *American Anthropologist*, **103**, 992–1013.

Henrich, J. (2004). Cultural group selection, coevolutionary processes and large-scale cooperation. *Journal of Economic and Behavioral Organization*, **53**, 3–35.

Henrich, J. (2006). Cooperation, punishment, and the evolution of human institutions. *Science*, **312**, 60–61.

Henrich, J. (2015). *The Secret of Our Success: How Culture Is Driving Human Evolution, Domesticating Our Species, and Making Us Smarter*. Princeton, NJ: Princeton University Press.

Henrich, J., & Gil-White, F. J. (2001). The evolution of prestige: freely conferred deference as a mechanism for enhancing the benefits of cultural transmission. *Evolution and Human Behavior*, **22**, 165–196.

Henrich, J., & Henrich, N. (2010). The evolution of cultural adaptations: Fijian food taboos protect against dangerous marine toxins. *Proceedings. Biological Sciences*, **277**, 3715–3724.

Henrich, J., Ensminger, J., McElreath, R., et al. (2010). Markets, religion, community size, and the evolution of fairness and punishment. *Science*, **327**, 1480–1484.

Hoppitt, W., & Laland, K. N. (2013). *Social Learning: An Introduction to Mechanisms, Methods, and Models*. Princeton, NJ: Princeton University Press.

Horowitz, M., Yaworsky, W., & Kickham, K. (2014). Whither the blank slate? A report on the reception of evolutionary biological ideas among sociological theorists. *Sociological Spectrum*, **34**, 489–509.

Hruschka, D. J., & Henrich, J. (2013). Institutions, parasites and the persistence of in-group preferences. *PLoS ONE*, **8**, e63642.

Ingold, T. (2007). The trouble with "evolutionary biology." *Anthropology Today*, **23**, 3–7.

Kempe, M., Lycett, S. J., & Mesoudi, A. (2014). From cultural traditions to cumulative culture: Parameterizing the differences between human and nonhuman culture. *Journal of Theoretical Biology*, **359**, 29–36.

Kendal, J., Giraldeau, L. A., & Laland, K. (2009). The evolution of social learning rules: Payoff-biased and frequency-dependent biased transmission. *Journal of Theoretical Biology*, **260**, 210–219.

Kirby, S., Dowman, M., & Griffiths, T. L. (2007). Innateness and culture in the evolution of language. *Proceedings of the National Academy of Sciences*, **104**, 5241–5245.

Kirby, S., Cornish, H., & Smith, K. (2008). Cumulative cultural evolution in the laboratory: An experimental approach to the origins of structure in human language. *Proceedings of the National Academy of Sciences*, **105**, 10681–10686.

Kitcher, P. (1985). *Vaulting Ambition: Sociobiology and the Quest for Human Nature*. Cambridge, MA: MIT Press.

Krasnow, M. M., Delton, A. W., Tooby, J., & Cosmides, L. (2013). Meeting now suggests we will meet again: Implications for debates on the evolution of cooperation. *Scientific Reports*, **3**, 1747.

Krebs, J. R., & Davies, N. B. (1984). *Behavioral Ecology: An Evolutionary Approach*. Oxford: Blackwell Scientific.

Kroeber, A. L. (1917). The superorganic. *American Anthropologist*, **19**, 163–213.

Kroeber, A. L., & Kluckohn, C. (1952). *Culture*. New York: Vantage.

Kurzban, R., & Haselton, M. G. (2006). Making hay out of straw? Real and imagined controversies in evolutionary psychology. In J. H. Barkow, ed., *Missing the Revolution: Darwinism for Social Scientists*. Oxford: Oxford University Press, pp. 149–162.

Laland, K. N., & Brown, G. R. (2011). *Sense and Nonsense*, 2nd ed. Oxford: Oxford University Press.

Lamba, S., & Mace, R. (2011). Demography and ecology drive variation in cooperation across human populations. *Proceedings of the National Academy of Sciences*, **108**, 14426–14430.

Lawson, D. W., Alvergne, A., & Gibson, M. A. (2012). The life-history trade-off between fertility and child survival. *Proceedings. Biological Sciences*, **279**, 4755–4764.

Lumsden, C. J., & Wilson, E. O. (1981). *Genes, Mind, and Culture: The Coevolutionary Process*. Cambridge, MA: Harvard University Press.

Lyle, H. F., & Smith, E. A. (2012). How conservative are evolutionary anthropologists? *Human Nature*, **23**, 306–322.

Mace, R., & Holden, C. J. (2005). A phylogenetic approach to cultural evolution. *Trends in Ecology & Evolution*, **20**, 116–121.

Matthews, L. J., Passmore, S., Richard, P. M., Gray, R. D., & Atkinson, Q. D. (2016). Shared cultural history as a predictor of political and economic changes among nation states. *PLoS ONE*, **11**, e0152979.

Maynard Smith, J., & Warren, N. (1982). Models of cultural and genetic change. *Evolution*, **36**, 620–627.

McElreath, R. (2016). *Statistical Rethinking: A Bayesian Course with Examples in R and Stan*. Boca Raton, FL: CRC Press.

McElreath, R., & Strimling, P. (2008). When natural selection favors imitation of parents. *Current Anthropology*, **49**, 307–316.

Mesoudi, A. (2009). The cultural dynamics of copycat suicide. *PLoS ONE*, **4**, e7252.

Mesoudi, A. (2011). *Cultural Evolution*. Chicago, IL: University of Chicago Press.

Mesoudi, A., & Whiten, A. (2004). The hierarchical transformation of event knowledge in human cultural transmission. *Journal of Cognition and Culture*, **4**, 1–24.

Mesoudi, A., & Whiten, A. (2008). The multiple roles of cultural transmission experiments in understanding human cultural evolution. *Philosophical Transactions of the Royal Society of London, Series B: Biological Sciences*, **363**, 3489–3501.

Mesoudi, A., Whiten, A., & Laland, K. N. (2004). Is human cultural evolution Darwinian? Evidence reviewed from the perspective of *The Origin of Species*. *Evolution*, **58**, 1–11.

Mesoudi, A., Whiten, A., & Dunbar, R. (2006). A bias for social information in human cultural transmission. *British Journal of Psychology*, **97**, 405–423.

Miton, H., Claidière, N., & Mercier, H. (2015). Universal cognitive mechanisms explain the cultural success of bloodletting. *Evolution and Human Behavior*, **36**, 303–312.

Morgan, L. H. (1877). *Ancient Society*. New York: Henry Holt.

Murdock, G. P. (1967). *Ethnographic Atlas*. Pittsburgh, PA: University of Pittsburgh Press.

Nettle, D. (2009). Ecological influences on human behavioral diversity: A review of recent findings. *Trends in Ecology & Evolution*, **24**, 618–624.

Nettle, D., Gibson, M. A., Lawson, D. W., & Sear, R. (2013). Human behavioral ecology: current research and future prospects. *Behavioral Ecology*, **24**, 1031–1040.

O'Brien, M. J., & Lyman, R. L. (2003). *Cladistics and Archaeology*. Salt Lake City, UT: University of Utah Press.

Open Science Collaboration (2015). Estimating the reproducibility of psychological science. *Science*, **349**, aac4716.

Pagel, M. (2009). Human language as a culturally transmitted replicator. *Nature Reviews: Genetics*, **10**, 405–415.

Pinker, S. (1994). *The Language Instinct*. New York: W. Morrow.

Pinker, S. (1997). *How the Mind Works*. New York: W. W. Norton & Co.

Pinker, S. (2010). The cognitive niche: Coevolution of intelligence, sociality, and language. *Proceedings of the National Academy of Sciences*, **107**, 8993–8999.

Plotkin, H. (1995). *Darwin Machines and the Nature of Knowledge*. London: Penguin.

Richerson, P. J., & Boyd, R. (2005). *Not by Genes Alone*. Chicago, IL: University Chicago Press.

Richerson, P. J., Baldini, R., Bell, A., et al. (2016). Cultural group selection plays an essential role in explaining human cooperation: a sketch of the evidence. *Behavioral and Brain Sciences*, **39**, e30.

Rogers, A. R. (1988). Does biology constrain culture? *American Anthropologist*, **90**, 819–831.

Rose, S., & Rose, H., eds. (2000). *Alas Poor Darwin: Arguments against Evolutionary Psychology*. London: Jonathon Cape.

Sahlins, M. (1976). *The Use and Abuse of Biology: An Anthropological Critique of Sociobiology*. Oxford: University of Michigan Press.

Schaller, M. (2006). Parasites, behavioral defenses, and the social psychological mechanisms through which cultures are evoked. *Psychological Inquiry*, **17**, 96–101.

Segerstråle, U. C. O. (2000). *Defenders of the Truth: The Battle for Science in the Sociobiology Debate and Beyond*. Oxford: Oxford University Press.

Servedio, M. R., Brandvain, Y., Dhole, S., et al. (2014). Not just a theory – The utility of mathematical models in evolutionary biology. *PLoS Biology*, **12**, e1002017.

Smaldino, P. E. (2016). Models are stupid, and we need more of them. In R. R. Vallacher, A. Nowak, & S. J. Read, eds., *Computational Models in Social Psychology*. Hove: Psychology Press, pp. 311–331.

Smith, E. A. (1998). Is Tibetan polyandry adaptive? Methodological and metatheoretical analyses. *Human Nature*, **9**, 225–261.

Sociobiology Study Group of Science for the People (1976). Sociobiology: Another biological determinism. *BioScience*, **26**, 182–186.

Sperber, D. (1996). *Explaining Culture: A Naturalistic Approach*. Oxford: Oxford University Press.

Tennie, C., Call, J., & Tomasello, M. (2009). Ratcheting up the ratchet: On the evolution of cumulative culture. *Philosophical Transactions of the Royal Society of London, Series B: Biological Sciences*, **364**, 2405–2415.

Tinbergen, N. (1963). On aims and methods of ethology. *Zeitschrift für Tierpsychologie*, **20**, 410–433.

Tooby, J. (2014). Learning and culture. Edge.org. What Scientific Idea Is Ready for Retirement? www.edge.org/response-detail/25343.

Tooby, J., & Cosmides, L. (1992). The psychological foundations of culture. In J. H. Barkow, L. Cosmides, & J. Tooby, eds., *The Adapted Mind*. Oxford: Oxford University Press, pp. 19–136.

Turchin, P. (2003). *Historical Dynamics: Why States Rise and Fall*. Princeton, NJ: Princeton University Press.

Turchin, P. (2015). *Ultrasociety: How 10,000 Years of War Made Humans the Greatest Cooperators on Earth*. Chaplin, CT: Beresta Books.

Turchin, P., Currie, T. E., Turner, E. A. L., & Gavrilets, S. (2013). War, space, and the evolution of Old World complex societies. *Proceedings of the National Academy of Sciences*, **110**, 16384–16389.

Tylor, E. B. (1871). *Primitive Culture*. London: John Murray.

Viveiros de Castro, E. (2014). Who is afraid of the ontological wolf? Some comments on an ongoing anthropological debate. Presented at: *CUSAS Annual Marilyn Strathern Lecture*, May 30, Cambridge, UK.

West, S. A., El Mouden, C., & Gardner, A. (2011). Sixteen common misconceptions about the evolution of cooperation in humans. *Evolution and Human Behavior*, **32**, 231–262.

Whiten, A., Goodall, J., McGrew, W. C., et al. (1999). Cultures in chimpanzees. *Nature*, **399**, 682–685.

Whiten, A., Caldwell, C. A., & Mesoudi, A. (2016). Cultural diffusion in humans and other animals. *Current Opinion in Psychology*, **8**, 15–21.

Williams, G. (1966). *Adaptation and Natural Selection*. Princeton, NJ: Princeton University Press.

Wilson, E. O. (1975). *Sociobiology: The New Synthesis*. Cambridge, MA: Harvard University Press.

Winterhalder, B., & Smith, E. A. (2000). Analyzing adaptive strategies: Human behavioral ecology at twenty-five. *Evolutionary Anthropology*, **9**, 51–72.

PART III

EVOLUTION AND NEUROSCIENCE

Late-twentieth-century evolutionary approaches to human behavior tended to shy away from explanations based on neuroscience. A justification for this was that evolutionary psychologists and behavioral ecologists dealt with ultimate rather than proximate levels of explanation (Cosmides & Tooby, 2000). This stance follows a prestigious heritage. Let us not forget that Darwin was able to formulate his theory of evolution by natural selection with no knowledge of the underlying material substrate (Darwin, 1859). Since the turn of the century, however, there has been a growing realization that, if evolutionary approaches are to succeed, then an understanding of the physical substrate of internal states and overt behavior will be required (Panksepp & Panksepp, 2000). Moreover, the development of credible neural perspectives clearly adds to the evolutionary methodological toolbox.

In Part III, we begin with Frederick M. Toates' consideration of the compatibility of evolutionary psychology with the neuroscience of motivation. Toates is one of a small number of academics (including Kent C. Berridge) who have dismantled old theories of motivation, in which the organism was *driven* by internal factors, and replaced these with the notion that learning and hormones modulate the power of the incentive to exert a pull (Toates, 2014). In his chapter, Toates poses (and seeks to answer) a number of questions such as: To what extent do motivational systems suggest adaptive design? And is there a mismatch between our motivational systems and our current environment?

Motivation also features in Eloise Stark, Kent C. Berridge, and Morten L. Kringelbach's contribution, as they consider the important question of whether humans are designed (by selective forces) to be happy. This is a timely question to ask, as well-being has become an increasingly important concept in positive psychology (and in neuroscience) in recent years (Kringelbach & Berridge, 2017; Stark, Vuust, & Kringelbach, 2018). Interestingly, Stark et al. suggest that well-being (or

"eudaimonia") may turn out to be an evolutionary by-product that happens to follow moments of bliss.

In our final contribution to Part III, Dwaipayan Adhya, Aicha Massrali, Arkoprovo Paul, Mark Kotter, Jason Carroll, Deepak Srivastava, Jack Price, and Simon Baron-Cohen consider epigenetic gene regulation during early developmental stages of cellular differentiation and its relationship to the development of autism. This highly technical, multiauthored chapter has been written by a team that includes world-leading authorities who bring together expertise from the fields of autism research, molecular genetics, and developmental neuroscience. In addition to presenting cutting-edge research into the neuroscientific bases of the development of autism, the team outlines the latest technological advances in this rapidly developing field.

REFERENCES

Cosmides, L., & Tooby, J. (2000). Evolutionary psychology and the emotions. In M. Lewis & J. Haviland, eds., *The Handbook of Emotions*, 2nd ed. New York. Guilford, pp. 91–116.

Darwin, C. (1859). *On The Origin of Species by Means of Natural Selection, or the Preservation of Favoured Races in the Struggle for Life*. London: John Murray.

Kringelbach, M. L., & Berridge, K. C. (2017). The affective core of emotion: Linking pleasure, subjective well-being and optimal metastability in the brain. *Emotion Review*, **9**, 191–199.

Panksepp, J., & Panksepp, J. B. (2000). The seven sins of evolutionary psychology. *Evolution and Cognition*, **6**, 108–131.

Stark, E., Vuust, P., & Kringelbach, M. L. (2018). Music, dance, and other art forms: New insights into the link between hedonia (pleasure) and eudaimonia (well-being). *Progress in Brain Research*, **237**, 129–152.

Toates, F. (2014). *How Sexual Desire Works: The Enigmatic Urge*. Cambridge, UK: Cambridge University Press.

7

Are Evolutionary Psychology and the Neuroscience of Motivation Compatible?

FREDERICK M. TOATES

7.1 INTRODUCTION

In evolutionary psychology (EP), motivation is the poor relation, not even earning an index entry in some of the principal textbooks (Buss, 2016; Campbell, 2013; Ray, 2013). By contrast, emotion sometimes justifies a full chapter (Ray, 2013). Of course, some phenomena linked to motivation, such as attachment and, in particular, sexual behavior, are very well described in the EP literature.

Motivational processes are the class that is responsible for selecting a course of action, organizing goal-directed behavior, and energizing behavior to meet the goal (Toates, 1986). Goals are selected on the basis of a combination of internal states and salient features of the environment and are pursued until the associated motivation is lowered to zero or a higher order priority emerges. Motivational processes are responsible for resolving competition and conflict between several potential goals that might simultaneously exist.

Under the heading of motivation, psychologists have traditionally studied feeding, drinking, exploratory behavior, sex, and drug-taking. Aggression has commonly been investigated as something distinct, as has social behavior. However, in recent years, some authors have extended the general heading of motivation to apply also to these forms of behavior. Fear can also be a powerful source of motivation.

The purpose of the present chapter is to see whether theories within EP are compatible with an understanding of the neuroscience of motivation. To answer this, a number of features of how motivational systems are constructed and operate need to be investigated, as follows: Is the array of motivational systems manifest in contemporary society indicative of universal problems of survival and reproduction in the environment of evolutionary adaptation (EEA)? Does each motivational system reveal a "design" suggestive of an adaptation to these environments? The evidence might include signs of evolutionary mismatch between what could have been an "optimal design" in the early hominin evolutionary environments and the outcome in the present very different environment.

A further question is: Are there any motivations revealed widely in the present environment, but were probably not present in the earlier environments? If so, might they still be understandable in terms of their exploitation of brain mechanisms that did serve an adaptive role in the earlier environments?

The essence of motivation is, of course, *action* in pursuit of a goal. This raises the issue of what constitutes the absence of such action: presumably passivity, and under some circumstances depression. A number of interesting questions follow: Is depression simply a failure of motivational mechanisms, or are there mechanisms that have evolved because depression has conferred adaptive value? If so, can the mechanisms underlying it manifest evolutionary mismatch in the contemporary environment?

A particular form of EP is sometimes known as the "Santa Barbara school," associated with Leda Cosmides, John Tooby, and their colleagues. According to this perspective, evidence would be expected to indicate that each motivational system is based upon a "module" dedicated to serving just the one motivation. This chapter will consider the evidence for such modules.

The next four sections consider four brain processes that embody the bases of motivational systems: (1) layered control; (2) incentive motivation; (3) inhibition; and (4) cost–benefit analysis.

7.2 LAYERED CONTROL

7.2.1 Basics

The terms "layered control," "levels of control," and "hierarchical control" refer to the assumption that the brain is organized as a hybrid of processes that have different properties and different responsibilities for behavior (Angus et al., 2016; Campbell, 2013; Maclean, 1990; Panksepp & Panksepp, 2000). In humans, brain regions that are evolutionarily old coexist with regions that are newer. In some studies of motivation and response production, the evolutionarily old processes are termed "System 1," whereas the newer processes are termed "System 2" (Toates, 1998, 2014).

In humans, subcortical processes underlying motivation share important features across mammalian species (Berridge & Kringelbach, 2013; Panksepp, 2014). These processes of emotion and motivation include seeking/wanting, fear, rage, panic/grief, maternal care, lust, and play (Panksepp, 2014).

The evidence suggests that affective states of pleasure and pain monitor the efficacy of these emotion–motivation systems. In their evolutionary origins, negative affect indicated a situation threatening to evolutionary fitness, whereas positive affect indicated a fitness-enhancing action. Thus, social loss, as in separation between caregiver and care-receiver, triggers an aversive state that the animal is motivated to reduce by regaining contact. Positive affect, as in play, indicates a fitness-enhancing activity. The fact that these basic processes have been conserved in evolution suggests their ubiquitous adaptive value, including throughout the course of human evolution. Much of the brain machinery of mammalian, including human, motivation (e.g., wanting and liking) is based in such subcortical brain regions.

It appears that in primates, especially humans, an evolutionarily newer region, the orbitofrontal cortex, recreates something of the functions of subcortical processes, but in a way that is less tied directly to behavior, thereby permitting more flexible control. Thus, regions of the orbitofrontal cortex embody the value of rewards and punishments, whereas the dorsolateral division of the prefrontal cortex exerts a role in the inhibition of behavior based upon information retrieved from long-term stores and held in working memory (Campbell, 2013).

In terms of both evolution and development, there is a transition in the weight of the control of behavior between such layers (Toates, 1998, 2005, 2014). Routine problems can be solved on the basis of largely subcortical processes, termed System 1. Newly emerging processes (System 2), both in the evolutionary and developmental sense, have a primary responsibility for the novel, creative features of behavior, where routine ("recurrent") solutions cannot solve the problem and indeed their expression needs to be inhibited (Chiappe & MacDonald, 2005). In terms of embodiment, this is perhaps most evident in the case of the prefrontal cortex, which occupies a larger percentage of the brain in humans than in any other species. It has rich connections to other brain regions.

The relative weight of control varies between layers as a function of such things as the formation of habits and current experience of emotion (Toates, 1998; see also Wagner & Heatherton, 2015). It is interesting to speculate on the adaptive significance of such a shift. As habits are formed, more weight is attributed to System 1, which presumably facilitates speedy responding and frees processing capacity within System 2 for engagement in nonroutine cognition. As emotional intensity increases, so it might be adaptive to fall back onto established solutions and to switch out control based upon future considerations.

7.2.2 Link to Evolutionary Psychology

In the study of EP, a school of thought proposes the existence of "domain-general" processes in the control of behavior, in addition to the "domain-specific" processes (Bolhuis et al., 2011; Chiappe & MacDonald, 2005; Panksepp & Panksepp, 2000; Toates, 2005). Domain-specific processes solve problems that were a recurrent feature of life in the EEA. They are something like, if not identical to, the modules of EP. They act fast and have evolved where a problem is solved by a recurrent mapping between stimulus input and response output. For example, a recurrent feature of the EEA would be the particularly strong triggering of male sexual desire by females with the right waist–hip ratio and at the optimal age for reproduction. If pursuit is successful, this would yield high levels of reward for the male and represent the most adaptive solution. It appears that control by domain-specific processes is mediated in large part by brain mechanisms outside the prefrontal cortex.

By contrast, the creativity and flexibility of behavior involve intentionality, goal-pursuit, and synthesizing disparate bits of information (Campbell, 2013). This is needed when confronting novel problems and is controlled by domain-general processes, with weight upon the prefrontal cortex. It involves what is normally termed "executive function," involving working memory, and it enables plans to be held in conscious memory in the face of distractions. Axiomatically, a solution based upon recurrent input–output links cannot solve such problems. Damage to the prefrontal cortex has a particularly disruptive effect on goal maintenance and the ability to inhibit irrelevant information.

By definition, motivation is what moves the animal. Cognition (e.g., cheater detection) without motivation (e.g., avoid or expose the cheater) would simply leave the animal buried in thought. According to Chiappe and MacDonald (2005), motivational processes are central to understanding how domain-specific and domain-general processes collaborate to maximize fitness. Thus, to return to the example of sexual motivation, domain specificity directs resources to a particular sexual target by triggering desire. However, particularly in humans, the necessary goal-directed appetitive activity leading to mating has probably always involved a wide and complex range of creative solutions, which are brought into conscious focus by the existence of sexual motivation. Which of these is chosen at a given time will depend upon the other individual's reaction, the features of an often changing physical and social context, and a history of learning.

Panksepp and Panksepp (2000) argue that the Santa Barbara school has placed too much explanatory weight upon specific modules that are said to have emerged in human evolution and are rooted in cortical structures. Rather, Panksepp and Panksepp suggest that much of human motivation is rooted in evolutionarily older subcortical structures shared broadly across mammalian species. Cortical embodiment could exemplify plasticity: a molding that is responsive to experience and input from subcortical mechanisms.

7.3 INCENTIVE MOTIVATION

7.3.1 Basics

The theory of incentive motivation (Berridge, 2004; Bindra, 1978; Toates, 1986) was developed in part as a reaction against earlier drive models. Its central premise is that motivational states arise from a combination of:

- Incentive stimuli and internal representations of them;
- Bodily states, such as nutrient and hormone levels.

For example, food and/or internal representations of it act together with nutrient levels to generate feeding motivation. Sexual incentives and internal representations of them act together with hormones to generate sexual motivation. In this way, a general principle can be seen to apply to both: (1) those motivations that have the function of regulating internal variables (e.g., fluid and nutrient levels); and (2) those that do not defend the internal environment of the body (e.g., sexual and social motivations).

In parallel with this theoretical development is another, which is entirely congruent: that of Jaak Panksepp and colleagues. Panksepp and Panksepp (2000) describe what they term an "expectancy/SEEKING," also known as a "general purpose foraging system." This employs the neurochemical dopamine and is energized in the range of situations in which the animal is moved to engagement toward such items as novel territory, food, or a mate. These theorists note that the system is subcortical, shared across mammalian species, and (p. 119):

... surely controls a diversity of human/animal aspirations and desires.

In humans, there are also cortically rooted desire systems, which work together with this system.

7.3.2 Wanting and Liking

It used to be assumed implicitly, if not explicitly, that the degree of wanting something corresponds to the degree of liking it once attained (Toates, 1986). In a landmark series of articles within the tradition of incentive motivation theory, Kent Berridge and colleagues discovered evidence for a dissociation between wanting and liking, based on different brain mechanisms (Berridge & Valenstein, 1991; Robinson & Berridge, 1993). For example, when the wanting brain mechanism is selectively stimulated, something might be pursued with vigor, but the degree of liking of it is relatively low. Conversely, when wanting is suppressed, an animal's motivation toward attaining a goal might be low, but a normal degree of liking is shown when in direct contact with the goal object.

This research established the role of the neurotransmitter dopamine in the wanting phase of motivation but not in the liking phase, which is based upon opioids, among other substances. A body of theory and experimentation that matches rather well to the liking–wanting distinction is described later. Cosmides and Tooby (1995, p. 57) argue:

There is simply no uniform element in sex, eating, drinking, staying warm (but not overheating), and so on, that could be used to build a general architecture that could learn to accomplish these behaviors.

In fact, evidence suggests a domain-general motivational process that employs dopamine and underlies each and every seeking behavior. Further evidence points to this process being shared with fear-based active avoidance (Berridge & Kringelbach, 2015; Blackburn et al., 1992). Opioidergic-based reward might also be domain general.

7.4 INHIBITION

From an evolutionary perspective, a process of active inhibition can be assumed to confer an adaptive advantage, in that under some conditions behavior needs to be restrained (Bjorklund & Kipp, 1996). This avoids suboptimal choices. For example, from an evolutionary perspective, it would be adaptive for females to show greater restraint and inhibition in a potential sexual encounter than it would for males (Chapter 27). Evidence points to sexual desire being a balance between a process of sexual excitation and inhibition, with the ratio of excitation to inhibition being greater in men (Carpenter et al., 2008). A domain-general system of behavioral inhibition is based in part within regions of the prefrontal cortex (Lopez et al., 2015).

7.5 COST–BENEFIT ANALYSIS

A major subject of investigation within behavioral ecology (e.g., in foraging for food) is that of behavioral analysis based upon cost–benefit calculation (Krebs & Davies, 1978). Similarly, evolutionary psychologists are attentive to the principle that evolutionary processes can only be understood as a trade-off between the costs and benefits of different courses of action. Within neuroscience, there is a considerable body of evidence on the brain mechanisms underlying cost–benefit analysis and decision-making. However, there has been little interaction between these bodies of theory.

In keeping with the principles of fitness maximization and optimal foraging, a study of the brain has revealed processes of cost–benefit analysis (Salamone et al., 2007). Given a choice of a reward of low value that is easily accessed or a reward of higher value that is accessed only by the exertion of effort, rats typically prefer the high-effort/high-reward option. When lesions are made to the dopamine terminal regions of the nucleus accumbens (N.Acc.), rats switch their choice to the low-reward/low-effort option.

The anterior cingulate cortex is involved in cost–benefit decision-making concerning economic decisions. A study by Rupp et al. (2009) on women found that this structure is also involved when the decision is one of weighing up the risk and benefits associated with potential sexual partners.

This chapter will now turn to some individual examples of motivational systems. It will first consider the uncontroversial cases of feeding and drinking. It will then go on to discuss social motivations and finally drug-taking. Sexual motivation, which tends to generate the most insight as well as heat in discussions of EP, deserves a chapter to itself (Chapter 27).

7.6 FEEDING AND DRINKING

7.6.1 Basics

Feeding and drinking feature rather little in discussions of EP, though the act of foraging is considered (Duchaine et al., 2001). This absence presumably represents the fact that they have a rather uncontroversial functional significance. However, from considerations of motivation, there are some interesting phenomena associated with feeding, which are relevant to EP.

7.6.2 Taste-Aversion Learning

For a variety of species, including humans, there appears to be a dedicated adaptive process underlying taste-aversion learning, also termed the "Garcia effect." When gastric-based nausea occurs, the animal is able to form an association with candidate tastes and smells that accompanied ingestion minutes or even hours earlier. The functional significance of this is clear in terms of protecting the animal from toxins in the diet. A general-purpose learning system that only associates events that are a second or so apart would be useless in solving this problem.

A formerly hedonically positive taste becomes aversive to rats following gastric upset (Itoga et al., 2016). Underlying this behavioral change is a change in the activity of neurons that code wanting and liking in a brain region known as the ventral pallidum. Prior to conditioning, the sweet taste excites their activity, but afterwards it inhibits activity.

7.6.3 Pregnancy Sickness

Rather than being an example of pathology, pregnancy sickness fits the criteria of being an adaptation and is one among several adaptations that act in a functionally coherent way (Profet, 1992). It protects the baby from toxins at a time of maximum vulnerability of the developing nervous system (i.e., usually starting at 2–4 weeks of pregnancy and ending at 14 weeks) (Profet, 1992). Some foods that were previously acceptable become intolerable.

A maximum advantage of such an adaptation would be obtained if ingestion were preempted by an aversion to the smell of risky foods, and this does indeed happen (Cameron, 2014; Patil et al., 2012). Sickness can be classically conditioned to environmental stimuli (Stockhorst et al., 2007), and this would be expected to increase the adaptive value.

In terms of the neuroscience of pregnancy sickness, a region in the brainstem is implicated, and it is termed the chemoreceptor trigger zone (Profet, 1992). It is known to be involved in taste aversion and is richly supplied with blood, and hence it is sensitive to toxins in the blood. Something associated only with the early stages of pregnancy must sensitize this area – a likely candidate is the elevated level of hormones: estrogens and progesterone (Patil et al., 2012).

7.6.4 Obesity

The current epidemic of obesity is associated with an abundance of energy-rich foods that can be obtained with minimal effort (Stice & Yokum, 2016). From a motivational perspective, this makes no sense in terms of the principles of homeostasis and a drive based upon nutrient deficits. However, from an incentive motivation perspective, it makes sense and illustrates evolutionary mismatch. High-energy foods, particularly refined sugars, obtained at low cost and with variety in their presentation, were obviously not part of humanity's early environments. These seem to be the qualities to which the feeding system is most sensitive. Variety in the diet can act to counter the satiety that would arise from a single food item, a phenomenon termed "sensory-specific satiety." This is reflected in heightened responsiveness to novel foods in various parts of the brain, including dopaminergic wanting processes (Papageorgiou et al., 2016; Rolls et al., 1986). Where low-energy food in a limited variety was available only with extensive foraging, one can speculate that a motivational process that is sensitive to variety was optimal in obtaining a balance of nutrients.

7.7 DOMINANCE AND AGGRESSION

7.7.1 Basics

Some authors consider an overarching motivation to be dominance-seeking, while aggression is triggered when other means of asserting dominance fail (Mazur & Booth, 1998). This is the terminology to be adopted here. Explicit violence can be very costly in adaptive terms. Getting a similar result without violence could be more adaptive, as in, for example, giving domineering glances or making malicious gossip about same-sex rivals.

Lorenz (1981) suggested that a motivational tendency to aggression simply builds up over time and is then discharged as aggressive behavior. However, few – if any – now believe in this model. If there is an internal variable that rises with deprivation from the opportunity to perform aggression and falls with satiation, no one has so far identified it (Toates & Archer, 1978).

There are various dichotomous expressions used to categorize aggression with a common underlying basis to them. Thus, in humans, and in keeping with hierarchical principles, the causes of aggression vary from triggering automatically by simple stimuli (e.g., one causing physical pain) to those arising only after complex cognitive processing (e.g., an assessment of being excluded socially) (MacDonald & Leary, 2005; see Section 7.9.3). In one dichotomy, aggression is classified as either reactive (i.e., involving anger in response to a current threat or frustration) or proactive (i.e., consciously planned) (Angus et al., 2016), though a given aggressive act can have features of both (Rosell & Siever, 2015). Another dichotomy very similar to this one is that between reactive and instrumental aggression, only with the latter being motivated to obtain some extrinsic goal, such as a victim's money (Glenn & Raine, 2009). Such dichotomies have echoes of control by System 1 and System 2, respectively (see Section 7.2). Something similar, if not identical, is expressed by Archer (2009, p. 206) as:

… I offer a different approach to moving from function to mechanism, one where the mechanisms underlying aggressive motivation are viewed as existing on a continuum from the inflexible to the flexible.

The motivation for reactive aggression against another individual is associated in humans with the conscious experience of anger (Denson, Schofield, & Fabiansson, 2015). It is commonly triggered by some kind of provocation (e.g., a challenge to resource ownership or other intrusion into the status and/or well-being of the individual, sometimes expressed as a threat to self-worth). An act of aggression against someone perceived to have caused offense can be affectively positive, pointing to features in common with unambiguously appetitive motivations (Ramírez et al., 2005). In keeping with incentive motivation, the presence of aggressive cues in the environment (e.g., a weapon) increases the tendency to show aggressive behavior (Berkowitz & LePage, 1967).

7.7.2 The Functional Level

Aggression can be very costly, and an assumption is that, in our EEA, it tended to be triggered only in situations in which there was some adaptive advantage to do so (e.g., competition over mates or territory or defense against threats) (Buss & Shackelford, 1997). As such, it would have been fine-tuned (e.g., escalating levels of threat) according to such things as the nature of the resources under threat, the size of the opponent, and social context. Indeed, across a number of species, including humans, there is evidence of such a cost–benefit analysis (Archer, 2009). By contrast, it would clearly be maladaptive for the system to behave in a homeostatic manner, with there to be regular discharges of energy via aggressive behavior irrespective of context and cost.

Although aggression can be very costly to both winners and losers, there clearly exists a certain amount of seemingly gratuitous ("maladaptive") violence that has both bodily and social consequences (Angus et al., 2016). This is exemplified by intermittent explosive disorder, characterized as repeated bouts of reactive aggression, the magnitude of which seems to be out of all proportion to the trigger (Gan et al., 2016). Whether such excesses were present in our EEA is a matter of speculation. Given the dopaminergic factor in the control of aggression (Gan et al., 2016), there presumably exists the possibility of this sometimes becoming freed from those triggers that would reflect adaptive operation of the system, as is the case with drug addiction and obesity.

7.7.3 Neuroscience

Often, explanations for aggression describe it as the result of a domain-general learning process. All of the arguments on why domain-general learning alone could not solve such adaptive problems apply here (Buss & Shackelford, 1997). Aggression involves dedicated brain processes of emotion and motivation (e.g., rage) serving an adaptive end, and these exist throughout mammalian species (Panksepp, 2014; Pinker, 2011). It is organized hierarchically, with newly evolved cortical regions modulating evolutionarily older subcortical regions (Angus et al., 2016). In taking costs and benefits into account, aggression therefore implies the use of the kind of neural process described in Section 7.5.

Anger/aggression is an approach motivation associated with relatively high activation in the front part of the left hemisphere (Angus et al., 2016; Carver & Harmon-Jones, 2009). It shows a number of the hallmarks of motivation (e.g., an appropriate direction of attention and narrowing of focus on particular targets).

Evidence points to a role of the dopaminergic behavioral activating system in aggression. The manic phase of bipolar disorder is sometimes associated with an increased frequency of violence, pointing to the role of increased activity by this system (Johnson et al., 2012). In this phase, people are prepared to exert high effort in order to earn rewards. The amygdala plays a primary role in computing emotional value to be attached to situations that are aggression related (Rosell & Siever, 2015).

Frustration is a major trigger to reactive aggression, and it involves a computation of actual and expected outcomes. Regions of the orbitofrontal cortex form the biological basis of the computation of violations of expectations (Blair, 2004) and thereby appear to be involved in frustration-induced aggression. The orbitofrontal cortex is also involved in the assessment of the value of such rewards as food.

The perception of an unfair offer triggers some of the same brain processes as those that underlie reactive aggression (White, 2014). Some authors argue that anger/aggression triggered by the perception of an unfair offer involves conventional incentive/reward processes (Buades-Rotger et al., 2016; De Quervain et al., 2004), though there is controversy on the exact structures involved (White et al., 2014).

The hormone testosterone acts to sensitize those areas of the brain that control aggression, and in the popular imagination there is a simple equation that more of it equals more aggression. However, the evidence suggests a more nuanced view (Archer, 2006; Dreher et al., 2016). For example, boosting levels of testosterone does not invariably increase aggression. Rather, in keeping with principles of evolutionary function, the secretion of testosterone usually reflects the cost–benefit calculation of aggression in the animal's present context (Archer, 2006).

Among young adult men, aggression toward other men is at its highest when competition for mates is at its most intense and testosterone levels are high. Testosterone increases aggression tendencies in men who are dominant and/or low in self-control (Carré et al., 2017). It would appear then that, if there is already a tendency to aggression, it is sensitized. This could represent a more adaptive solution than a universal increase in aggressive tendency. In men, boosting levels of testosterone increases the punishment given to wrongdoers, but also increases the reward given to a deserving other (Dreher et al., 2016). The authors interpret both of these actions as assertions of status.

Archer (2006) applied to humans what is termed "the challenge hypothesis." According to this, in situations of sexual competition (e.g., in the presence of apparently available females), adult males would be expected to exhibit: (1) heightened tendencies to aggression; and (2) a surge in the release of testosterone. Evidence points to an increase in testosterone levels in response to erotic stimuli. These tendencies would also appear when there is a dispute over status or resources. Sporting events are associated with increases in testosterone secretion in anticipation of the event and during it. Winning in a competition tends to increase testosterone relative to losing, particularly if the winning can be attributed to the skill of the winner. Just being a supporter of the winning side also tends to increase testosterone secretion. Conversely, after the birth of a child, when presumably competition for partners is absent or low, testosterone levels in males tend to fall.

Archer (2006) predicted that dominance achieved through aggression would be associated with relatively high levels of testosterone, and in general this is the case for both men and women. However, as noted, explicit aggression is only one means of asserting dominance, and there are other, more subtle ways of reacting to challenges.

A dichotomy can be drawn between male investment in: (1) short-term mating effort (i.e., relatively numerous partners and little investment in any of them); or (2) long-term mating effort associated with one or few partners and

paternal care. Archer (2006) predicts that testosterone levels will show a positive association with investment in short-term mating and a negative association with marital and parental investment. A number of pieces of evidence point to such associations, and Archer (2006, p. 337) argued that the studies reviewed show tentatively:

… an association between testosterone and a more extraverted, uninhibited, dominant personality. This is consistent with pursuing a shorter-term reproductive strategy, emphasizing mating rather than parental effort.

The amygdala and hypothalamus are two brain regions that are rich in testosterone receptors. Testosterone increases the reaction of these two regions to the presentation of angry faces (Carré et al., 2017), among other things increasing the attention directed to them. This points to the triggering of aggression-readiness by appropriate external stimuli, which is comparable to incentive sensitization.

7.7.4 Inhibition on Aggression

People commonly entertain emotionally hot fantasies of inflicting serious violence against selected others (e.g., love rivals), but this usually remains within the realm of fantasy (Crabb, 2000; Pinker, 2011). Clearly, it is functionally adaptive that much potential aggression should be inhibited. Social norms and pragmatic considerations are associated with inhibition on aggression (Denson, Schofield, & Fabiansson, 2015).

Modulation of reactive aggression arises from processing in the ventromedial region of the prefrontal cortex, and there are links from here to the amygdala (Angus et al., 2016; Blair, 2004, 2016). In nonhuman primates, this reflects adaptive cost–benefit (anticipated reward–punishment) considerations: the tendency to aggression is reduced when confronted by a dominant, threatening individual and increased in the presence of a threat from a subordinate. It seems logical to extrapolate this to humans. Where there is a conflict between the motivation to aggression and considerations of restraint, processes of cost–benefit analysis based within the anterior cingulate cortex also appear to be engaged.

According to subjective reports, women experience as much anger as do men (Archer, 2004), so how do we account for the sex difference in aggression? Campbell (2013) suggests that a simultaneous triggering of greater fear in women than in men might inhibit expression of anger as aggression. She notes that a fearful stimulus is associated with a relatively greater activation of the prefrontal region of the right hemisphere, a greater blink response, and greater sweating in women than in men. However, there is one situation in which females exhibit high levels of aggression: threat to her offspring. Campbell suggests that oxytocin secreted during suckling promotes maternal aggression toward threats.

7.8 EXPLORATION

As Litman (2005, p. 793) expresses it, with reference to humans:

Curiosity may be defined as a desire to know, to see, or to experience that motivates exploratory behavior directed towards the acquisition of new information . . .

Men tend to be more attracted to novelty-seeking and sensation-seeking than are women. This appears to be due to women's greater triggering of fear (anticipation of aversive consequences) in such situations (Cross et al., 2011).

Litman suggests that, as with other motivational systems, exploration exploits a dopamine-based wanting ("approach") system (e.g., toward novel items, sources of uncertainty) and an opioid-based liking system (triggered by resolution of uncertainty) (cf. Wittmann et al., 2008). In some situations, wanting can get out of alignment with liking, as in seeking disgusting or frightening events.

A problem faced by an animal is to resolve the conflict between exploiting a known resource and exploration of the environment, which might lead to other resources (Daw et al., 2006). In one study on human exploration, the weight of control appeared to rest upon frontal cortical structures, whereas in exploitation it moved to subcortical processes. Resolution of uncertainty is associated with activation of the N.Acc., among other structures (Jepma et al., 2012).

One can speculate that, in evolutionary history, rewards were generally unpredictable (e.g., past locations of prey or fruits would not invariably reward investigation), but the person needed to persist in exploration and foraging even in face of failure and fatigue. Hence, it would have been of adaptive value that: (1) unexpected rewards form associations with predictive cues; and (2) future engagement with cues predicting uncertain reward should strongly motivate and energize approach. This is indeed the case, as is reflected in the motivating role of uncertainty mediated via dopamine (Anselme, 2013). The casino and race course appear to capture something of this past environment, topics discussed further in Section 7.10.3.

7.9 SOCIAL MOTIVATION

7.9.1 General Principles

For a social-living species, such as humans, the adaptive value of establishing and maintaining regular social contact with one or more familiar individuals is clear (Over, 2016). A motivational system that underlies such positive interaction can be described. This would be accompanied by reward at attainment/renewal of contact and aversion at breaking contact. Indeed, a set of particular brain regions and the neurochemical dopamine are involved in motivating approach to social stimuli (Chevallier et al., 2012; Radke et al., 2016).

7.9.2 Infant–Mother Interaction

Across cultures, mothers typically spend much more time caring (holding, showing affection) for babies than do fathers (Campbell, 2013). Campbell writes (p. 73):

. . . through our evolutionary past from an infant's viewpoint, mothers have always been essential while fathers have been an optional extra.

What kind of motivational processes could form the basis of a child's attachment to a parent? What can trigger caring in the parent?

Darwin pointed out that there is something special about the face of an infant for it to trigger parental care (Kringelbach et al., 2008). Bowlby (1982) was the first to clearly articulate the existence of a distinct attachment system. This involved a reciprocal attraction between the caregiver – usually the mother – and the infant. Bowlby speculated that the infant forms an internal working model of the relationship with the mother. Under what might appear to be optimal conditions, this internal model is one of a mother who is responsive to the nutrient and comfort needs of the infant.

Panksepp (2014) describes a system of rapid-action social learning that forms an attachment bond. Furthermore, there is a "panic system" that triggers characteristic calls when isolated from the mother, with reinforcement coming in the form of reunion.

Evidence points to a process in the mother's brain that underlies an attraction to infant rather than adult faces (Kringelbach et al., 2008). Activity in the fusiform face area of the brain is triggered by both adult and infant faces, whereas the medial orbitofrontal cortex (mOFC) is triggered by infant but not by adult faces. This result was obtained whether the participants were parents or not. The result suggests that the mOFC is excited by the particular features of the infant face and is involved in drawing attention to them, presumably a feature of incentive salience attribution and caregiving. The reaction is so fast (130 ms) that the researchers concluded that it happens outside conscious awareness and has important features of the innate releasing mechanism described by early ethologists. Experiments point to a role of the mOFC in computing the reward value of other features of the environment, such as taste and olfaction (Kringelbach et al., 2008). The lateral OFC, by contrast, appears to encode aversive events.

Kim et al. (2016) suggest that the biological basis of the mother's bond formation to her infant is an increase in incentive salience on contact post-birth. This is triggered by the affective reaction to the baby, a process involving the amygdala.

7.9.3 Social Interaction Beyond the Mother–Infant Stage

Distinct from early attachment but building upon it, a later-emerging process of belonging to a larger social group with an associated motivational system has been identified (Baumeister & Leary, 1995; Over, 2016). Rewards come in the form of social acceptance and approval, whereas social exclusion triggers an aversive state, which the person is motivated to correct (Williams, 2007). Both motivate attempts to avoid social disapproval/ rejection and to restore social harmony. The functional significance is clear: exclusion from a social group would be maladaptive and a motivation to rejoin would be adaptive. Merely the threat of the psychological pain of exclusion could help to motivate continued staying in the social group.

MacDonald and Leary (2005) suggest that using the term "pain" to describe social exclusion is much more than just a colorful metaphor. Rather, social exclusion has important features in common with pain arising from tissue damage and exploits some of the same brain regions.

Concerning the incentive approach system, the dopaminergic pathway from the ventral tegmental area to the ventral striatum (a region that includes the N.Acc.) is involved in social reward (Gunaydin & Deisseroth, 2014; Radke et al., 2016). In humans, it is activated in anticipation of both social acceptance (an image of faces displaying this) and the possibility of avoiding social disapproval (an image of faces showing this) (Kohls et al., 2013). Cooperation with another human to solve a task triggers more activation of this region than cooperation with a computer (Fehr & Camerer, 2007). In one neuroimaging study, it was found that men but not women were more sensitive to monetary reward than social reward (a smiling face) (Spreckelmeyer et al., 2009). There is an interaction between dopaminergic neurotransmission and the neurohormone oxytocin such that oxytocin enhances activity in the ventral tegemental area–N.Acc. pathway (Groppe et al., 2013).

Turning now to social exclusion, MacDonald and Leary (2005) argue that brain processes that react to tissue damage were already in place when adaptations for social exclusion appeared, and hence they co-opted the existing processes. It might be somewhat surprising that tissue damage and social exclusion appear to share common brain processes since their inputs are very different. Nonetheless, neuroimaging has shown activation by a common set of brain structures in nociceptive and social pain.

The dorsal region of the anterior cingulate cortex (and possibly the periaqueductal gray [PAG]) is excited by both tissue damage and social exclusion. The PAG is closely involved in the generation of distress calls by rats (Panksepp, 1998). On a neurochemical level, opioids appear to be implicated in the alleviation of both nociceptive and social pain. The equivalent exogenous substance morphine reduces both nociceptive pain and distress-calling by the young of various vertebrate species (MacDonald & Leary, 2005).

7.9.4 Empathy

Empathy refers to the shared emotional experience of the mental state of another individual, the defining feature being the capacity to feel psychological pain in response to the pain and distress of another. This is a motivator of altruistic action toward the suffering individual. For a group-living species, such as humans, it is easy to appreciate the adaptive value of feeling empathy toward kin or group members. Indeed, the strength of empathy is to some extent associated with membership of an in-group (Hein et al., 2016).

There is a network of brain structures involving the anterior insula and anterior cingulate cortex that forms the basis of empathy (Hein et al., 2016; Wondra & Ellsworth, 2015). On exposure to a potential empathy-triggering situation, activation of this network depends upon in-group membership. Presumably, a generic process triggering altruism to all would not be adaptive. However, this reactivity is open to some fine-tuning in the light of a few repeated experiences. Altruistic gestures by members of an out-group trigger a disparity from expectation ("positive prediction error") registered in the insula and thereby a recalibration of the system to offer empathy to the formerly out-group individuals in future.

7.9.5 The Special Case of Autism

Autism has often featured in discussions of EP (Duchaine et al., 2001), since the social deficit seems to show modular properties (i.e., an island of disruption within a sea of normal functioning). The assumption is common that the fundamental deficit in autism is cognitive: a failure of a theory of mind mechanism (see also Chapter 9).

A different interpretation was given by Chevallier et al. (2012), who argue that "social motivation is a powerful force guiding human behavior and that disruption of social motivational mechanisms may constitute a primary deficit in autism" (p. 231). A neuroimaging study pointed to a particular deficit in reward-related brain regions in response to the presentation of what to controls would be social rewards (Scott-Van Zeeland et al., 2010). Hence, failures in theory of mind appear to be downstream failures attributable to a lack of social motivation.

7.10 ADDICTIVE ACTIVITIES

7.10.1 Drug-Taking

At first glance, drug-taking to the point of addiction might appear to be counterintuitive in terms of EP and fitness enhancement. However, a deeper consideration, involving both neuroscience and EP, makes it less puzzling.

The first consideration is to ask how it got started: What was reinforcing to our ancestors in ingesting those fruits and leaves that are at the base of modern addictive drugs? Anthropological evidence suggests that such things as ethanol from ripe fruits, the opium poppy, the coca leaf, and nicotine brought benefits in terms of pain relief, disease resistance, mood enhancement, facilitating social communication, and overcoming fatigue (Davis, 2014). The psychoactive components of the ingested substances would have been of low density and safe, so such consumption might well have been fitness enhancing. They could have become attractive either by their direct and rapid effect on the brain, by cultural transmission of knowledge on their health benefits, or both.

The equivalent substances that are available today and that lead to addiction are highly refined in their manufacture (i.e., they have greatly increased density of psychoactive ingredients). Furthermore, in some cases, the techniques of delivery that rapidly convey the substances to the bloodstream or lungs massively increase the speed with which psychoactive components arrive in the brain. Such substances as alcohol, amphetamine, cocaine, and heroin activate those same dopaminergic and opioidergic brain systems that evolved in mediating natural motivations such as those toward food and sex (Nesse & Berridge, 1997). Such rapid activation of these motivational systems is as if a natural reward of very high incentive value has just arrived. As Nesse and Berridge (p. 64) note:

Drugs of abuse create a signal in the brain that indicates, falsely, the arrival of a huge fitness benefit.

Hence, the brain sets up strong conditional associations with any candidate stimuli that are present (e.g., the sight of the delivery syringe).

Rewarding social contact is mediated in part through the release of natural opioids (see Section 7.9.3). In contemporary society, many people do not have fulfilling social contact in their lives. Therefore, drugs such as heroin accompanied by the social context of other drug users could help to "fill the vacuum" (Alexander, 2008).

7.10.2 Hyper-palatable Processed Food Addiction

These days, many people meet the criteria of food addiction, by analogy to drug addiction (i.e., withdrawal symptoms, craving, and continued intake in spite of adverse consequences and a wish to quit) (Davis, 2014). Given that food, unlike drugs, is necessary for survival, some prefer a term such as "hyper-palatable processed food addiction." Such high-energy foods produce excessive activation of the dopaminergic wanting process and, in some individuals, incentive sensitization similar to the effects of addictive drugs (Temple, 2016). One can speculate that, in our evolution, a mild form of sensitization to cues predictive of what were probably scarce palatable foods would have been adaptive. Finding them would have required considerable investment of energy.

7.10.3 Gambling

The core feature of gambling is to make some kind of investment through an activity that has an inherently *uncertain* outcome. Its existence extends way back in human history (Davis, 2014).

There are animal models of gambling: some species show a preference for instrumental activities for food that have a relatively uncertain outcome (even at some cost in gain) as compared to a predictable one (Anselme, 2013). When a reward that is relatively unpredictable arrives, it triggers a large release of dopamine. This causes the attribution of incentive salience to an available predictive cue, such as a light or sound that appeared just prior to the reward – a conditional stimulus (CS). Subsequently, such CSs trigger much more dopamine if there is only a 50 percent chance of the reward arriving following their presentation as compared to a 100 percent chance. Hence, the attraction of uncertainty appears explicable by the relatively high dopamine release under this condition, the inherent feature of gambling.

7.10.4 General Principles

For each of the three activities described in this section, a comparison of the present environment with that of early adaptation permits understanding of their addictive potential. A similar logic of evolutionary mismatch could be applied to gambling, shopping addiction, and kleptomania (in the case of the latter two, in terms of acquiring resources in the present environment of abundance), as well as sexual behavior (discussed in Chapter 27).

7.11 DEPRESSION

7.11.1 Introduction

Depression is, first and foremost, a disorder of motivation. Before we can fully understand depression, we must understand the systems that normally regulate motivation. We especially need to know if there are situations in which low mood is useful. (Nesse, 2001, p. 186)

Nesse describes what he calls the "clinician's illusion," "that aversive states are abnormal" (p. 181). By contrast, he argues that aversive states (e.g., pain, nausea, and fever) reflect the outputs of adaptive processes. They are all usually associated with pathological conditions, but are not in themselves pathological.[1] Rather, they are defense mechanisms against a pathological condition. Depression might be another such defense mechanism. A combined consideration of functional aspects and the neuroscience of depression could contribute to our understanding of this, which forms the topic of the present section.

---

[1] Pain can sometimes take on a life of its own, as in phantom limb pain. However, such chronic pain is best seen as a pathological reflection of an adaptive process. Natural selection cannot get every instance right.

7.11.2 The Nature of Depression

According to Beck and Bredemeier (2016), depression is precipitated by (p. 6, emphasis in original):

... the *perceived loss of the investment in a vital resource.*

This is typically the uncontrollable loss of a resource of adaptive significance (e.g., a partner, parent, child, home, or professional resource).

A defining feature of depression is a decreased motivation to engage with the world (e.g., the desires for food and sex are usually lowered) (Nesse, 2001). There is an unwillingness to exert effort to get rewards (Treadway, 2015). Tasks that call for initiative are particularly impaired. There is low self-esteem and pessimism about the future, and aggression toward dominants is inhibited (Price et al., 1994). The term "anhedonia" is commonly used to describe the patient's experience of depression, and it has traditionally been characterized by reduced motivation and pleasure in engaging with life's activities (Treadway, 2015).

7.11.3 Evolutionary Arguments

Panksepp (2014) took a comparative and developmental perspective on depression. He suggests that it represents the result of a chronic activation of a panic system that, in its evolutionary and developmental origins, would be triggered by breaking an infant's attachment to the mother. Presumably the panic reaction (e.g., distress calls) serves to help reestablish contact. However, why protracted failure of this system should trigger a state resembling depression and whether this has an adaptive value are topics of vigorous debate.

Turning to adults, depression could be the pathological outcome when there is a protracted experience of failure, with no perceived possibility of success at switching output, as exemplified by loss of a job or romantic relationship. It might be that a capacity to experience mild negative mood rather than full depression is an adaptation (Nettle, 2004). However, some ask: Could even major depression represent an adaptation (Durisko et al., 2015)?

Price et al. (1994) assemble evidence across various species that depression represents an adaptation. In the case of humans, they suggest that depression is a response to either lack of or loss of social status. For a number of species, by withdrawing from competition, certain adaptive advantages are achieved, such as conservation of resources and avoidance of further conflict by signaling submission.

Neese (2001) noted that behavioral ecologists have amassed evidence that animals are exquisitely sensitive to the outcomes of their behavior, in terms of, say, success at foraging (note the discussion in Section 7.5 of brain mechanisms that assess benefits and costs). When things are going well, the incentive engagement system is activated, associated in humans with the conscious experience of pleasure. Nesse argues that, reciprocally, a motivational system can only behave adaptively if a lack of success motivates quitting from a task and thereby being prepared to engage in another activity, such as foraging in a different environment. In these terms, negative affect is triggered by failing at a task and motivates quitting. Depression could be the product of such a process.

Analogies might be useful in understanding depression. As with addiction (see Section 7.10), depression could be a contemporary outcome of motivational systems that were adaptive in the early environment. Evolutionary mismatch could be relevant in both cases.

Consideration of pain is also insightful. It is easy to appreciate pain's adaptive value in terms of protecting the tissues of the body from damage. This comes in motivating a move of the body into a position to reduce pain, as well as a teaching process in the form of warnings of potential pain if certain behaviors are repeated. As Wittman (2014) argues, knowledge about pain helps to avoid it. One does not need to be mauled by a lion or fall from a tree to know that the experience would be painful. The human brain has the capacity to extrapolate and run simulations of the future. However, chronic pain (e.g., from cancer) would appear to serve no adaptive end. Rather, it is an inevitable outcome of possessing the adaptive mechanism of pain. Wittman suggests that similarly the knowledge of depression means that preemptive steps can be taken to avoid it (e.g., kin are cared for and nurtured). So, chronic depression, as with chronic pain, could well be maladaptive, but the threat of depression or short episodes of mild depression could be an adaptation.

Yet another behavioral system that might offer clues and has several features in common with depression is that of fever following infection (Hart, 1988). This involves an integrated physiological and psychological pattern that appears clearly to be an adaptation. It consists of the sick animal withdrawing from engagement with the world, being lethargic ("depressed"), sleeping, and showing little interest in food or sex until the underlying pathology is corrected. As in depression, corticosteroid hormone secretion is elevated. However, under extreme conditions, what is an otherwise adaptive response of fever can become pathological, suggesting another parallel with depression.

7.11.4 Neuroscience

As suggested by Nesse (2001), depression should be understood in the context of approach motivation, which at a biological level means the dopamine-based behavioral activation system (see Sections 7.3.1 and 7.7.3). Depression might simply reflect the default setting of this system. Even if this is the case, it could still be relevant to understanding motivation. Alternatively, depression could represent the opposite pole of this system, as a kind of mirror image. In such terms, the neutral crossover point between positive and negative could correspond to emotionless laziness (Wittman, 2014)!

Depression shows some coherence as an approximate mirror image of appetitive motivation. Appetitive motivation is associated with adaptive activation of dopamine, energization of behavior, incentive sensitization, and pleasure reactions to goal anticipation. Depression is associated with dopaminergic inactivation, inertia/lethargy, sensitization of future tendencies to depression, and low levels of anticipatory pleasure (Beck & Bredemeier, 2016; Wittman, 2014).

Let us revisit the distinction between wanting and liking (Robinson & Berridge, 1993). By analogy with appetitive behavior, as the term "anhedonia" has been traditionally used, it conflates wanting and liking in that lack of desire is subsumed along with lack of hedonism (Treadway, 2015). Could these be teased apart?

A fraction of depressed patients show a reduction in the effort that they are prepared to exert to obtain rewards (analogous to reduced wanting) (Treadway, 2015). There is some evidence pointing to impaired dopamine neurotransmission underlying this effect. Boosting dopaminergic neurotransmission can increase the amount of effort that nondepressed people are prepared to exert. Let us suppose that there is a fracture line between wanting and liking in depression and only wanting is a function of disrupted dopaminergic neurotransmission. Some other disruption must underlie the lack of pleasure, and a possible candidate is an opioid mechanism.[2]

Carver et al. (2008) adopted a hierarchical model in order to understand depression. According to this, the neurochemical serotonin has a role in establishing the balance between Systems 1 and 2 in controlling behavior. Low levels of serotonin give a bias toward control by System 1. However, in depression, this is accompanied by low dopaminergic activity, which causes a reduction of responsiveness in System 1. So, both systems show low responsiveness.

Concerning the possible adaptive value of depression, what is the relevance of the evidence from neuroscience pointing to relative inaction mediated by both Systems 1 and 2? Such passivity could serve well to keep the individual away from conflict and dominants, though whether it needs to be so protracted is a question for debate. A shorter period of passivity/withdrawal might seem a more adaptive strategy. The associated psychological pain could trigger empathy from others and thereby help.

Some argue that depression gives an opportunity for "time out" to reflect and to alter strategy (Durisko et al., 2015). However, depression appears to hinder the kind of adaptive controlled cognition that is associated with System 2 and to replace it with task-irrelevant and depression-related cognition (Carver et al., 2008; Hartlage et al., 1993). When finding a possible solution, any such planning needs to have the facility for energization of behavior,

and the evidence suggests that both systems are in a relatively passive mode and unable to take advantage of new strategies.

7.11.5 Epidemiology of Depression

Statistics strongly suggest that, in the affluent industrialized world during the course of the twentieth century, there was a large increase in the number of people suffering from depression. What kind of changes might have contributed to this? The breakdown of the closely knit family could be one such change. A factor suggested by Lambert (2006) is the switch from an active lifestyle to one of relative inactivity, exemplified by the large reduction in the number of farmworkers and increase in office/service-based jobs.

Extrapolation would suggest that the hunter–gatherer lifestyle of our ancestors involved very considerable muscular exertion, including extensive running (Schulkin, 2016). In other words, the modern brain is adapted to an early evolutionary environment that is very different from contemporary society. Lambert (2006, p. 500, emphasis in original) suggests that there has been a drastic reduction in "the opportunity to engage in physical activities that lead to desired outcomes (i.e. *effort-based rewards*)." It seems that a combination of possessing control and obtaining meaningful outcomes from physical actions is important. Lambert notes that, in a neuroimaging study, participants receiving earned rewards showed greater activation of the striatum, including the N.Acc., than when the rewards were unearned.

An important theoretical rationale for the use of behavioral activation therapy in treating depression (i.e., to encourage patients to engage again in rewarding activities) is that activation of the N.Acc. is lower in depression than in a control condition (Alexopoulos et al., 2016).

Extensive physical activity, as in regular jogging, offers some protection against depression (Schulkin, 2016), and the effect appears to be mediated by, among other things, the release of endorphins and endocannabinoids. Evidence suggests that these mediate the euphoric effects of exercise and its reinforcement value (Raichlen et al., 2013). There is also a cognitive factor of goal achievement. Various growth factors are released by exercise, and they contribute to neural plasticity, seen particularly in the hippocampus. This could explain, at least in part, the more long-term beneficial effects of exercise in raising low moods.

7.12 CONCLUSION AND DISCUSSION

The evidence reviewed suggests that a two-pronged approach integrating explanations in terms of causation and evolutionary function offers the most insight into motivation. This chapter set out to investigate whether EP and the neuroscience of motivation are compatible.

---

[2] There is some evidence for a wanting vs. liking distinction in the case of schizophrenia (Gard et al., 2007).

More specifically, the question translates as follows: Does a study of the brain processes that form the bases of motivation provide evidence of adaptations to early evolutionary environments?

A number of examples discussed point to the adaptive nature of brain processes, including those of cost–benefit analysis and inhibition. The process of incentive motivation is in itself an efficient one, in that the strength of motivation increases with incentive proximity, so that behavior tends to be engaged at the most optimal time corresponding to incentive availability. Evidence was discussed that disputes the conclusion of the Santa Barbara school that there is no common feature in the "design" of motivational systems.

Considering individual motivations, feeding perhaps provides the clearest examples of adaptations in the form of taste-aversion learning and pregnancy sickness (for examples of sexual motivation, see Chapter 27). Exploratory behavior reveals brain processes that would encourage some exploration of new environments.

A number of features of modern humans and their society suggest evolutionary mismatch. That is to say, motivational systems generate what appears to suboptimal behavior in terms of health and longevity, exemplified by obesity and addiction. However, these can be understood as reflecting the action of behavioral processes that would have been adaptive in an earlier environment (e.g., the contrast between the present-day low cost of gaining energy-rich food in an abundant variety and the assumed scarcity of earlier times). Addictive drug-taking "short-circuits" the reward pathways that form the basis of adaptive behavior.

A study of motivation reveals evolution as a "tinkerer." Evolutionarily new processes build upon older processes, and so a hybrid brain results (Maclean, 1990). A study of the neuroscience of motivation reveals that something like the dedicated modules suggested by the Santa Barbara school of EP do exist. However, they operate in conjunction with some general-purpose brain processes.

7.13 ACKNOWLEDGMENTS

I am very grateful to John Archer, Kent Berridge, and Olga Coschug-Toates for comments received on an early draft of this chapter.

REFERENCES

Alexander, B. (2008). *The Globalization of Addiction: A Study in Poverty of the Spirit*. Oxford: Oxford University Press.

Alexopoulos, G. S., Raue, P. J., Gunning, F., et al. (2016). "Engage" therapy: Behavioral activation and improvement of late-life major depression. *American Journal of Geriatric Psychiatry*, **24**(4), 320–326.

Angus, D. J., Schutter, D. J., Terburg, D., et al. (2016). A review of social neuroscience research on anger and aggression. In E. Harmon-Jones & M. Inzlicht, eds., *Social Neuroscience: Biological Approaches to Social Psychology*. London: Routledge, pp. 223–246.

Anselme, P. (2013). Dopamine, motivation, and the evolutionary significance of gambling-like behaviour. *Behavioural Brain Research*, **256**, 1–4.

Archer, J. (2004). Sex differences in aggression in real-world settings: A meta-analytic review. *Review of General Psychology*, **8**(4), 291–322.

Archer, J. (2006). Testosterone and human aggression: An evaluation of the challenge hypothesis. *Neuroscience & Biobehavioral Reviews*, **30**(3), 319–345.

Archer, J. (2009). The nature of human aggression. *International Journal of Law and Psychiatry*, **32**(4), 202–208.

Baumeister, R. F., & Leary, M. R. (1995). The need to belong: Desire for interpersonal attachments as a fundamental human motivation. *Psychological Bulletin*, **117**(3), 497–529.

Beck, A. T., & Bredemeier, K. (2016). A unified model of depression integrating clinical, cognitive, biological, and evolutionary perspectives. *Clinical Psychological Science*, **4**, 596–619.

Berkowitz, L., & LePage, A. (1967). Weapons as aggression-eliciting stimuli. *Journal of Personality and Social Psychology*, **7**(2), 202–207.

Berridge, K. C. (2004). Motivation concepts in behavioral neuroscience. *Physiology & Behavior*, **81**(2), 179–209.

Berridge, K. C., & Kringelbach, M. L. (2013). Neuroscience of affect: Brain mechanisms of pleasure and displeasure. *Current Opinion in Neurobiology*, **23**(3), 294–303.

Berridge, K. C., & Kringelbach, M. L. (2015). Pleasure systems in the brain. *Neuron*, **86**(3), 646–664.

Berridge, K. C., & Valenstein, E. S. (1991). What psychological process mediates feeding evoked by electrical stimulation of the lateral hypothalamus? *Behavioral Neuroscience*, **105**(1), 3–14.

Bindra, D. (1978). How adaptive behavior is produced: A perceptual–motivational alternative to response reinforcements. *Behavioral and Brain Sciences*, **1**(1), 41–52.

Bjorklund, D. F., & Kipp, K. (1996). Parental investment theory and gender differences in the evolution of inhibition mechanisms. *Psychological Bulletin*, **120**(2), 163–188.

Blackburn, J. R., Pfaus, J. G., & Phillips, A. G. (1992). Dopamine functions in appetitive and defensive behaviours. *Progress in Neurobiology*, **39**(3), 247–279.

Blair, R. J. R. (2004). The roles of orbital frontal cortex in the modulation of antisocial behavior. *Brain and Cognition*, **55**(1), 198–208.

Blair, R. J. R. (2016). The neurobiology of impulsive aggression. *Journal of Child and Adolescent Psychopharmacology*, **26**(1), 4–9.

Bolhuis, J. J., Brown, G. R., Richardson, R. C., et al. (2011). Darwin in mind: New opportunities for evolutionary psychology. *PLoS Biology*, **9**(7), e1001109.

Bowlby, J. (1982). *Attachment and Loss. Vol. 1 Attachment*. New York: Basic Books.

Buades-Rotger, M., Brunnlieb, C., Münte, T. F., et al. (2016). Winning is not enough: Ventral striatum connectivity during physical aggression. *Brain Imaging and Behavior*, **10**(1), 105–114.

Buss, D. M. (2016). *Evolutionary Psychology: The New Science of the Mind*. London: Routledge.

Buss, D. M., & Shackelford, T. K. (1997). Human aggression in evolutionary psychological perspective. *Clinical Psychology Review*, **17**(6), 605–619.

Cameron, E. L. (2014). Pregnancy and olfaction: A review. *Frontiers in Psychology*, **5**, 67.

Campbell, A. (2013). *A Mind of Her Own: The Evolutionary Psychology of Women*. Oxford: Oxford University Press.

Carpenter, D., Janssen, E., Graham, C., et al. (2008). Women's scores on the Sexual Inhibition/Sexual Excitation Scales (SIS/SES): Gender similarities and differences. *Journal of Sex Research*, **45**(1), 36–48.

Carré, J. M., Geniole, S. N., Ortiz, T. L., et al. (2017). Exogenous testosterone rapidly increases aggressive behavior in dominant and impulsive men. *Biological Psychiatry*, **82**(4), 249–256.

Carver, C. S., & Harmon-Jones, E. (2009). Anger is an approach-related affect: Evidence and implications. *Psychological Bulletin*, **135**(2), 183–204.

Carver, C. S., Johnson, S. L., & Joormann, J. (2008). Serotonergic function, two-mode models of self-regulation, and vulnerability to depression: What depression has in common with impulsive aggression. *Psychological Bulletin*, **134**(6), 912–943.

Chevallier, C., Kohls, G., Troiani, V., et al. (2012). The social motivation theory of autism. *Trends in Cognitive Sciences*, **16**(4), 231–239.

Chiappe, D., & MacDonald, K. (2005). The evolution of domain-general mechanisms in intelligence and learning. *The Journal of General Psychology*, **132**(1), 5–40.

Cosmides, L., & Tooby, J. (1995). From evolution to adaptations to behavior: Toward an integrated evolutionary psychology. In R. Wong, ed., *Biological Perspectives on Motivated Activities*. Norwood, NJ: Ablex, pp. 11–74.

Crabb, P. B. (2000). The material culture of homicidal fantasies. *Aggressive Behavior*, **26**(3), 225–234.

Cross, C. P., Copping, L. T., & Campbell, A. (2011). Sex differences in impulsivity: A meta-analysis. *Psychological Bulletin*, **137**(1), 976–130.

Davis, C. (2014). Evolutionary and neuropsychological perspectives on addictive behaviors and addictive substances: relevance to the "food addiction" construct. *Substance Abuse and Rehabilitation*, **5**, 129–137.

Daw, N. D., O'Doherty, J. P., Dayan, P., et al. (2006). Cortical substrates for exploratory decisions in humans. *Nature*, **441**(7095), 876–879.

Denson, T. F., Schofield, T. P., & Fabiansson, E. C. (2015). Aggressive desires. In W. Hofmann & L. F. Nordgren, eds., *The Psychology of Desire*. New York: Guilford Press, pp. 369–389.

De Quervain, D. J., Fischbacher, U., Treyer, V., et al. (2004). The neural basis of altruistic punishment. *Science*, **305**(5688), 1254–1258.

Dreher, J. C., Dunne, S., Pazderska, A., et al. (2016). Testosterone causes both prosocial and antisocial status-enhancing behaviors in human males. *Proceedings of the National Academy of Sciences*, **113**(41), 11633–11638.

Duchaine, B., Cosmides, L., & Tooby, J. (2001). Evolutionary psychology and the brain. *Current Opinion in Neurobiology*, **11**(2), 225–230.

Durisko, Z., Mulsant, B. H., & Andrews, P. W. (2015). An adaptationist perspective on the etiology of depression. *Journal of Affective Disorders*, **172**, 315–323.

Fehr, E., & Camerer, C. F. (2007). Social neuroeconomics: The neural circuitry of social preferences. *Trends in Cognitive Sciences*, **11**(10), 419–427.

Gan, G., Preston-Campbell, R. N., Moeller, S. J., et al. (2016). Reward vs. retaliation – The role of the mesocorticolimbic salience network in human reactive aggression. *Frontiers in Behavioral Neuroscience*, **10**, 179.

Gard, D. E., Kring, A. M., Gard, M. G., et al. (2007). Anhedonia in schizophrenia: distinctions between anticipatory and consummatory pleasure. *Schizophrenia Research*, **93**(1), 253–260.

Glenn, A. L., & Raine, A. (2009). Psychopathy and instrumental aggression: Evolutionary, neurobiological, and legal perspectives. *International Journal of Law and Psychiatry*, 32(4), 253–258.

Groppe, S. E., Gossen, A., Rademacher, L., et al. (2013). Oxytocin influences processing of socially relevant cues in the ventral tegmental area of the human brain. *Biological Psychiatry*, **74**(3), 172–179.

Gunaydin, L. A., & Deisseroth, K. (2014). Dopaminergic dynamics contributing to social behavior. *Cold Spring Harbor Symposia on Quantitative Biology*, **79**, 221–227.

Hart, B. L. (1988). Biological basis of the behavior of sick animals. *Neuroscience and Biobehavioral Reviews*, **12**(2), 123–137.

Hartlage, S., Alloy, L. B., Vázquez, C., et al. (1993). Automatic and effortful processing in depression. *Psychological Bulletin*, **113**(2), 247–278.

Hein, G., Engelmann, J. B., Vollberg, M. C., et al. (2016). How learning shapes the empathic brain. *Proceedings of the National Academy of Sciences*, **113**(1), 80–85.

Itoga, C. A., Berridge, K. C., & Aldridge, J. W. (2016). Ventral pallidal coding of a learned taste aversion. *Behavioural Brain Research*, **300**, 175–183.

Jepma, M., Verdonschot, R. G., Van Steenbergen, H., et al. (2012). Neural mechanisms underlying the induction and relief of perceptual curiosity. *Frontiers in Behavioral Neuroscience*, **6**, 5.

Johnson, S. L., Edge, M. D., Holmes, M. K., et al. (2012). The behavioral activation system and mania. *Annual Review of Clinical Psychology*, **8**, 243–267.

Kim, P., Strathearn, L., & Swain, J. E. (2016). The maternal brain and its plasticity in humans. *Hormones and Behavior*, **77**, 113–123.

Kohls, G., Perino, M. T., Taylor, J. M., et al. (2013). The nucleus accumbens is involved in both the pursuit of social reward and the avoidance of social punishment. *Neuropsychologia*, **51**(11), 2062–2069.

Krebs, J. R., & Davies, N. B. (1978). *Behavioural Ecology: An Evolutionary Approach*. Oxford: Blackwell Science.

Kringelbach, M. L., Lehtonen, A., Squire, S., et al. (2008). A specific and rapid neural signature for parental instinct. *PLoS ONE*, **3**(2), e1664.

Lambert, K. G. (2006). Rising rates of depression in today's society: Consideration of the roles of effort-based rewards and enhanced resilience in day-to-day functioning. *Neuroscience and Biobehavioral Reviews*, **30**(4), 497–510.

Litman, J. (2005). Curiosity and the pleasures of learning: Wanting and liking new information. *Cognition and Emotion*, **19**(6), 793–814.

Lopez, R. B., Wagner, D. D., & Heatherton, T. F. (2015). Neuroscience of desire regulation. In W. Hofmann & L. F. Nordgren, eds., *The Psychology of Desire*. New York: Guilford Press, pp. 146–160.

Lorenz, K. Z. (1981). *The Foundations of Ethology*. New York: Springer-Verlag.

MacDonald, G., & Leary, M. R. (2005). Why does social exclusion hurt? The relationship between social and physical pain. *Psychological Bulletin*, **131**(2), 202–223.

MacLean, P. D. (1990). *The Triune Brain in Evolution*. New York: Plenum Press.

Mazur, A., & Booth, A. (1998). Testosterone and dominance in men. *Behavioral and Brain Sciences*, **21**(3), 353–363.

Nesse, R. M. (2001). Motivation and melancholy: A Darwinian perspective. *Nebraska Symposium on Motivation*, **47**, 179–203.

Nesse, R. M., & Berridge, K. C. (1997). Psychoactive drug use in evolutionary perspective. *Science*, **278**(5335), 63–66.

Nettle, D. (2004). Evolutionary origins of depression: A review and reformulation. *Journal of Affective Disorders*, **81**(2), 91–102.

Over, H. (2016). The origins of belonging: social motivation in infants and young children. *Philosophical Transactions of the Royal Society B*, **371**(1686), 20150072.

Panksepp, J. (1998). *Affective Neuroscience: The Foundations of Human and Animal Emotions*. Oxford: Oxford University Press.

Panksepp, J. (2014). Seeking and loss in the ancestral genesis of resilience, depression, and addiction. In M. Kent, M. C. Davis, & J. W. Reich, eds., *The Resilience Handbook: Approaches to Stress and Trauma*. New York: Routledge, pp. 3–14.

Panksepp, J., & Panksepp, J. B. (2000). The seven sins of evolutionary psychology. *Evolution and Cognition*, **6**(2), 108–131.

Papageorgiou, G. K., Baudonnat, M., Cucca, F., et al. (2016). Mesolimbic dopamine encodes prediction errors in a state-dependent manner. *Cell Reports*, **15**(2), 221–228.

Patil, C. L., Abrams, E. T., Steinmetz, A. R., et al. (2012). Appetite sensations and nausea and vomiting in pregnancy: An overview of the explanations. *Ecology of Food and Nutrition*, **51**(5), 394–417.

Pinker, S. (2011). *The Better Angels of our Nature: The Decline of Violence in History and its Causes*. London: Penguin.

Price, J., Sloman, L., Gardner, R., et al. (1994). The social competition hypothesis of depression. *British Journal of Psychiatry*, **164**(3), 309–315.

Profet, M. (1992) Pregnancy sickness as adaptation: A deterrent to maternal ingestion of teratogens. In J. H. Barkow, L. Cosmides, & J. Tooby, eds., *The Adapted Mind: Evolutionary Psychology and the Generation of Culture*. New York: Oxford University Press, pp. 327–365.

Radke, S., Seidel, E. M., Eickhoff, S. B., et al. (2016). When opportunity meets motivation: Neural engagement during social approach is linked to high approach motivation. *NeuroImage*, **127**, 267–276.

Raichlen, D. A., Foster, A. D., Seillier, A., et al. (2013). Exercise-induced endocannabinoid signaling is modulated by intensity. *European Journal of Applied Physiology*, **113**(4), 869–875.

Ramírez, J. M., Bonniot-Cabanac, M. C., & Cabanac, M. (2005). Can impulsive aggression provide pleasure? *European Psychologist*, **10**(2), 136–145.

Ray, W. J. (2013). *Evolutionary psychology: Neuroscience Perspectives Concerning Human Behavior and Experience*. Los Angeles, CA: Sage.

Robinson, T. E., & Berridge, K. C. (1993). The neural basis of drug craving: An incentive-sensitization theory of addiction. *Brain Research Reviews*, **18**(3), 247–291.

Rolls, E. T., Murzi, E., Yaxley, S., et al. (1986). Sensory-specific satiety: Food-specific reduction in responsiveness of ventral forebrain neurons after feeding in the monkey. *Brain Research*, **368**(1), 79–86.

Rosell, D. R., & Siever, L. J. (2015). The neurobiology of aggression and violence. *CNS Spectrums*, **20**(3), 254–279.

Rupp, H. A., James, T. W., Ketterson, E. D., et al. (2009). The role of the anterior cingulate cortex in women's sexual decision making. *Neuroscience Letters*, **449**(1), 42–47.

Salamone, J. D., Correa, M., Farrar, A., et al. (2007). Effort-related functions of nucleus accumbens dopamine and associated forebrain circuits. *Psychopharmacology*, **191**(3), 461–482.

Schulkin, J. (2016). Evolutionary basis of human running and its impact on neural function. *Frontiers in Systems Neuroscience*, **10**, 59.

Spreckelmeyer, K. N., Krach, S., Kohls, G., et al. (2009). Anticipation of monetary and social reward differently activates mesolimbic brain structures in men and women. *Social Cognitive and Affective Neuroscience*, **4**(2), 158–165.

Stice, E., & Yokum, S. (2016). Neural vulnerability factors that increase risk for future weight gain. *Psychological Bulletin*, **142**(5), 447–471.

Stockhorst, U., Enck, P., & Klosterhalfen, S. (2007). Role of classical conditioning in learning gastrointestinal symptoms. *World Journal of Gastroenterology*, **13**(25), 3430–3437.

Temple, J. L. (2016). Behavioral sensitization of the reinforcing value of food: What food and drugs have in common. *Preventive Medicine*, **92**, 90–99.

Toates, F. (1986). *Motivational Systems*. Cambridge, UK: Cambridge University Press.

Toates, F. (1998). The interaction of cognitive and stimulus–response processes in the control of behaviour. *Neuroscience & Biobehavioral Reviews*, **22**(1), 59–83.

Toates, F. (2005). Evolutionary psychology – Towards a more integrative model. *Biology and Philosophy*, **20**(2–3), 305–328.

Toates, F. (2014). *How Sexual Desire Works: The Enigmatic Urge*. Cambridge, UK: Cambridge University Press.

Toates, F. M., & Archer, J. (1978). A comparative review of motivational systems using classical control theory. *Animal Behaviour*, **26**, 368–380.

Treadway, M. T. (2015). Liking little, wanting less. In W. Hofmann & L. F. Nordgren, eds., *The Psychology of Desire*. New York: Guilford Press, pp. 307–320.

Wagner, D. D., & Heatherton, T. F. (2015). Self-regulation and its failure: The seven deadly threats to self-regulation. In M. Mikulincer & P.R. Shaver, eds., *APA Handbook of Personality and Social Psychology: Vol. 1. Attitudes and Social Cognition*. Washington, DC: American Psychological Association, pp. 805–842.

White, S. F., Brislin, S. J., Sinclair, S., et al. (2014). Punishing unfairness: Rewarding or the organization of a reactively aggressive response? *Human Brain Mapping*, **35**(5), 2137–2147.

Williams, K. D. (2007). Ostracism. *Annual Review of Psychology*, **58**(1), 425–452.

Wittmann, B. C., Daw, N. D., Seymour, B., et al. (2008). Striatal activity underlies novelty-based choice in humans. *Neuron*, **58**(6), 967–973.

Wittman, D. (2014). Darwinian depression. *Journal of Affective Disorders*, **168**, 142–150.

Wondra, J. D., & Ellsworth, P. C. (2015). An appraisal theory of empathy and other vicarious emotional experiences. *Psychological Review*, **122**(3), 411–428.

Zeeland, S. V., Ashley, A., Dapretto, M., et al. (2010). Reward processing in autism. *Autism Research*, **3**(2), 53–67.

8 Are We Designed to Be Happy?

The Neuroscience of Making Sense of Pleasure

ELOISE STARK, KENT C. BERRIDGE, AND MORTEN L. KRINGELBACH

Man might be described fairly adequately, if simply, as a two-legged paradox. He has never become accustomed to the tragic miracle of consciousness. Perhaps, as has been suggested, his species is not set, has not jelled, but is still in a state of becoming, bound by his physical memories to a past of struggle and survival, limited in his futures by the uneasiness of thought and consciousness.

– John Steinbeck and Ed Ricketts

The science of happiness is still in its infancy; there is little consensus on how happiness might be best defined, let alone studied. Still, the pursuit of happiness was inscribed in the American Constitution as a fundamental right, and recently governments around the world have started to measure the happiness and well-being of people as a measure on par with gross national product.

Whatever it is, happiness is clearly desirable, yet worryingly fleeting. In this chapter, we will describe some of the scientific progress that has been made. We will take our lead from Aristotle, who was interested in the good life; a life lived embedded in meaningful values (Aristotle, 350 BC/1976). He divided the main ingredients into *hedonia* and *eudaimonia*. The former is perhaps best translated as "pleasure" (derived from *hedus*, the sweet taste of honey), while eudaimonia is often translated as "well-being," although it is probably better captured by "flourishing" or "meaningful pleasure." From this perspective, hedonia would seem to underlie eudaimonia, and therefore any insights into hedonia are valuable for an understanding of eudaimonia. As we will show here, the science of hedonia has made much exciting progress over the last decade, and as a result we are beginning to glimpse some of the mechanisms involved in eudaimonia. With time, perhaps science can come to provide novel insights into how best to enjoy the "tragic miracle of consciousness" rather than continue to merely struggle and survive.

Here, we first provide a brief overview of the neuroscience of pleasure, discussing its cyclical nature and the underlying brain networks involved in wanting, liking, and satiation. We show how disruptions to the pleasure cycle can lead to anhedonia, the lack of pleasure, which is a devastating main component of affective disorders.

Finally, we speculate on how hedonia might underpin eudaimonia through some of the shared brain networks and optimal metastability, which may help in making sense of pleasure.

8.1 A SCIENCE OF PLEASURE

Pleasure is not a unitary concept, and rewards involve a composite of several psychological components contained within the pleasure cycle (Berridge & Kringelbach, 2008). The first phase is characterized by "wanting," which is the motivational process of incentive salience where there is a desire to approach a reward and gain pleasure. This is followed by a phase dominated by "liking" or consummation, encompassing the core reactions to the hedonic impact of rewards. Lastly, the satiation phase follows when a reward is no longer liked and after which the cycle can begin anew. Throughout the pleasure cycle, "learning" is a crucial facet, as each approach and encounter with a stimulus updates its hedonic value, allowing for more accurate computations about the payoffs for future consummation.

The component processes of the pleasure cycle also have discriminable neural mechanisms. Although all three components may overlap, wanting processes usually dominate the initial appetitive phase, while liking processes dominate the subsequent consummatory phase that may include a peak in pleasure (such as an orgasm during sex), followed by satiety.

A successful strategy for studying pleasure in affective neuroscience has been to identify objective aspects of pleasure-elicited reactions and then triangulating toward underlying brain substrates. Crucially, this scientific strategy divides the concept of pleasure into two parts: the affective state, which can be measured objectively in behavior, physiology, and neural responses; and conscious affective feelings, which describe our subjective experience of pleasure (Kringelbach, 2004a). This conceptualization of pleasure allows for conscious, subjective feelings to play a central role in hedonic experiences, but maintains that the affective basis of a pleasure reaction is not limited to conscious feelings alone. The focus on objective measurements allows us to use animal models to explore the pleasure response, as they cannot report subjective sensations, and it is especially tractable to neuroscientific investigations involving brain manipulations.

Prominent affective reactions to stimuli, on a continuum from pleasure to displeasure, appear to be found in the behavior and brains of all mammals (Steiner et al., 2001), and they likely have important evolutionary functions (Kringelbach, 2009). On a very simple level, we approach stimuli that generate pleasure and we avoid stimuli that cause pain or displeasure. Through this simple behavioral response based upon valence, we learn rules that guide us to make adaptive choices. The fact that neural mechanisms for generating affective reactions are present and similar in most mammalian brains suggests that there is significant adaptive value that has been evolutionarily selected for and conserved across species throughout evolution (Kringelbach, 2010).

The study of pleasure has focused on finding objective hedonic reactions to taste in infants and in other animals (Berridge, 1996). Such hedonic reactions are found in newborn human infants on the first postnatal day (Steiner, 1973), where sugar water initiates "liking," modeled by an expression involving relaxed facial muscles and a contented licking of the lips. However, exposure to a bitter taste elicits an immediate "disgust" reaction, which is instantly recognizable to most adult humans by gaping of the mouth, headshakes, and scrunching of the nose and brow. Homologous "liking" and "disgust" orofacial expressions have been elicited in apes and monkeys, and even in rats and mice, characterized by rhythmic tongue protrusions and lateral lip licking in response to sweetness and gapes and headshakes in response to bitterness (Berridge, 2000; Grill & Norgren, 1978; Steiner et al., 2001). The basic sensorimotor circuitry underlying these instinctive affective expressions resides in the brainstem (Steiner, 1973), but such affective expressions are not mere brainstem reflexes, but rather are hierarchically controlled by forebrain structures. Forebrain circuitry exerts powerful descending control over brainstem and behavioral output. For example, states of satiety or an appetitive state may physiologically modulate the "liking" expression for a given taste via forebrain circuitry (Cabanac & Lafrance, 1990; Kaplan et al., 2000). Learned preferences and aversions may also associatively lead to modulations of "liking" responses.

Some have proposed a distinction between lower and higher pleasures. They like to point out that pleasure can be fundamental (e.g., for food or sexual pleasure) or it can be higher order (e.g., aesthetic pleasure, music, dance, altruism, monetary pleasure, or transcendent pleasures). A strict distinction between higher and lower pleasures does not, however, seem to hold from the point of view of neuroscience, given that the available evidence suggests that the same brain mechanisms are involved in both fundamental and higher-order pleasures (Crisp & Kringelbach, 2018; Kringelbach, 2010). The brain is designed to process all stimuli that give pleasure to an individual, and pleasure is likely to be generated by hedonic brain circuits that are distinct from the mediation of other features of the same events, such as sensory processing and cognitive processing (Kringelbach, 2005a). As a result, pleasure is a "hedonic gloss" on sensory stimuli that is generated by the brain through dedicated systems (Frijda, 2010). These dedicated hedonic brain systems may even be fundamental to eudaimonia.

8.2 HEDONIC HOTSPOTS AND INCENTIVE SALIENCE

The brain's reward-related circuitry is distributed across the brain, with some hedonic mechanisms found deep in the brain (brainstem, nucleus accumbens, ventral pallidum) and other regions in the cortex (orbitofrontal, cingulate, medial prefrontal and insular cortices). Brain networks that support *pleasure coding* are widespread, yet *pleasure causation*, which can be detected as increases in "liking" reactions as a result of brain manipulation, have been found only in a few connected hedonic hotspots (Berridge & Kringelbach, 2015). When stimulated with optogenetics or opioid, endocannabinoid, or other neurochemical modulators, hedonic hotspots are capable of generating enhancements in the "liking" response to a sensory pleasure such as the taste of sweetness (Smith et al., 2010). In rodents, each hedonic hotspot is approximately 1 mm^3 in volume, which would correspond to approximately 1 cm^3 in the human brain.

Hedonic hotspots have been found in rats in the nucleus accumbens shell, ventral pallidum, insula, and orbitofrontal cortex (Berridge & Kringelbach, 2015; Castro & Berridge, 2017), and also in deep brainstem regions, including the parabrachial nucleus in the pons. The pleasure-generating capacity of these hotspots has been revealed in part by animal studies in which microinjections of drugs stimulated neurochemical receptors on neurons within a designated hotspot, causing a doubling or tripling of the number of hedonic "liking" reactions normally elicited by a pleasant sucrose taste. Analogous to scattered islands that form a single archipelago, hedonic hotspots are anatomically distributed, but they structurally and functionally interact to form a functional integrated circuit. This circuit displays a hierarchical organization of control. The top levels function together as a cooperative heterarchy, such that simultaneous activity in hotspots in the nucleus accumbens and ventral pallidum is required for opioid stimulation in either forebrain site to enhance "liking" above normal.

It is also prescient to remember that the "liking" component of pleasure is translated into motivational processes in part by activating another component of reward termed "wanting" or incentive salience. "Wanting" makes stimuli attractive when attributed to them by mesolimbic brain systems (Berridge & Robinson, 2003). Incentive salience depends in particular on mesolimbic dopamine neurotransmission, although other neurotransmitters and structures are also involved.

It is important to separate incentive salience or "wanting" from hedonic impact or "liking" (Berridge, 2007), as an individual can "want" a reward without necessarily "liking" the same reward. Addiction is a special case where irrational "wanting" without "liking" can occur via incentive sensitization of the mesolimbic dopamine system and connected structures. At the extreme end of addiction, the individual may come to "want" or crave stimuli that are neither "liked" nor expected to be liked. This dissociation between "wanting" and "liking" is possible because incentive salience "wanting" mechanisms are mostly subcortical and separable from cortically mediated conscious planning and declarative expectation. This is one reason why addicts may compulsively "want" to take drugs, even if they do not want to do so at a more cognitive and conscious level, and this is surely a recipe for great unhappiness.

So while pleasure is generally associated only with the "liking" component or subjective sensations of pleasure that facilitate eudaimonia, it is clear that excessive or maladaptive "wanting" can generate unhappiness. These evaluative reactions – eudaimonia and unhappiness – are two sides of the same coin, as they are both generated by the pleasure system and components of the pleasure cycle.

8.3 HEDONIC COMPUTATIONS AT THE LEVEL OF THE CORTEX

The evidence suggests that hedonic computation of pleasure valence is separable from sensory computations. Human neuroimaging studies have shown that these hedonic computations involve brain regions such as the insula and orbitofrontal, medial prefrontal, and cingulate cortices. These regions are closely linked to the subcortical hedonic hotspots and are involved in the anticipation, appraisal, experience, and memory of pleasurable stimuli.

In humans, the encoding of pleasure appears to be cortically localized in a mid-anterior and roughly mid-lateral subregion of the orbitofrontal cortex of the prefrontal lobe. Neuroimaging activity in this specific region has been found to correlate with subjective pleasantness ratings of food varieties and other pleasures such as sexual orgasms, drugs, chocolate, and music. As a further demonstration of the importance of this region for pleasure encoding, activity in this special mid-anterior zone of the orbitofrontal cortex tracks changes in subjective pleasure, such as a decline in palatability (when the reward value of a food is reduced by eating it to satiety, compared to another food for which the subjective pleasure remains stable). This makes it a main candidate for the computation of the subjective experience of pleasure (Kringelbach, 2005b).

Furthermore, it has been shown that there is a medial–lateral distinction in the orbitofrontal cortex, combined with a concreteness–abstraction gradient in the posterior–anterior dimension. The medial part of the orbitofrontal cortex shows activity related to positive and negative valence (Kringelbach & Rolls, 2004), while the lateral orbitofrontal regions have been suggested to code unpleasant events, particularly when an action is needed (Kringelbach, 2004b; O'Doherty et al., 2001). Hedonic computations on sensory rewards such as taste are more likely to be found in posterior parts, while more complex or abstract reinforcers such as money are more likely to give rise to activity more anteriorly.

8.4 ANHEDONIA, THE LACK OF PLEASURE, AND AVOLITION, ITS MIMIC

With the pleasure system conceptualized as a cyclical flow of interacting processes, this gives us scope to describe anhedonia – or lack of pleasure – as the complete or partial breakdown of the interacting brain networks in the pleasure cycle. Anhedonia is important to the study of well-being, as someone who is experiencing anhedonia is unlikely to report high levels of well-being. This strongly suggests that the ability to experience pleasure, however fleeting, is important for our overall well-being. Anhedonia is a key feature of many (if not all) neuropsychiatric disorders (Whybrow, 1998).

We have previously proposed that anhedonia can be conceptualized as impairments in the components of the pleasure cycle – wanting, liking, and learning – that can lead to different expressions or subtypes of anhedonia (Rømer Thomsen et al., 2015). Anhedonia traditionally has been taken to imply an actual loss of "liking" for pleasures. However, in brief, there is scant evidence that most patients with neuropsychiatric illnesses have reduced liking, in that they consistently subjectively report normal levels of subjective pleasure. However, available data suggest that deficits in the motivational aspects of pleasure or wanting may play a crucial role in anhedonia, as patients with depression and schizophrenia are generally less willing to work for a reward compared to healthy controls (Fervaha et al., 2013; Gold et al., 2013; Treadway et al., 2012; Yang et al., 2014). They may reflect a loss of incentive salience rather than of pleasure, a motivational deficit that has been called avolition, anticipatory anhedonia, or motivational anhedonia.

There is also evidence that anhedonia is associated with a blunted or attenuated ability to learn to respond to feedback. Clinically depressed patients have demonstrated impairments in reward learning, correlated with self-reported anhedonia symptoms (Pizzagalli et al., 2008). Patients with schizophrenia have also consistently exhibited deficits in reward-driven learning (e.g., Waltz et al., 2007), although punishment-driven learning may display a different pattern.

Awareness of anhedonia as a trans-diagnostic symptom, with several subcomponents relating to complex psychological processes occurring on different levels of conscious awareness, is necessary in order to understand how this symptom occurs in psychiatric disorder. Patients with anhedonia are clearly not happy, suggesting that in order to be happy we have to be able to experience pleasure. Yet, the findings would also seem to suggest that intact liking is not sufficient for well-being. The fact that wanting is aberrant may suggest that well-being also depends on the anticipation of pleasure. As an example, it has been shown that patients suffering from depression and schizophrenia report reduced enjoyment when asked to rate past, future, or hypothetical experience (McFarland & Klein, 2009; Strauss & Gold, 2012; Watson & Naragon-Gainey, 2010).

8.5 LINKING HEDONIA AND EUDAIMONIA

Overall, the evidence suggests that hedonia is important to drive life, fulfilling the ancient evolutionary imperatives of survival and procreation. The animal kingdom appears to have the same mechanisms and networks underlying hedonia, but it would appear that humans may be uniquely capable of reporting on and making sense of pleasure.

Consciousness enables humans to consciously predict and anticipate the outcomes of choices and actions, which confers our species, at least, with a considerable evolutionary advantage. Our "liking" responses are often experienced consciously, and we are able to judge subjective pleasure. Consciousness enables us to experience pleasures, desires, and perhaps even eudaimonia.

Well-being, however, is a slippery concept not easily studied. As mentioned earlier, happiness is traditionally seen as consisting of two fundamental elements: hedonia (pleasure) and eudaimonia (a life well lived). In modern psychology, these two aspects are usually referred to as pleasure and meaning, and positive psychologists have recently proposed to add a third meaning-related component of engagement involving feelings of commitment and participation in life (Seligman et al., 2005). Yet Aristotle also wrote, "all agree our goal is eudaimonia, but people differ on what it is, some considering it pleasure…" (*Nicomachean Ethics* I, 4, 1095a14–25). Even though there is clearly a conceptual distinction between pleasure and meaning and engagement, hedonic and eudaimonic aspects empirically cohere together in happy people (Diener et al., 2006; Kahneman, 1999; Seligman et al., 2005) and may share overlapping mechanisms.

Surveys of happiness in societies have become commonplace and are seen as increasingly important measures. Asking people how happy they are does, however, offer little evidence as to the underlying neurobiology of happiness. Kahneman (1999) suggested that the best way to measure hedonics is to repeatedly measure an individual's hedonic state over time, thereby tracking their hedonic state during daily life. The study of hedonic *states* may be amenable to neuroscientific studies of the neural dynamic

supporting a happy brain and individual, also allowing us to look at how the networks underlying pleasure correspond to the networks underlying a eudaimonic state.

8.6 BRAIN NETWORKS AND METASTABILITY IN EUDAIMONIA

In terms of underlying brain networks, we have previously suggested that the so-called default mode network could be key to linking hedonia and eudaimonia (Kringelbach & Berridge, 2009). The default mode network is a network of brain regions that are active in the absence of any external task demands (Biswal et al., 1995; Gusnard & Raichle, 2001). This network includes hedonic regions such the anterior cingulate and orbitofrontal cortices, which have a relatively high density of opiate receptors and have been shown to undergo activity change in pathological states such as depression (Drevets et al., 1997). Intriguingly, one part of this network is the precuneus, which has been shown to engage in self-related mental representations during rest (Cavanna & Trimble, 2006; Lou et al., 1999) and where the activity correlates to self-reported questionnaire happiness scores (Sato et al., 2015). This might provide some insights into where the brain is computing the meaningful evaluations of pleasure related to the self and others, which is important for eudaimonia.

Equally, we have also proposed that the concept of metastability might be key to furthering our understanding (Kringelbach & Berridge, 2017). Metastability is a concept from the study of dynamic systems and is a measure of how the synchronization between the different regions fluctuates across time (Cabral et al., 2014). A system with optimal metastability has the potential for optimal exploration of the dynamic repertoire inherent in the brain's structural connectivity and thus achieving a balance between the fast and slow processing that characterize human cognition (Kringelbach et al., 2015). Conceptualized like this, such flexibility could underlie eudaimonia as a brain state with optimal flow of information in the pleasure system and other emotion-processing networks.

8.7 CONCLUSIONS

Evolution has shaped biological brains to be able to survive as individuals and as a species. Hedonia is a key driver in the underlying brain processing, constantly helping to make decisions that optimize the possibility of survival through wanting and seeking out food, sex, and conspecifics. The brain is disproportionally interested in rewards and spends much of its time in the cyclical pursuit and consummation of them. Learning is ubiquitous through this cycle and is essential for developing and updating the predictions upon which survival is predicated. In humans, our highly developed prediction mechanisms are a hallmark of our evolutionary success. Yet, this constant need for prediction is also potentially the root of much unhappiness, which is why the "tragic miracle of consciousness" can seem such an apt description of the human condition.

Nevertheless, the human condition also involves states of eudaimonia, which are fleeting, and the possible routes to entering them are not fully understood and not easily reproducible. It has become clear, however, that while evolution has clearly designed us to survive – for which pleasure is a necessary requirement – eudaimonia is what allows for moments of meaningful bliss. Yet the exploration of the links between hedonia and eudaimonia is important since it holds the promise of creating better treatments for affective disorders. In the long term, this might lead to the generation of not only more hedonia, but also potentially eudaimonia in the greatest number of people.

8.8 ACKNOWLEDGMENTS

Our research is supported by an MRC studentship to ES, as well as grants ERC Consolidator Grant: CAREGIVING (n. 615539) and the Center for Music in the Brain, funded by the Danish National Research Foundation (DNRF117) to MLK and from the NIH (MH63644 and DA015188) to KCB. This chapter is in part based on material in previously published articles (Berridge & Kringelbach, 2011).

REFERENCES

Aristotle (350 BCE/1976). *The Nicomachean Ethics, Book 10*, trans. by J. A. K. Thomson. London: Penguin Books.

Berridge, K. C. (1996). Food reward: Brain substrates of wanting and liking. *Neuroscience and Biobehavioral Reviews*, **20**, 1–25.

Berridge, K. C. (2000). Measuring hedonic impact in animals and infants: Microstructure of affective taste reactivity patterns. *Neuroscience and Biobehavioral Reviews*, **24**, 173–198.

Berridge, K. C. (2007). Brain reward systems for food incentives and hedonics in normal appetite and eating disorders. In T. C. Kirkham & S. J. Cooper, eds., *Appetite and Body Weight*. New York: Academic Press, pp. 191–216

Berridge, K. C., & Kringelbach, M. L. (2008). Affective neuroscience of pleasure: Reward in humans and animals. *Psychopharmacology*, **199**, 457–480.

Berridge, K. C., & Kringelbach, M. L. (2011). Building a neuroscience of pleasure and well-being. *Psychology of Well-Being*, **1**(1), 1–3.

Berridge, K. C., & Kringelbach, M. L. (2015). Pleasure systems in the brain. *Neuron*, **86**, 646–664.

Berridge, K. C., & Robinson, T. E. (2003). Parsing reward. *Trends in Neurosciences*, **26**, 507–513.

Biswal, B., Yetkin, F., Haughton, V., & Hyde, J. (1995). Functional connectivity in the motor cortex of resting human brain using echo-planar MRI. *Magnetic Resonance in Medicine*, **34**, 537–541.

Cabanac, M., & Lafrance, L. (1990). Postingestive alliesthesia: The rat tells the same story. *Physiology and Behavior*, **47**, 539–543.

Cabral, J., Kringelbach, M. L., & Deco, G. (2014). Exploring the network dynamics underlying brain activity during rest. *Progress in Neurobiology*, **114**, 102–131.

Castro, D. C., & Berridge, K. C. (2017). Opioid and orexin hedonic hotspots in rat orbitofrontal cortex and insula. *Proceedings of the National Academy of Sciences*, **114**, E9125–E9134.

Cavanna, A. E., & Trimble, M. R. (2006). The precuneus: A review of its functional anatomy and behavioural correlates. *Brain*, **129**, 564–583.

Crisp, R., & Kringelbach, M. L. (2018). Higher and lower pleasures revisited: Evidence from neuroscience. *Neuroethics*, **11**, 211–215.

Diener, E., Lucas, R. E., & Scollon, C. N. (2006). Beyond the hedonic treadmill: Revising the adaptation theory of well-being. *American Psychologist*, **61**, 305–314.

Drevets, W. C., Price, J. L., Simpson, J. R., Jr., et al. (1997). Subgenual prefrontal cortex abnormalities in mood disorders. *Nature*, **386**, 824–827.

Fervaha, G., Graff-Guerrero, A., Zakzanis, K. K., et al. (2013). Incentive motivation deficits in schizophrenia reflect effort computation impairments during cost–benefit decision-making. *Journal of Psychiatric Research*, **47**, 1590–1596.

Frijda, N. (2010). On the nature and function of pleasure. In M. L. Kringelbach & K. C. Berridge, eds., *Pleasures of the Brain*. New York: Oxford University Press, pp. 99–112.

Gold, J. M., Strauss, G. P., Waltz, J. A., et al. (2013). Negative symptoms of schizophrenia are associated with abnormal effort-cost computations. *Biological Psychiatry*, **74**, 130–136.

Grill, H. J., & Norgren, R. (1978). The taste reactivity test. I. Mimetic responses to gustatory stimuli in neurologically normal rats. *Brain Research*, **143**, 263–279.

Gusnard, D. A., & Raichle, M. E. (2001). Searching for a baseline: Functional imaging and the resting human brain. *Nature Reviews Neuroscience*, **2**, 685–694.

Kahneman, D. (1999). Objective happiness. In D. Kahneman, E. Diener, & N. Schwartz, eds., *Well-being: The Foundation of Hedonic Psychology*. New York: Russell Sage Foundation, pp. 3–25.

Kaplan, J. M., Roitman, M., & Grill, H. J. (2000). Food deprivation does not potentiate glucose taste reactivity responses of chronic decerebrate rats. *Brain Research*, **870**, 102–108.

Kringelbach, M. L. (2004a). Emotion. In R. L. Gregory, ed., *The Oxford Companion to the Mind*, 2nd ed. Oxford: Oxford University Press, pp. 287–290.

Kringelbach, M. L. (2004b). Learning to change. *PLoS Biology*, **2**, E140.

Kringelbach, M. L. (2005a). The human orbitofrontal cortex: linking reward to hedonic experience. *Nature Reviews Neuroscience*, **6**, 691–702.

Kringelbach, M. L. (2005b). The orbitofrontal cortex: Linking reward to hedonic experience. *Nature Reviews Neuroscience*, **6**, 691–702.

Kringelbach, M. L. (2009). *The Pleasure Center: Trust Your Animal Instincts*. New York: Oxford University Press.

Kringelbach, M. L. (2010). The hedonic brain: A functional neuroanatomy of human pleasure. In M. L. Kringelbach & K. C. Berridge, eds., *Pleasures of the Brain*. Oxford: Oxford University Press, pp. 202–221.

Kringelbach, M. L., & Berridge, K. C. (2009). Towards a functional neuroanatomy of pleasure and happiness in the brain. *Trends in Cognitive Sciences*, **13**, 479–487.

Kringelbach, M. L., & Berridge, K. C. (2017). The affective core of emotion: linking pleasure, subjective well-being and optimal metastability in the brain. *Emotion Review*, **9**, 191–199.

Kringelbach, M. L., & Rolls, E. T. (2004). The functional neuroanatomy of the human orbitofrontal cortex: Evidence from neuroimaging and neuropsychology. *Progress in Neurobiology*, **72**, 341–372.

Kringelbach, M. L., McIntosh, A. R., Ritter, P., Jirsa, V. K., & Deco, G. (2015). The rediscovery of slowness: Exploring the timing of cognition. *Trends in Cognitive Sciences*, **19**, 616–628.

Lou, H. C., Kjaer, T. W., Friberg, L., et al. (1999). A ^{15}O–H$_2$O PET study of meditation and the resting state of normal consciousness. *Human Brain Mapping*, **7**, 98–105.

McFarland, B. R., & Klein, D. N. (2009). Emotional reactivity in depression: Diminished responsiveness to anticipated reward but not to anticipated punishment or to nonreward or avoidance. *Depression and Anxiety*, **26**, 117–122.

O'Doherty, J., Rolls, E. T., Francis, S., Bowtell, R., & McGlone, F. (2001). Representation of pleasant and aversive taste in the human brain. *Journal of Neurophysiology*, **85**, 1315–1321.

Pizzagalli, D. A., Iosifescu, D., Hallett, L. A., Ratner, K. G., & Fava, M. (2008). Reduced hedonic capacity in major depressive disorder: Evidence from a probabilistic reward task. *Journal of Psychiatric Research*, **43**, 76–87.

Rømer Thomsen, K., Whybrow, P. C., & Kringelbach, M. L. (2015). Reconceptualising anhedonia: Novel perspectives on balancing the pleasure networks in the human brain. *Frontiers in Behavioral Neuroscience*, **9**, 49.

Sato, W., Kochiyama, T., Uono, S., et al. (2015). The structural neural substrate of subjective happiness. *Scientific Reports*, **5**, 16891.

Seligman, M. E., Steen, T. A., Park, N., & Peterson, C. (2005). Positive psychology progress: Empirical validation of interventions. *American Psychologist*, **60**, 410–421.

Smith, K. S., Mahler, S. V., Pecina, S., & Berridge, K. C. (2010). Hedonic hotspots: Generating sensory pleasure in the brain. In M. L. Kringelbach & K. C. Berridge, eds., *Pleasures of the Brain*. New York: Oxford University Press, pp. 27–49.

Steiner, J. E. (1973). The gustofacial response: Observation on normal and anencephalic newborn infants. *Symposium on Oral Sensation and Perception*, **4**, 254–278.

Steiner, J. E., Glaser, D., Hawilo, M. E., & Berridge, K. C. (2001). Comparative expression of hedonic impact: Affective reactions to taste by human infants and other primates. *Neuroscience and Biobehavioral Reviews*, **25**, 53–74.

Strauss, G. P., & Gold, J. M. (2012). A new perspective on anhedonia in schizophrenia. *American Journal of Psychiatry*, **169**, 364–373.

Treadway, M. T., Bossaller, N. A., Shelton, R. C., & Zald, D. H. (2012). Effort-based decision-making in major depressive disorder: a translational model of motivational anhedonia. *Journal of Abnormal Psychology*, **121**, 553–558.

Waltz, J. A., Frank, M. J., Robinson, B. M., & Gold, J. M. (2007). Selective reinforcement learning deficits in schizophrenia support predictions from computational models of striatal–cortical dysfunction. *Biological Psychiatry*, **62**, 756–764.

Watson, D., & Naragon-Gainey, K. (2010). On the specificity of positive emotional dysfunction in psychopathology: Evidence from the mood and anxiety disorders and schizophrenia/schizotypy. *Clinical Psychology Review*, **30**, 839–848.

Whybrow, P. C. (1998). *A Mood Apart: The Thinkers Guide to Emotion and its Disorder*. New York: Harper Perennial.

Yang, X. H., Huang, J., Zhu, C. Y., et al. (2014). Motivational deficits in effort-based decision making in individuals with subsyndromal depression, first-episode and remitted depression patients. *Psychiatry Research*, **220**, 874–882.

9

Environmental Pressures on Transgenerational Epigenetic Inheritance

An Evolutionary Development Mechanism Influencing Atypical Neurodevelopment in Autism?

DWAIPAYAN ADHYA, AICHA MASSRALI, ARKOPROVO PAUL, MARK KOTTER, JASON CARROLL, DEEPAK SRIVASTAVA, JACK PRICE, AND SIMON BARON-COHEN

9.1 INTRODUCTION

Research in developmental neuropsychiatric conditions has revealed morphological and functional divergences in the brain. In some cases, the divergences occur due to one or two highly penetrant genomic mutations. In case such as autism, mutations in varied sets of genes may produce a convergent autism behavioral phenotype. It is thus likely that there may be other forms of non-genomic regulation of gene expression during development affecting behavioral outcome. Epigenetic gene regulation is one such mechanism that can permanently switch on or switch off gene expression, and these epigenetic changes can be inherited from one cell stage to another during differentiation, mimicking the effects of genomic mutations. Epigenetic gene regulation occurring during early developmental stages of cellular differentiation, which are highly sensitive to environmental cues, is the primary mechanism responsible for the phenomenon known as *evolutionary development* or "evo-devo." This chapter discusses these mechanisms in the context of autism and the environmental factors that influence it. We also discuss the latest technologies that are used to study these mechanisms and the impacts they might have on our current understanding of neuropsychiatric conditions and their association with evolutionary development.

Autism spectrum disorders (ASDs), as defined by the fifth edition of the Diagnostic and Statistical Manual of Mental Disorders (DSM-5; APA, 2013), are neurodevelopmental conditions diagnosed on the basis of a dyad of behavioral impairments: impaired social communication, alongside unusually narrow and repetitive interests and activities. Atypical language development, which was a third behavioral impairment defined in DSM-IV, was removed in DSM-5, and is now classified as a co-occurring condition. The DSM-5 criteria also emphasize the dimensional nature of autism. Although the term ASD is frequently used, the term autism spectrum condition is sometimes preferred, as it also indicates a biomedical diagnosis requiring individual support and recognizes affected individuals as different, but without the negative connotations of the term "disorder" (Lai et al., 2014).

Many genes have been identified as having an association with an autism diagnosis (Bourgeron, 2015; O'Roak et al., 2011, 2012). The Simons Foundation Autism Research Initiative (SFARI) database catalogues the most relevant of these genes according to the type of genetic variations from whole-genome sequencing studies, rare genetic mutations, and mutations causing syndromic forms of autism. It was first published in 2009 (Basu et al., 2009), and an up-to-date reference for all known genes can be found at https://gene.sfari.org/autdb/HG_Home.do. This catalogue currently contains >800 genes with a significant association with autism, 99 of which are classified as "high-confidence" genes (being independently replicated, and most of which are rare genetic variants). Although common genetic variants are thought to contribute around 50 percent of the variance (Bourgeron, 2015), many common genetic variants have not been successfully replicated. Indeed, genome-wide association studies of autism have only replicated five common genetic variants linked to autism, even though the actual number is likely to be hundreds (Bourgeron, 2015). The large number of associated genes has made it a complicated task to identify pathways and mechanisms. One subset of autistic individuals[1] may carry a single mutation in a highly penetrant gene, while another subset may carry many common variants, each of small effect, while in a subset of typical individuals a deleterious mutation in the same gene(s) might not be enough to produce autism at all (Bourgeron, 2015). This extreme genetic heterogeneity has resulted in research into the environmental and hormonal associations of autism in an effort to understand if these environmental or epigenetic factors can help explain such diverse genetic mechanisms and if these implicate a common final pathway. Preliminary

---

[1] We use "identity-first" language (autistic individuals) rather than person-first language (individuals with autism) to reflect the preference of the autism community (http://autisticadvocacy.org/about-asan/identity-first-language).

estimates suggest environmental factors to account for up to 38.1 percent of all cases (Sassone-Corsi & Christen, 2015).

Although autism environmental research is still in its infancy, there have been many epidemiological studies demonstrating significant associations of various environmental agents with autism, and the timing of exposure (Figure 9.1) seems to be of importance to an autism outcome (Modabbernia et al., 2017; Rossignol et al., 2014). Preconception factors include advanced parental age (Wu et al., 2017) and exposure to toxic substances in the period prior to conception (McCanlies et al., 2012). Gestational factors include exposure to insecticides and pesticides (Rauh et al., 2006; Roberts & English, 2013), air pollution (Volk et al., 2011), drugs (Gentile, 2014; Kobayashi et al., 2016), high levels of fetal and maternal steroids (Baron-Cohen et al., 2015; Kosidou et al., 2016), and maternal stress and immune activity (Ronald et al., 2010; Masi et al., 2015). Events occurring at birth such as birth injury, trauma, and hypoxia are also strongly associated (Gardener et al., 2009, 2011; Modabbernia et al., 2016), while perinatal and childhood factors include exposure to inorganic mercury (Yoshimasu et al., 2014), heavy metals (Price et al., 2010), air pollution, and particulate matter (PM) (Windham et al., 2006). Many of the environmental and hormonal factors associated with autism have the ability to cause sweeping gene regulatory effects through transcriptional and epigenetic mechanisms (Abel & Zukin, 2008; Bao & Swaab, 2011; McCarthy & Nugent, 2013; Tsai et al., 2009). We should note that until these environmental factors are well replicated and can be established as being of "high confidence," we should keep an open mind that some of these may be spurious associations (http://tylervigen.com/spurious-correlations).

Human behavioral divergences associated with autism may involve developmental mechanisms causing permanent transgenerational inheritance during cellular differentiation in a phenomenon referred to as "evolutionary development" or evolution of development (or "evo-devo") (Carroll, 2008). This phenomenon is not well understood, as it involves several molecular mechanisms that produce wide-ranging phenotypes originating from a single genomic template (or genotype). It is a form of evolution that does not arise out of natural selection of genomic mutations, but by the selection of phenotypic features that mimic the effects of genomic mutations. This occurs as organisms undergo morphogenesis by differentiation of cells during development. Non-genomic factors such as regulatory proteins and RNA periodically switch on and switch off gene expression to produce heritable gene expression states analogous to the effects of genomic mutations. This selection of phenotypic features during development through the action of such non-genomic factors can also be triggered by atypical environmental stimuli. This brings about cellular divergences during development that are responsible for permanent phenotypic alterations during adulthood. The ability of the environment to alter and select cellular developmental mechanisms that result in permanent phenotypic divergences during adulthood is an example of how "evo-devo" mechanisms of transgenerational inheritance work in multicellular organisms during development.

One of the mechanisms of "evo-devo" involves epigenetic regulation that is later consolidated as heritable changes (Gerhart & Kirschner, 1997). This chapter reviews studies that have shown how environmental influences during development cause behavioral divergences with respect to autism and associated neurodevelopmental conditions, as well as providing a description of cutting-edge technologies that have started to investigate this heritable gene regulatory mechanism. This chapter also discusses these mechanisms in the context of autism and how the environment applies selection pressure to produce phenotypes that are missing heritability factors. As these processes are observable within a short time frame, they can be easily studied in humans using a laboratory model such as the induced pluripotent stem cell (iPSC) model, combining this with cutting-edge sequencing-based methods to study epigenetic regulation. The study of the role of the environment in bringing about

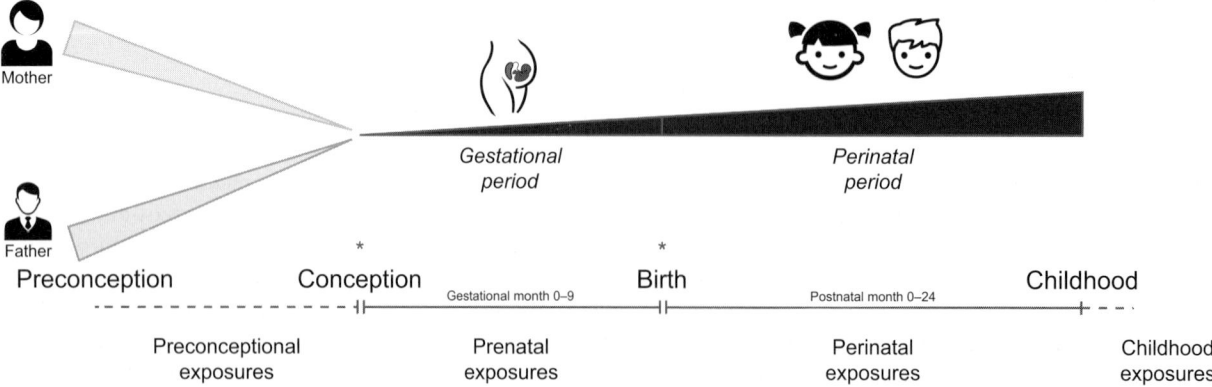

Figure 9.1 Timing of environmental exposures associated with autism. *Exposure to toxic substances or traumatic events around conception and birth are strongly associated with an autism outcome.

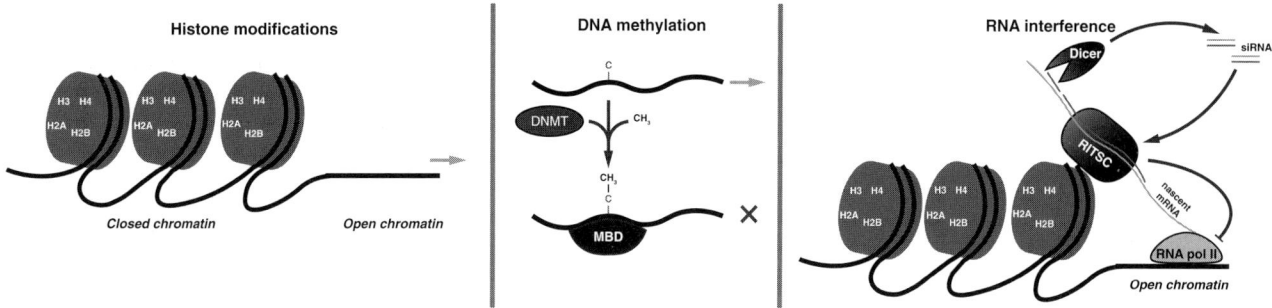

Figure 9.2 Major cellular epigenetic mechanisms responsible for transgenerational inheritance. siRNA: small interfering RNA.

permanent changes in brain development, especially with respect to autism, may provide novel insights into our understanding of the development of the human brain, a research theme that will likely predominate over the next decade. When studies using the above methods are performed in other animals, it will also give us a better understanding of the "evolution of function" (Carroll, 2008) of epigenetic proteins and the origin of epigenetic regulation as an agent of "evo-devo."

9.2 TRANSCRIPTIONAL AND EPIGENETIC REGULATORY MECHANISMS

Transcription factors are proteins that form complexes or may act alone to bind DNA and control the rate of gene expression (Latchman, 1997). They are also involved in the process of transcribing DNA into RNA, the first step in the central dogma of molecular biology (Crick, 1970). This transfer of sequential information from DNA to RNA then to protein is crucial to transforming genomic information into visible phenotypes. As a consequence, manipulation or disruption of this process can alter this sequential transfer of information, affecting the phenotype and often manifesting in a "disease" state. Most transcription factors act in a transient fashion, and removal of stimuli generally dismantles the transcription factor complex. However, certain classes of transcription factors, known as epigenetic factors, can produce a more long-term effect on gene regulation by a mechanism known as imprinting (Berger et al., 2009). Epigenetic factors can activate or repress genes permanently in a configuration that is often inherited into the next generation.

The first transcription factors discovered were polymerases between 1955 and 1961 (Hurwitz, 2005). Polymerases are the most fundamental component of any transcription factor complex, such as the DNA-dependent RNA polymerase, which transcribes DNA into RNA. Different transcription factors have dedicated functions in different tissues of the body. For example, MEF2, Pax6, CREB, and NF-κB are known to drive gene expression in the developing brain (Mao et al., 1999; O'Neill & Kaltschmidt, 1997; Rammes et al., 2000; Warren et al., 1999). NF-κB is one of the best-characterized transcription factors, which responds to changing levels of neurotransmitters as well as modulating the immune response against infection and injury in the brain (O'Neill & Kaltschmidt, 1997). Responses to different kinds of stimuli, including various environmental ones, are convergent functions of transcription factors. Epigenetic factors, which alter DNA conformation, act like regulating mechanisms for transcription factor activities.

This is an important step in determining cell fate and lineage, as it is epigenetic factors that determine which chromosomal regions are expressed based on stimuli received from the surrounding environment. This results in activation of a subset of genes, namely those that are required for determining a particular cell lineage. The combined activities of epigenetic factors and transcription factors are responsible for giving rise to the thousands of different cell types in the human body from a single set of genetic information. The difference between epigenetic and transcription factors is that the effects of the former are inherited in each sister cell after mitosis, while the latter act more on the instructions set by epigenetic factors depending on the lineage of the cells. It is believed that cancer cells show atypical epigenetic signatures, often derepressing lineage-nonspecific genes, causing genomic instability and cancer progression (Kanwal et al., 2015).

There are three types of epigenetic modifications: (1) DNA methylation; (2) histone regulation; and (3) gene silencing by noncoding RNA (Figure 9.2) (Handy et al., 2011). How some of these mechanisms are dysregulated to produce atypical phenotypes in autism is described in Section 9.3.

9.3 TIMING OF ENVIRONMENTAL EXPOSURE

Every cell in the body has a unique epigenetic signature. The epigenetic signatures of some cells may be more closely related than others depending on the tissues they form and the functions they carry out. Stem cells have an open epigenetic state, as opposed to differentiated cells (e.g., neurons, muscle, blood, and liver), which have an ON-memory epigenetic state that maintains the stability of the mechanisms necessary for cellular identity (Hormanseder et al., 2017). The temporal differences in transcriptional regulation mean that the stage of cellular development at which certain environmental exposures occurs is crucial in determining the phenotypic fate. This section discusses the timeline of exposures known to be associated with a diagnosis of autism. Possible cellular/molecular mechanisms resulting from these exposures are discussed in Section 9.4.

9.3.1 Preconception Exposure

Exposure of parents to toxic chemicals prior to conception is significantly associated with autism in children. A 1976 study found 26 percent of families with an autistic child (n = 78) had parents who were exposed to toxic chemicals at work, compared to only one percent of parents with typically developing (TD) children (Coleman, 1976). Parents of children with autism (n = 93) were more likely to have been exposed to lacquer, varnish, solvents, and asphalt from their workplace during the three months prior to conception, compared with parents of TD children (n = 81) (McCanlies et al., 2012). However, the overall significance of these studies is limited by their small sample sizes, and further investigation is required to these establish association in a wider population in case these are spurious correlations.

9.3.2 Prenatal Exposure

The fetus is in the most vulnerable stage of human development, and any chemical imbalances in the mother or atypical exposure to toxins can affect fetal development. Some of the most toxic substances for the fetus are pesticides, and prenatal exposure to organochlorine pesticides is associated with heightened likelihood of autism, with the highest likelihood in mothers who lived within 500 m of the source (Roberts et al., 2007).

Children with higher exposure to organophosphate insecticides are also significantly more likely to develop symptoms of pervasive developmental disorders (Rauh et al., 2006). Maternal exposure to air pollution during pregnancy is also highly associated with autism, and one study showed pregnant mothers living near freeways were more likely to have children with autism (Volk et al., 2011). Air pollutants associated with higher risk of autism include quinoline, styrene, ozone, nitrogen dioxide, and PM including mercury, lead, nickel, manganese, diesel particulates, and methylene chloride (Becerra et al., 2013; Kalkbrenner et al., 2010; Roberts et al., 2013; Volk et al., 2013). Again, which of these findings are true or spurious correlations needs to be investigated.

Steroids are important factors in brain development, and higher levels of circulating sex steroids in the male fetus are positively associated with autism (Baron-Cohen et al., 2011, 2015). Mothers with polycystic ovary syndrome, characterized by hyperandrogenism and elevated levels of androgens during pregnancy, are also at greater risk of having children with autism (Kosidou et al., 2016; Pohl et al., 2014). Prenatal stress is another trigger for

a range of postnatal outcomes associated with autism, including core symptoms such as hypersensitivity and repetitive behavior, cognitive deficits, and abnormalities in immune function (Kinney et al., 2008). The molecular mechanisms responsible for increased susceptibility to prenatal factors is not well understood.

9.3.3 Perinatal and Childhood Exposure

The period around birth and the first 24 months of a child's life are also susceptible periods for toxic environmental exposures, although to a lesser degree than the fetal stage. Exposures to pesticides such as phosphine (a fungicide) and dichlorodiphenyltrichloroethane (DDT) during childhood have been linked with increased prevalence of autism (Audouze & Grandjean, 2011; Garry et al., 2002). Childhood exposure to air pollution appears to be an especially important factor associated with autistic symptoms. A survey of children at ~24 months (2 years of age) identified cadmium, nickel, trichloroethylene, vinyl chloride, and diesel PM to be highly correlated with an autism diagnosis (Windham et al., 2006). Heavy metal exposure during childhood is also associated with autism prevalence. The role of mercury exposure in autism is not clear (Palmer et al., 2006), although some studies have found a strong positive association with autism (Blanchard et al., 2011; Palmer et al., 2009). However, childhood exposure to lead has been reported to have a high positive association with autism (Price et al., 2010). The perinatal period is also associated with the establishment and stabilization of healthy gut microbiota, and the disruption of this process either through birth by cesarean section, preterm birth, or the use of antibiotics can lead to atypical microbiota colonization, affecting brain development and behavior (Clarke et al., 2014; Diaz Heijtz et al., 2011). Epidemiological studies have revealed an association between autism and atypical perinatal microbiota (Finegold et al., 2002; Mittal et al., 2008), while altered microbiota colonization patterns have also been associated with studies in animal models of autism (de Theije et al., 2014b). Which of these putative environmental factors will emerge as "high confidence" (i.e., well replicated) remains to be seen.

The association of childhood toxic exposure with autism diagnosis demonstrates that the developing brain during the first two years of life is almost as susceptible to environmental insults as the fetal brain. The large body of epidemiological studies discussed in this section have identified an exposure window extending from preconception until two years after birth, demonstrating that the mechanisms responsible for the appearance of autism phenotypes can be triggered through exposures from parents, transmitted through the placenta, or transmitted through direct exposures after birth. At the prenatal, neonatal, and postnatal stages, the problem for scientists is to disentangle possible causal agents from mere correlational factors, since the epidemiological design that has revealed the majority of these agents and factors does not allow one to achieve such a separation. The strength of the iPSC model system is that causality can be directly tested by experimental manipulation.

9.4 MOLECULAR MECHANISMS TRIGGERED BY ENVIRONMENTAL AGENTS

The effect of environmental risk factors on mental health was not recognized until recent epidemiological studies demonstrated associations with neurodevelopmental conditions such as autism. The influence that these environmental agents have on the molecular machineries of the developing brain is still not fully understood. Some molecular mechanisms have been identified through pharmacological studies using animal models (Table 9.1), and they have revealed two major effects of environmental agents on animal cell biology, as discussed in this section: (1) activation of oxidative stress, the immune system, and inflammation; and (2) modulation of epigenetic gene regulation. It is interesting to note that the genes involved in both biological mechanisms are strongly associated with autism (Bourgeron, 2015); however, currently no evidence exists to suggest that environmental agents act via these mechanisms to trigger the condition. This section discusses the prominent molecular mechanisms activated by environmental risk factors, aiding understanding of their effects and their potential involvement in the triggering of neurological symptoms.

9.4.1 Pesticides

Pesticides are chemical compounds used to control, repel, or kill plants or animals that are considered to be "pests" (Corsini et al., 2013). They include herbicides, fungicides,

Table 9.1 Major environmental agents and mechanisms affected.

Environmental factor categories	Physiological system affected	Environmental agent	Mechanisms of action	References
Pesticides	Immune system	Organophosphates	Reduces T-cell proliferation, inhibits NK cells, LAK cells, and CTLs Dysregulates cell death	Galloway & Handy, 2003; Li, 2007
		Organochlorides	Immunosuppression Disrupts ability to inactivate or eliminate pathogens Aberrant T- and B-cell regulation Cytokine secretion	Cooper et al., 2004; Hermanowicz & Kossman, 1984; McConnachie & Zahalsky, 1992; Reed et al., 2004; Schaalan et al., 2012; Vine et al., 2000
	Cellular epigenetics	Organophosphates	Induces alkaline damage and DNA damage	Ray & Richards, 2001
		Organochlorines	Alters DNA methylation patterns Dysregulates histone acetylation Affects male and female reproductive systems Mimics epigenetic regulation by endogenous hormone Disrupts hypothalamic gene expression	Desaulniers et al., 2009; McLachlan et al., 2006; Shutoh et al., 2009; Song et al., 2010, 2011; Stouder & Paoloni-Giacobino, 2011; Zama & Uzumcu, 2009
Metals	Cellular oxidative stress	Cadmium	Replaces iron and copper in proteins Causes DNA damage in the CNS	Price & Joshi, 1983; Waalkes, 2000; Watjen & Beyersmann, 2004
		Lead	Inhibits trace element absorption Elevates oxidative stress	Gurer & Ercal, 2000; Hoffman et al., 2000;

Continued

Table 9.1 Cont.

Environmental factor categories	Physiological system affected	Environmental agent	Mechanisms of action	References
			Directly increases ROS Depletes antioxidant pools	Hunaiti & Soud, 2000; Patrick, 2006
		Nickel	Hydroxyl radical formation Oxidative DNA damage Elevate sister-chromatid exchanges during cell division Disrupts histone–DNA binding	Bal et al., 2000; Dally & Hartwig, 1997; M'Bemba-Meka et al., 2007
		Mercury	Accumulates as organomercuric compounds Crosses placental barrier and blood–brain barrier Hypoxia, oxidative stress, disruption of brain development	Guzzi & La Porta, 2008
	Cellular epigenetics	Cadmium	Increases DNMT activity Dysregulates DNA methylation Induces genomic imprinting	Doi et al., 2011; Somji et al., 2011; Takiguchi et al., 2003
		Nickel	Inhibits DNA repair Histone deacetylation Histone ubiquitination Disrupts histone–DNA binding Influences binding of MeCP2 to DNA	Borochov et al., 1984; Chen et al., 2006; Ji et al., 2008; Karaczyn et al., 2006, Ke et al., 2008; Martinez-Zamudio & Ha, 2011; Yan et al., 2003
		Mercury	Alters histone methylation and acetylation Reduces BDNF expression	Onishchenko et al., 2008
		Lead	Affects DNA methylation	Bollati & Baccarelli, 2010; Wright et al., 2010
Air pollution	Cellular oxidative stress and innate immunity system	UFPM	Neuro-inflammation Delivers toxic substances into the CNS Stimulates pro-inflammatory cytokines Induces microglial activation	Block & Calderon-Garciduenas, 2009; Long et al., 2007; Nemmar & Inuwa, 2008; Oberdorster et al., 2004; Tin Tin Win et al., 2008; Valavanidis et al., 2008; Wang et al., 2007, 2008
		Ground-level ozone	Oxidative stress on proteins and lipids Brain lipid peroxidation Dopaminergic neuron death Morphological damage in neurons Stimulates pro-inflammatory cytokines	Angoa-Perez et al., 2006; Araneda et al., 2008; Guevara-Guzman et al., 2009; Pereyra-Munoz et al., 2006
	Cellular epigenetics	PAH	Dysregulated methylation of genes associated with neural impairments	Bellavia et al., 2013; Breton et al., 2012; Liang et al., 2012;

Continued

Table 9.1 Cont.

Environmental factor categories	Physiological system affected	Environmental agent	Mechanisms of action	References
			Dysregulated cell death and survival mechanisms	Madrigano et al., 2012; Pavanello et al., 2010; Rossnerova et al., 2013; Tang et al., 2010
		PM	Decreased methylation of iNOS gene	Breton et al., 2012; Sofer et al., 2013
GI exposures	Immune system	Atypical bacterial gut flora	Maternal immune activation Affects neurogenesis Dysregulation of secondary messengers	Choi et al., 2016; Emanuele et al., 2010
		Microbe-released metabolites	Activates immune responses	Rieder et al., 2017
		Prolonged antibiotic treatment	Reduces monocyte numbers	Mohle et al., 2016
	Gut–brain axis	Prolonged antibiotic treatment	Atypical GABA receptor expression in PFC neurons Decreased neurogenesis	Mohle et al., 2016; Sharon et al., 2016
		Microbe-released metabolites	Autism symptoms Anxiety-like behaviors Bacteria-derived glutamate directly affect neurotransmission Atypical GABA signaling Dysregulation of synaptic transmission genes and neurophysiological mechanisms	Ding et al., 2017; Favre et al., 2013; Foley et al., 2014; Kuo & Liu, 2017; Mohle et al., 2016; Persico & Napolioni, 2013; Rieder et al., 2017; Sharon et al., 2016; Sheldon & Robinson, 2007; Wang et al., 2012

NK: natural killer; LAK: lymphokine-activated killer; CTL: cytotoxic T lymphocyte; CNS: central nervous system; ROS: reactive oxygen species; DNMT: DNA methyltransferase; MeCP2: methyl CpG binding protein 2; BDNF: brain-derived neurotrophic factor; UFPM: ultrafine particulate matter; PAH: polycyclic aromatic hydrocarbon; PM: particulate matter; iNOS: inducible nitric oxide synthase; GI: gastrointestinal; GABA: γ-aminobutyric acid; PFC: prefrontal cortex.

and insecticides. Pesticides have widespread uses, including heavy use in agriculture worldwide, and their pervasive presence in living and working environments is a major public health concern. Because of their widespread use, there are always low levels of pesticides prevalent in the circulation.

9.4.1.1 Immune System Modulation

Pesticide exposure has been associated with autism. Several studies have identified the immune system to be very susceptible to pesticide exposure, with children especially vulnerable because of their immature immune system (Corsini et al., 2008; Repetto & Baliga, 1997; Wigle et al., 2008). Pesticides can cause acute alterations in immune responses in children, making them susceptible to infection and immune disorders (Phillips, 2000). Organophosphates affect the immune system by reducing T-cell proliferation and inhibiting natural killer (NK) cells, lymphokine-activated killer cells, and cytotoxic T lymphocytes (Galloway &

Handy, 2003; Li, 2007). Several immune activation pathways are affected, and proteases responsible for inducing programmed cell death in target cells are suppressed (Galloway & Handy, 2003). Organochlorine pesticides have immunosuppressive effects. DDT can alter immunoglobulin levels (Cooper et al., 2004; Vine et al., 2000), disrupting its ability to inactivate or eliminate pathogens and thus altering resistance to infections (Hermanowicz & Kossman, 1984). Organochlorines are also responsible for aberrant peripheral T- and B-cell regulation (McConnachie & Zahalsky, 1992), decreased NK cell activation (Reed et al., 2004), cytokine secretion (Schaalan et al., 2012), and autoimmune activation (McConnachie & Zahalsky, 1992).

9.4.1.2 Epigenetic Modulation

In addition to immune dysregulation, pesticides are also able to modify epigenetic gene regulatory functions. DDT and other organochlorine pesticides are known to alter

DNA methylation patterns in the hippocampus in animal studies, and CpG islands have been found to be hypomethylated compared to typical controls (Desaulniers et al., 2009; Shutoh et al., 2009). One organochlorine pesticide, methoxychlor (MXC), is part of the environmental endocrine disruptor family, which mimic endogenous hormones and cause epigenetic dysregulation (McLachlan et al., 2006). MXC affects both the male and female reproductive systems (Stouder & Paoloni-Giacobino, 2011), disrupts expression of the hypothalamic genes responsible for normal reproductive function (Stouder & Paoloni-Giacobino, 2011), and downregulates estrogen receptor β expression in ovaries (Zama & Uzumcu, 2009). MXC also alters DNA methylation patterns of paternally as well as maternally imprinted genes (Stouder & Paoloni-Giacobino, 2011). Organochlorines Paraquat and Dieldrin have been associated with exacerbating Parkinson's disease symptoms. Exposure to Paraquat induces histone H3 acetylation and reduces histone deacetylase (HDAC) activity in N27 dopaminergic cells (Song et al., 2010, 2011). Dieldrin has also been associated with a time-dependent increase in histone H3 and H4 acetylation and upregulation of histone acetyltransferases, and prolonged Dieldrin exposure in rats has been shown to induce histone "hyperacetylation" in the striatum and substantia nigra (Song et al., 2010). Organophosphates are also able to cause alkaline damage and induce DNA methylation (Ray & Richards, 2001).

9.4.2 Metals

Metal ions are important elements that carry out a wide range of biological functions in the body. A delicate metal-ion homeostasis is maintained in cells with the help of protein transporters. However, as metals form strong ionic bonds, a breakdown of metal-ion homeostasis by the introduction of metals foreign to the body may result in strong binding of these metals to ligands designed for a different metal, and this may result in oxidative stress and deterioration of protein function, leading to cellular death and disease in the organism (Jomova & Valko, 2011).

9.4.2.1 Oxidative Stress

As discussed in Section 9.3.3, exposure to metals such as mercury, lead, nickel, manganese, and cadmium are associated with autism. Toxic exposure to cadmium can replace iron and copper in various proteins, thereby increasing free iron and copper ions, which can cause oxidative stress (Price & Joshi, 1983). Cadmium-induced free copper ions break down hydrogen peroxide into free radicals that are able to cause DNA damage in the central nervous system (Waalkes, 2000; Watjen & Beyersmann, 2004). Lead, a heavy metal, can damage cellular components by elevating oxidative stress, and it can competitively bind to protein enzymes, inhibiting trace element absorption (Patrick, 2006). Lead also deactivates certain classes of antioxidants (Patrick, 2006). Lead has a dual

mechanism of action to elevate oxidative stress: it directly increases reactive oxygen species (Gurer & Ercal, 2000; Hunaiti & Soud, 2000), while it also depletes antioxidant pools through inhibition of enzymes such as glutathione reductase and δ-aminolevulinic acid dehydrogenase (Hoffman et al., 2000).

Nickel is another metal that causes oxidative stress. Under different physiological conditions and ion concentrations, it can cause hydroxyl radical formation and oxidative DNA damage or induce elevated sister-chromatid exchanges during cell division (Dally & Hartwig, 1997; M'Bemba-Meka et al., 2007). Oxidation–reduction (redox) activity of nickel can disrupt histone–DNA binding (Bal et al., 2000).

Mercury accumulates inside the body in the form of organomercury compounds (Guzzi & La Porta, 2008). Ethylmercury and methylmercury can also cross the placental barrier and blood–brain barrier (BBB), but are not excreted through urine. The brain is the major organ affected by mercury compounds, where its accumulation causes hypoxic conditions, oxidative stress, cell death, and disruption of fetal brain development during gestation. Elemental mercury also accumulates in the brain.

9.4.2.2 Epigenetic Modulation

Metals ions can modulate transcriptional regulation through epigenetic modifications. Cadmium increases DNA methyltransferase (DNMT) activity and DNA hypermethylation-induced gene silencing in rats (Takiguchi et al., 2003). In the chicken embryo, short exposures of cadmium can have a hypomethylating effect (Doi et al., 2011). Cadmium can induce genomic imprinting, leading to heritable changes in chromatin structure (Somji et al., 2011). Nickel ions can modulate the DNA methylation status of cells. After exposure to nickel, cells that are transformed show silencing of DNA-repairing gene O6-methylguanine DNA methyltransferase (MGMT) expression by DNA hypermethylation of its promoter, in addition to histone H3 deacetylation and H3K9me2 enrichment (Ji et al., 2008). DNA–histone interactions can also be disrupted, leading to abnormal condensation of higher-order chromatin structures (Borochov et al., 1984) and global deregulation of gene expression (Martinez-Zamudio & Ha, 2011). Some studies have observed histone H2A/H2B ubiquitination after nickel exposure (Karaczyn et al., 2006; Ke et al., 2008). Nickel also influences the binding characteristics of MeCP2, an autism risk gene (Chen et al., 2006; Yan et al., 2003).

Mercury is the third metal showing an association with autism prevalence, and in its methylmercury form interacts with histone-modifying factors to increase levels of H3K27me3 and decrease levels of H3Ac in promoter regions of neurodevelopmental genes such as brain-derived neurotrophic factor (BDNF). BDNF, a gene associated with autism, is one of the promoters most affected, resulting in reduced expression in the dentate gyrus, with appearance of depression-like symptoms in mice (Onishchenko et al.,

2008). Lead has unknown effects on DNA methylation and epigenetic regulation; however, one study has shown lead to be associated with reduced DNA methylation levels (Wright et al., 2010). In vitro studies have also shown lead to affect DNA methylation (Bollati and Baccarelli, 2010).

9.4.3 Air Pollution

Air pollution is a heterogeneous mixture of suspended particles, gases, and fumes that have adverse effects on the human body. It is composed of PM, gases (e.g., carbon monoxide, sulfur oxides, nitrogen oxides, ground-level ozone), suspended organic compounds (e.g., aromatic hydrocarbons, endotoxins), and metals (e.g., mercury, lead, nickel, manganese, vanadium) (Block & Calderon-Garciduenas, 2009). Components of air pollution are strongly linked with autism occurrence (Becerra et al., 2013; Kalkbrenner et al., 2010; Roberts et al., 2013; Volk et al., 2013). Unsurprisingly, dominant molecular mechanisms triggered by air pollution include increased oxidative stress, altered innate immunity, and chronic neuro-inflammation (Block & Calderon-Garciduenas, 2009), and, in some cases, epigenetic modulation of gene expression (Ji et al., 2016; Lin et al., 2016, Silveyra & Floros, 2012).

9.4.3.1 Oxidative Stress and Innate Immune System Modulation

Unlike pesticides and metals, the effect of air pollution on the human brain has been the focus of many studies after it was found that air pollution could cause ischemic stroke (Thomson et al., 2007) and neurodegeneration (Calderon-Garciduenas et al., 2008a, 2008b). Air pollution can trigger the release of cytokines, which are major neuro-inflammatory agents and act as a communication links between the brain and peripheral immune system. Triggering of pro-inflammatory signals in peripheral organs such as the lungs, liver, and the cardiovascular system results in a systemic innate immune response. Blood cytokine concentrations of IL-1β, IL-6, and granulocyte-macrophage colony-stimulating factor are elevated, while also elevating immune cell populations of marrow-derived neutrophils and monocytes in both human and animal models (Mills et al., 2009).

Cytokines are able to cross the BBB to produce neuro-inflammation, neurotoxicity, and vascular damage (Ling et al., 2006; Manousakis et al., 2009; Perry et al., 2007; Qin et al., 2007; Rivest et al., 2000). Ultrafine PM (UFPM) can penetrate into neurons and the brain through several routes, producing direct air pollution-induced neuro inflammation by activation of the brain's immune system (Block & Calderon-Garciduenas, 2009; Valavanidis et al., 2008). Experimental inhalation of UFPM in rodents has revealed efficient translocation of these nano-sized particles in the systemic circulation and the brain (Nemmar & Inuwa, 2008; Oberdorster et al., 2004). UFPM is able to cross traditional barriers in the lungs, be carried by erythrocytes, cross the BBB, and finally end up in neurons (Block & Calderon-Garciduenas, 2009).

The nasal epithelium is another portal of entry for UFPM, and inhaled particles can be traced through the nasal olfactory pathway, trigeminal nerves, brainstem, and hippocampus (Wang et al., 2007, 2008). Once in the brain, UFPM may trigger an immune response in two ways: (1) being made up of nanocarbon-based particles, UFPM adsorbs toxic organic compounds that are easily delivered into the central nervous system, in the same way as designer nanoparticles deliver custom drugs. Metals, hydrocarbons, and lipopolysaccharides have been associated with UFPM-adsorbed chemical-induced neuro-inflammation; and (2) UFPM themselves may stimulate innate immunity in the brain, and rodent studies have shown that carbon black particles (model of UFPM without adsorbed compounds) elevate pro-inflammatory cytokines (IL-1β, TNF-α, IFN-γ) in the olfactory bulb (Tin Tin Win et al., 2008) and induce activation of microglial cells (Long et al., 2007).

Sometimes, emissions generated from industry and vehicles are able to react with each other to form ground-level ozone, which is a strong oxidizing agent when inhaled. Ozone reacts with proteins and lipids, inducing oxidative stress. It is unlikely that ozone physically reaches the brain, as it is a reactive molecule with a short half-life, but breakdown products can reach the brain, and studies in rats have shown that chronic exposure to ozone leads to brain lipid peroxidation, dopaminergic neuron death, and neuron morphological damage (Angoa-Perez et al., 2006; Guevara-Guzman et al., 2009; Pereyra-Munoz et al., 2006). Astrocytes located near brain capillaries in rats have also been shown to have enhanced ozone-induced expression of pro-inflammatory factors (Araneda et al., 2008).

9.4.3.2 Epigenetic Modulation

Some cellular/molecular changes brought about by air pollution are mediated through epigenetic regulation of transcription. Air pollutants such as polycyclic aromatic hydrocarbons (PAHs) including benzo[a]pyrene (B[a]P), by-products of incomplete fossil fuel combustion, and PM can induce hypomethylation and hypermethylation of specific genes associated with increased risk of neural impairments (Madrigano et al., 2012), cell death and survival mechanisms (Bellavia et al., 2013; Pavanello et al., 2010), and the immune system and inflammation (Breton et al., 2012). One study found air pollution induced hypomethylation in 58 CpG loci in different genes with a greater than 10 percent difference in children in a highly polluted area than in a non-polluted area (Rossnerova et al., 2013). PAH exposure was associated with hypomethylation in the promoters of MGMT, p53, and IL-6, genes involved in the cell death and survival mechanisms, and the immune system (Bellavia et al., 2013; Pavanello et al., 2010).

Another important gene that is able to detoxify PAHs, glutathione S-transferase Mu2 (GSTM2), is hypermethylated at its promoter region by PAHs, thus inhibiting its transcription, leading to suppressed detoxification processes (Tang et al., 2010). B[a]P induced promoter methylation of the retinoic acid receptor β2 (RAR-β2) gene (Ye & Xu, 2010), and also decreased acetylation of histone H3K14 associated with the steroid acute regulatory (StAR) protein (Liang et al., 2012). Increased levels of PM are also associated with decreased methylation within the CpG island of the inducible nitric oxide synthase (iNOS) gene, while increased methylation in endothelial nitric oxide synthase (eNOS) disrupted nitric oxide (a key biological messenger) production in the body (Breton et al., 2012). One study also found carbon black to be significantly associated with aberrant methylation patterns in 31 genes (Sofer et al., 2013).

9.4.4 Gastrointestinal Exposure

The gut microbiota play a primary role in gastrointestinal (GI) physiology and is able to modulate microbiota–gut–brain axis signaling (Vuong & Hsiao, 2017). Animal studies have demonstrated the importance of the establishment of symbiotic and commensal bacterial colonies in the gut, and germ-free rodents have been shown to exhibit altered behavior, most often exhibiting reduced social exploration (Arentsen et al., 2015; Crumeyrolle-Arias et al., 2014; Desbonnet et al., 2014) and abnormal transcriptomic profiles across the frontal cortex, striatum, amygdala, and hippocampus (Diaz Heijtz et al., 2011). Altered behavior can be restored by the manipulation of the microbiome by probiotic treatment (Coiro et al., 2015; Hsiao et al., 2013; Shi et al., 2003). In humans, the gut microbiota establish themselves within the first two years after birth (Diaz Heijtz et al., 2011), and disruption in this process can lead to atypical interaction of microbiota with gut-associated immune cells and the enteric nervous system (Quigley, 2016). Atypical gut microbiota may trigger an immune reaction, stimulating the body's immune cells to release inflammatory cytokines, or in certain cases, compounds released by microbiota may enter the bloodstream, and some of these compounds, such as γ-aminobutyric acid (GABA) and 5-hydroxytryptamine (5-HT), are neurotransmitters that can directly affect the central nervous system and alter behavior (Li et al., 2017; Quigley, 2016).

Large numbers of GI problems and microbiota anomalies have been reported in individuals with autism. The most consistent GI observations have been diarrhea, constipation, and abdominal pain, as well as the occurrence of "leaky gut." The Childhood Autism Risks from Genetics and Environment (CHARGE) study showed greater association of GI symptoms with social withdrawal, stereotypy, irritability, and hyperactivity (Chaidez et al., 2014). Fecal bacterial profiling in individuals with autism has revealed a greater abundance of certain bacterial species such as *Clostridium* (Finegold et al., 2002; Parracho et al., 2005; Song et al., 2004), *Sutterella* (Wang et al., 2013), *Lactobacillus*, and *Desulfovibrio* (Tomova et al., 2015), with reduced *Bacteroidetes/Firmicutes* ratio (Tomova

et al., 2015), *Prevotella*, *Coprococcus*, and *Veilonellaceae* (Kang et al., 2013; Williams et al., 2011). This section discusses the mechanisms and effects of gut-mediated immune system modulation and microbiota-released metabolites, as well as their relationship with neuropsychiatric conditions such as autism.

9.4.4.1 Immune System Modulation

The gut microbiota are able to interact with the immune system, altering circulating levels of brain inflammatory factors associated with autism. Elevated IFN-γ, IL-β, IL-6, IL-12p40, TNF-α, monocyte chemoattractant protein-1 (MCP-1), transforming growth factor-β (TGF-β), and chemokine (C–C motif) ligand 2 (CCL-2) have been shown to be associated with autism (El-Ansary & Al-Ayadhi, 2014; Jyonouchi et al., 2001; Suzuki et al., 2011; Xu et al., 2015). Elevated bacterial endotoxin levels have also been found in individuals with autism (Emanuele et al., 2010). Cellular immune response has also been shown to be elevated in autism. For example, an animal model study has shown that maternal immune activation, which is known to be associated with the development of autism, results in the production of Th17 immune cells and secretion of IL-17α, but treatment with the commensal bacteria species *Bacteroides fragilis* is able to correct many of its adverse effects (Choi et al., 2016).

Experiments in animals have also shown that gut microbiota can affect neurogenesis and essential processes inside neurons, including transcriptional regulation of secondary messengers, synaptic transmission genes including postsynaptic density 95 and synaptophysin, and neurophysiological mechanisms (Diaz Heijtz et al., 2011). Long-term antibiotic treatment of adult mice can induce decreased neurogenesis, which can be rescued by probiotic treatment and voluntary exercise (Mohle et al., 2016; Sharon et al., 2016). This occurs as a result of a reduced number of monocytes associated with prolonged antibiotic treatment (Mohle et al., 2016). Gut microbiota have also been associated with atypical GABA receptor expression in prefrontal cortex neurons (Bravo et al., 2011). Several of these mechanisms that are highly correlated with autism are yet to be studied in humans.

9.4.4.2 Gut Microbiota-Released Metabolites Activity

Gut microbiota are known to release metabolites that can directly affect brain function. These include short-chain fatty acids (SCFAs), phenol compounds, and free amino acids (FAAs), peptidoglycan, and lipopolysaccharide (Li et al., 2017; Sharon et al., 2016). For example, elevated SCFAs have been found in the fecal matter of children with autism (Wang et al., 2012). In mice, subcutaneous administration of a SCFA – propionic acid – was shown to result in anxiety-like behavior (Foley et al., 2014). Another SCFA – butyrate – is an inhibitor of HDACs (De Vadder et al., 2014), similar to valproic acid, which is a major autism risk factor (Favre et al., 2013; Kuo & Liu, 2017). Studies using a valproate animal model for autism have demonstrated dysregulation of gut microbiota (e.g., an increase in *Firmicutes/Bacteroidetes* species) and also elevation in butyrate levels (de Theije et al., 2014a, 2014b).

However, butyrate is also a ligand for certain G-protein-coupled receptors and is involved in ATP production. It may have a neuroprotective effect (Bourassa et al., 2016), and in one autism animal model study using BTBR mice, butyrate has been shown to improve autism-associated behavioral deficits (Kratsman et al., 2016). Gut bacteria also synthesize phenol compounds such as *p*-cresol from tyrosine, which is easily absorbed through the GI tract and has been postulated to worsen the symptoms of autism (Persico & Napolioni, 2013). High levels of FAAs are observed in fecal samples of children with autism, particularly glutamate, which is easily absorbed and acts as a neurotransmitter. An excess of glutamate may impact behavior and lead to neuronal death (Ding et al., 2017; Sheldon & Robinson, 2007). Peptidoglycans from bacterial cell walls and lipopolysaccharides, which are bacterial endotoxins, are also able to enter through the gut and activate immune responses and atypical neural pathways (Rieder et al., 2017).

Studies so far have identified numerous ways by which environmental pressures increase autism susceptibility and generally increase the propensity for mental health issues. The environmental agents act via multiple transcriptional and epigenetic mechanisms either through direct interaction with molecular machinery or through the activation of immune-responsive cellular pathways and oxidative stress. Most findings have unfortunately been made in animal models that do not properly recapitulate human neurodevelopment or neuropsychiatric conditions. To find out how transcriptional and epigenetic mechanisms imprint gene expression signatures on human cell lineages, research will need to be done on a comparable human model system.

Section 9.5 discusses the state of current research that includes animal-based model systems and the future of neuropsychiatric research using iPSC technology.

9.5 METHODS FOR STUDYING TRANSCRIPTIONAL AND EPIGENETIC REGULATION

Studying the interaction between protein and DNA has unique challenges. This is why we know very little about how proteins control gene expression regulation, which in some cases are also inherited from one cell generation to the next. Recent findings are, however, pointing toward a greater role of these gene-regulatory proteins, known as transcription and epigenetic factors. For example, there is greater acknowledgment of the fact that epigenetic gene regulation during development gives rise to the plethora of different cell types of the body from just one starting embryonic stem cell containing only one set of genomic information. The concerted regularity in which genes are "switched on" and "switched off" in different cell types during development, or the aberrant switching on/off of nonspecific genes in cancer cells, has also given rise to the question of how much of an impact the environment can play on this cellular machinery. This section deals with how genes are regulated inside cells and the methods that can capture this unique phenomenon (Table 9.2) to enable the study of typical and disease-state epigenetic gene expression regulation.

9.6 PROTEINS AND RNA INVOLVED IN EPIGENETIC REGULATION

The primary epigenetic proteins in the eukaryote cells are histones. There are four classes of histones: H1, H2 (H2A, H2B), H3, and H4. Histones H2A, H2B, H3, and H4 form

Table 9.2 Methods for studying epigenetic regulation.

Method	Epigenetic mechanism studied	References
Chromatin immunoprecipitation sequencing (ChIP-seq)	Protein–DNA interaction, especially histone–DNA	Hayashi, 1975; Johnson et al., 2007; Kuo & Allis, 1999; Madisen et al., 1998; Sen et al., 1976,
Chromosomal conformation capture (3 C)	Structural properties and spatial conformation of chromosomes	Dekker et al., 2002
Hi-C	Modification of 3 C method using deep sequencing to identify protein-linked DNA sequences	Belton et al., 2012; Hsieh et al., 2016; Lieberman-Aiden et al., 2009; Nagano et al., 2013; Ramani et al., 2017,
Chromatin interaction analysis by paired-end tag (ChIA-PET)	Combines ChIP and 3 C to enable detection of short- and long-range chromatin interactions by protein of interest	Davies et al., 2017; Fullwood & Ruan, 2009; Fullwood et al., 2009,

Continued

Table 9.2 Cont.

Method	Epigenetic mechanism studied	References
Chip-based methylation arrays	Targeted approach to studying methylated cytosines in the genome	Infinium BeadChip arrays from Illumina
Whole-genome bisulfite sequencing (WGBS)	Detects methylated CpG islands in the whole genome	Gu et al., 2011
Reduced representation bisulfite sequencing (RRBS)	Targeted detection of methylated CpG islands by restriction digestion using MspI	Gu et al., 2011; Guo et al., 2015; Nwankwo & Wilson, 1988; Xu et al., 2005,
MethylC-seq	All methylated cytosines, including non-CpG island methylated cytosines	Urich et al., 2015
Chromatin affinity purification with mass spectrometry (ChAP-MS)	Collective assembly of gene expression regulators at DNA regulatory regions	Byrum et al., 2012
Assay for transposase-accessible chromatin with high-throughput sequencing (ATAC-seq)	Open chromatin regions that can be accessed by the hyperactive Tn5 transposase	Buenrostro et al., 2013

an octamer around which DNA is packaged into structural units called nucleosomes (Vardabasso et al., 2014). DNA packaged in this way is inactive. However, histones are able to undergo covalent modifications to unpack the DNA in a precise, coordinated manner to bring about long-term changes in gene expression. Such changes are often inherited from one cell generation to the next (Feil & Fraga, 2012; Henikoff & Shilatifard, 2011; Law & Jacobsen, 2010). Another class of epigenetic proteins make direct modifications on the DNA to influence gene expression regulation. This class of proteins, called DNMTs, is known to methylate cytosine found as part of CpG dinucleotides (Law & Jacobsen, 2010). Methylated CpG islands (mCG) can then bind with methyl-CpG-binding domain proteins MBD1, MBD2, MBD4, and MeCP2, which mediate the repression of gene expression (Stroud et al., 2017). More recent studies have revealed a non-CG methylation site, a CA sequence that can also be methylated at the cytosine (mCA). CA and CG methylation by methyltransferases seems to occur at a high rate and at similar levels in adult neurons, and DNMT3A seems to be primarily responsible (He & Ecker, 2015; Lister et al., 2013; Stroud et al., 2017).

Other than epigenetic proteins, there are also RNA molecules that are able to regulate gene expression by inhibiting translation or degrading cytoplasmic mRNA (Castel & Martienssen, 2013). This form of gene regulation is known as RNA interference (RNAi) and involves microRNAs, small interfering RNAs, and PIWI protein-interacting RNAs. These regulatory RNA molecules are processed by enzymes such as Dicer and Drosha. These enzymes are known to interact with the environment and epigenetic factors to process this form of gene regulation through RNAi, which can be inherited from one cell generation to the next.

In mammals, many of these epigenetic factors have been found to be present at high levels during early embryonic stages of development, which is a possible explanation for cells at various stages of pluripotency being especially susceptible to environmental triggers. The stem cell differentiation process also involves systematic transitions from one epigenetic state to the next, and this most likely makes it vulnerable to epigenetic modifications from external cues (Feil & Fraga, 2012). The endocrine system and its disruptors are also able to influence germ line epigenetic modifications that can be perpetuated through transgenerational inheritance.

9.6.1 Molecular Methods for Epigenomic Studies

Studying epigenetic regulation of gene expression involves studying protein–DNA interactions, or more precisely, studying the interaction of epigenetic factors with the DNA, then identifying the binding DNA sequence to locate the gene that the epigenetic factor was regulating. But proteins such as histones have an extremely dynamic relationship with the DNA, and only transiently interact with the genome through weak chemical bonds. To be able to extract these proteins, they need to be cross-linked with the DNA, then identified using an antibody tag specific to the protein of interest. Cross-linking agents such as formaldehyde have been successfully used for this purpose, and this method of cross-linking protein–DNA to study epigenetic interactions is known as chromatin immunoprecipitation (ChIP) (Kuo & Allis, 1999). These types of ChIP cross-linking studies have been performed since the 1970s (Hayashi, 1975; Sen et al., 1976). However, the interest in ChIP waned in the 1980s due to the lack of quality antibodies. More selective antibodies were subsequently developed for looking at various configurations of histones (Kuo & Allis, 1999; Madisen et al., 1998). With the emergence of high-throughput sequencing, the sequence of the binding DNA fragment could be easily identified, and soon ChIP-seq became one of the most widely used methods for studying epigenetics (Johnson et al., 2007).

Using the same protein–protein and protein–DNA cross-linking abilities of formaldehyde, methods such as chromosomal conformation capture (3 C) have been devised to study the structural properties and spatial conformation of chromosomes (Dekker et al., 2002). While local chromosomal conformations (using ChIP) have been established as playing a big role in gene expression regulation, global structural features play lesser-known but equally important roles in gene expression and DNA recombination (Dekker et al., 2002). In this method, DNA fragments covalently linked together with proteins are purified and their sequences investigated using polymerase chain reaction (PCR), or more recently using deep sequencing in a high-throughput method known as Hi-C (Belton et al., 2012). Hi-C data can reveal the global activation state of the chromatin (regions of open or closed chromatin correspond to active or inactive states, respectively). Hubs of activation can then be identified and compared with other data sets to reveal how the genome is rearranged in response to a treatment, pathology, or developmental stage (Belton et al., 2012; Lieberman-Aiden et al., 2009). More recently, Hi-C conformations have been determined in single cells (Nagano et al., 2013; Ramani et al., 2017) and detection of long-range contacts has been considerably improved (Hsieh et al., 2016). A new method has been developed as an offshoot of Hi-C, known as chromatin interaction analysis by paired-end tag (ChIA-PET), which combines ChIP and 3 C to potentially enable detection of all kinds of sites bound by a protein of interest (Davies et al., 2017; Fullwood et al., 2009).

ChIA-PET can be used for the de novo identification of global chromatin interactions. This method is able to detect transcriptional regulation through long-range interaction of chromatin. Using this technique, it is possible to study transcription and epigenetic factors that bind to regulatory DNA elements that are located far away from the promoter of the genes they regulate. Highly specific monoclonal antibodies are used to pull down transcriptional epigenetic proteins of interest. The DNA bound to the protein complexes can then be digested enzymatically or using sonication, and then sequenced to reveal the nature of the protein–DNA interaction. The data output would be a combination of ChIP-seq and Hi-C, making this a strong tool for understanding global gene regulatory networks inside cells (Fullwood & Ruan, 2009; Fullwood et al., 2009).

The study of methylated cytosine residues in the genome, however, requires a different set of techniques compared to those described above. Cytosine methylation is quantified using sodium bisulfite treatment-based strategies, which convert unmethylated cytosines to uracil (which later on are converted to thymine using PCR amplification). Sodium bisulfite treatment does not affect methylated cytosines, allowing quantification of DNA methylation by the estimation of the cytosine-to-thymine ratio at known CpG sites (Ulahannan & Greally, 2015).

There are currently two approaches to isolating methylated cytosine as part of CpG sites (mCG). The first method involves a targeted approach using chip-based arrays such as the Infinium BeadChip arrays from Illumina. These arrays are designed to capture promoter CpG islands associated with RefSeq genes. The limitation of chip-based methods is that other biologically significant CpG regions on the genome cannot be detected. The second method involves taking a whole-genome bisulfite sequencing (WGBS) approach, or a modified version that allows improved capture of CpG islands, known as reduced representation bisulfite sequencing (RRBS) (Gu et al., 2011). In RRBS, CpG islands are isolated by briefly restriction digesting genomic DNA with the MspI restriction enzyme (Nwankwo & Wilson, 1988), which targets and cleaves CpG-rich regions (Xu et al., 2005). More recently, RRBS has been performed in single cells by reducing the loss of DNA associated with the purification steps. This is achieved by adjusting the buffer system and reaction volumes and combining the whole process into a single tube reaction (Guo et al., 2015).

The main caveat of the chip-based, WGBS-based, or RRBS-based methods is that they are not able to detect non-CpG island-methylated cytosines such as methylated CA sequences (mCA). This can be addressed using another method called MethylC-seq (Urich et al., 2015). In this method, complete genomic DNA is fragmented, processed and ligated to adapter sequences with methylated cytosines. The fragments are then treated with sodium bisulfite, followed by PCR amplification using primers specific for the adapters. The bisulfite-treated DNA fragments are sequenced then base-called against a reference genome, which reveals all of the thymine bases to be either a thymine or an unmethylated cytosine based on their location on the reference genome, while each cytosine base indicates the methylation status within the fragment. This method is more comprehensive than RRBS and DNA methylation arrays; however, it is also more costly, and thus is not a powerful method for studying large numbers of samples (Urich et al., 2015).

Additionally, there are some specialized methods for acquiring in-depth understanding of epigenetic gene regulation. A proteomics-based method known as chromatin affinity purification with mass spectrometry (ChAP-MS) can detect how gene expression regulators collectively assemble at DNA regulatory regions, something that cannot be detected by the ChIP-based methods described above, which can only detect single protein–DNA interactions (Byrum et al., 2012). By using mass spectrometric detection methods, ChAP-MS is also able to identify sequences of chromatin-associated proteins in an unbiased manner. Another method known as assay for transposase-accessible chromatin with high-throughput sequencing (ATAC-seq) takes a completely novel approach to epigenomic profiling by studying open chromatin regions. In ATAC-seq, accessible regions of the genome are scanned using a hyperactive Tn5 transposase. These genomic regions can then be sequenced to map regions of regulatory regions, transcription factor activity, and nucleosome positioning (Buenrostro et al., 2013). The main advantage of using ATAC-seq when compared to other epigenomic methods, is that it does not require millions of cells as the starting material, nor complex and time-consuming sample preparation steps. This makes it a powerful tool for studying epigenomics in clinical samples (Buenrostro et al., 2013).

From this section, we can see that epigenetic gene regulation is an extremely complex process that involves several protein and RNA factors. However, several methods have so far been developed to study this phenomenon from different angles. The first major method involves cross-linking DNA-binding proteins with their genomic targets and studying the protein sequence and DNA sequence. This reveals different conformations of epigenetic factors during gene expression regulation. The second major method involves sodium bisulfite treatment of genomic DNA to reveal methylation sites. This is able to reveal gene regulatory sequences across the whole genome. Different variations based on these basic principles balance depth with cost-effectiveness of experiments. These methods, when combined with the iPSC technology discussed in the Section 9.7, can transform our understanding of neurodevelopment.

9.7 ROLE OF IPSC-BASED MODELS IN THE FUTURE OF NEURO-EPIGENOMIC RESEARCH

The genome is a set of instructions that determine the fate of biological organisms. In the case of complex multicellular organisms such as humans, it is a challenge to understand how one set of instructions can yield hundreds of diverse types of cells that interact with each other to form dozens of different types of tissue. However, it is easier to understand that this might be made possible by expressing only a fraction of the genes in each type of tissue: those genes that are relevant to providing the tissue with its unique identity. But how do the cells determine which genes to express, and how is all of this done starting from a single cell – the embryo?

This phenomenon by which genes are selectively switched on or switched off, which are then inherited from one cell generation to the next, such that by studying the developmental stages one can observe the gradual evolution of cell types with evolving functions specific to a developmental stage until they reach their end states, has intrigued scientists for decades. The study of development, known as embryology, has until recently been performed exclusively in laboratory animals under stringent ethical guidelines, as embryos are the inception of all life-forms. But in the past decade there have been groundbreaking inventions now enabling researchers to grow embryo-like tissue in the lab from any individual by collecting skin biopsies, blood, or a few strand of hair. The cells that are "reprogrammed" as a result are known as iPSCs, and they share characteristics with the stem cells that form the embryo.

iPSCs, once generated in the lab, can be artificially induced to form any tissue of the body. This enables in-depth research into the development of different tissues. Nowhere has this method been more relevant than in the study of complex developmental conditions such as autism. During this first wave of research on developmental conditions, many research groups around the world have developed methods to differentiate iPSCs into neurons and three-dimensional brain tissue. Studies have been performed to characterize morphology, physiology, and gene expression in these neurons. These neurons recapitulate many of the characteristics of in vivo neurons. In the future, researchers are aiming to combine the neuronal differentiating methods with the epigenomic methods as described in Section 9.6 to unravel the cellular machinery that conducts the "switching on" and "switching off" of genes as they go through different stages of brain development. In this section, the iPSC technology is discussed in terms of how it is being utilized at present to understand brain development and the promise it holds for studying the epigenomics of development and complex developmental conditions in humans.

9.7.1 Role of the iPSC Technology in Developmental Neuroscience

The method of reprogramming adult cells to iPSCs was first described by Takahashi and Yamanaka in 2006 (Takahashi & Yamanaka, 2006). They first reported this in mouse fibroblasts that were reprogrammed into iPSCs by retroviral transduction using four transcription factors – OCT4, KLF4, c-MYC, and SOX2 – then subsequently applied the technology to human fibroblasts (Takahashi et al., 2007). The pluripotent nature of iPSCs meant that they could be differentiated into neurons by exposing them to the biochemical milieu that is known to push pluripotent cells into cells of neuronal lineage. Early neuronal differentiation studies involved the inhibition of bone morphogenetic protein and SMAD signaling to trigger neural induction (Chambers et al., 2009; Shi et al., 2012). The SMAD inhibition strategy has been used more recently to develop three-dimensional cerebral cortex-like structures of pyramidal neurons, known as human cortical spheroids (Pasca et al., 2015). Another independently developed method that recapitulates various stages of the developing brain as a whole does not use the SMAD inhibition strategy to direct differentiation toward a neuronal lineage. This method is able to generate complex heterogeneous tissues by improving growth conditions and providing the environment that is necessary to influence development through time-dependent intrinsic cues associated with the developing brain (Lancaster et al., 2013, 2017).

One of the main driving forces behind developing neural differentiation protocols has been enabling the study of neurodevelopmental conditions. Not surprisingly, syndromic forms of autism, including Fragile X syndrome (Bar-Nur et al., 2012; Sheridan et al., 2011), Rett syndrome (Amenduni et al., 2011; Ananiev et al., 2011; Cheung et al., 2011; Kim et al., 2011; Marchetto et al., 2010; Muotri et al., 2010), and Timothy syndrome (Krey et al., 2013; Pasca et al., 2011), were a few of the first neurological conditions to be explored using these methods. An iPSC model of idiopathic autism has also been established during the first wave of studies (DeRosa et al., 2012; Griesi-Oliveira et al., 2015; Marchetto et al., 2017; Mariani et al., 2015), with one of the studies (Mariani et al., 2015) having adopted the cerebral organoid method (Mariani et al., 2015). Genes such as MeCP2, mutations in which are associated with Rett syndrome, and DNMT3A (Sanders et al., 2015) are strongly associated with an autism outcome. MeCP2 and DNMT3A are also DNA methylation factors that play important roles in transgenerational epigenetic gene regulation. DNMT3A is also known to mediate methylation of non-CpG island cytosines as described in Section 9.6. Their mechanisms of action or downstream genes are not well characterized, and how mutations in these epigenetic factors might skew gene expression patterns toward a disease-causing state is not well understood. Whether the presence of mutations in these factors results in atypical responses to environmental insults during brain development is also not known. However, with the development of new epigenomic methods and the falling costs of next-generation sequencing, we could soon use iPSCs to start unraveling some of these mechanisms.

9.7.2 Current Research Shows Atypical Gene Expression Pathways with No Identifiable Genetic Association in Autism

An epigenomic study in autism is still cost prohibitive because autism is a complex condition, and to achieve statistical significance, a large sample size is recommended. Such studies have not been attempted using postmortem brain tissue, as it is a rare resource with highly variable tissue quality. Epigenomic studies using iPSCs have also been avoided, as it was not known whether the epigenetic signature of source tissue is completely reset after reprogramming into iPSCs. However, it has been established that iPSCs are functionally identical to embryonic stem cells, and there is accumulating evidence that their chromatin states are also fully recalibrated to that of a pluripotent cell

(Mascetti & Pedersen, 2016; Papp & Plath, 2013; Rouhani et al., 2014).

However, RNA sequencing and microarray-based methods have revealed significant differences in gene expression patterns in convergent pathways associated with autism. This suggests non-genomic factors influencing regulation of gene expression in autism. Voineagu et al. (2011) used autism postmortem brain tissue to undertake gene expression studies using microarrays. They discovered two major dysregulated mechanisms in the autism brain: (1) synaptic function, vesicular transport, and neuronal projection (APMB_asdM12); and (2) immune system and inflammatory, astrocyte, and microglia responses (APMB_asdM16). A 2016 study undertaking RNA-seq using autism postmortem brain tissue by Parikshak et al. (2016) demonstrated attenuated cortical patterning associated with the autism brain (gene modules: ACP_asdM5, ACP_asdM13, and ACP_asdM14). A third study using developing postmortem brain tissue showed enrichment of autism-associated pathways at more advanced stages of development compared to control (gene modules: dev_asdM2, dev_asdM3, dev_asdM13, dev_asdM16, and dev_asdM17) (Parikshak et al., 2013).

Furthermore, one study using autism iPSC-derived organoids identified significant differences in developmental trajectory between autism and control organoids (Mariani et al., 2015). A more recent study using iPSC-derived neurons also identified several autism-associated pathways (Marchetto et al., 2017). A study from our group undertook RNA-seq using iPSC-derived neurons and used similar analysis methods as the above postmortem and iPSC-based studies. We found a very high correlation of gene expression pathways in our iPSC-derived neurons with the gene expression pathways detected in those studies (Figure 9.3) (Adhya et al., 2018). We also found poor correlation of gene expression with exome single-nucleotide polymorphisms found in the participants.

The postmortem brain studies and autism iPSC studies, including one by our group, point toward a global non-genomic gene regulatory mechanism at play in autism, one that is responsible for producing the autism-associated phenotypes irrespective of genetic background. Study of the epigenomics of autism might thus provide crucial insights into why such neurodevelopmental conditions have such heterogeneous genetic backgrounds.

9.7.3 The Significance of Future Clinical Neuro-epigenomic Research Using iPSCs

At present, the epigenetics of neurodevelopment is not well understood. Many of the methods described in Section 9.5 have been developed in the last five years. It has also only been a decade since the Nobel Prize-winning iPSC technology was first published, and most

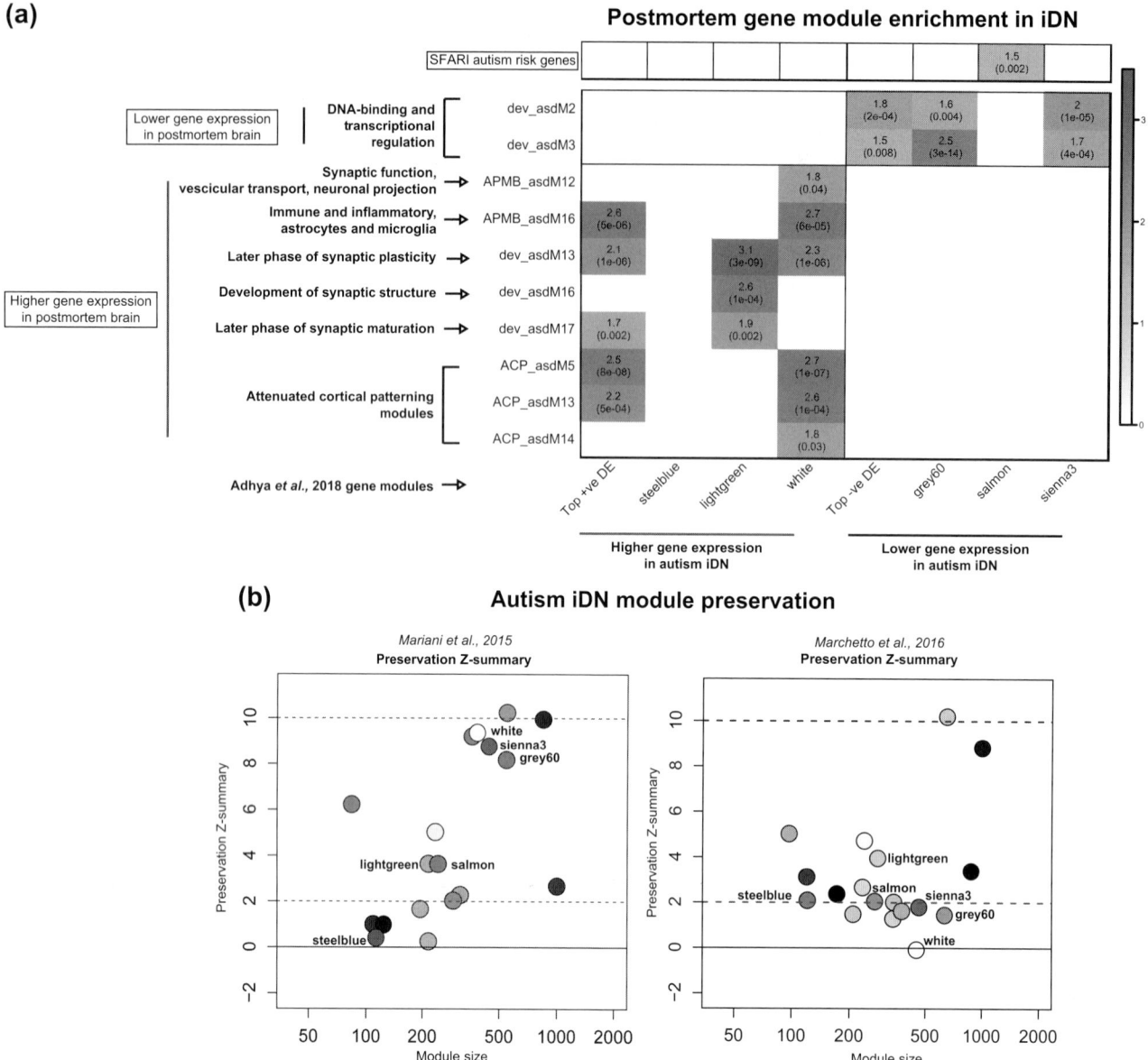

Figure 9.3 Enrichment analysis of autism-associated pathways and modules using induced pluripotent stem cell-derived neurons (iDN). (a) Enrichment of selected autism postmortem gene modules and SFARI autism risk gene list in iDN gene modules. Odds ratios (ORs) after logistic regression have been shown. Only ORs > 1.5 have been shown (*p*-value in parentheses). (b) Module preservation of autism minibrain gene modules in iDN gene modules from this study. DE = differential expression.

neuronal differentiation methods are still under various stages of development. However, as data from iPSC-derived neuronal tissue start accumulating, it is becoming clear that iPSCs are the starting material of choice for characterizing the wide array of phenotypes associated with neurodevelopmental conditions. Large sequencing studies have revealed hundreds of single-nucleotide polymorphisms with varying degrees of penetrance to be associated with autism (Bourgeron, 2015; O'Roak et al., 2011, 2012; Sanders et al., 2015). This has made research into a complex condition such as autism extremely challenging. Subsequent transcriptomic studies have reported quite a different story, and it has now been established that there are a set of

convergent gene expression pathways that are associated with autism (Marchetto et al., 2017; Mariani et al., 2015; Parikshak et al., 2013, 2016; Voineagu et al., 2011). These pathways have been repeatedly observed in postmortem brain tissue and iPSC-derived brain tissue, giving rise to additional questions on how genetic mutations interact with cellular gene regulatory systems to manifest a disease state. To learn how these cellular gene regulatory systems work, researchers need to take an in-depth look at the complicated epigenetic machinery of the developing brain. As primary brain tissue from sufferers cannot be obtained, and as postmortem brain tissue is a rare commodity and acquiring it involves many ethical dilemmas, the best solution for researchers is to model

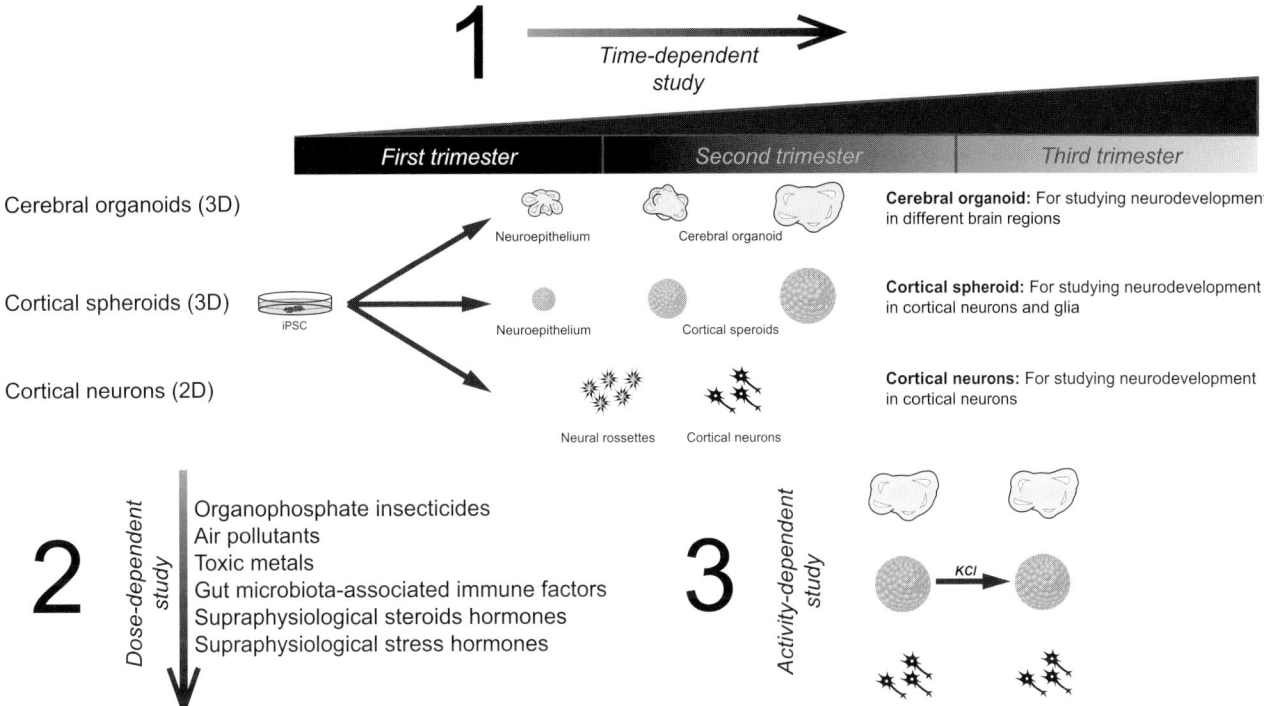

Figure 9.4 Two- and three-dimensional neuronal approaches that can be taken to study neurodevelopment using iPSCs. (1) Time-dependent study, to demonstrate cellular/molecular trajectories during neurodevelopment. (2) Dose-dependent study, to demonstrate effects of environmental agents during stages of neurodevelopment. (3) Activity-dependent study, to mimic the effect of neuronal activity and observe neuronal function during development.

neurodevelopmental conditions such as autism in vitro using iPSCs.

By using two- and three-dimensional neuronal differentiation protocols, early neurodevelopmental stages can easily be recapitulated (Figure 9.4). This allows the study of epigenetics of early stem cell differentiation, when it is most dynamic (Feil & Fraga, 2012). Three approaches can be taken to study neurodevelopment using iPSCs: (1) by undertaking a time-dependent study, one can track epigenetic transitions from one cell state to the next; (2) by adding environmental stimuli associated with autism, one can study epigenetic perturbations at these early stages to track divergences of epigenetic signatures; and (3) by artificially inducing neuronal activity in matured neurons (e.g., by administration of potassium chloride), one can track activity-dependent signaling in autism-associated epigenetic states, and this can mimic cellular differences in adult brain function in autism.

The iPSC technology is able to provide novel insights into human neurodevelopment, something that has not been accurately recapitulated in animal models. By studying epigenetics of transgenerational inheritance during neurodevelopment, we can understand the interaction between the genome and epigenetic factors, and how the environment might be able to cause perturbations to this interaction.

9.8 CONCLUSIONS

Over the last decade, there have been a growing number of studies showing a role of the environment in autism etiology, the development of the iPSC method for studying neurological conditions in vitro, and the development of various epigenomic methods to study transgenerational epigenetic inheritance. The convergence of research in these three fields has opened up a world of research opportunities to study how genes are regulated epigenetically during the progression of neurodevelopmental conditions.

Epigenomic studies in neuropsychiatric conditions have been performed using animals as well as human tissue. Blood and postmortem brain tissue from humans have been used so far. All of these tissue sources have major caveats for studying neuropsychiatric conditions. The animal brain is not able to recapitulate all of the stages of human brain development. Animals also do not share similar behavioral phenotypes with humans. Complex neuropsychiatric conditions such as autism and schizophrenia are difficult to model in animals, as they are heterogeneous conditions with hundreds of associated risk genes involved. Human tissue material such as blood and postmortem brain tissue can be used instead. However, there are questions over the validity of using blood as a biomarker for the brain. Postmortem brain tissue from individuals with neuropsychiatric conditions is a rare resource, and thus all such studies suffer from low sample sizes. In addition, it is not possible to study development in neuropsychiatric conditions with postmortem brain tissue, as most of the available brain tissue is from adult brains when it is fully mature. Furthermore, atypical gestational development cannot be studied in prenatal postmortem brain resources, as conditions like autism can only be diagnosed after birth (Bakulski et al., 2016). There have been a number of studies looking at typical epigenetic signatures of the brain and atypical epigenetic signatures in autism using animal models (Kim et al., 2016; McCormick et al., 2017), blood (Hannon et al., 2015; Naumova et al., 2018), and postmortem brain tissue (Nardone et al., 2017; Sun et al., 2016). However, all of these studies share the same caveats as described above.

iPSCs can be used to overcome many of the challenges discussed above. They can be differentiated using various neural differentiation protocols (two- or three-dimensional) to an early or late developmental stage of the brain, then used to investigate epigenetic signatures associated with healthy and diseased states associated with neuropsychiatric conditions. iPSCs also share the same genetic background with the individuals suffering from the particular condition, making it easier to correlate aberrant gene regulation with their genetic predispositions. iPSC-derived neural tissue can also be exposed to environmental toxins to gain insights into how environmental stimuli perturb transgenerational inheritance processes during neural development at very early stages when epigenetic gene regulation is most dynamic.

However, such studies are not without challenges. First, undertaking next-generation sequencing experiments, such as those required for RRBS, and MethylC-seq on a statistically significant sample size can be extremely resource intensive. Cellular reprogramming and neural differentiation also involve significant lab and reagent costs. Second, current neural differentiation methods have some issues that prevent their use as preclinical models of conditions. They can currently be developed until mid-second trimester, although theoretically they can be differentiated for longer. It is also not certain whether neurons developed in this manner show physiological activity similar to in vivo neurons or brain tissue. iPSC-derived three-dimensional neural tissue can also be quite heterogeneous, which compounds issues during analysis. Costs can also increase many times over as a result, as more experiments need to be done to achieve statistically significant observations.

Nevertheless, most issues related to iPSCs are technical in nature and are already being addressed. As the cost of next-generation sequencing falls, methylation sequencing costs will also fall. The scientific community is working closely with biotech companies to develop more efficient reprogramming methods. Neuronal differentiation methods are also improving, and various strategies are being implemented to reduce the heterogeneity of neuronal differentiation. It is understood that one of the main causes of heterogeneity is the health and cell-cycle stage of the starting stem cells. Studies using autism iPSC-derived neural tissue have already demonstrated electrophysiological activity (Marchetto et al., 2017; Mariani et al., 2015), and recent studies in three-dimensional cerebral organoids have shown the ability of rod cells to respond to light stimuli (Quadrato et al., 2017), thus demonstrating physiological activity in neurons derived from iPSCs.

Autism is already associated with environmental risk factors, although we have repeatedly noted throughout this chapter that these await independent replication. Autism is also associated with epigenetic mechanisms. The use of the iPSC technology with epigenomic methods for studying autism is expected to reveal different atypical patterns of transgenerational inheritance associated with autism. It is possible that neuropsychiatric conditions other than autism also have associated environmental risk factors. These kinds of studies would also provide information about the environmental risk factors associated with other neuropsychiatric conditions. Although gene therapy is often discussed as the future of personalized medicine, it is fraught with ethical issues, and it may be the case that a less invasive, lower-risk therapeutic strategy might constitute personalized epigenetic treatments that could recalibrate atypical epigenetic signatures and compensate for dysregulated gene expression without the need to make changes to the genome. Again, there is a need to start to discuss the ethical issues surrounding such approaches, and safety considerations remain of paramount importance.

9.9 ACKNOWLEDGMENTS

The authors were supported by the Autism Research Trust, the Chinese University of Hong Kong (CUHK), and EU-AIMS during the period of this work.

REFERENCES

Abel, T., & Zukin, R. S. (2008). Epigenetic targets of HDAC inhibition in neurodegenerative and psychiatric disorders. *Current Opinion in Pharmacology*, **8**, 57–64.

Adhya, D., Swarup, V., Nowosiad, P., et al. (2018). Shared gene co-expression networks in autism from induced pluripotent stem cell (iPSC) neurons. *bioRxiv*, doi.org/10.1101/349415.

Amenduni, M., De Filippis, R., Cheung, A. Y., et al. (2011). iPS cells to model CDKL5-related disorders. *European Journal of Human Genetics*, **19**, 1246–1255.

Ananiev, G., Williams, E. C., Li, H., & Chang, Q. (2011). Isogenic pairs of wild type and mutant induced pluripotent stem cell (iPSC) lines from Rett syndrome patients as in vitro disease model. *PLoS ONE*, **6**, e25255.

Angoa-Perez, M., Jiang, H., Rodriguez, A. I., et al. (2006). Estrogen counteracts ozone-induced oxidative stress and nigral neuronal death. *Neuroreport*, **17**, 629–633.

APA (2013). *Diagnostic and Statistical Manual of Mental Disorders (DSM-5®)*. Washington, DC: American Psychiatric Association Publishing.

Araneda, S., Commin, L., Atlagich, M., et al. (2008). VEGF over-expression in the astroglial cells of rat brainstem following ozone exposure. *Neurotoxicology*, **29**, 920–927.

Arentsen, T., Raith, H., Qian, Y., Forssberg, H., & Diaz Heijtz, R. (2015). Host microbiota modulates development of social preference in mice. *Microbial Ecology in Health and Disease*, **26**, 29719.

Audouze, K., & Grandjean, P. (2011). Application of computational systems biology to explore environmental toxicity hazards. *Environmental Health Perspectives*, **119**, 1754–1759.

Bakulski, K. M., Halladay, A., Hu, V. W., Mill, J., & Fallin, M. D. (2016). Epigenetic research in neuropsychiatric disorders: The "tissue issue". *Current Behavioral Neuroscience Reports*, **3**, 264–274.

Bal, W., Liang, R., Lukszo, J., et al. (2000). Ni(II) specifically cleaves the C-terminal tail of the major variant of histone H2A and forms an oxidative damage-mediating complex with the cleaved-off octapeptide. *Chemical Research in Toxicology*, **13**, 616–624.

Bao, A. M., & Swaab, D. F. (2011). Sexual differentiation of the human brain: relation to gender identity, sexual orientation and neuropsychiatric disorders. *Frontiers in Neuroendocrinology*, **32**, 214–226.

Bar-Nur, O., Caspi, I., & Benvenisty, N. (2012). Molecular analysis of FMR1 reactivation in fragile-X induced pluripotent stem cells and their neuronal derivatives. *Journal of Molecular Cell Biology*, **4**, 180–183.

Baron-Cohen, S., Lombardo, M. V., Auyeung, B., et al. (2011). Why are autism spectrum conditions more prevalent in males? *PLoS Biology*, **9**, e1001081.

Baron-Cohen, S., Auyeung, B., Norgaard-Pedersen, B., et al. (2015). Elevated fetal steroidogenic activity in autism. *Molecular Psychiatry*, **20**, 369–376.

Basu, S. N., Kollu, R., & Banerjee-Basu, S. (2009). AutDB: A gene reference resource for autism research. *Nucleic Acids Researchearch*, **37**, D832–D836.

Becerra, T. A., Wilhelm, M., Olsen, J., Cockburn, M., & Ritz, B. (2013). Ambient air pollution and autism in Los Angeles County, California. *Environmental Health Perspectives*, **121**, 380–386.

Bellavia, A., Urch, B., Speck, M., et al. (2013). DNA hypomethylation, ambient particulate matter, and increased blood pressure: findings from controlled human exposure experiments. *Journal of the American Heart Association*, **2**, e000212.

Belton, J. M., McCord, R. P., Gibcus, J. H., et al. (2012). Hi-C: A comprehensive technique to capture the conformation of genomes. *Methods*, **58**, 268–276.

Berger, S. L., Kouzarides, T., Shiekhattar, R., & Shilatifard, A. (2009). An operational definition of epigenetics. *Genes & Development*, **23**, 781–783.

Blanchard, K. S., Palmer, R. F., & Stein, Z. (2011). The value of ecologic studies: Mercury concentration in ambient air and the risk of autism. *Reviews in Environmental Health*, **26**, 111–118.

Block, M. L., & Calderon-Garciduenas, L. (2009). Air pollution: Mechanisms of neuroinflammation and CNS disease. *Trends in Neurosciences*, **32**, 506–516.

Bollati, V., & Baccarelli, A. (2010). Environmental epigenetics. *Heredity*, **105**, 105–112.

Borochov, N., Ausio, J., & Eisenberg, H. (1984). Interaction and conformational changes of chromatin with divalent ions. *Nucleic Acids Research*, **12**, 3089–3096.

Bourassa, M. W., Alim, I., Bultman, S. J., & Ratan, R. R. (2016). Butyrate, neuroepigenetics and the gut microbiome: Can a high fiber diet improve brain health? *Neuroscience Letters*, **625**, 56–63.

Bourgeron, T. (2015). From the genetic architecture to synaptic plasticity in autism spectrum disorder. *Nature Reviews Neuroscience*, **16**, 551–563.

Bravo, J. A., Forsythe, P., Chew, M. V., et al. (2011). Ingestion of *Lactobacillus* strain regulates emotional behavior and central GABA receptor expression in a mouse via the vagus nerve. *Proceedings of the National Academy of Sciences*, **108**, 16050–16055.

Breton, C. V., Salam, M. T., Wang, X., et al. (2012). Particulate matter, DNA methylation in nitric oxide synthase, and childhood respiratory disease. *Environmental Health Perspectives*, **120**, 1320–1326.

Buenrostro, J. D., Giresi, P. G., Zaba, L. C., Chang, H. Y., & Greenleaf, W. J. (2013). Transposition of native chromatin for fast and sensitive epigenomic profiling of open chromatin, DNA-binding proteins and nucleosome position. *Nature Methods*, **10**, 1213–1218.

Byrum, S. D., Raman, A., Taverna, S. D., & Tackett, A. J. (2012). ChAP-MS: A method for identification of proteins and histone posttranslational modifications at a single genomic locus. *Cell Reports*, **2**, 198–205.

Calderon-Garciduenas, L., Mora-Tiscareno, A., Ontiveros, E., et al. (2008a). Air pollution, cognitive deficits and brain abnormalities: A pilot study with children and dogs. *Brain and Cognition*, **68**, 117–127.

Calderon-Garciduenas, L., Solt, A. C., Henriquez-Roldan, C., et al. (2008b). Long-term air pollution exposure is associated with neuroinflammation, an altered innate immune response, disruption of the blood–brain barrier, ultrafine particulate deposition, and accumulation of amyloid beta-42 and alpha-synuclein in children and young adults. *Toxicologic Pathology*, **36**, 289–310.

Carroll, S. B. (2008). Evo-devo and an expanding evolutionary synthesis: a genetic theory of morphological evolution. *Cell*, **134**, 25–36.

Castel, S. E., & Martienssen, R. A. (2013). RNA interference in the nucleus: Roles for small RNAs in transcription, epigenetics and beyond. *Nature Reviews Genetics*, **14**, 100–112.

Chaidez, V., Hansen, R. L., & Hertz-Picciotto, I. (2014). Gastrointestinal problems in children with autism, developmental delays or typical development. *Journal of Autism and Developmental Disorders*, **44**, 1117–1127.

Chambers, S. M., Fasano, C. A., Papapetrou, E. P., et al. (2009). Highly efficient neural conversion of human ES and iPS cells by dual inhibition of SMAD signaling. *Nature Biotechnology*, **27**, 275–280.

Chen, H., Ke, Q., Kluz, T., Yan, Y., & Costa, M. (2006). Nickel ions increase histone H3 lysine 9 dimethylation and induce transgene silencing. *Molecular and Cellular Biology*, **26**, 3728–3737.

Cheung, A. Y., Horvath, L. M., Grafodatskaya, D., et al. (2011). Isolation of MECP2-null Rett syndrome patient hiPS cells and isogenic controls through X-chromosome inactivation. *Human Molecular Genetics*, **20**, 2103–2115.

Choi, G. B., Yim, Y. S., Wong, H., et al. (2016). The maternal interleukin-17a pathway in mice promotes autism-like phenotypes in offspring. *Science*, **351**, 933–939.

Clarke, G., O'Mahony, S. M., Dinan, T. G., & Cryan, J. F. (2014). Priming for health: Gut microbiota acquired in early life regulates physiology, brain and behaviour. *Acta Paediatrica*, **103**, 812–819.

Coiro, P., Padmashri, R., Suresh, A., et al. (2015). Impaired synaptic development in a maternal immune activation mouse model of neurodevelopmental disorders. *Brain, Behavior, and Immunity*, **50**, 249–258.

Coleman, M. (1976). *The Autistic Syndromes*. Amsterdam: North-Holland Publishing Company.

Cooper, G. S., Martin, S. A., Longnecker, M. P., Sandler, D. P., & Germolec, D. R. (2004). Associations between plasma DDE levels and immunologic measures in African-American farmers in North Carolina. *Environmental Health Perspectives*, **112**, 1080–1084.

Corsini, E., Liesivuori, J., Vergieva, T., Van Loveren, H., & Colosio, C. (2008). Effects of pesticide exposure on the human immune system. *Human & Experimental Toxicology*, **27**, 671–680.

Corsini, E., Sokooti, M., Galli, C. L., Moretto, A., & Colosio, C. (2013). Pesticide induced immunotoxicity in humans: A comprehensive review of the existing evidence. *Toxicology*, **307**, 123–135.

Crick, F. (1970). Central dogma of molecular biology. *Nature*, **227**, 561–563.

Crumeyrolle-Arias, M., Jaglin, M., Bruneau, A., et al. (2014). Absence of the gut microbiota enhances anxiety-like behavior and neuroendocrine response to acute stress in rats. *Psychoneuroendocrinology*, **42**, 207–217.

Dally, H., & Hartwig, A. (1997). Induction and repair inhibition of oxidative DNA damage by nickel(II) and cadmium(II) in mammalian cells. *Carcinogenesis*, **18**, 1021–1026.

Davies, J. O., Oudelaar, A. M., Higgs, D. R., & Hughes, J. R. (2017). How best to identify chromosomal interactions: A comparison of approaches. *Nature Methods*, **14**, 125–134.

de Theije, C. G., Koelink, P. J., Korte-Bouws, G. A., et al. (2014a). Intestinal inflammation in a murine model of autism spectrum disorders. *Brain, Behavior, and Immunity*, **37**, 240–247.

de Theije, C. G., Wopereis, H., Ramadan, M., et al. (2014b). Altered gut microbiota and activity in a murine model of autism spectrum disorders. *Brain, Behavior, and Immunity*, **37**, 197–206.

De Vadder, F., Kovatcheva-Datchary, P., Goncalves, D., et al. (2014). Microbiota-generated metabolites promote metabolic benefits via gut–brain neural circuits. *Cell*, **156**, 84–96.

Dekker, J., Rippe, K., Dekker, M., & Kleckner, N. (2002). Capturing chromosome conformation. *Science*, **295**, 1306–1311.

DeRosa, B. A., Van Baaren, J. M., Dubey, G. K., et al. (2012). Derivation of autism spectrum disorder-specific induced pluripotent stem cells from peripheral blood mononuclear cells. *Neuroscience Letters*, **516**, 9–14.

Desaulniers, D., Xiao, G. H., Lian, H., et al. (2009). Effects of mixtures of polychlorinated biphenyls, methylmercury, and organochlorine pesticides on hepatic DNA methylation in prepubertal female Sprague-Dawley rats. *International Journal of Toxicology*, **28**, 294–307.

Desbonnet, L., Clarke, G., Shanahan, F., Dinan, T. G., & Cryan, J. F. (2014). Microbiota is essential for social development in the mouse. *Molecular Psychiatry*, **19**, 146–148.

Diaz Heijtz, R., Wang, S., Anuar, F., et al. (2011). Normal gut microbiota modulates brain development and behavior. *Proceedings of the National Academy of Sciences*, **108**, 3047–3052.

Ding, H. T., Taur, Y., & Walkup, J. T. (2017). Gut microbiota and autism: Key concepts and findings. *Journal of Autism and Developmental Disorders*, **47**, 480–489.

Doi, T., Puri, P., McCann, A., Bannigan, J., & Thompson, J. (2011). Epigenetic effect of cadmium on global de novo DNA hypomethylation in the cadmium-induced ventral body wall defect (VBWD) in the chick model. *Toxicological Sciences*, **120**, 475–480.

El-Ansary, A., & Al-Ayadhi, L. (2014). GABAergic/glutamatergic imbalance relative to excessive neuroinflammation in autism spectrum disorders. *Journal of Neuroinflammation*, **11**, 189.

Emanuele, E., Orsi, P., Boso, M., et al. (2010). Low-grade endotoxemia in patients with severe autism. *Neuroscience Letters*, **471**, 162–165.

Favre, M. R., Barkat, T. R., Lamendola, D., et al. (2013). General developmental health in the VPA-rat model of autism. *Frontiers in Behavioral Neuroscience*, **7**, 88.

Feil, R., & Fraga, M. F. (2012). Epigenetics and the environment: Emerging patterns and implications. *Nature Reviews Genetics*, **13**, 97–109.

Finegold, S. M., Molitoris, D., Song, Y., et al. (2002). Gastrointestinal microflora studies in late-onset autism. *Clinical Infectious Diseases*, **35**, S6–S16.

Foley, K. A., Ossenkopp, K. P., Kavaliers, M., & Macfabe, D. F. (2014). Pre- and neonatal exposure to lipopolysaccharide or the enteric metabolite, propionic acid, alters development and behavior in adolescent rats in a sexually dimorphic manner. *PLoS ONE*, **9**, e87072.

Fullwood, M. J., & Ruan, Y. (2009). ChIP-based methods for the identification of long-range chromatin interactions. *Journal of Cellular Biochemistry*, **107**, 30–39.

Fullwood, M. J., Liu, M. H., Pan, Y. F., et al. (2009). An oestrogen-receptor-alpha-bound human chromatin interactome. *Nature*, **462**, 58–64.

Galloway, T., & Handy, R. (2003). Immunotoxicity of organophosphorous pesticides. *Ecotoxicology*, **12**, 345–363.

Gardener, H., Spiegelman, D., & Buka, S. L. (2009). Prenatal risk factors for autism: Comprehensive meta-analysis. *British Journal of Psychiatry*, **195**, 7–14.

Gardener, H., Spiegelman, D., & Buka, S. L. (2011). Perinatal and neonatal risk factors for autism: A comprehensive meta-analysis. *Pediatrics*, **128**, 344–355.

Garry, V. F., Harkins, M. E., Erickson, L. L., et al. (2002). Birth defects, season of conception, and sex of children born to pesticide applicators living in the Red River Valley of Minnesota, USA. *Environmental Health Perspectives*, **110**(Suppl. 3), 441–449.

Gentile, S. (2014). Risks of neurobehavioral teratogenicity associated with prenatal exposure to valproate monotherapy: A systematic review with regulatory repercussions. *CNS Spectrums*, **19**, 305–315.

Gerhart, J., & Kirschner, M. (1997). *Cells, Embryos, and Evolution: Toward a Cellular and Developmental Understanding of Phenotypic Variation and Evolutionary Adaptability*, Malden, MA: Blackwell Science.

Griesi-Oliveira, K., Acab, A., Gupta, A. R., et al. (2015). Modeling non-syndromic autism and the impact of TRPC6 disruption in human neurons. *Molecular Psychiatry*, **20**, 1350–1365.

Gu, H., Smith, Z. D., Bock, C., et al. (2011). Preparation of reduced representation bisulfite sequencing libraries for genome-scale DNA methylation profiling. *Nature Protocols*, **6**, 468–481.

Guevara-Guzman, R., Arriaga, V., Kendrick, K. M., et al. (2009). Estradiol prevents ozone-induced increases in brain lipid peroxidation and impaired social recognition memory in female rats. *Neuroscience*, **159**, 940–950.

Guo, H., Zhu, P., Guo, F., et al. (2015). Profiling DNA methylome landscapes of mammalian cells with single-cell reduced-representation bisulfite sequencing. *Nature Protocols*, **10**, 645–659.

Gurer, H., & Ercal, N. (2000). Can antioxidants be beneficial in the treatment of lead poisoning? *Free Radical Biology and Medicine*, **29**, 927–945.

Guzzi, G., & La Porta, C. A. (2008). Molecular mechanisms triggered by mercury. *Toxicology*, **244**, 1–12.

Handy, D. E., Castro, R., & Loscalzo, J. (2011). Epigenetic modifications: Basic mechanisms and role in cardiovascular disease. *Circulation*, **123**, 2145–2156.

Hannon, E., Lunnon, K., Schalkwyk, L., & Mill, J. (2015). Interindividual methylomic variation across blood, cortex, and cerebellum: Implications for epigenetic studies of neurological and neuropsychiatric phenotypes. *Epigenetics*, **10**, 1024–1032.

Hayashi, K. (1975). Distribution of histone F1 on calf thymus nucleohistone DNA. *Journal of Molecular Biology*, **94**, 397–408.

He, Y., & Ecker, J. R. (2015). Non-CG methylation in the human genome. *Annual Review of Genomics Human Genetics*, **16**, 55–77.

Henikoff, S., & Shilatifard, A. (2011). Histone modification: Cause or cog? *Trends in Genetics*, **27**, 389–396.

Hermanowicz, A., & Kossman, S. (1984). Neutrophil function and infectious disease in workers occupationally exposed to phosphoorganic pesticides: Role of mononuclear-derived chemotactic factor for neutrophils. *Clinical Immunology and Immunopathology*, **33**, 13–22.

Hoffman, D. J., Heinz, G. H., Sileo, L., et al. (2000). Developmental toxicity of lead-contaminated sediment in Canada geese (*Branta canadensis*). *Journal of Toxicology and Environmental Health A*, **59**, 235–252.

Hormanseder, E., Simeone, A., Allen, G. E., et al. (2017). H3K4 methylation dependent memory of somatic cell identity inhibits reprogramming and development of nuclear transfer embryos. *Cell Stem Cell*, **21**, 135–143.e6.

Hsiao, E. Y., McBride, S. W., Hsien, S., et al. (2013). Microbiota modulate behavioral and physiological abnormalities associated with neurodevelopmental disorders. *Cell*, **155**, 1451–1463.

Hsieh, T. S., Fudenberg, G., Goloborodko, A., & Rando, O. J. (2016). Micro-C XL: Assaying chromosome conformation from the nucleosome to the entire genome. *Nature Methods*, **13**, 1009–1011.

Hunaiti, A. A., & Soud, M. (2000). Effect of lead concentration on the level of glutathione, glutathione S-transferase, reductase and peroxidase in human blood. *Science of the Total Environment*, **248**, 45–50.

Hurwitz, J. (2005). The discovery of RNA polymerase. *Journal of Biological Chemistry*, **280**, 42477–42485.

Ji, W., Yang, L., Yu, L., et al. (2008). Epigenetic silencing of O6-methylguanine DNA methyltransferase gene in NiS-transformed cells. *Carcinogenesis*, **29**, 1267–1275.

Ji, H., Biagini Myers, J. M., Brandt, E. B., et al. (2016). Air pollution, epigenetics, and asthma. *Allergy, Asthma & Clinical Immunology*, **12**, 51.

Johnson, D. S., Mortazavi, A., Myers, R. M., & Wold, B. (2007). Genome-wide mapping of in vivo protein–DNA interactions. *Science*, **316**, 1497–1502.

Jomova, K., & Valko, M. (2011). Advances in metal-induced oxidative stress and human disease. *Toxicology*, **283**, 65–87.

Jyonouchi, H., Sun, S., & Le, H. (2001). Proinflammatory and regulatory cytokine production associated with innate and adaptive immune responses in children with autism spectrum disorders and developmental regression. *Journal of Neuroimmunology*, **120**, 170–179.

Kalkbrenner, A. E., Daniels, J. L., Chen, J. C., et al. (2010). Perinatal exposure to hazardous air pollutants and autism spectrum disorders at age 8. *Epidemiology*, **21**, 631–641.

Kang, D. W., Park, J. G., Ilhan, Z. E., et al. (2013). Reduced incidence of *Prevotella* and other fermenters in intestinal microflora of autistic children. *PLoS ONE*, **8**, e68322.

Kanwal, R., Gupta, K., & Gupta, S. (2015). Cancer epigenetics: An introduction. *Methods in Molecular Biology*, **1238**, 3–25.

Karaczyn, A. A., Golebiowski, F., & Kasprzak, K. S. (2006). Ni(II) affects ubiquitination of core histones H2B and H2A. *Experimental Cell Research*, **312**, 3252–3259.

Ke, Q., Ellen, T. P., & Costa, M. (2008). Nickel compounds induce histone ubiquitination by inhibiting histone deubiquitinating enzyme activity. *Toxicology and Applied Pharmacology*, **228**, 190–199.

Kim, K. C., Choi, C. S., Kim, J. W., et al. (2016). MeCP2 modulates sex differences in the postsynaptic development of the valproate animal model of autism. *Molecular Neurobiology*, **53**, 40–56.

Kim, K. Y., Hysolli, E., & Park, I. H. (2011). Neuronal maturation defect in induced pluripotent stem cells from patients with Rett syndrome. *Proceedings of the National Academy of Sciences*, **108**, 14169–14174.

Kinney, D. K., Munir, K. M., Crowley, D. J., & Miller, A. M. (2008). Prenatal stress and risk for autism. *Neuroscience & Biobehavioral Reviews*, **32**, 1519–1532.

Kobayashi, T., Matsuyama, T., Takeuchi, M., & Ito, S. (2016). Autism spectrum disorder and prenatal exposure to selective serotonin reuptake inhibitors: A systematic review and meta-analysis. *Reproductive Toxicology*, **65**, 170–178.

Kosidou, K., Dalman, C., Widman, L., et al. (2016). Maternal polycystic ovary syndrome and the risk of autism spectrum disorders in the offspring: A population-based nationwide study in Sweden. *Molecular Psychiatry*, **21**, 1441–1448.

Kratsman, N., Getselter, D., & Elliott, E. (2016). Sodium butyrate attenuates social behavior deficits and modifies the transcription of inhibitory/excitatory genes in the frontal cortex of an autism model. *Neuropharmacology*, **102**, 136–145.

Krey, J. F., Pasca, S. P., Shcheglovitov, A., et al. (2013). Timothy syndrome is associated with activity-dependent dendritic retraction in rodent and human neurons. *Nature Neuroscience*, **16**, 201–209.

Kuo, H. Y., & Liu, F. C. (2017). Valproic acid induces aberrant development of striatal compartments and corticostriatal pathways in a mouse model of autism spectrum disorder. *FASEB Journal*, **31**, 4458–4472.

Kuo, M. H., & Allis, C. D. (1999). In vivo cross-linking and immunoprecipitation for studying dynamic protein:DNA associations in a chromatin environment. *Methods*, **19**, 425–433.

Lai, M. C., Lombardo, M. V., & Baron-Cohen, S. (2014). Autism. *Lancet*, **383**, 896–910.

Lancaster, M. A., Renner, M., Martin, C. A., et al. (2013). Cerebral organoids model human brain development and microcephaly. *Nature*, **501**, 373–379.

Lancaster, M. A., Corsini, N. S., Wolfinger, S., et al. (2017). Guided self-organization and cortical plate formation in human brain organoids. *Nature Biotechnology*, **35**, 659–666.

Latchman, D. S. (1997). Transcription factors: An overview. *International Journal of Biochemistry & Cell Biology*, **29**, 1305–1312.

Law, J. A., & Jacobsen, S. E. (2010). Establishing, maintaining and modifying DNA methylation patterns in plants and animals. *Nature Reviews Genetics*, **11**, 204–220.

Li, Q. (2007). New mechanism of organophosphorus pesticide-induced immunotoxicity. *Journal of Nippon Medical School*, **74**, 70–73.

Li, Q., Han, Y., Dy, A. B. C., & Hagerman, R. J. (2017). The gut microbiota and autism spectrum disorders. *Frontiers in Cellular Neuroscience*, **11**, 120.

Liang, J., Zhu, H., Li, C., et al. (2012). Neonatal exposure to benzo[a]pyrene decreases the levels of serum testosterone and histone H3K14 acetylation of the StAR promoter in the testes of SD rats. *Toxicology*, **302**, 285–291.

Lieberman-Aiden, E., van Berkum, N. L., Williams, L., et al. (2009). Comprehensive mapping of long-range interactions reveals folding principles of the human genome. *Science*, **326**, 289–293.

Lin, V. W., Baccarelli, A. A., & Burris, H. H. (2016). Epigenetics – A potential mediator between air pollution and preterm birth. *Environmental Epigenetics*, **2**, dvv008.

Ling, Z., Zhu, Y., Tong, C., et al. (2006). Progressive dopamine neuron loss following supra-nigral lipopolysaccharide (LPS) infusion into rats exposed to LPS prenatally. *Experimental Neurology*, **199**, 499–512.

Lister, R., Mukamel, E. A., Nery, J. R., et al. (2013). Global epigenomic reconfiguration during mammalian brain development. *Science*, **341**, 1237905.

Long, T. C., Tajuba, J., Sama, P., et al. (2007). Nanosize titanium dioxide stimulates reactive oxygen species in brain microglia and damages neurons in vitro. *Environmental Health Perspectives*, **115**, 1631–1637.

M'Bemba-Meka, P., Lemieux, N., & Chakrabarti, S. K. (2007). Role of oxidative stress and intracellular calcium in nickel carbonate hydroxide-induced sister-chromatid exchange, and alterations in replication index and mitotic index in cultured human peripheral blood lymphocytes. *Archives of Toxicology*, **81**, 89–99.

Madisen, L., Krumm, A., Hebbes, T. R., & Groudine, M. (1998). The immunoglobulin heavy chain locus control region increases histone acetylation along linked c-myc genes. *Molecular and Cellular Biology*, **18**, 6281–6292.

Madrigano, J., Baccarelli, A., Mittleman, M. A., et al. (2012). Air pollution and DNA methylation: Interaction by psychological factors in the VA Normative Aging Study. *American Journal of Epidemiology*, **176**, 224–232.

Manousakis, G., Jensen, M. B., Chacon, M. R., Sattin, J. A., & Levine, R. L. (2009). The interface between stroke and infectious disease: Infectious diseases leading to stroke and infections complicating stroke. *Current Neurology and Neuroscience Reports*, **9**, 28–34.

Mao, Z., Bonni, A., Xia, F., Nadal-Vicens, M., & Greenberg, M. E. (1999). Neuronal activity-dependent cell survival mediated by transcription factor MEF2. *Science*, **286**, 785–790.

Marchetto, M. C., Carromeu, C., Acab, A., et al. (2010). A model for neural development and treatment of Rett syndrome using human induced pluripotent stem cells. *Cell*, **143**, 527–539.

Marchetto, M. C., Belinson, H., Tian, Y., et al. (2017). Altered proliferation and networks in neural cells derived from idiopathic autistic individuals. *Molecular Psychiatry*, **22**, 820–835.

Mariani, J., Coppola, G., Zhang, P., et al. (2015). FOXG1-dependent dysregulation of GABA/glutamate neuron differentiation in autism spectrum disorders. *Cell*, **162**, 375–390.

Martinez-Zamudio, R., & Ha, H. C. (2011). Environmental epigenetics in metal exposure. *Epigenetics*, **6**, 820–827.

Mascetti, V. L., & Pedersen, R. A. (2016). Human–mouse chimerism validates human stem cell pluripotency. *Cell Stem Cell*, **18**, 67–72.

Masi, A., Quintana, D. S., Glozier, N., et al. (2015). Cytokine aberrations in autism spectrum disorder: A systematic review and meta-analysis. *Molecular Psychiatry*, **20**, 440–446.

McCanlies, E. C., Fekedulegn, D., Mnatsakanova, A., et al. (2012). Parental occupational exposures and autism spectrum disorder. *Journal of Autism and Developmental Disorders*, **42**, 2323–2334.

McCarthy, M. M., & Nugent, B. M. (2013). Epigenetic contributions to hormonally-mediated sexual differentiation of the brain. *Journal of Neuroendocrinology*, **25**, 1133–1140.

McConnachie, P. R., & Zahalsky, A. C. (1992). Immune alterations in humans exposed to the termiticide technical chlordane. *Archives of Environmental Health*, **47**, 295–301.

McCormick, H., Young, P. E., Hur, S. S. J., et al. (2017). Isogenic mice exhibit sexually-dimorphic DNA methylation patterns across multiple tissues. *BMC Genomics*, **18**, 966.

McLachlan, J. A., Simpson, E., & Martin, M. (2006). Endocrine disrupters and female reproductive health. *Best Practice & Research: Clinical Endocrinology & Metabolism*, **20**, 63–75.

Mills, N. L., Donaldson, K., Hadoke, P. W., et al. (2009). Adverse cardiovascular effects of air pollution. *Nature Clinical Practice Cardiovascular Medicine*, **6**, 36–44.

Mittal, V. A., Ellman, L. M., & Cannon, T. D. (2008). Gene–environment interaction and covariation in schizophrenia: The role of obstetric complications. *Schizophrenia Bulletin*, **34**, 1083–1094.

Modabbernia, A., Mollon, J., Boffetta, P., & Reichenberg, A. (2016). Impaired gas exchange at birth and risk of intellectual disability and autism: A meta-analysis. *Journal of Autism and Developmental Disorders*, **46**, 1847–1859.

Modabbernia, A., Velthorst, E., & Reichenberg, A. (2017). Environmental risk factors for autism: An evidence-based review of systematic reviews and meta-analyses. *Molecular Autism*, **8**, 13.

Mohle, L., Mattei, D., Heimesaat, M. M., et al. (2016). Ly6 C(Hi) monocytes provide a link between antibiotic-induced changes in gut microbiota and adult hippocampal neurogenesis. *Cell Reports*, **15**, 1945–1956.

Muotri, A. R., Marchetto, M. C., Coufal, N. G., et al. (2010). L1 retrotransposition in neurons is modulated by MeCP2. *Nature*, **468**, 443–446.

Nagano, T., Lubling, Y., Stevens, T. J., et al. (2013). Single-cell Hi-C reveals cell-to-cell variability in chromosome structure. *Nature*, **502**, 59–64.

Nardone, S., Sams, D. S., Zito, A., Reuveni, E., & Elliott, E. (2017). Dysregulation of cortical neuron DNA methylation profile in autism spectrum disorder. *Cerebral Cortex*, **27**, 5739–5754.

Naumova, O. Y., Dozier, M., Dobrynin, P. V., et al. (2018). Developmental dynamics of the epigenome: A longitudinal study of three toddlers. *Neurotoxicology and Teratology*, **66**, 125–131.

Nemmar, A., & Inuwa, I. M. (2008). Diesel exhaust particles in blood trigger systemic and pulmonary morphological alterations. *Toxicology Letters*, **176**, 20–30.

Nwankwo, D. O., & Wilson, G. G. (1988). Cloning and expression of the MspI restriction and modification genes. *Gene*, **64**, 1–8.

O'Neill, L. A., & Kaltschmidt, C. (1997). NF-kappa B: A crucial transcription factor for glial and neuronal cell function. *Trends in Neurosciences*, **20**, 252–258.

O'Roak, B. J., Deriziotis, P., Lee, C., et al. (2011). Exome sequencing in sporadic autism spectrum disorders identifies severe de novo mutations. *Nature Genetics*, **43**, 585–589.

O'Roak, B. J., Vives, L., Girirajan, S., et al. (2012). Sporadic autism exomes reveal a highly interconnected protein network of de novo mutations. *Nature*, **485**, 246–250.

Oberdorster, G., Sharp, Z., Atudorei, V., et al. (2004). Translocation of inhaled ultrafine particles to the brain. *Inhalation Toxicology*, **16**, 437–445.

Onishchenko, N., Karpova, N., Sabri, F., Castren, E., & Ceccatelli, S. (2008). Long-lasting depression-like behavior and epigenetic changes of BDNF gene expression induced by perinatal exposure to methylmercury. *Journal of Neurochemistry*, **106**, 1378–1387.

Palmer, R. F., Blanchard, S., Stein, Z., Mandell, D., & Miller, C. (2006). Environmental mercury release, special education rates, and autism disorder: An ecological study of Texas. *Health & Place*, **12**, 203–209.

Palmer, R. F., Blanchard, S., & Wood, R. (2009). Proximity to point sources of environmental mercury release as a predictor of autism prevalence. *Health & Place*, **15**, 18–24.

Papp, B., & Plath, K. (2013). Epigenetics of reprogramming to induced pluripotency. *Cell*, **152**, 1324–1343.

Parikshak, N. N., Luo, R., Zhang, A., et al. (2013). Integrative functional genomic analyses implicate specific molecular pathways and circuits in autism. *Cell*, **155**, 1008–1021.

Parikshak, N. N., Swarup, V., Belgard, T. G., et al. (2016). Genome-wide changes in lncRNA, splicing, and regional gene expression patterns in autism. *Nature*, **540**, 423–427.

Parracho, H. M., Bingham, M. O., Gibson, G. R., & McCartney, A. L. (2005). Differences between the gut microflora of children with autistic spectrum disorders and that of healthy children. *Journal of Medical Microbiology*, **54**, 987–991.

Pasca, A. M., Sloan, S. A., Clarke, L. E., et al. (2015). Functional cortical neurons and astrocytes from human pluripotent stem cells in 3D culture. *Nature Methods*, **12**, 671–678.

Pasca, S. P., Portmann, T., Voineagu, I., et al. (2011). Using iPSC-derived neurons to uncover cellular phenotypes associated with Timothy syndrome. *Nature Medicine*, **17**, 1657–1662.

Patrick, L. (2006). Lead toxicity part II: The role of free radical damage and the use of antioxidants in the pathology and treatment of lead toxicity. *Alternative Medicine Review*, **11**, 114–127.

Pavanello, S., Pesatori, A. C., Dioni, L., et al. (2010). Shorter telomere length in peripheral blood lymphocytes of workers exposed to polycyclic aromatic hydrocarbons. *Carcinogenesis*, **31**, 216–221.

Pereyra-Munoz, N., Rugerio-Vargas, C., Angoa-Perez, M., Borgonio-Perez, G., & Rivas-Arancibia, S. (2006). Oxidative damage in substantia nigra and striatum of rats chronically exposed to ozone. *Journal of Chemical Neuroanatomy*, **31**, 114–123.

Perry, V. H., Cunningham, C., & Holmes, C. (2007). Systemic infections and inflammation affect chronic neurodegeneration. *Nature Reviews Immunology*, **7**, 161–167.

Persico, A. M., & Napolioni, V. (2013). Urinary *p*-cresol in autism spectrum disorder. *Neurotoxicology and Teratology*, **36**, 82–90.

Phillips, T. M. (2000). Assessing environmental exposure in children: Immunotoxicology screening. *Journal of Exposure Analysis and Environmental Epidemiology*, **10**, 769–775.

Pohl, A., Cassidy, S., Auyeung, B., & Baron-Cohen, S. (2014). Uncovering steroidopathy in women with autism: A latent class analysis. *Molecular Autism*, **5**, 27.

Price, C. S., Thompson, W. W., Goodson, B., et al. (2010). Prenatal and infant exposure to thimerosal from vaccines and immunoglobulins and risk of autism. *Pediatrics*, **126**, 656–664.

Price, D. J., & Joshi, J. G. (1983). Ferritin. Binding of beryllium and other divalent metal ions. *Journal of Biological Chemistry*, **258**, 10873–10880.

Qin, L., Wu, X., Block, M. L., et al. (2007). Systemic LPS causes chronic neuroinflammation and progressive neurodegeneration. *Glia*, **55**, 453–462.

Quadrato, G., Nguyen, T., Macosko, E. Z., et al. (2017). Cell diversity and network dynamics in photosensitive human brain organoids. *Nature*, **545**, 48–53.

Quigley, E. M. (2016). Leaky gut – Concept or clinical entity? *Current Opinion in Gastroenterology*, **32**, 74–79.

Ramani, V., Deng, X., Qiu, R., et al. (2017). Massively multiplex single-cell Hi-C. *Nature Methods*, **14**, 263–266.

Rammes, G., Steckler, T., Kresse, A., et al. (2000). Synaptic plasticity in the basolateral amygdala in transgenic mice expressing dominant-negative cAMP response element-binding protein (CREB) in forebrain. *European Journal of Neuroscience*, **12**, 2534–2546.

Rauh, V. A., Garfinkel, R., Perera, F. P., et al. (2006). Impact of prenatal chlorpyrifos exposure on neurodevelopment in the first 3 years of life among inner-city children. *Pediatrics*, **118**, e1845–e1859.

Ray, D. E., & Richards, P. G. (2001). The potential for toxic effects of chronic, low-dose exposure to organophosphates. *Toxicology Letters*, **120**, 343–351.

Reed, A., Dzon, L., Loganathan, B. G., & Whalen, M. M. (2004). Immunomodulation of human natural killer cell cytotoxic function by organochlorine pesticides. *Human & Experimental Toxicology*, **23**, 463–471.

Repetto, R., & Baliga, S. S. (1997). Pesticides and immunosuppression: The risks to public health. *Health Policy and Planning*, **12**, 97–106.

Rieder, R., Wisniewski, P. J., Alderman, B. L., & Campbell, S. C. (2017). Microbes and mental health: A review. *Brain, Behavior, and Immunity*, **66**, 9–17.

Rivest, S., Lacroix, S., Vallieres, L., et al. (2000). How the blood talks to the brain parenchyma and the paraventricular nucleus of the hypothalamus during systemic inflammatory and infectious stimuli. *Proceedings of the Society for Experimental Biology and Medicine*, **223**, 22–38.

Roberts, A. L., Lyall, K., Hart, J. E., et al. (2013). Perinatal air pollutant exposures and autism spectrum disorder in the children of Nurses' Health Study II participants. *Environmental Health Perspectives*, **121**, 978–984.

Roberts, E. M., & English, P. B. (2013). Bayesian modeling of time-dependent vulnerability to environmental hazards: An example using autism and pesticide data. *Statistics in Medicine*, **32**, 2308–2319.

Roberts, E. M., English, P. B., Grether, J. K., et al. (2007). Maternal residence near agricultural pesticide applications and autism spectrum disorders among children in the California Central Valley. *Environmental Health Perspectives*, **115**, 1482–1489.

Ronald, A., Pennell, C. E., & Whitehouse, A. J. (2010). Prenatal maternal stress associated with ADHD and autistic traits in early childhood. *Frontiers in Psychology*, **1**, 223.

Rossignol, D. A., Genuis, S. J., & Frye, R. E. (2014). Environmental toxicants and autism spectrum disorders: A systematic review. *Translational Psychiatry*, **4**, e360.

Rossnerova, A., Tulupova, E., Tabashidze, N., et al. (2013). Factors affecting the 27 K DNA methylation pattern in asthmatic and healthy children from locations with various environments. *Mutation Research*, **741–742**, 18–26.

Rouhani, F., Kumasaka, N., de Brito, M. C., et al. (2014). Genetic background drives transcriptional variation in human induced pluripotent stem cells. *PLoS Genetics*, **10**, e1004432.

Sanders, S. J., He, X., Willsey, A. J., et al. (2015). Insights into autism spectrum disorder genomic architecture and biology from 71 risk loci. *Neuron*, **87**, 1215–1233.

Sassone-Corsi, P., & Christen, Y. (2015). *A Time for Metabolism and Hormones*. New York: Springer Berlin Heidelberg.

Schaalan, M. F., Abdelraouf, S. M., Mohamed, W. A., & Hassanein, F. S. (2012). Correlation between maternal milk and infant serum levels of chlorinated pesticides (CP) and the impact of elevated CP on bleeding tendency and immune status in some infants in Egypt. *Journal of Immunotoxicology*, **9**, 15–24.

Sen, A., Sherr, C. J., & Todaro, G. J. (1976). Specific binding of the type C viral core protein p12 with purified viral RNA. *Cell*, **7**, 21–32.

Sharon, G., Sampson, T. R., Geschwind, D. H., & Mazmanian, S. K. (2016). The central nervous system and the gut microbiome. *Cell*, **167**, 915–932.

Sheldon, A. L., & Robinson, M. B. (2007). The role of glutamate transporters in neurodegenerative diseases and potential opportunities for intervention. *Neurochemistry International*, **51**, 333–355.

Sheridan, S. D., Theriault, K. M., Reis, S. A., et al. (2011). Epigenetic characterization of the FMR1 gene and aberrant neurodevelopment in human induced pluripotent stem cell models of fragile X syndrome. *PLoS ONE*, **6**, e26203.

Shi, L., Fatemi, S. H., Sidwell, R. W., & Patterson, P. H. (2003). Maternal influenza infection causes marked behavioral and pharmacological changes in the offspring. *Journal of Neuroscience*, **23**, 297–302.

Shi, Y., Kirwan, P., & Livesey, F. J. (2012). Directed differentiation of human pluripotent stem cells to cerebral cortex neurons and neural networks. *Nature Protocols*, **7**, 1836–1846.

Shutoh, Y., Takeda, M., Ohtsuka, R., et al. (2009). Low dose effects of dichlorodiphenyltrichloroethane (DDT) on gene transcription and DNA methylation in the hypothalamus of young male rats: Implication of hormesis-like effects. *Journal of Toxicological Sciences*, **34**, 469–482.

Silveyra, P., & Floros, J. (2012). Air pollution and epigenetics: Effects on SP-A and innate host defence in the lung. *Swiss Medical Weekly*, **142**, w13579.

Sofer, T., Baccarelli, A., Cantone, L., et al. (2013). Exposure to airborne particulate matter is associated with methylation pattern in the asthma pathway. *Epigenomics*, **5**, 147–154.

Somji, S., Garrett, S. H., Toni, C., et al. (2011). Differences in the epigenetic regulation of MT-3 gene expression between parental and Cd^{+2} or As^{+3} transformed human urothelial cells. *Cancer Cell International*, **11**, 2.

Song, C., Kanthasamy, A., Anantharam, V., Sun, F., & Kanthasamy, A. G. (2010). Environmental neurotoxic pesticide increases histone acetylation to promote apoptosis in dopaminergic neuronal cells: Relevance to epigenetic mechanisms of neurodegeneration. *Molecular Pharmacology*, **77**, 621–632.

Song, C., Kanthasamy, A., Jin, H., Anantharam, V., & Kanthasamy, A. G. (2011). Paraquat induces epigenetic changes by promoting histone acetylation in cell culture models of dopaminergic degeneration. *Neurotoxicology*, **32**, 586–595.

Song, Y., Liu, C., & Finegold, S. M. (2004). Real-time PCR quantitation of clostridia in feces of autistic children. *Applied and Environmental Microbiology*, **70**, 6459–6465.

Stouder, C., & Paoloni-Giacobino, A. (2011). Specific transgenerational imprinting effects of the endocrine disruptor methoxychlor on male gametes. *Reproduction*, **141**, 207–216.

Stroud, H., Su, S. C., Hrvatin, S., et al. (2017). Early-life gene expression in neurons modulates lasting epigenetic states. *Cell*, **171**, 1151–1164.e16.

Sun, W., Poschmann, J., Cruz-Herrera Del Rosario, R., et al. (2016). Histone acetylome-wide association study of autism spectrum disorder. *Cell*, **167**, 1385–1397.e11.

Suzuki, K., Matsuzaki, H., Iwata, K., et al. (2011). Plasma cytokine profiles in subjects with high-functioning autism spectrum disorders. *PLoS ONE*, **6**, e20470.

Takahashi, K., & Yamanaka, S. (2006). Induction of pluripotent stem cells from mouse embryonic and adult fibroblast cultures by defined factors. *Cell*, **126**, 663–676.

Takahashi, K., Tanabe, K., Ohnuki, M., et al. (2007). Induction of pluripotent stem cells from adult human fibroblasts by defined factors. *Cell*, **131**, 861–872.

Takiguchi, M., Achanzar, W. E., Qu, W., Li, G., & Waalkes, M. P. (2003). Effects of cadmium on DNA-(cytosine-5) methyltransferase activity and DNA methylation status during cadmium-induced cellular transformation. *Experimental Cell Research*, **286**, 355–365.

Tang, S. C., Sheu, G. T., Wong, R. H., et al. (2010). Expression of glutathione S-transferase M2 in stage I/II non-small cell lung cancer and alleviation of DNA damage exposure to benzo[a]pyrene. *Toxicology Letters*, **192**, 316–323.

Thomson, E. M., Kumarathasan, P., Calderon-Garciduenas, L., & Vincent, R. (2007). Air pollution alters brain and pituitary endothelin-1 and inducible nitric oxide synthase gene expression. *Environmental Research*, **105**, 224–233.

Tin Tin Win, S., Mitsushima, D., Yamamoto, S., et al. (2008). Changes in neurotransmitter levels and proinflammatory cytokine mRNA expressions in the mice olfactory bulb following nanoparticle exposure. *Toxicology and Applied Pharmacology*, **226**, 192–198.

Tomova, A., Husarova, V., Lakatosova, S., et al. (2015). Gastrointestinal microbiota in children with autism in Slovakia. *Physiology & Behavior*, **138**, 179–187.

Tsai, H. W., Grant, P. A., & Rissman, E. F. (2009). Sex differences in histone modifications in the neonatal mouse brain. *Epigenetics*, **4**, 47–53.

Ulahannan, N., & Greally, J. M. (2015). Genome-wide assays that identify and quantify modified cytosines in human disease studies. *Epigenetics & Chromatin*, **8**, 5.

Urich, M. A., Nery, J. R., Lister, R., Schmitz, R. J., & Ecker, J. R. (2015). MethylC-seq library preparation for base-resolution whole-genome bisulfite sequencing. *Nature Protocols*, **10**, 475–483.

Valavanidis, A., Fiotakis, K., & Vlachogianni, T. (2008). Airborne particulate matter and human health: Toxicological assessment and importance of size and composition of particles for oxidative damage and carcinogenic mechanisms. *Journal of Environmental Science Health, Part C: Environmental Carcinogenesis & Ecotoxicology Reviews*, **26**, 339–362.

Vardabasso, C., Hasson, D., Ratnakumar, K., et al. (2014). Histone variants: Emerging players in cancer biology. *Cellular and Molecular Life Sciences*, **71**, 379–404.

Vine, M. F., Stein, L., Weigle, K., et al. (2000). Effects on the immune system associated with living near a pesticide dump site. *Environmental Health Perspectives*, **108**, 1113–1124.

Voineagu, I., Wang, X., Johnston, P., et al. (2011). Transcriptomic analysis of autistic brain reveals convergent molecular pathology. *Nature*, **474**, 380–384.

Volk, H. E., Hertz-Picciotto, I., Delwiche, L., Lurmann, F., & McConnell, R. (2011). Residential proximity to freeways and autism in the CHARGE study. *Environmental Health Perspectives*, **119**, 873–877.

Volk, H. E., Lurmann, F., Penfold, B., Hertz-Picciotto, I., & McConnell, R. (2013). Traffic-related air pollution, particulate matter, and autism. *JAMA Psychiatry*, **70**, 71–77.

Vuong, H. E., & Hsiao, E. Y. (2017). Emerging roles for the gut microbiome in autism spectrum disorder. *Biological Psychiatry*, **81**, 411–423.

Waalkes, M. P. (2000). Cadmium carcinogenesis in review. *Journal of Inorganic Biochemistry*, **79**, 241–244.

Wang, B., Feng, W. Y., Wang, M., et al. (2007). Transport of intranasally instilled fine Fe_2O_3 particles into the brain: Micro-distribution, chemical states, and histopathological observation. *Biological Trace Element Research*, **118**, 233–243.

Wang, J., Liu, Y., Jiao, F., et al. (2008). Time-dependent translocation and potential impairment on central nervous system by intranasally instilled TiO_2 nanoparticles. *Toxicology*, **254**, 82–90.

Wang, L., Christophersen, C. T., Sorich, M. J., et al. (2012). Elevated fecal short chain fatty acid and ammonia concentrations in children with autism spectrum disorder. *Digestive Diseases and Sciences*, **57**, 2096–2102.

Wang, L., Christophersen, C. T., Sorich, M. J., et al. (2013). Increased abundance of *Sutterella* spp. and *Ruminococcus torques* in feces of children with autism spectrum disorder. *Molecular Autism*, **4**, 42.

Warren, N., Caric, D., Pratt, T., et al. (1999). The transcription factor, Pax6, is required for cell proliferation and differentiation in the developing cerebral cortex. *Cerebral Cortex*, **9**, 627–635.

Watjen, W., & Beyersmann, D. (2004). Cadmium-induced apoptosis in C6 glioma cells: Influence of oxidative stress. *Biometals*, **17**, 65–78.

Wigle, D. T., Arbuckle, T. E., Turner, M. C., et al. (2008). Epidemiologic evidence of relationships between reproductive and child health outcomes and environmental chemical contaminants. *Journal of Toxicology and Environmental Health, Part B: Critical Reviews*, **11**, 373–517.

Williams, B. L., Hornig, M., Buie, T., et al. (2011). Impaired carbohydrate digestion and transport and mucosal dysbiosis in the intestines of children with autism and gastrointestinal disturbances. *PLoS ONE*, **6**, e24585.

Windham, G. C., Zhang, L., Gunier, R., Croen, L. A., & Grether, J. K. (2006). Autism spectrum disorders in relation to distribution of hazardous air pollutants in the San Francisco Bay Area. *Environmental Health Perspectives*, **114**, 1438–1444.

Wright, R. O., Schwartz, J., Wright, R. J., et al. (2010). Biomarkers of lead exposure and DNA methylation within retrotransposons. *Environmental Health Perspectives*, **118**, 790–795.

Wu, S., Wu, F., Ding, Y., et al. (2017). Advanced parental age and autism risk in children: A systematic review and meta-analysis. *Acta Psychiatrica Scandinavica*, **135**, 29–41.

Xu, N., Li, X., & Zhong, Y. (2015). Inflammatory cytokines: Potential biomarkers of immunologic dysfunction in autism spectrum disorders. *Mediators of Inflammation*, **2015**, 531518.

Xu, Q. S., Roberts, R. J., & Guo, H. C. (2005). Two crystal forms of the restriction enzyme MspI-DNA complex show the same novel structure. *Protein Science*, **14**, 2590–2600.

Yan, Y., Kluz, T., Zhang, P., Chen, H. B., & Costa, M. (2003). Analysis of specific lysine histone H3 and H4 acetylation and methylation status in clones of cells with a gene silenced by nickel exposure. *Toxicology and Applied Pharmacology*, **190**, 272–277.

Ye, F., & Xu, X. C. (2010). Benzo[a]pyrene diol epoxide suppresses retinoic acid receptor-beta2 expression by recruiting DNA (cytosine-5-)-methyltransferase 3A. *Molecular Cancer*, **9**, 93.

Yoshimasu, K., Kiyohara, C., Takemura, S., & Nakai, K. (2014). A meta-analysis of the evidence on the impact of prenatal and early infancy exposures to mercury on autism and attention deficit/hyperactivity disorder in the childhood. *Neurotoxicology*, **44**, 121–131.

Zama, A. M., & Uzumcu, M. (2009). Fetal and neonatal exposure to the endocrine disruptor methoxychlor causes epigenetic alterations in adult ovarian genes. *Endocrinology*, **150**, 4681–4691.

PART IV

The Evolution of Social and Moral Behavior

After mate choice, social behavior is probably the largest area of study within the evolutionary social sciences. One reason for this is because cooperative behavior seemingly contradicts the selfish view of life that Darwinism advocates. As Darwin (1871) stated in *The Decent of Man*, 'He who was ready to sacrifice his life, as many a savage has been, rather than betray his comrades, would often leave no offspring to inherit his noble nature." Like Einstein, whose similarly misunderstood quotation that he "refuses to believe that God plays dice with the universe," Darwin was not denying the phenomenon of altruism, as is sometimes thought, but simply pointing out the inadequacies of his theory of natural selection to explain something that was obviously true.

The story goes that the first attempt to explain cooperation began with a comment made in a London pub by a suitably refreshed J. B. S. Haldane, who, following some inspired scribblings on an apocryphal napkin, announced that he would be "prepared to lay down his life for eight cousins or two brothers" (this witnessed by none other than John Maynard Smith). In some stories, this event inspired William Hamilton (1964) to develop the theory of inclusive fitness that shows there is no contradiction between Darwinism and altruism, so long as altruism is nepotistic. But assuming that Darwin's "comrades" are not relatives, his puzzle remained unsolved. Enter our third giant of evolutionary theory, Robert Trivers, who, in his 1971 book, showed how cooperation could evolve along strict Darwinian principles so long as favors were returned.

The story does not end there, of course, but the simple narrative does, and since Trivers and Hamilton, there has been an explosion of research on altruism, cooperation, and morality, which is sampled in this Part IV. It is worth pointing out how different this theorizing is to that in the traditional social sciences. For many social scientists, the problem that needs explaining is selfishness; witness Latané and Darley's classic work in the 1960s on the so-called "unresponsive bystander." "Why don't people help when others are in distress?" they asked. Evolutionary theory inverts this question and wishes to explain the

responsive bystander thus: "Why should people help when it is seemingly against their genetic interests?"

Colquhoun, Workman, and Fowler provide us with even more delicious historical perspective on the evolutionary attempt to solve the puzzle of altruism. Lance Workman, an evolutionary psychologist before the term existed, has, through his ability to see both sides of an argument, gently coaxed many into the subject area. As well as a historical perspective, Workman, with Luke Colquhoun and Jo Fowler, also discuss some of the more recent theories that have been advanced to explain the existence of altruism, including indirect reciprocity, signaling theory, and multilevel selection theory. They finish with a perspective that brings kin and nonkin altruism closer together by raising the possibility that emotional closeness – an important factor in nonkin cooperation – might have been an indicator in ancestral times of relatedness. Maybe Hamilton and Trivers are not so far apart at all.

Dennis L. Krebs has been studying social cognition and altruism since the 1980s and deserves a special mention for exploring the evolutionary foundations of morality before it attained its current status as one of the psychological questions du jour. As befitting his status as someone who arrived at the party before anyone else, his chapter covers practically all of the relevant moral phenomena and attempts to explain their evolution. If someone were to ask us to give them a summary of the field over the past 30 years, we would give them this chapter as essential reading.

Will Reader and Sara Hughes are recent arrivals at the morality party. The punch bowl has been drained, the nibbles tray emptied, and the sandwiches are curling at the edges; nevertheless, they find nourishment with a question that has been asked many times under different guises but has been subject to comparatively little evolutionary scrutiny. Whereas most questions of morality have asked why people do good, they ask why we morally judge others for acts that have no effect on us: third-party moral judgment. Their answer is that maybe it permits humans to emulate the superorganismic eusocial species (see

Chapter 4) at certain times, with morality being rooted in obligate interdependence and intergroup conflict.

Craig T. Palmer and Kathryn Coe explore the evolution of the human family. Both authors have long research careers in diverse areas of cultural anthropology and related disciplines that include the relationships between evolution, tradition, health, and kinship. Palmer also collaborated with Randy Thornhill (another contributor to Part IV) on the evolutionary origins of rape. In this fascinating chapter, they argue that common conceptions of the human family are misguided. This is possibly because they start with nonhuman primates, and possibly because many of the researchers are WEIRD (Western, Educated, Industrial, Rich, and Democratic) and have little knowledge of family systems in traditional societies. Their argument is that human families are better understood not by considering them as networks of kinship ties, but rather by considering families as "mechanism by which behaviors enabling many earlier generations of ancestors to leave descendants are transmitted to the next generation." The implications of this are manifest, not least that this view can place individuals in conflict with long-dead ancestors!

Both Randy Thornhill and Corey L. Fincher have conducted groundbreaking research of mate choice strategies and parasite resistance, and in their chapter they explain how parasites have influenced cultural values and mate choice preferences via the behavioral immune system. Unlike the biological immune system, which fights infection, the behavioral immune system aims to prevent infection in the first place by producing behaviors that avoid contamination. The disgust response is the classic example: things that contain harmful pathogens (rotten food, other people's effluvia) are harmful, so the disgust response evolved to help avoid them. Fascinatingly, the authors present evidence that liberal/individualist vs. conservative/collectivist worldviews might themselves be adaptations influenced by parasite load. In communities with high parasite load, people are more inward-looking and xenophobic, avoiding harmful contact with others. In low-load communities, people are more liberal and better able to reap the rewards of trade without suffering potential infection.

In their chapter, Alec T. Beall, who studies sexual attraction, and Jessica L. Tracy, a researcher on the evolutionary function of emotions, discuss shame. Until comparatively recently, shame was a neglected topic within psychology at least, perhaps because it is less visible than other emotions (anger, happiness, disgust) or possibly because it is something that we would rather not discuss. Whatever the reason, Beall and Tracy address the imbalance in their chapter and discuss shame alongside its opposite number, pride. Pride, they suggest, may function to motivate fitness-enhancing behavior, such as pursuing activities that increase status. Shame is an attempt to appease group members by showing humility and contrition following an event that lowers status, such as cheating or failing on a task.

Finally, Louise Barrett, an expert in social cognition and cognitive adaptation, begins her chapter by pondering why evolution has had such difficulty penetrating the social sciences. After a review of the usual suspects (evolutionary theory asks unhelpful questions, active hostility, ignorance, it is too revolutionary), she speculates that it might be that evolutionary psychology may not be radical enough. The principles laid down by the Santa Barbara school of evolutionary psychology have been read by some as evolutionary principles grafted onto good old-fashioned cognitive psychology (the brain is a computer, cognition happens inside the mind, there are conscious and unconscious processes) combined with Jerry Fodor's modularity of function. In its place, she proposes a view of cognition (or maybe of psychology itself) in which thought does not stop at the cranium's edge. Cognition can be distributed among individuals and among individuals and artifacts; cognition is social; cognition is embodied; and so on. Reading this chapter brings to mind Leonard Read's (1958) *I, Pencil*, a short story in which a pencil explains to the author how no single person knows how to make a pencil, but instead it is the result of many thousands of people who know only one thing (how to mine graphite, how to shape wood, etc.). If the success (if that is the word) of our species is dependent on obligate cooperation – the ability to specialize and delegate to others – then perhaps our current conception of thought warrants a rethink?

REFERENCE

Darwin, C. (1871). *The Descent of Man, and Selection in Relation to Sex*. London: John Murray.

Hamilton, W. D. (1964). The genetical evolution of social behaviour. II. *Journal of Theoretical Biology*, **7**(1), 17–52.

Read, L. (1958). *I, Pencil*. Atlanta, GA: Foundation for Economic Education.

Trivers, R. L. (1971). The evolution of reciprocal altruism. *Quarterly Review of Biology*, **46**(1), 35–57.

10 The Problem of Altruism and Future Directions

LUKE COLQUHOUN, LANCE WORKMAN, AND JO FOWLER

10.1 INTRODUCTION

Models that attempt to explain altruistic behavior in terms of natural selection are models designed to take the altruism out of altruism.

– Robert Trivers (1971, p. 35)

Augustes Comte first coined the term "altruism" as a cornerstone for his ethic doctrine Positivism, describing it as selfless concern for another's welfare (Sutton, 1982). The existence of altruism and basic human kindness has been heavily debated throughout history, permeating through psychological, philosophical, and theological fields. Early philosophical and religious figures discussed concepts such Saint Augustine's theory of conscience – that basic human kindness was an innate human feature (Fortin, 1996) – or Thomas Aquinas' synderesis rule, which states that humans desire to be good and can innately distinguish between right and wrong (Davies, 2014). While these speculations have little or no empirical basis, they do have something in common with more recent research: they view altruism as having an innate component. Augustine and Aquinas attributed God as the innate origin of human altruism. Evolutionary psychologists view altruism as genetically influenced, having arisen, in part, from the Darwinian selective forces of natural and sexual selection. Definitional confusion often occurs between evolutionary psychology, social science, and popular science (Bhogal, 2017; West, El Mouden, & Gardner, 2011). As such, it is important to note that altruism is distinct from mutual benefit or mutualism (behavior beneficial to both the actor and recipient; West, El Mouden, & Gardner, 2011; West, Griffin, & Gardner, 2007a, 2007b) and fairness (behavior beneficial to both the actor and recipient through the equal division of rewards; i.e., a 50/50 split between two parties; Bhogal, 2017; Chiang, 2010). However, all of these terms are often used interchangeably with cooperation. Modern research makes a distinction between evolutionary and psychological altruism. Evolutionary altruism is defined in terms of reproductive fitness; an act of true altruism would enhance the reproductive fitness of the recipient(s) and diminish the reproductive fitness of the altruist (Bell, 2008), whereas psychological altruism can be described as the promotion of the well-being of others without acting with self-interest (Sober & Wilson, 1998).

Altruism is of great interest to evolutionary psychologists because of the apparent dilemma it presents to Darwin's theory of natural selection, a dilemma he acknowledged:

if it could be proved that any part of the structure of any one species had been formed for the exclusive good of another species, it would annihilate my theory, for such could not have been produced through natural selection. (Darwin, 1859, p. 189)

To overcome this dilemma, evolutionary psychologists have engaged in the pursuit of explanations for apparent acts of altruism. From this pursuit, a number of possible explanations have been suggested. Two of the most widely accepted explanations are kin selection (Hamilton, 1964) and reciprocal altruism (Trivers, 1971). However, many other explanations have been proposed, from induced altruism (Trivers, 1985), to costly signaling theory (Waynforth, 2011; Zahavi & Zahavi, 1997), to more complex multilevel and group selection models (Boyd & Richerson, 1982; Feldman & Cavalli-Sforna, 1976).

As Darwin rightly states, the greatest threat to his theory of natural selection is any behavior that has evolved that, with thorough investigation, can be described as true altruism. As such, the goal of the theories discussed in this chapter is to demonstrate that true altruism, in an evolutionary sense, does not exist. From a gene-centered viewpoint, a gene mutation that encourages true altruistic behaviors will not be selected for. So while it may seem that natural selection and evolutionary explanations for apparent acts of altruism contradict Augustine's and Aquinas' views that humans are innately good, remember that the following evidence demonstrates that, while not altruistic from the genes' eye view, many species – humans included – have evolved to engage in pro-social and cooperative behaviors.

In this review of altruism, we consider the explanations that have been developed to explain apparent altruism from an evolutionary perspective, before moving on to suggest fruitful areas for future research. We begin with the highly influential concept of kin selection.

10.2 KIN SELECTION

By behaving in an apparently altruistic manner to their non-descendent kin, an individual can successfully pass on copies of their genes by boosting their kin's reproductive output. This is called indirect reproduction and, together with direct reproduction, it determines an individual's inclusive fitness (i.e., the proportion of genes, by common descent, passed into future generations; Hamilton, 1964a, 1964b). While the importance of kinship had previously been noted by researchers (Fisher, 1930; Haldane, 1932, 1955), Hamilton (1964a, 1964b) popularized the theory by providing a mathematical formula that describes the conditions in which what become known as "kin selection" can occur:

$$rB > C.$$

An altruistic act would increase an individual's inclusive fitness when the recipient's degree of relatedness (r) times the benefit to their fitness (B) outweighs the cost to the altruist's fitness (C); this has subsequently become known as "Hamilton's rule."

Examples of kin selection can be seen in a wide range of species and taking form as various different behaviors. Alarm call frequency in Belding's ground squirrels (*Urocitellus beldingi*) has been linked to the degree of relatedness to the endangered conspecific (Milius, 1998). This association is commonly found in many species, including tufted capuchin monkeys (*Cebus apella nigritus*; Wheeler, 2008), vervet monkeys (*Chlorocebus pygerythrus*; Cheney & Seyfarth, 1985), and yellow-bellied marmots (*Marmota flaviventris*; Blumstein et al., 1997). Alarm calls are not restricted to just animals; plants also have evolved the capacity to produce alarm calls. Lima beans (*Phaseolus lunatus*) release alarm pheromones when infested by spider mites of the Tetranychidae family; this allows conspecifics to activate defense genes and attracts spider mite predators. Kobayashi and Yamamura (2007) reported uninfected plants produce a secondary signal, but this costs resources and is seemingly unnecessary; they suggest that this behavior may benefit surrounding kin and increase the signalers' inclusive fitness. In numerous bird species, conspecifics help in other nesting attempts, from preparing the nest to caring for and defending the young. This behavior can be seen in Florida scrub jays (*Aphelocoma coerulescens*; Woolfenden, 1975) and white-fronted bee-eaters (*Merops bullockoides*; Emlen & Wrege, 1988), among others. Some species, such as the chestnut-crowned babbler (*Pomatostomus ruficeps*), are even

obligate cooperative breeders totally reliant on kin support (Browning et al., 2012). Kin-selected cooperative breeding can even be seen in fish such as daffodil cichlids (*Neolamprologus pulcher*; Dierkes et al., 2005) and mammals such as brown hyenas (*Hyaena brunnea*; Owens & Owens, 1984) and alpine marmots (*Marmota marmota*; Arnold, 1990a, 1990b). Kin selection can lead to a change of competition and strategy; the yellow jewelweed plant (*Impatiens pallida*) changes its growth pattern when surrounded by close kin by redirecting resources to branching and stem elongation as opposed to competitively directing resources to its leaves (Murphy & Dudley, 2009).

While these studies demonstrate the preferential treatment of kin, few studies have provided more than qualitative support for kin selection because of the difficulty in quantifying the benefits and costs of an altruistic act. Gorrell et al. (2010) delivers a quantitative example of Hamilton's rule. They found that in scenarios when asocial red squirrels (*Tamiasciurus hudsonicus*) adopt, the behavior followed Hamilton's rule. However, it should be noted that adoption is incredibly rare; only five cases were reported over 2,230 litters and 19 years of observation. Waibel, Floreano, and Keller (2011) overcome this problem by simulating 500 generations of artificial evolution with 200 groups of 8 foraging Alice robots (for initial development of the robots, see Waibel, Keller, & Floreano, 2009). They could manipulate costs, benefits, and "relatedness." The Alice robots had 33 "genomes" (neural network connection weights) that determined their sensory perception and behavior. The study reported Hamilton's rule always predicted the minimal relatedness required for altruism to evolve, even with the presence of pleiotropic (multiple phenotypic effects), epistatic (modifier gene needed to activate phenotypic display), and non-additive effects (dominant allele determines phenotypic display).

Kin selection can be used to explain cases where relatedness is very low, approximating 0; Refardt, Bergmiller, and Kümmerli (2013) investigated altruistic *Escherichia coli* strains that commit suicide upon phage T4rll infection in low-relation cultures, comparing them with non-altruistic *E. coli*. Phage infection would eradicate non-altruistic *E. coli* cultures, but altruistic *E. coli* would be unhindered. They suggest that because of the lethality of phage infection, the cost of suicide is low and the benefit to the culture is high, allowing for an inclusive fitness gain from altruistic behavior. Kin selection is also commonly used to explain the evolution of eusociality (Hamilton, 1964). Hamilton suggested that because eusocial Hymenoptera species engage in haplodiploidy (males are haploid and develop from unfertilized eggs, whereas females are diploid and develop from fertilized eggs), the nonbreeding females have a higher degree of relatedness with their sisters ($r = 0.75$) than any offspring they may have ($r = 0.5$). This explanation has been criticized, as there are species with multiple queens or queens taking multiple mates, therefore lowering the genetic relatedness of sisters (Andersson, 1984). Furthermore, this argument is dependent on haplodiploidy, and while eusociality has evolved more frequently in haplodiploid Hymenoptera (Andersson, 1984), other species engage in eusociality, including eusocial mammals, such as Damaraland mole-rats (*Fukomys damarensis*) and naked mole-rats (*Heterocephalus glaber*; Burda et al., 2000), and eusocial crustaceans, such as sponge-dwelling shrimps (*Synalpheus chacei, Synalpheus filitigitus,* and *Synalpheus regalis*; Duffy, Morrison, & Rios, 2000). Kin selection can still be used to counter these criticisms; ancestral monogamy has been determined in a number of cases in which eusociality has evolved (Hughes et al., 2008), potentially dismissing criticisms of multiple queens/mates, whereas inbreeding can lead to higher relatedness than is seen in haplodiploidy, as seen in naked mole-rats (Honeycutt, 1992), presenting options other than haplodiploidy that may encourage eusociality.

10.2.1 Kin Selection in Humans

Kin selection and Hamilton's rule can commonly be seen in many human behaviors. By analyzing Viking Orkneyinga and Njal's sagas, Dunbar, Clark, and Hurst (1995) considered the role of kinship, cost, and benefit played in murder and revenge. They demonstrated that when exacting revenge for fallen kin, if the cost was high (e.g., the aggressor is known as a superior fighter) or the kin's relatedness was low, then blood money settlements were more likely to occur. They also found alliances between kin were more likely to occur with fewer preconditions and that the murder of competitors for resources was inversely linked to relatedness. McCullough, Heath, and Fields (2006) found similar relationship dynamics in the English mid-millennial Cousins' War, demonstrating individuals were more likely to kill distant cousins opposed to close kin. Nepotism – the practice of favoring relatives usually by giving kin favorable roles and jobs – is a more common way in which kin selection can be observed. Nepotistic/kin selection tendencies can be seen in agricultural cooperation (Berte, 1988; Hames, 1987), cooperative breeding in humans (Crittenden & Marlowe, 2008), the sharing of food (Gurven et al., 2001), and selection for jobs (He, 2005; Lentz & Laband, 1989). Kin selection would be expected to be especially present in high-cost situations; this can be seen from studies examining disasters such as the Donner Party or Plymouth Colony disasters (Grayson, 1993; McCullough & Barton, 1991), or combat situations such as the Gulf War (Shavit, Fischer, & Koresh, 1994). Attempts to explore the effects of cost and degree of relatedness on helping behavior show that the share of a deceased's estate given to be proportional to beneficiaries' genetic relatedness (Smith, Kish, & Crawford, 1987). However, in a cross-cultural study where participants could gain monetary rewards for different family members by maintaining a painful skiing position with rewards increasing with time, as the beneficiaries' degree of relatedness increased, so did the length of time participants would maintain this cost (Madsen et al., 2007).

Beyond the effects of degree of relatedness on altruistic responses, research has suggested numerous possible factors that may influence kin selection. Burnstein, Crandall, and Kitayama (1994) conducted a large hypothetical scenario decision task comparing responses between everyday favors and life-or-death situations. Manipulating a number of factors that may influence kin selection, such as age or health, they found a number of differences between these costs of helping. In life-or-death scenarios, participants would help young, healthy, and/or wealthy individuals as opposed to very young or old, unhealthy, and/or poor individuals. They also found more help was given to premenopausal as opposed to postmenopausal women. But in the lower cost situation of everyday favors, the authors found a curvilinear effect for age, with the very young or old receiving more help, and the reverse or diminished effects for health, wealth, and many other factors. They attributed these results to Hamilton's rule, stating that these factors increase an individual's reproductive value; therefore, there would be greater inclusive fitness gain in helping kin with greater reproductive value, even if they are less close, especially in high-cost situations. In low-cost situations, morality and other social factors have more influence. While there was still found a preference to help close kin, the results demonstrate that it is not sufficient simply to state that kin selection should lead to favoring close relatives; rather, it should lead to the selection of the optimal indirect fitness gain. Similar results are reported by Essock-Vitale and McGuire (1985) using a sample of Los Angeles women. Perhaps the most researched factor is paternal uncertainty, which has been found to diminish altruism toward paternal relatives in nonhuman species (Gross & Shine, 1981). This can be seen as an influencing factor in human relationships, too; wealth is often distributed through maternal lineage (Gaulin & Schlegel, 1980). Studies exploring relationships between grandparents and grandchildren found evidence for this theory in the intensity of grief over the loss of grandchildren (Littlefield & Rushton, 1986), the emotional closeness to grandparents (Eisenberg, 1988; Pashos & McBurney, 2008; see also Grey & Brogdon, 2017; Pashos, Schwarz, & Bjorklund, 2016) or grandchildren (Littlefield & Rushton, 1986), and the amount of caregiving received from grandparents (Danielsbacka & Tanskanen, 2012, Danielsbacka, Tanskanen, Jokela, & Rotkirch, 2011; Euler & Weitzel, 1996; Laham, Gonsalkorale, & von Hippel, 2005).

10.3 RECIPROCAL ALTRUISM (RECIPROCATION)

By the early 1970s, it became clear that not all apparent altruistic behaviors could be explained by kin selection. In 1971, Robert Trivers published a groundbreaking article in which he outlined a way in which individuals of social species might aid each other without jettisoning Darwinian selective forces. In his concept of reciprocal altruism, Trivers (1971) explained that if the benefit to the recipient's fitness is greater than the detriment to the altruist's fitness, then, if a similar degree of help is reciprocated at a later date, both the altruist's and the recipient's fitness is increased and this behavior can be selected for. It should be noted that reciprocal altruism is more likely to evolve under a number of conditions. Species with long lifespans or that repeatedly interact with the same set of individuals are likely to have more altruistic events occur in their lifetimes. Species with the ability to recognize individuals are more capable of reciprocation and detecting cheats. Reciprocal altruism is often known simply as "reciprocation" today, and it has been reported in a wide range of species manifesting as a variety of different behaviors. Numerous bird species such as northern bald ibises (*Geronticus eremita*; Portugal et al., 2014; Voelkl et al., 2015), great white pelicans (*Pelecanus onocrotalus*; Weimerskirch et al., 2001), and pink-footed geese (*Anser brachyrhynchus*; Cutts & Speakman, 1994) engage in V-shaped flight formations, taking turns to utilize the upwash created by conspecifics at the front of the formation. Subordinate olive baboons (*Papio anubis*) assist fellow subordinates to outcompete a dominant conspecific for female attention (Packer, 1977), whereas vervet monkeys (*C. pygerythrus*) form reciprocal relationships and give predator alarm calls for nonkin (Seyfarth & Cheney, 1984). Reciprocal relationships can even be formed between different species; a cleaning symbiosis is an example of such a relationship. Feder (1966) reports a reciprocal relationship with the bluestreak cleaner wrasse (*Labroides dimidiatus*) cleaning parasites and dead skin off the larger predatory Nassau grouper (*Epinephelus striatus*). Reciprocation is the perfect example of how behavior may look altruistic at first glance, but we see that with time and patience the actor receives their payoff, increasing their chances of surviving and, in turn, reproducing and thus passing on copies of their genes. This may not be true altruism, but it does demonstrate the genetic advantage of such pro-social behaviors.

However, studies on reciprocation do not always clearly define reciprocal altruism; behaviors such as mutualism (immediate cooperative behavior) or manipulation (coercive or begging behavior) are generally not considered to be reciprocation (Clutton-Brock, 2009). Trivers (1971) did not include these behaviors in his explanation of reciprocal altruism, whereas others view these as related behaviors (Axelrod & Hamilton, 1981). Examples of this definitional dilemma include the cooperative hunting of African wild dogs (*Lycaon pictus*) and the territorial choruses and synchronized mobbing of southern pied babblers (*Turdoides bicolor*). Clutton-Brock (2009) points out there is no immediate cost–benefit asymmetry in the wild dogs' mutualistic behavior, and while the territorial behavior of one southern pied babbler may incur a cost, by working with the flock, the individual bird increases its own fitness. While vampire bats (*Desmodus rotundus*) have been observed regurgitating blood meals to starved conspecifics (DeNault & McFarlane, 1995; Wilkinson, 1984), Clutton-Brock (2009) suggests that manipulation may also account for this behavior. Vampire bats may increase their fitness by feeding begging conspecifics rather than persevering through their nuisance.

Further issues include behavior being misinterpreted as reciprocation. There is evidence that vampire bats' food sharing may be between relatives; if so, then this reciprocation falls under kin selection (Carter & Wilkinson, 2013). Seemingly reciprocal behavior may also be misinterpreted as altruism. Food-sharing relationships are traditionally viewed as reciprocal altruism (Moore, 1984). However, chimpanzees (*Pan troglodytes*) have been observed to only abandon food when they are already sated (Morris & Goodall, 1977; Nishida, Uehara, & Nyundo, 1979; Riss & Busse, 1977); this act does not cost the individual, and so does not constitute altruism. Warning calls in bird species have previously been used as an example of reciprocal altruism (Marler, 1955, 1957). However, saving a conspecific may limit the knowledge that the predator gains about the prey species and locale. Predatory vertebrates have been demonstrated to rely heavily on this learning to hone their skills (Rudebeck, 1950, 1951; Tinbergen, 1960); as such, the payoff for this costly behavior may not come from reciprocation at a later date, but by actively limiting a predator's hunting skills and preferences.

One theory – indirect reciprocity – suggests that in large groups, by demonstrating cooperative behavior, an individual increases their "reputation" as a cooperator. Other individuals may only provide help to those with a high "reputation" for such behavior; therefore, the cooperator gains a boost to fitness, and non-cooperators do not (Nowak & Sigmund, 2005; Panchanathan & Boyd, 2004; see Section 10.5).

10.3.1 Reciprocation in Humans

It has been suggested that reciprocal altruism rarely occurs in species outside of primates (Clutton-Brock, 2009). Trivers (1985) even suggested reciprocation may have been an important factor in human evolution. There is little doubt that humans engage in reciprocal relationships. Various examples exist, such as the "gift economy" of African !Kung San nomadic bands (Lee, 1979) and the South American Yanomami (Diamond, 2012), the dyadic food sharing of the Eurasian Dolgan and Nganasan (Ziker, 2005), and reciprocal constraint in warfare during World War I (Axelrod, 1984). In Axelrod's (1984) book, *The Evolution of Cooperation*, and Ashworth's (2000) *Trench Warfare, 1914–1918: The Live and Let Live System*, a collection of fascinating stories and accounts of apparent altruism are presented:

In one section the hour of 8 to 9 A.M. was regarded as consecrated to "private business," and certain places indicated by a flag were regarded as out of bounds by the snipers on both sides. (Ashworth, 2000, p. 33)

Some accounts are even rather humorous in hindsight, a stark contrast to the conditions and horrors we often hear regarding World War I and the accompanying human behavior:

I was having tea with A Company when we heard a lot of shouting and went out to investigate. We found our men and the Germans standing on their respective parapets. Suddenly a salvo arrived but did no damage. Naturally both sides got down and our men started swearing at the Germans, when all at once a brave German got on to his parapet and shouted out "We are very sorry about that; we hope no one was hurt. It is not our fault, it is that damned Prussian artillery." (Ashworth, 2000, p. 146)

A number of accounts speak of the punishment of cheats: "if the British shelled the Germans, the Germans replied" (Ashworth, 2000, p. 80). This represents a tit-for-tat-like system where the immediate cost of letting enemy combatants live benefits the actor at a later date, with relative peace and safety created through an uneasy truce. In contrast, the cost of cheating is swift and equal retaliation, appropriately referred to as a "live and let live" system by Axelrod and Ashworth. This behavior was noted by generals who, according to some accounts, made attempts to stop this behavior. Ashworth (2000, p. 37) comments that despite the severity of high command's reactions – "any attempt on his part to fraternize is to be instantly repressed" – truces still held, especially for the collection of fallen comrades.

Humans form complex reciprocal relationships, and it has been theorized that a number of methods are used to maintain these relationships. Frank (1993) suggests the use of a "commitment device," stating that human emotions and facial cues are used to demonstrate honest intentions. It should be noted that it has been suggested that human cheats could employ complex methods of hiding their intentions, such as Trivers' (2011) theory of self-deception, in which humans are not consciously aware that they will cheat, allowing them to hide dishonest cues. Singer (1981) suggests that blood donation may be an example of true altruism, as the donor cannot possibly expect that the recipient is likely to repay this behavior. However, there are a number of arguments against this example. Political scientist Robert Axelrod (1984) claims that the payoff for this behavior may come from demonstrating willingness to cooperate with others (indirect reciprocity). Altruism could serve as a valuable signal; not only is it attractive (see Section 10.5), but it could also alert potential cooperators and reciprocators to an actor's value as an "ally," thus benefiting the actor by increasing their reputation. Altruistic tendencies are shown to increase when behavior is made public, from altruism in economic games like public goods or dictator games (Andreoni & Petrie, 2004; Filiz-Ozbay & Ozbay, 2014; Rege & Telle, 2004), to greater engagement in volunteering

(Carpenter & Myers, 2010; Linardi & McConnell, 2011), or buying "green" products rather than cheaper alternatives (Griskevicius, Tybur, & Van den Bergh, 2010). Public rewards demonstrate this same influence; Low and Heinen (1993) reported greater donations when a fundraiser gave badges for involvement, and Lacetera and Macis (2010a, 2010b) reported participants were more willing to donate blood if they were promised a public reward. Furthermore, it is also possible that, while the recipient cannot repay the donator, should the donator need a blood transfusion in the future by supporting the continuation of blood donation, they would be rewarded with potentially lifesaving care.

10.3.2 Game Theory

Social scientists have long struggled to understand the circumstances under which people engage in apparently kind behavior. One way of exploring such a question is through "game theory." Game theory is the mathematical application of decision-making in situations of conflict or cooperation. Game theory can be tested via non–zero-sum games such as the "iterated prisoner's dilemma," in which both parties (A and B) can either cooperate or defect. The game is called the prisoner's dilemma because it is based on the scenario of two suspects being arrested by the police and placed in separate cells. Both are told that if they implicate the other (i.e., "defect") they will be set free, while the other will receive a harsh sentence (i.e., "sucker's payoff"). If neither talks (i.e., they "cooperate" with each other), however, then both will receive a light sentence. This leads to four possible outcomes: both cooperate; both defect; A cooperates while B defects; and B cooperates while A defects. The payoff for each is symbolized by one of four letters: T equals the temptation to defect; R equals the reward each receives should they cooperate; P equals the punishment received if both defect; and S equals the sucker's payoff (i.e., you cooperate and your partner defects). This means that the payoff decreases from T to R and P to S. Hence, mathematically, we can express the payoff in the prisoner's dilemma as $T > R > P > S$. In practice, the game is played for points, with the typical values being:

$$T = 5; R = 3; P = 1; S = 0.$$

In its simple form the game is played only once, but in the iterated version it is played a number of times, allowing each player to learn about the typical playing strategy of the other player (Axelrod, 1984; Deulofeu, 2014; Storey & Workman, 2013; Workman & Reader, 2014). When people play the iterated prisoner's dilemma, there are a number of strategies they can apply, such as always cooperate or always defect. Axelrod was curious to learn which strategy was the most successful, so he ran a tournament based around a large number of possible strategies in order to test them against each other. Axelrod (1984) discovered that the strategy known as "tit-for-tat" (cooperate first round, then select the option your opponent used last turn) came out on top against all others. Why should this be the case? Axelrod suggests it is a combination of three factors that leads to this success. First, the strategy is "nice" (always cooperates first turn); second, it is "forgiving" (cooperates so long as the opponent cooperated on the last turn, regardless of previous displays of defecting); and third, it is "retaliatory" (immediately punishes defection). This tit-for-tat strategy can be seen regularly in human interactions. In fact, a good example of this is the "live and let live" system reported in World War I (and in the retaliation that followed defection there). In the iterated prisoner's dilemma, we have a microcosm of real-life dilemmas that regularly occur in everyday life – from whether we should offer people a lift in our car, to tipping a waiter (or not), to deciding whether or not to remain faithful to our romantic partner. Do we cooperate or do we defect? In most cases, it will be to our advantage in the long run to cooperate – so long as this is reciprocated, that is.

10.4 COMPLEX GROUP SELECTIONIST MODELS

Today, most people who study the relationship between evolution and behavior consider that selection acts at the level of the individual, or even at the level of the gene. This has largely been the case ever since Hamilton developed kin selection theory in 1963. Prior to this, however, the concept of animals acting altruistically for the good of the group was widespread and particularly associated with Scottish ethologist Vero Wynne-Edwards (1962). In 1962, in his book *Animal Dispersion in Relation to Social Behavior*, Wynne-Edwards advocated this position, stating that animals were truly altruistic, as he believed was evident from behaviors such as alarm calls to controlled breeding in order to ensure the survival of the group. Due to prominent evolutionary theorists such as Bill Hamilton, Maynard-Smith, and George Williams, however, this concept was overturned due to the "subversion from within" problem (Maynard-Smith, 1964; Workman & Reader, 2014). The subversion from within problem simply states that, in a population where everybody only breeds when it is for the good of the group, individuals should arise that attempt to produce as many offspring as possible, selfishly ignoring the greater good of the group. Over a number of generations, the genes of the selfish individuals will lead to the extinction of the group altruists. Due to this problem, by the 1980s, Wynne-Edwards' notion of group selectionism became known as naive group selectionism.

10.4.1 New Group Selection

Despite the dismissal of naive group selectionism, since the 1980s, new forms of group selectionism have emerged, in part to help explain altruistic behavior (Laland & Brown, 2011). These have been labeled collectively as

"new group selection." There are three main overlapping forms of new group selection:

- Multilevel selection (MLS)
- Dual inheritance theory (DIT; or gene–culture coevolution)
- Cultural group selection (CGS)

Advocates of new forms of group selection generally either consider highly integrated species such as the social insects (Hymenoptera and Isoptera) and naked mole-rats (*Heterocephalus glaber*) and/or they consider human societies (Laland & Brown, 2011; Wilson & Brown, 1994). In the case of human societies, the argument is that since the invention of agriculture around 12,000 years before present humans have lived in increasingly complex societies that require increasingly complex forms of cooperation. This has become known as "ultrasociality." Human ultrasociality is believed to have arisen when we developed the concept of division of labor; that is, some individuals become experts, for example, at aspects of farming such as planting and tending crops, while others become experts at making tools or weapons to defend the tribe. According to new group selectionists, while cooperation is advantageous to the group of cooperators as a whole, it may also curb some individual freedoms. Cooperation increases the fitness of the cooperators when together they are able to collect more resources than the sum of resources that could be collected by each of them individually.

10.4.2 Multilevel Selection

In 1994, David Sloan Wilson and Elliott Sober began to argue that misgivings against of group selectionism had been overstated (Wilson, 2007; Wilson & Sober, 1994; Wilson & Wilson, 2007). They argued that, under some circumstances, the group can be perceived of as a single organism (sometimes called a superorganism). Groups where cooperation is high might outperform (in terms of reproductive output) those where cooperation is low, such that the individuals in that group pass on more copies of their genes. Wilson and Sober perceive this rather like a Russian doll, with each level contained within the one above. So we have genes within cells, cells within an organism, and organisms within a group. Each level has evolved by natural selection to maximize fitness or reproductive success. Furthermore, they claim that, under some circumstances, selection at the group level can outweigh selection at the individual level.

Wilson and Sober have argued that MLS may help to explain altruistic behavior for certain specific species, including social insects, naked mole-rats, and, importantly, humans. They suggest that, for humans, group selection can be a real force because we are so socially integrated, and they argue that, for many societies, we are encouraged to think altruistically about the group rather than the individual.

One supporter of MLS is E. O. Wilson (2005, 2012), who has done a volte-face and now argues that Hamilton's kin selection model does not explain the evolution of extreme sociality (i.e., for social insects). With coauthors Nowak and Tarnita, Wilson argued kin selection and inclusive fitness had little explanatory power for the behaviors of eusocial species, instead suggesting group-level selection held more sway. Nowak, Tarnita, and Wilson (2010) suggested Hamilton's model of:

$$rB > C$$

should be refined by the equation:

$$rB_k + B_e > C.$$

In this case, B_k is the benefit to kin (B in the original equation) and B_e is the benefit to the group. They also argue that, at least for social insects $B_e > rB_k$; that is, the benefit to the group outweighs the benefit to kin. This means that for both D. S. Wilson and E. O. Wilson, altruism should be explained in terms of selection at the colony level rather than at the kin level.

While it does have its supporters, this form of explanation for altruism remains controversial with mainstream evolutionists, because $rB > C$ works for most cases and $B_e > rB_k$ only works under some extreme circumstances. In fact, in response to Nowak, Tarnita, and Wilson (2010), 137 authors from various fields penned a defense of Hamilton's rule, describing their work as "a misunderstanding of evolutionary theory and a misrepresentation of the empirical literature" (Abbot et al., 2011). In summary, while MLS theories do recognize gene- and individual-level selection, staunch gene-level supporters doubt that MLS (or indeed group selection in any form) could occur as a consistent, reliable level of selection (Dawkins, 2016).

10.4.3 Dual Inheritance Theory

Another theoretical attempt to bring group selection back into our understanding of altruistic behavior is via DIT (also called gene–culture coevolution). DIT suggests that cultural transmission is possible in humans because, at some point in human evolutionary history, social learning (the ability to acquire beliefs, ideas, and strategies via imitation) became adaptive and was selected for (Baum et al., 2004; Efferson et al., 2008; Feldman & Cavalli Sforza, 1976; Henrich & McElreath, 2007; McElreath et al., 2005). The main idea here is that cultural evolution alters the physical and behavioral environment in which genetic changes are naturally selected for. Due to cultural transmission's bidirectional relationship and impact on genetic evolution, social learning limits the ability to track changes in the environment. As such, behaviors that are maladaptive, such as altruism, might still be naturally selected for because cultural evolution has altered the selective pressures affecting the selection of altruists (McElreath et al., 2005; Richerson & Boyd, 2005). Henrich and McElreath (2007) use early human farming as an example. If an individual were to imitate a successful farmer, some of the behaviors they would learn would be adaptive (e.g., knowing what, when, and where to plant), whereas some of the behaviors would be potentially maladaptive (e.g., making sacrifices to spirits).

In addition to explaining phenomena such as the rise of agriculture, dual inheritance theorists have been particularly concerned with providing answers as to why we see such high levels of altruism in our species. They suggest that our current propensity for altruistic behavior is due to our evolved tendency to gravitate toward group norms. One of these group norms has been labeled "strong reciprocity." Strong reciprocity is based on Trivers' notion of reciprocal altruism, but is more generalized and culture specific. Strong reciprocity refers to a combination of:

- Tendencies to reward others for cooperation
- Norm-abiding behavior
- Altruistic punishment (i.e., the propensity to impose sanctions on others who violate norms)

It is the norm-violation part that gives it its cultural specificity. What might be a norm in one culture may be abhorrent in another – such as wearing a bikini or a burka. The idea here is that those cultural groups that exhibited strong reciprocity in the past are considered to have outcompeted those cultural groups that did not exhibit strong reciprocity. This then results in genes that underlie prosocial behavior being selected for, leading ultimately to a form of universal pro-social behavior. Note that this differs from traditional evolutionary psychology explanations in that the selection of the genes follows the cultural change.

10.4.4 Cultural Group Selection

Another way of examining the concept of group-level selection in relation to altruistic tendencies is to argue that cultural transmission now outweighs genetic transmission. Such a stance is known as CGS (Boyd & Richerson, 1982, 1985). CGS differs from DIT in that the genes merely create a brain that allows for flexibility, with culture now having taken over as a force for change (Portin, 2015; Richerson et al., 2016). This means that CGS differs from gene–culture coevolution in that it does not see changes in gene frequencies as being a part of the explanation. CGS gains support in the form of game theory models (Soltis, Boyd, & Richerson, 1995) and population genetic formulization (Lehmann et al., 2007; Lehmann, Feldman, & Foster, 2008). CGS suggests that cultural transmission homogenizes the behaviors of individuals within a group; however, this is dependent on groups being distinct.

Laland and Brown (2011) provide an example of how this might work. Imagine a group where the members cooperate to build a stockade to protect themselves from other groups that might want to take their land or resources. During wartime, this group will suffer far fewer casualties than a group that does not build a stockade. This means that its membership will increase and new groups will be descended from it. Thus, to adherents of CGS, this is a form of group selection that does not directly involve specific genes, since there are no genes for stockade building. According to Laland and Brown, there are four features of CGS that increase a group's homogeneity and make this concept plausible:

- First, conformity helps maintain differences between groups (the process discriminates against nonconformers). It also reduces individual-level variation and competition.
- Second, selection of culturally selected traits is generally much faster than selection of genetic variants (this is very important when there is competition or even warfare between groups).
- Third, symbolic group marker systems (e.g., language or cultural icons) help groups to maintain group boundaries.
- Fourth, socially sanctioned punishments for cheating (an important part of strong reciprocity), when combined with cultural transmission of information (e.g., gossip) about such cheaters, removes the possible advantages of noncooperation.

Based on observations of pro-social behavior of various groups in New Guinea, Boyd and Richerson (2005) claim that CGS can work. There is, however, a downside to this, since CGS can lead to altruism within the group, but simultaneously promote selfishness between groups. In fact, Bowles (2009) looked at archaeological data on causes of death during the late Pleistocene and the early Holocene and concluded that between-group competition led to a large increase in mortality rates. In a sense, the claim that CGS can lead to altruism in humans really raises the question: Altruism toward whom? From this perspective, suicide bombers may be thought of as altruists to the cause of their group by other group members, but to non-group members they are anything but.

10.5 COSTLY SIGNALING THEORY: SEXUALLY SELECTED COMPETITIVE ALTRUISM

In 1975, Amotz Zahavi suggested that organisms may develop costly traits that appear to be a burden in order to signal quality. According to Zahavi, sexual selection may have led to the development of costly signals in males in particular due to their lower level of parental investment (Zahavi, 1975; see also Zahavi & Zahavi, 1997). Zahavi originally labeled this concept the "handicap principle," but it has subsequently been developed into "costly signaling theory" (Arnocky et al., 2016; Waynforth, 2011; Workman & Reader, 2015). Costly signaling theory states that maladaptive traits could possibly alert others (and, in particular, potential mates) that individuals have good genes because they have survived and functioned in spite of their disadvantage (Grafen, 1990). During the

development of costly signaling theory it became clear that the display must be "honest," otherwise it would quickly become ignored (Waynforth, 2011). In recent years, researchers have proposed that males are more likely to produce costly signals through altruistic behavior, either by aiding females or by aiding males to signal that they are of higher quality. In this way, both intra (within sexes) and inter (between sexes) components of sexual selection may be operating.

Costly signaling theory as an explanation for altruistic behavior in males seems to make sense on paper. The question is: Does it have empirical support? A number of studies suggest that it does in both humans and in other species. Barclay (2010) had both male and female participants read dating advertisement passages about the behavior of members of the opposite sex. Some of these descriptions suggested high levels of altruism while others did not. Barclay found that high levels of altruism were favored as showing potential for a long-term partner by both sexes, whereas when considering short-term partners, only females rated these more highly. Due to this and other findings, Barclay has described this phenomenon as "competitive altruism" (Barclay, 2011, 2013, 2016); that is, males in particular use apparent altruism to compete for the attention of the opposite sex. Following up on the themes explored by Barclay, Arnocky et al. (2016) found that "altruists" self-reported more sexual partners than "non-altruists." This was particularly the case for males. In a follow-up study, men who were more willing to donate money to a good cause reported a larger number of lifetime romantic partners (Arnocky et al., 2016). Both of these findings support the notion that altruism increases (male) mating success in humans. In addition to recent empirical studies, circumstantial evidence for this position arises from social anthropologists' studies of hunter–gatherer peoples. It is well established that, in forager societies, men who hunt and share meat enjoy greater reproductive success (Hawkes, 1991; Hill & Hurtado, 1996). In the words of Steve Stewart Williams:

The reason we're nice is that, throughout the course of our evolution, kind, loyal, and generous people attracted more romantic interest than their cruel, traitorous, or selfish peers, and therefore did better in the reproductive sweepstakes. (2018, p. 200)

As in kin altruism and reciprocity, sexually selected altruism has also been observed in nonhuman animals. A prime example of this is the case of the Arabian babbler (*Turdoides squamiceps*), a small brown bird nesting in communal groups in arid areas of Arabia, where males compete to act as sentinels for the group, risking their lives to look out for hawks (Zahavi & Zahavi, 1997). Males also compete to provide food for chicks, some of which are nonrelatives. According to the Zahavis, this ultra-helpfulness is a way of showing to both males and females in the group that "I have such great genes that, in addition to feeding and protecting myself, I can do this for others. If you're a female, I would be a good bet to recombine your genes with. If you're a male, don't bother competing with me." Barclay's term "competitive altruism" seems highly appropriate here.

10.6 INDUCED ALTRUISM

Induced altruism refers to an act of altruism that has been forced, usually through trickery (Trivers, 1985). One of the most frequently used examples is brood parasitism. Numerous bird species parasitize other bird species to rear their young, including the brown-headed cowbird (*Molothrus ater*), European magpie (*Pica pica*), and great spotted cuckoo (*Clamator glandarius*). To ensure the parasitized species altruistically rear their young, these parasites predate and mob noncompliant birds, thus enhancing the reproductive fitness of compliant birds (Soler, Soler, Martinez, & Moller, 1995). Numerous other behaviors may constitute induced altruism. Many species, such as the false cleanerfish (*Aspidontus taeniatus*) and the bluestriped fangblenny (*Plagiotremus rhinorhynchos*), mimic the behavior of cleanerfish to take advantage of the passive reactions of larger fish (Trivers, 1971). Members of the entomopathogenic fungal *Cordyceps* and *Ophiocordyceps* species, such as the zombie fungus (*Ophiocordyceps unilateralis*), infect insects and other arthropods, forcing them to seek suitable environments that encourage fungal growth and spread to other conspecifics (Mongkolsamrit et al., 2012). This morbid example is not unique in the biological world: our microbes, symbionts, and parasites can influence our emotions and our behaviors (Dobson, 1988). Numerous studies show that gut infections increase anxiety-like behaviors in mice (*Campylobacter jejuni*, Goehler et al., 2008; *Trichuris muris*, Bercik et al., 2010), probiotic treatment reduces/reverses anxiety- and depression-like behaviors in mice (*Bifidobacterium longum*, Bercik et al., 2010, 2011; *Lactobacillus rhamnosus*, Bravo et al., 2011; *T. muris*, Bercik et al., 2010), and infections influence social behaviors such as changes in sexual interest, as seen in *Plasmodium pediocetii*-infected greater sage-grouse (*Centrocercus urophasianus*; Johnson & Boyce, 1991) and Borna disease virus-infected common treeshrews (*Tupaia glis*; Sprankel et al., 1978). Human "autonomy" is not immune to these behavior changes (Allen et al., 2017), and it has been proposed that, like sexual behaviors and aggression, altruistic behaviors may also be influenced by our microbiome. Computational models demonstrate how microbes could promote this behavior within a species population. Lewin-Epstein, Aharonov, and Hadany (2017) argue that altruistic and cooperative behavior increases contact between hosts, and this contact increases the chance for the horizontal transfer of microbes, leading to selection of microbes that increase their host's altruistic tendencies.

10.7 FUTURE DIRECTIONS

It is clear that evolutionists have developed a number of explanations for the existence of apparent altruism. Despite criticism from some quarters, kin selection theory continues to provide a fruitful theoretical framework for understanding altruistic behavior. Moreover, a plethora of empirical studies have been developed that demonstrate how well kin selection theory stands up to scrutiny as an ultimate explanation for various forms of pro-social behavior. Arguably, however, this progress has not been matched by developments in understanding of the nature of the supporting proximate mechanisms. Back in 2001, Korchmaros and Kenny suggested that we may cultivate a better understanding of the proximate cues and mechanisms that support altruistic behavior by combining evolutionary psychology with developments in social and relationship psychology. Future research might follow this example by combining evolutionary and social/relationship principles, so that not only do we recognize why acts of apparent altruism have evolved, but also we develop a better understanding of the underlying proximate mechanisms.

One area that might prove fruitful in understanding the nature of adaptive proximate cues is the concept of emotional closeness. Korchmaros and Kenny (2001, 2006), along with many other researchers (Ackerman, Kenrick, & Schaller, 2007; Brown & Brown, 2006; Mills & Clark, 1994; Neyer & Lang, 2003) have suggested that emotional closeness serves as a proximate cue for kinship. They found that emotional closeness mediates the effect of relatedness and altruistic willingness to help for both low- and high-cost altruistic acts. Definitions of emotional closeness vary, but feelings of concern, trust, and enjoyment experienced in human relationships underpin most definitions (Lee, Mancini, & Maxwell, 1990). Similarly, the term "communal relationships" (relationships in which help is given out of care and concern, without the expectation or obligation of reciprocation; Clark & Mills, 1993) can be used to describe emotionally close interactions. Terms such as "empathic concern" and "oneness" (the self-perceived overlap of one's own identity and the identity of another individual; Cialdini et al., 1997) have also been linked to helping intentions. Kruger (2003), for example, reported that empathic concern and oneness are predictive of helping intentions. These definitions are rather vague and potentially overlap. Future research would do well to establish more distinct definitions. Regardless of this definitional overlap, the research into these proximate cues suggests that, in the human era of evolutionary adaptation, groups were composed chiefly of closely related individuals, meaning that those one was emotionally close to were most likely one's kin. Hence, it is likely that the kin selection heuristic rule of "help those you are emotionally close to" would have been selected for. This means that altruism toward emotionally close nonkin can be seen as hijacking a once-adaptive heuristic. Furthermore, in some studies, emotionally close

LUKE COLQUHOUN ET AL.

categories of nonkin friends have received equal or more help than kin (Cialdini et al., 1997; Essock & McGuire, 1980; Kruger, 2003).

Research on emotional closeness is advancing our understanding of the proximate cues and behaviors that are, arguably, derived from kin selection in humans. There is, however, still a lot more to learn about how this relationship operates. Bressan, Colarelli, and Cavalieri (2009) found that emotional closeness predicted altruism for nonrelatives but not relatives in high-cost scenarios. Advocates for the argument that reliable kinship cues and not proximate behavioral cues determine altruism in high-cost scenarios call this the "kinship premium" (Bressan, Colarelli, & Cavalieri, 2009; Curry, Roberts, & Dunbar, 2013; Pollet, Robert, & Dunbar, 2013; Stewart-Williams, 2007). The relationship between emotional closeness and kin selection is clearly a complicated one that is not well understood. Colquhoun, Davies, and Workman (2017) reported that, as emotional closeness increased, so did altruistic willingness to help across different degrees of relatedness. Furthermore, they found that cost of helping was an important consideration, as the effect size of emotional closeness was smaller for high-cost than low-cost helping in kin, but the reverse could be seen in nonkin. Further research is needed to determine the precise parameters of this emerging body of evidence concerning the roles of emotional closeness, kin selection, and kinship premium in altruistic responses.

Another area for future research is to examine the notion that altruism could serve as a valuable signal. Not only is apparently altruistic behavior seen as attractive, it can also act to boost status (see Section 10.5). Reputation and public visibility are possible proximate cues for decision-making regarding reciprocation. As human strategies are undoubtedly more complicated and flexible than game theory strategies, reputation may be important to securing aid and cooperation. Once again, incorporating social research into evolutionary explanations of altruism might help to provide clearer understanding. Laboratory and field experiments by Ariely, Bracha, and Meier (2009), Carpenter and Myers (2010), Griskevicius, Tybur, and Van den Bergh (2010), Karlan and McConnell (2014), Lacetera and Macis (2010a, 2010b), and Linardi and McConnell (2011) provide greater experimental and ecological validity than is usually present in the study of human altruism. However, there is no distinction in the literature between public visibility announcements, such as being seen to volunteer or donate (Carpenter & Myers, 2010; Griskevicius, Tybur, & Van den Bergh, 2010; Karlan & McConnell, 2014; Linardi & McConnell, 2011), and public visibility rewards, such as badges and awards (Lacetera & Macis, 2010a, 2010b; Low & Heinen, 1993). This distinction between announcements and rewards may prove to be a fruitful direction for studies of human reciprocation in relation to reputation. Finally, as our understanding of the proximate cues underlying the ultimate explanations of apparent altruism develops, so too does the potential to apply this new understanding. By fully understanding the reasoning behind these acts of apparent altruism, we might be able to unlock the potential here to encourage pro-social behaviors.

REFERENCES

Abbot, P., Abe, J., Alcock, J., et al. (2011). Inclusive fitness theory and eusociality. Nature, 471(7339), E1–E4.

Ackerman, J. M., Kenrick, D. T., & Schaller, M. (2007). Is friendship akin to kinship? Evolution and Human Behavior, 28, 365–374.

Allen, A. P., Dinan, T. G., Clarke, G., & Cryan, J. F. (2017). A psychology of the human brain–gut–microbiome axis. Social and Personality Psychology Compass, 11(4). e12309.

Andersson, M. (1984). The evolution of eusociality. Annual Review of Ecology and Systematics, 15, 165–189.

Andreoni, J., & Petrie, R. (2004). Public goods experiments without confidentiality: A glimpse into fund-raising. Journal of Public Economics, 88(7), 1605–1623.

Ariely, D., Bracha, A., & Meier, S. (2009). Doing good or doing well? Image motivation and monetary incentives in behaving prosocially. American Economic Review, 99(1), 544–555.

Arnocky, S., Piché, T., Albert, G., Ouellette, D., & Barclay, P. (2016). Altruism predicts mating success in humans. British Journal of Psychology, 108, 416–435.

Arnold, W. (1990a). The evolution of marmot sociality: I. Why disperse late? Behavioral Ecology and Sociobiology, 27(4), 229–237.

Arnold, W. (1990b). The evolution of marmot sociality: II. Costs and benefits of joint hibernation. Behavioral Ecology and Sociobiology, 27(4), 239–246.

Ashworth, T. (2000). Trench Warfare, 1914–1918: The Live and Let Live System, 2nd ed. Basingstoke: Pan Macmillan.

Axelrod, R. (1984). The Evolution of Co-operation. New York: Basic Books.

Axelrod, R., & Hamilton, W. D. (1981). The evolution of cooperation. Science, 211(4489), 1390–1396.

Barclay, P. (2010). Altruism as a courtship display: Some effects of third-party generosity on audience perceptions. British Journal of Psychology, 101(1), 123–135.

Barclay, P. (2011). Competitive helping increases with the size of biological markets and invades defection. Journal of Theoretical Biology, 281, 47–55.

Barclay, P. (2013). Strategies for cooperation in biological markets, especially for humans. Evolution and Human Behavior, 34(3), 164–175.

Barclay, P. (2016). Biological markets and the effects of partner choice on cooperation and friendship. Current Opinion in Psychology, 7, 33–38.

Baum, W. M., Richerson, P. J., Efferson, C. M., & Paciotti, B. M. (2004). Cultural evolution in laboratory micro-societies including traditions of rule-giving and rule-following. Evolution and Human Behavior, 25, 305–326.

Bell, D. C., & Richard, A. J. (2000). Caregiving: The forgotten element in attachment. Psychological Inquiry, 11, 69–83.

Bell, G. (2008). Selection: The Mechanism of Evolution, 2nd ed. Oxford: Oxford University Press.

Bercik, P., Verdu, E. F., Foster, J. A., et al. (2010). Chronic gastrointestinal inflammation induces anxiety-like behavior and alters central nervous system biochemistry in mice. Gastroenterology, 139(6), 2102–2112.

Bercik, P., Park, A. J., Sinclair, D., et al. (2011). The anxiolytic effect of *Bifidobacterium longum* NCC3001 involves vagal pathways for gut–brain communication. *Neurogastroenterology & Motility*, **23**(12), 1132–1139.

Berte, N. A. (1988). K'ekchi' horticultural labor exchange: Productive and reproductive implications. In L. Betzig, M. Borgerhofl Mulder, & P. Turke, eds., *Human Reproductive Behavior: A Darwinian Perspective*. Cambridge, UK: Cambridge University Press, pp. 83–95.

Bhogal, M. S. (2017). Physical Attractiveness, Altruism and Fairness in a Game-Theoretic Framework. Unpublished doctoral dissertation, University of Wolverhampton.

Blumstein, D. T., Steinmetz, J., Armitage, K. B., & Daniel, J. C. (1997). Alarm calling in yellow-bellied marmots: II. The importance of direct fitness. *Animal Behaviour*, **53**(1), 173–184.

Bowles, S. (2009). Did warfare among ancestral hunter–gatherers affect the evolution of human social behaviors? *Science*, **324** (5932), 1293–1298.

Boyd, R., & Richerson, P. J. (1982). Cultural transmission and the evolution of cooperative behavior. *Human Ecology*, **10**(3), 325–351.

Boyd, R., & Richerson, P. J. (1985). *Culture and the Evolutionary Process*. Chicago, IL: University of Chicago Press.

Boyd, R., & Richerson, P. J. (2005). *The Origin and Evolution of Cultures*. Oxford: Oxford University Press.

Bravo, J. A., Forsythe, P., Chew, M. V., et al. (2011). Ingestion of *Lactobacillus* strain regulates emotional behavior and central GABA receptor expression in a mouse via the vagus nerve. *Proceedings of the National Academy of Sciences*, **108**(38), 16050–16055.

Bressan, P., Colarelli, S. M., & Cavalieri, M. B. (2009). Biologically costly altruism depends on emotional closeness among step but not half or full sibling. *Evolutionary Psychology*, **7**(1), 118–132.

Brown, S. L., & Brown, R. M. (2006). Selective investment theory: Recasting the functional significance of close relationships. *Psychological Inquiry*, **17**, 1–29.

Browning, L. E., Patrick, S. C., Rollins, L. A., Griffith, S. C., & Russell, A. F. (2012). Kin selection, not group augmentation, predicts helping in an obligate cooperatively breeding bird. *Proceedings of the Royal Society of London B: Biological Sciences*, **279**(1743), 3861–3869.

Burda, H., Honeycutt, R. L., Begall, S., Locker-Grütjen, O., & Scharff, A. (2000). Are naked and common mole-rats eusocial and if so, why? *Behavioral Ecology and Sociobiology*, **47**(5), 293–303.

Burnstein, E., Crandall, C., & Kitayama, S. (1994). Some neo-Darwinian decision rules for altruism: Weighing cues for inclusive fitness as a function of the biological importance of the decision. *Journal of Personality and Social Psychology*, **67**(5), 773–789.

Carter, G., & Wilkinson, G. (2013). Does food sharing in vampire bats demonstrate reciprocity? *Communicative & integrative biology*, **6**(6), e25783.

Carpenter, J., & Myers, C. K. (2010). Why volunteer? Evidence on the role of altruism, image, and incentives. *Journal of Public Economics*, **94**(11), 911–920.

Cheney, D. L., & Seyfarth, R. M. (1985). Vervet monkey alarm calls: Manipulation through shared information? *Behaviour*, **94** (1), 150–166.

Chiang, Y. S. (2010). Self-interested partner selection can lead to the emergence of fairness. *Evolution and Human Behavior*, **31** (4), 265–270.

Cialdini, R. B., Brown, S. L., Lewis, B. P., Luce, C., & Neuberg, S. L. (1997). Reinterpreting the empathy–altruism relationship: When one into one equals oneness. *Journal of Personality and Social Psychology*, **73**(3), 481–494.

Clark, M. S., & Mills, J. (1993). The difference between communal and exchange relationships: What it is and is not. *Personality and Social Psychology Bulletin*, **19**(6), 684–691.

Clutton-Brock, T. (2009). Cooperation between non-kin in animal societies. *Nature*, **462**(7269), 51–57.

Colquhoun, L., Davies, J., & Workman, L. (2017). Kith or kin – Who do we favour? The relationship between emotional closeness and genetic relatedness in altruistic willingness to help. Presented at British Psychological Society Annual Conference.

Crittenden, A. N., & Marlowe, F. W. (2008). Allomaternal care among the Hadza of Tanzania. *Human Nature*, **19**(3), 249–262.

Curry, O., Roberts, S. G., & Dunbar, R. I. (2013). Altruism in social networks: Evidence for a "kinship premium." *British Journal of Psychology*, **104**(2), 283–295.

Cutts, C., & Speakman, J. (1994). Energy savings in formation flight of pink-footed geese. *Journal of Experimental Biology*, **189** (1), 251–261.

Danielsbacka M., & Tanskanen A. O. (2012). Adolescent grandchildren's perceptions of grandparents' involvement in UK: An interpretation from life course and evolutionary theory perspective. *European Journal of Ageing*, **9**, 329–341.

Danielsbacka, M., Tanskanen, A. O., Jokela, M., & Rotkirch, A. (2011). Grandparental child care in Europe: Evidence for preferential investment in more certain kin. *Evolutionary Psychology*, **9**, 3–24.

Darwin, C. (1859). *On The Origin of Species by Means of Natural Selection, or Preservation of Favoured Races in the Struggle for Life*. London: John Murray.

Davies, B. (2014). *Thomas Aquinas's Summa Theologiae: A Guide and Commentary*. Oxford: Oxford University Press.

Dawkins, R. (2016). *The Selfish Gene*. 4th ed. Oxford: Oxford University Press.

DeNault, L. K., & McFarlane, D. A. (1995). Reciprocal altruism between male vampire bats, *Desmodus rotundus. Animal Behaviour*, **49**(3), 855–856.

Deulofeu, J. (2014). *Prisioneros con dilemas y estrategias dominantes: Teoria de juegos. [Prisoners with dilemmas and dominant strategies: Game theory]*, trans. London: Windmill Books and Vespa Design, RBA Collectibles.

Diamond, J. (2012). *The World Until Yesterday: What Can We Learn from Traditional Societies?* London: Penguin UK.

Dierkes, P., Heg, D., Taborsky, M., Skubic, E., & Achmann, R. (2005). Genetic relatedness in groups is sex-specific and declines with age of helpers in a cooperatively breeding cichlid. *Ecology Letters*, **8**(9), 968–975.

Dobson, A. P. (1988). The population biology of parasite-induced changes in host behavior. *Quarterly Review of Biology*, **63**(2), 139–165.

Duffy, J. E., Morrison, C. L., & Ríos, R. (2000). Multiple origins of eusociality among sponge-dwelling shrimps (*Synalpheus*). *Evolution*, **54**(2), 503–516.

Dunbar, R. I., Clark, A., & Hurst, N. L. (1995). Conflict and cooperation among the Vikings: Contingent behavioral decisions. *Ethology and Sociobiology*, **16**(3), 233–246.

Efferson, C., Lalive, R., Richerson, P. J., McElreath, R., & Lubell, M. (2008). Conformists and mavericks: The empirics of frequency-dependent cultural transmission. *Evolution and Human Behavior*, **29**(1), 56–64.

Eisenberg, A. R. (1988). Grandchildren's perspectives on relationships with grandparents: The influence of gender across generations. *Sex Roles*, **19**, 205–217.

Emlen, S. T., & Wrege, P. H. (1988). The role of kinship in helping decisions among white-fronted bee-eaters. *Behavioral Ecology and Sociobiology*, **23**(5), 305–315.

Essock-Vitale, S. M., & McGuire, M. T. (1980). Predictions derived from the theories of kin selection and reciprocation assessed by anthropological data. *Ethology and Sociobiology*, **1** (3), 233–243.

Essock-Vitale, S. M., & McGuire, M. T. (1985). Women's lives viewed from an evolutionary perspective. II. Patterns of helping. *Ethology and Sociobiology*, **6**(3), 155–173.

Euler, H. A., & Weitzel, B. (1996). Discriminative grandparental solicitude as reproductive strategy. *Human Nature*, **7**, 39–59.

Feder, H. M. (1966). Cleaning symbioses in the marine environment. In S. M. Henry, ed., *Associations of Microorganisms, Plants and Marine Organisms*, Vol. 1. Cambridge, MA: Academic Press, pp. 327–380.

Feldman, M., & Cavalli-Sforna, L. (1976). Cultural and biological evolutionary processes, selection for a trait under complex transmission. *Theoretical Population Biology*, **9**, 238–259.

Filiz-Ozbay, E., & Ozbay, E. Y. (2014). Effect of an audience in public goods provision. *Experimental Economics*, **17**(2), 200–214.

Fisher, R. A. (1930). *The Genetical Theory of Natural Selection*. Wotton-under-Edge: Clarendon Press.

Fortin, E. L. (1996). *Classical Christianity and the Political Order: Reflections on the Theologico-political Problem*. Lanham, MD: Rowman & Littlefield.

Frank, R. H. (1993). The strategic role of the emotions reconciling over-and undersocialized accounts of behavior. *Rationality and Society*, **5**(2), 160–184.

Gaulin, S. J., & Schlegel, A. (1980). Paternal confidence and paternal investment: A cross cultural test of a sociobiological hypothesis. *Ethology and Sociobiology*, **1**(4), 301–309.

Goehler, L. E., Park, S. M., Opitz, N., Lyte, M., & Gaykema, R. P. (2008). *Campylobacter jejuni* infection increases anxiety-like behavior in the holeboard: Possible anatomical substrates for viscerosensory modulation of exploratory behavior. *Brain, Behavior, and Immunity*, **22**(3), 354–366.

Gorrell, J. C., McAdam, A. G., Coltman, D. W., Humphries, M. M., & Boutin, S. (2010). Adopting kin enhances inclusive fitness in asocial red squirrels. *Nature Communications*, **1**, 22.

Grafen, A. (1990). Biological signals as handicaps. *Journal of Theoretical Biology*, **144**, 517–546.

Grayson, D. K. (1993). Differential mortality and the Donner Party disaster. *Evolutionary Anthropology: Issues, News, and Reviews*, **2**(5), 151–159.

Grey, P. B., & Brogdon, E. (2017). Do step- and biological grandparents show differences in investment and emotional closeness with their grandchildren? *Evolutionary Psychology*, **15**(1), 1–9.

Griskevicius, V., Tybur, J. M., & Van den Bergh, B. (2010). Going green to be seen: Status, reputation, and conspicuous conservation. *Journal of Personality and Social Psychology*, **98**, 392–404.

Gross, M. R., & Shine, R. (1981). Parental care and mode of fertilization in ectothermic vertebrates. *Evolution*, **35**(4), 775–793.

Gurven, M., Allen-Arave, W., Hill, K., & Hurtado, A. M. (2001). Reservation food sharing among the Ache of Paraguay. *Human Nature*, **12**(4), 273–297.

Haldane, J. B. S. (1932). *The Causes of Evolution*, 3rd ed. Harlow: Longmans, Green & Co.

Haldane, J. B. S. (1955). Population genetics. *New Biology*, **18**, 34–51.

Hames, R. (1987). Garden labor exchange among the Ye'kwana. *Ethology and Sociobiology*, **8**(4), 259–284.

Hamilton, W. D. (1964a). The genetical evolution of social behaviour I. *Journal of Theoretical Biology*, **7**(1), 1–16.

Hamilton, W. D. (1964b). The genetical evolution of social behaviour II. *Journal of Theoretical Biology*, **7**(1), 17–52.

Hawkes, K. (1991). Showing off: Tests of an hypothesis about men's foraging goals. *Ethology and Sociobiology*, **12**(1), 29–54.

He, W. (2005). Introduction: Kinship and family in international context. *International Journal of Sociology and Social Policy*, **25** (3), 1–8.

Henrich, J., & McElreath, R. (2007). Dual-inheritance theory: The evolution of human cultural capacities and cultural evolution. In R. I. M. Dunbar & L. Barrett, eds., *Oxford Handbook of Evolutionary Psychology*. Oxford: Oxford University Press, pp. 555–570.

Hill, K., & Hurtado, A. M. (1996). *Ache Life History: The Ecology and Demography of a Foraging People*. London: Aldine.

Honeycutt, R. L. (1992). Naked mole-rats. *American Scientist*, **80** (1), 43–53.

Hughes, W. O., Oldroyd, B. P., Beekman, M., & Ratnieks, F. L. (2008). Ancestral monogamy shows kin selection is key to the evolution of eusociality. *Science*, **320**(5880), 1213–1216.

Johnson, L. L., & Boyce, M. S. (1991). Female choice of males with low parasite loads in sage grouse. In J. E. Loye & M. Zuk, eds., *Bird Parasite Interactions: Ecology, Evolution, and Behavior*. Oxford: Oxford University Press, pp. 377–388.

Karlan, D., & McConnell, M. A. (2014). Hey look at me: The effect of giving circles on giving. *Journal of Economic Behavior & Organization*, **106**, 402–412.

Kobayashi, Y., & Yamamura, N. (2007). Evolution of signal emission by uninfested plants to help nearby infested relatives. *Evolutionary Ecology*, **21**(3), 281–294.

Korchmaros, J. D., & Kenny, D. A. (2001). Emotional closeness as a mediator of the effect of genetic relatedness on altruism. *Psychological Science*, **12**(3), 262–265.

Korchmaros, J. D., & Kenny, D. A. (2006). An evolutionary and close-relationship model of helping. *Journal of Social and Personal Relationships*, **23**(1), 21–43.

Kruger, D. J. (2003). Evolution and altruism: Combining psychological mediators with naturally selected tendencies. *Evolution and Human Behavior*, **24**(2), 118–125.

Lacetera, N., & Macis, M. (2010a). Do all material incentives for pro-social activities backfire? The response to cash and non-cash incentives for blood donations. *Journal of Economic Psychology*, **31**(4), 738–748.

Lacetera, N., & Macis, M. (2010b). Social image concerns and prosocial behavior: Field evidence from a nonlinear incentive scheme. *Journal of Economic Behavior & Organization*, **76**(2), 225–237.

Laham, S. M., Gonsalkorale, K., & von Hippel, W. (2005). Darwinian grandparenting: Preferential investment in more certain kin. *Personality and Social Psychology Bulletin*, **31**, 63–72.

Laland, K. N., & Brown, G. R. (2011). *Sense and Nonsense: Evolutionary Perspectives on Human Behaviour*. Oxford: Oxford University Press.

Lee, R. B. (1979). *The !Kung San: Men, Women, and Work in a Foraging Society*. Cambridge, UK: Cambridge University Press.

Lee, T. R., Mancini, J. A., & Maxwell, J. W. (1990). Sibling relationships in adulthood: Contact patterns and motivations. *Journal of Marriage and the Family*, **52**(2), 431–440.

Lehmann, L., Rousset, F., Roze, D., & Keller, L. (2007). Strong reciprocity or strong ferocity? A population genetic view of the evolution of altruistic punishment. *American Naturalist*, **170**(1), 21–36.

Lehmann, L., Feldman, M. W., & Foster, K. R. (2008). Cultural transmission can inhibit the evolution of altruistic helping. *American Naturalist*, **172**(1), 12–24.

Lentz, B. F., & Laband, D. N. (1989). Why so many children of doctors become doctors: Nepotism vs. human capital transfers. *Journal of Human Resources*, **24**(3), 396–413.

Lewin-Epstein, O., Aharonov, R., & Hadany, L. (2017). Microbes can help explain the evolution of host altruism. *Nature Communications*, **8**, 14040.

Linardi, S., & McConnell, M. (2011). No excuses for good behavior. *Journal of Public Economics*, **95**(5–6), 445–454.

Littlefield, C. H., & Rushton, J. P. (1986). When a child dies: The sociobiology of bereavement. *Journal of Personality and Social Psychology*, **51**(4), 797–802.

Low, B. S., & Heinen, J. T. (1993). Population, resources, and environment: Implications of human behavioral ecology for conservation. *Population and Environment*, **15**(1), 7–41.

Madsen, E. A., Tunney, R. J., Fieldman, G., et al. (2007). Kinship and altruism: A cross-cultural experimental study. *British Journal of Psychology*, **98**(2), 339–359.

Marler, P. (1955). Characteristics of some animal calls. *Nature*, **176**, 6–8.

Marler, P. (1957). Specific distinctiveness in the communication signals of birds. *Behaviour*, **11**(1), 13–38.

Maynard Smith, J. (1964). Group selection and kin selection. *Nature*, **201**, 1145–1147.

McCullough, J. M., & Barton, E. Y. (1991). Relatedness and mortality risk during a crisis year: Plymouth colony, 1620–1621. *Ethology and Sociobiology*, **12**(3), 195–209.

McCullough, J. M., Heath, K. M., & Fields, J. D. (2006). Culling the cousins: Kingship, kinship, and competition in mid-millennial England. *History of the Family*, **11**(1), 59–66.

McElreath, R., Lubell, M., Richerson, P. J., et al. (2005). Applying evolutionary models to the laboratory study of social learning. *Evolution and Human Behavior*, **26**(6), 483–508.

Milius, S. (1998). The science of EEEEEK!: What a squeak can tell researchers about life, society, and all that. *Science News*, **154**(11), 174–175.

Mills, J., & Clark, M. S. (1994). Communal and exchange relationships: Controversies and research. In R. Erber & R. Gilmour, eds., *Theoretical Frameworks for Personal Relationships*. Mahwah, NJ: Lawrence Erlbaum, pp. 29–42.

Mongkolsamrit, S., Kobmoo, N., Tasanathai, K., et al. (2012). Life cycle, host range and temporal variation of *Ophiocordyceps unilateralis*/*Hirsutella formicarum* on Formicine ants. *Journal of Invertebrate Pathology*, **111**(3), 217–224.

Moore, J. (1984). The evolution of reciprocal sharing. *Ethology and Sociobiology*, **5**(1), 5–14.

Morris, K., & Goodall, J (1977). Competition for meat between chimpanzees and baboons of the Gombe National Park. *Folia Primatologica*. **28**, 109–121.

Murphy, G. P., & Dudley, S. A. (2009). Kin recognition: Competition and cooperation in *Impatiens* (Balsaminaceae). *American Journal of Botany*, **96**(11), 1990–1996.

Neyer, F. J., & Lang, F. R. (2003). Blood is thicker than water: Kinship orientation across adulthood. *Journal of Personality and Social Psychology*, **84**(2), 310–321.

Nishida, T., Uehara, S., & Nyundo, R (1979). Predatory behavior among wild chimpanzees of the Mahale Mountains. *Primates*, **20**, 1–20.

Nowak, M., & Sigmund, K. (2005) Evolution of indirect reciprocity. *Nature*, **437**, 1291–1298.

Nowak, M. A., Tarnita, C. E., & Wilson, E. O. (2010). The evolution of eusociality. *Nature*, **466**(7310), 1057–1062.

Owens, D., & Owens, M. (1984). Helping behaviour in brown hyenas. *Nature*, **308**(5962), 843–845.

Packer, C. (1977). Reciprocal altruism in olive baboons. *Nature*, **265**, 441–443.

Panchanathan, K., & Boyd, R. (2004). Indirect reciprocity can stabilize cooperation without the second-order free rider problem. *Nature*, **432**(7016), 499–502.

Pashos, A., & McBurney, D. H. (2008). Kin relationships and the caregiving biases of grandparents, aunts and uncles: A two generational questionnaire study. *Human Nature*, **19**, 311–330.

Pashos, A., Schwarz, S., & Bjorklund, D. F. (2016). Kin investment by step-grandparents – More than expected. *Evolutionary Psychology*, **14**(1), 1–11.

Pollet, T. V., Roberts, S. G., & Dunbar, R. I. (2013). Going that extra mile: Individuals travel further to maintain face-to-face contact with highly related kin than with less related kin. *PLoS ONE*, **8**(1), e53929.

Portin, P. (2015). A comparison of biological and cultural evolution. *Journal of Genetics*, **94**(1), 155–168.

Portugal, S. J., Hubel, T. Y., Fritz, J., et al. (2014). Upwash exploitation and downwash avoidance by flap phasing in ibis formation flight. *Nature*, **505**(7483), 399–402.

Refardt, D., Bergmiller, T., & Kümmerli, R. (2013). Altruism can evolve when relatedness is low: Evidence from bacteria committing suicide upon phage infection. *Proceedings of the Royal Society of London B: Biological Sciences*, **280**(1759), 20123035.

Rege, M., & Telle, K. (2004). The impact of social approval and framing on cooperation in public good situations. *Journal of Public Economics*, **88**(7), 1625–1644.

Richerson, P. J., & Boyd, R. (2005). *Not by Genes Alone: How Culture Transformed Human Biology*. Chicago, IL: University of Chicago Press.

Richerson, P. J., Baldini, R., Bell, A. V., et al. (2016). Cultural group selection plays an essential role in explaining human cooperation: A sketch of the evidence. *Behavioral and Brain Sciences*, **39**, 1–68.

Riss, D. C., & Busse, C. D (1977). Fifty-day observation of a free-ranging adult male chimpanzee. *Folia Primatologica*. **28**, 283–297.

Rudebeck, G. (1950). The choice of prey and modes of hunting of predatory birds with special reference to their selective effect. *Oikos*, **2**(1), 65–88.

Rudebeck, G. (1951). The choice of prey and modes of hunting of predatory birds with special reference to their selective effect. *Oikos*, **3**(2), 200–231.

Seyfarth, R. M., & Cheney, D. L. (1984). Grooming, alliances, and reciprocal altruism in vervet monkeys. *Nature*, **308**, 541–543.

Shavit, Y., Fischer, C. S., & Koresh, Y. (1994). Kin and nonkin under collective threat: Israeli networks during the gulf war. *Social Forces*, **72**(4), 1197–1215.

Singer, P. (1981). *The Expanding Circle: Ethics and Sociobiology*. New York: Farrar, Straus and Giroux.

Smith, M. S., Kish, B. J., & Crawford, C. B. (1987). Inheritance of wealth as human kin investment. *Ethology and Sociobiology*, **8**(3), 171–182.

Sober, E., & Wilson, D. S. (1998). *Unto Others: The Evolution and Psychology of Unselfish Behavior*. Cambridge, MA: Harvard University Press.

Soler, M., Soler, J. J., Martinez, J. G., & Moller, A. P. (1995). Magpie host manipulation by great spotted cuckoos: Evidence for an avian mafia? *Evolution*, **49**, 770–775.

Soltis, J., Boyd, R., & Richerson, P. J. (1995). Can group-functional behaviors evolve by cultural group selection?: An empirical test. *Current Anthropology*, **36**(3), 473–494.

Sprankel, H., Richarz, K., Ludwig, H., & Rott, R. (1978). Behavior alterations in tree shrews (*Tupaia glis*, Diard 1820) induced by Borna disease virus. *Medical Microbiology and Immunology*, **165**(1), 1–18.

Stewart-Williams, S. (2007). Altruism among kin vs. nonkin: Effects of cost of help and reciprocal exchange. *Evolution and Human Behavior*, **28**(3), 193–198.

Stewart-Williams, S. (2018). *The Ape That Understood the Universe: How the Mind and Culture Evolve*. Cambridge, UK: Cambridge University Press.

Storey, S., & Workman, L. (2013). The effects of temperature priming on cooperation in the iterated prisoner's dilemma. *Evolutionary Psychology*, **11**(1), 52–67.

Sutton, M. (1982). *Nationalism, Positivism, and Catholicism. The Politics of Charles Maurras and French Catholics 1890–1914*. Cambridge, UK: Cambridge University Press.

Tinbergen, L. (1960). *The Dynamics of Insect and Bird Populations in Pine Woods*. Leiden: Brill Archive.

Trivers, R. L. (1971). The evolution of reciprocal altruism. *Quarterly Review of Biology*, **46**, 35–57.

Trivers, R. L. (1985). *Social Evolution*. San Francisco, CA: Benjamin/Cummings.

Trivers, R. L. (2011). *The Folly of Fools: The Logic of Deceit and Self-Deception in Human Life*. New York: Basic Books.

Voelkl, B., Portugal, S. J., Unsöld, M., et al. (2015). Matching times of leading and following suggest cooperation through direct reciprocity during V-formation flight in ibis. *Proceedings of the National Academy of Sciences*, **112**(7), 2115–2120.

Waibel, M., Keller, L., & Floreano, D. (2009). Genetic team composition and level of selection in the evolution of cooperation. *IEEE Transactions on Evolutionary Computation*, **13**(3), 648–660.

Waibel, M., Floreano, D., & Keller, L. (2011). A quantitative test of Hamilton's rule for the evolution of altruism. *PLoS Biology*, **9** (5), e1000615.

Waynforth, D. (2011). Mate choice and sexual selection. In V. Swami, ed., *Evolutionary Psychology: A Critical Introduction*. Hoboken, NJ: Wiley-Blackwell, pp. 107–130.

Weimerskirch, H., Martin, J., Clerquin, Y., Alexandre, P., & Jiraskova, S. (2001). Energy saving in flight formation. *Nature*, **413**(6857), 697–698.

West, S. A., Griffin, A. S., & Gardner, A. (2007a). Evolutionary explanations for cooperation. *Current Biology*, **17**(16), R661–R672.

West, S. A., Griffin, A. S., & Gardner, A. (2007b). Social semantics: Altruism, cooperation, mutualism, strong reciprocity and group selection. *Journal of Evolutionary Biology*, **20**(2), 415–432.

West, S. A., El Mouden, C., & Gardner, A. (2011). Sixteen common misconceptions about the evolution of cooperation in humans. *Evolution and Human Behavior*, **32**(4), 231–262.

Wheeler, B. C. (2008). Selfish or altruistic? An analysis of alarm call function in wild capuchin monkeys, *Cebus apella nigritus*. *Animal Behaviour*, **76**(5), 1465–1475.

Wilkinson, G. S. (1984). Reciprocal food sharing in the vampire bat. *Nature*, **308**(5955), 181–184.

Wilson, D. S. (2007). *Evolution for Everyone: How Darwin's Theory Can Change the Way We Think About Our Lives*. New York: Delacorte Press.

Wilson, D. S., & Brown, J. M. (1994). *Poecilochirus carabi*: Behavioral and life-history adaptations to different hosts and the consequences of geographical shifts in host communities. In M. A. Houck, ed., *Mites: Ecological and Evolutionary Analyses of Life-History Patterns*. Berlin: Springer, pp. 1–22.

Wilson, D. S., & Sober, E. (1994). Reintroducing group selection to the human behavioral sciences. *Behavioural and Brain Sciences*, **17**, 585–654.

Wilson, D. S., & Wilson, E. O. (2007). Rethinking the theoretical foundation of sociobiology. *Quarterly Review of Biology*, **82**(4), 327–348.

Wilson, E. O. (2005). Kin selection as the key to altruism: Its rise and fall. *Social Research*, **72**(1), 159–166.

Wilson, E. O. (2012). *The Social Conquest of Earth*. New York: Norton.

Woolfenden, G. E. (1975). Florida scrub jay helpers at the nest. *The Auk*, **92**(1), 1–15.

Workman, L., & Reader, W. (2014). *Evolutionary Psychology*, 3rd ed. Cambridge: Cambridge University Press.

Workman, L., & Reader, W. (2015). *Evolution and Behavior*. Abingdon: Routledge.

Wynne-Edwards, V. C. (1962). *Animal Dispersion: In Relation to Social Behaviour*. Edinburgh: Oliver and Boyd.

Zahavi, A. (1975). Mate selection: A selection for a handicap. *Journal of Theoretical Biology*, **53**, 205–214.

Zahavi, A., & Zahavi, A. (1997). *The Handicap Principle: A Missing Piece of Darwin's Puzzle*. Oxford: Oxford University Press.

Ziker, J. (2005). Food sharing at meals: Kinship, reciprocity, and clustering in the Taimyr autonomous Okrug, northern Russia. *Human Nature*, **16**, 178–211.

11 Can Evolutionary Processes Explain the Origins of Morality?

DENNIS L. KREBS

11.1 INTRODUCTION

We humans have been called "the moral animal" (Wright, 1994). We help others, make sacrifices to uphold our groups, treat others fairly, respect authority, exert self-control, and seek to be pure of heart. We experience moral emotions such as sympathy, gratitude, forgiveness, guilt, and righteous indignation. We judge acts as right, wrong, fair, and unfair, and we judge people as good, bad, virtuous, and vicious. We harbor beliefs about the forms of conduct that we and others should and should not enact, and we justify these beliefs with moral reasons. Children react strongly to violations of moral rules at a very young age. We possess a conscience that bothers us when we violate our moral standards and principles. We express moral outrage at acts we consider unjust, and we are willing to make sacrifices – in some cases, even of our lives – to uphold our moral principles and to fight injustice. We praise moral exemplars such as Mother Theresa, Gandhi, and Martin Luther King Jr., and we condemn villains such as Hitler, Stalin, and Charles Manson.

Researchers have found that people from all cultures consider at least six forms of conduct moral: (1) altruism; (2) group solidarity; (3) respect for rules and authority; (4) fairness; (5) self-control; and (6) cleanliness and purity (Krebs, 2011). Such universals notwithstanding, cultures may differ in the salience of each of these moral norms, the relative value they place on them, and in the ways in which they instantiate them. For example, care and justice are valued highly in Western societies; purity and authority are valued highly in Orthodox Jewish and Hindu cultures (e.g., as manifest in dietary restrictions and severe punishments against those who disrespect prophets), and group solidarity is valued highly by the Japanese (Pinker, 2008). Within Western societies, liberals tend to place a higher value on altruism and fairness than conservatives do, whereas conservatives tend to value all universal moral norms roughly equally (Haidt & Joseph, 2004).

Although all people share basic moral values, no two people possess exactly the same sense of morality. Not only do people differ in the values they attach to different moral norms, they also differ in: (1) the particular acts and traits they consider altruistic, fair, pure, and so on; (2) the reasons that they consider such acts right and wrong; (3) their moral principles; and (4) the importance they place on morality compared to other values. Several theorists have likened the acquisition of moral values to the acquisition of language (Haidt, 2001; Hauser, 2006; Pinker, 2008; Richerson & Boyd, 2005). All people inherit mental mechanisms that endow them a species-specific capacity to develop moral values, but people from different cultures – and within the same culture – instantiate them in different ways.

How can we account for the moral aspect of human nature? How did it originate? How do people acquire dispositions to behave in moral ways, to experience moral emotions, to make moral judgments, and to acquire moral beliefs? How do people acquire a sense of morality?

11.2 EVOLUTIONARY EXPLANATIONS OF THE ORIGINS OF MORALITY

The most popular accounts of the origins of morality have been advanced by developmental psychologists, who have attributed it either to social learning (Bandura, 1991) or to the development of moral reasoning (Kohlberg, 1984). Although there is a wealth of evidence that social learning and cognitive development contribute to the acquisition of morality, these processes do not offer a complete account (Krebs, 2011; Krebs & Denton, 2006). For example, they fail to account for the origins of social learning mechanisms, moral reasoning mechanisms, moral intuitions, and moral emotions.

To fully explain the origins of morality, we must trace it back to its origins, not in children, but in the human species. Building on ideas introduced by Darwin (1871), contemporary evolutionary theorists have taken on this task. The basic assumptions of evolutionary accounts of morality are: (1) some members of early human groups inherited genes that, in interaction with environmental inputs, endowed them with moral traits; (2) these individuals passed more replicas of these genes on to future generations than other individuals did; and (3) the proportion of morality-producing genes increased in the population until moral traits became part of human nature. Evolutionary accounts expect moral traits to evolve either because they are adaptive, increasing individuals' fitness by helping them adapt to their (social) environments, or as by-products of other adaptive mental mechanisms, such as those that endow people with social learning abilities and the ability to reason.

11.3 THE PARADOX OF THE EVOLUTION OF MORALITY

At first glance, evolutionary accounts of the origins of morality seem implausible. How could early humans who inherited genes that disposed them to behave in moral ways have fared better in the struggle for existence than early humans who inherited genes that disposed them to behave in immoral ways? Don't animals that defeat others "red in tooth and claw" in competitions for resources and mates contribute more of their genes to ensuing generations than animals that behave in more moral ways? As expressed by the eminent evolutionary biologist, Richard Dawkins (1989), "if you look at the way natural selection works, it seems to follow that anything that has evolved by natural selection should be selfish. ... Much as we might wish to believe otherwise, universal love and welfare of the species as a whole are concepts that simply do not make evolutionary sense. ... Be warned that if you wish, as I do, to build a society in which individuals cooperate generously and unselfishly toward a common good, you can expect little help from biological nature" (pp. 3–4).

As brilliant and renowned as Richard Dawkins is, he reached the wrong conclusion about the evolution of morality, primarily because he conflated genetic selfishness, which pertains to the selection of genes, with biological and psychological forms of selfishness (a mistake that Dawkins corrected in revisions of his books). In a nutshell, behaving in biologically and psychologically altruistic and moral ways can increase fitness, and behaving in biologically and psychologically selfish and immoral ways can diminish fitness (Krebs, 2011). Contemporary evolutionary theorists have offered cogent accounts of the evolution of altruism and the other five forms of conduct that people from all cultures consider moral. Let us consider each in turn.

11.4 ORIGINS OF ALTRUISM

Although it might seem that dispositions to behave in biologically altruistic ways cannot evolve because, by definition, they diminish the survival prospects of altruists, and although it might seem that dispositions to behave in reproductively altruistic ways cannot evolve because they reduce the reproductive success of altruists, evolutionary theorists have identified three main routes to the evolution of biologically and reproductively altruistic dispositions: through sexual selection, kin selection, and group selection.

11.4.1 The Evolution of Altruism through Sexual Selection

To evolve, genetically influenced traits must be passed on to future generations. In the animal kingdom, the primary way of accomplishing this is through sexual reproduction. Because animals that survive without reproducing fail to pass their genes on through their offspring, it is in the genetic interest of animals that have not mated to be willing to risk costs to their prospects of surviving in order to increase their chances of reproducing. One of the ways in which members of many sexually reproducing species accomplish this is by making altruistic overtures to potential mates. Nuptial gift-giving is a common example. In effect, donors seek to exchange goods and services that contribute to recipients' chances of surviving in return for mating opportunities that contribute to donors' reproductive success.

The Expansion of Sexually Selected Altruism. In addition to behaving altruistically toward potential mates, individuals may attract mates by putting on costly displays that signal that they are willing and able to make sacrifices for the sake of other members of their groups. Such altruistic displays advertise that the individuals who emit them possess "good genes" that enable them to overcome handicaps (Miller, 2007). Research on mate choice in humans has found that members of both sexes rank altruistic traits such as kindness near the tops of their lists of desirable traits (Buss, 2004).

Mental Mechanisms Supporting Sexually Selected Altruism: Implications for Morality. Sexually selected mental mechanisms that induce individuals to sacrifice their survival interests in order to improve their chances of mating could be designed in any number of ways. To a large extent, attributions of morality are based on the ways in which such mechanisms are designed. Some sexually selected mechanisms induce individuals to behave in ways that observers consider altruistic and moral; others do not. For example, caring behaviors that stem from mechanisms that generate feelings of romantic love present as genuinely altruistic and good even though they increase individuals' reproductive success by inducing them to make long-term investments in their mates and offspring (Fessler & Haley, 2003). In contrast, superficially generous gestures that are aimed at short-term seduction seem much less altruistic and moral. Note that although sexually selected altruistic dispositions may be biologically altruistic, they are reproductively selfish.

11.4.2 The Evolution of Altruism through Kin Selection

Sexual reproduction is not the only way to transmit replicas of one's genes to future generations. Another way is to help those who possess copies of one's genes survive and reproduce. In particular, individuals could propagate "altruism genes" (i.e., genes that dispose individuals to behave in biologically and reproductively altruistic ways) by helping kin who carry copies of these genes. This insight was quantified by Hamilton (1964), who asserted that dispositions to behave in altruistic ways could evolve when the genetic costs (C) to donors were less than the genetic benefits (B) to recipients multiplied by the degree of relatedness (r) between donor and recipient ($C < rB$, or $rB > C$). Kin selection offers an explanation for the evolution of altruism proffered by parents to their offspring ($r = 0.5$), as well as altruistic acts observed in the animal kingdom directed toward siblings ($r = 0.5$) and other relatives, such as nephews and nieces ($r = 0.25$) and cousins ($r = 0.125$). There is a wealth of evidence that humans and other animals behave in accordance with Hamilton's rule (Bernstein, 2005).

It is important to recognize that Hamilton's rule does not imply that donors will allocate their altruism to recipients in terms of their degree of relatedness by, for example, allocating a quarter of their altruism to their nieces and nephews. It implies that, all else being equal, donors will allocate their altruism predominantly to those who are most likely to possess replicas of their genes, especially their offspring. Hamilton's rule also does not imply that individuals will not favor themselves; after all, the probability of them carrying replicas of their own genes is 100 percent. Hamilton's rule implies that donors will be disinclined to behave altruistically when the costs (C) of helping relatives are high and when the benefits (B) to relatives are low. Hamilton's rule also implies that donors should be more strongly inclined to behave altruistically toward recipients who possess the highest reproductive potential, because the genetic benefits to them (B) tend to be higher. Conversely, donors are not expected to be inclined to assist relatives who are unlikely to bear fecund offspring.

The Expansion of Kin-Selected Altruism. Strictly speaking, kin-selected altruistic dispositions should be triggered by kin. Fortunately for the evolution of altruism, however, members of most species, including our own, are not very good at determining the extent to which they are related to others, let alone the probability that potential recipients possess copies of their altruism genes. Like other animals, we infer the degree of genetic relatedness by attending to three main correlates of – or cues to – kinship: similarity (e.g., in appearance and smell), familiarity, and proximity. Because nonrelatives may look like relatives, be familiar to donors, and live in close proximity to them, nonrelatives may trigger kin-selected altruistic dispositions.

Mental Mechanisms Supporting Kin-Selected Altruism: Implications for Morality. Although the evolved mental mechanisms that render animals willing to sacrifice their biological and reproductive interests to assist their kin and those who are like their kin may be designed in different ways in different species, researchers have found that hormones such as oxytocin, vasopressin, prolactin, and endogenous opioids play an important role in regulating these mechanisms (Brown & Brown, 2006). The two emotional experiences that have been most closely associated with kin-selected forms of altruism in humans are empathy and sympathy (Batson, 1991).

Empathy and Sympathy. de Waal (2008) has reviewed research demonstrating that the mental mechanisms that evoke empathy and sympathy are activated by kinship cues such as "similarity, familiarity, social closeness, and positive experience with the other, [and that among humans] subjects empathize with a confederate's pleasure or distress if they perceive the relationship as cooperative." de Waal goes on to review research indicating that "seeing the pain of a cooperative confederate activates pain-related brain areas, but seeing the pain of an unfair confederate activates reward-related areas, at least in men," concluding that "the empathy mechanism is biased the way evolutionary theory would predict. Empathy is (a) activated in relation to those with whom one has a close or positive relationship, and (b) suppressed, or even turned into Schadenfreude, in relation to strangers and defectors" (p. 16).

Although empathy and sympathy can induce individuals to help others for selfish reasons, such as reducing their vicariously experienced distress or avoiding disapproval from victims and observers, a wealth of research by Batson (1991) and his colleagues indicates that empathy and sympathy may generate genuinely altruistic motivational states that dispose individuals to want to help those with whom they are empathizing as an end in itself, not a means to selfish ends. Helping others as an end in itself is commonly considered more moral than helping others as a means to a selfish end.

11.5 ORIGINS OF ALTRUISTIC "TRIBAL INSTINCTS"

The process of sexual selection helps explain altruism that increases donors' reproductive success, and the process of kin selection helps explain altruism toward kin and those who resemble kin, but what about altruistic acts that do not improve donors' reproductive success and are directed toward unrelated members of groups? There is a great deal of evidence – from social science research and from everyday life – that under some circumstances, humans and other animals may be willing to sacrifice their own welfare to promote the welfare of their groups. A prototypical example is the repeatedly documented willingness of soldiers to sacrifice their lives to protect members of their military groups. Brown and Brown (2006) cite the following passage from the US Army Field Manual: "while patriotism and sense of purpose will get American soldiers to the battlefield, the soldiers' own accounts (and many systematic studies) testify that what keeps them there amid the fear of death and mutilation is, above all, their loyalty to their fellow soldiers" (p. 38).

When Darwin (1871) was grappling with the problem that evidence of altruism in the animal kingdom presented for his theory of evolution, it occurred to him that altruistic dispositions that diminished the fitness of altruistic individuals could be selected if they increased the fitness of the altruists' groups by enabling them to outcompete more selfish groups in the struggle for existence. Perhaps, pondered Darwin, altruism that was disadvantageous to individuals could evolve through the selection of altruistic groups – a process that came to be called group selection. However, Darwin quickly detected a problem with this account of the evolution of altruism, namely that even if altruistic groups prevailed over more selfish groups, selfish members of the altruistic groups would be more likely than their more altruistic cohorts to survive and reproduce, eventually replacing them and rendering their groups selfish. Within-group selection for selfishness would outpace between-group selection for altruism.

11.5.1 The Evolution of Altruistic Tribal Instincts through Group Selection

The idea of group selection has not been received well by most evolutionary biologists, probably stemming from the flawed version of it popularized by Wynne-Edwards (1962). However, most evolutionary biologists accept the idea that it is theoretically possible for altruism to evolve through group selection in ideal circumstances. For example, Sober and Wilson (1998) argued that altruism can evolve through group selection if a population contains many groups composed of different proportions of altruists and selfish individuals, the altruistic groups are more fit than are selfish groups, the groups compete against one another and form new groups (e.g., by incorporating members of the groups they have defeated), and the altruistic groups reap greater gains in fitness in between-group competitions than selfish individuals reap in within-group competitions. Sober and Wilson adduced evidence that, during the course of human evolution, humans formed groups that competed against one another (not only physically, but also economically and culturally), and that altruistic groups outcompeted more selfish groups, expanded in size, and restructured themselves before within-group selection rendered them selfish. In the last decade, eminent biologists and mathematicians such as E. O. Wilson and Martin Nowak (Nowak, Tarnita, & Wilson, 2010) have concluded that group selection is as well – or better – equipped than kin selection to account for the evolution of altruism, and that models of group selection and models of kin selection are mathematically equivalent.

11.5.2 Cultural Group Selection

In addition to proposing that genes that code for altruism can evolve through biological forms of group selection, some theorists have suggested that altruistic "tribal instincts" have evolved in the human species through cultural group selection (Henrich, 2016; Richerson & Boyd, 2005). The main idea underlying cultural group selection is that units of culture – called by a variety of names, such

as culturgens and memes – can be selected, retained, replicated, and transmitted in ways analogous to the evolution of genes. Theorists such Richerson and Boyd (2005) have adduced evidence that groups guided by altruistic and cooperative cultural norms have prevailed over groups guided by more selfish cultural norms. These theorists have suggested that tribal instincts evolved through cultural group selection relatively recently in the human species, and that tribal instincts "are laid on top of more ancient social instincts rooted in kin selection and reciprocal altruism" (p. 191), which causes humans to be "simultaneously committed to tribes, family, and self, even though our simultaneous and conflicting commitments very often cause us … great anguish" (p. 191). Richerson and Boyd view tribal instincts as analogous to Chomskyan linguists' "principles and parameters … of language. … The innate principles furnish people with basic predispositions, emotional capacities, and social skills that are implemented in practice through highly variable cultural institutions, the parameters" (p. 191).

11.5.3 The Expansion of Group-Selected Altruism

Group-selected tribal instincts are triggered by cues to shared group identity and common culture. Although in the first instance individuals may define in-groups and out-groups on the basis of features such as race and nationality, social psychologists such as Tajfel and Turner (1985) have found that people quickly identify with new groups to which they have been assigned and favor members of these groups over members of other groups, even when the groups are defined in terms of such arbitrary criteria as eye color and art preferences.

11.5.4 Mental Mechanisms Supporting Group-Selected Altruism: Implications for Morality

As expressed by Richerson and Boyd (2001), "we are adapted to living in tribes, and the social institutions of tribes elicit strong – sometimes fanatical – commitment" (p. 215). According to Tajfel and Turner (1985), feelings of group solidarity stem from people's tendency to identify with in-groups and incorporate "social identities" into their self-concepts. Greene (2013) has reviewed evidence that the mental mechanisms that mediate moral judgments are triggered primarily by members of in-groups; people are not naturally inclined to make moral judgments about out-group members because moral adaptations evolved to uphold fitness-increasing social relations among members of in-groups.

11.6 ORIGINS OF RESPECT FOR AUTHORITY

To a significant extent, upholding the social order of one's in-groups entails respecting rules and the authority of those who preach and enforce the rules – people such as parents, teachers, judges, and police. Dispositions to respect authority are readily visible in many species, including primates, whose groups are structured in terms of dominance hierarchies. Deferential dispositions induce members of such groups to subordinate themselves to those who are more powerful or of higher status than they are and to conform to the rules, regulations, and norms of their groups.

Dispositions to respect the authority of dominant members of groups are adaptive for several reasons. To begin with, they induce subordinate members to avoid the wrath of bigger, stronger, and/or more powerful individuals. In addition, as expressed by Haslam (1997) with respect to primate groups, "low-ranking individuals ... typically benefit from protection, advice, leadership, and intervention in disputes" (p. 300). Further, upholding hierarchical social orders increases the effectiveness of groups in group-against-group competitions.

11.6.1 The Expansion of Respect for Authority

According to Haslam (1997), "there appear to be clear continuities between humans and nonhuman primates regarding the realization and representation of dominance" (p. 304), and by implication the realization and representation of respect for authority. Primitive dispositions to respect authority probably evolved in the human species in the same ways in which they evolved in other primate species. Humans possess primitive brain structures similar to those of other primates that, when activated, induce them to behave in primitively deferential ways. Like other social animals, humans who experience conflicts with members of their groups appraise the power and status of their opponents and adapt their social strategies accordingly, deferring to those who possess greater power and higher status. However, in addition, humans possess more sophisticated "new brain" structures that enable them to conceptualize authority in significantly more complex and flexible ways than other primates do, to make much finer distinctions between legitimate and illegitimate authority, to question authority and rules, to resolve conflicts between competing authorities, to control and reflect on their primitive deferential emotional reactions, and to make more rational decisions than other primates make about whether and when to respect and obey authority. Humans belong to many more groups than other primates do, so they are exposed to many more types of authority. Cultural artifacts such as weapons equalize power differentials.

11.6.2 Mental Mechanisms Supporting Respect for Authority: Implications for Morality

Researchers have found that hormones such as androgen and serotonin affect the activation of deferential dispositions in humans and other animals (Cummins, 2005). Two types of mental mechanisms, designed in quite different ways, may dispose humans and other animals to obey authority. The first type of mechanism induces individuals to obey authority in order to avoid punishment from those who are more powerful than they are. This mechanism is based primarily in fear. The second type of mechanism induces individuals to obey people of high status. This mechanism is based primarily in feelings of admiration and awe. Different mixtures of fear and awe may engender different kinds of deferential dispositions.

Developmental psychologists have found that respect for authority dominates young children's conceptions of morality (Kohlberg, 1984). Young children view morality primarily in terms of unilateral respect for rules and the adults who enforce them. Even though children tend to acquire more autonomous moral values as they grow older, most children and adults continue to believe that people have a moral obligation to obey rules and respect legitimate authority. The anthropologist Christopher Boehm (2000) has argued that "morality is based heavily on social pressure, punishment, and other kinds of direct social manipulation" (p. 82).

11.7 ORIGINS OF SELF-CONTROL

Animals make decisions about whether to indulge urges such as hunger, sexual desire, greed, and anger or whether to control their urges, delay gratification, and meet their needs in more temperate ways. In many cases, such decisions involve choosing between relatively certain short-term gains and less certain long-term gains. The hoarding behavior of squirrels offers evidence of the evolution of the capacity to delay gratification and invest in long-term security. Other examples involve refraining from reproduction until one is able to care for new offspring and constraining one's wrath in conflicts with members of one's group who could be useful in the future.

11.7.1 Mental Mechanisms Supporting Self-Control: Implications for Morality

We humans have inherited mental mechanisms that, even though they induced us to show optimal levels of self-control in ancestral environments, often dispose us to exert less self-control than is optimal in modern environments. Consider gluttony, for example. As expressed by Burnham and Phelan (2000):

For our ancestors ... saving through markets and money was not an option. Successful people would ram as much [food] as possible into their own stomachs and those of their genetic relatives ... In such an environment, the best way to save is, paradoxically, to consume. Rather than leave some precious energy lying around to mold or be stolen, put it in your stomach and have your body convert the food into an energy savings account. ... As we struggle to save money, our mammalian heritage lurks in the background. We know we ought to put some money in the bank, but consuming just feels so good. (p. 19)

The main reason that temperance and self-control are viewed as moral – even when manifest in extremes such as asceticism – is because they are usually beneficial to members of groups. Individuals who take more than their share because they are greedy, fail to do their share because they are slothful, fail to control their sexual desires, and injure or kill other members of their groups in anger diminish the fitness of their fellows. In addition, individuals who fail to exert optimal levels of self-control may jeopardize their own welfare, thus becoming a burden on the social system. Tobacco, drug, alcohol, and food consumption in modern society are cases in point.

11.8 ORIGINS OF FAIRNESS AND JUSTICE

In a seminal paper on social evolution, Trivers (1971) explained how a behavior he labeled "reciprocal altruism" could evolve. Invoking the prisoner's dilemma game as a model, Trivers demonstrated that two cooperative players who are willing to reciprocate can reap greater biological benefits in social exchanges than two selfish individuals bent on exploiting their partners by taking without giving (gross cheating) or taking more than they give (subtle cheating). Although "reciprocal altruism" may not be biologically altruistic because it produces return benefits, balanced reciprocity can be considered moral because it is fair. Game theorists have produced a great deal of evidence that fair strategies such as "Tit for Tat" and "Firm but Fair" can defeat selfish strategies such as "All Defect" in evolutionary games. An important implication of these findings is that fair strategies were selected not because animals valued fairness, but because they paid off better biologically than unfair strategies did.

Although evidence of reciprocal altruism has not been as prevalent in the animal kingdom as expected from theoretical work documenting its biological benefits (Hauser, 2006), it has been observed among primates. Negative forms ("getting even") are more prevalent that positive forms. According to de Waal and Brosnan (2006), "the squaring of accounts in the negative domain ... may represent a precursor to human justice, since justice can be viewed as a transformation of the urge for revenge, euphemized as retribution, in order to control and regulate behavior" (p. 88). On the positive side, "the memory of a received service, such as grooming, induces a positive attitude toward the same individual, a psychological mechanism described as 'gratitude' by Trivers" (p. 93). Behaving fairly pays off in relatively small groups because members tend to reward those who treat them and others fairly (by selecting them as exchange partners, enhancing their reputations, and treating them fairly), while punishing those who behave in selfish and unfair ways (by lashing out at them physically, rejecting them as exchange partners, and ostracizing them from their groups). It is in everyone except cheaters' interests to ensure that cheaters do not prosper.

11.8.1 The Expansion of Fairness

Some evolutionary theorists (e.g., Hauser, 2006) have accounted for the paucity of evidence of reciprocity and other forms of fair exchange in the animal kingdom by arguing that few species possess the mental abilities necessary to calculate the value of the goods and services, compare these values, and evaluate them in terms of appropriate standards of fairness. Such theorists have argued that dispositions to reciprocate and behave fairly have evolved only in species such as humans that possess large brains and advanced mental abilities. Other evolutionary theorists have disagreed, arguing that sophisticated cognitive abilities are not necessary to endow animals with a sense of fairness. In support of this position, Brosnan and de Waal (2003) cited research showing that capuchin monkeys expressed anger when offered a less desirable reward than other monkeys for performing the same task, and even refused to accept the less desirable reward. Brosnan and de Waal (2003) concluded that "capuchin monkeys thus seem to measure reward in relative terms, comparing their own rewards with those available and their own efforts with those of others" (p. 48). Such evidence and arguments notwithstanding, it is clear that humans stand alone in the animal kingdom in their ability to invoke the kinds of sophisticated conceptions of fairness necessary to uphold complex systems of cooperation.

11.9 MENTAL MECHANISMS SUPPORTING FAIR BEHAVIORS

Evolutionary theorists expect dispositions to behave fairly to be triggered in relatively egalitarian cooperative groups among individuals who are of similar status and who engage in repeated social exchanges. The evolved mental mechanisms that regulate these dispositions are expected to be sensitive to information pertaining to the probability of encountering exchange partners in the future, exchange partners' inclination to behave fairly, and the presence of observers. Researchers have found that people rarely invoke tit-for-tat types of fair strategies when engaging in exchanges with friends and loved ones; rather, people tend to support their friends over long periods of time without expecting them to return each and every favor (Tooby & Cosmides, 1996). Indeed, people may feel offended when their friends adopt an "exchange orientation" and insist on paying them back.

A suite of "moral emotions" motivate humans to behave fairly (Fessler & Haley, 2003; Trivers, 2006). Gratitude, a sense of indebtedness, and guilt dispose people to treat others fairly by repaying favors and righting the wrongs they have committed against them. Feelings of vindictiveness and righteous indignation dispose people to punish those who behave unfairly. Forgiveness disposes people to reestablish cooperative relations with those who have cheated them. In recent years, researchers have made a great deal of progress in mapping the design of the mental mechanisms that give rise to these moral emotions.

Gratitude and a Sense of Indebtedness. Feelings of indebtedness are experienced as negative emotional states that motive people to redress a perceived inequity, whereas feelings of gratitude are experienced as magnanimous positive emotional states that motivate people to help those who helped them and others. According to McCullough, Kimeldorf, and Cohen (2008), gratitude evolved "to stimulate not only direct reciprocal altruism, but also 'upstream reciprocity': passing benefits on to third parties instead of returning benefits to one's benefactors" (p. 283). Feelings of gratitude are affected by the value of the benefits that donors proffer, the costs of giving, the intentions that are attributed to them, and the extent to which the beneficial behaviors are viewed as voluntary, expected, and normative (McCullough et al., 2008).

Guilt. Moll et al. (2008) have found that guilt stems from "the blending of elementary subjective emotional experiences, which are ubiquitous in mammals, with emotional and cognitive mechanisms that are typically human" (p. 4). Feelings of guilt induce individuals to repair damaged social relations and restore mutually beneficial cooperative relations. Social psychological research has revealed that feelings of guilt are triggered when people feel responsible for wronging those to whom they are attached, especially when their behaviors violate moral norms.

Findings from research on the development of conscience in children are consistent with evolutionary analyses of its biological functions. For example, Thompson, Meyer, and McGinley (2006) have concluded:

[Y]oung children are motivated to cooperate with the expectations of parents ... to maintain the positive affectionate relationship that they enjoy. Viewed in this light, the parent–child relationship in early childhood can be conceived of as the young child's introduction into a relational system of reciprocity that supports moral conduct by sensitizing the child to the mutual obligations of close relationships. (p. 282)

Vindictiveness and Moral Indignation. Suffering unfair treatment from others may trigger feelings of vindictiveness, and observing others being treated unfairly may trigger feelings of moral indignation. Both of these emotional states motivate individuals to right wrongs by punishing those who have behaved in unfair ways.

Neuroscientists have found that areas of the brain that light up when people pursue rewards, such as the left prefrontal cortex and caudate nucleus, are activated when people feel vengeful, and areas of the brain that are associated with feelings of satisfaction, such as the nucleus accumbens, are activated when people take revenge or learn that those who have treated them unfairly have been punished. According to McCullough (2008), "revenge pays neurochemical dividends. ... Natural selection's logic here seems pretty easy to comprehend: by paying us back with pleasure, our brains ensure that we'll go to the trouble of seeking the social advantages that come from returning harm for harm" (p. 146).

Forgiveness. One of the biological disadvantages of holding grudges is losing potentially valuable exchange partners, and one of the disadvantages of getting even with those who have wronged you is that this strategy may lead to fitness-reducing blood feuds. Feelings of forgiveness evolved to counteract these problems. The biological function of forgiveness is to restore derailed fitness-increasing systems of cooperative exchange.

Forgiveness appears to be a culturally universal social instinct that mediates reconciliation in humans, chimpanzees, and other primates (McCullough, 2008). When individuals are wronged, they experience elevations in stress hormones such as cortisol that generate unpleasant feelings of anger and anxiety. Victims are able to allay these unpleasant states in two main ways: by forgiving those who have harmed them and by seeking revenge. Studies have found that forgiveness is activated when individuals: (1) feel attached to those who have wronged them; (2) value relationships with wrongdoers; (3) are confident that wrongdoers will not harm them again; and (4) believe that forgiving wrongdoers will increase their security. The mental mechanisms that generate feelings of forgiveness are regulated by such considerations as the intentions attributed to wrongdoers and judgments about the extent to which wrongdoers could have avoided inflicting harm, feel regretful, and are likely to reoffend.

11.10 ORIGINS OF PURITY

One of the disadvantages of social living is that members of groups may transmit diseases, spread germs, and contaminate one another. In addition, interacting with dirty individuals lowers the quality of life. For these reasons, members of all human groups value cleanliness and prefer to associate with members of their groups who practice proper hygiene.

11.10.1 Mental Mechanisms Upholding Purity: Disgust

Phenomena such as bodily fluids, dead animals, decaying food, and rancid smells that are associated with the risk of harm due to germs and pollution evoke feelings of disgust in humans and other primates (Rozin, Haidt, & McCauley, 2000). Sexual prohibitions such as those associated with menstruation, anal sex, and bestiality also often induce feelings of disgust (Rozin et al., 2000).

11.10.2 The Expansion of Cleanliness: Implications for Morality

Rozin et al. (2000) have suggested that, in humans, moral disgust is an "embodiment" of nonmoral disgust that originated in digestive repulsion, and that instinctive reactions to physical impurity and purity may be embodied in ideas about spiritual impurity and purity. There is a strong association between cleanliness and godliness in the Bible. For example, in Isaiah 1:16, it is written: "Wash you, make you clean; put away the evil of your doings from before mine eyes; cease to do evil." The adaptive function of disgust is to motivate people to withdraw from potentially noxious stimuli, whether physical or social in nature. It is in the adaptive interest of members of groups to steer clear of those who emit signs of impurity, to interact with those who seem pure, and to take measures to purify themselves.

Researchers have found that disgusting physical stimuli can intensify moral judgments. For example, in one study, participants housed in a dirty and messy room made harsher moral judgments than participants housed in a clean room (Schnall et al., 2008). Investigators also have found that the intensity of negative moral judgments diminishes after people wash their hands (Schnall et al., 2008).

11.11 THE NATURE OF MORALITY

What do the forms of conduct that people from all cultures consider moral have in common that earns them this honor? What is it about these types of behavior that renders them moral? In response to such questions, evolutionary theory directs us to ask what overriding biological functions these forms of conduct served when they evolved. How did they help early humans improve their fitness? The most obvious answer to this question is: by enabling members of groups to reap the biological benefits of sociality and cooperation. Altruistic forms of conduct induce members of groups to help one another meet their biological needs. Tribal instincts induce individuals to uphold the groups on which they are dependent. Deferential forms of conduct induce individuals to uphold the hierarchical moral orders of their groups, and fair forms of conduct induce individuals to uphold systems of cooperative exchange, both of which benefit them by increasing the effectiveness of the social systems that produce the resources they need to survive, reproduce, and propagate their genes. Self-controlling dispositions induce members of groups to constrain selfish and aggressive urges that would alienate potential exchange partners and threaten the sanctity of their groups, and forms of conduct that uphold purity enable individuals to avoid contaminating members of their groups and being contaminated by them. Behaving in moral ways pays off biologically when what goes around comes around.

11.12 ORIGINS OF A SENSE OF MORALITY

Accounting for the origins of dispositions to behave in ways that people from all cultures consider moral takes us a long way toward an evolutionary account of morality.

However, in addition, we must explain how the mental mechanisms that induce people to classify phenomena as moral originated. How were the mental mechanisms that endow people with a sense of morality selected?

11.12.1 The Moral Senses

Although the basic adaptive function of all aspects of a sense of morality is to uphold the moral orders of groups, different aspects of the moral sense – or different moral sentiments – contribute to this function in different ways. Some moral sentiments dispose individuals to induce other members of their groups to behave in moral ways, whereas other moral sentiments dispose individuals to behave in moral ways themselves.

11.12.2 Origins of Moral Sentiments Directed at Others

Moral sentiments that dispose individuals to induce others to behave morally probably originated in positive and negative emotional reactions to being treated in fitness-increasing and fitness-reducing ways by members of one's group and observing one's friends and relatives being treated in ways that diminished their fitness. Early humans would have reacted in positive ways when they and those with whom they were allied were treated in kind, loyal, respectful, fair, and uncontaminated ways because being treated in these ways would have helped them survive, reproduce, and propagate their genes. Early humans would have reacted negatively when they and those with whom they were allied were treated in cruel, disloyal, disrespectful, unfair, and unclean ways, just as modern humans do. Positive emotional reactions would have evoked a primitive evaluative sense that the fitness-increasing forms of conduct were right and that those who showed them were good, whereas negative reactions would have engendered the sense that the fitness-decreasing forms of conduct were wrong and that those responsible for perpetrating them were bad.

11.12.3 Origins of Moral Sentiments Directed at One's Self

People experience both forward-looking moral sentiments about what they ought to do and backward-looking sentiments about what they have done. People experience a sense of moral obligation when they believe they should – that is to say, have a duty to – behave in accordance with the moral norms of their groups. In contrast, people experience a guilty conscience when they believe they have done something wrong. As discussed, the biological function of conscience and guilt is to induce members of groups to regret behaving in ways that undermine the cooperative social orders of their groups, to motivate them to right the wrongs they have committed, and to induce them to reform.

11.12.4 Origins of a Sense of Fairness and Justice

According to Trivers, "a sense of fairness has evolved in the human species as the standard against which to measure the behavior of other people, so as to guard against cheating in reciprocal relationships" (1985, p. 388), and "such cheating is expected to generate strong emotional reactions, because unfair arrangements, repeated often, may exact a very strong cost in inclusive fitness" (2006, p. 77). From the perspective of evolutionary theory, it is obvious why individuals believe that others should treat them fairly; however, we would not expect people to feel equally strongly that they should treat others fairly. As asserted by Trivers (2006), "an attachment to fairness or justice is self-interested and we repeatedly see in life ... that victims of injustice feel the pain more strongly than do disinterested bystanders and far more than do the perpetrators" (p. 77).

In support of Trivers' observations, researchers have documented several self-serving biases in people's sense of fairness and justice (for reviews, see Haidt, 2001; Krebs, 2011). For example, people tend to underestimate how much they owe others and overestimate how much others owe them, undervalue the benefits of receiving and overvalue the costs of giving, and judge themselves less harshly when they treat others unfairly than they do when others treat them unfairly. However, this is not the full story. Evolutionary theory also leads us to expect significant constraints on self-serving biases in people's sense of justice because unconstrained biases do not wash in exchanges with others. Just as selfish strategies fail to pay off well against other selfish strategies in social interactions and economic games, self-serving biases tend to pay off poorly against other self-serving biases because they end up doing each other down. Indeed, Janicki (2004) found that people may actually invoke other-serving biases in exchanges with their friends. For example, people tend to feel worse when they fail to repay their friends than they do when their friends fail to repay them. The long-term fitness benefits of upholding friendships outweigh the short-term benefits of exploiting one's friends.

Like other primates, humans manifest a primitive sense of justice when they react in positive ways to being treated fairly and negative ways to being treated unfairly (de Waal, 2008). However, in addition, humans may react strongly to the ways in which third parties are treated – a phenomenon that has been labeled "strong reciprocity." As expressed by J. Q. Wilson (1993), "our sense of justice ... involves a desire to punish wrongdoers, even when we are not the victims, and that sense is a 'spontaneous' and 'natural' sentiment" (p. 40).

11.12.5 Emotional and Rational Aspects of a Sense of Morality

There is a long-standing debate among scholars about the extent to which people derive moral judgments from emotional and rational sources (Denton & Krebs, 2017).

Dual-process theorists such as Greene (2013) have attempted to resolve this dispute by demonstrating that two sets of mental mechanisms have evolved in humans: one that induces them to make moral judgments in emotional ways and another that induces them to make moral judgments in more rational ways. Denton and Krebs (2017) have accounted for evidence supporting dual-process models of moral decision-making by suggesting: (1) that early-evolved brain mechanisms that humans share with other primates give rise to emotional forms of moral decision-making; (2) that later-evolved brain mechanisms give rise to more rational forms of moral reasoning; and (3) that the later-evolved mechanisms increase in power in humans as they develop throughout their lives. According to Denton and Krebs (2017, p. 79):

[A]ttending to the fact that moral decision making mechanisms originated to solve adaptive problems in early human social environments that changed in more complex ways as humans evolved helps explain why people possess two types of moral decision-making mechanisms, and it supplies a basis for generating expectations about what the essential attributes of each type should be, the kinds of stimuli that trigger each type of mechanism, and how the two types interact. The first suite of moral decision-making mechanisms, originating in a set of primitive brain structures, evolved to solve the relatively simple moral problems faced by early humans. The second suite of moral decision-making mechanisms, controlled by more recently evolved brain structures, evolved to solve the increasingly complex moral problems experienced by humans as they evolved, including those that stemmed from membership in increasingly complex societies and participation in increasingly sophisticated forms of strategic social interaction.

In a similar vein, Haidt (2001) has concluded that "the affective system has primacy in every sense: It came first in phylogeny, it emerges first in ontogeny, it is triggered more quickly in real-time judgments, and it is more powerful and irrevocable when the two systems yield conflicting judgments" (p. 819). Moll et al. (2008) have suggested that "moral emotions might prove to be a key venue for understanding how phylogenetically old neural systems, such as the limbic system, were integrated with brain regions more recently shaped by evolution, such as the anterior PFC [prefrontal cortex], to produce moral judgment, moral reasoning, and behavior" (p. 17).

11.13 CONCLUSION

Morality is considered by many to be a uniquely human attribute instilled in children through learning. However, there is a great deal of evidence that members of other species behave in moral ways and even possess a primitive sense of morality, and there is a great deal of evidence that moral dispositions or instincts have evolved in the human species. Although accounts that explain the development of morality in children offer valuable insights into the proximate mechanisms that mediate moral judgments and moral behaviors, they offer little insight into the origins of these proximate mechanisms themselves. To fully understand the role that morality plays in human nature, we must understand how the evolved mental mechanisms that give rise to moral behaviors, moral emotions, moral judgments, and moral reasoning helped early humans adapt to their environments and foster their biological interests. The primary function of moral dispositions was to induce early humans to behave in ways that upheld social orders that supported fitness-increasing systems of cooperation and to induce them to resist the temptation to resolve conflicts of interest in maladaptively selfish ways. Mental mechanisms that disposed early humans to behave in moral ways were selected and shaped as humans evolved. Modern humans are naturally disposed to uphold culturally universal and culturally relative moral norms. We are moral by nature.

REFERENCES

Alexander, R. D. (1987). *The Biology of Moral Systems*. New York: Aldine de Gruyter.

Bandura, A. (1991). Social cognitive theory of moral thought and action. In W. M. Kurtines & J. L. Gewirtz, eds., *Handbook of Moral Behavior and Development*, Vol. 1. Hillsdale, NJ: Erlbaum, pp. 54–104.

Batson, C. D. (1991). *The Altruism Question: Toward a Social-Psychological Answer*. Hillsdale, NJ: Erlbaum.

Boehm, C. (2000). Conflict and the evolution of social control. *Journal of Consciousness Studies*, **7**, 79–101.

Boehm, C. (2012). *Moral Origins: The Evolution of Virtue Altruism and Shame*. New York: Basic Books.

Brosnan, S. F., & de Waal, F. B. (2003). Monkeys reject unequal pay. *Nature*, **425**, 297–299.

Brown, S. L., & Brown, M. (2006). Selective investment theory: Recasting the functional significance of close relationships. *Psychological Inquiry*, **17**, 30–59.

Burnham, T., & Phelan, J. (2000). *Mean Genes*. New York: Perseus Publishing.

Buss, D. (2004). *Evolutionary Psychology: The New Science of the Mind*, 2nd ed. Boston, MA: Allyn & Bacon.

Cummins, D. (2005). Dominance, status, and social hierarchies. In D. Buss, ed., *The Handbook of Evolutionary Psychology*. Hoboken, NJ: John Wiley & Sons, pp. 676–697.

Darwin, C. (1871). *The Descent of Man, and Selection in Relation to Sex*. London: John Murray.

Dawkins, R. (1989). *The Selfish Gene*. Oxford: Oxford University Press.

Denton, K. K., & Krebs, D. L. (2017). Rational and emotional sources of moral decision-making: An evolutionary-developmental account. *Evolutionary Psychological Science*, **3**, 72–85.

de Waal, F. B. M. (2008). Putting the altruism back in altruism. *Annual Review of Psychology*, **59**, 279–300.

de Waal, F. B. M., & Brosnan, S. F. (2006). Simple and complex reciprocity in animals. In P. M. Kappeler & C. P. van Schaik, eds., *Cooperation in Primates and Humans*. New York: Springer-Verlag, pp. 85–105.

Fessler, D. M. T., & Haley, K. J. (2003). The strategy of affect: Emotions in human cooperation. In P. Hammerstein, ed., *Genetic and Cultural Evolution of Cooperation*. Cambridge, MA: MIT Press, pp. 7–36.

Greene, J. D. (2013). *Moral Tribes: Emotion, Reason, and the Gap between Us and Them*. New York: Penguin Press.

Haidt, J. (2001). The emotional dog and its rational tail: A social intuitionist approach to moral judgment. *Psychological Review*, **108**, 814–834.

Haidt, J., & Joseph, C. (2004). Intuitive ethics: How innately prepared intuitions generate culturally variable virtues. *Daedalus*, **133**, 55–66.

Hamilton, W. D. (1964). The evolution of social behavior. *Journal of Theoretical Biology*, **7**, 1–52.

Haslam, N. (1997). Four grammars for primate social relations. In J. A. Simpson & D. T. Kenrick, eds., *Evolutionary Social Psychology*. Mahwah, NJ: Erlbaum, pp. 297–316.

Hauser, M. D. (2006). *Moral Minds: How Nature Designed Our Universal Sense of Right and Wrong*. New York: HarperCollins Publishers.

Henrich, J. (2016). *The Secret of Our Success*. Princeton, NJ: Princeton University Press.

Janicki, M. (2004). Beyond sociobiology: A kinder and gentler evolutionary view of human nature. In C. Crawford & C. Salmon, eds., *Evolutionary Psychology: Public Policy and Personal Decisions*. Mahwah, NJ: Erlbaum, pp. 51–72.

Kohlberg, L. (1984). *Essays in Moral Development: The Psychology of Moral Development*, Vol. 2. New York: Harper & Row.

Krebs, D. L. (2011). *The Origins of Morality*. New York: Oxford University Press.

Krebs, D. L., & Denton, K. (2006). Explanatory limitations of cognitive-developmental approaches to morality. *Psychological Review*, **113**, 672–675.

McCullough, M. E. (2008). *Beyond Revenge: The Evolution of the Forgiveness Instinct*. San Francisco, CA: Jossey-Bass.

McCullough, M. E., Kimeldorf, M. B., & Cohen, A. D. (2008). An adaptation for altruism? The social causes, social effects and social evolution of gratitude. *Current Directions in Psychological Science*, **17**, 281–285.

Miller, G. F. (2007). The sexual selection of moral virtues. *Quarterly Review of Biology*, **82**, 97–125.

Moll, J., di Oliveira-Sourza, R., Zahn, R., & Grafman, J. (2008). The cognitive neuroscience of moral emotions. In W. Sinnott-Armstrong, ed., *Moral Psychology: The Neuroscience of Morality: Emotion, Brain Disorders, and Development*. Cambridge, MA: MIT Press, pp. 1–18.

Nowak, M. (2006). Five rules for the evolution of cooperation. *Science*, **314**, 1560–1563.

Nowak, M. A., Tarnita, C. T., & Wilson, E. O. (2010). The evolution of eusociality. *Nature*, **466**, 1057–1062.

Pinker, S. (2008). The moral instinct. *New York Times*, January 13.

Richerson, P. J., & Boyd, R. (2001). The evolution of subjective commitment: A tribal instincts hypothesis. In R. Nesse, ed., *Evolution and the Capacity for Commitment*. New York: Russell Sage Foundation, pp. 186–219.

Richerson, P. J., & Boyd, R. (2005). *Not by Genes Alone: How Culture Transformed Human Evolution*. Chicago, IL: University of Chicago Press.

Rozin, P., Haidt, J., & McCauley, C. R. (2000). Disgust. In M. Lewis & J. M. Haviland-Jones, eds., *Handbook of Emotions*, 2nd ed. New York: Guilford Press, pp. 637–653.

Schnall, S., Haidt, J., Clore, G. L., & Jordan, A. H. (2008). Disgust as embodied moral judgment. *Personality and Social Psychology Bulletin*, **34**, 1096–1109.

Sober, E., & Wilson, D. S. (1998). *Unto Others: The Evolution and Psychology of Unselfish Behavior*. Cambridge, MA: Harvard University Press.

Tajfel, H., & Turner, J. C. (1985). The social identity theory of intergroup behavior. In S. Worchel & W. G. Austin, eds., *Psychology of Intergroup Relations*. Chicago, IL: Nelson-Hall, pp. 7–24.

Thompson, R., Meyer, S., & McGinley, M. (2006). Understanding values in relationships: The development of conscience. In M. Killen & J. Smetana, eds., *Handbook of Moral Development*. Mahwah, NJ: Erlbaum, pp. 267–298.

Tooby, J., & Cosmides, L. (1996). Friendship and the banker's paradox: Other pathways to the evolution of adaptations for altruism. *Proceedings of the British Academy*, **88**, 119–143.

Trivers, R. L. (1971). The evolution of reciprocal altruism. *Quarterly Review of Biology*, **46**, 35–57.

Trivers, R. (1985). *Social Evolution*. Menlo Park, CA: Benjamin Cummings.

Trivers, R. (2006). Reciprocal altruism: 30 years later. In P. M. Kappeler & C. P. Van Schaik, eds., *Cooperation in Primates and Humans*. New York: Springer-Verlag, pp. 67–84.

Wilson, J. Q. (1993). *The Moral Sense*. New York: The Free Press.

Wright, R. (1994). *The Moral Animal*. New York: Pantheon Books.

Wynne-Edwards, V. C. (1962). *Animal Dispersion in Relation to Social Behavior*. Edinburgh: Oliver and Boyd.

12 The Evolution and Function of Third-Party Moral Judgment

WILL READER AND SARA HUGHES

12.1 INTRODUCTION

There has been much recent research activity on the evolution of morality (Chapters 10 and 11, this volume; Haidt, 1993; Hauser, 2006; Krebs, 2011; Kurzban & DeScioli, 2009; Ridley, 1996; Wright, 1994), with most tending to focus upon the paradoxical behavior in situations in which a moral actor incurs a personal cost in order to help nonrelatives. This is paradoxical because, from an evolutionary point of view, any genes that produce a behavior that benefits nonrelatives at the expense of the individual in which those genes reside should find themselves at a fitness disadvantage and therefore would die out. In short, such behaviors should not evolve. This is, of course, a familiar problem, one that was discussed at length by Trivers (1971) under the guise of the evolution of altruism. Solutions to the altruism problem vary and include reciprocation, whereby doing good to others could evolve so long as others reciprocate, and indirect reciprocity, whereby doing good to others could evolve if it encourages third parties to cooperate with us (Alexander, 1987). Recent research on the evolution of morality has suggested that positive moral virtues such as sharing, reciprocity, and helping the group may be designed for the express purpose of enabling and facilitating cooperation (Curry, 2016; Curry, Mullins, & Whitehouse, 2019). In this volume, Dennis L. Krebs discusses the many possible mechanisms by which morality could have evolved (Chapter 11).

Another strand of research on morality is exploring the universality of specific morals (see the discussion introduced by Sloan-Wilson, Sloan, & Price, n.d.). These may be specific morals such as prohibitions against murder, rape, incest, or theft, as discussed by Brown (1986), or more general categories such as those proposed in Moral Foundations Theory: fairness, harm, in-group loyalty, authority, and purity (Haidt, 2012).

But there is another paradox to morality within the evolutionary literature that has received less discussion: third-party moral judgment, which is the subject of this chapter. Third-party moral judgment is the positive or (probably more usually) negative judgment of a behavior or trait in others that has *no material effect on the person judging*. For example, if an actor (first party) harms a recipient (second party) in some way, say by stealing money from them, an observer (third party) with no material interest in the second party's welfare seeks to condemn or punish the actor for their actions. "No material interest" here means that the third party is not a close relative of the recipient, therefore ruling out inclusive fitness as an explanation, and has no other investment in them – they are not an ally or potential ally and they do not owe the third party a favor.

There is a strong similarity between third-party moral judgment and two other phenomena that have been discussed in the literature: third-party punishment (e.g., Fehr & Fischbacher, 2004) and third-party moral condemnation (DeScioli & Kurzban, 2009). It is worth discussing the differences between these three phenomena.

First, third-party moral judgment is different from third-party punishment because while the act of judging may result in punishment, it is by no means inevitable. The person judging may decide that discretion is the better part of valor and choose not to punish, or they may be unable to punish. Similarly, it differs from third-party condemnation in that the third party may decide to keep their views private. Third-party moral judgment is therefore a precursor to and motivator of third-party moral condemnation or punishment.

A second difference is that in most considerations of third-party punishment there is an obvious *second* party. DeScioli and Kurzban (2009) describe moral condemnation as necessarily triadic, consisting of an agent (or transgressor), a patient (or victim), and an uninvolved witness. In third-party moral judgment, there need be no obvious second party. For example, people around the world and across history have been condemned and persecuted for their homosexuality, even if they only have consensual sex or are celibate.

Finally, although most instances where a third party judges some action or trait are likely to be negative, we admit the possibility that judgments may also be positive, perhaps leading to the actor being praised or rewarded. The Catholic Church, for example, has recognized many individuals as saints for (among other things) showing faith in the face of persecution, such as St. Thomas More, who was hanged, drawn, and quartered at the instruction of Henry VIII for refusing to accept Henry as the head of the church. More prosaically, there are the various medals and awards that are given out for acts of heroism or other good deeds, such the Victoria Cross, the Nobel Peace Prize, the Carnegie Medal, the Civil Courage Prize, and so on.

This brief discussion of differences is by no means to make a claim that third-party moral judgment is a novel concept that is qualitatively different from third-party punishment or third-party moral condemnation. The above discussion was purely to make clear what the properties of third-party morals are. If it turns out that the three concepts are the same, then this does not affect what we discuss in the rest of this chapter.

In this chapter, then, we explore possible evolutionary explanations for why people judge, condemn, punish, and praise each other in the ways described above, and we speculate that morality was an essential element in enabling the human species to become the dominant cultural force that it now is. We write this chapter very much in the spirit of inquiry rather than advocacy, and although much of the evidence supports our favored position, some of it does not.

12.2 WHAT NEEDS TO BE EXPLAINED?

For the purposes of clarity, we need to be clear on what we are trying to explain, and we do this using the following thought experiment: imagine a hypothetical situation in which people only cared about others whose lives had direct material value to their own. They may be relatives, owe the agent some obligation, or be part of an ongoing reciprocal relationship. Such individuals would show no compassion to a stranger, even if that person was part of their group, and they would show no compassion to someone known to them unless that person was perceived to be of some material value. Likewise, no one would care how other people in their group behaved or treated each other and how they directly or indirectly affected them. Today, we might consider such individuals as having psychopathic tendencies – only showing concern for others when it directly affected their own interests – and psychopaths are commonly considered to be without morals (Jonason et al., 2013). Assuming there is an adaptive explanation, what might the adaptive benefit be that marked a transition from not judging others to judging others? Before we can answer this question, we need to inspect more closely the characteristics of third-party moral judgment.

12.3 THE CHARACTERISTICS OF MORAL JUDGMENT

12.3.1 There Does Not Need to Be a Victim

In their classic paper, Haidt, Koller, and Dias (1993) demonstrated that participants will condemn actions that are entirely victimless: a family that eats their treasured dog who died from natural causes; a woman who cleans her house with the American flag; a brother and sister who have sex despite taking multiple precautions against pregnancy. Grey, Waytz, and Young (2012) disagree with this interpretation. They argue that the moral dyad of actor/agent/perpetrator and recipient/patient/victim is essential in all moral judgments (handily, they even examine many interpretations of "essential" and show the implications of these for their theory). They argue that in situations such as those explored by Haidt and other apparently "victimless crimes" such as masturbation, drug use, and suicide, the moral dyad remains intact. In such cases, the victim is God (masturbation is considered to be a sin in some religions), the perpetrator's relatives (e.g., suicide), or the person themselves (drug use). While the question remains as to whether it is possible to judge something as immoral without there being a victim or victims, two points must be noted. First, as DeScioli, Gilbert, and Kurzban (2012) point out, the fact that people might always be able to nominate a victim does not necessarily mean that that victim played a causal role in the moral decision-making process; the victims may have been created post hoc. Second, there is a difference between the intentional harm or exploitation of another individual (rape, murder, theft) and the unintended and potentially unforeseen consequences of one's actions on another person. Intention has repeatedly been shown to be an important ingredient in the moral calculus (e.g., Kant, 1999). This particular debate might, as so many have before it, spiral into inconsequentiality, but for the purposes of this chapter, let us define "victimless" as meaning that "no harm was intended to be meant to another person."

12.3.2 Many Judgments Are Not Based on the Consequences of an Action

The discussion about victimless crimes fits neatly with an observation made by DeScioli and Kurzban (2009) that, although moral condemnation often requires there to be negative consequences as a result of the action (e.g., a person suffers), it is not necessary (Haidt, 1993; Hauser, 2006). In Hauser's versions of the trolley problem, most people are happy to concede that a train that is about to kill five railway workers should be diverted down a branch line where it will kill only one worker, but they are far from happy sanctioning pushing a large man off a footbridge to stop the train. In both cases, the consequence is the same – one person is killed to save five – but the judgments are very different. Or take another classic moral dilemma: you are with a group of people hiding in an abandoned house. Soldiers are outside looking for you, but you are all safe as long as no one makes a sound. You notice a woman with a baby and the baby is about to cry. Do you: (1) kill the baby, thus preventing you from being discovered; or (2) not kill the baby, meaning that everyone in the house gets shot. There are many objections raised to this scenario. How do you kill the baby soundlessly? How do you know with 100 percent certainty that the baby will cry? How do you know that the soldiers will hear the cry? Are you certain that they will kill you all? Even once these doubts have been satisfied, most people are still uncomfortable with killing the baby (Cushman & Greene, 2012). This is a situation that is even more obvious than those in the trolley problem. You are killing one to save everyone, because if you do not act, everybody dies (including the baby). Interestingly, people high in dark triad traits (including psychopathy) are more likely to make the rational decision based on the outcomes (Arvan, 2013).

Like psychopaths, many nonhuman animals also appear to be consequentialists. Parents will not only kill their offspring if resources are severely limited (Hrdy, 1979), but many will also eat them to increase their own chances of survival. This is the familiar "red in tooth and claw" logic of natural selection, but we – and it appears nonhuman primates as well – buck this trend, at least some of the time (de Waal, 1996). Recent research even suggests that dogs (as well as capuchin monkeys) can make moral judgments and show moral condemnation by refusing to interact with people whom they consider unhelpful (Anderson et al., 2017; Nitzschner et al., 2012). The fact that all of the animals (including humans) that show moral reasoning live in complex social groups suggests that morality, including third-party moral judgment, is one of the answers to the questions posed by group living.

12.4 ADAPTATIONIST LOGIC

We have pointed out more than once that one of the properties of third-party moral judgment is that people reward or punish others even when their behavior has no material effect on the punisher. But, of course, if third-party moral judgment is an adaptation, then there must be some ultimate, average net benefit to the punisher's genes for it to evolve. It may, of course, not itself be an adaptation, but may instead be a by-product of something that *is* an adaptation. Alternatively, it may be a purely cultural phenomenon and nothing to do with evolution at all. Rather than defend against these latter two interpretations, we will simply assume that it is an adaptation and see how far this gets us. If our argument seems productive, then we can work on dismissing the alternatives later.

12.5 AN EXPLANATION FOR MORAL JUDGMENT: TAKING SIDES

DeScioli and Kurzban (2012) propose that third-party moral judgment (they use the word "condemnation" as discussed above, but the differences are not important here) evolved as a manner for managing intragroup conflict. Their account hinges on the principle that the more equally matched potential combatants are, the greater the possibility there is for full-scale and potentially ruinous conflict. In many animal species, males compete for access to females. Often the competition begins with displays in which each animal gives its best shot at intimidating the other. If one animal is demonstrably superior, the other will back down, done and dusted, and the winner takes all (Krebs & Davies, 2009). (In very unequally matched individuals, it is unlikely that it will get to this stage, as the inferior animal will not even chance his – or more rarely her – arm.) If the two animals cannot decide who is superior at this point, they may start to fight. Again, if one animal is obviously physically superior to the other, the fight will end with little damage to either; even a physically superior animal will not risk damage being sustained by a lucky blow. The problem arises when two animals are so equally matched that they have to go for the full 12 rounds; in such a situation, even the winner can end up damaged beyond repair. Equally matched combatants are a recipe for potential disaster.

DeScioli and Kurzban suggest that intragroup conflict was as common in our ancestors' time as it is today, and so long as conflict was small scale – between the disputants and perhaps immediate friends and family – it would pose little problem to the group as a whole. Problems arise when the dispute escalates and erstwhile bystanders stop standing by and enter the fray. In extreme cases, this might lead to the entire group being split down the middle into two warring factions. As for the nonhuman example given above, two equally matched groups create the potential for a long-lasting and injurious conflict, causing physical and psychological harm to both sides, depleting resources, and obstructing normal life (foraging, childrearing, hunting).

DeScioli and Kurzban discuss a number of possible solutions to this problem, finally suggesting that third-party moral condemnation (to use their term) is a potential solution. All that is needed is some way to favor one disputant over the other, and this can be done by the formation of (possibly arbitrary) moral principles that everyone in the group adheres to. The existence of a rule such as "stealing is wrong" means that when a potential theft-related conflict arises and starts to escalate, the majority sides with the victim rather than the perpetrator and costly conflict is avoided. The moral principles provide an easy, predetermined way to decide whose side they should take.

This theory could account for the evolution of third-party moral judgment if ancestral societies consisted of a small number of kinship or alliance networks, each of which comprised a large percentage of the group as a whole. If a society comprised two groups, A and B, then any slight to a member of group A by a member of group B would lead to members of group A retaliating, which would drag group B members in, with the resulting costly conflict described above.

It is important to note that these networks would need to be based on economic or kinship bonds as, remember, third-party moral judgment has not yet evolved (this is, after all, what we are trying to explain the evolution of), so, like the psychopaths above, people only care about the business of others to the extent that it impacts their own.

12.6 AN ALTERNATIVE EXPLANATION: THE SUPERORGANISM

There is an alternative that may be a direct competitor to the side-taking hypothesis or may supplement it. Whereas DeScioli and Kurzban focus on intragroup conflict to explain the evolution of moral judgment, we propose that the ultimate origins lie in interdependence and intergroup conflict. Unlike our closest primate relatives, ancestral humans became obligate collaborative foragers, meaning that not all adult individuals (or even families) could acquire enough calories and nutrients to sustain themselves, and they had to rely on the output of others, mainly younger, unmarried men (Hill & Hurtado, 2009). This was probably a result of the helplessness of human newborns who required greater investment from the father (Chapter 28), which in turn led to a decreased inter-birth interval in human females (Chapter 26). With many often helpless mouths to feed, many families were unable to obtain enough food for themselves without outside help (Hrdy, 2011). This increased interdependence prepared the way for social adaptations that enable sharing and collaboration, with concomitant mechanisms for preventing defection (Chapter 22; Tomasello et al., 2012). Tomasello et al. propose that, at this point in our evolutionary history, human group sizes were comparatively small, possibly corresponding to Dunbar's number of ~150 individuals. With groups of this size, it is likely that everyone knew everyone else and cooperation could be maintained by reputation-based means.

The next step in our social evolution, according to Tomasello et al., was an increase in average group size as a result of intergroup conflict – larger groups being, on average, more able to defend themselves against attack. Exactly why intergroup conflict increased is unknown, but at least one answer is that it was the result of increased population density: humans were becoming more numerous in favored parts of the world and conflict became inevitable (Kelly, 2000). Human history has been a story of warring tribes (Choi & Bowles, 2003; Dunbar et al., 1995; Langergraber et al., 2011; Wrangham, 1987). Conflicts between groups could include wars over territory, food, and potential mates (McAndrew, 2002; Wrangham, 1987). Such conflicts are commonly observed in current hunter–gatherer societies such as the Yanomamo (Chagnon, 2012). The increasingly large groups required as a result of this intergroup conflict made it difficult for in-group cooperation to be based upon personal knowledge and reputation, which led to the evolution of "group-mindedness" where people engage

in reciprocal relationships with people whom they see as their own kind whether they know each other or do not (Choi & Bowles, 2007; Gavrilets, 2015; Tomasello et al., 2012). "Own kind" here is a somewhat elastic concept that can include family, friends, near-neighbors, tribe, nation, and, if extraterrestrials were ever to threaten us, our entire species (Sherif, 1958). Groups naturally organize themselves around a common identity that includes food preferences, flags, clothing, initiation ceremonies, rites of passage, music, religious beliefs, body adornments, values, norms, architecture, art, and language. It has further been suggested that the human capacity for creating rites, rituals, and ceremonies evolved in order to act as identifiers for in-group members and to increase social cohesion (Legare & Watson-Jones, 2015)

The enforcement of these identifiers can be summed up in the phrase: "*We* don't do that, we do this." What this is all for, evolutionarily speaking, is to be of one flesh, the better to repel the attacks of those who wish to take our land, and the better to take the lands of others.

So this perspective suggests that moral condemnation promotes social cohesion in three ways. First, rehabilitation: to gently (or otherwise) encourage those that stray from the group values to return to the fold. Second, deterrence: by witnessing the condemnation of others, those who are tempted to stray might think again. Third, signaling: people condemn in order to show to everyone else that they are pious adherents to the party line, so much so that they are prepared to put their head above the parapet and risk harmful retaliation from those who they condemn (a modern interpretation of this is so-called internet virtue signaling; Bartholomew, 2015).

12.7 DOES IN-GROUP SIMILARITY OF CULTURAL PRACTICES PROMOTE COHESION?

The research on this point largely tries to answer the opposite question: Does diversity in cultural practices reduce group cohesion? This is partly due to Robert Putnam's (2007) suggestion that, in the face of diversity, we "hunker down – that is, to pull in like a turtle" and withdraw from social life (p. 149). This principle, known as the *constrict claim*, generally has some support from the literature. However, the principal effect seems to be that ethnically diverse areas drive down contact with and trust in neighbors. Increased ethnic diversity can actually increase interethnic trust, presumably because people are meeting real people from other ethnicities, rather than being presented with stereotypes. The patterns in the data are confusing, as the extent to which diversity negatively affects cohesion varies depending upon country, ethnicity, the measures of cohesion (behavioral, attitudinal), and a variety of other factors (Dinesen & Sønderskov, 2015; Van der Meer, 2015; Van der Meer & Tolsa, 2014).

One must be wary of drawing broad-brush conclusions from such research, as the societies that it investigates are doubtless very different from those of our ancestors. Our ancestors would have lived in smaller groups where each group member would have had personal knowledge of all other members, rather than in the densely packed, anonymous towns and cities that the above research focuses upon. The concept of "group" is also problematic here. Above, we discussed the nature of a group as being elastic and variable depending on which particular identity is activated. If we study an ethnically diverse neighborhood, does that constitute the group? And should we explore cohesion among all its members? Or should we see the neighborhood as comprising multiple groups that consist of the different ethnicities?

Overall, the real-world research suggests that increased diversity might, in some cases, reduce cohesion, which might be taken as suggesting that increased similarity might increase cohesion, but the picture is complex.

12.8 PROMOTING COHESION

Research based on laboratory studies and mathematical modeling is less ambiguous and shows how intergroup conflicts promote parochial altruism in groups (Bowles et al., 2003; Choi & Bowles, 2007) and how parochial altruism works advantageously for groups during conflicts. Groups whose members are more altruistic will have an evolutionary benefit and will outcompete less altruistic groups during intergroup conflicts. We propose that moral punishments partly evolved to increase success during intergroup conflicts by increasing similarities between individuals within groups. Increased similarities, such as sharing the same beliefs and behaving similarly, increase group cohesion. Groups that are more cohesive work cooperatively to achieve beneficial group goals (i.e., defending the group against an attack of an out-group). If one member of our group starts to develop different ideas and beliefs to us or starts acting differently, we become concerned that they may not be there to help us in the future when we need them (e.g., during conflicts). This may be because these individuals appear to be breaking ranks with the majority of the group. Individuals who act differently from the majority pose a risk of not contributing toward group goals due to holding different ideas. Third-party moral judgment, we suggest, is a mechanism designed to promote uniformity within a group so as to improve cohesion and thus success in intergroup conflicts. Given this, it should be the case that people are more likely to judge, condemn, and punish members of the in-group than those of the out-group for moral transgressions, because people should not be interested in promoting cohesion within a rival group.

There is some research that supports this position by establishment of a phenomenon known as the black sheep effect (Marques & Paez, 1994; Marques et al., 1992). The black sheep effect is a situation whereby in-group members are punished more severely for a transgression than is an out-group member; this is consistent with the hypothesis that third-party moral judgment has a function (possibly among others) to increase uniformity and cohesion. The term "black sheep" derives from the idiom that compares individuals within a group or family who stand out from the crowd by being a different color.

Although research into the black sheep effect supports our interpretation of third-party moral judgment, there is other research under the so-called moral hypocrisy that shows the opposite. For example, Valdesolo and DeSteno (2007) created an experimental situation in which participants who had transgressed had to judge their own fairness and the fairness of in- and out-group members whom the participant had observed making the same transgression. Participants judged in-group transgressions to be as (un)fair as their own, but judged out-group transgressions as significantly more unfair than those of both themselves and the in-group. This seems to show they are being *more* forgiving of the behavior of the in-group member than *less* forgiving toward the behavior of the out-group member because the judged fairness of the behavior of an unaffiliated member (neither in- nor out-group) was no different from that of the out-group.

The apparent contradiction between these findings might be resolved if we consider cohesion more carefully. In order to produce a group that can resist attacks from an out-group, we not only want the in-group to be cohesive, but it should also be of a large size. All else being equal, larger groups would be more effective in defense or attack than smaller groups (Dunbar, 1992). The best method for dealing with deviant behavior, therefore, would be to try to coax deviants back into the fold rather than killing, maiming, or ostracizing them. Only when it becomes clear that rehabilitation is a forlorn strategy might more punitive methods be necessary. Furthermore, the severity of in-group punishment might be moderated by the perceived threat of the deviant behavior to group stability. There is some evidence for both of these positions.

First, Gollwitzer and Keller (2010) investigated the role of group membership on punishment decisions as a result of the perpetrator hiding a book in a library so that other students would be unable to use it. The perpetrator was either part of the in- or out-group with respect to the participant who was making the judgment. Importantly, the researchers also portrayed the perpetrator as a first-time or repeat offender. They found that for first-time offenders, participants judged the out-group members more harshly; for repeat offenders, it was the in-group members who received the harsher judgment. This suggests the possibility of forgiveness for in-group transgressions, with minor punishments being meted out in the hope for rehabilitation. Second, van Prooijen (2006) found that in- and out-group punishments varied depending on certainty of guilt. In situations where guilt was certain, participants judged in-group transgressions more harshly – this is the black sheep effect – but where guilt was uncertain, it was the out-group members who were judged more harshly. This again shows that in-group transgressors are being given the benefit of the doubt. Finally, Iyer, Jetten, and Haslam (2012) found that the severity of punishment meted out for transgressions by in-group members increased with the perceived damage that their behavior caused to the group (see also Jettam & Hornsey, 2011). So not all deviance is equal – it is judged more harshly when it is habitual, when guilt is certain, and when it is seen to threaten in-group stability. (The key phrase here is "seen to"; whether or not it actually affects group stability is a matter for further research.)

Further research in support of our claim that third-party moral judgment arises as a result of intergroup competition can be found in a study by Rebers and Koopmans (2012). They used an *n*-person prisoner's dilemma task and measured the punishment of cooperators to in-group defectors. There were two situations: one in which the in-group was in direct competition with an out-group, and another in which there was no out-group present. Consistent with their predictions, Rebers and Koopmans found that in-group defection was punished more severely when there was intergroup competition than when there was not, exactly what we might expect if third-party moral judgment arose through intergroup competition.

12.9 BACK TO THE SUPERORGANISM

Our argument, therefore, is that third-party moral judgment is one mechanism among many that permits – or maybe it is better to say *enforces* – superorganismic behavior (Kroeber, 1917). It is the ability of humans to behave as a structured organism – to work together – that has undoubtedly led to their having the largest single-species biomass of any wild animal (we can dispute whether humans are actually wild, but to the extent that they are domesticated, they are domesticated by themselves, or other humans) and, of course, thriving in every continent bar Antarctica. Humans are different from that other example of a large-biomass animal, the termite (who could win the biomass competition were it not composed of many different species): termites' superorganismic properties are ensured by the relatively deterministic way that their behavior is controlled by pheromones; a termite has no individual agency, unless that individual agency benefits the greater good.

Irrespective of cultural differences, humans are always individual and social: even if they could master the complex smelting process, termites would not give medals for selfless bravery any more than humans would give a medal for selfishness. Neither would compute as a praiseworthy behavior. Humans' instincts for fulfilling their own interests are kept in check by their equally powerful instincts toward fulfilling the interests of the greater good.

Research suggests that there are, when painted with the broadest brush imaginable, two fundamentally distinct types of morality: individualizing morals and binding morals. This tells us something important. Unlike termites, humans should have – to a greater or lesser extent

and under certain conditions – some degree of freedom of self-determination. Unlike solitary animals such as bears or tigers, humans also have responsibilities beyond themselves and their kin in the dense and mutually beneficial reciprocal networks that permeate our social groups. Fiske's (1992) social relations theory sketches out the three or (more recently in our history) four basic compulsions that permit superorganismic behavior: community (emotions and morals that encourage us to operate as part of a group), hierarchy (we allow ourselves to be uncomplainingly influenced by those above us and influence those below us in some linear hierarchy), and reciprocity (we engage in often transient and complex exchange relationships with those with whom we have no personal attachment or involvement – although they will quite often be part of our group).

12.10 CONCLUSIONS AND OUTSTANDING QUESTIONS FOR FUTURE RESEARCH

Many will be disappointed by our suggestion that third-party moral judgment evolved to promote social cohesion and superorganismic behavior as a result of intergroup conflict. In some ways, we share this view. In response to our suggestion, one of the editors of this volume noted:

Sorry, this sounds like old-fashioned structural functionalism in which every aspect of a culture gets "explained" in terms of increasing within-group solidarity.

We do not wish to be tarred with the same brush as Dr. Pangloss, and therefore we feel it necessary to reiterate that we are by no means trying to explain "every aspect" of culture; there are surely many aspects of culture that have no survival or reproductive value. We are simple trying to explain one aspect, which is third-party moral judgment. The question is whether returning to an old-fashioned idea represents a failure of our imaginations or an indication that despite its advancing years there may be some explanatory power left in it, and the evidence above shows that this might be the case.

So far, this research shows only that people are likely (under certain circumstances) to punish the in-group more than the out-group, especially when the transgression threatens the stability of the group and when there is tangible intergroup rivalry or conflict. Although this is *likely* to induce cohesion, there is no evidence that it does. This is a missing piece from the jigsaw. There needs also to be more research on whether more cohesive groups are more successful in intergroup conflicts. The research that exists has focused more on the reverse – the so-called constrict problem – where increased diversity leads to decreased cohesion. As we have already pointed out, these situations are ambiguous and the measures of cohesion far from satisfactory. The main alternative theory – the side-taking hypothesis of DeScioli and Kurzban – may have a role to play. Although it is difficult to see how such a theory could have provided a selection pressure for third-party moral judgment in an environment where only material (genetic or resource-based) interests mattered, unless very specific conditions were obtained. The side-taking hypothesis could easily have amplified the effect once the behavior evolved.

REFERENCES

Alexander, R. D. (1987). *The Biology of Moral Systems*. New York: Aldine de Gruyter.

Anderson, J. R., Bucher, B., Chijiiwa, H., et al. (2017). Third-party social evaluations of humans by monkeys and dogs. *Neuroscience and Biobehavioral Reviews*, **82**, 95–109.

Arvan, M. (2013). Bad news for conservatives? Moral judgments and the Dark Triad personality traits: A correlational study. *Neuroethics*, **6**(2), 307–318.

Bartholomew, J. (2015). The awful rise of "virtue signalling". *Spectator*, April 18.

Bernhard, H., Fischbacher, U., & Fehr, E. (2006). Parochial altruism in humans. *Nature*, **442**(7105), 912–915.

Bowles, S., & Choi, J. K. (2003). The first property rights revolution. Presented at: "Workshop on the Co-evolution of Behaviors and Institutions," Santa Fe Institute.

Bowles, S., Choi, J. K., & Hopfensitz, A. (2003). The co-evolution of individual behaviours and social institutions. *Journal of Theoretical Biology*, **223**(2), 135–147.

Brown, R. (1986). *The Nature of Social Laws: Machiavelli to Mill*. Cambridge, UK: Cambridge University Press.

Chagnon, N. (2012). *The Yanomamo*. Scarborough, ON: Nelson Education.

Choi, J. K., & Bowles, S. (2007). The coevolution of parochial altruism and war. *Science*, **318**(5850), 636–640.

Curry, O. S. (2016). Morality as cooperation: A problem-centred approach. In T. K. Shackelford & R. D. Hansen, eds., *The Evolution of Morality*. New York: Springer International, pp. 27–51.

Curry, O. S., Mullins, D. A., & Whitehouse, H. (2019). Is it good to cooperate? Testing the theory of morality-as-cooperation in 60 societies. *Current Anthropology*, **60**, 1.

Cushman, F., & Greene, J. D. (2012). Finding faults: How moral dilemmas illuminate cognitive structure. *Social Neuroscience*, **7**(3), 269–279.

de Waal, F. B. M. (1996). *Good Natured: The Origins of Right and Wrong in Humans and Other Animals*. Cambridge, MA: Harvard University Press.

DeScioli, P., & Kurzban, R. (2009). Mysteries of morality. *Cognition*, **112**(2), 281–299.

DeScioli, P., & Kurzban, R. (2013). A solution to the mysteries of morality. *Psychological bulletin*, **139**(2), 477–496.

DeScioli, P., Gilbert, S. S., & Kurzban, R. (2012). Indelible victims and persistent punishers in moral cognition. *Psychological Inquiry*, **23**(2), 143–149.

Dinesen, P. T., & Sønderskov, K. M. (2015). Ethnic diversity and social trust: Evidence from the micro-context. *American Sociological Review*, **80**(3) 550–573.

Dunbar, R. I. (1992). Neocortex size as a constraint on group size in primates. *Journal of Human Evolution*, **22**(6), 469–493.

Dunbar, R. I., & Spoors, M. (1995). Social networks, support cliques, and kinship. *Human Nature*, **6**(3), 273–290.

Fehr, E., & Fischbacher, U. (2004). Third-party punishment and social norms. *Evolution and Human Behavior*, **25**, 63–87.

Fiske, A. P. (1992). The four elementary forms of sociality. *Psychological Review*, 99(4), 689–723.

Gavrilets, S. (2015). Collective action and the collaborative brain. *Journal of the Royal Society Interface*, **12**(102), 20141067.

Gollwitzer, M., & Keller, L. (2010). What you did only matters if you are one of us: Offenders' group membership moderates the effect of criminal history on punishment severity. *Social Psychology*, **41**(1), 20–26.

Gray, K., Young, L., & Waytz, A. (2012). Mind perception is the essence of morality. *Psychological Inquiry*, **23**(2), 101–124.

Haidt, J. (2012). *The Righteous Mind: Why Good People Are Divided by Religion and Politics*. New York: Pantheon.

Haidt, J., Koller, S. H., & Dias, M. G. (1993). Affect, culture, and morality, or is it wrong to eat your dog? *Journal of Personality and Social Psychology*, 65(4), 613–628.

Hardy, I. C. (1997). Possible factors influencing vertebrate sex ratios: An introductory overview. *Applied Animal Behaviour Science*, **51**(3–4), 217–241.

Hauser, M. (2006). *Moral Minds: How Nature Designed Our Universal Sense of Right and Wrong*. New York: Ecco/HarperCollins Publishers.

Hill, K., & Hurtado, A. M. (2009). Cooperative breeding in South American hunter–gatherers. *Proceedings of the Royal Society of London B: Biological Sciences*, **276**, rspb20091061.

Hrdy, S. B. (1979). Infanticide among animals: a review, classification, and examination of the implications for the reproductive strategies of females. *Ethology and Sociobiology*, **1**(1), 13–40.

Hrdy, S. B. (2011). *Mothers and Others*. Cambridge, MA: Harvard University Press.

Iyer, A., Jetten, J., & Haslam, S. A. (2012). Sugaring o'er the devil: Moral superiority and group identification help individuals downplay the implications of ingroup rule-breaking. *European Journal of Social Psychology*, **42**(2), 141–149.

Jetten, J., & Hornsey, M. J. (2011). The many faces of rebels. In J. Jetten & M. J. Hornsey, eds., *Rebels in Groups: Dissent, Deviance, Difference, and Defiance*. New York: Blackwell, pp. 1–13.

Jonason, P. K., Lyons, M., Bethell, E. J., & Ross, R. (2013). Different routes to limited empathy in the sexes: Examining the links between the Dark Triad and empathy. *Personality and Individual Differences*, **54**(5), 572–576.

Kant, I. (1999). *Critique of Pure Reason*. Cambridge, UK: Cambridge University Press.

Kelly, R. C. (2000). *Warless Societies and the Origin of War*. Ann Arbor: University of Michigan Press.

Krebs, D. L. (2011). *The Origins of Morality: An Evolutionary Account*. New York: Oxford University Press.

Krebs, J. R., & Davies, N. B., eds. (2009). *Behavioral Ecology: An Evolutionary Approach*. Hoboken, NJ: John Wiley & Sons.

Kroeber, A. L. (1917). The superorganic. *American Anthropologist*, **19**(2), 163–213.

Langergraber, K., Schubert, G., Rowney, C., et al. (2011). Genetic differentiation and the evolution of cooperation in chimpanzees and humans. *Proceedings of the Royal Society of London B: Biological Sciences*, **278**, rspb20102592.

Legare, C. H., & Watson-Jones, R. E. (2015). The evolution and ontogeny of ritual. In D. M. Buss, ed., *The Handbook of Evolutionary Psychology*. Hoboken, NJ: Wiley, pp. 1–19.

Marques, J. M., & Paez, D. (1994). The "black sheep effect": Social categorization, rejection of ingroup deviates, and perception of group variability. *European Review of Social Psychology*, **5**(1), 37–68.

Marques, J. M., Robalo, E. M., & Rocha, S. A. (1992). Ingroup bias and the "black sheep" effect: Assessing the impact of social identification and perceived variability on group judgements. *European Journal of Social Psychology*, **22**(4), 331–352.

McAndrew, F. T. (2002). New evolutionary perspectives on altruism: Multilevel-selection and costly-signaling theories. *Current Directions in Psychological Science*, **11**(2), 79–82.

McPherson, M., Smith-Lovin, L., & Cook, J. M. (2001). Birds of a feather: Homophily in social networks. *Annual Review of Sociology*, **27**, 415–444.

Meer, T. V. D., & Tolsma, J. (2014). Ethnic diversity and its effects on social cohesion. *Annual Review of Sociology*, **40**, 459–478.

Nitzschner, M., Melis, A. P., Kaminski, J., & Tomasello, M. (2012). Dogs (*Canis familiaris*) evaluate humans on the basis of direct experiences only. *PLoS ONE*, **7**(10), e46880.

Putnam, R. D. (2007). *E pluribus unum*: Diversity and community in the twenty-first century the 2006 Johan Skytte Prize Lecture. *Scandinavian Political Studies*, **30**(2), 137–174.

Rebers, S., & Koopmans, R. (2012). Altruistic punishment and between-group competition. *Human Nature*, **23**(2), 173–190.

Ridley, M. (1996). *The Origins of Virtue*. London: Penguin Books.

Schaller, M., & Duncan, L. A. (2007). The behavioral immune system: Its evolution and social psychological implications. In J. P. Forgas, M. G. Haselton, & W. von Hippel. *Evolution and the Social Mind: Evolutionary Psychology and Social Cognition*. New York: Psychology Press, pp. 293–307.

Sherif, M. (1958). Superordinate goals in the reduction of intergroup conflict. *American Journal of Sociology*, **63**, 349–356.

Sloan-Wilson, D., Sloan, M., & Price, M. (n.d.). Is there a universal morality? Introduction and overview of responses. *This View of Life*. The Evolution Institute. https://evolution-institute.org/is-there-a-universal-morality-introduction-and-overview-of-responses.

Tomasello, M., Melis, A. P., Tennie, C., et al. (2012). Two key steps in the evolution of human cooperation: The interdependence hypothesis. *Current Anthropology*, **53**(6), 673–692.

Trivers, R. L. (1971). The evolution of reciprocal altruism. *Quarterly Review of Biology*, **46**(1), 35–57.

Valdesolo, P., & DeSteno, D. (2007). Moral hypocrisy. *Psychological Science*, **18**(8), 689–690.

van der Meer, T. (2015). Care is required when making assertions about the relationship between diversity and social cohesion. *Democratic Audit Blog*. March 9. www.democraticaudit.com/2015/03/09/care-is-required-when-making-assertions-about-the-relationship-between-diversity-and-social-cohesion.

van Prooijen, J. W. (2006). Retributive reactions to suspected offenders: The importance of social categorizations and guilt probability. *Personality and Social Psychology Bulletin*, **32**(6), 715–726.

Wrangham, R. W. (1987). African apes: The significance of African apes for reconstructing human social evolution. In W. G. Kinzey, ed., *Evolution of Human Behavior: Primate Models*. New York: SUNY Press, pp. 51–71.

Wright, R. (1994). *The Moral Animal*. New York: Pantheon.

13 Evolution of the Human Family

CRAIG T. PALMER AND KATHRYN COE

13.1 INTRODUCTION

After describing common conceptions of the evolution of the human family as a unit, this chapter examines several problems with this conception and presents an alternative conception of the human family as a mechanism of tradition transmission instead of as a unit. It concludes with a discussion of some of the directions for future research generated by this reconception of the evolution of the human family.

The common conception of the family as a unit sees the evolution of the human family as a chain, in which each link in the chain is a successive form of family unit. Descriptions of this chain typically start with a hypothesized prehuman family unit that is largely based on comparisons with other primates and the fossil record, and end with the forms of family units found in the ethnographic record. The changes in each successive form of family unit are seen as resulting from selection favoring psychological mechanisms and patterns of behavior of the members of the unit that increased the number of copies of genes transmitted to the next generation. Specific explanations of this selection usually refer to Hamilton's rule (i.e., kin selection), paternity certainty, parent–offspring conflict, life history theory, and sexual selection. Debates usually focus on which nonhuman primate species is the best model for the structure of the initial human family unit, which selective forces acting on which psychological mechanisms and which behavior patterns are responsible for the changes in the family unit, and whether or not selection has operated at levels above the individual.

The problem with the conception of the human family as a unit is that it prevents an understanding of both the uniqueness of human kinship and the process by which this uniqueness has evolved. The uniqueness of human kinship is the dramatic increase during recent human evolution in the number of distantly related individuals identified as kin and the altruism directed toward these kin. To account for this unique aspect of human kinship, an alternative conception of the human family is

presented. This conception views a human family not as a link in a chain, but as the mechanism by which behaviors enabling many earlier generations of ancestors to leave descendants are transmitted to the next generation, which then replicates both these behaviors and their retransmission to the next generation, and thus potentially to an unlimited number of subsequent generations of descendants.

Conceptualizing the human family as a mechanism of tradition transmission sheds new light both on the causes of the interactions of the individuals constituting the fluid nexus of interaction labeled "a family" and on how the replication of this interaction over many generations constitutes a central part of the evolution of human social behavior. At the heart of this evolution was the transmission from parents to offspring of traditions identifying generation after generation of descendants as kin and encouraging altruism toward these individuals. The large webs of cooperating kin that resulted from the transmission of these traditions through the mechanism of the human family over many generations constituted a social environment unlike that experienced by any other species, thereby constituting a major selective force on recent human evolution (Palmer et al., 2016). This selection is best conceived of not as selection within or between groups, but as selection between different traditions transmitted from different ancestors. The traditions favored by natural selection were the traditions leaving the most descendants in distant generations, and thus being most likely to be included in the ethnographic record. This chapter concludes with a discussion of how future research based on this reconceptualization of the evolution of the human family may shed new light on many aspects of human social behavior.

13.2 THE CONVENTIONAL CONCEPTION OF THE EVOLUTION OF THE HUMAN FAMILY AS A UNIT

Conventional explanations of the evolution of the human family usually start with the existence of a family unit as it was postulated to have existed shortly after the divergence with the last common ancestor shared by humans and other living primates. Although it is acknowledged that this family unit would have differed to some extent from all of the forms of family units existing among contemporary primates and that there is variation among the contemporary family units found within these species, explanations of the evolution of the human family usually start with a family unit similar to those found among chimpanzees, bonobos, or gorillas.

The chimpanzee model usually posits communities consisting of between 30 and 60 adult males and females and their offspring, with the exact size and membership of these groups fluctuating as members come and go. Subgroups of various compositions form for various lengths of time within the larger community. In general, "Within these communities, groups of males define a territory that contains the smaller territories of adult females and their offspring … In other words, families are female-centered and nested within larger territories that are maintained by the group's males" (Geary et al., 2011, p. 370). Although the males of a chimpanzee community form coalitions, dominant "males aggressively achieve more matings with estrous females than do other males" (Geary et al., 2011, p. 371). Thus, chimpanzee family units consist of adult females and their offspring (Chapais, 2008; Goodall, 1986; Wrangham, 1980). Adult males are not included in chimpanzee family units because mating is generally promiscuous, with long-term pair-bonds and significant paternal investment in offspring being essentially absent. Female offspring generally leave the larger community upon reaching adulthood. The exact extent to which cooperation with other chimpanzees is based on more distant kinship ties extending beyond a family unit is still being debated, but it does not appear that interaction with more distantly related kin warrants the inclusion of distant kin within the chimpanzee family unit. Although bonobos differ from chimpanzees in having important female coalitions and "relatively weak" affiliative bonds between males (Stevens et al., 2007, p. 31), they are similar to chimpanzees in having neither paternal care nor interactions warranting the inclusion of distant kin within the family unit. For example, Palagi and Norscia (2013) found that kinship correlated with cooperative interaction among bonobos, but kin were limited to only individuals who were extremely closely related through females: "We considered as kin-related individuals belonging to grandmother/mother/offspring dyads and siblings ($r \geq 0.25$)" (p. 9).

In contrast to both chimpanzees and bonobos, the family in the gorilla model usually consists of "single-male harems with several females and their offspring" (Geary et al., 2011, p. 365). Males are thus included in the family, experiencing relatively high paternity certainty and contributing paternal investment to offspring. Female offspring typically leave the family unit upon reaching adulthood to attempt to form new families, while male offspring may stay after reaching adulthood in hopes of becoming the breeding male, or they may leave to form small "bachelor" groups with other males while attempting to form families of their own. As among chimpanzees and bonobos, the exact extent to which cooperation with other gorillas is based on more distant kinship ties extending beyond a family unit is still being debated, but it does not appear that interaction with more distantly related kin warrants the inclusion of distant kin within the gorilla family unit.

After selecting either the chimpanzee, bonobo, or gorilla model for their starting point, conventional explanations of the evolution of the human family attempt to account for the evolution of aspects of the family units found in the ethnographic record, but not present in the model chosen for their starting point. Thus, all models have to account for a life history of extended juvenile dependence and paternal investment greater than even among gorillas. Considerable attention is also usually given to the apparently universal existence of male coalitions in the ethnographic record. Attention is also sometimes given to the flexibility of residence found in the ethnographic record because although there is some variation in the typical pattern of daughters leaving the family upon reaching adulthood among chimpanzees, bonobos, and gorillas, in the ethnographic record, "postmarital residence patterns are remarkably diverse and flexible" (Chapais, 2008, p. 237; see also Kramer & Greaves, 2011; Palmer et al., 1997). Thus, explanations of the evolution of the human family usually focus on explaining a family unit featuring extended juvenile dependency, residential flexibility, and one adult male pair-bonded with one or several females, who has relatively high paternity certainty, engages in relatively high paternal investment, and forms male coalitions with adult males in other families, leading to the family being embedded in a larger community consisting of nonkin. Since bonobos lack pair-bonded males engaging in paternal investment and bonobo male coalitions are "very rare" (Stevens et al., 2007, p. 31), those using this model must account for both of these features of human kinship. Since male coalitions are already part of the chimpanzee model, those using this starting point focus on explaining the evolution of families that include an adult male pair-bonded with one or more females and engaging in paternal investment. Since single-male families are already part of the gorilla model, those using this starting point focus primarily on the formation of male coalitions between the adult males of different families, leading to the family being embedded within a larger community consisting of nonkin.

Although reasons for favoring one of these models over the others may include fossil evidence of the amount of sexual dimorphism of our ancestors after the split with chimpanzees, bonobos, and gorillas, arguments for the superiority of one of these models over the others are typically framed in terms of which evolutionary changes appear to be the most plausible. The extension of reproductive immaturity in the human life history is usually attributed to selection favoring increased social learning from parents (Palmer & Coe, 2010), although it has also been proposed to be related to sexually dimorphic mating tactics (Miller, 2000). Explanations of pair-bonding usually emphasize the reproductive benefits of male parental investment and/or mate guarding outweighing the reproductive costs of reduced mating effort as predicted by sexual selection/parental investment theory. Chapais (2008) emphasizes the ability of pair-bonds to identify fathers as kin, but the further benefit of the identification of kin *through* fathers resulting from pair-bonds, which greatly increases the number of individuals that an offspring could identify as kin (Palmer et al., 2006), has been largely overlooked. Explanations of male coalitions focus on the reproductive benefits of greater access to and defense of females and other resources explainable through mutualism, reciprocal altruism, or group selection. Explanations of differences in the amount of altruistic behavior toward the slightly more distant kin sometimes included in the "extended family" usually refer to kin selection and the influences of paternity certainty on the likely degree of relatedness (i.e., grandparents, aunts, and uncles related through females are expected to engage in more alloparenting and other altruistic acts than are those related through males). However, the key concept used to explain the behavioral interactions of the members of the "family unit" (i.e., the behavior between parents and offspring, as well as between siblings) is parent–offspring conflict (Trivers, 1974). As Mock (2011) explains, "The heart of Trivers's theory is couched in terms of how the two generations view the optimal division of parental investment. From the parent's vantage, each offspring is equally related to it (by the meiosis-conferred r of 0.50), but from an individual offspring's position, self is worth two siblings" (p. 57). Thus, parent–offspring conflict is often used to explain conflict between siblings and between parents and offspring, but also cooperation and alloparenting by siblings. One way of looking at this conflict is known as the parental manipulation explanation of altruism, which proposes that parents have evolved to encourage altruism among their offspring. More specifically, parents are predicted to try to influence their offspring to value each sibling as much as itself (Emlen et al., 1995; Geary & Flinn, 2001; Hrdy, 2005; Salmon & Malcolm, 2011; Wright, 1994, p. 166).

13.3 QUESTIONABLE ASSUMPTIONS IN CONVENTIONAL EXPLANATIONS OF THE EVOLUTION OF THE HUMAN FAMILY

The first feature of explanations of the evolution of the human family that warrants reconsideration is the assumption that "the family unit" is essentially coterminous with "kinship," and therefore an explanation of the interactions between individuals in a family constitutes an explanation of kinship. One of the reasons why this assumption is so commonly made in evolutionary explanations is that the altruism directed toward fellow members of a family unit *appears* to coincide with predictions based on kin selection. "A family" refers to the closely related individuals in a "nuclear family" (parents, offspring, siblings), but sometimes slightly more distantly related individuals (grandparents/grandchildren, aunts/uncles, nieces/nephews, and very rarely first cousins) may be included in a larger unit referred to as an "extended family." This is in accordance with the premise that kin selection is most likely to explain the altruistic behavior of parents toward their offspring and siblings toward each other, but might be applicable to the same slightly more distantly related other members who sometimes form an extended family. The problem with this assumption is that, unlike among chimpanzees, bonobos, and gorillas, human kinship extends far beyond even the "extended family," because traditions prescribe certain patterns of behavior toward distantly related individuals in specific ways. For example, among the Tiwi of Australia, complex traditional forms of interaction were followed by certain types of kin well outside the extended family in order to arrange marriages of "a MoFaSiDaSo to a MoMoBrDaDa" (Goodale, 1962, p. 457). Such traditionally prescribed patterns of behavior are particularly important because they are often extended to even far more distantly related kin and entail altruism toward kin far beyond the range of kin where altruism might be explained by kin selection. Reports of individuals traveling to a distant village for the first time and expecting "hospitality" from very distantly related kin living there are common in the ethnographic literature (Hughes, 1988, p. 77). For example, in a paper coauthored by Palmer and Coe, Lyle Steadman describes how a Hewa man living in the highlands of New Guinea visited a far-off area where he had heard he had distant relatives. Steadman reports that when the man arrived, he explained "he was a descendant of their clan ancestor. They welcomed him and he stayed with them for several weeks, re-turning with gifts for their kin" (Palmer et al., 2016, p. S185). Extremely distantly related kin are also seen as potential allies in warfare, and Evans-Pritchard (1940) reports how individual Nuer would unite for war "because their ancestors were sons of the same mother" (p. 143).

The failure of the human family to be coterminous with human kinship is closely related to a second feature of explanations of the evolution of the human family that is in need of reconsideration. This is the assumption that the evolution of the human family led to the human family being embedded in a larger community consisting of nonkin. Explanations starting with the chimpanzee and bonobo models see this as merely a continuation of the earlier social structure, while those using a gorilla model see the evolving "hominid family" becoming "embedded within expanding communities" (Geary et al., 2011, p. 368) of nonkin through the formation of alliances between unrelated families. Thus, proponents of the gorilla model must explain how the cooperation characterizing the larger communities of nonkin in which family units are imbedded came to exist. That is, how did "human groups overcome this inherent conflict between family units to form large, stable coalitions" (Flinn, 2011, p. 16)?

The assumption that "Kinship is embedded in human society" (Hepper, 2011, p. 211) is probably made because it accurately describes many contemporary nontraditional technological societies where families typically do exist within huge populations of individuals not identified as kin. However, this is not the end point that explanations of the evolution of the human family need to account for because these social environments are extremely recent, and the ethnographic record describes a very different social environment: "For most of human evolutionary history, our ancestors lived in small tribal societies based mostly on kinship ties" (Gorelik et al., 2011, p. 393). To account for how kinship in the form of "family units" could produce a much larger "tribal" society based on kinship, Gorelik et al. (2011) proposed that larger societies resulted from the *overlap* between one family unit and another: "If two individuals from the same society were unrelated, chances were high that both were related to a third individual" (p. 393). A key feature of this overlapping family unit hypothesis holds to the premise that all kinship is still found only within closely related family units, and thus the altruism associated with kinship is explainable by kin selection: "It is therefore likely that our moral and empathetic sentiments had their origins in adaptive acts of kindness directed toward individuals with whom genes 'for' that kindness were shared" (p. 393).

The fundamental problem with the assumption that human kinship is restricted to family units of closely related kin embedded within a larger society of nonkin, whether or not that explanation proposes that the larger society is held together by the overlap between family units, indirect reciprocity, group selection, or cultural group selection (Smaldino et al., 2013; West et al., 2011), is that this assumption does not match the following key aspects of kinship found in the ethnographic record:

First, humans favor closer kin over more distant kin even at genealogical distances where kin selection is essentially nonoperative. Second, this favoring of "close" distant kin over "distant" distant kin is prescribed by tradition and does not depend on the greater likelihood of reciprocal altruism from the closer kin. Finally, the favoring of "close" distant kin over "distant" distant kin does not depend on co-membership in the same social group. (Palmer & Steadman, 1997, p. 39)

The gulf between the view of human kinship confined to family units and the reality described in the ethnographic record is dramatically illustrated by Fortes' (1969) observation that the axiom of kinship amity "applies to all of the Tiv" (p. 237), where "the whole population of some 800,000 traces descent by traditional genealogical links

from a single founding ancestor" (Keesing, 1975, pp. 32–33).

13.4 AN ALTERNATIVE CONCEPTION OF THE EVOLUTION OF THE HUMAN FAMILY

To reconcile evolutionary explanations of the human family with the descriptions of kinship found in the ethnographic record requires a new conception of the relationship between kinship, "the family," and the larger "society" during human evolution. This new conception still starts with a chimpanzee-, bonobo-, or gorilla-like form of kinship and social organization and still holds that "The family is a key human evolutionary adaptation" (Flinn, 2011, p. 12). However, it rejects the claim that "The family has been the fundamental social *unit* throughout much of human evolutionary history" (Emlen et al., 1995, p. 148; emphasis added). In this new conception, the family is not seen as a unit, but as a mechanism that formed human societies through the process of replicating the transmission of certain crucial traditional behaviors from one generation to the next over centuries and millennia. The evolution of this human family mechanism involved selection favoring those forms of traditions that left the most descendants over vast numbers of generations. The first step in understanding this new conception is to understand how the transmission of traditional behaviors in a human family can do something no other process can: create a clan.

13.5 CLANS, TRADITIONS, AND THE TRANSFORMATION OF PREHUMAN FAMILY UNITS INTO HUMAN KINSHIP

Like families, clans are often talked about as if they too are a "unit" in the sense of a residential gathering. In actuality, clans are categories of individuals having the same unilineal descent name, which, by definition, occurs only as a result of inheriting the name from a common ancestor. Further, "all of a clan does not reside in the same locality" (Durkheim, 1961/1912, p. 123), and because clans are typically exogamous, no residential gathering ever includes only members of one clan as long as it includes at least one married couple (i.e., even a nuclear family includes members of two clans). The failure of a clan to form a "unit" raises the question of how a clear-cut category of kin (those individuals, and only those individuals, having the same clan name) could come to exist despite not being a residential gathering. The importance of answering this question becomes apparent when one considers the large sizes clans can grow to and the influence clan names potentially have on interactions between all of these individuals, regardless of whether or not they live together in the same residential gathering.

Shockingly, given the prominence of clans in the ethnographic record, one of the few explanations of how clans could have come to exist is the fictitious one described by

Kurt Vonnegut, who studied cultural anthropology as a graduate student at the University of Chicago. In his 1976 tongue-in-cheek novel *Slapstick: Or Lonesome No More!*, as Cronk (1999, p. 129) explains, Vonnegut's character "Dr. Swain" runs for president on the promise to use "the computers of the federal government" to assign large numbers of individuals the same new middle name, and thus "recreate kinship networks like those of our ancestors … [including] 190,000 cousins, all obligated to help fellow clan members." Assuming this tongue-in-cheek explanation is rejected, the only plausible scenario for the creation of large clans *requires* traditional behavior being transmitted from parents to offspring over many generations. More specifically, the identification of large numbers of kin requires a specific type of tradition where a parent gives his or her offspring some symbol that they are a descendant, such as a descent name, and then influences that offspring to copy that behavior and transmit that symbol to offspring of their own (Palmer & Steadman, 1997). When, *and only when*, this type of tradition persists over many generations can "large lineages or clans … grow up over time as the descendants of the original ancestor/ancestress accumulate" (Fox, 1967, p. 122). van den Berghe and Barash (1977) provide examples of huge numbers of individuals who share the same descent name, whose existence implies the traditional transmission of a descent name over many hundreds of generations: "large unilineal kin groups such as clans and lineages with memberships running into hundreds or thousands, or indeed in a few cases, millions" (p. 821). Explanations ignoring traditions and only focusing on evolved psychological mechanisms are obviously unable to account for clans. Explanations based on genetic or cultural group selection are also unable to account for the existence of clans because it is not membership in any "group" that leads to all 800,000 Tiv sharing "ties of reputed kinship" (Bohannan, 1958, p. 33). Instead, it is because of the tradition of transmitting genealogies: "A Tiv is a Tiv and can prove it. This proof consists in a genealogy through which every Tiv can trace his descent from Tiv himself" (Bohannan, 1958, p. 33).

As impressive as the number of individuals that can be identified as kin by virtue of having the *same* descent name transmitted from a common ancestor is, it is only a fraction of the number of non-clan kin that the transmission of descent names makes possible. Traditions not only allow individuals who share their clan name to be identified as kin; they allow any two people to be identified as kin as long as they can trace any combination of birth links to people who have the same clan name. For example, Bohannan (1958) described that among the patrilineal Tiv, in addition to all of the individuals with his own clan name, a man can identify as kin "patrilineages not his own," including his mother's, his father's mother's, his father's father's mother's, and his mother's mother's (p. 58). As Chapais (2008) explains, this is because " kinship is traced through both the mother and the father in all human societies" (p. 58). For example, mothers are identified as kin in patrilineal societies with clan exogamy, even though they are never of the same clan as their offspring, and fathers are identified as kin in matrilineal societies with clan exogamy, even though they are never of the same clan as their offspring. Thus, Chapais (2008) writes "kinship, because it is intrinsically bilateral, transcends unilineal descent fundamentally" (p. 59). Although this fact has been recognized, the implications of this identification of non-clan kin for the number of people an individual can identify as kin have gone largely unappreciated.

The time and effort put into faithfully transmitting names from ancestor to descendant over many generations in order to identify distantly related kin would provide no evolutionary advantage unless the naming tradition was accompanied by traditional behavior influencing descendants to behave in ways that increased evolutionary success. This accompanying traditional behavior is the encouragement of altruism toward kin (i.e., "the axiom of kinship amity"), which influences descendants to direct altruism toward kin, including kin far beyond the members of the family unit. It is this tradition of encouraging altruism toward kin that leads to thousands of distantly related co-descendants being "obligated to help each other," as in Vonnegut's fictional novel (Palmer et al., 2013). The altruism resulting from this traditional behavior extends far beyond the range of relatedness where altruism could have evolved through kin selection (Palmer & Palmer, 2016). Traditions identifying many individuals as kin – including individuals living in many different locations – and influencing those individuals to be altruistic toward kin are also what enable human forager postmarital residence patterns to be "remarkably diverse and flexible" (Chapais, 2008, p. 237; see also Kramer & Greaves, 2011; Steadman et al., 1996).

13.6 LONG-TERM EVOLUTIONARY SUCCESS: FROM REPRODUCTIVE SUCCESS TO DESCENDANT-LEAVING SUCCESS

An explanation of how traditions causing altruism toward very distant kin could have evolved requires a concept of evolutionary success (i.e., inclusive fitness) that takes into consideration a timescale compatible with the many generations required for large clans to form. The idea that evolutionary success is judged best not by the number of offspring, or even the number of grandchildren, but by the number of descendants in distant generations was noted by Alexander (1974), who posed the question: "What to measure, and what generation to measure it in, to determine which genetic line is winning (or what in fact constitutes 'winning')?" (p. 346). Elaborating on this point, Alexander (1974) asked: "What is to be measured and when should it be measured? Should we measure numbers of offspring produced, numbers reared, numbers breeding, numbers of grandchildren produced, reared, breeding, *etc.*?" (p. 374; emphasis added). The "*etc.*" is crucial because it indicates a realization that selection might be best measured as far into the future as possible, or at least that farther measurements would be superior to nearer ones. *This implies that the success of a behavior might be different when measured after more generations than it is in the one or two generations where it is typically measured.* West-Eberhard (1975) expanded on this crucial point in regard to the question of how parents can influence their offspring's behavior: "This ... raises the question of how far into future generations maternal control could be expected to operate ... [and] the further general question of just what it is that selection maximizes – whether number of children, grandchildren, great-grandchildren or *n*th descendants (see Alexander, 1974)" (p. 29). The answer reached was clearly stated by Dawkins (1982) when he wrote: "Ideally we might count the relative number of descendants alive after some very large number of generations" (p. 184). The concept of measuring evolutionary success by the number of descendants alive after many generations has been referred to as "descendant-leaving success" (Steadman & Palmer, 1995). This is in contrast to the conventional use of "reproductive success" when measuring evolutionary success in the next generation or, at the most, by counting grandchildren in the following generation. The distinction between the two types of "success" is important because it means that a trait that lowers reproductive success over one or two generations may lead to higher descendant-leaving success many generations later. Unfortunately, "researchers tend to forget that less is often more" (Mock, 2011, p. 60).

13.7 FROM PARENT–OFFSPRING CONFLICT TO ANCESTOR–DESCENDANT CONFLICT

Replacing the concept of reproductive success with descendant-leaving success facilitates an explanation of altruism toward very distant kin. This explanation is a multigenerational extension of the previously described parental manipulation explanation of altruism, and thus it is referred to as the ancestor manipulation explanation of altruism. The ancestor manipulation explanation holds that the parental manipulation of altruism can influence future generations when language and other behaviors allow that manipulation to become traditional (i.e., the encouragement of descendants to sacrifice for each other is replicated in successive generations). This multigenerational transmission of manipulation transforms parent–offspring conflict into ancestor–descendant conflict (Coe et al., 2010) and generates the prediction that selection favored ancestors who were successful in transmitting such traditions because they promoted the long-term descendant-leaving success of those ancestors. The view that the human family is where these traditions, essential to the existence of the patterns of social behavior found in the ethnographic record, are transmitted is consistent with the hypothesis that the extended immaturity that characterizes human life history allowed for social learning.

However, it focuses attention on the importance of the specific social learning involved in the transmission, replication, and retransmission of traditions. This new focus also adds a new dimension and level of importance to the conflict between parents and offspring. In addition to competition over resources, offspring are also in conflict with the attempts of their parents *and* earlier ancestors to influence (i.e., manipulate) their behavior in various ways, including engaging in forms of altruism that may be detrimental to that particular offspring's short-term fitness and, in some cases, that offspring's long-term descendant-leaving success. Further, to the extent that the offspring's behavior has been influenced by these traditions, the offspring will enter into similar conflict with their own offspring when they attempt to transmit those traditions to them.

The first step in understanding this extension is to recognize that "From a parental perspective, they are equally related to all their children *and grandchildren*, so parents have typically been under selective pressure to resist a particular offspring's [or grandchild's] demands, especially when offspring [or grandchildren] are trying to extract more than their fair share of resources" (Salmon & Shackelford, 2011, pp. 5–6, emphasis added). The next and crucial step is to realize that parents can continue to influence generations long after they are dead if they transmit traditions influencing each generation. An oversimplified example of how parent–offspring conflict could have been transformed into ancestor–descendant conflict is the following three-part exhortation by a parent to his or her offspring: "1) treat all of my other descendants as if they are as valuable to you as you are to yourself, 2) tell your offspring to also treat all of my descendants as if they are as valuable to them as they are to themselves, and 3) also tell your offspring to tell their own offspring these things" (Coe et al., 2010, p. 6). We propose that these traditions are known in the ethnographic record as the axiom of kinship amity. We also propose that they are apparently universal, as indicated in this observation by Turner (1968): the Ndembu concept of a good man – one "who bears no grudges, who is without jealousy, envy, pride, anger, covetousness, lust, greed, etc., and who honours his kinship obligations ... [and] respects and remembers his ancestors" (pp. 48–49) – is "an ethical code which would be recognized as valid by *all human groups*" (p. 48, emphasis added; see also Palmer & Begley, 2015). The reason for the apparent universality of such traditions is that they were favored by natural selection over alternative behaviors, including other traditions.

The prominent use of language – in addition to art and nonverbal communication – in the transmission of both descent names and the axiom of kinship amity suggests that language may explain why multigenerational traditions of ancestral manipulation did not evolve earlier in human evolution or in nonhuman primates. In any case, because of language and our extended period of parent–offspring interaction, the potential multigenerational consequences of traditional parental manipulation were particularly important among our human ancestors because "humans are parental manipulators par excellence" (Alexander, 1974, p. 367; see also Voland & Voland, 1995; West-Eberhard, 1975, p. 18). Indeed, the evidence of parents successfully manipulating the behavior of their offspring is so strong that it leads Mock (2011) to ask "whether offspring ever 'win' in nature" (p. 57). However, the segmentary opposition so often described in the ethnographic record indicates that parents were rarely, if ever, *totally* successful in influencing their offspring to follow traditions encouraging altruism toward co-descendants. Segmentary opposition occurs when "(descendants of) close kin stand together against more distant kin: (descendants of) brothers are allied against (descendants of) cousins, cousins against second cousins etc. Thus, even very distant kin will automatically put their conflicts to the side and unite against any threat from groups of non-kin" (Anthrobase, 2009). Sahlins (1961) points out that such a pattern of behavior, where the "closer the kin relation, the greater the sociability … the more distant, or more nearly unrelated, the less" (p. 331), is one of the "very widespread – nearly universal – features of human social organization" (p. 322). We propose that segmentary opposition results from the fact that the descendant-leaving success of each generation of ancestors would be increased by influencing their descendants to favor each other over those individuals only related through previous common ancestors. Hence, each ancestor's descendant-leaving interest would be in partial conflict with the traditions encouraging altruism toward a wider range of descendants that their parents are attempting to transmit to them. Segmentary opposition results whenever ancestors have only partial success in manipulating their offspring, causing the influence of each earlier generation of ancestor to decrease over time. Segmentary opposition would not exist if the traditions encouraging altruism toward all co-descendants of a particular ancestor were completely successful in influencing the following generation of descendants, as all of the co-descendants of that particular ancestor, no matter how distantly related, would value each other as much as they value themselves. On the other hand, human kinship would have been limited to the range of kin selection, as it is in other primates, if early ancestors had not been at least partially successful in traditionally manipulating generations of descendants. The ethnographic record clearly indicates that our ancestors had significant, but not complete, success in manipulating many generations of their descendants.

Ironically, Mock (2011) states that the evidence of successful parental manipulation "is important because the parent-wins solution does not require any theoretical framework beyond orthodox Darwinism" (p. 57). Recognition of the multigenerational consequences of the tradition replication of parental manipulation reverses this conclusion: evidence of the partial success of parental manipulation, and the tradition replication of that manipulation, requires a fundamental change in orthodox Darwinism. This, in turn, gives new meaning to Mock's conclusion that "The potential for parent–offspring conflict may be virtually ubiquitous, but its impact remains unknown" (p. 60).

13.8 DIRECTIONS FOR FUTURE RESEARCH

Reconceptualizing the human family as a mechanism of tradition transmission instead of as a unit has clear theoretical implications for evolutionary explanations of human behavior. This is because the ancestor manipulation explanation introduces a new dimension to the evolutionary studies of altruism and human cooperation. Although this is most directly applied to explaining patterns of behavior in traditional societies, like those making up the majority of the ethnographic record, it may also shed light indirectly on modern nontraditional societies. Although large categories of actual descendants of a common ancestor may no longer be significant in such societies, recognition of the importance of such categories during our recent evolutionary history provides new insights into the ubiquitous examples of kinship-like categories (e.g., modern religions, nations, common identity interest groups) that make up such a large aspect of both cooperation and conflict in modern nontraditional populations (Quirko, 2011). That is, providing a more complete and accurate understanding of actual human kinship and how this actual human kinship evolved to extend far beyond the "family unit" may lead to a better understanding of the similarities between kinship and more recent social categories.

A focus on the human family as a mechanism of tradition transmission may also be relevant to a variety of issues in contemporary populations, but it is important to avoid the naturalistic fallacy when considering this possibility. That is, it is important not to assume that traditional behaviors are necessarily good or necessarily bad. For example, some traditional food-eating rituals may have encouraged restraint, which both contributed to social cooperation and decreased the likelihood of obesity (Huntington & Hostetler, 2002, pp. 145–147), but many families may now be transmitting patterns of food consumption that contribute to obesity (Rollins et al., 2010).

13.9 CONCLUSION

Conventional explanations of the evolution of the human family assume a family unit of kin so closely related to each other that their altruistic behavior is explainable by kin selection, embedded inside a larger unit usually referred to as a community, tribe, or society of nonkin whose altruism and cooperation must be due to other processes such as reciprocal altruism or group selection. The acceptance of such a model may be related to the fact that most of the people trying to explain the evolution of the human family are from nontraditional modern technological societies, and for "most people in a modern technological society ... the significance of distinguishing relatives decreases beyond some level, such as that of first cousins, because of low relatedness" (Alexander, 1979, p. 148). Since the relationships among these kin are consistent with kin selection, evolutionary explanations based on this concept of the family unit appear to be "obviously consistent with a Darwinian model" (Alexander, 1979, p. 149). The problem with this conception of the evolution of the human family is that the ethnographic data from traditional societies are obviously inconsistent with an orthodox Darwinian model because not only are humans in every known traditional society able to identify kin far beyond first cousins, but also "extensive extra-familial nepotism" (Alexander, 1979, p. 211) appears to be universal. Quoting Murdock (1949, p. 14), Alexander (1979, p. 156) describes this pattern by stating that universally "some of the intimacy characteristic of relationships within the nuclear family tends to flow outward along the ramifying channels of kinship ties. ... [When an individual] needs assistance or services beyond what his family ... can provide, he is more likely to turn to his secondary, tertiary, or remoter relatives than to persons who are not his kinsmen." The reference to "traditional" societies is crucial because the transmission of traditions is what made the identification of distant kin possible. Traditions are also what enabled ancestors to influence their descendants to engage in altruism toward distant kin. As individuals transmitting such traditions experienced greater descendant-leaving success than those who did not, human kinship evolved from the "family units" found in other primates to the vast webs of kin that constituted the social environments of our ancestors as described in the ethnographic record.

REFERENCES

Alexander, R. D. (1974). The evolution of social behavior. *Annual Review of Ecology and Systematic*, **5**, 325–383.

Alexander, R. D. (1979). *Darwinism and Human Affairs*. Seattle, WA: University of Washington Press.

Anthrobase (2009). Segmentary Lineage/Segmentary Opposition. www.anthrobase.com/Dic/eng/def/segmentary-lineage.htm.

Bohannan, L. (1958). Political aspects of Tiv social organization. In J. Middleton & D. Tait, eds., *Tribes without Rulers: Studies in African Segmentary Systems*. London: Routledge & Kegan Paul, pp. 33–66.

Chapais, B. (2008). *Primeval Kinship: How Pair-Bonding Gave Birth to Human Society*. Cambridge, MA: Harvard University Press.

Coe, K., Palmer, A. L., Palmer, C. T., & DeVito, C. L. (2010). Culture, altruism, and conflict between ancestors and descendants. *Structure and Dynamics*, **4**(3), 1–17.

Cronk, L. (1999). *That Complex Whole: Culture and the Evolution of Human Behavior*. Boulder, CO: Westview Press.

Dawkins, R. (1982). *The Extended Phenotype*. Oxford: W. H. Freeman.

Durkheim, E. (1961/1912). *The Elementary Forms of the Religious Life*. New York: Collier.

Emlen, S. T., Wrege. P. H., & Demong, N. J. (1995). Making decisions in the family: An evolutionary perspective. *American Scientist*, **83**(2), 148–157.

Evans-Pritchard, E. E. (1940). *The Nuer*. London: Oxford University Press.

Flinn, M. V. (2011). Evolutionary anthropology of the human family. In C. Salmon & T. Shackelford, eds., *Evolutionary Family Psychology*. New York: Oxford University Press, pp. 12–32.

Fortes, M. (1969). *Kinship and the Social Order: The Legacy of Lewis Henry Morgan*. Chicago, IL: Aldine.

Fox, R. (1967). *Kinship and Marriage*. Harmondsworth: Penguin.

Geary, D. C., & Flinn, M. V. (2001). Evolution of human parental behavior and the human family. *Parenting: Science and Practice*, **1**(1 & 2), 5–61.

Geary, D. C., Bailey, D. H., & Oxford, J. (2011). Reflections on the human family. In C. Salmon & T. Shackelford, eds., *Evolutionary Family Psychology*. New York: Oxford University Press, pp. 365–385.

Goodale, J. C. (1962). Marriage contracts among the Tiwi. *Ethnology*, **1**(4), 452–466.

Goodall, J. (1986). *The Chimpanzees of Gombe: Patterns of Behavior*. Cambridge, MA: Harvard University Press.

Gorelik, B. K., Shackelford, T. K., & Salmon, C. A. (2011). Between conflict and cooperation: New horizons in the evolutionary science of the human family. In C. Salmon & T. Shackelford, eds., *Evolutionary Family Psychology*. New York: Oxford University Press, pp. 386–398.

Hepper, P. (2011). Kin recognition. In C. Salmon & T. Shackelford, eds., *Evolutionary Family Psychology*. New York: Oxford University Press, pp. 211–229.

Hrdy, S. B. (2005). *Mother Nature: A History of Mothers, Infants, and Natural Selection*. Cambridge, MA: Harvard University Press.

Hughes, A. L. (1988). *Evolution and Human Kinship*. New York: Oxford University Press.

Huntington, G. E., & Hostetler, J. A. (2002). *The Hutterites in North America*. Mason, OH: Cengage Learning.

Keesing, R. (1975). *Kin Groups and Social Structure*. New York: Holt, Rinehart and Winston.

Kramer, K., & Greaves, R. (2011). Postmarital residence and bilateral kin associations among hunter–gatherers: Pumé foragers living in the best of both worlds. *Human Nature*, **22**, 41–63.

Miller, G. (2000). *The Mating Mind: How Sexual Choice Shaped the Evolution of Human Nature*. New York: Random House.

Mock, D. W. (2011). The evolution of relationships in nonhuman families. In C. Salmon & T. Shackelford, eds., *Evolutionary Family Psychology*. New York: Oxford University Press, pp. 51–62.

Murdock, G. P. (1949). *Social Structure*. New York: The MacMillan Company.

Palagi, E., & Norscia, I. (2013). Bonobos protect and console friends and kin. *PLoS ONE*, **8**(11), e79290.

Palmer, C. T., & Begley, R. O. (2015). Costly signaling theory, sexual selection, and the influence of ancestors on religious behavior. In J. Slone, ed., *The Attraction of Religion: Connecting Religion, Sex, and Evolution*. London: Bloomsbury Press, pp. 93–107.

Palmer, C. T., Begley, R. O., Coe, K., & Steadman, L. B. (2013). Moral elevation and traditions: Ancestral encouragement of altruism through ritual and myth. *Journal of Ritual Studies*, **27**(2), 83–96.

Palmer, C. T., & Coe, K. (2010). Parenting, courtship, Disneyland and the human brain. *International Journal of Tourism Anthropology*, **1**(1), 1–14.

Palmer, C. T., Coe, K., & Steadman, L. B. (2016). Reconceptualizing the human social niche: How it came to exist and how it is changing. *Current Anthropology*, **57**(S13), S181–S191.

Palmer, C. T., & Palmer, A. L. (2016). Why traditional ethical codes prescribing self-sacrifice are a puzzle to evolutionary theory: The example of Besa. *Journal of the Association for the Study of Ethical Behavior/Evolutionary Biology in Literature*, **12**(1), 40–50.

Palmer, C. T., & Steadman, L. B. (1997). Human kinship as a descendant-leaving strategy: A solution to an evolutionary puzzle. *Journal of Social and Evolutionary Systems*, **20**(1), 39–51.

Palmer, C. T., Steadman, L. B., & Coe, K. (2006). More kin: An effect of the tradition of marriage. *Structure and Dynamics*, **1**(2), 1–16.

Quirko, H. N. (2011). Fictive kinship and induced altruism. In C. Salmon & T. Shackelford, eds., *Evolutionary Family Psychology*. New York: Oxford University Press, pp. 310–328.

Rollins, B. Y., Belue, R. Z., & Francis, L. A. (2010). The beneficial effect of family meals on obesity differs by race, sex, and household education: the national survey of children's health, 2003–2004. *Journal of the American Dietetic Association*, **110**(9), 1335–1339.

Sahlins, M. (1961). The segmentary lineage: An organization of predatory expansion. *American Anthropologist*, **63**(2), 322–345.

Salmon, C. A., & Malcolm, J. (2011). Parent–offspring conflict. In C. Salmon & T. Shackelford, eds., *Evolutionary Family Psychology*. New York: Oxford University Press, pp. 83–96.

Smaldino, P. E., Newson, L., Schank, J. C., & Richerson, P. J. (2013). Simulating the evolution of the human family: Cooperative breeding increases in harsh environments. *PLoS ONE*, **8**(11), e80753.

Steadman, L. B., & Palmer, C. T. (1995). Religion as an identifiable traditional behavior subject to natural selection. *Journal of Social and Evolutionary Systems*, **18**(2), 149–164.

Steadman, L. B., Palmer, C. T., & Tilley, C. F. (1996). The universality of ancestor worship. *Ethnology*, **35**, 63–76.

Stevens, J. M. G., Vervaecke, H., & Van Elsacker, L. (2007). The bonobo's adaptive potential: Social relations under captive conditions. In T. Furuichi & J. Thompson, eds., *The Bonobos: Behavior, Ecology, and Conservation*. New York: Springer, pp. 19–38.

Trivers, R. L. (1974). Parent–offspring conflict. *American Zoologist*, **14**(1), 249–264.

Turner, V. (1968). *The Drums of Affliction: A Study of Religious Processes among the Ndembu of Zambia*. Oxford: Oxford University Press.

van den Berghe, P. L., & Barash, D. P. (1977). Inclusive fitness and human family structure. *American Anthropologist*, **79**, 809–823.

Voland, E., & Voland, R. (1995). Parent–offspring conflict, the extended phenotype, and the evolution of conscience. *Journal of Social and Evolutionary Systems*, **18**(4), 397–412.

Vonnegut, K. (1976). *Slapstick, or Lonesome No More!* New York: Delacorte Press.

West, S. A., El Mouden, C., & Gardner, A. (2011). Sixteen common misconceptions about the evolution of cooperation in humans. *Evolution and Human Behavior*, **32**, 231–262.

West-Eberhard, M. J. (1975). The evolution of social behavior by kin selection. *Quarterly Review of Biology*, **50**, 1–33.

Wrangham, R. W. (1980). An ecological model of female-bonded primate groups. *Behaviour*, **75**, 262–300.

Wright, R. (1994). *The Moral Animal: Why We Are the Way We Are*. New York: Vintage.

14

The Parasite-Stress Theory of Cultural Values and Sociality

RANDY THORNHILL AND COREY L. FINCHER

14.1 INTRODUCTION

People are parasitically modified animals as a result of interactions with parasites both during evolutionary historical generations and during their lifetimes. By parasites we mean infectious disease agents of all types: viruses, bacteria, fungi, protozoa, helminths ("worms"), and arthropods. The evolutionary historical interactions with parasites gave rise to the evolutionary selection that crafted the functional design of the two human immune systems – the classical immune system and the behavioral immune system – as well as the functional integration of these two systems in defending against parasites. During the lifetimes of individual humans, their interactions with cues of actual or potential infectious diseases proximately cause the cultural values/preferences and associated emotions and cognitions that function to yield the personal social psychology and behavior that are optimal for the amount of infectious disease adversity locally (i.e., the optimal degree of behavioral immunity of a person). This new perspective on humans as being both ultimately (evolutionarily) and proximately modified by parasites and employing behavioral immunity adaptations as a defense against parasites is supported by a rich and growing scientific literature.

We have called the combination of the ecological and evolutionary concepts and the associated empirical evidence supporting this perspective "the parasite-stress theory of values and sociality." As an ecological and evolutionary theory of human cultural values (i.e., core human preferences), it applies widely across domains of human social life and human affairs in general. It is a general theory of human culture and sociality. In this chapter, we explain the conceptual basis of the theory and review several recent discoveries the theory's application has revealed about human social psychology and behavior. In novel ways, it informs and synthesizes knowledge of many categories of the social life and societal-level affairs of people, ranging from prejudice and egalitarianism to personality, economic patterns, core values, interpersonal and intergroup violence, governmental systems, gender relations, family structure, and the genesis and maintenance of cultural diversity across the world.

14.2 THE PARASITE-STRESS THEORY OF SOCIALITY/VALUES

14.2.1 Immunity

Established knowledge of the ecology and evolution of parasitic diseases provides part of the foundation for the parasite-stress theory of sociality or values. Infectious diseases are significant causes of evolutionary selection acting on all life. For humans, parasites appear to be a predominant cause of evolutionary change. Geneticists who study recent evolutionary changes in genes of the human genome report that parasites account for more evolutionary action across the genome than other environmental factors that are also sources of selection. In support of this, Fumagalli et al. (2011) reviewed much of the published evidence of recent evolution in the human genome in response to infectious diseases. They report that across 55 contemporary human populations, genes related to immunity exhibit significantly more allelic variation and hence evolutionary change across geographic regions compared to genes involved in dealing with 13 other environmental challenges (climatic and geographic factors, diet, metabolic traits, subsistence strategies). Thus, immunity genes are evolutionary hotspots, and forces of selection that are region specific act more strongly on those genes than other genes so far studied. Indeed, a sizeable portion of the current morbidity and mortality across the world and even across states of the USA appears to be attributable to parasitic diseases (Thornhill & Fincher, 2011).

In addition, infectious diseases also were a major source of morbidity and mortality, and hence of evolutionary selection, in human evolutionary history (Anderson & May, 1991; Dobson & Carper, 1996; Ewald, 1994; McNeill, 1998; Volk & Atkinson, 2013; Wolfe et al., 2007). Volk and Atkinson (2013) recently published a review of rates and causes of human juvenile mortality in ethnographic samples representative of ecological conditions in human evolutionary history. Infant mortality was about 25 percent and child mortality about 50 percent. The two largest mortality factors were infectious disease (especially gastrointestinal and respiratory illnesses) and infanticide, with the former greatly predominating. Thus, infectious disease was the chief cause of juvenile mortality in the evolutionary historical settings comprising the juvenile mortality data reviewed by Volk and Atkinson. Finally, the existence of complex, evolved human adaptations that are organized functionally to defend against parasites documents that evolutionary selection acting in the deep-time past of human evolutionary history directly favored individuals with defenses against infectious diseases.

Humans have two immune systems. One is the *classical immune system*: the biochemical, physiological, cellular, and tissue-based mechanisms of defense against parasites discussed in traditional textbooks for immunology courses taught at universities. The second is the *behavioral immune system*. This is composed of the psychological features and behaviors for infectious disease avoidance (Schaller, 2006; Schaller & Duncan, 2007), as well as for managing the effects of infectious diseases (Fincher & Thornhill, 2008a). The human behavioral immune system has only recently been researched in detail, and the resulting cornucopia of findings will require a major expansion of immunology textbooks. Furthermore, human behavioral immunity research is connecting knowledge of behavioral immunity, classical immunity, health, and sociality by anchoring the four topics in a shared proximate and ultimate causation of infectious disease adversity.

14.2.2 Host–Parasite Coevolution

Hosts and their parasites coevolve in antagonistic and perpetual races with adaptation, counter-adaptation, counter-counter-adaptation, and so on; there is no lasting adaptive solution that can be mounted by either side against the other (Ewald, 1994; Ridley, 1993; Thompson, 2005; Van Valen, 1973). In the human case, this dynamic, antagonistic interaction is illustrated by the fact that, despite the huge somatic allocation given to the classical immune system, people still get sick, and even small reductions in immunocompetence increase vulnerability to infectious disease.

Furthermore, host–parasite arms races are localized geographically across the range of a host species and its parasite, creating a coevolutionary mosaic involving genetic and phenotypic differences in host immune adaptation and corresponding parasite counter-adaptation (Thompson, 2005). An important outcome of the geographical localization of parasite–host coevolutionary races is that host defense works most effectively, or only, against the local parasite species, strains, or genotypes, and not against those evolving in nearby host groups. Hence,

out-groups may often harbor novel parasites that cannot be defended against very well or at all by an individual or his or her immunologically similar in-group members (Fincher & Thornhill, 2008a, 2008b; Fincher et al., 2008). Out-group individuals pose the additional infectious disease threat of lacking knowledge of local customs, manners, rituals, and norms in general, many of which, like sexual and other social customs and methods of hygiene or food preparation, may prevent infection from local parasites; also, individuals with out-group norms may carry out-group parasites (Fincher et al., 2008; Schaller & Neuberg, 2008). Norms of many types – culinary, linguistic, moral, sexual, nepotistic, religious, dress, and so on – are used by people both to display in-group affiliation and associated values and to distinguish in-group from out-group members. Norm differences between groups are often the basis of intergroup prejudice and hostility (i.e., xenophobia). Likewise, norm similarity is the basis of positive valuation of and altruism toward people (Murray et al., 2011; Norenzayan & Shariff, 2008; Park & Schaller, 2005).

The emotion of disgust is not only evoked in the context of perception of disease-laden cues, such as contaminated foods, sick people, parasites (e.g., worms), or parasite reservoirs (e.g., cockroaches), but also is commonly generalized to include: (1) groups of people who are perceived as harboring infectious disease; and (2) cultural behaviors that are different or unfamiliar (Curtis, 2007; Curtis et al., 2004). Thus, disgust directed toward out-group people appears to be a component of behavioral immunity that provides a boundary between the in-group and out-group and promotes out-group avoidance. This includes so-called moral disgust toward others in which others' beliefs, norms, values, manners, or behaviors are deemed morally undesirable or repugnant (Curtis et al., 2011; Inbar et al., 2012; Oaten et al., 2009; Schnall et al., 2008).

Evidence for geographically localized host–parasite coevolutionary races is abundant. On the parasite side of the race, a parasite geographical mosaic was found, for example, in research on the human protozoan parasite *Leishmania braziliensis*. Rougeron et al. (2009) described the high genetic diversity and subdivided population structure of this parasite across both Peru and Bolivia. They found high levels of microgeographic variation identifiable by at least 124 highly localized, physiologically and genetically distinct strains.

The extremely fine-grained geographic mosaic in *L. braziliensis* implies a similar microgeographic immunological genetic mosaic in human hosts. This type of spatial variation in host adaptation against local parasites, or, said differently, in host immune maladaptation against out-group-typical parasites, is a common pattern in the animal and plant infectious disease literature (e.g., Corby-Harris & Promislow, 2008; Dionne et al., 2007; Kaltz et al., 1999; Thompson, 2005; Tinsley et al., 2006). Specific human cases showing this include the caste-specific infectious diseases and associated caste-specific immunity among Indian castes in the same geographic locale (Pitchappan, 2002). Another case is found in the village-specific immune defenses against *Leishmania* parasites in adjacent Sudanese villages (Miller et al., 2007). In particular regions, the localization of host immunity to local parasites is so fine-grained that people inbreed, risking the potential costs of inbreeding depression, to maintain coadapted gene complexes important for coping with parasite infection in their offspring, as Denic and colleagues have shown for malaria across regions (Denic & Nicholls, 2007; Denic et al., 2008a, 2008b), and as we, in collaboration with colleagues, have shown for parasite stress in general across countries (Hoben, 2011; Hoben et al., 2010). On a broad scale, the localization of host–parasite coevolutionary races in humans is seen dramatically in the findings of the human genetic research conducted by Fumagalli and colleagues (2011) noted in Section 14.2.1: there is more regional variation in genes affecting classical immunity than in many other human genes affecting fitness.

There are other bodies of evidence of localized host immunity. One familiar type of evidence involves events where individuals from isolated groups interact with novel groups by conquest or trade and infectious disease transmission ensues, sometimes with drastic effects. This has occurred after the intra- and inter-continental movement of individuals brought about intergroup contact (Diamond, 1998; Good, 1972; McNeill, 1998). Other human examples of localized immunity are discussed in Thornhill and Fincher (2014).

14.2.3 Assortative Sociality: An Aspect of the Behavioral Immune System

In an ecological setting of high disease stress, philopatry (reduced dispersal), xenophobia, and ethnocentrism reflect evolved preferences/values and motivate behaviors for avoidance of novel parasites contained in outgroups and for the management of local infectious disease (Fincher & Thornhill, 2008a, 2008b, 2012a). Philopatry is the psychological preference for the natal locale. It is manifested in behaviors that reduce movements away from the natal location. In areas of high parasite adversity compared to areas of low parasite adversity, philopatry will be the optimal habitat preference because of the correspondent increase in social association with immunologically similar individuals and decreased contact with more distant, and differently parasitized, individuals and their habitats. Likewise, xenophobia – the avoidance and dislike of out-group members – discourages contact with out-groups and their likely different parasites. Neophobia – the dislike of new ideas, attitudes, and ways of doing – is a component of xenophobia; according to the parasite-stress theory of sociality, neophobia functions like xenophobia. Ethnocentrism is in-group favoritism entailing nepotism toward both the nuclear and extended family, as well as altruism toward other immunologically similar

in-group members. This support and loyalty toward in-group members defends against the morbidity and mortality effects of parasites (Navarrete & Fessler, 2006; Sugiyama, 2004; Sugiyama & Sugiyama, 2003). Sugiyama (2004) reported that in the Shiwiar, an indigenous society without ready access to modern medicine, health care in the forms of food and other assistance from in-group members to persons suffering from infectious disease is a major factor in lowering mortality. This parasite management benefit of embeddedness in the local in-group seems to characterize numerous traditional societies in the ethnographic record (Gurven et al., 2000; Sugiyama, 2004; Sugiyama & Sugiyama, 2003). To paraphrase Navarrete and Fessler (2006), in human evolutionary history, under high parasite stress, in-group members were the only health insurance one had, and it was adaptive to have always paid your premiums – in terms of social investment and loyalty toward in-group allies that buffer one and one's family against the morbidity and mortality of infectious disease.

Hence, philopatry, xenophobia (including neophobia), and ethnocentrism – the basic features of what we have called assortative sociality and important features of the behavioral immune system – are expected to be cultural values and normative behaviors predominantly in regions of high parasite stress (Fincher & Thornhill, 2008a; Fincher et al., 2008; Thornhill et al., 2009). Yet humans have experienced parasite gradients throughout history and continue to do so today (Dobson & Carper, 1996; Guernier et al., 2004; Lopez et al., 2006; Low, 1990; McNeill, 1998; Smith et al., 2007; Wolfe et al., 2007). Thus, we expect that the benefits and costs of assortative sociality will shift along the parasite-stress gradient such that in ecological settings of high parasite stress, high levels of assortative sociality will be more beneficial than in circumstances of low parasite stress. As parasite stress declines, the infectious disease risks to individuals of dispersal and interaction with out-groups decrease. Consequently, for individuals in ecological settings that are relatively low in parasite stress, out-group contacts and alliances can provide greater benefits than costs. The benefits of out-group interactions include gains through access to goods, services, and ideas of other groups, as well as through diversified and sometimes larger social networks for marriage and other social alliances (Fincher et al., 2008; Thornhill et al., 2009). Brown et al. (2016) have modeled the regional parasite gradient and its theoretical impact on the benefits and costs of different degrees of assortative sociality. They model their "opportunities–parasites trade-off hypothesis, according to which individuals place themselves at an optimal point on a trade-off between (a) the gains that may be achieved through accessing the resources of geographically or socially distant out-group members through openness to out-group interaction, and (b) the losses arising due to consequently increased risks of exotic infection to which immunity has not been developed" (p. 98). Research discussed later in this chapter has shown that the components of assortative sociality or behavioral immunity respond quantitatively to regional variation in parasite adversity, as predicted by Brown et al.'s and our reasoning.

14.2.4 Conditional Behavioral Immunity

The parasite-stress theory of sociality posits an ancestrally adaptive, condition-dependent adoption of in-group and out-group values and related social tactics by individuals, dependent on level of local parasite stress. This condition-dependent adaptation, like other condition-dependent adaptations, requires for its evolution local variation on a short timescale in the evolutionary selection pressures responsible for it. Hence, evolutionary historical selection due to morbidity and mortality from parasites varied locally in individual lifetimes and thereby favored contingent behavioral and psychological adaptations for assortative sociality.

The evolution of conditionality as an important feature of assortative sociality's functional design, rather than being a region-specific genetically distinct adaptation, is consistent with knowledge about infectious diseases. The dynamic nature of the host–parasite coevolutionary race generates considerable variation in the prevalence, transmissibility, and pathogenicity of the disease agent across the range of its host species, as well as on a fine-grained, local scale within a host individual's lifetime. Important factors affecting this variability at a single locale and in a single human generation are temporal changes in host group size, weather, disease vector abundance and behavior, and the number, virulence, and dynamics of the different diseases infecting hosts (Anderson & May, 1991; Corby-Harris & Promislow, 2008; Ewald, 1994; Guernier et al., 2004; Loker, 2012; Prugnolle et al., 2005). Thus, in-group assortative sociality is an example of evolved phenotypic plasticity within individuals. That is, the individual possesses a conditional strategy with multiple contingent tactics (Fincher et al., 2008; Schaller & Murray, 2008; Thornhill et al., 2009). Such plasticity in traits is favored by evolutionary selection when phenotypic change allows the individual to modify its phenotypic expression in directions that give greater net inclusive fitness benefit than that achieved by a single static phenotype.

A considerable body of research supports the hypothesis of an evolved contingent assortative sociality in people that functions against contagion. In some of the earliest studies of behavioral immunity, Faulkner et al. (2004) and Navarrete and Fessler (2006) provided evidence, based on numerous and diverse Western samples, that scores among individuals on scales that measure the degree of xenophobia and ethnocentrism correspond to chronic individual differences in worry about contracting or catching infectious disease (measured by the perceived vulnerability to disease scale; Duncan et al., 2009); those who perceive high infectious disease risk are more xenophobic and ethnocentric than those who perceive low disease risk. Importantly, this same research also showed that xenophobia and ethnocentrism within individuals increase under experimental primes of greater parasite salience in the current environment.

Other research has documented within-individual shifts in personality – toward greater introversion and avoidance of novelty – and in heightened classical immune responses as well as behavioral avoidance of strangers immediately after research participants view cues of infectious disease salience. Mortensen et al. (2010) reported that subjects viewing images with disease-salient cues immediately exhibited greater feelings promoting between-person avoidance (i.e., extraversion, openness to experience, and agreeableness were reduced) in comparison to these subjects' feelings upon viewing control images. These researchers also found that subjects with high scores on the scale of perceived vulnerability to disease showed greater feelings of interpersonal avoidance than did subjects with low scores on the same scale. Finally, these researchers reported that viewing disease-salient images resulted in increased avoidant arm movements when subjects viewed facial photos of strangers, especially for subjects high in perceived vulnerability to disease. Schaller et al. (2010) reported that participants who observed images of people with infectious disease symptoms (e.g., pockmarks, skin lesions, sneezing) immediately mounted a classical immune response. Their white blood cells produced elevated amounts of the inflammatory cytokine interleukin-6, which is produced when exposed to bacterial antigens. This immune response was not seen in participants who viewed control images, including those who viewed images depicting a person pointing a gun directly at them. Hence, the immune response was not a general reaction to danger or threat, but was specific to cues of other people with symptoms of parasitic infection. Research by Stevenson et al. (2011) compared salivary immune markers between research participants in whom disgust was induced by disease-relevant pictorial cues documented to be disgust elicitors (e.g., a dirty toilet, an eye infection) and other participants who were exposed to either negative but disease-irrelevant pictures or neutral pictures. The disgust-primed group showed an oral classical immune response, but the other groups did not.

When considered together, these studies by Mortensen et al. (2010), Schaller et al. (2010), and Stevenson et al. (2011) reveal that visually perceiving cues pertinent to risk of parasitic infection generates immediate cellular and biochemical classical immune responses, changes in perceptions of one's own personality, changes in disgust sensitivity, and behavioral actions that defend against contagion and motivate avoidance of infectious people. Hence, such cues activate markedly the classical immune system as well as the behavioral immune system, and the dual activation is functionally coordinated to defend against infectious disease threat.

The functional coordination of the two immune systems is seen in other research findings. There is a conditional increase in behavioral immunity upon activation of the classical immune system. Miller and Maner (2011) reported increased aversive reactions toward facial disfigurement by research participants recovering from an infectious disease. Also, the adaptive suppression of classical immunity in early pregnancy (to avoid damage to the nonself features of the fetus) activates an increase in behavioral immunity to compensate: more disgust sensitivity and ethnocentrism (Fessler et al., 2005, Navarrete, Fessler, & Eng, 2007). Finally, genetic factors that contribute to defective classical immunity activate an increase in two aspects of behavioral immunity: lower extraversion and lower openness to new experiences (MacMurray, Comings, & Napoloioni, 2014; Napolioni et al. 2014).

In sum, there is considerable evidence of both between-individual stable differences as well as within-individual conditionality in xenophobic and ethnocentric values and related personality features and behaviors, as well as evidence that both the between-individual consistency and within-individual contingency are caused by infectious disease problems in the local environment. The proximate means by which individuals assess local parasite stress – and thereby ontogenetically and contingently express the locally adaptive degree of assortative sociality – may include immune system activation (e.g., the frequency of infection; Stevenson et al., 2009) and social learning of local disease risks (Fincher et al., 2008). Both of these causes may act typically during an individual's development and account for the between-individual and within-individual variation in values affecting in- and out-group behavioral preferences and degree of philopatry. In regard to immune system activation, Stevenson et al. (2009) found that people with high contamination sensitivity (an individual difference variable related to perceived vulnerability to disease) reported a history of high infectious diseases (but not recency of infections), implying that an ontogeny of repeated activation of the classical immune system may underlie the adoption of assortative sociality values and associated behavior. These researchers also reported that people with high contamination sensitivity and disgust sensitivity had fewer recent infectious diseases than people with low sensitivities, providing evidence of a protective function of these emotions against these diseases.

Our emphasis on adaptive contingency in the adoption and use of assortative sociality tactics does not imply that we expect no variation across human groups in genetic adaptation for assortative sociality. As outlined by Durham (1991) and Blute (2010), culture–gene coevolution involves allelic frequency changes (i.e., evolution) that correspond to changes in cultural traits. Culture–gene coevolution may produce genetically differentiated cross-cultural variation in the values and behaviors of assortative sociality. For example, in areas of consistently high parasite prevalence, cultural practices of xenophobia,

philopatry, and ethnocentrism may select for alleles affecting psychological features that promote the learning and effective use of these values (Fincher et al., 2008). Our argument is that infectious disease problems are locally variable on a short timescale as a result of the temporal changes mentioned above, and hence significant conditionality will be favored and maintained by selection even in the presence of localized genetic adaptation functioning in adoption and use of local values and behaviors. There is some evidence that culture–gene coevolution may play a role in cross-national variation in assortative sociality, specifically in the value dimension collectivism–individualism (Chiao & Blizinsky, 2010; Way & Lieberman, 2010).

14.2.5 Behavioral Immunity Adaptively Manages False Positives

Only fairly recently has it been demonstrated scientifically that parasites, most of which are microscopic, cause disease. Natural selection in all animal species favors individuals with indirect knowledge of infectious disease risk and the avoidance of such risks. Hence, there are directly selected human psychological features that attend to and process information about environmental cues that, across generations of human evolutionary history, corresponded with risk of contagion. Moreover, given that an error in judging a contagion risk can be literally grave, selection has built behavioral immunity to adaptively accept many false positives (i.e., deduce contagion risk when it is absent; Curtis, 2007; Duncan & Schaller, 2009; Oaten et al., 2009). As a result, people's behavioral immune system sometimes overreacts to even the hint of contagion danger in our environment, including our social environment. This is why a person's encounter with a stranger who speaks a different dialect or possesses a different value system may evoke strong xenophobia toward the stranger. This, too, is a cause of the disgust and associated prejudice of many people toward sexual minorities (e. g., homosexuals), obese or very thin people, the elderly, people with noncontagious diseases, or physically or mentally challenged people who show behavior that deviates from the norm (for reviews, see Duncan & Schaller, 2009; Kouznetsova et al., 2012; Ryan et al., 2012; Terrizzi et al., 2010, 2012).

14.3 ADDITIONAL EVIDENCE SUPPORTING THE PARASITE-STRESS THEORY OF SOCIALITY

In Section 14.2, we mentioned several published empirical findings that support the parasite-stress theory of sociality. In this section, we discuss some additional supportive empirical results from the literature.

The cultural variable of collectivism–individualism is a major and much-researched variable in psychology and sociology for describing cross-cultural differences in values. This unidimensional variable corresponds closely with what political scientists call the rightist–leftist or conservatism–liberalism dimension, with high collectivism mapping onto high conservatism (low liberalism) and low collectivism mapping onto high liberalism/individualism (Fincher & Thornhill, 2012b). Fincher et al. (2008) hypothesized that regional differences in parasite adversity proximately cause variation in this dimension of values, with the following reasoning. The values and behaviors that define collectivism, such as ethnocentrism, xenophobia, and interdependent thinking dictated by the cognitive preferences of the in-group and related conformity with traditional ideas and ways, provide anti-parasite defenses and thus are optimal under conditions of high parasite adversity. Collectivist values and behaviors, we reasoned, function in behavioral immunity. In contrast, individualism confers benefits upon individuals such as personal autonomy and independent thinking, openness to new and nontraditional ideas and ways, and willingness to interact with a diversity of people. These individualistic/liberal traits, however, have the cost of an enhanced likelihood of contracting infectious disease. Thus, the lower the parasite stress, the greater the benefits of individualism relative to its costs. Specifically, Fincher et al. (2008) predicted a positive correlation between parasite stress and collectivism (negative with individualism) across cultures.

Across multiple intercorrelated measures of collectivism–individualism put together by cross-cultural researchers, Fincher et al. (2008) found, as hypothesized, that worldwide variation in parasite stress robustly predicted cross-national values of collectivism–individualism. Within regions with high severities of infectious diseases, human cultures are characterized by high collectivism, whereas cultures from regions of low parasite stress are highly individualistic. This pattern remained statistically significant when controlling for potential confounding variables, including societal wealth, population size, and latitude.

Subsequently, Thornhill et al. (2010) computed separate indices assessing the number of human parasitic diseases in each of two distinct categories – non-zoonotic and zoonotic – and examined the extent to which each index uniquely predicted cross-national differences in collectivism–individualism. The parasite-stress theory proposes that infectious diseases transmissible among humans (non-zoonotics) will be more important in predicting collectivism–individualism than those that humans can contract only from nonhuman animals and not subsequently pass to other humans (zoonotics). As hypothesized, the richness of non-zoonotic infectious disease agents predicted uniquely cross-national differences in collectivist–individualist values. In contrast, zoonotic parasite richness contributed little to cross-national relationships between parasite adversity and these values. Thus, worldwide variation in these values predicted by parasite adversity appears to be attributable almost entirely to the prevalence of non-zoonotic diseases. These cross-national results for numbers of diseases in the transmission categories in relation to collectivism–individualism were repeated with parasite severity measures (measures of number of infectious disease cases, not number of diseases). Non-zoonotic severity related much more strongly to collectivism–individualism than did zoonotic severity. Also, the measures of parasitic disease numbers were strongly and positively correlated with measures of parasite severity (Thornhill & Fincher, 2014).

Moreover, across the 50 states of the USA and 186 indigenous societies in the Standard Cross-Cultural Sample, collectivism correlated positively with parasite stress (and individualism correlated negatively; Cashdan & Steele, 2013; Fincher & Thornhill, 2012a). Furthermore, as with the cross-national results, collectivism across the US states correlated more strongly with non-zoonotic than with zoonotic human diseases (Thornhill & Fincher, 2014).

The "strength of family ties," a measure of collectivism developed by Fincher and Thornhill (2012a) and focused on extended family loyalty and support, also showed robust positive correlation with parasite stress across nations and across states of the USA. And, as predicted, in both the cross-national analysis and the analysis across US states, the strength of family ties was correlated more strongly with non-zoonotic infectious diseases than with zoonotic infectious diseases (Thornhill & Fincher, 2014).

The potential confounds examined in our analyses did not change any of the conclusions we have mentioned. Also, the basic relationships of values and parasite stress are robust at regional levels both cross-nationally (e.g., Murdock's six world regions) and across the USA (nine US census regions; Fincher & Thornhill 2012a; Fincher et al., 2008; Thornhill et al., 2010).

Guided by the parasite-stress theory of sociality, we have also investigated religiosity (religious commitment and participation) across countries and states of the USA. We hypothesized that religiosity is a collectivist/conservative value that functions to enhance in-group embeddedness and in-group boundary formation and maintenance and hence functions in defense against parasites. As hypothesized, the degree of religiosity at each of the two regional levels is strongly positively related to collectivism (negatively with individualism) and parasite adversity (Fincher & Thornhill, 2012a, 2012b). Thus, religiosity reflects and promotes behavioral immunity (see also Terrizzi et al., 2012).

As mentioned, we have hypothesized that absence of dispersal (philopatry) is a behavioral immunity defense against contact with novel parasites harbored in outgroups and the habitats they frequent, and that philopatry is the optimal habitat preference under high parasite stress. Evidence in support of this hypothesis is seen in movement patterns of people from both cross-national analysis and analysis across the 50 US states (Thornhill & Fincher, 2014), as well as analyses across indigenous societies (Cashdan & Steele, 2013; Fincher & Thornhill, 2008a).

With colleagues, we extended the parasite-stress theory to explain cross-national variation in degree of democratization, gender equality, sexual permissiveness, and property rights, as well as in personality (Fincher & Thornhill, 2012a; Thornhill et al., 2009, 2010; see also Gangestad et al., 2006; Murray et al., 2013; Schaller & Murray, 2008). Parasite stress and collectivism positively correlate with undemocratic (autocratic or authoritarian) governance, gender inequality, sexual restrictiveness (especially in women), property rights restrictions to elites, and the personality traits of introversion and closed-mindedness to new experiences. Hence, the behavioral immunity values characterizing collectivism give rise at the societal level to autocratic governance, undemocratic gender relations, illiberal sexuality of women, restricted property rights, and certain personality traits.

In separate studies with a colleague, we showed that parasite stress positively predicts frequencies of civil and other within-country warfare, revolutions, coups, and the absence of peace across countries of the world (Letendre et al., 2010, 2012). This work supported our hypothesis that the collectivism generated by parasite adversity, in particular the xenophobic component of collectivism, is a proximate cause of intergroup coalitional violence. Hence, this research indicates that certain features of behavioral immunity are causes of warfare and other intergroup coalitional violence.

Recent research on conformity in relation to infectious disease risk is another example of the generality of the parasite-stress theory of sociality and associated behavioral immunity. Conforming to the beliefs and values of the majority has benefits and costs. Benefits of socially navigating in a conformist group include the predictability of the way people think and behave. Moreover, when conformity is coupled with aversion and prejudice toward those who do not conform to the majority behavior, as it typically is, conformity will be protective against novel parasites in out-groups to which the conforming in-group is not immune (Fincher et al., 2008; Murray & Schaller, 2012; Murray et al., 2011; Wu & Chang, 2012). Costs of conformity include the low rate of generating and adopting ideas, especially ideas that are unfamiliar locally. However, preferring traditional ways of thinking and avoiding foreign ideas can be defenses against novel parasites in out-groups by way of reduced contact with out-groups. In line with this reasoning that conformity is an aspect of behavioral immunity and supporting the parasite-stress theory of sociality, Murray et al. (2011) showed that cross-national variation in conformity correlates positively with parasite adversity. Also, using a Canadian and a Chinese sample, respectively, Murray and Schaller (2012) and Wu and Chang (2012) examined individual differences in conformity values and found that scores on the scale that measures perceived vulnerability to disease correlate positively with conformity. Each of these two studies also included experiments that made infectious disease risk salient to research participants. The participants immediately became more conformist, but this change in values was not observed in control groups of participants, including controls presented with disease-irrelevant threat cues. In the Murray and Schaller (2012) study, the participants exposed to parasite-salient cues showed increased positivism toward conforming others. Murray and Schaller's (2012) and Wu and Chang's (2012) findings indicate that an individual's perception of threat of infectious disease, either arising from individual differences in perceived vulnerability to disease or due to immediate stimuli of parasite presence, causes the individual to adopt conformist values. The Murray and Schaller (2012) study also showed that individuals presented with cues of parasite presence in their immediate environment become prejudiced in favor of others with conformist values.

Other recent research also reveals the heuristic nature of the parasite-stress theory of sociality and associated behavioral immunity. Terrizzi et al. (2010, 2012) investigated individual differences in the relationship of disgust sensitivity with the conservative values of religiosity and prejudice against sexual minorities (homosexuals and bisexuals). They reported that disgust sensitivity positively predicts these values and argued that disgust, religiosity, and prejudice against sexual out-groups are components of the human behavioral immune system. In complementary work, Clay et al. (2012) showed that individual differences in disgust sensitivity and perceived vulnerability to disease positively correlate with collectivism and several other variables that reflect conservatism (e.g., traditionalism, conformity, and importance of societal stability and security). Moreover, Terrizzi et al. (2013) conducted a meta-analysis of 22 studies of individual differences in components of collectivism or of conservatism in relation to perceived vulnerability to disease or disgust sensitivity. They reported robust positive relationships among the variables. Consistent with our arguments and with the evidence presented above, Terrizzi et al. (2013) concluded that conservative values are defenses that reduce contact with infectious diseases.

Inspired by the parasite-stress theory of sociality, Reid et al. (2012) made a significant discovery for the scholarly discipline of linguistics. Reid and colleagues researched disgust sensitivity in relation to sound perception of dissimilarity to self's accent of foreign-accented English. Americans of high disgust sensitivity rated foreign-accented English as more dissimilar to their own accent than did Americans of low disgust sensitivity, even though the study participants were listening to the same speakers. The study also showed that research participants who viewed parasite-salient stimuli perceived a greater difference in foreign-accented English compared to their own accent, but participants viewing other threat stimuli (unrelated to parasite threat) did not. Given the positive relationship between conservatism and disgust (e.g., Terrizzi et al., 2013), these results imply that conservatives perceive greater differences between in-group and out-group spoken language than do liberals.

We hypothesize from the parasite-stress theory that the greater sensitivity of high-disgust people to differences between us and them, which was documented by Reid et al. (2012) for accents, may extend to many differences outside of language, such as the perception of differences in values, skin color, behavior, dress patterns, etc. Such perception biases may underlie the xenophobia sensitivity of conservatives.

Billing and Sherman (1998) and Sherman and Billing (1999) hypothesized that the value people place on the use of spices in cooking is a defense against foodborne human parasites. To test this, they investigated the types and numbers of spices used in recipes across many regions of the world. They found that temperature positively correlates with anti-pathogen spice use across regions. Regional temperature is a useful surrogate for parasite stress, with warmer temperature equating with more parasite adversity (Billing & Sherman, 1998; Guernier et al., 2004). Later research by Murray and Schaller (2010) reported a robust positive relationship across countries between spice use and parasite stress per se. Additional evidence that spicing foods is a form of behavioral immunity was reported by Prokop and Fancovicova (2011). They showed that individual differences in preference for and use of spiced food corresponded with concern about infectious diseases. Individuals who are high in worry about contagion (high scorers on the perceived vulnerability to disease scale) had stronger preferences for and greater consumption of spicy foods than individuals who are low on concern about contagion.

The parasite theory of sexual selection was proposed by Hamilton and Zuk (1982). It relates to the component of sociality involved in competition for mates and mate choice and hence is a sub-theory of the parasite-stress theory of sociality. The parasite theory of sexual selection argues that variation among individuals in genetic immunity to parasites and related phenotypic quality gives rise to sexual selection, which favors resistant individuals. Relative resistance of individuals is honestly depicted in traits such as the rooster's comb and the peacock's tail, and in humans in the traits affecting sexual attractiveness, especially developmental stability (bilateral symmetry) and the sexually dimorphic hormone markers on the face and body (estrogen markers in women and testosterone markers in men; Thornhill & Gangestad, 2008).

Gangestad and Buss (1993) and Gangestad et al. (2006) conducted cross-national research inspired by the parasite theory of sexual selection that empirically linked human mate choice, parasite stress, and behavioral immunity. They reported a positive correlation across countries between human parasite stress and the importance people place on physical attractiveness (good looks) in mate choice. This finding was hypothesized from their reasoning that physical attractiveness is a certification of genetic resistance to parasites – good genes for parasite resistance – and hence is expected to be valued more in high- than low-parasite-stress regions (for a review of the literature on the positive relationship between attractiveness and genetic quality in humans, see Thornhill & Gangestad, 2008).

Consistent with this finding, DeBruine et al. (2012) and Moore et al. (2013) showed that high parasite stress evokes in women an enhanced mate preference for facially masculine men and hence for men's facial markers of phenotypic and genetic quality. Women's preference for testosteronization/masculinity of male faces correlates significantly and positively with parasite stress across countries and across the US states. In related research, Jones et al. (2013a) linked masculinity preferences of women to pathogen disgust sensitivity. They reported that, in women, pathogen disgust positively correlates with their attractiveness ratings of masculinity in men's faces, bodies, and voices. In other research, Jones et al. (2013b) showed men's preferences for femininity in women's faces; also, apparent markers of phenotypic and genetic quality were positively correlated with men's pathogen disgust sensitivity.

Research findings reported by de Barra et al. (2013) reveal more about the mechanisms of the behavioral immune system that act during individuals' ontogeny and account for the regional differences in priority of good looks in mate selection. Compared to adults with a low infectious disease ontogenetic history, adults with a childhood background of high infectious disease incidence showed stronger attractiveness preference for mates with enhanced sex-typical facial hormone markers, and hence with relatively high phenotypic and genetic quality, including parasite resistance.

Little et al. (2010) experimentally presented to research participants pictures of high and absent parasite salience, after which they recorded the participants' attractiveness ratings of human faces that varied in symmetry and hormone markers. Symmetry, like sex-specific hormone markers in the face, is a likely marker of phenotypic and genetic quality. Little et al. found that people who were exposed to cues of high contagion risk, immediately showed increased attractiveness preferences for opposite-sex individuals with greater sex-specific hormone markers and symmetry compared to those seeing no contagion risk. Young et al. (2011) replicated this finding for symmetry.

Lee and Zeitsch's (2011) research indicates that women primed to perceive contagion in their current environment immediately adjusted their mate preferences for men with resources vs. men with high genetic quality. Showing women contagion cues activated the aspect of their behavioral immune system that increases their psychological preference for a mate of high genetic quality and reduces their preference for a mate with resources to provide.

Welling et al.'s (2007) research, like that of Gangestad and colleagues, DeBruine and colleagues, Jones and colleagues, de Barra and colleagues, Little and colleagues, and Lee and Zeitsch, indicates that physical attractiveness judgments are part of the behavioral immune system. Welling et al. reported that men and women who perceived themselves to be more vulnerable to infectious disease had stronger attractiveness preferences for healthy faces than did individuals who perceived themselves to be less vulnerable to disease.

As a theory of cultural diversity, the parasite-stress theory of sociality informs the processes causing new cultures to originate. We have proposed that, given the ecological localization of host defenses against parasites, the components of assortative sociality (limited dispersal, ethnocentrism, and xenophobia), by functioning in parasite avoidance and management, fractionate or segment and also factionalize an original culture's range and thereby contribute to the independence of the resulting segments (Fincher & Thornhill, 2008a, 2008b). Thus, the parasite-stress theory of sociality includes a hypothesis about the genesis of cultural or ethnic diversity.

Accordingly, infectious diseases can transform an ancestral culture into new cultures that arise side by side and without segmentation of the ancestral culture by a geographic barrier. We have called the process involved "the parasite-driven wedge" (Fincher & Thornhill, 2008a, 2008b; Thornhill & Fincher, 2013b). The geographically localized host–parasite races and corresponding localized behavioral immunity traits are the basic features of the wedge. The wedge subdivides an ancestral culture and thereby pushes segments apart, creating new cultures in the absence of the segmentation created by mountains or other geographic barriers.

Several predictions of the parasite-driven wedge hypothesis have been supported empirically. We have shown that the numbers of endemic religions (both major religions and ethnoreligions), as well as the numbers of endemic languages, across contemporary countries worldwide are related strongly and positively to parasite stress (Fincher & Thornhill, 2008a, 2008b). Also consistent with this aspect of the parasite-stress theory was the earlier finding by Cashdan (2001) that high-parasite-stress regions have more indigenous ethnic groups than low-parasite-stress regions. Our own and Cashdan's research on the diversity of human cultures across the world indicates that the behavioral immune system is a cause of the genesis of new cultures.

14.4 CONCLUDING COMMENTS

In this chapter, we have provided an introduction to the parasite-stress theory of values/sociality, including some of its empirical support. In our recent book, we give a much fuller treatment of the conceptual and empirical foundations of the theory, including its application to non-human animal sociality (Thornhill & Fincher, 2014). Evidence indicates that both a wide span of human affairs and major aspects of human cultural diversity can be understood in light of variable parasite stress and the range of value systems evoked by variable parasite stress. The same evidence supports the hypothesis that people have psychological adaptations that function to adopt values dependent upon local infectious disease adversity. Evidence also indicates that the parasite-stress theory of sociality informs other topics in ecology and evolutionary biology, such as family organization and speciation processes and biological diversity in general in nonhuman animals.

The overall goal of our book was to create a synthesis or unity, based on the parasite-stress theory, of many topics that traditionally have been viewed and studied as distinct and unconnected. The book shows the utility of the parasite-stress theory for the unification of areas of research and knowledge ranging from parasitology, immunology, moral systems, civil conflict, governmental systems, family life, sexual behavior, dispersal patterns, economics, personality, violence, religious commitment, biodiversity, and so on. We hope others will clarify and expand the unity we initiated. Many potential new research directions and associated hypotheses are discussed in our book.

The theory and its empirical support have been evaluated by scholars across a range of disciplines. The theory is far-reaching and thus open to comments and criticisms from the many research areas that investigate human affairs and other topics addressed by the theory. Our responses to criticisms and comments by other researchers are presented fully in our book (Thornhill & Fincher, 2014; also see Fincher & Thornhill, 2012b; Thornhill & Fincher, 2013a). We conclude that the criticisms to date do not falsify the theory, moderate its application to any of the topics we suggest it can explain, or discredit any body of its empirical support.

Our book summarizes evidence that emancipation from parasites is a key factor in the emancipation of people from prejudice, poverty, oppression, domestic conflict, and violence. Accordingly, cleaning up the parasites in a region will liberalize the values of the people in the region and result in increased out-group tolerance and amity and reduced authoritarianism, classism, and sexism. Hence, relief from infectious disease does more than improve health, increase longevity, and reduce child and overall morbidity and mortality rates – such relief produces people with an egalitarian priority. Those interested in making a more humane planet should study the ideas and findings reported in our book. We emphasize that the evidence we present in itself does not identify a more democratic planet as a morally correct goal because scientific facts do not and cannot identify moral correctness or incorrectness; instead, the evidence provides the way to achieve a more democratic world if people desire such a world. Yet, the same evidence points the way to making the Jim Crow South rise again or to turn the whole planet into only ultraconservative, totalitarian governmental polities. These undemocratic outcomes can be accomplished by promoting the welfare of human infectious diseases through practices that disenfranchise and impoverish people widely and limit or deny modern sanitation technology and health care.

REFERENCES

Anderson, R. M., & May, R. M. (1991). *Infectious Disease of Humans: Dynamics and Control*. Oxford: Oxford University Press.

Billing, J., & Sherman, P. W. (1998). Antimicrobial functions of spices: Why some like it hot. *Quarterly Review of Biology*, **73**, 3–49.

Blute, M. (2010). *Darwinian Sociocultural Evolution: Solutions to Dilemmas in Cultural and Social Theory*. Cambridge, UK: Cambridge University Press.

Brown, G. D. A., Fincher, C. L., & Walasek, L. (2016). Personality, parasites, political attitudes, and cooperation: A model of how infection prevalence influences openness and social group formation. *Topics in Cognitive Science*, **8**, 98–117.

Cashdan, E. (2001). Ethnic diversity and its environmental determinants: Effects on climate, pathogens, and habitat diversity. *American Anthropology*, **103**, 968–991.

Cashdan, E., & Steele, M. (2013). Pathogen prevalence, group bias, and collectivism in the standard cross-cultural sample. *Human Nature*, **24**, 59–75.

Chiao, J. Y., & Blizinsky, K. D. (2010). Culture–gene coevolution of individualism–collectivism and the serotonin transporter gene. *Proceedings of the Royal Society B*, **277**, 529–537.

Clay, R., Terrizzi, J. A., Jr., & Shook, N. J. (2012). Individual differences in the behavioral immune system and the emergence of cultural systems. *Journal of Social Psychology*, **43**, 174–184.

Corby-Harris, V., & Promislow, D. E. L. (2008). Host ecology shapes geographical variation for resistance to bacterial infection in Drosophila melanogaster. *Journal of Animal Ecology*, **77**, 768–776.

Curtis, V. A. (2007). Dirt, disgust and disease: A natural history of hygiene. *Journal of Epidemiology and Community Health*, **61**, 660–664.

Curtis, V., Aunger, R., & Rabie, T. (2004). Evidence that disgust evolved to protect from risk of disease. *Proceedings of the Royal Society B*, **271**, 17–31.

Curtis, V. A., de Barra, M., & Aunger, H. (2011). Disgust as an adaptive system for disease avoidance behaviour. *Philosophical Transactions of the Royal Society B*, **366**, 389–401.

de Barra, M., DeBruine, L., Jones, B., & Curtis, V. A. (2013). Illness in childhood predicts face preferences in adulthood. *Evolution and Human Behavior*, **34**, 384–389.

DeBruine, L. M., Little, A. C., & Jones, B. C. (2012). Extending parasite-stress theory to variation in human mate preferences. *Behavioral and Brain Sciences*, **35**, 86–87.

Denic, S., & Nicholls, M. G. (2007). Genetic benefits of consanguinity through selection of genotypes protective against malaria. *Human Biology*, **79**, 145–158.

Denic, S., Nagelkerke, N., & Agarwal, M. M. (2008a). Consanguineous marriages and endemic malaria: Can inbreeding increase population fitness? *Malaria Journal*, **7**, 150.

Denic, S., Nagelkerke, N., & Agarwal, M. M. (2008b). Consanguineous marriages: Do genetic benefits outweigh its costs in populations with alpha(+)-thalassemia, hemoglobin S, and malaria? *Evolution and Human Behavior*, **29**, 364–369.

Diamond, J. (1998). *Guns, Germs and Steel: The Fates of Human Societies*. New York: W.W. Norton.

Dionne, M., Miller, K. M., Dodson, J. J., Caron, F., & Bernatchez, L. (2007). Clinical variation in MHC diversity with temperature: Evidence for the role of host–pathogen interaction on local adaptation in Atlantic salmon. *Evolution*, **61**, 2154–2164.

Dobson, A. P., & Carper, E. R. (1996). Infectious diseases and human population history. *BioScience*, **46**, 115–126.

Duncan, L. A., Schaller, M., & Park, J. H. (2009). Perceived vulnerability to disease: Development and validation of a 15-item self-report instrument. *Personality and Individual Differences*, **47**, 541–546.

Durham, W. H. (1991). *Coevolution: Genes, Culture and Human Diversity*. Stanford, CA: Stanford University Press.

Ewald, P. W. (1994). *Evolution of Infectious Disease*. New York: Oxford University Press.

Faulkner, J., Schaller, M., Park, J. H., & Duncan, L.A. (2004). Evolved disease-avoidance mechanisms and contemporary xenophobic attitudes. *Group Processes and Intergroup Relations*, **7**, 333–353.

Fessler, D. M. T., Eng, S. J., & Navarrete, C. D. (2005). Elevated disgust sensitivity in the first trimester of pregnancy: Evidence supporting the compensatory prophylaxis hypothesis. *Evolution and Human Behavior*, **26**, 344–351.

Fincher, C. L., & Thornhill, R. (2008a). A parasite-driven wedge: Infectious diseases may explain language and other biodiversity. *Oikos*, **117**, 1289–1297.

Fincher, C. L., & Thornhill, R. (2008b). Assortative sociality, limited dispersal, infectious disease and the genesis of the global pattern of religion diversity. *Proceedings of the Royal Society of London, Biological Sciences*, **275**, 2587–2594.

Fincher, C. L., & Thornhill, R. (2012a). Parasite-stress promotes in-group assortative sociality: The cases of strong family ties and heightened religiosity. *Behavioral and Brain Sciences*, **35**, 61–79.

Fincher, C. L., & Thornhill, R. (2012b). The parasite-stress theory may be a general theory of culture and sociality response. *Behavioral and Brain Sciences*, **35**, 99–119.

Fincher, C. L., Thornhill, R., Murray, D. R., & Schaller, M. (2008). Pathogen prevalence predicts human cross-cultural variability in individualism/collectivism. *Proceedings of the Royal Society of London, Biological Sciences*, **275**, 1279–1285.

Fumagalli, M., Sironi, M., Pozzoli, U., et al. (2011). Signatures of environmental genetic adaptation pinpoint pathogens as the main selective pressure through human evolution. *PLoS Genetics*, **7**, e1002355.

Gangestad, S. W., & Buss, D. M. (1993). Pathogen prevalence and human mate preference. *Ethology and Sociobiology*, **14**, 89–96.

Gangestad, S. W., Haselton, M. G., & Buss, D. M. (2006). Evolutionary foundations of cultural variation: Evoked culture and mate preferences. *Psychological Inquiry*, **17**, 75–95.

Good, C. M. (1972). Salt, trade, and disease: Aspects of development in Africa's northern Great Lakes region. *International Journal of African Historical Studies*, **5**, 543–586.

Guernier, V., Hochberg, M. E., & Guegan, J. (2004). Ecology drives the worldwide distribution of human diseases. *PLoS Biology*, **2**, e141.

Gurven, M., Allen-Arave, W., Hill, K., & Hurtado, M. (2000). "It's a Wonderful Life": Signaling generosity among the Ache of Paraguay. *Evolution and Human Behavior*, **21**, 263–282.

Hamilton, W. D., & Zuk, M. (1982). Heritable true fitness and bright birds: A role for parasites? *Science*, **218**, 284–387.

Hoben, A. D. (2011). An evolutionary investigation of consanguineous marriages. Unpublished doctoral dissertation, University of Groningen.

Hoben, A. D., Buunk, A. P., Fincher, C. L., & Thornhill, R. (2010). On the adaptive origins and maladaptive consequences of human inbreeding: Parasite prevalence, immune functioning, and consanguineous marriage. *Evolutionary Psychology*, **8**, 658–676.

Inbar, Y., Pizarro, D. A., Iyer, R., & Raidt, J. (2012). Disgust sensitivity, political conservatism, and voting. *Social Psychological and Personality Science*, **5**, 537–544.

Jones, B. C., Feinberg, D. R., Watkins, C. D., et al. (2013a). Pathogen disgust predicts women's preferences for masculinity in men's voices, faces, and bodies. *Behavioral Ecology*, **24**, 373–379.

Jones, B. C., Fincher, C. L., Welling, L. L. M., et al. (2013b). Salivary cortisol and pathogen disgust predict men's preferences for feminine shape cues in women's faces. *Biological Psychology*, **92**, 233–240.

Kaltz, O., Gandon, S., Michalakis, Y., & Shykoff, J. A. (1999). Local maladaptation in the anther-smut fungus *Microbotryum violaceum* to its host plant *Silene latifolia*: Evidence from a cross-inoculation experiment. *Evolution*, **53**, 395–407.

Kouznetsova, D., Stevenson, R. J., Oaten, M. J., & Case, T. I. (2012). Disease-avoidant behaviour and its consequences. *Psychology and Health*, **27**, 491–506.

Lee, A. J., & Zietsch, B. P. (2011). Experimental evidence that women's mate preferences are directly influenced by cues of pathogen prevalence and resource scarcity. *Biology Letters*, **7**, 892–895.

Letendre, K., Fincher, C. L., & Thornhill, R. (2010). Does infectious disease cause global variation in the frequency of intrastate armed conflict and civil war? *Biological Reviews*, **85**, 669–683.

Letendre, K., Fincher, C. L., & Thornhill, R. (2012). Infectious disease, collectivism, and warfare. In T. Shackelford & V. Weekes-Shackelford, eds., *The Oxford Handbook on Evolutionary Perspectives on Violence, Homicide, and Warfare*. New York: Oxford University Press, pp. 351–371.

Little, A. C., DeBruine, L. M., & Jones, B. C. (2010). Exposure to visual cues of pathogen contagion changes preferences for masculinity and symmetry in opposite-sex faces. *Proceedings of the Royal Society of London B*, **278**, 2032–2039.

Loker, E. S. (2012). Macroevolutionary immunology: A role for immunity in the diversification of animal life. *Frontiers in Immunology*, **3**, 25.

Lopez, A. D., Mathers, C. D., Ezzati, M., Jamieson, D. T., & Murray, C. J. (2006). Global and regional burden of disease and risk factors, 2001: Systematic analysis of population health data. *Lancet*, **367**, 1747–1757.

Low, B. S. (1990). Marriage systems and pathogen stress in human societies. *American Zoologist*, **30**, 325–339.

MacMurray, J., Comings, D. E., & Napolioni, V. (2014). The gene–immune–behavioral pathway: Gamma-interferon (1FN-γ) simultaneously coordinates susceptibility to infectious disease and harm avoidance behaviors. *Brain, Behavior, and Immunity*, **35**, 169–175.

McNeill, W. H. (1998). *Plagues and Peoples*. Harpswell, ME: Anchor.

Miller, E. N., Fadl, M., Mohamed, H. S., et al. (2007). Y chromosome lineage- and village-specific genes on chromosomes 1p22 and 6q27 control visceral leishmaniasis in Sudan. *PLoS Genetics*, **3**, 679–688.

Miller, S. L., & Maner, J. K. (2011). Sick body, vigilant mind: The biological immune system activates the behavioral immune system. *Psychological Science*, **22**, 1467–1471.

Moore, F. R, Coetzee, V., Contreras-Garduño, J., et al. (2013). Cross-cultural variation in women's preferences for cues to sex- and stress-hormones in the male face. *Biology Letters*, **9**, 20130050.

Mortensen, C. R., Becker, D. V., Ackerman, J. M., Neuberg, S. L., & Kenrick, D. T. (2010). Infection breeds reticence: The effects of disease salience on self-perceptions of personality and behavioral avoidance tendencies. *Psychological Science*, **21**, 440–447.

Murray, D. R., & Schaller, M. (2010). Historical prevalence of infectious diseases within 230 geopolitical regions: A tool for investigating origins of culture. *Journal of Cross-Cultural Psychology*, **41**, 99–108.

Murray, D. R., & Schaller, M. (2012). Threat(s) and conformity deconstructed: Perceived threat of infectious disease and its implications for conformist attitudes and behavior. *European Journal of Social Psychology*, **42**, 180–188.

Murray, D. R., Trudeau, R., & Schaller M. (2011). On the origins of cultural differences in conformity: Four tests of the pathogen prevalence hypothesis. *Personality and Social Psychology Bulletin*, **37**, 318–329.

Murray, D. R., Schaller, M., & Suedfeld, P. (2013). Pathogens and politics: Further evidence that parasite prevalence predicts authoritarianism. *PLoS ONE*, **8**, e62275.

Napolioni, V., Murray, D. R., Cominngs, D. E., et al. (2014). Interaction between infectious diseases and personality traits: *ACP1* C* as a potential mediator. *Infection, Genetics and Evolution*, **26**, 267–273.

Navarrete, C. D., & Fessler, D. M. T. (2006). Disease avoidance and ethnocentrism: The effects of disease vulnerability and disgust sensitivity on intergroup attitudes. *Evolution and Human Behavior*, **27**, 270–282.

Navarrete, C. D., Fessler, D. M. T., & Eng, S. J. (2007). Elevated ethnocentrism in the first trimester of pregnancy. *Evolution and Human Behavior*, **28**, 60–65.

Norenzayan, A., & Shariff, A. F. (2008). The origin and evolution of religious prosociality. *Science*, **322**, 58–62.

Oaten, M., Stevenson, R. J., & Case, T. I. (2009). Disgust as a disease-avoidance mechanism. *Psychological Bulletin*, **135**, 303–321.

Park, J. H., & Schaller, M. (2005). Does attitude similarity serve as a heuristic cue for kinship? Evidence of an implicit cognitive association. *Evolution and Human Behavior*, **26**, 158–170.

Pitchappan, R. M. (2002). Castes, migration, immunogenetics and infectious diseases in south India. *Community Genetics*, **5**, 157–161.

Prokop, P., & Fačovičová, J. (2011). Preferences for spicy foods and disgust of ectoparasites are associated with reported health in humans. *Psihologija*, **44**, 281–293.

Prugnolle, F., Manica, A., Charpentier, M., et al. (2005). Pathogen-driven selection and worldwide HLA class I diversity. *Current Biology*, **15**, 1022–1027.

Reid, S. A., Zhang, J., Anderson, G. L., et al. (2012). Parasite primes make foreign-accented English sound more distant to people who are disgusted by pathogens (but not by sex or morality). *Evolution and Human Behavior*, **33**, 471–478.

Ridley, M. (1993). *The Red Queen: Sex and the Evolution of Human Nature*. New York: Macmillan Publishing Company.

Rougeron, V., De Meeus, T., Hide, M., et al. (2009). Extreme inbreeding in *Leishmania braziliensis*. *Proceedings of the National Academy of Sciences*, **106**, 10224–10229.

Ryan, S., Oaten, M., Stevenson, R. J., & Case, T. I. (2012). Facial disfigurement is treated like an infectious disease. *Evolution and Human Behavior*, **33**, 639–646.

Schaller, M. (2006). Parasites, behavioral defenses, and the social psychological mechanisms through which cultures are evoked. *Psychological Inquiry*, **17**, 96–101.

Schaller, M., & Duncan, L. (2007). The behavioral immune system: Its evolution and social psychological implications. In J. P. Forges, M. G. Haselton, & W. Von Hippel, eds., *Evolution and the Social Mind: Evolutionary Psychology and Social Cognition*. New York: Psychology Press, pp. 293–307.

Schaller, M., & Murray, D. (2008). Pathogens, personality, and culture: Disease prevalence predicts worldwide variability in sociosexuality, extraversion, and openness to experience. *Journal of Personality and Social Psychology*, **95**, 212–221.

Schaller, M., & Neuberg, S. L. (2008). Intergroup prejudices and intergroup conflicts. In C. Crawford & D. L. Krebs, eds., *Foundations of Evolutionary Psychology*. New York: Erlbaum, pp. 399–412.

Schaller, M., Miller, G. E., Gervais, W. M., Yager, S., & Chen, E. (2010). Mere visual perception of other people's disease symptoms facilitates a more aggressive immune response. *Psychological Science*, **21**, 649–652.

Schnall, S., Haidt, J., Clore, G.L., & Jordan, A. H. (2008). Disgust as embodied moral judgment. *Personality and Social Psychology Bulletin*, **34**, 1096–1109.

Sherman, P. W., & Billing, J. (1999). Darwinian gastronomy: Why we use spices. *BioScience*, **49**, 453–463.

Smith, K. F., Sax, D. F., Gaines, S. D., Guernier, V., & Guégan, J. F. (2007). Globalization of human infectious disease. *Ecology*, **88**, 1903–1910.

Stevenson, R. J., Case, T. I., & Oaten, M. J. (2009). Frequency and recency of infection and their relationship with disgust and contamination sensitivity. *Evolution and Human Behavior*, **30**, 363–368.

Stevenson, R. J., Hodgson, D., Oaten, M. J., Barouei, J., & Case, T. I. (2011). The effect of disgust on oral immune function. *Psychophysiology*, **48**, 900–907.

Sugiyama, L. S. (2004). Illness, injury, and disability among Shiwiar forager–horticulturalists: Implications of human life history. *American Journal of Physical Anthropology*, **123**, 371–389.

Sugiyama, L. S., & Sugiyama, M. S. (2003). Social roles, prestige, and health risk: Social niche specialization as a risk-buffering strategy. *Human Nature*, **14**, 165–190.

Terrizzi, J. A., Jr., Shook, N. J., & Ventis, W. L. (2010). Disgust: A predictor of social conservatism and prejudicial attitudes toward homosexuals. *Personality and Individual Differences*, **49**, 587–592.

Terrizzi, J. A., Jr., Shook, N. J., & Ventis, W. L. (2012). Religious conservatism: an evolutionarily evoked disease-avoidance strategy. *Religion, Brain and Behavior*, **2**, 105–120.

Terrizzi, J. A., Jr., Shook, N. J., & McDaniel, M. A. (2013). The behavioral immune system and social conservatism: A meta-analysis. *Evolution and Human Behavior*, **34**, 99–108.

Thompson, J. N. (2005). *The Geographic Mosaic of Coevolution*. Chicago, IL: University of Chicago Press.

Thornhill, R., & Fincher, C. L. (2011). Parasite stress promotes homicide and child maltreatment. *Philosophical Transactions of the Royal Society: Biological Sciences*, **366**, 3466–3477.

Thornhill, R., & Fincher, C. L. (2013a). Commentary on Hackman, J., & Hruschka, D. (2013). Fast life histories, not pathogens, account for state-level variation in homicide, child maltreatment, and family ties in the U.S. *Evolution and Human Behavior*, **34**, 314–315.

Thornhill, R., & Fincher, C. L. (2013b). The parasite-driven-wedge model of parapatric speciation. *Journal of Zoology*, **291**, 23–33.

Thornhill, R., & Fincher, C. L. (2014). *The Parasite-Stress Theory of Values and Sociality: Infectious Disease, History and Human Values Worldwide*. New York: Springer.

Thornhill, R., & Gangestad, S. W. (2008). *The Evolutionary Biology of Human Female Sexuality*. New York: Oxford University Press.

Thornhill, R., Fincher, C. L., & Aran, D. (2009). Parasites, democratization, and the liberalization of values across contemporary countries. *Biological Reviews*, **84**, 113–131.

Thornhill, R., Fincher, C. L., Murray, D. R., & Schaller, M. (2010). Zoonotic and non-zoonotic diseases in relation to human personality and societal values: Support for the parasite-stress model. *Evolutionary Psychology*, **8**, 151–169.

Tinsley, M. C., Blanford, S., & Jiggins, F. M. (2006). Genetic variation in *Drosophila melanogaster* pathogen susceptibility. *Parasitology*, **132**, 767–773.

Van Valen, L. (1973). A new evolutionary law. *Evolutionary Theory*, **1**, 1–30.

Volk, A. A., & Atkinson, J. A. (2013). Infant and child death in the human environment of evolutionary adaptation. *Evolution and Human Behavior*, **34**, 182–192.

Way, B. M., & Lieberman, M. D. (2010). Is there a genetic contribution to cultural differences? Collectivism, individualism and genetic markers of social sensitivity. *Social Cognitive and Affective Neuroscience*, **5**, 203–211.

Welling, L. L. M., Conway, C. A., DeBruine, L. M., & Jones, B. C. (2007). Perceived vulnerability to disease is positively related to the strength of preferences for apparent health in faces. *Journal of Evolutionary Psychology*, **5**, 131–139.

Wolfe, N. D., Dunavan, C. P., & Diamond, J. (2007). Origins of major human infectious diseases. *Nature*, **447**, 279–283.

Wu, B., & Chang, L. (2012). The social impact of pathogen threat: How disease salience influences conformity. *Personality and Individual Differences*, **53**, 50–54.

Young, S. G., Savvo, D. F., & Hugenberg, K. (2011). Vulnerability to disease is associated with a domain-specific preference for symmetrical faces relative to symmetrical non-face stimuli. *European Journal of Social Psychology*, **41**, 558–563.

15 The Evolution of Pride and Shame

ALEC T. BEALL AND JESSICA L. TRACY

What drives us to climb that highest mountain? And what do we subsequently feel when we realize we are extremely bad at mountain climbing and decide to fake an injury to get helicoptered back to base camp? The emotions that shape these events and our responses to them – pride and shame – play a central role in motivating and regulating many of people's thoughts, feelings, and behaviors (Tangney & Tracy, 2012). These self-conscious emotions drive people to work hard to succeed (Stipek, 1995; Weiner, 1985) and to behave in moral and pro-social ways in their relationships (Baumeister, Stillwell, & Heatherton, 1994; Leith & Baumeister, 1998; Retzinger, 1987). Yet despite their centrality to psychological functioning, pride and shame did not receive the same attention from early emotion researchers as the so-called basic emotions, such as joy, fear, and sadness (Campos, 1995; Fischer & Tangney, 1995).

There are both theoretical and methodological reasons for the relatively later arrival of research on self-conscious emotions. In the emotions literature, researchers have historically focused on studying those emotions that are biologically based, shared with other animals, pan-culturally experienced, and identifiable via discrete, universally recognized facial expressions – in other words, emotions that can be measured without reliance on verbal reports of internal experience (e.g., Davidson, 2001; Ekman, Levenson, & Friesen, 1983; LeDoux, 1996; Panksepp, 1998). These emotions have been labeled "basic" because of their presumed evolved origins and location (in most cases) at the basic level in a hierarchical classification of emotion terms (Johnson-Laird & Oatley, 1989; Shaver et al., 1987). Until the past decade, only a small subset of the vast number of emotions represented in the natural language – anger, fear, disgust, sadness, happiness, and surprise – were considered "basic" (Ekman, 1992; Izard, 1971). However, in recent years, a growing body of research has accumulated to suggest that pride and shame also meet at least the evolutionary-based criterion, and should therefore be considered "basic."

In this chapter, we review findings from many of these studies and highlight their central implication: pride and shame are likely to be adaptive parts of human nature. First, we review evidence supporting the claim that pride and shame are biologically innate, including studies demonstrating that both emotions are associated with nonverbal expressions that are reliably recognized and displayed by individuals across a wide range of cultures, and without visual learning (e.g., Tracy & Matsumoto, 2008; Tracy & Robins, 2008a). Next, we focus on the functionality of pride and shame, in terms of both their subjective experiences and their nonverbal displays. We argue that the subjective feelings associated with pride and shame aided human survival and reproductive success by facilitating behaviors oriented toward status-related goal pursuits. We further suggest that the prototypical nonverbal expressions associated with these emotions likely evolved from earlier nonhuman displays of dominance and submission, which became ritualized into the recognizable expressions that correspond to these emotions in humans today. Finally, we close with several proposed directions for future research. Our overarching goals for this chapter are to lay a foundation for continued programmatic research on pride and shame and to persuade readers that these emotions serve important and essential social functions, and thereby increase humans' adaptive fitness.

15.1 THE EVOLUTION OF PRIDE AND SHAME

All human societies reveal status differences among individuals that influence patterns of conflict, resource allocation, and mating (Fried, 1967), and often facilitate coordination on group tasks (Bales, 1950; Berger, Rosenholtz, & Zelditch, 1980; Ellis, 1995). Even the most egalitarian of human foragers show such status differences, despite the frequent presence of social norms that partially suppress them (Boehm, 1993; Lee, 1979; see also Henrich & Gil-White, 2001). High-status individuals tend to hold disproportionate influence, such that social status can be defined as the degree of influence one possesses over resource allocations, conflicts, and group decisions (Berger et al., 1980). In contrast, low-status individuals often passively give up these benefits, deferring to higher-status group members. As a result, high status tends to promote greater fitness than low status, and a large body of evidence attests to a strong relation between social rank and fitness or well-being (e.g., Barkow, 1975; Cowlishaw & Dunbar, 1991; Hill, 1984).

In evolutionary accounts, emotions are fitness-maximizing affective mechanisms that galvanize coordinated suites of physiological and psychological responses to recurrent events of evolutionary significance (Beall & Tracy, 2017; Cosmides & Tooby, 2000; Nesse & Ellsworth, 2009). The costs associated with being low status imposed large selective pressures on humans throughout history and led to the evolution of specialized affective mechanisms geared toward helping individuals successfully compete with other group members for status and signal their own (self-perceived) relative status. Functionalist accounts of emotions suggest that pride and shame are adaptive by virtue of helping individuals attain or maintain social status and acceptance in the eyes of their social group (Tracy & Robins, 2004a).

Pride may be a major part of the suite of affective mechanisms that motivate status-seeking efforts by supplying psychological rewards and recalibrating psychological systems to sustain attained status and providing the affective substrate for signaling (via pride displays) status achievements or self-perceived status increases (Tracy, Shariff, & Cheng, 2010). In contrast, feeling shame may function to inform the individual of a discrepancy between his or her current and desired state (Leary et al., 1995; Simon, 1967) or that the individual is not making sufficient progress toward his or her goals (Carver & Scheier, 1990; Larsen, 2000). These perceived discrepancies lead to an aversive emotional state (e.g., shame), which in turn provides a motivational push for individuals to halt their unsuccessful status-seeking strategies and minimize the potential negative consequences of their lowered status. In addition, the desire to avoid shame's unpleasant feelings may provide an anticipatory incentive to avoid engaging in future status-lowering actions (Fessler, 2007). Together, pride and shame may therefore represent psychological adaptations that guide the selection of strategies (including cognitions, subjective feelings, and behaviors) from an organism's repertoire, and thereby facilitate the acquiring, sustaining, and (through nonverbal displays) signaling of social status (Tracy et al., 2010).

Questions about the phylogenetic history of a particular human characteristic often beg speculation, given the difficulty of empirically tracing the path of evolution. Nonetheless, the question of whether a given faculty of the mind evolved to its present form in humans is an important one. For pride and shame, it is unclear whether their subjective feeling experiences are present, in any form, in nonhuman animals. Given that self-conscious emotions require complex self-evaluative processes, pride and shame may exist in a more primitive (less cognitively complex) form in other great apes and other animals in whom evidence of self-awareness (e.g., mirror self-recognition) has been documented (Hart & Karmel, 1996), but these emotions are unlikely to be experienced by animals that do not self-reflect or hold stable self-representations.

One of the most prominent gold-standard criteria used to determine whether a particular emotion is likely to be evolved (or "basic") is whether it has a distinct, cross-culturally recognized nonverbal expression (e.g., Ekman & Cordaro, 2011; Tracy & Randles, 2011). Although pride and shame were not included in the pantheon of emotions originally thought to meet this criterion (e.g., Ekman, 1992), studies conducted over the past decade demonstrate (quite conclusively in the case of pride) that both of these emotions are associated with reliably recognized nonverbal expressions that are spontaneously displayed in pride- and shame-eliciting situations.

15.1.1 The Pride Expression Is Universally Recognized and Displayed

The prototypical pride expression consists of an expanded and upright posture, head tilted slightly upward (about 20°), a small smile, and arms either akimbo with hands on the hips or raised above the head with hands in fists (Tracy & Robins, 2004b; 2007) (Figure 15.1).

Pride recognition rates in educated North American and Western European samples typically range from 80 to 90 percent, comparable to rates found for basic emotions such as happiness and sadness. Pride recognition tends to be somewhat lower in non-Western cultures, as is the case for all emotions originally documented in Western nations (Elfenbein & Ambady, 2002), but cross-cultural work has demonstrated reliable recognition of pride in two highly isolated, traditional small-scale societies in disparate parts of the world: Burkina Faso and Fiji (Tracy & Robins, 2008b; Tracy et al., 2013). Given that individuals in these samples are unlikely to have learned about the pride expression from cross-cultural transmission (in both groups, participants had never left the local community and had no access to any kind of media such as magazines, films, or television from beyond their community), these findings suggest that pride recognition is likely to be universal. Also like basic emotions, pride displays can be distinguished from other emotions quickly and efficiently from a single snapshot image (Tracy & Robins, 2008a), and children acquire the ability to recognize pride at the same age (four years old) at which they demonstrate accurate recognition (i.e., verbal labeling) of most other expressions (Tracy, Robins, & Lagattuta, 2005).

A number of authors have noted that the pride expression differs from other highly recognizable emotion expressions in that accurate recognition requires bodily and head movements as well as facial muscle movements (Tracy & Robins, 2004b). However, several researchers

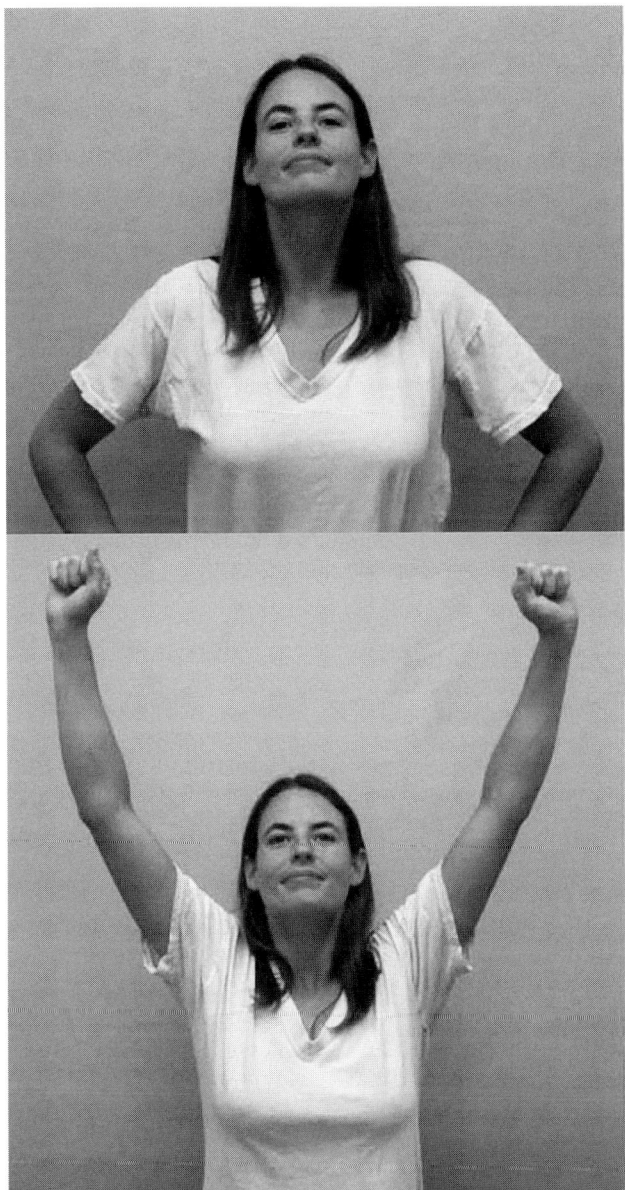

Figure 15.1 The prototypical pride expression.

that vocal bursts of achievement were fairly reliably identified as "achievement." Achievement recognition rates (M = 71 percent) were slightly lower than those typically found for visually observed pride nonverbal displays (i.e., M = 80 percent), but higher than those found for vocal bursts intended to convey contentment (M = 28 percent), relief (M = 57 percent), and pleasure (M = 37 percent; Sauter & Scott, 2007). Participants in this study were not given the option to identify vocal bursts of achievement as "pride," and in general, research on vocal expressions of emotion is still somewhat in its infancy, so further work is needed to determine whether feelings of pride – rather than merely the eliciting event of achievement – can be reliably conveyed through this medium.

In addition to substantial evidence indicating widespread reliable recognition for a visually observed pride expression, there is also evidence to suggest that this expression is spontaneously *displayed* in pride-eliciting situations of success. Behaviors such as head tilted upward, erect posture, and arms stretched upward and out from the body are reliably displayed by preschool children who have won a fight (Strayer & Strayer, 1976), high school students who have performed well on a class exam (Weisfeld & Beresford, 1982), and children as young as three years old in response to task success (Belsky, Domitrovich, & Crnic, 1997; Lewis, Alessandri, & Sullivan, 1992; Stipek, Recchia, & McClintic, 1992).

Moreover, this tendency to display pride in response to success generalizes across a wide range of cultures; the expression was found to be spontaneously displayed in response to victory at the 2012 Olympic Games by judo athletes from a wide range of nations, including individualistic countries such as Canada and Estonia, collectivistic countries such as China and Iran, countries with "secular–rational" values such as Belgium and Finland, and countries with more "traditional–religious" values such as Ireland and Poland (Tracy & Matsumoto, 2008). In all cases, individuals were substantially more likely to display pride if they won a judo match than if they lost, and winners tended to display all behaviors associated with the prototypical pride expression. Perhaps most important, these findings were replicated in an international sample of congenitally blind athletes participating in the Paralympic Games, who could not have learned to display pride through visual modeling (Tracy & Matsumoto, 2008).

15.1.2 The Shame Expression Is Universally Recognized and Spontaneously Displayed

The shame expression consists of essentially the opposite set of behaviors as pride: head tilted downward and lowered eye gaze, along with slumped posture (Izard, 1971; Keltner, 1995; Tracy & Matsumoto, 2008; Tracy, Robins, & Schriber, 2009) (Figure 15.2).

Like pride, shame is reliably recognized and, at least in educated Western populations, distinguished from similar

probing this distinction found that pride can be recognized at fairly high levels of accuracy from the face and head alone (i.e., without expanded posture) if shown as a dynamic display (i.e., via video; Nelson & Russell, 2011). This finding suggests that although static images of pride require expanded posture for accurate recognition, the observation of a head tilting upward removes this need, so in everyday interpersonal interactions pride displays are likely to be reliably recognized even when bodily movements (beyond the head) are not visible.

Studies of vocal displays of emotion have sought to identify a distinct pride vocal expression, but have produced somewhat mixed results. While one set of studies failed to find a recognizable vocal "burst" associated with pride (Simon-Thomas et al., 2009), another study found

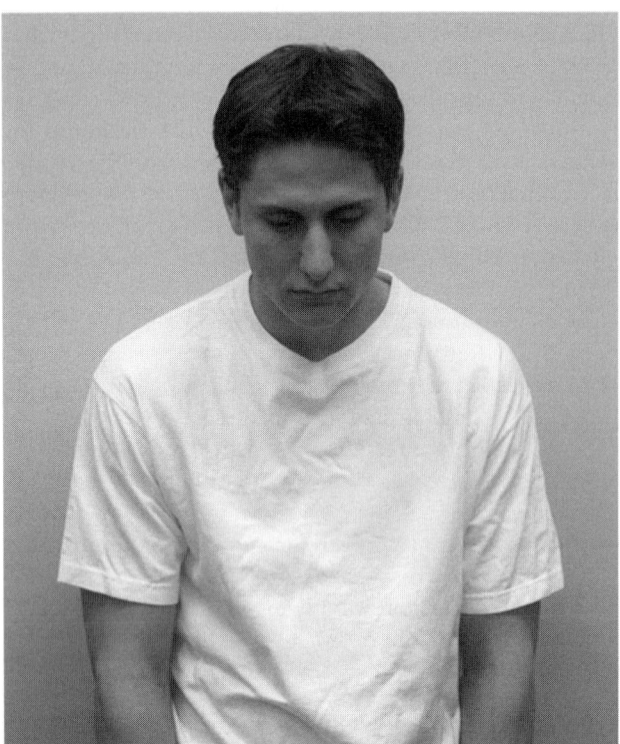

Figure 15.2 The prototypical shame expression.

emotions such as embarrassment and sadness (both of which share features with shame; Babcock & Sabini, 1990; Keltner, 1995). The discrimination of shame from most other emotions can also occur rapidly and efficiently (Tracy & Robins, 2008a). Shame recognition rates in North American samples are typically lower than those observed for pride, but, at 57 percent on average, they are not substantially lower than rates often found for certain basic emotions, such as fear (Haidt & Keltner, 1999; Keltner, 1995; Tracy et al., 2009). Furthermore, most of these studies included the eye gaze and head tilted downward components of shame only, and one study found that shame recognition rates become slightly, though not significantly, improved when the display includes slumped posture in addition to the downward head tilt. It is possible that future studies using a wider range of targets and judges will find additional improvements in shame recognition when such bodily features are added (Tracy et al., 2009). Shame recognition rates are considerably lower in non-Western cultures, but recognition was significantly greater than chance in the same two traditional small-scale societies where pride recognition was examined (in Burkina Faso and Fiji), providing at least initial evidence for the universality of shame displays (Tracy & Robins, 2008b; Tracy et al., 2012).

The recognizable shame expression is also spontaneously displayed in shame-eliciting situations of failure. Shame behaviors such as head tilted downward and slumped posture or narrowed shoulders have been documented in response to failure or loss of a fight in children

as young as 2.5–3.0 years old (Belsky et al., 1997; Lewis et al., 1992; Stipek et al., 1992), older children aged 3–10 (Ginsburg, 1980; Strayer & Strayer, 1976), high school students (Weisfeld & Beresford, 1982), and adult Olympic athletes from a wide range of countries (Tracy & Matsumoto, 2008). In that last study on Olympic athletes' responses to success and failure, athletes were found to reliably narrow their chests and slump their shoulders in response to defeat, but only if they were from countries *outside* of North America and Western Europe. This cultural difference – an absence of failure-based shame displays by individuals from the most individualistic and self-expression -valuing Western nations – is consistent with the strong devaluation of shame in these cultures (Tangney & Dearing, 2002; Tangney & Tracy, 2012). The finding that congenitally blind athletes across cultures, including several from Western nations, did reliably display shame in response to loss at the Paralympic Games suggests that the observed cultural difference in sighted athletes is likely to be a result of display suppression – sighted individuals from cultures where shame is devalued inhibiting the expression in accordance with local cultural norms (Tracy & Matsumoto, 2008).

Together, the accumulated evidence suggests that the pride and shame expressions are both likely to be universal and innate behavioral responses to success or failure. It is unlikely that these expressions would be recognized so consistently and robustly by individuals who could not have learned them through cross-cultural transmission, and be reliably and spontaneously displayed in pride- and shame-eliciting situations by individuals who have never seen others display them, if they were not innate human universals.

15.2 THE EVOLUTIONARY FUNCTION OF PRIDE

15.2.1 The Evolutionary Function of the Pride Subjective Experience

Pride feelings are reinforcing; there is no other emotion that not only makes individuals feel good, but feel good about *themselves*. Pride arises when individuals experience success in achievement or social contexts, and several lines of research are consistent with the suggestion that pride feelings are associated with the attainment of high status (Tracy, 2016). To take a few examples, individuals intuitively assume that people who feel pride hold high status (Tiedens, Ellsworth, & Moskowitz, 2000), agentic individuals (i.e., those who typically seek and possess power and control) tend to feel greater pride than those low in agency (Anderson & Berdahl, 2002), and individuals induced to feel pride tend to engage in high-status behaviors and be perceived by others as influential (Williams & DeSteno, 2009). In addition, pride experiences and the desire to attain these experiences motivate achievement, perseverance, and positive behavioral change in status-relevant domains (Verbeke, Belschak, & Bagozzi, 2004;

Weidman, Tracy, & Elliot, 2016; Williams & DeSteno, 2008). Consequent achievements – which some studies have traced directly back to pride (Weidman et al., 2016) – are, in turn, rewarded with social approval, acceptance, and high status.

Though it seems clear that pride functions to promote status, this conclusion is complicated by the numerous studies suggesting that pride is not a singular experience. Instead, it is best characterized as consisting of two distinct facets: a hubristic facet, marked by arrogance and conceit, and an authentic facet, fueled by feelings of accomplishment, confidence, and success (Tracy & Robins, 2007a). These two facets are conceptualized and experienced as distinct and independent and are associated with highly divergent personality profiles (Tracy & Robins, 2007a, 2014; Tracy et al., 2009). Hubristic pride is the more antisocial facet, associated with disagreeableness, aggression, and a lack of conscientiousness, as well as narcissism, problematic relationships, and poor mental health outcomes. In contrast, authentic pride is the more pro-social, achievement-oriented facet, associated with extraversion, agreeableness, conscientiousness, high self-esteem, satisfying interpersonal relationships, and positive mental health (Tracy et al., 2009). Cross-cultural studies have found evidence for the presence of both pride facets, along with their distinctive correlates, in two non-Western populations – China and South Korea – raising the possibility that both facets may be universal. Given the notably negative personality correlates of hubristic pride, it is not immediately evident why this facet would have evolved; however, there is evidence to suggest that both facets might effectively promote status, but along different avenues. While authentic pride may be ideally suited to promoting a *prestige*-based status, hubristic pride may also be functional by virtue of promoting *dominance*.

Henrich and Gil-White (2001) originally distinguished between dominance and prestige in their ethnographic review of rank attainment in small-scale traditional societies. They used the term "dominance" to refer to the attainment of rank via induced fear, typically through practices of intimidation and coercion. Prestige, in contrast, refers to social rank that is willingly granted to individuals who are recognized and respected for their skills, success, or knowledge. Experimental studies suggest that both dominance and prestige are effective means of attaining rank; individuals who wield both strategies successfully attain influence over their group (Cheng et al., 2013). Hubristic and authentic pride may therefore have separately evolved as the affective mechanisms that, respectively, underpin each of these systems (Cheng, Tracy, & Henrich, 2010; Tracy et al., 2010; Shariff et al., 2010).

More specifically, hubristic pride may facilitate the attainment of dominance by motivating individuals to behave in an aggressive and intimidating manner and providing them with a sense of grandiosity and entitlement that allows them to take power rather than earn it and to feel little empathy for those who get in the way.

Indeed, when individuals experience hubristic pride, they evaluate themselves as superior to others and experience a subjective sense of dominance and superiority and low empathy toward those who are different from them (Ashton-James & Tracy, 2012; Tracy et al., 2009).

In contrast, authentic pride may facilitate the attainment of prestige by motivating and reinforcing achievements and other indicators of competence and providing individuals with the feelings of genuine self-confidence that allow them to comfortably demonstrate both social attractiveness and generosity (Cheng et al., 2010). In order to retain subordinates' respect, prestigious individuals must avoid succumbing to feelings of power and superiority, and authentic pride may allow these individuals to focus on their achievements while maintaining some sense of humility. Indeed, findings show that authentic pride is associated with helping others or giving them advice, voluntary moral action, empathy toward out-group members, experiences of a pro-social "appreciative" form of humility, and a desire to work hard toward achievements (Ashton-James & Tracy, 2012; Hart & Matsuba, 2007; Tracy & Robins, 2007a; Tracy et al., 2009; Weidman et al., 2016; Weidman, Cheng, & Tracy, 2018).

In addition to these lines of work, several studies provide more direct evidence for the theorized unique associations between each pride facet and each status-attaining strategy (Cheng et al., 2010). First, in a study assessing dispositional levels of authentic and hubristic pride and dominance and prestige, individuals prone to authentic pride were found to rate themselves as highly prestigious, whereas those prone to hubristic pride rated themselves as more dominant. In a second study, this pattern was replicated using peer ratings of dominance and prestige; varsity athletes rated the extent to which fellow team members used each strategy. Individuals high in authentic pride were viewed as prestigious (but not dominant) by their peers, whereas those high in hubristic pride were viewed as dominant (but not prestigious). Follow-up analyses demonstrated that these effects could not be attributed to shared variance in positive affect; when controlling for authentic and hubristic pride, neither peer-rated prestige nor dominance was significantly related to generalized good feelings. These results therefore suggest that although individuals high in prestige are generally happy, likeable, and agreeable (Cheng et al., 2010), the emotion that accounts for their ability to attain high status is not generalized positive affect, but authentic pride. More broadly, these findings are consistent with the suggestion that both facets of pride function to facilitate status attainment, but through distinct mechanisms.

15.2.2 The Evolutionary Function of the Pride Nonverbal Expression

A large body of evidence suggests that, in humans, nonverbal displays of pride send a rapidly and automatically perceived message of high status to other group members

(Shariff & Tracy, 2009). The finding that pride displays lead to automatic perceptions of high status also generalizes to a small-scale traditional society on Fiji's outer islands, suggesting that pride may be a universal status signal (Tracy et al., 2013). Among educated Western samples, pride has been shown to signal high status more strongly than any other emotion expression examined, and the high-status message sent by the pride expression is powerful enough to override contradicting contextual information in shaping implicit judgments of status (Shariff, Tracy, & Markusoff, 2012). Together, these findings suggest that the pride expression may have evolved as a mechanism for informing other group members of self-perceived shifts in social status.

The presumed phylogenetic origins of pride displays are consistent with this account. The pride expression's features of expansive posture and head tilted upward create an overall appearance that is similar to the "inflated display" shown by dominant chimpanzees who have defeated a rival, or the "bluff display" documented in these animals prior to an agonistic encounter (presumably with the goal of intimidating a rival; de Waal, 1989). Other nonhuman dominance displays that are visually reminiscent of the pride expression include the chest-beating intimidation displays observed in mountain gorillas (Schaller, 1963) and the "strutting [and] confident air" that characterizes dominant Catarrhine monkeys (Maslow, 1936). Animals who show these behaviors typically receive high-status benefits such as greater attention and resources (e.g., Deaner, Khera, & Platt, 2005).

If the human pride display evolved from earlier nonhuman dominance displays that likely functioned to indicate a direct threat or power differential, then at some point in our evolutionary history the nonhuman bluff display became a more indirect communicative *signal* of deserved status, and was eventually ritualized into the recognizable pride expression (Eibl-Eisenfeldt, 1989; Shariff & Tracy, 2011). In other words, it is likely that these displays originated in a nonhuman ancestor as a way of intimidating rivals who threatened one's power, or of threatening others' power through the same intimidation. For high-status animals, it would be adaptive to respond to status threats with a quick display that is overtly intimidating, as this display alone could save resources that would otherwise need to be devoted to aggressive acts every time a new individual enters the social group.

The particular components of the pride display, and of nonhuman bluff displays, seem well suited for this function. The expanded posture and outstretched arms in humans and the generalized body expansion, shoulder raising, and fur piloerection in chimps make the animal appear larger, facilitating the assertion of dominance or power, and simultaneously attracting the attention of onlookers. In addition, the potentially "handicapping" open and expanded posture may indicate the sincerity of the display. Zahavi and Zahavi (1997) have argued that the veracity of a behavioral signal is established to conspecifics on the basis of whether it is handicapping – that is, costly to the sender. If individuals display such signals despite inherent risks (e.g., revealing oneself to a predator in the process of alerting others to the danger), onlookers can trust the message's sincerity. Thus, the potentially risky open posture associated with pride and bluff displays may have originated as an honest way of conveying one's dominance, success, or status.

Given how widely and reliably recognized the pride expression is, even among young children and individuals across highly diverse cultures, it is likely that recognizing pride has adaptive benefits for perceivers as well as expressers. In this view, the tendency to display pride in response to status attainment and success may have coevolved with a tendency to recognize the pride displays shown by high-status and successful others and to make functional inferences on that basis (Martens, Tracy, & Shariff, 2012). Specifically, observers may use others' pride displays to quickly and effortlessly determine which group members hold status, and thus are likely to possess knowledge or expertise that should be copied or followed. If this is the case, the ability to rapidly detect and understand the pride expression would benefit observers by biasing their social learning, such that individuals selectively copy or choose to learn from those displaying pride.

Two studies tested this account by examining whether financially motivated observers would choose to copy answers to difficult trivia questions provided by another group member (actually a confederate) if the other individual displayed pride (Martens & Tracy, 2013). Across both studies, participants copied the answers of pride-displaying confederates more frequently (approximately 80 percent of the time) than they copied the answers of confederates displaying neutral, shame, or – importantly – happy expressions. This finding further supports the claim that pride's functionality cannot be attributed to more generalized positive affect. It also suggests that, to the extent that pride displays are a reliable signal of knowledge or expertise, they are likely to be functional not only for those who display them and acquire higher status, but also for those who observe and automatically interpret pride in others.

15.3 THE EVOLUTIONARY FUNCTION OF SHAME

15.3.1 The Evolutionary Function of the Shame Subjective Experience

Shame is one of the most painful emotions to experience (Izard, 1971), probably because it signals a flawed self. Like physical pain, which is aversive but adaptive by virtue of promoting injury avoidance, shame experiences may have evolved as a kind of alarm system, warning individuals that a drop in social rank is imminent and that they should therefore change their behavior or depart from the situation to avoid the consequences (cf. Nesse, 1991).

Consistent with the argument that shame warns individuals of a drop in social rank, shame typically arises when individuals experience failure in achievement or social contexts (Tangney & Tracy, 2012; Tracy & Robins, 2004): when Westerners are asked to recount shame-eliciting events, some list situations involving a public failure in an achievement domain (e.g., losing an athletic competition), but most list situations involving a socially unacceptable transgression (e.g., being caught cheating on an exam). This same pattern has also been noted among members of a small-scale traditional fishing village, suggesting that the propensity to feel shame in response to a deviation from culturally normative behavior may be a human universal (Fessler, 2004). Shame may therefore be functional by virtue of informing experiencers of important social information such as downward shifts in rank (Tracy et al., 2010); the resultant unpleasant feelings provide a motivational push to avoid and minimize the negative consequences associated with their lowered status.

One way shame may function to help experiencers avoid the negative consequences associated with a downward shift in rank is by motivating them to act as trustworthy group members who behave in accordance with social norms (Fessler, 2007). Ancestral humans relied heavily on their social groups for protection from external threats (Lancaster, 1976). Those without strong interpersonal connections would have failed to secure the benefits of shared cultural knowledge and resources; as a result, group membership and belongingness were crucial to survival (e.g., Henrich & Boyd, 1998; Hill & Hurtado, 1989). Fessler (2007) has suggested that individuals behave as trustworthy group members and conform to the standards of their culture, in part, to avoid the painful shame experiences that arise from not behaving in these ways.

Furthermore, when individuals do violate social norms, they risk unpleasant reactions from others (e.g., anger, retaliation), which can be dangerous (Gilbert, 2007). Shame may have evolved in part to motivate the appeasement behaviors that reduce these unpleasant reactions after a social transgression has been committed. Appeasement is essential to the long-term survival of interpersonal relationships and to the maintenance of one's place within a social group (i.e., avoiding social rejection). Appeasing higher-status or more powerful others is a cost-efficient way of dealing with conflict; though it may come at the cost of social status, appeasing a more formidable opponent saves valuable resources that might be lost from fighting him or her (Keltner, Young, & Buswell, 1997). Furthermore, the time and energy saved by submitting and appeasing rather than fighting can be used for other pursuits that enhance fitness, such as resource and mate acquisition and retention (Gangestad & Simpson, 2000).

One study tested whether shame serves to appease by asking participants to read hypothetical scenarios about a fictitious CEO who apologized for a negative ecological incident (i.e., a chemical spill) caused by his company. Participants who learned that the CEO verbally expressed feelings of shame while apologizing were more satisfied with the apology than those who learned that he communicated guilt, or no emotion (Giner-Sorolla et al., 2008). Similarly, Proeve and Howells (2006) found that participants applied weaker penalties to fictitious sex offenders who were described as feeling ashamed than to offenders described as feeling sad and remorseful, or as feeling no emotion.

Indeed, under the assumption that shame can motivate positive behavioral change, governments have historically used public shaming as a punitive means of curbing bad behavior (Jacquet, 2011). Such institutionally sanctioned shaming practices remain common; examples include the statewide issuance of marked license plates for individuals convicted of driving under the influence (DUI) of alcohol (Nussbaum, 2006) and online lists of noncompliant taxpayers (Jacquet, 2011). Supporting the use of these practices, researchers have found that the experience of shame motivates individuals to improve their self-image (de Hooge, Zeelenberg, & Breugelmans, 2011) and that the threat of public shaming promotes greater contributions to a common good (Jacquet et al., 2011).

However, it is not clear that shame actually experienced about a particular wrongdoing promotes positive behavioral change relevant to that domain; in other words, whether the shame an individual feels as result of being forced to drive with a marked DUI license plate in fact reduces that individual's likelihood of future drinking and driving. In fact, despite the evidence suggesting that shame may serve an important adaptive function, past research also suggests that shame is associated with certain problematic consequences. Shame is strongly related to anger, both at the trait level (i.e., those who are dispositionally prone to shame also tend to experience anger) and the state level (i.e., experimental inductions of shame promote anger and blame; Heaven, Ciarrochi, & Leeson, 2009; Tangney, Stuewig, & Mashek, 2007). The anger associated with shame tends to lead to numerous forms of aggression (e.g., physical, verbal, and self-directed; Tangney et al., 1996). In addition, feeling shame tends to inhibit empathy (Leith & Baumeister, 1998) and perspective-taking (Yang, Yang, & Chiou, 2010) and typically promotes attempts to deny, hide, or escape shame-inducing situations (e.g., de Hooge, Zeelenberg, & Breugelmans, 2007). In short, though shame experiences may be evolutionarily adaptive, humans' desire to counteract, suppress, and avoid these painful experiences can lead to a number of social problems.

In a related vein, shame may also be a core emotion underlying the larger social problems and poor health associated with addiction. One study found that the degree to which recovering alcoholics demonstrated behavioral displays of shame while discussing their past drinking significantly and substantially predicted subsequent declines in their physical and mental health, their likelihood of relapsing, and the severity of that relapse. More specifically, newly sober alcoholics' shame displays strongly predicted whether they relapsed up to four months later, and the number of drinks they consumed (Randles & Tracy, 2013).

In sum, although there is good reason to think that shame evolved to serve adaptive functions related to appeasement, more research is needed to demonstrate shame's positive consequences because, at present, of all the emotions that are thought to be endemic to our species, shame appears to come with the most negative and even maladaptive social and psychological consequences (Randles & Tracy, 2015).

15.3.2 The Evolutionary Function of the Shame Nonverbal Expression

Nonverbal expressions of shame are automatically perceived as low status (Shariff & Tracy, 2009), and such perceptions can reduce an individual's fitness in a number of ways (e.g., Barkow, 1975; Cowlishaw & Dunbar, 1991; Leary et al., 1995). However, nonverbal displays of shame may also provide certain benefits to displayers by appeasing onlookers after a social transgression (Keltner & Buswell, 1997). Indeed, the nonverbal expression associated with shame may have evolved as a social signal that functions to inform onlookers of a transgressing individual's awareness that social norms have been violated and his or her respect for those norms. This communication likely increases perceptions of trustworthiness; the transgressor is choosing to openly acknowledge his or her error rather than pretending it did not happen, and thus indicating his or her sincere acknowledgment of, and respect for, the transgressed norm. This is an important message to send after a transgression, as those who break a social rule without communicating an admission of norm violation may be perceived as disrespectful of the group and likely to violate other norms in the future (Gilbert, 2007).

Consistent with this logic, displaying shame after a transgression may indirectly promote fitness by facilitating the formation of cooperative social ties (Barkow, 1989; Baumeister & Leary, 1995; Gilbert, 1997). Further, by openly deferring to higher-status individuals through the display of an expression readily recognized as conveying low status, individuals may effectively gain greater access to powerful and knowledgeable others who might otherwise be threatened by them (i.e., if they were perceived as potentially high-status competitors). Access to high-status prestigious individuals is crucial to social learning, given the important role that expert social models play in the transmission of cultural knowledge (Henrich & Gil-White, 2001).

A growing body of research is consistent with this account of shame displays as appeasement and deference signals. Behaviors associated with the human shame expression have been observed in a number of nonhuman animals during situations of submissive appeasement, suggesting that shame displays may have originated as appeasing submission displays in our nonhuman ancestors. Submissive chimpanzees have been observed to lower their bodies or crouch toward more dominant conspecifics during agonistic encounters (de Waal, 1989; van Hooff, 1973), and similar constricted body postures have been observed in the submissive displays of stump-tailed macaques and hamadryas and yellow baboons (Adams, 1981; Kummer, 1968; Leresche, 1976; Silk, 1987).

Several studies have more directly examined whether the shame display appeases observers after a failure or transgression has been committed. In one, Keltner and colleagues (1997) found that participants were more sympathetic toward hypothetical students who failed a class presentation when those students displayed shame than when they displayed embarrassment (unlike shame displays, embarrassment displays include a smile and face touching; Keltner, 1995). In a subsequent mock trial study, these researchers found that a hypothetical convicted drug dealer was judged as less guilty and given a weaker penalty when displaying shame and embarrassment compared to contempt or a neutral expression. Together, these results suggest that shame displays are more appeasing than an absence of emotional displays or contemptuous displays.

Other studies have examined the adaptive benefits of observing others' displays of shame. Given that shame displays are not as reliably or cross-culturally recognized as other emotions expressions (Haidt & Keltner, 1999; Tracy & Robins, 2008), it is possible that recognizing these displays is, in fact, ultimately less adaptive compared to recognizing displays of pride. However, there are several benefits that would likely accrue to those who can effectively infer shame in others. First, because shame displays communicate an individual's commitment to social norms and trustworthiness, observers may use these displays to quickly decipher which group members would make the best interaction partners. Observers who do so would reap the many benefits associated with cooperating with a trustworthy and committed group member (Fessler, 2007). Consequently, there may be survival-related benefits to effectively observing shame in others, using it to infer their level of commitment to the group, and choosing interaction partners on this basis. It may be for this reason that shame expressions are one of the most sexually attractive emotion expressions shown by both men and women (Tracy & Beall, 2011; but see Beall & Tracy, 2014).

Another benefit likely accrued by observers who reliably identify others' shame displays is conflict avoidance. Just as processes of appeasement reduce conflict for those who display shame, they may also reduce conflict for those who observe it. Even if observers are higher status than the shamed other, and thus likely to win an agonistic encounter, they will still save valuable resources by recognizing that the other is willing to submit and therefore that no fight is necessary.

Finally, reliably recognizing shame displays in others can be vitally beneficial for gaining knowledge about the social hierarchy. For group cooperation to be effective, individuals need to determine who has the power to make decisions; having a clearly delineated status hierarchy aids group coordination (Van Vugt, Hogan, & Kaiser, 2008). Thus, a quick and reliable means for higher-status leaders to identify subordinates who are likely to support their decisions rather than fight them – and follow orders rather than question them – allows for a streamlined decision-making process and a more effectively functioning group. Knowing who subordinates are may also facilitate leaders' ability to keep followers happy – an essential part of maintaining influence (Van Vugt et al., 2008).

15.4 FUTURE DIRECTIONS IN RESEARCH ON THE EVOLUTION OF PRIDE AND SHAME

15.4.1 Are There Gender Differences in Pride and Shame?

In general, most of the major findings discussed in this chapter hold across gender. For the research on emotion expressions, this means that similar patterns of results have emerged for both male and female displayers (as in the spontaneous display of pride by Olympic athletes, across cultures, after a success; e.g., Tracy & Matsumoto, 2008) and perceivers (as in the recognition of pride and shame by individuals across cultures; e.g., Tracy & Robins, 2008). However, exploring gender differences has not been a major focus of the bulk of this work, and for some previous studies, targets and/or perceivers of only one gender were included (e.g., Shariff, Tracy, & Markusoff, 2012). As a result, exploring possible gender differences in pride and shame experiences, expression, and recognition is a fruitful area for future research.

In fact, given gender differences in social status and rank-attainment processes (Henrich & Gil-White, 2001), there is reason to expect gender differences in pride and shame. For example, in certain contemporary cultures, men must constantly track their own status relative to those around them, whereas for women this is less essential (Tracy et al., 2013). In Fijian culture, each time men sit at a meal or socialize, they sit by rank, whereas women are considerably more flexible about status distinctions within their gender group (in fact, women often cram together so high-status men can spread out). As a result, Fijian men must be highly attuned to status differences. This may explain the finding that, among Fijians, gender has a significant effect on implicit associations between pride displays and high status, with women making relatively more errors and responding relatively more slowly than men (Tracy et al., 2013).

Other studies have documented gender differences in the frequency and intensity of pride and shame experiences; for example, men experience greater pride, at least of the hubristic variety (Tracy & Robins, 2007a), whereas women are somewhat more prone to shame – or at least are more willing to acknowledge their shameful experiences (Tangney, 1990). However, it is important to note that the magnitudes of these gender differences were relatively modest, and more work is needed before any strong conclusions can be drawn.

Finally, there is evidence to suggest that gender plays a role in the perceived attractiveness of pride-displaying others. In a series of four studies using different methods, men who displayed pride were found to be highly attractive to women compared to men who gave neutral, shame, or happy displays (male happy displays, in fact, were particularly unattractive). In contrast, women who displayed pride were perceived by male viewers as unattractive compared to women who displayed happy or shame expressions, and generally were less attractive than women who displayed neutral expressions (Tracy & Beall, 2011). These findings are consistent with the social status account of pride and with evolutionary theory, suggesting that women tend to seek male partners who seem to be reliable providers, whereas men place a higher value on a potential mate's youth, health, and apparent receptivity to sexual relationships (e.g., Buss, 2008). Interestingly, these studies also found that shame displays (which, as noted above, communicate commitment to social norms and trustworthiness) generally increased attractiveness for both men and women. However, subsequent research indicated that the results for shame were likely to be an artifact of Western cultural norms, as men displaying shame were found to be less attractive in a non-Western cultural context (i.e., India) and among women who were at highest risk of conception (Beall & Tracy, 2014). These additional results suggest that North American women's tendency to find shame-displaying men attractive may be due to local sociocultural norms, rather than being the result of an evolved mate choice.

15.4.2 What Are the Biological and Physiological Responses Associated with Pride and Shame?

Given that distinct emotions are thought to be critically involved in the orchestration of coordinated suites of biological and physiological responses (e.g., Al-Shawaf et al., 2015; Beall & Tracy, 2017), the subjective experiences of both pride and shame may be associated with measurable biological and physiological reactions that facilitate their respective adaptive goals. For example, some past research suggests that feeling shame is associated with increases in cortisol (Dickerson, Mycek, & Zaldivar, 2008) and pro-inflammatory cytokine activity (Dickerson et al., 2004), which may facilitate adaptive behaviors such as submission and withdrawal during shame-eliciting situations (Dickerson, Gruenewald, & Kemeny, 2009). Feeling pride, on the other hand, may be associated with physiological responses – such as increased testosterone – that facilitate gaining or maintaining prestige and/or dominance over peers. More work is needed that directly investigates the biological and/or physiological markers of pride and shame and how these associated responses may help to facilitate status-seeking and appeasement behaviors, respectively.

In addition to the subjective experiences of distinct emotions, a growing body of research also suggests that the prototypical nonverbal displays of distinct emotions may have adaptive biological functions (Shariff & Tracy, 2011). The fear expression provides a useful illustration: the widened eyes of individuals instructed to pose a fearful facial expression were found to increase the scope of their visual field and the speed of their eye movements, allowing expressers to better identify potentially threatening objects in their periphery (Susskind & Anderson, 2008). This line of reasoning has led some researchers to explore whether the physical configurations of the prototypical pride and shame nonverbal displays may serve adaptive biological and social functions beyond communication. For example, the open and expansive chest associated with the pride display may help to increase lung capacity, which, in turn, could prepare the displayer for success in status-driven agonistic encounters. Conversely, if shame's adaptive function is the reduction of harm by appeasing high-status others, its prototypical components (e.g., shoulders slumped, head tilted downward) may effectively hide bodily targets from potential attack in agonistic encounters (Shariff & Tracy, 2011). Though these hypotheses are somewhat speculative, examining the noncommunicative functions of the physical properties of pride and shame displays is an important direction for future research.

15.4.3 What Is the Evolutionary Function of the Shame Experience?

Although self-conscious emotions have received increased empirical attention over the past few decades, the majority of significant scholarly advances have concerned the evolution and function of pride; continued research is thus needed to further examine shame. One recently proposed sociocultural account might explain the evolutionary persistence of shame despite its seemingly maladaptive consequences. In their "sociodynamic model of emotions," Mesquita and Boiger (2014) suggest that emotions are more likely to emerge in certain sociocultural environments if they positively contribute to social cohesion within those contexts. An implication of this account is that emotions are experienced and displayed more frequently in certain social contexts where they produce better outcomes; in contrast, experiencing and displaying these emotions in social contexts where they are not highly valued could lead to various dysfunctions.

In the USA – a country with a highly individualistic culture that emphasizes independence and maintaining high self-esteem – shame is considered to be an undesirable emotional experience (Boiger et al., 2013); feelings of

shame are typically discouraged among American youth (e.g., Miller et al., 2002) and avoided in American literature (Cohen, 2003). However, in Japan – a country with a culture that emphasizes interdependence and monitoring/overcoming personal shortcomings – shame is valued (Boiger et al., 2013); feelings of shame are encouraged among Japanese youth (Lewis, 1995) and are considered to be important to self-improvement (Heine et al., 1999). Consistent with the sociodynamic model of emotions, shame is more frequently experienced in Japan than in the USA (Kitayama, Mesquita, & Karasawa, 2006). Furthermore, in the cross-cultural study of Olympic athletes' expressions of shame in response to loss mentioned earlier, participants from countries holding similar values to Japan (e.g., collectivism) were more likely to display shame than participants from more individualistic countries like the USA (Tracy & Matsumoto, 2008). This past work implies that shame may not be generally dysfunctional; instead, the dysfunction observed may be an artifact of a Western cultural emphasis on extreme individualism and self-enhancement at the expense of caring about one's group (Mesquita & Karasawa, 2004). That said, however, additional cross-cultural work is needed to further test this account.

15.4.4 What Is the Evolutionary Function of the Shame Nonverbal Expression?

Few studies have directly examined the functionality of shame's nonverbal display and whether it actually serves to appease observers after a transgression has been committed. As noted above, preliminary evidence suggests that shame displays are more appeasing than an absence of emotional displays or contemptuous displays (e.g., Keltner et al., 1997), but more research is needed. For example, it remains unclear whether shame displays in particular lead to reductions in punishment and whether these displays would more powerfully trigger forgiveness than other potentially appeasing displays, such as sadness. It is also unclear whether shame displays serve an appeasing function across cultures. Future studies are needed to manipulate the various appeasing emotion expressions separately, in a range of populations, to determine the extent to which shame uniquely and universally functions to appease.

15.5 CONCLUSION

This chapter reviewed considerable evidence suggesting that pride and shame are likely to be adaptations (i.e., evolved to serve particular functions relevant to enhancing fitness) and that their displays should also be considered evolved (e.g., Ekman, 1992). Research findings suggest that the prototypical pride and shame nonverbal expressions are recognized consistently and robustly by individuals who could not have learned them through cross-cultural transmission (i.e., films, television, magazines) and are reliably and spontaneously displayed in pride- and shame-eliciting situations by individuals who have never seen others display them. Pride and shame are therefore likely to be universal and innate affective responses to success and failure that each serve adaptive functions; pride may aid in reproductive fitness by motivating status-related goal pursuit and shame may aid in reproductive fitness by facilitating appeasement after a status-lowering event. However, there are still many questions that remain unanswered about the functions of these universal emotions, and there is much room for continued programmatic research regarding their evolution.

REFERENCES

Adams, D. B. (1981). Motivational systems of social behavior in male rats and monkeys: Are they homologous? *Aggressive Behavior*, **7**, 5–18.

Adler, N. E., Epel, E. S., Castellazzo, G., & Ickovics, J. R. (2000). Relationship of subjective and objective social status with psychological and physiological functioning: Preliminary data in healthy, white women. *Health Psychology*, **19**, 586–592.

Al-Shawaf, L., Conroy-Beam, D., Asao, K., & Buss, D. M. (2015). Human emotions: An evolutionary psychological perspective. *Emotion Review*, **8**, 173–186.

Anderson, C., & Berdahl, J. L. (2002). The experience of power: Examining the effects of power on approach and inhibition tendencies. *Journal of Personality and Social Psychology*, **83**, 1362–1377.

Ashton-James, C. E., & Tracy, J. L. (2012). Pride and prejudice: How feelings about the self influence judgments of others. *Personality and Social Psychology Bulletin*, **38**, 466–476.

Babcock, M. K., & Sabini, J. (1990). On differentiating embarrassment from shame. *European Journal of Social Psychology*, **20**, 151–169.

Bales, R. F. (1950). *Interaction Process Analysis: A Method for the Study of Small Groups*. Reading, MA: Addison-Wesley.

Barkow, J. H. (1975). Prestige and culture: A biosocial interpretation. *Current Anthropology*, **16**, 553–572.

Barkow, J. H. (1989). *Darwin, Sex, and Status*. Toronto: University of Toronto Press.

Bartlett, M. Y., & DeSteno, D. (2006). Gratitude and prosocial behavior: Helping when it costs you. *Psychological Science*, **17**, 319–325.

Baumeister, R. F., & Leary, M. R. (1995). The need to belong: Desire for interpersonal attachments as a fundamental human motivation. *Psychological Bulletin*, **117**, 497–529.

Baumeister, R. F., Stillwell, A. M., & Heatherton, T. F. (1994). Guilt: An interpersonal approach. *Psychological Bulletin*, **115**, 243–267.

Beall, A. T., & Tracy, J. L. (2014). The puzzling attractiveness of male shame. *Evolutionary Psychology*, **13**, 29–47.

Beall, A. T., & Tracy, J. L. (2017). Emotivational psychology: How distinct emotions facilitate fundamental motives. *Social and Personality Psychology Compass*, **11**, e12303.

Belsky, J., Domitrovich, C. E., & Crnic, K. (1997). Temperament and parenting antecedents of individual differences in 3-year-old pride and shame reactions. *Child Development*, **68**, 456–466.

Berger, J., Rosenholtz, S. J., & Zelditch, M. (1980). Status organizing processes. *Annual Review of Sociology*, **6**, 479–508.

Boehm, C. (1993). Egalitarian society and reverse dominance hierarchy. *Current Anthropology*, **34**, 227–254.

Boiger, M., Mesquita, B., Uchida, Y., & Barrett, L. F. (2013). Condoned or condemned the situational affordance of anger and shame in the United States and Japan. *Personality and Social Psychology Bulletin*, **39**, 540–553.

Buss, D. M. (2008). Human nature and individual differences: Evolution of human personality. In O. P. John, R. W. Robins, & L. A. Pervin, eds., *Handbook of Personality: Theory and Research*, 3rd ed. New York: Guilford Press, pp. 29–60.

Campos, J. J. (1995). Foreword. In J. P. Tangney & K. W. Fischer, eds., *Self-Conscious Emotions: The Psychology of Shame, Guilt, Embarrassment, and Pride*. New York: Guilford Press, pp. ix–xi.

Campos, J. J., Barrett, K. C., Lamb, M. E., Goldsmith, H. H., & Stenberg, C. (1983). Socioemotional development. In M. M. Haith & J. J. Campos, eds., *Handbook of Child Psychology (Vol. II): Infancy and Developmental Psychobiology*. New York: Wiley, pp. 783–915.

Carver, C. S., & Scheier, M. F. (1990). Origins and functions of positive and negative affect: A control-process view. *Psychological Review*, **97**, 19–35.

Carver, C. S., Sinclair, S., & Johnson, S. L. (2010). Authentic and hubristic pride: Differential relations to aspects of goal regulation, affect, and self-control. *Journal of Research in Personality*, **44**, 698–703.

Cheng, J. T., Tracy, J. L., & Henrich, J. (2010). Pride, personality, and the evolutionary foundations of human social status. *Evolution and Human Behavior*, **31**, 334–347.

Cheng, J. T., Tracy, J. L., Foulsham, T., Kingstone, A., & Henrich, J. (2013). Two ways to the top: Evidence that dominance and prestige are distinct yet viable avenues to social rank and influence. *Journal of Personality and Social Psychology*, **104**, 103.

Cohen, D. (2003). The American national conversation about (everything but) shame. *Social Research*, **70**, 1075–1108.

Cosmides, L., & Tooby, J. (2000). Evolutionary psychology and the emotions. *Handbook of Emotions*, **2**, 91–115.

Cowlishaw, G., & Dunbar, R. I. (1991). Dominance rank and mating success in male primates. *Animal Behaviour*, **41**, 1045–1056.

Davidson, R. J. (2001). The neural circuitry of emotion and affective style: Prefrontal cortex and amygdala contributions. *Social Science Information*, **40**, 11–37.

de Hooge, I. E., Zeelenberg, M., & Breugelmans, S. M. (2007). Moral sentiments and cooperation: Differential influences of shame and guilt, *Cognition and Emotion*, **21**, 1025–1042.

de Hooge, I. E., Zeelenberg, M., & Breugelmans, S. M. (2011) A functionalist account of shame-induced behavior. *Cognition and Emotion*, **25**, 939–946.

de Waal, F. (1989). *Peacemaking among Primates*. Cambridge, MA: Harvard University Press.

Deaner, R. O., Khera, A. V., & Platt, M. L. (2005). Monkeys pay per view: Adaptive valuation of social images by rhesus macaques. *Current Biology*, **15**, 543–548.

Dickerson, S. S., Kemeny, M. E., Aziz, N., Kim, K. H., & Fahey, J. L. (2004). Immunological effects of induced shame and guilt. *Psychosomatic Medicine*, **66**, 124–131.

Dickerson, S. S., Mycek, P. J., & Zaldivar, F. (2008). Negative social evaluation – but not mere social presence – elicits cortisol responses in the laboratory. *Health Psychology*, **27**, 116–121.

Dickerson, S. S., Gruenewald, T. L., & Kemeny, M. E. (2009). Psychobiological responses to social self threat: Functional or detrimental? *Self and Identity*, **8**, 270–285.

Eibl-Eisenfeldt, I. (1989). *Human Ethology*. New York: Aldine de Gruyter.

Ekman, P. (1992). An argument for basic emotions. *Cognition and Emotion*, **6**, 169–200.

Ekman, P., & Cordaro, D. (2011). What is meant by calling emotions basic? *Emotion Review*, **3**, 364–370.

Ekman, P., Sorenson, E. R., & Friesen, W. V. (1969). Pan-cultural elements in facial displays of emotion. *Science*, **164**, 86–88.

Ekman, P., Levenson, R. W., & Friesen, W. V. (1983). Autonomic nervous system activity distinguishes among emotions. *Science*, **221**, 1208–1210.

Ekman, P., Friesen, W. V., O'Sullivan, M., et al. (1987). Universals and cultural differences in the judgment of facial expressions of emotion. *Journal of Personality and Social Psychology*, **53**, 712–717.

Elfenbein, H. A., & Ambady, N. (2002). On the universality and cultural specificity of emotion recognition: A meta-analysis. *Psychological Bulletin*, **128**, 203–235.

Ellis, L. (1995). Dominance and reproductive success among non-human animals: A cross-species comparison. *Ethology and Sociobiology*, **16**, 257–333.

Ellsworth, P. C., & Smith, C. A. (1988). Shades of joy: Patterns of appraisal differentiating pleasant emotions. *Cognition and Emotion*, **2**, 301–331.

Fessler, D. M. T. (2004). Shame in two cultures: Implications for evolutionary approaches. *Journal of Cognition and Culture*, **4**, 207–262.

Fessler, D. M. T. (2007). From appeasement to conformity: Evolutionary and cultural perspectives on shame, competition, and cooperation. In J. L. Tracy, R. W. Robins, & J. P. Tangney, eds., *The Self-Conscious Emotions: Theory and Research*. New York: Guilford Press, pp. 174–193.

Fischer, J., Fischer, P., Englich, B., Aydin, N., & Frey, D. (2011). Empower my decisions: The effects of power gestures on confirmatory information processing. *Journal of Experimental Social Psychology*, **47**, 1146–1154.

Fischer, K. W., & Tangney, J. P. (1995). Self-conscious emotions and the affect revolution: Framework and overview. In J. P. Tangney & K. W. Fischer, eds., *Self-Conscious Emotions: The Psychology of Shame, Guilt, Embarrassment, and Pride*. New York: Guilford Press, pp. 3–24.

Fredrickson, B. L. (1998). What good are positive emotions? *Review of General Psychology*, **2**, 300–319.

Fried, M. H. (1967). *The Evolution of Political Society: An Essay in Political Anthropology*. New York: Random House.

Gangestad, S. W., & Simpson, J. A. (2000). The evolution of human mating: Trade-offs and strategic pluralism. *Behavioral and brain sciences*, **23**, 573–587.

Gilbert, P. (1997). The evolution of social attractiveness and its role in shame, humiliation, guilt and therapy. *British Journal of Medical Psychology*, **70**, 113–147.

Gilbert, P. (2007). The evolution of shame as a marker for relationship security: A biopsychosocial approach. In J. L. Tracy, R. W. Robins, & J. P. Tangney, eds., *The Self-Conscious Emotions: Theory and Research*. New York: Guilford Press, pp. 283–309.

Giner-Sorolla, R., Castano, E., Espinosa, P., & Brown, R. (2008). Shame expressions reduce recipient's insult from outgroup reparations. *Journal of Experimental Social Psychology*, **44**, 519–526.

Ginsburg, H. J. (1980). Playground as laboratory: Naturalistic studies of appeasement, altruism, and the omega child. In D. R. Omark, F. F. Strayer, & D. G. Freeman, eds., *Dominance Relations: An Ethological View of Human Conflict and Social Interaction*. New York: Garland, pp. 341–357.

Graham, S., & Weiner, B. (1986). From an attributional theory of emotion to developmental psychology: A round-trip ticket? *Social Cognition*, **4**, 152–179.

Greenwald, A. G., McGhee, D. E., & Schwartz, J. L. K. (1998). Measuring individual differences in implicit cognition: The implicit association test. *Journal of Personality and Social Psychology*, **74**, 1464–1480.

Gruber, J., & Johnson, S. L. (2009). Positive emotional traits and ambitious goals among people at risk for mania: The need for specificity. *International Journal of Cognitive Therapy*, **2**, 179–190.

Gruber, J., Culver, J. L., Johnson, S. L., et al. (2009). Do positive emotions predict symptomatic change in bipolar disorder? *Bipolar Disorders*, **11**, 330–336.

Gruber, J., Oveis, C., Keltner, D., & Johnson, S. L. (2011). A discrete emotions approach to positive emotion disturbance in depression. *Cognition and Emotion*, **25**, 40–52.

Haidt, J., & Keltner, D. (1999). Culture and facial expression: Open-ended methods find more expressions and a gradient of recognition. *Cognition and Emotion*, **13**, 225–266.

Harris, P. L., Olthof, T., Terwogt, M. M., & Hardman, C. E. (1987). Children's knowledge of the situations that provoke emotion. *International Journal of Behavioral Development*, **10**, 319–343.

Hart, D., & Karmel, M. P. (1996). Self-awareness and self-knowledge in humans, apes, and monkeys. In A. E. Russon, K. A. Bard, & S. T. Parker, eds., *Reaching into Thought: The Minds of the Great Apes*. New York: Cambridge University Press, pp. 325–347.

Hart, D., & Matsuba, M. K. (2007). The development of pride and moral life. In J. L. Tracy, R. W. Robins, & J. P. Tangney, eds., *The Self-Conscious Emotions: Theory and Research*. New York: Guilford Press, pp. 114–133.

Harter, S. (1983). Developmental perspective on the self-system. In E. M. Hetherington, ed., P. H. Mussen, series ed., *Handbook of Child Psychology: Vol 4. Socialization, Personality, and Social Development*, 4th ed. New York: Wiley, pp. 275–385.

Heaven, P. C. L., Ciarrochi, J., & Leeson, P. (2009). The longitudinal links between shame and increasing hostility during adolescence. *Personality and Individual Differences*, **47**, 841–844.

Heine, S. J., Lehman, D. R., Markus, H. R., & Kitayama, S. (1999). Is there a universal need for positive self-regard? *Psychological Review*, **106**, 766–794.

Henrich, J., & Boyd, R. (1998). The evolution of conformist transmission and between-group differences. *Evolution and Human Behavior*, **19**, 215–242.

Henrich, J., & Gil-White, F. J. (2001). The evolution of prestige: Freely conferred deference as a mechanism for enhancing the benefits of cultural transmission. *Evolution and Human Behavior*, **22**, 165–196.

Hill, J. (1984). Prestige and reproductive success in man. *Ethology and Sociobiology*, **5**, 77–95.

Hill, K., & Hurtado, A. M. (1989). Hunter–gatherers of the New World. *American Scientist*, **77**, 436–443.

Huang, L., Galinsky, A. D., Gruenfeld, D. H., & Guillory, L. E. (2010). Powerful postures versus powerful roles: Which is the proximate correlate of thought and behavior? *Psychological Science*, **22**, 95–102.

Izard, C. E. (1971). *The Face of Emotion*. East Norwalk, CT: Appleton-Century-Crofts.

Jacquet, J. (2011). Is shame necessary? In M. Brockman, ed., *Future Science: Essays from the Cutting Edge*. New York: Vintage Books, pp. 128–140.

Jacquet, J., Hauert, C., Traulsen, A., & Milinski, M. (2011). Shame and honor drive cooperation. *Biology Letters*, **7**, 899–901.

Johnson-Laird, P. N., & Oatley, K. (1989). The language of emotions: Analysis of a semantic field. *Cognition and Emotion*, **3**, 81–123.

Keltner, D. (1995). Signs of appeasement: Evidence for the distinct displays of embarrassment, amusement, and shame. *Journal of Personality and Social Psychology*, **61**, 441–454.

Keltner, D., & Buswell, B. N. (1997). Embarrassment: Its distinct form and appeasement functions. *Psychological Bulletin*, **122**, 250–270.

Keltner, D., Young, R. C., & Buswell, B. N. (1997). Appeasement in human emotion, social practice, and personality. *Aggressive Behavior*, **23**, 359–374.

Keltner, D., Haidt, J., & Shiota, M. N. (2006). Social functionalism and the evolution of emotions. In M. Schaller, J. A. Simpson, & D. T. Kenrick, eds., *Evolution and Social Psychology*. Madison, CT: Psychosocial Press, pp. 115–142.

Kitayama, S., Mesquita, B., & Karasawa, M. (2006). Cultural affordances and emotional experience: Socially engaging and disengaging emotions in Japan and the United States. *Journal of Personality and Social Psychology*, **91**, 890–903.

Kummer, H. (1968). *Social Organization of Hamadryas Baboons*. Chicago, IL: University of Chicago Press.

Lagattuta, K. H., & Thompson, R. A. (2007). The development of self-conscious emotions: Cognitive processes and social influences. In J. L. Tracy, R. W. Robins, & J. P. Tangney, eds., *The Self-Conscious Emotions: Theory and Research*. New York: Guilford Press, pp. 91–113.

Lancaster, J. B. (1976). *Primate Behavior and the Emergence of Human Culture*. New York: Holt.

Larsen, R. J. (2000). Toward a science of mood regulation. *Psychological Inquiry*, **11**, 129–141.

Leary, M. R., Tambor, E. S., Terdal, S. K., & Downs, D. L. (1995). Self-esteem as an interpersonal monitor: The sociometer hypothesis. *Journal of Personality and Social Psychology*, **68**, 518–530.

LeDoux, J. E. (1996). *The Emotional Brain: The Mysterious Underpinnings of Emotional Life*. New York: Simon & Schuster.

Lee, R. B. (1979). *The !Kung San: Men, Women and Work in a Foraging Society*. Cambridge, UK: Cambridge University Press.

Leith, K. P., & Baumeister, R. F. (1998). Empathy, shame, guilt, and narratives of interpersonal conflicts: Guilt-prone people are better at perspective taking. *Journal of Personality*, **66**, 1–37.

Leresche, L. A. (1976). Dyadic play in Hamadryas baboons. *Behaviour*, **57**, 190–205.

Levenson, R. W. (2011). Basic emotion questions. *Emotion Review*, **3**, 379–386.

Lewis, M. (2000). Self-conscious emotions: Embarrassment, pride, shame, and guilt. In M. Lewis & J. M. Haviland-Jones, eds., *Handbook of Emotions*, 2nd ed. New York: Guilford Press, pp. 623–636.

Lewis, M., Alessandri, S. M., & Sullivan, M. W. (1992). Differences in shame and pride as a function of children's gender and task difficulty. *Child Development*, **63**, 630–638.

Markus, H. R., & Kitayama, S. (1991). Culture and the self: Implications for cognition, emotion, and motivation. *Psychological Review*, **98**, 224–253.

Martens, J. P., & Tracy, J. L. (2013). The emotional origins of a social learning bias does the pride expression cue copying? *Social Psychological and Personality Science*, **4**, 492–499.

Martens, J. P., Tracy, J. L., Cheng, J. T., Parr, L. A., & Price, S. (2010). Do the chimpanzee bluff display and human pride expression share evolutionary origins? Poster presented at the 33rd Society for Personality and Social Psychology Pre-Conference on Evolutionary Psychology, Las Vegas, NV.

Martens, J. P., Tracy, J. L., & Shariff, A. F. (2012). Status signals: Adaptive benefits of displaying and observing the nonverbal expressions of pride and shame. *Cognition and Emotion*, **26**, 390–406.

Maslow, A. H. (1936). The role of dominance in the social and sexual behavior of infra-human primates: IV. The determination of hierarchy in pairs and in a group. *Pedagogical Seminary and Journal of Genetic Psychology*, **49**, 161–198.

Mazur, A. (1983). Hormones, aggression, and dominance in humans. In B. B. Svare, ed., *Hormones and Aggressive Behavior*. New York: Plenum Press, pp. 563–576.

Mazur, A., & Lamb, T. A. (1980). Testosterone, status, and mood in human males. *Hormones and Behavior*, **14**, 236–246.

Mehta, P. H., & Josephs, R. A. (2010). Testosterone and cortisol jointly regulate dominance: Evidence for a dual-hormone hypothesis. *Hormones and Behavior*, **58**, 898–906.

Mesquita, B., & Boiger, M. (2014). Emotions in context: A sociodynamic model of emotions. *Emotion Review*, **6**, 298–302.

Mesquita, B., & Karasawa, M. (2004). Self-conscious emotions as dynamic cultural processes. *Psychological Inquiry*, **15**, 161–166.

Miller, P. J., Wang, S.-H., Sandel, T., & Cho, G. E. (2002). Self-esteem as folk theory: A comparison of European American and Taiwanese mothers' beliefs. *Parenting: Science and Practice*, **2**, 209–239.

Morrison, D., & Gilbert, P. (2001). Social rank, shame and anger in primary and secondary psychopaths. *Journal of Forensic Psychiatry*, **12**, 330–356.

Nelson, N. L., & Russell, J. A. (2011). When dynamic, the head and face alone can express pride. *Emotion*, **11**, 990–993.

Nesse, R. M. (1991). What good is feeling bad? *Sciences*, **31**, 30–37.

Nesse, R. M., & Ellsworth, P. C. (2009). Evolution, emotions, and emotional disorders. *American Psychologist*, **64**, 129–139.

Newman, M. L., Sellers, J. G., & Josephs, R. A. (2005). Testosterone, cognition, and social status. *Hormones and Behavior*, **47**, 205–211.

Nussbaum, M. C. (2006). *Hiding from Humanity: Disgust, Shame, and the Law*. Princeton, NJ: Princeton University Press.

Orth, U., Robins, R. W., & Soto, C. J. (2008). Tracking the trajectory of shame, guilt, and pride across the lifespan. *Journal of Personality and Social Psychology*, **99**, 1061–1071.

Panksepp, J. (1998). *Affective Neuroscience: The Foundations of Human and Animal Emotions*. New York: Oxford University Press.

Panksepp, J., & Watt, D. (2011). What is basic about basic emotions? Lasting lessons from affective neuroscience. *Emotion Review*, **3**, 387–396.

Proeve, M. J., & Howells, K. (2006). Effects of remorse and shame and criminal justice experience on judgements about a sex offender. *Psychology, Crime and Law*, **12**, 145–161.

Randles, D., & Tracy, J. L. (2013). Nonverbal displays of shame predict relapse and declining health in recovering alcoholics. *Clinical Psychological Science*, **1**, 149–155.

Randles, D., & Tracy, J. L. (2015). Shame. In R. Segal & K. Stuckrad, eds., *Vocabulary for the Study of Religion*. Leiden: Brill, pp. 339–343.

Retzinger, S. M. (1987). Resentment and laughter: Video studies of the shame–rage spiral. In H. B. Lewis, ed., *The Role of Shame in Symptom Formation*. Hillsdale, NJ: Lawrence Erlbaum Associates, pp. 151–181.

Roberts, B. W., Wood, D., & Caspi, A. (2008). The development of personality traits in adulthood. In O. P. John, R. W. Robins, & L. A. Pervin, eds., *Handbook of Personality: Theory and Research*. New York: Guilford Press, pp. 375–398.

Robinson, R., Roberts, W. L., Strayer, J., & Koopman, R. (2007). Empathy and emotional responsiveness in delinquent and non-delinquent adolescents. *Social* Development, **16**, 555–579.

Rose, R. M., Gordon, T. P., & Bernstein, T. S. (1972). Plasma testosterone levels in the male rhesus: Influences of sexual and social stimuli. *Science*, **178**, 643–645.

Rosenberg, E. L., & Ekman, P. (1994). Coherence between expressive and experiential systems in emotion. *Cognition and Emotion*, **8**, 201–229.

Sauter, D. A., & Scott, S. K. (2007). More than one kind of happiness: Can we recognize vocal expressions of different positive states? *Motivation and Emotion*, **31**, 192–199.

Schaller, G. E. (1963). *The Mountain Gorilla: Ecology and Behavior*. Chicago, IL: University of Chicago Press.

Shariff, A. F., & Tracy, J. L. (2009). Knowing who's boss: Implicit perceptions of status from the nonverbal expression of pride. *Emotion*, **9**, 631–639.

Shariff, A. F., & Tracy, J. L. (2011). What are emotion expressions for? *Current Directions in Psychological Science*, **20**, 395–399.

Shariff, A. F., Tracy, J. L., Cheng, J. T., & Henrich, J. (2010). Further thoughts on the evolution of pride's two facets: A response to Clark. *Emotion Review*, **2**, 399–400.

Shariff, A. F., Tracy, J. L., & Markusoff, J. L. (2012). (Implicitly) judging a book by its cover: The power of pride and shame expressions in shaping judgments of social status. *Personality and Social Psychology Bulletin*, **38**, 1178–1193.

Shaver, P., Schwartz, J., Kirson, D., & O'Connor, C. (1987). Emotion knowledge: Further exploration of a prototype approach. *Journal of Personality and Social Psychology*, **52**, 1061–1086.

Silk, J. B. (1987). Correlates of agonistic and competitive interactions in pregnant baboons. *American Journal of Primatology*, **12**, 479–495.

Simon, H. A. (1967). Motivational and emotional controls of cognition. *Psychological Review*, **74**, 29–39.

Simon-Thomas, E. R., Keltner, D. J., Sauter, D., Sinicropi-Yao, L., & Abramson, A. (2009). The voice conveys specific emotions: Evidence from vocal burst displays. *Emotion*, **9**, 838–846.

Stipek, D. (1995). The development of pride and shame in toddlers. In J. P. Tangney & K. W. Fischer, eds., *Self-Conscious Emotions: The Psychology of Shame, Guilt, Embarrassment, and Pride*. New York: Guilford Press, pp. 237–254.

Stipek, D. (1998). Differences between Americans and Chinese in the circumstances evoking pride, shame and guilt. *Journal of Cross-Culture Psychology*, **29**, 616–629.

Stipek, D., Recchia, S., & McClintic, S. (1992). Self-evaluation in young children. *Monographs of the Society for Research in Child Development*, **57**, 100.

Strayer, F. F., & Strayer, J. (1976). An ethological analysis of social agonism and dominance relations among preschool children. *Child Development*, **47**, 980–989.

Susskind, J. M., & Anderson A. K. (2008). Facial expression form and function. *Communicative and Integrative Biology*, **1**, 148–149.

Takahashi, H., Matsuura, M., Koeda, M., et al. (2008). Brain activations during judgments of positive self-conscious emotion and positive basic emotion: Pride and joy. *Cerebral Cortex*, **18**, 898–903.

Takahashi, H., Yahata, N., Koeda, M., et al. (2004). Brain activation associated with evaluative processes of guilt and embarrassment: An fMRI study. *NeuroImage*, **23**, 967–974.

Tangney, J. P. (1990). Assessing individual differences in proneness to shame and guilt: Development of the Self-Conscious Affect and Attribution Inventory. *Journal of Personality and Social Psychology*, **59**, 102–111.

Tangney, J. P., & Dearing, R. (2002). *Shame and Guilt in Interpersonal Relationships*. New York: Guilford Press.

Tangney, J. P., & Fisher, K. W., eds. (1995). *Self-Conscious Emotions: The Psychology of Shame, Guilt, Embarrassment, and Pride*. New York: Guilford Press.

Tangney, J. P., Stuewig, J., & Mashek, D. J. (2007). Moral emotions and moral behavior. *Annual Review of Psychology*, **58**, 345–372.

Tangney, J. P., & Tracy, J. L. (2012). Self-conscious emotions. In M. Leary & J. P. Tangney, eds., *Handbook of Self and Identity*, 2nd ed. New York: Guilford Press, pp. 446–478.

Tangney, J. P., Wagner, P., & Gramzow, R. (1989). *The Test of Self-Conscious Affect*. Fairfax, VA: George Mason University.

Tangney, J. P., Wagner, P. E., Hill-Barlow, D., Marschall, D. E., & Gramzow, R. (1996). Relation of shame and guilt to constructive versus destructive responses to anger across the lifespan. *Journal of Personality and Social Psychology*, **70**, 797–809.

Thompson, R. A. (1989). Causal attributions and children's emotional understanding. In C. Saarni & P. L. Harris, eds., *Children's Understanding of Emotion*. New York: Cambridge University Press, pp. 117–150.

Tiedens, L., Ellsworth, P. C., & Mesquita, B. (2000). Sentimental stereotypes: Emotional expectations for high- and low-status group members. *Personality and Social Psychology Bulletin*, **26**, 560–575.

Tomkins, S. S. (1962). *Affect Imagery Consciousness: The Positive Affects*, Vol. 1. New York: Springer.

Tracy, J. L. (2016). *Take Pride: Why the Deadliest Sin Holds the Secret to Human Success*. Boston, MA: Houghton Mifflin Harcourt.

Tracy, J. L., & Beall, A. (2011). Happy guys finish last: The impact of emotional expressions on sexual attraction. *Emotion*, **11**, 1379–1387.

Tracy, J. L., Cheng, J. T., Robins, R. W., & Trzesniewski, K. H. (2009). Authentic and hubristic pride: The affective core of self-esteem and narcissism. *Self and Identity*, **8**, 196–213.

Tracy, J. L., & Matsumoto, D. (2008). The spontaneous display of pride and shame: Evidence for biologically innate nonverbal displays. *Proceedings of the National Academy of Sciences*, **105**, 11655–11660.

Tracy, J. L., & Prehn, C. (2012). The use of contextual knowledge to differentiate hubristic and authentic pride from a single non-verbal expression. *Cognition and Emotion*, **26**, 14–24.

Tracy, J. L., & Randles, D. (2011). Four models of basic emotions: A review of Ekman and Cordaro, Izard, Levenson, and Paksepp and Watt. *Emotion Review*, **3**, 397–405.

Tracy, J. L., & Robins, R. W. (2004a). Putting the self into self-conscious emotions: A theoretical model. *Psychological Inquiry*, **15**, 103–125.

Tracy, J. L., & Robins, R. W. (2004b). Show your pride: Evidence for a discrete emotion expression. *Psychological Science*, **15**, 194–197.

Tracy, J. L., & Robins, R. W. (2007a). The psychological structure of pride: A tale of two facets. *Journal of Personality and Social Psychology*, **92**, 506–525.

Tracy, J. L., & Robins, R. W. (2007b). Emerging insights into the nature and function of pride. *Current Directions in Psychological Science*, **16**, 147–150.

Tracy, J. L., & Robins, R. W. (2007c). The prototypical pride expression: Development of a nonverbal behavioral coding system. *Emotion*, **7**, 789–801.

Tracy, J. L., & Robins, R. W. (2008a). The nonverbal expression of pride: Evidence for crosscultural recognition. *Journal of Personality and Social Psychology*, **94**, 516–530.

Tracy, J. L., & Robins, R. W. (2008b). The automaticity of emotion recognition. *Emotion*, **7**, 789–801.

Tracy, J. L., Robins, R. W., & Lagattuta, K. H. (2005). Can children recognize the pride expression? *Emotion*, **5**, 251–257.

Tracy, J. L., Robins, R. W., & Schriber, R. A. (2009). Development of a FACS-verified set of basic and self-conscious emotion expressions. *Emotion*, **9**, 554–559.

Tracy, J. L., Shariff, A. F., & Cheng, J. T. (2010). A naturalist's view of pride. *Emotion Review*, **2**, 163–177.

Tracy, J. L., Shariff, A. F., Zhao, W., & Henrich, J. (2013). Cross-cultural evidence that the pride expression is a universal automatic status signal. *Journal of Experimental Psychology: General*, **142**, 163–180.

van Hooff, J. A. R. A. M. (1973). A structural analysis of the social behaviour of a semi-captive group of chimpanzees. In M. von Cranach & I. Vine, eds., *Social Communication and Movement: Studies of Interaction and Expression in Man and Chimpanzee*. New York: Academic Press, pp. 75–162.

Van Vugt, M., Hogan, R., & Kaiser, R. (2008). Leadership, followership, and evolution: Some lessons from the past. *American Psychologist*, **63**, 182–196.

Verbeke, W., Belschak, F., & Bagozzi, R. P. (2004). The adaptive consequences of pride in personal selling. *Journal of the Academy of Marketing Science*, **32**, 386–402.

Watson, D., Clark, L. A., & Tellegen, A. (1988). Development and validation of brief measures of positive and negative affect: The PANAS Scales. *Journal of Personality and Social Psychology*, **54**, 1063–1070.

Weidman, A. C., Cheng, J. T., & Tracy, J. L. (2018). The psychological structure of humility. *Journal of Personality and Social Psychology*, **114**, 153–178.

Weidman, A. C., Tracy, J. L., & Elliot, A. J. (2016). The benefits of following your pride: Authentic pride promotes achievement. *Journal of Personality*, **84**, 607–622.

Weiner, B. (1985). An attributional theory of achievement motivation and emotion. *Psychological Review*, **92**, 548–573.

Weisfeld, G. E., & Beresford, J. M. (1982). Erectness of posture as an indicator of dominance or success in humans. *Motivation and Emotion*, **6**, 113–131.

Williams, L. A., & DeSteno, D. (2008). Pride and perseverance: The motivational role of pride. *Journal of Personality and Social Psychology*, **94**, 1007–1017.

Williams, L. A., & DeSteno, D. (2009). Pride: Adaptive social emotion or seventh sin? *Psychological Science*, **20**, 284–288.

Yang, M. L., Yang, C. C., & Chiou, W. B. (2010). When guilt leads to other orientation and shame leads to egocentric self-focus: effects of differential priming of negative affects on perspective taking. *Social Behavior and Personality*, **38**, 605–614.

Zahavi, A., & Zahavi, A. (1997). *The Handicap Principle*. New York: Oxford University Press.

16 Thinking Outside the Head
Cognitive Ecologies and Evolutionary Psychology

LOUISE BARRETT

Man's nervous system does not merely enable him to acquire culture; it positively demands that he do so if it is going to function at all. Rather than culture acting only to supplement, develop, and extend organically based capacities logically and genetically prior to it, it would seem to be ingredient to those capacities themselves.

– Clifford Geertz (1973, p. 68)

Some of these pathways happen to be located outside the physical individual, others inside; but the characteristics of the system are in no way dependent upon any boundary lines which we may superpose upon the communicational map.

– Gregory Bateson (1960, p. 251)

That may sound like a strange idea. But it is hardly stranger, on reflection, than the commonplace idea that the activity of *brain-meat* realizes all that matters about human cognition.

– Andy Clark (2007, p. 164)

16.1 WHY HAS EVOLUTION NOT REVOLUTIONIZED THE SOCIAL SCIENCES?

Humans have colonized virtually every habitat on the planet and have transformed the Earth's environments to such an extent that there is a strong push to name the current geological epoch the "Anthropocene" (Smith & Zeder, 2013; Steffen et al., 2011). Understanding how humans have achieved such a feat – and how to deal with the unforeseen consequences of these actions – requires a massive, multidisciplinary effort, in which evolutionary-informed analyses of human behavior must surely play a part. Indeed, some researchers claim that an evolutionarily-informed and integrated social science is essential in order to make any headway on this issue (e.g., Mesoudi, 2011).[1] In particular, great claims have been made about the revolutionary promise of evolutionary psychology

(EP): how it will transform and unite the social sciences, providing us with a better, more complete explanation of the human mind (e.g., Tooby & Cosmides, 1992, 2005).

There is a complication here, however: Darwin's theory of natural selection was published over 150 years ago. If evolutionary theory were going to transform psychology, and the social sciences more broadly, surely it would have done so by now. The fact that it has not suggests that evolutionary theory may not be as useful to psychology and the other social sciences as we would like to think – non-evolutionary types may simply have different questions, interests, and explanatory targets; evolutionary ideas simply do no additional work, and thus add little value to their research projects. After all, many biochemists do not specifically test evolutionary theories, despite accepting the theory of evolution and recognizing the power of an evolutionary approach.

An alternative argument – one made more frequently – is that the social sciences are immune to the power of evolutionary thinking because of motivated resistance (Burke, 2014; Jonason & Dane, 2014). Many in the social sciences are alleged either to reject the notion that there is an evolved biological basis to our psychology – preferring instead a view of the human mind purely as a product of culture – or to reject science altogether, viewing it as just another mode of discourse and disputing any claims it might make to authority in explaining the natural world (Barkow, 2006; Ross, 1996). With respect to the latter, there is little one can do to convince the committed poststructuralist; any of the arguments one could put forward to defend a scientific stance are simply grist to their mill. The best we can do is to agree to disagree, and even that concession may stick in the craw of parties on both sides.

It would, however, be foolish to dismiss all of the work that casts a critical eye on science and its methods. There are a number of excellent, thoughtful scholars (some of whom were or are practicing scientists) who tackle issues in the history and sociology of science and provide a useful perspective on science as a human – and hence imperfect – social activity (e.g., Daston & Gallison, 2007; Marks, 2009;

---

[1] Although sympathetic to the idea that evolutionary approaches enrich the human sciences and are needed if we are to offer a complete explanation of human behavior, I am less enamored of the idea that all of the social sciences must be united and integrated under the banner of Darwinism. Rather, I prefer a pluralist stance: one that recognizes that an overt evolutionary framework is not essential to some of the interesting questions we wish to ask about human lifeways.

Pickering, 2010; Shapin, 2010; Shapin & Schaffer, 1985). Moreover, once understood in context, some of the arguments made against applying evolutionary theory to humans hold genuine currency: in anthropology, for example, it is readily apparent that early biological and evolutionary theorizing was both reductionist and racist, and linked strongly to colonialism (Marks, 2009; Sussman, 2014). Ongoing suspicion of biological explanations is therefore not difficult to understand given that, at certain times and places, the biological explanations on offer were indeed suspect.

At the same time, it is apparent that some social scientists have simply failed to register the shifts in evolutionary thinking that have taken place over the last century, often due to a lack of appropriate background (for discussions, see Burke, 2014; Park, 2007). The arguments raised against evolutionary thinking often describe positions that have been superseded both theoretically and empirically, including the idea that there are simple linear relationships between genes and behavior. Such outdated positions allow social scientists to persist with the argument that biologically-based explanations are both heavily reductionist and deterministic (see also Barkow, 2006). But this could not be further from the position now held by the majority of geneticists and biologists. As Fisher (2006, p. 270) puts it, referring expressly to cognition: "Genes do not specify behaviours or cognitive processes; they make regulatory factors, signaling molecules, receptors, enzymes, and so on, that interact in highly complex networks, modulated by environmental influences, in order to build and maintain the brain." In all fairness, however, it should also be mentioned that some evolutionary psychological studies do, in fact, display elements of simplistic evolutionary reasoning (e.g., Jonason et al., 2013). Moreover, its cause often can be traced to the same source – a reliance on outdated thinking, particularly from the 1970s – which West et al. (2011) refer to as the "disco problem." Not all of the criticism leveled at the human evolutionary sciences necessarily misses its mark.

Analyzing the tensions within and between the natural and social sciences suggests we are confronting two problems, one more tractable than the other (for a thoughtful and detailed discussion of these issues and possible solutions, see Barkow, 2006). As already noted, there is no hope of convincing the committed antiscientific poststructuralist, so we can abandon that as a nonstarter. The more tractable problem is whether other social scientists can be persuaded by modern evolutionary thinking. My suggestion – and the focus of this chapter – is that the latter might succeed if evolutionary social scientists engaged more fully with current thinking in philosophy of mind and cognitive science. From this, it may be possible, as Rowlands (2010) suggests, to build a new science of the mind – one that will appeal to both social and natural scientists alike – because it presents a different conception of what minds are.

16.2 REAL REVOLUTION

The real problem, then, is that evolutionary thinking in psychology is just not revolutionary enough. EP, after all, sticks pretty closely to standard cognitivist views of "Individualism, Intellectualism, and Internalism" (Hutto, 2013, p. 175), where cognitive processes are considered to be purely brain-bound individual computational–representational processes and that "the mind's main concern is not acting but thinking, and that paradigmatically thinking is directed at ascertaining truths" (Fodor, 2008, p. 8). In other words, EP is simply business as usual. The advantage of EP is that it provides a means to ground psychology as a discipline (at least in principle) by providing an explanation (or set of explanations) about what our minds are for and why our cognitive architecture takes the form that it does. What EP does not do, however, is question this model of the mind. The same is true of cognitive anthropology (e.g., Sperber, 1996) and gene–culture coevolutionary theory, which also accept a standard cognitivist approach to the mind, with culture often operationally defined as information inside people's head (although, of course, it does deal more explicitly with human culture as a system of inheritance distinct from our biology) (Henrich, 2016; Mesoudi, 2011; Richerson & Boyd, 2005).[2]

16.3 PICTURES OF THE MIND

The picture of the mind to which both EP and traditional cognitive psychology adhere is one that, as Rowlands (1999) puts it, "began life as a controversial philosophical thesis and then evolved into common sense." The philosophical thesis referred to was laid out by Descartes, who proposed that minds can be conceived as objects that possess certain properties, much like other organs of the body. These properties enable minds to carry out a particular function, and the function of the mind is to think. For Descartes, the difference between the mind and our other bodily organs was that the mind was a nonphysical substance. This came about because of Descartes' commitment to a mechanical universe and the atomism of the Scientific Revolution: the physical world was governed by purely mechanical principles, and the behavior of any macroscopic body could be explained in terms of the motions of the atoms of which it was made. In the context of the "body machine," Descartes could conceive of many functions that could be instituted mechanically, specifically "the reception by the external sense organs of light, sounds, smell, tastes, heat and other such qualities … the

---

[2] Human behavioral ecology, meanwhile, has tended to ignore the mind altogether, choosing instead to investigate the extent to which formal models developed to explain nonhuman animal behavior can account for the patterns seen in human populations; again, this is not revolutionary to behavioral ecologists and, for the most part, human behavioral ecology simply fails to address the concerns and interests of non-evolutionary anthropologists and sociologists. The latter are more interested in what people think and why, rather than the functional consequences of their behavior in terms of reproductive outcomes.

imprinting of the idea of these qualities in the organ of the 'common' sense and the imagination, the retention or stamping of these ideas in the memory, the internal movements of the appetites and passions and finally the external movements of all the limbs," but he could not conceive of a way in which rational thought could be explained mechanically: "it is for all practical purposes impossible for a machine to have enough different organs to make it act in all the contingencies of life in the way that reason makes us act"[3] (Descartes, 1637, the *Discourse*: Cottingham et al., 1985, p. 140). If thought could not be mechanized, then, by definition, it could not be physical. Hence, humans possessed a material brain that operated on mechanical principles, but a nonphysical mind that operated according to the principles of reason. This view of the mind as a "ghost in the machine" (Ryle, 1949) is one that has received fierce criticism and universal rejection by psychologists and neuroscientists alike; today, we consider the mind to have a material basis, and no one would admit to being a substance dualist in Descartes' sense. Moreover, the notion that the mind cannot be mechanized has also been roundly rejected with the rise of the computational theory of mind; none more so than within EP, with its emphasis on functionally specialized problem-solving algorithms.

As Rowlands (1999) points out, however, there is another facet to Ryle's expression: the ghostly mind is to be found *in* the machine. Although nonmaterial, Descartes nevertheless assumed that the mind was a wholly internal entity, to be found inside the skin of the organism that possessed it. While substance dualism was rejected, the assumption of internalism was subject to no criticism at all. Thus ignored, internalism has remained intact and unquestioned to this day. Exorcising the ghost from the machine has not, therefore, eliminated all elements of the Cartesian conception of the mind: although the mind itself is now material (and, to some, mechanical), it continues to remain bound securely within the skin and skull, and it is this picture that informs the human evolutionary sciences.

Other pictures are possible, however, and continuous efforts have been made to think differently about the mind over the course of the twentieth century. As Hutchins (2010) points out, early pioneers were the anthropologist, Gregory Bateson, who took the first steps to an "ecology of mind," the experimental psychologist, James Gibson, father of "ecological psychology," and the Russian developmental psychologist, Lev Vygotsky, who was instrumental in the rise of cultural–historical activity theory. John Dewey, the American pragmatist philosopher, should also be included as one of the "ancestors of a modern synthesis of cognitive ecology approaches" (Hutchins, 2010, p. 709), extending its provenance back into the nineteenth century. In each case, organism–environment relations were "seen in terms of coupling,

coordination, emergence and self-organization, rather than the transduction of information across a barrier" (Hutchins, 2010, p. 710). Consequently, the environment was seen not merely as the background against which behavior played out, but as an interface at which essential processes took place. In all cases, however, these ideas and research programs failed to gain sufficient traction to overcome the dominant Cartesian conception.

More recently, there has been a renewed effort to unseat the internalist view and demonstrate that the prevailing picture is not the only way to think about minds. These dissenting views fall under the umbrella term of 4E cognition (i.e., where the Es stand for embodied, embedded, enactive, and extended) (for a review, see Rowlands, 2010). Although there are some important differences between the different schools of thought encompassed by 4E cognition, all share in common the idea that the evolutionary origins of the mind and cognition are to be found in the control of perception and motor action in a dynamic environment. As a consequence, cognition cannot be divorced from its bodily and environmental contexts. Hence, it is a mistake to attribute cognitive processes to brains alone; rather, cognitive processes must be understood as contextual, relational phenomena, reflecting the nested arrangement of brains, bodies, and environments. Cognitive processes therefore spread beyond the brain, into the body, and out into the environment (e.g., Barrett, 2011; Chemero, 2009; Clark, 1997, 2008; Clark & Chalmers, 1998; Rowlands, 1999).

What this means is that, if cognitive processes have an environmental component, then we cannot understand such processes by focusing exclusively on what goes on inside the skin of the individual subject. Instead, they are positions that argue for what van Dijk (2016) calls a "horizontal attitude." By this, he means a strategy that "looks sideways" at the contexts of behavior and its environmental supports, rather than assuming the answer will be found by digging below the surface to identify a unique form of cognitive machinery (the standard "vertical" approach). I want to emphasize this point because, in previous work where my colleagues and I offered a 4E view as an alternative to the evolutionary psychological computational model of the mind (Barrett et al., 2014), critics and commentators argued that we had done no such thing: given that 4E cognition does not rule out the idea that minds and cognitive processes are computational, it does not offer an alternative to the dominant view of the mind within EP. This is accurate, but misses the point. What we were rejecting was not computationalism,[4] but internalism.

---

[3] Descartes, as should be apparent, would not have been a fan of massive modularity.

[4] Although I do, in fact, disagree with many aspects of computational–representational theories of mind and am sympathetic to radical enactivist views (e.g., Hutto & Myin, 2013) that argue that "basic minds" (i.e., pre- or non-linguistic minds) are nonrepresentational. For my purposes here, however, I can remain neutral on the issue: arguing that cognitive processes have a large environmental

A new science of the mind based on externalist principles will, I hope, be more conducive to engagement with the other social sciences; by resituating our understanding of the mental, we will naturally incorporate elements of their subject matter. Anthropology and sociology, in particular, are concerned with social processes as relational phenomena, and do not focus on the individual as a unit of explanation. Rather, the focus is on the relations and networks that help define people and the groups in which they live, the dense webs of meaning we create through social practices, as well as the manufacture and use of artifacts and the formation of various institutions (and other supra-individual structures) that are both created by and shape individual decision-making processes. In addition, the view of human nature that follows from a 4E approach helps undercut the distinction between culture and biology, nature and nurture, and may finally rid us of the ongoing, but sterile, debate regarding their relative importance.

16.4 A SURVEY OF COGNITIVE ECOLOGIES

Rejecting internalism is not, however, an all-or-nothing enterprise. There are a variety of positions one can take, some more conservative than others. Again, there is much debate over these distinctions and the relative plausibility of the various positions, but they all resemble each other in viewing all cognitive processes (and minds) as fundamentally and intimately tied to the manipulation of environmental resources, props, and scaffolds of various kinds, with various feedback, feedforward, and looping effects (Clark, 2008). Here, the environment is not simply a source of input to the cognitive system, but a necessary component.

Given this, a rejection of internalism is also a rejection of the idea that psychology can and should be a conventional laboratory science. Currently, we assume we can explain the mechanisms of psychology irrespective of the environment where the organism is located. Although we need the environment to construct evolutionary and developmental hypotheses concerning our capacities, the capacities themselves are viewed as purely internal. If cognitive processes are not fully internal, however, then standard laboratory setups may fail to capture the phenomenon of interest, or they may mistake behavior scaffolded by the environment as the product of internal processes. This does not mean psychology cannot be an experimental science, but it does mean that any such science must be informed by an understanding of the cultural practice of experimentation itself (Hutchins, 2008).

In what follows, I offer a brief survey of the externalist landscape in order to introduce these ideas to unfamiliar readers and to provide a sense of how thinking differently

about the mind leads to a new conception of how we can study the mind scientifically.

16.4.1 Distributed Cognition

One of the pioneers of what we could call the "second wave" of cognitive ecologists, following the earlier efforts of Bateson and Gibson, is Edwin Hutchins (1995, 2008, 2010). His form of cognitive ecology, which is often known by the term "distributed cognition," places the emphasis on collective activity rather than individual capacities, thus generating cognitive states that are socially distributed (i.e., the group as a whole is viewed as the entity that possesses a certain kind of knowledge or is attributed a particular belief or other mental state). Combined with the use of environmental resources and tools, cognitive states are thus spread over brain, body, and world. On Hutchins' account, cognition is also precisely defined with respect to symbol manipulation (something that is not true in general, where cognition and mind are often used interchangeably, and cognition itself is also defined vaguely), hence cognitive processes are strongly social and cultural in origin, and (seemingly) apply to humans alone. Indeed, one of Hutchins' most interesting points concerning the standard cognitivist, computational view is that cognitive scientists have not actually been modeling cognition itself, but rather the formal symbolic tools and systems that human cognition itself created and has now adapted to – consequently, computational models of the mind are themselves cultural symbolic tools.

Hutchins' (1995) classic work, *Cognition in the Wild*, deals with the art of navigation, providing a detailed analysis of the process in the Western military, along with a fascinating comparison to Micronesian navigation. His ethnographic study of an amphibious helicopter transport vessel revealed that navigation is a feat that no single individual can achieve alone. Instead, the task is accomplished via the layout of the ship, the way tasks are delegated across different individuals, the social hierarchy, and the way information is combined using specialized tools. This generates the knowledge needed to safely guide the vessel to harbor, but it is knowledge that resides in the entire sociocultural system of the ship, with no one individual possessing all of the knowledge and knowhow needed to complete the task.

Another classic study in "distributed cognitive ecologies" is Lynn Tribble's analysis of how Elizabethan actors in the Globe Theatre were able to learn a repertory of half a dozen plays, without full scripts or dedicated rehearsal time, by means of distributing the cognitive tasks involved across other players, the layout and organization of the stage, and the use of various external props and prompts (Tribble, 2011; see also Sutton, 2006). Distributed cognition, then, places most of its emphasis on how cognition is *socially* distributed – a feature I return to in Section 16.4.4. This distinguishes it from a number of the other externalist views, which are more concerned with the individual mind.

component does not require one to take any particular stand on the issue of computationalism.

16.4.2 Enactivism and Extensive Cognition

Possibly the most radical view on offer is the idea that the human mind or cognitive system[5] is unbounded, as John Dewey would put it (Dewey, 1916, 1981; see also Vaesen, 2014), or, as Hutto and Myin (2013) would say, "extensive." This view is most closely associated with the school of thought known as enactivism. This draws on work in dynamical systems theory and views cognitive processes as dynamic relations (or transactions) between organism and the environment. In this view, minds (or cognitive systems) are never confined to the head because, by definition, they are world-involving processes: there simply is no boundary to be drawn. This unboundedness is a source of continuity across the animal kingdom – all organisms exploit bodily and environmental sources in order to generate flexible, adaptive behavior in their particular niches (for a review, see Barrett, 2011) – and also a source of difference between humans and other animals: the sheer number and complexity of the external resources that humans have at their disposal is vast compared to those of other species.

Some in the classic enactivist camp make an even stronger argument, suggesting that any attempt to distinguish cognition from other kinds of processes is essentially meaningless, and they consider all life to be inherently cognitive (Varela et al., 1992). More specifically, Thompson (2007) offers us the position that "life and mind share a set of basic organizational properties, and the organizational properties of mind are an enriched version of those fundamental to life. Mind is life-like, and life is mind-like" (p. 128).

Although there are some potential problems with this argument,[6] from my perspective here it does help to make the broader point that enactivism, and indeed 4E views in general, are explicitly evolutionary, making use of what Rowlands (1999) refers to as the "barking dog principle" (based on the adage: Why buy a dog, and then bark yourself?). Specifically, if the environment already contains structures that are recruited and exploited to solve an organism's problems, then a thrifty evolutionary process will select for this solution, rather than the alternative of evolving expensive brain tissue to recreate these environmental structures as internal brain-based structures (but see Shapiro, 2010). As Brooks (1999) has shown, it often pays dividends simply to "use the world as its own best model."[7]

16.4.3 Extended Minds

An equally radical position is the extended mind hypothesis (Clark & Chalmers, 1998) where, again, the environmental resources we use and exploit in the service of adaptive behavior are argued to literally form part of an individual's cognitive system: humans are, in Andy Clark's (2003) words, "natural-born cyborgs" whose cognitive states are distributed across brain, body, and world.[8] As Clark (2001) puts it: "The intelligent process just *is* the spatially and temporally extended one which zig-zags between brain, body, and world." Importantly, external resources are only considered as parts of the mind when they are coupled in a robust, reliable, and persistent way to internal processes, and they count as part of the mind only during the execution of the relevant task (which is when this crucial form of coupling occurs). Thus, not all and every environmental resource we use forms part of our cognitive system.[9]

The classic example used by Clark and Chalmers (1998) is that of the fictional Otto, who has a mild form of Alzheimer's disease. Otto relies on a notebook containing all of the relevant information he needs to function. Otto wishes to visit the Museum of Modern Art (MoMA) in New York City, consults his notebook, and makes his way to 53rd Street. Clark and Chalmers (1998) argue that Otto undergoes precisely the same process as his friend, Inga, who uses her internal biological memory to retrieve MoMA's address: both act on the belief that it is located on 53rd Street. This being the case, we have no reason, bar a neurocentric prejudice, to treat the notebook as any

---

[5] As noted in the discussion of Hutchins, there is a distinction to be made between minds, mental states, and cognitive systems, but from here on I use the term interchangeably, following the lead of the authors whose work I discuss.

[6] One major problem with this position, as some see it, is that, if everything is cognitive, then nothing is (van Duijn et al., 2006); that is, we are left no means of getting to grips with why a bacterium differs from a baboon or a blowfly from a badger. Lack of space prevents me from addressing this further.

[7] Radical embodied and enactivist views are also distinguished by a rejection of the computational and representational mind that defines modern cognitive psychology (or at least for nonlinguistic or prelinguistic creatures like nonhuman animals and human babies). This position arises directly from a commitment to an unbounded mind, but for present purposes, we do not need to consider why this is in any further detail.

[8] Radical enactivists, however, argue that the extended mind thesis implicitly accepts the idea of an internal cognitive system: if cognitive processes extend out from the brain and body into the world, this suggests there is a fixed boundary to the mind beyond which extension occurs. Some extended mind theorists seem to consider this to be a terminological argument and that nothing of importance hangs on this distinction (e.g., Sutton, 2015). From my perspective here, the two positions are equally radical with respect to the pictures of the mind they offer.

[9] I emphasize this point because the extended mind thesis is the one that has come in for most criticism, with dissenters arguing that tools and other external resources are not literal constituents of our minds, but are only causally related to them – a stance known as the "coupling-constitution fallacy" (Adams & Aizawa, 2001, 2011). Part of the reason for offering such criticism is a fear of "cognitive bloat"; that is, of drawing the bounds too widely so that everything becomes a part of the human cognitive system, resulting in a motley assortment of external props and resources from which it will prove impossible to construct a useful, empirical science. This is a real concern and not one to be taken lightly, but for our purposes here, I do not need to get into the details of these philosophical discussions, not least because there is no consensus on what should count as a "mark of the cognitive" and so define where the boundary is drawn (i.e., what distinguishes between a constitutive vs. a merely causal contribution to cognitive processes). As Hurly (2010) has noted, this could simply be the source of another kind of error: if neither side in the debate can demonstrate conclusively where the bounds of the cognitive system should be drawn, why assume that any such boundary will favor internalism? Given this, many theorists have suggested that the issue is an empirical one to be decided by scientific efforts and not by a priori philosophical theorizing.

different from biological memory; that is, it is possible for a mental state to be externally located.

We can see this even more clearly by raising the possibility that Otto locates the address of MoMA via a chip implanted in his brain. Under these circumstances, the process of retrieval would take place internally – within Otto's brain – and we would have no hesitation including the chip as part of Otto's cognitive system. Why, then, should we think differently about Otto's external "chip," the notebook that he keeps in his pocket, rather than inside his head? Clark and Chalmers (1998) thus make the case that what matters is the role played by the various components of the cognitive system, not what these are made of or where they are located.

As the above example makes clear, Clark and Chalmers (1998) argue for what they called "the parity principle" between the use of external and internal resources (i.e., one directly replaces or stands in for the other), but other theorists have suggested that external resources can be complementary to internal functions in ways that augment and enhance internal processes, a position known as cognitive integration (Menary, 2007, 2010; Sutton, 2010). External resources can do this because they can store information in ways that are inherently more stable than biological processes, and also do not depend on a particular modality or the execution of a particular kind of task. A flash drive (memory stick) is more stable than our biological memory, for example, as the latter is altered by the process of retrieval. Consequently, external resources can also improve on our biological capacities rather than simply substitute for them (Menary, 2007, 2010; Sutton, 2010). In both first- and second-wave extended mind views, there is no explicit stance taken on the nature of internal cognitive processes (these can take an entirely conventional computational–representational form), and the question of interest is where the bounds of the cognitive system should be drawn.

16.4.4 Embedded or Scaffolded Minds

Perhaps the least radical of the 4E positions on offer (although still pretty radical from a standard internalist position) is the idea that minds and cognitive processes are "embedded": certain behaviors and cognitive skills are so strongly scaffolded by environmental resources that the processes in question would simply not be possible without such environmental supports (Rupert, 2009; Sterelny, 2010). In this view, there is an acceptance that cognitive processes are distributed across brain, body, and environment, but a rejection of the idea that one should therefore treat external resources as literal parts of the mind. This is because treating external resources as literal components of our minds/cognitive processes does no unique explanatory work: we can achieve the same insights by viewing cognitive processes as transcranial hybrids – part mental, part nonmental – because it is the transcranial nature of such processes that is crucial, not whether we consider

environmental resources as literal parts of our minds or cognitive processes (for a more detailed consideration of the claims of the extended vs. embedded mind, see Sprevak, 2010).

Sterelny (2010) justifies this position by referring to "extended digestion." Cooking and food preparation are widely acknowledged to have transformed hominin physiology by altering the selection pressures acting on jaws, teeth, and guts and, by extension, brains. We have reduced the demands of eating by chopping and slicing our food, and processes like heating, soaking, pounding, and fermentation have altered the chemical composition of food, improving its digestibility and nutritional value. This, in turn, is argued to have altered selection pressures on our gut and released a constraint on brain size evolution: cooking reduces the demands on digestion, with the result that the energy previously required to fuel expensive gut tissue could be diverted into brain growth. In addition, as Sterelny (2010) points out, we have also improved the nutritional value of our plant food resources via domestication and selective breeding.

The human gustatory niche is thus a heavily engineered one, and our digestion is environmentally scaffolded by all manner of technological props and supports. Indeed, we are now obligatory cooks: it is almost impossible for humans to subsist on a diet of raw food alone (Wrangham, 2009). Now Sterelny (2010) springs his trap, asking whether this means that "my soup pot, my food processor and my fine collection of choppers [are] part of my digestive system? As far as I know, no one has defended an extended stomach hypothesis, treating routine kitchen equipment as part of an agent's digestive system" (p. 468). As Sterelny (2010) goes on to point out, one reason why this might be is simply an accident of our intellectual history; that is, we could argue that digestion is extended in exactly the same way that Clark and others have argued that cognition is extended. But, Sterelny (2010) suggests, once we have acknowledged and appreciated the vital role that environmental scaffolds play in supporting our behavior, there is no further explanatory mileage to be had in supposing that "digestion takes place on my stove as well as in my body."[10] Personally, I have no

---

[10] Paul Thagard has offered a similar criticism in a *Psychology Today* blog post, where he introduces the notion of "extended breath" as a parody of the extended mind. Once again, the hapless Otto is pressed into service, this time suffering from emphysema and thus placed on a respirator. Otto's ability to continue breathing thus demonstrates that "breathing is not just a lung process, but can be extended into the world by machines such as ventilators." Thagard goes on, "Critics will undoubtedly complain that the extended breath hypothesis uses obvious observations about machines and interactions to obscure the fact that lungs are in fact the key organs for performing respiration," thus making the point that environmental props and resources need not and should not be considered constitutive parts of the process. As Paul Patton, a commentator on the piece, points out, however, Thagard makes a key blunder. Respiration is a cellular process that takes place in the mitochondria, and lungs are simply the sites of gas exchange. As such, lungs *are* part of a larger system, and do not perform respiration alone. Even if we assume Thagard is really talking about breathing, Patton suggests, his criticism is still wide of the mark,

trouble with a notion of an "extended stomach," especially given our increasing knowledge of the human microbiota – the bacterial communities that occupy the human gut (e.g., Nicholson et al., 2012). No one disputes that these form an integral part of our digestive processes, yet they are not an integral part of our bodies: we offload part of our digestive system onto these organisms. The difference between regarding cooking as part of digestion and the microbiota as part of digestion seems to rest purely on the fact that one operates outside and one inside the body, which is precisely the internalist prejudice that Clark and others are attempting to break down with respect to cognition. Indeed, this is partly why Clark (2008) makes the case for an extended view rather than a more conservative embedded view – only the former genuinely unseats our standard picture of the mind, whereas the embedded view allows us to keep that picture intact.

Sterelny's (2010) preference for scaffolding (embedding) over the extended mind is more subtle than it first appears, however. For Sterelny, environmental scaffolds occupy a multidimensional space that amplifies and augments an agent's cognitive capacity to varying degrees. On this view, the extended mind should not be opposed to the embedded mind, or given a special label, because this obscures the fact that extended minds are just a special case of a more general phenomenon; they occupy a particular niche in the overall landscape. To illustrate this, Sterelny (2010) discusses a number of these dimensions and how they vary. For example, certain environmental resources should inspire a higher level of trust than others – Otto's notebook, for example, strikes Sterelny as vulnerable to malicious alternation, such as the removal of vital information or the addition of falsehoods, whereas subway maps and other publicly shared resources can be treated as much more reliable precisely because they are shared (it is much easier to interfere with a single notebook than the signage of an entire subway system). Environmental resources can also vary from being interchangeable to highly individualized or personalized: a tennis player prefers a certain kind of racket, with specific grip and tension,

and a chef has a special set of personal knives. There is also a process of mutual modification: the person adjusts to fit the specific demands of the tools, just as the tools become shaped to the person, and the use of that tool becomes "entrenched," not just individualized: if forced to use an alternative to her preferred racket, the tennis player may well seem clumsy and less skilled. In contrast, we can usually swap between different pencils, or cutlery, without deficit – these tools are interchangeable. Finally, Sterelny (2010) discusses whether the skills in questions can be assigned to individuals or form part of a collective. In his view, certain kinds of skills are best seen as properties of collectives of people and their surroundings – that is, they are socially distributed in the way that Hutchins (1995, 2010) describes – and not of individuals alone, and in this sense it becomes difficult to fit them into the extended mind framework.

Thus, Sterelny's (2010) dimensional analysis suggests that extended processes occupy the corner of the landscape where there are high levels of trust, where there are high levels of personalization, and where the cognitive processes are individual achievements, not those of a collective. Moreover, Sterelny (2010) concludes that extended mind cases probably occupy a much smaller area than many might think because many of the environmental resources that extended mind theorists consider to be individual are actually collective. Much of the environment that humans exploit consists of the products of past cultural activity, and young humans are shaped by growing up in a highly engineered, cultural niche. The various resources that, in the extended view, become part of our individual minds are produced by a history of invention by others. In that sense, they are collective resources produced by and through cultural and social practices that existed prior to our use. Extended minds must therefore also be social and collective, at least partly, with cognitive states shared across other people through the tools we use and the manner in which we learn to engage with them. We have, in other words, come full circle back to Hutchins.

From this brief survey, it should be apparent that, if we acknowledge that our cultural–technological practices – the way people make and do things together – make a fundamental contribution to human cognition, we are faced with the rather large problem of how to make this fit with the dominant picture that privileges isolated, individual minds and internal, abstracted processes. As Hutchins (1995, p. 354) puts it, "[t]he early researchers in cognitive science placed a bet that the modularity of human cognition would be such that culture, context and history could be safely ignored at the outset, and then integrated in later. The bet did not pay off." Consequently, "some of what has been done in cognitive science must now be undone so that these things can be brought into the cognitive picture" (Hutchins, 1995, p. 354). There is, as they say, no time like the present.

---

because lungs also do not breathe on their own. They need to be situated in a chest cavity with a rib cage and a contractile diaphragm, plus the external atmosphere needs to contain an appropriate partial pressure of oxygen. As Patton argues, once explained correctly, breathing clearly is an embodied, embedded process, and there are real benefits to be had from adopting such a view: "Consider a respiratory doctor who knew only about gas exchange in the alveoli of the lungs, and was clueless about every other aspect of respiration. He or she wouldn't realize that polio, a disease that causes weakness of the diaphragm, could result in impaired breathing." The use of an iron lung in such cases is, clearly, an example of extended breath, because it does the work of the diaphragm. Similarly, a doctor ignorant of the role that the external atmosphere plays in respiration would not recommend supplemental oxygen to mountain climbers, an oversight that could also have fatal consequences. Thagard's parody actually makes the argument he set out to ridicule (www.psychologytoday.com/blog/hot-thought/201310/the-extended-breath).

16.5 MOVING FROM EVOLUTIONARY PSYCHOLOGY TO COGNITIVE ECOLOGIES

A 4E view of cognition, then, offers an evolutionary-grounded view of human cognitive abilities that treats human social and cultural practices and technologies as integral to those cognitive abilities, and not simply their product (Barrett et al., 2014). Once we recognize the role played by extra-neural bodily and environmental resources in our ongoing activity, it becomes harder to see where the mind ends and the rest of the world begins. Hopefully, it is also clear that an externalist picture of the mind denies neither the importance of the brain nor the existence of internal processes. Rather, it simply expands our conception of what it means to be a cognitive system. In other words, the contribution of e-cognition, and the extended–scaffolded view in particular, is entirely positive, expanding the explanatory resources we have available in our studies of psychology and cognitive science (Sprevak, 2010). This being the case, what implications does this view hold for how we actually study psychological processes? There are two answers to this. As far as "classic" cognitive psychologists are concerned, the answer is: not much. As hinted at above, adopting an extended (or embedded or scaffolded or distributed) perspective does not entail that we abandon the study of individual capacities. To argue that notebooks can (sometimes) function in the same way as biological memory – or whatever extended system we are interested in – is not to argue *against* investigating individual cognition in the traditional way, attending to the fine-grained specifics of individual performance.

Doing so, however, does require us to more closely scrutinize the strategy of methodological individualism (or solipsism) that has long dominated psychology. If our cognitive capacities extend to include parts of the environment (or even if they are only scaffolded), then bracketing off everything outside the individual subject may be counterproductive to a full understanding of human cognition and behavior. One can make an analogy with studies of nonhuman animals, where studies of natural behavior sometimes suggest the possession of cognitive capacities that cannot then be demonstrated under laboratory conditions. One explanation given for this is that a lack of experimental control encourages an overly liberal, and often anthropomorphic, interpretation of behavior under natural observational conditions. Another possibility, however, is that, under natural conditions, animals lean on the environment in ways that improve their problem-solving (or other) abilities (Barrett, 2012). If laboratory setups fail to include such environmental supports, under the assumption that the brain does all the heavy lifting, then the animal will obviously fail to manifest these abilities in the laboratory. The "classical" cognitivist might say, with justification, this is because the animal lacks the requisite cognitive abilities (because these are, by definition, brain-bound processes). What the cognitivist cannot do, however, is explain why the animal performs as it does in the wild, nor would there be any justification for the claim that natural observations are necessarily mistaken because the behavior cannot be reproduced in the laboratory (precisely because the behavior may now be crucially dependent on environmental supports). The extended/extensive mind advocate would say that, as the experiment (potentially) excludes part of the animal's cognitive system, we cannot say one way or another whether the animal possesses the ability in question. We can only do so by investigating the intact system. If we design experiments that can do so and reproduce the wild behavior in the laboratory, we will then have a deeper understanding of the animal's "cognition in the wild" – and this is true for both the cognitivist and the extended mind theorist. The same would seem to be true of human performance.

Thus, even the committed cognitivist may benefit from the extended/scaffolded view by becoming more aware of environmental structure and how it can be exploited within a cognitivist paradigm. For example, Landy and Goldstone (2007a) showed an effect of perceptual grouping on mathematical problem-solving where, for example, the physical spacing used in formal equations had a large impact on whether they were judged as valid or not (see also Landy & Goldtone, 2007b, 2010); that is, the physical format mattered to task success. A failure to recognize this, as Landy and Goldstone (2007a) note, could risk modeling this process as conceptual and interpretative, when aspects of it are actually perceptual. Similarly, Lock (2013), in her analysis of Alzheimer's disease, argues that more attention should be given to public health approaches to prevention, care, and coping – strategies that recognize the entanglement of minds and environments – rather than focusing purely on biomedical strategies aimed at studying the brain in isolation in an effort to find a cure. This, in turn, echoes a point made by Clark (2008), who describes how many people with Alzheimer's disease function much more effectively than one would predict thanks to the strategies they use to ensure their environment can compensate for deficits in their "onboard" memory, allowing them to remain independent (e.g., Post-it notes in strategic places, ensuring the layout remains stable and supports the memory functions they have lost).

One obvious response to the above would be to say, as Payne et al. (2001) do, that while the offloading of cognition and the use and exploitation of external tools are indeed widespread and useful, it is the human agent that, when all is said and done, determines the relative balance of internal vs. external strategies employed. This is similar to the objections leveled at the extended mind argument, which argue that the brain ultimately has the final say for much the same reason (e.g., Adams & Aizawa, 2001). The person – or brain – is the controller of action in a way that other things, like iPhones and notebooks, are not. Although stating that the person has agency may be more appropriate than saying the brain itself has agency (which would be to commit the so-called mereological fallacy; Bennett & Hacker, 2003), such a stance does seem to beg the question: it assumes we know what agency is and that it is endogenously generated, when both of these assumptions can be, and indeed are, contested. Another way to say this is that it is precisely this idea of a "privileged user" of our various non-neural and neural tools that the extended mind is designed to upend.

The problem with arguing that the brain must have the final say is that, in any given instance, there may be parts of the brain that are not actively involved. This being the case, does this mean that, for example, it is only the frontal lobes that have the final word? The problem with such an argument should be apparent, because what if, having done the necessary empirical work, we find that no unique subsystem has the final word in any given process? Would this mean that the thinking agent has disappeared (Clark, 2008)? The answer to this is obviously no: if we were to demonstrate such a thing, this would not mean that we as thinking agents would suddenly disappear in a puff of smoke. What it does means, however, is that, as Clark (2008, 2014) suggests, we need to think more deeply about what it means to be a thinking agent and what we mean when we speak of control, minds, and selves. We also need to consider more seriously the possibility that we just *are* a shifting coalition of tools, "soft selves" assembled and reassembled from the "grab-bag of neural, bodily and worldly elements" (Clark, 2014). In this sense, the case for the extended mind is a promissory note for a new science of the mind, not a new science of the mind in and of itself: it presents us with an opportunity to develop the kinds of methods and research designs that are needed in order to be able to test whether a given process shows, in the words of Sprevak (2010, p. 20) "internal self-sufficiency" ("mental processes are largely self-sufficient, and can be studied largely in isolation from environmental props") or is "externally dependent" ("mental processes depend intimately on environmental resources, and should be studied within the context of those resources").

The second answer regarding the implications of an extended/scaffolded approach concerns the evolutionary human sciences. In this case, an extended mind approach requires reconsideration of the notion that "our modern skulls house a stone age mind," where environmental resources serve purely as inputs to our evolved cognitive architecture (Cosmides & Tooby, 1997). This position leads to the argument that we are often mismatched to our present environments (i.e., that many aspects of modern society and our cultural practices are sources of disruption to our evolved cognitive programs), which raises the problem of how we create and cope with novelty, despite having minds apparently ill-equipped to cope with novel circumstances. Although more recent theoretical treatments of EP avoid this

conclusion, they do not allow environmental resources themselves to contribute to cognitive processes, not even as scaffolds (e.g., Barrett, 2015; Barrett & Kurzban, 2006). The picture remains fully internalist: cultural artifacts and practices are not picked out as being any different from other environmental inputs, nor do they exert any transformative influence on cognitive processes themselves. Hence, cultural activities remain distinct from, and firmly outside, the mind.

The same is true of gene–culture coevolutionary approaches, where culture is conceived of as internal mental representations (Mesoudi, 2011; Richerson & Boyd, 2005), and where the internalist mind sets limits on the kinds of cultural artifacts we produce and the practices in which we engage (Sperber, 1994). None of these positions recognize Clark's (2008) and Sterelny's (2010) point that cultural artifacts and practices can "turbocharge" our cognitive abilities via the way these outputs can serve as further inputs to our cognitive system (in much the same way that the exhaust flow in a turbocharged engine is used to drive more air into the cylinders and so burn more fuel). Refocusing efforts along externalist lines may not be too difficult to achieve given that cultural influences are integral to this research program – cultural practices need to span brain, body, and world and not simply be assumed to lie in the head alone. In some instances, this may simply require new operational definitions for the terms used in formal models. For example, the level of information assigned to individual A's memory could be increased by also adding information present in the memory of individuals in A's social network (e.g., Gintis, 2016, p. 101), while material artifacts could be modeled as amplifiers of the skills/knowledge possessed by demonstrators, or could themselves be assigned the ability to influence performance, decoupled from the presence of a skilled human actor. Classic ethnographic work along with more theoretically driven work, such as elements of Latour's (2005) actor-network theory, could also be mined for information on how cognitive processes are orchestrated by technology and cultural practices and provide a more detailed understanding of certain social contexts.

No doubt many current evolutionarily oriented researchers will argue that they are already focused on context, but for the most part, this represents an acknowledgment that environmental contexts vary, and that such variation leads to behavioral differences. Truly attending to context means working toward an understanding of how cultural practices and technologies contribute to the outcomes achieved and the impacts they have on internal processes themselves. As Wheeler and Clark (2008) put it, "[a] child whose early experience is shaped by the special environments provided by books and software programs, and whose own emerging cognitive profile favours certain elements within that culturally enabled nexus over other elements, will end up with a cognitive system that is not just superficially, but profoundly, different from that of a differently encultured child." It is promising to note some indications that a more externalist view of mind is beginning to infiltrate – Henrich (2016), for example, refers to the abacus as a "mental prosthesis" – but clearly we have a long way to go.

Interestingly, a leading proponent of EP has made the opposite criticism of gene–culture approaches, arguing that the inclusion of culture into evolutionary accounts is literally "mindless." Specifically, Cosmides (2016) has claimed that treating cultural processes as a second form of inheritance system, independent of the process of genetic evolution, "invites the inference that cultural inheritance can exist, be identified and understood without discovering the information processing architecture of the *many* evolved systems that generate, shape and create culture" (emphasis in original) and that this has a "long history in anthropology and the social sciences, where people have ignored [these mechanisms] entirely."

This criticism actually seems rather misplaced with respect to gene–culture coevolution theory and cognitive anthropology, both of which claim to have identified a number of internal cognitive mechanisms by which cultural practices are propagated. Leaving this aside, it is also apparent that, if we think of cultural resources as either scaffolds or constituents of our cognitive processes, this criticism simply has no purchase; indeed, it simply dissolves. Swapping an internalist picture for an externalist picture provides a way for cultural patterns to be brought into the psychological fold and, far from ignoring mechanisms completely, anthropological and ethnographic studies provide a rich source of insight and data that can be used to inform our psychological theories precisely because cultural practices are, depending on one's theoretical proclivities, essential scaffolds or literal parts of the human mind. Moreover, this approach also dissolves the nature–nurture dichotomy: if minds extend and incorporate environmental resources and tools (or are scaffolded and dependent on them), it makes no sense to speak of ancestral cognitive architecture, ill-suited to modern environments, because culture and biology are no longer opposed. There is, then, much to be gained from this (literally) mind-expanding view, and it would be narrow-minded indeed to remain clinging to a Cartesian conception of mind.

16.6 ACKNOWLEDGMENTS

This work was supported by the NSERC Canada Research Chairs Program and Discovery Grant Program. Many thanks to Gert Stulp for discussion on this issue and for helpful comments, and to Lance Workman and Will Reader for helpful and constructive comments on a previous draft.

REFERENCES

Adams, F., & Aizawa, K. (2001). The bounds of cognition. *Philosophical Psychology*, **14**(1), 43–64.

Adams, F., & Aizawa, K. (2011). *The Bounds of Cognition*. New York: John Wiley & Sons.

Barkow, J. H. (2006). Introduction: Sometimes the bus does wait. In J. H. Barkow, ed., *Missing the Revolution: Darwinism for Social Scientists*. Oxford: Oxford University Press, pp. 3–60.

Barrett, H. C. (2015). *The Shape of Thought: How Mental Adaptations Evolve*. New York: Oxford University Press.

Barrett, H. C., & Kurzban, R. (2006). Modularity in cognition: Framing the debate. *Psychological Review*, **113**, 628–647.

Barrett, L. (2011). *Beyond the Brain: How Body and Environment Shape Animal and Human Minds*. Princeton, NJ: Princeton University Press.

Barrett, L. (2012). Why behaviorism isn't satanism. In T. Shackelford & J. Vonk, eds., *The Oxford Handbook of Comparative Evolutionary Psychology*. New York: Oxford University Press, pp. 17–38.

Barrett, L., Pollet, T. V., & Stulp, G. (2014). From computers to cultivation: Reconceptualizing evolutionary psychology. *Frontiers in Psychology*, **5**, 867.

Bateson, G. (1960). Minimal requirements for a theory of schizophrenia. *AMA Archives of General Psychiatry*, **2**(5), 477–491.

Bennett, M. R., & Hacker, P. M. S. (2003). *Philosophical Foundations of Neuroscience*. Oxford: Blackwell Publishing.

Brooks, R. A. (1999). *Cambrian Intelligence: The Early History of the New AI*. Cambridge, MA: MIT Press.

Burke, D. (2014). Why isn't everyone an evolutionary psychologist? *Frontiers in Psychology*, **5**, 910.

Chemero, A. (2009). *Radical Embodied Cognitive Science*. Cambridge, MA: MIT Press.

Clark, A. (1997). *Being There: Putting Brain, Body, and World Together Again*. Cambridge, MA: MIT Press.

Clark, A. (2001). Reasons, robots and the extended mind. *Mind & Language*, **16**(2), 121–145.

Clark, A. (2003). *Natural-Born Cyborgs: Minds, Technologies, and the Future of Human Intelligence*. Oxford: Oxford University Press.

Clark, A. (2007). Curing cognitive hiccups: A defense of the extended mind. *Journal of Philosophy*, **104**(4), 163–192.

Clark, A. (2008). *Supersizing the Mind: Embodiment, Action, and Cognitive Extension*. New York: Oxford University Press.

Clark, A. (2014). *Mindware: An Introduction to the Philosophy of Cognitive Science*. Oxford: Oxford University Press.

Clark, A., & Chalmers, D. (1998). The extended mind. *Analysis*, **58**(1), 7–19.

Cosmides, L. (2016). Evolutionary psychology and the generation of culture. Invited presentation. Big Questions in Evolutionary Science and what they mean for Social-Personality Psychology. Presented at 17th Annual Convention of the Society for Personality and Social Psychology, San Diego, January 28–30.

Cosmides, L., & Tooby, J. (1997). Evolutionary Psychology: A Primer. www.cep.ucsb.edu/primer.html.

Cottingham, J., Stoothoff, R., & Murdoch, D., eds. (1985). *The Philosophical Writings of Descartes*. Vol. 1. Cambridge, UK: Cambridge University Press.

Daston, L., & Galison, P. (2007). *Objectivity*. New York: Zone Books.

Dewey, J. (1916). *Essays in Experimental Logic*. Chicago, IL: University of Chicago Press.

Dewey, J. (1981). *The Later Works: 1925–1953*. Chicago, IL: Southern Illinois University Press.

Fisher, S. E. (2006). Tangled webs: Tracing the connections between genes and cognition. *Cognition*, **101**(2), 270–297.

Fodor, J. A. (2008). *LOT 2: The Language of Thought Revisited*. Oxford: Oxford University Press.

Geertz, C. (1973). *The Interpretation of Cultures: Selected Essays*. New York: Basic books.

Gintis, H. (2016). *Individuality and Entanglement: The Moral and Material Bases of Social Life*. Princeton, NJ: Princeton University Press.

Henrich, J. (2016). *The Secret of Our Success: How Culture Is Driving Human Evolution, Domesticating Our Species, and Making Us Smarter*. Princeton, NJ: Princeton University Press.

Hurley, S. (2010). Varieties of externalism. In R. A. Menary, ed., *The Extended Mind*. Cambridge, MA: MIT Press, pp. 101–154.

Hutchins, E. (1995). *Cognition in the Wild*. Cambridge, MA: MIT Press.

Hutchins, E. (2008). The role of cultural practices in the emergence of modern human intelligence. *Philosophical Transactions of the Royal Society B: Biological Sciences*, **363** (1499), 2011–2019.

Hutchins, E. (2010). Cognitive ecology. *Topics in Cognitive Science*, **2**(4), 705–715.

Hutchins, E. (2011). Enculturating the supersized mind. *Philosophical Studies*, **152**(3), 437–446.

Hutto, D. D. (2013). Psychology unified: From folk psychology to radical enactivism. *Review of General Psychology*, **17**(2), 174–184.

Hutto, D. D., & Myin, E. (2013). *Radicalizing Enactivism: Basic Minds without Content*. Cambridge, MA: MIT Press.

Jonason, P. K., & Dane, L. K. (2014). How beliefs get in the way of the acceptance of evolutionary psychology. *Frontiers in Psychology*, **5**, 1212.

Jonason, P. K., Jones, A., & Lyons, M. (2013). Creatures of the night: Chronotypes and the Dark Triad traits. *Personality and Individual Differences*, **55**(5), 538–541.

Landy, D., & Goldstone, R. L. (2007a). How abstract is symbolic thought? *Journal of Experimental Psychology: Learning, Memory, and Cognition*, **33**(4), 720–733.

Landy, D., & Goldstone, R. L. (2007b). The alignment of ordering and space in arithmetic computation. *Proceedings of the Cognitive Science Society*, **29**, 437–442.

Landy, D., & Goldstone, R. L. (2010). Proximity and precedence in arithmetic. *Quarterly Journal of Experimental Psychology*, **63**, 1953–1968.

Latour, B. (2005). *Reassembling the Social: An Introduction to Actor-Network-Theory*. Oxford: Oxford University Press.

Lock, M. (2013). *The Alzheimer Conundrum: Entanglements of Dementia and Aging*. Princeton, NJ: Princeton University Press.

Marks, J. M. (2009). *Why I Am Not a Scientist: Anthropology and Modern Knowledge*. Oakland, CA: University of California Press.

Menary, R. A. (2007). *Cognitive Integration: Mind and Cognition Unbounded*. London: Palgrave Macmillan.

Menary, R. A. (2010). Cognitive integration and the extended mind. In R. A. Menary, ed., *The Extended Mind*. Cambridge, MA: MIT Press, pp. 227–243.

Mesoudi, A. (2011). *Cultural Evolution: How Darwinian Theory Can Explain Human Culture and Synthesize the Social Sciences*. Chicago, IL: University of Chicago Press.

Nicholson, J. K., Holmes, E., Kinross, J., et al. (2012). Host–gut microbiota metabolic interactions. *Science*, **336**(6086), 1262–1267.

Park, J. H. (2007). Persistent misunderstandings of inclusive fitness and kin selection: Their ubiquitous appearance in social psychology textbooks. *Evolutionary Psychology*, **5**(4), 860–873.

Payne, S. J., Howes, A., & Reader, W. R. (2001). Adaptively distributing cognition: A decision-making perspective on human–computer interaction. *Behaviour & Information Technology*, **20** (5), 339–346.

Pickering, A. (2010). *The Mangle of Practice: Time, Agency, and Science*. Chicago, IL: University of Chicago Press.

Richerson, P. J., & Boyd, R. (2005). *Not by Genes Alone: How Culture Transformed Human Evolution*. Chicago, IL: University of Chicago Press.

Ross, A., ed. (1996). *Science Wars*. Durham, NC: Duke University Press.

Rowlands, M. (1999). *The Body in Mind: Understanding Cognitive Processes*. Cambridge, UK: Cambridge University Press.

Rowlands, M. (2010). *The New Science of the Mind: From Extended Mind to Embodied Phenomenology*. Cambridge, MA: MIT Press.

Rupert, R. D. (2009). *Cognitive Systems and the Extended Mind*. Oxford: Oxford University Press.

Ryle, G. (1949). *The Concept of Mind*. London: Hutchinson.

Shapin, S. (2010). *Never Pure: Historical Studies of Science as if It Was Produced by People with Bodies, Situated in Time, Space, Culture, and Society, and Struggling for Credibility and Authority*. Baltimore, MD: Johns Hopkins University Press.

Shapin, S., & Schaffer, S. (1985). *Leviathan and the Air-Pump: Hobbes, Boyle, and the Experimental Life*. Princeton, NJ: Princeton University Press.

Shapiro, L. (2010). James Bond and the barking dog: Evolution and extended cognition. *Philosophy of Science*, **77**(3), 400–418.

Smith, B. D., & Zeder, M. A. (2013). The onset of the Anthropocene. *Anthropocene*, **4**, 8–13.

Sperber, D. (1994). The modularity of thought and the epidemiology of representations. In L. A. Hirschfeld & S. A. Gelman, eds., *Mapping the Mind: Domain Specificity in Cognition and Culture*, Cambridge, UK: Cambridge University Press, pp. 39–67.

Sperber, D. (1996). *Explaining Culture*. Oxford: Blackwell Publishers.

Sprevak, M. (2010). Inference to the hypothesis of extended cognition. *Studies in History and Philosophy of Science Part A*, **41**(4), 353–362.

Steffen, W., Grinevald, J., Crutzen, P., & McNeill, J. (2011). The Anthropocene: Conceptual and historical perspectives. *Philosophical Transactions of the Royal Society of London A: Mathematical, Physical and Engineering Sciences*, **369**(1938), 842–867.

Sterelny, K. (2010). Minds: Extended or scaffolded? *Phenomenology and the Cognitive Sciences*, **9**(4), 465–481.

Sussman, R. W. (2014). *The Myth of Race: The Troubling Persistence of an Unscientific Idea*. Cambridge, MA: Harvard University Press.

Sutton, J. (2006). Distributed cognition: Domains and dimensions. *Pragmatics & Cognition*, **14**(2), 235–247.

Sutton, J. (2010). Exograms and interdisciplinarity: History, the extended mind, and the civilizing process. In R. A. Menary, ed., *The Extended Mind*. Cambridge, MA: MIT Press, pp. 189–225.

Sutton, J. (2015). Remembering as public practice: Wittgenstein, memory, and distributed cognitive ecologies. In D. Moyal-Sharrock, V. Munz, & A. Coliva, eds., *Mind, Language, and Action: Proceedings of the 36th Wittgenstein Symposium*. Berlin: De Gruyter, pp. 409–443.

Thompson, E. (2007). *Mind in Life: Biology, Phenomenology, and the Sciences of Mind*. Harvard, MA: Harvard University Press.

Tooby, J., & Cosmides, L. (1992). The psychological foundations of culture. In J. Barkow, L. Cosmides, & J. Tooby, eds., *The Adapted Mind: Evolutionary Psychology and the Generation of Culture*. Oxford: Oxford University Press, pp. 19–136.

Tooby, J., & Cosmides, L. (2005). Evolutionary psychology: Conceptual foundations. In D. Buss, ed., *The Handbook of Evolutionary Psychology*. New York: John Wiley & Sons, pp. 5–67.

Tribble, E. (2011). *Cognition in the Globe: Attention and Memory in Shakespeare's Theatre*. New York: Springer.

Vaesen, K. (2014). Dewey on extended cognition and epistemology. *Philosophical Issues*, **24**(1), 426–438.

van Dijk, L. (2016). A horizontal attitude: Gibsonian psychology and an ontology of doing. Unpublished PhD thesis, University of Antwerp.

Varela, F. J., Thompson, E., & Rosch, E. (1992). *The Embodied Mind: Cognitive Science and Human Experience*. Cambridge, MA: MIT Press.

West, S. A., El Mouden, C., & Gardner, A. (2011). Sixteen common misconceptions about the evolution of cooperation in humans. *Evolution and Human Behavior*, **32**(4), 231–262.

Wheeler, M., & Clark, A. (2008). Culture, embodiment and genes: Unravelling the triple helix. *Philosophical Transactions of the Royal Society of London B: Biological Sciences*, **363**(1509), 3563–3575.

Wrangham, R. (2009). *Catching Fire: How Cooking Made Us Human*. New York: Basic Books.

PART V

EVOLUTION AND COGNITION

Cognition is the study of the mental processes that underlie behavior, such as thought, memory, reasoning, and perception. Since the 1950s, many psychologists have discussed these mental processes in terms of computation: information is taken in through the sensory systems, processed, and used to guide behavior, taking into account immediate and long-term goals. A common misconception is that the organism is aware of these mental processes or that they are making conscious decisions, for example, to reproduce. "When I go out on a date," people might say, "I'm not thinking of how my date's genes are." This is to miss the point. The point is not that people (or other animals) have some kind of explicit motive to find good genes; the claim is that we tend to find things attractive that correlate with good genes and act accordingly; we act *as if* we have this goal (Dennett, 2017).

Evolutionary psychology has probably not had the impact on the cognitive sciences that many of us thought it would following the publication in 1992 of *The Adapted Mind*, whose title surely points to a consilience of the evolutionary and cognitive sciences (Barkow, Cosmides, & Tooby, 1992). The reasons for this are obscure, but there is good evidence that a consideration of evolutionary theory can open up productive new research questions in the cognitive sciences.

The usefulness of the evolutionary approach is summarized by George C. Williams's famous quotation:

Is it not reasonable to anticipate that our understanding of the human mind would be aided greatly by knowing the purpose for which it was designed? (Williams, 1966, p. 16)

Without an understanding of what the mind is *for*, cognitive scientists are condemned to suffer the same fate as the blind men touching the elephant. They can hypothesize what each component (the tusks, the trunk, the tail) might do, but without an understanding of the elephant as a whole (and its ecology), they have no idea *why* an elephant has tusks, trunk, or tail, or what they are for.

We hope to elucidate the advantages of evolutionary cognitive science more thoroughly when discussing the individual chapters in Part V.

Most of the phenomena studied by cognitive scientists are purely mental processes and therefore are unobservable. Language is the exception. Michael Corballis, a psychologist who has spent many years studying cerebral asymmetry and language processing, turns his attention to the evolution of language (Chapter 19). Specifically, he discusses what he terms the *great leap forward*, but has elsewhere been described as the *big bang* (Pinker, 1994): that language appeared to have happened rather suddenly in our species. Traditionally minded psycholinguists would not consider this a problem – if, indeed, they would consider it at all – but armed with knowledge of how evolution works, we can see that this *is* a problem. Evolution tends not to trade in leaps and bangs, so if language did indeed evolve, then it must have evolved gradually, which means that some of the mechanisms underlying language must have had precursors in our prelinguistic ancestors and nonhuman relatives. We labor this point slightly because this is an example of the power of evolutionary thinking: it sets boundaries – "It could not have happened like that!" – that force deeper theorizing to find a solution.

Corballis' solution is an intriguing one. Some aspects of language (recursion, simple comprehension) may predate our own species by tens of millions of years, albeit being used for different purposes. As language evolved, it recruited these extant mechanisms for its own ends. Further, he proposes that the earliest human language might have been gestural rather than vocal, with the voice taking over only when our hands acquired other important things to occupy them, such as manipulating tools and weaponry. This also meant that you could communicate without seeing your interlocutors, which is likely to have been important when hunting.

When non-hearing people use sign language, we often credit their ingenuity in "making do" with a gestural system that was not designed for language. If Corballis is correct, they are not making do – they are merely defaulting to the earliest medium of human language. Maybe the fact that sign language comes so readily to non-hearing communities provides evidence that gestures were foundational in language evolution.

Among other things, Ádám Miklósi and Veronika Konok (Chapter 17), two Hungarian ethologists whose principal investigations have focused on dog cognition (or "dognition," as they call it), touch upon the vexed question of why language evolved: What was the ecological question to which language was the answer? Neuroanthropologist Terrance Deacon expressed the nature of this problem when he wrote

Looking for the adaptive benefits of language is like picking only one dessert in your favorite bakery: there are too many compelling options to choose from. What aspects of human social organization and adaptation wouldn't benefit from the evolution of language? (Deacon, 1997, p. 377)

Rather than looking for a single function, as many have before them, Miklósi and Konok embrace this embarrassment of riches and conclude that there may have been many different reasons that led to the evolution of language, some of them concerned with social bonding, some with coordinating hunting, and others with social manipulation and deception. They go further, however, and ask the question as to why people spend so much time engaged in linguistic activity. After all, the design of the human hand was as important in our evolution as language, yet we do not spend nearly so much time deriving pleasure from our opposable thumbs. Their answer is multifactorial, but rests upon their aforementioned claim that language evolved for many reasons and thus can be exercised in many different contexts. But they also see the importance of culture in the ubiquity of language. Runaway selection is where an initially arbitrary preference can select for a particular trait, leading to the trait becoming larger than it needs to be. Female choice, for example, drove the expansion of the peacock's tail to the extent that the peacock, no matter what he might prefer, cannot reverse the process.

Similarly, they argue, language became enmeshed in cultural practices such as rites and rituals, and cultural runaway selection ensured that it exploded from being used in a few rather specific contexts to its current omnipresence. Technology helped, too, with smartphones, social media, and online forums providing increasing outlets to exercise the language instinct. Like the poor peacock, even if people want to abandon their smartphones, they cannot for fear of becoming socially isolated.

In Chapter 18, developmental psychologists Peter J. LaFreniere and Rachelle M. Smith illustrate one of the main divergences between the traditional approach to cognitive science and one that is evolutionarily informed. Many adherents to the traditional approach assume that the evolution of the human mind led us to perceiving the external and internal world with ever greater fidelity. In short, it is all about our perceptions and reasoning more closely approximating the truth. This may be one of the reasons why early cognitive psychologists were so fixated

on logic, which can derive new truths from axioms by the application of logical rules (Pinker, 2005). Of course, it became pretty clear that we are not at all logical (Wason, 1968), and an industrial-scale research program was born in order to determine why. Surprisingly few asked the question, "Why should we be logical?" The evolutionary answer to this is that we need not be. The mind is not there to represent the world as it truly is – witness visual illusions – but to help us to make better decisions about what to do next, and if a false or distorted representation of the external world or our thoughts and feelings results in better decisions (ultimately, more surviving babies), then that is what we will get.

LaFreniere and Smith start from this premise in their explanation of the evolution of social intelligence. They argue that social intelligence is driven by tactical deception, or lying, in its various forms. Here again we see the evolutionary view turning conventional wisdom on its head. To most people (and many researchers), lying is a form of deviance and is therefore seen as a phenomenon worthy of explanation (presumably rounding up the usual suspects such as the parents or society at large). But from LaFreniere and Smith's perspective, lying is just a solution to the ecological circumstances of being human. We have to forge alliances with others, conceal important personal information, and manage multiple – often mutually antagonistic – relationships, none of which would be possible if we were obligate truth-tellers. Once lying became commonplace, there would be a selective pressure for lie detection, as it is in no one's interest to be duped. This in turn placed a premium on telling better lies, to the extent that we frequently believe our own lies: so-called self-deception. The lies need not be directly malign, and unlike psychopaths and other pathological liars, they are motivated to grease the wheels, pour oil on troubled water, and generally manage social situations, albeit often for the ultimate benefit of the deceiver. Their chapter shows how tactical deception emerges early in development, with the Rubicon being crossed at about four years of age when children's theory of mind (or mind reading) ability really starts to come onstream. We used to consider truth-telling to be the default position and deception to be deviant, but evolutionary research is increasingly demonstrating that deception is not only ubiquitous, but functional.

In Chapter 20, Andreas Wilke, a cognitive psychologist working at Clarkson University in New York State, explores how our evolutionary history of foraging might have shaped the way we make decisions in an uncertain world. In the natural world, resources such as food are not uniformly or randomly distributed; instead, they occur in "patches," meaning that they tend to occur in clumps. You will find berries on bushes, for example, so the bush is the patch, and between bushes you will find no berries. Anyone who has ever gone berry picking will know that early on the bush usually yields plenty of berries, but as time goes on the rate at which we can pick them will start to decline: the berries become harder to spot and will be less accessible, especially if the bush has thorns. The question that faces us is: When do we decide to give up on this particular bush and move on to another one? The mathematics of foraging theory (Stephens & Krebs, 1986) attempts to address these questions. One relevant factor is that the greater the travel time to the next patch, the longer you will tend to spend exploiting the current patch. This makes a great deal of sense if animals are designed to be optimal foragers. To return to the berry-picking example, if you have an hour to pick as many berries as you can, you do not want to waste time walking between distantly spaced bushes. On the other hand, if the bushes are very close together, you do not want to waste time struggling to find difficult-to-access berries – better to move on to the next one where pickings are easier. This is known as Charnov's marginal value theorem, and its mathematical predictions have been applied to a range of different foraging animals, such as bees, oystercatchers, and deer, as well as humans.

Wilke cleverly applies foraging theory to a range of behaviours that are not obviously related to foraging, including some classic cognitive fallacies. Of particular interest is the hot hand fallacy and the gambler's fallacy, respectively the belief that a success will be followed by another success and the belief that a losing streak will soon be followed by a win, assumptions that can often drive people to lose large sums of money at the roulette table or the slot machine. Wilke proposes that our evolved expectations from our history as a foraging species lead us to assume that wins or losses are like berries and occur in clumps, so when gamblers win, they keep gambling because they are certain they will win again (they assume they are in a patch and finding one berry surely predicts that they will find another). In the gambler's fallacy, people

assume that they are searching for a patch and one will turn up sooner or later if they just keep going, and when they do find it, the hot hand fallacy tells them that they will hit a rich seam of success. Unfortunately, the distributions of wins and losses at the casino are not patchy – they are as random as a coin flip, but our foraging brains tell us differently, often with a detrimental effect on our bank balance. On a more positive note, Wilke presents cross-cultural evidence that people with experience of coin flips are less likely than those with no experience to make the "forager's fallacy," so it may well be the case that problem gambling could potentially be reduced through education and experience. This is a nice example of a situation in which the sometimes arcane mathematics of foraging theory can be used to help understand and solve practical problems.

REFERENCES

Barkow, J. H., Cosmides, L., & Tooby, J., eds. (1992). *The Adapted Mind: Evolutionary Psychology and the Generation of Culture*. New York: Oxford University Press.

Deacon, T. W. (1997). *The Symbolic Species: The Co-evolution of Language and the Brain*. New York: W. W. Norton & Company.

Dennett, D. C. (2017). *From Bacteria to Bach and Back: The Evolution of Minds*. New York: W. W. Norton & Company.

Pinker, S. (1994). *The Language Instinct: How the Mind Creates Language*. New York: William Morrow.

Pinker, S. (2005). So how does the mind work? *Mind & Language*, **20**(1), 1–24.

Stephens, D. W., & Krebs, J. R. (1986). *Foraging Theory*. Princeton, NJ: Princeton University Press.

Wason, P. C. (1968). Reasoning about a rule. *Quarterly Journal of Experimental Psychology*, **20**(3), 273–281.

Williams, G. C. (1966). *Adaptation and Natural Selection*. Princeton, NJ: Princeton University Press.

17 Runaway Processes in Modern Human Culture

An Evolutionary Approach to Exaggerated Communication in Present Human Societies

ÁDÁM MIKLÓSI AND VERONIKA KONOK

17.1 INTRODUCTION

In biology, runaway selection has been proposed as an evolutionary mechanism to explain the emergence of exaggerated morphological and behavior features in animals. Such runaway processes usually manifest if there is a fundamental change in the constraints that control the emergence of a specific trait. Here, we argue that the exaggerated communication in modern humans is the expression of a biological feature that has been liberated from environmental control. Social evolution of humans has led to increased group size, which, in turn, selected for a biologically supported need for communication in order to continuously keep contact with an increasing number of group members. This selection process led eventually to the emergence of language a few tens of thousands of years ago. However, up to a few hundred years ago, spoken communication was limited by spatial and temporal constraints. The invention of modern, global communication networks (the Internet, radio networks) and new tools (mobile phones) removed these barriers that limited human communication in the past. As a result, present-day humans display strong tendencies for exaggerated communication based on their selected predisposition to keep in contact with group members.

17.2 RUNAWAY PROCESSES IN THE EVOLUTION OF EXAGGERATED TRAITS

One of the most famous and widely cited examples of runaway processes is the courtship behavior of the peacock male. Even one glimpse at a peacock's (*Pavo* spp.) showing off his adornments is enough to remember it for one's whole life. His tail feathers are unreasonably long, making him quite vulnerable to predation during walking or running. But he still seems to move them with relative ease up and down or to the side, and he shakes their upper parts gently or more rapidly in the vicinity of a female. The male's body seems to fade in front of the wall of these raised feathers, the color of which gives the whole perfor-

mance a quite mysterious appearance. So it is no wonder why many students of evolution, including Darwin (1871), speculated about how this creature is able to overcome such a handicap as it struggles for survival.

In nature, this display serves a clear function in mate choice. During the mating season, peacocks gather at a specific place (the lek) where the females also assemble. These encounters provide the best opportunity for the females to select the most promising male candidate for mating. This also saves a lot of time for them because they do not need to seek out males individually; they have them lined up at the same time, which also helps in making a well-founded choice. The male's task is to attract the female's attention and keep her in close range, and finally mate with her if possible. But in a ring of many other similarly decorated males, this task is not easy. The female also needs to have a good eye for choosing the best partner from among the many males. While the tail and the courtship behavior make perfect sense in courtship, this "masquerade" seemed to be quite out of context if one considers that the peacock has to wear it during all of his life.

The courtship of peacocks may be quite extraordinary, but it is by no means an isolated case. The naturalist can easily find many similarly astonishing modes of interaction in representatives of most animal clades.

For Darwin and his followers, it was clear that the peacock's tail was a product of evolution, despite assuming that in some respects it represents a huge disadvantage for the bearer. This seemed to be an insoluble paradox. The development and maintenance of the tail costs much energy in terms of consumed food, including specific food ingredients that are needed to keep the feathers in good condition and parasite free (Hill & Montgomerie, 1994). On top of such unavoidable investment, the tail seems to be rather a burden when the peacock is facing a potential predator. Why would evolution produce such a disadvantageous character? The possibility of it being a "by-product" can be excluded because there are so many other similar traits in other species. For example, the ability to produce a quite elaborate song evolved at least three times independently in birds (Mason, Shultz, & Burns, 2014). So rather than regarding these traits as evolutionary accidents, one needs to develop a theory that aims to explain their emergence and existence.

Darwin suggested that features like a peacock's tail were the result of sexual selection. He discriminated the process from natural selection because the selective agents were the members of the species – males or females – rather than the species' environment. Darwin's boldness in the formulation of this hypothesis was due to the fact that, at the time, little knowledge was available about the genetic mechanism of inheritance. While intra-sexual selection (e.g., male–male competition) was relatively easy to interpret, it was far from clear how females' choices could ensure that those males had more offspring than the less favorable ones. In Darwin's theory, females had a "free choice," which was not only difficult to accept in the Victorian era, but there was not much direct evidence for this theory either (Rosser, 1992).

Darwin was also aware of the close interrelation of natural and sexual selection, and that the former might facilitate or rather constrain the selection of a trait by the latter process. Although he was not able to provide an explanation for why "prettier" (preferred) partners have more offspring, he also generalized this theory to humans, explaining sexual dimorphism in height, female beauty, or musical talent as being substrates of sexual selection (Darwin, 1871).

17.2.1 Sexual Selection as a Runaway Process

For many years to come, the weighty debates about natural selection led researchers to ignore the seemingly unrealistic scenario of sexual selection. Fisher (1930) was one of those who took Darwin's idea seriously and aimed to provide an evolutionarily plausible explanation. By that time, the genetic concept of inheritance was more generally accepted, so for Fisher it was a more natural assumption that some phenotypic traits could be directly linked with genetic elements ("genes") and that naturally selected traits might become exaggerated by sexual selection. Thus, according to Fisher, female preference for some male traits might cause these traits to become more characteristic as a result of a runaway evolutionary process, which was based on positive feedback.

Let's take Haufe's (2007) description about the Fisherian idea. There is a starting population (G1) in which females show no preference for males, and two types of males (M1 and M2) exist in equal frequency. In the next generation (G2), a "mutant" female (F2) emerges who shows some preference for an observable male feature (e.g., longer tail) that is more typical of M2. The daughters of this female inherit with greater chance their mother's preference; the sons inherit their father's phenotype. Thus, in G3, there will be more F2 females showing a preference for M2, and step by step the frequencies of such females and males

increase. Note that this process is only possible if there is a specific genetic variability controlling the trait in males. If the mutant female prefers a nongenetic characteristic, no runaway process is initiated. If there is a genetic component, then both the genetic preference and the genes determining the characteristic are passed on to the offspring. The female preference could increase (become stronger) during subsequent generations, which could lead to exaggeration. The process may stop when the sexually selected characteristic becomes too costly for those individuals exhibiting such exaggerated features.

Importantly, the preference is expressed only in the females and the sexually dimorphic trait is expressed only in the males, but the individual genomes carry both types of genetic information; in other words, the female preference and the male trait become coupled at the genetic level, despite the differential expression in the two sexes. These traits become co-inherited. This is why females are not sensitive to the way in which their preference affects the males. They may continuously select for males even when the bearer of that trait pays a heavy cost, which may increase the morbidity of male individuals and the vulnerability of those males in the population with such extreme features. So in the end, males may pay a price for being attractive to the females.

Mutants preferring the opposite feature (e.g., short tails) have little effect in the long term because the other females do not show a preference for these offspring, so they eventually disappear from the population.

In summary, runway processes have three phases. The first phase starts when a female shows a preference for a male trait that is, however, quite accidental. In the second phase ("runaway"), the bearers of the trait increase their ratio in the population. In the third phase, the trait that emerges as a result of sexual selection may be subject to factors of natural selection. In some evolutionary scenarios, individuals of that species may have a selective disadvantage if natural selection acts too rapidly and there is "no way back" for the sexually selected trait.

17.2.2 Runway Processes in Evolution

Today, the fate of the peacocks seems to be safe; that is, the males seem to survive because there is a balance between producing the sexually selected trait and the selection on this trait in nature. It is very likely that many peacocks die very young, but there are still enough survivors to keep the population going. However, this equilibrium may change suddenly if, for example, a new predator emerges and the peacocks cannot find a possibility to escape.

The Fisherian model was described for many years as mere speculation, and the first mathematically based models emerged only in the 1960s. Lande (1981) argued that with specific conditions (large populations, males can mate with many females, repeated mutations maintain the variability of the trait) there could be an equilibrium between natural and sexual selection if there is a close relationship between the females' mean preference and the mean value of the males' character. There are also models that predict that populations consisting of such individuals may die out. It is also important to note that, in reality, sexual selection acts on multiple traits in concert. So these traits could mutually inhibit the "runaway" of the others. For example, if females select for both tail length and the dance of the males, then a too long tail would probably hamper dancing, so selection would act against longer tails.

17.2.3 An Alternative Approach to Sexual Selection

Back in 1975, Zahavi was not convinced by the Fisherian model, and he thought about alternative explanations of the emergence of extreme traits. He argued that the bearer of such traits must excel in other ways to be able to compensate for the costs. Thus, only peacocks in the best condition can have the "luxury" of developing an attractive tail. This means that the tail becomes a reliable ("honest") signal of the male's quality. The reliability of the signal is ensured by the costs involved; cheaters of lower quality do not have the necessary resources to develop an attractive tail. Although Zahavi's (1975) ideas were also received with some skepticism, Grafen (1990) provided mathematical underpinnings for the theory. This was followed by a surge of publications showing that traits preferred by the female are indeed good markers of male quality. For example, Pryke, Andersson, and Lawes (2001) have shown that in the highly polygynous red-collared widowbird (*Euplectes ardens*) females express a strong preference for males with longer tails. As expected, these males also have higher fitness in terms of reproductive success.

Others developed these ideas further and argued that the hormones controlling some secondary sexual traits have an inhibitory effect on the immune system (Folstad & Karter, 1992). Thus, males with higher testosterone levels may be more vulnerable to infections and parasites. This handicap can be compensated for only by males who are in good physical conditions. In one experiment, Verhulst, Dieleman, and Parmentier (1999) compared comb size in different domestic chicken (*Gallus domesticus*) lines that were selectively bred for low or high antibody response to sheep erythrocytes (one possible measure for immunocompetence). Cocks selected for low immunocompetence had larger combs, and they also had higher testosterone levels, as predicted. This is in good agreement with earlier research that found a strong positive correlation between comb size and testosterone level in males. Other similar results also suggest that hormonal status provides a connection between secondary sexual characteristics and selective factors. There is probably a maximum level of testosterone (and thus comb size) that can be tolerated or compensated for by other mechanisms in male chickens.

One can argue, however, that natural variation and environmental effects may obscure male differences for the females and make the latter less successful. Thus, it would be advantageous if the phenotypic marker would signal (predict) the "good" genotypes independently from environmental variations. The stalk-eyed fly (*Cyrtodiopsis dalmanni*) provides a nice example of this. Females of this species show a preference for males that possess longer eyestalks at the end of which the eyes sit (Wilkinson, Presgraves, & Crymes, 1998). If males are fed on low-quality food, then males with some genotypes develop long eyestalks independently of their body condition, while in others these features vary as a function of food (David et al., 2000). Thus, in the former case, female choice provides a benefit for the offspring.

The above ideas have been popularized in many books, and they are commonly referred to as the "handicap principle" (Zahavi, 1975). However, according to Getty (2006), instead of concentrating on the costs of having the preferred signal, we should acknowledge that for males of better quality it is worth investing in these traits to increase their chances of mating with the females. Thus, better males gain more from developing a high-quality sexually selected characteristic than their weaker counterparts.

17.2.4 The Role of Sexual Selection in Human Evolution

Darwin (1871) writes in the General Summary of *Descent of Man, and Selection in Relation to Sex* (p. 402): "Courage, pugnacity, perseverance, strength and size of body, weapons of all kinds, musical organs, both vocal and instrumental, bright colours, stripes and marks, and ornamental appendages, have all been indirectly gained by the one sex or the other, through the influence of love and jealousy, through the appreciation of the beautiful in sound, colour or form, and through the exertion of a choice." This is a brave sentence in the absence of any data, but characteristic of the synthesizing approach typical of Darwin.

The modern study of sexual selection in humans started 100 years later and was initiated under the flag of evolutionary psychology. Researchers have started to collect evidence for specific female preferences and for traits of males that may have been subject to female choice (and evolution) and may indeed indicate male quality. Research in humans is more constrained and lacks the possibility of constructing "invasive" experiments. Nevertheless, these ideas generated new experiments that would not have been performed otherwise.

Based on nonhuman research, it seemed rewarding to study which human morphological features are relied upon in partner choice. Looks, specifically facial attributes, are often mentioned as standing for beauty, intelligence, or effeminacy. Features of the eyes may have only an attention-grabbing function, but they may also reflect the quality of the person in terms of fitness. A recent review

(Little, Jones, & DeBruine, 2011) suggests that humans show some typical preferences toward men's or women's faces, and research has also provided evidence that many facial traits are associated with physiological control (e.g., testosterone; Penton-Voak & Chen, 2004), and thus indirectly are connected to fitness. In addition, European men and hunter–gatherers living in Tanzania found symmetrical faces more attractive (Little, Apicella, & Marlowe, 2007). Many studies in birds have revealed that bodily symmetry is a preferred cue for mate choice (e.g., Møller, 1992), and there is also indirect evidence in humans that more symmetrical men are stronger (Fink et al., 2014).

17.2.5 Runaway Processes in the Evolution of Nonsexual Traits

Since the original Fisherian proposal, the theory of runaway processes has been extended to phenomena unrelated to sexual selection. Sexual selection can be regarded as a subtype of natural selection when the interactions among conspecifics provide constraints in either sexual or nonsexual contexts (Lyon & Montgomerie, 2012; West-Eberhard, 1979). This theory expands Darwin's sexual selection theory by demonstrating that social competition in a variety of contexts unrelated to mating often favors the evolution of certain types of ornaments, weapons, and behaviors. For example, in American coot (*Fulica americana*) families, offspring display highly ornamented, bright orange feathers to their gray parents to get the food and nourishment that are critical for the offspring's survival (Lyon, Eadie, & Hamilton, 1994). Many male and female birds use ornamental plumage traits as badges of status during dominance interactions in winter flocks (Searcy & Nowicki, 2005). This kind of social selection differs from nonsocial (natural) selection in that an individual's fitness is not solely determined by its own phenotype, but also by the phenotypes (and genotypes) of the individuals with which it interacts (Wolf, Brodie, & Moore, 1999). As the interacting phenotype is both the agent and the target of selection, this can lead to feedback loops and trait exaggeration (Moore, Brodie, & Wolf, 1997). Thus, the trans-generational effect of sexual selection may give rise not only to feedback loops and runaway process affecting traits, but also to social interactions in general.

Runaway social selection can explain some human nonsexual phenomena. Individuals chose their social partners for cooperation and altruistic behavior, and this partner choice constitutes a selective force for certain traits (Noë & Hammerstein, 1994). Individuals try to outcompete each other in those aspects that are preferred by the choosing party (Noë & Hammerstein, 1994) or by the group as a whole. Nakamaru and Dieckmann (2009) demonstrated mathematically that cooperation and punishment of defectors in human groups underwent runaway selection, with evolution toward enhanced cooperation and ever stricter punishment norms mutually reinforcing each other.

In summary, Fisherian runaway selection may occur in many nonhuman species and in humans, although its effect on the phenotype may differ depending on the controlling mechanisms. Female preference can lead to positive selection for one or more morphological and behavioral features, some of which may be associated with the males' quality. Once such a process is unleashed, there is little chance for any "spontaneous" reversal unless there are other limiting factors outside of the interacting system.

17.3 RUNAWAY PROCESSES IN CULTURAL EVOLUTION

We suggest that runaway selection theory can be further expanded to analogous phenomena in human cultural evolution. Cultural evolution can be modeled through the same basic principles of variation and selection that are the bases of biological evolution (Cavalli-Sforza & Feldman, 1981). Instead of genes as replicating units of biological information, the replicating units of cultural information are memes (Dawkins, 1976). A meme is an information pattern in an individual's memory that is capable of being copied to another individual's memory. This includes anything that can be learned or remembered: ideas, knowledge, beliefs, habits, skills, etc. (Chielens & Heylighen, 2005). Memes are replicated through social learning, rule following, and other synchronizing behaviors.

In biological runaway processes, some genes spread despite the fact that they are costly in terms of survival. Similarly, some widespread memes in cultural evolution may also be disadvantageous for the individuals who hold them in their mind. For example, while in recent times thinness has become a popular meme (in women), individuals on whom this meme has a large effect (individuals with anorexia) suffer decreases in their fitness (mental and physical health). Similarly to biological runaway processes, where the selective agent (e.g., the female) and the owner of the phenotype (e.g., the male) are independent, being different individuals, in cultural runaway processes, the two interacting agents are independent because they occupy different organizational levels (e.g., individuals vs. groups/societies). Therefore, the runaway process (e.g., the cultural norm) can be insensitive to the fitness consequences of the runaway phenotype in the individuals.

Culture also has contributed to runaway processes by changing the equilibrium between our biological preferences and the environmental constraints. For example, one may assume that human preference for calorie-rich food (sweets and fat) could lead to obesity if the constraints of eating such valuable nourishment cease (for details, see Section 17.3.1).

Thus, the parallel here with biological runaway processes is that preferences increase some phenotypic traits beyond their optimal level, despite some disadvantages. In case of biological runaway processes, sexual/social selection often drives a trait in the opposite direction to natural selection (e.g., long tail/short tail of a peacock). In cultural runaway processes, cultural memes also shift a trait in a direction that is disadvantageous because of the presence of other constraints. It is often the disappearance or weakening of this (environmental) constraint that leads to the runaway process. For example, the disappearance of a predator in case of biological runaway processes or the disappearance/weakening of ecological, social, or cultural constraints in case of cultural runaway processes can lead to the exaggeration of a trait.

The selective agent in the case of cultural runaway processes is not always and solely the culture. The preference for a trait/behavior can be transmitted both genetically and by social or cultural learning. For example, in the case of food consumption, besides our biological preference for calorie-rich food (which is transmitted genetically), cultural norms and habits also facilitate the consuming of sweet and fatty food (see Section 17.3.1). Another example is the beauty ideals of cultures, which are based on biological sexual preferences, but are strengthened, exaggerated, or modified by norms, fashion, media stars, etc. Additionally, as mentioned above, culture weakens or erases those constraints that, in the past, limited the exaggeration of these biologically based traits. Drug consumption (or addiction) or sexual addiction and pornography are also examples of when a biological preference is liberated from constraints and runs away. Furthermore, there are often cultural compensating mechanisms that decrease the costs of a runaway trait. For example, medicines decrease the costs of runaway sugar consumption (diabetes) or drug consumption.

17.3.1 Runaway Processes in the Emergency of Obesity

Obesity is an increasingly prevalent phenomenon in modern human societies. Currently, nearly 70 percent of adults in the USA are classified as either overweight or obese, while this number was fewer than 40 percent just 40 years ago (Lavie, Milani, & Ventura, 2009). The proportion of obese children is also growing fast (Go et al., 2013). This is a huge problem, as obesity is associated with the prevalence of most cardiovascular diseases (Lavie et al., 2009) and accounts for nearly 20 percent of overall mortality (Masters et al., 2013). Health care costs are huge and growing, and may account for 18 percent of US health expenditures by 2030 (Go et al., 2013).

Humans show an innate preference for the flavor of sweet food (Mennella, 2014), and breast milk also contains oligosaccharides that strengthen the preference in the newborn baby in this direction.

In the preindustrial era, our biological preference for calorie-rich food (which served a function in survival) was constrained by the limited availability of food in the environment. There was a balance between the amount and type of food humans gained and consumed and their activity investment during the search (a balance between the burned calories and the consumed calories). Since the Industrial Revolution, however, we have been able to produce much more food than in the past, and these are much richer in sugar and fat. This led to a sharp rise in the intake of sugar and fat, and in parallel the eating of high-fiber foods decreased. The wide availability of food and industrialization in general markedly decreased physical activity. This led to an imbalance in energy management; most humans eat more calories than they need.

In traditional human societies, valuable food was regarded as a "treat" that played a special role in the community. This could explain why cultures created many memes that involved eating such precious nourishments only on such occasions (e.g., feasts for birthdays or for religious events).

Present-day culture also strengthens the preference for sweet and fatty flavors. Chocolate was once the privilege of the rich, which induced a positive concept of it. Today, chocolate and other sweets are available to everybody, and they are often used as tools of rewarding (especially in childhood) and as essential accessories in social events and celebrations (e.g., wedding cakes), and their consumption is encouraged by advertisements and their huge supply. The modern food industry also makes many types of "fast"/"junk" food widely available around the clock. Our culture is also characterized by a fetishizing of the enjoyment of flavor (e.g., gastronomy, etc.).

These habits and norms are spread by social learning and rule following. The influence of social learning in the development of obesity is illustrated by findings that a person's chance of becoming obese increases markedly (by 37–57 percent) if he or she had a friend, a sibling, or a spouse who becomes obese (Christakis & Fowler, 2007). Persons of the same sex have relatively greater influence on each other than those of the opposite sex, which also emphasizes the role of social learning. Eating habits also show a concordance between peers, especially spouses (Pachucki, Jacques, & Christakis, 2011).

Because of these biologically and culturally transmitted preferences, the trait of excessive sugar and fat consumption started to run away. The costs are obesity, diabetes, and other accompanying illnesses, such as cardiovascular diseases.

17.3.2 Runaway Processes May Affect Human "Beauty"

Regarding the face, preferences for certain facial features emerge early in development (e.g., Kramer et al., 1995), and people from different cultures agree on the features of attractiveness (e.g., Langlois et al., 2000); thus, there seems to be a biologically based preference for certain facial features. Averageness, symmetry, and sexual dimorphism are the main biologically based standards of beauty in both male and female faces (for a review, see Rhodes, 2006). Attractive female faces have more feminine features (e.g., smaller chin, fuller lips, and higher cheekbones) (Perrett, May, & Yoshikawa, 1994). Exaggeration of feminine features increases attractiveness (Johnston & Franklin, 1993), and when people generate beautiful female faces on a computer, they produce faces with more feminine traits (smaller chin, smaller lower face area, fuller lips) than average (Johnston & Franklin, 1993). Besides the face, an hourglass body shape (enlarged breasts, buttocks, and hips accompanied by a slender waist) are seen as attractive (Dixson et al., 2007).

In contrast, in male faces, characteristics such as thick brows, thin lips, and square chins are preferred (Thornhill & Gangestad, 1999), and women are more attracted to men with more muscular upper bodies (Fan et al., 2005).

Theorists have proposed that preferences for appearance may be advantageous for mate choice because attractive traits signal important aspects of mate quality, such as health (Rhodes, 2006; Rhodes et al., 2001). For example, a broad pelvic canal may reduce the birth difficulties associated with the large head of the human infant. A low waist/hip ratio indicates a high estrogen/testosterone ratio, which also favors reproductive function. The slender waist is a reasonably sensitive indicator of youth. Some attractive female facial features, like fuller lips, are affected by the female sexual hormone, estrogen (Thornhill & Møller, 1997), which plays an important role in the regulation of female reproduction.

Apart from a biologically supported preference in both sexes for these features, culture strengthens these preferences (e.g., by presenting beauty ideals in media, advertisements, and pornography). Ancient humans could manipulate their bodies and faces only in a limited way by using paints and jewels. More recently, possibilities have increased for modifying looks in women by applying makeup, earrings (which increase perceived symmetry), and certain types of clothes (e.g., push-up bras). However, all of these changes are quite transient and may be effective only in the short term.

A newer technology – plastic surgery – can change many parts of the face and body, and these modifications have a long-lasting effect. Women fill their lips and their wrinkles, enlarge their breasts, or get liposuction. These procedures are increasingly popular. For example, there were 15.9 million surgical and cosmetic procedures performed in the USA in 2015. Since 2000, overall procedures have risen 115 percent, and some types of procedures have been chosen at an even greater degree. For example, since 2000, in the USA, there has been considerable growth in breast lifts (89 percent), buttock lifts (252 percent), and breast augmentation (31 percent). Importantly, at present, the price of these procedures is a limiting factor (American Society of Plastic Surgery, 2015).

However, one may hypothesize that, based on the logic we applied to obesity in Section 17.3.1, human "beauty" could easily become subject to a runaway process because genetic and cultural preferences for such traits acting synergistically. This means that once the limiting factor (cost of plastic surgery) is removed, a runaway process might take over. Humans could easily modify their bodies even at a young age. This could weaken the relationship between genetic quality and facial and bodily markers and make mate choice less relevant from an evolutionary perspective.

17.4 THE EVOLUTION OF HUMAN COMMUNICATION

Although one risks being too anthropocentric and anthropomorphic, it is perhaps not an overstatement that human communicative behavior is exceptional in the animal kingdom. This trait in humans is exceptional with regard to both quality and quantity. We use language (in two modalities), as well as a broad range of nonlinguistic signaling systems. One study estimated that during workdays in an office, people spend between 50 and 80 percent of their time communicating (Klemmer & Snyder, 1972). Obviously, it is very difficult to get a precise estimation of this activity, but it seems to be a safe assumption that communication is the most frequent activity among humans. After much research on animals, there seems little chance that we will find a species that outperforms us in this area.

In ethological terms, communication is defined as a sender relying on a set of specifically evolved behaviors (cf. signals) throughout the course of repeated interactions to modify the behavior of a receiver in a manner that is advantageous to the sender (Dawkins & Krebs, 1978). One may add that receivers can also benefit from the interaction, but this is not a necessity. Using this definition clearly helps provide objectivity in the study of animal communication, rather than immediately focusing on the issue of content that is unavoidable if one refers to communication as "information transfer."

While most previous studies aimed to look for the evolutionary continuity of communicative behavior among primates, the first question should have been about the possible changes in the function of communication. Despite this, we give a very short overview of primate communication, and then provide an answer to the first question that should have been asked.

17.4.1 Trends in Communicative Behaviors in Primates

Scientific tradition and evolutionary arguments supported the idea that the understanding of human communication could be deepened by studying corresponding behaviors in our closest relatives – the primates. One of the first major efforts was to teach a few more or less enculturated apes some forms of human language (for a recent review, see Lyn, 2012). The main conclusion of this research was that these apes mastered a relatively complex level of communicating with the researchers by using specific symbols (up to 500), but they never reached the level of a human child of around four to five years old. The majority of the researchers also agree that, even at comparable levels, apes and humans rely on some distinct mental mechanisms, so it is likely that the apes' achievements are more akin to high-level simulations of the relevant human ability. Thus, they may rely on their generally complex cognitive abilities rather than using language-specific mental mechanisms. Moreover, parallel research showed that apes are not alone in having such a capacity. Dolphins (Herman, 1987) and parrots (Pepperberg, 2010) come very close to (or even outperform) the linguistic performance seen in apes.

A more interesting approach is to look for complex communication patterns by observing interacting primates in the wild (or semi-naturalistic conditions). In the early stages of this research, many observers were surprised at the relatively small amount of communication in the acoustic modality among primates compared to the use of visual signaling. A homology between nonhuman primate and human vocal communication seemed to be a natural assumption. The relatively constrained vocal communication and more expressive visual signaling in primates led to the alternative assumption that in humans vocal communication was based on a generalization of gestural signaling; that is, linguistic use of vocal signals evolved de novo in humans. This view has been generally strengthened by the similarity between vocal and sign languages at the level of both performance and mechanisms (e.g., Hickok, Bellugi, & Klima, 1996).

Interestingly, the functionally most "word-like" primate signals are vocal in nature. Calls indicating the presence of a specific type of predator (e.g., leopard, snake, bird of prey) are emitted by many species, including vervet monkeys (*Chlorocebus pygerythrus*) (Cheney & Seyfarth, 1990) and Campbell's mona monkeys (*Cercopithecus campbelli*) (Zuberbühler, 2012). In some calls, the sequence of the vocal units may also confer different messages. In putty-nosed monkeys (*Cercopithecus nictitans*), different sequences of the same vocal units initiate different reactions in the listeners (Arnold & Zuberbühler, 2006). Such detailed observations are difficult to make, so evidence is accumulating very slowly. If such complex communicative skills proved to be more widespread in nonhuman primates, one might reconsider the issue of homology in acoustic communication in nonhuman primates and language. Note, however, that an evolutionary relation does not necessarily explain the fate of a character in the descendant species.

17.4.2 The Human Behavior Complex

To explain the seemingly unrivaled role of communication in humans, we may need to take a different route from that used in nonhuman primates. While communication has an obvious general function (see Section 17.4.1), this behavior serves specific functions in many different social interactions. Thus, the first step is to review in what ways the human social group is different from those in other nonhuman primates. This approach was pioneered by Dunbar (2009), who was the first to emphasize that humans live in much larger groups and that language may have evolved for the function of keeping in contact (Dunbar, 1996), rather than being a medium of information transfer. Thus, in functional terms, language may have replaced mutual grooming, which plays an important role in interindividual relationships. We agree with Dunbar's view, but we take a broader approach to this question (see also David-Barrett & Dunbar, 2016).

Számadó and Szathmáry (2006) listed many possible scenarios for language evolution, including its role in hunting, tool-making, sexual selection, and mother–infant interaction. Although all of these scenarios are regarded as alternatives, there is a common approach of searching for "the only" evolutionary scenario that provided the selective environment for the emergence of language. Based on Csányi (2000), we would like to take a different view. Rather than searching for a single cause for the emergence of the complex and seemingly limitless communicative skills in humans, we suggest that the greater the number of the above scenarios that were significantly present in any ancestral human species, the more likely it would be that communication would become a general activity.

The story of human evolution started about six million years ago when some genera (*Ardipithecus*, *Paranthropus*), represented by a few species, diverted from the ancestors of present-day chimpanzees (Uzzel & Pilbeam, 1971; Wood & Harrison, 2011). This was followed by the emergence of *Australopithecus* spp. and finally *Homo* spp. Although this transition seems to have been a gradual one over a few million years, it is similarly likely that most of these species shared some behavioral and cognitive features. Taking a more holistic view, human evolution can be seen as representing different features of social behavior emerging in a mosaic fashion in one or other species. There is no "royal road" to the present condition of our species.

It may be more fruitful instead to trace the evolution of a single function rather to look at the ways in which human social behavior has changed in general, some or most of which were already emerging in earlier genera. Csányi (2000) considers three different aspects of social behavior.

1 *Sociality* refers to the mechanisms that maintain and control complex social interactions, including intimate relationships, such as attachment or group-level distribution of resources like food.

2 *Synchronization* deals with the capacity to conform to rules in synchrony (doing things together) or

independently (observing rules). These mechanisms include rhythmic activities, imitation, and the performance of rites.

3 *Construction* is the human capacity to organize complex actions and to create corresponding mental representations that would never emerge in single individuals acting more or less in isolation. Language, tools, and religions are examples of such constructive abilities.

These three factors give rise to the human behavior complex, which differs in many ways from the corresponding abilities of chimpanzees. It is the interaction of these factors that makes an important difference. Consider, for example, attachment behavior in humans (Bolwby, 1969). A similar behavior system may also be present in chimpanzees between the mother and her baby, but in humans attachment has a broader function. In monogamous humans, attachment also develops between the infant and both parents (Ainsworth, 1989); moreover, attachment has an important role in keeping the adult parents together (Fraley et al., 2005). This intimate relationship, which endures over many years, provides a good context for the emergence of synchronic behavior (doing things together) and constructive actions (learning/teaching language).

17.4.3 Functional Synergies May Facilitate the Evolution of Communicative Skills

The above scenario suggests a simple picture. While non-human primates may communicate in specific situations, the factors of the human behavior complex make humans communicate in almost all situations. Communicative signals that may have evolved for functioning in one or other context are more likely to be used in other contexts or be replaced by more general ones by selection. An interesting parallel to this can be observed in the dog. In wolves and other canines, acoustic communicative signals are context specific (Schassburger, 1993). It seems, however, that domestication has changed the role of barking, and it has become a general signal used in different contexts by correspondingly modified acoustic structures (Pongrácz et al., 2005).

Thus, it is likely that in the human-ancestral species communicative behavior was associated with a variable numbers of contexts, and a major change occurred when widespread context-specific signaling was replaced by a uniform mechanism of generating sounds and rules of sound sequences. It is important to stress that there is no need to suppose that any communicative interaction would have needed complex linguistic structures and/or expressed complicated thoughts. The limited linguistic performance of apes, one- to two-year-old human infants, or dogs (Pilley & Reid, 2011) shows that complex interactions can be structured even with a small number of words uttered in the absence of any grammatical structure.

17.4.4 Urge for Communicating in Humans

The uniform communication system could have been rapidly subject to selection in different ways. Note that selection can only act on genetic components, so we also have to assume that this "standardization" was supported by genetic changes. Darwin previously noted that sexual selection may have contributed to the evolution of language. Men might have used their voices in courtship and to compete with other men (Darwin, 1871). In addition, better communicative skills may have increased the individual's fitness in the group (e.g., through better maintenance of social relationships or through group members' preference for "talkative" persons).

If language became a general way of interacting, then it may have increased the divergence among isolated human populations. Language may have become a phenotypic barrier that restricted migration from and to human groups. Separated groups may have been subject to group selection (Sober & Wilson, 1994), giving advantage to groups that developed better communicative skills for organizing their actions. Such selection could have recruited further genes controlling humans' communicative behavior. At the simplest level, these genes had little to do with the more complex aspects of language (Progovac, 2016), perhaps only facilitating communicative signaling in general. This trend can be seen clearly in human infants who start to use a wide range of communicative signals for a very early stage and seem also to be very interested in talking to people well before they have the capacity to use complex linguistic expressions. This strong propensity for communicating is typical of humans, and thus may easily have become a substrate for cultural evolution.

17.5 RUNAWAY COMMUNICATION IN HUMANS

In Section 17.4, we have shown that humans have a biological tendency to communicate. Runaway processes in humans, although they are usually cultural in kind, are based on such biological preferences. Our biological preference to communicate may have formed the basis of the dramatic improvement of information and communication technology (ICT) in recent years.

However, biological preference is not enough. As we argued in Section 17.4, liberation from environmental control is the other unconditional factor for the emergence of runaway processes. The emergence of speech in ancient humans increased the effectiveness of communication in humans. It allowed for keeping contact with many humans at the same time compared to mutual grooming in primates (Dunbar, 1996). However, the speaking partners needed to be in close range, and interaction was dependent on mutual attention during waking hours.

These constraints were removed from the environment gradually during the last 100 years or so. First, the landline phone, then the mobile phone and the Internet (e.g., e-mail, chat rooms) increased dramatically the speed and the geographical distance of interaction (Dunbar, 2012), reducing both spatial and temporal limits. Today, social networking sites (SNSs) break down even more constraints: we are connected to hundreds (or thousands) of people, broadcasting to them at once and immediately what has happened to us. Furthermore, our relationships can be maintained and strengthened by sending short, shallow messages, emoticons, emojis, or stickers, or by just clicking on the "like" button (e.g., Derks, Bos, & Grumbkow, 2007; Scissors, Burke, & Wengrovitz, 2016). With the Internet, it is also possible to "speak" to millions of people. Copresence is neither spatially nor temporally necessary (we can read the message of a remote friend later as well). Additionally, it is also possible (and frequent; e.g., on Facebook) to interact with others online who we have never met before (Vodafone, 2013). Furthermore, communication can be maintained through more than one channel (e.g., it is possible to participate in several chat conversations at the same time while posting on Facebook and making phone calls, etc.).

The popularity of these technologies is huge and increasing. In 2015, three billion users (two-thirds of the global population) had access to the Internet, and this proportion is 87 percent for advanced economies (Perrin, 2015). Worldwide, there are over 1.79 billion monthly active Facebook users, from which 1.66 billion use their mobile phones to log in (Facebook, 2016). The number of active mobile subscriptions exceeds the total world population (Ericsson, 2014; Kemp, 2014). Therefore, these technologies not only make communication limitless, but also they are accessible to the majority of people. The lack of limits makes communication subject to runaway processes. As a result, present-day humans display strong tendencies for exaggerated communication that are based on their selected predisposition to keep in contact with group members.

17.5.1 Cultural Preferences in the Runaway Processes of Communication

The preference for communication is transmitted not only by genes, but also by culture. Besides our biological tendency to communicate, cultural factors can also give positive feedback and facilitate runaway effects. Humans look for ever better ways to communicate, and technological development exploits this, developing newer devices and ways of communicating, which in turn leads to even more desire in people to communicate. The economy is also a facilitator: in advanced economies people can afford to buy digital devices and most of them have access to the Internet, while in developing countries the penetration of devices and Internet access are still limited (Perrin, 2015).

Humans' synchronization capacity (e.g., social rule following) also affects the tendency to use ICT for communication. What others do in our surrounding (including the larger units – e.g., the country – and the smallest groups we belong to) affect us: norms, fashion, and

advertisements tell us that using ICT is trendy and desirable. Mobile phones can signal social status (Lycett & Dunbar, 2000) and can be used to compensate for social isolation (Stald, 2008). On Facebook or other SNSs, users develop a profile that shows the best picture of them ("impression management"; Manago et al., 2008; Strano, 2008; but see Back et al., 2010); therefore, this may also be capable of lifting one's social status. Additionally, advertisements and cultural beliefs suggest that ICT is necessary to be in contact with friends, partners, and family, and people often worry that they might miss out on social events if they do not use Facebook (Przybylski et al., 2013). Thus, the lack of involvement in these technologies threatens us with social exclusion.

17.5.2 Exaggeration of Communicative Behavior

The fact that communication has started to run away (to become exaggerated) is indicated by the amount of time we spend on these technologies and the effects we experience when we cannot use them. In the USA in 2014, people used their smartphones 3.3 hours a day on average, and in young adults aged between 18 and 24, this number is 5.2 hours a day (Salesforce Marketing Cloud, 2014). It seems that people are increasingly dependent on their mobile phones. For example, according to a survey in 2013, 79 percent of smartphone owners keep their mobile with them for all but two of their waking hours, and a quarter of smartphone users cannot even recall the last time their mobile was not within earshot (Levitas, 2013). About two-thirds of mobile users report distress on being separated from their mobile (left at home, run out of battery, etc.), a phenomenon that is called "nomophobia" (Bivin et al., 2013; King et al., 2013), and this proportion is even higher in young adults (Sharma et al., 2015). Separation from the mobile induces self-reported behavioral and physiological anxiety (Clayton, Leshner, & Almond, 2015; Konok, Pogány, & Miklósi, 2017) and an attentional bias to separation-related stimuli that is indicative of loneliness (Konok et al., 2017).

There is a similar tendency in connection with SNSs. There are 1.18 billion people worldwide who log on to Facebook every day (the number was 0.37 billion in 2011). Half of 18–24-year-olds go on Facebook right after they wake up, and 28 percent of them do this from their smartphone before getting out of bed (Zephoria, 2015). British teenagers spend 100 minutes per day on social media, 24 minutes on video calls, 21 minutes on e-mail, and 15 minutes on taking selfies that they would share on SNSs (Logicalis, 2016). Similarly to nomophobia, many people experience anxiety when they cannot use Facebook because they think that this leads to missing out on social life. This phenomenon is called "fear of missing out," characterized by a pervasive worry that others might be having rewarding experiences from which one is absent and a desire to stay continually connected with what others are doing (Przybylski et al., 2013). Additionally, around a fifth of 13–17-year-olds feel bad if they do not get a lot of "likes" for a photo they have posted (Rideout & Saphir, 2012).

17.5.3 Costs of Exaggerated Communication

Whether this exaggeration has a cost is still being debated. Exaggerated traits need extra resources, either in time, energy, money (in the case of humans), attention/cognitive resources, or others. Here, we review briefly what kinds of consequences this extra need for resources may have. In addition, exaggeration may be maintained if the costs are compensated for in the long term.

17.5.3.1 Shallow Relationships

Any human behavior that becomes excessive will take time away from other activities. For example, spending time with online communication may take time away from off-line interactions. Some argue that time spent on the Internet displaces time spent on socializing, particularly with family and in face-to-face activities (Mesch, 2006; Nie & Hillygus, 2002). Research suggests that use of social media can have a negative impact on social relationships and well-being (Schiffrin et al., 2010; van den Eijnden et al., 2008). In particular, the quality of communication over social media compared to face-to-face communication is poorer and does not support the building of strong, emotionally intense relationships (Cummings, Butler, & Kraut, 2002; Schiffrin et al., 2010).

There are also contrary findings suggesting that communication through the Internet only supplements other forms of communication (Wellman et al., 2001) and can have positive effects on building strong and satisfying social relationships (for a review, see Bargh & McKenna, 2004). In contrast, the above studies (Cummings, Butler, & Kraut, 2002; Schiffrin et al., 2010) suggest that the relationships that are maintained on the Internet are weaker (emotionally less close) than offline relationships.

Although by spending time on social media we can build a larger online social network, this does not lead to emotionally close relationships (Pollet, Roberts, & Dunbar, 2011). The social brain hypothesis (an explanation for the evolution of brain size in primates) may explain this (Dunbar, 2012). It assumes that humans typically cannot maintain more than 150 relationships at any one time (Dunbar, 2012), and that there are cognitive and time constraints on the number of relationships that can be maintained at particular levels of emotional intensity (Dunbar, 2008). In line with this, research has shown that network size is negatively associated with mean emotional closeness (Roberts et al., 2009).

Furthermore, extensive SNS usage is associated with decreased involvement in the real-life social community and with more frequent relationship problems (Kuss & Griffiths, 2011). Other studies (e.g., Ezoe & Toda, 2013; Kraut et al., 1998) showed that there is an association between extensive Internet/mobile use and depression and especially loneliness. Although the causal relationship is unclear (i.e., lonelier people may use ICT for social compensation), the results show that online communication does not decrease loneliness.

17.5.3.2 Distraction of Attention

Spending time in online communication may take away attentional resources (and also time) from learning, work, or other important cognitive tasks. This can lead to worse performance in school or the workplace, to attentional problems, and even to accidents.

For example, mobile usage was found to be associated with faster and less accurate responding to higher-level cognitive tasks (Abramson et al., 2009), and the use of a mobile phone in childhood/adolescence was found to be associated with later attention problems (Byun et al., 2013; Zheng et al., 2014). Although some of the authors of these studies explain this association through the effects of the electromagnetic radiation that the head is exposed to, it is also possible that the effect is behavioral and not physiological in kind (e.g., the usage of these devices may activate/require alternative cognitive styles or skills). For example, texting on a phone is helped by the autocorrect function (or "predictive texting"), which may train the user to favor speed over accuracy (Abramson et al., 2009), and this may lead to an impulsive response style. In addition to mobile phone usage, Facebook usage was also found to be associated with worse performance in a memory test (Frein, Jones, & Gerow, 2013), which can be attributed again to the fact that Facebook trains the user to look through large amounts of content without paying significant attention to them.

Another phenomenon that frequently accompanies the usage of ICT is multitasking (i.e., the consumption of more than one stream of content at the same time). In contrast to what people generally believe, heavy media multitaskers are worse at switching between tasks, which may be due to their reduced ability to filter out interference from irrelevant stimuli (e.g., from a previous task) (Ophir, Nass, & Wagner, 2009; but see Minear et al., 2013). Multitaskers are also more impulsive (Minear et al., 2013), but they are better at multisensory integration (Lui & Wong, 2012). Thus, multitasking may allow people to attend to many things, but rather superficially.

Perhaps because of attentional and time distraction and changes in cognitive processing, intensive use of ICT leads to lower levels of academic performance (e.g., Kirschner & Karpinski, 2010; Lepp, Barkley, & Karpinski, 2014). Multitasking has a negative effect on learning in school (Rosen et al., 2011). Students spend fewer than six minutes on a single task prior to switching, most often due to technological distractions, including social media and texting (Rosen et al., 2013). The educational system has to deal with the challenges caused by these potential changes in children's cognitive styles.

In addition, mobile use takes attention away from everyday situations (Nasar, Hecht, & Wener, 2008) and constricts visual awareness (Maples et al., 2008). This can cause risks in some cases. For example, mobile usage while driving leads to a higher prevalence of accidents (Redelmeier & Tibshirani, 1997), yet many people talk via mobile while driving (Pennay, 2006). Mobile phones also distract pedestrians' attention, and mobile phone-related injuries among pedestrians are increasing (Nasar & Troyer, 2013).

The use of ICT takes time away from and affects other important areas as well; for example, it negatively affects sleeping (which may have also an indirect negative effect on performance in school or the workplace) (e.g., Bruni et al., 2015) and leads to participation in fewer recreational sport activities and higher levels of obesity (Kautiainen et al., 2005). Therefore, this taking of time and attention away from other activities leads to costs at both the individual and social level.

17.6 CONCLUSIONS

In this chapter, we presented arguments that runaway processes can also be observed in humans in connection with different kinds of (sexual and nonsexual) behaviors. These runaway processes are cultural in kind, but biological preferences provide the basis for them. Communication is a good example, and studying the exaggerated features of communication in this framework may help in recognizing the potential adverse effects it exerts on society and individuals. Additionally, the theory of runaway processes can provide guidelines for solutions as well: as the disappearance of some of the limitations on communication caused the exaggeration, we may need to place new constraints on culture to prevent the negative consequences and compensate for the costs.

17.7 ACKNOWLEDGMENTS

The work was supported by the Hungarian Academy of Sciences (a grant to the MTA-ELTE Comparative Ethology Research Group [F01/031]), the János Bolyai Research Fellowship from the Hungarian Academy of Sciences, and the Research Grant NKFI K124458 The Good Mobile: Facilitating Human Relations Using an Ethologically Grounded Attachment Model.

REFERENCES

Abramson, M. J., Benke, G. P., Dimitriadis, C., et al. (2009). Mobile telephone use is associated with changes in cognitive function in young adolescents. *Bioelectromagnetics*, **30**(8), 678–686.

Ainsworth, M. S. (1989). Attachments beyond infancy. *American Psychologist*, **44**(4), 709–716.

American Society of Plastic Surgery. (2015). Plastic surgery procedural statistics. www.plasticsurgery.org/news/plastic-surgery-statistics.

Arnold, K., & Zuberbühler, K. (2006). The alarm-calling system of adult male putty-nosed monkeys, *Cercopithecus nictitans martini. Animal Behaviour*, **72**(3), 643–653.

Back, M. D., Stopfer, J. M., Vazire, S., et al. (2010). Facebook profiles reflect actual personality, not self-idealization. *Psychological Science*, **21**(3), 372–374.

Bargh, J. A., & McKenna, K. Y. A. (2004). The Internet and social life. *Annual Review of Psychology*, **55**(1), 573–590.

Bivin, J. B., Mathew, P., Thulasi, P. C., & Philip, J. (2013). Nomophobia – Do we really need to worry about? *Reviews of Progress*, **1**, 1–5.

Bowlby, J. (1969). *Attachment and Loss: Vol. 1. Attachment*. New York: Basic Books.

Bruni, O., Sette, S., Fontanesi, L., et al. (2015). Technology use and sleep quality in preadolescence and adolescence. *Journal of Clinical Sleep Medicine*, **11**(12), 1433–1441.

Byun, Y., Ha, M., Kwon, H., et al. (2013). Mobile phone use, blood lead levels, and attention deficit hyperactivity symptoms in children: A longitudinal study. *PLoS ONE*, **8**(3), e59742.

Cavalli-Sforza, L. L., & Feldman, M. W. (1981). Cultural transmission and evolution: A quantitative approach. *Monographs in Population Biology*, **16**, 1–388.

Cheney, D. L., & Seyfarth, R. M. (1990). *How Monkeys See the World*. Chicago, IL: University of Chicago Press.

Chielens, K., & Heylighen, F. (2005). Operationalization of meme selection criteria: Methodologies to empirically test memetic predictions. In *Proceedings of the Joint Symposium on Socially Inspired Computing (AISB'05)*. Hove: Society for the Study of Artificial Intelligence and the Simulation of Behaviour, pp. 14–20.

Christakis, N. A., & Fowler, J. H. (2007). The spread of obesity in a large social network over 32 years. *New England Journal of Medicine*, **357**(4), 370–379.

Clayton, R. B., Leshner, G., & Almond, A. (2015). The extended iSelf: The impact of iPhone separation on cognition, emotion, and physiology. *Journal of Computer-Mediated Communication*, **20**(2), 119–135.

Csányi, V. (2000). The "human behavior complex" and the compulsion of communication: Key factors of human evolution. *Semiotica*, **128**(3–4), 243–258.

Cummings, J. N., Butler, B., & Kraut, R. (2002). The quality of online social relationships. *Communications of the ACM*, **45**(7), 103–108.

Darwin, C. (1871). *The Descent of Man, and Selection in Relation to Sex*. London: John Murray.

David, P., Bjorksten, T., Fowler, K., & Pomiankowski, A. (2000). Condition-dependent signalling of genetic variation in stalk-eyed flies. *Nature*, **406**(6792), 186–188.

David-Barrett, T., & Dunbar, R. I. M. (2016). Language as a coordination tool evolves slowly. *Royal Society Open Science*, **3**(12), 160259.

Dawkins, R. (1976). *The Selfish Gene*. New York: Oxford University Press.

Dawkins, R., & Krebs, J. R. (1978). Animal signals: Information or manipulation. In J. R. Krebs & N. B. Davies, eds., *Behavioural Ecology: An Evolutionary Approach*. Oxford: Blackwell Publishing, pp. 282–309.

Derks, D., Bos, A. E. R., & Grumbkow, J. (2007). Emoticons and social interaction on the Internet: The importance of social context. *Computers in Human Behavior*, **23**(1), 842–849.

Dixson, B. J., Dixson, A. F., Li, B., & Anderson, M. J. (2007). Studies of human physique and sexual attractiveness: Sexual preferences of men and women in China. *American Journal of Human Biology*, **19**, 88–95.

Dunbar, R. I. M. (1996). *Gossip, Grooming and the Evolution of Language*. Cambridge, MA: Harvard University Press.

Dunbar, R. I. M. (2008). Cognitive constraints on the structure and dynamics of social networks. *Group Dynamics: Theory, Research, and Practice*, **12**(1), 7–16.

Dunbar, R. I. M. (2009). The social brain hypothesis and its implications for social evolution. *Annals of Human Biology*, **36**(5), 562–572.

Dunbar, R. I. M. (2012). Social cognition on the Internet: Testing constraints on social network size. *Philosophical Transactions of the Royal Society of London. Series B, Biological Sciences*, **367** (1599), 2192–2201.

Ericsson, J. (2014). Ericsson mobility report. https://edna.iea-4e .org/files/otherfiles/0000/0438/11_-_Ericsson_Mobility_ Report_-_Patrik_Cerwall.pdf.

Ezoe, S., & Toda, M. (2013). Relationships of loneliness and mobile phone dependence with Internet addiction in Japanese medical students. *Open Journal of Preventive Medicine*, **3**(6), 407–412.

Facebook. (2016). Facebook Reports Third Quarter 2016. https:// s21.q4cdn.com/399680738/files/doc_financials/2016/Q3/3.- Facebook-Reports-Third-Quarter-2016-Results.pdf.

Fan, J., Dai, W., Liu, F., & Wu, J. (2005). Visual perception of male body attractiveness. *Proceedings of the Royal Society of London B: Biological Sciences*, **272**(1560), 219–226.

Fink, B., Weege, B., Manning, J. T., & Trivers, R. (2014). Body symmetry and physical strength in human males. *American Journal of Human Biology*, **26**(5), 697–700.

Fisher, R. A. (1930). *The Genetical Theory of Natural Selection*. Oxford: Clarendon Press.

Folstad, I., & Karter, A. J. (1992). Parasites, bright males, and the immunocompetence handicap. *American Naturalist*, **139**(3), 603–622.

Fraley, R. C., & Shaver, P. R. (2005). The evolution and function of adult attachment: A comparative and phylogenetic analysis. *Journal of Personality and Social Psychology*, **89**, 731–746.

Frein, S. T., Jones, S. L., & Gerow, J. E. (2013). Computers in human behavior when it comes to Facebook there may be more to bad memory than just multitasking. *Computers in Human Behavior*, **29**(6), 2179–2182.

Getty, T. (2006). Sexually selected signals are not similar to sports handicaps. *Trends in Ecology & Evolution*, **21**(2), 83–88.

Go, A. S., Mozaffarian, D., Roger, V. L., et al. (2013). Heart disease and stroke statistics – 2013 update: A report from the American Heart Association. *Circulation*, **127**(1), e6–e245.

Grafen, A. (1990). Biological signals as handicaps. *Journal of Theoretical Biology*, **144**(4), 517–546.

Haufe, C. (2007). Sexual selection and mate choice in evolutionary psychology. *Biology & Philosophy*, **23**(1), 115–128.

Herman, L. M. (1987). Receptive competencies of language-trained animals. *Advances in the Study of Behavior*, **17**, 1–60.

Hickok, G., Bellugi, U., & Klima, E. S. (1996). The neurobiology of sign language and its implications for the neural basis of language. *Nature*, **381**(6584), 699–702.

Hill, G. E., & Montgomerie, R. (1994). Plumage colour signals nutritional condition in the house finch. *Proceedings of the Royal Society of London B: Biological Sciences*, **258**(1351), 47–52.

Johnston, V. S., & Franklin, M. (1993). Is beauty in the eye of the beholder? *Ethology and Sociobiology*, **14**(3), 183–199.

Kautiainen, S., Koivusilta, L., Lintonen, T., Virtanen, S. M., & Rimpelä, A. (2005). Use of information and communication technology and prevalence of overweight and obesity among adolescents. *International Journal of Obesity*, **29**(8), 925–933.

Kemp, S. (2014). Social, Digital and Mobile in 2015. http://zenith malaysia.com/infobank/Digital_social_and_mobile_in_ 2015_Executive_summary.pdf.

King, A. L. S., Valença, A. M., Silva, A. C. O., et al. (2013). Nomophobia: Dependency on virtual environments or social phobia? *Computers in Human Behavior*, **29**(1), 140–144.

Kirschner, P. A., & Karpinski, A. C. (2010). Facebook and academic performance. *Computers in Human Behavior*, **26**(6), 1237–1245.

Klemmer, E. T., & Snyder, F. W. (1972). Measurement of time spent communicating. *Journal of Communication*, **22**(2), 142–158.

Konok, V., Pogány, Á., & Miklósi, Á. (2017). Mobile attachment: Separation from the mobile induces physiological and behavioural stress and attentional bias to separation-related stimuli. *Computers in Human Behavior*, **71**, 228–239.

Kramer, S., Zebrowitz, L. A., San Giovanni, J. P., & Sherak, B. (1995). Infant preferences for attractiveness and babyfaceness. In B. G. Bardy, R. J. Bootsma, & Y. Guillard, eds., *Studies in Perception and Action III*. Hillsdale, NJ: Lawrence Erlbaum Associates, pp. 389–392.

Kraut, R., Patterson, M., Lundmark, V., et al. (1998). Internet paradox: A social technology that reduces social involvement and psychological well-being? *American Psychologist*, **53**(9), 1017–1031.

Kuss, D. J., & Griffiths, M. D. (2011). Online social networking and addiction – A review of the psychological literature. *International Journal of Environmental Research and Public Health*, **8**(12), 3528–3552.

Lande, R. (1981). Models of speciation by sexual selection on polygenic traits. *Evolution*, **78**(6), 3721–3725.

Langlois, J. H., Kalakanis, L., Rubenstein, A. J., et al. (2000). Maxims or myths of beauty? A meta-analytic and theoretical review. *Psychological Bulletin*, **126**(3), 390–423.

Lavie, C. J., Milani, R. V, & Ventura, H. O. (2009). Obesity and cardiovascular disease: Risk factor, paradox, and impact of weight loss. *Journal of the American College of Cardiology*, **53** (21), 1925–1932.

Lepp, A., Barkley, J. E., & Karpinski, A. C. (2014). The relationship between cell phone use, academic performance, anxiety, and Satisfaction with Life in college students. *Computers in Human Behavior*, **31**, 343–350.

Levitas, D. (2013). *Always Connected: How Smartphones And Social Keep Us Engaged*. Framingham, MA: International Data Corporation (IDC).

Little, A. C., Apicella, C. L., & Marlowe, F. W. (2007). Preferences for symmetry in human faces in two cultures: Data from the UK and the Hadza, an isolated group of hunter–gatherers. *Proceedings of the Royal Society of London B: Biological Sciences*, **274**(1629), 3113–3117.

Little, A. C., Jones, B. C., & DeBruine, L. M. (2011). The many faces of research on face perception. *Philosophical Transactions of the Royal Society of London B: Biological Sciences*, **366**(1571), 1634–1637.

Logicalis (2016). The age of digital enlightenment. www .uk.logicalis.com/globalassets/united-kingdom/microsites/real- time-generation/realtime-generation-2016-report.pdf.

Lui, K. F. H., & Wong, A. C.-N. (2012). Does media multitasking always hurt? A positive correlation between multitasking and multisensory integration. *Psychonomic Bulletin & Review*, **19**(4), 647–653.

Lycett, J. E., & Dunbar, R. I. M. (2000). Mobile phones as lekking devices among human males. *Human Nature*, **11**(1), 93–104.

Lyn, H. (2012). *Apes and the Evolution of Language: Taking Stock of 40 Years of Research*. Oxford: Oxford University Press.

Lyon, B. E., & Montgomerie, R. (2012). Sexual selection is a form of social selection. *Philosophical Transactions of the Royal Society of London B: Biological Sciences*, **367**(1600), 2266–2273.

Lyon, B. E., Eadie, J. M., & Hamilton, L. D. (1994). Parental choice selects for ornamental plumage in American coot chicks. *Nature*, **371**, 240–243.

Manago, A. M., Graham, M. B., Greenfield, P. M., & Salimkhan, G. (2008). Self-presentation and gender on MySpace. *Journal of Applied Developmental Psychology*, **29**(6), 446–458.

Maples, W. C., DeRosier, W., Hoenes, R., Bendure, R., & Moore, S. (2008). The effects of cell phone use on peripheral vision. *Optometry – Journal of the American Optometric Association*, **79**(1), 36–42.

Mason, N. A., Shultz, A. J., & Burns, K. J. (2014). Elaborate visual and acoustic signals evolve independently in a large, phenotypically diverse radiation of songbirds. *Proceedings of the Royal Society of London B: Biological Sciences*, **281**(1788), 20140967.

Masters, R. K., Reither, E. N., Powers, D. A., et al. (2013). The impact of obesity on US mortality levels: The importance of age and cohort factors in population estimates. *American Journal of Public Health*, **103**(10), 1895–1901.

Mennella, J. A. (2014). Ontogeny of taste preferences: Basic biology and implications for health. *American Journal of Clinical Nutrition*, **99**(3), 704S–711S.

Mesch, G. S. (2006). Family relations and the Internet: Exploring a family boundaries approach. *Journal of Family Communication*, **6**(2), 119–138.

Minear, M., Brasher, F., Mccurdy, M., Lewis, J., & Younggren, A (2013). Working memory, fluid intelligence, and impulsiveness in heavy media multitaskers. *Psychonomic Bulletin & Review*, **20**(6), 1274–1281.

Møller, A. P. (1992). Female swallow preference for symmetrical male sexual ornaments. *Nature*, **357**(6375), 238–240.

Moore, A. J., Brodie, E. D., & Wolf, J. B. (1997). Interacting phenotypes and the evolutionary process: I. Direct and indirect genetic effects of social interactions. *Evolution*, **51**(5), 1352–1362.

Nakamaru, M., & Dieckmann, U. (2009). Runaway selection for cooperation and strict-and-severe punishment. *Journal of Theoretical Biology*, **257**(1), 1–8.

Nasar, J., Hecht, P., & Wener, R. (2008). Mobile telephones, distracted attention, and pedestrian safety. *Accident Analysis & Prevention*, **40**(1), 69–75.

Nasar, J. L., & Troyer, D. (2013). Pedestrian injuries due to mobile phone use in public places. *Accident Analysis & Prevention*, **57**, 91–95.

Nie, N. H., & Hillygus, D. S. (2002). The impact of Internet use on sociability: Time-diary findings. *IT & Society*, **1**(1), 1–20.

Noë, R., & Hammerstein, P. (1994). Biological markets: Supply and demand determine the effect and mating mutualism of partner choice in cooperation. *Behavioral Ecology*, **35**(1), 1–11.

Ophir, E., Nass, C., & Wagner, A. D. (2009). Cognitive control in media multitaskers. *Proceedings of the National Academy of Sciences*, **106**(37), 15583–15587.

Pachucki, M. A., Jacques, P. F., & Christakis, N. A. (2011). Social network concordance in food choice among spouses, friends, and siblings. *American Journal of Public Health*, **101**(11), 2170–2177.

Pennay, D. (2006). *ATSB Research and Analysis Report. Road Safety Consultant Report CR 229, Community Attitudes to Road Safety – Wave 19*. Canberra: Australian Transport Safety Bureau.

Penton-Voak, I. S., & Chen, J. Y. (2004). High salivary testosterone is linked to masculine male facial appearance in humans. *Evolution and Human Behavior*, **25**(4), 229–241.

Pepperberg, I. M. (2010). Functional vocalizations by an African grey parrot (*Psittacus erithacus*). *Zeitschrift Für Tierpsychologie*, **55**(2), 139–160.

Perrett, D. I., May, K. A., & Yoshikawa, S. (1994). Facial shape and judgements of female attractiveness. *Nature*, **368**(6468), 239–242.

Perrin, A. (2015). Social Media Usage: 2005–2015 (October), 2005–2015. Pew Research Center. October 2015. www.pewinternet.org/2015/10/08/2015/Social-Networking-Usage-2005-2015.

Pilley, J. W., & Reid, A. K. (2011). Border collie comprehends object names as verbal referents. *Behavioural Processes*, **86**(2), 184–195.

Pollet, T. V., Roberts, S. G. B., & Dunbar, R. I. M. (2011). Use of social network sites and instant messaging does not lead to increased offline social network size, or to emotionally closer relationships with offline network members. *Cyberpsychology, Behavior, and Social Networking*, **14**(4), 253–258.

Pongrácz, P., Molnár, C., Miklósi, A., & Csányi, V. (2005). Human listeners are able to classify dog (*Canis familiaris*) barks recorded in different situations. *Journal of Comparative Psychology*, **119**(2), 136–144.

Progovac, L. (2016). A gradualist scenario for language evolution: Precise linguistic reconstruction of early human (and Neandertal) grammars. *Frontiers in Psychology*, **7**, 1714.

Pryke, S. R., Andersson, S., & Lawes, M. J. (2001). Sexual selection of multiple handicaps in the red-collared widowbird: Female choice of tail length but not carotenoid display. *Evolution*, **55**(7), 1452–1463.

Przybylski, A. K., Murayama, K., Dehaan, C. R., & Gladwell, V. (2013). Motivational, emotional, and behavioral correlates of fear of missing out. *Computers in Human Behavior*, **29**(4), 1841–1848.

Redelmeier, D. A., & Tibshirani, R. J. (1997). Association between cellular-telephone calls and motor vehicle collisions. *New England Journal of Medicine*, **336**(7), 453–458.

Rhodes, G. (2006). The evolutionary psychology of facial beauty. *Annual Review of Psychology*, **57**, 199–226.

Rhodes, G., Zebrowitz, L. A., Clark, A., et al. (2001). Do facial averageness and symmetry signal health? *Evolution and Human Behavior*, **22**, 31–46.

Rideout, V., & Saphir, M. (2012). *Social Media, Social Life: How Teens View Their Digital Lives*. San Francisco, CA: Common Sense Media.

Roberts, S. G. B., Dunbar, R. I. M., Pollet, T. V., & Kuppens, T. (2009). Exploring variation in active network size: Constraints and ego characteristics. *Social Networks*, **31**, 138–146.

Rosen, L. D., Carrier, L. M., & Cheever, N. A. (2013). Facebook and texting made me do it: Media-induced task-switching while studying. *Computers in Human Behavior*, **29**(3), 948–958.

Rosen, L. D., Lim, A. F., Carrier, L. M., & Cheever, N. A. (2011). An empirical examination of the educational impact of text message-induced task switching in the classroom: Educational implications and strategies to enhance learning. *Psicología Educativa* **17**, 163–177.

Rosser, S. V. (1992). *Biology and Feminism. A Dynamic Interaction*. New York: Twayne Publishers.

Salesforce Marketing Cloud (2014). 2014 mobile behavior report. https://brandcdn.exacttarget.com/sites/exacttarget/files/deliverables/etmc-2014mobilebehaviorreport.pdf.

Schassburger, R. M. (1993). *Vocal Communication in the Timber Wolf*, Canis lupus, *Linnaeus: Structure, Motivation, and Ontogeny*. Berlin: Paul Parey Scientific Publishers.

Schiffrin, H., Edelman, A., Falkenstern, M., & Stewart, C. (2010). The associations among computer-mediated communication, relationships, and well-being. *Cyberpsychology, Behavior, and Social Networking*, **13**(3), 299–306.

Scissors, L., Burke, M., & Wengrovitz, S. (2016). What's in a Like? Attitudes and behaviors around receiving Likes on Facebook. *Proceedings of the 19th ACM Conference on Computer-Supported Cooperative Work & Social Computing – CSCW*, **16**, 1499–1508.

Searcy, W. A., & Nowicki, S. (2005). *The Evolution of Animal Communication: Reliability and Deception in Signaling Systems*. Princeton, NJ: Princeton University Press.

Sharma, N., Sharma, P., Sharma, N., & Wavare, R. (2015). Rising concern of nomophobia amongst Indian medical students. *International Journal of Research in Medical Sciences*, **3**(3), 705–707.

Sober, E., & Wilson, D. S. (1994). A critical review of philosophical work on the units of selection problem. *Philosophy of Science*, **61**(4), 534–555.

Stald, G. (2008). Mobile identity: Youth, identity, and mobile communication media. In D. Buckingham, ed., *Youth, Identity, and Digital Media*. Cambridge, MA: MIT Press, pp. 143–164.

Strano, M. (2008). User descriptions and interpretations of self-presentation through Facebook profile images. *Journal of Psychosocial Research on Cyberspace*, **2**(2), 1–13.

Számadó, S., & Szathmáry, E. (2006). Selective scenarios for the emergence of natural language. *Trends in Ecology & Evolution*, **21**(10), 555–561.

Thornhill, R., & Gangestad, S. W. (1999). Facial attractiveness. *Trends in Cognitive Sciences*, **3**(12), 452–460.

Thornhill, R., & Møller, A. P. (1997). Developmental stability, disease and medicine. *Biological Reviews of the Cambridge Philosophical Society*, **72**(4), 497–548.

Uzzell, T., & Pilbeam, D. (1971). Phyletic divergence dates of hominoid primates: A comparison of fossil and molecular data. *Evolution*, **25**(4), 615–635.

van den Eijnden, R. J., Meerkerk, G.-J., Vermulst, A. A., Spijkerman, R., & Engels, R. C. (2008). Online communication, compulsive internet use, and psychosocial well-being among adolescents: A longitudinal study. *Developmental Psychology*, **44**(3), 655–665.

Verhulst, S., Dieleman, S. J., & Parmentier, H. K. (1999). A tradeoff between immunocompetence and sexual ornamentation in domestic fowl. *Proceedings of the National Academy of Sciences*, **96**(8), 4478–4481.

Vodafone (2013). Digital lives: How do teenagers in the UK navigate their digital world? www.vodafone.com/content/dam/vodafone/parents/assets_2013/pdf/digital_lives_report.pdf.

Wellman, B., Haase, A. Q., Witte, J., & Hampton, K. (2001). Does the Internet increase, decrease, or supplement social capital?: Social networks, participation, and community commitment. *American Behavioral Scientist*, **45**(3), 436–455.

West-Eberhard, M. J. (1979). Sexual selection, social competition, and evolution. *Proceedings of the American Philosophical Society*, **123**(4), 222–234.

Wilkinson, G. S., Presgraves, D. C., & Crymes, L. (1998). Male eye span in stalk-eyed flies indicates genetic quality by meiotic drive suppression. Nature, **391**(6664), 276–279.

Wilson, D. S., & Sober, E. (1994). Reintroducing group selection to the human behavioral sciences. *Behavioral and Brain Sciences*, **17**, 585–608.

Wolf, J. B., Brodie, E. D., & Moore, A. J. (1999). Interacting phenotypes and the evolutionary process. II. Selection resulting from social interactions. *American Naturalist*, **153**(3), 254–266.

Wood, B., & Harrison, T. (2011). The evolutionary context of the first hominins. *Nature*, **470**(7334), 347–352.

Zahavi, A. (1975). Mate selection – A selection for a handicap. *Journal of Theoretical Biology*, **53**(1), 205–214.

Zephoria (2015). The Top 20 Valuable Facebook Statistics – Updated October 2015. https://zephoria.com/top-15-valuable-facebook-statistics.

Zheng, F., Gao, P., He, M., et al. (2014). Association between mobile phone use and inattention in 7102 Chinese adolescents: A population-based cross-sectional study. *BMC Public Health*, **14**, 1022.

Zuberbühler, K. (2012). Communication in nonhuman primates. In J. Vonk & T. K. Shakelford, eds., *The Oxford Handbook of Comparative Evolutionary Psychology*. Oxford: Oxford University Press, pp. 320–338.

18 Ontogeny of Tactical Deception

PETER J. LAFRENIERE AND RACHELLE M. SMITH

Years ago, we began a research program in the development of deliberate tactical deception because we felt that deception was central to the evolution of human social intelligence. We were also convinced that it was central to a successful social life in the modern era, particularly the ability to unmask deception masquerading as benevolence or cooperation. This is the pith of the *Machiavellian Hypothesis* that came to the forefront of evolutionary thinking in the 1980s (e.g., Byrne & Whiten, 1988; Humphrey, 1976). Now, 40 years later, we review the empirical case for its progressive development in children.

We view this essential question regarding the ontogeny of deception within a broader framework of interrelated questions. In his influential paper, "On the Aims and Methods of Ethology" (1963), Tinbergen gave the discipline of ethology a strong and enduring paradigm. According to Tinbergen (1951), "broad descriptive reconnaissance of the whole system of phenomena is necessary in order to see each individual problem in its perspective, it is the only safeguard for a balanced approach in which analytical and synthetical thinking can cooperate" (p. 130). Tinbergen provides a succinct account of an ethological orientation to the study of behavior in his fourfold scheme of basic questions concerning (1) evolution, (2) development, (3) causal mechanisms, and (4) adaptive function. Within this framework, an ethologist seeks to understand behavior from a broad perspective that combines the questions of a developmental psychologist (How does a particular behavior develop? What causes it to occur?) with those of a zoologist (How did a particular behavior evolve? What are its adaptive functions?).

Today, many ethologists believe that signals evolved toward greater persuasiveness rather than toward increasingly accurate readouts of inner states and intentions, as was initially thought. For example, an animal might bluff in an agonistic encounter, or misdirect the attention of the receiver away from a food location. This position proposes that expressions that exaggerate or minimize internal emotional states may be strategically advantageous to their sender on some occasions, particularly if they are likely to be accepted as true (Dawkins & Krebs, 1978).

The individual's adaptation within the social group may depend upon its ability to manage its own signals and assess the veracity of signals directed to them. Owings and Morton (1998) argue that the tension between management and assessment issues drives the evolution of communicative systems to new heights by fostering successive adjustments in both the subtlety of displays and in the ability to interpret them. This "evolutionary arms race" is thought to have supported the evolution of our unique human social intelligence from its earlier primate roots. The cognitive skills underlying sophisticated tactical deception are quintessentially human, making human social life the most complex arena in which it is honed. While we focus our review of the research on the development of deliberate tactical deception on children, we recognize that deception itself is a much broader topic.

18.1 PHYLOGENY OF DECEPTION

Deception/detection as a strategy for the competitive struggle for survival, acquiring resources, securing mates, protecting offspring, etc., is ubiquitous throughout the animal world. Because of the enormous power of design via natural selection, deception in animals can be exceedingly clever. A phylogenetic analysis of deception reveals four qualitatively distinct levels (LaFreniere, 1988a; Mitchell, 1986).

1 Morphological (e.g., Batesian mimicry, or camouflage)
2 Genetic behavioral action patterns (e.g., broken wing display)
3 Conditioned behavior
4 Deliberate tactical deception

Distinguishing between the latter two forms of deception has proven difficult for researchers and has shifted their methods from anecdotal observations of clever deception in animals to more controlled study of deception in the lab. A key focus has been laboratory research with chimpanzees. In their landmark study entitled "Does the Chimpanzee Have a Theory of Mind," Premack and Woodruff explored whether captive chimpanzees could deceive their human trainers. Their ingenious experimental investigations of intentional communication in chimpanzees seemed to demonstrate that chimps were capable of clever deceptions that might be based upon their understanding of the desires, emotions, and intentions of others (Premack & Woodruff, 1978; Woodruff & Premack, 1979). They presented three- to four-year-old chimps with the following dilemma: the chimps knew the location of a hidden banana, but they could not reach it without the help of a trainer who could reach it but did not know where it was hidden. Over a number of trials, the chimps were exposed to two different trainers: a kind trainer who, if shown where the banana was hidden, would give it to them to eat; and a villainous trainer who, if shown the correct location, would eat the banana himself. Faced with such duplicity, one of the chimps, Sarah, learned to direct honest communications to the kind trainer, but to deceive the villainous trainer by pretending not to know where the banana was hidden, or even by pointing to the wrong container. Premack and Woodruff believed that Sarah consciously employed a deceptive strategy based upon her understanding of the intentions of the villainous trainer, though this interpretation remains controversial because it is difficult to distinguish such a strategy from other types of deception based on conditioning (for a review, see Call & Tomasello, 2008).

While chimpanzees appear to show some surprising abilities in deception, they make fewer attributions than humans, since they cannot attribute states of mind that they themselves do not possess. Naturalistic and experimental evidence suggests that apes make simple attributions of seeing, wanting, and expecting, rather than attributions about beliefs. Thus, they appear to possess only a limited theory of mind, similar to that of a three-year-old child. Premack's rule of thumb is that if a child of 3.5 years cannot do it, neither can the chimpanzee (Premack, 1988). Premack addressed this question by dividing states of mind into simple and complex states:

Simple states are those produced by processes that are hard-wired, automatic or reflex-like, and encapsulated … perception is the prototypic simple state, we may add others: first, certain basic motivational states; and secondly, somewhat more controversially, expectancy, a state that is produced by conditioning or simple learning. These three states – seeing, wanting, expecting – have in common a restricted and automatic production process that is independent of language both at the input level of the system and of internal representation … Complex states, of which belief is the prototype, are of course everything that simple states are not. Belief is not automatic, encapsulated, or hard-wired; moreover it definitely depends on language, most certainly at the level of internal representation though often also at the level of input to the system. (Premack, 1988, p. 172)

Current research indicates that this division between simple and complex states may provide a rough estimate of the abilities that chimpanzees and other nonhuman primates do and do not possess. Chimpanzees know what others can and cannot see, and they can use this information to compete more effectively for food (Call & Tomasello, 2008). However, it is less clear how far their visual perspective-taking skills go. Does this skill allow them to make correct inferences about what others know? Although chimpanzees possess a basic capacity to follow the gaze direction of others using body orientation as a cue rather than eyes, they seem to lack the visual perspective-taking skills of four-year-old children. Thus, chimpanzees do not appear to understand gaze as referential communication, and they clearly do not understand pointing as such (Povinelli & O'Neill, 2000). Nor do chimpanzees appear to discriminate between accidental and intentional behavior, nor between a knowledgeable and naive experimenter (Call & Tomasello, 2008). Chimpanzees have not been successful on nonverbal false-belief tests. In general, they appear to lack the human capacity for abstract causal reasoning, and to lack the ability to posit unobservable constructs to explain observable events that is central to a mature theory of mind (Bering & Povinelli, 2003). In a recent summary of 30 years of research (Call & Tomasello, 2008), growing consensus from experimental evidence seems to indicate that chimpanzees understand others in terms of a perception–goal psychology, but not in terms of a belief–desire psychology. Humans, of course, clearly do understand others in terms of their desires and beliefs, and they do so from an early age.

18.2 DEVELOPMENT OF TACTICAL DECEPTION IN CHILDREN

Tactical deception may be either active or passive. It can involve the active use of communicative skills in order to mislead another individual, but it can also be achieved when the deceiver withholds information. In humans, tactical deception is thought to rely upon various social cognitive skills, such as theory of mind – the understanding that other people have alternative representations of the world that may be true or false and may differ from one's own (Wellman, Cross, & Watson, 2001) – and an understanding of how the nonverbal cues one provides regarding one's intentions can influence the beliefs and actions of observers (Shultz & Cloghesy, 1981). Here, we review research on the development of such skills, including both false signaling and its detection.

How would young children react to the type of problem Premack presented to his chimps? In a series of early studies, we explored the development of tactical deception within a game context, in which the child had to try to conceal a hidden object from an adult confederate (LaFreniere, 1988b; LaFreniere & Menard, 1990). Our early results showed that young children, between three and four years of age, were very poor at concealing information or providing false information to their opponent. By about eight years of age, however, children showed greater awareness of the effects of their own behavior on the experimenter's beliefs and expectations, and greater success in their deception.

Subsequent research into the early development of children's deception has been lively, with numerous studies of verbal and nonverbal deception in a wide range of situations; both realistic (Josephs, 1993, 1994; Lewis, Stanger, & Sullivan, 1989) and highly artificial (Chandler, Fritz, & Hala, 1989; Hala, Chandler, & Fritz, 1991; Sodian et al., 1991). In general, this work demonstrates that the early forms of deception, up to the fourth year, are unsophisticated and consist of attempts to withhold information by not confessing to transgressions (Lewis et al., 1989) or by removing incriminating evidence (Chandler et al., 1989). In another variation on the hiding game used with chimps, Sodian and Schneider (1990) showed that only a few of the four-year-olds, half of the five-year-olds, but almost all of the six-year-olds were able to effectively use tactical strategies. Similarly, using a lying paradigm, Talwar, Gordon, and Lee (2007) demonstrated that six- and seven-year-olds engaged in strategies to avoid lie detection, including providing incorrect answers to questions designed to entrap them, while children between three and five years of age were more likely to blurt out correct answers, revealing their hidden intentions to a partner.

18.3 VERBAL LIES

Lying is an especially interesting form of tactical deception because verbal deception is uniquely human. Rudimentary lying emerges early and is relatively unsophisticated. Proficient lying relies on the coordination of cognitive abilities such as executive control of verbal and nonverbal expressions, and it requires a sophisticated understanding of social interactions, social expectations, and social skills (Talwar & Crossman, 2011). The liar must coordinate verbal and nonverbal expressions and fabricate displays that are congruent with the lie. The deceiver must inhibit verbal and nonverbal expressions, control for leakage of genuine information, and fabricate information that aligns with the deceptive intent. Successful lying requires consistency between the initial lie and follow-up behavior and expressions (Talwar & Crossman, 2011).

Beginning around age two, lying is typically self-serving and emerges in circumstances where the child is attempting to avoid punishment or to gain rewards (LaFreniere, 1988a). These first attempts are almost always egocentric and ineffective at this age. By age three, children start to offer verbal lies to conform to social expectations, to protect the feelings of others, and to smooth social interactions (Talwar & Crossman, 2011). Around four years of age, children begin to generate rudimentary intentional lies to influence the beliefs of others. These lies can be playful, defensive, aggressive, competitive, or protective (LaFreniere, 1988b), but they tend to be marked by inconsistency between verbal and nonverbal behavior or between verbal statements and follow-up remarks (Talwar, Gordon, & Lee, 2007), likely due to immature

executive functioning and self-assessment skills (Evans & Lee, 2013). By age six or seven, however, children are more able to effectively use tactical strategies to avoid lie detection (Talwar, Gordon, & Lee, 2007). Around seven or eight years of age, children start to demonstrate more sophisticated abilities and begin to coordinate verbal and nonverbal expressions to create overall more effective deceptive displays (Talwar & Crossman, 2011). These more sophisticated skills are honed across childhood, and lying proficiency continues to advance into young adulthood, likely due to increased executive control, theory of mind understanding, and inhibitory control (Debey et al., 2015).

18.4 NONVERBAL CUES TO DECEPTION

Of course, face-to-face lying is not just a verbal endeavor, but also requires nonverbal expressive control. To date, children's understanding of facial cues around the eyes and mouth has received the most attention from researchers. Monitoring the social partner's face, especially the eyes, is a prominent means of seeking information with respect to the person's intentions. A number of studies demonstrate that preschool children begin to use eye gaze to infer mental states such as desires, intentions, and eventually beliefs (Baron-Cohen, 2005; Wellman, Cross, & Watson, 2001) and are sensitive to duration and frequency of eye gaze when attempting to make inferences about these mental states. For example, Freire, Eskritt, and Lee (2004) found that in a deceptive task, five-year-olds, but not three-year-olds, were able to use eye-gaze cues to infer intentions, despite contradictory verbal cues.

Attention to other nonverbal facial cues, like smiling in game situations that could involve deception, has been shown in previous research to develop after four years of age. To investigate how well children could use expressive cues to detect ongoing deception in others, LaFreniere (1998b) modified an experimental task previously used by Shultz and Cloghesy (1981) to explore children's recursive awareness of intentionality. The card game was modified so that the truthfulness of the cue was contingent on the facial expression of the experimenter providing the cue. For example, every time the examiner sent a false signal, he smiled slightly. Children from four to eight years of age were instructed, "Try to guess the color of the next card." Preschoolers were rarely successful in solving the contingency detection task, with only 8 percent scoring significantly above chance levels across the 15 trials (more than 11 correct). First graders solved the contingency more often, with 50 percent scoring above chance. There was also a significant effect of condition in the older group. Three out of four of the older children solved the contingency task when the experimenter smiled while presenting a false cue. However, only one out of four was able to solve the task when the experimenter smiled while being truthful and kept a poker face while presenting a false cue (LaFreniere, 1998). Such *street smarts* seemed to indicate that they might have had some prior experience with simple deception in games with peers and had learned to associate certain facial cues like smiling with ongoing deception.

In a subsequent study (Smith & LaFreniere, 2013), we examined children from four to eight years of age in a competitive card game paradigm with an adult partner. Success required the child to recognize that the partner had the desire to gain access to the child's intention to deceive and could do so by observing the child's behavior. To be successful, the child needed to suppress behaviors that could reveal information to the adult or provide false cues to mislead the adult. By taking into account social cues that provide information to opponents (looking, verbalizations, leaked information, body movements), this study adds to our understanding of the development of children's deceptive skills beyond the age at which children typically pass classic false-belief tests (i.e., four years).

Our results revealed that children older than four years were significantly more likely to inhibit revealing nonverbal signals, use an irregular strategy when attempting to trick a partner, and use false or misleading nonverbal cues, all of which led to more success in the deception game (Smith & LaFreniere, 2013). Together with previous research, these results demonstrate that the skills required to successfully complete these tasks are not fully present in the average preschooler, but continue to develop in middle childhood.

To be successful at this task, the child needs to effectively conceal information from the adult opponent (Smith & LaFreniere, 2013). This required that the child be aware of what signals would be informative to a partner and have the ability to suppress or alter such signals effectively. This progression of understanding the opponent's intentions and expectations translated into the increase in success we witnessed over the three age groups, with only 11 percent of the four-year-olds reaching the success criterion on the task, 49 percent of the six-year-olds reaching the success criterion, and 76 percent of the eight-year-olds performing successfully at the task. Our experimental design (as was true of our earlier research on tactical deception) prioritizes ecological validity by placing the child in a game situation that required successfully deceiving the adult experimenter who was genuinely attempting to avoid being deceived. Although the experimental situation was artificial, the child's behavior and emotions were natural because how they achieved this success was left entirely up to the child.

A closer look at this increase in overall success reveals two specific behaviors (Smith & LaFreniere, 2013). First, the number of times a child changed his/her pointing strategy across trials significantly increased with age. Frequently changing the pointing strategy makes it more difficult for an opponent to gain a strategic advantage in the game and reveals that the child is aware of this dynamic aspect. On average, the 4-year-olds switched their pointing strategy fewer than 4 times per 20 card set,

the 6-year-olds switched 7.5 times, and the 8-year-olds switched more than 10 times. Thus, where four-year-olds were likely to persist in a predictable pointing strategy that could be readily exploited by their opponent, older children altered their pointing strategy frequently. Though six-year-olds were beginning to use a more flexible strategy by changing it after their opponent started using it to guess correctly, eight-year-olds were anticipating the response of the opponent and proactively shifting the veracity of their pointing clues.

A second important finding was children's insight about their strategy, as indicated by their comments about the task during the post-task interview (Smith & LaFreniere, 2013). As hypothesized, there was a significant stepwise increase in the percentage of children who verbalized a more mature understanding of how to effectively deceive an opponent. These trends demonstrate the emergence of a more complex understanding of tactical deception across this four-year period.

Other behaviors examined in this study did not show linear development with age, but examination of such behaviors nevertheless reveals clues about how tactical deception ability may develop (Smith & LaFreniere, 2013). While four-year-olds played the game with great interest and enthusiasm and demonstrated an understanding of the rules and object of the game, they explicitly revealed critical information much more frequently than the other two age groups. They also made fewer attempts to conceal information than the older groups and thus the opponent had access to all of the information needed to win the game.

Six-year-olds demonstrated an awareness of the need for a deceptive strategy, yet were still relatively unsuccessful in carrying it out compared to the eight-year-olds (Smith & LaFreniere, 2013). They showed obvious attempts at hiding information and superficially attempted to mislead their opponent, yet, unlike the eight-year-olds, the six-year-olds were still more likely to leak information by looking at, orienting to, or verbalizing the correct answer. Where the four-year-olds could be characterized by explicit revelation of information, the six-year-olds showed an increase in accidental revelations, attempts at hiding information, and generation of false information. This shows development in understanding the task demands as well as knowledge that they have the potential to influence a partner with their behavior. However, even at six years of age, children still had difficulty in carrying out these nascent strategies. For example, the six-year-olds showed the highest level of hiding behaviors, consisting of hiding their face with the card or hiding their body under the table. This demonstrates partial understanding of some of the cues a partner may use to gain access to private information, and they attempted to suppress or inhibit such cues. These tactics did make them more successful in the game than the four-year-olds, who did not demonstrate such attempts, but these behaviors were obvious deception cues. It is likely that this is a transitional strategy used when children first become aware that they are giving away information through their behavior. As they become adept at controlling their expressive behavior, children adopt more subtle and effective strategies. Thus, by eight years of age, hiding the face is replaced by controlling expressive facial cues. Since this more subtle "hiding" replaces the earlier naive hiding of their face or body, their level of hiding behaviors, as measured in this study, declines from its peak at age six. Instead, eight-year-old children adopt more mature strategies that demonstrate more confidence in their ability to conceal information through control and manipulation of their facial expression without needing to physically hide inadvertent cues from their face or body.

The six- and eight-year-old groups also demonstrated the strategy of providing false cues to their opponent, a low base rate behavior that was virtually absent in the four-year-old age group (Smith & LaFreniere, 2013). Such cues went beyond hiding and attempted to mislead an opponent by actively creating a false belief. This strategy is more effective because the mask of false information reduces the risk of leaking information through the eyes or face. The social interaction is smoother, with children looking at the examiner and both cards. It was originally hypothesized that eight-year-olds would provide significantly more false information than six-year-olds, but examination of the behaviors revealed no significant difference between the two age groups. It is possible that the specific demand characteristics of this particular task led to lower base rates of fabrication than in other tactical deception paradigms such as the hide-a-bear task (LaFreniere & Ménard, 1990), in which a much higher percentage of eight-year-olds showed this strategy. What has been established across a variety of tasks is that four-year-olds almost never use this strategy. This may be a by-product of increased practice with deceptive skills, but it is also likely to be reflective of increased cognitive understanding at these older ages.

As shown in previous deception research with children (Feldman, Jenkins, & Popoola, 1979; Gosselin, Warren, & Diotte, 2002; Josephs, 1994; LaFreniere, 1988b; Ruihe & Guoliang, 2006; Saarni, 1984), masking rather than neutralization of an emotional expression is generally a more effective strategy. Children who competed effectively in this paradigm shifted their strategy, provided occasional misleading cues, and actively engaged their partner rather than withdrawing in an attempt to suppress information. This is likely a product of increased insight into the partner's intentions and increased insight into how a partner may be interpreting one's own intentions, as well as increased confidence in one's own abilities to keep critical information private.

Although most four-year-olds are able to pass traditional false-belief understanding tasks, it becomes clear from experimental studies of tactical deception that this important step is a building block for an adult-like theory of mind rather than an end point. As with the study of tactical deception in primates, we feel that there is much to be gained in the interplay of field and laboratory research, in which insights from the field are translated into experimental techniques. The research findings regarding the development of deception and detection of deception reviewed thus far are drawn from Western samples, including the USA (e.g., Talwar & Lee, 2008; Smith & LaFreniere, 2013), Canada (e.g., LaFreniere & Ménard, 1990; Schultz & Cloghesy, 1981), the UK (e.g., Reddy, 2007), and the Netherlands (e.g., Debey et al., 2015). While similarities emerge across all of these groups, research in more diverse cultures needs to be conducted to confirm the universality of the development of deception/detection abilities.

What has been shown thus far is that to be successful in real-life deception (here mimicked as a competitive game) children must first be aware that the opponent has opposing intentions, and then be able to anticipate deceptive strategies that their opponent may use and alter their own behavior accordingly (Freire, Eskritt, & Lee, 2004; Shultz & Cloghesy, 1981). This more recursive level of awareness of intentions builds upon prior false-belief understanding and is part of a series of steps involving advances in cognitive tactics, expressive control, and experience in deception, toward the eventual acquisition of an adult-like theory of mind that is serviceable in the real world. Some scientists have speculated that tactical deception can become even more subtle than the conscious strategies that we have explored thus far.

18.5 SELF-DECEPTION

A number of evolutionary scholars have argued that self-deception evolved to facilitate interpersonal deception (Barkow, 1989; Lockard & Paulhus, 1988; Trivers, 1976, 2000, 2012; von Hippel & Trivers, 2011). According to this view, self-deception allows people to avoid the cues to conscious deception that might reveal their intention to deceive. It also eliminates the costly cognitive load that is typically associated with deceiving. Given the variety of methods for deceiving others, self-deception is also thought to manifest itself in a variety of different psychological processes. According to Trivers (2000), this may include denial of ongoing deception, self-inflation, ego-biased social theory, false narratives of intention, and a conscious mind that operates via denial and projection to create a self-serving world. For example, self-inflation allows people, mostly males, to display more confidence than is warranted, which has a host of social advantages (and disadvantages). However, Trivers is the first to admit that knowledge of self-deception remains highly speculative, consisting mostly of anecdotal and introspective reports, despite the fact that the concept was introduced to evolutionary scholars at least as early as 1976.

Of course, well before that, Freudian scholars explored self-deception in their clinical practice and developed a lexicon of concepts to explain human behavior, such as denial, rationalization, etc. This work, too, has been criticized as non-falsifiable in the Popperian sense. In the political arena, another form of self-deception was brilliantly analyzed by George Orwell. In his dystopian tale of future society, *Nineteen Eighty-Four*, Orwell (1949) explains that the Party could not protect its total power without constant propaganda and deception. Yet knowledge of this deception, even within the Inner Party itself, would be dangerous to the State. The solution to this paradox is "doublethink," the method of directly controlling thought.

To know and not to know, to be conscious of complete truthfulness while telling carefully constructed lies, to hold simultaneously two opinions which cancelled out, knowing them to be contradictory and believing in both of them, to use logic against logic, to repudiate morality while laying claim to it, to believe that democracy was impossible and that the Party was the guardian of democracy, to forget whatever it was necessary to forget, then to draw it back into memory again at the moment when it was needed, and then promptly to forget it again, and above all, to apply the same process to the process itself – that was the ultimate subtlety: consciously to induce unconsciousness, and then, once again, to become unconscious of the act of hypnosis you had just performed. Even to understand the word "doublethink" involved the use of doublethink. (Orwell, 1949, p. 32)

While the concept of self-deception has enduring interest to observers of human nature and society, it remains to be seen whether or not an empirical science of self-deception will be established. Thus far, the concept has not had much heuristic power. Despite this, we would agree with Freud, Orwell, Trivers, and others that it does seem to be part of our human nature.

18.6 CONCLUSION

From Machiavelli to the present, realists acknowledge that the subtleties and demands of adult social life call upon abilities to inhibit information and modify expressive behavior, especially in contexts of competition and conflict, and even, on occasion, to provide misinformation. The cognitive skills underlying sophisticated tactical deception in humans create a complex arena for honing adaptive social intelligence. This process begins in early childhood and, as we have shown, continues well past the age at which children pass false-belief tasks and acquire a rudimentary theory of mind. In adulthood, the arenas in which deception is most probable are those contexts that present the individual with competition or conflict. These are also likely to be situations that call for fine-grained assessments. Like a poker game or chess match, these situations give rise to strategies of bluff, exaggeration, and concealment. In human affairs, four key contexts strike us as the most likely to favor deception as a potential strategy:

1 Warfare
2 Commerce and trade
3 Law and politics

4 Love and romance

Sun Tzu proclaims "All warfare is based upon deception," which is likely the most quoted passage of his classic Chinese text, *The Art of War* (Griffith, 1963). But Sun Tzu further elaborates on deceptive tactics:

> Hence, when able to attack, we must seem unable;
> When using our forces, we must seem inactive;
> When we are near, we must make the enemy believe we are far away;
> When far away, we must make him believe we are near.
> Hold out baits to entice the enemy. Feign disorder, and crush him.
> If he is superior in strength, evade him.
> If your opponent is of choleric temper, seek to irritate him. Pretend to be weak, that he may grow arrogant.
> Attack him where he is unprepared; appear where you are not expected.
> These military devices, leading to victory, must not be divulged beforehand.

Throughout human history, from the Trojan horse to Pearl Harbor to modern encryption, the lessons of Sun Tzu have been apparent. As Winston Churchill famously said in the context of the Allied invasion of Europe, "In wartime, truth is so precious that she should always be attended by a bodyguard of lies." In our era, where naked military aggression is more likely to be punished by coalitionary action, secrecy and privacy of aggressive activities are paramount. For example, in the context of terrorism, the use of encryption in communication devices becomes an excellent means of coordinating activities away from the prying eyes and ears of the authorities who seek to expose and prevent such actions.

Similarly, subtle deception is a prominent strategy in the competitive business world from Wall Street to back-alley drug dealers. This is perhaps most apparent in the world of advertising, where products are always "new and improved" and side effects minor. This basic state of affairs has been well known for ages and needs no elaboration here, hence *caveat emptor*. Likewise in the world of politics, self-serving messages often take a form of loud and repetitive signaling akin to advertising any product on the market. This is so much the case that a rule of thumb for true signals is that they tend to be small and quiet, like the tell of a poker player or a brief micro-expression in the speaker's face or voice (Ekman & Friesen, 1975). Finally, crucial assessments that take place in human courtship in any society provide a means of detecting true from false signals from both the ardent suitor and the coy partner. For example, unlike a smile, a blush cannot be produced at will or easily inhibited, and thus is often taken to be a true signal. After courtship, sexual infidelity and attempts at its detection can be crucial for males to provide resources to offspring that they are confident are their own. In all of these domains, children gradually acquire much of what they will eventually understand about such subtleties after a prolonged period of relative innocence.

REFERENCES

Barkow, J. H. (1989). *Darwin, Sex and Status: Biological Approaches to Mind and Culture*. Toronto: University of Toronto Press.

Baron-Cohen, S. (2005). The empathizing system: A revision of the 1994 model of the mindreading system. In B. J. Ellis & D. F. Bjorklund, eds., *Origins of the Social Mind: Evolutionary Psychology and Child Development*. New York: Guilford Press, pp. 468–492.

Bering, J. M., & Povinelli, D. J. (2003). Comparing cognitive development. In D. Maestripieri, ed., *Primate Psychology*. Cambridge, MA: Harvard University Press, pp. 205–233.

Byrne, R. W., & Whiten, A. (1988). *Machiavellian Intelligence: Social Expertise and the Evolution of Intellect in Monkeys, Apes, and Humans*. New York: Clarendon Press/Oxford University Press.

Call, J., & Tomasello, M. (2008). Does the chimpanzee have a theory of mind? 30 years later. *Trends in Cognitive Sciences*, **12**(5), 187–192.

Chandler, M., Fritz, A., & Hala, S. (1989). Small-scale deceit: Deception as a marker of two-, three-, and four-year-olds' early theories of mind. *Child Development*, **60**, 1263–1277.

Dawkins, R., & Krebs, J. R. (1978). Animal signals: Information or manipulation? In J. R. Krebs & N. B. Davies, eds., *Behavioral Ecology*. Oxford: Blackwell Scientific Publishers, pp. 282–309.

Debey, E., De Schryver, M., Logan, G. D., Suchotzki, K., & Verschuere, B. (2015). From junior to senior Pinocchio: A cross-sectional lifespan investigation of deception. *Acta Psychologica*, **160**, 58–68.

Ekman, P., & Friesen, W. V. (1975). *Unmasking the Face*. Englewood Cliffs, NJ: Prentice Hall.

Evans, A. D., & Lee, K. (2013). Emergence of lying in very young children. *Developmental Psychology*, **49**(10), 1958–1963.

Feldman, R. S., Jenkins, L., & Popoola, O. (1979). Detection of deception in adults and children via facial expressions. *Child Development*, **50**, 350–355.

Freire, A., Eskritt, M., & Lee, K. (2004). Are eyes windows to a deceiver's soul? Children's use of another's eye gaze cues in a deceptive situation. *Developmental Psychology*, **40**(6), 1093–1104.

Gosselin, P., Warren, M., & Diotte, M. (2002). Motivation to hide emotion and children's understanding of the distinction between real and apparent emotions. *Journal of Genetic Psychology*, **163**(4), 479–495.

Griffith, S. B., trans. (1963). *The Art of War*. Oxford: Oxford University Press.

Hala, S., Chandler, M., & Fritz, A. S. (1991). Fledgling theories of mind: Deception as a marker of three-year-olds' understanding of false belief. *Child Development*, **62**(1), 83–97.

Humphrey, N. (1976). The social function of intellect. In P. P. G. Bateson & R. A. Hinde, eds., *Growing Points in Ethology*. Cambridge, UK: Cambridge University Press, pp. 303–317.

Josephs, I. E. (1993). *The Regulation of Emotional Expression in Preschool Children*. Münster, NY: Waxmann.

Josephs, I. E. (1994). Display rule behavior and understanding in preschool children. *Journal of Nonverbal Behavior*, **18**(4), 301–326.

LaFreniere, P. J. (1988a). L'evolution at la function de la tromperie. *Anthropologie et Societes*, **12**(3), 63–75.

LaFreniere, P. J. (1988b). The ontogeny of tactical deception in humans. In R. Byrne & A. Whiten, eds., *Evolution of Social Intelligence*. Oxford: Oxford University Press, pp. 238–252.

LaFreniere, P. J. (1998). Card sharks and poker faces: Links between developmental research on deception and evolutionary models. Presented at the International Society for Human Ethology, Vancouver, BC, Canada.

LaFreniere, P. J., & Ménard, J. M. (1990). Le développement de la tromperie tactique chez le jeune enfant. *Enfance*, **45**(4), 361–373.

Lewis, M., Stanger, C., & Sullivan, M. W. (1989). Deception in 3-year-olds. *Developmental Psychology*, **25**(3), 439–443.

Lockard, J. S., & Paulhus, D. S. (1988). *Self-Deception: An Adaptive Mechanism*. Englewood-Cliffs, NJ: Prentice Hall.

Mitchell, R. W. (1986). A framework for discussing deception. In R. W. Mitchell & N. S. Thompson, eds., *Deception: Perspectives on Human and Nonhuman Deceit*. Albany, NY: State University of New York Press, pp. 3–40.

Orwell, G. (1949). *Nineteen Eighty-Four*. London: Martin Secker & Warburg.

Owings, D. H., & Morton, E. S. (1998). *Animal Vocal Communication: A New Approach*. Cambridge, UK: Cambridge University Press.

Povinelli, D. J., & O'Neill, D. K. (2000). Do chimpanzees use their gestures to instruct each other? In S. Baron-Cohen, H. Tager-Flusberg, & D. J. Cohen, eds., *Understanding Other Minds: Perspectives from Developmental Cognitive Neuroscience*, 2nd ed. Oxford: Oxford University Press, pp. 459–487.

Premack, D. (1988). "Does the chimpanzee have a theory of mind" revisited. In R. W. Byrne & A. Whiten, eds., *Machiavellian Intelligence: Social Expertise and the Evolution of Intellect in Monkeys, Apes, and Humans*. Oxford: Clarendon Press, pp. 160–179.

Premack, D., & Woodruff, G. (1978). Does the chimpanzee have a theory of mind? *Behavioral and Brain Sciences*, **1**(4), 515–526.

Reddy, V. (2007). Getting back to the rough ground: Deception and 'social living'. In N. Emery, C. Nicola, & C. Frith, eds., *Social Intelligence: From Brain to Culture*. Oxford: Oxford University Press, pp. 219–244.

Ruihe, H., & Guoliang, Y. (2006). Children's understanding of emotional display rules and use of strategies. *Psychological Science (China)*, **29**(1), 18–21.

Saarni, C. (1984). An observational study of children's attempts to monitor their expressive behavior. *Child Development*, **55**(4), 1504–1513.

Schultz, T. R., & Cloghesy, K. (1981). Development of recursive awareness of intention. *Developmental Psychology*, **17**(4), 465–471.

Smith, R., & LaFreniere, P. (2013). Development of tactical deception from 4 to 8 years of age. *British Journal of Developmental Psychology*, **31**, 30–41.

Sodian, B., & Schneider, W. (1990). Children's understanding of cognitive cuing: How to manipulate cues to fool a competitor. *Child Development*, **61**(3), 697–704.

Sodian, B., Taylor, C., Harris, P. L., & Perner, J. (1991). Early deception and the child's theory of mind: False trails and genuine markers. *Child Development*, **62**(3), 468–483.

Talwar, V., & Crossman, A. (2011). From little white lies to filthy liars: The evolution of honesty and deception in young children. *Advances in Child Development and Behavior*, **40**, 139–179.

Talwar, V., & Lee, K. (2008). Social and cognitive correlates of children's lying behavior. *Child Development*, **79**, 866–881.

Talwar, V., Gordon, H. M., & Lee, K. (2007). Lying in the elementary school years: Verbal deception and its relation to second-order belief understanding. *Developmental Psychology*, **43**(3), 804–810.

Tinbergen, N. (1951). *The Study of Instinct*. Oxford: Clarendon Press.

Tinbergen, N. (1963). On aims and methods of ethology. *Ethology*, **20**(4), 410–433.

Trivers, R. (1976). Forward. In R. Dawkins, ed., *The Selfish Gene*. New York: Oxford University Press.

Trivers, R. (2000). The elements of a scientific theory of self-deception. *Annals of the New York Academy of Sciences*, **907**, 114–131.

Trivers, R. (2012). *The Folly of Fools: The Logic of Deceit and Self-Deception in Human Life*. Philadelphia, PA: Basic Books.

von Hippel, W., & Trivers, R. (2011). The evolution and psychology of self-deception. *Behavioral and Brain Sciences*, **34**, 16–56.

Wellman, H. M., Cross, D., & Watson, J. (2001). Meta-analysis of theory-of-mind development: The truth about false belief. *Child Development*, **72**(3), 655–684.

Woodruff, G., & Premack, D. (1979). Intentional communication in the chimpanzee: The development of deception. *Cognition*, **7**(4), 333–362.

19 The Evolution of Language

A Darwinian Approach

MICHAEL C. CORBALLIS

It is commonly assumed that expressive language is a secondary consequence of an abrupt change in human thought that emerged uniquely in humans within the past 100,000 years. This is difficult to reconcile with the theory of evolution by natural selection. An alternative view is that human thought evolved gradually over many millennia and has properties shared with other species. Some of these properties are critical to language; they include mental time travel, which underlies the linguistic properties of generativity and displacement, and theory of mind, which is critical to meaningful discourse. What does seem to be unique to humans is the ability to communicate our thoughts, rather than the nature of the thoughts themselves. Expressive language depended on the relatively late emergence of output systems flexible enough to map onto internal experience. I argue here that it was built first on manual gesture, with the gradual addition of vocalization.

19.1 BACKGROUND

Throughout history, language has been regarded as special, unique to humans, and the outcome of some "catastrophic" event, perhaps genetic, or perhaps delivered by a deity. According to the Bible, language was a gift to Adam, and for a time all people spoke the same language. When the people built the Tower of Babel to be closer to heaven, the Lord punished them for being disrespectful, scattered their language so that different peoples could no longer understand one another, and dispersed them around the world (Genesis 11: 1–9). In more recent times, Chomsky (2007, 2010) echoes the Biblical story by suggesting that language was bestowed in a single step on a single individual, whom he whimsically names Prometheus rather than Adam, giving rise to universal grammar (UG), common to all people, but again languages proliferated into the some 6,000 different languages that now exist on Earth. Chomsky further suggests that this transformation to UG occurred, as a "great leap forward," within the past 100,000 years.

The notion of a great leap forward is supported by some archaeologists and anthropologists on the basis of dramatic increases in such artifacts as manufactured tools,

cave drawings, symbolic representations, and other signs of increased acculturation. Tattersall (2012), for example, writes:

Our ancestors made an almost unimaginable transition from a non-symbolic, nonlinguistic way of processing information and communicating information about the world to the symbolic and linguistic condition we enjoy today. It is a qualitative leap in cognitive state unparalleled in history. Indeed ... the only reason we have for believing that such a leap could ever have been made, is that it *was* made. And it seems to have been made well *after* the acquisition by our species of its distinctive modern form. (p. 199)

Tattersall is not alone; similar views are expressed by several others (e.g., Hoffecker, 2005; Klein, 2008; Mellars, 2009).

Chomsky has frequently described language, or its embodiment, as "an organ of the body" (e.g., Chomsky, 2007, p. 5), comparable to the heart or lungs, so its sudden appearance in human evolution runs counter to Darwin's theory of evolution, and may even be considered to challenge the theory itself. Darwin (1859) himself wrote:

If it could be demonstrated that any complex organ existed, which could not possibly have been formed by numerous, successive, slight modifications, my theory would absolutely break down. But I can find no such case. (p. 158)

Could language signal the end of Darwinian theory?

Not all believe this to be so. Pinker and Bloom (1990) argued that the evolution of complex functions like language can only be explained in terms of incremental changes, distributed over time. They write that "there is every reason to believe that language has been shaped by natural selection as it is understood within the orthodox 'synthetic' or 'neo-Darwinian' theory of evolution" (p. 708). The archaeological notion of a great leap forward, also known as the "human revolution," has also been challenged (McBrearty & Brooks, 2000). It is hard to avoid the impression that discussions of human uniqueness are prompted in part by what has been termed the "human superiority complex" (Villa & Roebroeks, 2014).

Those who have argued for the sudden emergence of language have typically considered the transformation to involve the restructuring of thought, and not simply of communicative language. Thus, Chomsky (1995) writes of language as an internal system called I-language, which he considers to be the basis of human thought. This gives rise to a process of externalization as a secondary (and, to Chomsky, relatively uninteresting) mechanism creating the diversity of languages as actually spoken or signed. In his "Minimalist Program," Chomsky (1995) identified the principle component of I-language, or of UG, as what he termed "unbounded Merge," whereby abstract elements are merged recursively to create a potential infinity of structures. I-language, then, contains the common ingredient for UG, the mechanism common to all people that is fundamentally a property of thought rather than of communication, although it also underlies the vast diversity of all languages as spoken or signed.

This sheer diversity, though, has made it difficult to infer the nature of UG. Even if the concept of Merge seems to apply to the way language is structured, there is a question as to whether it can be truly said to be unbounded. Through Merge, phrases can be embedded within phrases in recursive fashion, but some languages, such as those of the Pirahã in Brazil (Everett, 2005) and the Iatmul of New Guinea (Evans, 2009), appear to have no recursive embedding at all, and even in a highly literate culture center-embedding only rarely goes beyond double embedding (a phrase within a phrase within a phrase) (Karlsson, 2007). Fujita (2009), among others, has proposed a gradual evolution of Merge through progressive levels of recursion, rather than a sudden arrival of unbounded Merge. Some have gone so far as to deny that UG can be said to exist at all. Evans and Levinson (2009), for instance, conclude that "the emperor of Universal Grammar has no clothes" (p. 438), and Tomasello (2009) has proclaimed that "Universal grammar is dead" (p. 470). Such pronouncements suggest that we need alternative accounts of language evolution and how language relates to thought.

19.2 TOWARD AN EVOLUTIONARY ACCOUNT

The burden of explaining the evolution of language is lessened if language as spoken or signed is differentiated from underlying thought processes. We can then consider language as a device for sharing our thoughts and experiences, rather than constituting those thoughts themselves. This more commonsense view of language, in the words of Dor (2015), "turns the Chomsky proposal on its head" (p. 2). We may then consider human thought to have evolved slowly, in incremental fashion, with at least some ingredients of thought shared with other species. It is primarily the capacity to share our thoughts rather than the thoughts themselves that may underlie the great leap forward, or the supposed human revolution. But even the

sharing of thoughts may have precursors to full-blown language, and is also evident in some nonhuman species.

Although language can be considered to be distinct from thought, the nature of language is shaped by our thought processes in the interests of effective communication. Let us therefore consider the structure of thought, how it evolved, and how it shapes language.

19.2.1 Mental Time Travel

A critical design feature of language is what has been termed "displacement" (Hockett, 1960), the capacity to refer to events at other places and other times, past or future. This in turn depends on the capacity to actually *imagine* such events, or what has been termed "mental time travel." In developing this notion, Suddendorf and Corballis (1997, 2007) also once proposed that it is unique to humans, and indeed may underlie the uniqueness of language. Behavioral studies have challenged this view in a number of species, including birds (e.g., Clayton, Bussey, & Dickinson, 2003) and chimpanzees (e.g., Janmaat et al., 2014). Behavioral evidence, though, can be ambiguous (e.g., Suddendorf & Corballis, 2010), and neurophysiology may bring us close to the nature of mental experience in nonhuman animals (Corballis, 2013; but see also Suddendorf, 2013). The hippocampus appears to play a critical role.

Brain imaging studies have shown that the hippocampus is activated when people imagine both past and future events (e.g., Addis, Wong, & Schacter, 2007; Buckner, 2010; Martin et al., 2011). Single-cell recordings suggest similar mental travels in the rat. So-called place cells in the rat hippocampus map out what has been termed a "cognitive map" (O'Keefe & Nadel, 1978), indicating where the animal is located in space. Hippocampal activity can persist in short-wave ripples after the animal has been removed from a spatial environment, such as a maze, and these ripples map out trajectories in that environment. The trajectories are sometimes "replays" of trajectories previously taken, sometimes the reverse of those trajectories (Foster & Wilson, 2006), sometimes trajectories the animal did not take (Gupta et al., 2010), some of which may be anticipations of future trajectories (Pfeiffer & Foster, 2013). Reviewing this evidence, Moser, Rowland, and Moser (2015) write that "the replay phenomenon may support 'mental time travel' … through the spatial map, both forward and backward in time" (p. 6).

Cells upstream from the hippocampus in the entorhinal cortex modulate and remap the firing of place cells. These include border cells that indicate proximity to geometric boundaries and grid cells that indicate spatial scale and head direction. The remappings may underlie such transformations as orientation in space, zooming into and out of different spatial scales, and location relative to landmarks. They operate in modular fashion, with vast numbers of possible combinations, which may well explain how even the humble rat can orient in different environments, real or imagined, and adjust for scale and viewpoint. Moser et al. (2015) note that just a few modules can produce a vast diversity of hippocampal firing patterns:

The mechanism would be similar to that of a combination lock in which 10,000 combinations may be generated with only four modules of 10 possible values each ... or that of an alphabet in which all words of a language can be generated by combining only 30 letters or less. (p. 11)

They acknowledge that this is "only a hypothesis," but it does seem to provide an explanation of how even the rat can imagine itself in different environments in different orientations, and at different spatial scales, in generative fashion. I can imagine myself in my office, or zoom out to my location in the building, on the campus, in the city, in the country – even on a world map. More importantly, the features of hippocampal mapping may account for the generativity of language itself, as we map words and sentences onto spatial and temporal understandings of events in the world. Generativity itself, then, may not be a function of language, but rather may be a function of the underlying events and ideas we want to convey. And even in the rat, imagined events seem to go beyond the memory of actual past events to include possible future events, and perhaps even purely imaginary ones with little or no reference to actuality. In humans, this capacity probably also includes fiction and storytelling, the generation of complex scenarios played out in different locations and at different times.

The suggestion here, then, is that thought itself is fundamentally spatiotemporal and a common feature of animals that move and need to know where they are, where they have been, and where they might go next. Imagined and remembered events include more than just locations and times; they may contain other ingredients, such as particular actions, individuals, objects, and emotions. Even in the humble rat, hippocampal cells respond to nonspatial as well as spatial information, such as odors, touch, and timing (Moser et al., 2015). There is of course a time span of some 66 million years since the common ancestor of rats and humans existed, and we might like to think that our mental lives have grown more complex and sophisticated than those of present-day rats, but the seeds for a generative mode of thought encompassing travels in space and time probably have ancient origins.

The origins of generative grammar, then, may lie in our understanding of the spatiotemporal world and our ability to create imaginary events. The capacity to share the experience of such events may well have evolved during the Pleistocene and the emergence of a hunter–gatherer society, where it would have been adaptive to share information about past expeditions and plans for future ones. But of course we share more than just events – we share ideas, opinions, theories, and emotions. The structure of such discourse, though, may still be fundamentally spatiotemporal. Lakoff (2014) suggests that even our nonspatial ideas are based on four basic metaphors: (1) thinking as moving (e.g., "Do you follow me?"); (2) understanding as seeing (e.g., "Do you see what I mean?"); (3) thinking as object manipulation (e.g., "Please turn it over in your mind"); and (4) thinking as eating (e.g., "Chew on that one for a while!"). Our understanding of space and time may well be the basis of all of our thinking, and it is a heritage of our peripatetic past.

19.2.2 Theory of Mind

There is, however, more to language than descriptions of events, whether imagined or real, metaphoric or actual. We also travel mentally into the minds of others. Our ability to do so depends on what has been termed "theory of mind," and it is the basis of a good deal of our storytelling. Indeed, humans are so addicted to stories – whether in the form of novels, TV soaps, or simple gossip – that it has been suggested that our species should be renamed *Homo narrans*, the storytellers (Niles, 2010). Understanding how other people think and feel is integral to cooperation in a society adapted to communal living.

Theory of mind, though, plays another role in language, since the sharing of experience depends on more than words and sentences. As the philosopher Paul Grice (1989) pointed out, I cannot have a meaningful conversation with you unless I know what you are thinking and know that you know that I know this. The words we actually use are seldom if ever sufficient to convey precise information; we rely extensively and often unconsciously on shared streams of thought – or as Chomsky (2007) put it, "communication relies on largely shared cognoscitive powers" (p. 10). The manner in which we use shared knowledge to extract information from linguistic utterances is explored by Sperber and Wilson (2002). Scott-Philips (2015) suggests that it is this *underdeterminacy* of language that makes it unique to humans, although studies of gestural communication among chimpanzees in the wild also suggest a lack of determinacy (Hobaiter & Byrne, 2011).

The broader question of whether nonhuman species are capable of theory of mind has been much discussed and disputed since Premack and Woodruff (1978) asked the explicit question, "Does the chimpanzee have a theory of mind?" As late as 2008, Penn, Holyoak, and Povinelli (2008) argued that even chimpanzees, our closest nonhuman relatives, have no theory of mind, describing such attributions as "Darwin's mistake," while Call and Tomasello (2008) conclude, more generously, that this 30 years of research has shown chimpanzees to have an understanding of the goals, intentions, perceptions, and knowledge of others, but no understanding of others' beliefs or desires. But even that claim may be too limited. A critical criterion for advanced theory of mind is that the individual shows understanding that another individual has a false belief. In a recent study, Krupenye et al. (2016) show that great apes, including chimpanzees, bonobos, and orangutans, look in anticipation of whether a human agent will falsely believe an object has been hidden. That is, they seem to pass the false-belief test, often regarded as the gold-standard test of theory of mind (Wimmer & Perner, 1983). This study joins a chorus of studies gradually showing greater mental continuity between humans and other species than is commonly assumed.

To summarize to this point, at least some of the properties of language are based on the nature of thought. Our thought processes are generative, and theory of mind also implies recursion – I know what you are thinking, and I know that you know that I think this. Of course, Chomsky also recognizes the relation between language and thinking, since his concept of I-language itself has to do with thought, and only secondarily with communication. The point of departure lies essentially in the manner in which thought evolved. Whereas Chomsky and others have viewed the symbolic, recursive nature of thought as having emerged exclusively and suddenly somewhat late in the history of our own species, the position taken here is that human thought evolved slowly over many millennia, originating as adaptations to space and time.

19.3 HOW EXPERIENCE IS SHARED

Compared to other species, humans show a remarkable capacity to share complex experiences, in which language plays a prominent (though not the only) role. Of course, other animals have special calls that alert others to such contingencies as danger or the discovery of food, or that signal dominance or territorial claims. Some songbirds have complex songs, but these are largely repetitive, if also pleasingly melodious. But as far as we know, no other species can map communicative signals onto the complex and ever-changing patterns of experience. As we saw in Section 19.2, that mapping is nonetheless far from precise and unambiguous, since it depends on shared thought, but it is nonetheless impressive for its diversity and plasticity. (That last sentence has probably never been written or uttered before, yet I am sure you understood it.)

The communication of experience requires the development of intentional signals that can be mapped onto our concepts about the world. At the simplest level, those concepts are objects and actions, and they are sufficient to describe simple events. The most basic symbols for describing events are nouns and verbs. At this level, some nonhuman animals appear easily able to understand symbolic communication. Savage-Rumbaugh, Shanker, and Taylor (1998) reported that Kanzi, a bonobo, was able to follow simple spoken instructions in English, made up of several words, at a level comparable to that of 2.5-year-old child. Kanzi is now said to understand some 3,000 spoken words (Raffaele, 2006). Even domestic dogs can rapidly learn the meanings of spoken words, even though they cannot themselves articulate them. A border collie called Rico responds accurately to spoken requests to fetch different objects from another room, and then either place the designated object in a box or bring it to a particular person (Kaminski, Call, & Fischer, 2004). This indicates some understanding of nouns and verbs at a receptive level, and also demonstrates displacement. Based on similar studies, another border collie called Chaser is said to respond meaningfully to the spoken names of 1,022 objects (Pilley & Reid, 2011).

These and other studies suggest that comprehension of abstract symbols is well within the capability of at least some other species, and was probably possible in mammals well before the emergence of hominins, and required no "great leap forward." But comprehension is not mirrored in production; one might give instructions to a dog or bonobo, but be unable to carry out a conversation with them. What was required for a true language of sharing was an intentional system that enabled the actual production of signals of sufficient diversity to map onto the diversity of our thoughts and experiences, so that others may then decode the message. Intentional systems with the requisite output as well as input properties seem to have emerged late in evolution.

The most obvious source of intentional signaling is bodily movement. Mammals and birds probably have the capacity to move intentionally about a terrain, but this of itself does not lend itself well to communication. In primates, adapted to an arboreal environment, a better source is movement of the hands, initially involved in such activities as climbing, grooming, catching insects, or plucking or picking up items of food for transport to the mouth. Intentional hand movements are represented in the mirror system, a network involving frontal, parietal, and temporal regions of the primate brain and specialized for such activities as reaching and grasping (Rizzolatti & Sinigaglia, 2010). The term "mirror" refers to the finding that the network is activated both when the animal makes an intentional hand movement and when it observes the same movement carried out by another individual, and in this respect it seems ideally preadapted for the evolution of communication (Rizzolatti & Arbib, 1998). Moreover, the system in humans that is homologous to the mirror system in primates encompasses the language circuit, including Broca's and Wernicke's areas. These considerations support the view that language evolved from manual gestures (e.g., Arbib, 2012; Corballis, 2002; Hewes, 1973).

Knott (2012) goes so far as to suggest that Chomsky's Minimalist Program and the concept of Merge can be applied to simple sensorimotor actions, such as grasping an object and bringing it to the mouth. He writes:

the universal, biologically given capacity for learning languages postulated by Minimalism exists, and its operation can be described using the kind of theoretical machinery which is posited by Minimalism. However, this capacity is not a language-specific one; Minimalism's syntactic analyses are in fact analyses of general properties of the human sensorimotor system, and of memory representations which reflect these properties. (p. 624)

Grammar itself, then, may not be beyond the grasp of the monkey!

The emergence of bipedalism, especially as it became obligatory rather than facultative with the emergence of the genus *Homo*, would no doubt have enhanced the possibilities of using the hands to generate signals of sufficient diversity to represent events, whether remembered or simply imagined. To convey sequences, gesture was probably elaborated as pantomime, during what Donald (1991) called the "mimetic culture" of the early Pleistocene. Pantomime involves whole-body action to represent events, but the essence of an event in space and time could be relayed more economically just using hands and arms, which were freed from any involvement in locomotion. Gestural language may well have developed to resemble modern sign languages invented by deaf communities.

Grammar itself may have originated in pantomime, with the sequencing of gestures following the sequences of the actions they convey. But further conventions were probably introduced to enable the embedding of information. In describing a hunting episode involving the slaughter of an antelope, say, it might have proved useful to interrupt the flow of action to explain the implement needed to effect the kill. Insertions into the thought processes themselves might then explain how subsidiary clauses are inserted into sentences. Conventions were also no doubt invented to refer to time – past, future, and "once upon a time" – as well as subtle nuances of meaning, such as negativity or conditionality.

Another medium for intentional and diverse output is voicing, giving rise to speech as a medium for sharing thoughts, although speech itself can be conceived as a gestural system comprising movements of the lips, the larynx, the velum, and the blade, body, and root of the tongue (Studdert-Kennedy, 2005). The vocal system does appear to offer the diversity necessary to convey complex messages, at least in humans. Nevertheless there are reasons to suppose that communicative language did not originate in vocal calls.

First, the ability to learn vocal sequences appears to be limited to only a few lineages of mammals and birds; among mammals, they include cetaceans, pinnipeds, bats, elephants, and humans, but not nonhuman primates (Pfenning et al., 2014). There is some evidence that great apes may be capable of limited vocal learning, but this more often seems to involve the modulation of existing calls (e.g., Slocombe & Zuberbühler, 2007) than the creation of new ones, or the production of sounds through vibrations of the tongue and lips rather than activation of the larynx (e.g., Hopkins, Taglialatela, & Leavens, 2007). Pfenning et al. (2014) show that the neurophysiological and genetic underpinnings of vocal control have more in common between humans and song-learning birds than between humans and nonhuman primates, implying convergent evolution. Learned speech-like vocalizations can be observed in some birds, such as African Grey parrots (e.g., Pepperberg, 1999), but attempts to teach great apes to speak were abandoned in favor of teaching language-like

behavior through gesture (e.g., Gardner & Gardner, 1969; Patterson & Gordon, 2001) or pointing to visual symbols on a keyboard (Savage-Rumbaugh et al., 1998).

Second, gesture offers the possibility of representations that mimic real-life events, including objects and actions, as in pantomime, or in simply using hand movements to represent such actions as the use of tools, or perhaps wiggling the fingers to represent walking or running. Spoken words, on the other hand, typically bear little relation to what they represent. The lack of correspondence between symbols and external reality is sometimes seen as a barrier to Darwinian accounts of how language evolved. Chomsky (2010), for example, writes: "Crucially, even the simplest words and concepts of human language lack the relation to mind-independent entities that appears to be characteristic of animal communication" (p. 57). These symbols, he supposes, are arbitrary, with nothing in their shape or sound to link them to anything in the natural world. They therefore cannot have been shaped by natural selection.

This problem is largely resolved if language originated from manual gestures, which probably began as iconic representations of what they referred to. This is well illustrated from signed languages, where many of the signs are iconic. It has been estimated, for example, that in Italian Sign Language some 50 percent of the hand signs and 67 percent of the bodily locations of signs stem from iconic representations (Pietrandrea, 2002), and Emmorey (2002) notes that in American Sign Language some signs are arbitrary, but many more are iconic. Over the course of time, though, signs become "conventionalized" (Burling, 1999), losing some of their iconicity in the interests of economy and stability. Spoken words may themselves be examples of conventionalization at a point in evolution where vocal learning became possible, perhaps relatively late in hominin evolution, so that speech gradually replaced manual gesture, perhaps because it is more economical in terms of energy consumption and because it allows communication at night or in the absence of visual contact. The shift from manual gesture to speech may be regarded as an early instance of miniaturization, reduced from movements of the hands and arms to the small confines of the mouth – although we still embellish and amplify our speech through the expressive movements of the hands. Tomasello (2008) writes that "it is possible that the human capacity [for language] evolved quite a long way in the service of gestural communication alone, and the vocal capacity is actually a very recent overlay" (p. 55).

Once conventionalized, words and signs are carried by linguistic cultures and are readily learned; we saw earlier that even dogs appear capable of learning the meanings of arbitrary spoken words, even if they cannot produce them. In any case, it is increasingly apparent that spoken words are not as arbitrary as is commonly assumed. In a study of words across nearly two-thirds of the world's spoken languages, basic vocabulary items were found to carry strong associations with specific speech sounds. For example, words for "nose" typically carried the nasal *n*, words for "tongue" the liquid *l*, words for "small" the short vowel *i*, words for "full" the plosives *p* or *b* (Blasi et al., 2016). Such associations recur frequently across different continents and lineages, suggesting that they are not inherited or borrowed. Rather, they seem to arise spontaneously and independently, perhaps reflecting a form of synesthesia (Ramachandran & Hubbard, 2001).

I suggest, then, that productive language originated in pantomime, but became increasingly conventionalized. Although there is some evidence for pantomime in great apes, it is likely that the system developed during the Pleistocene, when our hominin forebears adapted to a hunter–gatherer existence requiring increased social bonding and communication. The system became increasingly conventionalized, with symbols gradually losing the iconic aspect and becoming more arbitrary, or seemingly so. Grammatical conventions would also be gradually introduced, perhaps in an "expanding spiral" (Arbib, 2005), driven by pressures to relay more complex episodes and tell more complex stories, which in turn increased the complexity of the imagined events themselves. Although pantomime and gesture communication probably led the way, it is possible that intentional communication has always included elements of both sound and gesture, and normal speech is still habitually and meaningfully accompanied by gesture (Goldin-Meadow, 2014).

19.4 CONCLUSIONS

In this chapter, I have suggested that the evolution of language can be understood in terms of gradual evolution by natural selection, and not as a sudden and dramatic change restricted to our own species. An evolutionary account is rendered more plausible by differentiating thought from communicative language and supposing that human thought evolved gradually and in many respects parallels that in nonhuman species. The ability to communicate our thoughts, though, evolved relatively late, but that, too, was probably gradual rather than sudden. The critical period for the evolution of communicative language was probably the Pleistocene, dating from some 2.9 million years ago. The so-called great leap forward, if it can be considered to have taken place at all as a sudden event, may have been due to changes other than the emergence of language itself, and even those changes are likely to have been fairly slow rather than emerging in a single individual at a singular point in time. For example, a switch to a near-autonomous vocal mode may have freed the hands, enabling the explosion of tools and other cultural artifacts (Corballis, 2004). Again, the invention of ways of recording language beyond the moment of utterance, moving from cave drawings and markings to the invention of writing, created a ratchet effect, allowing material progress to build on previous advances. As Isaac Newton (1675) wrote: "If I have seen further it is by standing on the shoulders of giants."

Insofar as it recognizes that language is shaped by the nature of human thought, the account given here is not so very different from that proposed by Chomsky. The main difference is that Chomsky proposes that thought, in the form of I-language, is part of language itself and emerged in a single evolutionary step. Freeing language from thought allows an account in which thought can be considered to have evolved gradually over many millennia, with characteristics that can be found in other species and that may go back to our common ancestry with rats, and even with birds. One such characteristic may be generativity; hippocampal recordings suggest that even rats may generate imaginary sequences of events. Recursive thinking may have emerged later, perhaps in our great ape ancestors, although a recent report suggests that rhesus monkeys are capable of metacognition – knowing what they know (Rosati & Santos, 2016). The capacity to actually produce overt sentence-like sequences mapped onto experience was more recent still, and it depended on the emergence of flexible output systems. Communicate language, then, evolved long after our capacity for generative thought itself.

The account given here does not preclude interactions between language and thought, which are no doubt reciprocal. Indeed, many if not most of our thoughts are derived from sharing information with others. And words are so firmly associated with concepts that we can translate some of our internal thoughts into words, often as preparation for talk – or what Slobin (1960) called "thinking for speaking." But thought itself lies deeper, as when we cannot find the name of a person even if we know him or her quite well, or when we cannot find ways of expressing a complex or subtle thought. And when we do hear a story told by someone else, we put it into different words when we retell it. Thought itself is much more ancient and fundamental than the capacity to share it, and even we eloquent humans can probably do little more than scratch the surface. In his poem "Limitations," the English poet Siegfried Sassoon (1920) wrote:

> Words are fools
> Who follow blindly, once they get a lead.
> But thoughts are kingfishers that haunt the
>   pools
> Of quiet; seldom seen: And all you need
> Is just that flash of joy above your dream.

REFERENCES

Addis, D. R., Wong, A. T., & Schacter, D. L. (2007). Remembering the past and imagining the future: Common and distinct neural substrates during event construction and elaboration. *Neuropsychologia*, **45**, 1363–1377.

Arbib, M. A. (2005). From monkey-like action recognition to human language: An evolutionary framework for neurolinguistics. *Behavioral & Brain Sciences*, **28**, 105–168.

Arbib, M. A. (2012). *How the Brain Got Language: The Mirror System Hypothesis*. Oxford: Oxford University Press.

Blasi, D. E., Wichmann, S., Hammarström, H., Stadler, P. F., & Christiansen, M. H. (2016). Sound–meaning association biases evidenced across thousands of languages. *Proceedings of the National Academy of Sciences*, **113**, 10808–10823.

Buckner, R. L. (2010). The role of the hippocampus in prediction and imagination. *Annual Review of Psychology*, **61**, 27–48.

Burling, R. (1999). Motivation, conventionalization, and arbitrariness in the origin of language. In B. J. King, ed., *The Origins of Language: What Human Primates can Tell Us*. Santa Fe, NM: School of American Research Press, pp. 307–350.

Call, J., & Michael Tomasello, M. (2008). Does the chimpanzee have a theory of mind? 30 years later. *Trends in Cognitive Science*, **12**, 187–192.

Chomsky, N. (1995). *The Minimalist Program*. Cambridge, MA: MIT Press.

Chomsky, N. (2007). Biolinguistic explorations: design, development, evolution. *International Journal of Philosophical Studies*, **15**, 1–21.

Chomsky, N. (2010). Some simple evo devo theses: How true might they be for language? In R. K. Larson, V. Déprez, & H. Yamakido, eds., *The Evolution of Human Language*. Cambridge, UK: Cambridge University Press, pp. 45–62.

Clayton, N. S., Bussey, T. J., & Dickinson, A. (2003). Can animals recall the past and plan for the future? *Trends in Cognitive Sciences*, **4**, 685–691.

Corballis, M. C. (2002). *From Hand to Mouth: The Origins of Language*. Princeton, NJ: Princeton University Press.

Corballis, M. C. (2004). The origins of modernity: Was autonomous speech the critical factor? *Psychological Review*, **111**, 543–552.

Corballis, M. C. (2013). Mental time travel: A case for evolutionary continuity. *Trends in Cognitive Sciences*, **17**, 5–6.

Darwin, C. (1859). *On The Origin of Species by Means of Natural Selection, or Preservation of Favoured Races in the Struggle for Life*. London: John Murray.

Donald, M. (1991). *Origins of the Modern Mind*. Cambridge, MA: Harvard University Press.

Dor, D. (2015). *The Instruction of Imagination: Language as a Social Communication Technology*. New York: Oxford University Press.

Emmorey, K. (2002). *Language, Cognition, and Brain: Insights from Sign Language Research*. Hillsdale, NJ: Lawrence Erlbaum

Evans, N. (2009). *Dying Words*. Oxford: Wiley-Blackwell.

Evans, N., & Levinson, S. C. (2009). The myth of language universals: Language diversity and its importance for cognitive science. *Behavioral & Brain Sciences*, **32**, 429–492.

Everett, D. L. (2005). Cultural constraints on grammar and cognition in Pirahã. *Current Anthropology*, **46**, 621–646.

Foster, D. J., & Wilson, M. A. (2006). Reverse replay of behavioural sequences in hippocampal place cells during the awake state. *Nature*, **440**, 680–683.

Fujita, K. (2009). A prospect for evolutionary adequacy: Merge and the evolution and development of human language. *Biolinguistics*, **3**, 128–153.

Gardner, R. A. & Gardner, B. T. (1969). Teaching sign language to a chimpanzee. *Science*, **165**, 664–672.

Goldin-Meadow, S. (2014). Widening the lens: what the manual modality reveals about language, learning and cognition. *Philosophical Transactions of the Royal Society of London. Series B: Biological Sciences*, **369**, 20130295.

Grice, H. P. (1989). *Studies in the Ways of Words*. Cambridge, MA: Cambridge University Press.

Gupta, A. S., van der Meer, M. A. A., Touretzky, D. S., & Redish, A. D. (2010). Hippocampal replay is not a simple function of experience. *Neuron*, **65**, 695–705.

Hewes, G. W. (1973). Primate communication and the gestural origins of language. *Current Anthropology*, **14**, 5–24.

Hobaiter, C., & Byrne, R. W. (2011). Serial gesturing by wild chimpanzees: Its nature and function for communication. *Animal Cognition*, **14**, 827–838.

Hockett, C. F. (1960). The origins of speech. *Scientific American*, **203**, 88–96.

Hoffecker, J. F. (2005). Innovation and technological knowledge in the Upper Paleolithic. *Evolutionary Anthropology*, **14**, 186–198.

Hopkins, W. D., Taglialatela, J. P., & Leavens. D. A. (2007). Chimpanzees differentially produce novel vocalizations to capture the attention of a human. *Animal Behaviour*, **73**, 281–286.

Janmaat, K. R. L., Polansky, L., Ban, S. D., & Boesch, C. (2014). Wild chimpanzees plan their breakfast time, type, and location. *Proceedings of the National Academy of Sciences*, **111**, 16343–16348.

Kaminski, J., Call, J., & Fischer, J. (2004). Word learning in a domestic dog: Evidence for "fast mapping". *Science*, **304**, 1682–1683.

Karlsson, F. (2007). Constraints on multiple center-embedding of clauses. *Journal of Linguistics*, **43**, 365–392.

Klein, R. G. (2008). Out of Africa and the evolution of human behavior. *Evolutionary Anthropology*, **17**, 267–281.

Knott, A. (2012). *Sensorimotor Cognition and Natural Language Syntax*. Cambridge, MA: MIT Press.

Krupenye, C., Fumihiro, K., Hirata, S., & Call, J. (2016). At apes anticipate that other individual will act according to false beliefs. *Science*, **354**, 110–116.

Lakoff, G. (2014). Mapping the brain's metaphor circuitry: Metaphorical thought in everyday reason. *Frontiers in Human Neuroscience*, **9**, 958.

Martin, V. C. Schacter, D. L., Corballis, M. C., & Addis, D. R. (2011). A role for the hippocampus in encoding simulations of future events. *Proceedings of the National Academy of Sciences*, **108**, 13858–13863.

McBrearty, S., & Brooks, A. S. (2000). The revolution that wasn't: A new interpretation of the origin of modern human behavior. *Journal of Human Evolution*, **39**, 453–563.

Mellars, P. A. (2009). Origins of the female image. *Nature*, **459**, 176–177.

Moser, M. B., Rowland, D. C., & Moser, E. I. (2015). Place cells, grid cells, and memory. *Cold Spring Harbor Perspectives in Biology*, **7**, a021808.

Newton, I. (1675). Letter from Sir Isaac Newton to Robert Hooke. *Historical Society of Pennsylvania*. Retrieved from the Historical Society of Pennsylvania, August 7, 2016. https://discover .hsp.org/Record/dc-9792/Description#tabnav.

Niles, J. D. (2010). Homo narrans: *The Poetics and Anthropology of Oral Literature*. Philadelphia, PA: University of Pennsylvania Press.

O'Keefe, J., & Nadel, N. (1978). *The Hippocampus as a Cognitive Map*. Oxford: Clarendon Press.

Patterson, F. G. P., & Gordon, W. (2001). Twenty-seven years of project Koko and Michael. In B. M. F. Galdikas, N. E. Briggs, L. K. Sheeran, & J. Goodall, eds., *All Apes Great and Small, Vol. 1: African Apes*. New York: Kluver, pp. 165–176.

Penn, D. C., Holyoak, K. J., & Povinelli, D. J. (2008). Darwin's mistake: Explaining the discontinuity between human and non-human minds. *Behavioral & Brain Sciences*, **31**, 108–178.

Pepperberg, I. M. (1999). *The Alex Studies*. Cambridge, MA: Harvard University Press.

Pfeiffer, B. E., & Foster, D. J. (2013). Hippocampal place-cell sequences depict future paths to remembered goals. *Nature*, **497**, 74–79.

Pfenning, A. R., Hara, E., Whitney, O., et al. (2014). Convergent specializations in the brains of humans and song-learning birds. *Science*, **346**, 1333–1346.

Pietrandrea, P. (2002). Iconicity and arbitrariness in Italian Sign Language. *Sign Language Studies*, **2**, 296–321.

Pilley, J. W., & Reid, A. K. (2011). Border collie comprehends object names as verbal referents. *Behavioural Processes*, **86**, 184–195.

Pinker, S., & Bloom, P. (1990). Natural language and natural selection. *Behavioral & Brain Sciences*, **13**, 707–784.

Premack, D., & Woodruff, G. (1978). Does the chimpanzee have a theory of mind? *Behavioral & Brain Sciences*, **4**, 515–526.

Raffaele, P. (2006). Speaking bonobo. *Smithsonian Magazine*. www.smithsonianmag.com/science-nature/speaking-bonobo-134931541.

Ramachandran, V. S., & Hubbard, E. M. (2001). Synaesthesia – A window into perception, thought and language. *Journal of Consciousness Studies*, **8**, 3–34.

Rizzolatti, G., & Arbib, M. A. (1998). Language within our grasp. *Trends in Neurosciences*, **21**, 188–194.

Rizzolatti, G., & Sinigaglia, C. (2010). The functional role of the parieto-frontal mirror circuit: Interpretations and misinterpretations. *Nature Reviews Neuroscience*, **11**, 264–274.

Rosati, A. G., & Santos, L. R. (2016). Spontaneous metacognition in rhesus monkeys. *Psychological Science*, **27**, 1181–1191.

Sassoon, S. (1920). Limitations. In S. Sassoon, ed., *Picture-Show*. New York: E.P. Dutton.

Savage-Rumbaugh, S., Shanker, S. G., & Taylor, T. J. (1998). *Apes, Language, and the Human Mind*. New York: Oxford University Press.

Scott-Phillips, T. (2015). *Speaking Our Minds: Why Human Communication Is Different, and How Language Evolved to Make It Special*. Basingstoke: Palgrave Macmillan.

Slobin, D. I. (1960). From "thought and language" to "thinking for speaking". In J. Gumperz & S. C. Levinson, eds., *Rethinking Linguistic Relativity*. Cambridge: Cambridge University Press, pp. 70–96.

Slocombe, K. E., & Zuberbühler. K. (2007). Chimpanzees modify recruitment screams as a function of audience composition. *Proceedings of the National Academy of Sciences*, **104**, 17228–17233.

Sperber, D., & Wilson, D. (2002). Pragmatics, modularity and mind-reading. *Mind and Language*, **17**, 3–23.

Studdert-Kennedy, M. (2005). How did language go discrete? In M. Tallerman, ed., *Language Origins: Perspectives on Evolution*. Oxford: Oxford University Press, pp. 48–67.

Suddendorf, T. (2013). Mental time travel: Continuities and discontinuities. *Trends in Cognitive Sciences*, 17, 151–152.

Suddendorf, T., & Corballis, M. C. (1997). Mental time travel and the evolution of the human mind. *Genetic, Social, & General Psychology Monographs*, **123**, 133–167.

Suddendorf, T., & Corballis, M. C. (2007). The evolution of foresight: What is mental time travel, and is it unique to humans? *Behavioral & Brain Sciences*, **30**, 299–351.

Suddendorf, T., & Corballis, M. C. (2010). Behavioural evidence for mental time travel in nonhuman animals. *Behavioural Brain Research*, **215**, 292–298.

Tattersall, I. (2012). *Masters of the Planet: The Search for Human Origins*. New York: Palgrave Macmillan.

Tomasello, M. (2008). *The Origins of Human Communication*. Cambridge, MA: MIT Press.

Tomasello, M. (2009). Universal grammar is dead. *Behavioral & Brain Sciences*, **32**, 470–471.

Villa, P., & Roebroeks, W. (2014). Neandertal demise: An archaeological analysis of the modern human superiority complex. *PLoS ONE*, **9**, e96424.

Wimmer, H., & Perner, J. (1983). Beliefs about beliefs: Representation and constraining function of wrong beliefs in young children's understanding of deception. *Cognition*, **13**, 103–128.

20 The Adaptive Problem of Exploiting Resources

Human Foraging Behavior in Patchy Environments

ANDREAS WILKE

20.1 INTRODUCTION

Over evolutionary time, humans have had to solve problems regarding many important foraging activities, such as deciding where to find crucial resources, when to move on to more resource-rich locations with higher intake rates, and how well past foraging success might predict the future likelihood of return. This chapter will argue that these reoccurring foraging behaviors of our ancestral past left an eminent footprint in our evolved cognitive system – and specifically in our information processing mechanisms that deal with risk and uncertainty. This chapter on evolution, cognition, and decision-making will review empirical work from animal behavior, biological anthropology, and evolutionary psychology to show that our mind possesses various cognitive foraging adaptations that coevolved with the statistical regularities of natural resource environments. In nature, most resources are distributed in an aggregated manner, rather than randomly or in a dispersed manner, meaning that, on the whole, plants and animals are distributed in clumps or patches (i.e., areas with a high density of a given resource surrounded by areas with low resource density). In modern times, these evolved expectations of clumpy resources still influence our human decision-making behavior, and this can be observed in everything from searching for external physical objects, internally seeking information from memory, dealing with streaks of events, and facing random outcomes while gambling, as well as when making strategic hiding and searching decisions.

20.1.1 Exploiting Natural Resource Environments

How do humans and other animals maximize resource intake from the environment? Whenever resources are distributed in space and a local source can be depleted faster than it replenishes, it is important to decide whether one could do better by moving on to a different source elsewhere. Thus, the adaptive decision-maker not only faces a decision on *where* to forage, but also of *how long* they should forage in a particular spot (Charnov, 1976). In

other words, is it better for me to stay at the current site and continue to forage or to leave and travel to a new place? If nearly all of the resources at a site have already been found, choosing to stay longer at that location might be wasteful, as more and more time would be spent on finding the few remaining resources, when it would be better to move elsewhere, where the initial gain rate of resource intake is higher (Wilke, 2006). On the other hand, it would also be wasteful to leave a site too early, because then too much time might be taken up by traveling and searching for a new site. Like other animals, humans have faced decisions of this type in many domains – both presently, in modern decision environments, and in the past, in the ancestral living environments our minds evolved in.

20.1.2 Patchy Resource Distributions

In nature, resources are often distributed in clumps or patches – local areas with a high probability of a resource encounter that are surrounded by other areas where the probability of a resource encounter is much smaller or close to zero (Wilke, 2006). Travel distances within a patch are smaller than travel distances between patches, and patches are usually conceptualized as differing in quality (Hutchinson, Wilke, & Todd, 2008). Rich patches provide a higher initial return per time unit than poor patches. No matter the quality, when a forager enters a patch, the forager utilizes the resources and depletes them over time. Each resource typically has to be searched for, as resources are exhaustible and hidden (Wilke et al., 2009).

Classical optimal foraging theory in behavioral ecology addresses this adaptive patch-leaving decision problem in the marginal value theorem, which states that the optimal strategy for each individual is to leave the patch when the instantaneous rate of return from the current patch falls below the mean return rate from the environment under the optimal policy (Charnov, 1976). When an animal first enters a rich patch, resource gains from exploiting it are high, because there is a lot of food present and it is easy to find.

As time passes, however, food resources will be depleted and it will take longer and longer to find a food item. Hence, more time is spent searching and less is spent eating. This declining rate of energy gain can be represented by a graph of diminishing returns, because the "marginal value" of the patch decreases and the slope of the graph levels off (Figure 20.1). When the travel time between patches is taken into account, then the line giving the maximum long-term rate of energy gain is the line from point A that touches the gain curve at a tangent.

Whereas the marginal value theorem predicts general phenomena in optimal foraging in animal behavior quite well, it makes unrealistic assumptions about the information constraints that the foragers face, such as that the mean net return rate is hard (if not impossible) to infer when foraging in novel environments and that foraging, for many animals, is a succession of discrete events where resources are encountered stochastically, rather than via an energetic intake following a continuously curved function (McNamara, 1982; McNamara & Houston, 1985). Most importantly, though, the marginal value theorem does not link its functional optimality predictions with any proximate decision mechanisms that a forager can use (cf. Hutchinson & Gigerenzer, 2005). What proximate decision mechanisms could be utilized to guide an animal in deciding when to leave a patch? What informational cues could an animal obtain while foraging to adjust its patch residence time? And what rules for integrating these cues would perform efficiently?

20.1.3 Decision Mechanisms for Animal Patch Time Allocation

Behavioral ecologists have long studied this problem of patch time allocation and looked at the performance of simple heuristic patch-leaving strategies in varying environmental contexts (e.g., Bell, 1991; Iwasa, Higashi, & Yamamura, 1981). Most notably, research has focused on parasitoid insects and various bird species (e.g., Livoreil & Giraldeau, 1997; Wajnberg et al., 2003).

Waage (1979) described a model for how long insect parasitoids decide to remain on a particular patch. Female insect parasitoids (*Venturia canescens*) look for hosts into which they can lay their eggs. Here, the adaptive problem is that the desired hosts, typically other insects, are not evenly dispersed over the habitat, but rather come in groups of variable size – patchy environmental structures, as described in Section 20.1.2. In Waage's (1979) model, an initial tendency to stay in the patch is set by the intensity of a kairomone smell cue, which is a good indicator of the size of the patch and host density. Responsiveness decreases linearly with time spent at a patch, but each successful oviposition alters the waning of this response by adding an increment to the current level. When the level of responsiveness has dropped below a critical threshold, the parasitoid leaves the patch and looks for a new one. In a way, the decision mechanism operates like a piece of clockwork that is wound up a little

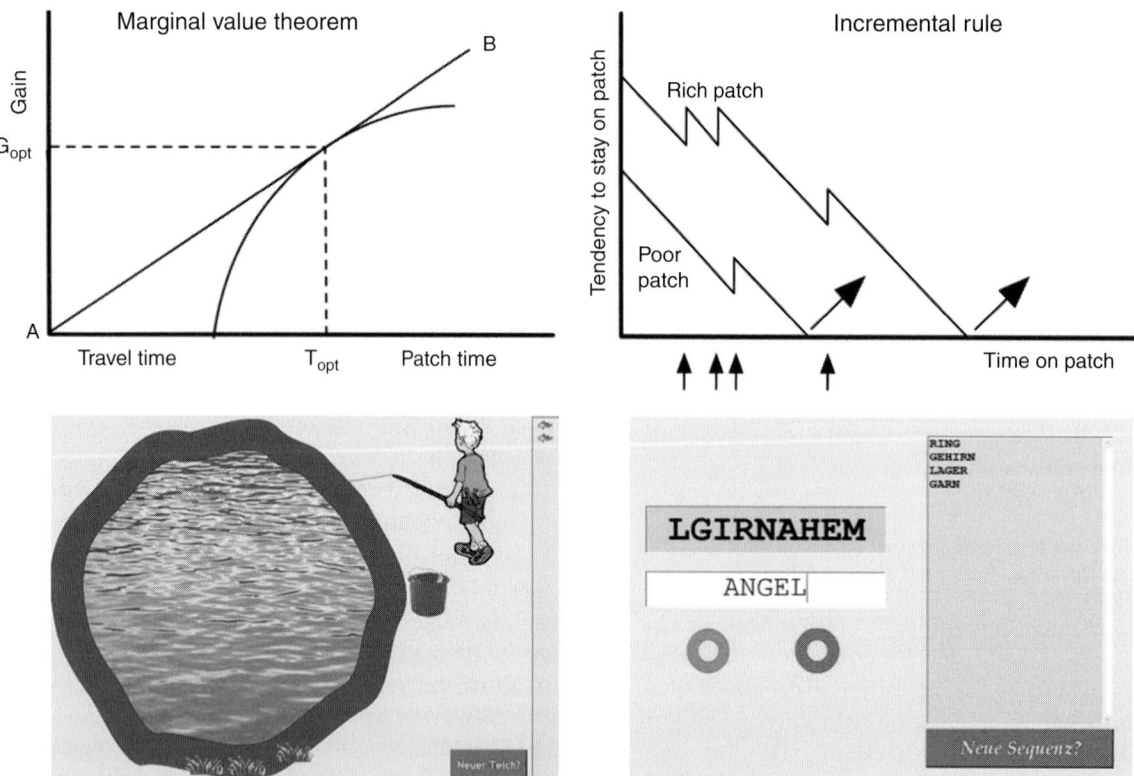

Figure 20.1 Upper left: The marginal value theorem. Upper right: The incremental rule. Lower left: The fishing task. Lower right: The word puzzle task. All images redrawn from Wilke (2006).

each time a free host is encountered – when the clock timer eventually expires, the animal leaves the patch. In the literature, such a mechanism has been termed an *incremental rule* (Figure 20.1). In research conducted on birds, Krebs, Ryan, and Charnov (1974) showed that the black-capped chickadee's (*Parus atricapillus*) search for meal-worms on artificial pinecones could be modeled with a *giving-up-time rule*. Here, similarly, the tendency to stay in a patch declines with unsuccessful searching, but is reset to the maximum with each resource item that is found.

These two decision rules perform surprisingly well, because the building blocks of the proximate mechanism are well adapted to the finer structure of the environment in which resource patches differ widely in quality. Some patches contain only a few resource items, while others contain a lot more; an ideal patch-leaving rule exploits such environmental patterns. Research on other animals foraging in patchy environments, for instance, has shown that natural selection fine-tunes these patch-leaving rules to the finer statistical structures in such environments (Alphen, Bernstein, & Driessen, 2003). Other parasitoid wasps, for instance, forage in patchy environments, too, but the amount of resources to be expected in their natural foraging landscape is much more similar across patches, with each patch more or less containing a very similar amount of exploitable resources. In such foraging contexts, a *decremental rule*, in which the initial tendency to stay in each patch is decreased step by step, outperforms any other patch-leaving rule – as now each new capture is more important in informing the forager that the patch has just gotten worse. In extreme cases, if each patch were to contain the exact same number of items in it, nothing outperforms a *fixed-number rule*, where the forager leaves after having found *n* items. In cases where the number of items is completely random, organisms do best by focusing on the time they have spent so far in the patch (a *fixed-time rule*) and ignoring their past, current, and future foraging success in determining their patch departure (Iwasa, Higashi, & Yamamura, 1981). Thus, these studies have revealed that the number of to-be-expected resource items in each patch is one of the most critical cues to determining the exact patch-leaving mechanism (Wajnberg, Fauvegue, & Pons, 2000).

20.1.4 Investigating Human Patch Time Allocation

Biological anthropologists investigate foraging decision-making among hunter–gatherer populations (e.g., Winterhalder & Smith, 1981). Most of their work, however, deals with specific environmental variables and how they systematically relate to a forager's diet (e.g., the dependence on gathered vs. hunted foods; Binford, 1990). Typical anthropological foraging models aim to explain which resource types should be incorporated into

the diet when foragers encounter them, what kind of resource patches should be included into the overall foray (Kelly, 1995), or how to resolve nutritional choice conflicts. A behavior that is well documented from observations among the Ache in Paraguay, for instance, occurs when a particular individual resource is sought after (e.g., a small mammal) but a different one is encountered during the foraging activity (e.g., honey). Ideally, a forager should switch to the resource encountered if it offers a higher energetic intake than the one that was originally sought after. This is what has been found in the Ache (Hill & Hurtado, 1996).

While these foraging decisions are related to the problems raised in this chapter, anthropological research on the topic has not fully addressed patch leaving in terms of actual cognitive mechanisms and the role specific environmental resource distributions play. Humans forage on a variety of different food sources coming in different qualities and distributions. Furthermore, given that humans are omnivores, one might expect that we possess the ability to adaptively and strategically select among a set of multiple foraging mechanisms for exploiting the food at hand (Wilke, 2006). To fill this gap, Wilke and colleagues conducted two controlled psychological laboratory experiments that tested whether the decision rules that evolved to direct animals on when to leave a food patch also underlie human decision-making. Both experiments differed in whether the search was external (e.g., finding physical objects) or internal (e.g., retrieving information from memory), but their environmental parameters (e.g., travel times, mean reward rates) were closely matched (Figure 20.1).

In the first experiment – the fishing task – participants were presented with a virtual landscape on a computer screen in which they had to monitor ponds, forage for fish, and decide how long to stay at each pond (for details, see Hutchinson, Wilke, & Todd, 2008; cf. Mata, Wilke, & Czienskowski, 2009). All ponds appeared equal, but the number of fish in each varied. Each participant experienced either a dispersed, an aggregated, or a random distribution of fish per patch. The rate at which fish appeared was proportional to the number of fish left in the pond, and participants saw only the number of fish caught at the current pond. Participants had no access to a clock to judge their staying times. The lab software recorded participants' foraging success, the number of patches they visited, and the time they spent at each of them (Czienskowski, 2005a). A performance-based payment at the end of the experiment rewarded them for the total number of fish they caught.

While foraging theory was originally developed in the context of food search, its formal assumptions were actually about the spatial and temporal distributions of items and their payoffs in any currency – not just calories. For this reason, psychologists have realized that models from foraging theory also apply well to problems of information search, such as when people seek information on the Internet (Pirolli, 2007) or when they search for information in memory (Hills, Todd, & Jones, 2015). Thus, in the second experiment, foraging for fish was replaced by cognitive information search, and participants were presented with a modified anagram search task in which they had to find word solutions using their memory lexicon (for details, see Wilke et al., 2009; cf. Mata, Wilke, & Czienskowski, 2013). Meaningful words had to be generated out of meaningless sequences of letter patches. As in the fishing task, participants experienced environments that differed in patch quality – with some sequences containing many possible word solutions and others only very few – and had to decide when to switch to the next sequence, resulting in a time delay before a new sequence was presented (Czienskowski, 2005b). Participants were again paid according to their overall foraging success, and the environmental parameters of this second experiment were closely matched to the first task.

The results showed that in both experiments people applied patch-leaving rules that matched the simple decision rules that are discussed in the animal behavior literature for aggregated environments (Wilke, 2006). One of these rules – the *incremental rule* – exploits the statistical pattern of aggregated environments by adding a time increment for each successful resource capture to an organism's patch-staying tendency. The other rule – the *giving-up-time rule* – adjusts the tendency to stay in a patch by utilizing the time since last capture. However, participants in both tasks did not adapt parameters of their patch-leaving rules much, if at all, to the types of environments they faced. Switching behavior after a fixed number of items (as predicted for the evenly dispersed environment) or switching after a fixed time (as predicted for the random environment) was not observed. On the contrary, participants used the same two patch-leaving rules that fit aggregated environments in the other types of environments (i.e., dispersed and random). Why was this the case?

Most species of plants and animals rarely, if ever, distribute themselves in a purely random manner in their natural environment, because individual organisms are not independent from one another: whereas mutual attraction leads to aggregation for some species, mutual repulsion leads to regularity (dispersed environments) in other species (Taylor, 1961; Taylor, Woiwod, & Perry, 1978). Most often, however, these deviations from randomness are in the direction of aggregation, since aggregation offers considerable ecological benefits (Krause & Ruxton, 2002). Such advantages can be responses to physical environments (e.g., resource availability, light, or temperature), reproductive behavior (e.g., clumped arrangements of eggs or offspring being in close parental contact), mutual attraction with other individuals of the same species (e.g., mating behavior), interactions with other species (e.g., parasites or predator avoidance), or benefits in foraging behavior (e.g., to capture prey types that would be too large, too agile, or too dangerous for a single individual). Since humans have been hunters and gatherers for much of their history (Tooby & DeVore, 1987), it could well be that our evolved psychology is adapted to assume such aggregated resource distributions as our default, because it would have offered a selective advantage under ancestral conditions. As we will see in Section 20.2, the idea that humans expect aggregation – autocorrelation in space and time – may also explain why apparent misconceptions of probabilistic thinking are hard to come by in our species.

20.2 STATISTICAL REGULARITIES IN NATURAL RESOURCE ENVIRONMENTS

Much research in psychology suggests that people have a difficult time properly thinking about randomness, often perceiving patterns that simply are not there. In particular, people seem to have difficulty conceiving of independent events, in which any given event has no influence on the outcomes of future ones (Falk, 1975; Nickerson, 2002). A classic paper by Gilovich, Vallone, and Tversky (1985) identified one such confusion, since then described as the *hot hand fallacy*. Gilovich et al. (1985) found that both basketball players and fans judged that a player's chance of sinking a shot was greater following a successful shot than a miss, with observers surmising that a player who had just made a basket was "on fire" and likely to continue making more. These judgments revealed an implicit assumption of streaks (or runs) in players' shooting success. This can be described as a positive recency effect: a successful shot, or hit, boosts the observers' subjective probability of predicting another hit. This means that the hot hand phenomenon reflects an implicit assumption on the part of the observer that hits are positively autocorrelated or clumped. However, when Gilovich et al. (1985) analyzed the actual data on which subjects' predictions were made, they found that the shots were, in fact, independent. In other words, a player's likelihood of making another successful shot was not correlated with whether or not they had made the last one. The hot hand assumption was therefore a mistake, at least as far as basketball was concerned. In recent years, psychologists have further delved into the *proximate mechanisms* of the hot hand phenomenon and its opposite, the gambler's fallacy, in which streaks are assumed to end (e.g., Oskarsson et al., 2009). In particular, studies have explored the role of the hot hand bias in specific sport disciplines other than basketball (e.g., Raab, Gula, & Gigerenzer, 2012) and its role in gambling behavior and finance (e.g., Croson & Sundali, 2005), as well as its occurrence in other age groups, such as older adults (Castel, Drolet Rossi, & McGillivray, 2012).

Similarly, people often perceive spatial clumps not only in random one-dimensional binary sequences – as in the hot hand phenomenon – but also in random two-dimensional resource patterns. Falk and Konold (1997) elegantly showed this by presenting subjects with a set of

10×10 resource grids in which half of the 100 squares were empty (colored white) and half had a resource (colored black). Each pattern was generated according to an alternation rate $p(A)$ that specified the probability that the next square would differ from the previous one (going from left to right and from top to bottom). Whereas grids with an alternation rate $p(A)$ near 0.5 are the least predictable (and hence most random), lower alternation rates create clusters of empty or full squares and higher alternation rates lead to more dispersion (Figure 20.2). But when asked to rate the randomness of the visual grid arrangements, participants failed to correctly give the highest ratings to grids with alternation rates near 0.5 – instead, they chose grids that were more dispersed (with $p(A)$ around 0.60–0.65). As in one-dimensional sequences, the least predictable random grid arrangements were perceived as having clusters of resources (Falk & Konold, 1997).

20.2.1 A Cognitive Adaptation for Detecting Clumped Resources

Recent research on the *ultimate function* of the hot hand phenomenon suggests that it is an adaptive human universal that is tied to our evolutionary history of foraging for patchy resources (Wilke & Barrett, 2009; Wilke & Mata, 2012; cf. Reifman, 2011). But how can one test whether hot hand thinking is culturally influenced or if it is more universally applied, as an evolutionary perspective would suggest?

Wilke and Barrett (2009) developed a computer game simulating a sequential search for resources to address this question, which was used to compare undergraduate subjects from the University of California, Los Angeles (UCLA) with Shuar hunter–horticulturalists from Amazonian Ecuador. During the simulated search, individuals were shown whether resources were present or absent in a series of locations and were asked to predict whether there would be resources in the next spot. In all experimental conditions, the distribution of resources was completely random.

However, different conditions used different types of resources. Some were natural resources, such as fruit trees and bird nests, while others were modern-day resources, such as parking spots and bus stops. Participants showed a high level of hot hand thinking across the board in both cultures, consistent with the idea that this is an evolved psychological default. Furthermore, two additional patterns emerged that support an evolutionary basis. First, more hot hand thinking was recorded for participants making predictions regarding natural resources compared to artificial, human-made resources, suggesting that hot hand thinking may have indeed evolved to aid our ancestors in their foraging pursuits. Second, when comparing decisions about coin tosses and foraged fruits, the authors found that Shuar hunter–horticulturalists showed equal levels of hot hand thinking for both, whereas UCLA students were at about the same level as Shuar subjects for fruits, yet lower for coin tosses. This suggests that familiarity arising from lifetime experience with the truly random nature of coin tosses might have helped the students learn to adjust away from their evolved default.

The hot hand bias is particularly adaptive in contexts where clumps often exist and dispersal or randomness is rare, such as in the foraging contexts described in Section 20.1.4 (cf. Haselton et al., 2009). If a forager is trying to predict how good a patch containing a certain resource is, then using a strong prior expectation for clumped resources is likely to provide better guesses than a random prior expectation. That is because the hot hand bias does not really decrease accuracy in random environments, as all strategies produce chance-level performance in this situation anyway (Scheibehenne, Wilke, & Todd, 2011). It might well be the case that what has been seen as a systematic error or fallacy in our modern decision-making apparatus may actually be a design feature of our ancestral cognitive system that evolved to reduce uncertainty in foraging situations. This possibility also highlights the role of ecological and evolutionary rationality – the principle that there is a match between the statistical

Clumpy distribution

$p(A) = 0.37$

Random distribution

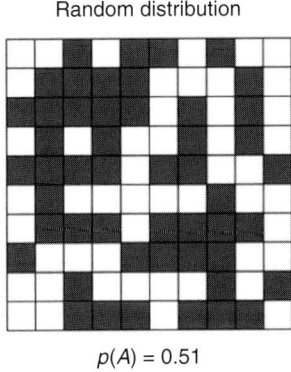

$p(A) = 0.51$

Dispersed distribution

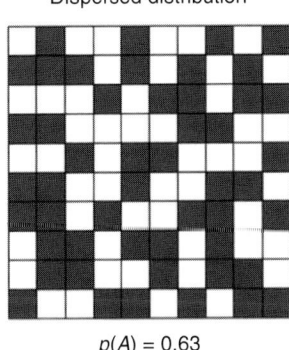

$p(A) = 0.63$

Figure 20.2 Three 10×10 resource grids of 50 white and 50 black squares that only differ in their underlying spatial distributions. Adapted from Falk and Konold (1997).

structure of objects, information regarding current (and past) environments, and the judgment and decision-making strategies of humans and other organisms (e.g., Fawcett et al., 2014; Todd & Gigerenzer, 2012).

20.2.2 New Directions for Hot Hand Research

To further explore the proposal that hot hand thinking is an evolved psychological adaptation to clumped resources, Wilke and collaborators developed a series of additional experimental tasks that simulated a sequential search for resources in various computer game environments and then asked subjects to make moment-to-moment predictions about the forthcoming presence or absence of resources as they navigated through these task environments. In recent experiments, they not only conducted cross-cultural research on small-scale foraging societies (as described in Section 20.2.1), but also conducted comparative studies by working with nonhuman primates, conducted clinical research by testing subjects with gambling disorders, and also tested the strategic interactions of multiple players in a behavioral decision-making study.

20.2.2.1 Other Species Share Our Proclivity to See Patterns in Random Data

An additional prediction resulting from investigations into the ultimate functions of the hot hand phenomenon is that nonhuman primate species with similar foraging histories might share our proclivity for positive recency in independent sequential events. Blanchard, Wilke, and Hayden (2014) directly tested for such hot hand biases in monkeys by setting up a task in which rewards in some conditions were positively correlated and in others negatively correlated. The authors hypothesized that if monkeys have an innate expectation for resources to be clumpy, we should see behavioral asymmetries in the adoption of the optimal behavior in positive vs. negative correlation conditions. The behavior of three monkeys was examined in a novel two-option risky choice task. In each trial, one option offered a reward and the other did not. The chance that the rewarded option would switch was set to 1 of 10 values and varied across days. Monkeys' patterns of choices were consistent with the hypothesis that they see more positive autocorrelation than there actually is across all conditions. These results are consistent with the data obtained from humans and suggest that monkeys' risky choices are determined by an inherent bias toward expectations of environmental patchiness as well. It seems as if the hot hand phenomenon is an ancient evolutionary bias that humans might share with other primate species.

20.2.2.2 Searching for Patterns in Places Where They Do Not Exist

Pathological gambling is a psychological and medical disorder identified in both the Diagnostic and Statistical Manual of Mental Disorders, 5th Edition (DSM-5) and the International Classification of Diseases 10th Revision (ICD-10) (American Psychiatric Association, 2013; World Health Organization, 1992). The lifetime prevalence rate varies worldwide, but has been estimated at around one percent of adults in the USA, with an additional two to three percent defined as problem gamblers (Shaffer, Hall, & Bilt, 1997). Problem gambling is on the rise in older populations and is often comorbid with other mental health problems (Hodgins, Stea, & Grant, 2011). Thus, unique approaches for studying gambling disorders could be helpful for the understanding and treatment of this disorder.

Could the hot hand effect play a role in human gambling behavior? The tendency to search for patterns in random data could explain part of the pleasure humans experience while gambling – for example, the experience of winning several times in a row could be highly compelling, leading one to believe that one is on a hot streak. But in addition to this universal propensity, there could very well be differences between individuals in just how "hot handed" they are – or just how prone they are to perceive streaks, even when they do not exist – possibly leading to differences in how much gambling on random outcomes is enjoyed. As for many evolved traits, individual differences in pattern perception could arise from a variety of factors, including genetic differences, environmental differences, or differences in individual experience. If such individual differences are predictive of the propensity to gamble, then this could have implications both for developing assessment tools to detect individuals' risks of developing a gambling problem and for developing interventions that might be effective, such as targeting gamblers' perceptions of randomness. Moreover, if the hypothesis that hot hand thinking is a universal human cognitive adaptation is right, then the risk of developing a gambling problem might also be a human universal, not restricted to those with a cultural history of gambling or those who have individual experience with it. Using a slot machine-like decision-making paradigm, Wilke et al. (2014) directly measured subjects' predictions of sequences. In their study, they found evidence that subjects who have a greater tendency to gamble also have a greater tendency to perceive illusionary patterns, as measured by their preferences for a random slot machine over a negatively autocorrelated one. Casino gamblers often played the random slot machine significantly more, even though a training phase and a history of outcomes was provided. Additionally, Wilke et al. (2014) found a significant group difference between gamblers and matched community members in their slot machine choice proportions. Performance on the behavioral choice task correlated with subjects' risk attitudes toward gambling and their frequency of play, as well as the selection of choice strategies in gambling activities. Thus, it is possible that interventions to teach subjects the properties of random devices might reduce their propensity to cognitive illusions – like hot hand thinking – that lead to an increased frequency of gambling.

20.2.2.3 Expectations of Clumpy Resources Influence Hiding and Searching Patterns

Our human ancestors evolved in environments with resources spread out in various patterns, with some quite difficult to find and others easier, depending on the adaptive goals of the agents involved in creating those resources. What spatial patterns do people expect when resources have been hidden so that they are either simple or difficult to find?

Wilke et al. (2015) addressed this question in terms of strategic foraging behavior via a sequential multiperson game in which participants hid resources for the next participant to try to find, either collaboratively or competitively (cf. Talbot et al., 2009). As predicted, the authors found that the type of interaction between players had a strong influence on hiding and searching behavior (see also Legge et al., 2012). When collaborating, resources were mostly hidden in clumpy distributions, as are commonly seen in the natural world, but when competing, resources were hidden in more dispersed and seemingly random patterns, to increase the searching difficulty for the other player. More dispersed resource distributions came at the cost of higher overall hiding (as well as searching) times, decreased payoffs, and increased difficulty when having to recall earlier hiding locations at the end of the experiment. In line with research on area-restricted search mechanisms in animals, participants' search strategies were affected by their underlying expectations for when resource distributions would be clumpy: when collaborating, participants appeared to use a win-stay, lose-shift strategy appropriate to clumpy resources (cf. Nowak & Sigmund, 1993). When competing, they moved similarly far after finding or not finding resources, as is more appropriate for a random (non-clumped) resource pattern. Thus, subjects' evolved expectations about two-dimensional resource patterns, whether clumpy or dispersed, may still drive aspects of human searching behavior to this day.

20.3 SUMMARY

Our ancestors evolved in environments in which resources were spread out in various patterns. These encountered resources were manifold, but their broader statistical structure was often either clumpy, random, or dispersed with regard to their distribution in space and time. Since detecting contingencies and patterns in your environment is an important adaptive challenge, our cognitive system seems to have evolved to expect the most common of these distributions – aggregated patchy resource distributions – to be our

default, as it most closely reflects the statistical pattern of many of the resources we were foraging for (e.g., plants and animals). Recently, Wilke et al. (2018) found support for this claim when investigating the exact ecological spatial patterns of different classes of resources in present environments. After observing and coding 15 different resources from both developed and natural domains – such as seats taken at a café and in a restaurant, occupied parking spots, group members of goose and cow groupings, and patterns of wilderness, wild forest, and water – the results showed that natural resource domains (e.g., animal distributions, habitat structures) indeed display more aggregated distribution patterns than those from human-developed resource domains and that many human-developed resource domains often contain aggregation. Random distributions occurred much less frequently than aggregated ones, and dispersed distributions were very rare. Thus, an evolutionarily informed study of our decision-making capacities should research our cognitive mechanisms in the ecological environments they evolved in and investigate the extent to which specific adaptive challenges – such as foraging for resources – may have left a distinct footprint on our evolved human cognition (cf. Barrett, 2018; Todd, Hills, & Robbins, 2012; Wilke & Todd, 2010).

REFERENCES

Alphen, J. J. M. van, Bernstein, C., & Driessen, G. (2003). Information acquisition and time allocation in insect parasitoids. *Trends in Ecology and Evolution*, **18**, 81–87.

American Psychiatric Association (2013). *Diagnostic and Statistical Manual of Mental Disorders, Fifth Edition*. Arlington, VA: American Psychiatric Association Publishing.

Barrett, H. C. (2018). The varieties of foraging experience. *Evolutionary Behavioral Sciences*, **12**, 135–138.

Bell, W. J. (1991). *Searching Behaviour: The Behavioural Ecology of Finding Resources*. New York: Chapman & Hall.

Binford, L. (1990). Mobility, housing, and environment: A comparative study. *Journal of Anthropological Research*, **46**, 119–152.

Blanchard, T. C., Wilke, A., & Hayden, B. Y. (2014). Hot hand bias in rhesus monkeys. *Journal of Experimental Psychology: Animal Learning and Cognition*, **40**, 280–286.

Castel, A. D., Drolet Rossi, A., & McGillivray, S. (2012). Beliefs about the "hot hand" in basketball across adult life span. *Psychology and Aging*, **27**, 601–605.

Charnov, E. L. (1976). Optimal foraging: The marginal value theorem. *Theoretical Population Biology*, **9**, 129–136.

Croson, R., & Sundali, J. (2005). The gambler's fallacy and the hot hand: Empirical data from Casinos. *Journal of Risk and Uncertainty*, **30**, 195–209.

Czienskowski, U. (2005a). *The Fishing Task Experiment* [computer software]. Berlin: Max Planck Institute for Human Development.

Czienskowski, U. (2005b). *The Word Puzzle Experiment* [computer software]. Berlin: Max Planck Institute for Human Development.

Falk, R. (1975). Perception of randomness. Unpublished doctoral dissertation (in Hebrew, with English abstract), Hebrew University of Jerusalem.

Falk, R., & Konold, C. (1997). Making sense of randomness: Implicit encoding as a basis for judgment. *Psychological Review*, **104**, 301–318.

Fawcett, T. W., Fallenstein, B., Higginson, A. D., et al. (2014). The evolution of decision rules in complex environments. *Trends in Cognitive Sciences*, **18**, 153–161.

Gilovich, T., Vallone, R., & Tversky, A. (1985). The hot hand in basketball: On the misperception of random sequences. *Cognitive psychology*, **17**, 295–314.

Haselton, M. G., Bryant, G. A., Wilke, A., et al. (2009). Adaptive rationality: An evolutionary perspective on cognitive bias. *Social Cognition*, **27**, 733–763.

Hill, K., & Hurtado, A. M. (1996). *Ache Life History: The Ecology and Demography of a Foraging People*. New York: Aldine de Gruyter.

Hills, T. T., Todd, P. M., & Jones, M. N. (2015). Foraging in semantic fields: How we search through memory. *Topics in Cognitive Science*, **7**, 513–534.

Hodgins, D. C., Stea, J. N., & Grant, J. E. (2011). Gambling disorders. *Lancet*, **378**, 1874–1884.

Hutchinson, J. M. C., & Gigerenzer, G. (2005). Simple heuristics and rules of thumb: Where psychologists and behavioral biologists might meet. *Behavioural Processes*, **69**, 97–124.

Hutchinson, J. M. C., Wilke, A., & Todd, P. M. (2008). Patch leaving in humans: Can a generalist adapt its rules to dispersal of items across patches? *Animal Behaviour*, **75**, 1331–1349.

Iwasa, Y., Higashi, M., & Yamamura, N. (1981). Prey distribution as a factor determining the choice of optimal foraging strategy. *American Naturalist*, **117**, 710–723.

Kelly, R. L. (1995). *The Foraging Spectrum*. Washington, DC: Smithsonian Institution Press.

Krause, J., & Ruxton, G. D. (2002). *Living in Groups: Oxford Series in Ecology and Evolution*. Oxford: Oxford University Press.

Krebs, J. R., Ryan, J. C., & Charnov, E. L. (1974). Hunting by expectation or optimal foraging? A study of patch use by chickadees. *Animal Behavior*, **22**, 953–964.

Legge, E. L. G., Spetch, M. L., Cenker, A., et al. (2012). Not all locations are created equal: Exploring how adults hide and search for objects. *PLOS ONE*, **7**, e36993.

Livoreil, B., & Giraldeau, L.-A. (1997). Patch departure decisions by spice finches foraging singly or in groups. *Animal Behavior*, **54**, 967–977.

Mata, R., Wilke, A., & Czienskowski, U. (2009). Cognitive aging and adaptive foraging behavior. *Journal of Gerontology: Psychological Sciences*, **64**, 474–481.

Mata, R., Wilke, A., & Czienskowski, U. (2013). Foraging across the life span: Is there a reduction in exploration with aging? *Frontiers in Neuroscience*, **7**, 53.

McNamara, J. M. (1982). Optimal patch use in a stochastic environment. *Theoretical Population Biology*, **21**, 269–288.

McNamara, J. M., & Houston, A. I. (1985). Optimal foraging and learning. *Journal of Theoretical Biology*, **117**, 231–249.

Nickerson, R. S. (2002). The production and perception of randomness. *Psychological Review*, **109**, 330–357.

Nowak, M. A., & Sigmund, K. (1993). A strategy of win-stay, lose-shift that outperforms tit-for-tat in the Prisoner's Dilemma game. *Nature*, **364**, 56–58.

Oskarsson, A. T., van Boven, L., McClelland, G. H., & Hastie, R. (2009). What's next? Judging sequences of binary events? *Psychological Bulletin*, **135**, 262–285.

Pirolli, P. (2007). *Information Foraging Theory: Adaptive Interaction with Information*. Oxford: Oxford University Press.

Raab, M., Gula, B., & Gigerenzer, G. (2012). The hot hand exists in volleyball and is used for allocation decisions. *Journal of Experimental Psychology: Applied*, **18**, 81–94.

Reifman, A. (2011). *Hot Hand: The Statistics behind Sports' Greatest Streaks*. Dulles, VA: Potomac Books.

Scheibehenne, B., Wilke, A., & Todd, P. M. (2011). Expectations of clumpy resources influence predictions of sequential events. *Evolution and Human Behavior*, **32**, 326–333.

Shaffer, H. J., Hall, M. N., & Bilt, J. V. (1997). *Estimating the Prevalence of Disordered Gambling Behavior in the United States and Canada: A Meta-analysis*. Boston, MA: Harvard Medical School.

Talbot, K. J., Legge, E. L. G., Bulkito, V., & Spetch, M. L. (2009). Hiding and searching strategies of adult humans in a virtual and a real-space room. *Learning and Motivation*, **40**, 221–233.

Taylor, L. R. (1961). Aggregation, variance and the mean. *Nature*, **189**, 732–735.

Taylor, L. R., Woiwod, I. P., & Perry, J. N. (1978). The density-dependence of spatial behaviour and the rarity of randomness. *Journal of Animal Ecology*, **47**, 383–406.

Todd, P. M., & Gigerenzer, G. (2012). What is ecological rationality? In P. M. Todd, G. Gigerenzer, & The ABC Research Group, eds., *Ecological Rationality: Intelligence in the World*. New York: Oxford University Press, pp. 3–32.

Todd, P. M., Hills, T. T., & Robbins, T. W., eds. (2012). *Cognitive Search: Evolution, Algorithms, and the Brain* (Strüngmann Forum Reports, Vol. 9). Cambridge, MA: MIT Press.

Tooby, J., & DeVore, I. (1987). The reconstruction of hominid behavioral evolution through strategic modeling. In W. G. Kinsey, ed., *The Evolution of Primate Behavior: Primate Models*. New York: State University of New York Press, pp. 183–237.

Waage, J. K. (1979). Foraging for patchily-distributed hosts by parasitoid, Nemeritis canescens. *Journal of Animal Ecology*, **48**, 353–371.

Wajnberg, E., Fauvegue, X., & Pons, O. (2000). Patch leaving decision rules and the marginal value theorem: An experimental analysis and a simulation model. *Behavioral Ecology*, **11**, 577–586.

Wajnberg, E., Gonsard, P.-A., Tabone, E., et al. (2003). A comparative analysis of patch-leaving decision rules in a parasitoid family. *Journal of Animal Ecology*, **72**, 618–626.

Wilke, A. (2006). Evolved responses to an uncertain world. Unpublished PhD thesis, Free University of Berlin.

Wilke, A., & Barrett, H. C. (2009). The hot hand phenomenon as a cognitive adaptation to clumped resources. *Evolution and Human Behavior*, **30**, 161–169.

Wilke, A., & Mata, R. (2012). Cognitive bias. In V. S. Ramachandran, ed., *Encyclopedia of Human Behavior*, 2nd ed. Maryland Heights, MO: Elsevier.

Wilke, A., & Todd, P. M. (2010). Past and present environments: The evolution of decision making. *Psicothema*, **22**, 4–8.

Wilke, A., Hutchinson, J. M. C., Todd, P. M., & Czienskowski, U. (2009). Fishing for the right words: Decision rules for human foraging behavior in internal search tasks. *Cognitive Science*, **33**, 497–529.

Wilke, A., Scheibehenne, B., Gaissmaier, W., McCanney, P., & Barrett, H. C. (2014). Illusionary pattern detection in habitual gamblers. *Evolution and Human Behavior*, **35**, 291–297.

Wilke, A., Minich, S., Panis, M., et al. (2015). A game of hide and seek: Expectations of clumpy resources influence hiding and searching patterns. *PLoS ONE*, **10**, e0130976.

Wilke, A., Lydick, J., Bedell, V., et al. (2018). Spatial dependency in local resource distributions. *Evolutionary Behavioral Sciences*, **12**, 163–172.

Winterhalder, B., & Smith, E. A. (1981). *Hunter–Gatherer Foraging Strategies: Ethnographic and Archaeological Analyses*. Chicago, IL: University of Chicago Press.

World Health Organization (1992). *ICD-10 Classifications of Mental and Behavioural Disorder: Clinical Descriptions and Diagnostic Guidelines*. Geneva: WHO Press.

PART VI

EVOLUTION AND DEVELOPMENT

The evolutionary approach is qualitatively different from all other academic attempts to study human behavior in that its focus on ultimate explanations problematizes the commonplace. When we ask our students what childhood is for, they quite often look perplexed. *"For?"* they seem to say. "Does it have to be *for* anything?" And we can sympathize; after all, a sapling does not have a different function from a fully grown tree – its function is the same. It is only the way it is due to the thermodynamic impossibility of going from seed to fully grown tree without passing – however briefly – through some intermediate size. So are children just thermodynamically constrained adults?

Alyson J. Myers, a developmental and comparative psychologist, and David F. Bjorklund, one of the founders of evolutionary developmental psychology, show us how children are so much more than human saplings. They show how, from birth, children are equipped with a variety of adaptive biases that enable them to negotiate the human and physical world. Some of these biases are concerned with negotiating their current world (*ontogenic adaptations*), which disappear like milk teeth when no longer needed, to be replaced by other ways of being. Interestingly, they propose that cognitive immaturity may itself be an adaptation. Other adaptations (*deferred adaptations*) prepare children to become more successful adults, with play being an example. Finally, *conditional adaptations* use the state of the childhood environment to work out what kind of adult it is best to become and so embark upon the most appropriate path. This chapter covers a lot of ground, and a lot of familiar (as well as less familiar) developmental phenomena are held up to evolutionary scrutiny, with fascinating results.

In Chapter 22, Sebastian Grüneisen and Emily Wyman, both experts in the development of cooperation, focus on the development of cooperative ability in childhood – specifically, using a problem first formulated by Rousseau known as the Stag Hunt. The bare bones of this dilemma is that by cooperating we can all reap greater rewards than by acting alone, but we can only obtain those greater rewards if we are able to recruit others to join in. Grüneisen and Wyman use comparative research exploring the differences between children and chimpanzees in their relative abilities to share attention, and they show that – as might be expected from an intensely cooperative species – children are better at representing the intentions of others, can switch roles within the collaborative action, and are better at representing joint goals. It seems that the origins of adult cooperation are already present at eight years or age, and even younger in some cases. As human societies grew larger, fueled by (among other things) intergroup conflict, there was a pressure for the establishment of behavioral norms that specified how each group member should behave. As before, it seems that young children develop these abilities early, including the identification of and preference for the in-group and the formation and enforcement of social norms.

Like blues singers, evolutionary psychologists frequently discuss the battle of the sexes; in Chapter 23, William M. Brown shows how this battle rages on in the bodies of our offspring. Brown, a long-time researcher in epigenetics in a diverse range of applications, discusses the head-spinning topic of genomic imprinting. In essence, the interests of paternal and maternal genes compete within the offspring: the father's genes want maximal investment in the offspring in terms of the mother's material and emotional provision, while the mother's genes would like to give their mother an easier ride. Why is this the case? One reason is that mothers are always certain that their children are biologically related to them, so the genes in their offspring want to share the mother's resources with their siblings. Fathers are never certain of paternity, so any offspring containing paternal genes wants to extract the maximum from their mother and keep it for him or herself. Brown shows how this form of intra-genomic conflict has implications for parental investment, language development, and the development of certain kinds of mental and physical illnesses.

Peter K. Smith's career has been largely devoted to conducting groundbreaking research into bullying, which is the topic Chapter 24. In an exhaustive review, Smith tackles traditional explanations for bullying – that it is the result of low self-esteem, poor social skills, and psychopathic tendencies – as well as more recent theories that see bullying as an adaptive way of attaining social standing, resources, and sexual partners. His conclusion is that this there is no clear single adaptive benefit to bullying. At least part of this is because research fails to distinguish in many cases between different kinds of bullies, methodological problems, and cultural differences. He also questions the notion that bullying is universal and ancient, suggesting that bullying may either be a facultative strategy that only reveals itself in certain environmental conditions or that it is merely a use of aggressive tendencies in situations when one can get away with it.

Frank J. Sulloway is a world-renowned psychologist and historian of science and is perhaps best known for two of his books, *Freud, Biologist of the Mind* (1992) and *Born to Rebel* (1996). It is the theme of the latter book that he discusses in Chapter 25. Research on birth order consistently shows that firstborns are more dominant, more conscientious, more conservative, and less agreeable than later-borns. Sulloway discusses the potential causes of these differences, including differential parental investment, sibling dominance hierarchies, and niche fitting. While some of these seem to have more effect than others, Sulloway argues that many different mechanisms are at work, and the challenge is to determine how they interact with each other.

REFERENCES

Sulloway, F. J. (1992). *Freud, Biologist of the Mind: Beyond the Psychoanalytic Legend*. Cambridge, MA: Harvard University Press.

Sulloway, F. J. (1996). *Born to Rebel: Birth Order, Family Dynamics, and Creative Lives*. New York: Pantheon Books.

more aggressive and risky behavior, become pregnant earlier, form unstable pair bonds, and provide less parental investment per offspring than children growing up in less harsh and more predictable environments (e.g., Ellis, 2004; Ellis et al., 2009; Nettle, 2010; Nettle & Cockerill, 2010; Placek & Quinlan, 2012; Wilson & Daly, 1997). For example, in a longitudinal study, children living in highly unpredictable environments (e.g., parental job changes, changes in residences, different adult males living in the household) during their first five years of life had earlier sexual debut, more sex partners, and higher levels of risk-taking and delinquent behavior at age 23 than children growing up in more predictable homes (Simpson et al., 2012). Although these are signs of maladjustment in contemporary society, engaging in these "risky" behaviors may be adaptive, from a reproductive perspective (or would have been adaptive for our ancestors), for children growing up in harsh and unpredictable environments (Ellis et al., 2009).

In an unstable environment, having a long-term investment in the development of a slow life history strategy does not optimize fitness. The energy that was invested in the future is wasted if the individual matures in an environment where life expectancy is short. Here, to respond to these environmental cues, it would serve the individual better to adopt a faster life history strategy.

21.2 DEVELOPING ADAPTATIONS: EVOLVED PROBABILISTIC COGNITIVE MECHANISMS

Most members of a species follow similar developmental trajectories. This is because they inherit not only a species-typical genome, but also a species-typical environment. For humans, a species-typical environment includes a womb during the prenatal period, a mother who nurtures and nurses infants post-birth, a social network of other people to help care for children, as well as other environmental factors such as temperature, humidity, gravity, and light cycles (Bjorklund & Ellis, 2014). These species-typical environmental factors constrain how a child's phenotype develops and the contexts in which genes are expressed.

Humans are not born as blank states. Rather, humans are prepared by natural selection to process some information more readily than others. According to Geary (1995, 2005), infants are born with *skeletal competencies*, information-processing biases or constraints that are fleshed out over the course of development, principally via play. Geary envisions these skeletal competencies to be the lowest levels of a hierarchically organized system depicted in Figure 21.1. Geary proposed two overarching domains – social (folk psychology) and ecological, the latter composed of the domains of folk biology and folk physics. Each of these domains, in turn, consists of more specific domains.

Our social, biological, and physical environments constrain how we are able to make sense of the world, and

having these skeletal competencies makes it easier to process some types of information more readily than others. These skeletal competencies occur because of selectively structured gene × environment × development interactions, which arise in each generation and are influenced by prenatal and postnatal environments, reflecting the inheritance of not only genes, but of entire developmental systems.

In order to better reflect the gene–environment interaction nature of development within an evolutionary framework, Bjorklund, Ellis, and Rosenberg (2007, p. 22) proposed the concept of *evolved probabilistic cognitive mechanisms* to complement the concept of evolved cognitive mechanisms of mainstream evolutionary psychology. Bjorklund et al. defined evolved probabilistic cognitive mechanisms as:

information-processing mechanisms that have evolved to solve recurrent problems faced by ancestral populations; however, they are expressed in a probabilistic fashion in each individual in a generation, based on the continuous and bidirectional interaction over time at all levels of organization, from the genetic through the cultural. These mechanisms are universal, in that they will develop in a species-typical manner when an individual experiences a species-typical environment over the course of ontogeny.

Evolved probabilistic cognitive mechanisms underlie all (or almost all) psychological adaptations. These evolved probabilistic cognitive mechanisms and their accompanying adaptations develop as a function of the interaction between all levels of an organism and its environment, ranging from genes through culture. Thus, prepared is not preformed (Bjorklund, 2003, 2016), with the eventual outcome being dependent not only on inherited genes, but also on the timing of relevant experiences. We provide four examples of how evolved probabilistic cognitive mechanisms can lead to adaptive outcomes, two from the realm of folk psychology (infant face perception, social learning), one from the realm of folk biology (fear of snakes), and one from the realm of folk physics (tool use).

21.2.1 Infant Face Perception

By 12 months of life, infants are able to better discriminate among faces of their own race (or species) than faces of other races (or species) (Lee, Quinn, & Pascalis, 2017). This specialization would seem to be adaptive, given that over a lifetime, most of our ancestors would rarely be required to differentiate the faces of people from other races (or animals from other species), but would need to make many such discriminations among members of their own race/tribe. This specialization develops following the tenets proposed by evolved evolutionary cognitive mechanisms.

Early in development, infants have a general tendency to be attentive to faces, or at least to perceptual features associated with faces. For example, early in infancy, babies are attentive to curvilinear, top-heavy (i.e., more

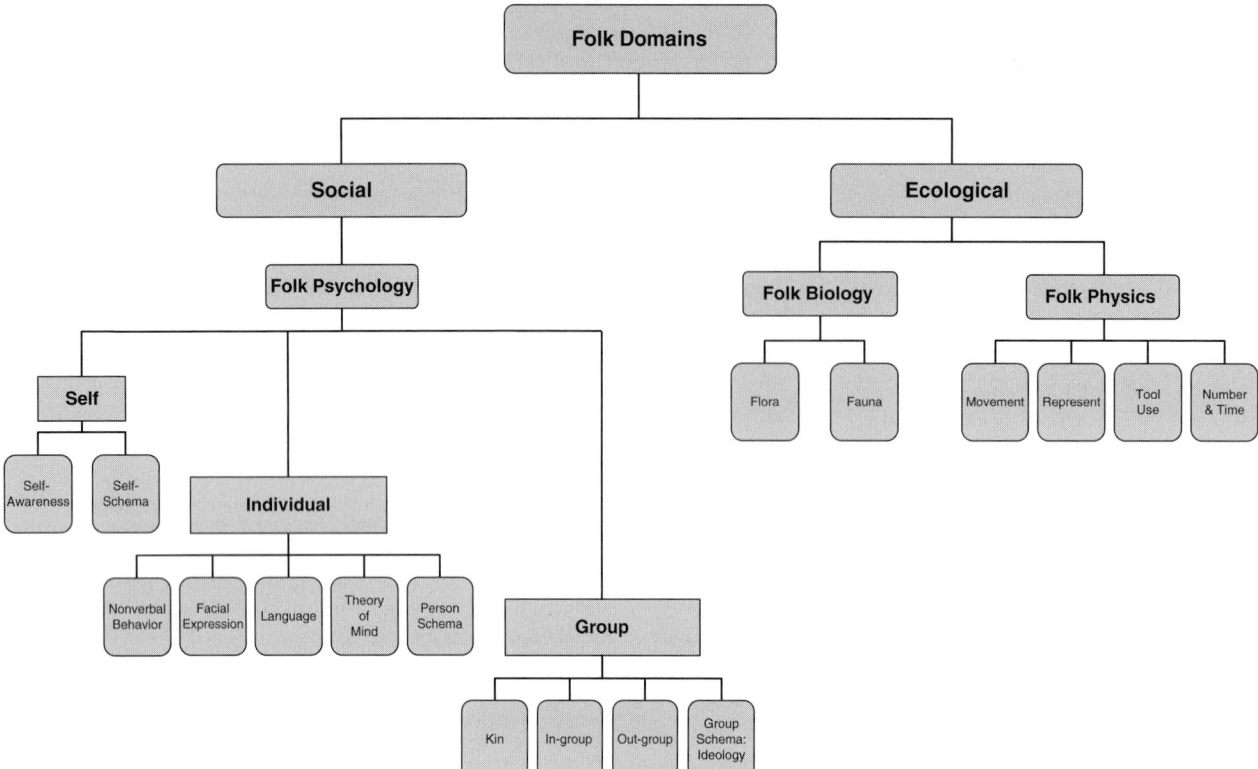

Figure 21.1 Geary's hierarchical mode.

information on the top half of an image than the bottom half), vertically symmetrical (left and right sides are similar) stimuli with high contrast (Bjorklund, 2015). These all happen to be features characteristic of the primate face, accounting for infants' bias to be attentive to faces without having to propose an innate "face module." Newborns (Di Georgio et al., 2012), one-month-olds (Wakako et al., 2014), and six-month-olds (Pascalis, de Haan, & Nelson, 2002) are also able to discriminate among upright (and thus top-heavy) faces of both humans and monkeys better than inverted faces, and at three months of age infants can discriminate among faces from their own race and other races equally well (Kelly et al., 2007, 2009). However, by nine months of age, infants are increasingly able to discriminate among human faces, but no longer among monkey faces (Pascalis et al., 2002; unless they have regular exposure to monkey faces – Pascalis et al., 2005), and among faces from their own race, but not faces from other races (Kelly et al., 2007, 2009; unless they are exposed to faces from other races – Anzures et al., 2012).

These findings demonstrate the importance of experience in shaping the development of the infants' face-processing systems and the interaction of inherited biases with species-typical experiences, as reflected by evolved probabilistic cognitive mechanisms. Infants enter the world with some perceptual biases and low-level processing abilities and are initially able to make discriminations for faces of different species and ethnicities equally well.

However, with experience, neural reorganization occurs and infants' processing narrows, increasing their abilities to discriminate among faces they experience frequently (humans and faces from their own race) and decreasing their ability to discriminate among faces they do not experience frequently (monkeys and faces from other races).

21.2.2 Social Learning

Social learning is a mechanism that allows us to acquire behaviors by observing other individuals without engaging in trial-and-error learning. Social learning increases ecological success by avoiding the costs and energy expenditure used in individual learning. Human culture is transmitted from individual to individual, community to community, and generation to generation. Typically, the transmission of culture is from adults to children, with each new generation beginning with the knowledge of the previous generation. This allows the rapid expansion of knowledge and information over time. Consistent with the concept of evolved probabilistic cognitive mechanisms, there are important social biases that promote the development of social learning, beginning in infancy (see also Legare, 2017).

Neonates are biased to orient toward social stimuli. Infants are biased to attend to biological motion (Bardi, Regolin, & Simion, 2014), to faces (Mondloch et al., 1999), and more specifically to eyes (Gava et al., 2008). As discussed previously, newborns tend to match facial expressions (Meltzoff & Moore, 1977), which also helps to foster bonds and relationships between the infant and caregiver, increasing infants' chances of survival (Simpson & Belsky, 2008; Thompson, 2006). Being attentive to social others may be a necessary condition to efficient social learning, but only a beginning. In order to learn from observing others, children must have a basic understanding of others as goal-directed individuals who can cause things to happen.

Viewing Others as Intentional Agents. One of the most important and basic abilities of infants' understanding of other people, which serves as a basis for the development of social learning, is viewing others as intentional agents. This means understanding that other individuals cause things to happen and that their behaviors are goal directed. Young infants are socially oriented, but they lack the ability to view and treat others as intentional agents until the latter part of the first year (Tomasello, 2009; Tomasello & Carpenter, 2007). *Shared attention* is early evidence of this. Shared attention involves a triadic interaction between an infant, another person, and an object. For example, a mother may point or direct her gaze toward a toy while getting the infant's attention. Rudimentary signs of shared attention begin in the early days of an infant's life, but it takes until approximately nine months of age before an infant actively engages in shared attention. During shared attention, the two participants experience the same thing at the same time and know that they are experiencing this together (Tomasello & Carpenter, 2007).

In order to have shared attention, infants must be able to demonstrate gaze following. They must be able to follow a social partner's eye gaze toward something or someone. Gaze-following and shared-attention abilities improve over the first year, with 10-month-old infants, but not 9-month-old infants, being able to understand social gaze and expecting partners in conversation to look at one another (Beier & Spelke, 2012). Between 12 and 18 months of age, infants will use another person's gaze to direct their own attention (Brooks & Meltzoff, 2002). At 12 months of age, infants are able to point to objects in order to inform others (Liszkowski, Carpenter, & Tomasello, 2007). They will also point to objects to direct an adult's attention to something the adult is searching for (Liszkowski et al., 2006). At approximately 12 months of age, infants will begin to engage in *social referencing*, using a parent's facial expression, gestures, tone of voice, or a combination of these things to determine their actions in an uncertain situation (e.g., "Should I laugh or cry?") (Hornik, Risenhoover, & Gunnar, 1987; Vaish & Striano, 2004).

Having the capability to gaze-follow and engage in shared attention provides important biases that set the stage for the development of social learning.

Social Learning: Imitation and Emulation. There are various forms of social learning, some more effective than others, and some seemingly well suited for children to acquire the knowledge and skills of their culture. Children two years of age and younger typically engage in *emulation*, in which they recognize the goal of a model but do not copy the exact behaviors the model exhibited to attain the goal (Nielson, 2006). In contrast, in true *imitation*, the learner understands the model's goal and uses the same or similar behaviors to attain said goal (Tomasello, 2000). Beginning around three years of age, children are more apt to engage in imitation rather than emulation, and in fact often do so to the point of copying all actions of the model, even if those actions are irrelevant to attaining the goal (Lyons, Young, & Kiel, 2007; Nielson, 2006). This is referred to as *overimitation*. For example, in a study by Lyons et al. (2007), children watched an adult perform a series of actions to open a box to retrieve a toy locked inside. When the children were given the same box, they copied all of the actions, relevant and irrelevant, to open the box, even after being warned to avoid unnecessary actions. Children do not always imitate blindly. Preschool-aged children will imitate more selectively if they have some awareness of the intentions of the model (Gardiner, 2014; Gardiner, Greif, & Bjorklund, 2011) or if they know the specific goal of a task beforehand (Williamson & Markman, 2006). Yet, overimitation seems to be the primary social-learning strategy for preschoolers, and persists into adulthood in some contexts (McGuigan, Makinson, & Whiten, 2011). Overimitation is not limited to children from Western cultures, but has also been seen in two- to six-year-old Kalahari Bushman children (Nielsen & Tomaselli, 2010).

A number of researchers have proposed that overimitation reflects an evolved adaptation. For example, overimitation may reflect an evolved heuristic for learning about cultural artifacts and their uses (Whiten et al., 2009). Unlike other species, humans have thousands of artifacts that they must learn how to use. An effective way to learn about a cultural object is to copy modeled behaviors exactly. While this may lead to some irrelevant actions being acquired, such actions can be "weeded out" through individual learning. In support of this argument, children tend to think that modeled actions are normative. For example, three- to five-year-old children corrected a puppet that omitted unnecessary actions that were previously performed by an adult because the puppet was "doing it wrong" (Kenward, 2012). Children's indiscriminate imitation may be a human adaptation that permits quick and accurate transmission of information between individuals (Csibra & Gergely, 2011; Gergely & Csibra, 2005). It also eliminates the costs in time and effort associated with trial-and-error and individual learning.

There is much diversity in human social environments, yet social-cognitive development follows a species-typical trajectory across cultures. Consistent with the concept of evolved probabilistic cognitive mechanisms, low-level biases to attend to faces, to share attention, and to copy the actions of others, in interaction with a broadly defined species-typical environment over the course of early development, produces a social-learning system with sufficient plasticity to permit children to acquire the skills and values associated with their culture.

21.2.3 Fear of Snakes

Evolved probabilistic cognitive mechanisms underlie some emotions, in particular fear, especially for evolutionarily relevant stimuli. For example, ancestral primates had reasons to fear snakes. However, neither children nor monkeys show an inherent fear of snakes, although both can easily acquire such fear through social learning. For example, monkeys who have never seen a snake before easily learn to respond fearfully to one after observing another monkey react fearfully to a snake. In contrast, monkeys do not act fearfully to a rabbit or flower after viewing another monkey reacting fearfully to these "neutral" stimuli (Cook & Mineka, 1989). Thus, monkeys do not have an innate fear of snakes, but develop one easily via observation.

Humans demonstrate a similar preparation for acquiring fear of snakes (and also spiders) (Lobue & Rakison, 2013). While many humans in developed countries do not interact with snakes frequently, snakes still kill more humans worldwide than any other vertebrate (Kasturiratne et al., 2008), and humans seem biased to be especially attentive to slithering serpents. For example, adults (Öhman, Flykt, & Esteves, 2001), children (LoBue, 2010), and infants (LoBue & DeLoache, 2010) more readily identify snakes and spiders in a background of flowers and mushrooms than vice versa, and nine-month-olds are more attentive to photographs of snakes, as reflected by patterns of evoked potentials, than to photographs of neutral stimuli (Hoehl & Pauen, 2017), suggesting a "preparation" to attend to these potentially dangerous evolutionarily relevant stimuli.

However, like monkeys, human infants are not necessarily innately fearful of snakes. For example, after demonstrating that 9- to 10-month-old infants showed no bias to respond to videos of snakes vs. other animals, DeLoache and LoBue (2009) showed 7- to 18-month-old infants videos of a snake and another exotic animal, such as a hippopotamus or elephant, on adjacent video screens. On different trials, a nonsense phrase was spoken in either a fearful or a pleasant tone, sometimes paired with the snake and sometimes paired with the other animal. Infants looked significantly longer at the snake when it was accompanied by a fearful voice than by the pleasant voice. The researchers did not find a significant difference in looking times for the pleasant and fearful voice when paired with the other animals. These findings indicate that infants are biased to associate snakes, but not other potentially dangerous animals, with fearful vocalizations. A subsequent experiment replaced the video displays with photographs of snakes and other animals and failed to find any bias to associate fearful voices with snakes. This result suggests that it is not the snake itself, but rather the sinusoidal movement of the snake, that is the feature producing special attention to them (and perhaps spiders: see Rakison, 2005) and a tendency to associate the perception of a snake's movement with fearful vocalizations.

It is clear that infants are not innately born with a fear of snakes or spiders, but rather, they are prepared to acquire such fears. Consistent with the concept of evolved probabilistic cognitive mechanisms, primate infants appear to possess perceptual biases increasing the probability that fearful responses to these evolutionarily relevant stimuli will be learned. These perceptual biases precede the fearful responses (LoBue, 2013), meaning that experience plays a vital role in the acquisition of these adaptive behaviors.

21.2.4 Tool Use

Infants around the world manually explore artifacts ("What can the object do?") and around nine months of age begin to manually play with objects ("What can *I* do with the object?") (Belsky & Most, 1981). Such object exploration and play result in infants discovering objects' *affordances* (i.e., functional relationships between objects and the environment), which may facilitate subsequent tool use (Lockman & Kahrs, 2017). For instance, five- and six-year-old children who had the opportunity to explore the properties of a pipe cleaner were subsequently more likely to appropriately modify pipe cleaners (bend them so they can be used as hooks) to solve a problem (retrieve an out-of-reach object) than children not permitted to manipulate the materials beforehand (Cutting et al., 2014). In other research, discussed earlier with respect to sex differences in play style, the amount of object-oriented play three-year-old boys (but not girls) engaged in during a free-play session predicted their performance on a later toy-retrieval task (Gredlein & Bjorklund, 2005), with the authors speculating that boys may have evolved to be more sensitive to such environmental experiences than girls.

In addition to being motivated to interact with objects, children are biased to believe that an artifact used to solve a problem was in fact designed for a specific function, termed the *design stance*. For instance, when three-year-old children were shown a function for a new tool (e.g., a box-like object used for catching bugs), they were less likely to see an alternative use for the object (e.g., collecting raindrops), but rather associated it with its originally designed purpose (German & Johnson, 2002). This reflects a lack of flexibility in tool use, termed *functional fixedness*, which is usually interpreted as a hindrance to problem-solving. However, most tools were indeed constructed for a specific purpose, and given young children's need to master the use of many different tools, the design stance likely reflects an adaptive mechanism fostering children's mastery of many of the tools prevalent in their culture.

Coupled with the design stance is young children's tendency to readily learn about tool use from watching other people use them (e.g., Flynn & Whiten, 2012; Gardiner et al., 2012). As discussed in Section 21.2.2 when examining the development of social learning, beginning around the age of three years, children tend to engage in overimitation, copying *all* behaviors associated with a model, whether relevant or not (Lyons et al., 2007). Children assume (usually correctly) that all of a model's behavior is normative, facilitating the acquisition of culturally specific use of tools (Kenward, 2012). Consistent with the concept of evolved probabilistic cognitive mechanisms, this pattern – from manipulation through the design stance to overimitation of actions on artifacts – makes effective tool use in humans almost inevitable, but rather than being based on a "tool-use instinct," it has its origins in early-developing biases to manipulate and play with objects and later to see purpose in the artifacts surrounding them, making it highly likely that they will discover culturally prescribed uses for objects.

21.3 CONCLUSION

Human brains and minds have experienced natural selection just as human bodies have, producing cognitive and behavioral adaptations to handle recurrent problems that our ancestors encountered (Bjorklund, 2015). Infants enter this world not as blank slates, but with biases, and they are prepared to attend to and process certain types of information and stimuli more readily than others. These evolved cognitive mechanisms are probabilistic and develop in a species-typical manner when children experience a species-typical environment. Children inherit lower-level biases that prepare them to attend to or process information more effectively, such as a fear of snakes or spiders. Natural selection has provided these initial biases, and a species-typical childhood environment eventually produces species-typical adult behavior. Adaptations play a central role in evolutionary psychological theory; however, adaptations have developmental histories, and an understanding of how such adaptations develop provides greater insights for evolutionary psychological explication.

Evolutionary developmental psychologists recognize the importance of childhood and the adaptations natural selection have shaped to survive childhood and prepare children for adulthood. These evolved characteristics develop via continuous and reciprocal bidirectional gene–environment interactions that emerge over time. Evolutionary developmental psychology emphasizes the importance of examining development through an evolutionary lens and of considering the interaction between a child and all levels, from genes to culture. From this perspective, we are able to see that not only does "nothing in biology make sense except in the light of evolution" (Dobzhansky, 1973), but that the same applies for psychology and human development.

REFERENCES

Abravanel, E., & Sigafoos, A. D. (1984). Exploring the presence of imitation during early infancy. *Child Development*, **55**, 381–392.

Anzures, G., Wheeler, A., Quinn, P. C., et al. (2012). Brief daily exposures to Asian females reverses perceptual narrowing for Asian faces in Caucasian infants. *Journal of Experimental Child Psychology*, **112**(4), 484–495.

Bardi, L., Regolin, L., & Simion, F. (2014). The first time ever I saw your feet: Inversion effect in newborns' sensitivity to biological motion. *Developmental Psychology*, **50**(4), 986.

Beier, J. S., & Spelke, E. S. (2012). Infants' developing understanding of social gaze. *Child Development*, **83**(2), 486–496.

Belsky, J., & Most, R. K. (1981). From exploration to play: A cross-sectional study of infant free play behavior. *Developmental Psychology*, **17**(5), 630.

Belsky, J., Steinberg, L., & Draper, P. (1991). Childhood experience, interpersonal development, and reproductive strategy: An evolutionary theory of socialization. *Child Development*, **62**(4), 647–670.

Belsky, J., Steinberg, L. D., Houts, R. M., et al. (2007). Family rearing antecedents of pubertal timing. *Child Development*, **78**(4), 1302–1321.

Bjorklund, D. F. (1987). How age changes in knowledge base contribute to the development of children's memory: An interpretive review. *Developmental Review*, **7**(2), 93–130.

Bjorklund, D. F. (1997). The role immaturity in human development. *Psychological Bulletin*, **122**(2), 153–169.

Bjorklund, D. F. (2003). Evolutionary psychology from a developmental systems perspective: comment on Lickliter and Honeycutt (2003). *Psychological Bulletin*, **129**, 836–841.

Bjorklund, D. F. (2007). *Why Youth Is* Not *Wasted on the Young: Immaturity in Human Development*. Oxford: Blackwell.

Bjorklund, D. F. (2015). Developing adaptations. *Developmental Review*, **38**, 13–35.

Bjorklund, D. F. (2016). Prepared is not preformed: Comment on Witherington and Lickliter. *Human Development*, **59**, 235–241.

Bjorklund, D. F., & Ellis, B. J. (2014). Children, childhood, and development in evolutionary perspective. *Developmental Review*, **34**(3), 225–264.

Bjorklund, D. F., Gaultney, J. F., & Green, B. L. (1993). "I watch therefore I can do": The development of meta-imitation over the preschool years and the advantage of optimism in one's imitative skills. In R. Pasnak & M. L. Howe, eds., *Emerging Themes in Cognitive Development*, Vol. 1. New York: Springer Verlag, pp. 79–102.

Bjorklund, D. F., & Green, B. L. (1992). The adaptive nature of cognitive immaturity. *American Psychologist*, **47**, 46–54.

Bjorklund, D. F., & Pellegrini, A. D. (2002). *The Origins of Human Nature: Evolutionary Developmental Psychology*. Washington, DC: American Psychological Association.

Bjorklund, D. F., Ellis, B. J., & Rosenberg, J. S. (2007). Evolved probabilistic cognitive mechanisms: An evolutionary approach to gene × environment × development interactions. In R. V. Kail, ed., *Advances in Child Development and Behavior*, Vol. 35. Oxford: Elsevier, pp. 1–39.

Bjorklund, D. F., Hernández Blasi, C., & Ellis, B. J. (2016). Evolutionary developmental psychology. In D. M. Buss, ed., *Evolutionary Psychology Handbook*, 2nd ed., Vol. 2. New York: Wiley, pp. 904–925.

Boyce, W. T., & Ellis, B. J. (2005). Biological sensitivity to context: I. An evolutionary-developmental theory of the origins and functions of stress reactivity. *Development and Psychopathology*, **17**, 271–301.

Brooks, R., & Meltzoff, A. N. (2002). The importance of eyes: How infants interpret adult looking behavior. *Developmental Psychology*, **38**, 958–966.

Buss, D. M., Haselton, M. G., Shackelford, T. K., Bleske, A. L., & Wakefield, J. C. (1998). Adaptations, exaptations, and spandrels. *American Psychologist*, **53**, 533–548.

Byrne, R. W. (2005). Social cognition: Imitation, imitation, imitation. *Current Biology*, **15**, R489–R500.

Cook, M., & Mineka, S. (1989). Observational conditioning of fear to fear-relevant versus fear-irrelevant stimuli in rhesus monkeys. *Journal of Abnormal Psychology*, **98**, 448–459.

Costello, E. J., Sung, M., Worthman, C., & Angold, A. (2007). Pubertal maturation and the development of alcohol use and abuse. *Drug and Alcohol Dependence*, **88**(Suppl. 1), S50–S59.

Csibra, G., & Gergely, G. (2011). Natural pedagogy as evolutionary adaptation. *Philosophical Transactions of the Royal Society of London. Series B: Biological Sciences*, **366**, 1149–1157.

Cutting, N., Apperly, I. A., Chappell, J., & Beck, S. R. (2014). The puzzling difficulty of tool innovation: Why can't children piece their knowledge together? *Journal of Experimental Child Psychology*, **125**, 110–117.

DeLoache, J. S., & LoBue, V. (2009). The narrow fellow in the grass: Human infants associate snakes and fear. *Developmental Science*, **12**, 201–207.

Di Giorgio, E., Leo, I., Pascalis, O., & Simion, F. (2012). Is the face-perception system human-specific at birth? *Developmental Psychology*, **48**, 1083–1090.

Dobzhansky, T. (1973). *Genetic Diversity and Human Equality*. New York: Basic Books.

Dunlosky, J., & Rawson, K. A. (2012). Overconfidence produces underachievement: Inaccurate self-evaluations undermine students' learning and retention. *Learning and Instruction*, **22**, 271–280.

Ellis, B. J. (2004). Timing of pubertal maturation in girls: An integrated life history approach. *Psychological Bulletin*, **130**, 920–958.

Ellis, B. J., & Graber, J. (2000). Psychosocial antecedents of variation in girls' pubertal timing: Maternal depression, stepfather presence, and marital and family stress. *Child Development*, **71**, 485–501.

Ellis, B. J., Figueredo, A. J., Brumbach, B. H., & Schlomer, G. L. (2009). Fundamental dimensions of environmental risk: The impact of harsh versus unpredictable environments on the evolution and development of life history strategies. *Human Nature*, **20**, 204–268.

Ellis, B. J., McFadyen-Ketchum, S., Dodge, K. A., Pettit, G. S., & Bates, J. E. (1999). Quality of early family relationships and individual differences in the timing of pubertal maturation in girls: A longitudinal test of an evolutionary model. *Journal of Personality and Social Psychology*, **77**, 387–401.

Flynn, E., & Whiten, A. (2012). Experimental "microcultures" in young children: Identifying biographic, cognitive, and social predictors of information transmission. *Child Development*, **83** (3), 911–925.

Furlow, F. B. (1997). Human neonatal cry quality as an honest signal of fitness. *Evolution and Human Behavior*, **18**, 175–193.

Gardiner, A. K. (2014). Beyond irrelevant actions: Understanding the role of intentionality in children's imitation of relevant actions. *Journal of Experimental Child Psychology*, **119**, 54–72.

Gardiner, A., Bjorklund, D. F., Greif, M. L., & Gray, S. K. (2012). Choosing and using tools: Prior experience and task difficulty influence preschoolers' tool-use strategies. *Cognitive Development*, **27**, 240–254.

Gardiner, A., Greif, M., & Bjorklund, D. F. (2011). Guided by intention: Preschoolers' imitation reflects inferences of causation. *Journal of Cognition and Development*, **12**, 355–373.

Gava, L., Valenza, E., Turati, C., & de Schonen, S. (2008). Effect of partial occlusion on newborns' face preference and recognition. *Developmental Science*, **11**, 563–574.

Geary, D. C. (1995). Reflections of evolution and culture in children's cognition: Implications for mathematical development and instruction. *American Psychologist*, **50**, 24–37.

Geary, D. C. (2002). Sexual selection and human life history. In R. V. Kail, ed., *Advances in Child Development and Behavior*, Vol. 30. San Diego, CA: Academic Press, pp. 41–101.

Geary, D. C. (2005). *The Origin of Mind: Evolution of Brain, Cognition, and General Intelligence*. Washington, DC: American Psychological Association.

Geary, D. C. (2010). *Male, Female: The Evolution of Human Sex Differences*, 2nd ed. Washington, DC: American Psychological Association.

Geary, D. C., & Bjorklund, D. F. (2000). Evolutionary developmental psychology. *Child Development*, **71**, 57–65.

Gergely, G., & Csibra, G. (2005). The social construction of the cultural mind: Imitative learning as a mechanism of human pedagogy. *Interaction Studies*, **6**, 463–481.

German, T. P., & Johnson, S. C. (2002). Function and the origins of the design stance. *Journal of Cognition and Development*, **3**, 279–300.

Graber, J. A., Brooks-Gunn, J., & Warren, M. P. (1995). The antecedents of menarcheal age: Heredity, family environment, and stressful life events. *Child Development*, **66**, 346–359.

Gredlein, J. M., & Bjorklund, D. F. (2005). Sex differences in young children's use of tools in a problem-solving task: The role of object-oriented play. *Human Nature*, **16**, 211–232.

Heimann, M. (1989). Neonatal imitation gaze aversion and mother–infant interaction. *Infant Behavior & Development*, **12**, 495–505.

Hernández Blasi, C., & Bjorklund, D. F. (2003). Evolutionary developmental psychology: A new tool for better understanding human ontogeny. *Human Development*, **46**, 259–281.

Hill, K. (1993). Life history theory and evolutionary anthropology. *Evolutionary Anthropology*, **2**, 78–88.

Hill, K., & Kaplan, H. (1999). Life history traits in humans: Theory and empirical studies. *Annual Review of Anthropology*, **28**, 397–430.

Hoehl, S., & Pauen, S. (2017). Do infants associate spiders and snakes with fearful facial expressions? *Evolution and Human Behavior*, **38**, 404–413.

Hornik, R., Risenhoover, N., & Gunnar, M. (1987). The effects of maternal positive, neutral, and negative affective communications on infant responses to new toys. *Child Development*, **58**, 937–944.

Jacobson, S. W. (1979). Matching behavior in the young infant. *Child Development*, **50**, 425–430.

Kaplan, H., & Lancaster, J. B. (2003). An evolutionary and ecological analysis of human fertility, mating patterns and parental investment. In K. W. Wachter & R. A. Bulatao, eds., *Offspring: Fertility Behavior in Biodemographic Perspective*. Washington, DC: National Research Council, National Academies Press, pp. 170–223.

Kaplan, H., Hill, K., Lancaster, J., & Hurtado, A. M. (2000). A theory of human life history evolution: Diet intelligence, and longevity. *Evolutionary Anthropology*, **9**, 156–185.

Kasturiratne, A., Wickremasinghe, A. R., de Silva, N., et al. (2008). The global burden of snakebite: A literature analysis and modelling based on regional estimates of envenoming and deaths. *PLoS Medicine*, **5**, 1591–1604.

Keeley, L. H. (1996). *War before Civilization: The Myth of the Peaceful Savage*. New York: Oxford University Press.

Kelly, D. J., Liu, S., Lee, K., et al. (2009). Development of the other-race effect in infancy: Evidence toward universality? *Journal of Experimental Child Psychology*, **104**, 105–114.

Kelly, D. J., Quinn, P. C., Slater, A. M., et al. (2007). The other-race effect develops during infancy. *Psychological Science*, **18**, 1084–1089.

Kenward, B. (2012). Over-imitating preschoolers believe unnecessary actions are normative and enforce their performance by a third party. *Journal of Experimental Child Psychology*, **112**, 195–207.

Konner, M. (2010). *The Evolution of Childhood: Relationships, Emotions, Mind*. Cambridge, MA: Belknap Press.

Lee, K., Quinn, P. C., & Pascalis, O. (2017). Face race processing and racial bias in early development: A perceptual-social linkage. *Current Directions in Psychological Science*, **26**, 256–262.

Legare, C. H. (2017). Cumulative cultural learning: Development and diversity. *Proceedings of the National Academy of Sciences*, **114**(30), 7877–7883.

Legerstee, M. (1991). The role of person and object in eliciting early imitation. *Journal of Experimental Child Psychology*, **51**, 423–433.

Liszkowski, U., Carpenter, M., Striano, T., & Tomasello, M. (2006). 12- and 18-month-olds point to provide information for others. *Journal of Cognition and Development*, **7**, 173–187.

Liszkowski, U., Carpenter, M., & Tomasello, M. (2007). Pointing out new news, old news, and absent referents at 12 months of age. *Developmental Science*, **10**, F1–F7.

LoBue, V. (2010). And along came a spider: An attentional bias for the detection of spiders in young children and adults. *Journal of Experimental Child Psychology*, **107**, 59–66.

LoBue, V. (2013). What are we so afraid of? How early attention shapes our most common fears. *Child Development Perspectives*, **7**, 38–42.

LoBue, V., & DeLoache, J. S. (2010). Superior detection of threat-relevant stimuli in infancy. *Developmental Science*, **13**, 221–228.

LoBue, V., & Rakison, D. H. (2013). What we fear most: Developmental advantage for threat-relevant stimuli. *Developmental Review*, **33**, 285–303.

Lockhart, K. L., Chang, B., & Tyler, S. (2002). Young children's belief about the stability of traits: Protective optimism. *Child Development*, **73**, 1408–1430.

Lockman, J. J., & Kahrs, B. A. (2017). New insights into the development of human tool use *Current Directions in Psychological Science*, **26**, 330–334.

Lyons, D. E., Young, A. G., & Keil, F. C. (2007). The hidden structure of overimitation. *Proceedings of the National Academy of Sciences*, **104**, 19751–19756.

McGuigan, N., Makinson, J., & Whiten, A. (2011). From over-imitation to super-copying: Adults imitate causally irrelevant aspects of tool use with higher fidelity than young children. *British Journal of Psychology*, **102**, 1–18.

Meaney, M. J. (2007). Environmental programming of phenotypic diversity in female reproductive strategies. *Advances in Genetics*, **59**, 173–215.

Meltzoff, A. N., & Moore, M. K. (1977). Imitation of facial and manual gestures by human neonates. *Science*, **198**, 75–78.

Meltzoff, A. N., & Moore, M. K. (1992). Early imitation within a functional framework: The importance of person identity, movement, and development. *Infant Behavior and Development*, **15**(4), 479–505.

Menard, J. L., & Hakvoort, R. M. (2007). Variations of maternal care alter offspring levels of behavioural defensiveness in adulthood: Evidence for a threshold model. *Behavioural Brain Research*, **176**, 302–313.

Mondloch, C. J., Lewis, T. M., Budreau, D. R., et al. (1999). Face perception during early infancy. *Psychological Science*, **10**, 419–422.

Mood, D. W. (1979). Sentence comprehension in preschool children: Testing an adaptive egocentrism hypothesis. *Child Development*, **50**: 247–250.

Nagy, E., & Molnar, P. (2004). *Homo imitans* or *Homo provocans*? Human imprinting model of neonatal imitation. *Infant Behavior and Development*, **27**(1), 54–63.

Nettle, D. (2010). Dying young and living fast: Variation in life history across English neighborhoods. *Behavioral Ecology*, **21**, 387–395.

Nettle, D., & Cockerill, M. (2010). Development of social variation in reproductive schedules: A study of an English urban area. *PLoS ONE*, **5**, e12690.

Nielsen, M. (2006). Copying actions and copying outcomes: Social learning through the second year. *Developmental Psychology*, **42**, 555–565.

Nielsen, M., & Tomaselli, K. (2010). Over-imitation in the Kalahari Desert and the origins of human cultural cognition. *Psychological Science*, **5**, 729–736.

Öhman, A., Flykt, A., & Esteves, F. (2001). Emotion drives attention: Detecting the snake in the grass. *Journal of Experimental Psychology: General*, **130**, 466–478.

Oostenbroek, J., Suddendorf, T., Nielsen, M., et al. (2016). Comprehensive longitudinal study challenges the existence of neonatal imitation in humans. *Current Biology*, **26**(10), 1334–1338.

Oppenheim, R. W. (1981). Ontogenetic adaptations and retrogressive processes in the development of the nervous system and behavior. In K. J. Connolly & H. F. R. Prechtl, eds., *Maturation and Development: Biological and Psychological Perspectives*. Philadelphia, PA: International Medical Publications, pp. 73–108.

Parent, C. I., & Meaney, M. J. (2008). The influence of natural variations in maternal care on play fighting in the rat. *Developmental Psychobiology*, **50**, 767–776.

Pascalis, O., de Haan, M., & Nelson, C. A. (2002). Is face processing species-specific during the first year of life? *Science*, **296**, 1321–1323.

Pascalis, O., Scott, L. S., Kelly, D. J., et al. (2005). Plasticity of face processing in infancy. *Proceeding of the National Academy of Sciences*, **102**, 5297–5300.

Pellegrini, A. D. (2013). Play. In P. Zelazo, ed., *Oxford Handbook of Developmental Psychology*. New York: Oxford University Press, pp. 276–299.

Pellegrini, A. D., & Smith, P. K. (1998). Physical activity play: The nature and function of neglected aspect of play. *Child Development*, **69**, 577–598.

Placek, C. D., & Quinlan, R. J. (2012). Adolescent fertility and risky environments: A population-level perspective across the lifespan. *Proceedings of the Royal Society B*, **279**, 4003–4008.

Quinlan, R. J. (2007). Human parental effort and environmental risk. *Proceedings of the Royal Society B*, **274**, 121–125.

Rakison, D. H. (2005). Infant perception and cognition: An evolutionary perspective on early learning. In B. J. Ellis & D. F. Bjorklund, eds., *Origins of the Social Mind: Evolutionary Psychology and Child Development*. New York: Guilford, pp. 317–353.

Ross, J., Anderson, J. R., & Campbell, R. N. (2011). I remember me: Mnemonic self-reference effects in preschool children. *Monographs of the Society for Research in Child Development*, **76**(3), 300.

Saxbe, D. E., & Repetti, R. L. (2009). Brief report: Fathers' and mothers' marital relationship predicts daughters' pubertal development two years later. *Journal of Adolescence*, **32**, 415–423.

Shin, H. E., Bjorklund, D. F., & Beck, E. F. (2007). The adaptive nature of children's overestimation in a strategic memory task. *Cognitive Development*, **22**, 197–212.

Simpson, J. A., & Belsky, J. (2008). Attachment theory within a modern evolutionary framework: Theory, research, and clinical applications. In J. Cassidy & P. R. Shaver, eds., *Handbook of Attachment: Theory, Research, and Clinical Applications*, 2nd ed. New York: Guilford Press, pp. 131–157.

Simpson, J. A., Griskevicius, V., Kuo, S., Sung, S., & Collins, W. A. (2012). Evolution, stress, and sensitive periods: The influence of unpredictability in early versus late childhood on sex and risky behavior. *Developmental Psychology*, **48**, 674–686.

Stearns, S. (1992). *The Evolution of Life Histories*. Oxford: Oxford University Press.

Thompson, R. A. (2006). The development of the person: Social understanding, relationships, conscience, self. In W. Damon & R. M. Lerner, gen. eds.; N. Eisenberg, vol. ed., *Handbook of Child Psychology: Vol. 3, Social, Emotional, and Personality Development*, 6th ed. New York: Wiley, pp. 24–98.

Tomasello, M. (2000). Culture and cognitive development. *Current Directions in Psychological Science*, **9**, 37–40.

Tomasello, M. (2009). *Why We Cooperate*. Cambridge, MA: MIT Press.

Tomasello, M., & Carpenter, M. (2007). Shared intentionality. *Developmental Science*, **10**, 121–125.

Tooby, J., & Cosmides, L. (1992). The psychological foundations of culture. In L. Cosmides, J. Tooby, & J. H. Barkow, eds., *The Adapted Mind*. New York: Oxford University Press, pp. 19–136.

Trickett, P. K., Noll, J. G., & Putnam, F. W. (2011). The impact of sexual abuse on female development: Lessons from a multigenerational, longitudinal research study. *Development and Psychopathology*, **23**, 453–476.

Vaish, A., & Striano, T. (2004). Is visual reference necessary? Contributions of facial versus vocal cues in 12-month-olds' social referencing behavior. *Developmental Science*, **7**, 261–269.

Volk, A. A., & Atkinson, J. A. (2013). Infant and child death in the human environment of evolutionary adaptation. *Evolution and Human Behavior*, **34**, 182–192.

Wakako, S., Wada, K., Yamamoto, T., Mohri, I., & Taniike, M. (2014). Development of preference for conspecific faces in human infants. *Developmental Psychology*, **50**, 979–985.

Whiten, A., McGuigan, N., Marshall-Pescini, S., & Hopper, L. M. (2009). Emulation, imitation, over-imitation and the scope of culture for child and chimpanzee. *Philosophical Transactions of the Royal Society of London. Series B: Biological Sciences*, **364**, 2417–2428.

Williams, G. C. (1966). *Adaptation and Natural Selection*. Princeton, NJ: Cambridge University Press.

Williamson, R. A., & Markman, E. M. (2006). Precision of imitation as a function of preschoolers' understanding of the goal of the demonstration. *Developmental Psychology*, **42**, 723–731.

Wilson, M., & Daly, M. (1997). Life expectancy, economic inequality, homicide, and reproductive timing in Chicago neighborhoods. *British Medical Journal*, **314**, 1271–1274.

22 The Ontogeny and Evolution of Cooperation

SEBASTIAN GRÜNEISEN AND EMILY WYMAN

Striking examples of human cooperation include people donating blood, paying their taxes, and helping total strangers on the street. These are acts of altruistic cooperation – behaviors that benefit the collective at a cost to the individual. To many researchers, explaining altruistic behaviors is central to understanding human cooperative uniqueness (Fehr & Fischbacher, 2003; Gintis et al., 2005), with the central question being how the fruits of cooperation can be enjoyed without being exploited by individuals who free-ride on the benevolent actions of others while not contributing themselves. Over recent decades, substantial advances have been made in identifying the factors that sustain cooperation in this context (Camerer, 2011; Hammerstein, 2003; Milinski, Semmann, & Krambeck, 2002). Here, we take a different approach and argue that an equally fundamental challenge of cooperation is for individuals to coordinate their behavior in order to generate mutual benefits (the "forgotten problem of cooperation"; Calcott, 2008). We propose that a defining human characteristic, which in its scope and flexibility is unparalleled in the animal kingdom, is the ability to coordinate actions with others – even with total strangers – in the pursuit of shared goals.

To investigate these issues, idealized models of cooperative challenges from economic game theory are commonly utilized to uncover people's underlying motivations in different types of interaction. For instance, the models used to investigate altruistic cooperation (such as tax compliance) are typically social dilemmas (such as the prisoner's dilemma) in which individuals choose to either cooperate or defect. Mutual cooperation offers both players a better payoff than mutual defection, but a dilemma arises since defecting against a cooperating partner yields an even higher gain.

In contrast to social dilemmas, coordination problems model situations in which multiple individuals have a common interest in coordinating, there are several actions they can choose from, and they are interdependent such that each person's best choice depends on others' choices, and vice versa (Lewis, 1969). A classic example

is the so-called Stag Hunt problem (first formally described by Rousseau), which models situations in which individuals choose between acting individually and launching cooperation. Specifically, individuals can decide to hunt for a "hare," which yields a small but certain reward, or to jointly hunt for a "stag," which provides a large reward but only if everyone joins the hunt. Importantly, hunting a stag is risky and requires coordination, since a lone attempt will leave that individual empty-handed (real-life examples include starting a joint business, joining a protest movement, or any activity in which one's willingness to participate is conditional upon others' participation).

Other important coordination problems include "pure coordination games" in which there are multiple, equally functional ways of cooperating and individuals have to converge on the same one to be successful (as when deciding which side of the road to drive on) or situations in which individuals have divergent interests as to how to coordinate but coordination failure is worse for everyone than any other outcome (e.g., when one of two drivers has to yield for the other at an intersection, a situation known as a Chicken Game).

In this chapter, we argue and present evidence that humans have evolved skills and motivations that allow them to flexibly solve these types of coordination problems. We approach this topic from a developmental-comparative perspective and review behavioral research with young children to investigate cooperative tendencies that develop reliably across human childhood. Throughout, we compare this approach to studies with chimpanzees, human's closest living evolutionary relative, to address questions regarding the evolution of these capacities.

In Section 22.1, we present evidence showing that already by infancy human children, but not chimpanzees, possess abilities for joint attention, as well as distinct social motivations for joint engagement. Over the first years of life, children reliably begin to participate in simple joint interactions that are cognitively rich in the sense that they involve representing joint goals, commitments to joint activities, and a conceptualization of the task from a shared interpersonal perspective. In Section 20.2, we show how these abilities provide the basis for more complex forms of mutualistic collaboration, particularly capacities for joint decision-making in coordination problems. In contrast to chimpanzees, whose coordination strategies tend to be unstable and are based on individual goals, preschool children solve these problems by using the knowledge they share with others (a cognitive descendent of joint attention capacities), and they even learn to coordinate when they have conflicting interests to those of their interaction partners. In Section 22.3, we demonstrate that human children display skills and motivations for collaboration in groups based on conventional cultural practices and normative rules. This enables them to reliably collaborate even with in-group strangers. Together, these findings contribute to our understanding of the key foundations of human forms of cooperation.

22.1 COLLABORATING WITH JOINT ATTENTION AND SHARED GOALS

By their first months of life, infants are already strongly motivated to coordinate their engagements with others. In simple face-to-face interactions, they participate in intimate bouts of mutual eye contact involving smiling and vocalizing in discernable turn-taking sequences with caregivers, often referred to as "proto-conversations" (Trevarthen, 1979). Around their first birthday, however, there is something of a "revolution" in infants' social development, as they begin to attend jointly with others to aspects of the external environment, "triangulating" attention between themselves, the partner, and objects or events of mutual interest (Bakeman & Adamson, 1984; Carpenter, Nagell, & Tomasello, 1998; Tomasello, 1995).

The term "revolution" is applied here because, during these bouts of "joint attention," the infant and another individual attend not just to a specific referent, but simultaneously to each other's attention. That is, they know together that they are attending to the same thing such that the simple activities they perform (such as rolling a ball back and forth) are occurring in a cognitively "shared frame" based on mutual awareness (Carpenter & Call, 2013; Tomasello, 1995).

Infants' motivation for actively sharing attention can be seen particularly clearly in their declarative pointing gestures: having pointed to something new in the environment, infants are only satisfied and so cease pointing when an adult both looks to the correct referent and looks back to the infant to acknowledge that it has been shared in view (Liszkowski et al., 2004). Relatedly, infants also closely track what they know together with other individuals through joint experience. After infants as young as 14 month old have shared one of three objects with an adult by playing with it with them, and the adult then ambiguously gestures toward those objects, infants naturally give her the one they had previously shared. In contrast, after observing the adult play with the object alone or if a different adult made the request, infants choose randomly between the three objects (Moll et al., 2008).

The ability of infants to participate in joint attention and to track what they know together with others represents a critical foundation for the emergence of joint action. Specifically, it enables infants to jointly attend to entities or events in the environment that can then become the focus of a shared goal. For example, it allows them not only to attend to a box and each other's relation to the box, but also to jointly focus on their goal of opening the box, on the basis of which they can adopt complementary roles such as "stabilizer" and "opener" to get to the contents inside.

Shared goals are in fact part of a wider suite of sociocognitive capacities that characterize much of human cooperation, generally known as "shared intentionality" (Tomasello et al., 2005). A pair or group of individuals have a shared intention when their reasons for acting cannot be reduced to their individual intentions. For instance, when two individuals share an intention to take a walk together, this is not reducible to each separately intending to stroll down the same path. It is because they intend to walk together that either intends to walk at all (Searle, 1990).

A hallmark of shared intentional cooperation is that it entails not just shared goals, but also commitments to those goals: one person cannot abandon the walk without explanation or they may risk reprimand, and any reprimand will be recognized as legitimate (Gilbert, 1996).

Children as young as two years of age appear to conceive of collaborative activities as involving shared intentions. In one study paradigm, for instance, 18- and 24-month-olds first successfully completed simple collaborative tasks with an adult partner that either required the action of the partner (Warneken, Chen, & Tomasello, 2006) or did not (Warneken, Gräfenhain, & Tomasello, 2012). When the adult unexpectedly reneged on the joint activity, children did not simply abandon the task or try to complete it alone (as might be expected had they acted according to individual goals). Instead, they attempted to make eye contact and reengage the adult partner by gesturing or vocalizing, and they did this regardless of whether or not they needed the partner to continue. This suggests that children were collaborating with the adult partner not simply to complete the task, but rather in virtue of the shared goal of performing the task together.

From around age three, young children also start to demonstrate a basic notion of the commitments involved in collaborative activities. When an adult verbally declares that they will play a game with the child by saying "let's play" and subsequently breaks away from the activity, children seem surprised and try to reengage the adult. However, if the adult abandons the activity without previously having issued a verbal commitment, children see this as relatively unproblematic (Gräfenhain et al., 2009).

Conversely, children who have agreed to play with an adult partner are not only less likely to abandon the activity when enticed with an alternative play option than children who have made no verbal commitment, but those who do abandon the game often "take leave"; for instance, by announcing their departure. Lastly, children who have collaborated with a peer take care to ensure that their peer receives equal payoffs (e.g., by helping the partner or by sharing rewards) in ways that peers who have acted individually do not (Hamann et al., 2011; Hamann, Warneken, & Tomasello, 2012; Warneken et al., 2011).

Another important aspect of shared intentional cooperation is the ability to adopt distinct and mutually responsive roles toward the joint goal (Bratman, 1992). Correspondingly, children attend to their partner's as well as their own role in joint action, as evidenced by their ability to swiftly engage in role reversals: children who first perform one role in a collaborative activity are subsequently more efficient in executing the opposite role than naive children (Carpenter, Tomasello, & Striano, 2005; Fletcher, Warneken, & Tomasello, 2012). This role interchangeability may indicate that children conceptualize joint activities from something like "a bird's eye view" or a "we perspective" (Tomasello, 2016; see also Milward, Kita, & Apperly, 2014).

Given that very young children engage in joint attention, form shared goals, coordinate actions with others, and commit to joint activities, questions arise about whether chimpanzees collaborate in similar ways. Chimpanzees do indeed possess several cognitive capacities considered crucial to human cooperation. They understand other's intentions (Call et al., 2004) and they know what others can see or know and predict other's behaviors accordingly (Hare, Call, & Tomasello, 2001; Kaminski, Call, & Tomasello, 2008; Karg et al., 2015; Schmelz, Call, & Tomasello, 2011). However, chimpanzees' joint action does not seem to be based on shared intentions. For example, unlike children who attempt to reengage collaborative partners who suddenly cease collaborating, chimpanzees try to solve the task alone, suggesting that they act on individual goals and do not sense any commitments to those goals (Warneken et al., 2006). Furthermore, unlike children who are adept at role reversals in joint action, chimpanzees are slow at performing a partner's role, despite having seen the partner perform it (Fletcher et al., 2012). This suggests that that they may not attend to both roles simultaneously as children do. More generally, there seem to be divergent social preferences for cooperating in humans and chimpanzees: when given the opportunity to obtain food alone or collaboratively, chimpanzees typically prefer to act alone, and they switch to cooperation only when this is materially advantageous. Children, by contrast, prefer to cooperate overall (Butler & Walton, 2013; Bullinger, Melis, & Tomasello, 2011; Rekkers, Haun, & Tomasello, 2012). These differences indicate that many complex cooperative activities in chimpanzees, such as group hunting of monkeys, may be best understood in terms of multiple acts of individual intentionality, rather than a collective act with shared intentions (Tomasello et al., 2012).

Multiple reasons may account for this lack of shared intentionality. While chimpanzees are motivated to attend to what others are attending to (Bräuer, Call, & Tomasello, 2005) and even direct other's attention to objects (Call & Tomasello, 2007; Leavens, Hopkins, & Thomas, 2004), there is currently no convincing evidence that they engage in joint attention: they do not attempt to share attention with others (e.g., by showing objects to others or by using declarative gestures; Tomasello & Carpenter, 2005; Tomonaga et al., 2004) in ways that children reliably do (Liszkowski et al., 2004), and they show no evidence of comprehension of having been in joint attentional engagement (e.g., that they know together that they were looking at the same thing; Carpenter & Call, 2013). This apparent lack of joint attention skills may preclude them from forming truly joint goals.

Furthermore, chimpanzee cooperation appears fundamentally constrained by competitive motivation. For example, while chimpanzee dyads successfully managed to draw in a wooden platform containing pre-divided food by simultaneously pulling on each of two ends of a rope (Hirata & Fuwa, 2007), cooperation quickly fell apart if the food was clumped in one pile (Melis, Hare, & Tomasello, 2006). In these cases, dominant individuals typically monopolized access to the food and subordinates quickly lost motivation to collaborate on subsequent trials. Even if the food was dispersed, however, success in the task was closely linked to social tolerance outside the experimental setting, and intolerant dyads predominantly failed to collaborate. In contrast, three-year-olds were successful over many trials in similar situations (Warneken et al., 2011).

Overall, while children's early cooperative interactions may appear simple on the surface, they are cognitively rich: they involve joint attention and shared goals that they understand as entailing commitments and mutually responsive roles. Collaborative action in chimpanzees, by contrast, apparently lacks this basic interpersonal frame of joint attention, which may in part explain why they do not seem to form truly shared goals. Hence, while chimpanzees are skilled collaborators in some contexts (e.g., when their goals happen to coincide, when interacting with tolerant partners, and when dominant individuals are prevented from monopolizing the spoils), they do not show the basic building blocks of shared intentional cooperation.

As children grow older, however, they increasingly face more complex cooperative challenges; namely, to strategically coordinate decisions (e.g., in choosing whether or not to cooperate when cooperation entails a risk of some kind, when there are multiple ways of cooperating and one must be adopted by all parties concerned, or when collaborative partners have conflicting preferences as to how to cooperate). These cooperative contexts are motivationally and cognitively more complex than those dealt with so far.

However, joint attention and shared intentionality constitute key building blocks as, over the course of the preschool years, children develop the skills for tackling these new challenges.

22.2 COORDINATING JOINT DECISIONS WITH SHARED KNOWLEDGE

The Stag Hunt parable, originally made famous by Rousseau, symbolizes the challenge of establishing cooperation when this is in everyone's interest but individuals need assurance that others will join, and it is a classic example of a coordination problem. Coordination problems are situations in which several individuals have a common interest in cooperating, but have multiple actions to decide between. Crucially, they are interdependent, such that each person's best decision depends on others' decisions, and vice versa (Lewis, 1969). Examples of the Stag Hunt variety include any collaborative activity that entails risk of loss, such as literally going hunting for food, but also choosing to start a joint business with others, or deciding to join a protest movement (where one's willingness to participate is conditional upon others' participation). The key psychological challenge is to accurately work out what others are likely to do. The problem arises because others will be reasoning similarly about what we will do, so, in principle, one needs to reason about others' reasoning about our reasoning, and so on, in a potentially infinite regress (for which there is no logical cutoff point; Gilbert, 1989). To coordinate successfully, individuals therefore require some kind of "meeting of minds" (as Schelling, 1960, metaphorically put it); that is, some kind of shared knowledge.

What exactly shared knowledge is and how it gets established has long been a matter of debate. Classic accounts posit that extensive recursive mental state reasoning (where I calculate what you know that I know that you know, etc.) is necessary for people to assume shared knowledge exists (Lewis, 1969; Schelling, 1960). We will examine some alternative proposals to this over the course of this section. But we begin here with the most straightforward case: one sure way to establish shared knowledge is by actively sharing things via explicit communication (Carpenter & Liebal, 2011; Clark, 1996), and this is precisely what young children naturally do when trying to coordinate. In a recent study, dyads of four-year-olds were presented with a playful version of the Stag Hunt game (Duguid et al., 2014). Each individual had a box from which they could extract a low-value reward ("hare"). For brief periods, a high-value reward ("stag") became available that required cooperation for capture, so that children had to decide between either sticking with the safe option or attempting risky cooperation. For the purposes of comparison, a directly analogous setup was provided to chimpanzee dyads in a parallel study.

One key finding was that, in a baseline condition, both children and chimpanzees cooperated almost all of the time, with the only difference being that children attempted to guarantee success by using imperative communication (e.g., "Come quick") once they were at the stag. Chimpanzees rarely communicated. However, when the risk levels were raised in another condition (partners were prevented from seeing each other and the value of the hare increased so that abandoning it became costlier), cooperative attempts by chimpanzees declined. Children, in contrast, compensated and achieved cooperation rates close to 100 percent by communicating prior to their decision to leave the hare (e.g., by shouting, "The stag is here!"). So while chimpanzees seemed to use a leader–follower strategy (such that one runs and the other follows) that became unstable once the risks were raised, children ensured that the arrival of the stag was shared knowledge between them, which reduced the risk of making failed cooperative attempts.

The mechanism by which communication can be used to build shared knowledge becomes particularly clear in another Stag Hunt study, this time run only with children (Wyman, Rakoczy, & Tomasello, 2013). Children sat beside an adult partner and played a game requiring each player to insert a tool into one of two holes. While a hare prize was available on all rounds to both players, on some rounds, a high-value stag briefly became available too, so that players could either safely retrieve the hare alone (using hole 1) or try to coordinate to retrieve the stag (using hole 2). On these rounds, when the adult simply looked to the stag, children tended to go for the hare (even though they could see the adult seeing the stag). But when, upon seeing the stag, the adult also made eye contact with the child, children chose to cooperate more often. Although the communicative look here was entirely without content, it seemed to function to make mutually manifest that the newly arrived stag was in shared attention, thus enabling coordination.

This is in line with the view that joint attention is, in fact, a rudimentary form of shared knowledge: just as a person can "know that you know that I know," in joint attention, they may "see that you see that I see." But sharing things in joint attention or via communication may enable individuals to bypass this complex inferential reasoning process altogether, since they can simply see another individual attend to a target and themselves, making shared knowledge to a degree self-evident (for discussions, see Peacocke, 2005; Tollefsen, 2005; Tomasello, 1995).

Now, let us consider a different type of coordination problem in which the question is not whether or not to cooperate, but rather how to cooperate when there are multiple options (so-called pure coordination problems). Imagine two friends who have agreed to meet in a large town square at a certain time but have not specified an exact location. Despite multiple potential meeting places – cafes, bars, monuments, etc. – they manage to meet simply by going to the landmark that each identifies as the most prominent (e.g., a statue in the center of the square; for evidence demonstrating adults' surprising proficiency at solving similar problems, see Mehta, Stramer, & Sugden, 1994; Schelling, 1960). What this example illustrates is that people can often derive and use shared knowledge in the complete absence of communication – in this case, by using salience (i.e., aspects of the environment that are prominent and can be assumed to be prominent to everyone; Schelling, 1960).

In a direct test of children's abilities to do this, they were presented with the following coordination task: two children were each given a ball that they both had to insert into the same four identical boxes in order to receive a treat. If they picked different boxes, they would both get nothing. The problem was that they chose sequentially – only one of them was in the room at a time – so they could not communicate or see each other's choices. However, three of the boxes were marked, somewhat oddly, with identical pictures of ice cream cones, while the fourth box displayed a picture of celery. Children aged five and older typically chose the saliently marked box in order to coordinate. In contrast, in a control condition in which children received two balls and thus were not required to coordinate, they disregarded the salient option and chose the attractive ice cream boxes instead (Grueneisen, Wyman, & Tomasello, 2015a). Children thus proficiently used their shared knowledge that one option was salient to jointly converge on a solution.

However, assessing shared knowledge is not always a simple exercise. We frequently face ambiguous situations and have to deal with potential discrepancies regarding the extent to which we share knowledge with others (e.g., "Does my partner know the meeting location?" "Does my partner know I've seen the stag?"). It is in these situations in particular that recursive mental state reasoning appears crucial in order for individuals to coordinate their decisions successfully (Thomas et al., 2014). And this, too, appears to come naturally to children.

Using the same basic paradigm with the four boxes, six year-olds first learned that one box contained a larger reward than the others and this box was clearly marked, making it the obvious solution (Grueneisen, Wyman, & Tomasello, 2015b). At test, however, children were informed separately that the largest reward was mistakenly placed in a different box. In addition, children either learned that their partner did not know this or that their partner knew about the location of the highest reward but not that they themselves knew about it, too. In both cases, children adjusted their choices to their partner's belief state: if their partner did not know where the largest reward was hidden, they chose the incorrectly marked box (i.e., the one their partner believed the high payoff to be in). Even more impressively, even if the partner also knew where the highest payoff was, they still did not choose that box if they thought their partner did not know that they knew this, too.

So when children sense that the information held between themselves and a coordination partner is discrepant, they engage in complex mental state reasoning to ensure successful coordination. As children reach school age, they are thus highly proficient at aligning their decisions with others based on the knowledge they share, which provides them with a central tool for solving coordination problems.

Thus far, however, we have only considered situations in which the interests of both partners are perfectly aligned. But individuals often have divergent preferences as to how a joint activity should be performed, and they need to coordinate despite conflicting motives (e.g., when one of two colleagues has to do an undesirable part of a joint project or one of two drivers has to yield for the other at an intersection). This type of situation is commonly modeled as a "Chicken Game" (also known as a "Snowdrift Game"; Doebeli & Hauer, 2005; Rapoport & Chammah, 1966). In the Chicken Game story, two drivers drive toward each other on course for collision. If they both continue straight they will crash, so at least one driver needs to swerve for the catastrophe to be avoided. The driver who swerves, however, loses face for being a coward (and so will be called "the chicken"). Hence, despite their opposing interests (each prefers the other to swerve), they share a strong incentive to align their decisions since coordination failure is the worst outcome for all involved.

Relatively little is known about children's behavior in these situations, but one recent study speaks directly to the issue (Grueneisen & Tomasello, 2017): two five-year-olds each steered an automated toy train carrying a reward. The trains simultaneously moved toward each other so that in order to avoid a crash – which would leave both children empty-handed – one train had to swerve. By swerving, however, the trains lost a portion of the rewards, so that it was in each child's interest to go straight and for their partner to swerve. Most dyads solved the problem by spontaneously developing a turn-taking strategy (i.e., by alternating who swerves), and this was often based on explicit joint agreements where children would propose, for example, "First I swerve, then you, OK?" (see also Melis et al., 2016). It appears that while chimpanzees can also coordinate their behavior in situations of conflict, they do not engage in turn-taking compromises (Melis et al., 2016), and the primary mechanism by which they coordinate seems to be that dominant individuals are willing to pay a cost (equivalent to swerving), knowing they can monopolize a good portion of the reward afterwards (Schneider, Melis, & Tomasello, 2012).

The emerging picture indicates fundamental species differences in how humans and chimpanzees coordinate decisions. While chimpanzees are efficient coordinators in some contexts, they appear to be mainly successful if one individual launches and another sees this and follows. When risk levels increase, and individuals can no longer see each other, their success rates drop. Children, by contrast, communicate strategically to ensure coordination, and this communication can be so minimal that it simply makes mutually manifest that important new events (e.g., the arrival of the stag) are in joint attention. They jointly recognize that salient features of the environment can be used as focal points for coordination, and when they realize that shared knowledge is lacking, they use recursive mental state reasoning to guide their decisions. Lastly, children spontaneously develop joint compromises (e.g., turn-taking) to enable continued coordination even when they have conflicting interests. This rich suite of behavioral and psychological capacities indicates that solving coordination problems comes fairly naturally to humans.

22.3 COLLABORATION IN LARGE GROUPS

Modern human life requires people to align their actions with many different partners, often complete strangers. This means that in order to coordinate, people cannot rely on shared knowledge derived from direct personal experiences, as, for example, when engaging in economic exchanges with strangers, when navigating through traffic, or when working in large-scale companies with hundreds of employees. In what might be termed "societal interactions," commonly known social rules typically prescribe how individuals in the relevant in-group are to behave. These rules guide a wide range of different behaviors, from how to greet others and what to wear, to which side of the road to drive on or which currency to use in trade.

Rules of this kind are typically conventionalized solutions to recurrent coordination problems (Lewis, 1969) that groups collectively converge on over repeated interactions (e.g., driving on the right or shaking hands to great each other). Over time, everyone comes to expect everyone else to conform to these conventions, and, furthermore, preferences for conformity to the convention develop, particularly when nonconformity comes at a cost: driving on the wrong side of the road can kill people and failing to greet in the appropriate manner can cause offense, with serious consequences in some contexts. Thus, rather than simply being a statistical regularity, conventional rules frequently have normative force such that they specify how people should behave (Gilbert, 1989) and sanctions are applied in response to violations (Bicchieri, 2006).

Over development, children need to swiftly pick up the various rules of their culture and, indeed, from a young age, children show a strong motivation to copy common behaviors in their social environment. For example, after observing alternative ways of extracting a reward, children as young as two years of age tend to adopt the method used by the majority of demonstrators. In fact, chimpanzees copy majority behaviors similarly. However, when chimpanzees know about the solution to a problem, they ignore majority behavior unless adopting it becomes materially advantageous, whereas children reliably copy even elements they know to be causally irrelevant, and favor majority behaviors for no other reason than to conform (Haun, Rekers, & Tomasello, 2012, 2014; Horner & Whiten, 2005; Lyons, Young, & Keil, 2007).

This sensitivity to majority information combined with a motivation to socially conform enables children to work out the conventional coordination solutions that their group commonly uses. In a variant of the coordination game described in Section 22.2 (requiring children to insert a ball into the same one of four boxes without communicating), five-year-olds first observed other pairs of children engage with the problem before they themselves played. Critically, they witnessed a majority of pairs using one box to retrieve treats, while a minority of pairs used a different box to the same effect. When it was their own turn to choose, children tended to adopt the majority strategy themselves in order to coordinate with a partner (they followed the majority less often when they had both balls and so no coordination was necessary; Grueneisen, Wyman, & Tomasello, 2015c). Hence, preschoolers can use majority information to infer conventional behavior and adapt their coordination decisions accordingly.

Moreover, children appear to understand rules as socially constructed practices that are established by some form of collective agreement or at least consent (Gilbert, 2008) and not as simply inalterable facts handed down by authorities (Hardecker, Schmidt, & Tomasello, 2017; for a detailed discussion, see Wyman, 2014). For example, five-year-olds sometimes spontaneously establish arbitrary game rules in peer groups, readily transmit them to novices, and treat them as objective normative standards in the sense that they demand conformity from group members (Göckeritz, Schmidt, & Tomasello, 2014).

Even more impressively, children can even devise and agree on rules when they have conflicting interests (Grueneisen & Tomasello, 2019): using the train setup mentioned in Section 22.2 (Grueneisen & Tomasello, 2017) groups of four children were repeatedly presented with a Chicken Game. In each round, two children were randomly selected to play while the other two children watched, with the inevitable conflict on display for all to see. In occasional game breaks, the group of four assembled and was encouraged to establish additional game rules. This situation thus mimicked, in some respects, societal interactions in which interaction partners constantly change and groups create rules for individuals to apply in subsequent encounters. Eight-year-olds (but not five-year-olds) readily managed to agree on rules that mostly consisted of impartial procedures to determine who would be advantaged. Crucially, playing with these rules substantially improved children's coordination success, reduced interpersonal conflict, and made the process of coordinating considerably more efficient (by removing the need for repeated negotiations). This demonstrates children's increasing sophistication at regulating their own interactions in spite of conflicting motives and underlines the functional properties of social rules in human groups: namely, that they can substantially facilitate cooperative outcomes in recurrent coordination problems even among changing partners or strangers.

But from a very young age, children do not merely understand social rules as indicating what people typically do, but as having normative force, specifying what *ought* to be done (Bicchieri, 2006; Ostrom, 2000). Indeed, they appear to view rules as objective standards that apply to everyone, as is evidenced by their willingness to police third-party violations: if a puppet deviates from a rule, three-year-olds protest using normative language (e.g., "No, that's not how it's done"; Rakoczy, Warneken, & Tomasello, 2008; Schmidt, Rakoczy, & Tomasello, 2011). And they do not apply norms indiscriminately, but understand their context specificity. For example, when an individual declares that they will join a normatively governed activity (such as participating in a game), children protest if the individual violates the relevant norms, but not if the person did not declare their participation (Schmidt et al., 2016; Wyman, Rakoczy, & Tomasello, 2009). Relatedly, three-year-olds are more likely to protest and demand rule conformity when they witness violations of rules by in-group members than by out-group members (Schmidt, Rakoczy, & Tomasello, 2012). Children thus understand the basic notion of a normative community – that individuals belong to different groups and are therefore subject to the conventions and norms of those particular groups.

However, children do not simply distinguish between different groups, but show distinct preferences for their in-group (Dunham, Baron, & Carey, 2011). While even young infants use group markers such as language, accent, and race to assign individuals to in-group and out-group categories (Cameron et al., 2001; Kinzler, Dupoux, & Spelke, 2007), preschool children are heavily influenced by these as well as totally arbitrary categories in their social decisions, such as preferring in-group members as friends (Kinzler et al., 2007) or treating them favorably when dividing resources (Benozio & Diesendruck, 2017; Fehr, Bernhard, & Rockenbach, 2008).

A recent study shows that children also attend to group membership in their cooperative decision-making (Wyman, Einav, Over, & Starmer, unpublished data): children aged between 8 and 11 were first randomly assigned either to the same or to a different group as an adult partner, after which they played a series of cooperative games – both coordination games as well as social dilemmas. In social dilemmas, which embody a conflict between collective and individual interests (as in paying one's taxes), children who played with an in-group partner were more likely to cooperate than children who played with an out-group partner. In coordination problems, by contrast, which provide a mutualistic context, children cooperated equally often regardless of their partner's group membership.

The second result relates to the reasoning processes that children used in the study. "Team reasoning" involves thinking as a collective unit as opposed to an individual unit (as if asking oneself, as it were, not "what's best for me," but rather "what's best for us"; Bacharach, 2006). Children participating in social dilemmas with in-group members were more likely to explain themselves in plural terms (such as "we," "us," and "our") than children playing with out-group members. In coordination problems, they used these terms equally often when playing with in-group and out-group members (and at levels similar to those seen in the in-group players in social dilemmas). This suggests that engaging with coordination problems may induce children to think from a collective perspective in ways that boost cooperation, whereas in social dilemmas this is only the case when individuals share a group identity.

One way in which in-group membership does foster coordination, however, is through the epistemic inferences that one can make about individuals from one's own cultural group. Any modern-day human belongs to a large and complex set of social groups such that a person can, for example, simultaneously identify as "male," "German," "left wing," "carpenter," "protestant," "football fan," and so on. These categories point to sets of common cultural knowledge shared among all members of particular communities due to their shared cultural background (e.g., as we are both fans of the same football team, I can assume you know the result of the last game). It is the knowledge we can expect everyone to share simply by virtue of being a member of our group (Clark, 1996).

This type of shared knowledge greatly facilitates interpersonal coordination among group members, and this can already be seen among children. In one study, for instance, one child was asked to hide a sticker in one of four boxes, knowing that a second child would try to guess which box it was in. If the guesser was correct, they would both win a price. When playing with a partner from their own culture, children mostly hid the sticker in a box marked with a cue presumed to be commonly known (e.g., their national flag). When playing with a child from a different culture from theirs, however, children tended to hide the sticker in a box marked with culturally neutral cues (Goldvicht-Bacon & Diesendruck, 2016). This shows that even preschool children have a notion of the common knowledge they share with members of their cultural community, and they actively use this to coordinate their decisions with in-group strangers.

The cooperative activities explored in this section represent cooperative cultural practices. Here, children acquire the conventional processes that their community uses to cooperate, for instance, by attending to majority behavior. They look for – and even themselves invent – social rules to regulate their interactions, and this improves the success and efficiency of coordinated action while reducing conflict. They perceive the normative force behind social rules, actively enforcing the rules themselves, and they grasp the basic principle of a normative and conventional cultural community, cooperating widely when it is in their interest, but preferably with in-group members when cooperation comes at a cost. The shared knowledge that children use to coordinate in these situations is now not gained from interpersonal experience, but inferred based on a shared cultural background. At this point, children understand and engage with the basic ingredients of "societal interactions."

22.4 CONCLUSION

Whereas chimpanzees understand what others attend to, human children reliably engage in joint attention to share interest and coordinate emotional states with others. While chimpanzees understand that others have goals and predict their behavior based on these goals, human children form joint goals to collaborate with others. And while chimpanzees track what others know, human children form shared knowledge based on common personal experiences or by virtue of a shared cultural background, and they use this knowledge to coordinate decisions in coordination problems or in societal interactions with cultural group members. It is these capacities for sharing psychological states and knowledge with others that enable ever more flexible ways of interpersonal coordination and thus represent a central building block of human forms of cooperation and culture.

But what can account for the emergence of these capacities since humans last shared a common ancestor with chimpanzees about six to eight million years ago? The findings presented in this chapter are consistent with a theoretical model proposing two key steps in the evolution of human cooperation (Tomasello et al., 2012). In a first step, according to this model, humans became obligate collaborative foragers such that collaboration became compulsory for survival and procreation. This created new forms of interdependence in the sense that individuals fundamentally relied on one another for foraging success. This gave rise to the various coordination problems requiring individuals to put their heads together in order to collectively generate benefits.

Early coordination problems likely included situations of the Stag Hunt variety in which individuals had to coordinate their cooperative efforts since lone attempts were costly or entailed serious risks (Alvard & Nolin, 2002), but also pure coordination problems in which individuals had to converge on one and the same coordination solution (e.g., complementary hunting strategies).

This pressure to coordinate behavior for subsistence thus resulted in the emergence of skills and motivations for mutualistic collaboration in small-scale interactions with stable cooperative partners, which – as we have shown in Sections 22.1 and 22.2 – reliably develop in contemporary human children over the first years of life. Supposedly, this process was also driven by partner choice such that good collaborators (i.e., individuals who engaged in joint attention, formed shared goals and shared knowledge, and who were committed to collaborative activities) were selected as cooperative partners, thus further facilitating the emergence of precisely these skills (Tomasello et al., 2012).

Presumably, early humans also had to coordinate despite having conflicting interests that required them to compromise (e.g., in Chicken Game scenarios, as when one of two hunters had to incur the cost of assuming a risky position for the hunt to be successful). Reaching joint compromises was also crucial when dividing the spoils of cooperation and, as we have seen, the inability to do this appears to be a major limiting factor of sustaining collaboration in chimpanzees (Melis et al., 2006). Partner choice in the context of interdependent collaboration presumably played a critical role as well by selecting against despotic individuals who attempted to dominate access to resources, thus leading to relatively egalitarian motives in humans (Boehm, 1999; Hare, 2017). Indeed, recent evidence suggests that partner choice may have specifically led to dispositions toward fairness in humans (Debove, André, & Baumard, 2015; Debove, Baumard, & André, 2015), which corresponds to the reported findings that children, but not chimpanzees, are prone to establish equal resource divisions and fair compromises (Grueneisen & Tomasello, 2017; Hamann et al., 2011).

Hence, according to this story, interdependence and partner choice can account for the first milestone in the evolution of human cooperation, namely the development of capacities for mutualistic collaboration – particularly skills for sharing psychological states and knowledge in small-scale interactions.

In a second evolutionary step, these cooperative capacities were scaled up to group life such that individuals now interacted with many different partners, and even relative strangers. Population expansions as well as competition between groups required individuals to accurately identify group members who have the requisite skills and the trustworthiness to be good collaborative partners. This resulted in various markers of group membership and group identity such that individuals who engaged in conventional cultural practices (e.g., who spoke and dressed in certain ways) could be assumed to possess the relevant skills, knowledge, and dispositions to be trusted in collaborative endeavors (Tomasello et al., 2012). As we have demonstrated in Section 22.3, children develop these requisite skills early on and distinguish between individuals based on group membership cues, draw inferences about the knowledge they share with others in virtue of being from the same group, and, in accordance, adapt their cooperative decisions (Dunham et al., 2011; Goldvicht-Bacon & Diesendruck, 2016; Wyman et al., unpublished data).

Moreover, recurrent coordination problems with multiple individuals resulted in conventionalized coordination solutions (Lewis, 1969) and the emergence of numerous social rules regulating interactions among members of cultural communities. Crucially, these rules frequently attained normative status such that everyone had mutual expectations in shared knowledge of how everyone in the relevant group ought to behave in specific situations (Bicchieri, 2006; Gilbert, 1989). This generated pressure to conform to the rules to ensure effective coordination with changing partners, but also because nonconformity was seen as potentially harmful to group life and was punished via various sanctioning mechanisms (Boyd, Gintis, & Bowles, 2010; Fehr & Fischbacher, 2004; Raihani & Bshary, 2011). Young children also come equipped to deal with these challenges, as they not only swiftly pick up the rules specifying "how *we* do things in our group," but also develop a refined understanding of the normative nature of much of human interaction.

In modern humans, these various social rules culminate in social institutions, which describe cultural practices that are based on collective group goals and involve norms specifying standardized roles and procedures (e.g., for producing and dividing resources; Turner, 1997). Social institutions further assign statuses to individuals playing particular roles (e.g., "police officer" or "president") that are typically symbolically marked and involve entitlements and obligations. Crucially, these new statuses that everyone must recognize and respect exist because – and only because – everyone agrees in shared knowledge that they do (Searle, 1995). Hence, this key aspect of modern human cooperative life and culture also exists by virtue of human-specific abilities for interpersonal sharing, which, as we have seen, have their origin in our species' long history of mutualistic collaboration and coordination.

REFERENCES

Alvard, M., & Nolin, D. (2002). Rousseau's whale hunt? Coordination among big-game hunters. *Current Anthropology*, **43**, 533–559.

Bacharach, M. (2006). *Beyond Individual Choices: Teams and Frames in Game Theory*. Princeton, NJ: Princeton University Press.

Bakeman, R., & Adamson, L. B. (1984). Coordinating attention to people and objects in mother infant and peer infant interaction. *Child Development*, **55**, 1278–1289.

Benozio, A., & Diesendruck, G. (2017). Parochial compliance: Young children's biased consideration of authorities' preferences regarding intergroup interactions. *Child Development*, 88, 1527–1535.

Bicchieri, C. (2006). *The Grammar of Society: The Nature and Dynamics of Social Norms*. New York: Cambridge University Press.

Boehm, C. (1999). *Hierarchy in the Forest: The Evolution of Egalitarian Behavior*. Cambridge, MA: Harvard University Press.

Boyd, R., Gintis, H., & Bowles, S. (2010). Coordinated punishment of defectors sustains cooperation and can proliferate when rare. *Science*, **328**, 617–620.

Bratman, M. (1992). Shared cooperative activity. *Psychological Review*, **101**, 327–341.

Bräuer, J., Call, J., & Tomasello, M. (2005). All great ape species follow gaze to distant locations and around barriers. *Journal of Comparative Psychology*, **119**, 145–154.

Bullinger, A. F., Melis, A. P., & Tomasello, M. (2011). Chimpanzees, *Pan troglodytes*, prefer individual over collaborative strategies towards goals. *Animal Behaviour*, **82**, 1135–1141.

Butler, L. P., & Walton, G. M. (2013). The opportunity to collaborate increases preschoolers' motivation for challenging tasks. *Journal of Experimental Child Psychology*, **116**, 953–961.

Calcott, B. (2008). The other cooperation problem: Generating benefit. *Biology & Philosophy*, **23**, 179–203.

Call, J., & Tomasello, M. (2007). *The Gestural Communication of Apes and Monkeys*. Manhaw, NJ: LEA.

Call, J., Hare, B., Carpenter, M., & Tomasello, M. (2004). "Unwilling" versus "unable": Chimpanzees' understanding of human intentional action. *Developmental Science*, **7**, 488–498.

Camerer, C. F. (2011). *Behavioral Game Theory: Experiments in Strategic Interaction*. Princeton, NJ: Princeton University Press.

Cameron, J. A., Alvarez, J. M., Ruble, D. N., & Fuligni, A. J. (2001). Children's lay theories about ingroups and outgroups: Reconceptualizing research on prejudice. *Personality and Social Psychology Review*, **5**, 118–128.

Carpenter, M., & Call, J. (2013). How joint is the joint attention of chimpanzees and human infants? In H. S. Terrace & J. Metcalfe, eds., *Agency and Joint Attention*. Oxford: Oxford University Press, pp. 49–61.

Carpenter, M., & Liebal, K. (2011). Joint attention, communication, and knowing together in infancy. In A. Seemann, ed., *Joint Attention: New Developments*. Cambridge, MA: MIT Press, pp. 159–181.

Carpenter, M., Nagell, K., & Tomasello, M. (1998). Social cognition, joint attention, and communicative competence from 9 to 15 months of age. *Monographs of the Society for Research in Child Development*, 63, 1–143.

Carpenter, M., Tomasello, M., & Striano, T. (2005). Role reversal imitation and language in typically developing infants and children with autism. *Infancy*, **8**, 253–278.

Clark, H. H. (1996). *Using Language*. Cambridge, UK: Cambridge University Press.

Debove, S., André, J. B., & Baumard, N. (2015). Partner choice creates fairness in humans. *Proceedings of the Royal Society B: Biological Sciences*, **282**, 20150392.

Debove, S., Baumard, N., & André, J. B. (2015). Evolution of equal division among unequal partners. *Evolution*, **69**, 561–569.

Doebeli, M., & Hauer, C. (2005). Models of cooperation based on the prisoner's dilemma and the snowdrift game. *Ecology Letters*, **8**, 748–766.

Duguid, S., Wyman, E., Bullinger, A. F., Herfurth-Majstorovic, K., & Tomasello, M. (2014). Coordination strategies of chimpanzees and human children in a Stag Hunt game. *Proceedings of the Royal Society B: Biological Sciences*, **281**, 20141973.

Dunham, Y., Baron, A. S., & Carey, S. (2011). Consequences of "minimal" group affiliations in children. *Child Development*, **82**, 793–811.

Fehr, E., & Fischbacher, U. (2003). The nature of human altruism. *Nature*, **425**, 785–791.

Fehr, E., & Fischbacher, U. (2004). Third-party punishment and social norms. *Evolution and Human Behavior*, **25**, 63–87.

Fehr, E., Bernhard, H., & Rockenbach, B. (2008). Egalitarianism in young children. *Nature* **454**, 1079–1084

Fletcher, G. E., Warneken, F., & Tomasello, M. (2012). Differences in cognitive processes underlying the collaborative activities of children and chimpanzees. *Cognitive Development*, **27**, 136–153.

Gilbert, M. (1989). *On Social Facts*. London: Routledge.

Gilbert, M. (1996). *Living Together: Rationality, Sociality, and Obligation*. Lanham, MD: Rowman & Littlefield.

Gilbert, M. (2008). Social convention revisited. *Topoi*, **27**, 5–16.

Gintis, H., Bowles, S., Boyd, R., & Fehr, E., eds. (2005). *Moral Sentiments and Material Interests*. Cambridge, MA: MIT Press.

Göckeritz, S., Schmidt, M. F. H., & Tomasello, M. (2014). Young children's creation and transmission of social norms. *Cognitive Development*, **30**, 81–95.

Goldvicht-Bacon, E., & Diesendruck, G. (2016). Children's capacity to use cultural focal points in coordination problems. *Cognition*, **149**, 95–103.

Gräfenhain, M., Behne, T., Carpenter, M., & Tomasello, M. (2009). Young children's understanding of joint commitments. *Developmental Psychology*, **45**, 1430–1443.

Grueneisen, S., & Tomasello, M. (2017). Children coordinate in a social dilemma by taking turns and along dominance asymmetries. *Developmental Psychology*, 53, 265–273.

Grueneisen, S., & Tomasello, M. (2019). Children use rules to coordinate in a social dilemma. *Journal of Experimental Child Psychology*, 179, 362–374.

Grueneisen, S., Wyman, E., & Tomasello, M. (2015a). Children use salience to solve coordination problems. *Developmental Science*, **18**, 495–501.

Grueneisen, S., Wyman, E., & Tomasello, M. (2015b). Conforming to coordinate: Children use majority information for peer coordination. *British Journal of Developmental Psychology*, **33**, 136–147.

Grueneisen, S., Wyman, E., & Tomasello, M. (2015c). "I know you don't know I know …" Children use second-order false-belief reasoning for peer coordination. *Child Development*, **86**, 287–293.

Hamann, K., Warneken, F., Greenberg, J. R., & Tomasello, M. (2011). Collaboration encourages equal sharing in children but not in chimpanzees. *Nature*, **476**, 328–331.

Hamann, K., Warneken, F., & Tomasello, M. (2012). Children's developing commitments to joint goals. *Child Development*, **83**, 137–145.

Hammerstein, P., ed. (2003). *Genetic and Cultural Evolution of Cooperation*. Cambridge, MA: MIT Press.

Hardecker, S., Schmidt, M. F. H., & Tomasello, M. (2017). Children's developing understanding of the conventionality of rules. *Journal of Cognition and Development*, **18**, 163–188.

Hare, B. (2017). Survival of the friendliest: Homo sapiens evolved via selection for prosociality. *Annual Review of Psychology*, 68, 155–186.

Hare, B., Call, J., & Tomasello, M. (2001). Do chimpanzees know what conspecifics know and do not know? *Animal Behaviour*, **61**, 139–151.

Haun, D. B. M., Rekers, Y., & Tomasello, M. (2012). Majority-biased transmission in chimpanzees and human children, but not orangutans. *Current Biology*, **22**, 727–731.

Haun, D. B. M., Rekers, Y., & Tomasello, M. (2014). Children conform to the behavior of peers; other great apes stick with what they know. *Psychological Science*, **25**, 2160–2167.

Hirata, S., & Fuwa, K. (2007). Chimpanzees (*Pan troglodytes*) learn to act with other individuals in a cooperative task. *Primates*, **48**, 13–21.

Horner, V., & Whiten, A. (2005). Causal knowledge and imitation/emulation switching in chimpanzees (*Pan troglodytes*) and children (*Homo sapiens*). *Animal Cognition*, **8**, 164–181.

Kaminski, J., Call, J., & Tomasello, M. (2008). Chimpanzees know what others know but not what they believe. *Cognition*, **109**, 224–234.

Karg, K., Schmelz, M., Call, J., & Tomasello, M. (2015). Chimpanzees strategically manipulate what others can see. *Animal Cognition*, **18**, 1069–1076.

Kinzler, K. D., Dupoux, E., & Spelke, E. S. (2007). The native language of social cognition. *Proceedings of the National Academy of Sciences*, **104**, 12577–12580.

Leavens, D. A., Hopkins, W. D., & Thomas, R. K. (2004). Referential communication by chimpanzees (*Pan troglodytes*). *Journal of Comparative Psychology*, **118**, 48–57.

Lewis, D. (1969). *Convention: A Philosophical Study*. Cambridge, MA: Harvard University Press.

Liszkowski, U., Carpenter, M., Henning, A., Striano, T., & Tomasello, M. (2004). Twelve-month-olds point to share attention and interest. *Developmental Science*, **7**, 297–307.

Lyons, D. E., Young, A. G., & Keil, F. C. (2007). The hidden structure of overimitation. *Proceedings of the National Academy of Sciences*, **104**, 19751–19756.

Mehta, J., Starmer, C., & Sugden, R. (1994). The nature of salience: An experimental investigation. *American Economic Review*, **84**, 658–673.

Melis, A. P., Hare, B., & Tomasello, M. (2006). Engineering cooperation in chimpanzees: Tolerance constraints on cooperation. *Animal Behaviour*, **72**, 275–286.

Melis, A. P., Grocke, P., Kalbitz, J., & Tomasello, M. (2016). One for you, one for me: Humans' unique turn-taking skills. *Psychological Science*, 27, 987–996.

Milinski, M., Semmann, D., & Krambeck, H. J. (2002). Reputation helps solve the "Tragedy of the Commons". *Nature*, **415**, 424–426.

Milward, S., Kita, S., & Apperly, I. A. (2014). The development of co-representation effects in a joint task: Do children represent a co-actor? *Cognition*, **132**, 269–279.

Moll, H., Richter, N., Carpenter, M., & Tomasello, M. (2008). Fourteen-month-olds know what "we" have shared in a special way. *Infancy*, **13**, 90–101.

Ostrom, E. (2000). Collective action and the evolution of social norms. *Journal of Economic Perspectives*, **14**, 137–158.

Peacocke, C. (2005). Joint attention: Its nature, reflexivity, and relation to common knowledge. In N. Eilan, C. Hoerl, T. McCormack, & J. Roessler, eds., *Joint Attention: Communication and Other Minds*. Oxford: Clarendon Press, pp. 298–324.

Raihani, N. J., & Bshary, R. (2011). The evolution of punishment in n-player games: A volunteer's dilemma. *Evolution*, **65**, 2725–2728.

Rakoczy, H., Warneken, F., & Tomasello, M. (2008). The sources of normativity: Young children's awareness of the normative structure of games. *Developmental Psychology*, **44**, 875–881.

Rapoport, A., & Chammah, A. M. (1966). The game of chicken. *American Behavioral Scientist*, **10**, 10–28.

Rekers, Y., Haun, D. B. M., & Tomasello, M. (2011). Children, but not chimpanzees, prefer to collaborate. *Current Biology*, **21**, 1756–1758.

Schelling, T. (1960). *The Strategy of Conflict*. Cambridge, MA: Harvard University Press.

Schmelz, M., Call, J., & Tomasello, M. (2011). Chimpanzees know that others make inferences. *Proceedings of the National Academy of Sciences*, **108**, 3077–3079.

Schmidt, M. F. H., Rakoczy, H., & Tomasello, M. (2011). Young children attribute normativity to novel actions without pedagogy or normative language. *Developmental Science*, **14**, 530–539.

Schmidt, M. F. H., Rakoczy, H., & Tomasello, M. (2012). Young children enforce social norms selectively depending on the violator's group affiliation. *Cognition*, **124**, 325–333.

Schmidt, M. F. H., Butler, L. P., Heinz, J., & Tomasello, M. (2016). Young children see a single action and infer a social norm: Promiscuous normativity in 3-year-olds. *Psychological Science*, **27**, 1360–1370.

Schneider, A. -C., Melis, A. P., & Tomasello, M. (2012). How chimpanzees solve collective action problems. *Proceedings of the Royal Society B: Biological Sciences*, **279**, 4946–4954.

Searle, J. R. (1990). Collective intentions and actions. In P. R. Cohen, J. Morgan, & M. E. Pollack, eds., *Intentions in Communication*. Cambridge, MA: MIT Press, pp. 227–241.

Searle, J. R. (1995). *The Construction of Social Reality*. London: Penguin.

Thomas, K. A., DeScioli, P., Haque, O. S., & Pinker, S. (2014). The psychology of coordination and common knowledge. *Journal of Personality and Social Psychology*, **107**, 657–676.

Tollefsen, D. (2005). Let's pretend! Children and joint action. *Philosophy of the Social Sciences*, **35**, 75–97.

Tomasello, M. (1995). Joint attention as social cognition. In C. Moore & P. J. Dunham, eds., *Joint Attention: Its Origins and Role in Development*. Hillsdale, NJ: Erlbaum, pp. 103–130.

Tomasello, M. (2016). *A Natural History of Human Morality*. Cambridge, MA: Harvard University Press.

Tomasello, M., & Carpenter, M. (2005). The emergence of social cognition in three young chimpanzees. *Monographs of the Society for Research in Child Development*, **70**, vii–132.

Tomasello, M., Carpenter, M., Call, J., Behne, T., & Moll, H. (2005). Understanding and sharing intentions: The origins of cultural cognition. *Behavioral and Brain Sciences*, **28**, 675–735.

Tomasello, M., Melis, A., Tennie, C., Wyman, E., & Herrmann, E. (2012). Two key steps in the evolution of human: The interdependence hypothesis. *Current Anthropology*, **53**, 673–692.

Tomonaga, M., Tanaka, M., Matsuzawa, T., et al. (2004). Development of social cognition in infant chimpanzees (*Pan troglodytes*): Face recognition, smiling, gaze, and the lack of triadic interactions. *Japanese Psychological Research*, **46**, 227–235.

Trevarthen, C. (1979). Communication and cooperation in early infancy. A description of primary intersubjectivity. In M. Bullowa, ed., *Before Speech: The Beginning of Human Communication*. Cambridge, UK: Cambridge University Press, pp. 321–347.

Turner, J. (1997). *The Institutional Order: Economy, Kinship, Religion, Polity, Law, and Education in Evolutionary and Comparative Perspective*. New York: Longman.

Warneken, F., Chen, F., & Tomasello, M. (2006). Cooperative activities in young children and chimpanzees. *Child Development*, **77**, 640–663.

Warneken, W., Lohse, K., Melis, A. P., & Tomasello, M. (2011). Young children share the spoils after collaboration. *Psychological Science*, **22**, 267–273.

Warneken, F., Gräfenhain, M., & Tomasello, M. (2012). Collaborative partner or social tool? New evidence for young children's understanding of joint intentions in collaborative activities. *Developmental Science*, **15**, 54–61.

Wyman, E. (2014). Language and collective fiction: From children's pretense to social institutions. In J. Lewis, C. Knight, & D. Dor, eds., *The Social Origins of Language*. Oxford: Oxford University Press, pp. 171–183.

Wyman, E., Rakoczy, H., & Tomasello, M. (2009). Normativity and context in young children's pretend play. *Cognitive Development*, **24**, 146–155.

Wyman, E., Rakoczy, H., & Tomasello, M. (2013). Non-verbal communication enables coordination with others in a children's "Stag Hunt" game. *European Journal of Developmental Psychology*, **10**, 597–610.

23 Genomic Imprinting Is Critical for Understanding the Development and Adaptive Design of Psychological Mechanisms in Humans and Other Animals

WILLIAM M. BROWN

Genomic imprinting is genotype-independent parent-of-origin gene expression. It is a well-understood example of epigenetic inheritance, as the environment (i.e., whether a gene resided in the egg or sperm in the previous generation) leaves a reversible mark on offspring DNA. These marks are erased when the offspring produces its own gametes (i.e., sperm or egg), so that new parent-of-origin marks can be placed. The phenotypic effects of imprinted genes are significant, especially in terms of growth and neurobiology. Imprinted genes are rare in mammalian genomes, but here I make the case for the idea that they are critically important for our understanding of how natural selection shaped functional adaptations. There are approximately 100 imprinted genes in humans that have differentially methylated regions by parent of origin. The effects of imprinted genes are species, isoform (i.e., alternative transcripts), tissue, and developmentally dependent. Haig (2000) has provided the best explanation for the mammalian expression patterns of imprinted genes, called *kinship* or *parental antagonism* theory. Kinship theory posits that multiple paternity and sex-biased dispersal caused parental gene conflicts within offspring. Specifically, paternal gene expression favors increased maternal investment to the benefit of patrilineal inclusive fitness at a cost to matrilineal inclusive fitness, while maternal gene expression patterns are designed to minimize these costs. *In utero*, imprinted genes are required for healthy development and the actions are antagonistic, whereby maternal genes restrict growth and paternal genes enhance growth. At birth, the actions of paternally expressed genes change when offspring behaviorally acquire resources from the mother (e.g., via breastfeeding and demand vocalizations). Maternal genes within offspring are expected to lower physiological and psychological resource demands. As offspring are weaned, they increase their home range and their ability to feed themselves. This in turn causes parentally derived imprinted genes within offspring to engage in conflicts over foraging for their own food vs. relying on maternal or asymmetric relative provisioning. Language faculties also involve an inherent conflict (Brown, 2011), whereby paternally derived genes favor escalating verbal demands of the mother and the filtering of maternal instruction, while maternal genes encourage the maintenance of cooperation among maternal relatives (which may include punishment of asymmetrically related sibling defectors). Given the fact that one critical aspect of the environment (whether a gene was in an egg or sperm in the previous generation) has been encoded to adaptively alter gene expression in offspring, this raises the strong possibility of other environmentally transmitted epigenetic effects across generations. I briefly point out how epigenetic transmission biases (e.g., DNA methylation) may hitchhike upon imprinting machinery for new vistas of evolutionary conflict (e.g., deprivation-linked loss in maternal investment during ontogeny) and its resolution.

> Perhaps the most interesting thing to come out of the realization of possible conflict within the genome is a philosophical one. We see that we are not even in principle the consistent wholes that some schools of philosophy would have us be. (Hamilton, 1987/2001, pp. 354–355)

In the above passage, the late W. D. Hamilton highlighted the importance of intra-genomic conflict for our philosophical understanding of human psychology (Hamilton, 1987/2001). Despite Hamilton's early insight, evolutionary psychology has proceeded much like behavioral ecology before it, specifically by positing the adaptive function for a trait (morphological or psychological) and the specific adaptive problem it was designed by selection to solve from the replicator's "point of view" of optimizing inclusive fitness. However, as Hamilton (1987/2001) was deeply aware, matrilineal and patrilineal inclusive fitness interests can diverge, causing genetic conflicts within offspring (Haig, 2000). This chapter will introduce genomic imprinting and the importance of understanding the subsequent within-organism conflicts that ensue.

Genomic imprinting is genotype-independent parent-of-origin gene expression. For most genes, we inherit two identical working parental copies. However, for genomic imprinting, epigenetic tags (i.e., methyl groups) are placed on either parental copy, rendering the other parental copy inactive (Abramowitz & Bartolomei, 2012). Imprinted genes are the best example of "epigenetic inheritance" because environmental information (i.e., whether a gene was present in an egg or sperm in the previous generation) is placed on DNA. When offspring produce gametes, imprints are erased. Approximately one percent of the human genome is imprinted (Morison, Paton, & Cleverley, 2001). Despite the rarity of imprinted genes, they are of great importance due to their large phenotypic effects on physical and psychological development (Buiting, 2010; Eggermann, 2009; Kong et al., 2009; Mackay & Temple, 2010; Monk et al., 2009; Small et al., 2011). Furthermore, epigenetic dysregulation in imprinted genes – which often have growth-enhancing and tumor-suppressing functions – predict disease and cancer outcomes (Portela & Esteller, 2010). More information is provided in this chapter on the primary epigenetic mechanism behind imprinting, namely DNA methylation. DNA methylation allows for phenotypic modification and the possibility of multiple sources of environmental information beyond parental origin (e.g., stress) to be transmitted to the next generation.

23.1 DNA METHYLATION, GROWTH, AND HEALTH: INTRODUCING PARENTAL ANTAGONISM THEORY

One of the most well-studied epigenetic mechanisms mediating imprinting is DNA methylation. DNA methylation refers to a process where methyl groups are placed on a genetic region, thereby altering expression (e.g., often suppressing gene expression like a car brake). DNA methylation refers to the adding of a methyl group on a cytosine base. It occurs primarily in the context of dinucleotides consisting of a cytosine base followed immediately by a guanine base (CpG), which cluster in regions called CpG islands (Deaton & Bird, 2011). CpG dinucleotides are rare in mammals (~1 percent), but ~50 percent of gene promoters are linked to CpG islands, which are often unmethylated in healthy cells. Cells become methylated in a tissue- and age-specific manner during development. Where DNA methylation occurs can be critical for its effect. DNA methylation at a gene's promoter is linked to silencing (i.e., less gene expression); in contrast, DNA methylation outside the promoter region (e.g., gene body) is sometimes associated with increased gene expression. In the case of genomic imprinting, hypermethylation of one of the two parental alleles leads to monoallelic expression.

Kinship theory (Haig, 2000) is currently the best theory to explain the phenotypic effects of imprinted genes. Specifically, kinship theory posits that the paternal genome within offspring fosters growth to the benefit of patrilineal inclusive fitness at a cost to the maternal genome, while maternal genomes attempt to minimize costs by suppressing growth. Haig's (2000) approach proposes that imprinting evolved due to conflicts of interest between parental genomes over resource acquisition and delivery.

In polygamous species, patrigenes (genes expressed within offspring inherited from the father) favor fetal growth at the expense of depleting maternal resources and disadvantaging future offspring. Meanwhile, matrigenes (genes expressed within offspring inherited from the mother) will oppose selfish paternal effects and conserve resources to optimize the inclusive fitness of the mother and future offspring, such as sharing or delivering resources equally to her offspring. Haig's (2000) theory predicts that paternally expressed imprinted genes will often promote growth to the detriment of future asymmetrically related siblings, while maternally expressed genes will have opposite effects of reducing costs on matrilineal inclusive fitness. Some of the well-known imprinted genes and imprinting disorders related to growth are discussed in the following paragraphs.

IGF2 – an imprinted gene and paternally expressed in humans – regulates muscle development. *IGF2* is upregulated early in *MyoD*-induced myocyte differentiation, and *IGF2* inhibition leads to reduced expression of *MyoD* target genes, which suggests that *IGF2* is essential for amplifying and maintaining *MyoD* efficacy (Wilson & Rotwein, 2006). *IGF2*'s role as a paternally derived skeletal muscle growth enhancer is consistent with kinship theory (Haig, 2000, 2002).

Angelman syndrome (AS) and Prader–Willi syndrome (PWS) are imprinting disorders. PWS is caused by an overexpression of maternal genes on chromosome 15, while AS is due to an overexpression of paternal genes. Muscle biopsies of 11 PWS children have been investigated using histochemical and morphometric methods (Sone, 1994). The phenotypic abnormalities included: (1) fiber size variation of both type 1 and 2 fibers; (2) type 2 fiber atrophy; (3) increased numbers of type 2 C fibers; and (4) decreased numbers of type 2B fibers. These findings are consistent with the overexpression of maternal genes suppressing skeletal muscle growth.

DNA methylation and imprinted loci influence muscle hypertrophy – extremely muscled hindquarters – in callipyge sheep (Murphy et al., 2006). Hypomethylation of *Clpg1* causes muscle hypertrophy in part due to the overexpression of *Dlk1* (Murphy et al., 2006), a paternally expressed imprinted gene associated with muscle precursor cell (myoblast) differentiation (Bassel-Duby & Olson, 2006). DNA demethylation promotes skeletal myotube maturation (Hupkes et al., 2011). Early experiments in the 1970s showed that DNA methyltransferase inhibitors (e.g., 5-azacytidine) induced transdifferentiation of fibroblasts into myoblasts (Taylor & Jones, 1979). More recently, in C2C12 cultures, Hupkes et al. (2011) noticed that upon treatment with a methylation inhibitor (i.e., 5-azacytidine), myotubes spontaneously acquired repetitive membrane activity, intracellular calcium transients, and contractility. Hupkes et al. (2011) suggested that DNA methylation may pose an epigenetic barrier to C2C12 myotubes reaching maturity. However, when imprinted genes are involved in skeletal muscle development, the so-called

DNA methylation barrier will likely be parent-of-origin dependent.

It is important to note that not all growth is positive. For example, the early growth of young humans is beneficial (e.g., developing secondary sexual characteristics), but growth among older humans is costly in terms of survival (e.g., cancers). DNA methylation has been implicated in aging and cancer.

Specifically, cancer cells are characterized by DNA methylation loss among growth enhancers (i.e., increasing expression of growth-related genes) and a coordinated DNA methylation gain near tumor suppressors, which silences their cancer-preventing function (Cruickshanks & Tufarelli, 2009). Cancer cells are characterized by a global loss of DNA methylation among growth enhancers and the coordinated acquisition of hypermethylation at the CpG islands of tumor-suppressor genes. Global hypomethylation occurs primarily at parasitic DNA regions of the genome. For example, the Long Interspersed Nuclear Element (LINE) family member L1 is hypomethylated in a variety of cancers, such as breast and colon (Cruickshanks & Tufarelli, 2009).

23.2 DISCRIMINATIVE GRANDPARENTAL SOLICITUDE: REWINDING THE EPIGENETIC CLOCK

It is expected that maternal genes in most cases will benefit from increased longevity to deliver investment to daughters' grandchildren. Paternal genes are designed to extract rather than share or deliver resources to matrilineal kin due to paternity uncertainty. Therefore, it is expected that the growth-suppressor effects of maternal genes will be responsive to condition (e.g., being physically able to invest in grand-offspring). The maternal genome may be responsible for "rewinding" the epigenetic clock to deliver grandparental investment. Consistent with this hypothesis is the fact that being physically capable of and giving social support to kin are positive correlates of health and possibly longevity (Brown et al., 2005). More recently, evidence has been shown to support the hypothesis that exercise interventions may reduce epigenetic age (Brown, 2015). A notable correlate of disease is aging, and DNA methylation is one of the best predictors of age. Research on this new area refers to the "epigenetic clock." The epigenetic clock is defined as the DNA methylation profile predictive of chronological age. Notably, DNA methylation profiles are excellent at detecting age acceleration due to disease (Horvath, 2013, 2015). DNA methylation age has the following properties germane to exercise epigenetics: (1) it is near zero for embryonic and induced pluripotent stem cells; (2) it correlates with cell passage number; and (3) it provides a heritable measure of age acceleration. A total of 6,000 cancer samples (from 20 cancer types) exhibited age acceleration of decades (Horvath, 2013, 2015) compared to healthy cells. DNA methylation age fulfills the biomarker criteria, but whether it is a cause or consequence of aging itself is not known.

Exercise interventions among patients could help ameliorate the degeneration associated with epigenomic age acceleration. For example, one study of older people (aged 58–90 years) with cerebrovascular disease suggests that physical function improvements during hospitalization covaried with sub-telomeric methylation of long telomeres (Maeda et al., 2011). In a landmark study (Zeng et al., 2012), beneficial effects of exercise were found among 12 females with breast cancer exposed to a 6-month walking regime. The key finding was that females with breast cancer lived longer after exercise, and this was associated with decreased DNA methylation of the tumor-suppressor gene *L3MBTL1*, notably a paternally expressed imprinted gene. These associations were from a single study of peripheral blood leukocytes after 6 months of moderate-intensity walking in a sample of 12 females with breast cancer who were approximately 57 years old (Zeng et al., 2012). Of those six genes, only *PPP2R3A* showed a relationship with cancer. After exercise, *PPP2R3A* showed increased DNA methylation, but this was correlated with increased *PPP2R3A* gene expression and better survival among patients. I briefly summarize the findings showing epigenetic changes after exercise (six months of moderate-intensity walking in a sample of postmenopausal females with breast cancer; Zeng et al., 2012). The key finding of the paper was that females with breast cancer lived longer after exercise, and this was associated with decreased DNA methylation of the tumor-suppressor gene *L3MBTL1*, which is an imprinted gene (paternally expressed). From an evolutionary perspective, a relaxing of silencing of this gene could occur when female grandparental investment benefits patrilineal inclusive fitness. Of note, the following genes showed changes in DNA methylation after exercise but had no impact on gene expression or cancer outcomes in this sample: *CXCL10, DCC, RASA1, SULF1*, and *TMEM100*. However, *PPP2R3A* was shown to increase DNA methylation in peripheral blood leukocytes after 6 months of moderate0intensity walking in a sample of 12 females with breast cancer who were approximately 57 years old. Increased *PPP2R3A* DNA methylation was associated with increased gene expression and breast cancer survival. This highlights that increased DNA methylation is not always associated with decreased gene expression. Specifically, DNA methylation (i.e., adding methyl groups to a cytosine base) at a gene's promoter is linked to silencing (i.e., less gene expression); in contrast, DNA methylation outside the promoter region (e.g., gene body) is sometimes associated with increased gene expression. It is important to note that DNA methylation changes with tissue and age in addition to the region of the gene with the marks. We need to elucidate the region of the gene in question before hypotheses can be reliably drawn regarding a direct link between exercise-associated DNA methylation change and cancer. Also of note is tissue type. In the case of blood samples, which are heterogeneous with regard to cell type, additional caution is needed for those studying behavioral epigenetics. Zeng et al.'s (2012)

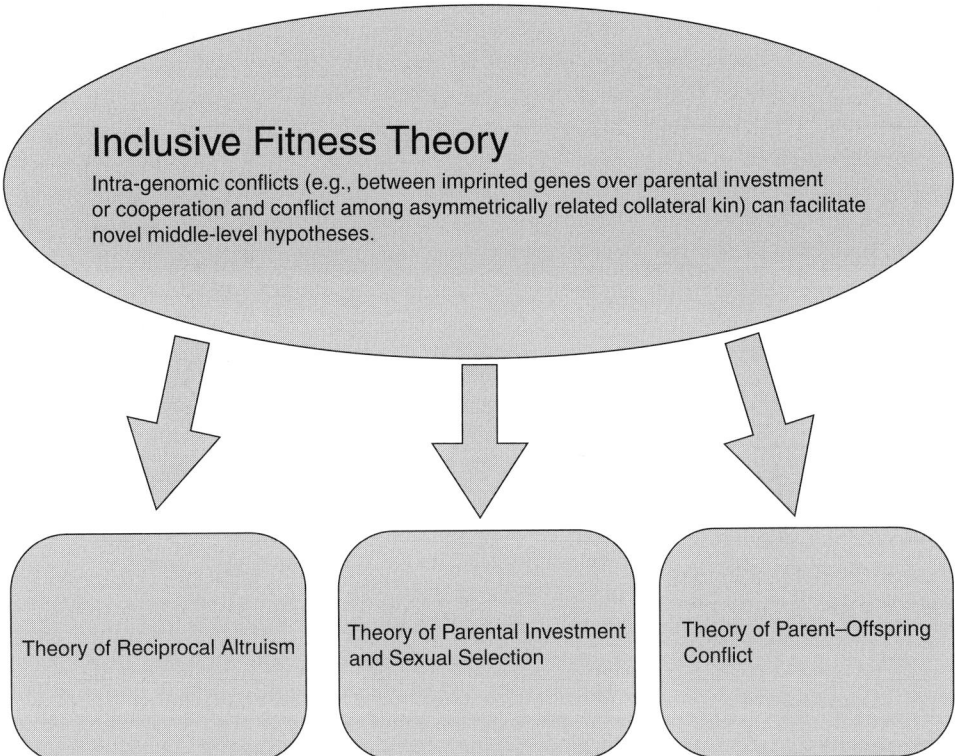

Figure 23.1 Three core middle-level evolutionary theories and the general evolutionary theory that motivated each.

observation of exercise-induced DNA methylation change in *L3MBTL1* was based on blood samples. Fortunately, their subsequent analyses showed that the changes were related to gene expression in tumor samples, and gene expression was associated with patient survival. So at least for this one imprinted gene epigenetic changes in the blood may be clinically relevant. Gillman et al. (2018) have suggested that epigenetic modification of inflammatory genes may provide a mechanistic link between exercise and breast cancer. Specifically, Gillman et al. (2018) recently presented empirical evidence that DNA methylation of inflammatory genes mediates the relationship between cardiorespiratory fitness and methylation changes in genes increasing breast cancer risk. Beyond these specific studies with small samples of patients, a meta-analysis (Brown, 2015) and recent review (Santos, Elliott-Sale, & Sale, 2017) suggest that exercise modifies epigenomes, and that the effects are more pronounced among older people. If these epigenetic alterations due to exercise primarily target older people, it may be an adaptive design feature for the delivery of grandparental investment to matrilineal relatives.

To further develop the study of human imprinted genes, an evolutionary psychological perspective should be taken (e.g., life history theory). Brown and Olding (2017) have suggested that the epigenetic changes occurring across the lifespan may reflect adaptive shifts in life history strategy (somatic efforts to inclusive fitness efforts) and correlate with condition-dependent ecological circumstances (e.g.,

deprivation). Below is an example of how evolutionary psychologists can help explicate and increase of our understanding of imprinted genes.

Evolutionary psychology can help elucidate imprinting conflicts and the adaptive design of epigenetic machinery. To understand how this is achievable, it is worth revisiting Buss's (1995) conceptual framework of evolutionary theory for helping to generate evolutionary psychological hypotheses (Figure 23.1). The three middle evolutionary theories derived from inclusive fitness theory were: (1) theory of reciprocal altruism; (2) theory of parental investment and sexual selection; and (3) theory of parent–offspring conflict. For all three middle-level evolutionary theories, imprinted genes likely play a significant role given their conceptual origins and links to inclusive fitness theory. Here, I adopt Buss's (1995) approach in order to outline the importance of imprinted genes for understanding evolved psychological mechanisms.

23.3 THEORY OF RECIPROCAL ALTRUISM: INTRA-GENOMIC CONFLICT RESOLUTION

How can a knowledge of imprinted genes change how we think about reciprocal altruism, which was posited to involve the social exchange between two individual organisms? If we take the metaphor of selfish genetic elements seriously, individuals can easily be reconfigured to individual genes, such as paternally derived and maternally derived alleles within a developing organism. Since their inclusive fitness interests diverge, genomic cooperation may be governed by the logic of reciprocity. Haig (2003) was the first to detail this idea, pointing out that intra-genomic cooperation may be more easily explained than cooperation between different organisms. I think it is fair to ask how imprinted genes can inform our study of reciprocal altruism in between-organism contexts. For example, parental alleles cause asymmetric kin relations to be treated as nonkin social exchanges. Breakdowns of trust among kin are common and could potentially be restored through reciprocity.

Recently, Micheletti, Ruxton, and Gardner (2017) have suggested that human intergroup violence could be further elucidated by incorporating an intra-genomic conflict perspective. Specifically, in a mathematical model, Micheletti, Ruxton, and Gardner (2017) show that warfare drives the evolution of sex-biased dispersal and sex-biased dispersal modulates intra-genomic conflicts in relation to warfare. It remains to be seen if such complex social evolutionary dynamics can be mediated or resolved via intra-personal reciprocity (Haig, 2003).

23.4 THEORY OF PARENTAL INVESTMENT AND SEXUAL SELECTION

The hypothesis that the sex investing the most in offspring development will be choosier when selecting a mate while the least-investing sex will expend greater effort on secondary sexual characteristic display traits could well be one of the most elegant in behavioral science. At first glance, it may be unclear how imprinted genes could influence sexual selection itself. Theoretically, it is becoming increasingly apparent that sexual and parental antagonisms may be linked in important ways. For example, Faria, Varela, and Gardner (2017) have showed how mating ecology (e.g., sex-biased dispersal causing relatedness asymmetry) can drive the evolution of intra-genomic conflict in contexts of sexual conflict.

Imprinted genes in humans are critically involved in the development of secondary sexual characteristics (e.g., skeletal muscle; Brown, 2015) and degree of maternal investment (amount extracted vs. delivered; Haig, 1993). Rather than viewing the degree of maternal investment and sexual characteristics as a constant, imprinted gene networks may cause fluctuations in the relative sex differences in parental investment and investment in secondary sexual characteristics. This may explain why parental investment varies across time and culture. Intra-genomic conflicts are expected to occur over incestuous sexual desires (Haig, 1999) and the inability to remain religiously celibate (Brown, 2001). More recently, evidence has shown that the loss of paternally expressed genes in the case of a family was associated with early puberty and the development of secondary sexual characteristics (Dauber et al., 2017). For example, in a case study sample, breast development occurred at 4.6–5.9 years of age. Early puberty could be interpreted as reducing demands on mothers

(e.g., leaving natal area/becoming independent). Dauber et al. (2017) make a persuasive case that their evidence of loss-of-function mutations in two paternally expressed imprinted genes (*MKRN3*, *DLK1*) associated with central precocious puberty may be explained by kinship theory. Specifically, they hypothesized that maternally expressed genes favor earlier puberty to reduce maternal demands, while paternally expressed imprinted genes will delay puberty, prolonging the maternal-dependence phase of human childhood.

23.5 THEORY OF PARENT–OFFSPRING CONFLICT

Parent–offspring conflict is quite well understood from a genetic conflict approach, so I will not go into much detail on the importance of imprinted genes. Many phenotypes associated with imprinted genes are likely to be consequences of underlying parent–offspring conflicts. For example, the well-known (fetal conflicts during pregnancy) and less well-known examples (sex-biased dispersal and discordant sibling relations) have been summarized elsewhere (Haig, 2000). Conflicts within offspring are not restricted to physiological growth in humans. For example, paternal genes may favor nonverbal expressions, such as smiling or crying, to extract resources from mothers (Brown & Consedine, 2004). Likewise, language faculties also involve an inherent conflict (Brown, 2011), whereby paternally derived genes favor verbal demands of the mother and the filtering of maternal instructions, while maternal genes encourage the maintenance of cooperation among maternal relatives (which may include punishment of asymmetrically related sibling defectors). More recently, Mehr et al. (2017) found in a sample of 39 people with PWS – a rare genomic imprinting disorder whereby paternally expressed genes on chromosome 15 are unexpressed – increased music contentment (e.g., increased motion and reduced heart rate while listening to music) compared to controls. The authors argue that this is evidence that intra-genomic conflict may have shaped the psychology of music. Specifically, when paternal genes are less active in developing offspring, we expect to see fewer demands of their mothers' investment. It may be worth noting that motion while listening to music could be a cooperative gesture and in the context of maternal compliance and sibling cooperation we may well expect increased levels of such cooperative gestures among individuals with PWS compared to controls.

Perhaps less appreciated is how the degree of parental investment early in development may cause changes in the epigenetic regulation of evolved psychological mechanisms. For example, if ecological constraints (e.g., *in utero* exposure to stressful, deprived environments) cause mothers to invest less, counteradaptations by paternal genes may ensue via decreasing the DNA methylation of paternal genes (increasing internal conflict) designed to increase maternal investment. Counteradaptations are expected to evolve in response to decreasing maternal care during development.

One example is that the neural areas responsible for rapid eye movement (REM) sleep and maternal–neonate co-sleeping behavior are in part controlled by imprinted genes (McNamara, 2014). It is hypothesized that the adaptive function of REM sleep in humans was to escalate or resolve intra-genomic conflicts between asymmetrically related kin. Specifically, nightmares cause offspring to wake, increasing maternal provisioning when there are risks to decreased maternal investment. In contrast, maternal genes within offspring mediate secure attachments, leading to less frequent nightmares and nighttime feeds. Bruni et al. (2004) found that children with AS (an overexpression of paternal genes) exhibit more nightmares and night wakings. Presumably, this functions to accrue more maternal provisioning (McNamara, 2014; Morales-Lara, De-La-Peña, & Murillo-Rodríguez, 2018; Tucci, 2016). Tucci's (2016) review of the imprinting and sleep regulation literature makes it clear that overexpression of maternal genes (or underexpression of paternal genes) at chromosomal region 15q11–13 is characterized by excessive sleepiness and REM sleep deficits. In contrast, an overexpression of paternal genes (or underexpression of maternal genes) at chromosomal region 15q11–13 is characterized by reductions in sleep (e.g., frequent and prolonged night wakings). The pattern of the involvement of imprinted genes in sleep regulation appears consistent with kinship theory. Degree of insecure attachment is a positive correlate of nightmare frequency (Brown, 2018), presumably to increase hypervigilance so as to monitor, retain, or regain parental investment. However, from a parental antagonism perspective, this raises the possibility that maternal genes could be selected to "erase" awareness of nightmares as a person regains consciousness in order to counteract the effects of paternal genes' sleep time tactics designed to enhance daytime vigilance regarding investment loss.

It is important to note that the epigenetic responses to investment loss may not be adaptive at high levels of stress. For example, recent work by Pepper and Nettle (2017) provoked Brown and Olding (2017) to hypothesize that there may be behaviorally adaptive epigenetic responses to low to moderate levels of deprivation. Excessive deprivation (e.g., caloric restriction, lack of parental investment, overcrowding, high mortality, etc.) is toxic. However, in low doses, it may be beneficial, which is known as epigenetic-based hormesis (Chalk & Brown, 2014). Hormesis is when low doses of a toxin or stressor are beneficial (e.g., improved health, stress tolerance, growth, or longevity), but higher doses are toxic or lethal. There is no reason to presume that the biological constraints of hormesis responses to stress discovered in mice do not translate to human stress. Brown and Olding (2017) pointed out that mice and human epigenomes are different and therefore different epigenetic machinery is likely to be involved, but the underlying logic is the same.

Consider extreme psychopathy: such responses (characterized by extreme selfishness and impulsivity) are surely not adaptive given their low frequency in population and the lack of beneficial outcomes. Evidence of higher reproductive success is weak among extreme psychopaths. A moderate level of impulsivity and psychopathy is likely an adaptive response to deprivation, but the highest levels of psychopathy in response to extreme deprivation are likely to be disastrous for the epigenome and adaptive outcomes. It is a maladaptation. A hormesis perspective helps to reconcile the adaptation with pathology perspectives, which have been battling for decades. The hypothesized epigenetic loci can be clarified (Bernal et al., 2013; Park et al., 2017; Swartz, Hariri, & Williamson, 2017; Waterland & Jirtle, 2003), and they can be used as a barometer of maladaptive responses to unreasonably high deprivation not systematically encountered by our ancestors. In summary, epigenetic adaptive responses can help distinguish between when a condition-dependent adaptive response switches to a maladaptation due to exorbitantly high levels to stress (e.g., loss of maternal investment due to deprivation).

23.6 SUMMARY AND FUTURE DIRECTIONS

It is now accepted that the same DNA machinery that is important for genomic imprinting is also responsible for how the environment shapes our epigenomes – the sum of molecular structures residing "above" the genome that can switch genes on and off during our lifetimes. Epigenetics is the study of how environmental factors alter heritable variation in gene expression without changing the underlying DNA sequence. Epigenetic effects are neatly illustrated by the visible differences ("phenotypic differences") between identical twins. One epigenetic pathway to phenotypic difference is the maternal environment *in utero*. Changes to the fetal epigenome during pregnancy can have lifelong consequences on an offspring's cardiovascular and metabolic health (Gluckman et al., 2009). There is good evidence that maternal stress causes epigenetic changes in offspring *in utero* (Chalk & Brown, 2014).

Imprinted genes and kinship theory have important implications for evolutionary psychologists interested in adult humans (Ubeda & Gardner, 2011). However, evolutionary scientists are needed to help understand the strategic nature of imprinting in the context of ecology, life, and natural history. More human studies are needed – despite the causal rigor of nonhuman model organism research – since the key epigenetic elements responsible for regulating hunger, energy expenditure, adiposity, glucose homeostasis, and exercise-associated DNA methylation are differentially marked and read differently between humans and other organisms (Dent & Isles, 2014). Unfortunately, molecular researchers interested in model organisms and cross-species generalization of the molecular machinery mediating imprinting (SanMiguel &

Bartolomei, 2018) may not have an appreciation of the species and ecological diversity of functional adaptations. Thus, there may be fruitful collaborations to be had between evolutionary and molecular researchers interested in furthering the study of imprinting in humans.

Future work using next-generation techniques for precisely quantifying DNA methylation (and other epigenetic modifications such a microRNA and histone modifications) will help to elucidate the causes and consequences of functional adaptations. The key mechanism for organisms to receive environmental feedback during their development occurs via epigenetic modifications in imprinted gene networks. However, caveats are required. The evolutionary histories of imprinted genes can differ by gene and species. Furthermore, there is an obvious constraint on our causal interpretations due to our inability to studying imprinting in healthy human brain tissue. These caveats should give us pause. Nonetheless, I can imagine a future science of molecular epigenetics where the nuanced and complex social world of selfish genetic elements will become theoretically and empirically richer with the input of behavioral scientists.

REFERENCES

Abramowitz, L. K., & Bartolomei, M. S. (2012). Genomic imprinting: Recognition and marking of imprinted loci. *Current Opinion in Genetics and Development*, **22**, 72–8.

Bassel-Duby R., & Olson, E. N. (2006). Signaling pathways in skeletal muscle remodeling. *Annual Reviews Biochemistry*, **75**, 19–37.

Bernal, A. J., Dolinoy, D. C., Huang, D., et al. (2013). Adaptive radiation-induced epigenetic alterations mitigated by antioxidants. *FASEB Journal*, **27**(2), 665–671.

Brown, W. M. (2001). Genomic imprinting and the cognitive architecture mediating human culture. *Journal of Cognition and Culture*, **1**, 251–58.

Brown, W. M. (2011). The parental antagonism theory of language evolution: Preliminary evidence for the proposal. *Human Biology*, **83**(2), 213–245.

Brown, W. M. (2015). Exercise-associated DNA methylation change in skeletal muscle and the importance of imprinted genes: A bioinformatics meta-analysis. *British Journal of Sports Medicine*, **49**(24), 1567–1578.

Brown, W. M. (2018). Conflict within: The epigenetics of troubled sleep. *Clinical Psychiatry*, **4**, 24.

Brown, W. M., & Consedine, N. S. (2004). Just how happy is the happy puppet? An emotion signalling and kinship theory perspective on the behavioral phenotype of children with Angelman syndrome. *Medical Hypotheses*, **63**, 377–385.

Brown, W. M., & Olding, R. J. (2017). Epigenetic-based hormesis and age-dependent altruism: Additions to the behavioural constellation of deprivation. *Behavioral and Brain Sciences*, **40**, e320.

Brown, W. M., Consedine, N. S., & Magai, C. (2005). Altruism relates to health in an ethnically diverse sample of older adults. *Journals of Gerontology*, **60B**(3), 143–152.

Bruni, O., Ferri, R., Dagostino, G., et al. (2004). Sleep disturbances in Angelman syndrome: A questionnaire study. *Brain and Development*, **26**(4), 233–240.

Buiting, K. (2010). Prader–Willi syndrome and Angelman syndrome. *American Journal of Medical Genetics Part C: Seminars in Medical Genetics*, **154C**(3), 365–376.

Buss, D. M. (1995). Evolutionary psychology: A new paradigm for psychological science. *Psychological Inquiry*, **6**(1), 1–30.

Chalk, T. E. W., & Brown, W. M. (2014). Exercise epigenetics and the fetal origins of disease. *Epigenomics*, **6**(5), 469–472.

Cruickshanks, H. A., & Tufarelli, C. (2009). Isolation of cancer-specific chimeric transcripts induced by hypomethylation of the LINE-1 antisense promoter. *Genomics*, **94**(6), 397–406.

Dauber, A., Cunha-Silva, M., Macedo, D. B., et al. (2017). Paternally inherited *DLK1* deletion associated with familial central precocious puberty. *Journal of Clinical Endocrinology & Metabolism*, **102**(5), 1557–1567.

Deaton, A. M., & Bird, A. (2011). CpG islands and the regulation of transcription. *Genes & Development*, **25**(10), 1010–1022.

Dent, C. L., & Isles, A. R. (2014). Brain-expressed imprinted genes and adult behaviour: The example of Nesp and Grb10. *Mammalian Genome*, **25**(1–2), 87–93.

Eggermann, T. (2009). Silver–Russell and Beckwith–Wiedemann syndromes: Opposite (epi)mutations in 11p15 result in opposite clinical pictures. *Hormone Research in Paediatrics*, **71**(2), 30–35.

Faria, G. S., Varela, S. A., & Gardner, A. (2017). Sexual selection modulates genetic conflicts and patterns of genomic imprinting. *Evolution*, **71**(3), 526–40.

Gillman, A. S., Gardiner, C. K., Koljack, C. E., & Bryan, A. D. (2018). Body mass index, diet, and exercise: Testing possible linkages to breast cancer risk via DNA methylation. *Breast Cancer Research and Treatment*, **168**(1), 241–248.

Gluckman, P. D., Hanson, M. A., Buklijas, T., Low, F. M., & Beedle, A. S. (2009). Epigenetic mechanisms that underpin metabolic and cardiovascular diseases. *Nature Reviews Endocrinology*, **5**(7), 401–408.

Haig, D. (1993). Genetic conflicts in human pregnancy. *Quarterly Review of Biology*, **68**(4), 495–532.

Haig, D. (1999). Asymmetric relations. *Evolution and Human Behavior*, **20**(2), 83–98.

Haig, D. (2000). Genomic imprinting, sex-biased dispersal, and social behavior. *Annals of the New York Academy of Sciences*, **907**(1), 149–163.

Haig, D. (2002). *Genomic Imprinting and Kinship*. Piscataway, NJ: Rutgers University Press.

Haig, D. (2003). On intrapersonal reciprocity. *Evolution and Human Behavior*, **24**, 418–425.

Haig, D. (2010). Transfers and transitions: Parent–offspring conflict, genomic imprinting, and the evolution of human life history. *Proceedings of the National Academy of Sciences*, **107** (Suppl. 1), 1731–1735.

Haig, D., & Wharton, R. (2003). Prader–Willi syndrome and the evolution of human childhood. *American Journal of Human Biology*, **15**(3), 320–329.

Hamilton, W. D. (1987). Discriminating nepotism: Expectable, common, overlooked. In D. J. C. Fletcher & C. D. Michener, eds., *Kin Recognition in Animals*. New York: Wiley, pp. 417–438. Reprinted in Hamilton, W. D. (2001). *Narrow Roads of Gene Land: Volume 2, Evolution of Sex*. Oxford: W.H. Freeman, pp. 345–366.

Horvath, S. (2013). DNA methylation age of human tissues and cell types. *Genome Biology*, **14**(10), R115.

Horvath, S. (2015). Erratum to: DNA methylation age of human tissues and cell types. *Genome Biology*, **16**, 96.

Hupkes, M., Jonsson, M. K., Scheenen, W. J., et al. (2011). Epigenetics: DNA demethylation promotes skeletal myotube maturation. *FASEB Journal*, **25**(11), 3861–3872.

Kong, A., Steinthorsdottir, V., Masson, G., et al. (2009). Parental origin of sequence variants associated with complex diseases. *Nature*, **462**(7275), 868–874.

Mackay, D. J., & Temple, I. K. (2010). Transient neonatal diabetes mellitus type 1. *American Journal of Medical Genetics Part C: Seminars in Medical Genetics*, **154C**(3), 335–342.

Maeda, T., Oyama, J., Higuchi, Y., et al. (2011). The physical ability of Japanese female elderly with cerebrovascular disease correlates with the telomere length and subtelomeric methylation status in their peripheral blood leukocytes. *Gerontology*, **57**(2), 137–143.

McNamara, P. (2014). Comment on David Haig's troubled sleep: Implications for functions of infant sleep. *Evolution, Medicine, and Public Health*, **2014**(1), 54–56.

Mehr, S. A., Kotler, J., Howard, R. M., Haig, D., & Krasnow, M. M. (2017). Genomic imprinting is implicated in the psychology of music. *Psychological Science*, **28**(10), 1455–1467.

Micheletti, A. J., Ruxton, G. D., & Gardner, A. (2017). Intrafamily and intragenomic conflicts in human warfare. *Proceedings of the Royal Society B: Biological Sciences*, **284**(1849), 201z2699.

Monk, D., Arnaud, P., Frost, J., et al. (2009). Reciprocal imprinting of human *GRB10* in placental trophoblast and brain: Evolutionary conservation of reversed allelic expression. *Human Molecular Genetics*, **18**(16), 3066–3074.

Morales-Lara, D., De-La-Peña, C., & Murillo-Rodríguez, E. (2018). Dad's snoring may have left molecular scars in your DNA: The emerging role of epigenetics in sleep disorders. *Molecular Neurobiology*, **55**(4), 2713–2724.

Morison, I. M. (2001). The imprinted gene and parent-of-origin effect database. *Nucleic Acids Research*, **29**(1), 275–276.

Murphy, S. K. (2006). Callipyge mutation affects gene expression in cis: A potential role for chromatin structure. *Genome Research*, **16**(3), 340–346.

Park, S. S., Skaar, D. A., Jirtle, R. L., & Hoyo, C. (2017). Epigenetics, obesity and early-life cadmium or lead exposure. *Epigenomics*, **9**(1), 57–75.

Pepper, G. V., & Nettle, D. (2017). The behavioural constellation of deprivation: Causes and consequences. *Behavioral and Brain Sciences*, **40**, e346.

Portela, A., & Esteller, M. (2010). Epigenetic modifications and human disease. *Nature Biotechnology*, **28**(10), 1057–1068.

SanMiguel, J. M., & Bartolomei, M. S. (2018). DNA methylation dynamics of genomic imprinting in mouse development. *Biology of Reproduction*, **99**, 252–262.

Santos, L., Elliott-Sale, K. J., & Sale, C. (2017). Exercise and bone health across the lifespan. *Biogerontology*, **18**(6), 931–946.

Small, K. S., Hedman, Å. K., Grundberg, E., et al. (2011). Identification of an imprinted master trans regulator at the *KLF14* locus related to multiple metabolic phenotypes. *Nature Genetics*, **43**(6), 561–564.

Sone, S. (1994). Muscle histochemistry in the Prader–Willi syndrome. *Brain & Development*, **16**, 183–188.

Swartz, J. R., Hariri, A. R., & Williamson, D. E. (2017). An epigenetic mechanism links socioeconomic status to changes in depression-related brain function in high-risk adolescents. *Molecular Psychiatry*, **22**, 209–214.

Taylor, S. M., & Jones, P. A. (1979). Multiple new phenotypes induced in 10T1/2 and 3T3 cells treated with 5-azacytidine. *Cell*, **17**, 771–779.

Tucci, V. (2016). Genomic imprinting: A new epigenetic perspective of sleep regulation. *PLoS Genetics*, **12**(5), e1006004.

Ubeda, F., & Gardner, A. (2011). A model for genomic imprinting in the social brain: Adults. *Evolution*, **65**, 462–475.

Waterland, R. A., & Jirtle, R. L. (2003). Transposable elements: Targets for early nutritional effects on epigenetic gene regulation. *Molecular and Cellular Biology*, **23**(15), 5293–5300.

Wilson, E. M., & Rotwein, P. (2006). Control of MyoD function during Initiation of muscle differentiation by an autocrine signaling pathway activated by insulin-like growth factor-II. *Journal of Biological Chemistry*, **281**(40), 29962–29971.

Zeng, H., Irwin, M. L., Lu, L., et al. (2012). Physical activity and breast cancer survival: An epigenetic link through reduced methylation of a tumor suppressor gene *L3MBTL1*. *Breast Cancer Research and Treatment*, **133**(1), 127–135.

24 Evolutionary Explanations for Bullying Behavior

PETER K. SMITH

Research on bullying, mostly focusing on children of school age, has been active since the 1970s. Paralleling earlier work on aggression, bullying has often been described as maladaptive and dysfunctional behavior, and this has informed some intervention efforts. However, and again as for aggression generally, this view has been challenged in the 2000s (Ellis et al., 2012; Hawley, Little, & Rodkin, 2007; Kolbert & Crothers, 2003; Volk et al., 2012). It has been argued that bullying behavior is universal (historically and culturally, as well as in contemporary urban societies); that it is heritable, perhaps in part via temperament; and that it can have advantages for those who bully. The advantages would ultimately be for reproductive success, but via physical resources and social status, as well as attractiveness to the opposite sex. These arguments acknowledge that there are variations among those who bully (e.g., the subset of bully–victims may indeed be acting in dysfunctional ways) and among different types of bullying. It is also acknowledged that bullying can bring costs as well as benefits, and it is argued that it is a facultative adaptation. Several implications for anti-bullying interventions have been discussed, arising from these views. This chapter will describe these views and consider them in the light of available evidence. First, bullying will be defined and some relevant research findings about bullying will be described.

24.1 WHAT BULLYING IS AND WHAT WE KNOW ABOUT IT

Bullying is generally defined as a subset of aggressive behavior – so, behavior with an intent to harm another person. It is a subset of aggression, but to be called bullying, most researchers agree that two further criteria need to be satisfied – namely repetition and an imbalance of power. This definition stems from the early work of Olweus (1999) and is widely although not universally accepted (Smith, 2014). Bullying is thus distinguished from more commonplace acts of aggression that occur infrequently or between children of roughly equal status and power.

The main types of traditional or offline bullying described are physical (hitting, kicking), taking or damaging belongings, extortion, verbal threats or harassment, social exclusion, and spreading nasty rumors. While physical bullying and damaging belongings are offline, the other forms can also occur online in cyberbullying, via mobile phones and the Internet (Kowalski et al., 2014). Cyberbullying can include quite sophisticated procedures such as masquerading as someone else or sending compromising images that might or might not be genuine.

Various roles have been delineated in bullying (Salmivalli et al., 1996). These roles may be identified via anonymous questionnaires or by peer nomination procedures from classmates (Smith, 2014). The role of the bully has been split into ringleader bullies (who organize the bullying and initiate it), followers (who join in after the ringleader), and reinforcers (who laugh and encourage bullying without joining in further). In addition to victims of bullying, some children may be identified as bully–victims – nominated as both bullies and as victims to a significant extent. These children may sometimes be labeled as "provocative victims," in that their behavior may provoke attacks or retaliation that they or others interpret as bullying. Some studies therefore delineate "pure bullies" and "pure victims" as those not appreciably nominated for the other role, so as to avoid overlap with the "bully–victim" category; this may be especially important when considering possible correlates or outcomes of behavior in these roles.

Bullying happens in a social context and usually some other children see it happening. Besides the roles associated directly with bully and victim, many studies have looked at roles of bystanders (who see the bullying but stay out of it entirely) and defenders (who help the victim or challenge the bully), whereas the outsider role may refer to those who have not seen the bullying happen.

The prevalence of these roles naturally depends very much on the criteria used for role assignation. It is also affected by age, gender, school context, societal context, and other factors (Hong & Espelage, 2012). As a rough estimate, and using reasonably stringent criteria, one might expect around 10 percent of pupils to be victims, 5 percent to be bullies, and 1–2 percent to be bully–victims. Also, passive bystanders are likely to considerably outnumber defenders.

Being in one of these roles matters. Generally, and especially if the bullying is severe or prolonged, victims of bullying have low self-esteem and feel depressed. Their schoolwork is likely to suffer and there are risks of self-harm, suicidal ideation, and, in some cases, actual suicide. These effects are well documented and the causal influence strengthened by longitudinal studies (Hellfeldt et al., 2018; Zwierzynska et al., 2013). In addition, they may be long term, with lack of trust in relationships and depression more likely in adult life (Bowes et al., 2015; Ttofi et al., 2011).

However, the correlates and outcomes of being a bully are more debated. Some bullies are popular, at least in the sense of perceived popularity; they (more likely the ringleader bullies) are seen as powerful in the peer group and able to command social support, even if not necessarily being liked by many children (as in sociometric popularity; e.g., Peeters et al., 2010; Veenstra et al., 2005). Findings about self-esteem are varied, but may depend on which aspects of self-esteem are measured; children who bully others may recognize that their behavior is not approved of by the school and may not rate themselves highly in terms of academic self-worth, but may score at normal or high levels in others aspects of perceived competence or self-esteem in regard to athletic skills or peer-group relations (O'Moore, 2000; Salmivalli et al., 1999). Findings on depression and other aspects of physical and mental health are also mixed, as will be discussed in detail in Section 24.6.3. What is well established, however, is that bullying children are likely to be aggressive or rule-breaking in other ways, and to be at greater risk of offending and of violent behavior in later life (Ttofi et al., 2012).

The effects of bullying are not limited to those who are directly involved. Others (bystanders and outsiders) are likely to see it or at least know about it, and if no action is taken, the implicit lesson will be that this kind of abuse of power is tolerated. It affects the climate of the school (Rivers et al., 2009). Because of all of these likely consequences of bullying, and out of concern for the rights of all children not to be subject to such kinds of abuse (Greene, 2006), there have been efforts in many countries to take actions against it.

Actions against bullying cover a range of options. Although causal factors involve societal factors (e.g., socio-economic inequality; Elgar et al., 2009) and family factors (e.g., parent and sibling relations; Lereya et al., 2013; Wolke et al., 2015), most interventions have focused on the school itself. They range through school policies, curriculum work, playground supervision, peer support systems, assertiveness training, disciplinary sanctions, restorative approaches, and counseling-based approaches, as well as e-safety and similar measures specifically for cyberbullying.

Many schools pick and choose these options, but there are also many programs that package options in a systematic way. Some notable ones are the Olweus program in Norway, KiVa in Finland, ViSC in Austria, Steps to Respect in the USA, and Friendly Schools in Australia. These programs have had variable success; some have had little impact, while some have shown success in reducing bully and victim prevalence rates by up to 50 percent (Smith, 2019). A meta-analysis by Ttofi and Farrington (2011) suggested average success rates of about 20 percent; in other words, worthwhile, but with much more still to be done.

There are well-documented age trends in bullying. Although aggression is common in preschool and infant school ages, bullying as a systematic kind of behavior against a victim is not so obvious, and roles are poorly defined (Monks, 2011). However, bullying roles are relatively clear and stable by middle childhood, from seven or eight years of age. The incidence of reports of being a victim are relatively high at these younger ages, but this may be because younger children define bullying rather broadly, including one-off attacks or fights between equals. There is usually some decrease in victim rates through the adolescent years; however, those who are victims in adolescence may be in a more difficult position, since they may have been in the role for some time, and attitudes to victims are less sympathetic in adolescence. In contrast to victim rates, rates of bullying do not show a consistent decrease with age through adolescence (Smith, Madsen, & Moody, 1999).

There are also well-established gender differences in bullying. Boys are more involved in physical forms of bullying, but there is little gender difference in verbal forms. As regards social exclusion, rumor spreading, and cyberbullying (often on social networking sites), findings are mixed, with some studies finding girls to be more involved, but others not; however, at least relative to physical bullying, girl bullies seem to prefer these more relational forms. Physical strength is more important in boys' peer groups and reputation is more important in girls' peer groups, so this gender difference has been explained as each gender choosing the most effective method of displaying power and/or damaging a victim (Archer, 2009; Guerra et al., 2011).

Both age and gender differences appear when assessing attitudes to bullying, as found in a number of countries. Most children may say they feel some sympathy for victims, with this being more so for girls than boys. However, such pro victim attitudes decrease with age, reaching a low point around 14–15 years, before slowly climbing again (Rigby, 1997). This age trend is similar in both boys and girls. Given the heightened importance of peer-group status and conformity in adolescence, this would be in line with the status power of bullies and a greater unwillingness to side with less powerful victims in mid-adolescence.

24.2 (MAL)ADAPTIVENESS OF AGGRESSIVE BEHAVIOR

Aggressive behavior is commonplace in all species of living things and serves a number of obvious functions for the individual. It is commonplace in human behavior, and even though the extent and inevitability of aggression is debated (Fry, 2004), it can have benefits for individuals or groups in terms of safeguarding or acquiring resources useful for survival and reproduction, even at the expense of others (Archer, 1988). Nevertheless, religious and moral principles and ethical values tend to devalue or forbid aggressive behavior, perhaps as a kind of "social preaching" (Campbell, 1975) that exhorts individuals to behave in ways that may not suit best their individual or immediate family interests, but promote a more harmonious larger society.

In child development research especially, aggressive behavior has been seen as socially undesirable. This is perhaps relatively uncontentious, although there could and should be more debate about the extent to which all aggressive behavior should be socially sanctioned (Smith, 2007). But many researchers in developmental psychology have gone further and seen aggressive behavior as a sign of the aggressive child lacking social skills or being socially incompetent. In particular, the social skills model of Crick and Dodge (1994, 1996) was used to suggest that aggressive children had some deficits at various stages of the interactional social information processing model. The implication was that training aggressive children in skills such as perspective taking and in conflict resolution strategies would be a good approach to changing their behavior.

This perspective changed radically by the 2000s, as both the weight of evidence and different disciplinary perspectives brought about a reassessment of children's aggressive behavior. This was encapsulated in the edited collection by Hawley, Little, and Rodkin (2007). Much evidence showed that some aggressive children had high status in their peer groups and had some measure of popularity. Kaukiainen et al. (1999) demonstrated that social intelligence correlated positively with relational aggression in particular. Some supposed deficits in social skills that aggressive children had might simply reflect different evaluations of aggressive and nonaggressive behavior choices. Thus, "socially undesirable" did not necessarily mean "socially incompetent." Indeed, Hawley's research with young children suggested that what she called "bistrategic controllers," who could flexibly use both pro-social and aggressive behaviors with peers, were the most successful by several criteria (Hawley, 2003). Hawley specifically invoked evolutionary theory as a background to her research. An overview of this perspective on aggression is provided by Bjorklund and Hawley (2014).

24.3 (MAL)ADAPTIVENESS OF BULLYING BEHAVIOR

Some aspects of the above critique also apply to bullying behaviors. Bullying is a severe and unfair form of aggressive behavior and appears even more socially undesirable than aggressive behavior generally. But bullying need not be socially incompetent. An early demonstration of this was provided by Sutton, Smith, and Swettenham (1999b), who found that in a sample of English 8–11-year-olds, ringleader bullies scored highest on theory-of-mind tasks; they argued that ringleader bullies in particular might deploy skills such as knowing how to choose vulnerable victims, how to choose the right social context to initiate bullying, how best to put down the victim in a way that demonstrated their own power, and how to avoid detection by adults. An exchange on the significance of these findings (Crick & Dodge, 1999; Sutton et al., 1999a, 1999c) exemplified different and shifting views at the turn of the century.

Since then, considerable evidence has come to light that some children who bully others have high status and are perceived as popular in their peer groups. There have also been a number of publications invoking evolutionary theory as a backdrop to explaining bullying behavior. An early example was Kolbert and Crothers (2003). Subsequent work by Volk, Ellis, and others has developed this line of reasoning further.

24.4 ARGUMENTS FOR THE ADAPTIVENESS OF BULLYING BEHAVIOR AND A POSSIBLE EVOLUTIONARY EXPLANATION

Kolbert and Crothers (2003) used an evolutionary psychology perspective to examine bullying behavior. Their main hypothesis was that social dominance in children and early adolescence was established through bullying. They argued that this was consistent with the nature of bullying behaviors (demonstrating power over a weaker opponent), the perceived popularity of many bullies, and the age trends and gender differences. An interesting subsidiary hypothesis was that physical bullying in boys and relational bullying in girls were mainly used to assert or establish dominance, while verbal bullying was mainly used to maintain dominance. Kolbert and Crothers also suggested implications for interventions against bullying, notably questioning the value of assertiveness training for victims (as potentially dangerous) and advocating ways of encouraging bullies to maintain their status in pro-social rather than aggressive ways.

Another hypothesis presented by Kolbert and Crothers was that bullying might have advantages for victims. While acknowledging the damage caused to victims, they suggested that it would be valuable for them to learn their place in the dominance hierarchy and stay safe from harm. However, this hypothesis seems much more questionable. It is true that in many nonhuman species the dominance hierarchy effectively functions as a way of avoiding very costly physical fights between individuals; an animal lower in the hierarchy, if it flees an encounter, will escape further attacks, although at the cost (for the time being at least) of losing out on reproductive opportunities. But human victims of bullying are often in a different situation. They cannot easily "escape" in the same way, and attacks may continue – indeed, crying and other signs of submissiveness seem to encourage bullies rather than causing the bullying to stop (Smith, Shu, & Madsen, 2001). It is the perpetrators of bullying who initiate such behaviors and may benefit from them, and it is the possible evolutionary adaptiveness of bullying perpetration that has been developed subsequently and will be considered further here.

The evolutionary approach was developed further by Volk et al. (2012). They provided a broad-ranged argument that bullying, especially in adolescence, was a facultative evolutionary adaptation. They argued that bullying was a universal phenomenon, found in hunter–gatherer societies and throughout different historical epochs. They also argued that it was prevalent – not, of course, that all children were bullies, but that it was found at appreciable levels rather than being a very occasional phenomenon.

In fact, the extent of bullying in hunter–gatherer societies is disputed. Volk et al. (2012, p. 233) concede that "overt/direct forms of bullying are rare in hunter–gatherers," although they point out that it can happen in conditions of extreme food scarcity. Many authors consider hunter–gatherers (unlike settled agricultural communities) to be quite egalitarian and not concerned with gaining status (Fry, 2004). However, bullying is certainly common in contemporary societies, although it varies considerably in prevalence (Smith, 2014).

Volk et al. argued that this provided a prima facie case for considering bullying as an adaptation. The argument was then developed by arguing that: (1) bullying had a heritable basis; and (2) bullying had advantages for survival and reproductive success. The advantages were considered under three aspects: gaining dominance and social status (and hence access to resources); individual growth, strength, and health; and individual reproductive success (using measures such as attractiveness to and dating or sexual opportunities with the opposite sex). More succinctly, these were termed as reputation, resources, and reproduction (Volk et al., 2015).

Volk et al. argued that bullying was a facultative adaptation, not only because most children were not bullies, but also because levels of bullying varied considerably in different societies. A decision to bully would be based (largely unconsciously) on a cost/benefit analysis of the consequences. This naturally led to a discussion of how anti-bullying interventions might be more successful, emphasizing more pro-social status routes for bullies rather than social skills training.

This approach was also expounded in a more general discussion of the evolutionary basis of risk-taking behavior by Ellis et al. (2012). It has been developed in later papers by Volk and other colleagues. Volk et al. (2016) provide an overview, and they also develop the implications of this approach for intervention work (considered in Section 24.8). Next, the evidence base for various assumptions of the evolutionary approach are examined, starting with the evidence for a genetic basis.

24.5 GENETICS OF BULLYING AND VICTIMIZATION

Behavioral genetic research in this area has used the classic twin method, which compares concordance in a characteristic for monozygotic (genetically identical) and dizygotic (nonidentical) twins. The variance in a measured trait can then be allocated to genetic factors, shared environment (common to the twins), or non-shared environment (experienced differently by the twins). Data concerning bullying and victimization have come from two major studies: the Quebec Newborn Twin Study, a population of 648 twin pairs born in the greater Montreal area of Canada between 1995 and 1998; and the Environmental Risk (E-Risk) Longitudinal Twin Study of 1,116 families with same-sex twins born in England and Wales during 1994–1995.

Boivin et al. (2012) reported data from the Quebec study on peer victimization using peer-, teacher-, and self-report data at kindergarten, grade 1 and grade 4 (about 6, 7 and 10 years of age). The estimates of genetic variance varied considerably by informant; however, for peer nominations, there was a clear increase with age in proportion of genetic variance, with this being 21 percent at 6 years, 29 percent at 7 years, and 66 percent at 10 years. This study did not report data on bullying others.

Ball et al. (2008) reported data from the E-Risk study, where being victimized by peers was assessed at 9–10 years by mother reports. Their best-fit model suggested that 73 percent of the variance was genetic. They also assessed bullying others, from a combination of mother and teacher reports; here, they found 61 percent of the variance was genetic, with 39 percent ascribed to non-shared environment (and zero to shared environment).

There are limitations to twin studies, one being how readily one can generalize from a twin population to the majority non-twin population. Also, the proportions of variance do depend on the models used. Nevertheless, it seems to be reasonably well established that there are genetic factors involved in being a victim, with the Ball et al. (2008) study indicating that this is so for being a bully. The genetic contribution to being a victim increases through the primary school years, probably reflecting the more transient nature of peer victimization in the four- to six-year period and its more stable nature by middle childhood.

24.5.1 Possible Factors Underlying Heritability

Any genetic factors must operate through various mechanisms. Ball et al. (2008) suggest a number of these: for being a victim, they mention introverted personality, social cognitive deficits, and emotional regulation and displays. For bullying, they mention impulsivity and sensation-seeking, biases in social cognitions, and low emotionality and poor emotional regulation. They also found a limited correlation between the bully and victim measures of r = 0.25; they suggested that this modest linkage was largely genetic and that it might be due to a common factor of emotional dysregulation (also found to be highly heritable). This would apply to the bully–victim category. So far as pure bullies are concerned, Volk et al. (2012) have suggested aspects of temperament and personality and general aggressiveness as being partially heritable traits that could facilitate bullying.

It is quite plausible that some of the personality characteristics suggested above could underlie the findings of heritability regarding bullying behavior. Early research had indicated that bullying others was associated with having a hot-tempered personality (Olweus, 1993), readily attributing hostile motives to others, and having defensive egotism – reacting angrily to minor threats to self-esteem (Salmivalli et al., 1999). Jolliffe and Farrington (2011) found impulsivity to be the most important predictor of bullying from a range of factors examined in English adolescents, including cognitive and affective empathy and parental supervision.

Another relevant dimension of personality has been Machiavellianism, generally defined as thinking that other people are untrustworthy and can be manipulated in interpersonal situations. Links to bullying have been found in a Scottish sample of 9–12-year-olds (Sutton & Keogh, 2000) and a Greek sample of 9–12-year-olds (Andreou, 2004). More recent work has focused on the "Dark Triad," comprising Machiavellianism, narcissism (sense of entitlement and superiority), and psychopathy (impulsive, thrill-seeking, low empathy, and low anxiety). Of these, psychopathy has been the most strongly implicated in cyberbullying in US college students (Gibb & Devereux, 2014). The related concept of callous–unemotional traits, and particularly the uncaring subscale of this, was associated with bullying in English 11–12-year-olds by Muñoz, Qualter, and Padgett (2011), even when taking empathy deficits into account.

The Big Five personality factors (friendliness, emotional instability, intellectual openness, energy, and conscientiousness) have also been linked to bullying. Tani et al. (2003) related pro-bully roles (bully, assistant, reinforcer) in 8–10-year-old Italian children to high levels of emotional instability and low levels of friendliness. A similar personality battery is the HEXACO measure, which assesses six factors (honesty–humility, emotionality, agreeableness, conscientiousness, extraversion, and openness). In a study with Canadian 10–18-year-olds, Book, Volk, and Hosker (2012) found bullies to be particularly low on honesty–humility, and also low on emotionality, agreeableness, and conscientiousness. Subsequently, Farrell et al. (2014) examined this relationship separately for five subtypes of bullying; all five correlated negatively with honesty–humility.

These personality traits have some degree of heritability and may facilitate bullying. A heritable trait would be maintained in a population if (at least under some circumstances) it brings benefits. Next, we will examine the various benefits (and costs) proposed for bullying behavior.

24.6 ARE THERE BENEFITS AND COSTS TO BEING A BULLY?

In considering this, and especially any possible advantages or benefits to bullying behavior, it is important to consider subtypes of bullies and especially the bully–victim category. The evolutionary arguments of Volk et al. (2012) acknowledged the very considerable empirical evidence that bully–victims score poorly on many measures of adaptive functioning. Unfortunately, quite a number of studies on correlates of bullying, which show bullies scoring high on measures such as depression, did not separate out bully–victims from bullies. A meta-analysis by Gini and Pozzoli (2009) found bullies to have a higher risk of psychosomatic symptoms, but did not clearly separate out studies that distinguished bully–victims from the bully category. Similarly, the extensive meta-analysis by Cook et al. (2010) ostensibly separated out correlates for bully, victim, and bully–victim based on 153 independent samples, but only 31 included bully–victims as a category, and these studies were not separated out from the remainder. Furthermore, the meta-analysis by Ttofi, Farrington, and Lösel (2011) on bullies and victims included many studies where the pure bully and bully–victim categories were considered together.

This is not such an important issue when positive outcomes are reported for bullies generally. For example, if bullies are found to have high perceived popularity, even if the study did not separate out bully–victims, it is very likely that bully–victims are not perceived as popular, so if anything the finding for pure bullies would be strengthened. But this is not the case for negative findings. If bullies generally are found to be more depressed than noninvolved children (as in Cook et al., 2010; Ttofi et al., 2011), this could conceivably be entirely due to the inclusion of bully–victims (who, when considered separately, do score high on depression; Cook et al., 2010).

In considering the following studies, control for the bully–victim category is not considered essential when advantages to being a bully are considered. However, for costs to being a bully, only studies that separate pure bullies and bully–victims are considered worth reporting. There is no current meta-analysis of such studies; included here are those studies that are cited frequently in the relevant literature.

24.6.1 Bullies, Reputation, and Resource Control

A number of studies consistently show that many bullies score high in perceived popularity – they are seen as popular. They are also seen by classmates as having high status and being influential. This does not necessarily mean that they are strongly liked by many children (which is measured by sociometric popularity).

For example, Vaillancourt, Hymel, and McDougall (2003) surveyed Canadian pupils aged 11–17 years (grades 6–10). They obtained peer nominations of bullying and social status. They delineated subgroups of pupils nominated as bullying others in terms of perceived social power – high power, medium power, and low power. The high-power bullies were high on perceived popularity and also better liked compared to low-power bullies, and they were also high on assets such as being physically attractive, wearing stylish clothing, and being good athletes. Juvonen, Graham, and Schuster (2003) found corresponding results in a US sample, with bullies being highest on social status, although not particularly liked in terms of peers saying they might avoid them.

Olthof et al. (2011) examined peer-nominated participant roles, including ringleader, assistant, and reinforcer bullies, victims, and bully–victims, among Dutch children aged 10–12 years. Rather consistent differences were found for the three main measures of perceived popularity, resource control, and desired dominance. Ringleader bullies scored highest on all of these, followed by reinforcers and assistants (with similar scores). Bully–victims scored next highest on desired dominance and resource control but lower on perceived popularity, where defenders scored next highest.

In another Dutch study, Reijntjes et al. (2013a, 2013b) distinguished three trajectories of pupils followed in three waves over two years (ages 10–12 years); they used peer nominations of bullying scores. Most pupils (82 percent) had consistently low bullying scores, another 11 percent showed consistently moderate scores, and a third group (7 percent) had consistently high scores. Teachers rated the high-bullying group as being high in resource control (2013a), and peers rated them as high in perceived popularity (2013b).

24.6.2 Bullies and Social Competence and Theory of Mind Skills

Studies in this area have yielded mixed findings. Besides the issue of separating out bully–victims from pure bullies, the nature of the measures used needs to be carefully considered. Consider the term "social competence." This was assessed by Haynie et al. (2001) in a study that did separate out bully–victims. The sample item given for their measure of social competence was "Compared with other kids, resisting dares from other kids is (much harder to much easier)" (p. 36). Presumably answering that for you it is "much easier" to resist dares gives the highest social competence score, but in terms of social status this could well be problematic – rising to a dare might be a necessary part of maintaining high status. In fact, many such social competence or social problem-solving measures may confound what is socially desirable from an adult viewpoint (e.g., selecting nonaggressive response options) with actual social skill (Hawley, 2003; Smith, 2007). Another example (also from Haynie et al., 2001) is "self-control." Their sample item for this is "I lose my temper and 'let people have it' when I am angry (never, some of the time, most of the time, always)" (p. 36). Losing one's temper easily may not be socially desirable, but it is not clear

that it adversely affects status, and perhaps the opposite is true. Given this pervasive problem, studies on these kinds of measures of social competence are not considered further.

Such difficulties do not apply to assessing theory of mind, which depends on more value-neutral measures such as understanding double-bluff stories or reading emotions from the eyes. Following the report from Sutton et al. (1999b) that ringleader bullies scored highest among the participant roles on advanced theory of mind skills, there have been further studies using similar measures. In a narrative review, Smith (2017) located nine studies in total that assessed bully role. Of these, five reported some evidence linking bullying to better theory of mind scores; one found an interaction effect of bullying with levels of narcissism; two gave nonsignificant results; and one reported a negative correlation, which disappeared after accounting for socioeconomic status and child maltreatment. The review concluded that "Any assumption that bullies are generally poor in theory of mind skills seems strongly counter-indicated by the available evidence" (p. 94). Three of the five positive studies specifically assessed ringleader bullies, for whom the association with high theory of mind scores seems best supported.

24.6.3 Bullies and Physical and Mental Health

This is by far the most contentious area of research. Challenging the earlier consensus of bullies as showing maladaptive behavior, some researchers have since written about the "superior mental health (lack of psychological stress) of bullies" (Juvonen et al., 2003, p. 1235), or that "being a bully promotes better mental health" (Koh & Wong, 2017, p. 14). Volk et al. (2012) concluded that "bullies tend to have some health benefits" and "no more total health problems than neutral, uninvolved children," and that "pure bullies (excluding bully–victims) report equal or better mental health than uninvolved adolescents" (p. 225).

However, these conclusions drew on specific studies rather than a broad review of the evidence. It is certainly important to only look at studies that separate pure bullies from bully–victims. In the following, only such studies are considered. Findings (in chronological order), together with the main aspects of each study, are summarized in Table 24.1. The focus is on how pure bullies compare with noninvolved children. In the great majority of studies, bully–victims and victims scored the worst on health measures; a few exceptions are mentioned in the narrative review that follows.

Kumpulainen, Räsänen, and Henttonen (1999) carried out a four-year longitudinal study. They assessed pupils on the Rutter A2 and B2 (parents; teachers) scales of psychological disturbance (which include "aggressive" items, so of less interest here) and the Children's Depression Scale. Bullies scored less well than noninvolved controls at both time points, although this specific difference was not tested for significance.

Table 24.1 Comparisons of pure bullies and noninvolved children on measures of physical and mental health. Only studies with control for bully–victim status are included.

Study	Country	Date of data collection	N pupils; age	N (%): bullies	Bully data method	Internalizing problems	Physical health
Kumpulainen et al. (1999)	Finland	1989/1993	1,268; 8–12 longitudinal	108 (8.5); 100 (7.9)	Self-report	Bullies worse on depression but not tested for significance	
Forero et al. (1999)	Australia	1996	3,918; mean 11.8	928 (23.7)	Self-report	NS for mental health measure	Bullies worse on composite psychosomatic symptoms
Kaltiala-Heino et al. (1999)	Finland	1997	16,410; 14–16	901 (5.5)	Self-report	Bullies worse on depression and suicidal ideation	
Haynie et al. (2001)	USA	NA	4,263; grades 6–8	142 (25.8)	Self-report	Bullies worse on depressive symptoms	
Wolke et al. (2001)	UK	1996/1998	1,982; 6–9	Direct 70 (4.3), indirect 18 (1.1)	Self-report		NS for physical health NS for psychosomatic symptoms
Juvonen et al. (2003)	USA	2000/2001	1,985; mean 11.5	139 (7)	Peer nomination	Bullies better on depression, social anxiety, and loneliness	
Vaillancourt et al. (2003)	Canada	NA	555; 11–17	51 high-power (9.2)	Peer nomination	NS for measures of depression and loneliness	
Fekkes et al. (2004)	Netherlands	1999	2,766; 9–12	96 (3.5)	Self-report	NS for measure of depression	Bullies worse for headache, bed-wetting; NS for ten other symptoms
Nansel et al. (2004)	25 countries	1997/1998	113, 200, average 4,528 per country; 11, 13, 15	Average 10% by country, range 3–20%	Self-report	Bullies worse on "emotional adjustment"	Bullies worse on "health problems"
Ivarsson et al. (2005)	Sweden	NA	183; 13–16	33 (18.0)	Self-report	Bullies worse on withdrawn and anxious/depressed, but both NS	Bullies higher on somatic symptoms, but NS
Nordhagen et al. (2005)	5 Nordic countries	1996	10,664; 2–17	294 (4.3)	Parent report	Bullies worse on psychiatric/nervous problems	Bullies worse on hearing impairment
Stein et al. (2007)	USA	2003	1,312 boys; 11–18	299 (22.8)	Self-report	Bullies worse on purpose of life and control of life, but NS	Bullies higher on injuries, but NS
Schnohr and Niclasen (2006)	Greenland	2002	891; 11, 13, 15	59 (6.6)	Self-report		Bullies worse on three health measures, significant for stomach ache; NS for self-rated health

Continued

Study	Country	Year	Sample (N; age)	Measure	Bullies better on mental health measure	Bullies worse on health problems
Volk et al. (2006)	Canada	1998	About 6,500; 12–19	About 396 (6.1)	Self-report		
Espelage and Holt (2007)	USA	NA	684; mean 14.5	102 (14.9)	Self-report	Bullies worse on anxiety/depression	
Liang et al. (2007)	South Africa	NA	5,074; 14 and 17	418 (3.2)	Self-report	Bullies worse on suicidal ideation and suicidal attempt, but NS in regression model	
Gini (2008)	Italy	2006	565; mean 9.6	63 (11.2)	Self-report		Bullies worse on six psychosomatic symptoms, only significant for sleeping problems, feeling tense
Wang et al. (2011)	USA	2005/2006	7,313; mean 14.2	Frequent bullies 1.9–5.2% by type	Self-report	Bullies worse on depression scale	
Winsper et al. (2012)	UK	2002/2003	6,043; 10–13	At 8 years 55 (1.1), at 10 years 52 (0.9)	Self-, mother, and teacher report	Bullies consistently worse on suicidal ideation and behavior, some comparisons significant	
Copeland et al. (2013)	USA	1993/2010	1,420; 9, 11, 13	112 (5.0) or 100	Self- and parent report	Bullies higher on depression, anxiety, suicidality in childhood, but all NS. No effects as adults	
Kowalski and Limber (2013)	USA	2007	931; 11–19	156 (16.8) traditional; 54 (5.8)	Self-report	Traditional bullies worse on anxiety (NS) but better on depression and worse on suicidal ideation; cyberbullies very slightly worse on anxiety (NS), worse on depression (NS), and worse on suicidal ideation	
Yang et al. (2016)	Finland	NA	17,586; 9–15	733 (4.2) self-report; 1,499 (8.5) peers; 332 (1.9) profile	3 methods	Bullies worse on composite internalizing symptoms	
Koh and Wong (2017)	Canada	2014	133; 13–16	15 (11.3) or 11	Self-report	Bullies better on depression; worse on social anxiety, but NS	
Le et al. (2017)	Vietnam	2014/2015	1,424; mean 14.7	94 (6.6)	Self-report	NS for depression and psychological distress; worse for suicidal ideation	

NA = not available; NS = nonsignificant.

Forero et al. (1999) used a World Health Organization survey on health behaviors, including eight psychosomatic symptoms, and a four-item mental health measure (on happiness and loneliness). Bullies scored significantly higher (i.e., worse) on composite psychosomatic symptoms, but were no worse on mental health.

Kaltiala-Heino et al. (1999) used the Beck Depression Inventory, which also includes an item (separately analyzed) on suicidal ideation. Bullies scored higher on depression (odds ratio 4.5) and suicidal ideation (odds ratio 8.7); in fact, taking other factors into account, bullies had the highest risk of all roles for suicidal ideation.

Haynie et al. (2001) assessed depressive symptoms using a subscale from the Weinberger Adjustment Inventory; on this, bullies scored significantly higher than comparisons (noninvolved).

Wolke et al. (2001) obtained parental ratings of seven items of physical health (headaches, feeling sick, etc.) and seven items of psychosomatic health (bed-wetting, poor appetite, etc.). They also distinguished direct bullies and relational bullies. Direct bullies had the fewest total physical health problems and total psychosomatic problems, but neither were significantly different from noninvolved; and the superior health of direct bullies was slightly reversed in regression analyses accounting for gender, school year, and ethnic minority.

Juvonen et al. (2003) used the Children's Depression Inventory, Social Anxiety Scale for Adolescents and a 16-item loneliness measure. Bullies scored lowest on both of these.

Vaillancourt et al. (2003) used the Child Depression Inventory and a Loneliness and Social Dissatisfaction Scale to compare bullies of high, moderate, or low power (rather than separating out bully–victims). High-power bullies scored higher on depression but lower on loneliness than low-power bullies, but these differences were not significant. No direct comparison was made to noninvolved pupils, but the overall correlations of the two scale measures with bullying score were both small and nonsignificant.

Fekkes, Pijpers, and Verloove-Vanhorick (2004) assessed depression using the Depression Questionnaire for Children and 12 health symptoms. Bullies scored slightly but not significantly higher on depression and about the same as noninvolved children in most health symptoms, but significantly worse on headaches and bed-wetting.

Nansel et al. (2004) used Health Behaviour of School-aged Children (HBSC) survey data to examine correlates with psychosocial adjustment using data from 25 countries. Correlates were obtained with health problems and emotional adjustment (details of these measures are not given in the report). Bullies scored worse than noninvolved children in health problems (all 25 countries) and emotional adjustment (all 25 countries).

Ivarsson et al. (2005) compared groups on the Achenbach Youth Self-Report Scale. Findings for an internalizing subscale of this (withdrawn, somatic complaints, anxious/depressed) found both girl and boy bullies scoring higher on each of these than noninvolved pupils, although not to a statistically significant extent.

Nordhagen et al. (2005) reported data from a comparative study in Norway, Sweden, Denmark, Finland, and Iceland. A variety of parent-reported health and welfare measures were compared across groups. Bullies were generally reported as not significantly different from controls, but they did show significantly higher levels of hearing impairment and psychiatric/nervous problems.

Stein, Dukes, and Warren (2007) gathered data on psychosocial health, including a purpose in life scale (my life is empty/exciting), a control of life scale (in my hands/out of my hands), and an injury scale. Bullies scored worse than neutrals (noninvolved) on all of these scales, but differences were not statistically significant.

Schnohr and Niclasen (2006) used HBSC survey data to assess frequencies of headache, stomachache, and sleeping difficulties, as well as self-rated health. Bullies scored worse on all three health measures (unusually, worse than victims and bully–victims as well as noninvolved children), but only significantly for stomachache. They did not score any higher than noninvolved children on self-rated health, however.

Volk et al. (2006) used HBSC data on a health problems index (headache, stomachache, etc.), on which bullies scored higher. However, they were also the only group scoring lower (better) on mental health problems (feeling nervous, sleeping difficulties, etc.).

Espelage and Holt (2007) found that bullies scored significantly worse than noninvolved children on the Achenbach Youth Self-Report Scale of anxiety/depression.

Liang, Flisher, and Lombard (2007) used a self-report risk behavior questionnaire, including questions on suicidal ideation and suicidal attempt. Bullies had higher risk ratios for these (1.87 and 1.91, respectively) than controls, but these were not specifically tested for significance, and they became smaller (1.15 and 0.98, respectively) in a logistic regression analysis of multiple risk factors.

Gini (2008) obtained self-reports for six psychosomatic complaints. Bullies scored worse on all of these, but only significantly so for two: sleeping problems and feeling tense.

In more use of HBSC data, Wang, Nansel, and Iannotti (2011) distinguished eight mutually exclusive groups: non-involved; occasional or frequent for each of bully, victim, or bully–victim; and mixed. They assessed each on a six-item depression scale. Bullies were more depressed than noninvolved children. Also, frequent bullies scored higher for depression than occasional bullies (for physical, verbal, and relational bullying, but this was not significant for cyberbullying).

Winsper et al. (2012) assessed bullying roles from a combination of self-, mother, and teacher reports at 8 and 10 years and obtained self-reports of suicidal ideation and suicidal/self-injurious behavior (hurt yourself, tried to kill yourself) at 11–12 years. Bullies scored worse on both measures on all comparisons, especially based on measures at 8 years and for self- and teacher reports of bullying (nonsignificant for 10 years and for mother reports for ideation). This remained true after controlling for potentially confounding factors, although now this was only significant for self-reported bullying at 8 years.

Copeland et al. (2013) carried out a longitudinal study from ages 9–16 through to young adulthood (twenties). Depression, anxiety, and suicidality were assessed as part of a Child and Adolescent Psychiatric Assessment, and later a Young Adult Psychiatric Assessment. Bullies scored worse on all of these measures in childhood, but not significantly. Adult outcomes were not significantly different from noninvolved.

Kowalski and Limber (2013) assessed correlates of both traditional bullying and cyberbullying, including measures of physical health (headache, poor appetite, sadness, etc.), depression (Beck Youth Depression Scale), and anxiety (Beck Youth Anxiety Scale), including one item on suicidal ideation. For traditional bullying, bullies scored higher on anxiety (not significantly) but lower on depression, and higher on suicidal ideation. For cyberbullying, bullies scored very slightly higher on anxiety (not significantly), higher on depression (not significantly), and higher on suicidal ideation. Some gender interactions were reported, notably for cyberbullying, where girl bullies only scored higher on anxiety.

Yang, Li, and Salmivalli (2016) gathered data using three identification methods (self, peer, and profile method). They assessed internalizing problems, including self-esteem, as well as depression by the Beck Depression Inventory and a nine-item social anxiety scale. Only a combined score of these was used in analysis, so effects of self-esteem cannot be separated out from depression. Bullies scored worse on internalizing problems than the noninvolved pupils. This was a consistent finding whichever identification method was used, but it was not tested specifically for statistical significance (the focus being on bully–victims).

Koh and Wong (2017) gathered self-report data from a sample of 15 bullies (the sample was apparently smaller, with 11 bullies, when comparisons were made: their table 2). Bullies scored significantly lowest on depression (and highest on global self-esteem and on social status). However, they scored highest on social anxiety, but this difference was not significant. The authors interpreted the findings from the perspective of evolutionary psychology, gaining considerable publicity (Blackwell, 2015), despite the limitations (small sample, all self-report data).

Le et al. (2017) gathered data over a six-month period using the Centre for Epidemiological Studies Depression Scale, the Kessler Psychological Distress Scale, and three items on suicidal ideation. The pure bullies did not show higher levels of depression than noninvolved students, nor did they show higher levels of psychological distress (such as "feeling tired for no good reason"); however, they did

report significantly higher levels of suicidal ideation than noninvolved students.

Examining Table 24.1, a clear majority of studies find pure bullies to be worse on measures of internalizing problems (usually depression), suicidal ideation, and other measures of physical and mental health. Many of these differences are statistically significant, and in cases where they are not, the direction of effect is usually for bullies to have worse scores. Only a few studies find bullies scoring better. However, a possible caveat is that most of the studies relied on self-report data to identify bullies, victims, bully–victims, and noninvolved. Juvonen et al. (2003), commenting on how their findings contrasted with some earlier findings, suggested that the peer nomination method they used was more accurate than self-report data. Vaillancourt et al. (2003) also used peer nominations and got nonsignificant findings. Yang et al. (2016) presented data comparing self and peer methods; self-report yielded a substantial difference for internalizing problems (odds ratio for bullies 1.25, noninvolved 1.03), but this was much less for peer nominations (odds ratio for bullies 1.09, noninvolved 1.07).

In summary, self-report data from a wide range of different countries overwhelmingly suggest that pure bullies score worse on measures of mental and physical health, although they do not score as badly as do victims and bully–victims. However, peer nomination data, which are available from very few studies, are inconclusive.

24.6.4 Bullies and Dating Success

Connolly et al. (2000) reported on dating experiences of Canadian adolescents aged 9–15 years (grades 5–8). Altogether, 196 bullies (134 boys and 62 girls; these included bully–victims) were compared with a matched comparison group who were not bullies or victims. Significant differences were found for age at which they started dating, with bullies starting about five months earlier; bullies also reported more types of dating activities and more time in other-sex contact. For boy bullies only, they more often had a current girlfriend. Findings on puberty were under-analyzed, but it was reported that "bullying was linked to advanced dating activities, independent of the adolescents' reports of their pubertal development" (p. 306). The quality of these dating relationships was reported by bullies as having less intimacy, affection, and commitment, and characterized by more aggression and less equitable power relationships. This latter finding was confirmed by Espelage and Holt (2007), who found that bullies reported more physical dating violence and emotional abuse in dating relationships compared to noninvolved individuals, although bully–victims scored by far the highest on these measures.

Arnocky and Vaillancourt (2012) assessed peer aggression, bullying, and dating status ("going out with someone now") in Canadian adolescents aged 11–14 years (grades 6–9). There were approximately equal numbers of boys and girls among the 310 (in the abstract) or 350 (in the main text) participants. This was a longitudinal study over one year. Both bivariate correlations and longitudinal analysis found that indirect aggression significantly predicted dating success (for both boys and girls); neither physical aggression nor bullying status did so. Unfortunately, the bullying variable would have included bully–victims, and victim status was a significant negative predictor of dating success; it is thus uncertain whether pure bully status would have predicted dating success.

Volk et al. (2015) reported on two studies, one with 334 adolescents aged 11–18 years and a second with 143 (in the main text; 144 in the abstract) college students with a mean age 18.5 years. In both studies, self-reported bullying predicted having dated, number of dating partners, having had sexual activity, and number of sexual partners. The conclusion in this study that "bullying is associated with a 1.5–2× greater likelihood of having had sexual intercourse" (p. 7) is overstated, as what was assessed was "sexual activity of any kind" (p. 4), and the likelihood "should read 1.5×" (Volk, 2016, pers. comm.). Again the bullying variable would have included bully–victims, but this seems less important when positive findings are reported, as if bully–victims are less successful, then this might increase the actual advantage to bullies.

The last conclusion is complicated by a subsequent report by Dane et al. (2017), apparently further analyzing study 1 from Volk et al. (2015) but distinguishing bullies, victims, and bully–victims using either physical or relational bullying in relation to dating or sexual partners. Physical bullying was related to number of dating and sexual partners, especially for boys. Unexpectedly, associations were just as strong for physical bully–victims. Relational bullying was not associated with the number of dating partners and was not significantly associated with number of sexual partners. However, being a relational bully–victim was related significantly to number of dating partners, especially for girls.

24.6.5 Summary of Costs and Benefits of Bullying

Some analyses of this issue have been (and continue to be) compromised by the confounding of what is socially desirable with what is socially skillful or advantageous to the individual in measures such as social competence. However, some conclusions can be made. First, there is considerable evidence that pure bullies (but not bully–victims) are often seen as powerful in the peer group; this does not mean that they are liked a lot, but it does mean that they are seen as exercising resource control. Second, there is evidence that this extends to dating success from a number of studies, although such relationships are characterized by more aggression and abuse. Third, ringleader bullies in particular seem to have good theory of mind abilities, and certainly the stereotype of the bully as socially incompetent seems radically incorrect.

This does not mean that there is no cost to being a bully. Contrary to some suggestions made by some evolutionary psychologists, the majority of studies that have examined the mental and physical health status of pure bullies show that (while they are better off than victims and bully–victims) they have poorer health than noninvolved children. This certainly applies to self-reported bullies; more evidence is needed for peer-nominated bullies.

It is not surprising that there are costs to being a bully. It is a risky strategy, especially if or when schools or the wider social milieu disapprove of bullying. However, the possibility of adopting a risky strategy might be selected for (i.e., a facultative adaptation).

24.7 IS BULLYING REALLY AN EVOLUTIONARY ADAPTATION?

Bullying is a particular form of aggression. The capability for being aggressive is clearly an evolutionary adaptation. But what about bullying?

There is evidence that bullying is heritable, but it rests on one study, which used a less-than-optimal measure of bullying (mother plus teacher report), and we do not know how much of this heritability was carried by general aggressiveness.

We know that bullying has been prevalent across many societies, but it is less clear that it was common in the hunter–gatherer environment that characterized much of human evolution. The ethos here is generally described as egalitarian. Children's peer groups would have been relatively small and multi-aged, and it has been argued that this would encourage cooperation and sharing rather than bullying (Gray, 2011). Such peer groups in hunter–gatherer societies contrast greatly with the large, same-age peer groups found in contemporary schools, where bullying could have fewer costs and greater benefits.

In contemporary urban societies, we have good evidence that bullying can have both benefits and costs for fitness. It can bring power and resource control, including dating opportunities in adolescence, but it can have costs for physical and mental health. It may also bring sanctions from adults.

It may be too early to judge whether bullying is a specific evolutionary adaptation. Conceivably, it is a by-product of the capacity for aggression and of using aggressive behavior in certain circumstances where pursuing dominance by displaying power against weaker peers may bring payoffs. Such behavior might be a relatively new feature in human cultural evolution, facilitated by the large same-age peer groups found in urban schools. Such large groups from which one cannot easily escape, whether in schools, prisons, workplaces, or the armed forces, can provide an opportunity for bullying that is not so prevalently found in other settings (Monks et al., 2009).

24.7.1 Cyberbullying as a New, Virtual Environment

Just as modern schools have provided a new peer group environment, so mobile phones and the Internet have provided a new virtual environment. While these tools are generally beneficial, cyber-aggression, harassment, and cyberbullying are not uncommon and can provide experiences that are just as harmful for victims as offline or traditional bullying (Livingstone & Smith, 2014).

The forms of cyberbullying differ somewhat from traditional bullying. Generally, cyber-victimization and perpetration seem more intertwined in the novel environment of cyberbullying (e.g., Festl et al., 2017; Mishna et al., 2012). Cyberbullying is typically not face-to-face, and it is often done via social networking sites. Much of it is in the nature of nasty forms of gossip that demean and damage someone's reputation. Gossip itself has been described as having evolutionary origins (Dunbar, 2004), and cyberbullying can be seen as a proliferation of this in a new environmental context (Bertolotti & Magnani, 2013). However, it would not be helpful to describe cyberbullying itself as an evolutionary adaptation; rather, it exploits the potential for gossip in a new environment.

24.8 IMPLICATIONS FOR INTERVENTIONS

Arguments about whether bullying is truly an evolutionary adaptation do not matter too much when it comes to implications for intervention. Here, the main message is that bullies can be quite socially skilled, they can gain status and resources through bullying, they may be motivated to do so, and it is not useful to characterize this as pathological behavior. However, it is likely to be risky behavior, and the benefits and costs will depend on context.

Attempts to reduce bullying behavior could work with the individual bullies, the victims, and the wider context. A challenge for working with bullies will be to change behavior in those bullies for whom the bullying is providing benefits and low costs. An analysis of the KiVa program – one of the more successful interventions – found that while it reduced bullying rates in bullies who were less popular or of average popularity in the peer group, it did not do so with bullies who were more popular (Garandeau et al., 2014). Ellis et al. (2016) argue that, especially in secondary schools, the status benefits from bullying are such that alternative, pro-social, status-enhancing activities need to be provided for such popular bullies. They describe a "meaningful roles" approach: all pupils, but especially those who are inclined to bully, are given pro-social activities with some responsibility that helps them have respect in the peer group, such as helping run activities, organize sporting competitions, or acting as peer tutors. However, the feasibility and success of such an approach in reducing bullying behavior still need to be ascertained.

Bullies need victims, and many typical victims lack friendship skills and coping strategies. Assertiveness training and buddying schemes may help victims and potential victims be less rewarding targets for bullies. Peer defenders can be helpful, although they are only likely to be effective if they themselves are of high status in the group. More widely, the climate of the peer group will be important. If bullies do not have at least perceived popularity in the peer group, motivation for continued bullying will be reduced. Clear sanctions against bullying and effective school policies are widely recognized to be important.

The distinctive input from the evolutionary approach here is to suggest that, when acting to change the cost/benefit nature of the environment, we must be fully cognizant of what these costs and benefits are for the bullying child. Simple adult sanctions that "bullying is wrong" might be counterproductive. Effective action needs to acknowledge the social competence and perceived popularity of many bullies and the status gains they seek and may get from bullying.

REFERENCES

Andreou, E. (2004). Bully/victim problems and their association with Machiavellianism and self-efficacy in Greek primary school children. *British Journal of Educational Psychology*, **74**, 297–309.

Archer, J. (1988). *The Behavioural Biology of Aggression*. Cambridge, UK: Cambridge University Press.

Archer, J. (2009). Does sexual selection explain human sex differences in aggression? *Behavioral and Brain Sciences*, **32**, 249–311.

Arnocky, S., & Vaillancourt, T. (2012). A multi-informant longitudinal study on the relationship between aggression, peer victimization, and dating status in adolescence. *Evolutionary Psychology*, **10**, 253–270.

Ball, H. A., Arsenault, L., Taylor, A., et al. (2008). Genetic and environmental influences on victims, bullies and bully–victims in childhood. *Journal of Child Psychiatry and Psychiatry* **49**, 104–112.

Bertolotti, T., & Magnani, L. (2013). A philosophical and evolutionary approach to cyber-bullying: social networks and the disruption of sub-moralities. *Ethics and Information Technology*, **15**, 285–299.

Bjorklund, D. F., & Hawley, P. H. (2014). Aggression grows up: Looking through an evolutionary developmental lens to understand the causes and consequences of human aggression. In T. K. Shackelford & R. D. Hansen, eds., *The Evolution of Violence*. New York: Springer, pp. 159–186.

Blackwell, T. (2015). Provocative new study finds bullies have highest self esteem, social status, lowest rates of depression. https://nationalpost.com/health/provocative-new-study-finds-bullies-have-highest-self-esteem-social-status-lowest-rates-of-depression.

Boivin, M., Brendgen, M., Vitaro, F., et al. (2012). Strong genetic contribution to peer relationship difficulties at school entry: Findings from a longitudinal twin study. *Child Development*, **84**, 1098–1114.

Book, A. S., Volk, A. A., & Hosker, A. (2012). Adolescent bullying and personality: An adaptive approach. *Personality and Individual Differences*, **52**, 218–223.

Bowes, L., Joinson, C., Wolke, D., & Lewis, G. (2015). Peer victimisation during adolescence and its impact on depression in early adulthood: Prospective cohort study in the United Kingdom. *British Medical Journal*, **350**, h2469.

Campbell, D. T. (1975). On the conflicts between biological and social evolution and between psychology and moral tradition. *American Psychologist*, **30**, 1103–1126.

Connolly, J., Pepler, D., Craig, W., & Taradash, A. (2000). Dating experiences of bullies in early adolescence. *Child Maltreatment*, **5**, 299–310.

Cook, C. R., Williams, K. R., Guerra, N. G., Kim, T. E., & Sadek, S. (2010). Predictors of bullying and victimization in childhood and adolescence: A meta-analytic investigation. *School Psychology Quarterly*, **25**, 65–83.

Copeland, W. E., Wolke, D., Angold, A., & Costello, J. (2013). Adult psychiatric outcomes of bullying and being bullied by peers in childhood and adolescence. *Journal of the American Medical Association: Psychiatry*, **70**, 419–426.

Crick, N. R., & Dodge, K. (1994). A review and reformulation of social-information-processing mechanisms in children's social adjustment. *Psychological Bulletin*, **115**, 74–101.

Crick, N., & Dodge, K. (1996). Social information processing mechanisms in reactive and proactive aggression. *Child Development*, **67**, 993–1002.

Crick, N. R., & Dodge, K. A. (1999). "Superiority" is in the eye of the beholder: A comment on Sutton, Smith and Swettenham. *Social Development*, **8**, 128–131.

Dane, A. V., Marini, Z. A. Volk, A. A., & Vaillancourt, T. (2017). Physical and relational bullying and victimization: Differential relations with adolescent dating and sexual behavior. *Aggressive Behavior*, **43**, 111–122.

Dunbar, R. I. M. (2004). Gossip in an evolutionary perspective. *Review of General Psychology*, **8**, 100–110.

Elgar, F. J., Craig, W., Boyce, W., Morgan, A., & Vella-Zarb, R. (2009). Income inequality and school bullying: Multilevel study of adolescents in 37 countries. *Journal of Adolescent Health*, **45**, 351–359.

Ellis, B. J., del Guidice, M., Dishion, T. J., et al. (2012). The evolutionary basis of risk taking behavior: Implications for science, policy, and practice. *Developmental Psychology*, **48**, 598–623.

Ellis, B. J., Volk, A. A., Gonzalez, J.-M., & Embry, D. D. (2016). The meaningful roles intervention: An evolutionary approach to reducing bullying and increasing prosocial behavior. *Journal of Research on Adolescence*, **26**, 622–637.

Espelage, D., & Holt, M. K. (2007). Dating violence and sexual harassment across the bully–victim continuum among middle and high school students. *Journal of Youth and Adolescence*, **36**, 799–811.

Farrell, A. H., della Cioppa, V., Volk, A. A., & Book, A. S. (2014). Predicting bullying heterogeneity with the HEXACO model of personality. *International Journal of Advances in Psychology*, **3**, 30–39.

Farrington, D. P., Lösel, F., Ttofi, M. M., & Theodorakis, N. (2012). *School Bullying, Depression and Offending Behaviour Later in Life: An Updated Systematic Review of Longitudinal Studies*. Stockholm: Swedish National Council for Crime Prevention.

Fekkes, M., Pijpers, F. I. M., & Verloove-Vanhorick, S. P. (2004). Bullying behaviour and associations with psychosomatic complaints and depression in victims. *Journal of Pediatrics*, **144**, 17–22.

Festl, R., Vogelgesang, J., Scharkow, M., & Quandt, T. (2017). Longitudinal patterns of involvement in cyberbullying: Results from a latent transition analysis. *Computers in Human Behavior*, **66**, 7–15.

Forero, R., McLellan, L., Rissel, C., & Bauman, A. (1999). Bullying behaviour and psychosocial health among school students in New South Wales, Australia: Cross sectional survey. *British Medical Journal*, **319**, 344–348.

Fry, D. P. (2004). *The Human Potential for Peace: Challenging the War Assumption*. New York: Oxford University Press.

Garandeau, C. F., Lee, I. A., & Salmivalli, C. (2014). Differential effects of the KiVa antibullying program on popular and unpopular bullies. *Journal of Applied Developmental Psychology*, **35**, 44–50.

Gibb, Z. G., & Devereux, P. G. (2014). Who does that anyway? Predictors and personality correlates of cyberbullying in college. *Computers in Human Behavior*, **38**, 8–16.

Gini, G. (2008). Associations between bullying behaviour, psychosomatic complaints, emotional and behavioural problems. *Journal of Paediatrics and Child Health*, **44**, 492–497.

Gini, G., & Pozzoli, T. (2009). Association between bullying and psychosomatic symptoms: A meta-analysis. *Pediatrics*, **123**, 1059–1065.

Gray, P. (2011). The special value of children's age-mixed play. *American Journal of Play*, **3**, 500–522.

Greene, M. B. (2006). Bullying in schools: A plea for a measure of human rights. *Journal of Social Issues*, **62**, 63–79.

Guerra, N., Williams, K. R., & Sadek, S. (2011). Understanding bullying and victimization during childhood and adolescence: A mixed methods study. *Child Development*, **82**, 295–310.

Hawley, P. (2003). Strategies of control, aggression, and morality in preschoolers: An evolutionary perspective. *Journal of Experimental Child Psychology*, **85**, 213–235.

Hawley, P., Little, T. D., & Rodkin, P., eds. (2007). *Aggression and Adaptation: The Bright Side to Bad Behavior*. Mahwah, NJ: Lawrence Erlbaum.

Haynie, D. L., Nansel, T., Eitel, P., et al. (2001). Bullies, victims, and bully/victims: Distinct groups of at-risk youth. *Journal of Early Adolescence*, **21**, 29–49.

Hellfeldt, K., Gill, P. E., & Johansson, B. (2018). Longitudinal analysis of links between bullying victimization and psychosomatic maladjustment in Swedish schoolchildren. *Journal of School Violence*, **17**, 86–98.

Hong, J. S., & Espelage, D. L. (2012). A review of research on bullying and peer victimization in school: An ecological systems analysis. *Aggression and Violent Behavior*, **17**, 311–322.

Houser, J. J., Mayeux, L., & Cross, C. (2015). Peer status and aggression as predictors of dating popularity in adolescence. *Journal of Youth and Adolescence*, **44**, 683–695.

Ingram, G. P. D. (2014). From hitting to tattling to gossip: An evolutionary rationale for the development of indirect aggression. *Evolutionary Psychology*, **12**, 343–363.

Ivarsson, T., Broberg, A. G., Arvidsson, T., & Gillberg, C. (2005). Bullying in adolescence: Psychiatric problems in victims and bullies as measured by the Youth Self Report (YSR) and the Depression Self-Rating Scale (DSRS). *Nordic Journal of Psychiatry*, **59**, 365–373.

Jolliffe, D., & Farrington, D. P. (2011). Is low empathy related to bullying after controlling for individual and social background variables? *Journal of Adolescence*, **34**, 59–71.

Juvonen, J., Graham, S., & Schuster, M. A. (2003). Bullying among young adolescents: The strong, the weak, and the troubled. *Pediatrics*, **112**, 1231–1237.

Kaltiala-Heino, R., Rimpelä, M., Marttunen, M., Rimpelä, A., & Rantanen, P. (1999). Bullying, depression and suicidal ideation in Finnish adolescents: School survey. *British Medical Journal*, **319**, 348–351.

Kaukiainen, A., Björkqvist, K., Lagerspetz, K., et al. (1999). The relationship between social intelligence, empathy, and three types of aggression. *Aggressive Behavior*, **25**, 81–89.

Koh, J.-B., & Wong, J. S. (2017). Survival of the fittest and the sexiest: Evolutionary origins of adolescent bullying. *Journal of Interpersonal Violence*, **32**, 2668–2690.

Kolbert, J. B., & Crothers, L. M. (2003). Bullying and evolutionary psychology: The dominance hierarchy among students and implications for school personnel. *Journal of School Violence*, **2**, 73–91.

Kowalski, R. M., & Limber, S. P. (2013). Psychological, physical, and academic correlates of cyberbullying and traditional bullying. *Journal of Adolescent Health*, **53**, S13–S20.

Kowalski, R. M., Giumetti, G. W., Schroeder, A. N., & Lattanner, M. R. (2014). Bullying in the digital age: A critical review and meta-analysis of cyberbullying research among youth. *Psychological Bulletin*, **140**, 1073–1137.

Kumpulainen, K., Räsänen, E., & Henttonen, I. (1999). Children involved in bullying: Psychological disturbance and the persistence of the involvement. *Child Abuse & Neglect*, **23**, 1253–1262.

Le, H. T. H., Nguyen, H. T., Campbell, M. A., et al. (2017). Longitudinal associations between bullying and mental health among adolescents in Vietnam. *International Journal of Public Health*, **62**(Suppl. 1), 51–61.

Lereya, S. T., Samara, M., & Wolke, D. (2013). Parenting behavior and the risk of becoming a victim and a bully/victim: A meta-analysis study. *Child Abuse & Neglect*, **37**, 1091–1108.

Liang, H., Flisher, A. J., & Lombard, C. J. (2007). Bullying, violence, and risk behavior in South African school students. *Child Abuse & Neglect*, **31**, 161–171.

Livingstone, S., & Smith, P. K. (2014). Research review: Harms experienced by child users of online and mobile technologies: The nature, prevalence and management of sexual and aggressive risks in the digital age. *Journal of Child Psychology & Psychiatry*, **55**, 635–654.

Mishna, F., Khoury-Kassabrib, M., Gadallaa, T., & Daciuka, J. (2012). Risk factors for involvement in cyber bullying: Victims, bullies and bully–victims. *Children and Youth Services Review*, **34**, 63–70.

Monks, C. P. (2011). Peer-victimisation in preschool. In C. P. Monks & I. Coyne, eds., *Bullying in Different Contexts*. Cambridge, UK: Cambridge University Press, pp. 12–35.

Monks, C. P., Smith, P. K., Naylor, P., et al. (2009). Bullying in different contexts: commonalities, differences and the role of theory. *Aggression and Violent Behavior*, **14**, 146–156.

Muñoz, L. C., Qualter, P., & Padgett, G. (2011). Empathy and bullying: Exploring the influence of callous–unemotional traits. *Child Psychiatry & Human Development*, **42**, 183–196.

Nansel, T. R., Craig, W., Overpeck, M. D., Saluja, G., Ruan, W. J., & The Health Behaviour in School-aged Children Bullying Analyses Working Group. (2004). Cross-national consistency in the relationship between bullying behaviors and psychosocial adjustment. *Archives of Pediatrics and Adolescent Medicine*, **158**, 730–736.

Nordhagen, R., Nielsen, A., Stigum, H., & Köhler, L. (2005). Parental reported bullying among Nordic children: A population-based study. *Child: Care, Health & Development*, **31**, 693–701.

Olthof, T., Goossens, F. A., Vermande, M. M., Aleva, E. A., & van der Meulen, M. (2011). Bullying as strategic behavior: Relations with desired and acquired dominance in the peer group. *Journal of School Psychology*, **49**, 339–359.

Olweus, D. (1993). *Bullying at School: What We Know and What We Can Do*. Oxford: Blackwell.

Olweus, D. (1999). Sweden. In P. K. Smith, Y. Morita, J. Junger-Tas, et al., eds., *The Nature of School Bullying: A Cross-National Perspective*. London/New York: Routledge, pp. 7–27.

O'Moore, M. (2000). Critical issues for teacher training to counter bullying and victimisation in Ireland. *Aggressive Behavior*, **26**, 99–111.

Peeters, M., Cillessen, A. H. N., & Scholte, R. H. J. (2010). Clueless or powerful? Identifying subtypes of bullies in adolescence. *Journal of Youth and Adolescence*, **39**, 1041–1052.

Reijntjes, A., Vermande, M., Goossens, F. A., et al. (2013a). Developmental trajectories of bullying and social dominance in youth. *Child Abuse and Neglect*, **37**, 224–234.

Reijntjes, A., Vermande, M., Olthof, T., et al. (2013b). Costs and benefits of bullying in the context of the peer group: A three wave longitudinal analysis. *Journal of Abnormal Child Psychology*, **41**, 1217–1229.

Rigby, K. (1997). Attitudes and beliefs about bullying among Australian school children. *Irish Journal of Psychology*, **18**, 202–220.

Rivers, I., Poteat, V. P., Noret, N., & Ashurst, N. (2009). Observing bullying at school: The mental health implications of witness status. *School Psychology Quarterly*, **24**, 211–223.

Salmivalli, C. (2010). Bullying and the peer group: A review. *Aggression and Violent Behavior*, **15**, 112–120.

Salmivalli, C., Kaukiainen, A., Kaistaniemi, L., & Lagerspetz, K. (1999). Self-evaluated self-esteem, peer-evaluated self-esteem, and defensive egotism as predictors of adolescents' participation in bullying situations. *Personality and Social Psychology Bulletin*, **25**, 1268–1278.

Salmivalli, C., Lagerspetz, K., Björkqvist, K., Österman, K., & Kaukiainen, A. (1996). Bullying as a group process: Participant roles and their relations to social status within the group. *Aggressive Behavior*, **22**, 1–15.

Schnohr, C., & Niclasen, B. V.-L. (2006). Bullying among Greenlandic school-children: Development since 1994 and relations to health and health behaviour. *International Journal of Circumpolar Health*, **65**, 305–312.

Smith, P. K. (2007). Why has aggression been thought of as maladaptive? In P. Hawley, T. D. Little, & P. Rodkin, eds., *Aggression and Adaptation*. Mahwah, NJ: Lawrence Erlbaum, pp. 65–83.

Smith, P. K. (2014). *Understanding School Bullying: Its Nature & Prevention Strategies*. Thousand Oaks, CA: Sage Publications.

Smith, P. K. (2017). Bullying and theory of mind: A review. *Current Psychiatry Reviews*, **13**, 90–95.

Smith, P. K., ed. (2019). *Making an Impact on School Bullying: Interventions and Recommendations*. London: Routledge.

Smith, P. K., Madsen, K., & Moody, J. (1999). What causes the age decline in reports of being bullied at school? Towards a developmental analysis of risks of being bullied. *Educational Research*, **41**, 267–285.

Smith, P. K., Shu, S., & Madsen, K. (2001). Characteristics of victims of school bullying: Developmental changes in coping strategies and skills. In J. Juvonen & S. Graham, eds., *Peer Harassment at School: The Plight of the Vulnerable and Victimised*. New York: Guildford, pp. 332–352.

Stein, J. A., Dukes, R. L., & Warren, J. I. (2007). Adolescent male bullies, victims, and bully–victims: A comparison of psychosocial and behavioral characteristics. *Journal of Pediatric Psychology*, **32**, 272–282.

Sutton, J., & Keogh, E. (2000). Social competition in school: Relationships with bullying, Machiavellianism and personality. *British Journal of Educational Psychology*, **70**, 443–456.

Sutton, J., Smith, P. K., & Swettenham, J. (1999a). Bullying and "theory of mind": A critique of the "social skills deficit" view of anti-social behaviour. *Social Development*, **8**, 117–127.

Sutton, J., Smith, P. K., & Swettenham, J. (1999b). Social cognition and bullying: Social inadequacy or skilled manipulation? *British Journal of Developmental Psychology*, **17**, 435–450.

Sutton, J., Smith, P. K., & Swettenham, J. (1999c). Socially undesirable need not be incompetent: A response to Crick and Dodge. *Social Development*, **8**, 132–134.

Tani, F., Greenman, P. S., Schneider, B. H., & Fregoso, M. (2003). Bullying and the Big Five: A study of childhood personality and participant roles in bullying incidents. *School Psychology International*, **24**, 131–146.

Ttofi, M. M., Farrington, D. P., & Lösel, F. (2011). Do the victims of school bullies tend to become depressed later in life? A systematic review and meta-analysis of longitudinal studies. *Journal of Aggression, Conflict and Peace Research*, **3**, 63–73.

Ttofi, M. M., Farrington, D. P., & Lösel, F. (2012). School bullying as a predictor of violence later in life: A systematic review and meta-analysis of prospective longitudinal studies. *Aggression and Violent Behavior*, **17**, 405–418.

Vaillancourt, T., Hymel, S., & McDougall, P. (2003). Bullying is power: Implications for school-based intervention strategies. *Journal of Applied School Psychology*, **19**, 157–176.

Veenstra, R., Lindenberg, S., Oldehinkel, A.J., et al. (2005). Bullying and victimization in elementary schools: Victims, bully/victims, and uninvolved adolescents. *Developmental Psychology*, **41**, 672–682.

Volk, A., Craig, W., Boyce, W., & King, M. (2006). Adolescent risk correlates of bullying and different types of victimization. *International Journal of Adolescent Medicine and Health*, **18**, 375–386.

Volk, A. A., Camilleri, J. A., Dane, A. V., & Marini, Z. A. (2012). Is adolescent bullying an evolutionary adaptation? *Aggressive Behavior*, **38**, 222–238.

Volk, A. A., Dane, A. V., Marini, Z. A., & Vaillancourt, T. (2015). Adolescent bullying, dating, and mating: Testing an evolutionary hypothesis. *Evolutionary Psychology*, **13**, 1–11.

Volk, A. A., Farrell, A. H., Franklin, P., Mularczyk, K. P., & Provenzano, D. A. (2016). Adolescent bullying in schools: An evolutionary perspective. In D. C. Geary & D. B. Berch, eds., *Evolutionary Perspectives on Child Development and Education*. Basel: Springer, pp. 167–191.

Wang, J., Nansel, T. R., & Iannotti, R. J. (2011). Cyber and traditional bullying: Differential association. *Journal of Adolescent Health*, **48**, 415–417.

Winsper, C., Lereya, T., Zanarini, M., & Wolke, D. (2012). Involvement in bullying and suicide-related behavior at 11 years: A prospective birth cohort study. *Journal of the American Academy of Child & Adolescent Psychiatry*, **51**, 271–282.

Wolke, D., & Lereya, S. T. (2015). Long-term effects of bullying. *Archives of Disease in Childhood*, **100**, 879–885.

Wolke, D., Tippett, N., & Dantchev, S. (2015). Bullying in the family: Sibling bullying. *Lancet Psychiatry*, **2**, 917–929.

Wolke, D., Woods, S., Bloomfield, I., & Karstadt, L. (2001). Bullying involvement in primary school and common health problems. *Archives of Disease in Childhood*, **85**, 197–201.

Yang, A., Li, X., & Salmivalli, C. (2016). Maladjustment of bully–victims: validation with three identification methods. *Educational Psychology*, **36**, 1390–1407.

Yeager, D. S., Fong, C. J., Lee, H. Y., & Espelage, D. L. (2015). Declines in efficacy of anti-bullying programs among older adolescents: Theory and a three-level meta-analysis. *Journal of Applied Developmental Psychology*, **37**, 36–51.

Zwierzynska, K., Wolke, D., & Lereya, T. S. (2013). Peer victimization in childhood and internalizing problems in adolescence: A prospective longitudinal study. *Journal of Abnormal Child Psychology*, **41**, 309–323.

25 Birth Order and Evolutionary Psychology

FRANK J. SULLOWAY

25.1 INTRODUCTION

Birth order has long been thought to have a lasting influence on people's lives through its effects on social customs and by fostering individual differences in personality and social behavior. Historically, birth order has been linked with well-documented differences in professional opportunities and achievement, emigration patterns, likelihood of reproduction, mortality rates, inheritance practices, and the politics of royal succession (Altus, 1966; Boone, 1986; Bu, 2016; Duby, 1977; Galton, 1874; Herlihy, 1977; Hrdy & Judge, 1993; Sulloway, 1996). An analysis of birth order and the social customs in 39 non-Western societies found that firstborns generally receive more extensive birth ceremonies than do their younger siblings, are allotted special privileges, and, even in adulthood, exert authority over their brothers and sisters (Rosenblatt & Skoogberg, 1974). Additionally, firstborns in these 39 societies received a greater share of parental property than did laterborns. Firstborns are also more likely than laterborns to be named after a parent, a practice that in turn tends to increase parental investment (MacAndrew, King, & Honoroff, 2002)

Alfred Adler (1870–1937) was one of the earliest psychologists to call attention to birth order as a source of individual differences in personality and social outlook. His theories drew heavily on his own personal experiences as a secondborn. Two years his senior, Adler's older brother, Sigmund, was favored by their parents, and he also presented a formidable rival by being physically accomplished, domineering, and financially successful. In his writings about birth order, Adler (1927, 1928) argued that firstborns undergo a traumatic "dethronement" at the birth of a younger sibling, an experience they subsequently try to overcome by emulating parents and by emphasizing the importance of law and order, which may drive them to become "power-hungry conservatives." Adler believed that middleborns, in an effort to surpass older siblings, try harder, and that middle children are also more easygoing and cooperative. Youngest children, Adler maintained, suffer from feelings of inferiority, and

because they never undergo a dethronement experience, they often become spoiled and lazy.

In contrast to Adler's approach to the consequences of birth order, which was based on personal and clinical observations rather than systematic research, the first notable efforts to study birth order and personality in a rigorous fashion were conducted by University of Chicago psychologist Helen Koch. Between 1954 and 1960, Koch published a series of 10 studies that were distinguished by their thoughtful research design. Koch selected 384 Chicago-area children between the ages of five and six who had all grown up in white, intact, two-child families. She divided her study participants into 24 matched subgroups to control for variations in birth order, sex, sex of sibling, and age spacing between siblings. The study included 58 behavioral measures, which were based on teachers' ratings of the 384 participants. Relative to laterborns, firstborns were judged as being more competitive, self-confident, emotionally intense, concerned about status, and upset by defeat (Koch, 1955, 1956).

One particularly interesting aspect of Koch's findings was the remarkable number of interaction effects she documented among the various behavioral measures included in her study. Directly or indirectly, birth order participated in 31 significant effects, but only 5 of these were main effects, with the other 26 being interactions involving sex, sibling's sex, and age spacing. Even though Koch's findings showed that firstborns of both sexes were generally less agreeable than laterborns, these birth-order effects were expressed differently by gender, largely in accordance with gender-based norms of behavior. For example, teachers judged firstborn boys to be more angry, vengeful, and cruel than laterborn boys, but none of these traits were significantly associated with birth order among girls, although teachers did rate firstborn girls as being more quarrelsome. Subsequent reanalysis of Koch's findings underscores the conclusion that attributing specific personality traits to birth order can be misleading without taking moderator variables into account (Brim, 1958; Sulloway, 1996).

In the wake of Koch's pioneering studies, an extensive literature – consisting of more than 2,000 publications – has documented significant birth-order differences in intellectual performance, personality, and social behavior. These findings have nevertheless been the subject of continuing controversy (Bleske-Rechek & Kelley, 2014; Damian & Roberts, 2015; Ernst & Angst, 1983; Harris, 1998, 2000; Jefferson, Herbst, & McCrae, 1998; Marini & Kurtz, 2011; Michalski & Shackelford, 2002; Rohrer, Egloff, & Schmukle, 2015; Saroglou & Fiasse, 2003; Schooler, 1972). Most of the empirical discrepancies fueling this continuing debate can be attributed to four problems: (1) inadequate sample sizes, resulting in numerous studies that lack sufficient statistical power to provide adequate tests of the hypotheses being considered; (2) absence of controls for important covariates, such as sibship size and social class, which covary with birth order and can confound its relationship to measures of human behavior; (3) failure to treat birth order as a multivariate construct that sometimes entails quadratic and other nonlinear trends; and (4) insufficient appreciation of the fact that birth order is an imperfect proxy for the real causes that lie behind sibling differences in personality and social behavior – namely, differences in age, physical size, relative status, and differing roles and relationships within the family.

It is important to distinguish between biological and functional birth order. What is generally influential about birth order is how children are raised, not how they are born, as Kristensen and Bjerkedal (2007) have shown with respect to differences in IQ. Divorce and remarriage, along with the early death of a sibling, can change a child's functional birth order and hence lead to different family experiences as the child is growing up. Similarly, age spacing, as Helen Koch (1955, 1956) convincingly showed, mediates the effects of birth order. Given a particularly large age gap, for example, functional birth order can be quite different from biological birth order, causing a younger sibling to grow up more like an only child than a typical laterborn.

The magnitude of birth-order differences is often underestimated in studies owing to other common methodological problems, which include measurement error and, more importantly, the fact that some birth-order differences are sensitive to situational factors rather than expressing themselves in a simple, invariant manner under all circumstances (Salmon, 1998; Sulloway, 1996: Sulloway, John, Gosling, & Potter, unpublished data). One would anticipate, for example, that firstborns would seek to dominate their inferiors, but to act in a gracious and subordinate manner when interacting with superiors – a behavioral style that has been termed a "pecking order personality" (Block, 1995). Most personality surveys are not designed to capture such subtle person-by-situation variations in the expression of behavior (Mischel & Shoda, 1995).

25.2 THE BIOLOGY OF SIBLING COMPETITION

From a Darwinian perspective, one would expect birth order to be a reasonably important consideration among the various influences that shape people's lives. At the heart of this evolutionary argument lie the two closely related phenomena of sibling competition and sibling rivalry, which are widespread among animal species, including insects, fish, reptiles, birds, and mammals. Sibling competition stems from a simple Darwinian fact. On average, full siblings in sexually reproducing species share only half their genes, unless they are identical twins. For this reason, siblings are twice as related to themselves as they are to a full brother or sister. William Hamilton (1964) drew on this fundamental genetic insight to argue that siblings ought to compete for scarce resources whenever the benefits to another sibling are less than twice the costs to the self. Seen from this Darwinian perspective, sibling competition is an important source of parent–offspring conflict. Offspring are twice as related to themselves as they are to a parent or sibling, but parents are equally related to all of their offspring. Parents are therefore inclined to invest in other offspring to a greater extent than is desired by the offspring themselves, which can lead to conflicts over the extent and timing of any resources allocated to a sibling. Weaning conflicts represent a prime example of such parent–offspring disagreements, as weaning curtails the resources being invested in one offspring, allowing mothers to become sexually fertile again and to invest in other offspring (Trivers, 1974).

Competitive behavior among siblings is particularly common among seabirds and predatory birds, and such competition can lead to siblicide. "In all siblicidal species studied to date," one research team has noted, "there is a striking tendency for the victim to be the youngest member of the brood" (Mock, Drummond, & Stinson, 1990, p. 445). Because it is not in the parents' interests to do so, they almost never interfere in these lethal conflicts.

Siblicide takes two general forms: obligate and facultative. Among avian species in which siblicide is obligate – occurring in almost every instance – parents are ecologically incapable of raising more than a single offspring. Females nevertheless lay and incubate a second egg because this additional egg requires only a small physiological investment, one that ensures that time is not unnecessarily wasted during the limited breeding season if the first chick becomes a victim of predation or otherwise dies in the nest. Among species in which siblicide is facultative, sibling competition is regulated by the available food supply. For example, female blue-footed boobies (*Sula nebouxii*) lay two or three eggs. When the body weight of the eldest chick drops below 80 percent of normal, this chick engages in aggressive pecking directed at its younger siblings, killing them outright or forcing them from the nest, where they eventually die of starvation and exposure (Drummond & García-Chevalas, 1989).

Even plant species sometimes engage in siblicide. A species of black plum (*Syzygium cuminii*) produces seeds that have 25–30 ovules, each of which is a full sibling. After the first of these ovules is fertilized, it secretes a "death chemical" that suppresses the metabolism of sucrose, killing off all the other ovules (Krishnamurthy, Shaanker, & Ganeshaiah, 1997).

25.3 PERSONALITY AND SOCIAL BEHAVIOR

Research on birth order and personality can be usefully subdivided into two different classes of studies: those based on within-family comparisons (e.g., when parents rate their own children or when siblings rate one another); and those based on between-family studies, in which the findings are based on single participants drawn from different families. The results obtained from between-family studies have often been criticized because they tend to confound differences in birth order that may actually be due to disparities in family size and social class, which covary with birth order (Ernst & Angst, 1983). More specifically, lower-class families are biased for larger sibships, so studies failing to control for social class and/or sibship size can produce unreliable results. For this reason, researchers have generally given greater credence to the results of within-family studies (Rodgers et al., 2000; Wichman, Rodgers, & McCallum, 2006). Although within-family studies eliminate the possibility of confounds arising from differences in sibship size, social class, and most other aspects of family background, they have also been criticized by researchers who argue that any reported birth-order differences may simply reflect widely held stereotypes about birth order, including the kinds of personality differences that are expected because siblings differ by age (Rohrer et al., 2015).

Although the published literature on birth order and personality contains many studies that show few or no differences, one sees a clearer pattern when these studies are amalgamated using the technique of meta-analysis (Sulloway, 1995, 1996, 2010). Analyzing multiple studies by this method, which benefits from greater statistical power, reveals a reasonably consistent pattern of small to moderate birth-order differences. It is useful to review these findings in terms of the "Five Factor Model" of personality (Costa & McCrae, 1992), although it is important to note that pooling findings in terms of just five broad personality dimensions sometimes misses important differences that are found when the same data are analyzed at the level of specific traits. These "Big Five" dimensions include conscientiousness, agreeableness, extraversion, openness to experience, and neuroticism.

25.3.1 Conscientiousness

In within-family studies of personality, firstborns consistently emerge as being more conscientious than their younger siblings. Eldest children, for example, are judged by parents and siblings as being more organized, self-disciplined, planful, and deliberate compared with their younger brothers and sisters. They are also generally perceived as being the "achiever" of the family (Bu, 2016; Healey & Ellis, 2007; Paulhus, Trapnell, & Chen, 1999; Sulloway, 1996, 2001, 2010). These empirical findings are consistent with the view that firstborns are more likely than laterborns to act as a surrogate parent, which entails the responsibility of caring for younger siblings.

25.3.2 Agreeableness

Studies have repeatedly shown that laterborns are more agreeable than firstborns (Koch, 1955, 1956; Jefferson, Herbst, & McCrae, 1998; Michalski & Shackelford, 2002; Paulhus et al., 1999; Sulloway, 1996, 2001, 2010). One possible reason for this difference is that firstborns are able to use their advantages in age and physical size to achieve dominance over their younger siblings, whereas laterborns tend to cultivate lower-power strategies, which include cooperation, bargaining, pleading, and requesting assistance from adults. In an experimental study involving an investment game, laterborns returned 1.4 times more money to their game partners than did firstborns (Courtiol, Raymond, & Faurie, 2009). In this same study, greater generosity entailed greater risk-taking and trust, and the altruism manifested by laterborns was usually not reciprocated. Middleborns tend to score higher than either firstborns or lastborns on most measures of agreeableness (Sulloway, 2001). This difference appears to reflect the unique family status of middle children, sandwiched as they are between firstborns, who have superiority in age, power, and status, and lastborns, who tend to receive greater attention and emotional investment from parents.

25.3.3 Extraversion

At the trait level, birth-order differences associated with the Big Five personality dimension of extraversion do not generally exhibit consistent outcomes, even though laterborns often score higher on aggregate measures. Disparities within this personality domain occur because firstborns tend to be more dominant and assertive than their younger siblings, whereas laterborns tend to be more fun-loving, affectionate, and sociable (Beck, Burnet, & Vosper, 2006; Dixon et al., 2008; Sulloway, 1996, 2001, 2010; Sulloway & Zweigenhaft, 2010). All of these varied attributes represent different aspects of extraversion. The fact that this personality dimension does not produce outcomes that are altogether consistent in terms of birth order underscores the argument that these sibling differences in personality arise from the pursuit of differing strategies within the family. There is no a priori reason why sibling strategies should always align themselves with the Big Five dimensions in the same consistent manner as do, for example, the biological underpinnings of personality, which, as research in behavioral genetics has shown, provide much of the cohesion of these five dimensions (Jang et al., 2002; Loehlin, 1992).

Risk-taking is another aspect of extraversion that has shown reasonably consistent birth-order differences. From a Darwinian viewpoint, laterborns are expected to take more risks than their older siblings in an effort to compensate for lesser parental investment (Daly & Wilson, 1988). One form of risk-taking that has received extensive study in relation to birth order involves participation in dangerous sports. Sulloway and Zweigenhaft (2010) undertook a meta-analysis of 24 previous studies on this subject (n = 8,340). They found that younger siblings, compared with firstborns and only children, were 1.5 times more likely to play dangerous sports such as football, soccer, and rugby, whereas firstborns and only children tended to engage in safer sports, which included tennis, track sports, and swimming. The same study examined the careers of 700 brothers who played Major League baseball. Relative to their older brothers, younger brothers were 1.7 times more likely to attempt stolen bases, which is associated with one of the highest injury rates in baseball. Younger brothers were also more likely to be hit by pitches, which occurs more often among players who are tenacious about guarding the strike zone. Of particular interest in this study is that these findings were confined to players whose ages were within five years of each other and who also played in similar positions (being either pitchers or field players). In other words, younger brothers took more risks, but only when they were in direct competition with older brothers whose baseball careers were usually concurrent with their own and whose performance statistics were directly comparable.

Another between-family study of risk-taking during adolescence, using data from 9,859 participants in the National Longitudinal Survey of Youth – 1997, found that younger siblings were more likely than firstborns to engage in a broad series of risk-taking behaviors (Argys et al., 2005). These risky behaviors included smoking, alcohol consumption, marijuana use, sexual activity, and being involved in delinquent behaviors (including theft, destruction of property, and carrying a handgun). These effects were directly proportional to birth rank. For example, a secondborn was 4 percent more likely than a firstborn to use marijuana, but a fifthborn was 14 percent more likely to do so. Similarly, in a study conducted in Italy, firstborns were found to be more "conservative" by a factor of 1.6, as measured by the desire to live in safe and secure environments (Barni, Vieno, & Alfieri, 2014).

25.3.4 Openness to Experience

Like the findings for extraversion, those for openness to experience exhibit significant heterogeneity by birth order, depending on what types of openness are being considered (Sulloway, 1996). Firstborns tend to score higher than laterborns in connection with the "intellectual" aspects of openness, including those directly associated with intelligence. Laterborns, by contrast, tend to score higher on aspects of openness that reflect imagination, unconventionality, and openness to innovation (Saad, Gill, & Nataraajan, 2005; Skinner & Fox-Francoeur, 2010; Sulloway, 1996, 2001, 2010). For example, Paulhus et al. (1999) found that laterborns were twice as likely as firstborns to be described as "the rebel" of the family, a result replicated by Rohde et al. (2003) in four different countries (Israel, Norway, Russia, and Spain) and by Kobayashi et al. (2001) in a Japanese sample. Healey and Ellis (2007) also found that laterborns, compared with firstborns, were more "rebellious" and "nonconformist." Similarly, in an analysis of 6,566 participants involved in 121 revolutions and reform movements in world history, Sulloway (1996) found that laterborns generally supported the "radical" alternative.

25.3.5 Neuroticism

Like the findings for extraversion and openness to experience, those for neuroticism are significantly heterogeneous at the trait level when analyzed according to birth order. Firstborns tend to score higher on those measures that involve feeling tense and being concerned about status. By contrast, laterborns score higher on those aspects of neuroticism that reflect self-consciousness and low self-esteem (Kidwell, 1982; Sulloway, 2001, 2010).

25.3.6 Intellectual Performance

Intellectual performance, including intelligence, is generally considered as being a separate psychological domain from personality, although it is moderately correlated with some of the aspects of openness to experience – particularly openness to ideas (Costa & McCrae, 1992). Considerable research has shown that firstborns tend to have higher IQs than their younger siblings. In addition, IQ scores generally decline by about two to three points with each successive birth rank (Belmont & Marolla, 1973; Kristensen & Bjerkedal, 2007; Sulloway, 2007; Zajonc & Mullally, 1997; Zajonc & Sulloway, 2007). These findings about intellectual ability are consistent with resource-dilution theories about parental investment (Downey, 2001), including Zajonc's (1976) confluence model, which posits that IQ is influenced by the collective intelligence of the family as a whole. Because a newborn has an effective IQ of zero, each new addition to the family lowers the family's collective intelligence.

According to Zajonc, firstborns also benefit from a "teaching" effect, which helps them to overcome the negative effects of intellectual resource dilution. Teaching is thought to promote intelligence owing to the cognitive effort required to master information being taught to younger siblings. The confluence model is also consistent with the well-documented tendency for IQ scores to increase slightly among lastborns growing up in large families. For these children, the intellectual environment may actually be richer relative to that experienced by their older siblings, because the lastborns in such families benefit from interactions with the many older quasi-adults whose overall cognitive abilities are nearing adulthood levels.

25.3.7 Social Attitudes

Birth order exerts an influence on social attitudes. In the largest study to date, which included 193,422 high school students and was controlled for sex, sibship size, social class, race, and religious background, Theroux (1993) found that laterborns were more likely than firstborns to support a liberal perspective with regard to premarital sex and legalization of abortion. Laterborns in this same study were also less likely to attend church. In a noteworthy real-life study, Zweigenhaft and Von Ammon (2000) analyzed the arrest records for people involved in a labor dispute that took place at a Kmart in Greensboro, North Carolina. They found that laterborns were 2.2 times more likely than firstborns to have multiple arrests for their protest activities. In another study conducted by Zweigenhaft (2002), laterborns growing up between 1969 and 1982 were more likely to use marijuana than were firstborns. Although there were no overall differences in rates of political activism in this same sample, laterborns were more likely than firstborns to engage in political activism during those years when activism was not a popular activity. Given that "activism" can take many different forms, this last finding was the relevant test of the hypothesis, as only some forms of activism are liberal, controversial, or unconventional (and hence likely to be supported by younger siblings).

A series of other studies have nevertheless failed to find consistent support for the relationship between birth order and social and political attitudes. Freese, Powell, and Steelman (1999) analyzed participants drawn from the General Social Survey and found only 3 significant birth-order effects out of the 33 relevant measures – none of which were in the predicted direction. Similarly, Førland, Korsvik, and Christophersen (2012) failed to find a significant relationship between birth order and involvement in radical political activism in Norway during the 1960s, although the trend was in the expected direction. Despite these and other failures to document a relationship between birth order and social attitudes, a meta-analysis of 20 different studies (n = 11,240), including 8 studies that did not support the hypothesis, produced a significant mean-weighted correlation of 0.09 between being laterborn and supporting the liberal alternative (Sulloway, 2001). An effect of this magnitude means that laterborns were 44 percent more likely than firstborns to support a liberal position (the relative risk ratio) – a finding that is equivalent to the gender gap in social attitudes in the USA (Clark & Clark, 2008).

Although meta-analysis supports the existence of a modest tendency for laterborns to prefer liberal points of view, there is significant heterogeneity among these previously published results. This heterogeneity strongly suggests that the relationship between birth order and social attitudes is being moderated by other important influences that sometimes introduce substantial variations in the reported findings. One plausible moderating factor is the role played by parents, an influence that in turn is likely affected by varying cultural and historical considerations. In an analysis of participants in 28 scientific revolutions during the last four centuries, Sulloway (1996) found that firstborns were twice as likely as laterborns to share their parents' social attitudes, a difference also reported by Kulik (2004) in 294 Israeli families, although this effect was smaller.

The magnitude of birth-order differences in social attitudes may also depend on the importance that parents personally place on transmitting their own social attitudes to the next generation. Among immigrant families, for example, parents often encourage their offspring to preserve certain family values that are threatened with being lost during the family's assimilation into a new cultural setting. Skinner (1992) analyzed data on social attitudes among the members of 649 Chinese Tokok families living in Indonesia. Parents in these immigrant families almost always favored older siblings, who received more of the family's resources and whose arranged marriages were generally more advantageous. Parents expected all of their children – especially eldest sons – to accept an arranged marriage bringing social and financial benefits to the family, as well as to provide for the parents in their old age. Eldest sons were also given responsibility for observing the rituals associated with ancestor worship and, above all, were expected to conform to Confucian ideas about filial obedience. In response to these parental pressures, Skinner found, Tokok eldest sons were more socially and politically conservative than their younger siblings, and they were also more likely to display "filial loyalty."

Similarly, Manaster and colleagues (1998) studied the transmission of social attitudes from parents to offspring among 1,844 first- and second-generation Japanese Americans. The firstborn children of immigrants were more likely than laterborn children to maintain their Japanese cultural ties (e.g., by living in Japanese neighborhoods, by continuing to read and speak Japanese, by endorsing traditional Japanese values, and by worshiping in the Shinto or Buddhist faith of their parents). Politically, Japanese-American firstborns were also 1.4 times more likely to support conservative political candidates in their adopted country. Because the laterborns in this two-generational study were more likely to reject their parents' conservative values, they proved to be better at assimilating themselves into American society. As a result, Japanese-American laterborns ended up achieving greater socioeconomic status than did their more conservative elder siblings. Another noteworthy finding of this study was that birth-order effects were systematically stronger for males than females, underscoring the special role that eldest sons have in the patriarchal family system that is characteristic of Japanese families.

The pressure felt among some immigrant families to preserve family values during cultural assimilation is analogous to the greater pressure to conform to parental social values that prevailed in prior centuries (Turner, 1990). It was not uncommon, in the past, for children who lost their religious faith to be ostracized by their scandalized families and, in some cases, to be banished from their homes – as were the fates of novelist George Eliot and social reformer Francis Power Cobbe in the mid-nineteenth century. Based on such evidence, it is likely that the magnitude of birth-order differences in social attitudes is sensitive to cultural and historical variations and may have dissipated somewhat over the last few centuries.

25.4 MECHANISMS RESPONSIBLE FOR BIRTH-ORDER DIFFERENCES

At least seven different mechanisms have been suggested as causing birth-order differences in personality, intellectual performance, and social behavior. These mechanisms include: (1) disparities in parental investment; (2) differences in parental treatment; (3) sibling dominance hierarchies; (4) niche picking or niche specialization within the family; (5) de-identification – or the tendency for siblings to try to differentiate themselves from their brothers and sisters; (6) prenatal effects; and (7) birth-order stereotypes. It seems likely that most birth-order effects reflect the simultaneous influence of multiple mechanisms.

25.4.1 Parental Investment

One clear source of sibling differences is disparities in parental investment. Like the offspring of many animal species, human offspring are heavily dependent on parental investment, especially during the first few years after birth, but also continuing through much of adult life. Based on evolutionary theory, differences are expected in the amount of investment that parents allocate to their offspring, and parents should invest more in offspring who are most likely to transmit their parents' genes to the next generation. Trivers (1972) has defined parental investment as any investment in time, energy, or resources that benefits one offspring at a cost to another offspring. Among humans, biases in parental investment are well documented and have often been based on differences in birth order, gender, and offspring quality (Daly & Wilson, 1988; Hrdy & Judge, 1993).

Before 1800 or so, half of all children died from the diseases of childhood. Research has shown that parental discrimination between offspring differing in age, gender, and phenotypic quality often influenced who survived this vulnerable period of life (Boone, 1986; Voland, 1988, 1990). Historically, older children were typically favored by parents because, having already survived many of the potentially lethal diseases of childhood, these children represented better Darwinian bets for attaining adulthood and having children of their own. Infanticide, for example, is an accepted practice among many traditional societies, occurring when offspring are too closely spaced or when parental resources are very limited. Never, however, do traditional societies sanction the killing of the elder of two children (Daly & Wilson, 1988).

One noteworthy form of parental investment involves inheritance customs, which historically have tended to entail some degree of discrimination by birth order. At least three different systems of inheritance may be distinguished: primogeniture – a policy leaving all parental property to the eldest child or son; secundogeniture, which allows for inheritance by the second and other younger sons (although the legal title to land was often retained by the firstborn); and ultimogeniture, or inheritance by the youngest child. These different systems of inheritance have almost always been associated with differing economic and local geographic circumstances that tend to make one system advantageous over the others (Hrdy & Judge, 1993).

Primogeniture has been common whenever there are constraints on available land and economic opportunity. By leaving all or most of the family assets to the eldest son or child, parents avoided subdividing the family lands while also favoring perpetuation of the family's patronymic. Secundogeniture and other forms of equal inheritance tend to occur when risk and skill are especially important for economic success (e.g., in Renaissance Venice, where speculative trade was a major source of economic prosperity). By dividing family assets among offspring, parents increased the chances that one or more of these offspring would ultimately be successful (Herlihy, 1977). Ultimogeniture has been most common in regions where the rates of death taxes are high, as this strategy lengthens the interval between taxations.

Typically, disparities in parental investment tend to involve U-shaped distributions of family resources, with middleborns receiving the short end of the stick (Salmon, 1999; Salmon & Daly, 1998; Salmon, Shackelford, & Michalski, 2012). These U-shaped distributions can arise from what has been called the "equity heuristic" (Hertwig, Davis, & Sulloway, 2002), which is a version of the "resource-dilution" hypotheses (Downey, 2001; Lehmann, Nuevo-Chiquero, & Vidal-Fernandez, 2018; Price, 2008; Zajonc, 1976; Zajonc & Mullally, 1997). The equity heuristic consists of treating children equally, but it leads counterintuitively to an unequal cumulative distribution of resources. Unlike middle children, firstborns and lastborns both experience a time when they are the only child growing up within the home. As long as resources are always divided equally among all of the children living at home, middle children inevitably receive less parental investment, as they never enjoy a period of exclusive investment. Although the equity heuristic predicts U-shaped trends in cumulative parental resource allocation, this heuristic also posits linear trends whenever the importance of a particular resource (e.g., being vaccinated) is limited to the first few years of life. A linear outcome occurs because lastborns cannot play catchup in adolescence if a resource at that time is no longer pertinent to life outcomes.

Some lastborns benefit from a common variation in parental investment patterns that occur as mothers age. As mothers approach the end of their childbearing years, their youngest child is increasingly likely to be the last child of the family. Inasmuch as the last child cannot be replaced, it makes Darwinian sense for mothers to invest more substantially in this child relative to older offspring (Rohde et al., 2003; Salmon & Daly, 1998). Consistent with this prediction, older mothers report that they feel more emotionally attached to their lastborn children than to other offspring (Suitor & Pillemer, 2007). Like the effects of the equity heuristic, the allocation of greater parental investment to youngest children augments U-shaped trends in birth rank.

Numerous studies have confirmed these theoretical expectations about the allocation of parental investment (Hertwig et al., 2002; Price, 2008). Lindert (1977) tallied the amount of childcare time devoted to siblings in 1,296 families living in the USA. Parental investment was assessed in terms of total childcare hours during the first 18 years of life. In two-child families there was a 15 percent reduction in childcare time compared with what an only child received. In families with three or more children, middleborns received 10 percent less childcare than did firstborns or lastborns.

Differences in parental investment can influence physical stature as well as overall health. Horton (1988) analyzed the height and weight of 1,903 children living in the Philippines. She found that laterborns were shorter and weighed less than firstborns, a result pointing to reduced nourishment among children growing up in larger sibships. In another study, laterborn girls were found to reach menarche at a later age than firstborns, a difference that also appears to reflect nutritional differences (Surbey, 1998). Studies of vaccination rates have revealed that compliance declines by about 20–30 percent with each increase in birth rank (Hertwig et al., 2002). In the light of this collective evidence, it is not surprising that birth rank is correlated with mortality. In a Swedish study that included 14,192 children, third- and fourth-born offspring were twice as likely to die before the age of 10 compared with firstborns (Modin, 2002; see also Barclay & Kolk, 2015).

25.4.2 Parental Treatment Effects

Findings about differences in parental investment are paralleled by disparities in the ways in which parents treat their children as they are growing up. New parents tend to more anxious about childrearing compared with parents who have already had at least one offspring. Claxton (1994) has shown that parents provide differing degrees of feedback to their children according to birth order. Firstborns are more likely to receive "process" feedback, which guides the child in the actual execution of tasks. Firstborns are also more likely to receive "outcome" feedback, which informs the child about performance after the fact. Another study found that mothers were more interfering, extreme, and inconsistent with their firstborn children, who in turn manifested significantly more dependent behavior than their younger siblings (Hilton, 1967). These differences in parental treatment may help to account for the common finding that firstborns are more parent oriented than their younger siblings, and are more likely to conform to parental expectations (Ernst & Angst, 1983; Sulloway, 1996; Ziv & Hermel, 2011).

25.4.3 Sibling Dominance Hierarchies

It is natural for siblings to establish dominance hierarchies based on differences in age, size, power, and status. Older children tend to dominate their younger siblings, and they generally use higher-power tactics to get their siblings to do what they want (Sutton-Smith & Rosenberg, 1970). Sibling dominance hierarchies explain the common finding from personality surveys that, relative to their younger siblings, firstborns score lower on most aspects of agreeableness (Michalski & Shackelford, 2002; Paulhus et al., 1999; Sulloway, 1996, 2001, 2010).

25.4.4 Niche Partitioning

In biology, an ecological niche is defined as the fit between an organism and those aspects of its environment that typify the organism's adaptedness to that environment and its resources. In a similar manner, siblings tend to develop personal niches within the family so as to encourage an optimal level of parental investment, as well as to reduce competition with other siblings. As Darwin (1859) argued in *On the Origin of Species*, competition between species typically leads to specialization and divergence, because diversification reduces costly competition over the same resources. In human families, divergence tends to benefit younger siblings. This is because cultivating a differing family niche, or set of skills, makes it harder for parents to make direct comparisons between their offspring – comparisons that usually favor elder children because of their greater age and ability.

Diversification into different family niches is also a natural consequence of the fact that children growing up together are at different ages. Age, in turn, is linked with differing opportunities to engage in age-specific tasks and activities. Owing to their greater age, firstborns tend to occupy the niche of a surrogate parent, which leads them to emulate parental behavior and to become the "responsible" child in the family. Given the closer relationship that firstborns, in their role as surrogate parents, have with their parents, it is not surprising that mothers are more likely to turn to their eldest children when facing a personal difficulty or crisis (Salmon & Daly, 1998; Suitor & Pillemer, 2007). Because laterborns are less mature than their older siblings, they are not as well suited to taking on the role of a surrogate parent. For example, laterborns are rarely asked to babysit an older sibling. Similarly, if parents ask two of their children to buy something at a nearby store, they are unlikely to entrust money for the purchase to the younger of the two children.

Differences in family niches appear to underlie sibling disparities in at least four of the Big Five personality dimensions. Disparities in conscientiousness appear to arise, in part, through differences in surrogate parenting (Sulloway et al., unpublished data). Surrogate parenting also appears to promote differences in extraversion, through the tendency of surrogate parents to be more assertive than their younger siblings, including engaging in bossy behavior (which leads to lower agreeableness). The cultivation of differing family niches is also likely responsible for the tendency for laterborns to score higher than firstborns on aspects of openness to experience that reflect imagination and unconventionality. These two attributes may help younger siblings to find a unique and unoccupied niche in the family system.

25.4.5 De-identification

Siblings often seek to differentiate themselves, especially from other siblings who are close in age – a process that has been dubbed "de-identification" (Schachter et al., 1978). A secondborn, for example, may seek to develop different interests from a firstborn, and a thirdborn may do the same thing with respect to a secondborn. This process can lead to zigzag trends within the family, whereby adjacent siblings end up being more different from one another than nonadjacent siblings (Skinner, 1992). Such trends have been shown to extend to patterns of identification with parents, with one child having a closer relationship with one parent based on the preference exhibited by another child in the family (Rohde et al., 2003; Schachter, 1982).

25.4.6 Prenatal Effects

Differences in intrauterine environments are known to influence at least one well-documented birth-order difference. Among males (but not females), higher birth rank is correlated with an increasing likelihood of homosexual orientation (Blanchard, 1997, 2004). The explanation for this finding is that a small percentage of mothers (about 15 percent of the population) have a tendency to develop antibodies to the antigens associated with the male-specific histocompatibility complex. These antibodies make it increasingly difficult for mothers to masculinize successive male fetuses, which increases the rate of homosexuality by about 33 percent for each older brother present within the family.

According to Beer and Horn (2000), prenatal hypomasculinization may account for some of the personality differences that are typically found in studies of birth order, with firstborn males exhibiting higher levels of competitive achievement and with laterborn males expressing more "feminine" traits, including higher scores on agreeableness. This hypothesis leads to the prediction that personality differences by birth order should be systematically larger for males than females. This prediction was tested by Sulloway (2001) in a large within-family study that included 4,510 participants. No birth-order differences were found to vary in a significant manner, by sex, for any of the Big Five personality dimension. Additionally, after being controlled for birth order (first/not first), birth rank among males was not significantly associated with any of more than 30 personality traits, including those that typically exhibit gender differences. It therefore appears that intrauterine hypomasculinization, if it does contribute to birth-order differences in personality, provides only a small additional effect, since the previously mentioned tests had 80 percent power to detect disparities amounting to just 1/40th of the overall birth-order differences found in this study. Healey and Ellis (2007) reached a similar conclusion in their within-family study of 325 sibling pairs.

25.4.7 Birth-Order Stereotypes

Research has documented the existence of birth-order stereotypes. Not surprisingly, these stereotypes are nearly identical to the kinds of sibling differences that have been reported in the literature (Baskett, 1985; Musun-Miller, 1993; Nyman, 1995). It is widely believed, for example, that firstborns are more intelligent than laterborns, that they work harder in their studies, and that they tend to be more successful professionally (Herrera et al., 2003). For their own part, laterborns are generally thought to be more easygoing and extraverted than firstborns. The fact that stereotypes exist about birth order does not necessarily invalidate the results of birth-order studies any more than the widespread existence of gender stereotypes calls into question the well-established findings about sex differences in personality (Feingold, 1994). Sulloway (2001) conducted several formal tests suggesting that birth-order stereotypes contribute only minimally to self-reported personality differences.

25.5 FUTURE RESEARCH DIRECTIONS

One of the greatest challenges for future research on birth order is to sort out the relative influence exerted by the various disparate mechanisms proposed to explain the observed effects. Depending on the traits, behaviors, and abilities being considered, it seems likely that multiple mechanisms are generally at work, making it difficult to identify causation in a clear-cut manner. Additionally, a better understanding of these varied mechanisms, including various mediators and moderators of birth-order effects more generally, is needed to understand why studies sometimes produce inconsistent results.

Most of the mechanisms that underlie birth-order effects are expected to vary according to behavioral context. Behaviors that are documented in within-family contexts, for example, are not necessarily expected to generalize to behaviors outside the family milieu because different mechanisms may apply. As a case in point, firstborns have consistently emerged as being more "conscientious" than laterborns in within-family contexts, a finding that appears to reflect the typical firstborn role as a surrogate parent. By contrast, laterborns – particularly middle children – tend to be more peer oriented than firstborns (Salmon, 1999, 2003; Salmon & Daly, 1998). For this reason, they ought to be more conscientious than firstborns in honoring commitments to close friends, and evidence supports such a conclusion (Sulloway et al., unpublished data). Hence, studies that fail to consider the social contexts in which specific behaviors are being expressed run the risk of producing superficially conflicting findings that are properly attributable to person-by-situation interaction effects.

Similarly, laterborns ought to score higher in some aspects of motivation for conscientious behavior (e.g., in academic, athletic, and workplace settings) because they often seek to surpass the achievements of their elder siblings. In a study of Dutch undergraduates, Carette, Anseel, and Van Yperen (2011) found that laterborns were 1.6 times more likely than firstborns to score higher in "performance goals" – namely, those judged by reference to another person's achievements – whereas firstborns were 1.5 times more likely than laterborns to score higher with regard to "mastery goals," or those judged by reference to a person's own abilities. A less discerning study that had not thought to distinguish between these two disparate sources of motivation would likely have concluded that no birth-order differences exist in conscientiousness-related motivation. Thus, firstborns and laterborns may sometimes manifest similar personality traits and behaviors, but they may do so for different reasons.

In a related vein, birth-order effects in openness to experience are often sensitive to the prevailing social context, as can be seen from previous research on openness to innovation. Historically, firstborns have supported many revolutionary innovations in science, some of which laterborns have opposed. In addition, firstborn scientists have earned a proportionately larger share of Nobel Prizes (Clark & Rice, 1982). But the nature of these firstborn innovations, which have tended to be highly technical, have typically contrasted with the kinds of scientific innovations championed by younger siblings. These laterborn-led innovations have often entailed dangerous materialistic and atheistic implications, including Copernican and Darwinian theory (Numbers, 1998; Sulloway, 1996). Both of these revolutionary theories engendered heated controversies that transcended the normal boundaries of scientific

debate and endured for more than a century (Kuhn, 1962). Interestingly, the kinds of person-by-situation interaction effects that have determined whether firstborns or laterborns have supported major revolutions in science involve the same ideological factors that moderate other important predictors of support for innovation, including age, disciplinary and national differences, and mentoring influences (Sulloway, 2009, 2014).

Context matters for much of human behavior, and it often matters a lot (Mischel & Shoda, 1995; Ross & Nisbett, 1991). Unfortunately, a consideration of situational factors has generally been missing from previous research on birth order. It seems likely that birth-order effects are sometimes being underestimated – or overlooked entirely – owing to a failure to consider the critical role of person-by-situation interactions in how this family influence is manifested in real-life behavior. In a Darwinian world in which behaviors are usually adapted to specific contexts, we should expect such nuanced findings. Future studies will hopefully address this and other unresolved issues, including the diverse mechanisms that appear to underlie birth-order effects.

REFERENCES

Adler, A. (1927). *Understanding Human Nature*. New York: Greenberg.

Adler, A. (1928). Characteristics of the first, second, and third child. *Children*, **3**, 14–52.

Altus, W. D. (1966). Birth order and its sequelae. *Science*, **151**, 44–49.

Argys, L. M., Rees, D. I., Averett, S. L., & Witoonchart, B. (2005). Birth order and risky adolescent behavior. *Economic Inquiry*, **44**, 215–233.

Barclay, K. J., & Kolk, M. (2015). Birth order and mortality: A population-based cohort study. *Demography*, **52**, 613–639.

Barni, D., Roccato, M., Vieno, A., & Alfieri, S. (2014). Birth order and conservatism: A multilevel test of Sulloway's "born to rebel" thesis. *Personality and Individual Differences*, **66**, 58–63.

Baskett, L. M. (1985). Sibling status: Adult expectations. *Developmental Psychology*, **21**, 441–445.

Beck, E., Burnet, K. L., & Vosper, J. (2006). Birth-order effects on facets of extraversion. *Personality and Individual Differences*, **40**, 953–959.

Beer, J., & Horn, J. M. (2000). The influence of rearing order on personality development within two adoption cohorts. *Journal of Personality*, **68**, 789–819.

Belmont, L., & Marolla, F. A. (1973). Birth order, family size, and intelligence. *Science*, **182**, 1096–1101.

Blanchard, R. (1997). Birth order and sibling sex ratio in homosexual versus heterosexual males and females. *Annual Review of Sex Research*, **8**, 27–67.

Blanchard, R. (2004). Quantitative and theoretical analyses of the relation between older brothers and homosexuality in men. *Journal of Theoretical Biology*, **230**, 173–187.

Bleske-Rechek, A., & Kelley, J. A. (2014). Birth order and personality: A within-family test using independent self-reports from both firstborn and laterborn siblings. *Personality and Individual Differences*, **56**, 15–18.

Block, J. (1995). A contrarian view of the Five-Factor approach to personality description. *Psychological Bulletin*, **117**, 187–215.

Boone, J. L. (1986). Parental investment and elite family structure in preindustrial states: A case study of late medieval-early modern Portuguese genealogies. *American Anthropologist*, **88**, 859–878.

Brim, O. G., Jr. (1958). Family structure and sex role learning by children: A further analysis of Helen Koch's data. *Sociometry*, **21**, 1–16.

Bu, F. (2016). Examining sibling configuration effects on young people's educational aspiration and attainment. *Advances in Life Course Research*, **27**, 69–79.

Carette, B., Anseel, F., & Van Yperen, N. W. (2011). Born to learn or born to win? Birth order effects on achievement goals. *Journal of Research in Personality*, **45**, 500–503.

Clark, C., & Clark, J. (2008). *Women at the Polls: The Gender Gap, Cultural Politics, and Contested Constituencies in the United States*. Newcastle upon Tyne: Cambridge Scholars Publishing.

Clark, R. D., & Rice, G. A. (1982). Family constellations and eminence: The birth orders of Nobel Prize winners. *Journal of Psychology*, **110**, 281–287.

Claxton, R. (1994). Empirical relationships between birth order and two types of parental feedback. *Psychological Record*, **44**, 475–487.

Costa, P. T., & McCrae, R. R. (1992). *NEO PI-R Professional Manual*. Odessa, FL: Psychological Assessment Resources.

Courtiol, A., Raymond, M., & Faurie, C. (2009). Birth order affects behaviour in the investment game: Firstborns are less trustful and reciprocate less. *Animal Behaviour*, **78**, 1405–1411.

Daly, M., & Wilson, M. (1988). *Homicide*. New York: Aldine de Gruyter.

Damian, R. I., & Roberts, B. W. (2015). The associations of birth order with personality and intelligence in a representative sample of U.S. high school students. *Journal of Research in Personality*, **58**, 96–105.

Darwin, C. (1859). *On The Origin of Species by Means of Natural Selection, or Preservation of Favoured Races in the Struggle for Life*. London: John Murray.

Dixon, M. M., Reyes, C. J., Leppert, M. F., & Pappas, L. M. (2008). Personality and birth order in large families. *Personality and Individual Differences*, **44**, 119–128.

Downey, D. (2001). Number of sibling and intellectual development: The resource dilution explanation. *American Psychologist*, **56**, 497–504.

Drummond, H., & García-Chevelas, C. (1989). Food shortage influences sibling aggression in the blue-footed booby. *Animal Behavior*, **37**, 806–818.

Duby, G. (1977). *The Chivalrous Society*, trans. by C. Poston. Berkeley, CA: University of California Press.

Ernst, C., & Angst, J. (1983). *Birth Order: Its Influence on Personality*. Berlin/New York: Springer-Verlag.

Feingold, A. (1994). Gender differences in personality: A meta-analysis. *Psychological Bulletin*, **116**, 429–456.

Førland, T. E., Korsvik, T. R., & Christophersen, K-A. (2012). Brought up to rebel in the sixties: Birth order irrelevant, parental worldview decisive. *Political Psychology*, **33**, 825–838.

Freese, J., Powell, B., & Steelman, L. C. (1999). Rebel without a cause or effect: Birth order and social attitudes. *American Sociological Review*, **64**, 207–231.

Galton, F. (1874). *English Men of Science*. London: Macmillan.

Hamilton, W. (1964). The genetical evolution of social behavior. Parts 1 and II. *Journal of Theoretical Biology*, **7**, 1–16; 17–52.

Harris, J. R. (1998). *The Nurture Assumption: Why Children Turn Out the Way They Do*. New York: Free Press.

Harris, J. R. (2000). Context-specific learning, personality, and birth order. *Current Directions in Psychological Science*, **9**, 174–177.

Healey, M., & Ellis, B. (2007). Birth order, conscientiousness, and openness to experience: Tests of the family-niche model of personality using a within-family methodology. *Evolution and Human Behavior*, **38**, 55–59.

Herlihy, D. (1977). Family and property in Renaissance Florence. In D. Herlihy & A. L. Udovitch, eds., *The Medieval City*. New Haven, CT: Yale University Press, pp. 3–24.

Herrera, N. C., Zajonc, R. B., Wieczorkowska, G., & Cichomski, B. (2003). Beliefs about birth rank and their reflection in reality. *Journal of Personality and Social Psychology*, **85**, 142–150.

Hertwig, R., Davis, J. N., & Sulloway, F. J. (2002). Parental investment: How an equity motive can produce inequality. *Psychological Bulletin*, **128**, 728–745.

Hilton, I. (1967). Differences in the behavior of mothers toward first- and later-born children. *Journal of Personality and Social Psychology*, **7**, 282–290.

Horton, S. (1988). Birth order and child nutritional status: evidence from the Philippines. *Economic Development and Cultural Change*, **36**, 341–354.

Hrdy, S., & Judge, D. S. (1993). Darwin and the puzzle of primogeniture. *Human Nature*, **4**, 1–45.

Jang, K. L., Livesley, W. J., Angeleitner, A., Riemann, R., & Vernon, P. A. (2002). Genetic and environmental influences on the covariance of facets defining the domains of the five-factor model of personality. *Personality and Individual Differences*, **33**, 83–101.

Jefferson, T., Herbst, J. H., & McCrae, R. R. (1998). Associations between birth order and personality traits: Evidence from self-reports and observer ratings. *Journal of Research in Personality*, **32**, 498–509.

Kidwell, J. S. (1982). The neglected birth order: Middleborns. *Journal of Marriage and the Family*, **44**, 225–235.

Kobayashi, T., Hasegawa, T., Hiraiwa-Hasegawa, M., & Kurashima, O. (2001). The effect of birth order on personality and familial sentiment in Japan. Poster presented at the 13th conference of the Human Behavior and Evolution Society, University College of London, London.

Koch, H. (1955). Some personality correlates of sex, sibling position, and sex of sibling among five- and six-year-old children. *Genetic Psychology Monographs*, **52**, 3–50.

Koch, H. (1956). Attitudes of young children toward their peers as related to certain characteristics of their siblings. *Psychological Monographs*, **70**, 1–41.

Krishnamurthy, K. S., Shaanker, R. U., & Ganeshaiah, K. N. (1997). Seed abortion in an animal dispersed species, *Syzgium cuminii* (L.). Skeels (Myrtaceae): The chemical basis. *Current Science*, **73**, 869–873.

Kristensen, P., & Bjerkedal, T. (2007). Explaining the relation between birth order and intelligence. *Science*, **316**, 1717.

Kuhn, T. S. (1962). *The Structure of Scientific Revolutions*. Chicago, IL: University of Chicago Press.

Kulik, L. (2004). The impact of birth order on intergenerational transmission of attitudes from parents to adolescent sons: The Israeli case. *Journal of Youth and Adolescence*, **33**, 149–157.

Lehmann, J.-Y. K., Nuevo-Chiquero, A., & Vidal-Fernandez, M. (2018). The early origins of birth order differences in children's outcomes and parental behavior. *Journal of Human Resources*, **53**, 123–156.

Lindert, P. H. (1977). Sibling position and achievement. *Journal of Human Resources*, **12**, 198–219.

Loehlin, J. C. (1992). *Genes and Environment in Personality Development*. Newbury Park, CA: Sage Publications.

MacAndrew, F. T., King, J. C., & Honoroff, L. R. (2002). A sociological analysis of namesaking patterns in 322 American families. *Journal of Applied Social Psychology*, **32**, 851–864.

Manaster, G. J., Rhodes, C., Marcus, M. B., & Chan, J. C. (1998). The role of birth order in the acculturation of Japanese Americans. *Psychologia*, **41**, 155–170.

Marini, V. A., & Kurtz, J. E. (2011). Birth order differences in normal personality traits: Perspectives from within and outside the family. *Personality and Individual Differences*, **51**, 910–914.

Michalski, R. L., & Shackelford, T. K. (2002). An attempted replication of the relationship between birth order and personality. *Journal of Research in Personality*, **36**, 182–186.

Mischel, W., & Shoda, Y. (1995). A cognitive–affective system theory of personality: Reconceptualizing situations, dispositions, dynamics, and invariance in personality structure. *Psychological Review*, **102**, 246–268.

Mock, D. W., Drummond, H., & Stinson, C. H. (1990). Avian siblicide. *American Scientist*, **78**, 438–449.

Modin, B. (2002). Birth order and mortality: A life-long follow-up of 14,200 boys and girls born in early 20th century Sweden. *Social Science & Medicine*, **54**, 1051–1064.

Musun-Miller, L. (1993). Sibling status effects: Parents' perceptions of their own children. *Journal of Genetic Psychology*, **154**, 189–198.

Numbers, R. L. (1998). *Darwinism Comes to America*. Cambridge, MA: Harvard University Press.

Nyman, L. (1995). The identification of birth order personality attributes. *Journal of Psychology*, **129**, 51–59.

Paulhus, D. L., Chen, D., & Trapnell, P. D. (1999). Birth order and personality within families. *Psychological Science*, **10**, 482–488.

Price, J. (2008). Parent–child quality time: Does birth order matter? *Journal of Human Resources*, **43**, 240–265.

Rodgers, J. L., Cleveland, H. H., van den Oord, E., & Rowe, D. C. (2000). Resolving the debate over birth order, family size, and intelligence. *American Psychologist*, **55**, 599–612.

Rohde, P. A., Atzwanger, K., Butovskaya, M., et al. (2003). Perceived parental favoritism, closeness to kin, and the rebel of the family: The effects of birth order and sex. *Evolution and Human Behavior*, **24**, 261–276.

Rohrer, J. M., Egloff, B., & Schmukle, S. (2015). Examining the effects of birth order on personality. *Proceedings of the National Academy of Sciences*, **112**, 14224–14229.

Rosenblatt, P. C., & Skoogberg, E. L. (1974). Birth order in cross-cultural perspective. *Developmental Psychology*, **10**, 48–54.

Ross, L., & Nisbett, R. E. (1991). *The Person and the Situation: Perspectives on Social Psychology*. New York: McGraw-Hill.

Saad, G., Gill, T., & Nataraajan, R. (2005). Are laterborns more nonconforming and open to innovation than firstborns? A Darwinian perspective. *Journal of Business Research*, **58**, 902–909.

Salmon, C. A. (1998). The evocative nature of kin terminology in political rhetoric. *Politics and the Life Sciences*, **17**, 51–57.

Salmon, C. A. (1999). On the impact of sex and birth order on contact with kin. *Human Nature*, **10**, 183–197.

Salmon, C. A. (2003). Birth order and relationships. *Human Nature*, **14**, 73–88.

Salmon, C. A., & Daly, M. (1998). Birth order and familial sentiment: Middleborns are different. *Evolution and Human Behavior*, **19**, 299–312.

Salmon, C. A., Shackelford, T. K., & Michalski, R. L. (2012). Birth order, sex of child, and perceptions of parental favoritism. *Personality and Individual Differences*, **52**, 357–362.

Saroglou, V., & Fiasse, L. (2003). Birth order, personality, and religion: A study among young adults from a three-sibling family. *Personality and Individual Differences*, **35**, 19–29.

Schachter, F. F. (1982). Sibling deidentification and split-parent identifications: A family tetrad. In M. E. Lamb & B. Sutton-Smith, eds., *Sibling Relationships: Their Nature and Significance across the Lifespan*. Hillsdale, NJ: Lawrence Erlbaum, pp. 123–152.

Schachter, F. F., Gilutz, G., Shore, E., & Adler, M. (1978). Sibling deidentification judged by mothers: Cross-validation and developmental studies. *Child Development*, **49**, 543–546.

Schooler, C. (1972). Birth order effects: Not here, not now! *Psychological Bulletin*, **78**, 161–175.

Skinner, G. W. (1992). Seek a loyal subject in a filial son: Family roots of political orientation in Chinese society. In Institute of Modern History, Academia Sinica, ed., *Family Process and Political Process in Modern Chinese History*, Vol. 2. Taipei: Chiang Ching-kou Foundation for International Scholarly Exchange, pp. 943–993.

Skinner, N. F., & Fox-Francoeur, C. A. (2010). Personality implications of adaption-innovation: V. Birth order as a determinant of cognitive style. *Social Behavior and Personality*, **38**, 237–240.

Suitor, J. J., & Pillemer, K. (2007). Mother's favoritism in later life: The role of children's birth order. *Research on Aging*, **29**, 32–55.

Sulloway, F. J. (1995). Birth order and evolutionary psychology. *Psychological Inquiry*, **6**, 75–80.

Sulloway, F. J. (1996). *Born to Rebel: Birth Order, Family Dynamics, and Creative Lives*. New York: Pantheon.

Sulloway, F. J. (2001). Birth order, sibling competition, and human behavior. In H. R. Holcomb III, ed., *Conceptual Challenges in Evolutionary Psychology: Innovative Research Strategies*. Dordrecht/Boston, MA: Kluwer Academic Publishers, pp. 39–83.

Sulloway, F. J. (2007). Birth order and intelligence. *Science*, **317**, 1711–1712.

Sulloway, F. J. (2009). Sources of scientific innovation: A meta-analytic approach. *Perspectives on Psychological Science*, **4**, 455–459.

Sulloway, F. J. (2010). Why siblings are like Darwin's Finches: Birth order, sibling competition, and adaptive divergence within the family. In D. M. Buss & P. H. Hawley, eds., *The Evolution of Personality and Individual Differences*. Oxford: Oxford University Press, pp. 86–119.

Sulloway, F. J. (2014). Openness to scientific innovation. In D. K. Simonton, ed., *The Handbook of Genius*. Oxford: Wiley-Blackwell, pp. 546–563.

Sulloway, F. J., & Zweigenhaft, R. L. (2010). Birth order and risk taking in athletics: A meta-analysis and study of major league baseball players. *Personality and Social Psychology Review*, **14**, 402–416.

Surbey, M. K. (1998). Parent and offspring strategies in the transition at adolescence. *Human Nature*, **9**, 67–94.

Sutton-Smith, B., & Rosenberg, B. G. (1970). *The Sibling*. New York: Holt, Rinehart and Winston.

Theroux, N. L. (1993). Birth order and its relationship to academic achievement and selected personality traits.

Unpublished doctoral dissertation, University of California, Los Angeles.

Trivers, R. L. (1972). Parental investment and sexual selection. In B. Campbell, ed., *Sexual Selection and the Descent of Man, 1871–1971*. Chicago, IL: Aldine-Atherton, pp. 136–170.

Trivers, R. L. (1974). Parent–offspring conflict. *American Zoologist*, **14**, 249–264.

Turner, F. M. (1990). The Victorian crisis of faith and the faith that was lost. In R. J. Helmstadter & B. Lightman, eds., *Victorian Faith in Crisis: Essays on Continuity and Change in Nineteenth-Century Religious Belief*. London: Macmillan, pp. 9–38.

Voland, E. (1988). Differential infant and child mortality in evolutionary perspective: Data from the late 17th to 19th century Ostfriesland (Germany). In L. Betzig, M. Borgerhoff Mulder, & P. Turke, eds., *Human Reproduction Behaviour: A Darwinian Perspective*. Cambridge, UK: Cambridge University Press, pp. 253–262.

Voland, E. (1990). Differential reproductive success within the Krummhorn population (Germany, 18th and 19th centuries). *Behavioral Ecology and Sociobiology*, **26**, 54–72.

Wichman, A. L., Rodgers, J. L., & MacCallum, R. C. (2006). A multilevel approach to the relationship between birth order and intelligence. *Personality and Social Psychology Bulletin*, **32**, 117–127.

Zajonc, R. B. (1976). Family configuration and intelligence. *Science*, **192**, 227–236.

Zajonc, R. B., & Mullally, P. R. (1997). Birth order: Reconciling conflicting effects. *American Psychologist*, **52**, 685–699.

Zajonc, R. B., & Sulloway, F. J. (2007). The confluence model: Birth order as a within-family or between-family dynamic? *Personality and Social Psychology Bulletin*, **33**, 1187–1194.

Ziv, I., & Hermel, O. (2011). Birth order effects on the separation process in young adults: An evolutionary and dynamic approach. *American Journal of Psychology*, **124**, 261–273.

Zweigenhaft, R. L. (2002). Birth order effects and rebelliousness: Political activism and involvement with marijuana. *Political Psychology*, **23**, 219–233.

Zweigenhaft, R. L., & Von Ammon, J. (2000). Birth order and civil disobedience: A test of Sulloway's "born to rebel" hypothesis. *Journal of Social Psychology*, **140**, 624–627.

PART VII

SEXUAL SELECTION AND HUMAN SEX DIFFERENCES

If you have ever walked too close to a goose, then you will probably have noticed her tendency to back away. In marked contrast, if you have ever walked too close to a gander, you will no doubt have noticed his tendency to puff up his feathers and hiss in a clear threat display. This response is designed to make you want to back away. Why should this be the case? According to Darwin's theory of natural selection, what is good for the goose should also be good for the gander. In order to account for differences between the sexes in behavior (and in form), Darwin introduced a second selection mechanism, which he termed sexual selection (Darwin, 1859, 1871). To him, there were two components to this form of selection – competition for access to the opposite sex and the choosiness of one sex in selecting members of the other. The former was largely a male concern, whereas the latter was of concern to the females of a population. Ever since its inception, the concept of sexual selection has been a controversial one. In Darwin's time, other naturalists considered the notion of female choice to be absurd, as in most cases males were larger and more powerful than their female counterparts (even Wallace rejected the notion – see Cronin, 1991). Twentieth-century ethological research, however, demonstrated how correct Darwin was in predicting the selective power of female choice (Fisher, 1915; Jones & Ratterman, 2009). Despite this, Darwin never quite worked out why females were the choosier sex. This puzzle was finally resolved a century later when Robert Trivers developed parental investment theory (Trivers, 1972). This postulated that the sex that invests most in offspring will be maximally choosy, whereas the sex that invests least is the one most likely to engage in promiscuous behavior. When applied to nonhuman species, a combination of sexual selection and parental investment theory became a well-accepted explanation for sex differences in behavior (Cronin, 1991). When, however, human behavioral ecologists and evolutionary psychologists turned their attention to sex differences in our own species, the use of sexual selection and parental investment theory led to a great deal of (often acrimonious) debate. The debate concerning the extent to which sexual selection and parental investment theory can explain sex/gender differences persists today. In Part VII, we explore aspects of this debate from a variety of informed positions.

We begin with the late Anne Campbell's examination of the relationship between sexual selection and sex differences in levels of fear. Anne Campbell was a rare academic in this field, in that her research is well respected by evolutionary scientists and feminist theorists alike. Even those who took issue with her conclusions accepted her as an international authority on gender differences in behavior. The majority of research into fear and aggression has focused on factors that affect these states in males. Although women are clearly capable of violent, aggressive behavior (Campbell, 2013), cross-culturally, when placed in agonistic situations, they are more likely to gravitate to dovelike than to hawklike strategies when compared with men. Campbell suggests that women are more prone to fear than men because, in contrast to men, "living slow and dying late" enhance their inclusive fitness. Interestingly, she also points out that, in cases of defending the young, females then have a reduction in levels of fear. Hence, when considering sex differences in behavior, we should always consider contextual factors.

This is followed by Frederick M. Toates' discussion of the relationship between evolution and sexual desire. Here, Toates distills a number of the arguments laid out in his award-winning book, *How Sexual Desire Works* (Toates, 2014). Building on his chapter on motivation theory in Chapter 7, Toates argues that we need to bring causal factors into evolutionary psychology if we are to understand many otherwise inexplicable features of sexual desire. Such features include differences in sexual behavior between men and women (Barkow, 1989; Buss, 2015). Although such differences are compatible with sexual selection theory, Toates explains how our understanding of sex differences can be more fully fleshed out by applying incentive motivation theory.

Species vary in their degree of sexual dimorphism, both in terms of physical attributes and in terms of behavior. Taking birds as an example, peacocks are far more elaborate than peahens, whereas for European robins the sexes are indistinguishable, even to the trained observer. Such differences are explained by the level of male parental investment. Peacocks provide no parental investment beyond the sperm they dispense (to as many hens as possible). Since all they provide is their genes, they are highly competitive and signal the quality of said genes via their obvious "sexiness." A male robin feeds the broodlings with as much vigor as his mate up to and beyond the point of fledging. Most evolutionists have suggested men and women are rather similar to peacocks and peahens, at least with respect to males' propensity (given the opportunity) toward polygyny. Since 2014, Steve Stewart-Williams has been questioning this assumption (Stewart-Williams & Thomas, 2013; see also Stewart-Williams, 2018). In his contribution to Part VII, Stewart-Williams presents evidence that, as a species, humans are more like the monomorphic pair-bonding robin than we are the highly dimorphic polygynous peacock. The main reason for this is the fact that human babies are born relatively altricial, meaning that it pays both parents to invest in offspring – hence high male parental investment. Similar levels of parental investment lead, in turn, to robinlike low levels of sexual dimorphism. Stewart-Williams calls this the "mutual mate choice" model, in contrast to the often accepted "males compete, females choose" model.

David M. Buss is one of the major figures in the development of evolutionary psychology, and Daniel Conroy-Beam is a rapidly up-and-coming researcher in this field. In their contribution to Part VII, Conroy-Beam and Buss point out that evolutionary approaches have had major successes in unraveling the nature of human mate choice preferences and the factors such as gender and cultural background that predict such preferences. Despite this progress, we know far less about the gap between ideal and actual mating outcomes. In their mathematical treatment, Conroy-Beam and Buss present a multidimensional model of mate selection to explain a number of counterintuitive findings in the current mate choice literature.

The existence of homosexuality has long been a paradox for evolutionary scientists. Given that both male and female individuals who form enduring same-sex romantic relationships produce significantly fewer offspring than heterosexual individuals, this widespread state of affairs is a conundrum for evolutionists (Schwartz et al., 2010). Does this mean that human sexuality is free from evolutionary concerns, or are homosexual individuals adopting an alternative adaptive strategy? In the contribution by Paul L. Vasey, Lanna J. Petterson, Scott W. Semenyna, Francisco R. Gómez, and Doug P. VanderLaan, this second explanation is considered for male homosexuality (technically "male androphilia"). Vasey – who is a world-leading expert in this field – and his team reexamine the notion that such men are adopting a kin-selected, avuncular role to pass on their genes indirectly via their nephews and nieces. This cross-cultural anthropological review provides an insight into the current state of play in this fascinating area of research.

To return to our avian analogies, it might be said that the final contribution to Part VII certainly puts the cat among the pigeons, as Maryanne L. Fisher, Justin R. Garcia, and Rebecca L. Burch consider the possibility that evolutionary psychology might benefit from an integration of feminist perspectives. They begin by suggesting that sexual selection theory has been used by evolutionary scientists to reinforce sexual inequalities. In their feminist theorist critique, Fisher et al. also examine the charges that evolutionary approaches reinforce sexual double standards, ignore sociocultural and gendered factors, and reify heteronormativity. Fisher, Garcia, and Burch are all experienced scholars in the fields of both feminist theory and evolutionary psychology. Hence, their critique of evolutionary approaches also considers areas where feminist scholars have, at times, presented misinterpretations of the aims and findings of evolutionists. Such feminist scholars have frequently been accused of creating straw man arguments and of simplifying evolutionary interpretations of behavior. Are they the pigeons or the statue? You decide.

REFERENCES

Barkow, J. H. (1989). *Darwin, Sex, and Status: Biological Approaches to Mind and Culture*. Toronto: University of Toronto Press.

Buss, D. M. (2015). *Evolutionary Psychology: The New Science of the Mind*. New York: Routledge.

Campbell, A. (2013). *A Mind of Her Own: The Evolutionary Psychology of Women*. Oxford: Oxford University Press.

Cronin, H. (1991). *The Ant and the Peacock: Altruism and Sexual Selection from Darwin to Today*. Cambridge, UK: Cambridge University Press.

Darwin, C. (1859). *On The Origin of Species by Means of Natural Selection, or Preservation of Favoured Races in the Struggle for Life*. London: John Murray.

Darwin, C. (1871). *The Descent of Man, and Selection in Relation to Sex*. London: John Murray.

Fisher, R. A. (1915). The evolution of sexual preference. *Eugenics Review*, **7**, 184–192.

Jones, A. G., & Ratterman, N. L. (2009). Mate choice and sexual selection: What have we learned since Darwin? *Proceedings of the National Academy of Sciences*, **106**, 10001–10008.

Schwartz, G., Kim, R. M., Kolundziji, A. B., Rieger, G., & Sanders, A. R. (2010). Biodemographic and physical correlates of sexual orientation in men. *Archives of Sexual Behavior*, **39**, 93–109.

Stewart-Williams, S. (2018). *The Ape That Understood the Universe: How the Mind and Culture Evolve*. Cambridge, UK: Cambridge University Press.

Stewart-Williams, S., & Thomas, A. G. (2013). The ape that thought it was a peacock: Does evolutionary psychology exaggerate human sex differences? *Psychological Inquiry*, **24**, 137–168.

Toates, F. (2014). *How Sexual Desire Works: The Enigmatic Urge*. Cambridge, UK: Cambridge University Press.

Trivers, R. L. (1972). Parental investment and sexual selection. In B. Campbell, ed., *Sexual Selection and the Descent of Man, 1871–1971*. Chicago, IL: Aldine-Atherton, pp. 136–179.

26 Survival, Selection, and Sex Differences in Fear

ANNE CAMPBELL

A cursory glance at human behavior suggests that men are considerably less careful with their lives than women. They are up to four times more likely to die from external causes (Kruger & Nesse, 2006). Many of these deaths are the result of car accidents, and men commit 97 percent of drink-driving offenses in Britain (Social Issues Research Centre, 2004). Men also die at a higher rate from nonvehicle accidents, such as drowning, falling, and electrocution (Pampel, 2001). They are the victims in 77 percent of homicides and the perpetrators in 88 percent of them (Federal Bureau of Investigation, 2012). They participate more often in extreme sports such as skydiving and mountain climbing (Robinson, 2008). The evolutionary explanation for men's seemingly careless attitude to survival is that such behaviors are men's way of conspicuously showing off their appetite for risk, the better to elicit respect from other men and sexual interest from women. In this chapter, I will argue that this account is a partial one, and that a fuller understanding of sex differences in risk-taking requires closer examination of women as well.

26.1 SEXUAL SELECTION ACTS ON FEMALES, TOO

The study of sexual selection has been, in the main, the study of male rather than female behavior. There is a theoretical (rather than sexist) reason for this. In a sexually reproducing species, the contributions of the two parents are rarely equal. The egg produced by the female is more costly than male sperm. Indeed, while it takes about 28 days for a woman's reproductive system to ready itself for a possible conception, sperm can be produced at the rate of 1,500 per second (Nakagawa et al., 2010). Male biologists such as George Williams (1996, p. 118) seem rather shamefaced about the inequity: "A sperm is not a contribution to the next generation; it is a claim on contributions put into the egg by another individual. Males of most species make no investments in the next generation, but merely compete with one another for the opportunity to exploit investments made by females." This inequity continues with gestation and lactation, during which time a female is effectively out of action

(reproductively speaking) while the male is technically free to continue inseminating other women. This is what creates the ubiquitous male-biased operational sex ratio (OSR; the ratio of reproductively competent and available males to females). There are never enough females to go round because a proportion of them are always in a nonreproductive state, making them a limiting resource for which males compete. This competition (involving male combat and female choice) means that some males will mate more than their fair share and others will not mate at all. But all females, according to the male-centered theoretical perspective, will be able to find an obliging male to copulate with, so, as a sex, they will be reproducing at close to their maximum rate (Daly & Wilson, 1988). This means that almost all females will contribute genes to the next generation, but the same is not true of males. Only the "best" males will reproduce, which means there is stronger selection pressure on male traits. Since females do not need to compete for copulation opportunities, it was thought that they had little to compete about. In this male-centered view, the contribution of females to the selection process is to act as genetic quality controllers of male genes.

Many sex differences have been attributed to sexual selection acting on males, including aggression (more aggressive males gain rank and so reproduce more), upper body strength (the better to intimidate rival males), spatial ability (better hunters catch more prey and attract more mates), and sex drive (to take advantage of all mating possibilities). Another sex difference ascribed to sexual selection acting on males is risk-taking. Wilson and Daly (1985) proposed that men's willingness to fight over apparently trivial causes is a manifestation of men's competition for status, respect, and reputation ultimately deriving from sexual selection. This competition, they argued, extends beyond physical domination into the realms of risk-taking more generally. "We furthermore expect an evolved inclination toward the social display of one's competitive risk-taking skills, and again this should be especially a masculine trait. Just why the maintenance of reputation should require incurring risk can be answered in terms of pressures favoring 'honest advertising'" (Wilson & Daly, 1985, p. 66). People imagine (incorrectly as it turns out) that male risk-takers are larger, stronger, and more violent than risk-avoiders (Fessler et al., 2014). So according to traditional sex role accounts, men were selected to seek out risks because it increased their reproductive success – either by intimidating rivals or by impressing prospective mates.

But this viewpoint is not without problems. It is premised on the assumption that sexual selection acting on traits such as male aggression will operate as a positive-feedback loop with consistent selection for ever-greater male aggression. However, if the competition becomes too dangerous or if some males lack the ability to win such same-sex aggressive encounters, a better strategy would be to remain with a single female contributing to the care of joint offspring (Gangestad & Simpson, 2000; Kokko & Jennions, 2008). Consistent with this, a cross-species meta-analysis (Weir, Grant, & Hutchings, 2011) demonstrated that aggression rises with an increasingly male-biased OSR until it approaches 2:1, when aggression rates begin to decrease due to the intensity of competition. As the OSR increases, courtship rates decrease and mate guarding increases, reflecting males' attempts to retain their current mates rather than competing for new ones. Although the selective forces that led to widespread human pair-bonding are debated, Gavrilets (2012) found support for a plausible model that showed that, if females were willing to forego extra-pair matings and preferred males who provided resources, male provisioning would increase driven chiefly by low-ranking males who would otherwise fail to reproduce at all. Because these males outnumbered elite males, selection would have acted more strongly on them and their strategy, causing it to spread through the male population.

Notice that the traditional male-focused argument rests on a characterization of humans as a fundamentally polygynous species, or at least a species in which the variance in men's reproductive success substantially exceeds that of women. Yet there is considerable evidence that humans have been fundamentally monogamous for over a million years. True, even under monogamy, men's reproductive success remains slightly higher than women's because men are more likely than women to remarry and produce further children (Jokela et al., 2010). But a history of human pair-bonding is supported by our species' relatively modest sexual dimorphism compared to that of polygynous primates such as gorillas (Stewart-Williams & Thomas, 2013). Archaeological findings, relevant to the "obstetric crisis" and the "premature" birth of infants, date biparental care to the start of the *Homo* line, 1.5–2.0 million years ago (Eastwick, 2009). Other archaeological findings place it even earlier with *Australopithecus afarensis*, the predecessors of *Homo* (Reno et al., 2003). Contemporary evidence of men's adaptation to paternal investment can be seen in the downregulation of men's testosterone levels following pair-bonding and fatherhood (Gray & Anderson, 2010) and the rise in oxytocin levels in fathers (as well as mothers) after the birth of a baby and during interaction with them (Gordon et al., 2010).

The effects of biparental care and pair-bonding on relations between the sexes are profound. Under polygyny, males can exploit females quite ruthlessly without suffering any costs themselves, but monogamy means that anything that hurts a female (prevents her from achieving her reproductive potential) hurts her male partner just as much. Monogamy reduces the sexual conflict that drives antagonistic coevolution (Holland & Rice, 1999). Under monogamy, males behave in a less exploitative way toward their female partners and so females do not need to evolve counterstrategies of resistance. Humans' basically monogamous mating system, through its effects on individual behavior, brings with it societal rewards in terms of reduced crime rates and stronger within-group alliances. For women, it reduces gender inequality and domestic conflict while increasing child survival (Henrich, Boyd, & Richerson, 2012). Monogamy also means that mate selection becomes a two-way street: when men constrain their (potentially) high reproductive output to that of a single woman, they become much choosier. Now women as well as men must compete for the best partners. Marketplace forces mean that this often results in assortative mating. Men's mate choice exerts selection pressure on women much more strongly than under a polygynous system, and this means that there are more marked differences between women in their reproductive success.

Whichever mating system is in operation, there is an area of a woman's life that is ruthlessly subject to selection: the production and survival of children. In examining this, we need to step back a little and consider the rather male-centered conceptualization of reproductive success and sexual selection. In some studies, male reproductive success has been equated with number of copulations. This occurred historically in primate studies because DNA analysis to establish paternity was not readily available (Ellis, 1995) and because in humans the use of contraception interrupts the usual relationship between intercourse and pregnancy so that frequency of the former has sometimes been used as a proxy for the latter (e.g., Perusse, 1993). At a conceptual level, the equation of intercourse with reproductive success is likely attributable to a selective reading of Darwin (1859), who defined sexual selection as "a struggle between the males for possession of the females" (van Wyhe, 2002). However, he later offered a more inclusive definition (Darwin, 1871): "the advantage which certain individuals have over others of the same sex and species solely in respect of reproduction" (van Wyhe, 2002). In this definition, sexual selection encompasses the entire reproductive process and so includes the contribution made by women to the survival of their offspring. As Hrdy (1999, p. 81) bluntly put it: "Unless mating results in the production of offspring who themselves survive infancy and the juvenile years and position themselves so as to reproduce, sex is only so much sound and undulation signifying nothing." The action of sexual selection on women means that women who gestated, birthed, nursed, fed, and protected a greater number of offspring left more copies of their genes in future generations than less capable mothers. We are the recipients of those successful maternal genes.

I have noted that although monogamy dramatically reduces the sex difference in reproductive variance, it does not eliminate it completely. With this in mind, let us revisit the male-centered argument that greater prizes warrant greater risks; or, to translate this proposal into a reciprocal female version: since women vary less in their reproductive success than men, there is correspondingly less incentive for them to compete reproductively with one another (no implication of conscious intent or rivalry intended). Remember that women as a sex are not competing with men: since every child has a mother and a father, the mean output of the two sexes is identical. The fact that there is greater variance in male reproductive success is irrelevant to the action of selection on females' reproductive success. When variance is constrained as it is among women, even small differences between individuals are crucial. It is analogous to a race where the competitors are very closely bunched and differ little in maximum speed. Even a slight advantage is critical to improving one's position, just as the birth or survival of even one "extra" child can have a real impact on the genetic contribution that a woman leaves behind. When variance is tight, a small genetic adjustment that may appear to offer only a slight reproductive advantage can be disproportionately important – and therefore visible to selection. Any trait whose net effect is to improve mothering will be favored, and the genes that underlie it will be copied into more bodies in the next generation.

26.1.1 Mothers Matter Most

There are time and energy constraints on the number of offspring women can produce. Their output is limited by lengthy infertile periods of gestation and lactation and truncated by menopause long before somatic senescence. Their investment in each offspring is measured in years as a result of children's long and demanding period of dependency. Despite this, women are unusual among primates in their short inter-birth intervals, resulting in the need to care simultaneously for more than one dependent child. The energetic demands on mothers and the replacement costs of each child mean that children are extremely precious. Yet childhood is the time in the human life cycle when natural selection acts most strongly (Jones, 2009). In hunter–gatherer societies, on average, only 57 percent of children survive to age 15. In these societies, the infant mortality rate is over 30 times greater than in the USA, and the early child mortality rate is 100 times greater (Gurven & Kaplan, 2007).

The key person standing between a young child and death is their chief caretaker – their mother. In a review

of studies spanning 28 societies lacking access to contraception and medical care, Sear and Mace (2008) examined the impact of parental survival on offspring survival. The populations sampled ranged from eighteenth-century China to twentieth-century Nepal; from Burkina Faso to New York State. In every case, a mother's survival increased the likelihood of her children surviving. The percentage of children surviving a mother's death ranged from 2 to 50 percent. The beneficial effects were more marked at younger ages, before children are weaned. In rural Gambia, a mother's death multiplies the odds of her child's death by 6.2 in infancy, 5.2 in toddlerhood, and 1.4 in childhood (Sear et al., 2002). Using data from seventeenth-century Quebec, Pavard et al. (2005) cross-tabulated the age of the child when the mother died and the age at which the child itself died, controlling for a range of variables, including the possibility of transmitted infection and shared genetic vulnerability. If the mother died while the infant was still a neonate, the odds of the child dying in the neonatal period were multiplied by 5.52, dropping to 1.27 when the child was aged 5–15 years.

Perhaps the most striking finding from Sear and Mace's (2008) review is that in 68 percent of cases a father's death had no impact on the survival chances of his child. In 32 percent of cases, it actually improved the child's odds of surviving. Even among the Tsimane of Bolivia, who have high levels of paternal provisioning and low divorce rates, the early death of a father had no impact on children's age of first reproduction, completed fertility, or number of surviving offspring (Winking et al., 2011). It seems that the presence of a male partner was advantageous to women not because it increased offspring survival, but because it decreased the workload on mothers. A synthesis of data from 10 studies of gatherer societies (Kaplan et al., 2000) suggests that between 60 percent (among the Nukak) and 84 percent (among the Ache) of the calories consumed are contributed by men. With men relieving women of the full burden of provisioning, women were better able to feed their dependent children, sustain pregnancy and lactation, and return to normal cycling more quickly, thus shortening inter-birth intervals (Worthman et al., 1993). But the presence of a father did not dilute the central role of the mother in keeping her children alive.

So we have a situation where women invest more heavily than men in offspring, small differences between women in their reproductive output are highly visible to selection, early childhood is a time of heightened mortality risk, and mothers play a key role in ensuring their children's survival. How might the confluence of these factors adaptively sculpt the female brain? Women with a heightened ability to detect danger would have left more surviving children – both because they were better able to protect their children's safety and because by avoiding risks to themselves they stayed alive to care for them. In Section 26.2, I examine two areas where women's sensitivity to fear works to enhance their survival: their greater avoidance of risk in general and aggression in particular.

26.2 THE USES OF FEAR

26.2.1 Avoiding Risk

Men's greater engagement in risky behavior, noted at the outset of this chapter, has been confirmed by a meta-analysis of controlled behavioral observations, although the overall effect size was quite modest ($d = 0.19$; Byrnes, Miller, & Schaffer, 1999). To isolate and manipulate the degree of risk involved, decision-making tasks have been used. Economists have asked respondents to choose between entering lotteries with different objective probabilities of winning (some hypothetical and some with real payoffs), where women have been found to be more risk averse (Croson & Gneezy, 2009). Decisions about investment choices similarly show women to be more conservative, although studies of entrepreneurs and fund managers show a smaller sex difference, perhaps reflecting self-selection or enculturation into risk acceptance (see also Nelson, 2015). The smallest sex differences in risk-taking are found when participants were asked to make hypothetical third-party decisions (e.g., indicating what the odds of a successful outcome would have to be before recommending that a patient undergo a risky operation). These decisions carry no personal consequences and are cognitively "cold."

It has been suggested that the role of cognition has been overemphasized in decision-making generally and especially in risky decision-making. Gut reactions (the "affect heuristic") often guide behavior (Slovic et al., 2002), and risk-taking has been proposed to have a strong affective component (Loewenstein et al., 2001). Indeed, Byrnes et al. (1999, p. 378) suggested that differences in fear "may explain gender differences in risk taking more adequately than the cognitive processes involved in the reflective evaluation of options." There is considerable evidence that women are more alert to the potential negative consequences of risky decisions than men. In one study (Harris, Jenkins, & Glaser, 2006), women reported that they were less likely than men to take risks in the areas of gambling, recreation, and health (although not socially), and the sex differences were significantly mediated by women's higher ratings of the likelihood of negative consequences and the severity of those consequences were they to occur.

This hypervigilance for negative outcomes is called punishment sensitivity and has been measured by questionnaire responses to items such as "I feel pretty worried or upset when I think or know somebody is angry at me" and "I have very few fears compared to my friends" (reverse scored) (Carver & White, 1994). Punishment sensitivity shows a significant sex difference favoring women, with an effect size of $d = -0.33$ over 17 studies (Cross, Copping,

& Campbell, 2011), although its companion scale measuring sensitivity to reward or positive outcomes shows no sex difference ($d = 0.01$). Even among women who engage in high-risk activities, it is their low sensitivity to punishment that predicts their differential willingness to take risks. Maher, Thomson, and Carlson (2015) used measures of reward and punishment sensitivity to predict sport-related risk-taking among a group of skiers and snowboarders. In women, punishment sensitivity was strongly negatively predictive of risk taking ($r = -0.64$), while the correlation for men was significantly lower ($r = -0.36$). In men, reward sensitivity was positively associated with risk-taking, but it showed no association in women.

Greater female sensitivity to negative outcomes is also seen in a decision-making task (Powell & Ansic, 1997) in which participants choose between definitely receiving a sum of money (ranging between $50 and $120) or gambling (with a 50 percent probability of winning $200 and a 50 percent probability of winning nothing). Women are more risk averse, preferring the certain option. However, when the task is inverted so that it is framed in terms of potential losses, women take the "risky" (gambling) option more than men. This outcome is congruent with women being more loss sensitive: they avoid the potential loss in the first version of the task (by taking the "safe" option) and avoid the certain loss in the second version (by taking the "risky" option). In a hypothetical lottery study, women's significantly greater reluctance to buy a lottery ticket disappeared when the intensity of emotional reactions to losing was statistically controlled (Eriksson & Simpson, 2010).

The Balloon Analogue Risk Task (BART) is another widely used test of risky decision-making. Participants pump up a computer-simulated balloon, with each inflation increasing their prize points. If the balloon bursts, they lose all of the points they have accumulated in that round. They can choose to "bank" (protect) their point winnings at any point and move on to the next round. When punishing outcomes were made salient by framing the instruction negatively to emphasize that money could be lost if the balloons burst, women but not men showed a reduction in risk-taking (Gabriel & Williamson, 2010). If women's heightened fear serves an adaptive function, we would expect women to be especially risk avoidant when they sense potential threat. This turns out to be the case. Stress has opposite effects on the two sexes: in women it decreases risk-taking on BART while enhancing it in men (Daughters et al., 2013; Lighthall, Mather, & Gorlick, 2009). We would also expect that making infants salient would decrease risk-taking in women. Using the BART paradigm, Fischer and Hills (2012) had participants play in a regular baseline condition and when a photograph of a same-sex adult, an opposite-sex adult, or a baby (with whom they would share their winnings) was displayed during the task. In line with the hypothesis that women's risk aversion is associated with their maternal roles, women but not men were significantly more risk averse when an image of a baby was displayed on the screen (Fischer & Hills, 2012). The only condition affecting men was the photo of the same-sex adult, which significantly increased risk-taking.

It is paradoxical that activities that generate fear and thus should be aversive can also be attractive. Funfair rides, horror movies, and extreme sports all capitalize on this and are generally more appealing to men than women. Campbell et al. (2016) reasoned that fear may become aversive at higher subjective levels. We examined how well men's lower level of fear could explain their taste for risk by presenting 668 participants with descriptions of 27 risky situations. For each situation, they rated their fear and also the extent to which they would find the experience alarming vs. exhilarating (hedonic tone). There was a significant sex difference, with men rating the situations as both less frightening and more exhilarating, although the significant association between fear and exhilaration was equally negative for both sexes (men $r = -0.43$, women $r = -0.42$). We used mediation analysis to investigate whether the sex difference in hedonic tone was explained by the sex difference in fear. Approximately 60 percent of the variance in the sex difference in enjoyment of risky situations was explained by men's lower levels of fear in these same situations. This suggests that a considerable portion of men's appetite for risk-taking may be a result of their lower fear. The level of fear at which a risky experience tips over from being enjoyable to alarming is lower in women than men. Men enjoy risky situations in part because they are experienced as less fear-inducing.

26.2.2 Avoiding Direct Aggression

Hundreds of studies have examined sex differences in aggression using a range of techniques, including laboratory experiments, observation, personality assessment, and self- and peer-reported behavior. The results fit a clear pattern: the more dangerous and risky the form of aggression, the larger the sex difference. This appears to be true cross-culturally (Archer, 2009; Campbell, 2006). Criminal statistics show the strongest sex difference, with men committing over seven times more homicides than women and over three times more aggravated assaults (Federal Bureau of Investigation, 2012). Psychological studies show that for physical acts such as hitting, punching, and kicking, the effect size lies between $d = 0.59$ and $d = 0.91$, while for verbal acts such as abuse and threats, the effect size is between $d = 0.28$ and $d = 0.46$ (Knight, Fabes, & Higgins, 1996; Knight et al., 2002).

This pattern of sex differences might be mediated by men's stronger anger (an *approach* motivation, with anger acting as an accelerator pedal) or by women's greater fear (an *avoidant* motivation, with fear acting as a braking mechanism). Men and women are capable of experiencing both emotions, but the relative balance between them is tipped more strongly in favor of overt aggression in men. But is this because men experience greater anger or lower fear?

Stereotypes notwithstanding, men do not experience anger more frequently or more intensely than women. This conclusion was reached by Kring (2000) and confirmed by Archer (2004), who reported an effect size of $d = -0.003$ in his meta-analysis of 46 samples. In a poll of over 2,000 Americans, women reported more episodes of anger during the last 7 days than men, even after controlling for sex differences in emotional expressiveness (Mirowsky & Ross, 1995). Using national data from the US General Social Survey, Simon and Nath (2004) also found no difference in the frequency of anger, although women exceeded men significantly in the intensity and duration of the emotion.

There is, however, much evidence that the sexes differ in the frequency and intensity of fear (reviewed more fully in Section 26.3). Among children, girls express fear earlier and more frequently than boys (Cote et al., 2002; Else-Quest et al., 2006; Gartstein & Rothbart, 2003). Among adults, women experience fear more intensely and frequently than men (Brody & Hall, 1993; Fischer, 1993; Gullone, 2000). Crucially, women's lower threshold for experiencing fear has direct consequences for aggressive behavior. Two independent meta-analyses concluded that fear was a powerful mediator of sex differences in aggression (Bettencourt & Miller, 1996; Eagly & Steffen, 1986). In experimental studies, women evaluate the same objective situation as more dangerous and more fear-provoking than men, and these appraisals significantly explain the sex difference in aggressive behavior.

One way that women can compete without compromising their safety or risking their lives is through acts that ostracize, stigmatize, and otherwise exclude others from social interaction without direct physical confrontation. Such acts do not eliminate or physically injure the target, nor do they demonstrate the greater size, strength, or belligerence of the attacker. They do, however, inflict stress and diminish the opponents' social standing and support. The target is attacked circuitously so the aggressor can remain unidentified and safe from retaliation, at least in the short term. This set of behaviors is referred to as *indirect or relational aggression* (Björkqvist, Lagerspetz, & Kaukiainen 1992; Crick & Grotpeter, 1995). These stigmatizing and excluding strategies can have devastating effects on the victim (McDougall & Vaillancourt, 2015). While early studies suggested that females exceeded males in their use of indirect aggression, Archer's (2004) meta-analysis found a negligible sex difference ($d = -0.02$). Regardless of whether women exceed men or are simply equal to them, the basic pattern remains intact: sex differences are magnified for risky forms of aggression and minimized (sometimes reversed) for less physically dangerous forms such as indirect aggression.

A key component of indirect aggression is the use of gossip to undermine an opponent's reputation and decrease her social capital. Because physical attractiveness is especially important to men's choice of mates, women should not only compete with one another to meet men's preference, but could use gossip as a way to derogate rivals. Physical appearance is the only topic of gossip to show a significant sex difference (Nevo, Nevo, & Derech-Zehavi, 1993) and is more often mentioned in relation to female targets (Hall, 2002). Of the 28 tactics that participants reported using to make same-sex rivals undesirable to the opposite sex, "derogate competitor's appearance" was used significantly more often by women than by men (Buss & Dedden, 1990).

Another trait that men find unattractive in a potential long-term partner is infidelity (Buss & Schmitt, 1993). Paternity is never certain for a man, and the costs of investing in another man's child are great. Because one of the best predictors of future behavior is past behavior, women can benefit from undermining their rivals' sexual reputation while defending their own. That is why terms such as "slag," "tart," or "whore" are powerful sources of reputation challenge among women (Campbell, 1982, 1995; Duncan, 1999; Lees, 1993; Marsh & Patton, 1986). Girls actively collude in enforcing the double standard through gossip and rumor-spreading (Baumeister & Twenge, 2002). These tactics are as visible among university students (Milhausen & Herold, 1999) as among gang members in deprived inner-city areas (Campbell, 1984; Joe Laidler & Hunt, 2001). In a rare experimental study of indirect aggression, Vaillancourt and Sharma (2011) asked female friendship dyads to wait in a room where they were briefly interrupted by a female confederate. She was dressed either conservatively or provocatively. The reactions of the young women during and after the interruption were blind coded. Women in the provocative condition were rated as "bitchier" in their reactions and, at a behavioral level, engaged in more negative verbal comments, "once-overs," and laughter following the confederate's exit.

26.3 WHERE DO THE SEX DIFFERENCES IN FEAR LIE?

Fear is an evolved psychological and biological state triggered by threat. It involves activation of specific neural circuitry, the autonomic nervous system, and the hypothalamic–pituitary axis (HPA). This gives rise to emotional, cognitive, and behavioral responses designed to promote safety. Sex differences have been examined in many of these areas and the results can provide clues as to whether women have a lower threshold for fear (a more sensitive system) or whether they show a more intense fear response than men (a more reactive system).

26.3.1 Self-Report

The sex differences in risky behavior and decision-making reviewed in Section 26.2.1 are strongly supported by men's and women's reports of their experience of fear. Sex differences in reported intensity and duration of fear are found

cross-culturally (Brebner, 2003; Fischer & Manstead, 2000; Fischer et al., 2004). A recently developed 27-item sex-invariant fear inventory (Campbell et al., 2016) showed an effect size favoring women that ranged between $d = -1.04$ (UK sample) and $d = -1.23$ (Romanian sample). There is a large effect size of $d = -0.78$ favoring women for the Harm Avoidance scale of the Multidimensional Personality Questionnaire (Cross, Copping, & Campbell, 2011), and a similar effect size ($d = 0.76$) favoring men is found on the Fearlessness scale of the Psychopathic Personality Inventory (Lee & Salekin, 2010). The most widely used measures of fear are the Fear Survey Schedules (FSS-II and FSS-III), which assess phobias – marked and persistent fears of specific objects or situations that the sufferer recognizes to be excessive or unreasonable. Both measures show sex differences ranging between $d = -0.32$ and $d = -0.76$ (Arrindell et al., 2003; Bernstein & Allen 1969). In a more recent inventory (McCraw & Velentiner, 2015), women scored significantly higher than men on each of five scales designed to assess phobic reactions (desire to escape or avoid; feelings of fear or anxiety; physiological symptoms; fear of losing control; risk analysis).

A concern about self-report measures is the extent to which apparent sex differences are artifacts of gendered self-presentation resulting from men's reluctance to admit fear. However, studies that control for gender role adherence find a significant effect of biological sex (Arrindell et al., 1993, 1999; Campbell et al., 2016; Dillon et al., 1985), and informing participants that the honesty of their responses are verifiable by physiological measures does not eliminate the sex difference in reported fear (McLean & Hope, 2010; Pickersgill & Arrindell, 1994; Stoyanova & Hope, 2012).

26.3.2 Inducing Fear in the Laboratory

Within ethical constraints, researchers have developed a number of ways of inducing fear under controlled conditions to examine behavioral, reflexive, autonomic, and cognitive reactions. The *Behavioral Approach Task*, as its name implies, provides a behaviorally based ordinal score of a participant's willingness to approach and make contact with a feared stimulus. Because aversion to spiders is very common, they have been used to induce fear among nonselected samples of undergraduates. Participants' approach behavior is scored on a stepwise Guttman-type scale from 1 (entering the room where the spider is) to 12 (allowing it to crawl on their hand or touching it for 3 seconds with 2 fingers). Compared to men, women rated their fear higher before and during the task and took significantly fewer approach steps (McLean & Hope 2010; Stoyanova & Hope, 2012). In children aged 9–12 years (Klein, Becker, & Rinck, 2011), girls reported higher fear and proceeded through fewer steps on the Behavioral Approach Task. However, no differences in heart rate have been found during the task.

The *potentiated startle reflex* measures the well-established amplifying effect of fear on the automatic startle reflex. While viewing fear-inducing images or under threat of shock, a sudden-onset auditory tone is delivered. Electrodes above the orbicularis oculi record the magnitude of the facial startle. This is compared with baseline startle or the startle during neutral images to quantify the degree of enhancement caused by the fear-inducing stimuli. The neural circuitry of potentiated startle is well established: auditory pathways carry information to the caudal pontine reticular nucleus, where it meets afferent information from the amygdala. Damage to this incoming pathway eliminates fear modulation of the startle response while leaving baseline startle intact (Davis et al., 1993). Potentiated startle is stronger in women than men (Bradley et al., 2001; Lang et al., 1993) and in girls compared to boys (McManis et al., 2001). A cross-sectional study of children (aged between three and nine years) and adults found that females had a higher overall startle magnitude and higher overall response probabilities, although this was found across all film stimuli (Quevedo et al., 2010). A longitudinal study of 10–13-year-old adolescents found that fear potentiation increased with pubertal stage in girls who showed stronger fear potentiation at all pubertal stages. In young adulthood, women showed a stronger enhanced startle to fearful and angry faces than men (Anokhin & Golosheykin, 2010). In light of this, the role of testosterone has been examined. A placebo-controlled within-subjects study found that administration of testosterone to women reduced fear-potentiated startle to a threatened electric shock while leaving baseline startle unaffected (Hermans et al., 2006). However, a follow-up study found these testosterone effects only in a subgroup of highly anxious women (Hermans et al., 2007).

Inhalation of CO$_2$-enriched air is an effective way of inducing brief but intense fear. During normal breathing, CO$_2$ composes only 0.04 percent of inhaled air. This is experimentally increased to as high as 35 percent CO$_2$ in some studies. Inhalation of CO$_2$ produces a dose-dependent increase in fear and panic symptomatology, which is thought to be an evolved response to possible suffocation. Studies have examined self-reported fear and autonomic responses. There is some evidence that women find the experience more alarming than men. They reported less control over their bodily responses, more fear, greater experiential panic, and more panic symptoms and were less willing to repeat the experiment than men (Kelly, Forsyth, & Karekla, 2006; Nillni et al., 2012). Bunaciu et al. (2012) found that it was only in the most severe condition (20 percent CO$_2$) that women reported greater anxiety. On the other hand, two studies which tested participants over a wide range of conditions (0, 9.0, 17.5 and 35.0 percent CO$_2$) found no sex differences in reported anxiety or panic symptoms (Griez et al., 2007; Leibold et al., 2013). Sex differences in physiological measures have been examined in three studies with mainly null results. During the task itself, neither Leibold et al.

(2013) nor Kelly et al. (2006) found a sex difference in heart rate. Although the task triggered a rapid rise in systolic and diastolic blood pressure, there was no sex difference (Leibold et al., 2013), nor have differences been found for skin conductance levels or responses (Kelly et al., 2006; Nillni et al., 2012) or frontalis muscle tension (Kelly et al., 2006).

The *dot probe task* exploits the fact that fear-inducing stimuli preferentially capture attention (Bar-Haim et al., 2007; Yiend, 2010). Two stimuli (one of which is threatening) appear simultaneously at different locations on a monitor (e.g., right and left). After a short exposure time, the stimuli disappear and a probe appears at one of the two locations. Participants are asked to indicate as quickly as possible the spatial location of the probe. A bias toward threat is inferred when the reaction time on congruent trials (when the probe appears at the same location as the threatening stimulus) is faster than incongruent trails (when the probe appears at the location of the neutral stimulus). Typically, the bias index is computed by subtracting congruent trials from incongruent trials such that a positive value indicates bias toward threat (attentional vigilance) and a negative value indicates bias away from threat (attentional avoidance). The direction of bias depends on the threat intensity of the stimuli because it has been proposed that individuals show avoidance to mild threats but vigilance to more dangerous threats. In a meta-analysis (Bar-Haim et al., 2007), stimuli used in laboratory studies did not produce a bias in nonclinical and low-vulnerability samples ($d = -0.01$), but a moderate vigilance bias was found in anxious samples ($d = 0.45$). Given that women are more prone to anxiety than men, the dot probe technique might be expected to reveal greater threat vigilance in women.

Despite the large number of studies that use the technique, very few have explicitly addressed or reported sex differences (eight in the last five years), and those that do have mainly used clinical or selected samples. Because of the relative uniformity of the paradigm (which increases the comparability of studies), we decided to undertake a "mini" meta-analysis investigating sex differences in nonselected samples. A Web of Science search using the terms "attention* bias* AND fear AND probe" identified 108 studies since 2011. Studies using nonclinical male and female samples were selected and, in studies of patients with clinical anxiety or preselected high-anxiety groups, authors were contacted for control group information (n = 35). Data on sex differences were received from 15 authors, providing data on 23 studies (Table 26.1). All data were coded such that a positive value indicated threat vigilance. The female mean was subtracted from the male mean so that a positive effect size indicates a higher mean score in men. While this sample is neither comprehensive nor random, it provides some information regarding sex differences in attentional bias. Although effect sizes across studies varied in valence and magnitude (from $g = -0.75$ to $g = 1.22$), the Q measure of heterogeneity

Table 26.1 Meta-analysis of sex differences in the dot probe task.

Study	Stimulus	Condition reported	Male n	Male mean (SD)	Female n	Female mean (SD)	Total n	g	Q
Facial stimuli									
Tian and Smith (2011)	Unpleasant faces	After moderate exercise	17	−2.00 (8.00)	17	−12.00 (8.00)	34	1.22	9.33
Schofield et al. (2012)	Angry faces		16	−8.09 (49.84)	23	−2.17 (45.80)	39	−0.12	0.28
Schofield et al. (2012)	Fearful faces		16	10.58 (45.69)	23	0.07 (62.27)	39	0.18	0.17
Carlson and Mujica-Parodi (2015)	Fearful faces	Supraliminal presentation	25	10.58 (9.73)	30	13.78 (14.45)	55	−0.25	1.23
Carlson and Mujica-Parodi (2015)	Fearful faces	Subliminal (masked) presentation	25	10.77 (10.45)	30	6.61 (16.56)	55	0.29	0.78
Carlson et al. (2012)	Fearful faces (masked)		29	4.39 (19.40)	21	4.48 (15.99)	50	0.00	0.04
Carlson et al. (2013)	Fearful faces (masked)		6	9.72 (10.63)	9	10.71 (14.97)	15	−0.07	0.05
Threat stimuli									
Bardeen and Orcutt (2011)	General threat pictures	150-ms stimulus-onset asynchrony	41	a	56	a	97	−0.02	0.12
Bardeen and Orcutt (2011)	General threat pictures	500-ms stimulus-onset asynchrony	41	a	56	a	97	0.22	0.66
Vogt et al. (2013)	Threatening pictures	Experiment 1	3	−7.09 (87.63)	6	8.56 (39.88)	9	−0.24	0.17
Dittmar et al. (2011)	Social threat words		13	0.01 (0.03)	14	0.00 (0.03)	27	0.32	0.49
Illness/pain stimuli									
Jasper and Whitthoft (2011)	Health threat pictures	175-ms exposure duration	23	−2.43 (21.29)	60	1.33 (21.64)	83	−0.17	0.82
Jasper and Whitthoft (2011)	Health threat pictures	500-ms exposure duration	23	−3.12 (15.20)	60	−4.15 (22.81)	83	0.05	0.00

Continued

Study	Stimulus	Condition	n	Mean (SD)	n	Mean (SD)	n		
Yang et al. (2013)	Health catastrophe words	High fear of pain group	3	50.09 (53.86)	10	−6.06 (45.35)	13	1.11	2.21
Yang et al. (2013)	Health catastrophe words	Low fear of pain group	1	−10.70 (39.41)	10	25.49 (39.41)	11	−0.84	0.68
Van Ryckeghem et al. (2012)	Pain words		11	−5.32 (23.24)	42	2.26 (17.80)	53	−0.39	1.68
Dittmar et al. (2011)	Pain-related words		13	−0.01 (0.02)	14	0.01 (0.03)	27	−0.75	3.98
McDermott et al. (2013)	Headache-related pictures		72	−3.56 (21.59)	152	−1.69 (42.82)	224	−0.05	0.49
Spider stimuli									
Van Bockstaele et al. (2011a)	Spider pictures		11	8.50 (42.01)	42	−8.33 (47.47)	53	0.36	0.81
Van Bockstaele et al. (2011b)	Spider pictures	Pretraining data from controls	13	−9.57 (17.58)	52	−7.24 (28.05)	65	−0.09	0.20
Participant stress manipulated									
Vogt et al. (2013)	Cue signaling threat (aversive noise)	Experiment 3	5	5.95 (21.26)	22	20.83 (34.60)	27	−0.44	0.95
Lee et al. (2014)	Faces (salient) vs. places (nonsalient)	High arousal (tone predicts shock)	14	20.34 (19.23)	38	13.95 (16.32)	52	0.37	1.01
Lee et al. (2014)	Faces vs. places	Low arousal (tone predicts no shock)	14	12.50 (20.26)	38	4.99 (20.32)	52	0.36	0.99
All studies							**1,260**	**0.05**	**27.13**

[a] The g value was estimated from reported correlations.

was not significant (χ^2 = 27.12, df = 22). The overall value of g = 0.05 indicates no appreciable sex difference. In Table 26.1, the studies have been grouped according to the type of threat stimulus used, but there was no evidence of a consistent sex difference in any of these clusters. However, the task itself has been criticized in terms of internal consistency and test–retest reliability (Schmukle, 2005), and hence it necessarily shows poor convergent validity (typically r = <0.10) with other tasks measuring attention bias (Van Bockstaele et al., 2014).

26.3.3 Neural Correlates of Threat Images

In neuropsychological studies of emotional processing, facial images of different emotions are presented in order to examine regional activation in the viewer's brain. However, the processes that are assumed to be taking place are rarely made explicit. It is assumed that viewers are "decoding" the emotion that may (but need not) involve duplicating the subjective experience of the expressed emotion (Wood et al., 2016), or viewers may experience their own emotional reaction to the facial expression (Davis et al., 2011). Hence, emotional processing of fear may involve: the identification of facial features associated with fear; the subjective experience of fear; or another emotion (such as sympathy) elicited by the sight of a frightened face. If emotional faces elicit a reactive (rather than a "mirroring") response, then an angry face would be more effective in stimulating fear than a fearful face, and this may be especially true for women (McClure et al., 2004). Some neuroimaging studies use "emotional" pictures from the international affective picture system (IAPS; Lang, Ohman, & Vaitl, 1988) collection that have been pre-rated as neutral, positive ("pleasant"), and negative ("unpleasant"), which means the analysis is based on relatively gross categories of emotion. Even where discrete emotions have been examined, meta-analyses often collapse them into binary groupings such as positive vs. negative (Stevens & Hamann, 2012) or approach related vs. avoidance related (Wager et al., 2003). This means that studies of fear are merged with studies of anger, disgust, guilt, and sadness. Meta-analyses of sex differences in emotional processing are further compromised because the majority of studies use participants of only one sex so that, in making a cross-sex comparison of regional activations to "negative" emotions, men and women have experienced different stimuli, task instructions, and experimental paradigms.

Despite these concerns, quantitative and narrative reviews have generally concluded that women show a stronger response than men to negative stimuli, particularly fear-inducing stimuli (Cahill, 2006; Hamann, 2005; Stevens & Hamann, 2012; Whittle et al., 2011). Although the fear circuit remains poorly understood, several key structures are consistently implicated, of which the amygdala is the most prominent. Because it responds to emotions other than fear (albeit more weakly and less

consistently), some have proposed that it operates as an alerting mechanism that prioritizes – but does not respond exclusively to – signals of threat or danger. The "salience network" (Seeley et al., 2007) includes another three key areas that are active in a number of fear studies. The anterior insula cortex and the orbital frontoinsula cortex situated at the junction of the anterior insula and posterior orbitofrontal cortex (OFC) are thought to give rise to subjective emotion by interoceptive monitoring of bodily states. The role of the dorsal anterior cingulate cortex is less clear but may involve appraisal of negative emotion and attention switching. These three regions also contain von Economo neurons found chiefly in humans and great apes, which have been implicated in empathy, social awareness, and self-control. These areas have projections to the OFC that, among other things (Stalnaker, Cooch, & Schoenbaum, 2015), may be involved in registering, valuing, and modulating emotional experience.

Regarding sex differences in fear specifically, boys showed more efficient habituation of amygdala activation than girls to repeated presentation of fearful faces (Thomas et al., 2001). This was replicated in adults by Williams et al. (2005), with men showing attenuation of right amygdala activation between the early and late stages of the experiment, while women's right amygdala activation increased both in terms of extent (voxel number) and effect size (standardized magnitude difference between control and fear conditions). Women, unlike men, demonstrated a persistent amygdala response to negative stimuli even when they had been familiarized with them (Andreano, Dickerson, & Barrett, 2014). McClure et al. (2004) used neutral, happy, fearful, and angry emotional expressions, with a particular interest in the contrast between fearful faces ("ambiguous threat" because it signals the indirect presence of threat in the locality of the actor) and angry faces ("unambiguous threat" because the gaze is directed unambiguously at the viewer). Women showed a pattern of enhanced activity specifically in those contrasts involving the angry face condition. To angry vs. neutral faces, women showed greater activity than men in the right amygdala and right OFC. The contrast between angry and fearful faces in women showed greater bilateral amygdala and right OFC activity to the angry faces, suggesting that for women an angry face is more fear-provoking than a fearful face. While men showed increased activity in at least one region to all except happy faces, women's responses were specific to the angry face condition, which activated the OFC, amygdala, and anterior cingulate. The sexes were distinguished not by the strength of the blood-oxygen-level-dependent imaging signal, but by the selectivity of circuit activation.

Several data-based hypotheses have been offered about sex differences in emotion processing. In response to highly arousing negative images, Lungu et al. (2015) replicated the key role of the right amygdala and established its functional connectivity with the dorsomedial prefrontal cortex (dmPFC) implicated in social cognition and action selection. This amygdala–dmPFC connectivity was stronger in men, correlated positively with testosterone, and correlated negatively with subjective ratings of emotional intensity. This suggests that men may have a more evaluative and action-oriented orientation to emotion, while women attend to the feelings elicited by the stimuli. It has also been suggested that men and women experience emotion differently in terms of orientation or focus. Moriguchi et al. (2014) found that in response to emotional pictures women's reported arousal was accompanied by increased activity in the anterior insula, which is implicated in interoception and subjective emotional experience, suggesting an interior focus. Men's arousal was correlated with activity in the primary visual cortex and fusiform gyrus, which are involved in face processing, suggesting an exterior focus. The functional connection between the anterior insula and anterior cingulate cortices (elements of the "salience network" responsible for regulating attention; Craig, 2009) was also stronger in men, suggesting that they may be more capable of switching from an internal to an external orientation in response to threat.

Contrary to my suggestion that women experience greater fear than men in response to threat, Taylor et al. (2000) proposed that women react to threat with a calming "tend-and-befriend" response mediated by the anxiolytic neuropeptide oxytocin. Oxytocin was proposed to adaptively soothe women and increase bonding with one another and their offspring. Early research on male participants confirmed that oxytocin administration was associated with a reduction in amygdala activation to fearful faces and threat-related scenes (Lischke et al., 2012). However, two studies (Domes et al., 2010; Lischke et al., 2012) have now demonstrated that in women intranasal oxytocin significantly *increases* rather than decreases amygdala activation in response to fearful faces and threatening scenes. Contrary to the tend-and-befriend hypothesis, I argue that enhanced rather than diminished reactivity to threat has been adaptive for women. In women, noradrenaline (normally released in the face of imminent danger) increased both subjective fear ratings and amygdala activity to fearful faces while decreasing OFC activity (Schwabe et al., 2013). In men, it had the opposite effect, decreasing their fear ratings and amygdala reactivity. Noradrenaline appears to enhance the strength of women's immediate reactions to fear-provoking stimuli, while having the opposite effect in men.

26.3.4 Peripheral Indices of Fear

Fear conditioning studies (in which a neutral conditioned stimulus acquires fear-inducing properties through its association with an unconditioned stimulus such as electric shock) typically measure skin conductance response as an index of successful conditioning. Sex differences are rarely found (Lebron-Milad et al., 2012). Similarly, heart

rate measures in response to fear typically fail to show sex differences, although there is some evidence that women's heart rates are more closely associated with (reduced) parasympathetic reactivity and men's with (increased) sympathetic reactivity (Dart, Du, & Kingwell, 2002; Nugent et al., 2011).

Studies of sex differences in HPA responses typically manipulate stress rather than fear, and sex differences depend on the type of stressor used. Men show a greater cortisol response to social performance challenges (such as the Trier Test), while women show a stronger response to interaction challenges such as interpersonal rejection (Stroud, Salovey, & Epel, 2002). However, between puberty and menopause, HPA responses are generally lower in women than men (Goel et al., 2014; Ordaz & Luna, 2012). This timing strongly implicates gonadal hormones. In the luteal phase, women show an enhanced salivary cortisol response to stress that is not significantly different from men, suggesting that the follicular phase, with its higher levels of estrogen, has a dampening effect on HPA reactivity (Kudielka & Kirschbaum, 2005). In line with this, cortisol levels show the expected changes as a function of oral contraceptive use (reduced), pregnancy (reduced), and menopause (enhanced). Kajantie and Phillips (2006) offer an evolutionary interpretation of the estrogen–cortisol link, suggesting that it reduces maternal stress reactivity that might be damaging to the fetus. High levels of stress during pregnancy are associated with early spontaneous abortion, premature delivery, smaller offspring, HPA and sympathetic nervous system hypersensitivity, and reduced activity in several neurotransmitter systems. During pregnancy, estradiol peaks at up to 100 times higher levels than during the menstrual cycle. The less dramatic but still significant effects of estrogen during normal menstrual cycles may be an epiphenomenal artifact of its role in pregnancy.

26.3.5 Summary

Looking at the fear system overall, sex differences in autonomic reactivity to threat appear weak and inconsistent. Although there is some evidence of an attenuated hypothalamic–pituitary response in women resulting from the close interaction between the HPA and the hypothalamic–gonadal axis, this tentative conclusion is based on studies of psychosocial stress, which are not equivalent to fear studies in terms of threat immediacy, intensity, or duration. In perhaps the majority of real-world instances, the HPA response may be activated only very briefly because the fear stimulus is transient. Sex differences are also not apparent for selective attention to threat (dot probe paradigm). This is a rough measure of system sensitivity. If women identified the presence of threat earlier and were engaged by it for longer, then this would suggest a more sensitive threat detection system, but there is currently little evidence of this. Nonetheless, the distinction between hypervigilance and selective attention deserves further attention in terms of possible sex differences (Richards et al., 2014).

Rather, the neuroimaging evidence points to stronger corticolimbic system activation in response to threat, with greater and more persistent activation in areas associated with interoception and the subjective experience of emotion. At a reflexive level, this can be seen in the greater extent to which fear enhances women's automatic startle response. At a conscious level, it results in greater sensitivity to negative outcomes, whether measured as a personality trait (punishment sensitivity) or in risk-averse choices. At an experiential level, women find fear to be a more distressing experience and report more intense fear even when peripheral measures show no sex difference. Their internal experience is manifested behaviorally in laboratory tasks such as the behavioral approach paradigm, as well as in the world beyond where they take fewer risks with their lives.

26.4 CONCLUSIONS

I have proposed that, because offspring survival depends more on the presence and care of a mother than a father, there has been stronger selection pressure on women than men to protect themselves from external threats to their lives. I have proposed that the target of selection has been the fear system, and the evidence suggests that women show a stronger reaction to threatening stimuli than men rather than a lower threshold for threat detection. This stronger reactivity is most marked in corticolimbic regions implicated in registering the salience of external events, interoception (monitoring of internal states), and the subjective experience of fear.

It bears repeating that any sex difference is a matter of degree, not kind. For example, even with a large effect size of $d = 1.00$ (as found by Campbell et al., 2016), 62 percent of the male and female fear distributions overlap. Men and women are, after all, morphs of the same species, and selection pressures "nudge" rather than "shove." Nonetheless, differences between men and women become more noticeable at the extreme ends of the fear distribution. While women's enhanced response to threat evolved because it was advantageous, there can be such a thing as too much fear. Post-traumatic stress disorder (PTSD) was reclassified in the fifth edition of the *Diagnostic and Statistical Manual of Mental Disorders* as a "trauma and stressor-related disorder" rather than an anxiety disorder. The emotional intensity accompanying the trauma (defined as experiencing or witnessing threatened or actual death, serious injury, or sexual violation) remains at the heart of the disorder and is predictive of later PTSD symptomatology. Women are twice as likely as men to develop PTSD during their lifetimes (10.4 vs. 5.0 percent; Kessler et al., 1995). This difference is not a function of sex differences in rate of exposure, type of trauma, or history of traumatization (Olff et al., 2007). Rather, it arises from sex differences in the intensity of fear engendered by the traumatic event. Among these peri-traumatic variables, it is not the objective facts of the trauma itself, but rather how it is experienced by the survivor that is critical (e.g., Bovin & Marks, 2011; Olff et al., 2005; Ozer et al., 2003). Even when men and women experience the same type of trauma, women are twice as likely to develop PTSD (e.g., DeLisi et al., 2003; Holbrook et al., 2002; Stein, Walker, & Forde, 2000). For example, a study of an explosion in a Danish fireworks factory showed that women were 2.4 times more likely to experience PTSD than men (Spindler, Elkit, & Christiansen, 2010). This sex difference became nonsignificant when perceived danger and trait anxiety were entered, underlining the importance of women's stronger subjective experience of fear as a risk factor for PTSD symptomatology.

At the other end of the fear spectrum we find psychopathy – a pathology in which men outnumber women by four to one (Coid et al., 2009; Wynn, Høiseth, & Pettersen, 2012). A key characteristic of primary psychopathy is the absence of fear and anxiety, manifested in deficiencies in passive avoidance learning, potentiated startle, and electrodermal response (Patrick, Bradley, & Lang, 1993). The fearlessness theory (Lykken, 1957) has been extended to encompass weak behavioral avoidance systems resulting in punishment insensitivity and a failure of somatic markers to guide decision-making away from punishing outcomes (Van Honk, & Schutter, 2006). A large study of undergraduates (Lee & Salekin, 2010) found an effect size of $d = 0.67$ favoring men on the widely used the Psychopathic Personality Inventory – Short Form (PPI-SF) screening device, with the largest sex difference emerging on the Fearlessness subscale ($d = 0.76$).

Fear serves a crucial evolutionary function in alerting organisms to danger. There are benefits and costs associated with recalibrating it: to achieve a net advantage, the benefits must exceed the costs. The downregulation of fear in men frees them to compete via combat and risk-taking with the associated benefits of enhanced kudos and mate choice. The price is potential injury and earlier death, but these costs are mitigated in the currency of reproductive success because women bear the brunt of childrearing. It appears that the costs of upregulating the fear system in women have been minimized by increasing the strength of the fear response rather than incurring the "false alarm" costs associated with increasing the system's sensitivity. The benefits for women lie in lower mortality and greater longevity, which enhance mothering and enable grandmothering. Among women, living slow and dying late has been a winning strategy.

REFERENCES

Andreano, J. M., Dickerson, B. C., & Barrett, L. F. (2014). Sex differences in the persistence of the amygdala response to negative material. *Social Cognitive and Affective Neuroscience*, **9**, 1388–1394.

Anokhin, A. P., & Golosheykin, S. (2010). Startle modulation by affective faces. *Biological Psychology*, **83**(1), 37–40.

Archer, J. (2004). Sex differences in aggression in real world settings: A meta-analytic review. *Review of General Psychology*, **8**, 291–322.

Archer, J. (2009). Does sexual selection explain human sex differences in aggression? *Behavioral and Brain Sciences*, **32**, 249–311.

Arrindell, W. A., Eisemann, M., Richter, J., et al. (2003). Phobic anxiety in 11 nations. Part I: Dimensional consistency of the five factor model. *Behaviour Research and Therapy*, **41**, 461–479.

Arrindell, W. A., Kolk, A. M., Pickersgill, M. J., & Hageman, W. J. (1993). Biological sex, sex role orientation, masculine sex role stress, dissimulation and self reported fears. *Advances in Behaviour Research and Therapy*, **15**, 103–146.

Arrindell, W. A., Mulkens, S., Kok, J., & Vollenbroek, J. (1999). Disgust sensitivity and the sex difference in fears to common indigenous animals. *Behavior Research and Therapy*, **37**(3), 273–280.

Bardeen, J. R., & Orcutt, H. K. (2011). Attentional control as a moderator of the relationship between posttraumatic stress symptoms and attentional threat bias. *Journal of Anxiety Disorders*, **25**(8), 1008–1018.

Bar-Haim, Y., Lamy, D., Pergamin, L., Bakermans-Kranenburg, M. J., & van IJzendoorn, M. H. (2007). Threat-related attentional bias in anxious and nonanxious individuals: A meta-analytic study. *Psychological Bulletin*, **133**, 1–24.

Baumeister, R. F., & Twenge, J. M. (2002). Cultural suppression of female sexuality. *Review of General Psychology*, **6**, 166–203.

Bernstein, D. A., & Allen, G. J. (1969). Fear Survey Schedule (II): Normative data and factor analyses based on a large college sample. *Behaviour Research and Therapy*, **7**, 403–407.

Bettencourt, B. A., & Miller, N. (1996). Gender differences in aggression as a function of provocation: A meta-analysis. *Psychological Bulletin*, **119**, 422–447.

Björkqvist, K., Lagerspetz, K., & Kaukiainen, A. (1992). Do girls manipulate and boys fight? Developmental trends in regard to direct and indirect aggression. *Aggressive Behavior*, **18**, 117–127.

Bovin, M. J., & Marx, B. P. (2011). The importance of the peritraumatic experience in defining traumatic stress. *Psychological Bulletin*, **137**(1), 47–67.

Bradley, M. M., Codispoti, M., Sabatinelli, D., & Lang, P. J. (2001). Emotion and motivation II: Sex differences in picture processing. *Emotion*, **1**(3), 300–319.

Brebner, J. (2003). Gender and emotions. *Personality and Individual Differences*, **34**, 387–394.

Brody, L. R., & Hall, J. A. (1993). Gender and emotion. In M. Lewis & J. M. Haviland, eds., *Handbook of Emotions*. New York: Guildford Press, pp. 447–460.

Brown, L. M. (1998). *Raising Their Voices: The Politics of Girls' Anger*. London: Harvard University Press.

Bunaciu, L., Feldner, M. T., Babson, K. A., Zvolensky, M. J., & Eifert, G. H. (2012). Biological sex and panic-relevant anxious reactivity to abrupt increases in bodily arousal as a function of biological challenge intensity. *Journal of Behavior Therapy and Experimental Psychiatry*, **43**, 526–531.

Buss, D. M., & Dedden, L. A. (1990). Derogation of competitors. *Journal of Personal and Social Relationships*, **7**, 395–422.

Buss, D. M., & Schmitt, D. (1993). Sexual strategies theory: An evolutionary perspective on human mating. *Psychological Review*, **100**, 204–232.

Byrnes, J. P., Miller, D. C., & Schafer, W. D. (1999). Gender differences in risk taking: A meta-analysis. *Psychological Bulletin*, **125**, 367–383.

Cahill, L. (2006). Why sex matters for neuroscience. *Nature Reviews Neuroscience*, **7**(6), 477–484.

Campbell, A. (1982). Female aggression. In P. Marsh & A. Campbell, eds., *Aggression and Violence*. Oxford: Blackwell, pp. 137–150.

Campbell, A. (1984). *The Girls in the Gang*. Oxford: Blackwell.

Campbell, A. (1995). A few good men: Evolutionary psychology and female adolescent aggression. *Ethology and Sociobiology*, **16**, 99–123.

Campbell, A. (2006). Sex differences in direct aggression: What are the psychological mediators? *Aggression and Violent Behavior*, **11**, 237–264.

Campbell, A., Coombes, C., David, R., et al. (2016). Sex differences are not attenuated by a sex-invariant measure of fear: The Situated Fear Questionnaire. *Personality and Individual Differences*, **97**, 210–219.

Carlson, J. M., Cha, J., & Mujica-Parodi, L. R. (2013). Functional and structural amygdala–anterior cingulate connectivity correlates with attentional bias to masked fearful faces. *Cortex*, **49**(9), 2595–2600.

Carlson, J. M., & Mujica-Parodi, L. R. (2015). Facilitated attentional orienting and delayed disengagement to conscious and nonconscious fearful faces. *Journal of Nonverbal Behavior*, **39**(1), 69–77.

Carlson, J. M., Mujica-Parodi, L. R., Harmon-Jones, E., & Hajcak, G. (2012). The orienting of spatial attention to backward masked fearful faces is associated with variation in the serotonin transporter gene. *Emotion*, **12**(2), 203–207.

Carver, C. S., & White, T. L. (1994). Behavioral inhibition, behavioral activation, and affective responses to impending reward and punishment: The BIS/BAS scales. *Journal of Personality and Social Psychology*, **67**, 319–333.

Coid, J., Yang, M., Ullrich, S., Roberts, A., & Hare, R. D. (2009). Prevalence and correlates of psychopathic traits in the household population of Great Britain. *International Journal of Law and Psychiatry*, **32**(2), 65–73.

Cote, S., Tremblay, R. E., Nagin, D., Zoccolillo, M., & Vitaro, F. (2002). The development of impulsivity, fearfulness and helpfulness during childhood: Patterns of consistency and change in the trajectories of boys and girls. *Journal of Child Psychology and Psychiatry and Allied Disciplines*, **43**, 609–618.

Craig, A. D. (2009). How do you feel – Now? The anterior insula and human awareness. *Nature Reviews Neuroscience*, **10**, 59–70.

Crick, N. R., & Grotpeter, J. K. (1995). Relational aggression, gender and social-psychological adjustment. *Child Development*, **66**, 710–722.

Croson, R., & Gneezy, U. (2009). Gender differences in preferences. *Journal of Economic Literature*, **47**(2), 448–474.

Cross, C. P., Copping, L. T., & Campbell, A. (2011). Sex differences in impulsivity: A meta-analysis. *Psychological Bulletin*, **137**(1), 97–130.

Daly, M., & Wilson, M. (1988). *Homicide*. Piscataway, NJ: Transaction.

Dart, A. M., Du, X. J., & Kingwell, B. A. (2002). Gender, sex hormones and autonomic nervous control of the cardiovascular system. *Cardiovascular Research*, **53**, 678–687.

Darwin, C. (1859). *On The Origin of Species by Means of Natural Selection, or the Preservation of Favoured Races in the Struggle for Life*. London: John Murray.

Darwin, C. (1871). *The Descent of Man, and Selection in Relation to Sex*. London: John Murray.

Daughters, S. B., Gorka, S. M., Matusiewicz, A., & Anderson, K. G. (2013). Gender specific effect of psychological stress and cortisol reactivity on adolescent risk taking. *Journal of Abnormal Child Psychology*, **41**, 749–758.

Davis, M., Falls, W. A., Campeau, S., & Kim, M. (1993). Fear-potentiated startle: a neural and pharmacological analysis. *Behavioural Brain Research*, **58**(1–2), 175–198.

Davis, F. C., Somerville, L. H., Ruberry, E. J., et al. (2011). A tale of two negatives: Differential memory modulation by threat-related facial expressions. *Emotion*, **11**(3), 647–655.

DeLisi, L. E., Maurizio, A., Yost, M., et al. (2003). A survey of New Yorkers after the Sept. 11, 2001, terrorist attacks. *American Journal of Psychiatry*, **160**, 780–3.

Dillon, K. M., Wolf, E., & Katz, H. (1985). Sex roles, gender, and fear. *Journal of Psychology*, **119**, 355–359.

Dittmar, O., Krehl, R., & Lautenbacher, S. (2011). Interrelation of self-report, behavioural and electrophysiological measures assessing pain-related information processing. *Pain Research and Management*, **16**(1), 33–40.

Domes, G., Lischke, A., Berger, C., et al. (2010). Effects of intranasal oxytocin on emotional face processing in women. *Psychoneuroendocrinology* **35**, 83–93.

Duncan, N. (1999). *Sexual Bullying: Gender Conflict and Pupil Culture in Secondary Schools*. London: Routledge.

Eagly, A. H., & Steffen, V. J. (1986). Gender and aggressive behavior: A meta-analytic review of the social psychological literature. *Psychological Bulletin*, **100**, 309–330.

Eastwick, P. W. (2009). Beyond the Pleistocene: Using phylogeny and constraint to inform the evolutionary psychology of human mating. *Psychological Bulletin*, **135**, 794–821.

Ellis, L. (1995). Dominance and reproductive success among non-human animals: A cross-species comparison. *Ethology and Sociobiology*, **16**(4), 257–333.

Else-Quest, N. M., Hyde, J. S., Goldsmith, H. H., & Van Hulle, C. A. (2006). Gender differences in temperament: A meta analysis. *Psychological Bulletin*, **132**, 33–72.

Eriksson, K., & Simpson, B. (2010). Emotional reactions to losing explain gender differences in entering a risk lottery. *Judgement and Decision Making*, **5**, 159–163.

Federal Bureau of Investigation (2012). Crime in the United States 2011. www.fbi.gov/about-us/cjis/ucr/crime-in-the-u.s/2011/crime-in-the-u.s.-2011/tables/table-33.

Fessler, D. M. T., Tiokhin, L. B., Holbrook, C., Gervais, M. M., & Snyder, J. K. (2014). Foundations of the Crazy Bastard Hypothesis: Nonviolent physical risk-taking enhances conceptualized formidability. *Evolution and Human Behavior*, **35**, 26–33.

Fischer, A. H. (1993). Sex differences in emotionality: Fact or stereotype? *Feminism and Psychology*, **3**, 303–318.

Fischer, A. H., & Manstead, A. S. R. (2000). Gender and emotions in different cultures. In A. H. Fischer, ed., *Gender and Emotion: Social Psychological Perspectives*. Cambridge, UK: Cambridge University Press, pp. 71–94.

Fischer, A. H., Mosquera, P. M. R., van Vianen, A., & Manstead, A. S. R. (2004). Gender and culture differences in emotion. *Emotion*, **4**, 87–94.

Fischer, D., & Hills, T. T. (2012). The baby effect and young male syndrome: Social influences on cooperative risk-taking in women and men. *Evolution and Human Behavior*, **33**, 530–536.

Gabriel, K. A., & Williamson, A. (2010). Framing alters risk-taking behavior on a modified Balloon Analogue Risk Task (BART) in a sex-specific manner. *Psychological Reports*, **107**(3), 699–712.

Gangestad, S. W., & Simpson, J. A. (2000). The evolution of human mating: Trade-offs and strategic pluralism. *Behavioral and Brain Sciences*, **23**(4), 573–587.

Gartstein, M. A., & Rothbart, M. K. (2003). Studying infant temperament via the Revised Infant Behavior Questionnaire. *Infant Behavior and Development*, **26**, 64–86.

Gavrilets, S. (2012). Human origins and the transition from promiscuity to pair-bonding. *Proceedings of the National Academy of Sciences*, **109**, 9923–9928.

Goel, N., Workman, J. L., Lee, T. T., Innala, L., & Viau, V. (2014). Sex differences in the HPA axis. *Comprehensive Physiology*, **4**, 1121–1155.

Gordon, I., Zagoory-Sharon, O., Leckman, J. F., & Feldman, R. (2010). Prolactin, oxytocin, and the development of paternal behavior across the first six months of fatherhood. *Hormones and Behavior*, **58**, 513–518.

Gray, P. B., & Anderson, K. G. (2010). *Fatherhood: Evolution and Human Paternal Behavior*. Cambridge, MA: Harvard University Press.

Griez, E. J., Colasanti, A., van Diest, R., Salamon, E., & Schruers, K. (2007). Carbon dioxide inhalation induces dose-dependent and age-related negative affectivity. *PLoS ONE*, **2**, e987.

Gullone, E. (2000). The development of normal fear: A century of research. *Clinical Psychology Review*, **20**, 429–451.

Gurven, M., & Kaplan, H. (2007). Longevity among hunter–gatherers: A cross-cultural examination. *Population and Development Review*, **33**(2), 321–365.

Hall, K. (2002). Who do men and women gossip about and what is discussed about them? Unpublished dissertation, Durham University.

Hamann, S. (2005). Sex differences in the responses of the human amygdala. *Neuroscientist*, **11**, 288–293.

Harris, C. R., Jenkins, M., & Glaser, D. (2006). Gender differences in risk assessment: Why do women take fewer risks than men? *Judgment and Decision Making*, **1**(1), 48–63.

Henrich, J., Boyd, R., & Richerson, P. J. (2012). The puzzle of monogamous marriage. *Philosophical Transactions of the Royal Society B*, **367**, 657–669.

Hermans, E. J., Putman, P., Baas, J. M., Koppeschaar, H. P., & van Honk, J. (2006). A single administration of testosterone reduces fear-potentiated startle in humans. *Biological Psychology*, **59**(9), 872–874.

Hermans, E. J., Putman, P., Baas, J. M., et al. (2007). Exogenous testosterone attenuates the integrated central stress response in healthy young women. *Psychoneuroendocrinology*, **32**(8–10), 1052–1061.

Holbrook, T. L., Hoyt, D. B., Stein, M. B., & Sieber, W. J. (2002). Gender differences in long-term posttraumatic stress disorder outcomes after major trauma: Women are at higher risk of adverse outcomes than men. *Journal of Trauma*, **53**, 882–888.

Holland, B., & Rice, W. R. (1999). Experimental removal of sexual selection reverses intersexual antagonistic coevolution and removes a reproductive load. *Proceedings of the National Academy of Sciences*, **96**, 5083–5088.

Hrdy, S. B. (1999). *Mother Nature: Natural Selection and the Female of the Species*. London: Chatto & Windus.

Jasper, F., & Whitthoft, M. (2011). Health anxiety and attentional bias: The time course of vigilance and avoidance in light of pictorial illness information. *Journal of Anxiety Disorders*, **25** (8), 1131–1138.

Joe Laidler, K., & Hunt, G. (2001). Accomplishing femininity among the girls in the gang. *British Journal of Criminology*, **41**, 656–678.

Jokela, M., Rotkirch, A., Rickard, I. J., Pettav, J., & Lummaa, V. (2010). Serial monogamy increases reproductive success in men but not in women. *Behavioral Ecology*, **21**, 906–912.

Jones, J. H. (2009). The force of selection on the human life cycle. *Evolution and Human Behavior*, **30**, 305–314.

Kajantie, E., & Phillips, D. I. W. (2006). The effects of sex and hormonal status on the physiological response to acute psychosocial stress. *Psychoneuroendocrinology*, **31**, 151–178.

Kaplan, H., Hill, K., Lancaster, J., & Hurtado, A. M. (2000). A theory of human life history evolution: Diet, intelligence, and longevity. *Evolutionary Anthropology*, **9**, 156–185.

Kelly, M. M., Forsyth, J. P., & Karekla, M. (2006). Sex differences in response to a panicogenic challenge procedure: An experimental evaluation of panic vulnerability in a non-clinical sample. *Behaviour Research and Therapy*, **44**, 1421–1430.

Kessler, R. C., Sonnega, A., Bromet, E., Hughes, M., & Nelson, C. B. (1995). PTSD in the National Comorbidity Survey. *Archives of General Psychiatry*, **52**, 1048–1060.

Klein, A. K., Becker, E. S., & Rinck, M. (2011). Approach and avoidance tendencies in spider fearful children: The Approach–Avoidance Task. *Journal of Child and Family Studies*, **20**, 224–231.

Knight, G. P., Fabes, R. A., & Higgins, D. A. (1996). Concerns about drawing causal inferences from meta-analyses: An example in the study of gender differences in aggression. *Psychological Bulletin*, **119**, 410–421.

Knight, G. P., Guthrie, I. L., Page, M. C., & Fabes, R. A. (2002). Emotional arousal and gender differences in aggression: A meta-analysis. *Aggressive Behavior*, **28**, 366–393.

Kokko, H., & Jennions, M. D. (2008). Parental investment, sexual selection and sex ratios. *Journal of Evolutionary Biology*, **21**, 919–948.

Kring, A. M. (2000). Gender and anger. In A. H. Fischer, ed., *Gender and Emotion: Social Psychological Perspectives*. Cambridge, UK: Cambridge University Press, pp. 211–231.

Kruger, D. J., & Nesse, R. M. (2006). An evolutionary life-history framework for understanding sex differences in human mortality rates. *Human Nature*, **17**, 74–97.

Kudielka, B. M., & Kirschbaum, C. (2005). Sex differences in HPA axis responses to stress: A review. *Biological Psychology*, **69**, 113–132.

Lang, P. J., Greenwald, M. K., Bradley, M. M., & Hamm, A. O. (1993). Looking at pictures: Affective, facial, visceral, and behavioral reactions. *Psychophysiology*, **30**(3), 261–273.

Lang, P. J., Ohman, A., & Vaitl, D. (1988). *The International Affective Picture System (Photographic Slides)*. Gainesville, FL: The Center for Research in Psychophysiology, University of Florida.

Lebron-Milad, K., Abbs, B., Milad, M. R., et al. (2012). Sex differences in the neurobiology of fear conditioning and extinction: A preliminary fMRI study of shared sex differences with stress-arousal circuitry. *Biology of Mood & Anxiety Disorders*, **2**, 7.

Lee, T. H., Sakaki, M., Cheng, R., Velasco, R., & Mather, M. (2014). Emotional arousal amplifies the effects of biased competition in the brain. *Social Cognitive and Affective Neuroscience*, **9**(12), 2067–2077.

Lee, Z., & Salekin, R. T. (2010). Psychopathy in a noninstitutional sample: Differences in primary and secondary subtypes. *Personality Disorders: Theory, Research, and Treatment*, **1**(3), 153–169.

Lees, S. (1993). *Sugar and Spice: Sexuality and Adolescent Girls*. London: Penguin.

Leibold, N. K., Viechtbauer, W., Goossens, L., et al. (2013). Carbon dioxide inhalation as a human experimental model of panic: The relationship between emotions and cardiovascular physiology. *Biological Psychology*, **94**, 331–340.

Lighthall, N. R., Mather, M., & Gorlick, M. A. (2009). Acute stress increases sex differences in risk seeking in the Balloon Analogue Risk Task. *PLoS ONE*, **4**, e6002.

Lischke, A., Gamer, M., Berger, C., et al. (2012). Oxytocin increases amygdala reactivity to threatening scenes in females. *Psychoneuroendocrinology*, **37**(9), 1431–1438.

Loewenstein, G. F., Weber, E. U., Hsee, C. K., & Welch, N. (2001). Risk as feelings. *Psychological Bulletin*, **127**(2), 267–286.

Lungu, O., Potvin, S., Tikasz, A., & Mendrek, A. (2015). Sex differences in effective fronto-limbic connectivity during negative emotion processing. *Psychoneuroendocrinology*, **62**, 180–188.

Lykken, D. T. (1957). A study of anxiety in the sociopathic personality. *Journal of Abnormal and Social Psychology*, **55**, 6–10.

Maher, A. M., Thomson, C. J., & Carlson, S. R. (2015). Risk-taking and impulsive personality traits in proficient downhill sports enthusiasts. *Personality and Individual Differences*, **79**, 20–24.

Marsh, P., & Paton, R. (1986). Gender, social class and conceptual schemas of aggression. In A. Campbell & J. Gibbs, eds., *Violent Transactions: The Limits of Personality*. Oxford: Blackwell, pp. 59–85.

McClure, E. B., Monk, C. S., Nelson E. E., et al. (2004). A developmental examination of gender differences in brain engagement during evaluation of threat. *Biological Psychiatry*, **55**, 1047–1055.

McCraw, K. S., & Valentiner, D. P. (2015). The Circumscribed Fear Measure: Development and initial validation of a trans-stimulus phobia measure. *Psychological Assessment*, **27** (2), 403–414.

McDermott, M. J., Peck, K. R., Walters, A. B., & Smitherman, T. A. (2013). Do episodic migraineurs selectively attend to headache-related visual stimuli? *Headache*, **53**(2), 356–364.

McDougall, P., & Vaillancourt, T. (2015). Long-term adult outcomes of peer victimization in childhood and adolescence pathways to adjustment and maladjustment. *American Psychologist*, **70**(4), 300–310.

McLean, C. P., & Hope, D. A. (2010). Subjective anxiety and behavioral avoidance: Gender, gender role, and perceived confirmability of self-report. *Journal of Anxiety Disorders*, **24**, 494–502.

McManis, M. H., Bradley, M. M., Berg, W. K., Cuthbert, B. N., & Lang, P. J. (2001). Emotional reactions in children: Verbal, physiological, and behavioral responses to affective pictures. *Psychophysiology*, **38**(2), 222–231.

Milhausen, R. R., & Herold, E. S. (1999). Does the sexual double standard still exist? Perceptions of university women. *Journal of Sex Research*, **36**, 361–368.

Mirowsky, J., & Ross, C. E. (1995). Sex differences in distress: Real or artifact? *American Sociological Review*, **60**, 449–468.

Moriguchi, Y., Touroutoglou, A., Dickerson, B. C., & Barrett, L. F. (2014). Sex differences in the neural correlates of affective experience. *Social, Cognitive and Affective Neuroscience*, **9**, 591–600.

Nakagawa, T., Sharma, M., Nabeshima, Y., Braun, R. E., & Yoshida, S. (2010). Functional hierarchy and reversibility within the murine spermatogenic stem cell compartment. *Science*, **328**(5974), 62–67.

Nelson, J. (2015). Are women really more risk-averse than men? A re-analysis of the literature using expanded methods. *Journal of Economic Surveys*, **29**(3), 566–585.

Nevo, O., Nevo, B., & Derech-Zehavi, A. (1993). The development of the Tendency to Gossip Questionnaire: Construct and concurrent validity for a sample of Israeli college students. *Educational and Psychological Measurement*, **53**, 973–981.

Nillni, Y. I., Berenz, E. C., Rohan, K. J., & Zvolensky, M. J. (2012). Sex differences in panic-relevant responding to a 10% carbon dioxide-enriched air biological challenge. *Journal of Anxiety Disorders*, **26**, 165–172.

Nugent, A. C., Bain, E. E., Thayer, J. F., Sollers, J. J., & Drevets, W. C. (2011). Sex differences in the neural correlates of autonomic arousal: A pilot PET study. *International Journal of Psychophysiology*, **80**, 182–191.

Olff, M., Langeland, W., & Gersons, B. P. R. (2005). Effects of appraisal and coping on the neuroendocrine response to extreme stress. *Neuroscience and Biobehavioral Reviews*, **29**, 457–467.

Olff, M., Langeland, W., Draijer, N., & Gersons, B. P. (2007). Gender differences in posttraumatic stress disorder. *Psychological Bulletin*, **133**(2), 183–204.

Ordaz, S., & Luna, B. (2012). Sex differences in physiological reactivity to acute psychosocial stress in adolescence. *Psychoneuroendocrinology*, **37**(8), 1135–1157.

Ozer, E. J., Best, S. R., Lipsey, T. L., & Weiss, D. S. (2003). Predictors of posttraumatic stress disorder and symptoms in adults: A meta-analysis. *Psychological Bulletin*, **129**(1), 52–73.

Pampel, F. C. (2001). Gender equality and the sex differential in mortality from accidents in high income nations. *Population Research and Policy Review*, **20**, 397–421.

Patrick, C.J., Bradley, M.M., & Lang, P.J. (1993) Emotion in the criminal psychopath: Startle reflex modulation. *Journal of Abnormal Psychology*, **102**, 89–92.

Pavard, S., Gagnon, A., Desjardins, B., & Heyer, E. (2005). Mother's death and child survival: The case of early Quebec. *Journal of Biosocial Science*, **37**, 209–227.

Perusse, D. (1993). Cultural and reproductive success in industrial societies: Testing the relationship at the proximate and ultimate levels. *Behavioral and Brain Sciences*, **16**(2), 267–283.

Pickersgill, M. J., & Arrindell, W. A. (1994). Men are innocent until proven guilty: A comment on the examination of sex differences by Pierce and Kirkpatrick (1992). *Behavior Research and Therapy*, **32**(1), 21–28.

Powell, M., & Ansic, D. (1997). Gender differences in risk behaviour in financial decision-making: An experimental analysis. *Journal of Economic Psychology*, **18**, 605–628.

Quevedo, K., Smith, T., Donzella, B., Schunk, E., & Gunnar, M. (2010). The startle response: Developmental effects and a paradigm for children and adults. *Developmental Psychobiology*, **52**(1), 78–89.

Reno, P. L., Meindl, R. S., McCollum, M. A., & Lovejoy, C. O. (2003). Sexual dimorphism in *Australopithecus afarensis* was similar to that of modern humans. *Proceedings of the National Academy of Sciences*, **100**, 9404–9409.

Richards, H. J., Benson, V., Donnelly, N., & Hadwin, J. (2014). Exploring the function of selective attention and hypervigilance for threat in anxiety. *Clinical Psychology Review*, **34**, 1–13.

Robinson, V. (2008). *Everyday Masculinities and Extreme Sports: Male Identity and Rock Climbing*. Oxford: Berg.

Schmukle, S. C. (2005). Unreliability of the dot probe task. *European Journal of Personality*, **19**, 595–605.

Schofield, C. A., Johnson, A. L., Inhoff, A. W., & Coles, M. E. (2012). Social anxiety and difficulty disengaging threat: Evidence from eye-tracking. *Cognition & Emotion*, **26**(2), 300–311.

Schwabe, L., Höffken, O., Tegenthoff, M., & Wolf, O. T. (2013). Opposite effects of noradrenergic arousal on amygdala processing of fearful faces in men and women. *NeuroImage*, **73**, 1–7.

Sear, R., & Mace, R. (2008). Who keeps children alive? A review of the effects of kin on child survival. *Evolution and Human Behavior*, **29**, 1–18.

Sear, R., Steele, F., McGregor, A. A., & Mace, R. (2002). The effects of kin on child mortality in rural Gambia. *Demography*, **39**, 43–63.

Seeley, W. W., Menon, V., Schatzberg, A. F., et al. (2007). Dissociable intrinsic connectivity networks for salience processing and executive control. *Journal of Neuroscience*, **27**(9), 2349–2356.

Simmons, R. (2002). *Odd Girl Out: The Hidden Culture of Aggression in Girls*. London: Harcourt.

Simon, R. W., & Nath, L. E. (2004). Gender and emotion in the United States: Do men and women differ in self-reports of feelings and expressive behavior? *American Journal of Sociology*, **109**, 1137–1176.

Slovic, P., Finucane, M. L., Peters, E., & MacGregor, D. G. (2002). The affect heuristic. In T. Gilovich, D. Griffin, & D. Kahneman, eds., *Heuristics and Biases: The Psychology of Intuitive Judgment*. Cambridge, MA/New York: Cambridge University Press, pp. 397–420.

Social Issues Research Centre (2004). Sex differences in driving and insurance risk: An analysis of the social and psychological differences between men and women that are relevant to their driving behaviour. www.sirc.org/publik/driving.pdf.

Spindler, H., Elkit, A., & Christiansen, D. (2010). Risk factors for posttraumatic stress disorder following an industrial disaster in a residential area: A note on the origin of observed gender differences. *Gender Medicine*, **7**(2), 156–165.

Stalnaker, T. A., Cooch, N. K., & Schoenbaum, G. (2015). What the orbitofrontal cortex does not do. *Nature Neuroscience*, **18** (5), 620–627.

Stein, M. B., Walker, J. R., & Forde, D. R. (2000). Gender differences in susceptibility to posttraumatic stress disorder. *Behavior Research and Therapy*, **38**, 619–628.

Stevens, J. S., & Hamann, S. (2012). Sex differences in brain activation to emotional stimuli: A meta-analysis of neuroimaging studies. *Neuropsychologia*, **50**, 1578–1593.

Stewart-Williams, S., & Thomas, A. G. (2013). The ape that thought it was a peacock: Does evolutionary psychology exaggerate human sex differences? *Psychological Inquiry*, **24**(3), 137–168.

Stoyanova, M., & Hope, D. A. (2012). Gender, gender roles, and anxiety: Perceived confirmability of self report, behavioral avoidance, and physiological reactivity. *Journal of Anxiety Disorders*, **26**, 206–214.

Stroud, L. R., Salovey, P., & Epel, E. S. (2002). Sex differences in stress responses: Social rejection versus achievement stress. *Biological Psychiatry*, **52**, 318–327.

Taylor, S. E., Klein, L. C., Lewis, B. P., et al. (2000). Biobehavioral responses to tress in females: Tend-and-befriend, not fight-or-flight. *Psychological Review*, **107**, 411–429.

Thomas, K. M., Drevets, W. C., Whalen, P. J., et al. (2001). Amygdala response to facial expressions in children and adults. *Biological Psychiatry*, **49**, 309–316.

Tian, Q., & Smith, J. C. (2011). Attentional bias to emotional stimuli is altered during moderate- but not high-intensity exercise. *Emotion*, **11**(6), 1415–1424.

Vaillancourt, T. & Sharma, A. (2011). Intolerance of sexy peers: Intrasexual competition among women. *Aggressive Behavior*, **37**(6), 568–576.

Van Bockstaele, B., Verschuere, B., Koster, E. H. W., et al. (2011a). Differential predictive power of self report and implicit measures on behavioral and physiological fear responses to spiders. *International Journal of Psychophysiology*, **79**(2), 166–174.

Van Bockstaele, B., Verschuere, B., Koster, E. H. W., et al. (2011b). Effects of attention training on self-reported, implicit, physiological and behavioral measures of spider fear. *Journal of Behavior Therapy and Experimental Psychiatry*, **42**, 211–218.

Van Bockstaele, B., Verschuere, B., Tibboel, H., et al. (2014). A review of current evidence for the causal impact of attentional bias on fear and anxiety. *Psychological Bulletin*, **140**(3), 682–721.

Van Honk, J., & Schutter, D. J. L. G. (2006). Unmasking feigned sanity: A neurobiological model of emotion processing in primary psychopathy. *Cognitive Neuropsychiatry*, **11**(3), 285–306.

Van Ryckeghem, D. M. L., Crombez, G., Van Hulle, L., & Van Damme, S. (2012). Attentional bias towards pain-related information diminishes the efficacy of distraction. *Pain*, **153**(12), 2345–2351.

Van Wyhe, J., ed. (2002). *The Complete Work of Charles Darwin Online*. http://darwin-online.org.uk.

Vogt, J., De Houwer, J., Crombez, G., & Van Damme, S. (2013). Competing for attentional priority: Temporary goals versus threats. *Emotion*, **13**(3), 587–598.

Wager, T. D., Phan, K. L., Liberzon, I., & Taylor, S. F. (2003). Valence, gender, and lateralization of functional brain anatomy in emotion: a meta-analysis of findings from neuroimaging. *NeuroImage*, **19**, 513–531.

Weir, L. K., Grant, J. W. A., & Hutchings, J. A. (2011). The influence of operational sex ratio on the intensity of competition for mates. *American Naturalist*, **177**(2), 167–176.

Whittle, S., Yucel, M., Yap, M. B. H., & Allen, N. B. (2011). Sex differences in the neural correlates of emotion: evidence from neuroimaging. *Biological Psychology*, **87**, 319–333.

Williams, G. C. (1996). *Plan and Purpose in Nature*. London: Phoenix.

Williams, L. M., Barton, M. J., Kemp, A. H., et al. (2005). Distinct amygdala–autonomic arousal profiles in response to fear signals in healthy males and females. *NeuroImage*, **28**(3), 618–625.

Wilson, M., & Daly, M. (1985). Competitiveness, risk taking and violence: The young male syndrome. *Ethology and Sociobiology*, **6**, 59–73.

Winking, J., Gurven, M., & Kaplan, H. (2011). Father death and adult success among the Tsimane: Implications for marriage and divorce. *Evolution and Human Behavior*, **32**(2), 79–89.

Wood, A., Rychlowska, M., Korb, S., & Niedenthal, P. (2016). Fashioning the face: Sensorimotor simulation contributes to facial expression recognition. *Trends in Cognitive Sciences*, **20**(3), 227–240.

Worthman, C. M., Jenkins, C. L., Stallings, J. F., & Daina, N.L. (1993). Attenuation of nursing-related ovarian suppression and high fertility in well-nourished, intensively breast-feeding Amele women of lowland Papua New Guinea. *Journal of Biosocial Science*, **25**, 425–443.

Wynn, M. H., Høiseth, M. H., & Pettersen, G. (2012). Psychopathy in women: Theoretical and clinical perspectives. *International Journal of Women's Health*, **4**, 257–263.

Yang, Z., Jackson, T., & Chen, H. (2013). Effects of chronic pain and pain-related fear on orienting and maintenance of attention: An eye movement study. *Journal of Pain*, **14**(10), 1148–1157.

Yiend, J. (2010). The effects of emotion on attention: A review of attentional processing of emotional information. *Cognition & Emotion*, **24**, 3–47.

27 The Enigmatic Urge
How Sexual Desire Works

FREDERICK M. TOATES

27.1 INTRODUCTION

This chapter explores how motivation theory (Chapter 15) set into an evolutionary framework can illuminate sexual desire. As implied by its title, some features of sexual desire appear to be enigmatic. However, a study of the combination of causal processes and evolutionary psychology can make them less so.

Examples of enigmatic and seemingly inexplicable features of sexual desire are to be found at numerous points along the wide spectrum of sexual responses that people exhibit. These range from something all-embracing and even addictive, including homicidal sex-linked violence, through to the absence of desire: asexuality. Some men spend hours absorbed in the biologically nonfunctional activity of watching child pornography, thereby risking their freedom, marriages, and careers. Others require a fetish object, such as rubber, to be present in order to feel sexual desire.

Some women claim little desire for sex but, for a fraction of these, if they actually get round to engaging in it, they find this pleasurable (Meana, 2010). Women can show sexual arousal as indexed by their genital response even when they find the encounter unattractive (Laan & Janssen, 2008) or abhorrent, as in rape (Levin & van Berlo, 2004). Meston and Buss (2007) discovered 237 different reasons for why people engage in sexual behavior, with a desire for sexual pleasure and/or babies being only two of these. Can these enigmatic phenomena be linked to motivational and evolutionary considerations?

Differences between men and women in their sexual behavior provide strong evidence in favor of evolutionary adaptations and feature at center stage in evolutionary psychology (Barkow et al., 1992; Buss, 2015; Workman & Reader, 2014). Therefore, a principal task of the chapter is to explore how a model of motivation can accommodate sex differences.

Concerning causal processes, several theoretical assumptions and models lie behind the arguments to be presented here, as described in Section 27.2.

27.2 BASIC DESIGN PRINCIPLES UNDERLYING DESIRE

27.2.1 Introduction

This section integrates three processes and the associated bodies of theory that are sometimes discussed separately: (1) hierarchical control, (2) incentive motivation and (3) control by a balance between excitation and inhibition.

27.2.2 Hierarchical Control

In humans, evolutionarily old ("low-level") brain structures, termed "System 1," coexist with newer species-specific structures that have more recently evolved, termed "System 2" (Maclean, 1990). At a low level, something like the modules of evolutionary psychology can mediate the control of desire outside conscious awareness, but this interacts with consciously accessible processes (Chapter 15). Newly emerging structures are added to older structures. This organization appears to be evident in both excitation and inhibition, described in Sections 27.2.3 and 27.2.4.

Low-level controls, largely subcortically based, give an automatic ("magnet-like") quality to behavior, drawing the individual forward toward those stimuli that have incentive salience. Higher-level controls, largely mediated by the prefrontal cortex, have more rationally conscious and goal-directed qualities to them (Berridge, 2001; Toates, 2006).

Layers of control can act in cooperation or competition (Toates, 2014). Both levels can trigger forward engagement. However, under other circumstances, the high level might offer restraint on behavior, whereas the low level might pull one toward engagement. Conversely, the high level might produce sexual engagement in the absence of any low-level triggered desire, as in "voluntary" sexual behavior in the face of disgust.

27.2.3 The Excitatory Role of Incentive Stimuli

Most theorists no longer postulate a diffuse drive arising somewhere in the body tissues that *drives* behavior.

Models based upon incentive principles (Chapter 15) have now largely replaced the older drive models (Berridge, 2004; Bindra, 1978; Toates, 1986). In incentive terms, sexual desire is triggered by a combination of: (1) sexual incentives in the environment (sexually attractive individuals, cues associated with them and portrayals of them); (2) contextual information triggered in memory by the incentive image; (3) sexual fantasy; and (4) sensory information arising at the genitals (Toates, 2014).

When presented rapidly at only a subliminal level in both men and women, sexual images rapidly trigger indices of sexual desire, approach motivation, associated positive sexual memories, and positive affect (Gillath & Collins, 2016). This appears to represent the effect of an uncomplicated "pure" incentive process. When these stimuli are presented for a longer duration and enter conscious awareness, the effects are more mixed and can take an aversive quality. Presumably, given time, a wider range of associations – some involving inhibitions – are recruited.

A distinction between wanting and liking was introduced in Chapter 7 and applies to sexual desire (Toates, 2014). Wanting appears to be closely associated with dopamine and liking with opioids.

The physical features associated with sexual attractiveness include, for both sexes, facial symmetry (Buss, 2015). For women, factors include the male's facial masculinity, height, muscular condition, and physique, and for men's desire, factors include the woman's facial femininity, hair quality, skin, and waist-to-hip ratio (Buss, 2015). Men tend to assess sexual attractiveness (and experience the associated desire) mainly on the basis of physical attributes. Setting opposite-sex targets into a context that suggests social status has a greater effect on men's attractiveness to women than on women's attractiveness to men (Townsend, 1998).

Some qualities (e.g., intelligence) can be assessed as a first guess based upon physical appearance. Whereas physical features impact the sensory systems immediately, such qualities as capability, dependability, kindness, and loyalty can only be assessed over longer periods of time. A principle that forms a foundation of evolutionary psychology is that a woman's parental investment is much higher than that of a man. Therefore, it is of adaptive value for the female to perform more extensive and protracted assessment of mate value (Buss, 2015). The value of such assessment can presumably account in part for women's greater sexual reserve. Women tend to put weight on the man's potential for resource-holding and commitment, assessed by such things as extrapolation from appearance, the man's behavior, and evidence of social status (Townsend & Wasserman, 1997). Evidence of positive interactions with children increases the attraction that women feel toward a man (Buss, 2015).

A sample of men are closer together in their ranking of sexual attraction of target models than are women, an effect that is amplified when the models are in bathing suits (Townsend & Wasserman, 1997). This points to universals in the relatively narrow range of the male taste regarding high female attractiveness.

Men tend to have an optimal age of attraction for sexual partners at around 25 years (Antfolk et al., 2015; see also Kenrick & Keefe, 1992). Men younger than this therefore desire an older sex partner, whereas men older than this desire a younger partner. By contrast, women's optimal desire is for a male slightly older than themselves, and the desired age increases in parallel with the female's age. Again, this points to the relatively fixed physical attributes of optimal attraction in males but the female's accommodation to changing circumstances.

This section will develop the foundations of a model of human sexual desire that incorporates sex differences. Imagine the physical features of another individual being detected and analyzed by sensory and perceptual processes. Presumably, any direct contribution that these features make to sexual desire will arise rather rapidly for both men and women. However, as just described, this direct contribution to sexual excitation is typically more powerful in men than in women. As a result of early development, the features that excite desire tend to be more restricted by the individual's sexual orientation in men than is the case for women (see Section 27.6.2).

Evolutionary psychology involves intelligent speculation. So, let us assume that, as a result of development, there exists something like a template ("internal model") for the necessary physical features of the other individual that potentially trigger sexual desire.[1] This is similar to the ideas of search images, sexual imprinting, and "love-maps," as suggested by Money (1986). The template might normally be formed in part as a result of early exposure to conspecifics. Masturbation to images could serve to consolidate the template (Bogaert, 2015). In addition to the triggers that are well articulated in the literature of evolutionary psychology, idiosyncratic features can be assimilated into the template developmentally (discussed in Section 27.10).

Consider the link between physical features and the associations that they trigger. Let us assume that, in the case of humans, activation of the template by the physical presence of the other individual or in the imagination can excite associations and meanings in memory, which act reciprocally with the template. These associations can excite or inhibit activation of the template. The level of excitation that is generated, if any, constitutes sexual desire. It is further suggested that a sex difference exists in the relative weight attached to this template and its associations as determinants of sexual desire. In triggering desire, there are two types of association that can be formed with physical features, as follows: idiosyncratic enduring associations are sometimes rapidly and automatically formed between the target individual ("incentive sexual object") and a particular other object(s) and/or

[1] This must surely be the case where fantasy in the absence of physical stimulation from the other individual is concerned.

event(s). These associations are formed at a low level and can be persistent. Physically present things paired with excitation of the template lock into association with it more strongly in men than in women, hence fetishes and partialism (being excited by only part of a body) being predominantly male phenomena (discussed in Section 27.10.1).

There are also longer-term and more flexible, consciously accessible associations formed. In sexual desire, extensive cognitive processing over relatively long periods of time involving the behavior of the partner plays a much bigger role in women's desire than men's (Buss, 2015). This either increases or inhibits the excitation arising from sensory features of the individual impacting upon the template.

Although men's greater desire for sex partners outside any established bond forms a bedrock of evolutionary psychology (Workman & Reader, 2014), women nonetheless sometimes seek so-called short-term mating, including extra-pair mating (Gangestad & Simpson, 2000). Although controversial, evidence suggests that, when this happens, the female seeks a particularly attractive male, thereby placing less weight on associations. In other words, the physical features represented in the template assume greater weight as compared to the kind of mate accepted for a long-term commitment.

After time in an established relationship, women tend to place more weight upon associations with their regular partner, such as to receive confirmation of their desirability (Basson et al., 2003).

The nature of inhibition will now be described.

27.2.4 The Inhibition of Desire

In evolutionary terms, mating incurs enormous potential costs, particularly for women. For example, a male's rival might be in the vicinity or, for a woman, a suboptimal choice could tie up her reproductive capacity for over nine months. It would seem that the costs have been such that a loss of excitation alone is insufficient to prevent suboptimal sexual behavior (Janssen & Bancroft, 2007). As a very rough generalization, three different forms of inhibition can be identified as that exerted by (Toates, 2014):

1 Factors arising from stimuli that are physically present before or at the time of potential sexual activity (e.g., perception of erectile failure, triggers to disgust or fear)
2 Changes that arise immediately at the end of sexual activity and inhibit future sexual desire
3 Consideration of long-term consequences, which might exert inhibition at any stage (e.g., thoughts of a pledge of chastity until marriage or of compromising a pair-bond)

These three forms of inhibition are not mutually exclusive and there can be overlap between them.

Factor 1 would seem to include what Janssen and Bancroft (2007) term "inhibition due to threat of performance failure." Factor 3 seems to correspond to what they term "inhibition due to the threat of performance consequences." However, the questionnaire from which these two factors were derived relied upon reports by people of their real or hypothetical sexual experience. Factors not open to conscious introspection were necessarily excluded, and these will also be considered here.

Sexual activity is a time of vulnerability (e.g., risk of predators, jealous conspecifics). This would be expected to exert some restraint on sexual desire outside an established bond and/or under suboptimal environmental conditions. Compared to men, women show elevated fear of novelty (Campbell, 2013). The amygdala plays a crucial role in the fear reaction to physically present stimuli, an effect mediated in part at a subcortical level (LeDoux, 1999; Victor & Hariri, 2016).

Guilt and regret might only be experienced after a sexual encounter. However, the human capacity for projection into the future permits such experience in an anticipatory fashion at the time of making a decision on whether to engage in sexual behavior.

The evolutionary roots of disgust ("basic disgust") appear to lie in protecting animals from sources of pathogens, such as ulcers, wounds, rotten food, and saliva (Tybur et al., 2013). Prior exposure to stimuli triggering basic disgust reduces subsequent sexual arousal in women (Fleischman et al., 2015).

In addition, in humans, the reaction has been coopted to serve other functions in two other classes of disgust: sexual disgust and moral disgust (Tybur et al., 2013). All three forms would be expected to be potential sources of inhibition on either sexual desire itself or putting it into effect in behavior.

Tybur et al. (2013) argue that the function of sexual disgust is to offer a defense against suboptimal mating (e.g., for a 16-year-old individual, this might be the prospect of having sex with an 85-year-old). Passivity through disgust is not necessarily the only vehicle for avoiding suboptimal mating. Rather, for a strong emotion of disgust, the individual exhibiting desire could trigger both *active* rejection and avoidance by the intended target.

Depending upon the individual, moral disgust is triggered by various activities (e.g., incest, prostitution, pedophilia, and bestiality). Compared to men, women show higher levels of basic, moral, and sexual disgust (Tybur et al., 2009).

Different forms of disgust might make different processing demands. Basic disgust would appear to be directly triggered by the stimulus properties of the offending object with minimal processing demand. Moral disgust requires some processing in terms of meaning, though this might not be accessible to conscious awareness (Haidt, 2001). For example, a father might trigger no disgust in a daughter – quite the contrary – until he makes what she interprets to be sexual gestures toward her.

We now turn to consider inhibitory factors that arise immediately following sexual activity. In men, orgasm/ejaculation generally exerts more inhibition on desire and arousal than does orgasm in women. There is an active process of inhibition mediated by serotonin, among other neurochemicals (Rubio-Casillas et al., 2015). The functional significance would appear to be to restrain mating until the seminal fluids have had time to replenish.

Following their first intercourse, women students are more likely than men to report "sadness, guilt, nervousness, tension, embarrassment and fear" (Guggino & Ponzetti Jr., 1997, p. 189). Such feelings are present shortly after the sexual interaction and are thereby likely to devalue the sexual incentive. Conversely, men tend to have a higher frequency of regret concerning their *inaction* (i.e., lost opportunities; Galperin et al., 2013). Men and women do not show a general difference in regret, with the sex difference appearing specifically in a *sexual* context (Campbell, 2013; Dickson et al., 1998; Galperin et al., 2013). This reflects the relative cost–benefits of engaging in casual sex when comparing the sexes.

From a functional perspective, short-term mating opportunities might need to be passed over in the interests of more promising longer-term gains (Bjorklund & Kipp, 1996). This involves projection into the future and calling up information on possible scenarios (e.g., family disapproval).

Evidence suggests that women are better at such inhibition than are men, which fits with the notion of a greater cost to women of suboptimal mating. Regions of the prefrontal cortex are implicated in the exertion of such inhibition. Bjorklund and Kipp (1996, p. 167) suggest that this is partly domain-specific:

women will hold an advantage over men on most tasks involving sexual or social content when it comes to inhibiting specific behaviours, including impulse control, resistance to temptation, delay of gratification, and control of emotional responding.

However, there are suggestions that women's greater inhibitory capacity is only evident in the phase of the menstrual cycle when they are most fertile (Hosseini-Kamkar & Morton, 2014). Other researchers fail to find enhanced inhibition in women, but find that males' greater tendency to succumb to temptation is due to elevated excitation in the presence of sexual stimuli (Tidwell & Eastwick, 2013).

In a culture in which most people know about the link between sexual intercourse and producing babies (as well as sexually transmitted diseases), it might be expected that this would contribute greater inhibitory weight in girls than boys when contemplating casual sex (Baldwin & Baldwin, 1997).

Having introduced the bases of a model, the discussion now turns to a range of different phenomena that can be better understood in such terms.

27.3 SEX DIFFERENCES IN INCENTIVE MOTIVATION

27.3.1 Behavioral Evidence

Various authors have noted men's higher interest in sex relative to that of women (Baldwin & Baldwin, 1997; Baumeister et al., 2001; Buss, 2015; Knoth et al., 1988; Schmitt, 2003; Symons, 1979). This section considers the evidence.

Boys typically start their sexual activity at a younger age than girls. In a new relationship, typically women hold out longer in agreeing to sex than do men. Initiation of sex is usually because of persuasion by the male. Women decline sexual invitations much more frequently than do men.

Men express a wish for a higher frequency of sexual outlets than do women and for a larger number of partners over a lifetime than do women. Men express a desire for a wider range of sexual activities with these novel partners than do women. Men have a higher actual number of sex partners than women. Of course, logically, for every consensual sexual act by a man, there must be an equivalent for a woman. However, this is misleading because of the factor of prostitution, which is typically not motivated by sexual desire. Men show a higher frequency of masturbation than do women. Women more commonly seek therapy for low sexual interest

Note that these are all measures of sexual *wanting*, rather than *liking*. It is difficult, if not conceptually impossible, to assess whether women like sex as much as or even more than do men.

These differences do not represent an absolute dichotomy. There is much intrasexual variation and some intersexual overlap, but they establish average values.

27.3.2 Bases of the Difference

A sex difference appears when the attention-grabbing capacity of subliminal sexual images is measured (Snowden et al., 2016). Men are specifically more captured by images of women than of other men, whereas women are more diffuse in allocating attention roughly equally to the images of the two sexes.

The behavioral phenomena described in Section 27.3.1 are in line with what would be expected from evolutionary considerations (i.e., a relatively higher ratio of excitation to inhibition in men, manifest as higher levels of expression in men with greater female reserve and caution). This range of differences found universally is suggestive of a biologically adaptive basis. This is not to ignore the role of sociocultural factors that universally act in the same direction (Baldwin & Baldwin, 1997). However, these cultural factors might well themselves be reflective of biological adaptation (Schmitt, 2003).

At least one additional factor might also be included: when women engage in short-term mating, as in casual sex, there is often a wish that this could turn into something more durable (Buss, 2015).

27.4 DEVELOPMENT OF DESIRE AND ITS EXPRESSION IN BEHAVIOR

27.4.1 Similarities in the Sexes

The age of around 10 years is normally when children first experience sexual desire. This is true of both boys and girls, and is equally true irrespective of their sexual orientation (McClintock & Herdt, 1996). This age corresponds to maturation of the adrenal glands and secretion of hormones from there. Even by an age of four to five years, children exhibit a preference for facial attractiveness as judged by the criteria shown by adults (Boothroyd et al., 2014). In terms of the present chapter, this represents the formation of a template underlying later incentive engagement.

27.4.2 Differences between the Sexes

A developmental history lies behind the emergence of the sex differences in the determinants of desire (described in Section 27.2.3), in which biological, individual learning, and sociocultural influences intertwine in complex ways (Baldwin & Baldwin, 1997). Most obviously, the male penis is much larger than the female clitoris as a source of self-generated pleasure, deliberately induced or accidentally.

Boys masturbate more than girls (and engage in more sex play) and have more frequent sexual fantasies and episodes of spontaneous sexual arousal (Knoth et al., 1988). One can speculate that, during masturbation, images of attractive others lock into association with the pleasures of sexual stimulation. Boys have more erotic dreams associated with orgasm than do girls, another potential route for the assimilation of hedonically potent representations in memory. In one sample, girls' first awareness of sexual arousal came in the presence of a boy, often one for whom they held feelings, whereas boys' first awareness of sexual arousal tended to occur with impersonal visual stimuli, such as nude pictures (Knoth et al., 1988). Menstruation can bring an association of the female genitals with a lack of cleanliness and thereby a potential source of loss of sexual self-confidence in girls. First intercourse is more likely to be associated with pain in girls than in boys.

27.4.3 Development: A Hierarchical Perspective

The brain regions contributing to the layers of control introduced in Section 27.3 develop at different rates (Shulman et al., 2016; Steinberg, 2008). Expressed in the terms of the present chapter, the largely subcortical System 1 develops faster than System 2, which is mainly embodied in prefrontal cortical structures. Adolescence is a time when there is disproportionate weight upon System 1.

This relative overactivity of System 1 is usually discussed in terms of high risk-taking at adolescence compared to childhood and adulthood. This is manifest by a number of risky activities, including experimenting with drugs and unsafe sex, resulting in a disproportionate incidence of sexually transmitted diseases and unwanted pregnancies (Victor & Hariri, 2016). Evidence points to a relatively weak prefrontal restraint process in the case of those adolescents who engage in risky sexual behavior (Goldenberg et al., 2013).

Traditionally, the discussion has concerned harmful aspects of risk-taking. However, more recently, attention has been drawn to the possible adaptive benefits of risk-taking (Ellis et al., 2012). Researchers have described the "adaptive role" of the relatively high subcortical dopaminergic (i.e., System 1) activity at adolescence (Telzer, 2016).

It can be speculated that, in our early evolution, such activity and the forward engagement that was triggered by it would have motivated exploration of new territories, seeking competitive advantages, new mating opportunities, etc. (Ellis et al., 2012). Within contemporary society, such high activity can be associated with "adaptive" roles, such as high persistence in college and with prosocial activities. Whether dangerous risk-taking or stability is achieved depends upon the social and cultural context. Some of the dangers present in our environment (e.g., fast cars, injecting drugs) were not around in our early evolution.

This dual role can speculatively be linked to underlying brain processes. Under some conditions, System 2 will tend to offer inhibition on System 1 (e.g., when contemplating a risky option). Heightened System 1 activity might then override the goal-directed inhibitory tendencies arising in System 2. Under other circumstances (e.g., pursuing academic success), Systems 1 and 2 could be acting in the same direction in triggering forward engagement.

The unpredictability and harshness of the child's environment is associated with relatively early appearance of "risk-taking" behaviors, including risky sexual activity (Doom et al., 2016; Ellis et al., 2012). Based upon incentive motivation principles, there could be an animal model of this. Rats reared under conditions of social deprivation (Lomanowska et al., 2011) or reward uncertainty (Anselme, 2010) show enhanced strengths of incentive salience.

Some evidence suggests that the developing brain monitors environmental indices of abuse, stress, and uncertainty, such that sexual maturation is accelerated under adverse conditions (Belsky et al., 2010; Ryan, Mendle, & Markowitz, 2015). This is said to be of adaptive significance since it increases the chances of early reproduction when life prospects are poor. Under stressful rearing conditions, there could be an amplification of the difference in development of prefrontal cortical restraint processes and subcortical excitatory ones. The extent to which this contributes to early sexual maturation and whether it reflects an evolutionary adaptation or prefrontal cortical psychopathology is perhaps something of a philosophical issue.

27.5 EROTIC PLASTICITY

Women show greater *erotic plasticity* than men (Baumeister, 2000; Diamond, 2012). This term refers to the extent to which sexual behavior alters with changes in social context. For example, women are better able to honor a pledge of chastity in a religious cause than are men.

A consideration of the bases of sexual desire and sex differences can illuminate this issue. As noted in Section 27.2.3, men are more strongly driven by physical features and women are more strongly driven by relationship meanings. In most environments, males have exposure to an abundance of sexual incentives to keep their incentive processes sensitized over time. This array is unlikely to change much in fundamental properties. By contrast, meanings can change drastically and almost instantaneously (e.g., one person walks out on a relationship and another walks in, or a much-loved partner is arrested for pedophilia).

Perhaps the best evidence in favor of plasticity comes from a study of sexual orientation, as described in Section 27.6.

27.6 BEYOND HETEROSEXUALITY

Sexual orientation is indexed by the *target* of sexual desire rather than any particular form of behavior, which is entirely compatible with an incentive motivation interpretation. Evidence suggests that prenatal hormonal influences play an important role in determining sexual orientation (Meyer-Bahlburg et al., 1995).

27.6.1 Homosexuality

Interestingly, one sees the gender difference even more clearly when considering homosexual as compared to heterosexual desire (Townsend, 1998). The number of partners of gay men is typically much higher than that of lesbian women. In heterosexuals, female restraint would normally curb male promiscuity, whereas in gay males there is no such brake and individual desires are likely to be more evenly matched. Reciprocally, the lesbian's lower desire for novelty and greater emphasis upon intimacy is mirrored by those of her partner.

27.6.2 Bisexuality

For nominal heterosexuals, women are more likely than men to find same-sex sexual contact attractive and to engage in this (Baumeister, 2000; Diamond, 2012), exemplifying the wider range of acceptable physical stimuli shown by women. That is to say, women's sexual orientation is much more "plastic" than that of men. Women have more homosexual themes in their fantasies than do men (Binter et al., 2012). More commonly than men, women pass through cycling phases of heterosexual and homosexual desire according to personal circumstances (e.g., meeting a new friend;

Diamond, 2012). A possible adaptive significance of women's greater bisexuality comes in the suggestion that those who bonded with other women were at an advantage in raising children and sexual contact is a good means of bonding (Radtke, 2013).

Depending upon their sexual orientation, hypersexual men tend to show desire toward *either* men *or* women but not both, whereas for women the tendency to show bisexual attraction increases with intensity of desire (Lippa, 2006). Increasing desire amplifies the strengths of men's templates, but broadens those of women.

27.6.3 Asexuality

Some people show no sexual desire directed toward other individuals, such that asexuality is a kind of identity to them (Bogaert, 2015). They sometimes engage in masturbation, with associated pleasure. This suggests that the developmental process that attributes sexual incentive salience to other individuals and links this to pleasure processes is not functioning normally. It also points to a dissociation between wanting and liking. Asexuality is more common among women than men, which is perhaps not surprising given the normally weaker links between raw physical stimuli and desire in women.

27.7 INTERACTION WITH OTHER MOTIVATIONS

27.7.1 A Range of Interacting Motivations

As documented by conscious report, a vast range of different motivations underlie a given consensual sexual interaction (Meston & Buss, 2007). For example, reasons given for engaging in sexual activity include those such as "I felt sorry for him" that have the characteristic of altruism, as well as "to make me feel wanted by someone." The feminist scholar, Tiefer (1991, p. 18), writes:

men and women are raised with different sets of sexual values – men towards varied experience and physical gratification, women towards intimacy and emotional communion … Socioeconomic subordination, threats of pregnancy, fear of male violence and society's double standard reduces women's power in heterosexual relationships.

By definition, putting into effect both pro-social and anger-related motivations involves forward engagement with the world (Chapter 15), which is associated with activation of brain dopamine systems (Carver & Harmon-Jones, 2009; Rademacher et al., 2017). Thus, it might be speculated that there is an added effect of sexual and non-sexual goals in their incentive value. Two primary interactions are described in Sections 27.7.2 and 27.7.3.

27.7.2 Link with Love, Commitment, and Romance

Women are more likely than men to emphasize love as a basis for sexual behavior, whereas men tend to value physical "release" (Hyde, 1996; Meston & Buss, 2007). The sexual desire that women feel for men is increased by prior priming with romantic imagery, an effect not seen in men (Dewitte, 2015). Guggino and Ponzetti Jr. (1997) concluded that an association between love and sexual pleasure "does not appear to apply to men in this study in the same way it does to women" (p. 199). This is another example of where women's sexual desire is contextualized in terms of meaning.

For women's sexuality, Basson et al. (2003, p. 222) write:

the reasons motivating sexual interaction remain highly varied and include many that are partially or totally non-sexual, for example, a wish to experience tenderness/appreciation for the partner, or a need to confirm one's desirability.

To some extent, extramarital affairs tend to mirror the sex difference seen prior to marriage (Glass & Wright, 1985). Adulterous husbands tend to be motivated by physically triggered sexual novelty, and sexual intercourse plays the dominant role. Adulterous wives tend to be dissatisfied with their marriages and thereby seek additional qualities such as a recognition of self-worth outside the marriage (Townsend, 1998). Emotional attachment to the extra-pair male tends to play an important role in women's infidelity.

The fact that asexual people can still feel and show romantic desires toward others is evidence of the potential decoupling between sex and romance (Bogaert, 2015).

27.7.3 Sexual Domination and Violence

Violence is more a male than a female behavior, and sexual violence is a particularly strong example of this general pattern (Hyde, 1996). An incentive motivation model would reject dichotomies of the kind that sexual violence is *either* motivated by sexual desire *or* by dominance/aggression. Rather, it would see sexual desire as invariably being present, and these underlying motivations would merge their effects when a sexual target is pursued aggressively (cf. Palmer, 1988; Thornhill, 1996). There appears to be a varying contribution from dominance/aggression depending upon the individual and ranging from the use of instrumental aggression with no desire to injure the victim to where violence is an essential ingredient (Palmer, 1988). As with sexual desire, power motivation is sensitized by testosterone (Hall, Stanton, & Schultheiss, 2010). For those males who have a desire to dominate, the subliminal presentation of dominance-related words tends to increase the attractiveness rating of the female being viewed (Bargh et al., 1995).

Statistics on the age of female rape victims indicate a tendency for these to concentrate at around 15 years (Felson & Cundiff, 2012, 2014), pointing to the role of sexual desire in this behavior as triggered by particular physical features. Vulnerability at this age might also be a contributing factor.

27.8 REPRESENTATIONS

This section looks at representations of sexual behavior, both those generated in fantasy and those in the form of pictures, films, and stories.

27.8.1 Fantasy

Fantasy tends to reflect reality of desire, whether or not such desire is ever realized in action. Men have a higher frequency of sexual fantasies than do women (Knoth et al., 1988; Leitenberg & Henning, 1995). In their fantasies, men typically switch between partners and engage in a wider range of activities than do women, who tend to run a meaningful narrative (Leitenberg & Henning, 1995). It appears that even boys' dreams tend to be more explicitly sexual than those of girls (Money, 1986).

27.8.2 Pornography, Erotica, and Romance Literature

Men tend to find explicit hard-core pornography more attractive than do women. Women prefer erotica with a story line attached (Laan & Both, 2008). Romance novels tend to attract more women than men readers (Money, 1986; Townsend, 1998). The hard-core material is visual with little or no meaning or personal significance attached, whereas what women find most attractive is implicitly meaning related and any explicit sexuality is strongly contextualized.

27.9 SEXUAL ADDICTION

The term "sexual addiction" refers to the situation in which sexual behavior shows some features in common with chemical addictions (e.g., distress, escalation, tolerance, and ambivalence; Dhuffar & Griffiths, 2015; Phillips et al., 2015). It is mainly a male phenomenon, particularly where it concerns viewing impersonal hard-core pornography. Where women are involved, the tendency is for the addictive activity to have at least some modicum of meaning and interpersonal connection, as obtained in sex chat rooms (McKeague, 2014; Schneider, 2000), sometimes taking more of the quality of love addiction (Reynaud et al., 2010).

Sexual addiction illustrates the principle of incentive motivation as it is not manifest as a generally heightened sexuality, but rather as pursuit of a specific goal often associated with novelty, such as pornographic images or pickups in a bar (Money, 1986). A regular partner might get ignored in the process. Addicted men show a particularly strong reactivity of the incentive salience pathway to cues of sexual novelty (Banca et al., 2016; Brand et al., 2016).

Sexual addiction would seem to be a good example of evolutionary mismatch, in that the presence of sexual stimuli (e.g., in the form of moving images) clearly is very different from what was available in our early evolution.

Neither did our early environment probably contain an abundance of young females dressed in a provocative way and available in exchange for some kind of material offering. The ethological notion of supernormal stimuli (Tinbergen, 1951) can be offered as a kind of animal model of this process.

27.10 SOME PARTICULAR TASTES

27.10.1 Paraphilias

A paraphilia is something outside the range of conventional activity, such as voyeurism, exhibitionism, and fetishes. These are overwhelmingly male phenomena (Money, 1986; Townsend, 1998) and are defined in terms of simple physical stimuli (e.g., approach to a visual stimulus) or reactions to anonymous women.

Fetishes exemplify triggering of desire by the *physical properties* of an object or situation. Partialism refers to a particular sexual attraction to a part of the body (Money, 1986). Occasionally, sexual desire is aroused especially strongly by the perception of a missing body part, as in an attraction to amputees.

It is uncertain how fetishes arise in the first place. Some evidence points to the fetish object or event being present at an early age in association with arousal of a sexual or nonsexual nature (Money, 1986; reviewed by Toates, 2014). Where the early event is traumatic, as in a boy who has been dressed in girls' clothes, it can later get transformed into something erotically attractive (Money, 1986). In other cases, one can speculate that the fetish might have been present only in the imagination.

Strictly speaking, a fetish is something *necessary* for sexual desire and arousal to occur. A milder form representing a similar phenomenon would appear to be the amplification of "normal" sexual desire by objects paired with sexual targets (Money, 1986). The ubiquitous appeal of stockings, suspenders, and high heels in adult stores and pornography exemplifies this. Boys show inordinate fascination with girls' underwear, with little evidence of any reciprocal attraction.

Similarly, a mild form of partialism is represented by the male tendency to be aroused by the sight of part of the female body and, from this, to extrapolate to the whole woman (Townsend, 1998). Tales of the Victorian English covering table legs to preempt desire might be simply apocryphal, but for there to have been any rationale in the procedure it would have needed to be men's rather than women's desire that they had in mind.

Fetishes and partialism could represent the aberrant outcome of an otherwise adaptive process: a form of imprinting plus classical conditioning to the whole sexual incentive. It is not hard to appreciate the functional value of such a process and that selection might sometimes "get it wrong."

27.10.2 Pedophilia

Pedophilia is a striking example of where sexual desire is triggered by particular physical features, and it is overwhelmingly (almost exclusively) a male phenomenon (Quinsey, 2003). In the terms advanced here, the template does not change with the advancing years of development. In keeping with an incentive motivation interpretation, compared to controls, pedophiles can be distinguished at an early stage of automatic processing of erotic stimuli, where adult images trigger a diminished response (Knott et al., 2016).

27.10.3 Groupies

The term "groupie" refers to girls who follow famous men such as pop stars or sportsmen (Townsend, 1998). A percentage of these engage in one-night stands with their heroes. This would appear to be an example of women engaging in the kind of "basic behavior" that men normally desire.

However, a moment's consideration reveals that the physical attributes of the male are strongly contextualized in terms of meaning. He usually has an enormous resource-holding potential, and any contact with a given groupie can be interpreted by the girl as an index of her special – if not unique – self-worth. Sportsmen who attract groupies in a social context where they are known fail to attract such female attention in a social context where they are not known (Townsend, 1998).

27.10.4 Prostitution and Romance Tourism

In all societies studied so far, prostitution is overwhelmingly a service offered to men, whether by women or other men (Townsend, 1998). It usually epitomizes casual sex offered with no hint of romance or complications, though the occasional male client seeks a girlfriend experience (Pruitt & Krull, 2011). The physical stimulation offered by the woman is exaggerated by the use of sexualized attire.

The nearest equivalent that men offer women is generally not described as prostitution but rather as "romance tourism," and it is exemplified by Western women traveling to Jamaica and other exotic destinations for liaisons (Pruitt & LaFont, 1995). Although sexual relations frequently occur, they tend to be set in a broader context of courtship, flattery of the woman, emotional involvement, and sightseeing together, and often with the woman's wish for durability in the relationship.

27.11 SUMMARY AND CONCLUSIONS

This chapter has argued the case for integration across some traditional boundaries, in particular the study of evolutionary psychology and that of causal processes. A fuller understanding requires consideration of several causal processes in order to understand sexual desire and how it links to sexual behavior: those of incentive motivation, hierarchical control, and excitation/inhibition.

The principle of incentive motivation accords with theories that involve both essentialist and sociocultural factors (Tolman & Diamond, 2001). Common across sexual desires and orientations is a universal incentive motivation process. However, what comes to form an incentive can be dependent to some extent upon social experience, exemplified by such things as female erotic plasticity.

The incentive motivation model has been extended to incorporate the strength of incentives being associated not only with the incentive's physical properties, but also the associations that it triggers. Such associations, formed by both men and women, can be simple and physical, as in fetishes, or complex, as in links to social status acquisition, trustworthiness, and resource-holding potential. They can be either excitatory or inhibitory.

Both excitation and inhibition can arise automatically ("low level") as a result of basic stimulus properties (e.g., physical appearance, excitatory; basic disgust, inhibitory), but also as a consequence of complex cognitive processing ("high level"), as in a favorable assessment of commitment (excitatory) or anticipatory regret concerning consequences of sexual action (inhibitory).

It was argued that the relative weight attached to physical stimuli and associations can vary with various factors. Women appear to place most weight on complex associations and men to place most weight on physical stimuli (Symons, 1979). The evidence to support this comes from such things as women's greater sexual reserve, even in the presence of physically attractive stimuli, and their greater weight upon meaningful connections, such as romantic associations. Men tend to show a greater attraction to casual sex, impersonal hard-core pornography, and prostitution, as well as a greater proneness to fetishes and partialism. Men's sexual orientation tends to be more clearly dichotomous than that of women. Pedophilia, defined by age/appearance of the desired individual, is overwhelmingly a male phenomenon.

Tables 27.1 and 27.2 summarize some sex difference in sexual desire and behavior.

Sexual behavior can be motivated by not only "pure" sexual desire, but also by a variety of other social motivations (Meston & Buss, 2007). Evidence is accumulating that there exists a common currency underlying forward engagement and that this is based upon dopamine. It seems logical to suppose a summation of sexual desire and other motivations in the pursuit of sexual incentives. Women tend to place more weight on such incentives as gaining love, whereas for men dominance seems more often to be linked with sexual desire.

Much of what is described in this chapter concerning sex differences in desire makes perfectly good sense in terms of adaptations to an earlier environment. However, some things would appear to be enigmatic and need to be addressed. The existence of pedophilia seems anomalous, but might be explicable in terms of errors in the process of incentive formation. Women's greater bisexuality can make sense in

Table 27.1 Comparing relative contributions to the strength of sexual desire in women and men.

Stimulus driven or meaning	Women Weight on meaning	Men Weight on physical stimulus
Intensity/frequency	Much change over lifetime with social circumstances	Relatively little change with social circumstances
Sexual orientation	Change with circumstances	More stable with changing circumstances
	Pedophilia very rare	Pedophilia less rare
Ideal type of partner	Not so tightly defined by physical characteristics	Defined largely by physical characteristics
	Inconsistent across individuals	Relatively consistent across individuals
	Sensitive to context (e.g., assessment of commitment)	Less sensitive to context (e.g., social status)
Fetishes/partialism	Very rare	Not so rare
Experience in the virtual world (e.g., pornography, erotica, fantasy)	Erotica: meaning often involved, soft	Pornography: visual, anonymous, hard-core
	Fantasy: relatively infrequent, narratively linked, not spontaneous	Fantasy: frequent, impersonal and apparently spontaneous, with sudden transitions of imagined players
Sex received in exchange for goods/money	Usually romance linked, as in sex tourism	Usually straightforward prostitution
Hypersexuality	Increased bisexuality	Orientation specific
Addiction	Personal dimension favored/romance	Anonymous contact and to pornography

terms of broader notions of adaptation. The attraction of pornography and the existence of sexual addiction make sense in terms of evolutionary mismatch, as possibly does sexual risk-taking.

The model presented here, in which men are more strongly stimulus sensitive and women more strongly

Table 27.2 Comparing sources of inhibition in women and men.

	Women	Men
Orgasmic	Relatively low	Relatively high
Fear/disgust (current and anticipatory)	Relatively high	Relatively low
Regret	Relatively high	Relatively low

meaning sensitive, might also apply to sex-related phenomena other than desire. These could be based upon similar if not identical components of information processing. First, there is the phenomenon of the genital reaction to sexual stimuli (Chivers, 2017). For men, this correlates rather closely with their subjective desire and sexual orientation. By contrast, in women there is a weaker association, whereby a range of sexual stimuli are effective. Toates (2017) suggested that women's reactions are relatively unresponsive to male sexual stimuli per se, but rather involve extraction of sexual meaning from them. As noted in Section 27.1, women's genital arousal can be associated with little or no desire or even with sexual abhorrence, as in rape (Levin & van Berlo, 2004).

Second, there is the phenomenon of sexual jealousy. Buss et al. (1992) found that sexual infidelity per se is more distressing to men, whereas emotional infidelity is more distressing to women. It could be argued that, in evolutionary terms, the threat posed by infidelity is reliably triggered by straightforward sensory information in men, whereas women extract more long-term meaning in their assessments and emotional triggering. Emotional infidelity might sometimes be more difficult to detect than sexual infidelity, involving extrapolation beyond the immediate sensory input. For example, as a group-living species, many people maintain strong platonic relationships with the opposite sex, involving close cooperation and some bodily contact but no sexual or romantic association. On exposure to a range of jealousy-triggering scripts, men showed activation of subcortical regions involved in extracting sexual and aggressive salience. Women showed activation in cortical regions involved in extracting intentions and violations of social norms (Takahashi et al., 2006).

27.12 ACKNOWLEDGMENTS

I am grateful to Kent Berridge and Olga Coschug-Toates for their comments on this chapter.

REFERENCES

Anselme, P. (2010). The uncertainty processing theory of motivation. *Behavioural Brain Research*, **208**(2), 291–310.

Antfolk, J., Salo, B., Alanko, K., et al. (2015). Women's and men's sexual preferences and activities with respect to the partner's age: Evidence for female choice. *Evolution and Human Behavior*, **36**(1), 73–79.

Baldwin, J. D., & Baldwin, J. I. (1997). Gender differences in sexual interest. *Archives of Sexual Behavior*, **26**(2), 181–210.

Banca, P., Morris, L. S., Mitchell, S., et al. (2016). Novelty, conditioning and attentional bias to sexual rewards. *Journal of Psychiatric Research*, **72**, 91–101.

Bargh, J. A., Raymond, P., Pryor, J. B., & Strack, F. (1995). Attractiveness of the underling: An automatic power → sex association and its consequences for sexual harassment and aggression. *Journal of Personality and Social Psychology*, **68**(5), 768–781.

Barkow, J. H., Cosmides, L., & Tooby, J. (1992) *The Adapted Mind: Evolutionary Psychology and the Generation of Culture*. New York: Oxford University Press.

Basson, R., Leiblum, S., Brotto, L., et al. (2003). Definitions of women's sexual dysfunction reconsidered: Advocating expansion and revision. *Journal of Psychosomatic Obstetrics & Gynecology*, **24**(4), 221–229.

Baumeister, R. F. (2000). Gender differences in erotic plasticity: The female sex drive as socially flexible and responsive. *Psychological Bulletin*, **126**(3), 347–374.

Baumeister, R. F., Catanese, K. R., & Vohs, K. D. (2001). Is there a gender difference in strength of sex drive? Theoretical views, conceptual distinctions, and a review of relevant evidence. *Personality and Social Psychology Review*, **5**(3), 242–273.

Belsky, J., Steinberg, L., Houts, R. M., & Halpern-Felsher, B. L. (2010). The development of reproductive strategy in females: Early maternal harshness → earlier menarche → increased sexual risk taking. *Developmental Psychology*, **46**(1), 120–128.

Berridge, K. C. (2001). Reward learning: Reinforcement, incentives, and expectations. In D. L. Medin, ed., *Psychology of Learning and Motivation: Advances in Research and Theory*, Vol. 40. New York: Academic Press, pp. 223–278.

Berridge, K. C. (2004). Motivation concepts in behavioral neuroscience. *Physiology & Behavior*, **81**(2), 179–209.

Bindra, D. (1978). How adaptive behavior is produced: A perceptual–motivational alternative to response reinforcements. *Behavioral and Brain Sciences*, **1**(1), 41–52.

Binter, J., Leongómez, J. D., Moyano, N., et al. (2012). Sex differences in the incidence of sexual fantasies focused on evolutionary relevant objects. *Anthropologie*, **50**(1), 83–93.

Bjorklund, D. F., & Kipp, K. (1996). Parental investment theory and gender differences in the evolution of inhibition mechanisms. *Psychological Bulletin*, **120**(2), 163–188.

Bogaert, A. F. (2015). Asexuality: What it is and why it matters. *Journal of Sex Research*, **52**(4), 362–379.

Boothroyd, L. G., Meins, E., Vukovic, J., & Burt, D. M. (2014). Developmental changes in children's facial preferences. *Evolution and Human Behavior*, **35**(5), 376–383.

Brand, M., Snagowski, J., Laier, C., & Maderwald, S. (2016). Ventral striatum activity when watching preferred pornographic pictures is correlated with symptoms of Internet pornography addiction. *NeuroImage*, **129**, 224–232.

Buss, D. (2015). *Evolutionary Psychology: The New Science of the Mind*. Hove: The Psychology Press.

Buss, D. M., Larsen, R. J., Westen, D., & Semmelroth, J. (1992). Sex differences in jealousy: Evolution, physiology, and psychology. *Psychological Science*, **3**(4), 251–255.

Campbell, A. (2013). *A Mind of Her Own: The Evolutionary Psychology of Women*. Oxford: Oxford University Press.

Carver, C. S., & Harmon-Jones, E. (2009). Anger is an approach-related affect: Evidence and implications. *Psychological Bulletin*, **135**, 183–204.

Chivers, M. L. (2017). The specificity of women's sexual response and its relationship with sexual orientations: A review and ten hypotheses. *Archives of Sexual Behavior*, **46**(5), 1161–1179.

Dewitte, M. (2015). Gender differences in liking and wanting sex: Examining the role of motivational context and implicit versus explicit processing. *Archives of Sexual Behavior*, **44**(6), 1663–1674.

Dhuffar, M. K., & Griffiths, M. D. (2015). Understanding conceptualisations of female sex addiction and recovery using interpretative phenomenological analysis. *Psychology Research*, **5**(10), 585–603.

Diamond, L. M. (2012). The desire disorder in research on sexual orientation in women: Contributions of dynamical systems theory. *Archives of Sexual Behavior*, **41**(1), 73–83.

Dickson, N., Paul, C., Herbison, P., & Silva, P. (1998). First sexual intercourse: Age, coercion, and later regrets reported by a birth cohort. *British Medical Journal*, **316**(7124), 29–33.

Doom, J. R., Vanzomeren-Dohm, A. A., & Simpson, J. A. (2016). Early unpredictability predicts increased adolescent externalizing behaviors and substance use: A life history perspective. *Development and Psychopathology*, **28**(4 Pt 2), 1505–1516.

Ellis, B. J., & Symons, D. (1990). Sex differences in sexual fantasy: An evolutionary psychological approach. *Journal of Sex Research*, **27**(4), 527–555.

Ellis, B. J., Del Giudice, M., Dishion, T. J., et al. (2012). The evolutionary basis of risky adolescent behavior: Implications for science, policy, and practice. *Developmental Psychology*, **48**(3), 598–623.

Felson, R. B., & Cundiff, P. R. (2012). Age and sexual assault during robberies. *Evolution and Human Behavior*, **13**, 10–16.

Felson, R. B., & Cundiff, P. R. (2014). Sexual assault as a crime against young people. *Archives of Sexual Behavior*, **43**, 273–284

Fleischman, D. S., Hamilton, L. D., Fessler, D. M., & Meston, C. M. (2015). Disgust versus lust: Exploring the interactions of disgust and fear with sexual arousal in women. *PLoS ONE*, **10**(6), e0118151.

Galperin, A., Haselton, M. G., Frederick, D. A., et al. (2013). Sexual regret: Evidence for evolved sex differences. *Archives of Sexual Behavior*, **42**(7), 1145–1161.

Gangestad, S. W., & Simpson, J. A. (2000). The evolution of human mating: Trade-offs and strategic pluralism. *Behavioral and Brain Sciences*, **23**(4), 573–587.

Gillath, O., & Collins, T. (2016). Unconscious desire: The affective and motivational aspects of subliminal sexual priming. *Archives of Sexual Behavior*, **45**(1), 5–20.

Glass, S. P., & Wright, T. L. (1985). Sex differences in type of extramarital involvement and marital dissatisfaction. *Sex Roles*, **12**(9–10), 1101–1120.

Goldenberg, D., Telzer, E. H., Lieberman, M. D., Fuligni, A., & Galván, A. (2013). Neural mechanisms of impulse control in sexually risky adolescents. *Developmental Cognitive Neuroscience*, **6**, 23–29.

Guggino, J. M., & Ponzetti Jr., J. J. (1997). Gender differences in affective reactions to first coitus. *Journal of Adolescence*, **20**(2), 189–200.

Haidt, J. (2001). The emotional dog and its rational tail: A social intuitionist approach to moral judgment. *Psychological Review*, **108**(4), 814–834.

Hall, J. L., Stanton, S. J., & Schultheiss, O. C. (2010). Biopsychological and neural processes of implicit motivation. In O. Schultheiss & J. C. Brunstein, eds., *Implicit Motives*. Oxford: Oxford University Press, pp. 279–307.

Hosseini-Kamkar, N., & Morton, J. B. (2014). Sex differences in self-regulation: An evolutionary perspective. *Frontiers in Neuroscience*, **8**, 233.

Hyde, J. S. (1996). Where are the gender differences? Where are the gender similarities? In D. M. Buss & N. M. Malamuth, eds., *Sex, Power, Conflict: Evolutionary and Feminist Perspectives*. New York: Oxford University Press, pp. 107–118.

Janssen, E., & Bancroft, J. (2007). The dual-control model: The role of sexual inhibition and excitation in sexual arousal and behavior. In E. Janssen, ed., *The Psychophysiology of Sex*. Bloomington, IN: Indiana University Press, pp. 197–222.

Kenrick, D. T., & Keefe, R. C. (1992). Age preferences in mates reflect sex differences in human reproductive strategies. *Behavioral and Brain Sciences*, **15**(1), 75–91.

Kenrick, D. T., Groth, G. E., Trost, M. R., & Sadalla, E. K. (1993). Integrating evolutionary and social exchange perspectives on relationships: Effects of gender, self-appraisal, and involvement level on mate selection criteria. *Journal of Personality and Social Psychology*, **64**(6), 951–969.

Knoth, R., Boyd, K., & Singer, B. (1988). Empirical tests of sexual selection theory: Predictions of sex differences in onset, intensity, and time course of sexual arousal. *Journal of Sex Research*, **24**(1), 73–89.

Knott, V., Impey, D., Fisher, D., Delpero, E., & Fedoroff, P. (2016). Pedophilic brain potential responses to adult erotic stimuli. *Brain Research*, **1632**, 127–140.

Laan, E., & Both, S. (2008). What makes women experience desire? *Feminism & Psychology*, **18**(4), 505–514.

Laan, E., & Janssen, E. (2007). How do men and women feel? Determinants of subjective experience of sexual arousal. In E. Janssen, ed., *The Psychophysiology of Sex*. Bloomington, IN: Indiana University Press, pp. 278–290.

LeDoux, J. (1999). *The Emotional Brain*. London: Phoenix.

Leitenberg, H., & Henning, K. (1995). Sexual fantasy. *Psychological Bulletin*, **117**(3), 469–496.

Levin, R. J., & van Berlo, W. (2004). Sexual arousal and orgasm in subjects who experience forced or non-consensual sexual stimulation – A review. *Journal of Clinical and Forensic Medicine*, **11**, 82–86.

Lippa, R. A. (2006). Is high sex drive associated with increased sexual attraction to both sexes? It depends on whether you are male or female. *Psychological Science*, **17**(1), 46–52.

Lomanowska, A. M., Lovic, V., Rankine, M. J., et al. (2011). Inadequate early social experience increases the incentive salience of reward-related cues in adulthood. *Behavioural Brain Research*, **220**(1), 91–99.

MacLean, P. D. (1990). *The Triune Brain in Evolution*. New York: Plenum Press.

McClintock, M. K., & Herdt, G. (1996). Rethinking puberty: The development of sexual attraction. *Current Directions in Psychological Science*, **5**(6), 178–183.

McKeague, E. L. (2014). Differentiating the female sex addict: A literature review focused on themes of gender difference used to inform recommendations for treating women with sex addiction. *Sexual Addiction & Compulsivity*, **21**(3), 203–224.

Meana, M. (2010) Elucidating women's (hetero)sexual desire: Definitional challenges and content expansion. *Journal of Sex Research*, **47**, 104–122.

Meston, C. M., & Buss, D. M. (2007). Why humans have sex. *Archives of Sexual Behavior*, **36**(4), 477–507.

Meyer-Bahlburg, H. F., Ehrhardt, A. A., Rosen, L. R., et al. (1995). Prenatal estrogens and the development of homosexual orientation. *Developmental Psychology*, **31**(1), 12–21.

Money, J. (1986). *Lovemaps*. New York: Irvington Publishers.

Palmer, C. T. (1988). Twelve reasons why rape is not sexually motivated: A skeptical examination. *Journal of Sex Research*, **25**(4), 512–530.

Phillips, B., Hajela, R., & Hilton Jr., D. L. (2015). Sex addiction as a disease: Evidence for assessment, diagnosis, and response to critics. *Sexual Addiction & Compulsivity*, **22**(2), 167–192.

Pruitt, D., & LaFont, S. (1995). For love and money: Romance tourism in Jamaica. *Annals of Tourism Research*, **22**(2), 422–440.

Pruitt, M. V., & Krull, A. C. (2011). Escort advertisements and male patronage of prostitutes. *Deviant Behavior*, **32**(1), 38–63.

Quinsey, V. L. (2003). The etiology of anomalous sexual preferences in men. *Annals of the New York Academy of Sciences*, **989**(1), 105–117.

Rademacher, L., Schulte-Rüther, M., Hanewald, B., & Lammertz, S. (2017). Reward: From basic reinforcers to anticipation of social cues. *Current Topics in Behavioral Neurosciences*, **30**, 207–221.

Radtke, S. (2013). Sexual fluidity in women: How feminist research influenced evolutionary studies of same-sex behavior. *Journal of Social, Evolutionary, and Cultural Psychology*, **7**(4), 336–343.

Reynaud, M., Karila, L., Blecha, L., & Benyamina, A. (2010). Is love passion an addictive disorder? *American Journal of Drug and Alcohol Abuse*, **36**(5), 261–267.

Rubio-Casillas, A., Rodríguez-Quintero, C. M., Rodríguez-Manzo, G., & Fernández-Guasti, A. (2015). Unraveling the modulatory actions of serotonin on male rat sexual responses. *Neuroscience & Biobehavioral Reviews*, **55**, 234–246.

Ryan, R. M., Mendle, J., & Markowitz, A. J. (2015). Early childhood maltreatment and girls' sexual behavior: The mediating role of pubertal timing. *Journal of Adolescent Health*, **57**(3), 342–347.

Schmitt, D. P. (2003). Universal sex differences in the desire for sexual variety: Tests from 52 nations, 6 continents, and 13 islands. *Journal of Personality and Social Psychology*, **85**(1), 85–104.

Schneider, J. P. (2000). A qualitative study of cybersex participants: Gender differences, recovery issues, and implications for therapists. *Sexual Addiction & Compulsivity*, **7**, 249–278.

Shulman, E. P., Smith, A. R., Silva, K., et al. (2016). The dual systems model: Review, reappraisal, and reaffirmation. *Developmental Cognitive Neuroscience*, **17**, 103–117.

Snowden, R. J., Curl, C., Jobbins, K., Lavington, C., & Gray, N. S. (2016). Automatic direction of spatial attention to male versus female stimuli: A comparison of heterosexual men and women. *Archives of Sexual Behavior*, **45**(4), 843–853.

Steinberg, L. (2008). A social neuroscience perspective on adolescent risk-taking. *Developmental Review*, **28**(1), 78–106.

Symons, D. (1979). *The Evolution of Human Sexuality*. New York: Oxford University Press.

Takahashi, H., Matsuura, M., Yahata, N., et al. (2006). Men and women show distinct brain activations during imagery of sexual and emotional infidelity. *NeuroImage*, **32**(3), 1299–1307.

Telzer, E. H. (2016). Dopaminergic reward sensitivity can promote adolescent health: A new perspective on the mechanism of ventral striatum activation. *Developmental Cognitive Neuroscience*, **17**, 57–67.

Thornhill, N. W. (1996). Psychological adaptation to sexual coercion in victims and offenders. In D. M. Buss & N. M. Malamuth, eds., *Sex, Power, Conflict: Evolutionary and Feminist Perspectives*. New York: Oxford University Press, pp. 90–104.

Tidwell, N. D., & Eastwick, P. W. (2013). Sex differences in succumbing to sexual temptations: A function of impulse or control? *Personality and Social Psychology Bulletin*, **39**, 1620–1633.

Tiefer, L. (1991). Historical, scientific, clinical and feminist criticisms of "the human sexual response cycle" model. *Annual Review of Sex Research*, **2**(1), 1–23.

Tinbergen, N. (1951). *The Study of Instinct*. New York: Oxford University Press.

Toates, F. (1986). *Motivational Systems*. Cambridge: Cambridge University Press.

Toates, F. (2006). A model of the hierarchy of behaviour, cognition, and consciousness. *Consciousness and Cognition*, **15**(1), 75–118.

Toates, F. (2014). *How Sexual Desire Works: The Enigmatic Urge*. Cambridge, UK: Cambridge University Press.

Toates, F. (2017). A hierarchical model might cast some light on the anomaly. *Archives of Sexual Behavior*, **46**(5), 1203–1205.

Tolman, D. L., & Diamond, L. M. (2001). Desegregating sexuality research: Cultural and biological perspectives on gender and desire. *Annual Review of Sex Research*, **12**, 33–74.

Townsend, J. M. (1998). *What Women Want – What Men Want*. Oxford: Oxford University Press.

Townsend, J. M., & Wasserman, T. (1997). The perception of sexual attractiveness: Sex differences in variability. *Archives of Sexual Behavior*, **26**(3), 243–268.

Tybur, J. M., Lieberman, D., & Griskevicius, V. (2009). Microbes, mating, and morality: Individual differences in three functional domains of disgust. *Journal of Personality and Social Psychology*, **97**(1), 103–122.

Tybur, J. M., Lieberman, D., Kurzban, R., & DeScioli, P. (2013). Disgust: Evolved function and structure. *Psychological Review*, **120**(1), 65–84.

Victor, E. C., & Hariri, A. R. (2016). A neuroscience perspective on sexual risk behavior in adolescence and emerging adulthood. *Development and Psychopathology*, **28**(2), 471–487.

Workman, L., & Reader, W. (2014). *Evolutionary Psychology*, 3rd ed. Cambridge, UK: Cambridge University Press.

28 Are Humans Peacocks or Robins?

STEVE STEWART-WILLIAMS

Sociobiological approaches have made great inroads into psychological science over the last few decades. This has not come without a fight. One of the main fronts on which the battle has been fought is the origins of human sex differences. Evolutionary psychologists have made a strong case that many basic sex differences in our species have an evolutionary origin; the case is now so strong, in fact, that it seems unreasonable to deny a significant evolutionary contribution. A question mark remains, however, over the relative magnitude of the evolved differences. Are we highly dimorphic, polygynous animals like peacocks? Or are we relatively monomorphic, pair-bonding animals like robins? In this chapter, I argue that we are closer to the latter than the former – a fact that makes us somewhat anomalous among the animals. In many species, the males alone compete for mates and the females alone choose from among the males on offer. In our species, in contrast, both sexes compete for mates and both are choosy about their mates. Certainly, males compete more fervently and females are choosier, at least in early courtship and for low-commitment relationships. But the most conspicuous feature of the human mating system is mutual mate choice, coupled with relatively modest levels of overall dimorphism.

At first glance, this might seem to clash with predictions from evolutionary psychology. On closer inspection, though, the pattern makes good Darwinian sense. One of the main driving forces behind the evolution of sex differences is parental investment. Across species, larger sex differences in parental investment are associated with greater levels of dimorphism. In most species, the females invest a great deal, the males little or nothing, and thus sex differences are substantial. But in our species, males often invest in offspring as well (albeit less reliably and to a lesser extent than females). As a result, most sex differences in *Homo sapiens* are comparatively muted.

Contrary to popular opinion, the evolutionary psychological literature strongly supports this assertion. To illustrate, this chapter surveys three main lines of research conducted by evolutionary psychologists, namely sex differences in the desire for casual sex, sex differences in the strength of certain mate preferences, and sex differences in proneness to sexual vs. emotional jealousy. In each case, I argue that although the differences are real, and although they have an evolutionary origin, they turn out to be relatively modest compared to the differences found in most other species. There is little doubt at this point that humans exhibit meaningful dimorphism in a range of psychological attributes. It is easy, however, to overstate the level of dimorphism, and thus easy to blur the emerging picture of our evolved nature.

28.1 THE EVOLUTION OF SEX DIFFERENCES

Even a cursory glance at the animal kingdom reveals a widespread trend: males and females in most species differ from one another in characteristic ways. Biologists refer to this as *sexual dimorphism: di* means two; *morph* means form; thus, dimorphism means "two forms." (The antonym is *sexual monomorphism*.) Among the most common sex differences are the following. First, males in many species have a stronger, less discriminating sex drive than females and a greater appetite for multiple mates. Second, among the "higher" vertebrates, and particularly the mammals, males are often larger than females. Third, males are typically more physically aggressive than females and possess a range of built-in weapons, including antlers, tusks, and oversized canines. Fourth, males are commonly more ornamented than females: in some species, they're more colorful, for instance; in others, they have ornamental tail feathers or head crests. Fifth, females are generally choosier than males about their sexual partners. Sixth, females usually contribute more than males to the rearing of offspring. And seventh, females tend to live longer than males.

Where do these differences come from? This is a question that biologists have wrestled with since Darwin's 1871 book, *The Descent of Man and Selection in Relation to Sex*. The biggest single step toward an answer came in 1972, when Robert Trivers unveiled his *parental investment theory*. According to Trivers, most sex differences trace back to a single "master" difference – namely that, in most species, one sex invests more into offspring than the other. Wherever this is the case, the maximum number of offspring that the higher-investing sex can produce is curtailed relative to the lower-investing sex (Clutton-Brock & Vincent, 1991). This simple fact has important and far-reaching consequences (Stewart-Williams, 2018).

To see how, imagine a species in which the sex difference in parental investment is especially large: one sex invests a great deal in offspring, the other very little. The first thing to notice is that members of the low-investing sex can potentially have a very high number of offspring, and any trait that increases their chances of doing so stands a good chance of being selected. This includes, most obviously, a strong and undiscriminating sex drive and a desire for multiple, novel sexual partners. It also includes larger body size, greater strength and aggressiveness, and more fearsome weaponry, all of which help their owners to vanquish same-sex rivals and acquire larger numbers of mates.

Meanwhile, members of the high-investing sex typically evolve to be choosy about their sexual partners, preferentially mating with individuals exhibiting signs of good health or good genes. This is because the high investors can have relatively few offspring in their lifetimes, and thus any suboptimal partner choice can potentially deal a much greater blow to their lifetime fitness. Once in place, the mate-choice criteria of the high-investing sex operate as a new selection pressure on the low-investing sex, often resulting in the evolution of sexual ornamentation. To take a hypothetical example, if the high investors prefer to mate with low investors with large noses, then over the generations, the low investors' noses will grow larger and larger: a multigenerational Pinocchio effect. Finally, the high-investing sex typically evolves to live longer. This is partly because members of this sex spend less time engaged in risky competition with same-sex rivals, and partly because as the sex that looks after the young – their fitness is dependent not simply on siring offspring, but on staying alive to care for them as well (Campbell, 2002).

In most species, the females invest considerably more than the males in offspring. As such, sexual selection often produces what Andrew Thomas and I (2013b) dubbed "MCFC species": species in which *males compete* for mates and/or *females choose* from among the males on offer.

The MCFC schema applies well to many species, but does it apply to humans? Initially, it certainly seems to. Like other mammals, the obligatory biological expenditure required to produce a single viable offspring is notably higher for women than for men. Most obviously, women bear the biological burden of pregnancy, parturition, and lactation. On top of that, in every culture on record, women provide more hands-on care of offspring than men (Wood & Eagly, 2002). As an alien scientist would predict for a species with such a profile, men are larger, more aggressive, and more inclined to pursue

multiple mates than women. Women, in contrast, are smaller, choosier about their sexual partners, and live longer. In other words, for the most part, humans fit the pattern that describes most sexually dimorphic species. This constitutes a strong argument that the sex differences in our species have an evolutionary origin, rather than being wholly a product of learning or culture (Stewart-Williams, 2018; Stewart-Williams & Thomas, 2013a, 2013b).

At first glance, then, the MCFC model seems to fit well. On closer inspection, however, the fit starts to seem less comfortable. The differences found in our species are not nearly as large or as strongly polarized as those found in most other mammals. Rather than females exercising mate choice and males competing for females, both sexes have species-typical mate preferences (Buss, 1989) and both compete for desirable members of the other sex (Buss, 1988; Campbell, 2002). Rather than males being ornamented and females drab, both sexes have prominent secondary sexual characteristics, including men's beards and V-shaped torsos and women's breasts and "hourglass" figures (Barber, 1995). These are not trivial qualifications; they make human beings a striking exception to the MCFC rule, and extremely unusual within the wider animal kingdom.

The human pattern might initially appear to clash with parental investment theory. For many, Trivers' theory is identified with the claim that, when one sex invests more in offspring than the other, the higher-investing sex evolves to be choosy about its mates, whereas the lower-investing sex evolves to compete for access to the choosier sex. As such, parental investment theory is widely viewed as a theory of sex differences. It is, however, equally a theory of sex similarities, for a natural implication of Trivers' (1972) theory – one which Trivers himself spelled out in some detail – is that when both sexes make comparable investments in offspring, sexual dimorphism is reduced. (More precisely, this happens when the sex difference in maximum reproductive rate is reduced for any of a number of reasons.) This is the case in a wide range of species, including around 90 percent of birds (Griffith, Owens, & Thuman, 2002). Could it be the case for human beings?

If we focus on the minimum biological investment required from each sex to produce a single offspring, it might seem not. But although the *minimum* investment is much lower for men than women, the *typical* investment is not. Human males generally provide at least some post-coital parental input (Gray & Anderson, 2010; Marlowe, 2003), often in the context of relatively durable pair-bonds (Marlowe, 2004). As such, the sex difference in parental investment is diminished, and we should expect a corresponding diminution in the level of sexual dimorphism in reproductively relevant aspects of human psychology.

Most mammals have not taken this path, so why would we? One possibility is as follows. As brain size increased in the hominin lineage, our young became increasingly costly to rear. They were born in a relatively underdeveloped and helpless state (Martin, 1990) and had an extended childhood (Kaplan, 1994). Consequently, human young required additional care from individuals other than the mother. Often, this came from grandparents (especially maternal grandmothers), siblings (especially older sisters), aunts and grandaunts, and unrelated friends (Hawkes, O'Connell, & Blurton Jones, 1989; Hrdy, 2009; Sear & Mace, 2008). However, it also often came from the father (Marlowe, 2000). As a result of the high cost of rearing young, pair-bonding and biparental care became important elements in humans' reproductive repertoire. Pair-bonds rarely last for life, and there are individual and cross-cultural differences in the extent to which fathers invest in offspring (Hrdy, 2009). Nonetheless, there is good evidence that men everywhere have the capacity to fall in love and form pair-bonds (Jankowiak & Fisher, 1992), and to bond with their offspring (Gettler et al., 2011). Because ancestral men and women both invested heavily in offspring, both evolved to be choosy about their long-term mates, and both evolved to compete for the most desirable mates available. In short, the evolution of large, clever brains turned us into a relatively monomorphic animal. (See Stewart-Williams & Thomas, 2013b, for a more detailed discussion.)

If this is correct, then the lesson of parental investment theory for our species is not that women invest a lot in offspring and men very little, and thus that there must be large sex differences. The lesson is roughly the opposite: that both sexes invest a great deal, and thus that we would expect to find somewhat modest sex differences.

Certainly, in most societies, men have somewhat higher reproductive variability than women – that is, they exhibit greater variability in the number of offspring they sire and thus have a higher maximum reproductive rate (Brown, Laland, & Borgerhoff Mulder, 2009). As a result, men are presumably more sexually selected. Nonetheless, human sex differences are nowhere near as large as those found in classic exemplars of sexual selection, such as peacocks and elephant seals, and the most striking thing about our species is not the differences, but the fact that *both* sexes are quite strongly sexually selected.

This is not the impression one often gets from popular media depictions of evolutionary psychology, however, or even sometimes from the evolutionary psychological literature itself. In the following sections, I'll examine the relative magnitude of human sex differences in each of three important domains and consider whether they're as large as people sometimes claim. I'll begin with sex differences in the desire for casual sex.

28.2 KEEPING IT CASUAL

One of the central topics in evolutionary psychology is male–female differences in sexual strategies – that is, the extent to which individuals pursue long-term versus short-

term relationships. To many laypeople, and many psychologists in other areas, evolutionary psychology's position on this issue is fully encapsulated in William James's nitrous oxide-fueled rhyme:

> Higgamous Hoggamous,
> Woman's monogamous.
> Hoggamous Higgamous,
> Man is polygamous!

Evolutionary psychology is often chided for this (supposed) claim. Furze et al. (2011), for instance, ask: "Is it true that men are promiscuous and women are not? The data tell a different story" (p. 55). The data do indeed tell a different story – but so do evolutionary psychologists. According to prominent theories in the field, such as Sexual Strategies Theory (Buss & Schmitt, 1993) and Strategic Pluralism Theory (Gangestad & Simpson, 2000), men as well as women often pursue long-term relationships, and women as well as men sometimes pursue short-term relationships. The sex difference is one of degree rather than kind, and applies largely to short-term mating. This is a point evolutionary psychologists have made repeatedly over the course of several decades. In fact, as David Buss (2003) notes, "Given our explicitness on this issue, when a critic describes the theory as proposing that 'men are promiscuous, women are monogamous,' one can only wonder about the person's scholarship, training, or eyesight" (p. 225). Nonetheless, the case can be made that evolutionary psychologists have sometimes inadvertently given the impression that the differences in mating strategies are much larger than their own data indicate. In this section, I'll discuss some examples and consider how large the difference really is.

28.2.1 Reproductive Variability

Let's start with the sex difference in reproductive variability. This difference is, as mentioned, closely linked to the level of sexual dimorphism in a species: the bigger the sex difference in reproductive variability, the more dimorphic the species. How large is this difference in *Homo sapiens*?

The impression one often gets is that the difference is extremely large. It is common to hear, for instance, that for men the maximum number of offspring is virtually unlimited or runs into the thousands, whereas for women it is unlikely to stretch much above a dozen (e.g., Dawkins, 1989, p. 142; Miller, 2000, p. 86). The natural inference is that sex differences in human sexuality must be correspondingly large. This inference is reinforced by one of the most famous statistics in evolutionary psychology: the official world record for the number of offspring sired by any one man. The record holder is Ismail the Bloodthirsty, the Sharifian emperor of Morocco from 1672 to 1727, who had hundreds of wives and concubines and reputedly sired 888 children (Gould, 2000). In contrast, the official world record for a woman is 69, which, although perhaps more surprising than the male record, illustrates the fact that the reproductive ceiling for women is considerably lower than that for men.

All of this fosters the impression that men have vastly greater reproductive variability than women, and thus that men must have evolved a vastly greater desire for multiple sexual partners. This would be a shaky conclusion, however. Although it is true that men can, in principle, sire hundreds or even thousands of offspring, in practice this is vanishingly rare. Ismail and his ilk are outliers among outliers; that's why their stories are so attention-grabbing. Furthermore, the extreme form of harem polygyny practiced by these despotic leaders was only possible in early large-scale civilizations, which allowed small numbers of individuals to monopolize extreme levels of power. For most of our evolutionary history, we lived in small-scale forager societies. Consequently, more representative levels of reproductive variability can be found in the ethnographies of these groups. A comprehensive survey by Brown and colleagues (2009) revealed that, among foragers, there is often relatively little difference in men and women's reproductive variability. Averaging across groups, men did have somewhat greater reproductive variability than women, as we would expect. But the size of the sex difference was often small, and sometimes there was no difference at all. Indeed, in a handful of groups, there was slightly more female than male variability. Thus, in the type of environment in which we spent most of our evolution, the sex difference in reproductive variability was nowhere near the magnitude suggested by Ismail the Bloodthirsty and his ilk (see also Betzig, 2012; Labuda et al., 2010). Because the difference in reproductive variability was relatively small, we might predict that sex differences in sexual psychology would be relatively small as well.

28.2.2 Opportunistic Males

This seems like a reasonable expectation; there is, however, at least one reason to question it. In his important 1979 book *The Evolution of Human Sexuality*, Donald Symons argued that, even if opportunities for low-investment couplings or harem polygyny were rare throughout most of human evolution, the inclusive fitness benefits of these activities would have been so great for males that males may have evolved to take advantage of any such opportunities, just in case they ever arose. As a result, men may have a much stronger desire than women for short-term sex and multiple partners, even though most men have few opportunities to satisfy such desires, and some men have none.

What should we make of Symons' conjecture? To begin with, it is worth noting that the basic argument would not just apply to humans. In most species, the obligatory physiological investment in offspring is much smaller for males than females; as such, males in most species could, in principle, increase their reproductive success through opportunistic mating. It is not the case, though, that males

in every species are equally opportunistic. This is because there are costs and benefits to any strategy, and the balance of costs and benefits for an opportunistic strategy differs from species to species. For male chimpanzees and bonobos, there is little to gain from paternal care because of high levels of promiscuous mating and low paternity certainty. For male owl monkeys, on the other hand, investing in offspring has clearly paid reproductive dividends, as this is their obligate strategy (Huck et al., 2014).

How does the Darwinian cost–benefit analysis come out for humans? It is easy to see the advantage of short-term mating for men (i.e., increasing offspring number at little biological cost), and this is often emphasized in discussions of the topic. But it is also important to factor in the selection pressures running in the opposite direction. In many forager societies, infants and young children are much less likely to survive without two investing parents (Dwyer & Minnegal, 1993; Hill & Hurtado, 1996). Assuming – as seems reasonable – that this was the case throughout our evolutionary history, it would have made it less profitable for most of our male ancestors to spend all their time pursuing new sexual conquests. To the extent that male parental care was necessary to bring a child to nutritional independence, this would have weakened the selection pressure for a polygynous male psychology. Clearly, it did not eliminate it altogether. It seems probable, however, that the average level of polygynous desire found in men is lower than we would expect if we focused only on the benefits of polygynous mating and overlooked the costs.

28.2.3 Sizing Up the Effect Size

The above argument is a theoretical one. The final court of appeal, though, is the evidence. What does the evidence tell us? First, a number of studies have found that, on average, men and women are similarly interested in forming long-term relationships and have comparably high standards for their long-term mates (Buss & Schmitt, 1993; Kenrick et al., 1994; Stewart-Williams, Butler, & Thomas, 2017). We are therefore notably monomorphic in a long-term mating context. In a short-term context, in contrast, we exhibit a higher level of dimorphism. Various lines of evidence point to this conclusion. In one large cross-cultural survey, David Schmitt (2005) looked at sex differences in *sociosexual orientation* (SO): people's willingness to engage in sex outside the confines of a committed relationship. In every one of more than 50 nations examined, the mean SO score was higher for men than for women. In the same survey, Schmitt and colleagues (2003) found that, in all the major world regions, the average number of sexual partners desired across the lifespan was higher for men than for women (although note that the averages for both sexes varied a lot across regions, and that the average for women in some regions was higher than the average for men in others). As well as self-report data, the sex difference in short-term interest has been documented in studies looking at people's responses to apparently real sexual solicitations (Clark & Hatfield, 1989) and in observations of real-world consumer behavior (e.g., the consumption of pornography vs. romance novels; Symons, 1979).

These data are the tip of a large iceberg, in light of which it is no longer reasonable to deny that the sexes differ in this domain. It *is* still reasonable to worry, though, that the magnitude of the difference is sometimes overstated. First, it is important to emphasize that the data do not support some of the more extravagant claims made by evolutionary psychologists. This includes Symons' (1979) claim that, "With respect to human sexuality, there is a female human nature and a male human nature and these natures are extraordinarily different" (p. 11). Claims like this gloss over the fact that the distributions for men and women on virtually every trait strongly overlap (Hyde, 2005). To suggest that men and women have distinct and discrete sexual psychologies, based on average differences in overlapping distributions, is to misdescribe the data.[1]

But even some less extreme claims may require reining in. Evolutionary psychologists often point out that the sex difference in the willingness to engage in casual sex is one of the largest sex differences known to psychological science. This is true. It is possible, though, that even our largest differences are relatively modest when human beings are considered shoulder to shoulder with other animals. One of the largest cross-cultural studies of sex differences in sexuality supports this suggestion. In an online survey of 53 nations, Lippa (2009) reported an average d value of 0.74 for the sex difference in SO, which is close to the customary cutoff point for a large effect size: 0.8. As a point of comparison, however, consider a morphological sex difference: the sex difference in height. This is a difference we all have an intuitive grasp of; we know that it's there, but we also know that there is a great deal of overlap between the sexes, and that the difference is nothing like that found in highly dimorphic animals such as gorillas or elephant seals. Importantly, the sex difference in height in Lippa's study ($d = 1.63$) was *more than twice the magnitude* of the SO difference in the same sample. Furthermore, the human SO difference is comparable to that of the size difference in gibbons (0.8; calculated from data in Schultz, 1941, reported by Geissmann, 1993). This means that, if humans are highly dimorphic for SO, gibbons must be highly dimorphic for size. Gibbons, however, are the archetypal example of

a sexually monomorphic primate. Certainly, the SO difference is large *for a human sex difference*. But this is not because it's a large difference in any absolute sense; it's because this relatively small difference is being compared with differences that are even smaller still.

In sum, sex differences in human mating strategies are real but relatively modest. This is exactly what we would expect on the assumption that, although ancestral humans engaged in some polygynous and short-term mating, our primary reproductive pattern was pair-bonding and biparental care.

28.3 MATE PREFERENCES

Another arena in which evolutionary psychologists have explored – and sometimes overstated – sex differences is mate preferences. The most famous findings here concern differences in the importance placed on physical attractiveness in a mate (the average is higher for men) and differences in the importance placed on wealth and status (the average is higher for women). These differences are found not only in university students, but in people of varying walks of life and ages (Sprecher, Sullivan, & Hatfield, 1994), and not only in self-report studies, but in studies of real-world behavior (Feingold, 1992). Furthermore, the differences have been found in a diverse array of cultures, included modern industrialized nations (Buss, 1989; Lippa, 2007), preindustrialized nations, and small-scale bands and tribes (Gottschall et al., 2004). This cross-cultural convergence is just what we would expect if the differences had an evolutionary origin.

How large are the differences, though? People often assume that, according to evolutionary psychologists, they are extremely large: that women are only interested in a man's social standing and the size of his wallet, for instance, and that men are only interested in "physically beautiful but dumb women" (McCaughey, 2007, p. 118). The actual claims of evolutionary psychologists are much more measured and moderate. Nonetheless, even they sometimes talk as if there are discrete sex differences in mate criteria. Cartwright (2008), for instance, wrote that "Females are predicted to look for high status males who are good providers, whereas males are predicted to look for young, healthy and fertile females who are good child bearers" (p. 270). Similarly, Kanazawa (2003) observed that, "In every society, men prefer young and attractive women for mates, and women prefer wealthy and powerful men" (p. 292). Such descriptions suggest substantial, dichotomous differences. The evidence for these differences, however, invariably consists of relatively modest differences in the central tendencies of highly variable and overlapping distributions. There is thus a mismatch between the data and the verbal descriptions of the data: we are comparing apples but concluding oranges.

---

[1] The strongest response to this argument comes from Marco Del Giudice and colleagues (2012), who claim that, although the sex difference in any given trait may be modest, when we consider several, related traits simultaneously, the resulting multivariate differences are notably larger. For a critique of this claim, see the appendix in Stewart-Williams and Thomas (2013b, pp. 167–168), and for a response to this critique, see Del Giudice (2013). Note that, regardless of the utility of the multivariate approach, it is fair to say that humans are less dimorphic than most mammals and than most nonhuman species.

Of course, when it comes to preferences for specific physical features, the differences often are genuinely dichotomous. It is not the case, for example, that heterosexual men and women both like protruding breasts in a mate, but that men like them more; instead, men like them and women do not. However, when it comes to traits such as good looks and social standing, the differences are nowhere near as stark. Let's consider each of these cases in term.

28.3.1 Looking Good

One of the most famous findings in evolutionary psychology concerns the importance people place on good looks in a prospective mate. As mentioned, in most samples, men rate looks as more important than do women (Buss, 1989; Feingold, 1990; Lippa, 2007). Although this is a consistent and well-replicated finding, there is still some question about how it should fit into our picture of the species. The quotations above give the impression that a mate's physical appearance is of the utmost importance to men but relatively unimportant to women. This is not what the data indicate, however.

First, although the average level of importance placed on physical attractiveness in a mate is higher for men than for women, this does not necessarily mean that attractiveness is the undisputed central concern of most men. Indeed, some of the most famous research in evolutionary psychology suggests otherwise. Buss (1990) conducted a large and extremely influential cross-national study examining what young adults want in a long-term partner. Averaging across the study's 37 samples, men ranked good looks 10th out of 18 traits, after love, dependable character, emotional stability/maturity, pleasing disposition, good health, education/intelligence, sociability, desire for home and children, and refinement/neatness. In contrast, women ranked good looks 13th. Certainly, in some studies, good looks appear higher on the agenda for men (e.g., Li, 2002; Lippa, 2007). Nonetheless, it is interesting and instructive that in Buss's research – the research that first attracted widespread attention to the idea that men evolved to put more weight than women on good looks – this trait appeared so far down the list, and yet many people concluded that good looks are men's primary concern. The findings did not support the generalization.

On the flip side of this coin, the common claim that men are interested in good looks whereas women are interested in resources and status seems to imply that a mate's looks are relatively unimportant to women. Again, however, this is not what the data tell us. Although on average good looks are *more* important to men, both sexes commonly place a fair amount of weight on good looks in a partner, and thus the sex difference is not especially large. Buss et al. (1990) had respondents rate the importance of good looks on a 0–3 scale, with the anchors *irrelevant*, *desirable*, *important*, and *indispensable*. Collapsing across the 37 samples, the average for both sexes fell between *desirable* and *important*. The male average was close to *important*

(1.91), whereas the female average was right in the middle (1.46). Notably, there was less than a half-point difference between the averages for each sex.

Looking at the literature more broadly, meta-analyses tend to yield reliable but modest effects. Feingold (1990), for instance, found an average effect size of $d = 0.54$ for the sex difference in importance placed on physical attractiveness. This is conventionally described as a medium effect. To put it into perspective, though, an effect size of this magnitude is around two-thirds that of the size difference in monomorphic gibbons, and implies around 80 percent overlap between the male and female distributions. If you were to select pairs of men and women at random, the woman would be more interested than the man in a mate's looks in around a third of pairs. This is a minority, certainly, but not a trivial one; it is, after all, closer to 50 percent than to zero (Stewart-Williams & Thomas, 2013a).

One might object that the evidence cited thus far consists solely of self-report data, and that such data cannot always be trusted. But when we look at actual mating behavior, the same pattern emerges. It is not the case that physical attractiveness has no currency for men on the mating market. Attractive men have more sexual partners than less attractive men, clearly implying that a mate's looks are important to women (Gangestad & Thornhill, 1997). Furthermore, this is not the case only for short-term sexual relationships. Men and women mate assortatively on attractiveness even in long-term pairings (Feingold, 1988). This suggests that, although there is an average sex difference in the importance placed on physical attractiveness, both sexes nonetheless base their long-term mating decisions to an important extent on this attribute.

The modest sex difference in the preference for physical attractiveness actually makes good sense when placed under a Darwinian microscope. A strong case has been made that the facial and bodily attributes that humans evolved to find attractive correlate with qualities such as youthfulness, fertility, and good health (Henderson & Anglin, 2003; Shackelford & Larsen, 1997). Good health, in turn, signals good genes and/or an absence of potentially transmissible diseases. These things are relevant to both sexes, not just to men. Recall that, across species, the more that members of a given sex invest in offspring, the choosier they tend to be about their long-term sexual partners. If we were a species in which only one sex invested heavily in offspring, the expectation would be that good looks would be a concern only to that sex (although note that this would be the females, not the males). Given, though, that both sexes commonly invest in offspring in our species, the expectation would be that *both* sexes would consider good looks important.

Indeed, given that human females have typically invested more, the puzzle is that it is *males* that put more weight on this trait. The adaptive rationale for this apparent anomaly is that women have a narrower window of fertility than men, as a result of menopause, and thus that a partner's youthfulness is more important to men's reproductive success than it is to women's (Buss, 1989; Symons, 1979). This asymmetry would definitely lead us to expect a difference. However, it would not necessarily lead us to expect a large one. Although men do not experience the complete cessation of fertility in middle age that women do, their fertility declines more precipitously than was once believed. To begin with, men's testosterone levels tend to nosedive with age (Harman et al., 2001), which may lead to a reduction in courtship effort and intrasexual competition. In addition, the quality and quantity of men's sperm declines throughout adulthood (Kong et al., 2012). The implication is this: the sex difference in age-related fertility is smaller than previously thought, which provides a theoretical rationale for thinking that the sex difference in the importance of physical attractiveness will be smaller than previously thought. Once again, a small sex difference is not inconsistent with an evolutionary perspective. It is exactly what an evolutionary perspective predicts.

28.3.2 Wealth and Status

Next, consider the claim that women evolved to seek wealthy, high-status mates. Again, a mountain of research suggests that there is indeed a sex difference in this domain. Buss (1989), for instance, found that women placed more weight on good financial prospects in a long-term mate in 36 of his 37 cross-cultural samples; Pérusse (1993) found that high-status males in a French Canadian city had more sexual partners than their lower-status compatriots; and Betzig (1989) found that one of the most common causes of divorce across cultures was the husband's failure to provide resources. In addition, historical and anthropological data indicate that men who have more power, more status, and more wealth tend also to have younger wives, more wives, more affairs, and more offspring than men lower on the totem pole (Betzig, 1986; Hawkes, 1991). The effects of women's status and wealth on men are much less pronounced.

It is hard to deny, then, that sex differences in the preference for resources and status are real. As some of the first discoveries in the newly minted field of evolutionary psychology, they assumed a central position in our picture of women's mating predilections. But even if resources and status do tend to be more important to women than men, we need to ask again how important they actually are. Often, they are fairly *un*important. Starting with resources, in Buss's classic study, women ranked good financial prospects 12th out of 18 traits, after love, dependable character, emotional stability/maturity, pleasing disposition, education/intelligence, sociability, good health, desire for home and children, ambition/industriousness, refinement/neatness, and similar education. Men, in contrast, placed it just one rung lower, at number 13. Likewise, in Lippa's (2007) survey, women ranked "money" 20th out of 23 traits and "prosperity" 22nd (men ranked these traits 21st and 23rd, respectively). Importantly, this put these pecuniary variables *below* good looks for women: good looks were in eighth place, challenging the common assumption that wealth always trumps good looks for women. Real-world studies bolster the challenge. In an analysis of Lonely Hearts Personal Ads, for instance, Pawlowski and Dunbar (2001) found that, although more women than men requested resources, fewer women requested resources than requested physical attractiveness (24 vs. 33 percent). Note that none of this would have surprised Darwin, who once wrote of a society in which "very ugly, though rich men, have been known to fail in getting wives" (1871, p. 667).

All of these arguments apply with equal force to the preference for social status in a mate. In Buss et al.'s 1990 paper, women ranked status 13th out of 18 traits. Importantly, this put status *in equal place* with good looks, again challenging the notion that a mate's looks are vastly less important to women than less tangible assets. Corroborating this conclusion, Lippa (2007) found that women ranked status 21st out of 23 traits, but good looks 8th. Of course, the fact that status is not the be-all and end-all for most women does not imply that there is no sex difference in this domain. There demonstrably is. In one meta-analysis, Feingold (1992) found an effect size of $d = 0.69$ for socioeconomic status, conventionally described as a medium to large difference. Putting this into perspective, however, it is worth noting, first, that this difference is still somewhat smaller than the size difference in monomorphic gibbons; second, that it represents around 70 percent overlap between the male and female distributions; and third, that if one were to take pairs of men and women at random, the man would be more interested than the woman in a mate's socioeconomic status in nearly one in three pairs – again, a minority, but a nontrivial one.

It seems, then, that as with wealth, the strong focus on women's preference for status in a mate is somewhat misleading. Remember: my argument for this conclusion does not involve selectively citing the work of opponents of evolutionary psychology or unrepresentative findings. It rests instead on some of the largest studies ever conducted on the subject. The Buss study in particular is a seminal work in evolutionary psychology, conducted by one of the field's main proponents. Nonetheless, the data suggest much more modest sex differences in mate preferences than are sometimes claimed.

28.4 JEALOUSY

Let's consider one final example: romantic jealousy. According to evolutionary psychologists, the emotion of jealousy was "designed" by natural selection to motivate mate guarding. The best-known idea associated with this approach focuses on sex differences in the triggers of jealousy, and stems from the notion that the primary function of mate guarding differs by sex. For men, the primary function is avoiding being cuckolded; for women, the primary function is avoiding the loss of an investing mate to another woman (Buss et al., 1992; Daly, Wilson, & Weghorst, 1982; Symons, 1979). This leads to the prediction that concerns about *sexual* infidelity will loom larger for men, whereas for women concerns about *emotional* infidelity will loom larger. The sexually differentiated jealousy hypothesis is usually contrasted with non-evolutionary explanations, according to which jealousy is an immature emotion, a sign of insecurity, or an arbitrary invention of specific cultures (summarized in Buss, 2000).

Let me say first that I think it is much more plausible that jealousy is a product of natural selection than an invention of culture, that it is part of the basic design of human nature, and that its evolutionary function is to motivate mate guarding (Stewart-Williams, 2018). Aside from anything else, other socially monogamous animals, including gibbons and many pair-bonding birds, attempt to prevent their partners from fraternizing with other-sex individuals, just as jealous humans do (Reichard, 2003). This suggests that ensuring partner fidelity is an important selection pressure in pair-bonding species, and one that can and does lead to the evolution of psychological adaptations designed to counter the threat. There also appear to be genuine sex differences in this realm, as we will see in the following sections. Nonetheless, the magnitude of these differences, as with those we have already considered, is easily and often overstated.

28.4.1 Sophie's Choice

The most common method of testing the sexually differentiated jealousy hypothesis is to ask participants to imagine that their long-term partner is involved with someone else, and then to ask which would upset them more: their partner having sex with the other person or their partner developing a deep emotional attachment to the other person. This is sometimes known as the *Sophie's choice dilemma*. Buss and colleagues' (1992) first foray into this area yielded fairly typical results: 60 percent of young men nominated sexual infidelity as the more upsetting option as opposed to only 17 percent of young women. This basic pattern has been found in a range of countries in the West, in Asia, and in South America (Bendixen, Kennair, & Buss, 2015; Buss, Shackelford, & Kirkpatrick, 1999; Buunk et al., 1996; de Souza et al., 2006; Geary et al., 1995) and in real-world responses to revelations of infidelity (Kuhle, 2011). It has also been found in at least one small-scale, natural fertility society (Scelza, 2014). A recent meta-analysis concluded that the sex difference is real and robust (Sagarin et al., 2012).

How, though, should we characterize this difference? One way would be to say that heterosexual men generally worry more about their partner having sex with another man than forming a close bond with him, whereas heterosexual women worry more about their partner forming a close bond with another woman than having sex with her. Buunk et al. (1996), for instance, noted that men "exhibit greater psychological and physiological distress

to sexual than to emotional infidelity of their partner, and women . . . exhibit more distress to emotional than to sexual infidelity (p. 139). To many, this might appear to be a direct description of the data.

But it's not. This is most obvious when we focus on men. In the initial Buss et al. (1992) study, 60 percent of men nominated sexual infidelity as the more upsetting option. We cannot conclude that men in general are more concerned about sexual than emotional infidelity when only a little more than half of men fit this description. Granted, we would be on safer ground saying that women exhibited a clear preference: 83 percent chose emotional infidelity as the most upsetting option. However, given that the proportion of men choosing sexual infidelity was closer to 50 percent than to 100 percent, it is closer to the truth to say that men were evenly split than to say that they were more worried about sexual infidelity.

Given these results, to claim that men are more worried about sexual than emotional infidelity is to commit what I call the *majority rules fallacy*. This is the fallacy of treating the majority tendency as if it characterizes the group as a whole. It is, in effect, a misplaced democratic process: a majority of men – a small majority – "votes" for sexual infidelity as the more upsetting option, and thus that tendency is "elected" to the position of human nature: men in general are more upset by sexual infidelity. This is not accurate, however; some men are, but many are not.

We can push the point further. Several studies explicitly *contradict* the claim that most men are more upset by sexual infidelity. In one study, a majority of German men (around 75 percent) reported that they would be more upset by *emotional* infidelity (Buunk et al., 1996). Similar results have been found in China (Geary et al., 1995) and Japan (Buss et al., 1999). It was still the case, in all of these studies, that *more men than women* chose sexual infidelity as the most upsetting option. However, *most* men did not. One cannot legitimately conclude that men in general are more concerned about sexual infidelity when, in some samples, a greater number of men report that they would be more concerned about *emotional* infidelity.

Again, this is not to deny that there is a sex difference or to deny that this difference has an evolutionary origin. The point is simply that the difference is not accurately described by statements such as "men find sexual infidelity more upsetting, women emotional infidelity." A more accurate description would be that "a *larger proportion* of men than women find sexual infidelity more upsetting, whereas a larger proportion of women are more upset by emotional infidelity." If these statements look like two different ways of saying the same thing, we need to look more closely. The first formulation exaggerates the sex difference by ignoring the variation within each sex. Furthermore, even if we treat it as a statement about the majority preference, it is false for some samples (e.g., Buunk et al.'s German sample). The second formulation, in contrast, is a weaker but more accurate claim, which acknowledges the distribution of preferences within groups of individuals, rather than projecting a singular preference onto most individuals within each of those groups.

28.4.2 Stacking the Deck

One might respond that, even if we characterize the difference in terms of the relative proportions of men versus women who are more upset by each option, there is still a rather sizeable sex difference. The problem, though, is that using a forced-choice methodology, although it does not create a sex difference out of thin air, is likely to inflate whatever difference there is. As Buss (2000) himself noted, the original reason for adopting the Sophie's choice paradigm was that, if participants are asked to rate how upset they would be about sexual versus emotional infidelity using continuous, free-response measures, there is often a ceiling effect; virtually everyone reports being extremely upset by both, and thus it is difficult to detect any sex differences. However, this in itself is surely an important fact about human nature: *most* men and women are upset by *both* aspects of infidelity. Consistent with this interpretation, Lishner and colleagues (2008) reran the basic Buss et al. forced-choice dilemma, but added a third option: "Both of the above options would upset me equally." A majority of men *and* women chose this option.

To be clear, sex differences in jealousy *can* often be detected using continuous measures (Bendixen et al., 2015; Pietrzak et al., 2002; Sagarin et al., 2012). Importantly, though, these differences are usually much more modest than the Sophie's choice differences. In a meta-analysis of 45 studies using continuous self-report measures, Sagarin et al. (2012) found an average sex difference of $g^* = 0.258$. (g^* is an effect size measure comparable to Cohen's *d*.) Similarly, in studies looking at people's physiological reactions to thoughts of a partner's sexual versus emotional infidelity, including changes in heart rate, blood pressure, and corrugator brow contraction, the sex differences tend to be much more modest than those derived from the standard forced-choice items (e.g., Baschnagel & Edlund, 2016; Buss et al., 1992).

A reasonable conclusion, then, is that although jealousy was very probably favored by natural selection for somewhat different reasons in each sex (i.e., avoiding being cuckolded vs. avoiding being left holding the baby), at a proximate and phenomenological level, it is overwhelmingly similar in both sexes (Harris, 2013; Stewart-Williams, 2018). There *are* average differences, but these are swamped by the cross-sex commonality – namely that sexual infidelity and emotional infidelity are both highly upsetting to most members of both sexes.

28.5 SUMMING UP

It is widely assumed that, because the obligatory biological investment of human females is so much greater than that of males, parental investment theory implies that there will be large sex differences in our species. If parental investment theory really *did* predict that men and women will be radically different, then the data gathered by evolutionary psychologists would falsify this prediction and falsify the theory, because the data show that there are generally *not* large sex differences. The data do not falsify the theory, however, because parental investment theory does not predict large sex differences for our species. Although the obligatory male investment in offspring is low, the typical investment is much higher. As such, we should generally expect modest sex differences in sexuality in our species. Certainly such differences do exist, and certainly these differences suggest a long history of mild effective polygyny (i.e., greater male than female reproductive variability). However, the best data and theory in evolutionary psychology suggest that these differences are relatively minor, as a result of the fact that our species is relatively biparental.

REFERENCES

Barber, N. (1995). The evolutionary psychology of physical attractiveness: Sexual selection and human morphology. *Ethology and Sociobiology*, **16**, 395–424.

Baschnagel, J. S., & Edlund, J. E. (2016). Affective modification of the startle eyeblink response during sexual and emotional infidelity scripts. *Evolutionary Psychological Science*, **2**, 114–122.

Bendixen, M., Kennair, L. E. O., & Buss, D. M. (2015). Jealousy: Evidence of strong sex differences using both forced choice and continuous measure paradigms. *Personality and Individual Differences*, **86**, 212–216.

Betzig, L. L. (1986). *Despotism and Differential Reproduction: A Darwinian View of History*. Hawthorne, NY: Aldine.

Betzig, L. (1989). Causes of conjugal dissolution: A cross-cultural study. *Current Anthropology*, **30**, 654–676.

Betzig, L. (2012). Means, variances, and ranges in reproductive success: Comparative evidence. *Evolution and Human Behavior*, **33**, 309–317.

Brown, J., Laland, K. N., & Borgerhoff Mulder, M. (2009). Bateman's principles and human sex roles. *Trends in Ecology and Evolution*, **24**, 297–304.

Buss, D. M. (1988). The evolution of human intrasexual competition: Tactics of mate attraction. *Journal of Personality and Social Psychology*, **54**, 616–628.

Buss, D. M. (1989). Sex differences in human mate preferences: Evolutionary hypotheses tested in 37 cultures. *Behavioral and Brain Sciences*, **12**, 1–49.

Buss, D. M. (2000). *The Dangerous Passion: Why Jealousy Is as Necessary as Love and Sex*. New York: Free Press.

Buss, D. M. (2003). Sexual strategies: A journey into controversy. *Psychological Inquiry*, **14**, 219–226.

Buss, D. M., Abbott, M., Angleitner, A., et al. (1990). International preferences in selecting mates: A study of 37 cultures. *Journal of Cross-Cultural Psychology*, **21**, 5–47.

Buss, D. M., Larsen, R. J., Westen, D., & Semmelroth, J. (1992). Sex differences in jealousy: Evolution, physiology, and psychology. *Psychological Science*, **3**, 251–255.

Buss, D. M., & Schmitt, D. P. (1993). Sexual strategies theory: An evolutionary perspective on human mating. *Psychological Review*, **100**, 204–232.

Buss, D. M., Shackelford, T. K., & Kirkpatrick, L. A. (1999). Jealousy and the nature of beliefs about infidelity: Tests of competing hypotheses about sex differences in the United States, Korea, and Japan. *Personal Relationships*, **16**, 125–150.

Buunk, B. P., Angleitner, A., Oubaid, V., & Buss, D. M. (1996). Sex differences in jealousy in evolutionary and cultural perspective: Tests from the Netherlands, Germany, and the United States. *Psychological Science*, **7**, 359–363.

Campbell, A. (2002). *A Mind of Her Own: The Evolutionary Psychology of Women*. Oxford: Oxford University Press.

Cartwright, J. (2008). *Evolution and Human Behaviour: Darwinian Perspectives on Human Nature*, 2nd ed. Basingstoke: Palgrave Macmillan.

Clark, R. D., & Hatfield, E. (1989). Gender differences in receptivity to sexual offers. *Journal of Psychology and Human Sexuality*, **2**, 39–55.

Clutton-Brock, T. H., & Vincent, A. C. J. (1991). Sexual selection and the potential reproductive rates of males and females. *Nature*, **351**, 58–60.

Daly, M., Wilson, M., & Weghorst, S. J. (1982). Male sexual jealousy. *Ethology and Sociobiology*, **3**, 11–27.

Darwin, C. (1871). *The Descent of Man, and Selection in Relation to Sex*. London: Murray.

Dawkins, R. (1989). *The Selfish Gene*, 2nd ed. Oxford: Oxford University Press.

de Souza, A. A. L., Verderane, M. P., Taira, J. T., & Otta, E. (2006). Emotional and sexual jealousy as a function of sex and sexual orientation in a Brazilian sample. *Psychological Reports*, **98**, 529–535.

Del Giudice, M. (2013). Multivariate misgivings: Is D a valid measure of group and sex differences? *Evolutionary Psychology*, **11**, 1067–1076.

Del Giudice, M., Booth, T., & Irwing, P. (2012). The distance between Mars and Venus: Measuring global sex differences in personality. *PLoS ONE*, **7**, e29265.

Dwyer, P. D., & Minnegal, M. (1993). Are Kubo hunters "show-offs"? *Ethology and Sociobiology*, **14**, 53–70.

Feingold, A. (1988). Matching for attractiveness in romantic partners and same-sex friends: A meta-analysis and theoretical critique. *Psychological Bulletin*, **104**, 226–235.

Feingold, A. (1990). Gender differences in effects of physical attractiveness on romantic attraction: A comparison across five research paradigms. *Journal of Personality and Social Psychology*, **59**, 981–993.

Feingold, A. (1992). Gender differences in mate selection preferences: A test of the paternal investment model. *Psychological Bulletin*, **112**, 125–139.

Furze, B., Savy, P., Brym, R. J., & Lie, J., eds. (2011). *Sociology in Today's World*, 2nd ed. South Melbourne: Cengage Learning.

Gangestad, S. W., & Simpson, J. A. (2000). The evolution of human mating: Trade-offs and strategic pluralism. *Behavioral and Brain Sciences*, **23**, 573–587.

Gangestad, S. W., & Thornhill, R. (1997). Human sexual selection and developmental stability. In J. A. Simpson & D. T. Kenrick, eds., *Evolutionary Social Psychology*. Hillsdale, NJ: Erlbaum, pp. 169–196.

Geary, D. C., Rumsey, M., Bow-Thomas, C. C., & Hoard, M. K. (1995). Sexual jealousy as a facultative trait: Evidence from the pattern of sex differences in adults from China and the United States. *Ethology and Sociobiology*, **16**, 355–383.

Geissmann, T. (1993). Evolution of communication in gibbons (Hylobatidae). Unpublished doctoral dissertation, Zürich University.

Gettler, L. T., McDade, T. W., Feranil, A. B., & Kuzawa, C. W. (2011). Longitudinal evidence that fatherhood decreases testosterone in human males. *Proceedings of the National Academy of Sciences*, **108**, 16194–16199.

Gottschall, J., Martin, J., Quish, H., & Rea, J. (2004). Sex differences in mate choice criteria are reflected in folktales from around the world and in historical European literature. *Evolution and Human Behavior*, **25**, 102–112.

Gould, R. G. (2000). How many children could Moulay Ismail have had? *Evolution and Human Behavior*, **21**, 295–296.

Gray, P. B., & Anderson, K. G. (2010). *Fatherhood: Evolution and Human Paternal Behavior*. Cambridge, MA: Harvard University Press.

Griffith, S. C., Owens, I. P. F., & Thuman, K. A. (2002). Extra pair paternity in birds: A review of interspecific variation and adaptive function. *Molecular Ecology*, **11**, 2195–2212.

Harman, S. M., Metter, E. J., Tobin, J. D., Pearson, J., & Blackman, M. R. (2001). Longitudinal effects of aging on serum total and free testosterone levels in healthy men. *Journal of Clinical Endocrinology and Metabolism*, **86**, 724–731.

Harris, C. R. (2013). Humans, deer, and sea dragons: How evolutionary psychology has misconstrued human sex differences. *Psychological Inquiry*, **24**, 195–201.

Hawkes, K. (1991). Showing off: Tests of an hypothesis about men's foraging goals. *Ethology and Sociobiology*, **12**, 29–54.

Hawkes, K., O'Connell, J. F., & Blurton Jones, N. (1989). Hardworking Hadza grandmothers. In V. Standen & R. Foley, eds., *Comparative Socioecology*. London: Basil Blackwell, pp. 341–266.

Henderson, J. J. A., & Anglin, J. M. (2003). Facial attractiveness predicts longevity. *Evolution and Human Behavior*, **24**, 303–364.

Hill, K., & Hurtado, A. M. (1996). *Demographic/Life History of Ache Foragers*. Hawthorne, NY: Aldine de Gruyter.

Hrdy, S. B. (2009). *Mothers and Others: The Evolutionary Origin of Mutual Understanding*. Cambridge, MA: Harvard University Press.

Huck, M., Fernandez-Duque, E., Babb, P., & Schurr, T. (2014). Correlates of genetic monogamy in socially monogamous mammals: Insights from Azara's owl monkeys. *Proceedings of the Royal Society of London B: Biological Sciences*, **281**, 20140195.

Hyde, J. S. (2005). The gender similarities hypothesis. *American Psychologist*, **60**, 581–592.

Jankowiak, W., & Fisher, E. (1992). Cross-cultural perspective on romantic love. *Ethnology*, **31**, 149–156.

Kanazawa, S. (2003). Can evolutionary psychology explain reproductive behavior in the contemporary United States? *Sociological Quarterly*, **44**, 291–302.

Kaplan, H. (1994). Evolutionary and wealth flow theories of fertility: Empirical tests and new models. *Population and Development Review*, **20**, 753–791.

Kenrick, D. T., Neuberg, S. L., Zierk, K. L., & Krones, J. M. (1994). Evolution and social cognition: Contrast effects as a function of sex, dominance, and physical attractiveness. *Personality and Social Psychology Bulletin*, **20**, 210–217.

Kong, A., Frigge, M. L., Masson, G., et al. (2012). Rate of de novo mutations and the importance of father's age to disease risk. *Nature*, **488**, 471–475.

Kuhle, B. X. (2011). Did you have sex with him? Do you love her? An in vivo test of sex differences in jealous interrogations. *Personality and Individual Differences*, **51**, 1044–1047.

Labuda, D., Lefebvre, J.-F., Nadeau, P., & Roy-Gagnon, M.-H. (2010). Female-to-male breeding ratio in modern humans: An analysis based on historical recombinations. *American Journal of Human Genetics*, **86**, 353–363.

Li, N. P., Bailey, J. M., & Kenrick, D. T. (2002). The necessities and luxuries of mate preferences: Testing the tradeoffs. *Journal of Personality and Social Psychology*, **82**, 947–955.

Lippa, R. A. (2007). The preferred traits of mates in a cross-national study of heterosexual and homosexual men and women: An examination of biological and cultural influences. *Archives of Sexual Behavior*, **36**, 193–208.

Lippa, R. A. (2009). Sex differences in sex drive, sociosexuality, and height across 53 nations: Testing evolutionary and social structural theories. *Archives of Sexual Behavior*, **38**, 631–651.

Lishner, D. A., Nguyen, S., Stocks, E. L., & Zillmer, E. J. (2008). Are sexual and emotional infidelity equally upsetting to men and women? Making sense of forced-choice responses. *Evolutionary Psychology*, **6**, 667–675.

Marlowe, F. (2000). Paternal investment and the human mating system. *Behavioural Processes*, **51**, 45–61.

Marlowe, F. W. (2003). A critical period for provisioning by Hadza men: Implications for pair bonding. *Evolution and Human Behavior*, **24**, 217–229.

Marlowe, F. W. (2004). Mate preferences among Hadza hunter–gatherers. *Human Nature*, **15**, 365–376.

Martin, R. D. (1990). *Primate Origins and Evolution: A Phylogenetic Reconstruction*. Princeton, NJ: Princeton University Press.

McCaughey, M. (2007). *The Caveman Mystique: Pop Darwinism and the Debates over Sex, Violence, and Science*. New York: Routledge.

Miller, G. F. (2000). *The Mating Mind: How Sexual Choice Shaped the Evolution of Human Nature*. New York: Vintage Books.

Pawlowski, B., & Dunbar, R. I. M. (2001). Human mate choice decisions. In R. Noe, P. Hammerstein, & J. A. R. A. M. van Hooff, eds., *Economic Models of Human and Animal Behaviour*. Cambridge, UK: Cambridge University Press, pp. 187–202.

Pérusse, D. (1993). Cultural and reproductive success in industrial societies: Testing the relationship at the proximate and ultimate levels. *Behavioral and Brain Sciences*, **16**, 267–322.

Pietrzak, R., Laird, J. D., Stevens, D. A., & Thompson, N. S. (2002). Sex differences in human jealousy: A coordinated study of forced-choice, continuous rating-scale, and physiological responses on the same subjects. *Evolution and Human Behavior*, **23**, 83–95.

Reichard, U. H. (2003). Monogamy: Past and present. In U. H. Reichard & C. Boesch, eds., *Monogamy: Mating Strategies and Partnerships in Birds, Humans and Other Mammals*. Cambridge, UK: Cambridge University Press, pp. 3–26.

Sagarin, B. J., Martin, A. L., Coutinho, S. A., et al. (2012). Sex differences in jealousy: A meta-analytic examination. *Evolution and Human Behavior*, **33**, 595–614.

Scelza, B. A. (2014). Jealousy in a small-scale, natural fertility population: The roles of paternity, investment and love in jealous response. *Evolution and Human Behavior*, **35**, 103–108.

Schmitt, D. P. (2005). Sociosexuality from Argentina to Zimbabwe: A 48-nation study of sex, culture, and strategies of human mating. *Behavioral and Brain Sciences*, **28**, 247–275.

Schmitt, D. P., & 118 Members of the International Sexuality Description Project (2003). Universal sex differences in the desire for sexual variety: Tests from 52 nations, 6 continents, and 13 islands. *Journal of Personality and Social Psychology*, **85**, 85–104.

Sear, R., & Mace, R. (2008). Who keeps children alive? A review of the effects of kin on child survival. *Evolution and Human Behavior*, **29**, 1–18.

Shackelford, T. K., & Larsen, R. J. (1997). Facial asymmetry as an indicator of psychological, emotional, and physiological distress. *Journal of Personality and Social Psychology*, **72**, 456–466.

Sprecher, S., Sullivan, Q., & Hatfield, E. (1994). Mate selection preferences: Gender differences examined in a national sample. *Journal of Personality and Social Psychology*, **66**, 1074–1080.

Stewart-Williams, S. (2018). *The Ape That Understood the Universe: How the Mind and Culture Evolve*. Cambridge, UK: Cambridge University Press.

Stewart-Williams, S., Butler, C. A., & Thomas, A. G. (2017). Sexual history and present attractiveness: People want a mate with a bit of a past, but not too much. *Journal of Sex Research*, **54**, 1097–1105.

Stewart-Williams, S., & Thomas, A. G. (2013a). The ape that kicked the hornet's nest: Response to commentaries on "The Ape That Thought It Was a Peacock." *Psychological Inquiry*, **24**, 248–271.

Stewart-Williams, S., & Thomas, A. G. (2013b). The ape that thought it was a peacock: Does evolutionary psychology exaggerate human sex differences? *Psychological Inquiry*, **24**, 137–168.

Symons, D. (1979). *The Evolution of Human Sexuality*. Oxford: Oxford University Press.

Trivers, R. L. (1972). Parental investment and sexual selection. In B. Campbell, ed., *Sexual Selection and the Descent of Man: 1871–1971*. Chicago, IL: Aldine Press, pp. 136–179.

Wood, W., & Eagly, A. (2002). A cross-cultural analysis of the behavior of women and men: Implications for the origins of sex differences. *Psychological Bulletin*, **128**, 699–727.

29 Human Mate Selection

A Multidimensional Approach

DANIEL CONROY-BEAM AND DAVID M. BUSS

Ask a few people to describe their ideal partner and you will get the same response every time: a thoughtful pause followed by a deluge. Our partners must be kind, intelligent, and physically attractive. They should be in good health, good with people, good parents, and good providers. Some of our ideals are very specific: straight hair, freckled, with a gap between their front teeth. Others are more abstract: loyal, creative, passionate, and driven. Many ideals are almost contradictory: spontaneous but reliable; flexible but strong-willed; youthful but responsible. Humans seem picky when it comes to mate choice.

And this pickiness makes great sense from an evolutionary perspective. For humans, a mate represents a potential reproductive partner, a parenting partner, a cooperation partner, and more. Throughout human evolutionary history, who an individual selected as their mate would have had profound impacts on their reproductive success. So selection historically would have favored the evolution of adaptations capable of motivating individuals to preferentially pursue mates who increased their reproductive success and to avoid mates who decreased it.

This rationale suggests that modern humans will have inherited an array of preference adaptations that focus our mating ideals on traits that would have historically conveyed fitness benefits to our ancestors. A large body of research in evolutionary psychology supports this hypothesis. Humans across cultures are known to express ideal mate preferences for many traits, including kindness (Buss, 1989; Hatfield & Sprecher, 1995), health (DeBruine et al., 2010), specific age ranges (Kenrick & Keefe, 1992), physical attractiveness (Sugiyama, 2004), and resource acquisition potential (Townsend & Levy, 1990). Mates possessing such features would have provided a variety of fitness benefits to ancestors insofar as they signaled fecundity, reproductive value, survivability, parenting ability, cooperative ability, and so on.

These preferences also vary across sex (Buss, 1989), context (Kenrick et al., 1990), and ecology (Gangestad & Buss, 1993) in ways that are consistent with evolutionary functional hypotheses. Further, there exists a cottage industry dedicated to documenting previously undiscovered mate preferences. New, seemingly functional mate preferences are continually reported in the evolutionary psychological literature (e.g., exploitability: Goetz et al., 2012; lumbar curvature: Lewis et al., 2015).

29.1 WHAT DO PREFERENCES MATTER?

Overall, research attempting to document human mate preferences has been extraordinarily successful. But this research is undergirded by one key assumption: people are actually motivated to pursue mates who embody their preferences. This assumption is taken for granted in much mate preference research. But studies that attempt to bridge the gap between stated ideals on the one hand and attraction and mate selection on the other often come up short.

For instance, Kurzban and Weeden (2007) analyzed mate preferences and mate choices from a large sample of speed-dating events. They collected measures of each participant's stated preferences and computed choice scores for each participant reflecting the average trait value they selected from their potential mates. The rationale was straightforward: if a woman says she prefers short partners, she should tend to select partners who are short, and the average height of all of her selected partners should be relatively short. A woman who prefers tall partners should instead tend to select tall partners and have a tall choice score for height. But surprisingly, participants' advertised preferences in a potential mate had little to no power to predict the participants' corresponding choice scores.

Todd et al. (2007) similarly found that stated ideals were poor predictors of choices in a speed-dating context across several dimensions: wealth, family commitment, physical attractiveness, health, and overall desirability. In this study, preferences generally did not predict the traits of chosen partners; for men, preferences *negatively* predicted choices, when they predicted them at all. Another speed-dating study found that stated ideal preferences were no help in predicting the attractions *or* choices of speed-daters (Eastwick & Finkel, 2008). For instance, people who reported a stronger ideal preference for financial prospects were not more likely to select or even report stronger desire for potential mates whom they judged to have better financial prospects. Finally, outside of a speed-dating context, Campbell, Chin, and Stanton (2016) surveyed the ideal mate preferences of a community sample of initially single participants. They then followed up with these participants after they formed romantic relationships and collected measures of their new partners' traits. In this sample, ideal preferences did significantly predict the traits of later chosen partners, but the effect size was small: an r of just 0.14 overall. In general, a person's ideal mate preferences appear to be at best of modest utility in predicting their attraction to or selection of potential mates.

29.2 THE PUZZLE OF MATE CHOICE AND SOME RESOLUTIONS

From these two sets of findings, the mating literature poses a puzzle. On the one hand, humans across cultures express an array of ideal mate preferences. These ideal mate preferences appear functional in that they target traits that would have provided fitness benefits to human ancestors. Ideal mate preferences also shift across individuals, contexts, and cultures in ways that are consistent with evolutionary functional hypotheses. And crucially, functional mate preferences could only have evolved if they historically motivated people to select mates who embodied them. On the other hand, research on attraction and actual mate choice commonly finds that ideal mate preferences have weak if any power to predict attraction and actual mate selection. How can this puzzling contradiction be resolved?

29.2.1 Possible Resolution 1: Ideal Mate Preferences Are Ineffectual

The first possible way to explain the puzzle of mate choice is to conclude that ideal mate preferences simply do not affect attraction and mate selection. This could occur because ideals apply in reasoning about abstract but not actual partners (e.g., Eastwick et al., 2014). If this were the case, attraction and mate selection would need to be governed by other contextual factors, such as proximity. Others have proposed that attraction and mate selection are simply inherently random (Lykken & Tellegen, 1993) – an unlikely scenario given the incredible importance of mate selection to fitness. More likely, people could be unwilling or unable to express the true preferences that guide their attractions and mate choice behavior (Todd et al., 2007). No matter the cause, according to this possible resolution, ideal mate preferences have poor power in predicting attraction and mate selection simply because stated ideals are not part of the causal pathway leading to mate choice.

This resolution is logically possible. However, adopting this resolution requires the assumption that the large array of mate preferences, documented by psychologists and anthropologists across cultures using methods ranging from self-report (e.g., Buss, 1989), to surveying personal advertisements (e.g., Wiederman, 1993), to analyses of cultural artifacts (e.g., Gottschall et al., 2004) and that correspond strongly to evolutionary functional principles, are simply nonfunctional psychological delusions – a unique and universal bug in the design of human mating psychology. This strikes us as a bizarrely non-parsimonious conclusion, and as we show in Section 29.3, it is unlikely to be correct.

29.2.2 Possible Resolution 2: Research on Attraction and Mate Choice Is Methodologically Limited

A second potential resolution of the puzzle of mate choice is to conclude that mate preferences do actually drive mate selection, but studies of mate selection have been beset by methodological limitations. Indirect evidence in favor of this resolution comes from work linking preferences and choices at the group level. For example, women on average express a preference for men who are older than them (e.g., Buss, 1989; Conroy-Beam & Buss, 2019). According to marriage records, women on average also tend to marry older men (e.g., Kenrick & Keefe, 1992). Swedish historical records indicate that this trend applied at least as far back as the nineteenth century (Low, 1991). This work shows that preferences are linked to mate choices at least at the group average level, even if preferences and choice are difficult to link at the individual level. Given this, it is possible that the difficulty in linking individual preferences to attraction and choice is due primarily to methodological limitations of the studies exploring individual choice.

Many of the limitations of mate choice research appear to stem from inherent limitations of speed dating, a common venue for attraction and mate choice research. First, interactions in speed dating are brief by design. These short meetings may not give participants enough time to learn the information that is relevant to their many mate preferences (Kurzban & Weeden, 2007). Second, a more fundamental limitation concerns the typical participants encountered at speed-dating events. People preferentially attend speed-dating events where they are likely to encounter potential mates who fulfill their mate preferences (Kurzban & Weeden, 2007). Additionally, people who are highly undesirable might avoid speed-dating events if they anticipate that they will be rejected (Li et al., 2013). These processes combine to limit attendance at speed-dating events to people who possess at least moderately desirable traits, suggesting a substantial restriction of range.

This restriction of range could limit the apparent predictive power of ideal mate preferences for two reasons. First, statistically, if participants are exposed only to a narrow range of trait values, they can express only a narrow range of attractions, thereby limiting the measured predictive power of preferences even if they would have stronger effects on attraction and mate choice across a less restricted range of trait values. Second, mate preferences may act as much or more to motivate aversion of undesirable partners as to motivate pursuit of desirable partners (Li et al., 2013). If speed-dating events are not attended by people with undesirable traits, this aversion effect might be missed by speed-dating studies. Indeed, studies that experimentally increase variance on the low end of traits such as physical attractiveness and status – two traits hypothesized by evolutionary psychologists to be important to mate choice – find substantially stronger effects of mate preferences in predicting mate choice (Li et al., 2013).

These methodological limitations could explain why speed-dating studies commonly find no or weak predictive power of mate preferences and are important cautions for future research. However, they provide little explanation of the weak effects of preferences found in studies such as Campbell et al.'s (2016) community sample of naturally formed relationships. A qualitatively different resolution to the mate choice puzzle appears necessary to explain these effects.

29.2.3 Possible Resolution 3: Prior Theoretical Models of Mate Choice Are Flawed

A third potential resolution to the puzzle of mate choice concerns the theoretical models researchers use to make predictions about attraction and mate choice. Most mating research treats mate selection as a linear, univariate process. For instance, Campbell et al. (2016) attempted to predict mate choices from mate preferences using multi-level models that separately predicted each partner trait value from the participant's corresponding preference value. Similarly, Kurzban and Weeden (2007), Todd et al. (2007), and Eastwick and Finkel (2008) constructed linear models that attempted to predict each partner trait value separately from participants' corresponding preference values.

These models implicitly model a potential mate as a collection of qualities, each of which has an effect on the potential mate's overall desirability dependent on the evaluator's mate preferences. Within a linear model, a crucial assumption (typically implicit) is that these effects are independent and additive: a potential mate's desirability in terms of physical attractiveness has no bearing on their desirability in terms of status, kindness, or intelligence. Their overall desirability is simply the sum of the desirability they earn from each of their trait dimensions.

But potential mates do not occur as sums of traits. Rather, each potential mate represents a point at the intersection of several preference dimensions. A mate who is physically attractive might also be cruel and unintelligent; a mate who is lovely interpersonally may not be lovely physically. And even comparatively small differences at the unidimensional level, such as between ideals and potential mates, can yield surprisingly large differences at the multidimensional level (Conroy-Beam et al., 2015; Del Giudice, Booth, & Irwing, 2012). Models that can better account for the multidimensional nature of mates and mate selection may find that preferences are more powerfully motivating than linear models make them appear.

29.3 A MULTIDIMENSIONAL MODEL OF HUMAN MATE SELECTION

Here, we detail the elements of a novel theoretical model of human mate selection that takes into account three critical elements of mate choice: (1) the multidimensional nature of human preference integration; (2) the multiplicity of human mate preferences; and (3) the constraints on mate choice imposed by mating markets that possess multiple individuals acting on their preferences in a multidimensional manner. We combine extant data on human mating with computer simulations to show how accounting for these elements in models of mate choice is essential for observing the true effects of mate preferences on downstream mating outcomes such as attraction and mate choice.

29.3.1 The Multidimensionality of Preference Integration

Consider building a robot that is able to reliably select fitness-beneficial mates from among a larger pool of potentials. This is no easy task. Potential mates vary along an infinite number of dimensions. Some are taller, some have more wrinkly knees, some are healthier, and some have moles arranged in shapes that more closely approximate classical Greek constellations. Only a tiny subset of the dimensions along which potential mates vary would actually impact the fitness benefits these mates offered. Within each of these dimensions, only a small region of the trait space would be fitness beneficial rather than fitness neutral or costly. A mating robot therefore first needs to know which trait dimensions are relevant to fitness and what values of those traits signal maximum benefits. That is, the robot needs mate preferences.

With mate preferences in hand, the robot next needs to compare its preference value on each trait dimension to the corresponding trait value offered by each potential mate. These comparisons will provide the mating robot with a list of trait-level desirabilities for each potential mate it encounters. Finally, in order to select a mate, the robot next needs to apply an algorithm to these lists that integrates each list down into some decision variable. This integration process crucially must produce a decision variable that is correlated with the total fitness benefits each mate offers, such that a mate who is high in mate value according to the decision value offers fitness benefits, whereas a mate who is low in mate value offers fitness costs.

The nature of the algorithm that the mating robot uses to integrate its preferences will have profound implications for how these preferences relate statistically to the robot's attractions and choices. The most commonly assumed model for human mate preference integration, implicitly or explicitly, is a linear model equivalent to linear regression. For instance, Eastwick et al. (2014) argue that for preferences to show predictive validity one must be able to demonstrate an interaction between the evaluator's preference value and the potential mate's trait value in predicting "romantic evaluations," such as overall attractiveness. Implicit in this argument is an assumption that mate preferences are integrated in a linear fashion: that a strong preference for, say, physical attractiveness exerts its effect on mate selection by applying a large slope relating mate physical attractiveness values to overall mate attractiveness. In this case, a physical attractiveness preference would manifest statistically as an interaction between preference and trait values in predicting overall attraction ratings.

This linear model of preference integration *could*, in principle, capture the nature of human preference integration. But other algorithms are possible, and there are good reasons for supposing that preference integration is not linear. Linear models make great sense for calculating outcomes such as the total force acting on an object in a given direction. This is because total force truly is the sum of many independent forces (the predictors) weighted by the direction of those forces (the slopes). But a potential mate is not a stream of disembodied trait values waiting to be summed by a mate selector. Nor are the fitness benefits a mate offers necessarily equal to the sum of benefits implied by each trait dimension. A mate who is beautiful, intelligent, or outgoing cannot offer many net fitness benefits if they are also sadistic, infertile, or lethally infectious (e.g., Li et al., 2002).

Potential mates are not linear combinations but multidimensional objects: constellations of trait values that present simultaneously and must be accepted or rejected as a whole. Computer simulations show that integration algorithms that better reflect the multidimensional nature of actual potential mates lead to more fitness-beneficial mate choices and greater reproductive success (Conroy-Beam & Buss, 2016b). In this study, we constructed an agent-based model in which agents evolve mate preferences and integrate them according to inherited algorithms in order to select mates who vary along invisible dimensions of fitness. The linear combination algorithm commonly assumed in the mate selection literature performed poorly and was almost immediately driven to extinction by alternative algorithms in all model runs.

The best-performing algorithm in this model of mate choice evolution was a Euclidean algorithm. The Euclidean algorithm is multidimensional because it represents preferences and potential mates not as weighted sums, but as points within an n-dimensional preference space. Imagine a person with just two mate preferences, such as for kindness and intelligence. One could form for this person a two-dimensional preference space, where one direction through the space represents increasing kindness and the other direction represents increasing intelligence. Each point in this preference space represents a possible set of mate preferences; so, if this hypothetical person prefers a kind and intelligent mate, their preference point would fall somewhere in the first quadrant of the preference space.

Any potential mate this person encounters can also be placed within the preference space. For instance, a mate who is very kind and intelligent would fall in the first quadrant of the preference space as well, perhaps near the person's preference point. A mate who is kind but not very intelligent would fall somewhere in the second quadrant, whereas an intelligent but cruel mate would fall in the fourth quadrant. With this two-dimensional preference space, the hypothetical person can integrate their preferences into an overall evaluation of any mate they encounter by simply calculating the straight-line distance between their own preference point and the potential mate's trait point through the preference space. When this Euclidean distance is large, value as a mate is low, and so proportionally should be motivation to pursue that mate.

The Euclidean algorithm performs well in computer simulations of mate choice evolution, but more importantly, it also performs well in describing real human relationships. People's actual long-term mates tend to fall close to their ideal preferences through the multidimensional preference space, exactly as they would if mate preferences were integrated according to a Euclidean algorithm (Conroy-Beam & Buss, 2016b). Additionally, people who are themselves higher in mate value according to Euclidean calculations – that is, who are closer to the opposite sex's aggregate preference point in the preference space – tend to be mated to partners closer to their own mate preferences. This mate value effect only emerges in computer simulations wherein agents integrate their preferences with a Euclidean algorithm; alternative integration algorithms, including linear algorithms, do not reproduce the mate value effect observed in real relationships.

Moreover, Euclidean mate value variables also have power in predicting relationship satisfaction (Conroy-Beam, Goetz, & Buss, 2016). People are dissatisfied with partners who are lower in Euclidean mate value than themselves. However, this effect disappears if those mates are closer to the person's preferences within the preference space than the majority of alternative mates. Altogether, humans appear to use something similar to a Euclidean algorithm to integrate their preferences when evaluating and selecting mates.

This insight into the nature of human mate preference integration is critical because the Euclidean algorithm has a number of features that are important for understanding how mate preferences are related to attraction and mate choice – and for understanding why it sometimes appears as though they are not. First of all, the Euclidean algorithm can be used to integrate any number of mate preferences. We introduced the algorithm with a two-dimensional example for simplicity, but the Euclidean algorithm can equivalently integrate two, two dozen, or two hundred mate preferences into a single mate value variable.

Second, the Euclidean algorithm is nonlinear in that its outputs are not additive functions of its inputs. Consider a simple scenario in which a person with a vector of preferences $p = (5, 5)$ encounters a potential mate with a vector of traits $t = (7, 4)$. With a linear model, the desirability of this potential mate along each dimension would be:

$$d_1 = 5 * 7 = 35$$

$$d_2 = 5 * 4 = 20.$$

The overall attractiveness of the potential mate would be:
$$d_O = d_1 + d_2 = 5 * 7 + 5 * 4 = 55.$$

The output of the linear integration algorithm across all inputs is equivalent to the sum of the outputs of the algorithm when applied to each input. This is not true when preferences are integrated according to a Euclidean algorithm. With Euclidean integration, the desirability of the potential mate on each dimension would instead be:

$$d_1 = \sqrt{(5-7)^2} = 2,$$

$$d_2 = \sqrt{(5-4)^2} = 1.$$

Here, lower values are associated with higher desirability. According to the Euclidean algorithm, overall desirability would be:

$$d_O = \sqrt{(5-7)^2 + (5-4)^2} = 2.24.$$

With Euclidean integration, d_O (2.24) is not equivalent to $d_1 + d_2$ (3); in fact, because lower values mean greater desirability, this potential mate is higher in overall mate value than the sum of the individual desirabilities would suggest.

If human mate preferences are actually integrated according to a Euclidean algorithm, as appears to be the case, this nonlinearity would cause mate preferences to have counterintuitive relationships with perceptions of mate value and thereby feelings of attraction. Here, we review two such counterintuitive consequences of nonlinear integration: preferences have interactive effects on attraction; and these effects make attraction difficult to predict linearly from ideal mate preferences.

Mate Preferences Have Interactive Effects on Attraction.
Lee et al. (2014) used multivariate analyses to explore attraction to potential mates as a function of facial attractiveness, sexual dimorphism, and intelligence. They found that, for men, intelligence and physical attractiveness interacted to predict attraction: increasing intelligence had a larger effect on the overall attractiveness of facially attractive women. Farrelly, Clemson, and Guthrie (2016) found interactive effects of women's preferences on attraction: women report greater attraction to altruistic men, but the effect of altruism is stronger for physically attractive men. These interactive effects are surprising assuming a linear model of mate preference integration, where each preference affects just its corresponding trait dimension. They will also tend to make the effects of preferences difficult to observe as long as researchers use models that do not incorporate interaction effects. A Euclidean model, on the other hand, both predicts and accounts for interactive effects of mate preferences.

Consider a toy model in which mate selection is based on five traits, each of which can range in value from $t = 1$ to $t = 5$ (Table 29.1). Suppose an individual, Emma, has ideal mate preferences $p = 5$ for all five traits. Further, suppose Emma encounters two potential mates: Charles and Alfred. Both Charles and Alfred have a value of $t = 1$ for traits 1–4 – very far from Emma's ideal. Charles and Alfred only differ from one another in their fifth trait: Alfred has a value of $t_5 = 1$ for trait 5, whereas Charles has a value of $t_5 = 5$.

Emma can compute Alfred and Charles' mate values to her by calculating the Euclidean distance between her preference vector and each of their trait vectors. If this Euclidean distance is transformed such that shorter distances yield higher mate values, this single difference in trait value between Charles and Alfred is sufficient to cause a small difference in mate value: Alfred's mate value will be 0, whereas Charles' mate value will be 1.07 (Table 29.1). In this case, variation between Charles and Alfred on trait 5 does not substantially impact their mate values because Charles and Alfred are far from Emma's ideals on the other four traits.

However, if the remaining four traits had been closer to Emma's preferred value, the difference between Charles and Alfred would have had a larger effect on their overall mate values. To see this, consider two new mates, George and Bill. These two new mates differ in the same way as Charles and Alfred did on trait 5: George has a value of $t_5 = 1$ for trait 5 and the Bill has a value of $t_5 = 5$. But unlike Charles and Alfred, George and Bill have values of $t = 5$ for traits 1–4. Because of this, the same difference in trait value on trait 5 has a much larger impact on mate values to Emma: George's overall mate value to Emma is 4, whereas Bill's overall mate value is 8. The same change in trait value between two potential mates had 3.73 times the impact on overall mate value depending on the value of all other traits (Table 29.1).

Table 29.1 A toy model of mate selection using a Euclidean algorithm to integrate five preferences.

	Trait 1	Trait 2	Trait 3	Trait 4	Trait 5	Mate value
Emma's preference	5	5	5	5	5	
Alfred	1	1	1	1	1	0
Charles	1	1	1	1	5	1.07
George	5	5	5	5	1	4
Bill	5	5	5	5	5	8

Note: Euclidean distances were transformed by multiplying each distance by −1 and adding the maximum possible distance value (8).

These interactive effects are directly analogous to interactions observed between mate preferences in prior research (e.g., Farrelly et al., 2016; Lee et al., 2014). They will also tend to make the effects of preferences on attraction and mate choice more difficult to observe. Because the impact of any one trait on mate value depends on the value of all others, models that incorporate only the direct effects of each mate preference independently will be considerably noisy. This noise will occur to the extent that, for instance, mates with non-preferred trait values are relatively high in mate value because their other traits are near ideal or mates with preferred trait values are low in mate value because their other traits are far from ideal. But because each preference can, in principle, interact with all other preferences, accounting for this apparent noise would require incorporating more attraction terms than most psychological studies have the statistical power to detect. The interactive effects of mate preferences created by Euclidean integration are simply unwieldy in accounting for using linear models of mate preferences. But models that predict attraction and mate selection from Euclidean distances, and therefore better represent the multidimensional nature of mate preference integration, will instead easily account for these interaction effects.

The Effects of Mate Preferences Are Difficult to Observe with Linear Models. The interactive effects of mate preferences on attraction are just one consequence of Euclidean mate preference integration. If mate preferences are actually integrated in mate choice according to a Euclidean algorithm, the fundamentally nonlinear nature of this algorithm will also make the effects of preferences on attraction generally difficult to observe using purely linear models. To illustrate this, we constructed a simple agent-based model. In this model, 1,000 agents take turns evaluating the same potential mate. Each agent has five mate preferences, drawn from random normal distributions and constrained to values between one and seven. The potential mate is generated randomly at the start of the model run and has five traits generated in the same way as the agents' preferences.

We repeated this model for 1,000 iterations in total. In each iteration of the model, each agent simply reports how attractive they find the random potential mate using one of two preference integration algorithms. The first is linear: the summed product of their preferences and their potential mate's traits. The second is Euclidean: attraction is computed as the Euclidean distance between their preferences and the potential mate's traits, transformed such that higher values indicate higher mate value. Under both algorithms, attraction is fully determined by mate preferences. Across iterations, the model saves the preferences of each agent and the attraction value given to the potential mate according to both preference integration algorithms. Within each model run, we used linear regression to predict attraction to the potential mate from the main effects of agent preferences. We saved the multiple R^2 from each model within iterations and calculated the average multiple R^2 across iterations.

When mate preferences were integrated according to a linear algorithm, a linear model perfectly predicted attraction: the multiple R^2 of all models across iterations was $R^2 = 1.0$, 95 percent confidence interval (CI) 1.00–1.00. But when mate preferences were integrated according to a Euclidean algorithm, preferences' effects on attraction were much more difficult to observe: $R^2 = 0.51$, 95 percent CI 0.50–0.52. Incorporating the interaction between preferences increased the variance explained by the linear model, $R^2 = 0.55$, 95 percent CI 0.54–0.56. Nonetheless, even the model with the interaction term included explained a relatively small proportion of the variance considering that attraction was fully systematic in this agent-based model. Overall, this model demonstrates that when mate preferences are integrated in a nonlinear fashion, linear models are dramatically less effective in discovering their effects. Researchers who analyze the relationship between mate preferences, attraction, and mate choice using only linear models risk missing much of mate preferences' true effects.

Consideration of the algorithm by which people integrate their mate preferences in making mating decisions is vital for understanding how preferences relate to attraction and mate choice. Mating researchers often implicitly assume a linear model of preference integration (e.g., Eastwick et al., 2014). But computer simulations and studies of real couples suggest that human mate preference integration is nonlinear and Euclidean (Conroy-Beam & Buss, 2016b; Conroy-Beam, Goetz, & Buss, 2016). The nonlinearity of this integration algorithm will cause mate preferences to have interactive effects on attraction and will make linear models poor predictors of attraction overall. If human mate preference integration is truly nonlinear, researchers hoping to fully capture the effects of mate preferences on attraction and mate choice need to use statistical models that reflect this nonlinearity.

29.3.2 The Multiplicity of Human Mate Preferences

The nature of human mate preference integration contributes to the difficulty in observing the effects of mate preferences on attraction and mate selection. But further complications come from one important truism about preferences themselves: human mate preferences are numerous. Sexually reproducing species commonly base their mate choices on multiple preferences (Candolin, 2003). For instance, peahens famously prefer peacocks with elaborate trains, but they also prefer to mate with peacocks that perform more frequent behavioral displays (Loyau, Jalme, & Sorci, 2005). Male fruit flies (*Drosophila melanogaster*) have several documented mate preferences, including those for female pheromones (Billeter et al., 2009), age (Lüpold et al., 2011), and body size (Byrne & Rice, 2006).

The use of multiple mate preferences can make good functional sense because variation along different trait dimensions can reflect variation along different dimensions or components of fitness. A mate that is more fecund offers more fitness benefits, as does a mate who commands better territory. But the mate who is the most fecund is not necessarily the mate who commands the best territory. Selection can consequently favor the evolution of multiple preferences for multiple cues of mate value.

This rationale applies to humans as well, and people across cultures readily express a large number of mate preferences. Buss (1989) had people from 37 different cultures around the world express their preferences in a mate across 18 different preference dimensions. These dimensions came from a questionnaire that had been administered to American samples since the 1940s (Hill, 1945). Fletcher et al. (1999) used participant nomination to develop a self-report scale of ideals that contains 69 items, which they reduce to five factors of ideals. This multiplicity of preferences is also not limited to industrialized societies. Marlowe (2004) interviewed Hadza hunter–gatherers for their preferences in an ideal partner and found that the nominated traits could broadly be grouped into seven categories: good character, fertility, fidelity, foraging ability, good looks, intelligence, and youth.

Using multiple mate preferences to guide mate choice presumably would have allowed ancestral humans to discriminate among potential mates on multiple dimensions of fitness. But the existence of these many preferences also increases the complexity of mate choice. This is largely because ideal mates represent small points in the comparatively enormous and rapidly growing space of possible mates. Imagine a situation in which potential mates can only take on one of seven trait values for each preferred trait dimension, as in a common Likert scale. With just one mate preference, ideal mates would represent 1/7 of all possible mates. With two mate preferences, ideal mates represent just 1/49 of all possible mates. The total population of the Hadza is around 1,000 people (Marlowe, 2004), but with seven preferences, an ideal mate would be just one out of 823,543 possible partners. A person living in California would do well to limit themselves to nine mate preferences; then their ideal would represent 1 out of 40 million possible mates relative to California's population of 38 million. Just 12 mate preferences are enough to make the space of possible mates nearly twice the global human population.

Actual potential mates are of course not distributed uniformly throughout the trait space, but trait spaces also do not neatly vary in increments of sevenths. Regardless of the exact size of the space of potential mates, it is clear that the odds of any given person encountering their ideal mate, across all of their many preference dimensions, are slim to none. Rather, each individual encounters an array of imperfect mates, each of whom will fulfill some preferences but not others. The mate who is ideally physically attractive may also be too cruel or too careless; the mate who is just the right height might also be hairy and unhealthy. These imperfections necessitate trade-offs in mate choice: people are often forced to forgo partners who are ideal on some dimensions for alternatives who are closer to ideals overall. Mate selection is therefore less about finding one's ideal partner and more about finding the best compromise among a set of ideals.

These trade-offs, inherent to real human mating markets, are not obvious from the analyses mating researchers commonly use to study attraction and mate choice. Mating researchers commonly assume that if attraction and mate selection are governed by mate preferences, then preferences should be correlated with corresponding trait values in desired or selected mates. For example, if physical attractiveness preferences powerfully motivate mate selection in a speed-dating environment, then people who express strong physical attractiveness preferences should tend to select mates who are in general more physically attractive. Weak correlations between physical attractiveness preferences and the physical attractiveness of chosen partners are taken to mean that physical attractiveness preferences have little motivational power (e.g., Eastwick & Finkel, 2008; Todd et al., 2007).

But this rationale overlooks the fact that each participant's physical attractiveness preference is being actively traded off against the participants' other preferences. The speed-dating partner who was most physically attractive could be far from ideal on other dimensions – perhaps kindness, status, or financial prospects. A participant who rejects a highly physically attractive but rude and low-status potential mate in favor of a moderately attractive but kind and high-status mate would be considered by typical analyses not to be acting on their preferences. But this person is actually acting simultaneously on three preferences! The trade-offs required by basing mate selection on multiple preferences not only limit people's ability to find their ideal mates, but also limit the apparent effects of individual mate preferences on attraction and mate choice. This will be particularly true when the effects of preferences are measured in terms of the predictive power of individual preferences in isolation as opposed to more appropriate, multivariate measures of preference fulfillment.

We constructed another agent-based model to illustrate the effects of trade-offs on: (1) the ability of individuals to fulfill their preferences; and (2) the predictive power of individual mate preferences. In each iteration of this model, 100 agents are each given 100 random potential mates to select from. Each agent possesses a set of mate preferences that correspond to a set of traits in their potential mates. Preferences and traits are drawn randomly from normal distributions and are constrained to values between one and seven.

In order to manipulate trade-offs, we manipulated the number of preferences agents used to make their mating decisions. The number of preferences used in mate selection ranged in value from 1 preference to 70 preferences across model runs. Each agent used their mate preferences to calculate the mate value of each of their 100 potential mates using a Euclidean algorithm and simply selected the mate they found most attractive. There was no random noise in agent attraction values. Additionally, there was no mutual mate choice in this model. As in speed dating, agents simply nominated the mate they most preferred to mate with. Therefore, attraction and mate selection within this model were based entirely on the agent's mate preferences.

After all agents completed mate selection, we calculated two values. The first was the extent to which the agents fulfilled their mate preferences overall. To calculate mate preference fulfillment, we calculated the average Euclidean distance between each agent's set of mate preferences and their mate's corresponding traits. These distances were scaled to adjust for the number of preferences used in mate selection and then transformed such that higher values indicated more preference fulfillment. The minimum preference fulfillment after these transformations was 0 and the maximum was 10.

The second value we calculated was the average overall predictive power of mate preferences. To calculate this value, we calculated the correlation between each agent preference and the corresponding partner trait across the agent couple population from that model run. Higher values indicate that the agent's preference was more predictive of their partner's trait value. When mate selection was based on multiple preferences, the preference–trait correlation was averaged across preference dimensions.

We ran this model 50 times in total for each number of mate preferences. Figure 29.1(a) shows the effect of trade-offs on overall mate preference fulfillment. When mate selection is based on just a single preference, agents on average have an easier time finding preference-fulfilling mates; preference fulfillment across the model runs was $M = 9.96$ out of a maximum possible 10. Preference fulfillment declines exponentially thereafter and drops to a value of $M = 9.05$ across model runs when selection is based on just five preferences. With 20 mate preferences, average preference fulfillment across the model runs was $M = 8.12$, or 81.2 percent of maximum. The effect of additional preferences continually diminishes, however, and preference fulfillment approaches an asymptote near 7.5, or 75 percent of maximum. Therefore, although preference fulfillment does decline as agents are forced to make trade-offs among greater numbers of preferences, agents integrating their preferences according to a Euclidean algorithm are always able to find mates who are reasonably good fits to their preferences overall.

Trade-offs between preferences had a much more dramatic effect on the predictive power of individual preferences. When mate selection is based on just one preference, that preference is very predictive on average: the average correlation between preference and trait value was $r = 0.99$ across model runs. But with five preferences, the average correlation between preferences and traits drops to $r = 0.83$. Predictive power plummets quickly thereafter, dropping to $r = 0.45$ with 20 preferences. When agents based their mate choices on 70 preferences, the average correlation between preferences and traits is $r = 0.24$, and the slope relating preference number to predictive power is still substantially negative.

It is worth noting that these correlations are based on a scenario in which attraction is determined entirely by mate preferences. Further, in this model, attraction, preferences, and traits are measured completely noise-free. That trade-offs can cause such low predictive power even in this idealistic scenario demands a reinterpretation of the weak predictive power of preferences found in previous research. The essential trade-offs inherent to realistic mate choice powerfully constrain the power any individual preference can have in terms of predicting actual mate choices. But this weak predictive power does not mean that preferences have little behavioral effect. In fact, it means quite the opposite: individual mate preferences can ironically have poor power in predicting mate choice precisely because they have meaningful behavioral effects. When people pursue mates based on multiple mate preferences, each mate preference will act to pull behavior against the others, turning mate selection into a compromise between several

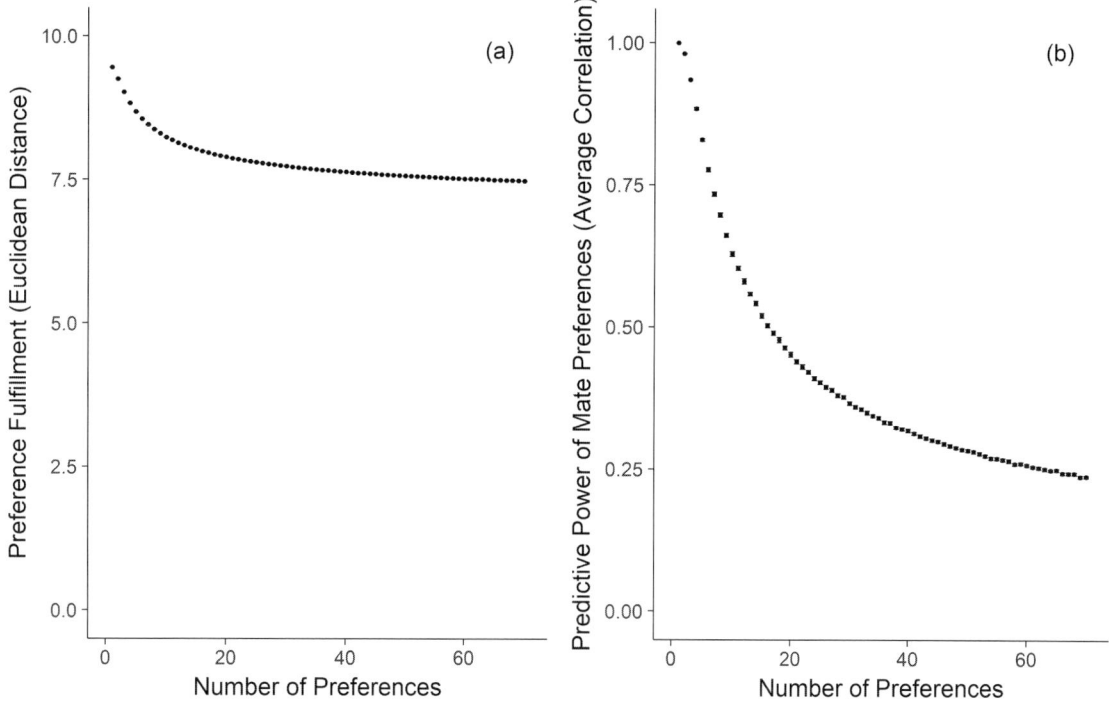

Figure 29.1 (a) Mate preference fulfillment and (b) predictive power of mate preferences as a function of number of mate preferences used in mate selection. The ability of agents to fulfill their mate preferences declines as agents are forced to trade off among an increasing number of mate preferences. These trade-offs also dramatically decrease the power of individual mate preferences to predict the traits of chosen partners. Error bars represent 95 percent confidence intervals.

conflicting ideals. The result is that mate preferences can be strongly fulfilled overall, but no single mate preference dimension in isolation can have consistent power to predict mating outcomes.

29.3.3 The Constraints of Realistic Human Mating Markets

The model in Section 29.3.2 showed that limited preference fulfillment and weak predictive power of mate preferences are the norm when mate selection is based on multiple mate preferences that must be traded off against one another. However, these correlations are likely still to be optimistic estimates of the magnitude of effect preference researchers can expect to observe when studying actual mate choices. For instance, Campbell et al. (2016) asked their community sample to report on 38 characteristics in both their ideal and actual partners. Our trade-off model would predict an average predictive power of $r = 0.33$, whereas they found an overall effect size of $r = 0.14$. What explains the difference?

A key difference between the agents in our model and people in real relationships is that our trade-off model was based on a scenario akin to speed dating, in which each individual simply determines which mate they most prefer. This is an essential step of the mate choice process, but mating is more than just preferring. In order to form an actual relationship, an individual, like the participants in

Campbell et al.'s (2016) sample, must navigate a mating market full of potential mates and rivals, each of whom is acting on their own unique set of preferences and possesses a unique set of traits. This collection of preferring and pursuing individuals creates a host of dynamics that will constrain the ways in which mate preferences relate to actual mate choice (Conroy-Beam & Buss, 2016a; Lykken & Tellegen, 1993).

One dynamic, already discussed in Section 29.3.2, is what we will call *availability*. In selecting a mate, a person is limited to selecting from among the mates available in their local mating market. The upper bound of preference fulfillment consequently is the most preference-fulfilling partner a given person can expect to actually encounter. A person surrounded by preference-fulfilling potential mates has more power to act on their preferences than a person surrounded by potential mates who are far from their preferences.

Availability of preference-fulfilling potential mates is necessary for forming a preference-fulfilling mateship, but it is not sufficient. One reason for this is a second important dynamic of mate choice: *human mating markets are inherently competitive*. For temporal, energetic, financial, and normative reasons, each person in a mating market can ultimately enter only a finite number of relationships. This limit is strictest in monogamous societies, where each person can only mate with one other person at a time. But even in polygamous societies, the maximum number of mates is most commonly just 4–10 partners; further, in most societies, less than 20 percent of men actually engage in polygamous relationships (Low, 1988). Those who are already mated to others are largely unavailable as mates, although mate poaching (Schmitt & Buss, 2001) and mate switching (Buss et al., 2017) render this constraint not absolute. Forming a preference-fulfilling mateship requires not only encountering preference-fulfilling partners, but also initiating relationships with them before rivals do.

Finally, one reason why rivals *can* beat a person to their ideal partner is a third and critical dynamic of human mating markets: *human mate choice is mutual.* It is not enough for a person to simply select their ideal partner; to form a mateship, that person's ideal partner must also select them in return. Attraction is not guaranteed to be symmetrical. Any person's ideal partner may ideally prefer a rival; a person who fails to secure the mutual attraction of their ideal partner will be forced to pursue a less ideal alternative. The need to secure mutual attraction limits people's ability to fulfill their preferences for all but the most desirable individuals (Conroy-Beam & Buss, 2016b).

Each of these three key dynamics constrains mate preference fulfillment and thereby necessarily reduces the predictive power of individual mate preferences. They also likely interact with each other. The need to secure mutual attraction, for example, would be a more severe constraint in mating markets that are competitive than in theoretical mating markets where each person can have an unlimited number of mates. These dynamics naturally co-occur in real human mating markets, so it is difficult for researchers to measure their independent and interactive effects. However, with agent-based models, we can experimentally manipulate the constraints imposed on simulated mating markets in order to estimate their effects on mate preference fulfillment and the predictive power of mate preferences.

To implement these realistic constraints, we expanded on our trade-off model of mate choice. In each iteration of this model, we separately generated 50 male and 50 female agents, each of whom had a set of 40 mate preferences drawn from random normal distributions. These agents selected one another as mates under varying conditions of constraint. We manipulated availability by changing how agent traits were initially generated. When the availability constraint was imposed, agent traits were generated by drawing trait values from random normal distributions independent of the preference distribution of the opposite sex. When the availability constraint was absent, we generated agent traits using the matrix of preference values of the opposite sex. Each agent's trait values were copied from the preference values of one unique opposite-sex agent. In this way, each agent was guaranteed at least one potential mate who perfectly fulfilled their ideal preferences.

In order to manipulate the competitive nature of the simulated mating market, we systematically varied the maximum number of mates each agent could acquire. When the markets were constrained by competitiveness, each agent was allowed only one partner. As soon as an agent was selected by a mate, they were removed from the mating market and no other agents were allowed to mate with them. In uncompetitive markets, however, each agent was allowed an unlimited number of mates. Each agent selected only one mate, but they were allowed to be selected by as many potential mates as were interested in them. This way, an agent's chances of mating with their ideal mate would not be affected even if a rival paired with them first.

Finally, we manipulated the mutual attraction constraint by varying how attraction related to mate choice. In all model runs, each agent calculated how attracted they were to all opposite-sex agents by computing the Euclidean distance between their own preferences and the traits of each potential mate. These differences were transformed such that higher values indicated a potential mate was closer to the agent's preferences. This process generated two matrices: a matrix of values indicating how attractive each male agent found each female agent; and a second matrix of values indicating how attractive each female agent found each male agent.

When markets were not constrained by mutual attraction, mate selection was based on these two matrices independently. For male mate choices, the model paired each male with whatever available female he was most attracted to. Female attraction values did not influence this pairing process. The reverse occurred for female choices: each female was paired with the male she was most attracted to, regardless of those males' reciprocal attractions. When markets were constrained by mutual attraction, mate selection was based on the product of male and female attraction values. Each male was paired with the female with whom he had the strongest mutual attraction (his attraction value for her multiplied by her attraction value for him) and vice versa for females. When the competition constraint was also present, the model formed relationships by first selecting the most mutually attracted pair in the population, pairing them, and removing both agents from the mating market; this process was iterated until all agents had a partner.

We manipulated each of these market dynamics independently across model runs, yielding eight markets constrained by different combinations of dynamics. We repeated the model 100 times for each market and in each we calculated average preference fulfillment and the predictive power of mate preferences for the resulting relationships. For each agent, these values were computed only for relationships that the agent chose to enter based on their mate preferences and not for relationships where the agent was simply selected by an opposite-sex agent.

Figure 29.2(a) shows the effect of market constraints on preference fulfillment and Figure 29.2(b) shows the

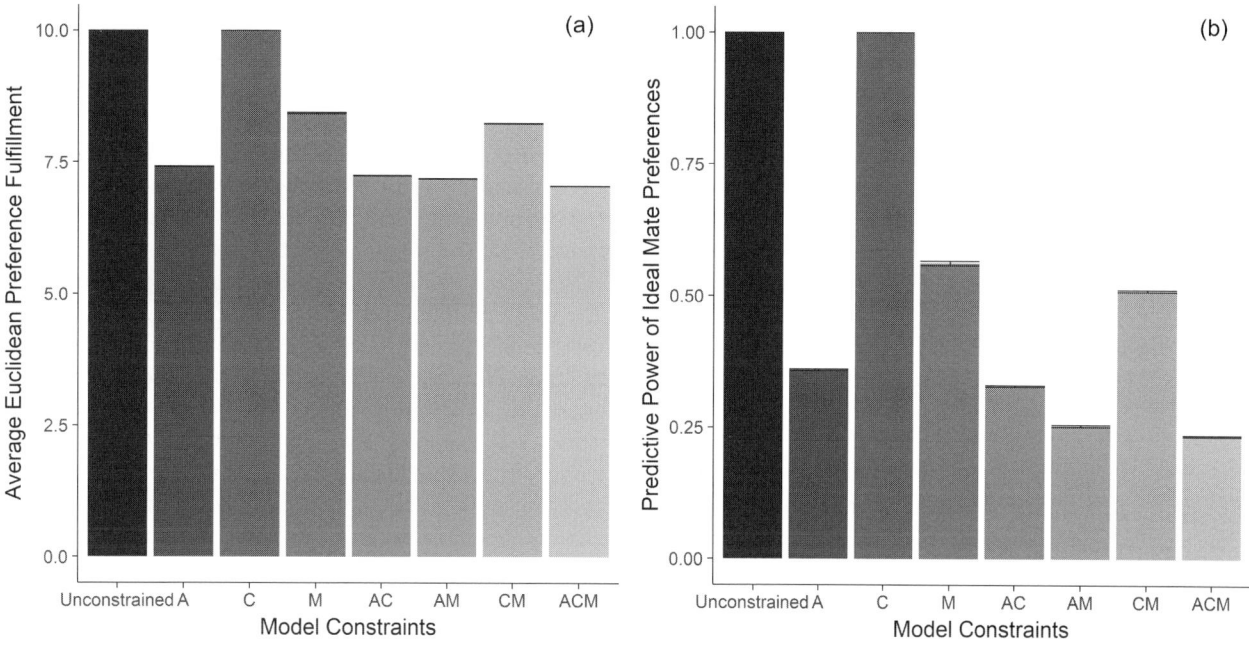

Figure 29.2 (a) Mate preference fulfillment and (b) predictive power of mate preferences as a function of mating market constraints. A = availability; C = competitiveness; M = mutual attraction. Error bars represent 95 percent confidence intervals.

predictive power of preferences. Intuitively, preference fulfillment and predictive power are strongest on average when mating markets are completely unconstrained and are weakest when mating markets are fully constrained. Overall, as in the trade-off model, the predictive power of preferences was more strongly impacted by constraints than was mate preference fulfillment. The single constraint with the largest impact was availability. Competitiveness did not have a main effect on preference fulfillment or predictive power, but it did interact with the other constraints, such that markets that were competitive and constrained by another dynamic showed significantly lower preference fulfillment and predictive power than markets that were constrained but uncompetitive. Finally, the fully constrained market showed an average predictive power of mate preferences of $r = 0.24$, more similar to Campbell et al.'s (2016) value of $r = 0.14$. This similarity is particularly striking given the absence of perceptual error, decision error, or measurement error in our agent-based model – each of which would act to further attenuate the relationship between individual mate preferences and the traits of chosen partners.

This model provides a first step toward allowing researchers to quantify the impact of mating market features on individual mating decisions and relationships. There are a number of ways in which this model could be improved. For example, the competitiveness of a mating market is a function of both the number of mates each person can accommodate and the extent to which the same mates are pursued by larger pools of potentials. Competition will be fiercer when attraction is concentrated on a small subset of mates than when it is spread more equivalently throughout the mating market.

The magnitude of the availability constraint can also vary continuously and is likely to be somewhat less strict than the availability constraint we imposed in this model. In real human mating markets, the mate preferences of one sex are the selection pressures of the opposite sex. Mate preferences therefore can and do shape the trait distributions of the opposite sex across generations (Buss & Barnes, 1986; Keller et al., 2013). This means that the trait distributions of one sex are not random with respect to the preference distributions of the opposite sex, as they were in our model. Models that incorporate such evolutionary processes alongside manipulations of market constraints will give more realistic estimates of the impacts markets can have on mating decisions.

Nonetheless, our model demonstrates empirically that realistic mating market dynamics can constrain the apparent effects of mate preferences on mating decisions, *even in markets where attraction and mate selection are entirely governed by preferences*. Individual agents acting on their preferences will trade off their preferences based on the mates available to them, outcompete one another for access to scarce preference-fulfilling mates, and turn away suitors who do not sufficiently fulfill their preferences. These dynamics are produced by the behavioral effects of preferences and ironically result in preferences having limited power in predicting the traits of chosen partners. This means that the weak effects of preferences observed in the prior literature (e.g., Campbell et al., 2016) do not indicate a small role of mate preferences in mating decisions. Rather, the weak predictive power of individual mate preferences considered one at a time suggest that mate preferences play an important role in motivating the mating behavior of individuals embedded within complex, dynamic mating markets. Researchers must use models that are capable of accounting for the three key constraints imposed by these markets – *mate availability, competitive mating markets*, and *mutual mate choice* – when interpreting the relationships between ideal mate preferences and downstream mating outcomes, such as attraction and mate selection.

29.4 CONCLUSIONS

Evolutionary psychologists have had notable success in documenting human mate preferences and their predictable variability across individuals, genders, contexts, and cultures. Research attempting to bridge the gap between these ideal preferences and actual mating outcomes, however, commonly comes up short. Mate preferences frequently show low power in predicting attraction to potential mates, as well as the traits of actually chosen partners (e.g., Campbell et al., 2016; Kurzban & Weeden, 2007). This shortcoming could emerge: (1) because ideal preferences actually play no role in mating behavior; (2) because research on mating outcomes is systematically methodologically limited; or (3) because mating researchers have conceptualized the nature of human mate preferences and mate selection inaccurately.

We propose that accurately appraising the role of mate preferences in downstream outcomes such as attraction and mate selection requires accounting for several realities of human mate selection. First, mate preferences appear to be integrated according to a nonlinear Euclidean algorithm (Conroy-Beam & Buss, 2016b; Conroy-Beam, Goetz, & Buss, 2016). This algorithm has several nonintuitive consequences for how mate preferences relate to estimates of mate value and feelings of attraction. This algorithm also explains several surprising findings in the extant literature, including interactions between mate preferences (e.g., Lee et al., 2014) and the poor power of preferences to predict attraction (e.g., Eastwick & Finkel, 2008). Second, the fact that humans hold many mate preferences inevitably forces trade-offs between preferences in mate choice. Accounting for these trade-offs in computer simulations causes mate preferences to yield predictive power similar to that documented in the mate choice literature. Third, the existence of multiple individuals attempting to select mates by acting on their preferences produces mating markets that impose several constraints when the ideal rubber hits the mate choice road. Experimentally manipulating these constraints within simulated markets where mate choice is fully determined by mate preferences yields precisely the relationships between preferences and mate choice that have been incorrectly interpreted to mean preferences have little behavioral power.

Altogether, we show that our multidimensional model of mate selection, which more fully matches the complexity of mate choice, is able to account for the puzzling findings in the mate choice literature. Individual mate preferences do not have strong power to predict mating outcomes such as attraction and mate selection. However, this low predictive power does not signify a lack of motivational power. Individual preferences have limited power in predicting mating outcomes because they are integrated with one another nonlinearly, force trade-offs against one another, and constrain one another when they guide behavior within larger mating markets. Mate preferences appear to do so little precisely because they do so much.

Our model has proven useful for conceptualizing and generating predictions about mate choice. Nonetheless it could be improved in several ways. Models of mating markets must include evolutionary forces in order to account for the effects of intersexual selection and intrasexual competition on the nature of preferences, traits, and their interrelationships. The Euclidean algorithm we employed for modeling preference integration also needs to be further refined. For example, at present, the Euclidean algorithm treats all preference dimensions as though they are equally important – an implausible assumption. The algorithm could be improved by allowing dimensions that are more important in mate choice to extend further in the multidimensional preference space.

Nonetheless, our novel model of mate selection both accounts for puzzling findings within the extant mating literature and provides researchers with tools for further understanding the links between ideal and real mating outcomes. Continuing to apply agent-based modeling and multidimensional analyses of mate preferences could prove useful for researchers hoping to understand the links between preferences and attraction, the roles preferences play in regulating ongoing relationships, and the ways in which human mating behavior shapes the nature of human mating markets and human evolution.

REFERENCES

Billeter, J. C., Atallah, J., Krupp, J. J., Millar, J. G., & Levine, J. D. (2009). Specialized cells tag sexual and species identity in Drosophila melanogaster. *Nature*, **461**(7266), 987–991.

Buss, D. M. (1989). Sex differences in human mate preferences: Evolutionary hypotheses tested in 37 cultures. *Behavioral and Brain Sciences*, **12**(1), 1–14.

Buss, D. M., & Barnes, M. (1986). Preferences in human mate selection. *Journal of Personality and Social Psychology*, **50**(3), 559–570.

Buss, D. M., Goetz, C., Duntley, J. D., Asao, K., & Conroy-Beam, D. (2017). The mate switching hypothesis. *Personality and Individual Differences*, **104**, 143–149.

Byrne, P. G., & Rice, W. R. (2006). Evidence for adaptive male mate choice in the fruit fly Drosophila melanogaster. *Proceedings of the Royal Society of London B: Biological Sciences*, **273**(1589), 917–922.

Campbell, L., Chin, K., & Stanton, S. C. (2016). Initial evidence that individuals form new relationships with partners that more closely match their ideal preferences. *Collabra*, **2**(1), 2.

Candolin, U. (2003). The use of multiple cues in mate choice. *Biological Reviews*, **78**(4), 575–595.

Conroy-Beam, D., & Buss, D. M. (2016a). Do mate preferences influence actual mating decisions? Evidence from computer simulations and three studies of mated couples. *Journal of Personality and Social Psychology*, **111**(1), 53–66.

Conroy-Beam, D., & Buss, D. M. (2016b). How are mate preferences linked with actual mate selection? Tests of mate preference integration algorithms using computer simulations and actual mating couples. *PLoS ONE*, **11**(6), e0156078.

Conroy-Beam, D., & Buss, D. M. (2019). Why is age so important in human mating? Evolved age preferences and their influences on multiple mating behaviors. *Evolutionary Behavioral Sciences*, **13**(2), 127–157.

Conroy-Beam, D., Buss, D. M., Pham, M. N., & Shackelford, T. K. (2015). How sexually dimorphic are human mate preferences? *Personality and Social Psychology Bulletin*, **41**(8), 1082–1093.

Conroy-Beam, D., Goetz, C. D., & Buss, D. M. (2016). What predicts romantic relationship satisfaction and mate retention intensity: Mate preference fulfillment or mate value discrepancies? *Evolution and Human Behavior*, **37**(6), 440–448.

DeBruine, L. M., Jones, B. C., Crawford, J. R., Welling, L. L., & Little, A. C. (2010). The health of a nation predicts their mate preferences: Cross-cultural variation in women's preferences for masculinized male faces. *Proceedings of the Royal Society of London B: Biological Sciences*, **277**(1692), 2405–2410.

Del Giudice, M., Booth, T., & Irwing, P. (2012). The distance between Mars and Venus: Measuring global sex differences in personality. *PLoS ONE*, **7**(1), e29265.

Eastwick, P. W., & Finkel, E. J. (2008). Sex differences in mate preferences revisited: Do people know what they initially desire in a romantic partner? *Journal of Personality and Social Psychology*, **94**(2), 245–264.

Eastwick, P. W., Luchies, L. B., Finkel, E. J., & Hunt, L. L. (2014). The predictive validity of ideal partner preferences: A review and meta-analysis. *Psychological Bulletin*, **140**(3), 623–665.

Farrelly, D., Clemson, P., & Guthrie, M. (2016). Are women's mate preferences for altruism also influenced by physical attractiveness? *Evolutionary Psychology*, **14**(1), 1–6.

Fletcher, G. J., Simpson, J. A., Thomas, G., & Giles, L. (1999). Ideals in intimate relationships. *Journal of Personality and Social Psychology*, **76**(1), 72–89.

Gangestad, S. W., & Buss, D. M. (1993). Pathogen prevalence and human mate preferences. *Ethology and sociobiology*, **14**(2), 89–96.

Goetz, C. D., Easton, J. A., Lewis, D. M., & Buss, D. M. (2012). Sexual exploitability: Observable cues and their link to sexual attraction. *Evolution and Human Behavior*, **33**(4), 417–426.

Gottschall, J., Martin, J., Quish, H., & Rea, J. (2004). Sex differences in mate choice criteria are reflected in folktales from around the world and in historical European literature. *Evolution and Human Behavior*, **25**(2), 102–112.

Hatfield, E., & Sprecher, S. (1995). Men's and women's preferences in marital partners in the United States, Russia, and Japan. *Journal of Cross-Cultural Psychology*, **26**(6), 728–750.

Hill, R. (1945). Campus values in mate selection. *Journal of Home Economics*, **37**, 554–558.

Keller, M. C., Garver-Apgar, C. E., Wright, M. J., et al. (2013). The genetic correlation between height and IQ: Shared genes or assortative mating? *PLoS Genetics*, **9**(4), e1003451.

Kenrick, D. T., & Keefe, R. C. (1992). Age preferences in mates reflect sex differences in human reproductive strategies. *Behavioral and Brain Sciences*, **15**(1), 75–91.

Kenrick, D. T., Sadalla, E. K., Groth, G., & Trost, M. R. (1990). Evolution, traits, and the stages of human courtship: Qualifying the parental investment model. *Journal of Personality*, **58**(1), 97–116.

Kurzban, R., & Weeden, J. (2007). Do advertised preferences predict the behavior of speed daters? *Personal Relationships*, **14**(4), 623–632.

Lee, A. J., Dubbs, S. L., Von Hippel, W., Brooks, R. C., & Zietsch, B. P. (2014). A multivariate approach to human mate preferences. *Evolution and Human Behavior*, **35**(3), 193–203.

Lewis, D. M., Russell, E. M., Al-Shawaf, L., & Buss, D. M. (2015). Lumbar curvature: A previously undiscovered standard of attractiveness. *Evolution and Human Behavior*, **36**(5), 345–350.

Li, N. P., Bailey, J. M., Kenrick, D. T., & Linsenmeier, J. A. (2002). The necessities and luxuries of mate preferences: Testing the tradeoffs. *Journal of Personality and Social Psychology*, **82**(6), 947–955.

Li, N. P., Yong, J. C., Tov, W., et al. (2013). Mate preferences do predict attraction and choices in the early stages of mate selection. *Journal of Personality and Social Psychology*, **105**(5), 757–776.

Low, B. S. (1988). Measures of polygyny in humans. *Current Anthropology*, **29**(1), 189–194.

Low, B. S. (1991). Reproductive life in nineteenth century Sweden: An evolutionary perspective on demographic phenomena. *Ethology and Sociobiology*, **12**(6), 411–448.

Loyau, A., Saint Jalme, M., Cagniant, C., & Sorci, G. (2005). Multiple sexual advertisements honestly reflect health status in peacocks (*Pavo cristatus*). *Behavioral Ecology and Sociobiology*, **58**(6), 552–557.

Lüpold, S., Manier, M. K., Ala-Honkola, O., Belote, J. M., & Pitnick, S. (2011). Male *Drosophila melanogaster* adjust ejaculate size based on female mating status, fecundity, and age. *Behavioral Ecology*, **22**(1), 184–191.

Lykken, D. T., & Tellegen, A. (1993). Is human mating adventitious or the result of lawful choice? A twin study of mate selection. *Journal of Personality and Social Psychology*, **65**(1), 56–68.

Marlowe, F. W. (2004). Mate preferences among Hadza hunter-gatherers. *Human Nature*, **15**(4), 365–376.

Schmitt, D. P. (2004). Patterns and universals of mate poaching across 53 nations: The effects of sex, culture, and personality on romantically attracting another person's partner. *Journal of Personality and Social Psychology*, **86**(4), 560–584.

Schmitt, D. P., & Buss, D. M. (2001). Human mate poaching: Tactics and temptations for infiltrating existing mateships. *Journal of Personality and Social Psychology*, **80**(6), 894–917.

Sugiyama, L. S. (2004). Is beauty in the context-sensitive adaptations of the beholder?: Shiwiar use of waist-to-hip ratio in assessments of female mate value. *Evolution and Human Behavior*, **25**(1), 51–62.

Todd, P. M., Penke, L., Fasolo, B., & Lenton, A. P. (2007). Different cognitive processes underlie human mate choices and mate preferences. *Proceedings of the National Academy of Sciences*, **104**(38), 15011–15016.

Townsend, J. M., & Levy, G. D. (1990). Effects of potential partners' physical attractiveness and socioeconomic status on sexuality and partner selection. *Archives of Sexual Behavior*, **19**(2), 149–164.

Wiederman, M. W. (1993). Evolved gender differences in mate preferences: Evidence from personal advertisements. *Ethology and Sociobiology*, **14**(5), 331–351.

30 Kin Selection and the Evolution of Male Androphilia

PAUL L. VASEY, LANNA J. PETTERSON, SCOTT W. SEMENYNA, FRANCISCO R. GÓMEZ JIMÉNEZ, AND DOUG P. VANDERLAAN

30.1 MALE ANDROPHILIA IS AN EVOLUTIONARY PARADOX

Androphilia refers to sexual attraction and arousal to adult males, whereas *gynephilia* refers to sexual attraction and arousal to adult females. Male androphilia is considered one of the outstanding paradoxes of evolutionary biology because its very existence flouts our expectations concerning what constitutes an evolutionarily viable trait (Bailey & Zuk, 2009). In humans, male androphilia is heritable, as evinced by twin studies (Alanko et al., 2010; Bailey et al., 2000; Kendler et al., 2000; Långström et al., 2010), as well as research in the area of molecular genetics (Hamer et al., 1993; Mustanski et al., 2005; Sanders et al., 2015). Despite the heritability of this trait, androphilic males reproduce at *far* lower rates when compared to gynephilic males, if they reproduce at all, which, very often, they do not (e.g., Bell & Weinberg, 1978; King et al., 2005; Saghir & Robins, 1973; Schwartz et al., 2010). According to the logic of Darwinian theory, a heritable trait that reduces reproduction should quickly become extinct when it exists side by side with alternative phenotypes that facilitate significantly higher reproduction. In this sense, male androphilia represents a profound evolutionary conundrum.

This claim might be dismissed if it could be demonstrated that male androphilia is a historically recent phenomenon that did not exist in the evolutionary past. However, ancient texts suggest that male androphilia has existed for well over two millennia (e.g., Crompton, 2003; Deakin, 1965; Sweet & Zwilling, 1993). Prehistoric rock art, which some scholars interpret as depictions of male homosexual activity, dates back even further to the Mesolithic era (15,000–5,000 BP) (e.g., Taylor, 1996, p. 145). Although male same-sex sexual behavior is not necessarily indicative of male androphilia, it is suggestive of the fact.

By sampling statistically independent societies, Hames et al. (2017) demonstrated that male androphilia is present in most cultural regions of the world. In those cultural regions where it has not been documented, there is – more often than not – an absence of evidence as opposed to any explicit statements on the part of ethnographers

that male androphilia is truly absent. On the basis of this evidence, we suggest that male androphilia is a *context-independent universal* (Brown, 1991) that occurs regardless of sociocultural context so long as the population in question is large enough for this low-frequency trait – which occurs in fewer than six percent of males – to be expressed. No matter how rigorously a society attempts to eliminate it, male androphilia emerges (e.g., "G," 1980; Schuvaloff, 1976). Accordingly, claims that male androphilia is entirely absent in a population (e.g., Hewlett & Hewlett, 2010) should be viewed with skepticism. This is in no way meant to impugn the field skills of the ethnographers in question, but merely to underscore that establishing facts about sexual orientation is complicated enough in the relatively open contemporary Western world, where representative surveys and even objective measures can be employed. Doing so in non-Western cultures where same-sex sexuality is often taboo can be much more difficult. In light of these considerations, it is important to recognize the limitations of the data and to have a healthy skepticism regarding surprising claims that would seem to contradict otherwise well-established patterns.

In sum, male androphilia is influenced by genetic factors, yet it reduces reproduction. Nonetheless, it appears to have persisted over evolutionary time and reliably occurs across diverse cultural contexts. The persistence of the genetic factors underlying male androphilia from one generation to the next requires explanation when viewed from the perspective of Darwinian selection, a process that favors the evolution of reproductively viable traits.

30.2 MALE ANDROPHILIA IN CROSS-CULTURAL PERSPECTIVE

Any attempt to understand male androphilia from an evolutionary perspective requires that we recognize that the manner in which this trait is publically expressed varies cross-culturally. Cross-cultural elaborations of this trait are not, however, infinite. Rather, they take on only two primary forms, which are related to gender role enactment and gender identity. These two forms include *cisgender* and *transgender* male androphilia. *Cisgender male androphiles* usually occupy the gender role that is typical for their sex and they behave in a relatively masculine manner. In contrast, *transgender androphilic males* often occupy gender roles that are atypical for their sex and they behave in an effeminate manner.

It is important to note that regardless of how male androphilia is expressed, it is associated with gender atypicality in both childhood and adulthood. The strength of this association varies, however, depending on the form that male androphilia takes. Cisgender androphilic males are relatively feminine as boys compared to their gynephilic counterparts (Bailey & Zucker, 1995), but they behaviorally defeminize to varying degrees as they develop. It has been suggested that this behavioral defeminization probably occurs in response to culturally specific gender role expectations, which hold that male-bodied individuals should behave in a masculine manner (Bailey, 2003; Rieger & Savin-Williams, 2012). In contrast, in cultures where transgender male androphilia is the norm, feminine boys develop into feminine adult males. Consequently, adult cisgender androphilic males are relatively masculine when compared to adult transgender androphilic males. Conversely, on average, both are relatively feminine when compared to adult male gynephiles (Bailey, 2003; Lippa, 2005). The effect sizes associated with male sexual orientation differences in childhood and adulthood gender atypicality tend to be large by the standards of most behavioral science research. This holds true regardless of whether one compares gynephilic males with transgender androphilic males or cisgender ones (Bailey, 2003; Bailey & Zucker, 1995; Lippa, 2005).

Both cisgender and transgender male androphilia may occur within a given culture, but typically one or the other tends to predominate (Whitam & Mathy, 1986). For example, the cisgender form is more common in many Western cultures, where they are known as *gay men*. When the transgender form does occur in Western cultures, such individuals are often identified – and often self-identify – as *transwomen*.

Outside the West, the transgender form appears to be more common. In some of these cultures, androphilic males are recognized by themselves and others as being neither "men" nor "women," but rather as members of a "third" gender (Nanda, 2014). Contemporary examples include, but are by no means limited to, Navajo *nádleeh* in the southwestern USA (Thomas, 1997), the *kathoey* of Thailand (Totman, 2003), the *hijra* of India (Nanda, 1998), the *xanith* of Oman (Wikan, 1977), the *muxe* of Mexico (Mirandé, 2017), the *'yan daudu* of Nigeria (Gaudio, 2009), and the *fa'afafine* of Samoa (Vasey & VanderLaan, 2014). Although cisgender androphilic males are recognized in some cultures as third-gender males, the typical pattern seen cross-culturally is that third-gender categories are occupied by transgender androphilic males.

30.3 THE SEXUAL PARTNERS OF MALE ANDROPHILES

Most cisgender androphilic males express aversion toward sexual activity with feminine males (Bailey et al., 1994; Petterson et al., 2018). Consequently, cisgender androphilic males typically seek out each other for sexual interactions and romantic relationships. These relationships have been called "egalitarian" because the partners, who are both postpubertal, tend to not be markedly different in age (i.e., age differences generally do not exceed one generation). In addition, partners in egalitarian relationships both tend to be relatively masculine and, as such, do not differ from each other in terms of gender-related characteristics. Within such relationships, partners do not adopt special social roles and they treat each other as social equals (Adams, 1986). The classic example, which most readers will be familiar with, is, of course, gay male couples.

In contrast to gay men, it is exceedingly rare that transgender androphilic males seek out each other for sexual interactions or romantic relationships. Instead, these individuals engage in sexual/romantic relationships with masculine males who self-identify – and are identified by others – as "men" (Petterson et al., 2015, 2016, 2018; Whitam, 1992). These relationships are sometimes referred to as "gender stratified" or "heterogendered." As such, transgender androphilic males and their cisgender male sexual partners are differentiated by gender-related characteristics. Furthermore, they sometimes adopt special social roles relative to each other and consequently they do not treat each other as social equals (Adams, 1986).

Very little work has been conducted on the masculine male sexual partners of transgender androphilic males. However, using both subjective (i.e., self-report) and objective (i.e., viewing time) measures, Petterson et al. (2015) found that Samoan men who engage in sexual interactions with *fa'afafine* demonstrated a unique, relatively bisexual pattern of sexual attraction intermediate to the androphilic one demonstrated by *fa'afafine* and the gynephilic one demonstrated by men who only engage in sexual interactions with women. Further, Petterson et al. (2016) found that activity role during oral sex distinguishes different degrees of bisexuality in these men, with those who fellate their *fa'afafine* partners falling more toward the androphilic extreme of bisexuality and those who are only fellated falling more toward the gynephilic extreme.

Although both bisexual and androphilic males exhibit same-sex sexual attraction, the two should not be conceptualized as synonymous from either a theoretical or empirical standpoint. From an evolutionary perspective, the former is much less of a conundrum than the latter, because bisexual males can and do reproduce, whereas exclusively androphilic males typically do not. This chapter focuses on elucidating the larger evolutionary paradox that is *exclusive* male androphilia.

30.4 CROSS-CULTURALLY INVARIANT PROPERTIES OF MALE ANDROPHILIA

Highlighting similarities between androphilic natal males who are transgender and those who are cisgender has become opprobrium in the West, where the latter are known as "transgender women" and the former are known as "gay men" (for those readers interested in an introduction to the politics of research on sexual orientation and gender diversity, see Bailey & Triea, 2007; Berling, 2001; Dreger, 2008). Nonetheless, the existing data – and they are substantial – indicate that these two forms of male androphilia are cultural variants of the same trait.

Cisgender and transgender androphilic males share numerous biodemographic correlates cross-culturally. For example, both tend to be laterborn among their siblings, have a greater number of older biological brothers (the "fraternal birth order effect"[1]; e.g., Blanchard, 2018; Blanchard & VanderLaan, 2015; Bozkurt et al., 2015; Semenyna et al., 2017c; VanderLaan & Vasey, 2011; VanderLaan et al., 2017a; Vasey & VanderLaan, 2007), cluster within families (e.g., Gómez et al., 2018; Gómez-Gil et al., 2011; Schwartz et al., 2010; Semenyna et al., 2017b; VanderLaan et al., 2013a, 2013c; Winter, 2006), occur at similar prevalence rates across different populations (e.g., approximately two to six percent: Gates, 2011; Gómez et al., 2018; Semenyna et al., 2017b; VanderLaan et al., 2013a; Whitam & Mathy, 1986), and exhibit little or no reproductive success (e.g., Nila et al., 2018; Schwartz et al., 2010; Vasey et al., 2014). In addition, the odds ratios associated with the fraternal birth order effect in various populations of cisgender and transgender male androphiles are remarkably consistent, suggesting that the manner in which older brothers influence the development of male androphilia is constant across diverse populations (e.g., Cantor, 2002; VanderLaan & Vasey, 2011; VanderLaan et al., 2017a).

Cisgender and transgender androphilic males also share various psychodevelopmental correlates cross-culturally. For example, compared to their gynephilic male counterparts, both tend to exhibit greater levels of childhood female-typical behavior (e.g., nurturing play with dolls), lower levels of childhood male-typical behavior (e.g., rough-and-tumble play), and elevated cross-sex beliefs and wishes in childhood (e.g., "I wish I was a girl"; e.g., Bailey & Zucker, 1995; Bartlett & Vasey, 2006; Besharat et al., 2016; Cardoso, 2005, 2009; Green, 1987; Petterson et al., 2017; Rieger et al., 2008; Steensma et al., 2013; Vasey & Bartlett, 2007; Whitam & Mathy, 1986). In adulthood, both forms tend to exhibit elevated female-typical occupational interests (e.g., Hart, 1968; Lippa, 2005; Semenyna & Vasey, 2016; Whitam & Mathy, 1986; Zheng et al., 2011).

Finally, experimental studies indicate that transgender and cisgender male androphiles exhibit virtually identical viewing time patterns when presented with images of men and women (Petterson et al., 2018). Taken together, these cross-culturally universal biodemographic, psychodevelopmental, and behavioral–cognitive correlates of male androphilia indicate that cisgender and transgender androphilic males share a common biological etiology despite being superficially different in appearance. This in no way diminishes the important differences that exist between cisgender and transgender androphilic males in terms of appearance, comportment, and self-concept. However, on the basis of the substantial evidence presented above, we contend that such differences are due to distinct cultural processes, not biological ones.

30.5 MALE ANDROPHILIA IN THE ANCESTRAL ENVIRONMENT

Given that there are two forms of male androphilia that occur cross-culturally, the question arises as to which one – cisgender or transgender – is ancestral to the other. This question is a critical one for anyone who seeks to accurately test hypotheses pertaining to the evolution of this trait. More derived forms of this trait may reflect historically recent cultural influences, which may obscure the outcome of evolutionary process and, as such, represent suboptimal models when testing evolutionary hypotheses.

Numerous researchers have presented evidence indicating that the ancestral human sociocultural environment was likely characterized by hunter–gatherers living in small groups with relatively egalitarian sociopolitical structures and shamanistic belief systems (e.g., Binford, 2001; Hill et al., 2011; Marlowe, 2005; McBrearty & Brooks, 2000; Sanderson & Roberts, 2008; Smith, 1999; Winkelman, 2010; Woodburn, 1982). If it could be demonstrated that these conditions are more often associated with one form of male androphilia or the other, then this would furnish important clues as to which of the two forms was the ancestral one.

With these concerns in mind, VanderLaan et al. (2013b) compared societies in which transgender androphilic males predominate with those in which they do not using information derived from the Standard Cross-Cultural Sample (SCCS). The SCCS provides data related to a subset of the world's nonindustrial societies. These societies are relatively independent from one another in terms of linguistic relationship, cultural resemblance, and geographic continuity. As such, the SCCS provides a sample of world societies that reduces the problem of nonindependence due to cultural diffusion or common cultural derivation (i.e., Galton's problem).

Compared to non-transgender societies, transgender societies were characterized by a significantly greater presence of ancestral sociocultural conditions (i.e., smaller group size, hunting and gathering, egalitarian political structure, and shamanistic religious beliefs). Given this, VanderLaan et al. (2013b) concluded that the ancestral

---

[1] The *fraternal birth order effect* refers to the well-established finding that the number of older biological brothers increases the odds of androphilia in laterborn males (Blanchard, 2018).

form of male androphilia was likely the transgender one. As such, compared to their cisgender counterparts, transgender androphilic males likely represent better models for testing hypotheses about the evolution of male androphilia. This is because by using such models we are more apt to capture the evolutionary processes that gave rise to this trait.

30.6 *FA'AFAFINE*: THIRD-GENDER MALES IN SAMOA

Given that the ancestral form of male androphilia is likely the transgender form, the remainder of this chapter emphasizes our research on third-gender males. Since 2003, we have conducted research annually (or more often than not biannually) in Samoa, a Polynesian island nation in the south Pacific where male androphilia is predominately expressed in the transgender form by individuals who are known locally as a third gender: *fa'afafine*. *Fa'afafine* enjoy a relatively high level of acceptance in Samoa that, while not absolute, stands in stark contrast to the situation experienced by Western transgender male androphiles (e.g., Namaste, 2000). They are highly visible and active members of their society. For example, *fa'afafine* occupy all manner of positions, from stay-at-home caregivers to Assistant Chief Executive Officers in the government. The Prime Minister of Samoa, the Honorable Tuilaepa Sailele Malielegaoi, is Patron of the National *Fa'afafine* Association and has spoken publically on many occasions about the value of *fa'afafine* for Samoan society.

30.7 KIN SELECTION AND THE EVOLUTION OF MALE ANDROPHILIA

The kin selection hypothesis has been proposed as one solution to the evolutionary conundrum of male androphilia. It holds that genes for male androphilia could be maintained in a population if enhancing one's indirect fitness offset the cost of not reproducing directly (Wilson, 1975). Indirect fitness is a measure of an individual's impact on the fitness of kin (who share some identical genes by virtue of descent), weighted by the degree of relatedness (Hamilton, 1963). Theoretically speaking, androphilic males can increase their indirect fitness by channeling altruistic behavior toward kin, which, in principle, then allows kin to increase their reproductive success.

The most basic prediction that flows from the kin selection hypothesis is that androphilic males should exhibit elevated kin-directed altruism compared to gynephilic males and androphilic females. No support for this basic prediction has been garnered among cisgender androphilic male ("gay men") participants. This is true for both Western (i.e., USA, UK, Canada, Spain, and Italy: Abild et al., 2014; Bobrow & Bailey, 2001; Camperio Ciani et al., 2016; Forrester et al., 2011; Rahman & Hull, 2005; VanderLaan et al., 2016) and non-Western cultures (i.e., Japan: Vasey & VanderLaan, 2012).

In light of this evidence, it is striking that the kin selection hypothesis has been repeatedly supported in Samoa, where *transgender* androphilic males predominate and are recognized as a third gender (i.e., *fa'afafine*). Multiple studies have shown that the avuncular tendencies (i.e., uncle-like willingness to allocate resources toward nieces and nephews) of *fa'afafine* are significantly elevated compared to those of Samoan gynephilic men (VanderLaan & Vasey, 2012; Vasey & VanderLaan, 2010b; Vasey et al., 2007). In addition, *fa'afafine* exhibit significantly elevated avuncular tendencies compared to the materteral (aunt-like) tendencies of Samoan mothers (VanderLaan et al., 2017b; Vasey & VanderLaan, 2009) and childless women (Vasey & VanderLaan, 2009), although this latter effect requires further replication.

Research demonstrates that *fa'afafine*'s elevated avuncularity – as opposed to the lack thereof among gay men in other cultures – does not appear to be attributable to any of the following: socially desirable responding on the part of *fa'afafine* (Vasey & VanderLaan, 2010a); greater social acceptance of same-sex sexuality in Samoa compared to other cultures (Forrester et al., 2011); lack of direct parental responsibilities (Vasey & VanderLaan, 2009, 2010b); close geographic proximity to family (Abild et al., 2014); a collectivistic cultural context (Vasey & VanderLaan, 2012); taking on the childcare role of women as part of their "third-gender" status (Vasey & VanderLaan, 2009); having more time to allocate to nieces and nephews due to a lack sexual/romantic relationship involvement (VanderLaan & Vasey, 2012); and, (trans)gender role expectations that *fa'afafine* will contribute more than other family members to caring for nieces and nephews (VanderLaan et al., 2015a).

It should be clear from the research described above that much of our work has focused on falsifying the kin selection hypothesis for male androphilia by examining alternative explanations that might account for *fa'afafine*'s elevated avuncularity. It should be equally clear that none of the alternative explanations we have tested to date have been supported. Taken together, this body of work is consistent with the conclusion that elevated avuncularity by androphilic males is an adaptation that evolved via kin selection. That being said, establishing that a given trait is an adaptation involves repeatedly satisfying adaptive design criteria empirically while simultaneously ruling out alternatives (Buss et al., 1998). Adaptive design implies complexity, economy, efficiency, reliability, precision, and functionality (Williams, 1966).

30.8 EVIDENCE OF ADAPTIVE DESIGN IN THE AVUNCULAR COGNITION OF SAMOAN *FA'AFAFINE*

In addition to examining alternative explanations for the unique Samoan findings, a number of our studies have tested more refined predictions derived from the kin selection hypothesis to assess whether the avuncular cognition of *fa'afafine* shows evidence of having been shaped by kin selection. These studies have produced a number of findings consistent with the conclusion that *fa'afafine*'s avuncular cognition exhibits hallmarks of adaptive design.

First, whereas Samoan men and women decrease their willingness to invest in nieces and nephews when they are in a sexual/romantic relationship, the cognition of *fa'afafine* appears to protect against this tendency by maintaining a high level of willingness to invest in nieces and nephews regardless of their relationship status (VanderLaan & Vasey, 2012).

Second, in light of the issue of paternity certainty, one can always be more confident in the genetic relatedness of sisters' children than brothers' children, and so androphilic males should favor investing in the former to maximize indirect fitness. Compared to Samoan men and women, *fa'afafine* show a bias to invest in their sisters' children, particularly when the consequences of investment are nontrivial (e.g., paying for school or medical expenses; VanderLaan & Vasey, 2014). Additional research demonstrates that *fa'afafine*'s avuncularity is more contingent on the existence of older sisters than it is on that of older brothers (VanderLaan & Vasey, 2013).

Third, because younger children are especially vulnerable to mortality, an efficient means of maximizing indirect fitness would be to bias investment toward younger nieces and nephews. Compared to Samoan men and women, *fa'afafine* are more likely to show such a bias (VanderLaan & Vasey, 2014), and this also translates into behavioral differences in terms of the allocation of monetary resources (Vasey & VanderLaan, 2010c).

Fourth, as interest in investing in nieces and nephews increases, so too does interest in investing in children more generally, including nonkin children. Consequently, dissociating one's interest toward investing in nieces and nephews versus nonkin children would allow one to maximize investment in kin. Compared to Samoan men and women, *fa'afafine* show a greater dissociation between these domains (Vasey & VanderLaan, 2010a). It is noteworthy that, compared to heterosexual men, gay men in Canada also show this greater dissociation (Forrester et al., 2011). This suggests that the avuncular cognition of cisgender androphilic males might also exhibit hallmarks of adaptive design due to kin selection, even though more direct support for the kin selection hypothesis in the form of elevated avuncular tendencies is lacking among this group.

30.9 A PROXIMATE MECHANISM FOR ELEVATED KIN-DIRECTED ALTRUISM BY SAMOAN *FA'AFAFINE*

Williams (1992) hypothesized that third-gender males in many non-Western cultures excel at various labor practices as a way of striving for prestige within their families and communities. He argued that one consequence of this pattern of prestige acquisition was that third-gender males sometimes behave in a competitive manner when engaged in labor. Consistent with this possibility, Williams (1992) presented numerous examples from various native North American cultures. Other examples can be found throughout the ethnographic literature in which third-gender males are described by themselves and by the cisgender members of their own groups as "better" than men, women, or both when performing particular tasks. For example, Balinese *bantong banci* (i.e., feminine, same-sex-attracted males) are described as especially skilled at making religious offerings (Duff-Cooper, 1985). Similarly, the Hausa *'yan dandu* of Nigeria are seen as particularly adept intermediaries between men and women, because it is thought that their status as feminine males allows them to interact and relate well with *both* cisgender women and men (Pittin, 1983).

In light of this literature, it is possible that *fa'afafine*'s elevated avuncular tendencies are motivated by a greater desire to strive for prestige within their families compared to men and women. Semenyna and Vasey (2019) tested this possibility and demonstrated that Samoan cisgender men and women do not differ in the degree to which they seek familial recognition of their altruistic behavior toward nieces and nephews. However, compared to men, *fa'afafine* placed greater importance on acknowledgment by kin of their avuncularity. These results illustrate one potential proximate cognitive mechanism for the elevated kin-directed altruism of *fa'afafine*.

30.10 CHILDHOOD SEPARATION ANXIETY AND THE DEVELOPMENT OF ELEVATED KIN-DIRECTED ALTRUISM

In addition to the psychodevelopmental correlates of male androphilia outlined in Section 30.4, research indicates that both types of androphilic males exhibit elevated indicators of childhood separation anxiety. Childhood separation anxiety refers to developmentally inappropriate and excessive anxiety related to separation (or impending separation) from major attachment figures, such as the primary caretaker or a close family member (American Psychiatric Association, 2013). This condition appears to be a female-typical trait as it is more commonly manifested among girls compared to boys (Bowen et al., 1990; Shear et al., 2006).

Clinical research conducted in Canada indicates that highly feminine boys – the majority of whom will grow up to be androphilic men (Green, 1987; Singh, 2012) – exhibit more indicators of separation anxiety relative to controls (Coates & Person, 1985; Zucker et al., 1996). Retrospective research conducted on nonclinical populations in different cultures corroborates previous prospective, clinical findings. In Canada, adult cisgender androphilic males recalled elevated indicators of childhood separation anxiety when compared to their gynephilic counterparts (VanderLaan et al., 2011a, 2015b, 2016). Similarly, in Samoa, adult *fa'afafine* recalled elevated indicators of childhood separation anxiety when compared to Samoan gynephilic men and androphilic women (VanderLaan et al., 2017b; Vasey et al., 2011).

Evidence for elevated recalled indicators of childhood separation anxiety in androphilic males across two highly disparate cultures such as Canada and Samoa suggests that this type of anxiety, when it occurs at nonclinical levels in childhood, is a cross-culturally invariant aspect of psychosocial development in androphilic males. This relationship exists regardless of whether male androphilia manifests in the cisgender or the transgender form. Once again, this suggests that the two forms of male androphilia share a similar etiology despite being different in terms of gender role presentation and identity. It is interesting to note in the context of this discussion that links between male same-sex sexual preference and anxiety have even been reported in other species. Male rats who prefer to engage in sexual activity with same-sex partners exhibit heightened anxiety in elevated maze and forced swim tests compared to their opposite-sex-attracted counterparts (García-Cárdenas et al., 2015; Hernández & Fernández-Guasti, 2018).

Because separation anxiety is characterized by distress experienced during periods of separation from major attachment figures (who are invariably close kin such as parents and siblings), elevated separation anxiety might reflect heightened kin attachment among androphilic males. Indeed, retrospective research conducted in Canada and in Samoa is consistent with this possibility and demonstrates that androphilic males' childhood separation anxiety is driven primarily by a concern about the well-being of kin during periods of separation (e.g., worrying kin will be hurt in an accident) as opposed to other circumstances involving separation (e.g., having to go to school) (VanderLaan et al., 2015b, 2016, 2017b). This pattern is consistent with the kin selection hypothesis in that worrying about the well-being of kin would prompt behaviors to secure their well-being, and thus improve the ability of kin to contribute to one's indirect fitness. In light of this literature, the question arises as to why Canadian cisgender androphilic males would exhibit elevated concern for the well-being of kin in childhood but then fail to do so in adulthood.

30.11 THE ADAPTIVE FEMININE PHENOTYPE MODEL

The *adaptive feminine phenotype model* attempts to reconcile the mixed cross-cultural and developmental research bearing on the kin selection hypothesis (VanderLaan et al., 2016). According to this model, in childhood, concern for one's kin manifests as elevated separation anxiety in (pre) androphilic boys and is part of an overall pattern of feminine behavior. Later, in adulthood, concern for one's kin

is expressed as elevated kin-directed altruism by adult male androphiles. As such, childhood separation anxiety that is motivated by concern for the well-being of kin is conceptualized as a developmental precursor of elevated kin-directed altruism, such as avuncularity, in adulthood (VanderLaan et al., 2011b). However, the model stipulates that adult patterns of elevated kin-directed altruism are contingent on the continued expression of femininity. As such, elevated traits of childhood separation anxiety are predicted to occur in all (pre)androphilic boys, regardless of their cultural milieu. In contrast, elevated kin-directed altruism in adulthood is predicted to occur in transgender androphilic males such as *fa'afafine*, who are feminine, but not in cisgender androphilic males such as gay men, who present publicly in a relatively masculine manner.

Diverse lines of evidence are consistent with the adaptive feminine phenotype model. In childhood, elevated cross-gender behavior and identification characterizes both cisgender (e.g., Bailey & Zucker, 1995) and transgender androphilic males (e.g., Bartlett & Vasey, 2006; Whitam, 1997) compared to their gynephilic counterparts. Consistent with the adaptive feminine phenotype model, both types of androphilic males exhibit elevated childhood separation anxiety when compared to gynephilic males (Gómez et al., 2017; VanderLaan et al., 2011a; Vasey et al., 2011). In both types, this anxiety is primarily motivated by concern about the well-being of kin (VanderLaan, 2015b, 2016, 2017b). Moreover, in both cisgender and transgender androphilic males, childhood separation anxiety is positively associated with childhood female-typical behavior (VanderLaan et al., 2011a, 2015b, 2017b).

In adulthood, Samoan *fa'afafine* maintain their marked femininity. In contrast, cisgender androphilic "gay" men behaviorally defeminize as they transition into adulthood (Bailey, 2003) and become much more masculine in comparison to *fa'afafine*. Consistent with the predictions of the adaptive feminine phenotype model, Samoan *fa'afafine* exhibit elevated avuncularity in adulthood (Vasey & VanderLaan, 2009; 2010b, 2010c; Vasey et al., 2007), but Canadian gay men do not (Abild et al., 2014; Forrester et al., 2011; VanderLaan et al., 2016). Among Samoan *fa'afafine*, childhood concern about the well-being of siblings is positively associated with adulthood avuncularity (VanderLaan et al., 2017b), which is consistent with the suggestion by VanderLaan et al. (2011b) that the former is a developmental precursor of the later.

Although Canadian gay men do not exhibit elevated kin-directed altruism, research by VanderLaan et al. (2015b, 2016) demonstrates that their lifespan expression of kin-directed altruism is linked with their lifespan expression of feminine gender expression. Feminine gender expression is not only associated with childhood separation anxiety in Canadian gay men (see Section 30.10), but it is also associated with their willingness in adulthood to invest in nieces and nephews (VanderLaan et al., 2016). Moreover, kin-directed altruism, as measured by degree of childhood concern for parents' well-being and adulthood avuncular tendencies, decreases from childhood to adulthood more dramatically in androphilic men compared to opposite-sex-attracted men and women, and this decrease corresponds to a concomitant decrease in femininity (VanderLaan et al., 2016).

The plausibility of the adaptive feminine phenotype model is further supported by additional lines of research. To begin with, if kin selection produced an androphilic male phenotype that exhibits elevated kin-directed altruism, but only in combination with feminine gender expression, then there should be a close genetic association between these traits. Indeed, twin studies have indicated that common genetic factors underlie same-sex attraction and gender-atypical behavior (Alanko et al., 2010), as well as neuroticism, which is conceptually similar to anxiety (Zietsch et al., 2011). Furthermore, findings from a mouse model showed that Xq28, a region of the X chromosome linked to human male sexual orientation (Hamer et al., 1993; Sanders et al., 2015), is associated with anxious behavioral patterns (i.e., less exploratory behavior) using open field, elevated maze, and light–dark box paradigms (Samaco et al., 2012). Thus, there appears to be some cursory evidence to suggest that common genetic factors underlie the psychological characteristics investigated in the current study.

In addition, if this model is accurate, then androphilic males would have had to exhibit feminine gender expression in the ancestral environment. As described in Section 30.5, transgender androphilic males were likely present in ancestral human societies (VanderLaan et al., 2013b). According to the adaptive feminine phenotype model, the femininity characterizing these males would have facilitated elevated kin-directed altruism.

Although the proximate mechanism(s) by which behavioral femininity facilitates such a disposition toward kin-directed altruism is presently unclear, a meta-analytic review of the cross-cultural and historical literature indicates that female kin are more important than male kin, including fathers, to the survival of offspring (Sear & Mace, 2008). As such, being a female-typical individual may in some way increase an androphilic male's propensity to engage in behaviors that ultimately benefit indirect fitness by enhancing the survival and subsequent reproductive success of nieces and nephews.

30.12 LIMITATIONS AND FUTURE DIRECTIONS

The primary criticism of the kin selection hypothesis concerns its ability to account for the persistence of genetic factors underpinning male androphilia. Some have argued that kin-directed altruism on the part of androphilic males cannot be sufficient to offset the reproductive costs associated with male androphilia (Bailey, 2003; Dickemann, 1995; LeVay, 2016). Specifically, because nieces and nephews are only half as related to oneself as direct offspring, kin-directed altruism would have to result in two nieces and nephews for every offspring foregone. The idea

that this level of kin-directed altruism is possible has been viewed with considerable skepticism, casting doubt on the viability of the kin selection hypothesis.

That said, this critique may be overly simplistic. It is possible that kin selection processes are not sufficient to completely explain the persistence of the genetic factors underpinning male androphilia, but they may contribute to their persistence, at least in part. Indeed, research by Nila et al. (2018) suggests that kin selection reduces the direct reproductive costs of androphilic males by approximately 20 percent, thus contributing to the genetic maintenance of this trait, but additional processes must also be at work.

In addition, an important consideration is that population growth was only slightly above replacement in the human ancestral environment, wherein the average person produced only two offspring who survived and reproduced in turn (Belovsky, 1988). As such, for the kin selection hypothesis to have been viable ancestrally, androphilic males' kin-directed altruism would only have had to result in four nieces and/or nephews who survived and reproduced. A bias toward investing in younger nieces and nephews – like that exhibited by Samoan fa'afafine – might improve the survival of those kin, thus allowing androphilic males to achieve this mark (VanderLaan & Vasey, 2012, 2013, 2014; Vasey & VanderLaan, 2010a).

Lastly, if the adaptive design features characterizing androphilic male avuncular cognition – such as those described in Section 30.8 for Samoan fa'afafine and Canadian gay men – are indeed the product of past selection, then kin selection processes contributed, at least in part, to the persistence of the genetic factors underpinning male androphilia ipso facto (VanderLaan & Vasey, 2012, 2013, 2014; Vasey & VanderLaan, 2010a). Additional demonstrations that androphilic males' avuncular cognition is characterized by hallmarks of adaptive design would further challenge critiques about the viability of the kin selection hypothesis.

A primary limitation of the research literature concerning this hypothesis is the lack of replication studies outside of Samoa. As such, additional tests in other populations where transgender androphilic males predominate would be valuable. To date, only one such study has been conducted among the Urak Lawoi people inhabiting Ko Lipe island in the Andaman Sea off the coast of Thailand (Camperio Ciani et al., 2016). In that population, transgender androphilic males are known locally as a third gender: na-ning (Vasey et al., 2016). Camperio Ciani et al. (2016) found no evidence that na-ning (whom the authors referred to using the Thai term, kathoey) exhibited elevated avuncularity when compared to Urak Lawoi gynephilic men. Unfortunately, Camperio Ciani et al.'s (2016) study of the Urak Lawoi is compromised by the fact that the sample sizes were small and the statistical analyses the authors employed had several limitations (Vasey et al., 2016).

The study by Nila et al. (2018) also deserves mention. These authors conducted research in Java, Indonesia, and found that androphilic males expressed elevated avuncular tendencies compared to gynephilic males. These tendencies manifested behaviorally in the form of increased transfer of money from androphilic uncles to their nieces and nephews. Nila et al.'s (2018) sample consisted of 82 androphilic males, some of whom (n = 11) declared themselves to be members of a traditional third gender, *waria*. In light of the adaptive feminine phenotype model, it would have been interesting to be shown the extent to which Nila et al.'s (2018) sample of androphilic males were appreciably feminine compared to their gynephilic counterparts. Presumably, all of the *waria* participants would have been considerably more feminine than the gynephilic male participants, but the degree to which this would have been the case for the other androphilic male participants is uncertain.

To further address this issue, in 2015, we initiated fieldwork among the Zapotec people in the Istmo region of Oaxaca, Mexico (which encompasses the Tehuantepec and Juchitán regions), where we have conducted biannual research ever since. In this area, androphilic males are recognized as a third gender, known locally as *muxe*. In contemporary usage, the term *muxe* refers to any androphilic male who routinely adopts the receptive position during anal intercourse. *Muxes* vary in terms of the degree to which they present publically in a feminine manner (Mirandé, 2017), and consequently, Istmo Zapotec recognize two types of *muxes: muxe gunaa* and *muxe nguiiu* (i.e., Zapotec for *muxe* woman and *muxe* man, respectively). *Muxe gunaa* are transgender androphilic males, comparable to the Samoan *fa'afafine*. In contrast, *muxe nguiiu* are cisgender androphilic males, comparable to Western "gay" men, who dress in men's clothes and present publically in a relatively masculine manner. Like the Samoan *fa'afafine*, *muxe* enjoy a relatively high level of acceptance that, while not absolute, stands in stark contrast to the situation experienced by Western transgender male androphiles (e.g., Namaste, 2000). Perhaps the most prominent public display of acceptance is *La Vela de las Autenticas Intrepidas Buscadoras del Peligro* (The Festival of the Authentic Intrepid Seekers of Dangers), a four-day festival in honor of the *muxes* and celebrated each November in Juchitán. More than 10,000 community members and visitors attend this festival, including representatives from the local Catholic Church (Mirandé, 2017).

Our initial research on *muxes*, as it pertains to the kin selection hypothesis, is promising in that we demonstrated that *muxes* exhibit elevated indicators of recalled childhood separation anxiety compared to their gynephilic male counterparts, but they do not differ in this regard from androphilic women (Gómez et al., 2017). This is true for both *muxe gunaa* and *muxe nguiiu*, who do not differ from each other in terms of this measure. This pattern is precisely what one would predict on the basis of the adaptive feminine phenotype model. Future research on *muxes*

will address the relationship between femininity and indicators of recalled separation anxiety in childhood, as well as tendencies for avuncularity in adulthood.

There is also a need to further examine androphilic males in cultures where they identify as gay men to test whether they show the same hallmarks of adaptive design that characterize the avuncular cognition of *fa'afafine* (VanderLaan & Vasey, 2012, 2013, 2014; Vasey & VanderLaan, 2010c). Even though gay men do not report elevated willingness to invest in kin, they may evince some of the same avuncular cognition design features as *fa'afafine*, which would help clarify the role of kin selection processes in the evolution of male androphilia across populations (e.g., Forrester et al., 2011).

A major gap in the research literature concerns the effects of androphilic males' kin-directed altruism on kin. If androphilic males' kin investments have positive influences on indirect fitness, then one would expect to observe increases in the quantity and/or quality (i.e., physical, psychological well-being) of their kin. To date, no such research has been undertaken, and such tests are key to evaluating the kin selection hypothesis.

Another aspect of research that has yet to be investigated empirically is the relationship, if any, between kin selection and alternative processes hypothesized to contribute to the evolution of male androphilia. Some alternative processes, such as sexually antagonistic selection, have garnered considerable support (e.g., Camperio Ciani & Pellizzari, 2012; Camperio Ciani et al., 2004; Iemmola & Camperio Ciani, 2009; Semenyna et al., 2017a; VanderLaan et al., 2012). A thorough description of these alternatives is beyond the scope of this chapter, but suffice to say that the way forward in solving the evolutionary conundrum that is male androphilia will require viewing these alternative processes as complimentary, rather than mutually exclusive (Vasey & VanderLaan, 2014). If kin selection provides only a partial explanation for the evolutionary persistence of the genes underpinning male androphilia – as work by Nila et al. (2018) suggests – then we will need to explore how alternative evolutionary process, like sexually antagonistic selection, interact with kin selection to maintain male androphilia over evolutionary time (Vasey & VanderLaan, 2014). Further advances in theory and research will be required to address this issue.

REFERENCES

Abild, M. L., VanderLaan, D. P., & Vasey, P. L. (2014). Does geographic proximity influence the expression of avuncular tendencies in Canadian androphilic males? *Journal of Cognition and Culture*, **14**, 41–63.

Adams, B. D. (1986). Age, structure and sexuality. *Journal of Homosexuality*, **11**, 19–33.

Alanko, K., Santtila, P., Harlaar, N., et al. (2010). Common genetic effects of gender atypical behavior in childhood and sexual orientation in adulthood: A study of Finnish twins. *Archives of Sexual Behavior*, **39**, 81–92.

American Psychiatric Association (2013). *Diagnostic and Statistical Manual of Mental Disorders*, 5th ed. Arlington, VA: American Psychiatric Publishing.

Bailey, J. M. (2003). *The Man Who Would Be Queen: The Science of Gender-Bending and Transsexualism*. Washington, DC: Joseph Henry Press.

Bailey, J. M., & Triea, K. (2007). What many transgender activists don't want you to know: And why you should know it anyway. *Perspectives in Biology and Medicine*, **50**, 521–534.

Bailey, J. M., & Zucker, K. J. (1995). Childhood sex-typed behavior and sexual orientation: A conceptual analysis and quantitative review. *Developmental Psychology*, **31**, 43–55.

Bailey, J. M., Gaulin, S., Agyei, Y., & Gladue, B. A. (1994). Effects of gender and sexual orientation on evolutionarily relevant aspects of human mating psychology. *Journal of Personality and Social Psychology*, **66**, 1081–1093.

Bailey, J. M., Dunne, M. P., & Martin, N. G. (2000). Genetics and environmental influences on sexual orientation and its correlates in an Australian twin sample. *Journal of Personality and Social Psychology*, **78**, 524–536.

Bailey, N. W., & Zuk, M. (2009). Same-sex sexual behavior and evolution. *Trends in Ecology and Evolution*, **24**, 439–446.

Bartlett, N. B., & Vasey, P. L. (2006). A retrospective study of childhood gender-atypical behavior in Samoan fa'afafine. *Archives of Sexual Behavior*, **35**, 559–566.

Bell, A. P., & Weinberg, M. S. (1978). *Homosexualities: A Study of Diversity among Men and Women*. New York: Simon and Schuster.

Belovsky, G. E. (1988). An optimal foraging-based model of hunter–gatherer population dynamics. *Journal of Anthropological Archaeology*, **7**, 329–372.

Berling, T. (2001). *Sissyphobia: Gay Men and Effeminate Behavior*. Binghamton, NY: Harrington Park Press.

Besharat, M. A., Karimi, S., & Saadati, M. (2016). A comparison of childhood gender nonconformity and fertility rate in a lineage in male homosexuals and heterosexuals. *Contemporary Psychology*, **10**, 3–14.

Binford, L. R. (2001). *Constructing Frames of References: An Analytical Method for Archaeological Theory Building Using Hunter–Gatherer and Environmental Data Sets*. Berkeley, CA: University of California Press.

Blanchard, R. (2018). Fraternal birth order, family size, and male homosexuality: Meta-analysis of studies spanning 25 years. *Archives of Sexual Behavior*, **47**, 1–15.

Blanchard, R., & VanderLaan, D. P. (2015). Commentary on Kishida & Rahman (2015), including a meta-analysis of relevant studies on fraternal birth order and sexual orientation in men. *Archives of Sexual Behavior*, **44**, 1503–1509.

Bobrow, B., & Bailey, J. M. (2001). Is male homosexuality maintained via kin selection? *Evolution and Human Behavior*, **22**, 361–368.

Bowen, R. C., Offord, D. R., & Boyle, M. H. (1990). The prevalence of overanxious disorder and separation anxiety disorder: Results from the Ontario Child Health Study. *Journal of the American Academy of Child and Adolescent Psychiatry*, **29**, 753–758.

Bozkurt, A., Bozkurt, O. H., & Sonmez, I. (2015). Birth order and sibling sex ratio in a population with high fertility: Are Turkish male-to-female transsexuals different? *Archives of Sexual Behavior*, **44**, 1331–1337.

Brown, D. E. (1991). *Human Universals*. New York: McGraw Hill.

Buss, D. M., Haselton, M. G., Shackelford, T. K., Bleske, A. L., & Wakefield, J. C. (1998). Adaptations, exaptations, and spandrels. *American Psychologist*, **53**, 533–548.

Camperio Ciani, A., & Pellizzari, E. (2012). Fecundity of paternal and maternal non-parental female relatives of homosexual and heterosexual men. *PLoS ONE*, **7**(12), e51088.

Camperio Ciani, A., Corna, F., & Capiluppi, C. (2004). Evidence for maternally inherited factors favoring male homosexuality and promoting female fecundity. *Proceedings of the Royal Society of London. Series B: Biological Sciences*, **271**, 2217–2221.

Camperio Ciani, A., Battaglia, U., & Liotta, M. (2016). Societal norms rather than sexual orientation influence kin altruism and avuncularity in tribal Urak-Lawoi, Italian, and Spanish adult males. *Journal of Sex Research*, **53**, 137–148.

Cantor, J. M., Blanchard, R., Paterson, A. D., & Bogaert, A. F. (2002). How many gay men owe their sexual orientation to fraternal birth order? *Archives of Sexual Behavior*, **31**, 63–71.

Cardoso, F. L. (2005). Cultural universals and differences in male homosexuality: The case of a Brazilian fishing village. *Archives of Sexual Behavior*, **34**, 103–109.

Cardoso, F. L. (2009). Recalled sex-typed behavior in childhood and sports' preferences in adulthood of heterosexual, bisexual, and homosexual men from Brazil, Turkey, and Thailand. *Archives of Sexual Behavior*, **38**, 726–736.

Coates, S., & Person, E. S. (1985). Extreme boyhood femininity: Isolated behavior or pervasive disorder? *Journal of the American Academy of Child Psychiatry*, **24**, 702–709.

Crompton, L. (2003). *Homosexuality and Civilization*. Cambridge, MA: Belknap Press.

Deakin, T. (1965). Evidence for homosexuality in ancient Egypt. *International Journal of Greek Love*, **1**, 31–38.

Dickemann, M. (1995). Wilson's panchreston: The inclusive fitness hypothesis of sociobiology re-examined. *Journal of Homosexuality*, **28**, 147–183.

Dreger, A. D. (2008). The controversy surrounding *The Man Who Would Be Queen*: A case history of the politics of science, identity, and sex in the internet age. *Archives of Sexual Behavior*, **37**, 366–421.

Duff-Cooper, A. (1985). Notes about some Balinese ideas and practices connected with sex from Western Lombok. *Anthropos*, **80**, 403–419.

Forrester, D. L., VanderLaan, D. P., Parker, J. L., & Vasey, P. L. (2011). Male sexual orientation and avuncularity in Canada: Implications for the kin selection hypothesis. *Journal of Cognition and Culture*, **11**, 339–352.

"G" (1980). The secret life of Moscow. *Christopher Street*, June, pp. 15–22.

García-Cárdenas, N., Olvera-Hernández, S., Gómez-Quintanar, B. N., & Fernández-Guasti, A. (2015). Male rats with same-sex preference show high experimental anxiety and lack of anxiogenic-like effect of fluoxetine in the plus maze test. *Pharmacology, Biochemistry and Behavior*, **135**, 128–135

Gates, G. J. (2011). How many people are lesbian, gay, bisexual, and transgender? *The Williams Institute*. https://escholarship .org/uc/item/09h684x2.

Gaudio, R. C. (2009). *Allah Made Us: Sexual Outlaws in an Islamic African City*. Hoboken, NJ: Wiley-Blackwell.

Gómez, F. R., Semenyna, S. W., Court, L. & Vasey, P. L. (2017). Recalled childhood separation anxiety in Istmo Zapotec men, women, and *muxes*. *Archives of Sexual Behavior*, **46**, 109–117.

Gómez, F. R., Semenyna, S. W., Court, L., & Vasey, P. L. (2018). Familial clustering of male androphilia among Istmo Zapotec men and *muxe*. *PLoS ONE*, **13**(2), e0192683.

Gómez-Gil, E., Esteva, I., Carrasco, R., et al. (2011). Birth order and ratio of brothers to sisters in Spanish transsexuals. *Archives of Sexual Behavior*, **40**, 505–510.

Green, R. (1987). *The "Sissy Boy Syndrome" and the Development of Homosexuality*. New Haven, CT: Yale University Press.

Hamer, D. H., Hu, S., Magnuson, V. L., Hu, N., & Pattatucci, A. M. L. (1993). A linkage between DNA markers on the X chromosome and male sexual orientation. *Science*, **261**, 321–327.

Hames, R., Garfield, Z. H., & Garfield, M. J. (2017). Is male androphilia a context-dependent cross-cultural universal? *Archives of Sexual Behavior*, **46**, 63–71.

Hamilton, W. D. (1963). The evolution of altruistic behavior. *American Naturalist*, **97**, 354–356.

Hart, D. V. (1968). Homosexuality and transvestism in the Philippines: The Cebuan Filipino *bayot* and *lakin-on*. *Behavior Science Notes*, **3**, 211–248.

Hernández, A., & Fernández-Guasti, A. (2018). Male rates with same-sex preference show higher immobility in the forced swim test, but similar effects of fluoxetine and desipramine than males that prefer females. *Pharmacology, Biochemistry and Behavior*, **171**, 39–45.

Hewlett, B., & Hewlett, B. L. (2010). Sex and searching for children among Aka foragers and Ngandu farmers of Central Africa. *African Studies Monographs*, **31**, 107–125.

Hill, K. R., Walker, R. S., Bozicevic, M., et al. (2011). Co-residence patterns in hunter–gatherer societies show unique human social structure. *Science*, **331**, 1286–1289.

Iemmola, F., & Camperio Ciani, A. (2009). New evidence of genetic factors influencing sexual orientation in men: Female fecundity increase in the maternal line. *Archives of Sexual Behavior*, **38**, 393–399.

Kendler, K. S., Thronton, L. M., Gilman, S. E., & Kessler, R. C. (2000). Sexual orientation in a US national sample of twin and non-twin sibling pairs. *American Journal of Psychiatry*, **157**, 1843–1846.

King, M. D., Green, J., Osborn, D. P. J., et al. (2005). Family size in white gay and heterosexual men. *Archives of Sexual Behavior*, **34**, 117–122.

Långström, N., Rahman, Q., Calström, E., & Lichtenstein, P. (2010). Genetic and environmental effects on same sex-sexual behavior: A population study of twins in Sweden. *Archives of Sexual Behavior*, **39**, 75–80.

LeVay, S. (2016). *Gay, Straight and the Reason Why: The Science of Sexual Orientation*. Oxford: Oxford University Press.

Lippa, R. A. (2005). *Gender, Nature, Nurture*, 2nd ed. Mahwah, NJ: Erlbaum.

Marlowe, F. W. (2005). Hunters–gatherers and human evolution. *Evolutionary Anthropology*, **14**, 54–67.

McBrearty, S., & Brooks, A. S. (2000). The revolution that wasn't: A new interpretation of the origins of modern human behavior. *Journal of Human Evolution*, **39**, 453–563.

Mirandé, A. (2017). *Behind the Mask: Gender Hybridity in a Zapotec Community*. Tucson, AZ: University of Arizona Press.

Mustanski, B. S., Dupree, M. G., Nievergelt, C. M., et al. (2005). A genomewide scan of male sexual orientation. *Human Genetics*, **116**, 272–278.

Namaste, V. (2000). *Invisible Lives: The Erasure of Transsexual and Transgender People*. Chicago, IL: University of Chicago Press.

Nanda, S. (1998). *Neither Man nor Woman: The Hijras of India*. Belmont, CA: Wadsworth Publishing Company.

Nanda, S. (2014). *Gender Diversity: Crosscultural Variations*. Long Grove, IL: Waveland Press.

Nila, S., Barthes, J., Crochet, P.-A., Suryobroto, B., & Raymond, M. (2018). Kin selection and male homosexual preference in Indonesia. *Archives of Sexual Behavior*, **47**(8), 2455–2465.

Petterson, L. J., Dixson, B. J., Little, A. C., & Vasey, P. L. (2015). Viewing time measures of sexual orientation in Samoan cisgender men who engage in sexual interactions with *fa'afine*. *PLoS ONE*, **10**(2), e0116529.

Petterson, L. J., Dixson, B. J., Little, A. C., & Vasey, P. L. (2016). Reconsidering male bisexuality: Sexual activity role and sexual attraction in Samoan men who engage in sexual interactions with fa'afafine. *Psychology of Sexual Orientation and Gender Diversity*, **3**, 11–26.

Petterson, L. J., Wrightson, C., & Vasey, P. L. (2017). A retrospective study of childhood gender-atypical behavior in Japanese androphilic males. *Archives of Sexual Behavior*, **46**, 119–127.

Petterson, L. J., Dixson, B. J., Little, A. C., & Vasey, P. L. (2018). Viewing time and self-report measures of sexual attraction in Samoan cisgender and transgender androphilic males. *Archives of Sexual Behavior*, **47**(8), 2427–2434.

Pittin, R. (1983). Houses of women: A focus on alternative lifestyles in Katsina City. In C. Oppong, ed., *Female and Male in West Africa*. London: George Allen & Unwin, pp. 291–302.

Rahman, Q., & Hull, M. S. (2005). An empirical test of the kin selection hypothesis for male homosexuality. *Archives of Sexual Behavior*, **34**, 461–467.

Rieger, G., & Savin-Williams, R. C. (2012). Gender nonconformity, sexual orientation, and psychological well-being. *Archives of Sexual Behavior*, **41**, 611–621.

Rieger, G., Linsenmeier, J. A., Gygax, L., & Bailey, J. M. (2008). Sexual orientation and childhood gender nonconformity: Evidence from home videos. *Developmental Psychology*, **44**, 46–58.

Saghir, M. T., & Robins, E. (1973). *Male and Female Homosexuality: A Comprehensive Investigation*. Baltimore, MD: Williams & Wilkins.

Samaco, R. C., Mandel-Brehm, C., McGraw, C. M., et al. (2012). Crh and Oprm1 mediate anxiety-related behavior and social approach in a mouse model of MECP2 duplication syndrome. *Nature Genetics*, **44**, 206–211.

Sanders, A. R., Martin, E. R., Beecham, G. W., et al. (2015). Genome-wide scan demonstrates significant linkage for male sexual orientation. *Psychological Medicine*, **45**, 1379–1388.

Sanderson, S. K., & Roberts, W. W. (2008). The evolutionary forms of the religious life: A cross-cultural, quantitative analysis. *American Anthropologist*, **110**, 454–466.

Schuvaloff, G. (1976). Gay life in Russia. *Christopher Street*, September, pp. 14–22.

Schwartz, G., Kim, R. M., Kolundziji, A. B., Rieger, G., & Sanders, A. R. (2010). Biodemographic and physical correlates of sexual orientation in men. *Archives of Sexual Behavior*, **39**, 93–109.

Sear, R., & Mace, R. (2008). Who keeps children alive? A review of the effects of kin on child survival. *Evolution and Human Behavior*, **29**, 1–18.

Semenyna, S. W., & Vasey, P. L. (2016). The relationship between adult occupational preferences and childhood gender nonconformity among Samoan women, men, and *fa'afafine*. *Human Nature*, **27**, 283–295.

Semenyna, S. W., & Vasey, P. L. (2019). Prestige striving in Samoa: A comparison of men, women and *fa'afafine*. *Journal of Homosexuality*, **66**(11), 1535–1545.

Semenyna, S. W., Petterson, L. J. & VanderLaan, D. P., & Vasey, P. L. (2017a). A comparison of the reproductive output among the relatives of Samoan androphilic *fa'afafine* and gynephilic men. *Archives of Sexual Behavior*, **46**, 87–93.

Semenyna, S. W., Petterson, L. J. & VanderLaan, D. P., & Vasey, P. L. (2017b). Familial patterning and prevalence of male androphilia in Samoa. *Journal of Sex Research*, **54**, 1077–1084.

Semenyna, S. W., VanderLaan, D. P., & Vasey, P. L. (2017c). Birth order and recalled gender nonconformity in Samoan men and *fa'afafine*. *Developmental Psychobiology*, **59**, 338–347.

Shear, K., Jin, R., Ruscio, A. M., Walters, E. E., & Kessler, R. C. (2006). Prevalence and correlates of estimated DSM-IV child and adult separation anxiety disorder in the National Comorbidity Survey replication. *American Journal of Psychiatry*, **163**, 1074–1083.

Singh, D. (2012). A follow-up of boys with gender identity disorder. Unpublished doctoral dissertation, University of Toronto.

Smith, A. B. (1999). Archaeology and the evolution of hunter–gatherers. In R. B. Lee & R. Daly, eds., *The Cambridge Encyclopedia of Hunters and Gatherers*. Cambridge, UK: Cambridge University Press, pp. 384–390.

Steensma, T. D., van der Ende, J., Verhulst, F. C., & Cohen-Kettenis, P. T. (2013). Gender variance in childhood and sexual orientation in adulthood: A prospective study. *Journal of Sexual Medicine*, **10**, 2723–2733.

Sweet, M., & Zwilling, L. (1993). The first medicalization: The taxonomy and etiology of queerness in classical Indian medicine. *Journal of the History of Sexuality*, **3**, 590–697.

Taylor, T. (1996). *The Prehistory of Sex: Four Million Years of Human Sexual Culture*. New York: Bantam Books.

Thomas, W. (1997). Navajo cultural constructions of gender and sexuality. In S. E. Jacobs, W. Thomas, & S. Lang, eds., *Two-Spirit People: Native American Gender Identity, Sexuality and Spirituality*. Chicago, IL: University of Illinois Press, pp. 156–173.

Totman, R. (2003). *The Third Sex – Kathoey: Thailand's Ladyboys*. London: Souvenir Press.

VanderLaan, D. P., & Vasey, P. L. (2011). Male sexual orientation in Independent Samoa: Evidence for the fraternal birth order and maternal fecundity effects. *Archives of Sexual Behavior*, **40**, 495–503.

VanderLaan, D. P., & Vasey, P. L. (2012). Relationship status and elevated avuncularity in Samoan *fa'afafine*. *Personal Relationships*, **19**, 326–339.

VanderLaan, D. P., & Vasey, P. L. (2013). Birth order and avuncular tendencies in Samoan men and *fa'afafine*. *Archives of Sexual Behavior*, **42**, 371–379.

VanderLaan, D. P., & Vasey, P. L. (2014). Evidence of enhanced cognitive biases for maximizing indirect fitness in Samoan *fa'afafine*. *Archives of Sexual Behavior*, **43**, 1009–1022.

VanderLaan, D. P., Gothreau, L. M., Bartlett, N. H., & Vasey, P. L. (2011a). Recalled separation anxiety and gender atypicality in childhood: A study of Canadian heterosexual and homosexual men and women. *Archives of Sexual Behavior*, **40**, 1233–1240.

VanderLaan, D. P., Gothreau, L. M., Bartlett, N. H., & Vasey, P. L. (2011b). Separation anxiety in feminine boys: Pathological or prosocial? *Journal of Gay and Lesbian Mental Health*, **15**, 30–45.

VanderLaan, D. P., Forrester, D. L., Petterson, L. J. & Vasey, P. L. (2012). Offspring production among the extended relatives of Samoan men and *fa'afafine*. *PLoS ONE*, **7**(4), e36088.

VanderLaan, D. P., Forrester, D. L., Petterson, L. J., & Vasey, P. L. (2013a). The prevalence of *fa'afafine* relatives among Samoan men and *fa'afafine*. *Archives of Sexual Behavior*, **42**, 353–359.

VanderLaan, D. P., Ren, Z., & Vasey, P. L. (2013b). Male androphilia in the ancestral environment: An ethnological analysis. *Human Nature*, **24**, 375–401.

VanderLaan, D. P., Vokey, J. R., & Vasey, P. L. (2013c). Is male androphilia familial in non-Western cultures? The case of a Samoan village. *Archives of Sexual Behavior*, **42**, 361–370.

VanderLaan, D. P., Petterson, L. J., Mallard, R. W., & Vasey, P. L. (2015a). (Trans)gender role expectations and childcare in Samoa. *Journal of Sex Research*, **52**, 710–720.

VanderLaan, D. P., Petterson, L. J., & Vasey, P. L. (2015b). Elevated childhood separation anxiety: An early developmental expression of heightened concern for kin in homosexual men? *Personality and Individual Differences*, **81**, 188–194.

VanderLaan, D. P., Petterson, L. J., & Vasey, P. L. (2016). Femininity and kin-directed altruism in androphilic men: A test of an evolutionary developmental model. *Archives of Sexual Behavior*, **45**, 619–633.

VanderLaan, D. P., Blanchard, R., Zucker, K. J., et al. (2017a). Birth order and androphilic male-to-female transsexualism in Brazil. *Journal of Biosocial Science*, **49**, 527–535.

VanderLaan, D. P., Petterson, L. J., & Vasey, P. L. (2017b). Elevated kin-directed altruism emerges in childhood and is linked to feminine gender expression in Samoan *fa'afafine*: A retrospective study. *Archives of Sexual Behavior*, **46**, 95–108.

Vasey, P. L., & Bartlett, N. H. (2007). What can the Samoan *fa'afafine* teach us about the Western concept of "gender identity disorder in childhood"? *Perspectives in Biology and Medicine*, **50**, 481–490.

Vasey, P. L., & VanderLaan, D. P. (2007). Birth order and male androphilia in Samoan *fa'afafine*. *Proceedings of the Royal Society B: Biological Sciences*, **274**, 1437–1442.

Vasey, P. L., & VanderLaan, D. P. (2009). Materteral and avuncular tendencies in Samoa: A comparative study of women, men and *fa'afafine*. *Human Nature*, **20**, 269–281.

Vasey, P. L., & VanderLaan, D. P. (2010a). An adaptive cognitive dissociation between willingness to help kin and non-kin in Samoan *fa'afafine*. *Psychological Science*, **21**, 292–297.

Vasey, P. L., & VanderLaan, D. P. (2010b). Avuncular tendencies in Samoan *fa'afafine* and the evolution of male androphila. *Archives of Sexual Behavior*, **39**, 821–830.

Vasey, P. L., & VanderLaan, D. P. (2010c). Monetary exchanges with nieces and nephews: A comparison of Samoan men, women, and *fa'afafine*. *Evolution and Human Behavior*, **31**, 373–380.

Vasey, P. L., & VanderLaan, D. P. (2012). Male sexual orientation and avuncularity in Japan: Implications for the kin selection hypothesis. *Archives of Sexual Behavior*, **41**, 209–215.

Vasey, P. L., & VanderLaan, D. P. (2014). Evolving research on the evolution of male androphilia. *Canadian Journal of Human Sexuality*, **23**, 137–147.

Vasey, P. L., Pocock, D. S., & VanderLaan, D. P. (2007). Kin selection and male androphilia in Samoan *fa'afafine*. *Evolution & Human Behavior*, **28**, 159–167.

Vasey, P. L., VanderLaan, D.P., Gothreau, L., & Bartlett, N. H. (2011). Traits of separation anxiety in childhood: A comparison of Samoan men, women and *fa'afafine*. *Archives of Sexual Behavior*, **40**, 511–517.

Vasey, P. L., Parker, J. L., & VanderLaan, D. P. (2014). Comparative reproductive output of androphilic and gynephilic males in Samoa. *Archives of Sexual Behavior*, **43**, 363–367.

Vasey, P. L., VanderLaan, D. P., Hames, R., & Jaidee, A. (2016). A problematic test of the kin selection hypothesis among the Urak-Lawoi of Ko Lipeh, Thailand: Commentary on Camperio Ciani, Battaglia, & Liotta (2015). *Journal of Sex Research*, **53**, 149–152.

Whitam, F. L. (1992). *Bayot* and callboy in the Philippines. In S. O. Murray, ed., *Oceanic Homosexualities*. New York: Garland, pp. 231–248.

Whitam, F. L. (1997). Culturally universal aspects of male homosexual transvestites and transsexuals. In B. Bullough, V. Bullough, & J. Elias, eds., *Gender Blending*. Amherst, NY: Prometheus, pp. 189–203.

Whitam, F. L., & Mathy, R. M. (1986). *Male Homosexuality in Four Societies: Brazil, Guatemala, the Philippines, and the United States*. New York: Praeger.

Wikan, U. (1977). Man becomes woman: Transsexualism in Oman as a key to gender roles. *Man*, **12**, 304–319.

Williams, G. C. (1966). *Adaptation and Natural Selection*. Princeton, NJ: Princeton University Press.

Williams, W. (1992). *The Spirit and the Flesh: Sexual Diversity in American Indian Culture*. Boston, MA: Beacon.

Wilson, E. O. (1975). *Sociobiology: The New Synthesis*. Cambridge, MA: Belknap Press.

Winkelman, M. (2010). *Shamanism: A Biopsychosocial Paradigm of Consciousness and Healing*. Santa Barbara, CA: Praeger.

Winter, S. J. (2006). Thai transgenders in focus: Demographics, transitions and identities. *International Journal of Transgenderism*, **9**, 15–27.

Woodburn, J. (1982). Egalitarian societies. *Man*, **17**, 431–451.

Zheng, L., Lippa, R. A., & Zheng, Y. (2011). Sex and sexual orientation differences in personality in China. *Archives of Sexual Behavior*, **40**, 533–541.

Zietsch, B. P., Verweij, K. J. H., Bailey, J. M., Wright, M. J., & Martin, N. G. (2011). Sexual orientation and psychiatric vulnerability: A twin study of neuroticism and psychoticism. *Archives of Sexual Behavior*, **40**, 133–142.

Zucker, K. J., Bradley, S. J., & Lowry Sullivan, C. B. (1996). Traits of separation anxiety in boys with gender identity disorder. *Journal of the American Academy of Child and Adolescent Psychiatry*, **35**, 791–798.

31 Evolutionary Psychology
Thoughts on Integrating Feminist Perspectives

MARYANNE L. FISHER, JUSTIN R. GARCIA, AND REBECCA L. BURCH

31.1 INTRODUCTION

The discussion of evolutionary theory and feminist ideology has existed for decades and has been obstructed by assumptions, generalizations, misunderstandings, and omissions from both points of view. Terminology, in particular, has had important consequences for comprehension. We note, like Barkow (2006), that there is no term that accurately captures the variety of work of those applying Darwinian theory to the study of human behavior. We apply "evolutionary psychology" here as it is a prevalent term that is used currently, but it also reinforces the goal of focusing on human nature as an outcome of biological evolution (Barkow, 2006). To provide as much clarity and simplicity as possible in this chapter, we will refer to evolutionary scientists as "evolutionists" and feminist scholars as "feminists" as they were in Hrdy (1981/1999).

Thus, we note that as evolutionists we rely on the distinction proposed by Unger (1979) that *sex* refers to biological designations of being female or male and *gender* refers to the social construction of one's femininity, masculinity, or androgenity, which is often designated by societies based on one's sex. This distinction relies on Money and colleagues (1955), who argued that *sex* was best suited to examining individual physical traits and *gender* for psychological traits. Per feminists Muehlenhard and Peterson (2011), while many behavioral scientists recognize such a distinction, current term definitions and uses vary widely and at times inconsistently. In this chapter, we use these terms distinctly to mark sex and gender as separate constructs, but we acknowledge that this is predicted to be a controversial decision (Radtke, 2017). There is an enduring uneasy relationship between evolutionists and feminists when it comes to the study of human behavior. In general, feminists tend to view work incorporating an evolutionary perspective as being somewhat anti-feminist, unnecessarily reductionist, or at least neglecting to be inclusionary of sociocultural and gendered factors. This unease has been written about at length for nearly four decades (Fausto-Sterling, 1997; Fedigan, 1986; Gowaty, 1997a, 1997b; Hager, 1997; Hrdy, 1981/1999; Kelly, 2014; Nier & Campbell, 2013; for a review

of the tension between psychology and feminisms in general, see Gavey & Braun, 2008; Morgan et al., 2011).

The key arguments tend to center around the appropriate level of analysis for issues in which biological and cultural factors that shape sexuality collide. For instance, evolutionists tend to view sexual inequalities in the context of parental investment differentials, leading to the criticism that this evolutionist approach to sexuality reinforces sexual double standards but does little to ameliorate social inequalities and injustices, especially toward women and minority populations. Likewise, the areas collide over sex differences in major traits (such as cognitive abilities and emotional reactions) that reify heteronormativity/heterosexualism, the propagation of stereotypical and ethnocentric beauty standards for women, or what is interpreted as the rationalization of harmful behaviors of men toward women (e.g., speculation about the adaptive function of rape: Thornhill & Palmer, 2001; family conflict: Daly & Wilson, 1999; or domestic violence: Wilson & Daly, 1998).

These concerns are perhaps understandable given some of the existing academic and popular literature on human nature. As one example, renowned evolutionary biologist E. O. Wilson once wrote, "In hunter–gatherer societies, men hunt and women stay at home. This strong bias persists in most agricultural and industrial societies, and on that grounds, appears to have a genetic origin. ... Even with identical education and equal access to all professions, men are likely to continue to play a disproportionate role in political life, business and science" (1978, p. 133). While this claim was written nearly four decades ago, such a perspective continues to undermine the multiple roles women (and men) have played in human evolution. It is important to note, however, that such claims have changed considerably since this time, with new research and more dynamic evolutionary biocultural theoretical models leading to new data and new insights (e.g., Hawkes, O'Connell, & Blurton Jones, 2018). In fact, as we will review, more recent work has begun to challenge these assumptions, integrating evolutionary and feminist scholarship to provide a more holistic understanding of how sex and gender manifest in different societies. Indeed, there is a growth of scholars who label themselves (or at least identify to some extent) as feminist evolutionists, as evidenced by the formation of the Feminist Evolutionary Perspectives Society in 2009 (Sokol-Chang & Fisher, 2013).

While there is a wide variety of mid-level evolutionary theoretical models, these theories of human cognition and behavior are grounded in Darwinian theories of natural and sexual selection. This epistemological cornerstone is what most of evolutionary psychological research has emerged from, with the substantial amount of work focused on human sex differences generally deriving from Darwinian models of sexual selection. Despite both the grandeur and parsimony of his insights on evolutionary theory, Darwin's views on women were exceedingly narrow and plagued by Victorian stereotypes. For example, he proposes that the "most able men will have succeeded best in defending and providing for themselves, their wives and offspring" (1871, p. 383). Darwin wrote about how males compete with other males to impress potential mates, such that females will simply – and passively – accept the winner and devote their lives to taking care of his offspring (1871). Moreover, in *The Descent of Man, and Selection in Relation to Sex* (1871), he wrote:

The chief distinction in the intellectual powers of the two sexes is shown by man's attaining to a higher eminence, in whatever he takes up, than can woman – whether requiring deep thought, reason, or imagination, or merely the use of the senses and hands ... if men are capable of a decided pre-eminence over women in many subjects, the average of mental power in man must be above that of women ... [men have had] to defend their females, as well as their young, from enemies of all kinds, and to hunt for their joint subsistence. But to avoid enemies or to attack them with success, to capture wild animals and to fashion weapons requires the aid of the higher mental faculties, namely, observation, reason, invention, or imagination. These various faculties will thus have been continually put to the test and selected during manhood. (p. 311)

Scholars generally accept the view that Darwin was not a feminist. Even while Darwin was alive and publishing, scholars such as Antoinette Brown Blackwell (1875) were questioning the veracity and androcentric views within his writings. His responses to feminist critiques later in his life echoed his original writings and the Victorian era he lived in: "I certainly think that women though generally superior to men [in] moral qualities are inferior intellectually" (correspondence to Caroline Kennard, 1882, as recounted in Saini, 2017). This androcentric view spurred Eliza Burt Gamble to write a detailed data-driven response entitled *The Evolution of Woman, an Inquiry into the Dogma of Her Inferiority to Man* in 1894. Despite these arguments, there is speculation that other personal communications reflect an effort to help female scientists gain credibility in a male-dominated academic world (www.cam.ac.uk/research/news/darwins-women). Furthermore, Darwin emphasized the importance of female choice in sexual selection, which

was a controversial idea (for a review, see Hosken & House, 2011).

Those who followed Darwin's work and built upon it, such as E. O. Wilson, were likewise often androcentric in their intellectual worldviews. While no longer a justifiable position today, this heritage is not altogether surprising given the history and philosophy of the natural sciences. Fisher (1930) extended Darwin's work on sexual selection by focusing on intrasexual competition among males for mates, while overlooking the potential roles that females play in selective processes beyond choosing the most fit male possible as a mate (if Fisher had studied Gamble, 1894, as well as Darwin, this omission would have been pointed out to him). The ubiquitous example of the peacock's train feathers paints this picture vividly; males with the most elaborate trains are chosen more often as mates than males with less elaborate trains. Females, as selectors, look drab in comparison, as their plumage has evolved in response to the natural selection pressure to blend in and avoid predators. Yet, more recent theoretical work, including on humans, shows that females do indeed compete for mates as well, though often in less direct ways (for an overview, see Fisher, 2017).

Even as feminist scholars like Blackwell and Gamble offered rebuttals well over a century ago, much evolutionary psychological research has been historically misled due to assumptions of women being less aggressive or competitive with respect to mating, and it has too often neglected topics such as mothering, female social behavior, female reverse dominance hierarchies, women's formations of alliances, aggression, intrasexual competition, and, generally, women's role in human evolution at large (Fisher, Sokol Chang, & Garcia, 2013)

Axiomatic to some of the early work in evolutionary psychology was that male–male competition was the driving force of sexual selection in humans, and with it the evolution of adaptive traits (be they morphological or behavioral). One example is the *Man the Hunter* hypothesis, in which many of humanity's intellectual achievements, including language and complex cooperation, were thought to be the result of the selective pressure enacted on men to hunt large game (Lee & Devore, 1968). This hypothesis was countered by the *Woman the Gatherer* hypothesis (Tanner & Zihlman, 1976), positing that much of what we consider fundamental to human nature could arguably be explained better by the gathering of nutrients and preparation of food for social consumption mostly by women, but also by men. However, the situation is far from clear-cut; in some hunter–gatherer societies, men do contribute significantly in terms of total calories (e.g., Hill & Kaplan, 1999), while in others, men's contributions are far less (e.g., Hawkes, O'Connell, & Blurton Jones, 1997). This inconsistency led to the model being modified, reducing the focus on men supporting women and instead emphasizing the role of female kin assisting each other, as is seen in allomothering. This assistance, primarily in terms of food acquisition and preparation, underpins humans' relatively fast rate of reproduction (for a review, see Mace, 2013). Once one delves into the topic more deeply, the significant role women are playing in evolution becomes apparent, and in many cases it has been female evolutionary scholars (Tanner, Zihlman, Hawkes, and Mace) who have provided this research.

This oversight is not limited to evolutionary studies. In comparison to men, shockingly little is known about women's evolved physiology and individual differences in women's biology. Reproductive physiology is an exception, but only as women endure conditions that men do not (for a review of women's reproductive physiology from an evolutionary perspective, see Vitzthum, 2008). Recently, some have noted a decline in male bias in human studies, but it remains entrenched (and has even increased) in other non-human animal work (Beery & Zucker, 2016). In realms of medical science such as clinical research, including animal model drug trials, research tends to more often be based on male subjects in order to avoid the complexities of and individual differences in female reproductive physiology (Schmucker, O'Mahony, & Vesell, 1994), even though such beliefs are without foundation (Beery, 2018). When these topics are examined, they are often viewed in terms of how they affect women specifically, rather than how observed patterns directly affect human evolution overall. In contrast, theorizing about men has often been stated in terms of humans more generally. We note, though, that in the past decade there have been some small steps toward correcting this bias (e.g., Robles & Kane, 2014). This pattern is not only applicable to evolutionary psychology, but also to those studying many aspects of human biology; indeed, even mass media magazines still contain stories about how physicians are wrongfully treating women as though they are men (e.g., Adler, 2017).

Despite these issues, the tide is turning, and like many others working within the evolutionary framework, we argue that it is time for (at least some) feminist and evolutionary perspectives to become more integrated. It is critical to point out that there is no singular view of feminism (and hence no unitary, integrated structure that can be simply and accurately called feminism). There is no central, general theory that holds feminist views together (Eagly & Wood, 2011), and indeed, even the terms "feminist" and "theory" are debated within the feminist community (for a review, see Radtke, 2017). Vandermassen (2005) clearly demonstrates the array of feminisms, which vary substantially according to how they view equality, as well as the emphasis they place on, for example, power dynamics, identity, and political activism. Likewise, Campbell (2006) points out that feminism is highly diversified and composed of many differing feminisms. She finds that academic feminists can be broadly characterized as social constructionists (i.e., rejecting the scientific method) or liberals (i.e., accepting the scientific method but taking issue with the studied topics and conclusions). Liberalists can then be classified as environmental (i.e., gender differences result

from culture) or evolutionary (i.e., accepting a role of natural and sexual selection). As various forms of feminist scholarship, much like evolutionary sciences, advance, learn, and respond to itself and cross-disciplinary critiques, these definitions and categories become more nuanced and/or blurred. Indeed, while we openly advocate the need to merge at least some parts of feminist thought with evolutionary perspectives, it is unrealistic to expect such a merger with all of the forms of feminism. (As an aside, we further note that we are hardly espousing a new idea because there have been biologists who are feminists [e.g., Haraway, 1978], as well as those working directly toward a feminist evolutionary biology [e.g., Fausto-Sterling, Gowaty, & Zuk, 1997; Smuts, 1995].)

It is unlikely that these intellectual arenas will fully merge because research questions, methodologies, and units of analysis are often quite different across evolutionary perspectives and the array of feminisms. However, we propose that an *integration*, even if limited, will lead to interesting research questions, correct past assumptions, allow for more accurate theory formation and testing, encourage the application of evolutionary research to address real-world problems, and allow feminists to further expand their focus beyond sociocultural influences to include biological principles.

While it could be argued that we are proposing a feminist evolutionary psychology, our view is not limited to only this perspective. Our aim is to see feminists better incorporate evolutionary principles, primarily by way of including biological influences and the rigors of the scientific method. We are not proposing that the feminisms, and in particular feminist epistemology, must leave behind their view of gender, for example, or that gender is embedded in social relations that may be hierarchical, but rather that there is "value added" by including evolutionary views. In other words, both feminist studies and evolutionary perspectives of humans may benefit, and consequently we propose that in order to move forward, an integration between these two perspectives is warranted and needed. Indeed, there have been many researchers who have tried to integrate these perspectives, beginning with Blackwell (1875) and Gamble (1894) well over a century ago. Obstacles have not only prevented further integration, but have kept the innovative research that does exist from getting the attention it deserves.

One of the main sticking points that divides feminist scholarship from evolutionary psychological viewpoints is the role of social constructionism, which is at the heart of much feminist writing (e.g., Heywood, 2013). Feminists often argue that evolutionary psychology relies on essentialism and is plagued with biological determinism, and instead they favor social and cultural influences (e.g., Eagly & Wood, 2011). Social constructionism generally refers to "the belief that rather than our living in a readily knowable "out there" reality, we dwell in a world that is socially constructed, constructed by our experiences with others, and validated consensually and communally"

(Barkow, 2006, p. 24). Rather than being juxtaposed against evolutionary psychology, we (and others, such as Heywood, 2013) agree that the latter may be informed by the former. As Barkow (2006) reviews, we are a culture-bearing species, meaning humans have an incredible reliance on socially transmitted information. This capacity to socially transmit and receive information itself has evolved, which is key for any argument about social construction. Moreover, it is this capacity for the human brain and mind to adapt to social environments – built into our evolved physiology – that is so important to the natural history of our species. Humans rely on information to learn about local environmental (i.e., geographic, social, familial, cultural) conditions to help respond and thrive, meaning that we have evolved mechanisms that consequently permit social construction. In this way, social construction is not *against* biological explanations of behavior, but rather can be synergistic with a biological explanation; both are constrained by our evolved brains and bodies. As Janicki and Krebs (1998, p. 202) wrote, "purely biological models will not be able to supply a full account of human behavior without reference to cultural process."

In summary, evolutionary theory has existed for well over a century, and feminist responses to it commenced shortly after its emergence. Additionally, feminisms have long studied the effect of culture, which in turn is created by evolved minds. However, both disciplines have historically failed to see where integration was obvious, necessary, or beneficial. These impediments must be identified, reviewed, and eliminated. Therefore, we will next provide an overview of ongoing obstacles to moving forward with integration between feminist scholarship and evolutionary psychological research. We then explore how gender, as an isolated, socially constructed variable, has been viewed historically within evolutionary psychology (and related disciplines). This discussion will require considering both the social construction of gender and sexuality and the role of the biological material body in shaping some aspects of gender and sexual expression. Last, we briefly discuss views of sex differences, as this area in particular has caused tension between evolutionary psychology and feminist studies.

31.2 OBSTACLES TO MOVING FORWARD

One obstacle may have been that the intellectual timing was not "right" for fields relying on evolutionary perspectives or feminist scholarship. Indeed, there has been evolutionary research lauding the role of women for decades, but little of it has gained traction. More recently, it seems a zeitgeist for the topic has occurred. Special issues of peer-reviewed journals have been devoted to the topic: *Sex Roles* (2011 and 2013); a section of the *Journal of Social, Evolutionary, and Cultural Psychology* (2013); and books such as Vandermassen's (2005) *Who's Afraid of Charles*

Darwin?: Debating Feminism and Evolutionary Theory and Fisher, Garcia, and Sokol-Chang's (2013) edited volume, *Evolution's Empress: Darwinian Perspectives on the Nature of Women*. It is critical to note, though, that elsewhere feminists have called for strong movement toward developing empirical rejoinders and ideological rebuttals to evolutionary psychology, including a special issue of *Hypatia* (2012; Meyers, 2012) and *Dialectical Anthropology* (2014; Grossi et al., 2014).

The time has arrived to have meaningful discussions and debates about incorporating feminist perspectives within evolutionary behavioral research, and hopefully evolutionary perspectives within feminism. Although feminism can be defined in many ways, for the sake of moving forward with the chapter, we view it as largely a "social movement and political program aimed at ameliorating the position of women in society" (Campbell, 2006, p. 63). Campbell points out that many within the field of evolutionary psychology (including human behavioral ecology, sociobiology, ethology, etc.) have written about feminist issues, but when we review the feminist literature, these works have received little attention. The attention that does exist is generally negative (e.g., Kelly, 2014), although there are some refreshing exceptions (e.g., Fehr, 2012).

31.2.1 Problems with Integration

"Reflection is fundamental to all feminist practice," according to Macleod, Marecek, and Capdevila (2014, p. 3). Thus, we begin by reflecting on the long-standing issues that have prevented feminist theories from being more fully integrated with evolution-based theories. Overall, evolutionary behavioral studies and feminism(s) are fields with vastly different intellectual approaches. As Eagly and Wood (2013) note, feminist perspectives are diverse but traditionally and typically emphasize sociocultural influences with the goal of promoting gender equality and altering patriarchal structures. In contrast, feminists argue that evolutionists usually emphasize genetic adaptations, meaning that issues such as gender inequality are seen as inevitable consequences that result from a long history of human adaptations (for a review, see Weaver, 2017). The veracity of these viewpoints is not the focus here, but rather that this perception is prevalent. Therefore, these two broad areas are rooted in entirely distinct histories and epistemologies that typically lead to conclusions that, when juxtaposed, seem to have minimal overlap. Each perspective rests on a set of assumptions regarding the forces and mechanisms that have shaped (and continue to impinge upon) the human condition. Evolutionists immerse themselves in the scientific method and evolutionary theory, asking questions they believe arise from past, objective research. Feminists raise questions about that very objectivity, pointing to the ways in which bias in research and action can have real-world social and political effects (Fisher, Sokol Chang, & Garcia, 2013).

This unease has been written about at length, perhaps most recently by Weaver (2017) and Kelly (2014), but also elsewhere (Fausto-Sterling, 1997; Fedigan, 1986; Gowaty, 1997a, 1997b; Hager, 1997; Hrdy, 1981/1999; Nier & Campbell, 2013; for a review of the tension between psychology and feminisms in general, see Gavey & Braun, 2008; Morgan et al., 2011). Weaver (2017, p. 33) points out that historically the two areas have been at odds, given that "feminists tak[e] issue with poorly supported and harmful conclusions evolutionists make about humans, and evolutionists often [are] up in arms about feminists bringing their 'ideology' into science." Problematic androcentric views, often relying on Western heteronormative ideals (as alluded to earlier when reviewing Wilson's view, for example), continue to undermine the multiple roles women (and men) have played in human evolution.

One potential concern is that there is a misunderstanding of what evolutionary psychology is and is not. Barkow (2006, p. 19) posits that evolutionary psychology is essential for understanding the problems our species face, but it is not a moral guide for action. That is, one can use evolutionary psychology to understand how social problems such as inequality function, or why they become issues, but it is not a commentary on morality (a discussion on the naturalistic fallacy can be found later in this section). There seems to also be a misunderstanding of evolutionary psychology as focused solely on biology, as is easily seen in a statement that opened the 2013 special issue of *Sex Roles*. Konik and Smith (2013) state, "This view of biological and social forces engaging in an interplay to shape the brain and behavior is critically important, as overly simplistic essentialist biological models can have negative repercussions for women" (p. 482). There is no discussion of what work within evolutionary psychology is entirely biological, or what the actual repercussions for women would be, if this were actually correct.

Gowaty (2012) speaks directly to the misunderstandings between feminists and evolutionists. She points out that feminists at times assume that evolutionary interpretations of behavior are arguing for fixed, invariant traits; that evolutionists speak in absolutes regarding human behavior and sex differences. This perception of genetic determinism is also found in sociology texts. Leahy (2012) found sociology textbooks often rejected explicit evolutionary theory as "biological," and thereby incompatible with the flexibility of human behavior. This perspective ignores research that leads to the opposite conclusions regarding flexibility, behavioral plasticity, epigenetics, and other evolutionary work. While it is possible this misinterpretation is intentional (purposely creating straw man arguments, for example), much of this tension can be diffused if both disciplines focus on newer and more nuanced literature.

Most recently, Winegard, Winegard, and Deaner (2014) examined cognitive errors, such as biological determinism and straw man arguments, in 15 widely used sex and gender textbooks. In total, 12 of the 15 discussed evolutionary psychology, and those contained multiple errors. The straw man error and the biological determinism error were the most frequently occurring, but the texts also included the naturalistic fallacy and the intentionalistic fallacy (making judgments based on the assumed intent of the author). This last fallacy is an important one, as some academics see biology-based research as a "form of apologetics for an unjust social system and for myriad other social evils (e.g., sexism, racism, classism)" (p. 475). Interestingly, evolutionary psychologists are no more likely to be conservative than academics in other fields (Tybur, Miller, & Gangestad, 2007).

As an alternative, we argue that evolutionary psychology is quite demonstrably already integrating nonbiological theory. The early influence of Darwinian evolution drew some attention to the psychology of women, such as Shield's (1975) now classic article on the functionalist paradigm and the influence of evolutionary theory on the study of the psychology of women. However, at that time, evolutionary theory was still viewed as part of the "biological determinism" paradigm so widely critiqued by feminists (for examples, see the essays within Hubbard, Henifin, & Fried, 1982).

As we have reviewed elsewhere (Fisher, Sokol Chang, & Garcia, 2013), often non-evolutionist researchers and those who use research findings may conflate "what is" with "what ought to be," which is often called the *naturalistic fallacy*. Using the example of research into the evolution of rape (Thornhill & Palmer, 2001), we argued that this work (on "what is") was seen as applicable for social views and punishment of rape ("what ought to be"). One analysis of this early argument, however, noted that Thornhill and Palmer themselves committed the naturalistic fallacy nine times in their original thesis on rape (Sloan-Wilson, Dietrich, & Clark, 2003). This illustrates well the logical complications of navigating these theoretical waters.

Another example that could be taken out of context is work on the evolution of parenting. Parental investment theory (Trivers, 1972) posits that the sex that has the most obligatory investment is the sex that provides investment, while the other sex can (and often does) flee after mating. For humans, women at the minimum must carry the neonate and then breastfeed, meaning that their obligate investment exceeds that of men. Yet, humans are complex, and selection pressures have led to trends for both maternal and paternal investment. At the very least, there are strong pressures for men to supplement a mother's ability to gather resources during the period she is breastfeeding, the most sensitive period of investment (Marlowe, 2000). It is easy to conflate the obligatory investment provided by mothers (i.e., what is) with an argument for mothers to remain the primary caregiver (i.e., what ought to be), as occurred following World War II and the return of men to the workforce to retake positions held down by women (Badinter, 1982).

We are aware that at times these issues result from research findings being taken out of context, particularly by the news media and at times perhaps social media. Thus, it is imperative that researchers engage with these disseminators in order to present their findings accurately, precisely, and with care so that misunderstandings are limited (Fisher, Kruger, & Garcia, 2011). This attention to detail is especially important given that evolutionary psychological findings receive attention from mass media outlets, including shows such as *The Today Show*, CNN, and NPR's *Talk of the Nation*, and are publically talked about by celebrities (e.g., Gwyneth Paltrow; Weaver, 2017). This is particularly true of more sensational or controversial claims. However, as Konik (2013) points out, "evolutionary research has the potential to be used for political aims, regardless of how thoughtfully the results are communicated" (p. 546). The potential misuse of research then becomes a problem; should certain research topics be avoided because they may yield (politically, socially, or culturally) unsavory findings? Such a dilemma comes face to face with the explicit goal of feminist scholarship of addressing social justice or engaging with scientific results that may be harmful (e.g., Weaver, 2017). Indeed, there are numerous issues to consider, including how feminist praxis can be achieved when research findings may be used to further reinforce sexual double standards. Politicization of research on sex differences is particularly problematic (for a brief review, see Fisher, Sokol Chang, & Garcia, 2013). As Fehr (2012) states, feminists do not simply deny scientific conclusions but instead explore the justifications that led to those conclusions, such as the social context for assumptions like androcentrism or sexism. In this vein, Konik (2013) comments that by focusing on politicization, researchers have failed to "acknowledge that research emphasizing sex differences in evolution has been used to marginalize women" (p. 547). For support, she cites Eagly and Wood (2011, p. 759), who claim that evolutionists' views of sex differences as inevitable consequences of adaptations have led them to be largely unresponsive to socioeconomic and political changes in society. But the arguments go deeper than simply weighing the potential consequences of the work or social contexts; the very theoretical foundation of what inspires the work is questioned. Easterlin (2013) is against the feminist "dogmatic insistence on constructionism, the perspective that [sex] differences are socially produced rather than the result of innate predispositions" (p. 391). She proposes that comprehending sex differences from an evolutionary view leads to a more accurate understanding of how sex differences result in inequality. As Weaver (2017) reviews, feminists disagree, partly because of the foundation on which the evolutionary-based study of sex differences rests. The evolution of anisogamy (i.e., sexual reproduction with unequal sizes of gametes [sperm and ova]) and the resulting male variance in reproduction (Bateman, 1948) that leads to males pursuing a mating strategy of quantity and females of quality are the reasons for feminists' criticism. Parental investment theory (Trivers, 1972; see also Bateman, 1948) leads to the assumptions that women are the more investing sex because of disproportionate costs related to their gametes, gestation, and postpartum childcare, that males profit more readily than females from repeated matings via increased reproductive success, that males are more eager and less discriminating in mating than females, and that male reproductive success is more variable than female reproductive success. As Kokko and Jennions (2008) report, these conjectures lead to committing the Concorde fallacy, particularly as applied to females, which is when animals (including humans) defend an investment where the defense costs more than desertion or abandonment in favor of an alternative strategy. Humans, like animals, must make decisions depending on future potential payoffs rather than past costs. Past reproductive decisions may affect future payoffs (e.g., a woman may have limited her ability to invest time and proper resources into her future children), but Kokko and Jennions show that the argument remains weak and that theory based on anisogamy must be applied with caution (see also Ellingsen & Robles, 2012). It is indisputable that the applicability of Trivers' and Bateman's work to the study of humans is problematic; the assumption of greater female investment in reproduction and greater male variation in reproductive success does not apply to all species, and primates are a particular exception (e.g., Hrdy, 1981/1999, 1986).

Liesen (2013) proposes an easy solution: those interested in women's behavior should look at evolutionary biology and behavioral ecology, as evolutionary psychology relies on old models derived from those fields that do not incorporate recent findings on phenotypic and behavioral plasticity, the flexibility of mating behaviors in both men and women (e.g., Gowaty & Hubbell, 2005), and corrections to female passivity that were initially documented by Bateman (1948) and by Trivers (1972). Essentially, many of the tensions between evolutionists and feminists are based on research that is either outdated, sensationalized and politicized, or (in the case of the naturalistic fallacy) misunderstood by both parties. The charge to both evolutionists and feminists is to examine the most recent, in-depth, and nuanced research findings when building foundations for their arguments.

Interestingly, Gowaty (2013) argues that Bateman's findings were flawed and proposes that instead of examining sex differences, one should redirect the focus to individual differences. This shift removes the possibility of politicizing sex differences. Heywood (2013) makes a similar case, contending that the use of genetic theory (and biological essentialism more broadly) leads to the belief that sex differences are hardwired. Consequently, women are persistently seen as caretakers and mothers, given that theory surrounding reproductive success dictates that their primary concern is to keep infants alive. Men, in contrast, are seen as needing to compete for mates and resources, which allows them to be more versatile and to occupy more valuable roles. In general, feminists are concerned that biological explanations for behavior, particularly with regard to psychological sex differences, may be seen as innately part of human nature, and hence hardwired and not changeable (Eagly & Wood, 2013). Feminists who rely on social constructionism typically reject determinism, meaning that there is a reluctance to say that there is a specific cause for issues such as gender inequality. That is, for them there is no higher-order theory that can be tested. Consequently, their explanations tend to be circular, answerable only to themselves, and lacking a completeness that incorporates multiple lines of evidence or perspectives.

A final obstacle worth mentioning pertains to views of feminists held by those working within evolutionary science. Some, like Richardson (2010), argue that feminist perspectives of science are often reduced to the extent that they are caricatured and overly simplified. She reports that there is a spirit of villainizing feminists, with the charge that feminists are in some way anti-science in that they may diminish the influence of science, deny findings that are incongruent with feminist ideals, or reject science outright by espousing that it is biased and fraudulent in terms of objectivity.

Indeed, even the thought of trying to integrate evolutionary psychology with feminism can lead to some unpleasant personal experiences. For example, Keller (1982) writes about her nervousness and defensiveness with respect to the potential conflict between science (she is a mathematical biologist) and feminism. While perhaps not as large an impediment to merging evolutionary psychology with feminism as theoretical concerns, the negative experiences of feminists and the trepidation of those trying to merge conflicting areas are important and need acknowledgment.

31.2.2 Small and Slow Progress Is Still Progress

New developments in evolutionary theory, from the modern synthesis to the extended synthesis of evolutionary biology, show significant promise for integrating feminist and evolutionary approaches to the study of gender and sexuality in behavioral sciences (Garcia & Heywood, 2016). These frameworks highlight features of biology such as epigenetics, where the social environment is as much a part of adaptive responses as material biology, and how the two are co-constitutive, one responding to the other, often in heritable ways. Similar frameworks in behavioral ecology, cultural evolution, and gene–culture coevolution are built on a consideration of biological species in a social environmental context.

While strides have been made, we argue that there remains much work to be done. This work needs to come

from both feminist scholarship and evolutionary psychology. We previously noted that incorporating feminist views within evolutionary psychology has made inroads, but there is resistance for the reverse transaction. As scholars working within the evolutionary perspective, we can speak most readily to this direction of integration and, in particular, document examples of ways in which our areas need to address areas of potential conflict to be able to move ahead. One strong example would be Gowaty's reexamination of Bateman and Trivers, reviewed in the previous section: investigation into previously held assumptions, not surprisingly, is a goal of feminist evolutionists (Sokol-Chang & Fisher, 2013).

One obvious gain is that researchers can reexamine theories that are widely conceived to be the pillars of the field and start to question their accuracy, which may lead to novel topics for investigation. Most importantly, there is now actually discourse about integrating various perspectives with evolutionary psychology, and researchers are challenging existing theoretical models, as we have reviewed. These steps toward integration may be small, but they still represent critical and needed progress.

31.3 AN EVOLUTIONARY VIEW OF GENDER

One area where there has been minimal progress is an evolutionary psychology of gender. Indeed, the study of gender is still often reduced to being a debate about whether "nature" or "nurture" has the larger influence (e.g., Eagly & Wood, 2013). A quick literature search (conducted in the fall of 2018) for scientific articles or academic books reveals only a small handful of articles for the keywords "evolutionary psychology" and "gender" where the articles actually address gender as a social construct rather than as a synonym for biological sex. Here, we examine how gender, as a socially constructed variable, has been viewed within evolutionary psychology and the role of the biological material body in shaping various aspects of gender.

Social construction explanations of gender are often founded on the idea of the mind as a tabula rasa (blank slate) where infants are shaped through rewards, learning, social models, and so on, until they meet the societal expectations for sex-appropriate behavior (for a review and criticism, see Campbell, 2006). Eagly and Wood (2013) point out that models of socialization have become increasing complex over the past 25 years, such that a wide range of influences (e.g., peers, teachers, media, interaction of traits, developmental processes, sociocultural contexts) and the processes that lead to socialization of gender (e.g., self-regulation, emulation, conformity) have been identified. However, much of the existing social research suggests that as gender roles change, beliefs about gender, including stereotypes, should also change, but this has yielded little supportive evidence (Campbell, 2006). While there can be no serious opposition to the conclusion that environment and culture influence the expression of sex differences, we should be cautious of propositions that then conclude they have no biological basis. As Campbell (2006) points out, if sex/gender differences were arbitrary, "it is a curious coincidence that they follow such a similar pattern around the world" (p. 80), and further, Barkow (2006) makes it clear that culture is created by human minds that in turn have been shaped by evolutionary forces.

As behavioral and social scientists have shown time and again, the issues of the magnitude of gender differences and similarities remain unresolved (for reviews, see Eagly & Wood, 2013; Hyde, 2014). Moreover, when there are statistically significant gender differences observed in research, one can ask whether the effect sizes are meaningful and robust. That is, while a small or medium aggregate percentage difference may exist between men and women in a given study, if there is significant overlap between the genders, we can and should ask whether that difference is significant enough to intuit that the difference has resulted from selection on distinct sex-specific patterns (Hyde, 2014). Conversely, we should question whether such a difference is attributable to cultural factors that shape masculinity and femininity or is a result of both biology and culture. These questions are perhaps best asked in the domain of evolutionary biology, in terms of mechanisms of selection, trait distribution, and frequency selection, but they nonetheless are pivotal to consider in evolutionary psychology.

We propose that evolutionary psychology focuses too often on sex differences, rather than addressing potential similarities; this discussion is rarely about gender itself. One valuable way for evolutionary psychology to contribute to this area of research is through its reliance on both natural and sexual selection. Sexual selection is the process in which members of one sex (and presumably gender) compete among themselves to gain mating access, while members of the other sex (and gender) show preferences for particular characteristics in a potential mate. In contrast, natural selection addresses the same adaptive pressures on both sexes and should lead to similarities. By examining characteristics in light of the potential selective pressures that lead to their formation, an evolutionary perspective could easily be used to examine both sex and gender differences and similarities.

Research on gender differences and similarities are important for several practical reasons. First, findings from psychological research may be used to guide the creation and implementation of policy. One obvious example is how research findings have led to the creation of single-sex classrooms (Eckes & McCall, 2014); another is whether public institutions will allow trans individuals to use bathrooms congruent with their self-reported gender or sex (e.g., Luecke, 2011). Second, research results may lead to conclusions that then become limiting stereotypes that may influence behavior, either by biasing perceptions or by causing the creation of a self-fulfilling prophesy. The classic example is the belief that girls are not able to perform well in mathematics, a stereotype that develops before there are actual differences in performance (e.g., elementary school students; Cvencek, Meltzoff, & Greenwald, 2011).

Mate choice in particular must be discussed, as it clearly demonstrates a difference between evolutionary psychology and feminist scholarship, as examined within a context of gender roles. Eagly and Wood (2013), among others, argue that evolutionary psychology's reliance on self-report measures is problematic. While results show a sex difference, for example, in mate preferences or sexual behavioral preferences (e.g., desired number of mates, sexual variety, preference for casual sex), these differences might at least partly stem from social desirability and adherence to gender role norms. One study demonstrates this issue very clearly. Using the bogus pipeline methodology (i.e., fake lie detector), Alexander and Fisher (2003) found that American men report more diverse and earlier sexual experiences than American women, adhering to gender norms for the USA, particularly if they believed their responses would be viewed (with attribution to the participant's identity) by a peer. However, when they believed lying could be detected, the responses showed minimal sex differences. Likewise, Eastwick et al. (2011) have pointed out that implicit measures (such as attitudinal views, spontaneous associations between concepts) may produce more accurate findings than explicit measures (e.g., self-report data as revealed on surveys), as they also tend to show less of a noticeable sex difference. We additionally note that psychology, like other social sciences, is facing problems of replicability of published findings (Camerer et al., 2018), and hence further work is needed.

Gender roles also exist within the sociocultural context in which they occur. Given that much of evolutionary theorizing has taken place within WEIRD cultures (Western, Educated, Industrialized, Rich, and Democratic) and that the data used to test evolutionary hypotheses often come from WEIRD populations, cultures that do not fit the norms of these groups are important to include (Barkow, 1989, 2006; Henrich, Heine, & Norenzayan, 2010). For example, evolutionists theorize that men more than women have a preference for sexual variety and extramarital sex. However, as Eagly and Wood (2013) report, some non-industrial societies reveal prevalent extramarital sex by women, which may be even supported by social structures. Thus, the exciting challenge for evolutionary psychology is to understand how the heterosexual reproductive motive interacts with sexual differentiation and the cultural construction and regulation of gender roles, which themselves may be cultural (if not genetic) adaptations to environments.

What is truly missing, though, is empirical or even theoretical research into socially constructed gender (i.e., femininity, masculinity, androgenity, and undifferentiated individuals; e.g., Bem, 1974) from an evolutionary perspective. How do recent findings from gender and sexuality studies relate to findings from evolutionary psychology? Searching the literature reveals a mammoth gap, with only a small handful of exceptions. For example, VanderLaan, Petterson, and Vasey (2016) examined adulthood gender independently of genetic sex and found femininity positively correlated with uncle/aunt-like tendencies among androphilic (i.e., sexually attracted and aroused by men) males and females. Among the *fa'afafine* in Samoa (Vasey & Bartlett, 2007), there is a culturally recognized third gender of androphilic males who adopt a unique feminine gender role. The evolutionarily informed work of Vasey, VanderLaan, and colleagues on the *fa'afafine* reminds us that gender categories, and their expression, are fundamentally tied to varying cultural systems around the world. For instance, this team found that families containing *fa'afafine* have better survival and reproduction outcomes, suggesting a possible role for kin selection in the presence and maintenance of a third gender in this population (VanderLaan et al., 2012).

We see a small yet promising step forward with respect to acknowledging sexual diversity, which may be useful to note here. Women's sexual fluidity is now a topic of research within evolutionary psychology (Apostolou, 2016; Khule & Radtke, 2013), although there is limited theorizing on "what sexual orientation orients" (Diamond, 2009). There also remain many other forms of sexual expression and orientation to examine, and their interplay with gender or biological sex will hopefully become a topic of interest to those working within the evolutionary perspective.

31.4 MOVEMENT TOWARD RECONCILIATION

There are some criticisms within the feminist scholarship regarding evolutionary psychology that are unlikely to be reconcilable (and vice versa). There is criticism that evolutionary psychology has been functioning too much in isolation and not interacting constructively with other disciplines in the social sciences and humanities, and hence there are limited opportunities for evolutionary-based scholars to be exposed to alternative theories about human behavior (e.g., Grasswick, 2010). The androcentrism and inclusion of sexist assumptions at the core of evolutionary psychological theory are also problematic (Hubbard, 1990; Martin, 2003). Feminists such as Fausto-Sterling (1997) have argued that human culture is simply too complex to be adequately captured by evolutionary psychology. Last, feminists have also stated that evolutionary psychology is morally irresponsible and may be used to compellingly justify harmful stereotypes, such as those prevalent regarding gender (Weaver, 2017).

Despite a blatantly biased starting point to evolutionary psychology, we propose that there have been evolutionary-based feminists contributing work even prior to its inception, like Blackwell (1875) and Gamble (1894). Whether they have been heard is another matter. These contributions have continued with research such as Tanner and Zihlman's work on gathering (1976), Sarah Hrdy's work on infanticide (1980) and cooperative breeding (e.g., 2009), Patricia Gowaty's work on mate choice (1993), and social relationships among female chimpanzees amid much earlier research solely focused on male–male relations (for a review, see Baker & Smuts, 1994). Within the last few decades, evolutionary-based scholars have been increasingly noticing the need to include feminist scholarship in their theory development (Fisher, Garcia, & Sokol Chang, 2013; Vandermassen, 2005). We are increasingly concerned that those who are dismissing the field of evolutionary studies in broad brushstrokes, on the grounds that it fails to include feminist thinking, may not have noticed this decades-long body of research and may be too easily tossing away a very useful framework that bridges much of the natural and social sciences. Similarly, however, those dismissing feminist approaches, including contributions of queer and trans theory to the study of sexuality, are no less uninformed and unaware of new integrative scholarship.

It is unsettling that some working within a feminist viewpoint, such as Kelly (2014), view the potential integration of feminisms and evolutionary psychology as infeasible and unproductive. Kelly asserts that any attempts by those working within an evolutionary perspective to integrate the two views are doing so on the grounds of showing their field is "politically progressive … a springboard for a politicized feminism" (p. 3; see also Weaver, 2017). She further proposes that the attempts that have been made, such as Fehr (2012), Fisher, Garcia, and Sokol-Chang (2013), and Vandermassen (2005), do not adequately capture the achievements of feminist scholarship. She dismisses these works outright and comments that some of these contributions (e.g., Fisher et al., 2013) are outdated in their attempts, in that they are starting with the "Where are the women?" question that spurred much of modern feminist thinking. The inference Kelly makes is that once these scholars get knee-deep in more advanced feminist thought, they will need to abandon their evolutionary perspective.

There are at least two immediately identifiable problems with this line of argument. First, it assumes that feminist thinking will remain unusually stable rather than continue to change, while in fact feminist scholars working on the topics of gender and sexuality are more than ever before integrating information from the biological sciences into their writing (for a review, see Radtke, 2017). Second, if evolutionary perspectives continue to be altered and move in novel directions by examining and integrating feminisms, it is simply not possible to try to accurately predict outcomes. For example, while it is highly unlikely that those integrating feminist scholarship into their evolutionary theorizing will become entirely social constructionist, it is also highly unlikely they will not consider the important role of dynamic social environments on behavior (Fehr, 2012). As theoretical models and empirical evidence in both "camps" advance, so too does the thinking.

We staunchly propose that it is possible for an informed and productive integration between feminist thinking and evolutionary psychology, albeit a partial one. This thought is not novel or even new; the need to integrate feminisms and evolutionary-related ideas has been written about extensively elsewhere. We do note, however, that the integration is slow. There have been repeated efforts to integrate the two perspectives by those using the evolutionary framework (e.g., Campbell, 2006; Fedigan, 1986; Fisher, Garcia, & Sokol Chang, 2013; Gowaty, 1997a, 1997b; Hrdy, 1981/1999; Liesen, 2007; Vandermassen, 2005), but these works seemed to have been largely ignored by those working in feminist scholarship. We therefore propose that the slow growth of integration is in part due to feminists lamenting what they see as past insults. The same can be said for evolutionists, who readily dismiss the contributions of feminists. As Hrdy (1981/1999, p. xxvii) writes, although it may be cathartic to respond to the other view (whether it be evolutionary psychology or feminist theory) without proper analysis, it postpones much needed dialogue "that has to happen if we are to build current experiment[ation] in women's rights on a more secure foundation, based on a deeper understanding of why issues like 'women's reproductive rights' are so charged in the first place." Science, scholarship, and ideas evolve. It is important to reject oversimplified models of human behavior that fail to stay abreast of the tensions and developments within fields.

31.4.1 Evidence of Change

To date, attempts to integrate feminist and evolutionary perspectives have tended to embrace the philosophy that it "means paying equivalent attention to selection pressures on females as well as those acting on males. 'Feminism' becomes political only when countervailing biases deny females equal consideration ... Rather than introducing new sources of bias, or seducing researchers into politically correct positions unsupported by evidence, feminist critiques led many of us to revise incorrect starting assumptions" (Hrdy, 2013, pp. xviii–xix). Thus, although some researchers who apply evolutionary theory to their work may suggest that they are feminists simply because they look at women, we agree with others (e.g., Gavey & Braun, 2008) that this is not sufficient. We also note that a recent review of the evolutionary psychology literature finds *less* of a male bias in authorship than in social psychology, but that is a rather discouraging barometer (Meredith, 2013; Schmitt, 2015).

The goal of ameliorating the position of women in society is supported by a large number of evolutionists (e.g., Campbell, 2006), but it was noted previously that "the women were missing" at evolutionary-themed conferences (Sokol Chang et al., 2013). There were very few presentations about mothering, how women have shaped human language development, or women's contributions to the human ancestral diet, for example. "Instead, like many evolutionary-themed conferences, much of the research presented about women ... reflected a greater societal focus on women's mate attractiveness and value" (Sokol Chang et al., 2013, p. 286). This observation led to a call for work that either addressed the specific selective pressures faced by women, empirical reconsiderations of prior androcentric theories and findings, or work that integrated feminist and evolutionary perspectives; this call resulted in an edited volume (Fisher, Garcia, & Sokol Chang, 2013). That volume, like much of the feminist evolutionary psychology research before it, primarily investigated women's behavior with an aim of correcting past biases to create new hypotheses.

There are other feminist-related topics that have garnered recent attention. There has been an increasing amount of work in the last *months* on issues such as allomothering (Kramer & Veile, 2018), the importance of grandmothers (Hawkes & Finlay, 2018), and topics in women's health such as menopause (Mattison et al., 2018), just to list a few. Similarly, important research on male roles has broadened the scope from inter- and intrasexual competition to topics such as parenting (for a review, see Gray & Anderson, 2012). Another example of a topic that has received significant attention within the past decade is women's engagement in competition with other women (Campbell, 2006; Fisher, 2013). While much of this work has been conducted within a heteronormative framework with an emphasis on competing to win mates, more recent work is examining how mothers compete for limited resources that affect their children outside a mating context (Fisher & Moule, 2013). This work positions women as active agents in the mating world, rather than the passive role previously assigned to women (beginning with Darwin, 1871). A last example is Falk's research (2004) on the evolution of language resulting from the vocalizations used by mothers to calm and reassure their infants.

There is movement away from simply listing the prominent women in evolutionary psychology (though this still happens; e.g., Buss, 2013; Schmitt, 2015) toward highlighting research on the unique and shared roles women have played in human evolution. We believe that current sociopolitical and physical environments are important for understanding human cognition and behavior, and they play a prominent role in correcting biases in evolutionary psychology. Yet two truisms ring clearly: "feminism is the radical notion that women are people" (Shear, 1986, p. 6), and "evolutionary psychology is the radical notion that human behavior is part of the natural world" (Geher, 2013). We must acknowledge that humans are biological beings, as well as the importance of history, including evolutionary history, on influencing human behavior and cognition. Eating illustrates this relationship. We can view eating habits as being caused by the food choices of individuals given available resources and individual preferences. However, we can also see how human diets and digestion have been shaped by ancient practices, such as the ability to create fire in order to cook food (Wrangham, 2009). In between the two, historic practices, such as raising cattle for milk, have further shaped the ability of some groups of people to digest lactose (Tishkoff et al., 2007). Such a multilevel approach is not limited to physical traits, but also relates to psychological traits. An integration of feminist and evolutionary perspectives can likewise enhance our understanding of human psychology.

While we remain optimists that an integration of feminist and evolutionary theory can lead to productive, novel, and interesting questions, we fully acknowledge that it will not be a smooth process. There are a number of ways to move toward integration. Fehr (2012) outlines a few of these, with one having strong resonance with us as authors of this chapter. She reviews the influence of Jeanne Altmann in primatology and suggests that part of her ability to reshape the field (to use systematic sampling of behavior) was to embody a non-adversarial approach. Fehr writes, "the possibility of entering a research community with entrenched and problematic methods and theories, and improving it from the inside through constructive epistemic interactions, does exist" (p. 62). She also points out that it takes an enormous investment of time and energy to create and maintain the relationships that facilitate meaningful interactions.

Another starting point, proposed by Tate (2013), is to recognize both genetic influences (favored by evolutionists) and experiential and psychosocial influences (favored by feminist and cultural theorists). Regardless of how one proceeds, an integration of these two frameworks is possible. Dismissing all work by evolutionists is simply shortsighted, limiting, and counterproductive. We have argued that although such dismissal is understandable based on historical androcentric views, there have been considerable developments in understanding human evolution more fully due to the use of feminisms. With more input from feminist scholars, these developments could grow substantially.

REFERENCES

Adler, K. W. (2017). Women are dying because doctors treat us like men. *Marie Claire*. www.marieclaire.com/health-fitness/a26741/doctors-treat-women-like-men.

Alexander, M. G., & Fisher, T. D. (2003). Truth and consequences: Using the bogus pipeline to examine sex differences in self-reported sexuality. *Journal of Sex Research*, **40**(1), 27–35.

Apostolou, M. (2016). The evolution of female same-sex attractions: The weak selection pressures hypothesis. *Evolutionary Behavioral Sciences*, **10**(4), 270–283.

Badinter, E. (1982). *The Myth of Motherhood: A Historical View of the Maternal Instinct*. London: Souvenir Press Ltd.

Baker, K. C., & Smuts, B. B. (1994). Social relationships of female chimpanzees. In R. W. Wrangham, ed., *Chimpanzee Cultures*. Cambridge, MA: Harvard University Press, pp. 227–242.

Barkow, J. H. (1989). *Darwin, Sex, and Status: Biological Approaches to Mind and Culture*. Toronto: University of Toronto Press.

Barkow, J. H. (2006). Introduction: Sometimes the bus does wait. In J. H. Barkow, ed., *Missing the Revolution: Darwinism for Social Scientists*. New York: Oxford University Press, pp. 3–62.

Bateman, A. J. (1948). Intra-sexual selection in *Drosophila*. *Heredity*, **2**, 349–368.

Beery, A. K. (2018). Inclusion of females does not increase variability in rodent research studies. *Current Opinion in Behavioral Sciences*, **23**, 143–149.

Beery, A. K., & Zucker, I. (2016). Sex bias in neuroscience and biomedical research. *Neuroscience and Biobehavioral Reviews*, **35**(3), 565–572.

Bem, S. L. (1974). The measurement of psychological androgyny. *Journal of Consulting and Clinical Psychology*, **42**(2), 155–162.

Blackwell, A. B. (1875). *The Sexes throughout Nature*. New York: Putnam.

Buss, D. M. (2013). Feminist evolutionary psychology: Some reflections. *Journal of Social, Evolutionary, and Cultural Psychology*, **7**(4), 295–296.

Camerer, C. F., Dreber, A., Holzmeister, F., et al. (2018). Evaluating the replicability of social science experiments in *Nature* and *Science* between 2010 and 2015. *Nature Human Behavior*, **2**, 637–644.

Campbell, A. (2006). Feminism and evolutionary psychology. In J. H. Barkow, ed., *Missing the Revolution: Darwinism for Social Scientists*. New York: Oxford University Press, pp. 63–99.

Cvencek, D., Meltzoff, A. N., & Greenwald, A. G. (2011). Math–gender stereotypes in elementary school children. *Child Development*, **82**(3), 766–779.

Daly, M., & Wilson, M. (1999). *The Truth about Cinderella: A Darwinian View of Parental Love*. Princeton. NJ: Yale University Press.

Darwin, C. (1871). *The Descent of Man, and Selection in Relation to Sex*. London: John Murray.

Dawkins, R., & Carlisle, T. R. (1976). Parental investment, mate desertion and a fallacy. *Nature*, **262**(5564), 131–133.

Diamond, L. M. (2009). *Sexual Fluidity: Understanding Women's Love and Desire*. Cambridge, MA: Harvard University Press.

Eagly, A., & Wood, W. (2011). Feminism and the evolution of sex differences and similarities. *Sex Roles*, **64**, 758–767.

Eagly, A. H., & Wood, W. (2013). Feminism and evolutionary psychology: Moving forward. *Sex Roles*, **69**, 549–556.

Easterlin, N. (2013). From reproductive resource to autonomous individuality? Charlotte Brontë's *Jane Eyre*. In M. L. Fisher, J. R. Garcia, & R. Sokol-Chang, eds., *Evolution's Empress: Darwinian Perspectives on the Nature of Women*. New York: Oxford University Press, pp. 390–405.

Eastwick, P. A., Eagly, A. H., Finkel, E. J., & Johnson, S. E. (2011). Implicit and explicit preferences for physical attractiveness in a romantic partner: A double dissociation in predictive validity. *Journal of Personality and Social Psychology*, **101**(3), 571–574.

Eckes, S. E., & McCall, S. D. (2014). The potential impact of social science research on legal issues surrounding single-sex classrooms and schools. *Educational Administration Quarterly*, **50**(2), 195–232.

Ellingsen, T., & Robles, J. (2012). The evolution of parental investment: Re-examining the anisogamy argument. *Journal of Theoretical Biology*, **299**, 113–119.

Falk, D. (2004). Prelinguistic evolution in early hominins: Whence motherese? *Behavioral and Brain Sciences*, **27**, 491–534.

Fausto-Sterling, A. (1997). Feminism and behavioral evolution: A taxonomy. In P. Gowaty, ed., *Feminist and Evolutionary Biology: Boundaries, Intersections, and Frontiers*. New York: Chapman and Hall, pp. 42–60.

Fausto-Sterling, A., Gowaty, P. A., & Zuk, M. (1997). Evolutionary psychology and Darwinian feminism. *Feminist Studies*, **23**, 403–418.

Fedigan, L. M. (1986). The changing role of women in models of human evolution. *Annual Review of Anthropology*, **15**, 25–66.

Fehr, C. (2012). Feminist engagement with evolutionary psychology. *Hypatia*, **27**(1), 50–72.

Fisher, M. (2013). Women's intrasexual competition. In M. Fisher, J. Garcia, & R. Chang, eds., *Evolution's Empress: Darwinian Perspectives on the Nature of Women*. New York: Oxford University Press, pp. 19–42.

Fisher, M., ed. (2017). *The Oxford Handbook of Women and Competition*. New York: Oxford University Press.

Fisher, M., & Moule, K. (2013). A new direction for intrasexual competition research: Cooperative versus competitive motherhood. *Journal of Social, Evolutionary and Cultural Psychology*, **7**, 318–325.

Fisher, M., Kruger, D., & Garcia, J. (2011). Understanding and enhancing the role of the mass media in evolutionary psychology education. *Evolution: Education and Outreach*, **4**(1), 75–82.

Fisher, M., Garcia, J., & Sokol Chang, R., eds. (2013). *Evolution's Empress: Darwinian Perspectives on the Nature of Women*. New York: Oxford University Press.

Fisher, M., Sokol Chang, R., & Garcia, J. (2013). Introduction to *Evolution's Empress*. In M. Fisher, J. Garcia, & R. Chang, eds., *Evolution's Empress: Darwinian Perspectives on the Nature of Women*. New York: Oxford University Press, pp. 1–16.

Fisher, R. A. (1930). *Genetical Theory of Natural Selection*. Oxford: Clarendon Press.

Gamble, E. B. (1894). *The Evolution of Woman: An Inquiry into the Dogma of Her Inferiority to Man*. New York: G. P. Putnam's Sons.

Garcia, J. R., & Heywood, L. L. (2016). Moving toward integrative feminist evolutionary behavioral sciences. *Feminism & Psychology*, **26**(3), 327–334.

Gavey, N., & Braun, V. (2008). Editorial. *Feminism and Psychology*, **18**(1), 5–12.

Geher, G. (2013). Feminism and evolutionary psychology: Complementary. *Psychology Today*. www.psychologytoday.com/ca/blog/darwins-subterranean-world/201305/feminism-and-evolutionary-psychology-complementary.

Gowaty, P. A. (1993). Differential dispersal, local resource competition, and sex ratio variation in birds. *American Naturalist*, **141**(2), 263–280.

Gowaty, P. A. (1997a). Darwinian feminists and feminist evolutionists. In P. A. Gowaty, ed., *Feminism and Evolutionary Biology*. New York: Chapman, pp. 1–7.

Gowaty, P. A., ed. (1997b). *Feminism and Evolutionary Biology*. New York: Chapman.

Gowaty, P. (2012). *Feminism and Evolutionary Biology: Boundaries, Intersections and Frontiers*. Berlin: Springer Science & Business Media.

Gowaty, P. A. (2013). A sex-neutral theoretical framework for making strong inferences about the origins of sex roles. In M. L. Fisher, J. R. Garcia, & R. Sokol Chang, eds., *Evolution's Empress: Darwinian Perspectives on the Nature of Women*. New York: Oxford University Press, pp. 85–112.

Gowaty, P. A., & Hubbell, S. P. (2005). Chance, time allocation, and the evolution of adaptively flexible sex role behavior. *Integrative and Comparative Biology*, **45**(5), 931–944.

Grasswick, H. (2010). Scientific and lay communities: Earning epistemic trust through knowledge sharing. *Synthese*, **177**(3), 387–409.

Gray, P. B., & Anderson, K. G. (2012). *Fatherhood: Evolution and Human Paternal Behavior*. Cambridge, MA: Harvard University Press.

Grossi, G., Kelly, S., Nash, A., & Parameswaran, G. (2014). Challenging dangerous ideas: A multi-disciplinary critique of evolutionary psychology. *Dialectical Anthropology*, **38**(3), 281–285.

Hager, L. D., ed. (1997). *Women in Human Evolution*. London: Routledge.

Haraway, D. (1978). Animal sociology and a natural economy of the body politic, part I: A political physiology of dominance. *Signs*, **4**(1), 21–36.

Hawkes, K., & Finlay, B. L. (2018). Mammalian brain development and our grandmothering life history. *Physiology & Behavior*, **193**, 55–68.

Hawkes, K., O'Connell, J. F., & Blurton Jones, N. G. (1997). Hadza women's time allocation, offspring provisioning, and the evolution of long postmenopausal life spans. *Current Anthropology*, **38**, 551–577.

Henrich, J., Heine, S. J., & Norenzayan, A. (2010). The weirdest people in the world? *Behavioral and Brain Science*, **33**(2–3), 61–83.

Heywood, L. L. (2013). The quick and the dead: Gendered agency in the history of Western science and evolutionary theory. In M. L. Fisher, J. R. Garcia, & R. Sokol-Chang, eds., *Evolution's Empress: Darwinian Perspectives on the Nature of Women*. New York: Oxford University Press, pp. 439–461.

Hill, K., & Kaplan, H. (1999). Life history traits in humans: Theory and empirical studies. *Annual Review of Anthropology*, **28**, 397–430.

Hosken, D. J., & House, C. M. (2011). Sexual selection. *Current Biology*, **21**(2), R62–R65.

Hrdy, S. B. (1980). *The Langurs of Abu: Female and Male Strategies of Reproduction*. Cambridge, MA: Harvard University Press.

Hrdy, S. B. (1981/1999). *The Woman That Never Evolved*. Cambridge, MA: Harvard University Press.

Hrdy, S. B. (1986). Empathy, polyandry, and the myth of the coy female. In R. Bleier, ed., *Feminist Approaches to Science*. New York: Teachers College Press, pp. 119–146.

Hrdy, S. B. (2009). *Mothers and Others: The Evolutionary Origins of Mutual Understanding*. Cambridge, MA: Harvard University Press.

Hrdy, S. B. (2013). Overdue dialogue: Forward to *Evolution's Empress*. In M. L. Fisher, J. R. Garcia, & R. Sokol-Chang, eds., *Evolution's Empress: Darwinian Perspectives on the Nature of Women*. New York: Oxford University Press, pp. xv–xix.

Hubbard, R. (1990). *The Politics of Women's Biology*. Piscataway, NJ: Rutgers University Press.

Hubbard, R., Henifin, M. S., & Fried, B., eds. (1982). *Biological Woman: The Convenient Myth*. Rochester, VT: Schenkman Books.

Hyde, J. S. (2005). The gender similarities hypothesis. *American Psychologist*, **60**, 581–592.

Hyde, J. S. (2014). Gender similarities and differences. *Annual Review of Psychology*, **65**, 373–398.

Janicki, M., & Krebs, D. L. (1998). Evolutionary approaches to culture. In C. Crawford & D. L. Krebs, eds., *Handbook on Evolutionary Psychology: Ideas, Issues and Applications*. Mahwah, NJ: Erlbaum, pp. 163–207.

Keller, E. F. (1982). Feminism and science. *Signs*, **7**(3), 589–602.

Kelly, S. (2014). Tofu feminism: Can feminist theory absorb evolutionary psychology? *Dialectical Anthropology*, **38**(3), 287–304.

Khule, B., & Radtke, S. (2013). Born both ways: The alloparenting hypothesis for sexual fluidity in women. *Evolutionary Psychology*, **11**(2), 304–323.

Kokko, H., & Jennions, M. D. (2008). Parental investment, sexual selection and sex ratios. *Journal of Evolutionary Biology*, **21**(4), 919–948.

Konik, J. (2013). Book review: Will we make it to the alter? Seeking to unite feminist and evolutionary psychology. *Sex Roles*, **69**, 546–548.

Konik, J., & Smith, C. (2013). In search of complexity: Seeking to integrate feminist and evolutionary perspectives in psychology. *Sex Roles*, **69**, 481–483.

Kramer, K. L., & Veile, A. (2018). Infant allocare in traditional societies. *Physiology & Behavior*, **193**, 117–126.

Leahy, T. (2012). The elephant in the room: Human nature and the sociology textbooks. *Current Sociology*, **60**(6), 806–823.

Lee, R. B., & DeVore, I., eds. (1968). *Man the Hunter*. New York: Aldine.

Liesen, L. (2007). Women, behavior, and evolution: Understanding the debate between feminist evolutionists and evolutionary psychologists. *Politics and the Life Sciences*, **26**(1), 51–70.

Liesen, L. (2013). The tangled web she weaves: The evolution of female–female aggression and status seeking. In M. L. Fisher, J. R. Garcia, & R. Sokol-Chang, eds., *Evolution's Empress: Darwinian Perspectives on the Nature of Women*. New York: Oxford University Press, pp. 43–62.

Luecke, J. C. (2011). Working with transgender children and their classmates in pre-adolescence: Just be supportive. *Journal of LGBT Youth*, **8**(2), 116–156.

Macleod, C., Marecek, J., & Capdevila, R. (2014). *Feminism & Psychology* going forward. *Feminism & Psychology*, **24**(1), 3–17.

Mace, R. (2013). Cooperation and conflict between women in the family. *Evolutionary Anthropology*, **22**(5), 251–258.

Marlowe, F. (2000). Paternal investment and the human mating system. *Behavioural Processes*, **51**(1), 45–61.

Martin, E. (2003). What is "rape"?—Toward a historical ethnographic approach. In C. Travis, ed., *Evolution, Gender, and Rape*. Cambridge, MA: MIT Press, pp. 363–381.

Mattison, S., Moya, C., Reynolds, A., & Towner, M. C. (2018). Evolutionary demography of age at last birth: integrating approaches from human behavioural ecology and cultural evolution. *Philosophical Transactions of the Royal Society B*, **373** (1743), 20170060.

Meredith, T. (2013). A journal of one's own. *Journal of Social, Evolutionary, and Cultural Psychology*, **7**(4), 354–360.

Meyers, D. T. (2012). Feminist critiques of evolutionary psychology. *Hypatia*, **27**(1), 1–2.

Money, J., Hampson, J. G., & Hampson, J. L. (1955). An examination of some basic sexual concepts: The evidence of human hermaphroditism. *Bulletin of the Johns Hopkins Hospital*, **97**, 301–319.

Morgan, M., Coombes, L., Neill-Weston, F., & Weatherley, G. E. (2011). Shaping feminist psychologies in Aotearoa: History, paradox, transformation. In A. Rutherford, R. Capdevila,

V. Undurti, & I. Palmary, eds., *Handbook of International Feminisms: Perspectives on Psychology, Women, Culture, and Rights*. New York: Springer, pp. 195–218.

Muehlenhard, C. L., & Peterson, Z. D. (2011). Distinguishing between sex and gender: History, current conceptualizations, and implications. *Sex Roles*, **64**(11), 791–803.

Nier, J. A., & Campbell, S. D. (2013). Two outsiders' view on feminism and evolutionary psychology: An opportune time for adversarial collaboration. *Sex Roles*, **69**, 503–506.

Radtke, H. L. (2017). Feminist theory in *Feminism & Psychology* [part 1]: Dealing with differences and negotiating the biological. *Feminism & Psychology*, **27**(3), 357–377.

Richardson, S. S. (2010). Feminist philosophy of science: History, contributions, and challenges. *Synthese*, **177**(3), 337–362.

Robles, T. F., & Kane, H. S. (2014). Normative processes, individual differences, and implications for health. *Journal of Personality*, **82**(6), 515–527.

Rosenberg, K., & Trevathan, W. (2002). Birth, obstetrics and human evolution. *British Journal of Obstetrics and Gynaecology*, **109**, 1199–1206.

Saini, A. (2017). *Inferior: How Science Got Women Wrong and the New Research That's Rewriting the Story*. Boston, MA: Beacon Press.

Schmitt, D. P. (2015). On accusations of exceptional male bias in evolutionary psychology: Placing sex differences in citation counts in proper evidentiary contexts. *Evolutionary Behavioral Sciences*, **9**(2), 69–72.

Schmucker, D. L., O'Mahony, M. S., & Vesell, E. S. (1994). Women in clinical drug trials: An update. *Clinical Pharmacokinetics*, **27** (6), 411–417.

Shear, M. (1986). A review of *A Feminist Dictionary*. *New Directions for Women*, **15**(3), 6.

Shields, S. A. (1975). Functionalism, Darwinism, and the psychology of women: A study in social myth. *American Psychologist*, **30** (7), 739–754.

Sloan-Wilson, D., Dietrich, E., & Clark, A. (2003). On the inappropriate use of the naturalistic fallacy in evolutionary psychology. *Biology and Philosophy*, **18**(5), 669–681.

Smuts, B. (1995). The evolutionary origins of patriarchy. *Human Nature*, **6**(1), 1–32.

Sokol-Chang, R., & Fisher, M. L. (2013). Letter of purpose of the Feminist Evolutionary Psychology Society. *Journal of Social, Evolutionary, and Cultural Psychology*, **7**(4), 286–294.

Tanner, N., & Zihlman, A. (1976). Women in evolution part I: Innovation and selection in human origins. *Signs: Journal of Women in Culture and Society*, **1**(3 Part 1), 558–608.

Tate, C. C. (2013). Addressing conceptual confusions regarding evolutionary theorizing: How and why evolutionary psychology and feminism do not oppose each other. *Sex Roles*, **69**, 491–502.

Thornhill, R., & Palmer, C. T. (2001). *A Natural History of Rape: Biological Bases of Sexual Coercion*. Cambridge, MA: MIT Press.

Tishkoff, S. A., Reed, F. A., Ranciaro, A., et al. (2007). Convergent adaptation of human lactase persistence in Africa and Europe. *Nature Genetics*, **39**(1), 31–40.

Trivers, R. (1972). Parental investment and sexual selection. In B. Campbell, ed., *Sexual Selection and the Descent of Man: 1871–1971*. Chicago, IL: Aldine, pp. 136–179.

Tybur, J. M., Miller, G. F., & Gangestad, S. W. (2007). Testing the controversy. *Human Nature*, **18**(4), 313–328.

Unger, R. K. (1979). Toward a redefinition of sex and gender. *American Psychologist*, **34**, 1085–1094.

VanderLaan, D. P., Forrester, D. L., Petterson, L. J, & Vasey, P. L. (2012). Offspring production among the extended relatives of Samoan men and *fa'afafine*. *PLoS ONE*, **7**(4), e36088.

VanderLaan, D. P., Petterson, L. J., & Vasey, P. L. (2016). Femininity and kin-directed altruism in androphilic men: A test of an evolutionary development model. *Archives of Sexual Behavior*, **45**(3), 619–633.

Vandermassen, G. (2005). *Who's Afraid of Charles Darwin? Debating Feminism and Evolutionary Theory*. Lanham, MD: Rowman and Littlefield.

Vasey, P. L., & Bartlett, N. H. (2007). What can the Samoan "*fa'afafine*" teach us about the Western concept of gender identity disorder in childhood? *Perspectives in Biology and Medicine*, **50**(4), 481–490.

Vitzthum, V. J. (2008). Evolutionary models of women's reproductive functioning. *Annual Review of Anthropology*, **37**, 53–73.

Weaver, S. (2017). A constructive critical assessment of feminist evolutionary psychology. Unpublished dissertation, University of Waterloo.

Wilson, E. O. (1978). *On Human Nature*. Cambridge, MA: Harvard University Press.

Wilson, M., & Daly, M. (1998). Lethal and nonlethal violence against wives and the evolutionary psychology of male sexual proprietariness. In R. E. Dobash & R. P. Dobash, eds., *Rethinking Violence against Wives*. Thousand Oaks, CA: Sage, pp. 199–230.

Winegard, B. M., Winegard, B. M., & Deaner, R. O. (2014). Misrepresentations of evolutionary psychology in sex and gender textbooks. *Evolutionary Psychology*, **12**(3), 474–508.

Wrangham, R. (2009). *Catching Fire: How Cooking Made Us Human*. New York: Basic Books.

PART VIII

ABNORMAL BEHAVIOR AND EVOLUTIONARY PSYCHOPATHOLOGY

Life on Earth has been evolving for some 3.8 billion years. The last common ancestor of *Pan* and *Homo* can be traced back to some 7–10 million years ago (White et al., 2009). Given all of this time for evolutionary change, you would think that people would be in pretty good shape by now. And yet current estimates suggest that around one in four of us will suffer from a mental health problem at some point in our lives (Ray, 2018). Given that mental health issues are, by definition, maladaptive, this state of affairs appears incongruous (Del Giudice, 2018). How might we explain this apparent proneness in our species to "abnormal" psychological states? In recent years, evolutionists have developed a number of hypotheses to explain this conundrum. These include the hypothesis that there is a mismatch between our current environment and the one in which our species arose and the notion that some traits increase reproduction but at a cost to health and longevity. In Part VIII, building on Part III, we examine a number of these recently developed hypotheses. We first present two reviews of the relationship between evolution and abnormal psychology (also known as psychopathology), before moving on to consider two specific areas of recent research.

William J. Ray has a long-standing interest in the relationship between abnormal psychology and evolutionary psychology (he has written well-received textbooks on both). In his contribution, Ray demonstrates how a number of prominent researchers, from John Hughlings Jackson to Jaak Panksepp, have used evolutionary theory to help explain the existence of psychopathology. He then goes on to examine the proposition that psychological adaptations appropriate to our ancestral past may no longer be functional due to the rapid change in our social environment. He considers, in particular, the hypothesis that a large number of specific human abilities are geared to social interactions that many people no longer engage in. Ray suggests that these social changes can lead to a sense of loss and resultant states of depression and anxiety.

Alfonso Troisi is a rare breed indeed, being both a psychiatrist and an expert in evolutionary psychology. Alongside the late Michael McGuire of UCLA, he was key to the development of Darwinian psychiatry during the 1990s (McGuire & Troisi, 1998; see also Nesse & Williams, 1995). In his contribution to Part VIII, Troisi first clarifies the nature of Darwinian psychiatry, before examining a number of mental health issues that evolutionary principles might be brought to bear on. These include the notion that schizophrenia may arise as a by-product of recent positive selection for language, creative thinking, and cognitive abilities. Also considered is the conception of depressive symptoms as adaptive mechanisms that may have evolved to elicit comfort from social contacts. Finally, Troisi argues that, if psychiatric symptoms can be shown to serve adaptive functions, then this will have implications for the treatment of mental disorders.

In the third of our contributions to Part VIII, Paul Gilbert considers the evolution of caring and compassion in our species. Gilbert, who is a recognized leading authority in this field, traces current human caring and compassionate behavior back to the period around two million years ago when hominin ancestors began to develop forms of social intelligence based around language and symbolism. Although Gilbert's chapter is largely concerned with positive social adjustment, it also highlights how developmental processes can lead to outcomes that are generally considered pathological. An evolutionary perspective might suggest, however, that such pathologies are the consequences of specific survival and reproductive strategies given current contexts (Gilbert, 2009).

Rebecca Brewer and Geoffrey Bird consider a fascinating and, until recently, little-understood condition – alexithymia – in our last contribution to Part VIII. Individuals suffering from alexithymia experience emotions but find it very difficult to identify which emotion they are feeling. Brewer and Bird have made significant progress in developing our understanding of alexithymia (Brewer, Cook, & Bird, 2016). In their chapter, they consider the relationship between alexithymia and interoception (the ability to perceive and recognize the internal state of one's body), before considering the possibility that alexithymia may be an adaptation. This reminds us once again that perceived "abnormal" states of mind might, under some circumstances, be either adaptations or derived from adaptive responses that would have been selected during our evolutionary history.

REFERENCES

Brewer, R., Cook, R., & Bird, G. (2016). Alexithymia: A general deficit of interoception. *Royal Society Open Science*, **3**(10), 150664.

Del Giudice, M. (2018). *Evolutionary Psychopathology: A Unified Approach*. New York: Oxford University Press.

Gilbert, P. (2009). *The Compassionate Mind: A New Approach to the Challenge of Life*. London: Constable & Robinson.

McGuire, M. T., & Troisi, A. (1998). *Darwinian Psychiatry*. New York: Oxford University Press.

Nesse, R. M., & Williams, G. C. (1995). *Evolution and Healing: The New Science of Darwinian Medicine*. London: Weidenfeld and Nicolson.

Ray, W. J. (2018). *Abnormal Psychology: Neuroscience Determinants of Human Behavior and Experience*, 2nd ed. Thousand Oaks, CA: Sage.

White, T. D., Asfaw, B., Beyene Y., et al. (2009). *Ardipithecus ramidus* and the paleobiology of early hominids. *Science*, **326** (5949), 75–86.

32 Psychopathology from an Evolutionary Perspective

WILLIAM J. RAY

32.1 INTRODUCTION

The purpose of this chapter is to consider the manner in which evolutionary perspectives offer an additional level of understanding to the field of psychopathology. This perspective offers both long-term and short-term considerations of psychological difficulties in everyday life. For all organisms, one of the main themes of evolution is the manner in which organisms are in close connection with their environment. It is this close connection that allows for change – including the turning on and off of genetic processes – to take place. In psychopathological disorders, this close connection with both the external and internal environment of the person may be dysfunctional.

When a person loses contact with the current environment and applies strategies that worked perhaps in an earlier time, then unsuccessful adaptation is the result. This lack of connectedness to our environment may take place on both external and internal levels. On an external level, the person finds him or herself to be different from the group, or even seeks to be separate from others. This is not our historical experience – humans have never lived as isolated individuals.

As a species, we have always lived in close contact with other humans, leading to the development of societies and cultures. In fact, many of the specific abilities of humans are geared toward social interactions on a variety of levels. When we no longer have the connection with the group, we experience a sense of loss. This loss often carries with it the experience of negative affect and depression and a need to withdraw. On an internal level, humans often have the need to explain to themselves the events that have just occurred, which may include anger, distorted perceptions, or a genuine plan for recovery. We refer to the extreme cases as psychopathology.

Although there is no one single definition of what represents abnormal processes, five ideas have been critical (Ray, 2018):

1 The processes involved in psychopathology are maladaptive and not in the individual's best interest.
2 The processes cause personal distress.
3 The processes represent a deviance from both cultural and statistical norms.
4 The person has difficulty connecting with his or her environment and also with him or herself.
5 There is the inability of an individual with a mental disorder to fully consider alternative ways of thinking, feeling, or doing. That is to say, they often do not see, feel, or think that there are alternatives. This results in their psychological processes being rigid and patterned. Having fewer alternatives also suggests that they have less freedom in any given situation.

Historically, psychology has focused on the signs and symptoms of mental disorders. The emphasis has been on developing systems of diagnosis, such as the *International Classification of Diseases* (ICD), published by the World Health Organization (WHO), and the *Diagnostic and Statistical Manual of Mental Disorders* (DSM), published by the American Psychiatric Association. Although these systems have been updated and informed by current research, evolution has not played a prominent role in diagnostic criteria for psychopathology. Evolutionary perspectives have, however, played an important role in considering how psychopathology might develop.

32.2 MODELS OF PSYCHOPATHOLOGY FROM AN EVOLUTIONARY PERSPECTIVE

Overall, we can ask critical questions concerning psychopathology that relate to other evolutionary processes.

1 First, we ask if the experience of mental illness is universal. If it were not universal, then it would be difficult to argue that we should study psychopathology from an evolutionary perspective. If it is a universal process, such as emotionality or language, then we can begin to

ask what the nature of mental illness is and how its existence fits into our history as humans.

2 Second, we ask if there is an adaptive value to the behaviors and experiences displayed in psychopathology. It is easy to see that there is a value in not trusting what someone tells you some of the time, but is there any adaptive value in not trusting what anyone tells you all of the time or thinking that everyone is always out to get you?

3 Third, we look for evidence of psychopathology across human history. This includes the question of whether we see signs of psychopathology in nonhuman species.

4 Fourth, we seek to understand the nature of psychopathology. That is to say, should we consider psychopathology to be qualitatively different from normal functioning, or is it a situation in which normal processes have been taken to the extreme? We know, for example, that allergic reactions are situations in which our immune system is overreactive. We also know that fever is the process in which body temperature is raised to fight infection; however, the fever uses energy and can damage the body.

5 Fifth, we ask if it is protective in some manner. Like sickle cell anemia, does having schizophrenia or depression, for example, make you less likely to experience another disorder?

6 Sixth, we ask if psychopathology is a recent process. That is, should we consider psychopathology as the result of a mental system designed in the Stone Age interacting with a fast-paced, modern environment?

These questions are not mutually exclusive. Overall, they also represent some of the ways in which scientists and others have sought to understand psychopathology. From an evolutionary perspective, the study of psychopathology begins with the three instincts of survival, sexuality, and socialness suggested by Charles Darwin. From this perspective, psychopathology becomes a disruption of these instinctual processes.

32.2.1 Historical Perspective of Psychopathology

The modern study of psychopathology is generally dated as beginning in the late 1700s. As with the move toward experimentation and empiricism in physics and chemistry, the study of psychopathology initially began with careful description. There was also a shift away from external mechanisms, such as possession by spirits, toward explanations in terms of natural processes. With this shift came a differentiation between patients in mental hospitals who experienced mental disorders from those who were just misfits in their society. Marquis de Sade, who wrote novels of a sexual nature, was released by the director of a mental hospital in Paris during this period with the statement, "He is not mad, his only madness is vice" (Szasz, 1970).

The difference between vice and madness is one of many dichotomies that have plagued the study of psychopathology

over the centuries and continue to this day as larger intellectual and societal questions. For example, is drug addiction a pleasure-seeking mechanism that is being overused? It was only at the end of the twentieth century that addictions such as alcoholism were officially viewed by the federal government as physiological disorders rather than problems of will.

Underlying many of these dichotomies is the dualism associated with the French philosopher Rene Descartes, who, in the 1600s, described human beings as being influenced by both physiological processes, such as those of the body, which could be measured scientifically, and mental processes, including the soul, which follow different laws. Even without the religious questions of Descartes' time, the mind–body dichotomy has influenced psychopathology research. A separate but related question is the extent to which psychopathological symptoms are to be viewed as representing underlying brain pathology or an exaggeration of everyday behavior. "Both" is the answer underlying current therapeutic approaches. This implies that psychosis is the result of a brain disorder, whereas affective disorders such as generalized anxiety disorder are seen as exaggerations of normal processes. Evolutionary perspectives offer the opportunity to move beyond these current dichotomies by focusing on the manner in which psychopathology developed within the context of our evolutionary history.

32.2.1.1 John Hughlings Jackson and Paul MacLean

For an evolutionary perspective on the nervous system, let us turn to the work of John Hughlings Jackson in the 1800s and Paul MacLean in the 1900s concerning the brain and evolution. Hughlings Jackson was interested in brain evolution and its relationship with mental disorders (Ploog, 2003). He introduced the idea of different functional levels in the nervous system. The higher levels of the cortex were seen to inhibit the lower, subcortical ones. Hughlings Jackson further saw the lower levels as more structured, in the sense that they performed fixed processes. One example of such a fixed and automatic process is the startle reflex. We all jump when we unexpectedly hear a loud noise. The higher levels, on the other hand, are more flexible in how they deal with information and include voluntary components in their response.

For Hughlings Jackson, evolution was the development over time of these higher and more complex centers. Evolution was the movement from the more automatic centers of lower brain processes to the more voluntary ones. In mental disorders such as epilepsy, Hughlings Jackson viewed the higher systems as no longer inhibiting the lower ones. Without inhibition, lower brain processes can propagate unchecked. This results in the large electroencephalogram waves seen in epilepsy. Hughlings Jackson called this process "dissolution of the nervous system."

In his view of a hierarchical representation of the brain, Hughlings Jackson was a forerunner to Paul MacLean and his concept of the triune brain. The idea of the triune brain suggested that, from an evolutionary perspective, the brain could be seen as consisting of three major components. The first system was involved in the regulation of basic systems such as temperature, breathing, and sleep, as well as fixed types of displays seen in relation to self-preservation and sexuality. This involves a large part of the forebrain in reptiles and birds. The second system, related to emotionality, came with the development of the limbic system. The transition from reptiles to mammals included three important processes: first, nursing of the infant in the context of maternal care; second, oral communication for maintaining mother–child contact; and third, play and its role in social development. Finally, the third system for MacLean was the neo-mammalian brain. It is this aspect of our human nature that allows us to consider future possibilities, have extensive knowledge of and relationships with our external environment, and perform a large variety of cognitive and social processes. Given these separate levels and types of information, there is the possibility that conflict within the brain might result. For example, what may be important for you on a higher level may be at odds with the initial basic reaction. The basic idea that ties the triune brain to psychopathology is this potential conflict between cognitive, emotional, and reptilian responses.

32.2.1.2 Sigmund Freud

Sigmund Freud was greatly influenced by Darwin. In fact, Freud quotes Darwin in many of his written works (Bailey, 1987; Sulloway, 1979). On some topics, such as the development of emotions, including fears, in children, Darwin and Freud followed parallel tracks. One important theme in both of their works was the role of instinctual processes in child development. Darwin emphasized two main instinctual processes: self-preservation and sexual selection. Similarly, Freud said that he began his studies with the idea that hunger and love are what move the world. However, in his writings, Freud emphasized sexual processes over those of self-preservation. As Freud described in his early work, he viewed instinctual energy as a form of biological energy that could build up unless it was in some way expressed. The lack of expression of sexual energy, for example, was seen to be the basis of various types of psychopathology. Likewise, the experience of sexual and self-preservation instincts, when met with the restrictive rules of a particular society, were seen to lead to anxiety. Overall, Freud saw neurosis as an attempt by individuals to treat the problem they were experiencing. For example, anxious individuals often worry too much about problems or events that could happen in the future. Although consideration of future problems may be helpful in problem-solving, constant consideration or worry is not. Although Freud did not focus on empirical research, more recent researchers have sought to integrate his ideas into the neurosciences (Carhart-Harris & Friston, 2010).

32.2.1.3 Carl Jung

Carl Jung was in many ways a forerunner of the modern evolutionary psychology perspective. He understood that humans come into the world with a variety of instinctual processes and that there is an important genetic component to human behavior. Like Paul MacLean, Jung viewed humans as being influenced by an evolutionary history that could be manifested on a variety of levels, which were capable of conflicting with one another. One important emphasis of Jung's work was his suggestion that all humans are similar in their psychological processes and that these processes have a basis in our evolutionary history. To demonstrate this, Jung collected dreams and stories from around the world that, he suggested, showed similar patterns of relationships. Jung was also interested in symbols and the manner in which humans have used symbols to represent and bring forth artistic and religious meaning. Jung was also important for his conceptualization of psychological development across the lifespan. He described the manner in which the stage of life brings forth a particular pattern of functioning, which he called an archetype.

32.2.1.4 John Bowlby

John Bowlby carefully observed children and recorded his observations in a series of books and articles focused on secure attachment as well as loss and separation anxiety (Bowlby, 1951, 1961, 1969, 1982, 1988). From a development perspective, Bowlby laid out in great detail the manner in which evolutionarily significant processes such as attachment could, in certain circumstances, lead to anxiety rather than comfort. Attachment is the situation in which a caregiver and an infant establish a close psychological and physical relationship, which protects the infant and allows for optimal development.

In terms of the general characteristics of attachment, Bowlby reported that children who develop a secure bond, or attachment, with a caregiver or parent, who is usually their mother, display patterns of activity that are especially strong from the end of the first year of life until about three years of age in relation to that caregiver. First, the infant shows distress when the caregiver leaves. Second, the infant smiles, makes noises, or shows other signs of pleasure when the caregiver returns. Third, the infant shows distress when approached by a stranger, unless the caregiver encourages the interaction. And fourth, the infant shows more exploratory behaviors in an unfamiliar situation when the caregiver is present.

The model that Bowlby articulated was that if an emotional caregiver were not present or a close connection could not be established, then the result would be distress. Initial distress can be seen as a means of attempting to reestablish the connection. Continued distress is seen to lead to fear and anxiety of an enduring nature. Current research with affective disorders such as anxiety has shown a relationship between the degree of anxiety and the type of attachment relationship in childhood. Other researchers have suggested that apparently "abnormal" types of attachment may represent strategies that offer infants a way to obtain resources from their caregivers in difficult situations (Belsky, 2005).

32.2.1.5 Jaak Panksepp

Jaak Panksepp (1998, 2004) suggested that there are seven basic systems from which emotions develop. These systems are anger, fear, sexual lust, maternal care, and separation distress, as well as playfulness and a resource acquisition system. He further suggests that these can also be seen as related to particular psychopathologies. In the following list, I will review the emotional systems and the psychopathologies that develop from them.

1 Anger or rage is evoked when there is stiff competition for resources. This system can also be aroused by restraint, frustration, and other irritations. Out of this system can emerge irritability, contempt, and hatred. Psychopathologies associated with this system would be those involving aggression, such as conduct disorders in children, psychopathic tendencies, and personality disorders.

2 Fear is evoked when the organism is in the presence of danger. This system will evoke freezing at low levels of arousal and flight at higher levels. Research suggests that external stimuli may be processed at fast but less conscious levels through low-level brain circuits or slower but more accurately at high cognitive levels. Out of this system can emerge simple anxiety, worry, and psychic trauma. Psychopathologies associated with this system include generalized anxiety disorder, phobias, and various forms of post-traumatic stress disorder.

3 Sexual lust becomes manifest during puberty, although the basic components of the system exist early in the development of the organism. Out of this system can emerge erotic feelings, as well as jealousy. Psychopathologies associated with this system include fetishes and sexual addictions.

4 Care systems are designed to allow us to nurture one another. Out of this system can emerge nurturance, love, and attraction. Psychopathologies associated with this system include dependency disorders, attachment disorders, and aloofness.

5 Separation distress is seen when a young organism is separated from its mother. In a variety of species, the infant will cry out in these situations. Extreme cases become more of a panic situation. Panksepp suggests that these feelings of abandonment may build on early pain circuits. Out of this system can emerge sadness, guilt, shame, and shyness. Psychopathologies associated with this system include panic attacks, pathological grief, depression, agoraphobia, and social phobia.

6 Play is seen in a variety of species and is often accompanied by the expression of joy and laughter. Play is considered an important preparation for later social life. Out of this system can emerge joy and happy

playfulness. Psychopathologies associated with this system include mania and disorders of hyperactivity, such as attention deficit hyperactivity disorder.

7 The seeking system controls our desire to find and harvest the resources of the world. Across species, it is related to the motivation to obtain resources from the environment. In humans, it is connected with goal-directed urges and positive expectations concerning the world, as well as with awareness and appraisals of the world. Out of this system can emerge interest, frustration, and caring. Psychopathologies associated with this system include obsessive–compulsive disorder (OCD), paranoid schizophrenia, and addictive disorders.

32.3 PSYCHOPATHOLOGY IN TERMS OF BRAIN FUNCTIONING

Networks have been studied in terms of a variety of cognitive and emotional tasks (Bressler & Menon, 2010; Raichle, 2015). These include separate networks involved in the processing of visual information, auditory information, sensorimotor processes, attentional processes, executive control, salience, and the default mode. Three of these networks have been examined in terms of psychopathology (Menon, 2011). These are the baseline or default network (also called the intrinsic network), the central executive network, and the salience network. The default network is active when an individual is not performing a particular task, such as when one's mind wanders or is processing internal information. The central executive network is involved in higher-order cognitive and attentional tasks. The salience network is important for monitoring critical external events as well as internal states. Psychopathological disorders such as schizophrenia, depression, anxiety, dementia, and autism have been shown to have problems in turning networks on or off, as well as problems in the connections within the network itself.

32.4 UNIVERSAL PSYCHOPATHOLOGY

Is the experience of mental illness universal? If psychopathology is part of our human makeup, then we would expect to see similar manifestations of it worldwide. One classic study in this regard was performed by Jane Murphy of Harvard University (Murphy, 1976). It dates from the 1970s, when mental illness was seen to be related to learning and the social construction of norms. In fact, some suggested that mental illness was just a myth developed by Western societies. In this perspective, neither the individual nor his or her acts are abnormal in an objective sense. One important implication of this view was that what would be seen as mental illness in a Western industrial culture might be interpreted very differently in a less developed rural culture. That is to say, mental illness in this perspective was viewed as a social construction of the society. The alternative to this perspective is more similar to what we saw with human processes, such as emotionality, in which humans throughout the world show similar expression of the basic emotions.

If mental illness is part of our human history, as evolutionary psychologists would suggest, then we would expect to find it across a variety of cultures. Dr. Murphy first studied two geographically separate and distinct non-Western groups: the Eskimos of northwest Alaska and the Yorubas of rural tropical Nigeria. Although many researchers of that time would have expected to find the conceptions of normality and abnormality to be very different in the two cultures, this is not what she found. What she found was that these cultures were well acquainted with processes in which a person was said to be "out of their mind." This included doing strange things, as well as hearing voices. Dr. Murphy concluded that processes of disturbed thought and behavior similar to schizophrenia are found in most cultures, and that most cultures have a distinct name in their language for these processes. Additionally, she reported that these cultures had a variety of words for what traditionally is referred to as neurosis, although today we would refer to these as *affective disorders*. Affective disorders include feeling anxious, tense, or fearful of being with others, as well as being troubled and not able to sleep. One Eskimo term was translated as worrying too much until it makes the person sick. Thus, it appears that most cultures have words for what has been called "psychosis," "neurosis," and "normalcy." What is also interesting is that many cultures also have words for describing people who are out of their mind, but not crazy. That is, there are those who perceive alternative versions of reality or invoke non-normal states. These individuals include witch doctors, shamans, and artists. Violation appears to be an important distinction for this concept.

To add evidence to her argument that psychopathology is indeed part of our human nature, Murphy also reviewed a large variety of studies conducted by others that looked at how common mental illness was in different cultures. The suggestion here is that if its prevalence is similar in cultures across the world, then it is more likely to be part of the human condition rather than culturally derived. These studies suggest that common forms of mental illness, such as schizophrenia, are found in similar rates around the world. Overall, this research established that mental illness was not a concept created by a given culture, but rather was part of the human condition in both its recognition and its prevalence. This set the stage for a development that came to be known as *evolutionary psychopathology* or *Darwinian psychiatry*.

32.4.1 Adaptive Value

Psychopathology from an evolutionary perspective also goes beyond the traditional psychological and physiological considerations (Ray, 2013, 2018). One question might be how long, in terms of our human history, a particular psychopathology has existed. Let us take schizophrenia as an example. A WHO study examined the presence of schizophrenia in a number of countries with very different racial and cultural backgrounds (Sartorius et al., 1986). If schizophrenia had an important environmental component, then one would expect to see different manifestations of the disorder in different cultures. What these authors found was that, despite the different cultural and racial backgrounds surveyed, the experience of schizophrenia was remarkably similar across countries. Likewise, the risk of developing schizophrenia was similar in terms of total population presence – about one percent. Further, the disorder had a similar time course in its occurrence, with its characteristics first being seen in young adults.

If you put these facts together, it suggests that schizophrenia is a disorder that has always been part of the human experience. Because it is found throughout the world in strikingly similar ways, this suggests that it existed before humans migrated out of Africa. The genes related to schizophrenia were carried by early humans who migrated from Africa and thus its presence is equally likely throughout the world. Given these estimates on the history of the disorder, one might ask why schizophrenia continues to exist. We know that individuals with schizophrenia tend to have fewer children than individuals without the disorder. Thus, we might assume that schizophrenia would have disappeared over evolutionary time in that it reduces reproductive success and has a genetic component (Huxley et al., 1964). However, this is not the case.

This creates a mystery for evolutionary psychologists to solve. In order to answer this question, we can draw on many of the considerations presented in the discussion of health. Perhaps, in the same way that sickle cell anemia is associated with protection against malaria, schizophrenia protects the person from another disorder. Or perhaps, like the reaction of rats to stress that results in depression-like symptoms, the symptoms seen in schizophrenia are the result of a long chain of stressful events in which the organism breaks down in its ability to function. Psychopathology could even go in a more positive direction and be associated with creative and nontraditional views of the world. For example, there are a variety of accounts that have noted greater creativity in families of individuals with schizophrenia (Andreasen, 2005).

The evolutionary perspective helps us to ask questions such as what function a disorder might serve, as well as how it came about. In the same way that pain can be seen as a warning system to the body to protect it from tissue damage, anxiety may have evolved to protect the person from other types of potential threats. For example, many of the outward expressions of social anxiety parallel what is seen in dominance interactions in primates. Submissive monkeys avoid contact with most dominant ones, as do individuals experiencing social anxiety. This suggests the possibility that anxiety may have its evolutionary origins in dominance structures. If this were the case, then we might expect to see some relationship to sexual instinctual processes, as is the case with dominance. Indeed, social anxiety begins to be manifested just prior to the onset of puberty, at around eight years of age. The evolutionary perspective can also help us think about how psychopathology should be treated.

What function does psychopathology play? One possibility is that it could be an exaggeration of a normal process. For example, consider basic attentional factors. Clearly, in a variety of situations, it is protective for humans to scan the environment for danger. However, if this behavior is exaggerated, as is seen in some forms of anxiety disorders with constant screening of the environment, then this takes protection to the extreme and limits full human functioning. Developmentally, there is some suggestion that the scanning of the environment seen in some individuals with generalized anxiety disorder is associated with being required prematurely as children to play a responsible parental role for others. That is, if your mother was disorganized when you were a child, you may have needed to watch out for your own needs and maybe even those of your mother as well.

Using this perspective, it has been suggested that the auditory hallucinations seen in schizophrenia result from a breakdown in a mechanism that identifies whether an action was produced by the self or another (Frith, 1992). How do we actually know whether what we experience as a voice in our heads came from ourselves or from someone in the external world? We have a variety of clues, such as whether we produced it and whether we can control the voice. However, if the mechanisms that give us these experiences were not functioning, what would our experience be? Perhaps, like people with schizophrenia, we would not be able to tell the difference between external voices and internal voices.

Two additional possibilities are present when seemingly positive traits become associated with negative outcomes. For example, there is a suggestion that genetic processes associated with creativity and intelligence may also be related to psychopathology. This is seen in the case of bipolar disorder, in which there is a statistically greater number of artists who experience this order than would be expected by chance (Andreasen, 2005; Jamison, 1993). The other model uses the analogy of sugar and obesity as its basis. Humans and many other organisms enjoy sweet-tasting foods. Thousands of years ago this was not a problem, since sweet foods were not easy to come by. However, today, sugar is readily available in a large variety of foods, which in turn may lead to obesity. Drug addictions may follow a similar model. Overall, the evolutionary perspective helps us to better understand psychopathology by creating innovative predictions to be tested.

Another model suggests that we consider psychopathology as reflecting a lack of functioning in a normal process,

in the same way as we think of color-blindness as a special case of color vision. In this situation, psychopathology results when the physiological networks are not functioning correctly or missing some critical element.

Another disorder that has been approached from an evolutionary perspective is the category of personality disorders. Personality disorders reflect a rigid approach to dealing with social relationships. Two commonly discussed personality disorders are psychopathic personality and hysteria. Psychopaths are described as manipulative, callous, dishonest, and self-centered. They are antisocial in the sense that they display no need to follow the traditional rules of a society, and they display little remorse or guilt for their actions. For example, they would contract and collect money for a job they would never do. They would clearly qualify as what evolutionary psychologists refer to as "cheaters." On the other hand, individuals with a histrionic personality disorder overly seek the attention of others and are very emotional in their reactions. They can be manipulative in their interpersonal relationships.

Harpending and Sobus (1987) suggest that the psychopath and the hysteric represent different adaptive strategies in relation to sexuality. Both of these personality types are viewed by Harpending and Sobus as "cheaters." Given that it is more common to see male psychopaths and female hysterics, these researchers suggest that this results from different reproductive strategies. A male cheater in a sexual relationship should be able to persuade a female to copulate with him while deceiving her about his commitment to her and his willingness to offer resources for the offspring. A female cheater, on the other hand, would exaggerate her need for the male and make herself appear helpless and in need so that he would give her additional attention and resources. She would also be willing to put her own needs ahead of her offspring, even to the extent of abandoning them. The work of Harpending and Sobus shows how evolutionary thinking can help to explain both the motivational factors of a particular disorder and the demonstrated gender differences.

In humans, there is another layer of complexity involved in the process. Part of this complexity comes from the fact that humans are born less fully developed at birth than many other species, and thus are sensitive to changes in their environment as they continue to develop. This includes our relationships with our families and others we initially come into contact with. As humans, we develop societal and cultural perspectives. These perspectives become the backdrop of our environment. Unlike animals that live within nature, we as humans largely live within the backdrop of our culture.

32.5 PSYCHOPATHOLOGY IN THE CONTEXT OF NATURAL FUNCTIONING

Mental disorders can be considered as a failure of psychological adaptations to perform their natural function. In thinking about psychopathology from this evolutionary perspective, Wakefield (1992, 2015, 2016) suggests that we first need to understand natural function as well as its opposite, dysfunction. Natural function is what a particular process has evolved to do. For example, the heart is designed to pump blood. Other processes, such as the sounds that the heart makes, are by-products of this process. Thus, it would not be suggested that the heart evolved to make sounds. In this example, dysfunction would be the situation in which the heart could not adequately pump blood. On a perceptual level, the visual system has the function of processing external stimulation, whereas a dysfunction is this system could lead to visual hallucinations. Likewise, fear serves a natural function by protecting an organism in dangerous situations. However, when the fear is manifested in situations in which there is no clear danger, then there is a dysfunction. In terms of traditional psychopathology, the person in such a situation would be described as displaying a phobia. Phobias, such as the case in which a person refuses to leave his or her house, would be considered dysfunctional. Wakefield concluded that a condition is a mental disorder if two criteria are met: first, the condition causes harm to the person; and second, the condition results from the inability of some mental mechanism to perform its natural function. The emphasis is on what the mechanism evolved to do and if there is a problem in accomplishing this task.

This approach can be further clarified by asking three types of questions (Nesse, 2005). The first question asks whether cognitive and brain mechanisms are normal or defective. If the brain mechanisms are defective, as might result from accidents or trauma, then an evolutionary analysis of psychopathology would be inappropriate. The second type of question asks whether symptoms arise from novel aspects of the environment. As we have seen, there are novel aspects of modern environments that may bring forth responses that would have been appropriate in an earlier time, but not within modern society. The third type of question asks whether the symptoms are in the interest of the individual, his or her genes, or neither.

From these three types of questions, Nesse (2005, 2015) develops a categorization system to consider psychopathology. He first looks at emotional, cognitive, and behavioral responses that arise from normal systems. He suggests that there are five conditions in which these responses may arise:

1 There are normal responses that may be useful to the person, such as anxiety or anger, but can be experienced as aversive.
2 There are normal responses that would benefit the individual's genes, but may be at odds with the individual's own interest.
3 There are normal responses that may not be useful in a particular case.
4 There are normal responses that may not be useful presently, but would have been useful in a previous time.

5 There are normal responses that are not harmful to the individual, but may be viewed as abnormal by a person's group or culture.

From this more global perspective, it is now possible to consider specific types of mental disturbances.

32.5.1 Developing Fears

From an evolutionary perspective, to be fearful in the presence of dangerous situations would be adaptive (Öhman, 2009). One scientific aspect of this is the question of how fixed or plastic these fears are. From research, we know that certain phobias run in families, suggesting a genetic component. However, not everyone has exactly the same fears, suggesting that fear may be the outcome of a developmental process. One classic study in fear development is that of Susan Mineka and her colleagues (for an overview, see Öhman & Mineka, 2001). It had been observed that primates in the wild show a fear of snakes. Since a similar fear was seen in lab monkeys, it was assumed that this fear was somehow innate. However, Mineka asked whether early experience could influence this. In particular, she wanted to know whether observational learning could play a role. What she and her colleagues did was to compare wild-reared rhesus monkeys with those who had been reared in the lab. The wild-reared monkeys who had been brought to the lab some 24 years earlier showed a fear of snakes. This fear existed even though they would have had no experience with snakes during their time in the lab. The lab-reared monkeys, on the other hand, did not show any fear of snakes. If fact, they would reach over the snake to grab food. How did monkeys develop the fear of snakes? What Mineka did next was pair a wild-reared monkey with a young lab-reared one. A snake was then presented, and the wild-reared monkey showed fear. The young lab-reared monkey was able to observe this. After this, the lab-reared monkey also showed fear. Clearly, the lab-reared monkey had the ability to quickly acquire the fear, but required an experience in which another monkey showed fear for this to happen. The next question Mineka and her colleagues asked was regarding the importance of the feared object itself. In a very clever study, she showed some of the young monkeys a videotape of a wild monkey showing fear toward a snake. As expected, they acquired the fear of snakes. However, with another group of young monkeys, she edited the tape so that what the young monkeys saw was the original fear reaction of the older monkey, but this time to a flower. If fear was acquired by a simple associative learning situation in which the stimulus did not matter, then you would expect the young monkeys to acquire a fear of flowers. This was not the case. From this and a variety of other studies, it appears that fear can be learned through observation only to evolutionarily important objects. At this point, I want to describe a variety of disorders that have been considered from an evolutionary perspective.

32.5.2 Depression

Depression is characterized by depressed mood in which one feels sad or empty without any sense of pleasure in one's activities. All individuals experience depressed mood for brief periods, which is usually accompanied by feelings of sadness, loss of energy, social withdrawal, and often negative thoughts about oneself. With a depressive disorder, the individual may also experience sleep problems and weight changes. Included with the disorder is a sense of worthlessness and self-blame. Clinical depression is seen when the majority of these symptoms last for a period of time. There is a gender difference in that, over the course of a lifetime, about 1 in 4 females and 1 in 10 males experience a major depressive episode (Rutter, 2006). Research suggests that depression is equally influenced by genetic and environmental factors (Rutter, 2006). In one set of studies, monkeys with a genetic risk for depression were raised by either highly responsive or less responsive foster mothers (Suomi, 1997, 1999). In this situation, the mothers determined the outcome, with more responsive mothers having less depressed infants.

In terms of an evolutionary perspective, Andrews (2007) suggests that there are only a limited number of hypotheses that might explain the existence of depression. One hypothesis would view depression as resulting from a novel reaction of the nervous system to the modern environment. This hypothesis would suggest that changes in our environment have come quickly, whereas evolutionary changes in our nervous system are much slower. However, Andrews rejects this hypothesis, since depression is seen in a variety of animals such as rats, cats, and primates, whose environment has not changed drastically in recent times. A second hypothesis views depression as a process that evolved because it was connected to another trait that was extremely adaptive. This connection has yet to be found.

Another hypothesis suggests that depression evolved because it had a useful function. Gilbert (2005) asks whether depression might serve such an advantage. Such an advantage would be found in situations in which positive affect and drive should be toned down. One simple answer is that a reduction of positive affect could make one more sensitive to threat. Gilbert then takes this question a step further and asks: At what point does the increase in depressive mood become maladaptive?

Allen and Badcock (2003) begin with the role of depressed mood in our evolutionary history. They see depressed mood as having evolved in relation to social processes. As with other researchers, they suggest that a depressed state represents a risk management strategy in response to a situation that has a low probability of success and a high probability of risk. The emphasis is on social situations in which an individual would be at risk for being excluded either from groups or from a relationship with another individual. Basically, they see depressed mood as the result of a computational problem on the part of the individual. That is to say, on some level, the individual evaluates the situation. In a situation in which there is a high risk of being excluded, this evaluation leads to depressed mood. Intrinsic to this way of thinking is the assumption that depressed mood was adaptive in our evolutionary history. That is to say, feeling depressed would help the individual solve a problem faced by humans from the earliest times. Overall, there are three questions that need to be approached from an evolutionary perspective in relation to depression (Nesse, 1990):

1 What are the situations that occur over and over again in our environment of evolutionary adaptiveness that are responsible for depressive states?
2 What are the selection pressures in these situations? Said in other terms: What reproductive goals would have been threatened?
3 What are the characteristics of depressed mood that would have enabled the organism to cope with these threats?

These questions have been answered by evolutionary psychologists in terms of three broad models: (1) theories of resource conservation; (2) theories of social competition; and (3) theories of attachment.

32.5.2.1 Theories of Resource Conservation

Theories of resource conservation suggest that depressive mood protects the organism by conserving energy. By reducing energy expenditure, the organism can both protect itself in the present situation and conserve energy that can be used in future productive situations. When organisms experience stress, as illustrated in the work of Jay Weiss (1977) and others, uncontrollable events can affect the organism and produce a depression-like state with similar characteristics found in clinical depression. In these theories, clinical depression results when an individual does not move on to the next situation, but rather continues in a situation in which there are few positive payoffs.

32.5.2.2 Theories of Social Competition

In discussing dominance hierarchies across species, it was noted that the most powerful individual has a greater chance of mating and passing on his genes. Typically, two males fight to determine which will be higher in power. It has also been observed that when one of the animals loses such a competition, this animal begins to make submissive gestures. David Buss, in a variety of writings, has extended this type of thinking to humans, and has suggested that there exists a powerful motivation to acquire rank and status, especially among human males (for an overview, see Buss, 2015). Price suggested that there is a connection between depressive mood and losing a fight for status and resources (Price, 1996). In particular, he suggested that the losing organism adopts a strategy in which he signals a desire to not continue the competition and withdraw. The winner, on the other hand, tends to escalate the competition and increasingly displays threatening behaviors. From this viewpoint, depression is seen as an involuntary deescalating strategy that signals to the other individual that he has won.

32.5.2.3 Theories of Attachment

John Bowlby developed a theoretical understanding of interpersonal relationships based on the interactions of a child with his or her parents. This type of bonding, of course, has great survival value for a human infant who cannot take care of himself or herself. As we also discussed, attachment patterns can also be seen in later human interpersonal and sexual relationships. A disruption of the relationship, either in terms of infant–caregiver or later interpersonal ones, is experienced by the person as a loss. It is this loss that results in the withdrawal and depressed mood. On the infant level, the value of such a depression would be to protect the child by reducing the desire to explore or leave the present environment, which would protect the child from potential danger. A similar process would work on the adult human level, in which an attachment and the resulting depression would protect the individual from further hurt until mental healing could take place.

32.5.2.4 Social Risk Hypothesis

Allen and Badcock (2003) begin to integrate the three previous broad models of depression with the social risk hypothesis. This hypothesis suggests that when significant interpersonal relationships are disrupted, including social humiliation or defeat, depressed mood is the outcome. For them, depressed mood is a signal to the individual. It is associated with risk aversion, which has evolved over human history. In this sense, it is a protective mechanism that prevents further critical losses. It is protective in two ways: first, depressed mood reduces the desire of the individual to immediately enter a social relationship in which there could be an adverse outcome; and second, the outward signs of depressed mood, including changes in voice tone, reaction time, eye contact, and facial expression, signal to others signs of submission and helplessness.

32.5.3 Obsessive–Compulsive Disorder

OCD is characterized by repetitive thoughts and feelings, usually followed by behaviors in response to these. The thoughts are usually perceived as unpleasant and not wanted. A distinction is made between obsessions and compulsions. Obsessions are generally unwelcomed thoughts that come into one's head. In studies examining these thoughts in patients with OCD, they involve avoiding contamination, aggressive impulses, sexual content, somatic concerns, and the need for order. Compulsions are the behaviors that one uses to respond to these thoughts. Some behaviors, like cleaning or placing objects

in order, reflect a desire to respond to the obsessions. Other compulsions, such as handwashing, are more avoidant in nature for fear of what one might say, do, or experience in a particular situation. Often individuals with OCD will constantly check to see if they have performed a particular behavior, such as turning off the stove or unplugging an iron. Interestingly, individuals with OCD may be aware that their thoughts and actions may seem bizarre to others, but they cannot dismiss the thoughts or need for action.

There is clearly a parallel between the themes found in OCD and concerns expressed by those without the disorder. Most individuals naturally avoid contamination or express concerns when they experience unusual bodily sensations. On a society level, there are often rituals concerned with health and success in the world. Tribal cultures perform rituals to avoid evil spirits or to bring in the good ones. Most modern societies have a variety of rituals, including not walking under a ladder, stepping on sidewalk cracks, or other behaviors as ways of avoiding bad luck. Sports teams also have rituals for how to prepare for an important game. Not performing these rituals may result in a feeling of anxiety for many individuals.

A variety of studies have suggested that OCD is found throughout the world in similar rates, although it is more frequent in females than males (for an overview, see Feygin, Swain, & Leckman, 2006). Feygin et al. (2006) suggest that OCD results from the exaggeration of normal traits, which can be mapped onto a developmental trajectory. In particular, they discuss four developmental themes as a response to stress. Each of these themes could be tied to a normal developmental stage in which fear or threatening situations were overemphasized.

1 The first theme is loss, similar to anxious attachment. The obsession is that someone could be lost to the person. There are a wide variety of situations in which this could happen. A friend, lover, or child could be killed in an accident, for example. To prevent this, the compulsion is to check on the person to make sure that he or she is still safe or to create situations in which the person will not be able to be in a situation with risk.
2 The second theme is physical security in one's own environment. A common manifestation is that the person checks to make sure everything is in its place.
3 The third theme is environmental cleanliness. The fear is that objects, or the person him or herself, are dirty and that this will result in disease or other negative events. The behavior, of course, is to clean obsessively.
4 The fourth theme is that the person will be deprived of resources or objects that are important to him or her. A person who experiences these obsessions will either hoard objects or resources or try to prevent any situation in which he or she could come experience loss.

Another model of OCD draws from animal behavior and the development of fixed action patterns. One of the most familiar of these was Lorenz's geese that would follow whatever moving object interacted with them during the critical period of 16–36 hours after birth. Similar stereotypical behavior is seen in social and sexual displays. For example, some lizards move their heads up and down when they greet another lizard, and the male peacock displays its features in the presence of a female. The basic idea is that these patterns are triggered by particular stimuli and result in behaviors that are very similar each time they are manifested. Overall, these behaviors help the organism survive and reproduce. Thus, they are viewed as adaptive.

It also has been observed that these fixed patterns of behavior can be seen being repeated frequently in conditions of trauma, stress, or non-natural conditions such as a zoo. Such a frequent manifestation of the same behavior would not be appropriate for the situation. Clearly, you would find it strange if someone kept greeting you over and over again, even after you acknowledged their presence. Given that the frequent repeating of patterned behaviors is generally associated with highly aroused organisms, one view is that this is the attempt of the organism to reduce stress. In many species, including humans, repetitive patterns, such as rocking movements, grooming, vocalizations, and pacing, are seen in periods of stress. These repetitive processes are seen by some to lie at the basis of OCD and to involve the basal ganglia, which is a more primitive brain area.

OCD appears to run in families, with first-degree relatives of a person with OCD having a higher risk of developing OCD themselves. This would lead one to expect a genetic relationship to exist. However, this has been difficult to demonstrate. Another line of research involves the search for underlying similarities in the brain processing that is associated with OCD. In one study, individuals with OCD, their first-degree relatives without OCD, and a matched control group were given a task that required the individual to reverse what was learned previously (Chamberlain et al., 2008). On each side of a screen, the participant was shown a face superimposed on top of a building. One of the images was considered the target. The faces and buildings were different, and the person had to guess which image was considered correct. After each guess, feedback was given as to whether the person was correct or not. Thus, after a few trails, the person learned the "correct" image. Once an individual was correct on six trials, either the "correct" image was changed or a new set of images was presented. Either way, new learning was required. Using brain imaging techniques (functional magnetic resonance imaging), it was shown that individuals with OCD and their unaffected relatives had similar patterns of brain activation as compared to controls. In particular, controls showed greater brain activation in the orbitofrontal cortex than the OCD individuals and their relatives. Other studies have shown this area of the frontal cortex to be related to reversal learning, flexibility, and decision-making, which reflects the difficulties of those with OCD.

32.5.4 Schizophrenia

Schizophrenia is one of the most debilitating of the mental disorders. It affects one's ability to express oneself clearly, to have close social relationships, and to express and experience positive emotions (for reviews, see Andreasen, 2001, 2005; Walker et al., 2004). Some individuals with schizophrenia also hear voices, see images not seen by others, or believe that others wish to harm or control them. It affects about one percent of the population. It is seen throughout the world with similar symptoms, regardless of culture or geographic location. Typically, two types of symptoms can be demonstrated in the disorder. One type is called negative, since what is important is what is not present. Lack of affect in situations that call for it, poor motivation, and social withdrawal are all examples of negative symptoms. The other type of symptom is called positive, and these symptoms relate to what is experienced. Visual or auditory hallucinations, delusions, and bizarre thoughts are all examples of positive symptoms. In general, the onset of schizophrenia occurs in the late teens or early 20s. It more often affects males than females. Since it tends to run in families and is seen throughout the world, it is thought to have a genetic component, but there does not appear to be a simple genetic relationship, as is seen in Huntington's disease or cystic fibrosis. Twin studies show a higher concordance rate, of approximately 40 percent, in monozygotic twins, compared to 10 percent in dizygotic twins if the other twin also developed schizophrenia. A variety of genetic studies suggest that genetic factors account for about 82 percent of the variance for any one individual developing schizophrenia, suggesting that environmental factors are less critical (Rutter, 2006). Adopted children from families with schizophrenia show a similar rate to those raised with their natural families, also suggesting that rearing factors per se are not related to its development.

It has been noted that highly gifted and creative individuals manifest schizophrenic-like traits, referred to as "schizotypal" traits, without having the disorder. However, it is not uncommon for these individuals to have a first-degree relative with schizophrenia, suggesting a genetic component. Andreasen (2001, 2005) suggests that there may be a connection with scientific creativity and schizophrenia within one's family. She notes that a number of Nobel Laureates had family members who were thought to have had schizophrenia, including Albert Einstein, Bertrand Russell, and John Nash, who had the condition himself. (As you may remember, John Nash's story was described in the film A Beautiful Mind.) Although a variety of research studies suggest a genetic component, it is clearly not a simple one involving a single gene.

A number of researchers have examined schizophrenia from an evolutionary perspective (for an overview, see Burns, 2004). Tim Crow, similarly to many other researchers, begins with the paradox of how a genetically related disorder can exist without a reproductive advantage

(Crow, 2000; Mitchell & Crow, 2005). In order to consider this question, Crow notes that schizophrenia is seen in all cultures, suggesting that its primary determinant is more independent of the environment than other disorders. Multiple sclerosis, for example, occurs in greater frequency the farther from the equator one grew up, suggesting an important role for environmental conditions in the development of the disorder. Given that schizophrenia is found throughout the world, this suggests that the environment plays less of a direct role. Further, given that it is found in all groups, Crow suggests that its origin in relation to humans must have occurred before the mass movement from Africa less than 100,000 years ago. What was the event in the entire human population that led to schizophrenia? Crow suggests it was the development of the capacity for language. The anatomical event associated with this was the development of the independence of the two cerebral hemispheres of the human brain. This asymmetry allowed for language to be more specialized in the left hemisphere and for spatial abilities to be more represented in the right hemisphere of the human brain. However, it has been shown that understanding certain emotional aspects of language, as well as understanding humor, sarcasm, and metaphor, may involve the right hemisphere. Crow notes that individuals with schizophrenia display symptoms that represent problems with language. He suggests that schizophrenia is related to problems of development in relation to language that involve the inability of the person to distinguish his or her thoughts from speech itself, including what others say to the person.

Jonathan Burns suggests that schizophrenia is related to the evolution of the social brain in humans (Burns, 2004). In particular, he sees it as being related to frontal areas and their connections to other parts of the human brain that are involved in our ability to represent our relationships. This includes the ability to know about others' intentions. The potential for schizophrenia is the result of trade-offs at two distinct times in our long evolutionary history. The first trade-off happened between 2 and 16 million years ago. This was the time in which complex cerebral interconnectivity and specialized circuits were being evolved in the brain in relation to the cognitive and intellectual demands of living in social groups. To perform the tasks required for social living, a higher level of cognitive functioning was required. In order for the brain to develop the circuits required, brain maturation was lengthened. That is, given the physical constraints of the developing brain in the human fetus, the period for brain development needed to be lengthened. This trade-off meant that the developing brain experienced a long period of time in which complex gene interactions or accidents could happen.

For Burns, the second trade-off happened more recently, about 100,000–150,000 years ago. This date is important. Since schizophrenia is seen in all cultures with similar symptoms, it is assumed that the genes involved in its manifestation would have evolved before

humans migrated out of Africa. What happened at this point was that some individuals experienced non-normal connections in the frontal areas of the brain. These connections resulted in some individuals being especially creative and thinking in different ways. These individuals may have been able to make important contributions to culture, much as our present-day artists and creative thinkers do. However, some individuals demonstrated a more severe version of these connections, which resulted in psychopathological experiences. Burns further suggests that this different way of experiencing the world, in either its mild or severe form, did not have any reproductive advantage. However, since the genes that controlled these experiences evolved as a part of the larger cortical networks needed for the cognitive and intellectual demands of social life, these genes continued to be passed on through their connections with adaptive mechanisms. Thus, according to Burns, schizophrenia represents one of the prices paid for evolving complex cognitive and social abilities.

32.5.5 Abnormal Social Psychology

As clinicians have moved from more learning theory-directed theories of psychopathology to those derived from the neurosciences, and thus consistent with the evolutionary perspective, a variety of disorders have been reexamined. These include autism, Williams syndrome, psychopathy, and social phobia. In the same way that we can speak of affective disorders such as depression or anxiety, we can also speak of social disorders.

From an evolutionary perspective, these social disorders are related to problems in function seen as part of the larger social instinct rather than sexual or self-preservation functions. Humans have always lived in groups, which has supported the development of social functions such as empathy and predicting the behaviors of others. One set of disorders in which this ability is interrupted are autism spectrum disorders.

32.5.5.1 Autism Spectrum Disorders
Individuals with autism display problems in three areas (Baron-Cohen & Belmonte, 2005): social functions, communications, and restrictions in behaviors and interests. In particular, individuals with autism look less at other individuals and show reduced responses in the mirror neuron system. They also show difficulty in interpersonal communication and tend to repeat activities in a stereotyped manner. One evolutionary perspective suggests that autism is an extreme case of males developing abilities to understand, control, and develop systems rather than develop empathy (Baron-Cohen, 2002). Empathy is seen as an ability that has developed in a greater degree in females.

Previously, those on the autism spectrum with a mild form of the condition were described in terms of Asperger syndrome. Those with Asperger syndrome show social impairments and restricted behaviors, but fewer problems in terms of language and communication. One current theory of autism centers on problems related to empathy. In particular, it is suggested that individuals with autism fail to develop a theory of mind. Theory of mind refers to one's ability to infer the mental states of others in relation to their actions or situations. In a variety of studies, individuals with autism were able to describe what was going on regarding someone else's behavior on the perceptual level, but showed difficulties when asked to describe the social/emotional processes that would be expected to accompany the behaviors. This appears to be a general lack of ability, since they also have problems reflecting on these aspects of their own behavior. Another characteristic of autism is the desire to have a stable set of routines, which results in problems shifting attention. In terms of cortical areas involved, individuals with autism display a variety of differences in those areas previously described as the social brain. Briefly, these are the amygdala, specific areas of the frontal lobes, and areas of the temporal lobe. There is also some suggestion that infants who develop autism show an overgrowth of brain volume followed by deceleration in the first years of life. Amaral and his colleagues have described the neuroanatomy of autism (Amaral, Schumann, & Nordahl, 2008).

Current neuroscience studies show that by the time of full brain development, a person with autism shows deficits in the areas that make up the social brain (Barak & Feng, 2016; Harris, 2016). For example, whereas individuals without autism tend to scan the eyes when looking at another person, those with autism focus on the mouth. Since various features of the eyes give another person both emotional and social information, those with autism miss out on this information. Coupled with the inability to empathize in autism is the superior ability to systemize. Systemizing is the ability to analyze objects or events in terms of their structures and future behaviors. These deficits and abilities make up the empathizing–systemizing theory of autism (Baron-Cohen & Belmonte, 2005). In terms of genetics, twin studies of individuals with these social difficulties show 92 percent concordance for monozygotic twins vs. 10 percent concordance for dizygotic pairs.

32.5.5.2 Williams Syndrome

Whereas those individuals with autism spectrum disorders show a lack of socialness, those with Williams syndrome show increased socialness (Barak & Feng, 2016). Williams syndrome is a fairly rare genetic condition that is related to chromosome 7 and leads to a unique pattern of both strengths and weaknesses. Weaknesses include delayed developmental milestones and problems with certain organ systems, such as the heart, and problems with spatial abilities and fine motor coordination. Strengths include good language skills and social behaviors. These individuals are overly social, especially interested in other people, and able to recognize the mental states of others.

In fact, these individuals are socially fearless and will eagerly engage with strangers. If you were to show individuals with Williams syndrome and those without it a series of threatening faces and threatening scenes, you would see that individuals with Williams syndrome would show amygdala activation only to threatening scenes, whereas those without Williams syndrome would respond to both threatening faces and threatening scenes (Meyer-Lindenberg et al., 2005; Meyer-Lindenberg, Mervis, & Berman, 2006).

An interesting evolutionary perspective suggests a link between sociability and music in general and those with Williams syndrome in specific (Schulkin & Raglan, 2014). Music has been part of our evolutionary history for at least 50,000 years and is connected with social events. Those with Williams syndrome show more engagement and liking of music than normal controls. On the other hand, those individuals on the autism spectrum show decreased perception of emotions in music. This is seen as an evolved set of features that include socialness, empathy, and musical engagement, which is emphasized in those with Williams syndrome.

32.6 SUMMARY

Psychopathology is present is all societies. Five ideas have been critical in the definition of psychopathology. First, the processes involved in psychopathology are maladaptive and not in the individual's best interest. Second, the processes cause personal distress. Third, the processes represent deviance from both cultural and statistical norms. Fourth, the person has difficulty connecting with his or her environment and also with him or herself. Fifth, there is the inability of an individual with a mental disorder to fully consider alternative ways of thinking, feeling, or doing.

Evolutionary perspectives can inform an understanding of psychopathology. In this chapter, six questions were raised. First, is psychopathology universal, given that it is allows us to examine its nature and development? Second, questions of adaptive values can be considered. Third, seeing signs of psychopathology in other species allows us to see psychopathology from an even broader evolutionary standpoint. Fourth, we can consider how psychopathologies can develop from normal processes. Fifth, we can ask whether psychopathologies are protective in some manner. And sixth, we can ask whether psychopathologies are recent and have existed throughout our human history.

Historically, a variety of researchers have seen psychopathology as related to basic evolutionary mechanisms. In this chapter, the works of John Hughlings Jackson, Paul MacLean, Sigmund Freud, Carl Jung, John Bowlby, and Jack Panksepp were considered. Further specific disorders were described. Overall, psychopathologies represent situations in which adaptive processes do not develop or function appropriately.

REFERENCES

Allen, N., & Badcock, P. (2003). The social risk hypothesis of depressed mood: Evolutionary, psychosocial, and neurobiological perspectives. *Psychological Bulletin*, **129**, 887–913.

Andreasen, N. (2001). *Brave New Brain*. New York: Oxford University Press.

Andreasen, N. (2005). *The Creating Brain: The Neuroscience of Genius*. New York: Dana Press.

Andrews, P. (2007). Reconstructing the evolution of the mind is depressingly difficult. In S. Gangestad & J. Simpson, eds., *The Evolution of Mind*. New York: Guilford Press, pp. 45–52.

Bailey, K. (1987). *Human Paleopsychology*. Hillsdale, NJ: Lawrence Erlbaum Associates.

Barak, B., & Feng, G. (2016). Neurobiology of social behavior abnormalities in autism and Williams syndrome. *Nature Neuroscience*, **19**, 647–655.

Baron-Cohen. S. (2002). The extreme male brain theory of autism. *Trends in Cognitive Sciences*, **6**, 248–254.

Belsky, J. (2005). Differential susceptibility to rearing influence: An evolutionary hypothesis and some evidence. In B. Ellis & D. Bjorklund, eds., *Origins of the Social Mind: Evolutionary Psychology and Child Development*. New York: Guilford Press, pp. 139–163.

Bowlby, J. (1951). *Maternal Care and Mental Health*. Geneva: World Health Organization Press.

Bowlby, J. (1961). Childhood mourning and its implications for psychiatry. The Adolf Meyer Lecture. *American Journal of Psychiatry*, **118**, 481–497.

Bowlby, J. (1969). *Attachment and Loss: Vol. 1. Attachment*. London: Hogarth.

Bowlby, J. (1982). Attachment and loss: Retrospect and prospect. *American Journal of Orthopsychiatry*, **52**, 664–678.

Bowlby, J. (1988). *A Secure Base: Clinical Applications of Attachment Theory*. London: Routledge.

Bressler, S., & Menon, V. (2010). Large scale brain networks in cognition emerging methods and principles. *Trends in Cognitive Sciences*, **14**, 277–290.

Burns, J. (2004). An evolutionary theory of schizophrenia: Cortical connectivity, metarepresentation, and the social brain. *Behavioral and Brain Sciences*, **27**, 831–885.

Buss, D., ed. (2015). *The Handbook of Evolutionary Psychology*. Hoboken, NJ: Wiley.

Carhart-Harris, R. L., & Friston, K. J. (2010). The default-mode, ego-functions and free-energy: A neurobiological account of Freudian ideas. *Brain*, **133**(4), 1265–1283.

Chamberlain, S., Menzies, L., Hampshire, A., et al. (2008). Orbitofrontal dysfunction in patients with obsessive–compulsive disorder and their unaffected relatives. *Science*, **321**, 421–422.

Crow, T. (2000). Schizophrenia as the price that Homo sapiens pays for language: A resolution of the central paradox in the origin of the species. *Brain Research Reviews*, **31**, 118–129.

Feygin, D., Swain, J., & Leckman, J. (2006). The normalcy of neurosis: Evolutionary origins of obsessive–compulsive disorder and related behaviors. *Progress in Neuropsychopharmacology & Biological Psychiatry*, **30**, 854–864.

Frith, C. (1992). *The Cognitive Neuropsychology of Schizophrenia*. London: LEA.

Gilbert, P. (2005). Evolution and depression: Issues and implications. *Psychological Medicine*, **36**, 287–297.

Harpending, H., & Sobus, J. (1987). Sociopathy as an adaptation. *Ethology and Sociobiology*, **8**, 63–72.

Harris, J. C. (2016). The origin and natural history of autism spectrum disorders, *Nature Neuroscience*, **19**, 1390–1391.

Huxley, J., Mayr, E., Osmond, H., & Hoffer, A. (1964). Schizophrenia as a genetic morphism. *Nature*, **204**, 220–221.

Jamison, K. (1993). *Touched with Fire*. New York: Simon & Schuster.

Menon, V. (2011). Large scale brain networks and psychopathology: A unifying triple network model. *Trends in Cognitive Sciences*, **15**, 483–506.

Meyer-Lindenberg, A., Hariri, A. R., Munoz, K. E., et al. (2005). Neural correlates of genetically abnormal social cognition in Williams syndrome. *Nature Neuroscience*, **8** (8), 991–993.

Meyer-Lindenberg, A., Mervis, C. B., & Berman, K. F. (2006). Neural mechanisms in Williams syndrome: A unique window to genetic influences on cognition and behaviour. *Nature Reviews. Neuroscience*, **7**(5), 380–393.

Mitchell, R., & Crow, T. (2005). Right hemisphere language functions and schizophrenia: The forgotten hemisphere? *Brain*, **128**, 963–978.

Murphy, J. (1976). Psychiatric labeling in cross-cultural perspective. *Science*, **191**, 1019–1028.

Nesse, R. M. (1990). Evolutionary explanations of emotions. *Human Nature*, **1**, 261–289.

Nesse, R. M. (2005). Natural selection and the regulation of defenses: A signal detection analysis of the smoke detector principle. *Evolution and Human Behavior*, **26**, 88–105.

Neese, R. M. (2015). Evolutionary psychology and mental health. In D. Buss, ed., *The Handbook of Evolutionary Psychology*. Hoboken, NJ: Wiley, pp. 903–927.

Öhman, A. (2009). Of snakes and faces: An evolutionary perspective on the psychology of fear. *Scandinavian Journal of Psychology*, **50**, 543–552.

Öhman, A., & Mineka, S. (2001). Fears, phobias, and preparedness: Toward an evolved module of fear and fear learning. *Psychological Review*, **108**, 483–522.

Panksepp, J. (1998). *Affective Neuroscience: The Foundations of Human and Animal Emotions*. New York: Oxford University Press.

Panksepp, J., ed. (2004). *Textbook of Biological Psychiatry*. New York: Wiley.

Ploog, D. (2003). The place of the triune brain in psychiatry. *Physiology and Behavior*, **79**, 487–493.

Price, P. (1996). *Biological Evolution*. Pacific Grove, CA: Brooks/Cole.

Raichle, M. E. (2015). The brain's default mode network. *Annual Review of Neuroscience*, **38**, 413–427.

Ray, W. J. (2013). *Evolutionary Psychology: Neuroscience Determinants of Human Behavior and Experience*, Thousand Oaks, CA: Sage.

Ray, W. J. (2018). *Abnormal Psychology: Neuroscience Determinants of Human Behavior and Experience*, 2nd ed. Thousand Oaks, CA: Sage.

Rutter, M. (2006). *Genes and Behavior: Nature–Nurture Interplay Explained*. Malden, MA: Blackwell Publishing.

Sartorius, N., Jablensky, A., Korten, A., et al. (1986). Early manifestations and first-contact incidence of schizophrenia in different cultures. A preliminary report on the initial evaluation phase of the WHO Collaborative Study on determinants of outcome of severe mental disorders. *Psychological Medicine*, **16**, 909–928.

Schulkin, J., & Ragian, G. (2014). The evolution of music and human social capability. *Frontiers in Neuroscience*, **8**, 292.

Sulloway, F. (1979). *Freud, Biologist of the Mind*. New York: Basic Books.

Suomi, S. (1997). Early determinants of behavior: Evidence from primate studies. *British Medical Bulletin*, **53**, 170–184.

Suomi, S. (1999). Attachment in rhesus monkeys. In J. Cassidy & P. Shaver, eds., *Handbook of Attachment: Theory, Research, and Clinical Applications*. New York: Guilford Press, pp. 181–197.

Szasz, T. (1970). *The Manufacture of Madness: A Comparative Study of the Inquisition and the Mental Health Movement*. New York: Harper & Row.

Wakefield, J. (1992). The concept of mental disorder: On the boundary between biological facts and social values. *American Psychologist*, **47**, 373–388.

Wakefield, J. (2015). Biological function and dysfunction: Conceptual foundations of evolutionary psychopathology. In D. Buss, ed., *The Handbook of Evolutionary Psychology*. Hoboken, NJ: Wiley, pp. 988–1006.

Wakefield, J. (2016). Diagnostic issues and controversies in DSM-5: Return of the false positives problem. *Annual Review of Clinical Psychology*, **12**, 105–132.

Walker, E., Kestler, L., Bollini, A., & Hochman, K. (2004). Schizophrenia: Etiology and course. *Annual Review of Psychology*, **55**, 401–430.

Weiss, J. (1977). Psychosomatic disorders: Ulcers. In J. Maser & M. Seligman, eds., *Psychopathology: Experimental Models*. San Francisco, CA: W. H. Freeman and Company.

33 Are We on the Verge of Darwinian Psychiatry?

ALFONSO TROISI

33.1 INTRODUCTION

There are many general definitions of Darwinian psychiatry in academic textbooks (e.g., "Psychiatric disorders viewed in evolutionary context"; McGuire & Troisi, 1998, p. vii) and journal articles (e.g., "Any attempt to make sense of mental disorders within the general framework of contemporary evolutionary theory"; Adriaens & De Block, 2010, p. 132). The common theme of these definitions is that Darwinian psychiatry originates from the hybridization of two distant fields of research: psychopathology and evolutionary biology. For those who are not familiar with the concepts and methods of contemporary evolutionary biology (which is true for most mental health professionals), these definitions are unlikely to stimulate the curiosity to learn more about Darwinian psychiatry. The reason for this is that the theoretical and practical utility of the evolutionary approach to psychopathology is not self-evident and becomes clear only if one takes on two intellectual exercises: first, avoiding clichés that misrepresent the evolutionary approach to psychiatric disorders (e.g., "mental illness is a regression to primitive stages of human evolution" or "mental disorders are diseases of civilization"); and second, acquiring advanced knowledge of the theoretical framework that underlies contemporary evolutionary biology, including its most important sub-theories (e.g., life history theory, attachment theory, sexual selection theory, niche construction theory).

This chapter is organized around the two intellectual exercises sketched above. The first part is directed to those readers who are not familiar with the concepts and methods of contemporary evolutionary biology. It has the twofold aim of clarifying what Darwinian psychiatry is and what it is not. The emphasis will be not only on the original features that distinguish the Darwinian approach to mental disorders from traditional methods of research and clinical practice in psychiatry, but also on some misleading descriptions of Darwinian psychiatry that need to be amended. In this first part, the discussion of theories and data simulates an imaginary dialogue with a mental health professional who is skeptical of Darwinian psychiatry and asks the basic questions "Why should I take an interest in evolutionary biology? How can it improve my understanding of mental disorders and treatment strategies?"

The second part of the chapter is directed to those students who are expert in evolutionary sciences of human behavior and are interested in knowing how Darwinian psychiatry integrates with its allied disciplines. The emphasis will be on the contributions of evolutionary psychology, ethology, behavioral ecology, and evolutionary anthropology to Darwinian psychiatry and on the methodological problems that need to be addressed in order to improve the scientific validity and clinical utility of the evolutionary approach to mental disorders. I feel it is fair to warn the reader that this second part of the chapter largely reflects my own vision of Darwinian psychiatry, a vision that other evolutionary psychiatrists might not endorse.

I started this introduction by questioning the scarce appeal of general definitions of Darwinian psychiatry due to the difficulty of understanding at first sight what is new and useful in the evolutionary study of mental disorders. An alternative definition that may arouse the interests of mental health professionals focuses directly on the questions posed by Darwinian psychiatry rather than on its theoretical foundations. Darwinian psychiatry can be defined as the branch of psychiatry that deals with the phylogenetic origin and adaptive significance of mental disorders. In my opinion, such a definition has the merit of targeting the most original aspect of the evolutionary approach in biology and medicine: asking questions related to ultimate causation that non-evolutionary disciplines never ask. Questions are the core of scientific investigation, and discovering that a clinical condition can be examined in the light of new and original questions is a powerful incentive to raise interest in Darwinian psychiatry.

33.2 DARWINIAN PSYCHIATRY: THE BASICS

33.2.1 Evolutionary Questions

In the 1960s, two papers were published that had a massive influence on how contemporary biologists understand causation in biological systems, including the behavior of living organisms. In November 1961, Ernst Mayr, an evolutionary biologist who participated in the Modern Synthesis of the 1930s and 1940s that emerged as neo-Darwinism, published a paper entitled "Cause and Effect in Biology" (Mayr, 1961). In that article, Mayr distinguished two currents of biological thought that differ by the questions they ask. Functional biology deals with the mechanisms controlling the functionality of organic elements, from molecules to individuals. The questions a functional biologist asks are proximate questions and are preceded by "how." For example, applied to psychopathology, proximate questions are as follows: How does brain serotonin regulate impulsivity? How does upbringing environment exert epigenetic effects? How does early trauma increase the risk of eating disorders? Evolutionary biology, the second current, focuses on the phylogenetic history and adaptive significance of biological traits. The questions an evolutionary biologist asks are ultimate questions and are preceded by "why": (1) Why are kids afraid of the dark but not of electric sockets? (2) Why are men more promiscuous than women? (3) Why has natural selection not eliminated genetic vulnerability to psychotic disorders?

In 1963, Niko Tinbergen, an ethologist who would win a Nobel Prize a decade later, wrote a classic paper entitled "On Aims and Methods in Ethology" (Tinbergen, 1963). Tinbergen pointed out that four fundamentally different types of problem are raised in biology, which he listed as survival value, ontogeny, evolution, and causation. These problems can be expressed as four questions about any feature of an organism, including psychological processes and behavior: (1) What is it for? (2) How did it develop during the lifetime of the individual? (3) How did it evolve over the history of the species? (4) How does it work? Questions about ontogeny and causation (numbers 2 and 4 above) are proximate questions. Questions about survival value (or adaptive significance in current language) and evolutionary history (or phylogeny; numbers 1 and 3 above) are ultimate questions.

Although both Mayr's and Tinbergen's classifications have been criticized in recent years (Bateson & Laland, 2013; Laland et al., 2011), they remain excellent starting points to understand what makes the evolutionary approach so different from traditional ways of reasoning in biology and medicine. Medicine and psychiatry are strongly settled in the territory of functional biology. Therefore, when clinicians speak of the cause of a disease or symptom, they mean the proximate events that set in motion the dysfunctional processes we call pathogenesis. But this analysis omits ultimate causes, those evolutionary forces that lead to or away from conditions that favor or prevent the advent of proximate cause (Childs, 1999). The mechanistic mentality is drawn only to the immediate reasons for illness and to what needs fixing. Yet a deeper understanding of a manifestation of disease, including the discovery of its possible adaptive significance, can be attained only by taking into account those evolutionary factors that made the proximate causes possible.

Having explained the distinction between proximate and ultimate questions, I should clarify through selected examples what would be the added value in psychiatry of asking questions about phylogeny and adaptive significance.

33.2.2 Phylogeny

There is evidence that the prevalence rates of psychiatric disorders have changed over time (Healy, 2014). Yet, these epidemiological data span a short time period (at best, hundreds of years) when compared with the duration of the evolutionary history of anatomically modern *Homo sapiens* (circa 200,000 years). What do we know of the prevalence of mental illness in prehistory? Very little indeed. Nevertheless, some empirical data related to human evolutionary history can improve our understanding of psychiatric disorders observed in contemporary populations.

A recent study (Srinivasan et al., 2016) has hypothesized that schizophrenia is a by-product of the complex evolution of the human brain and a compromise for humans' language, creative-thinking, and cognitive abilities. The authors arrived at such a conclusion by applying a polygenic statistical approach and analyzing data from recent large genome-wide association studies. They found that gene loci associated with schizophrenia are significantly more prevalent in genomic regions that are likely to have undergone recent positive selection in humans. The measure used to assess recent positive selection was the Neanderthal selective sweep score, which is a likelihood index of phylogenetic divergence between modern humans and Neanderthals. Three implications of this study are worth discussing. First, the results go exactly in the opposite direction of the cliché often applied in the past to characterize the Darwinian approach to psychiatric disorders (i.e., that mental illness reflects a condition of phylogenetic regression and atavism). The evolution of novel cognitive abilities, not the regression to a more primitive stage, seems to be the evolutionary reason for the existence of psychosis. Second, the sophisticated and empirical methodology of this study clearly fends off the accusations of evolutionary psychiatry promulgating "just-so" stories. Third, the results are an excellent demonstration of how the biology of proximate questions can integrate profitably with the biology of ultimate questions. The identification of the genes predisposing to schizophrenia belongs to the realm of functional biology, but the explanation of their persistence in contemporary populations is a task for evolutionary biologists.

The phylogenetic perspective is also important for recognizing that much of the environment we inhabit is increasingly out of the optimal range for which our body's physiological and psychological mechanisms were designed by the processes of evolution (the mismatch hypothesis). Many aspects of the physical and social environment have been dramatically altered by human action, particularly in our recent past (Gluckman & Hanson, 2006). An innovative and original contribution that Darwinian psychiatry can offer to the prevention of mental disorders is the identification of those modifiable risk factors that were absent in the ancestral environment but are common in modern environments.

My favorite example is the revolution in the procedures of pediatric hospitalization caused by the pioneering work of the British psychiatrist John Bowlby (Troisi, 2012). Misguided by the wrong assumptions of psychoanalysis and learning theory, for a long time professionals and childcare workers downplayed the significance of mother–child separation and ignored the emotional needs of hospitalized children. If a young child needed to be hospitalized, it was standard practice to prevent or severely restrict parental visitation. In the UK during the 1940s and 1950s, parents were allowed to visit their sick children in the hospital for only one hour per week (Karen, 1994). The physiologic needs of the children, particularly for food and warmth, were placed ahead of the children's need for an affectionate relationship with their mothers. The prevailing view of those times was that if a child was fed by a variety of caregivers, the relationship with the mother would hold no special significance for the child. Inspired by his evolutionary and ethological studies, Bowlby showed that the attachment system is genetically "wired" into human nature through intense directional selection during our phylogenetic history. The reason why the attachment system evolved and remains so deeply ingrained in the human brain and behavior is that it provided a good solution to one of the most daunting adaptive problems our ancestors faced: how to survive through the most perilous years of social and physical development (Simpson & Belsky, 2008). The growing acceptance of the attachment theory in the 1960s eventually altered hospital practice and signaled with unmistakable clarity the mismatch between professionals' prejudices and the real needs of young children. Thanks to Bowlby's work, successful prevention of emotional disorders related to pediatric hospitalization has become feasible.

The mismatch hypothesis teaches two important lessons to clinicians: first, the greater the degree of match between an individual's evolutionary constraints and his or her environment, the more likely the individual is to enjoy physical and mental health; and second, when diagnosing a condition of emotional suffering, consider the possibility that mismatch is playing a role in symptom causation (Troisi, 2018).

The "dark side" of the phylogenetic perspective is the risk of inventing explanations for mental disorders that are likely to be "phylogenetic fantasies" (Adriaens & De Block, 2010, p. 139). A detailed reconstruction of what happened throughout our evolutionary history is virtually impossible, especially when behavior and social relationships are the focus of investigation. The difficulty of validating hypotheses against empirical data is fertile ground for uncontrolled speculations, and it is also the major reason fomenting accusations of promulgating "just-so" stories by the critics of Darwinian psychiatry. Two possible examples of "phylogenetic fantasies" are the group-splitting hypothesis of schizophrenia and the hypothesis of the evolutionary origin of bipolar disorder.

Steven and Price (1996) argued that our ancestors lived in groups that would have grown to a point at which they outran available resources, whereupon the group would have had to split up. Schizoid individuals would have been regarded as charismatic and hence able to draw followers away with them to start a new group; this would have been good for the group. Steven and Price contend that natural selection has fixed a predisposition to schizophrenia as an enduring component of the human genome because it was an adaptation to facilitate group splitting.

Sherman (2012) argued that bipolar behaviors might have developed among Neanderthals as seasonal adaptations. Neanderthals were the most cold-adapted of all hominins, which would be consistent with the idea that they developed bipolar behaviors as a climatic adaptation. Sherman's hypothesis suggests that, especially for women in their reproductive years, gorging occurred in preparation for winter, which was followed by depressive behaviors during the winter: desires declined, including desires for food, activity, social contact, and sex. Initiative was minimal. Not only would depressive behaviors have reduced activity levels and so helping to conserve energy, but the lack of desire for food and sexual activity decreased conflict within the group. With the departure of winter, depression was replaced by hypomania: optimism, self-confidence, and lots of energy to accomplish the tasks set aside during the winter. This was a time for socializing, mating, travel, and getting things done. During the Ice Ages, there was sometimes only two months of reliable weather in the areas where Neanderthals lived. Under these conditions, hypomanic behavior contributed importantly to survival.

Although imaginative and original, these evolutionary hypotheses are difficult or impossible to test and unlikely to gain the interests of those who have the practical task of diagnosing and treating people with schizophrenia or bipolar disorder.

33.2.3 Adaptive Significance

For most psychiatrists and other mental health professionals, evolution means evolutionary change. However, evolutionary theory is a theory of adaptation through selection, not just a theory of change over evolutionary time. Adaptationist reasoning is a crucial feature of Darwinian psychiatry. This may appear incongruous because, by definition, pathological conditions are maladaptive. Yet, there are several reasons that make it useful to look for features of adaptation when analyzing psychiatric symptoms and syndromes.

Contemporary medicine conceptualizes symptoms as dysfunctional manifestations of an underlying pathology (Sharpe & Walker, 2009). Yet, not all medical symptoms are the same: some manifestations of disease are sophisticated adaptations, not just epiphenomena of a broken machine (Nesse, 2005). Empirical evidence that some symptoms are useful defenses that evolved to counteract pathologic processes comes mostly from the field of infectious diseases (e.g., Evans et al., 2015). However, preliminary data suggest that the distinction between dysfunctional symptoms and adaptive symptoms can be important in psychiatry, too.

Keller and Nesse (2006) introduced and tested a new framework for understanding the adaptive significance of depressive symptoms. Their hypothesis (the "situation–symptom congruence" hypothesis) predicts that, if different depressive symptoms serve different functions, then different events that precipitate a depressive episode should give rise to different symptom patterns that increase the ability to cope with the adaptive challenges specific to each situation. The hypothesis was tested by asking 445 participants to identify depressive symptoms that followed a recent adverse situation. Guilt, rumination, fatigue, and pessimism were prominent following failed efforts; crying, sadness, and desire for social support were prominent following social losses. These significant differences were replicated in an experiment in which 113 students were randomly assigned to visualize a major failure or the death of a loved one. The results of the study confirmed the prediction that symptoms eliciting comfort (e.g., crying) should be especially prominent when social bonds are threatened, lacking, or lost, whereas symptoms dissuading the individual from pursuing current and potential goals (e.g., pessimism and fatigue) should arise when the environment is unpropitious and future efforts are unlikely to succeed.

The possibility that some psychiatric symptoms may serve an adaptive function has practical implications. The therapeutic elimination of defensive symptoms could make the prognosis worse and interfere with recovery (Troisi, 2012), as happens with some symptoms of infectious diseases. In addition, the reinterpretation of some symptoms as defensive responses may help to identify the environmental circumstances that play a role in the etiology of the disorder. On the other hand, a major risk of the adaptationist approach is the invention of outlandish explanations to elucidate the selection pressures that forged psychopathological traits (Troisi, 2006). I agree with the criticism "against those evolutionary thinkers who assume that adaptive forces are the only possible

explanations for common, heritable polymorphisms such as mental disorders" (Keller & Miller, 2006, p. 403). It is a mistake to equate Darwinian psychiatry with adaptationist psychopathology. Most psychiatric disorders reflect a failure of psychological and behavioral adaptations to perform their evolved functions. Yet, dysfunction is difficult to infer without an understanding of function, and this is the reason why the study of adaptive significance is important in psychiatry.

33.3 DARWINIAN PSYCHIATRY: A NEW LOOK AT OLD PROBLEMS

More than other fields of medicine, psychiatry deals with some theoretical problems that remain unsolved. Discouraged by the difficulty of finding unanimous solutions, clinical psychiatrists tend to ignore these basic problems, considering them more appropriate for philosophical discussion than scientific inquiry. Darwinian psychiatry can offer a different perspective on these old questions and renew clinicians' interest because the evolutionary approach is based on empirical data and has clear implications for clinical decisions. Here, I will discuss two of these unsolved problems: the definition of mental disorder and mind–brain dualism.

33.3.1 Definition of Mental Disorder

There is a paradox in contemporary psychiatry. Clinicians diagnose and treat hundreds of different disorders, but they do not have a general definition of mental disorder. Allen Frances, an American psychiatrist who was chairman of the DSM-IV (the fourth edition of the *Diagnostic and Statistical Manual of Mental Disorders* of the American Psychiatric Association) Task Force, described the paradox cogently: "The concept of mental disorder is so amorphous, protean, and heterogeneous that it inherently defies definition. This is a hole at the center of psychiatric classification" (Frances, 2012, p. 24).

The problem of defining mental disorder is not just an academic exercise. Without a clear definition, the diagnostic process incurs the double risk of false positives (i.e., making a psychiatric diagnosis in people who are healthy) and false negatives (i.e., not diagnosing sick people). Considering the stigma that accompanies most psychiatric diagnoses, false positives are a major problem and foment accusations of medicalizing normal human conditions. On the other hand, false negatives are persons needing professional care who are erroneously deprived of treatment. For those psychiatrists who believe that mental disorders are real medical entities and not social or ideological constructs, the solution of the problem consists in identifying the pathophysiologic mechanisms that underlie psychiatric conditions (Pierre, 2012). They argue that, throughout the history of medicine, diagnostic constructs of dubious validity gained the status of real diseases when technology-driven discoveries allowed the identification of somatic lesions. Sooner or later, this will also happen in psychiatry. Based on the terminology explained in Section 33.2.1, they argue for a definition of mental disorder grounded on proximate mechanisms.

I disagree. To formulate a scientific and value-free definition of mental disorder, we should focus on ultimate consequences, not proximate mechanisms. There is no factual or value-free feature in the concept of pathophysiology viewed as an inner mechanism underlying a condition whose morbidity status is surrounded by controversy. For example, in type 2 diabetes, insulin resistance is a pathophysiologic mechanism because there is general agreement that diabetes is a disease. Yet, the larger volume of left-handers' corpus callosum is not a pathophysiologic mechanism because left-handedness is considered a natural variant. It is worth noting that diabetes has been considered a disease since antiquity, but the status of left-handedness has changed over time: until well into the twentieth century it was considered a disease (Gutwinski et al., 2011). Personal and social values orient the decision as to whether a condition is a disorder or not. If it is agreed as being a disorder, the discovered mechanism underlying the condition is considered a dysfunction or a lesion. If not, the mechanism is considered a correlate of a normal variant.

Unlike proximate mechanisms, ultimate consequences are criteria of morbidity that are objective and immune to the perils of cultural relativism. I have proposed an evolutionary definition of mental disorder based on ultimate consequences (Troisi, 2015). The definition includes five criteria. Criterion A defines mental disorder as a maladaptive syndrome; that is, a psychological or behavioral syndrome that impacts negatively on the individual's inclusive fitness. Criteria B and C specify that: (1) individuals who have a mental disorder make choices that penalize their inclusive fitness, relative to all feasible alternative strategies in a specific environment; and (2) at the individual level, the inability to achieve short-term biological goals is a valid proxy indicator of mental disorder when estimates of inclusive fitness cannot be made. Criteria D and E make it clear that neither the demonstration of a pathophysiology underlying the syndrome nor the demonstration of distress associated with the syndrome are necessary or sufficient for a diagnosis of mental disorder. I refer the reader to my original paper (Troisi, 2015) for a methodological discussion of the scientific validity of each criterion. Here, I would mention an aspect that shows how being a Darwinian psychiatrist does not mean one must ignore sociocultural aspects of diagnosis.

The definition I have proposed reveals a contradiction that, in my opinion, is the real reason why decades of debate and hundreds of academic publications have produced no general consensus. A purely scientific definition of mental disorder is possible, but it cannot be used in clinical practice because, unlike other biological sciences, medicine and psychiatry are strongly influenced by social values and public expectations. Understandably, clinicians and their patients want to eliminate suffering, not to promote biological adaptation. Yet, evolution did not shape our minds for well-being and social harmony, and the correlation between well-being and adaptation is tenuous (Nesse, 2004; Troisi, 2011). In addition, the history of medicine demonstrates that often answers to the question "what is a disorder?" are identified by presuming the desired outcome and then adjusting one's interpretation of the research data to guarantee arrival at that outcome. The inflexible application of the criterion of maladaptation would produce unwanted effects on psychiatric classification. Behaviors that we value should be considered as mental disorders and conditions that we dislike and want to change should be viewed as sophisticated adaptations. There is nothing in the concept of mental disorder that inherently defies definition. Rather, the hole at the center of psychiatric classification reflects the irresoluble tension between scientific evidence and cultural preferences.

33.3.2 Mind–Brain Dualism

Mental health professionals continue to employ a mind–brain dichotomy when reasoning about clinical cases (Gabbard, 2005). In clinical discourse, references to "mind" and "brain" have become a form of code for different ways to think about the etiology of psychiatric disorders and their treatment. The etiology and pathogenesis of "biological" disorders would depend mainly on genetic predisposition and neural dysfunction, whereas environmental factors and interpersonal problems would be the main causal factors of "psychological" disorders. Somatic therapies (e.g., medication) would be best indicated for biological disorders and psychotherapy would be the treatment of choice for psychological disorders. The negative consequences of mind–brain dualism are many, including quarrels over the boundaries between neurology and psychiatry, patients' confusion about their diagnostic status, and biased attribution of intentionality and responsibility (Miresco & Kirmayer, 2006).

Despite countless attempts to promote an integrative model (i.e., the so-called biopsychosocial model), the dualistic view continues to dominate psychiatric theory and practice. In my opinion, one major reason for psychiatrists' resistance to abandoning the dualistic way of thinking is that the models proposed to replace the mind–brain dichotomy are based on causal factors reflecting proximate mechanisms. If the biological, psychological, and social determinants of psychiatric disorders are conceptualized as distinct categories acting as proximate causes, the search for the relative role of each factor in the etiology of psychiatric disorders is inevitable (Baillie et al., 2009). The integrative model rapidly turns into an additive, eclectic framework that does not explain the conceptual relationship between its components (Henningsen, 2015).

The divide between mind and brain fades away if one reasons in terms of adaptive function and employs the concept of behavioral systems. A behavioral system (e.g., the attachment system) can be defined as an integrated group of functionally related components consisting of specific psychological processes, physiological mechanisms, anatomical structures, and genetic influences. To serve the adaptive function that increases (or increased in the past) the likelihood of survival and/or reproduction, a behavioral system needs: (1) a set of activating stimuli; (2) a set of behaviors that are enacted to attain the system's goal; (3) a specific set goal (i.e., a change in the person–environment relationship that terminates the system's activation); (4) genetic, anatomical, and physiological substrates that allow it to function; and (5) cognitive and emotional processes that motivate and guide the system's activity. The fact that, in the course of evolution, all of these components were selected to serve a common adaptive function explains the conceptual relationship between them. No component has an intrinsic priority over the others and malfunctioning can originate anywhere in the system. When this happens, malfunctioning propagates to all of the components. From such a perspective, it does not make any sense to say that a dysfunctional style of insecure attachment is either a biological or a psychological disorder. And, more generally, it does not make any sense to say that a psychiatric disorder is "organic" or "functional" (where functional is used to indicate that there is no demonstrable somatic lesion; Kanaan et al., 2012).

33.4 DARWINIAN PSYCHIATRY AND ITS ALLIED DISCIPLINES

Darwinian psychiatry shares with other disciplines the objective of explaining the human mind and behavior from an evolutionary perspective. A brief review of the theoretical framework and methodological features of these allied disciplines allows us to define better the place of Darwinian psychiatry in contemporary behavioral biology.

33.4.1 Evolutionary Psychology

Evolutionary psychology conceptualizes the human mind as consisting of a mosaic of species-specific mental mechanisms developed through the process of natural selection that allowed our human ancestors to adapt to the hunter–gatherer environment in which our species evolved (Crawford & Krebs, 2008). This environment is referred to as the environment of evolutionary adaptedness (EEA), which is defined as a statistical composite of the adaptation-relevant properties of the ancestral environments encountered by members of ancestral populations, weighted by their frequency and fitness consequences.

The application to psychiatric disorders of some hypotheses advanced by evolutionary psychologists has produced interesting results that have clinical implications. One example is the mismatch hypothesis that

I have mentioned above. Vulnerability to some mental disorders can be explained by focusing on the differences between the EEA and modern environments. In medicine, the mismatch hypothesis has proved to be a valid explanation for the epidemics of obesity and its complications, including diabetes, atherosclerosis, and hypertension, in modern and otherwise increasingly healthy societies (Genné-Bacon, 2014). The explanation is that natural selection shaped eating preferences that maximize intake of calorically dense foods that were rarely available to our ancestors, but are available in abundance now. In psychiatry, the same explanation may apply to drug use and addiction: purified psychoactive substances were unavailable in the ancestral environment, but are now prevalent in many societies (Durrant et al., 2009).

Mind modularity is another hypothesis that has generated interesting lines of research in psychopathology. The empathizing–systemizing (E–S) theory suggests that humans have evolved two parallel and complementary cognitive–affective systems. According to this theory, "empathizing" involves the motivation and skills required to understand and interact appropriately with the social world, and "systemizing" describes the drive to analyze, understand, and manipulate the physical world. On average, males demonstrate a stronger drive to systemize, whereas females tend toward empathizing. Under the E–S model, autism would represent an extreme expression of male-typical cognition involving a strongly skewed profile of enhanced systemizing and reduced empathizing (Baron-Cohen, 2010). By contrast, an overdevelopment or overexpression of the female psychological phenotype would manifest with clinical profiles that are currently classified as borderline personality, depression, and psychotic–affective disorders. The diametrical hypothesis of autism and psychosis is based on a large body of anatomical, physiological, and behavioral data and implies genomic imprinting (i.e., an epigenetic mechanism by which certain genes are expressed in a parent of origin-specific manner) in the pathogenic process (Crespi & Go, 2015).

Here, I will not discuss the several aspects of evolutionary psychology that can be criticized from a biological perspective (Panksepp & Panksepp, 2000). I will limit my analysis to a point that will be further developed in Section 33.4.2 dealing with ethology: behavioral analysis. Evolutionary psychologists maintain that human behavior itself does not contribute to fitness and that the psychological mechanisms that produce behavior should be the focus of evolutionary analysis (Symons, 1989). Ironically, such an emphasis on psychological processes to the detriment of behavioral analysis is in line with mainstream psychiatric thinking. In my view, Darwinian psychiatrists should reject the idea that psychological processes should be the exclusive focus of evolutionary analysis for two reasons: first, it distracts research and clinical observation from assessing the actual behavior of persons with mental disorders; and second, it ignores the fact that natural selection cannot operate on psychological processes unless they manifest through physiology and behavior.

33.4.2 Ethology

Historically, ethology was the first evolutionary discipline to study behavior. Studies of animals and humans in unconstrained conditions played a major role in the development of powerful methods for observing and measuring behavior (Martin & Bateson, 2007). The discipline has important implications for Darwinian psychiatry because it pushes the collection of clinical data away from the artificial settings of psychiatrists' offices and hospital wards and encourages observation of patients in the environments where they live. Psychiatry has computers full of clinical data collected through clinician-rated scales, self-rated scales, projective tests, neuropsychological tests, psychophysiological measures, brain imaging, hormonal assays, and genetic testing (just to name a few), but with few exceptions it does not have data documenting what people with psychiatric conditions do in their everyday lives.

Insistence on observing individuals in their natural environments differentiates ethology from other disciplines studying human behavior, including evolutionary psychology. Ethology emphasizes the importance of naturalistic observation because the functionality of a single behavior or an entire behavioral repertoire can be assessed only by viewing how the actor of that behavior copes with specific environmental challenges and constraints. The artificial settings where mental health assessments generally occur (e.g., the hospital ward, the consulting room, the psychiatrist's office) may give a distorted picture of what happens to patients in their natural environments. For example, depressed individuals can engender negative mood and rejection in those with whom they interact. Roommates of depressed college students like their roommates less and tend to reject them more than do roommates of nondepressed subjects. Spouses of depressed partners report feeling more depressed, hostile, and critical following interactions with their partners. However, studies focusing on interactions of depressed people with strangers have failed to replicate these findings (Nezlek et al., 2000). Thus, the negative quality of depressed persons' interactions appears to emerge fully only in specific social environments; namely, those in which there is a close relationship among participants (Hale et al., 1997).

The full potential of ethological studies of people with mental disorders is yet to be realized. Despite these limitations, those ethological studies that have been published offer a fresh perspective on different aspects of psychiatric conditions and improve the definitions of several clinical phenotypes (Troisi, 1999).

33.4.3 Behavioral Ecology

Behavioral ecology is the study of the evolutionary basis for behavior due to ecological pressures (Davies et al., 2012). Behavioral ecology has brought the evolutionary study of behavior to a level of complexity and sophistication that was unimaginable when ethology was founded in the first half of the twentieth century. For example, mathematical modeling – one of behavioral ecology's trademarks – has become routine in the analysis of behavioral patterns. Theoretical advances have also led to new evolutionary concepts, including life history theory, kin selection, inclusive fitness, parent–offspring conflict, and alternative strategies. Among the evolutionary behavioral sciences, behavioral ecology probably offers most original contributions to Darwinian psychiatry. Here, I will briefly discuss the psychiatric implications of the concept of alternative strategies.

The approach of medicine and psychiatry to individual differences reflects typological thinking (i.e., postulating homogeneity in a population as the natural state and variation as the result of some sort of interference; Chung, 2003). When this way of thinking is applied to clinical practice, most somatic and psychological traits are assumed to fall into a normal distribution, with most of the cases in the middle and a few at the extremes. These extremes, which constitute only a small percentage of the total population, are arbitrarily lopped off and labeled "abnormal" or "pathological" and the far larger percentage clustering around the middle is arbitrarily called "normal." Unlike clinicians, evolutionary biologists are aware that variation within natural populations is widespread at every level, from overt behavior to DNA sequences, and that such variation is a prerequisite for evolution. A hallmark of modern evolutionary models is the capacity to integrate explanations focusing on species-typical patterns with explanations focusing on individual differences that diverge from these modal patterns (Del Giudice, 2018).

The focus on adaptive individual differences is exemplified by studies of alternative strategies, defined as the presence of two or more discrete behavioral variants among adults of one sex and one population when those variants serve the same functional end, such as more than one way of foraging, or attracting mates, or nesting (Brockmann, 2001). Alternative strategies may evolve as genetically based programs and be maintained within a population because the environment is heterogeneous and what works in one ecological niche does not necessarily do as well as in another. A second account of alternative strategies emphasizes their "frequency-dependent" origins. A given strategy is only advantageous if it is displayed by no more than a certain proportion of the population. If the variant strategy becomes too common, its advantages are reduced and the proportion of individuals enacting it declines.

When reconsidered from the perspective of behavioral ecology, some conditions that have been diagnosed as personality disorders are best understood as evolved alternative strategies (Troisi, 2005). For example, individuals with primary psychopathy are deceitful and manipulative in order to gain personal profit. They lack empathy and tend to be callous, cynical, and contemptuous of the feelings, rights, and suffering of others. In an evolutionary context, primary psychopathy is thought to represent a high-risk strategy of social defection associated with resource acquisition and reproduction. In a society made up primarily of cooperators, genes for cheaters could enter the population and remain, provided that the benefits of cheating outweigh the costs. As long as selective pressures for cooperative strategies coexist with counterpressures for cheating, a mixture of phenotypes will result, such that some sort of statistical equilibrium will be approached. Primary psychopathy should thus be expected to be maintained as a low-level, frequency-dependent alternative strategy (Krupp et al., 2013).

Individual differences in personality can be conceptualized as either categories (e.g., personality types) or dimensions. Categorical diagnoses of personality disorders fit well with the concept of alternative strategies. Empirical data recorded in large clinical samples have provided preliminary evidence for the adaptive significance of some pathological personality profiles. However, it is likely that the full potential of the evolutionary approach to individual differences will be revealed by dimensional studies aimed at demonstrating that normal-range personality variation has adaptive functions rather than being random noise around a behavioral optimum (Vall et al., 2016).

33.4.4 Evolutionary Anthropology

Most of the research and clinical data published in the psychiatric literature derive from studies of WEIRD (Western, Educated, Industrialized, Rich, and Democratic) populations (Henrich et al., 2010). The extent to which these data are generalizable to people living in other cultural contexts is dubious. Adopting a comparative perspective, cross-cultural psychiatry analyzes the influence of culture on psychopathology. Yet, the comparative perspective can be widened to include the evolutionary history of psychopathology in our phylogenetic lineage, with particular reference to the impacts of different ecological contexts on psychiatric vulnerability.

The comparative perspective is the framework of reference for evolutionary anthropology, defined as the interdisciplinary study of the evolution of human physiology and behavior and the relationship between nonhuman primates, early humans, and modern humans. When focusing on behavior, evolutionary anthropologists study how ecological differences impact on social organization and interpersonal relationships (Gibson & Lawson, 2015). Considering that obtaining a direct picture of psychopathology among our prehistoric ancestors is impossible, most evolutionary anthropologists suggest that contemporary hunter–gatherers offer the best glimpse of what life might have been like during the later phases of human evolution. In anthropology, societies of hunter–gatherers are classified within the wider category of "traditional societies." Evolutionary anthropology can contribute to Darwinian psychiatry by defining which psychiatric conditions described in current classifications are diagnosable in traditional societies and how other group members in these communities react to individuals with psychopathology.

Studies of psychopathology in traditional societies are complicated by many methodological problems, including the "category fallacy," defined as the reification of a nosological category developed for a particular cultural group that is then applied to members of another culture for whom it lacks coherence and its validity has not been established (Kleinman, 1987). In spite of the complexity of applying standardized diagnostic procedures in different ethnographic contexts, there is preliminary evidence that the basic manifestations of psychiatric disorders (i.e., anxiety, depression, somatization, and psychosis) are present in traditional societies (Fabrega, 1997; Pridmore, 2009). However, this is a neglected area of research, and in the absence of systematic data (e.g., prevalence rates, ages of onset, precipitating factors), any generalization is impossible.

Studies of abnormal behaviors in nonhuman primates are another potential source of information for exploring the evolutionary history of psychopathology in our phylogenetic lineage. Unfortunately, because of methodological problems, knowledge of the occurrence and phenomenology of psychiatric disorders in nonhuman primates has progressed slowly. For example, many abnormal behaviors have been observed only after exposure to the extremely severe and highly artificial stressors employed in experimental studies, and the search for spontaneous psychopathology in natural or quasi-natural environments has been limited to gross behavioral abnormalities, often ignoring subtle individual differences. Despite these limitations, there is evidence that monkeys and apes can suffer from behavioral disorders similar to those that afflict human subjects (e.g., maternal abuse of infants in macaques and fatal depression following the death of the mother in adolescent chimpanzees; Troisi, 2003).

33.5 CONCLUSIONS

The interrogative title of this chapter was chosen by the editors of this volume probably because they were expecting the author to formulate clear predictions on the future of Darwinian psychiatry. Yet, they did not consider that the author is a psychiatrist, and psychiatrists can be vague if they want to be. (Do you remember the dialog in the 1999 movie *Analyze This* with Billy Crystal and Robert De Niro? "Doc, if you gotta talk, try to be vague. Can you do that?" "I'm a psychiatrist. Believe me, I can be vague.") Therefore, my answer to the question "are we on the verge of Darwinian psychiatry?" is yes and no.

The answer is yes because the scope of contemporary psychiatry is much wider than it was in the past. Only a few decades ago, psychiatrists dealt mostly with hallucinations, delusions, suicidal behavior, agitation, mania, severe depression, obsessions, or sexual perversions. Today, they also focus on theory of mind, alexithymia, Machiavellianism, dysfunctional attachment, and pathological altruism. To explore these new areas of research, psychiatry needs new theoretical models that explain the normal and abnormal functioning of human mind and behavior (Troisi, 2017). Thus, my prediction is that psychiatrists will look with growing interest at the insights of the evolutionary behavioral sciences to get a better understanding of the complexity of clinical syndromes.

The answer is no because psychiatry is a clinical science, and clinicians pay little attention to those disciplines that apparently have scarce utility for their everyday practice. When evolutionary psychiatrists suggest ultimate explanations of vulnerability to psychopathology based on phenomena such as evolutionary trade-offs, genome lag, and historical constraints, certainly they are offering a deeper understanding of the origins of psychiatric disorders. But let us be practical – when patients with psychosis or panic attacks arrive at the hospital, all that matters is relieving their symptoms. Thus, my prediction is that psychiatrists will continue to ignore Darwinian psychiatry unless evolutionary explanations are tied to effective solutions of clinical problems.

REFERENCES

Adriaens, P. R., & De Block, A. (2010). The evolutionary turn in psychiatry: A historical overview. *History of Psychiatry*, **21**(82 Pt 2), 131–143.

Baillie, D., McCabe, R., & Priebe, S. (2009). Aetiology of depression and schizophrenia: Current views of British psychiatrists. *Psychiatric Bulletin*, **33**, 374–377.

Baron-Cohen, S. (2010). Empathizing, systemizing, and the extreme male brain theory of autism. *Progress Brain Research*, **186**, 167–175.

Bateson, P., & Laland, K. N. (2013). Tinbergen's four questions: An appreciation and an update. *Trends in Ecology and Evolution*, **28**, 712–718.

Belsky, J. (2016). The differential susceptibility hypothesis: Sensitivity to the environment for better and for worse. *JAMA Pediatrics*, **170**(4), 321–322.

Brockmann, H. J. (2001). The evolution of alternative strategies and tactics. *Advances in the Study of Behavior*, **30**, 1–51.

Childs, B. (1999). *Genetic Medicine. A Logic of Disease*. Baltimore, MD: Johns Hopkins University Press.

Chung, C. (2003). On the origin of the typological/population distinction in Ernst Mayr's changing views of species, 1942–1959. *Studies in History and Philosophy of Biology & Biomedical Sciences*, **34**, 277–296.

Crawford, C., & Krebs, D. (2008). *Foundations of Evolutionary Psychology*. New York: Lawrence Erlbaum.

Crespi, B. J., & Go, M. C. (2015). Diametrical diseases reflect evolutionary–genetic tradeoffs: Evidence from psychiatry, neurology, rheumatology, oncology and immunology. *Evolution, Medicine, and Public Health*, **1**, 216–253.

Davies, N. B., Krebs, J. R., & West, S. A. (2012). *An Introduction to Behavioural Ecology*, 4th ed. New York: Wiley-Blackwell.

Del Giudice, M. (2018). *Evolutionary Psychopathology: A Unified Approach*. New York: Oxford University Press.

Durrant, R., Adamson, S., Todd, F., & Sellman, D. (2009). Drug use and addiction: Evolutionary perspective. *Australian New Zealand Journal of Psychiatry*, **43**(11), 1049–1056.

Evans, S. S., Repasky, E. A., & Fisher, D. T. (2015). Fever and the thermal regulation of immunity: The immune system feels the heat. *Nature Review Immunology*, **15**(6), 335–349.

Fabrega, H., Jr. (1997). *Evolution of Sickness and Healing*. Berkeley, CA: University of California Press.

Frances, A. (2012). There is a time and place for every umpire. *Philosophy, Ethics, and Humanities in Medicine*, **7**, 24–26.

Gabbard, G. O. (2005). Mind, brain, and personality disorders. *American Journal of Psychiatry*, **162**(4), 648–655.

Genné-Bacon, E. A. (2014). Thinking evolutionarily about obesity. *Yale Journal of Biology and Medicine*, **87**(2), 99–112.

Gibson, M. A., & Lawson, D. W. (2015). Applying evolutionary anthropology. *Evolutionary Anthropology*, **24**(1), 3–14.

Gluckman, P., & Hanson, M. (2006). *Mismatch. Why Our World No Longer Fits Our Bodies*. Oxford: Oxford University Press.

Gutwinski, S., Löscher, A., Mahler, L., et al. (2011). Understanding left-handedness. *Deutsches Ärzteblatt International*, **108**, 849–853.

Hale, W. W., 3rd, Jansen, J. H., Bouhuys, A. L., Jenner, J. A., & van den Hoofdakker, R. H. (1997). Non-verbal behavioral interactions of depressed patients with partners and strangers: The role of behavioral social support and involvement in depression persistence. *Journal of Affective Disorders*, **44**, 111–122.

Healy, D. (2014). Psychiatric "diseases" in history. *History of Psychiatry*, **25**(4), 450–458.

Henningsen, P. (2015). Still modern? Developing the biopsychosocial model for the 21st century. *Journal of Psychosomatic Research*, **79**(5), 362–363.

Henrich, J., Heine, S. J., & Norenzayan, A. (2010). The weirdest people in the world? *Behavioral and Brain Sciences*, **33**(2–3), 61–83.

Kanaan, R. A., Armstrong, D., & Wessely, S. C. (2012). The function of "functional": A mixed methods investigation. *Journal of Neurology Neurosurgery and Psychiatry*, **83**(3), 248–250.

Karen, R. (1994). *Becoming Attached*. New York: Warner.

Keller, M. C., & Miller, G. (2006). Resolving the paradox of common, harmful, heritable mental disorders: Which evolutionary genetic models work best? *Behavioral Brain Sciences*, **29**, 385–404.

Keller, M. C., & Nesse, R. M. (2006). The evolutionary significance of depressive symptoms: Different adverse situations lead to different depressive symptom patterns. *Journal of Personality and Social Psychology*, **91**, 316–330.

Kleinman, A. (1987). Anthropology and psychiatry. The role of culture in cross-cultural research on illness. *British Journal of Psychiatry*, **151**, 447–454.

Krupp, D. B., Sewall, L. A., Lalumière, M. L., Sheriff, C., & Harris, G. T. (2013). Psychopathy, adaptation, and disorder. *Frontiers in Psychology*, **4**, 139.

Laland, K. N., Sterelny, K., Odling-Smee, J., Hoppitt, W., & Uller, T. (2011). Cause and effect in biology revisited: Is Mayr's proximate-ultimate dichotomy still useful? *Science*, **334**(6062), 1512–1516.

Martin, P., & Bateson, P. (2007). *Measuring Behaviour. An Introductory Guide*. Cambridge, UK: Cambridge University Press.

Mayr, E. (1961). Cause and effect in biology. *Science*, **134**(3489), 1501–1506.

McGuire, M. T., & Troisi, A. (1998). *Darwinian Psychiatry*. New York: Oxford University Press.

Miresco, M. J., & Kirmayer, L. J. (2006). The persistence of mind–brain dualism in psychiatric reasoning about clinical scenarios. *American Journal of Psychiatry*, **163**(5), 913–918.

Nesse, R. M. (2004). Natural selection and the elusiveness of happiness. *Philosophical Transactions of the Royal Society B: Biological Sciences*, **359**(1449), 1333–1347.

Nesse, R. M. (2005). Natural selection and the regulation of defenses: A signal detection analysis of the smoke detector principle. *Evolution and Human Behavior*, **26**, 88–105.

Nezlek, J. B., Hampton, C. P., & Shean, G. D. (2000). Clinical depression and day-to-day social interaction in a community sample. *Journal of Abnormal Psychology*, **109**, 11–19.

Panksepp, J., & Panksepp, J. B. (2000). The seven sins of evolutionary psychology. *Evolution and Cognition*, **6**(2), 108–131.

Pierre, J. (2012). Commentary. *Philosophy, Ethics, and Humanities in Medicine*, **7**, 21–22.

Pridmore, S. (2009). Australian Aboriginal stories and psychopathology. *Asian Journal of Psychiatry*, **2**(4), 139–142.

Sharpe, M., & Walker, J. (2009). Symptoms: A new approach. *Psychiatry*, **8**, 146–148.

Sherman, J. A. (2012). Evolutionary origin of bipolar disorder-revised: EOBD-R. *Medical Hypotheses*, **78**(1), 113–122.

Simpson, J. A., & Belsky, J. (2008). Attachment theory within a modern evolutionary framework. In J. Cassidy & P. R. Shaver, eds.) *Handbook of Attachment. Theory, Research, and Clinical Applications*, 2nd ed. New York: Guilford Press, pp. 131–157.

Srinivasan, S., Bettella, F., Mattingsdal, M., et al. (2016). Genetic markers of human evolution are enriched in schizophrenia. *Biological Psychiatry*, **80**(4), 284–292.

Stevens, A., & Price, J. (1996). *Evolutionary Psychiatry: A New Beginning*. New York: Routledge.

Symons, D. (1989). A critique of Darwinian anthropology. *Ethology and Sociobiology*, **10**, 131–144.

Tinbergen, N. (1963). On aims and methods of ethology. *Zeitschrift für Tierpsychologie*, **20**, 410–433.

Troisi, A. (1999). Ethological research in clinical psychiatry: The study of nonverbal behavior during interviews. *Neuroscience and Biobehavioral Review*, **23**, 905–913.

Troisi, A. (2003). Psychopathology. In D. Maestripieri, ed., *Primate Psychology*. Cambridge, MA: Harvard University Press, pp. 451–470.

Troisi, A. (2005). The concept of alternative strategies and its relevance to psychiatry and clinical psychology. *Neuroscience and Biobehavioral Review*, **29**(1), 159–168.

Troisi, A. (2006). Adaptationism and medicalization: The Scylla and Charybdis of Darwinian psychiatry. *Behavioral and Brain Sciences*, **29**, 422–423.

Troisi, A. (2011). Psychoactive drug use: Expand the scope of outcome assessment. *Behavioral Brain Sciences*, **34**, 324–325.

Troisi, A. (2012). Mental health and well-being: Clinical applications of Darwinian psychiatry. In S. Roberts, ed., *Applied Evolutionary Psychology*. New York: Oxford University Press, pp. 277–289.

Troisi, A. (2015). The evolutionary diagnosis of mental disorder. *Wiley Interdisciplinary Reviews Cognitive Science*, **6**(3), 323–331.

Troisi, A. (2017). *The Painted Mind: Behavioral Science Reflected in Great Paintings*. New York: Oxford University Press.

Troisi, A. (2018). Psychotraumatology: What researchers and clinicians can learn from an evolutionary perspective. *Seminars in Cell & Developmental Biology*, **77**, 153–160.

Vall, G., Gutiérrez, F., Peri, J. M., et al. (2016). Seven dimensions of personality pathology are under sexual selection in modern Spain. *Evolution and Human Behavior*, **37**, 169–178.

34 The Evolution of Pro-social Behavior

From Caring to Compassion

PAUL GILBERT

Some of the roots of pro-social behavior, of which caring and compassion are forms, are from the evolution of parental investment and caring (Brown & Brown, 2015; Fogel, Melson, & Mistry, 1986; Gilbert, 1989/2016, 2009; Mayseless, 2016; Preston, 2013; Seppälä, Simon-Thomas, Brown, Worline, Cameron, & Doty. (2017)). There are a number of different dictionary definitions of caring. A typical one is "the provision of what is needed for the well-being or protection of a person or thing" (www .dictionary.com/browse/caring). Fogel, Melson, and Mistry (1986) suggested that the core elements of care-nurturance are "The provision of guidance, protection and care for the purpose of fostering developmental change congruent with the expected potential for change of the object of nurturance" (p. 55). They also suggested that nurturance involves *awareness* of the need to be nurturing, *motivation* to nurture, *expression* of nurturing feelings, *understanding* what is needed to be nurturing, and an ability to match nurturing with *feedback* from the nurtured. Nurturing then needs to be skillfully enacted. Importantly, with human cognitive insight, this basic motivation can be used with animate and inanimate objects, such as caring for one's children, family pet, rose garden, or prized car. Compassion, however, is reserved for sentient beings and (for the most part) their conscious subjective experiences. It can be defined as *a sensitivity to suffering in self and others with a commitment to try to alleviate and prevent it* (Gilbert & Choden, 2013). Hence, compassion requires the capacity for empathic insights into the experiences of others and concern to have an impact on aversive and distressing experiences.

On the journey into the evolution of caring and compassion, we can begin with Tinbergen (1963), who suggested that an evolutionary analysis of any psychological process requires at least four domains of focus. Two address "why" questions and two address "how" questions. So, first is exploring why certain physical and psychological traits (like caring) evolve. These relate to issues of: (1) phylogenetic origin, the evolutionary pressures and challenges the adaptation solved, and cross-species comparisons; and (2) the adaptive value of evolved traits and how they contribute to an animal's reproductive fitness over a lifetime. The second set of issues relate to *how* certain traits (like caring) function at any point in time. These are related to: (3) ontogenetic development and how traits (which can mature and function in different ways according to the context in which they mature) are patterned via lifetime contextual influences (phenotype); and (4) proximate influences in terms of the mechanisms (e.g., physiological systems, acquired beliefs) that are triggered by certain stimuli and contexts (Buss, 2015; Tinbergen, 1963). Evolution gives rise to physiological processes by which environmental stimuli are translated into behavioral outputs. Hence, evolutionary approaches attend to the evolution of both the behavioral and the physiological infrastructures by which different motivations emerge within, and are expressed by, organisms (Dawkins, 1976). A core reason for this is because these are the mechanisms that both evolution over the long term and ontogenetic influence over the short term work on, building and changing circuits "here" and rewiring them "there." This chapter will address these themes in regard to pro-social and caring behavior.

34.1 EVOLUTIONARY FUNCTIONAL ANALYSIS OF CARING AND PRO-SOCIAL BEHAVIOR

It is now well understood that the mechanisms of evolution are via algorithmic replication (Buss, 2015; Dawkins, 1976; Dunbar, 2016). What is passed from generation to generation is nothing physical, but rather information; algorithms for organizing molecules into proteins, into cells, into cell assemblies, into specific physical systems, and into physically integrated living beings (Dawkins, 1976). Hence, genes are information stores that build organisms with physiochemical and physiological systems that are oriented to the life tasks necessary for (their algorithmic) survival and reproduction. Central is the notion that physical brains and bodies are motivated to behave in certain ways that increase the likelihood of genetic (information-carrying) replication. Different emergent motivations can roughly be distinguished between those that evolved from the *challenges of survival* and those that evolved from the *challenges of reproduction*. The two challenges of reproduction (for sexually reproducing organisms) are, first, to ensure finding and sexually engaging with a breeding partner and, second, to ensure that resultant offspring have a chance of becoming breeding partners themselves. The evolution of caring originated primarily but not solely to aid offspring survival. As indicated in Figure 34.1, a range of strategies for survival and reproduction (gene replication) give rise to motivations that become choreographed according to the niche the organism grows within and hence the orchestration of the organism's phenotypes.

First, survival motives involve both harm-avoidance motives (injuries, diseases, and losses) and "acquisition of resources" motives. The most common defenses here are fight, flight, freeze, and faint. Second, animals need to be able to discriminate species-specific food from nonfood and avoid poisons with expelling (vomiting and diarrhea) if toxic substances have been ingested. Third are shelter-seeking strategies (below ground, on the ground, above the ground). The evolution of mammalian caring enabled parents to provide all of these to the infant; that is, protection, food, and shelter.

Reproduction motives underpin intersexual and intrasexual competition for access to conspecifics' genes and post-birth investment strategies. Males and females tend to follow different strategies both for sexual competition and for parental investment (Geary, 2000; Sapolsky, 2017). On making it to reproductive age, and once successful copulation has taken place, interest switches to *survival* of the offspring. In fact, it is their survival and reproduction that counts. Here, there are two fundamental strategies for offspring survival, called *r* and *k* selection. *r* selection is associated with high numbers with low parental investment typical of many egg-laying species, such as most fish, turtles, and reptiles. Their brains are hardwired to be functioning at birth and to disperse rapidly after birth to avoid predation, including, at times, from their own parents (MacLean, 1985). What ensures genetic replication in the next generation is large numbers. As ecologies change, however, so do the strategies for

reproduction. For example, crocodiles produce far fewer eggs than turtles and are responsive to their hatchlings' calls as they clamber from the nest. They pick them up in their mouths and carry them into the water, but thereafter the hatchlings are on their own. The parent neither feeds them nor guards/protects them nor provides shelter. Different ecologies call forth different reproductive strategies, and one major set of changes was with the evolution of warm-bloodedness, live birth, and small numbers of young with post-birth parental/caring investment, sometimes referred to as *k* selection. The evolution of caring then emerges from *k* selection.

With a few exceptions, it was mainly mammalian evolution that was to advance caring behavior in the form of complex forms of parental investment (Geary, 2000; Mayseless, 2016). This required substantial changes to the physiology of threat detection and defense. For example, offspring no longer dispersed into hiding for safety, but evolved to seek proximity to the secure, protective, provisioning base of the parent (Bowlby, 1969; Gilbert, 1989/2016; Mikulincer & Shaver, 2017). This enables close interpersonal contact, connection, and complex physiological co-regulation (Hofer, 1994; Porges, 2007). Mammalian young, then, do not have to fear being purposefully harmed or eaten by the parent, nor to be self-protective or self-sustaining, because of parental care provision and investment.

So, first, parent and offspring needed to be attracted in order to stay close to each other rather than disperse (proximity-seeking mechanisms). In species like cows, horses, and elephants, infants are mobile within the first minutes of life and can be partly responsible for staying close to the parent. Primate infants, however, are nonmobile from birth, requiring the parent to hold and maintain closeness to the infant. For primates, and especially humans, this physical closeness and skin-to-skin contact has implications for (neuro)physiological maturation (Dunbar, 2010; Hofer, 1994), including of the autonomic nervous system (Bergman, Linley, & Fawcus, 2004). Second, resources (including quantity and quality of food) and social stimuli provided by the parent have major impacts on the growth and physiological regulation of the infant (Hofer, 1994). Third, in some species in which parents provide food and protection, especially primates, (immobile) infants are required to be quiescent for much of the time as they grow and develop (e.g., infants sleep a lot). Fourth, because primate infants cannot move themselves, the parent is the major source of the stimuli that impact on the physiological maturation and regulation of the infant (Bergman et al., 2004; Hofer, 1994); they are a source of stimulation curiosity, excitement, and also soothing, calming, and (dis)stress regulation (Bowlby, 1969; Mikulincer & Shaver, 2017; Music, 2014). The evolution of attachment is the basis for the experience of early warmth, affiliation, and emotional soothing. Attachment disruption and loss impact on these emotion systems, giving rise to anxiety and depression (Bowlby, 1980; Harlow

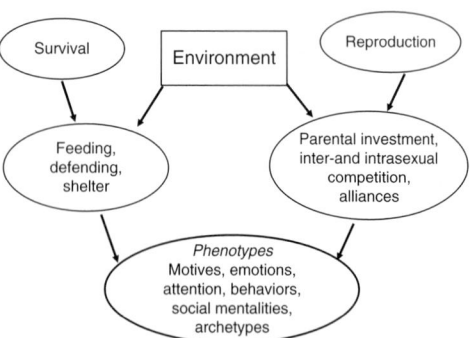

Figure 34.1 Evolution, strategies, phenotypes, social mentalities, and archetypes.

& Mears, 1979; Mikulincer & Shaver, 2017). The evolved defensive strategies to disruptions in attachment and affiliation are protest (anger, anxiety, and crying) and despair (retardation, loss of positive emotions, and hiding).

34.2 FUNCTIONS OF CARING

Over 50 years ago, Harlow (Harlow & Mears, 1979) and Bowlby (1969, 1973, 1980; Mikulincer & Shaver, 2017) highlighted the evolution of attachment mechanisms as providing sources of safety and soothing. Harlow showed that monkeys preferred to seek comfort, especially when threatened or distressed, from a terry cloth wire surrogate rather than from a wire one that gave food. Bowlby highlighted three dimensions of attachment that he labeled *proximity-seeking*, a *secure base* from which infants could explore and return, and a *safe haven*, indicating the capacity for the parent to soothe and regulate the infant's arousal, especially if distressed. Since then, studies have explored a variety of interpersonal styles that are believed to be linked to the experience of these pro-social caring functions in childhood. The most basic of these patterns have been labeled as *secure*, where children are able to trust others and turn to them for comfort and various forms of help when needed. Insecure, anxious children are fearful of abandonment and at times mistrusting. Avoidant children are also distrusting, but disengaged and self-reliant (Holmes & Slade, 2018). Even if children are provisioned, fed, sheltered, and kept out of danger, a lack of affectionate, caring interpersonal stimuli and interactions can have profound effects on the developing infant, as was sadly confirmed in studies of orphaned Romanian children (Rutter et al., 2007). As infants become more mobile, parents act as a *secure base*, facilitating and encouraging gradual exploration of their environment (Bowlby, 1969; Mikulincer & Shaver, 2017).

All of these inputs will be utilized by the developing child as indicators of what kind of environment they are growing up in and hence what kind of "brain" and interpersonal strategies are to be choreographed (Siegel, 2011). To facilitate these interactional sequences between parent and child, k selected regulation processes evolve and operate a sequence of adaptations. One was the orchestration of a set of neurophysiological processes that involved oxytocin (Carter, 2014; Taylor, 2006). Others were the major adaptations and evolution of the myelinated parasympathetic system – the vagal nerve that links a range of internal organs to central control systems (Thayer et al., 2009, 2012). Indeed, the vagus nerve is connected to a range of organs, including the heart and gut, and importantly it plays a role in the activation and orchestration of the frontal cortex (Thayer et al., 2012). In addition, these systems, along with the amygdala, are very sensitive to interpersonal stimuli such as voice tones, facial expressions, and touch (Dunbar, 2010; Petrocchi, & Cheli (2019) Porges, 2007). These adaptations did a number of things, such as enabling mammals to recognize their own kin, be oriented to move closer rather than away, suppress the fight-or-flight threat responses when doing so, and provisioning and protecting/defending them. Importantly, too, one of the functions of parental care and investment is *sensitivity to distress and preparedness to act appropriately to relieve that distress*. This is also the basic sentiment and *core of compassion*, and compassion utilizes the same evolved physiological pathways as basic caring behavior (Gilbert, 2005, 2009, 2014, 2017 Seppälä, Simon-Thomas, Brown, Worline, Cameron, & Doty. (2017)).

In regard to humans specifically, many researchers have highlighted necessary evolved functions, other than those of protecting and soothing, as part of caring in early close relationships. Hrdy (2009) highlighted the fact that we are the only primate that allows our babies to be touched and held by others from birth. Touch is extremely important for a range of processes (Dunbar, 2010). So right from birth humans can be exposed to multiple carers (e.g., relatives), which sensitizes them to different types of people and social interaction. Early life environments also provide opportunities for children to learn how they "exist in the minds of others," particularly the degree to which they are valued, liked, and wanted (likely and able to elicit investment from others); or, to put it another way, they learn about their social influence and their ability to stimulate positive emotions about themselves in the minds of others (Barkow, 1989; Gilbert, 1989/2016, 1997). Children learn about their emotions and how to express them appropriately. In addition, they experience empathic validation of their emotions from parents and relatives and receive encouragement and reinforcement, along with moral training. These inputs help the child to create internal working models of themselves and others that allow them to enter the competitive world of alliances and sexual partners with confidence, utilizing socially attracting tactics in contrast to (say) threatening, bullying, and narcissistic social strategies. To put this another way, human parental care has evolved to function in a myriad of complex ways not only to aid physiological maturation, but also to choreograph and support newly evolved human cognitive systems that are essential for social intercourse, the creation of a self-identity, and emotion regulation. Caring behavior therefore is among the most profoundly important of all human motives. Seppälä, Simon-Thomas, Brown, Worline, Cameron, & Doty. (2017).

34.3 CARING AS A MOTIVATION

Animals have many life tasks that underpin a variety of motives, such as harm avoidance, exerting control over resources, status seeking, sexual reproduction, and caring with the intent of fostering the development of others, particularly – but not only – kin. Although caring involves emotion, it is a fundamental motivational process; it can be enacted even in the absence of emotion. A core aspect of caring of sentient beings has been labeled *altruistic responding* (Preston, 2013). Given that different forms of altruism (kin and reciprocal) are often seen as underpinning the evolution of caring motivation, Preston (2013) suggests a definition of altruistic responding:

Altruistic responding is defined as any form of helping that applies when the giver is motivated to assist a specific target *after perceiving their distress or need* ... Altruistic responding implies an active behavioral response initiated by the perception of need, which is differentiated from cooperative, diffuse, or unintentional forms of altruism that likely derive from other evolutionary and mechanistic origins ... Altruistic responding further narrows these classifications to only include cases where the motivation to respond is fomented by direct or indirect perception of the other's distress or need ... This excludes cases that emerged later in time or include diverse processes, such as cooperation or helping influenced by strategic goals, social norms, display rules, or mate signaling. (p. 1307, emphasis added)

She identifies passive and active forms of caring. Passive forms are providing soothing and comforting, whereas active forms are specific behaviors designed to rescue, alleviate distress, and feed, which require motor activation and appropriate behaviors. This definition excludes concepts of sharing or acts that focus on the flourishing and well-being of others or general caring; this makes altruistic caring a subclass of caring motivation in general. Others, however, argue that human altruism also evolved with – and is central to – cooperation, which is about sharing (Warneken & Tomasello, 2009).

Common to many approaches to caring behavior, Preston (2013) locates the origins of altruistic responding in the evolution of detecting *and* responding to (retrieving/rescuing) distress calls in infants – coming to their aid. Indeed, caring and compassion, like all motivational systems, have to have at least two core aspects to their evolution: the classic "stimulus–response" algorithm process. First is stimulus detection, such that evolution builds physiological systems that are able to detect and react to particular stimuli that indicate information relating to the fulfillment of social goals (Gilbert, 1989/2016). Sometimes these are regarded as modularized systems (Geary & Huffman, 2002). For example, many threat stimuli stimulate automatic algorithms for defense (hatchling birds will crouch in the nest at the shadow of a hawk passing over; animals may freeze and attend to the sound of a predator; individuals may become alarmed at an angry or aggressive face and voice tone). Second, there are menus of appropriate behaviors to different classes of stimulus (e.g., animals run from a threat stimulus, approach and try to eat a food stimulus, and orient for courting and sexual behavior to a sexual stimulus). Caring behavior also clearly begins with stimulus detection. First is kin recognition (Burnstein, Crandall, & Kitayama, 1994; Buss, 2015; MacLean, 1985). Animals rarely care for just any offspring and are kin selective, although of course they can make mistakes and be cuckolded. In addition, some

primates will adopt and care for orphaned infants (Goodall, 1990), and humans are also well-known to do this. Following identification of the target comes selective attentiveness to the actual type of signals emanating from the target (e.g., offspring as the object of caring). Obvious signals are those that stimulate action such as hunger calls (go feed), distress calls (go rescue/sooth), or awareness of proximity distance (go find). Others include signals of security and safeness, where (say) the infant is resting or sleeping and no specific action is required other than some form of vigilance for any change that calls for action.

34.3.1 Caring as a Social Mentality

As noted in Section 34.3, caring and altruistic behaviors did not evolve as one-way streets, but as dynamic, reciprocal behavioral interactions. So caring behavior evolves with appropriate stimulus-detection and response repertoires in dynamic reciprocal interactions – interpersonal dances. For example, as a parent responds to the infant's distress or need, this will impact on the physiology and behavioral outputs of the infant, which in turn will become signals activating physiological and behavioral changes in the parent. The infant's changing behavior signals (is a stimulus) to the parent to do more, less, or change their behavior in a repeating sequence of interactions, changing slightly with each repetition.

Motivations that involve competencies for dynamic, socially reciprocal interactions have been called *social mentalities* (Gilbert, 1989/2016, 2005, 2017). Part of the motivation, therefore, is not just to influence and organize one's own mind, but to behave in a way that impacts on the mind/body of the other – to stimulate the behavioral algorithms in another. These processing systems are in dynamic fluctuating states, adjusting moment by moment to changing stimulus processing and responding and communicating in a kind of interpersonal dance of dynamic, reciprocal interactions. Social mentalities emerge from basic social motives (e.g., status acquisition, sexual opportunity, caring for offspring, building alliances), but complex social processing systems (e.g., empathy) are required in order to engage in complex, dynamic reciprocal interpersonal dances and to create states of mind in the minds of all those engaged in that interpersonal dance (Gilbert, 1989/2016, 2017).

Although caring is clearly linked to distress sensitivity/awareness, the origins of an empathic recognition of distress calls was likely in the service of threat processing rather than caring. Indeed, many animals, on hearing another in distress, may produce flight behavior (a distress call indicates danger), and signals of disease (hence possible contamination) certainly create flight behavior (Panksepp & Panksepp, 2013). In the parent–child relationship, however, exactly the opposite has to occur, such that a distress signal activates approach and engagement (rescue), not flight or disengagement behavior, and signs of illness trigger caring. This is an important dimension to keep in mind because signals of disease and distress, if they are identified as originating in nonkin and or out-group members, may activate harm-avoidance behaviors such as disengagement, avoidance, or even fight/attack. This is why we admire the courage of individuals who are prepared to try and help people who are diseased and potentially infectious, such as those traveling to West Africa to address the Ebola crisis. In stark contrast, when out-group members are attacked, they are often described in "disease and infection" terms, as Hitler did with the Jewish people (Gilbert, 2005).

34.3.2 The Flow of Caring and Compassion

What this means is that evolution works on the nature of social interactions. Signs of vulnerability, weakness, need, or disease can trigger caring, rejection, or hostile behavior depending on the nature of the relationship. For example, in reciprocating relationships, these kinds of signals may make one unattractive as a friend, employee, or sexual choice. Hence, is it useful to note the flow of compassion such that we can be compassionate to others, we can be observant of, receptive to, and grateful for the compassion from others, and, in humans, who can have feelings about themselves, we can be compassionate to ourselves (in contrast to hostile and rejecting, as in being harshly self-critical). Exploring the evolution of social mentalities is a way of thinking about evolution working on dynamic interactions rather than on single, socially decontextualized individuals. Both the giving and receiving of caring compassion operate through overlapping processes; for example, both are affected by oxytocin and the vagus nerve (Carter, 2014; Petrocchi, & Cheli, 2019).

34.4 CARING FOR NONKIN

Helping and caring for nonkin others has sometimes being referred to as reciprocal altruism in the sense that one is more likely to care for others if that investment is likely to be reciprocated (Buss, 2015). Indeed, there is considerable evidence that the emotions involved in caring (like sympathy and affection) are influenced by the nature of the relationship (close–distant, like–dislike) (Loewenstein & Small, 2007). Importantly, if caring behavior to nonkin has survival and reproductive payoffs, then it is likely to spread through populations. Indeed, it turns out that caring and altruistic behavior can have payoffs in many domains.

As evolution modifies physiological systems that facilitate different competencies, these competencies can begin to be utilized for a range of different purposes. For example, it is well known that human empathy can be used to be caring or for Machiavellian self-advancement or to cause harm (Gilbert et al., 2017b; Zaki, 2014). Similarly, caring behavior can be used for a whole variety of purposes, one being to advance one's reputation (Catarino et al., 2014; Goetz et al., 2010). There is also good evidence

that, when it comes to long-term mating partners (but not short-term ones or flings), signals of caring, empathy, and altruism are high on the list of attractive qualities in mates; we are looking for helpful and trustful others to commit to (Farrelly, Clemson, & Guthrie, 2016; Jensen-Campbell, Graziano, & West, 1995). Similar evaluations pertain when it comes to developing alliances and close friendships. We are more attracted to individuals we perceive to be empathic and caring than non-empathic and non-caring. Emotions like guilt, with associated feelings of remorse and sadness, can arise in contexts in which, by acts of either omission or commission, we have hurt people, especially (but not only) those we were supposed to care for. This makes guilt and the avoidance of harming others central to the evolution of caring. Indeed, caring comes with mechanisms for avoiding causing harm and, to some extent, these mechanisms must be turned off when we deliberately choose to cause harm, as, for example, in vengeance or in war (Gilbert, 1989/2016, 2009, 2018). Individuals can also be shamed and stigmatized (damaging their reputation) for not demonstrating caring or compassion in contexts when it was expected (Gilbert, 2007). Given that humans evolved in small, isolated groups of reciprocal altruistic relating, caring for each other and ensuring the prosperity of others had considerable advantages for the long-term survival of the individuals in those groups, and if they survived and prospered, so would you. Importantly, there is no a priori reason for these algorithms to be conscious (Bargh, 2017).

34.4.1 Creating an Interpersonal Field

When using a social mentalities approach and focusing on reciprocal dynamic interactions, it is useful to keep in mind that different strategies and algorithms will try to create states of mind in their hosts *and their targets* that enable those strategies to function. Hence, they are operating in – and arising out of – an "interacting field of other minds," rather than just being completely self-contained. For example, individuals who utilize dominant, hostile leadership strategies will need to be successful in creating forms of followership and, to some degree, fearful, submissive states of mind in those around them rather than igniting aggressive competitors that simply retaliate (Gilbert & McGuire, 1998). In other words, they need to emit signals/behaviors that stimulate the amygdala and hypothalamic–pituitary axis systems in those they are aimed at, turning on submissive subordinate strategies and algorithms in them. Or a sexual signal needs to stimulate the pituitary of the target such that the target becomes sexually interested in the original sender. It is the same with caring and altruistic behavior. In order for (the strategies for) altruistic and caring behaviors to flourish, they will have to impact on the minds of those who receive them. In the case of the infant, this is to promote healthy development. In the case of allies or others in one's local group, these behaviors seek to reduce their aggressiveness and increase their preparedness for resource cooperation, sharing, and helpfulness. Those minds must not only be biologically built to detect and respond to the stimulus, but also be choreographed by local conditions to be receptive to it and then respond in a way that is appropriate to maintain or develop that role formation. Only if two or more individuals are (more or less) algorithmically attuned can a role be formed, otherwise conflict arises. There are many factors that will stop algorithmic attunement. For example, we may drop out of caring or cooperative interactions if we perceive the other is cheating or exploiting us. It is easier for us to both give care to and accept care from people we like and know than people we do not like and do not know. Hauser, Preston, and Stansfield (2014) found that we are more likely to help people who seem happier than people who seem more miserable and distressed. And we are more likely to help and to want to help individuals who show gratitude and appreciation than those who do not.

Reputations for caring and altruistic behavior, and for being appreciative of receiving care in contrast to being seen as selfish, aggressive, ungrateful, or exploitative, would advantage individuals within those groups. They are more likely to be targets for sharing and reciprocation – a form of prestige and reputation competition (Barkow, 1989). However, people can exploit this in that they can behave in caring ways to ensure social acceptance and to be liked and develop self-images that translate into social performances of being "a likeable person" (Catarino et al., 2014; Goetz et al., 2010). Nonetheless, altruistic, caring, and compassionate behavior probably evolved first as a selected reproductive strategy that became increasingly adapted for a range of social behaviors, particularly in the context of close-knit, isolated human groups (Dunbar, 2016). Caring builds alliances and creates a sense of safeness and security in interpersonal relating. This is advantageous not only in terms of practical resource sharing, but also in the co-regulation of physiological systems. Individuals who care for each other and who share friendly and pro-social interpersonal exchanges have higher levels of oxytocin, healthier heart rate variability, lower levels of stress and cortisol, more robust immune systems, and generally better health than individuals who are relating in competitive, critical, intrusive, hostile, or neglectful ways (Frederickson et al., 2008; Seppälä, et al., (2017).) – a fact that many business and Dark Triad leaders have failed to recognize, preferring instead to use threats and insecurity to motivate people (Sachs, 2012).

Of special note, too, is the archaeological evidence suggesting that early humans cared for individuals who were disadvantaged by injury and disease (Spikins, 2017; Spikins, Rutherford, & Needham, 2017). This is unusual, and indeed, for many (but not all) species, diseased and injured animals are either avoided or even at times attacked. Goodall (1990) discussed the tragedy of when the chimpanzee group she was studying was struck by polio. Individuals who were paralyzed were often avoided.

But whether or not prehuman examples can be found, it is definitely the case that caring for the injured and sick can be seen as a fundamental human trait that has flourished in modern-day medical practices and professions and numerous rescue and teaching services. Caring for the sick and injured may also have gained impetus because older people played important roles in many domains of life, including childcare. Postmenopausal survival and grandparenting offers distinct advantages in a species like humans who share childcare and who require extensive education into emotional and social awareness of group-appropriate behavior (Hrdy, 2009). The point is that the desire to be caring and helpful to others and to address people's suffering, especially when injured, sick, or aged, is well established in the archaeological and historical records. As we will note in Section 34.5.1, however, there are boundaries to this, and humans can be equally cruel, sadistic, and destructive to fellow humans.

34.5 PHYSIOLOGIES AND CARING

One of Tinbergen's questions is concerned with how traits evolved, including their physiological infrastructures. Recent years has revealed quite a lot about these in relation to caring. Caring behavior evolved with a series of neurophysiological adaptations and processes, such as oxytocin and vasopressin networks and receptors (Carter, 2014) and the myelinated vagal nerve of the parasympathetic system (Mayseless, 2016; Music, 2014; Porges, 2007; Thayer et al., 2012). Humans are highly adapted to needing caring environments to provide their attachment needs, to enable their psychological and physiological maturation, and later for reciprocal sharing. How those needs are met (or not, or how they are abused), especially early in life, has a huge impact on the phenotypes for personality development and maturing dispositions for caring, pro-social, or antisocial behavior (Mikulincer & Shaver, 2017; Music, 2014). Indeed, epigenetic studies show that genes can be turned on or off in response to environmental challenges of poor care, thus influencing phenotypic expressions (Cowan et al., 2016). Over recent years, there has been increasing research exploring these evolved physiological mediators for caring behavior and how they link to pro- and anti-social behavior and health outcomes. It is beyond the scope of this chapter to give a comprehensive review, but there are a few key issues to highlight.

34.5.1 Oxytocin

Although various of the catecholamines and other neurotransmitters like endorphins are linked to caring behavior and its consequences, probably the most specifically linked to the evolution of caring and social bonding is oxytocin (Carter, 2014; Taylor, 2006). There is now good research suggesting that oxytocin is associated with a range of pro-social behaviors that underpin mammalian and especially human sociality and caring, including parent–infant recognition and caring, pair-bonding, friendship formation, and social memory (Carter, 2014; MacDonald, 2012). Variations in the oxytocin receptor gene impact pro-social behavior and empathic accuracy (Laursen et al., 2014). Research also indicates that variations in the oxytocin gene may be detectable by observers in that individuals carrying the G allele are rated by others as more helpful and pro-social than those carrying the A allele (Kogan et al., 2011). However, oxytocin is by no means a "love all" hormone, for it also influences stimulus detection for (particularly) kin and in-group members. Oxytocin can increase *hostility to outsiders* and maternal aggression to potential threats to their infants (Carter, 2014; De Dreu et al., 2011; MacDonald, 2012). In fact, given that care is an expensive resource to dispense, it would make sense that evolution would create algorithms that were highly discriminatory (Burnstein et al., 1994), and oxytocin may well fit the bill of increasing dispositions for caring and altruistic behavior to some and reducing it to others (Shamay-Tsoory & Abu-Akel, 2016). Importantly, then, context sensitivity plays a role in caring. Oxytocin is not only associated with care provision, but also with seeking care when distressed and when care is available for downregulating stress (Taylor, 2006). In a recent study, Ebert et al. (2018) found that, in borderline patients, fears of giving and receiving compassion and caring were linked to blood levels of oxytocin.

34.5.2 The Vagus Nerve

In addition to rescuing and protecting, a core aspect of mammalian caring behavior, particularly to infants, is to facilitate soothing, calming, and safeness (Bowlby, 1969; Gilbert, 1989/2016; Porges, 2007). It is now known that a range of inputs from the parent, such as touch, food, provisioning, warmth, and protection, impact on the maturation of a number of different neurophysiological systems (Hofer, 1994). Put simply, the sympathetic system tends toward activation for defense and resource seeking (and joyful excitement states, which can also be part of attachment), whereas the parasympathetic system is associated with "rest and digest" and has a soothing, downregulating effect on the sympathetic system. Branches of the parasympathetic system, especially the myelinated vagus nerve, have been utilized and modified through affiliative relating, especially attachment. When balanced, such that individuals have good parasympathetic tone and heart rate variability, this is a marker for pro-social and helping behavior. Kogan et al. (2014) provide a major overview of the link between this branch of the parasympathetic system and pro-social behavior.

Importantly, then, the vagus nerve played a fundamental role in the evolution of mammalian caring behavior (Brown & Brown, 2016; Mayseless, 2016; Music, 2014; Petrocchi, & Cheli, 2019; Porges, 2007). It is linked to a range of central anatomical structures, including the frontal cortex. Good vagal tone also indicates adaptive integrative processing in the frontal cortex (Thayer et al., 2009). Indeed, there is a direct link between vagal tone as measured by heart rate variability and caring and compassionate orientation (Keltner et al., 2014; Kirby et al., 2017a). Bornemann et al. (2016) show that deliberate acts of helpfulness and helping improved heart rate variability. Importantly, however, the operation of the autonomic nervous system can change according to whether individuals are engaging with the suffering of others and experiencing distress themselves, are able to cope with that distress without engaging in avoidance, and also are able to initiate appropriate helping behavior. There is no one-size-fits-all in terms of response profile.

34.5.3 Paternal Caring and Testosterone

Another hormone recently implicated in changes in human caring behavior, particularly male caring behavior, is testosterone, which shares complex relationships with vasopressin, oxytocin, and reward-sensitive neurotransmitters like dopamine (Carter, 2014). Changes in the facial morphology of evolving humans linked to changes in testosterone in human evolution suggest a trend toward feminization, associated with increased caring and social affiliation behavior for children, within groups, and in pair-bonding (Cieri et al., 2014). Although shared care and paternal investment are central to bird reproductive strategies (Wesołowski, 2004), these are actually uncommon in live-birth mammals. Indeed, in many mammals, such as horses, sheep, elephants, and so forth, males take no interest in the provision of care to their offspring. Among primates, while gibbons show paternal caring and investment, these are relatively unique characteristics to humans. Most primate males do not know who their offspring are, and so there was probably a relationship between paternal certainty and the evolution of paternal investment.

There are two types of strategies of paternal investment. One involves males seeking to provide resources to support females who then provide most of the emotional and physical care to offspring. The other involves much more social interaction with offspring themselves. These are very culturally variant. In certain ecologies that are dangerous, males are required to be fearless. This creates rituals of demonstrations of toughness and competitiveness, with clear gendered roles and with males taking little interest in children, even treating them harshly. In contrast, in secure and safe environments, males tend spend time with and are caring and affectionate with (their) children (for a review, see Gilmore, 1990).

Muller et al. (2009) compared two African groups, showing that high paternal investment groups differed in testosterone levels in predicted directions. Laboratory studies with human fathers have also shown complex relationships between oxytocin, testosterone, and paternal caring (Weisman, Zagoory-Sharon, & Feldman, 2014). As noted by Muller et al. (2009), testosterone tends to promote competitive behavior and is very sensitive to winning and losing competitions. This may be important when we think about how different social and economic contexts may stimulate either competitive or cooperative behaviors, with consequences for hormone profiles like testosterone (Kemper, 1990) and paternal infant caring. However, as recently reviewed by Eisenegger, Haushofer, and Fehr (2011) and Sapolsky (2017), testosterone is also context sensitive in terms of its actual behavioral outputs. For example, testosterone is linked to status seeking, but the methods of status seeking, whether aggressive or prosocial, depend upon the social context in which that status is sought. Kirby and Kirby (2017) also review evidence of the complex interactions between testosterone and oxytocin in terms of compassion and caregiving behavior. For example, some status-seeking behavior in males can be related to protection (e.g., as in the rescue and protect professions). Hence, in highly altruistic environments, status seeking can be compassionate and altruistic, with demonstrations of helpfulness. (Petrocchi, & Cheli, 2019; Sapolsky, 2017)

34.5.4 Gender Variation

It is often suggested that male caring tends to be more practical and instrumental, whereas female caring is more emotional and physical, although this strict demarcation has been questioned because context plays a huge role (Gilmore, 1990; Jaffee & Hyde, 2000). Males can also be very caring, supportive, and rescuing–protecting of each other, particularly in dangerous situations. Indeed, armies play on this fact in their training, and this was particularly important for the Spartans. However, women can be equally brave in rescuing colleagues when provided with opportunities. Nonetheless, women are regarded as using emotionally focused caring as a means of stress regulation more than men (Taylor, 2006). So caring can be expressed in different ways, in different contexts, and within different types of relationship. The caring that takes place between bonded couples can be very pronounced in humans, being much less so in other primates. This can be associated with proximity seeking, wanting to be together, and with affectionate caring and closeness after sexual encounters. Most primates separate after short sexual encounters. In addition, humans are capable of a range of caring behaviors, even to strangers, as in the medical, caring, and rescuing professions. So our capacities for caring and the forms and functions of caring depend on social contexts, the type of relationship, and the participants involved. While some forms of caring can involve love, many forms (e.g., doctor–patient) do not.

We tend to assess caring and compassion as a single dimension, but it is likely that there are different dimensions, underpinned by different strategies and algorithms, such that male-on-male, male-on-female, female-on-female, female-on-male, male-on-child, female-on-child, friend-on-friend, and stranger-on-stranger caring may all fall under overlapping but also slightly different processing systems. We already know that oxytocin is not a general compassion hormone because it can actually increase aggressiveness to outsiders. Future research needs to be more subtle in its focus.

34.6 ONTOGENETIC DEVELOPMENT: EPIGENETICS

Another core theme in an evolutionary approach is the *ontogenetic and developmental* aspects of motivational systems such as caring. There is good evidence that the autonomic nervous system and, in particular, vagal tone and heart rate variability are very influenced and to some extent programmed by the quality of the attachment relationships and experiences of early caring (Bergman et al., 2004; Keltner et al., 2014; Kirby et al., 2017a). Recent work has shown that oxytocin, and its functional properties, are also very influenced by early life histories (Alves et al., 2015). Similarly, the impact of testosterone on behavior is influenced by early life history (Eisenegger, Haushofer, & Fehr, 2011), as are threat-processing systems. The frontal cortex, amygdala, and hypothalamic–pituitary–adrenal axis are highly susceptible to early life experiences, increasing threat sensitivity and impulsivity in the context of early life stress (Cowan et al., 2016; Shonkoff et al., 2012). Indeed, there seem to be few physiological systems underpinning variations in social behavior that are not heavily influenced by early life experiences. Importantly, these are all sources of variations in caring behavior to self and others (Brown & Brown, 2015; Mayseless, 2016).

Early life experiences also impact on gene expression itself. Recent years have seen major changes in how we understand the process of maturation and the development of phenotypes. One way this happens is through methylation of DNA. This is a process by which methyl groups are added to the DNA molecule. They act as switches on the outside of the double helix that enable the gene to be turned on or off. Hence, research in epigenetics has revealed considerable capacity for variation in gene expression due to the process of methylation. Some of these methylated changes may be passed across the generations (Cowan et al., 2016; Shonkoff et al., 2012).

Allied to these developments in epigenetics are life history approaches that suggest that early environments that are threatening, stressful, and, in humans, lack empathic caring, orient individuals for relatively fast, high-risk /high-gain survival and reproductive strategies that are self-focused, narcissistic, threat sensitive, impulsive, and relatively lacking in cooperative and caring motivations. In contrast, individuals who grow up in caring, supportive environments choreograph phenotypes adapted to that environment, where reciprocal caring, supporting, and sharing are more adaptive than self-centered impulsivity (Del Giudice, 2016). As noted, there are many potential candidates for the environment-linked physiological orchestration of these strategies. Trait dispositions to being caring, sensitive or insensitive to the distress of others, and motivated or unmotivated to do something about it are therefore partly linked to early life experiences. Indeed, individuals who mature in threatening, stressful, and hostile environments are considerably more vulnerable to subsequent mental and physical health problems, criminality, unstable relationships, and unreliable pro-social behavior (Del Giudice, 2016). These can develop into problems when they themselves become parents or leaders, perpetrating cycles of disadvantage and environmental and social toxicity. Nonetheless, rather than seeing these as pathologies, they can be seen as the maturation of specific survival and reproductive strategies to fit specific social contexts. The more we see these as pathologies rather than natural consequences of context, the less likely we are to address context and the more likely we are to try and address and "treat" individuals, which is to miss the point of an evolutionary analysis.

However, even people who come from supportive, loving environments can, in certain contexts, behave in very callous, hostile, even sadistic ways, particularly when orchestrated for tribal conflicts or obedience to authority (Zimbardo, 2007). Interestingly, although friendliness and agreeableness are personality traits commonly associated with compassion, people who are friendly and agreeable may not necessarily be the most courageous and caring in certain threatening contexts. Bègue et al. (2015) found that individuals high on agreeableness may try to avoid conflict, can be submissive to avoid conflict, and may be complicit in Milgram-like obedience experiments. Crucially, then, kind people, "looking after one's elderly neighbor or trying to treat others with kindness" may be caring, but may struggle to overcome fears of standing up against hostile authority. In contrast, individuals who may appear to have less agreeable temperaments may be more courageous in fighting injustice. As already mentioned, caring behavior has boundaries, with many inhibitors as well as facilitators (for a review, see Gilbert & Mascaro, 2017). We are more likely to save two of our own children than 50 children in difficulty down the road; we are more likely to help people we know and like than strangers and those we do not like. Bomber pilots are not psychopaths, even though they know perfectly well the suffering they are causing by their actions. So while we have evolved extraordinary competencies for caring and compassion, these represent one strategic orientation or one motivational system that competes with others, both inside our own minds and interpersonally (Huang & Bargh, 2014). This, in a competition of "what to feel and how to act with whom and in which context," highlights the importance of our context sensitivity.

This raises the issue of how distress sensitivity and concern for others can be turned off in certain contexts, specifically those related to threats to the self, tribal conflicts, and predation. Indeed, some argue that our capacity for sadism is the result of predator mechanisms in which predators are impervious to the distress they are causing their prey (Nell, 2006). Regardless of the mechanisms for cruel and destructive behaviors, the key focus is that caring behavior has facilitators and inhibitors and can be turned off. These sources of turning off distress sensitivity and caring motivation can reside both within the person in terms of genetic predisposition, epigenetic change, previous traumas, or current (political) beliefs, but also link to sociocultural contexts and triggers. It is unfortunately quite easy to create contexts that turn off compassion and concern for others in favor of cruelty and causing harm (Zimbardo, 2007). Certainly, the evolution of caring therefore needs to be viewed against the backdrop of the darker sides to humanity and the ease with which we can engage in sadistic cruelty (Gilbert, 1994, 2005, 2015, 2018).

34.7 PROXIMATE INFLUENCES: NEW BRAINS, NEW CONTEXTS, AND WAYS OF IMAGINING THE WORLD

Finally, we can explore proximate influences for caring and how mammalian caring motives can become human compassion. Proximate influences are situated in current ecological and social contexts and the ways in which minds interpret and respond to them. To explore this requires recognition of the fact that human minds have evolved unique competencies for interpreting and responding to social contexts. Moreover, our interpretations of – and responses to – the contents of our own minds create complex feedback loops that influence mental and physical states and behaviors. Humans constitute the only species that can intentionally, with an aware purpose, change their phenotypic expressions – or, break their programming.

Again, we can turn to evolution for how this extraordinary flexibility has come about. Humans are unique in that, around two million years ago, some of our ancestors began to develop forms of reasoning intelligence rooted partly in language and symbol use, social intelligence linked to empathy and also different forms of insightful consciousness that do not depend on language, reasoning or social intelligence (Dunbar, 2016, Siegel, 2016). One process the human mind clearly does is integrate information from these different competencies in very complex ways. For example, driving a car requires complex forms conscious intentions, of attention, changing gears, being aware of speed, other drivers, road conditions, and direction of travel, and having a conversation with somebody next to you. These everyday human acts are highly focused, yet they require multiple domains of attention that integrate multiple domains of information to pursue a variety of tasks (e.g., driving and having a conversation), sometimes for hours on end. Although many behaviors require integration of different competencies, such as in hunting or flying, this level of attention focusing, concentration, and complexity of integration is likely to be way beyond any other mammal. No animal could learn to play a Rachmaninov piano concerto from memory. These are extraordinary feats of information integration and flexibility (Gilbert, 2009, 2017; Siegel, 2016). Clearly, these competencies to *flexibly integrate* information will have a major impact on motivational and emotion processing because we can balance and understand conflicting processes within us, although some of these conflicting motives and emotions can be unconscious (Bargh, 2017; Huang & Bargh, 2014). Importantly, this is not just integration, but also transformation, because this level of integration of complexity gives rise to the potential emergence of new forms of knowledge and awareness that may be the basis for *insight*. New patterns of organization can emerge from the relationships between subcomponents, but only if the subcomponents are interacting and affecting each other (Johnson, 2002).

Additionally, our competencies for language and symbol use have clearly been important in the evolution of human intelligence (Dunbar, 2016). As Dunbar notes, however, language evolved as a communication process, and you need some degree of mentalization to be able to match words with the meaning of what the communicating person is trying to communicate. And of course, we can communicate good or bad information about other people by language, including information of whether they are worthy or not of compassion. What is especially important about language is the communication of meanings, motives, and feelings. For example, in some (e.g., political or male-dominated) groups, words and concepts like compassion dependency, vulnerability, or caring are textured with concepts like indulgent, weak, or soft. These groups prefer terms for strength, independence, and toughness that emphasize male–male competition. So individuals following certain strategies will try to create mental social niches via language that will suppress some strategies in favor of others. Looked at in this way, linguistic discourses create the seas of meaning for how to fit in with certain social groups: "if I want to prosper in this group then I must see myself as *x*, behave as *y*, and adopt the values of *z*." These will sometimes be conducive to compassion, but at other times will seriously inhibit it (Gilbert & Mascaro, 2017).

A third set of competencies are those for imagination, fantasy, and pretend. Imagination activates sensorimotor patterns and memory and is nonlinguistic (Singer, 2006). By running imaginary scenarios in the mind, we can create inner virtual realities and anticipate the consequences of our behavior without actually engaging in behaviors ("Suppose I did *X*. What would happen if I did *Y*?"). Imagining consequences before we act offers huge evolutionary advantages. And we can imagine chains of connectedness such as, "I think that Sally thinks that James thinks that Jim thinks *X*" (Suddendorf & Whitten, 2001). These second-order representations are extremely important for complex social behaviors and relational development. Their role in caring behavior seems clear. While language may aid this type of thinking, it may also simply label processes that have already been "thought about" or imagined. Hence, conscious awareness and verbal labeling are often quite late in information processing chains of neural events, and conscious verbal articulation of our thinking can be inaccurate and simply a justification for what we are already feeling (Bargh, 2017; Haidt, 2001).

In addition, our imaginary and fantasy worlds can powerfully recruit physiological systems. For example, lying alone in bed and fantasizing a sexual scene, knowingly and on purpose, is to stimulate a set of cells in one's pituitary to change the body to prepare for action and, of course, pleasure. Again, this is not using language – try using verbal stimuli without imagery to get sexually aroused! Crucially, we can also choose to practice fantasizing about caring- and compassion-focused ideas, feelings, and motives precisely to stimulate processes such as heart rate variability and neurophysiological systems. There is now considerable evidence that regular practices of compassion impact on a range of beneficial physiological systems and pro-social behaviors (e.g., Goleman & Davidson, 2017; Singer & Bolz, 2012; Weng et al., 2013, 2018). To some degree, caring that relates to various forms of empathy relies on imagery. It relies on us imagining ourselves in the shoes of the other, seeing the world through their eyes. It takes us into the world of theory of mind and mentalization and what has been called "mindsight" (Siegel, 2011). Whitten (1999) argues, "Reading others' minds makes minds deeply social in that those minds interpenetrate each other" (p. 177). This is particularly important when it comes to perspective taking rather than mere emotional contagion, as it requires effort and a desire to do it.

A fourth new competency is for objective self-awareness and to create and imagine a *sense of self as object* and as we exist for others (Sedikides & Skowronski, 1997). We can form images of ourselves as we want to be and as we do not want to be, and these can be sources of pride and shame that then become motives in themselves to seek or avoid (Gilbert, 2007). Crucial here is the choreography of our motives such that in some contexts we may wish to become an individual who is seen as tough, fearless, and potentially aggressive, yet in a different context as an individual who is compassionate, kind-hearted, and helpful (Gilbert, 2009).

Crucially, our experience of the kind of self we are is also socially constructed through our interactions and discourses with others. For example, had I been kidnapped as a three-day-old baby into a violent drug gang and experienced abuse, this version of Paul Gilbert as a professor writing this chapter would not exist. A totally different epigenetic version with different brain wiring, who would be potentially quite violent and impulsive, would exist. Although we naturally are wedded to versions of ourselves, believing them to be in some sense "real," in fact they are illusions created by the brain to facilitate coherence of action (Taylor, 1989). What is important is that the inner patterning of a sense of self and the inner monitoring process supports a self-presentation that fits the social niche. In a symbolizing, self-aware mind that has theory of mind and metacognition competencies and can regulate a multitude of desires and possible roles, not having a clear self-identity can create *too much flexibility* and competing possibilities for thought and actions (McGregor & Marigold, 2003). Many psychologists now see "the self" and self-identity like this, as an organizing system that coordinates memories, emotions, beliefs, and other processes for a cohesive securing of goals in specific contexts/social niches. Taylor (1989) puts it clearly when he says:

To know who I am is a species of knowing where I stand on key issues. My identity is defined by the commitments and identifications which provide the frame or horizon within which I can try to determine from case to case what is good, or valuable, or what ought to be done, or what I endorse or oppose. In other words, it is the horizon within which I am capable of taking a stand. (p. 27)

Importantly, although we are of course niche sensitive, we can also to some extent choose self-identities. There is evidence that creating a care-focused and compassionate self-identity has significant benefits for social relating and health, in comparison to self-focused, self-defensive, shame-avoidant presentations (Crocker & Canevello, 2012). Reed and Aquino (2003) suggest that caring, kindness, and honesty attributes can become important for self-identity, which they call moral identity. Having such attributes at the core of "one's identity" is associated with less hostile behavior to out-groups and increased probability of forgiveness in harmful situations. So, the creation of self-identities around compassionate motives is important for internal physiological organization and external social behavior.

Fifth are complex intersubjectivity and empathic (Theory of Mind) competencies that allow insight into the minds of others and, particularly importantly, other people's intentionality (Trevarthen & Aitken, 2001). Intersubjectivity is slightly different from empathy and involves a sharing of perspectives; for example, when the parent points, the child looks to where the parent is pointing, not to the tip of the finger. Compassion motives are clearly linked to intersubjectivity (Gilbert, 2005, 2017). Empathic competencies that involve emotional contagion, the ability to feel as others do, and having a knowing awareness of this, as well as a cognitive perspective on why the other may be feeling as they are feeling, are some of humanity's important new competencies (Decety & Ickes, 2011). Siegel (2011) calls this "mindsight" – literally, the ability to see into other minds. To some extent, this is again influenced by early life history. However, like other competencies, it can be used in the service of many different motivations, both pro- and anti-social (Zaki, 2014). Various forms of hostile motivation and emotion, where we see others as threats or as undeserving of compassion, can turn off empathy and certainly turn off caring

interest (Decety & Cowell, 2014; Gilbert, 2015; Gilbert & Mascaro, 2017).

Sixth, we have a highly self-monitoring mind, sometimes linked to what has been called the "default mode," which allows us to monitor our own thoughts, behaviors, and body states moment by moment, form judgments about them, and attribute causes to them. We can them be content, angry, or frightened by them, and make efforts to change them – or even avoid them, as in experiential avoidance. Many psychotherapies pay particular attention to these self-monitoring processes and the degree to which they are accepted, repressed, or avoided. Self-monitoring is usually goal directed and can also be a source of pride or self-criticism and shame. There is now good evidence that hostile forms of self-monitoring and self-criticism can be very damaging to mental health and that switching to a more compassionate orientation is very beneficial to mental and physical well-being (Goleman & Davidson, 2017; Kirby, 2017).

Self-monitoring can also underpin fixed rumination. We are able to hold things in mind and repeatedly bring them into mind. This can be very useful when one is voluntarily repeatedly thinking about a problem and then begins to see ways to solve it. It can enable creativity and the emergence of new insights. But as with so many of our competencies, it can have a dark side. Our minds sometimes get involuntarily stuck on a threat-focused theme with repetitive intrusions and rumination. For example, repeatedly remembering an argument will also repeatedly stimulate a range of physiological systems, including the amygdala and stress hormones. We can ruminate on vengeance rather than compassion, and ruminative cycles can become locked into emotions that drive each other. Grief states are typically highly ruminative, as are certain anxious and depressed states. This is why mindfulness training along with breath training to stimulate the parasympathetic system can help people to notice ruminative cycles and learn how to switch out of them and into focusing on compassion, which will stimulate different physiological systems that counteract threat system physiology (Gilbert & Choden, 2013; Goleman & Davidson, 2017; Kirby et al., 2017a; Weng et al., 2013, 2018).

A seventh competency is the nature of consciousness itself, which may be different in humans from that of other animals since we have a consciousness of being conscious, a kind of "knowing awareness" that is also the basis for mindfulness. We can mindfully observe our minds, observing our thoughts and behaviors from this "knowing awareness" position. We know that we know; we are aware that we are aware. We can then choose to learn to stand back and take an observer orientation regarding our minds and the contents of our consciousness (Siegel, 2016). For many thousands of years, this competency has been at the center of mind training in which individuals knowingly and purposefully pay attention to the moment-by-moment flow of the contents of

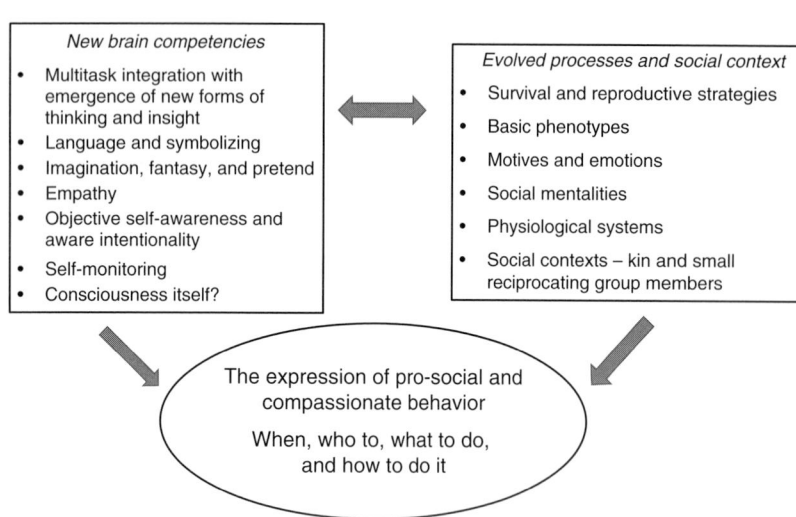

Figure 34.2 The multiple influences on compassionate phenotypes.

mind without overly identifying with them, treating them as if one were watching a movie (Goleman & Davidson, 2017). The idea is that as attention becomes less captured by the moment-by-moment undulating processing of the motivations and emotions churning away, it settles into its observer mode and can gain new insights into the nature of the mind itself. In some traditions, such as the Theravada Buddhist tradition, this is the practice of shifting to an observer mind that, over time, provides clarity of understanding regarding the nature of the mind. This then gives rise to an understanding of the interdependency from which compassion can flow (Nhat Hanh, 2009). So, in this tradition, caring and compassion for others are naturally arising consequences of a settled, insightful mind (Gilbert & Choden, 2013).

These are only some of the recent human competencies, but their implications are profound because all motivational systems are subject to potential regulation through these new brain processes (see Figure 34.2).

34.7.1 Making the Choices

With all of these competencies, we can take just about any motivational system and create extraordinarily novel ways of fulfilling that motive. For example, in the case of sexuality, we can invent contraception such that we do not have to experience the consequences of our acts; we can create a whole pornography industry. When we knowingly understand the impacts of our behaviors, we may choose not to act out what we feel or want to do. Even though driving a car is inherently dangerous and can make us anxious, our desire to travel means that we override that anxiety. There are many areas in life where we need to learn to override anxiety and develop courage. Even though we are attracted to sweet, high-fat, and high-salt foods, we can choose to override our desires in order to be fit. We can take exercise with a knowing and precise intention. No animal can knowingly and intentionally engage in a behavior – no zebra, as far as we know,

chooses to do laps around the savanna to become super-fit and be ready for any lions that might chase them.

Our human competencies make morals and ethics possible. For example, we can choose not to have an affair with somebody we really like because we know it would harm our partners and (more self-interestedly) our relationships. Loewenstein and Small (2007) have highlighted the fact that caring and compassion are relatively bounded by evolutionary concerns (caring is focused on kin and friends), and it is these that guide our emotions and motives. Importantly, however, we can override these when we develop a more mindful and rational morality that is linked to a particular kind of self-identity. In fact, as they indicate in their review, relying on evolutionarily informed emotion to drive compassion may produce very non-compassionate results.

Many animals obviously care for each other and certainly their young, but we would not necessarily call that compassion. While caring is basically an attentional sensitivity to the needs of self or another followed by the appropriate actions to address them, compassion has that same orientation, but is recruited and textured by recently evolved human cognitive competencies, including ones such as perspective taking and mentalization. Compassion then arises when we knowingly seek to utilize and cultivate our caring motives, including developing the wisdom of how to best deal with distress (Gilbert, 2017). This is demonstrated in Figure 34.3.

Although there are slightly different definitions for compassion (Gilbert, 2017), there are now measures for the compassionate engagement and actions that are rooted in evolutionarily informed and competencies-based approaches (Gilbert et al., 2017a).

34.7.2 Harnessing and Cultivating Compassion

Like any motivational system, caring and pro-social motives can be regulated, cultivated, accentuated, or

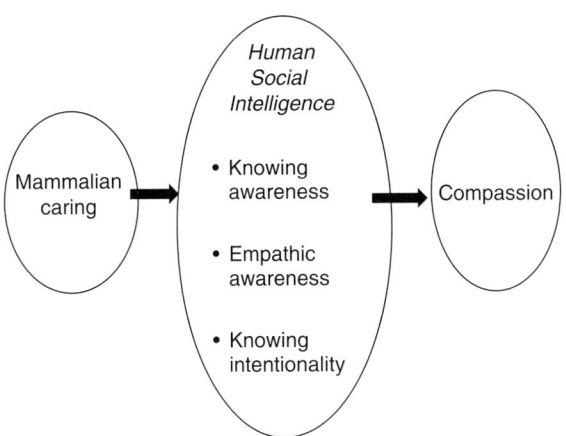

Figure 34.3 Evolution: from caring to compassion.

deactivated through new brain processes and the deliberate creation of social contexts that offer opportunities for practice and reward. Given the evolved functions of caring and the physiological mediators of caring, there would be good reason to believe that cultivating caring motives would have a range of beneficial physiological effects. Indeed, this is the case. The last 20 years have seen much research demonstrating that compassion training beneficially alters the autonomic nervous system, threat-processing systems, immune systems, cardiovascular systems, and, via neuroplasticity, cortical processing systems (for reviews, see Frederickson et al., 2008; Goleman & Davidson, 2017; Singer & Bolz, 2012). A meta-analysis of compassionate interventions and trainings that examined 21 randomized controlled trials found significant improvements for well-being and reductions in suffering, such as depression and anxiety symptoms (Kirby et al., 2017b). Indeed, compassion cultivation (*bodhicitta*) has been the basis of many spiritual paths, particularly Mahayana Buddhism (Gilbert & Choden, 2013; Tsering, 2005, 2008). There are a growing number of different compassion models and training programs to promote physiological change, ethical and pro-social behavior, and well-being (Goleman & Davidson, 2017). Some have been developed particularly for psychotherapy with complex mental health problems (Gilbert, 2010).

In addition, various compassion mind trainings are becoming increasingly important in a range of social contexts. There are concerns as to whether we should allow our schools to drift further and further into narcissistic, competitive self-interest, with consequences for vulnerability to mental health and reducing community values, or whether schools should be the grounding for the teaching of ethical and compassionate values and behavior (Weissbourd & Jones, 2014). Businesses and economists are also beginning to recognize that simply allowing our minds to do what they want means they will gravitate toward competitive self-interest (Sachs, 2012). Training guides and books are constantly encouraging leaders to become more mindful and more compassionate.

Crucially, then, we are not slaves to our nature, but have an opportunity to create contexts both inside our own minds and in our social relationships that hone certain motives and their physiological profiles at the expense of others. This is important because we cannot simply rely on the emergence of contexts to create the kinds of justice, fairness, ecological sensitivity and compassion we may wish to see. Indeed, self-interest and tribalism can equally recruit new brain competencies, and they have done so for thousands of years, with some very horrific consequences. We may be vehicles for our genes and for different algorithms and strategies that are "battling" for the control of the minds. The key issue is the degree to which we can choose to tilt this competition toward pro-social behavior.

Currently, many commentators recognize that we have not really come to terms with the advent of agriculture and the creation of surplus, which have allowed group sizes to expand and for the advantages of accumulation to become enormous. Privilege was protected through the use of violence, with dominant (mostly) males using other males as enforcers, secret police, and so forth. For the most part, submission to authority was the primary cohesive factor, rather than compassion, fairness, and sharing. Within 10,000 years, we have created cultures with a range of mismatches between social context and evolved motives (Li, van Vugt, & Colarelli, 2017). This mismatch is responsible for both the science that powers medicine and compassionate behavior, but equally antisocial behavior and the intense individual and group-based competitiveness that can undermine compassion and sharing (Gilbert, 2018). Hence, when taking an evolutionary lens to caring and compassionate responses, part of our science must be to consider how to boost some of its facilitators and address some of its inherent inhibitors (Gilbert & Mascaro, 2017).

34.7.3 Sharing and Caring versus Controlling and Holding

What we may be witnessing in different social and "resource-sparse versus resource-plenty" contexts is the playing out of different algorithms and strategies. For example, in low-resource hunter–gatherer groups, the strategies employed may well be for caring and sharing (Dunbar, 2016). This promotes mutual benefit, particularly where reputations for sharing can be important and being reliant on each other is central to survival. In high-resource environments with expanding group sizes, abilities to store resources, and with the advent of trading, survival and reproductive strategies change to 'control and hold' – an earlier, more territorial, reptilian-like strategy for survival and reproduction. Here, it becomes more adaptive not to share resources, but to "grow them" and defend them for self or small kin networks. Indeed, there is evidence that caring and sharing tend to be greater in low-resource rather than high-resource environments. In a number of studies, Piff et al. (2010) found that low-resource, lower-social class individuals were more charitable, trusting, and helpful compared to high-class individuals, even though the latter could afford to be more charitable. There is now growing evidence that as people become wealthier and more powerful they become less empathic, less sensitive to the suffering of others, and less compassionate (Keltner, 2016; Van Kleef, et al., 2008). Piff (2014) has shown that increasing wealth also correlates with an increasing narcissistic sense of entitlement and a reduced orientation to sharing.

So while we can think of compassion as a strategy that probably evolved to support kin survival and was adaptive in small-group contexts. We are now faced with the problem of how to root compassion in mega-groups that excessive resources and wealth, societies of strangers, and nations, particularly when there are political movements that are stimulating algorithms for competitive, self-focused, resource-accumulating, protectionist, anti-immigration, and tribal conflict. Increasing concern has been raised at the disparities of wealth across countries and within countries and the hardship this causes (Sachs, 2012). Understanding the evolved processes underpinning compassion and contextual and multifactorial nature of compassion, rather than viewing it as a single process, may help in addressing these issues.

34.8 CONCLUSIONS

This chapter addressed the evolution of caring and compassion through the four questions raised by Tinbergen (1963). Although obviously selective, it has attempted to address phylogenetic origins, adaptive function as a reproductive and alliance-supporting strategy, its underlying physiology, and how caring is turned into compassion via recently evolved, complex cognitive competencies. The chapter has also highlighted that caring is a social mentality rooted in the processing of dynamic reciprocal and changing interactions. Moreover, the evolution of caring simultaneously operates in mechanisms within the giver and receiver. Furthermore, similar physiological systems such as oxytocin and the vagus nerve are involved in both giving and being receptive to – and affected by – caring and compassion

Although Machiavellian motives and competencies have played a role in the evolution of human social intelligence, it is now recognized that it was pro-social behavior rather than aggressive behavior that drove much of human evolution, particularly human intelligence and capacities for information sharing via language (Carter, 2014; Dunbar, 2010, 2016; Porges, 2007). Indeed, there is now good evidence that when humans feel embedded in supportive, caring, and validating relationships, when they themselves can be caring and helpful and valued by others, these relating patterns are associated with a wide range of physiological and mental health-promoting processes (Frederickson et al., 2008; Goleman & Davidson, 2017; Singer & Bolz, 2012). Having more robust immune systems helps with infections, and affiliative, generated heart rate variability

benefits cardiovascular and other physiological systems conducive to health. Good vagal tone, as measured by heart rate variability, also supports empathy, cognitive processing, and general problem-solving. In contrast, environments that are highly competitive, self-focused, and potentially hostile are associated with high levels of stress and poorer organ functioning (Cowan et al., 2016).

We stand at a point in history when we are beginning to recognize the evolved functional systems, strategies, and algorithms of the brain and the extraordinary importance of caring early in life, but also the degree to which caring has natural (kin, alliance, and tribal) boundaries. In addition, for modern societies, we are recognizing the dangers of unregulated competitiveness in terms of creating huge disparities of resources and ecological damage. The need to facilitate ethical, compassionate motives for ourselves and others in the world we live in is becoming increasingly urgent.

The challenge is that just as we came to understand the importance of infection and hygiene and that diseases flourish in toxic environments, we must promote the understanding that the dark side of humanity flourishes in toxic social environments. It is very easy to inspire humans to violence and enjoyment of narcissistic self-interest with a toning down of empathic concern for the misfortune of others. Hence, to use evolutionary science for the betterment of humanity requires us to understand both the facilitators and the inhibitors of caring and compassion (Gilbert & Mascaro, 2017). Although many might like to see humanity as a compassionate species – and we can be – we can also be one of the nastiest, most brutal, and most exploitative species to have ever walked this planet. An evolutionary analysis helps us understand the inherent competition between different motives, strategies, and algorithms as information flows within us and between us. Our challenges are to try to create a more compassionate mind in a more compassionate and pro-social world. Currently, most science fiction writers paint a very dim and dark view of the future; scientific understanding of what promotes, cultivates, and supports (and inhibits) compassion could do much to offset this. That compassion and pro-social behavior is good for us in so many ways is beyond doubt. Building a more compassionate world, however, poses a complex scientific question of how to do so!

REFERENCES

Alves, E., Fielder, A., Ghabriel, N., Sawyer, M., & Buisman-Pijlman, F. T. (2015). Early social environment affects the endogenous oxytocin system: A review and future directions. *Frontiers in Endocrinology*, **6**, 32.

Bargh, J. (2017). *Before You Know It: The Unconscious Reasons We Do What We Do*. New York: Touchstone.

Barkow, J. H. (1989). *Darwin, Sex, and Status*. Toronto: Toronto University Press.

Bègue, L., Beauvois, J. L., Courbet, D., et al. (2015). Personality predicts obedience in a Milgram paradigm. *Journal of Personality*, **83**(3), 299–306.

Bergman, N. J., Linley, L. L., & Fawcus, S. R. (2004). Randomized controlled trial of skin-to-skin contact from birth versus conventional incubator for physiological stabilization in 1200-to 2199-gram newborns. *Acta Paediatrica*, **93**(6), 779–785.

Bornemann, B., Kok, B. E., Böckler, A., & Singer, T. (2016). Helping from the heart: Voluntary upregulation of heart rate variability predicts altruistic behavior. *Biological Psychiatry*, **119**, 54–63.

Bowlby, J. (1969). *Attachment: Attachment and Loss*, Vol. 1. London: Hogarth Press.

Bowlby, J. (1973). *Separation, Anxiety and Anger. Attachment and Loss*, Vol. 2. London: Hogarth Press.

Bowlby, J. (1980). *Loss: Sadness and Depression. Attachment and Loss*, Vol. 3. London: Hogarth Press.

Brown, S. L., & Brown, R. M. (2015). Connecting prosocial behavior to improved physical health: Contributions from the neurobiology of parenting. *Neuroscience and Biobehavioral Reviews*, **55**, 1–17.

Burnstein, E., Crandall, C., & Kitayama, S. (1994). Some neo-Darwinian rules for altruism: Weighing cues for inclusive fitness as a function of biological importance of the decision. *Journal of Personality and Social Psychology*, **67**, 773–807.

Buss, D. (2015). *Evolutionary Psychology: The New Science of the Mind*. Hove: Psychology Press.

Carter, C. S. (2014). Oxytocin pathways and the evolution of human behavior. *Annual Review of Psychology*, **65**, 17–39.

Carter, C. S. (2014). Oxytocin pathways and the evolution of human behavior. *Annual Review Psychology*, **65**, 17–39.

Catarino, F., Gilbert, P., McEwan, K., & Baião, R. (2014). Compassion motivations: Distinguishing submissive compassion from genuine compassion and its association with shame, submissive behavior, depression, anxiety and stress. *Journal of Social and Clinical Psychology*, **33**, 399–412.

Cieri, R. L., Churchill, S. E., Franciscus, R. G., Tan, J., & Hare, B. (2014). Craniofacial feminization, social tolerance, and the origins of behavioral modernity. *Current Anthropology*, **55**, 419–443.

Cowan, C. S. M., Callaghan, B. L., Kan, J. M., & Richardson, R. (2016). The lasting impact of early-life adversity on individuals and their descendants: Potential mechanisms and hope for intervention. *Genes, Brain and Behavior*, **15**(1), 155–168.

Crocker, J., & Canevello, A. (2012). Consequences of self-image and compassionate goals. In P. G. Devine & A. Plant, eds., *Advances in Experimental Social Psychology*. New York: Elsevier, pp. 229–277.

Dawkins, R. (1976). *The Selfish Gene*. London: Cape.

Decety, J., & Ickes, W., eds. (2011). *The Social Neuroscience of Empathy*. Cambridge, MA: Bradford Press.

Decety, J., & Cowell, J. M. (2014). Friends or foes: Is empathy necessary for moral behavior? *Perspectives on Psychological Science*, **9**(5), 525–537.

De Dreu, C. K. W., Greer, L. L., Van Kleef, G. A., Shalvi, S., & Handgraaf, M. J. J. (2011). Oxytocin promotes human ethnocentrism. *Proceedings of the National Academy of Sciences*, **108**, 1262–1266.

Del Giudice, M. (2016). The life history model of psychopathology explains the structure of psychiatric disorders and the emergence of the *p* factor: A simulation study. *Clinical Psychological Science*, **4**(2), 299–311.

Dunbar, R. I. M. (2007). Mind the bonding gap: Or why humans aren't just great apes. *Proceedings of the British Academy*, **154**, 403–433.

Dunbar, R. I. M. (2010). The social role of touch in humans and primates: Behavioral function and neurobiological mechanisms. *Neuroscience and Biobehavioral Reviews*, **34**, 260–268.

Dunbar, R. (2016). *Human Evolution*. Oxford: Oxford University Press.

Ebert, A., Edel, M. A., Gilbert, P., & Brüne, M. (2018). Endogenous oxytocin is associated with the experience of compassion and recalled upbringing in borderline personality disorder. *Depression and Anxiety*, **35**(1), 50–57.

Eisenegger, C., Haushofer, J., & Fehr, E. (2011). The role of testosterone in social interaction. *Trends in Cognitive Sciences*, **15**(6), 263–271.

Farrelly, D., Clemson, P., & Guthrie, M. (2016). Are women's mate preferences for altruism also influenced by physical attractiveness? *Evolutionary Psychology*, **14**(1), 1–6.

Fogel, A., Melson, G. F., & Mistry, J. (1986). Conceptualising the determinants of nurturance: A reassessment of sex differences. In A. Fogel & G. F. Melson, eds., *Origins of Nurturance: Developmental, Biological and Cultural Perspectives on Caregiving*. Hillsdale, NJ: Lawrence Erlbaum Associates, Inc., pp. 53–67.

Fredrickson, B. L., Cohn, M. A., Coffey, K. A., Pek, J., & Finkel. S. A. (2008). Open hearts build lives: Positive emotions, induced through loving-kindness meditation, build consequential personal resources. *Journal of Personality and Social Psychology*, **95**, 1045–1062.

Geary, D. C. (2000). Evolution and proximate expression of human parental investment. *Psychological Bulletin*, **126**, 55–77.

Geary, D. C., & Huffman, K. J. (2002). Brain and cognitive evolution: Forms of modularity and functions of the mind. *Psychological Bulletin*, **128**, 667–698.

Gilbert, P. (1989/2016). *Human Nature and Suffering*. Hove: Lawrence Erlbaum Associates.

Gilbert, P. (1994). Male violence: Towards an integration. In J. Archer, ed., *Male Violence*. London: Routledge and Kegan Paul, pp. 352–389.

Gilbert, P. (1997). The evolution of social attractiveness and its role in shame, humiliation, guilt and therapy. *British Journal of Medical Psychology*, **70**, 113–147.

Gilbert, P. (2005). Compassion and cruelty: A biopsychosocial approach. In P. Gilbert, ed., *Compassion: Conceptualisations, Research and Use in Psychotherapy*. London: Routledge, pp. 3–74.

Gilbert, P. (2007). The evolution of shame as a marker for relationship security. In J. L. Tracy, R. W. Robins, & J. P. Tangney, eds., *The Self-Conscious Emotions: Theory and Research*. New York: Guilford, pp. 283–309.

Gilbert, P. (2009). *The Compassionate Mind: A New Approach to the Challenge of Life*. London: Constable & Robinson.

Gilbert, P. (2010). *Compassion Focused Therapy: The CBT Distinctive Features Series*. London: Routledge.

Gilbert, P. (2014). The origins and nature of compassion focused therapy. *British Journal of Clinical Psychology*, **53**, 6–41.

Gilbert, P. (2015). The evolution and social dynamics of compassion. *Social and Personality Psychology Compass*, **9**, 239–254.

Gilbert, P. (2017). Compassion as a social mentality: An evolutionary approach. In P. Gilbert, ed., *Compassion: Concepts, Research and Applications*. London: Routledge, pp. 31–68.

Gilbert, P. (2018). *Living Like Crazy*. York: Annwyn House.

Gilbert, P., & Choden (2013). *Mindful Compassion*. London: Constable & Robinson.

Gilbert, P., & Mascaro, J. (2017). Compassion: Fears, blocks, and resistances: An evolutionary investigation. In E. Sappla & J. Doty, eds., *Handbook of Compassion*. New York: Oxford University Press, pp. 399–420.

Gilbert, P., & McGuire, M. (1998). Shame, status and social roles: The psychobiological continuum from monkeys to humans. In P. Gilbert & B. Andrews, eds., *Shame: Interpersonal Behavior, Psychopathology and Culture*. New York: Oxford University Press, pp. 99–125.

Gilbert, P., Catarino, F., Duarte, C., et al. (2017a). The development of compassionate engagement and action scales for self and others. *Journal of Compassionate Health Care*, **4**, 4.

Gilbert, P., Catarino, F., Sousa, J., et al. (2017b). Measuring competitive self-focus perspective taking, submissive compassion and compassion goals. *Journal of Compassionate Health Care*, **4**, 5.

Gilmore, D. D. (1990). *Manhood in the Making: Cultural Concepts of Masculinity*. New Haven, CT: Yale University Press.

Goetz, J. E., Keltner, D., & Simon-Thomas, E. (2010). Compassion: An evolutionary analysis and empirical review. *Psychological Bulletin*. **136**, 351–374.

Goleman, D., & Davidson, R. J. (2017). *Altered Traits: Science Reveals How Meditation Changes Your Mind, Brain, and Body*. London: Penguin.

Goodall, J. (1990). *Through a Window: My Thirty Years with the Chimpanzees of Gombe*. London: Penguin.

Haidt, J. (2001). The emotional dog and its rational tail: A social intuitionist approach to moral judgment. *Psychological Review*, **108**(4), 814–834.

Harlow, H. F., & Mears, C. (1979). *The Human Model: Primate Perspectives*. New York: Winston & Sons.

Hauser, D. J., Preston, S. D., & Stansfield, R. B. (2014). Altruism in the wild: When affiliative motives to help positive people overtake empathic motives to help the distressed. *Journal of Experimental Psychology: General*, **143**(3), 1295–1305.

Hofer, M. A. (1994). Early relationships as regulators of infant physiology and behavior. *Acta Paediatiricia Supplement*, **397**, 9–18.

Holmes, J., & Slade, A., eds. (2018). *Attachment in Therapeutic Practice*. London: Routledge.

Hrdy, S. B. (2009). *Mothers and Others. The Evolutionary Origins of Mutual Understanding*. Boston, MA: Harvard University Press.

Huang, J. Y., & Bargh, J. A. (2014). The selfish goal: Autonomously operating motivational structures as the proximate cause of human judgment and behavior. *Brain and Behavioral Sciences*, **37**, 121–175.

Jaffee, S., & Hyde, J. S. (2000). Gender differences in moral orientation: A meta-analysis. *Psychological Bulletin*, **126**, 703–726.

Jensen-Campbell, L. A., Graziano, W. G., & West, S. G. (1995). Dominance, prosocial orientation, and female preferences: Do nice guys really finish last? *Journal of Personality and Social Psychology*, **68**(3), 427–440.

Johnson, S. (2002). *Emergence: The Connected Lives of Ants, Brains, Cities and Software*. London: Penguin.

Keltner, D. (2016). *The Power Paradox: How We Gain and Lose Influence*. London: Allen Lane.

Keltner, D., Kogan, A., Piff, P. K., & Saturn, S. R. (2014). The sociocultural appraisals, values, and emotions (SAVE) framework of prosociality: Core processes from gene to meme. *Annual Review of Psychology*, **65**, 425–460.

Kemper, T. D. (1990). *Social Structure and Testosterone: Explorations of the Socio-Bio-Social Chain*. New Brunswick, NJ: Rutgers University Press.

Kirby, J. N. (2017). Compassion interventions: The programmes, the evidence, and implications for research and practice. *Psychology and Psychotherapy: Theory, Research and Practice*, **90**(3), 432–455.

Kirby, J. N., & Gilbert, P. (2017). The emergence of the compassion focused therapies. In P. Gilbert, ed., *Compassion: Concepts, Research and Applications*. London: Routledge, pp. 258–285.

Kirby, J. N., & Kirby, P. G. (2017). An evolutionary model to conceptualise masculinity and compassion in male teenagers: A unifying framework. *Clinical Psychologist*, **21**(2), 74–89.

Kirby, J. N., Doty, J., Petrocchi, N., & Gilbert, P. (2017a). The current and future role of heart rate variability for assessing and training compassion. *Frontiers in Public Health*, **5**, 40.

Kirby, J. N., Tellegen, C. L., & Steindl, S. (2017b). A systematic review and meta-analysis of compassion-based interventions: Current state of knowledge and future directions. *Behavior Therapy*, **48**(6), 778–792.

Kogan, A., Saslow, L. R., Impett, E. A., et al. (2011). Thin-slicing study of the oxytocin receptor (OXTR) gene and the evaluation and expression of the prosocial disposition. *Proceedings of the National Academy of Sciences*, **108**(48), 19189–19192.

Kogan, A., Oveis, C., Carr, E. W., et al. (2014). Vagal activity is quadratically related to prosocial traits, prosocial emotions, and observer perceptions of prosociality. *Journal of Personality and Social Psychology*, **107**(6), 1051–1063.

Laursen, R. F., Siebner, H. R., Haren, T., et al. (2014). Variation in the oxytocin receptor gene is associated with behavioral and neural correlates of empathic accuracy. *Frontiers in Behavioral Neuroscience*, **8**, 423.

Li, N. P., van Vugt, M., & Colarelli, S. M. (2017). The evolutionary mismatch hypothesis: Implications for psychological science. *Current Directions in Psychological Science*, **27**, 38–44.

Loewenstein, G., & Small, D. A. (2007). The scarecrow and the tin man: The vicissitudes of human sympathy and caring. *Review of General Psychology*, **11**, 112–126.

MacDonald, K. S. (2012). Sex, receptors, and attachment: A review of individual factors influencing response to oxytocin. *Frontiers in Neuroscience*, **6**, 194.

MacLean, P. (1985). Brain evolution relating to family, play and the separation call. *Archives of General Psychiatry*, **42**, 405–417.

Mayseless, O. (2016). *The Caring Motivation: An Integrated Theory*. Oxford: Oxford University Press.

McGregor, I., & Marigold, D. C. (2003). Defensive zeal and the uncertain self: What makes you so sure? *Journal of Personality and Social Psychology*, **85**, 838–852.

Mikulincer, M., & Shaver, P. R. (2017). *Attachment in Adulthood: Structure, Dynamics, and Change*, 2nd ed. New York: Guilford.

Muller, M. N., Marlowe, F. W., Bugumba, R., & Ellison, P. T. (2009). Testosterone and paternal care in East African foragers and pastoralists. *Proceedings of the Royal Society of Biological Sciences*, **276**, 347–354.

Music, G. (2014). *The Good Life: Wellbeing and the Neuroscience of Altruism, Selfishness and Immorality*. London: Routledge.

Nell, V. (2006). Cruelty's rewards: The gratifications of perpetrators and spectators. *Behavioral and Brain Sciences*, **29**, 211–257.

Nhat Hanh, T. (2009). *The Blooming of a Lotus: Guided Meditation for Achieving the Miracle of Mindfulness*. London: Beacon Press.

Park, G., & Thayer, J. F. (2014). From the heart to the mind: Cardiac vagal tone modulates top-down and bottom-up visual perception and attention to emotional stimuli. *Frontiers in Psychology*, **5**, 278.

Panksepp, J., & Panksepp, J. B. (2013). Toward a cross-species understanding of empathy. *Trends in Neurosciences*, **36**, 489–496.

Petrocchi, N, & Cheli, S. (2019). The social brain and heart rate variability: Implications for psychotherapy. *Psychology and Psychotherapy*: Special edition: Building an Integrative Science for Psychotherapy for the 21st-Century.

Piff, P. K. (2014). Wealth and The inflated self: Class, entitlement, and narcissism. *Personality and Social Psychology Bulletin*, **40**, 34–43.

Piff, P. K., Kraus, M. W., Côté, S., Cheng, B. H., & Keltner, D. (2010). Having less, giving more: The influence of social class on prosocial behavior. *Journal of Personality and Social Psychology*, **99**, 771–777.

Porges, S. W. (2007). The polyvagal perspective. *Biological Psychology*, **74**, 116–143.

Preston, S. D. (2013). The origins of altruism in offspring care. *Psychological Bulletin*, **139**, 1305–1041.

Reed, A., & Aquino, K. F. (2003). Moral identity and the expanding circle of moral regard toward out groups. *Journal of Personality and Social Psychology*, **64**, 1270–1286.

Rutter, M., Beckett, C., Castle, J., et al. (2007). Effects of profound early institutional deprivation: An overview of findings from a UK longitudinal study of Romanian adoptees. *European Journal of Developmental Psychology*, **4**(3), 332–350.

Sachs, J. (2012). *The Price of Civilization: Economics and Ethics after the Fall*. New York: Vintage.

Sapolsky, R. M. (2017). *Behave: The Biology of Humans at Our Best and Worst*. London: Penguin.

Sedikides, C., & Skowronski, J. J. (1997). The symbolic self in evolutionary context. *Personality and Social Psychology Review*, **1**, 80–102.

Seppälä, E. M., Simon-Thomas, E., Brown, S. L., Worline, M. C., Cameron, C. D., & Doty, J. R. (Eds.). (2017). The Oxford handbook of compassion science. New York: Oxford University Press.

Shamay-Tsoory, S. G., & Abu-Akel, A. (2016). The social salience hypothesis of oxytocin. *Biological Psychiatry*, **79**(3), 194–202.

Shonkoff, J. P., Garner, A. S., Siegel, B. S., et al.; Committee on Early Childhood, Adoption, and Dependent Care (2012). The lifelong effects of early childhood adversity and toxic stress. *Pediatrics*, **129**(1), e232–e246.

Siegel, D. J. (2016) *Mind: A Journey to the Heart of Being Human*. New York: Norton.

Siegel, D. J. (2011). *Mindsight: Transform Your Brain with the New Science of Kindness*. New York: One World Publications.

Singer, J. L. (2006). *Imagery in Psychotherapy*: New York: American Psychological Association.

Singer, T., & Bolz, M., eds. (2012). Compassion: Bridging practice and science. www.compassion-training.org.

Singer, T., & Klimecki, O. M. (2014). Empathy and compassion. *Current Biology*, **24**(18), r875–r878.

Spikins, P. (2017). Prehistoric origin: The compassion of distant strangers. In P. Gilbert, ed., *Compassion: Concepts, Research and Applications*. London: Routledge, pp. 16–30.

Spikins, P., Rutherford, H. E., & Needham, A. P. (2010). From homininity to humanity: Compassion from the earliest archaics to modern humans. *Journal of Archaeology, Conscious and Culture*, **3**, 303–325.

Suddendorf, T., & Whitten, A. (2001). Mental evolutions and development: Evidence for secondary representation in children, great apes and other animals. *Psychological Bulletin*, **127**, 629–650.

Taylor, C. (1989). *Sources of the Self: The Making of the Modern Identity*. Cambridge, UK: Cambridge University Press

Taylor, S. E. (2006). Tend and befriend: Biobehavioral bases of affiliation under stress. *Current Directions in Psychological Science*, **15**, 273–277.

Thayer, J. F., Hansen, A. L., Saus-Rose, E., & Johnsen, B. H. (2009). Heart rate variability, self-regulation and the neurovisceral model of health. *Annals of Behavioral Medicine*, **37**, 141–153.

Thayer, J. F., Åhs, F., Fredrikson, M., Sollers, J. J., & Wager, T. D. (2012). A meta-analysis of heart rate variability and neuroimaging studies: Implications for heart rate variability as a marker of stress and health. *Neuroscience & Biobehavioral Reviews*, **36**(2), 747–756.

Tinbergen, N. (1963). On the aims and methods of ethology. *Zeitschrift für Tierpsychologie*, **20**, 410–433.

Trevarthen, C., & Aitken, K. (2001). Infant intersubjectivity: Research, theory, and clinical applications. *Journal of Child Psychology and Psychiatry*, **42**, 3–48.

Tsering, G. T. (2005). *The Four Noble Truths: The Foundation of Buddhist Thought*, Vol. 1. Boston, MA: Wisdom Publications.

Tsering, G. T. (2008). *The Awakening Mind: The Foundation of Buddhist Thought*, Vol. 4. London: Wisdom Publications.

Warneken, F., & Tomasello, M. (2009). The roots of human altruism. *British Journal of Psychology*, **100**, 455–471.

Van Vugt, M., & Park, J. H. (2009). Guns, germs, and sex: How evolution shaped our intergroup psychology. *Social and Personality Psychology Compass*, **3**, 927–938.

Weisman, O., Zagoory-Sharon, O., & Feldman, R. (2014). Oxytocin administration, salivary testosterone, and father–infant social behavior. *Progress in Neuro-Psychopharmacology and Biological Psychiatry*, **49**, 47–52.

Weissbourd, R., & Jones, S. (2014). The children we mean to raise: The real messages adults are sending about values. Making Caring Common Project, Graduate School of Education, Harvard University. https://mcc.gse.harvard.edu/reports/children-mean-raise.

Weng, H. Y., Fox A. S., Shackman, A. J., et al. (2013). Compassion training alters altruism and neural responses to suffering. *Psychological Science*, **24**, 1171–1180.

Weng, H. Y., Lapate, R. C., Stodola, D. E., Rogers, G. M., & Davidson, R. J. (2018). Visual attention to suffering after compassion training is associated with decreased amygdala responses. *Frontiers in Psychology*, **9**, 771.

Wesołowski, T. (2004). The origin of parental care in birds: A reassessment. *Behavioral Ecology*, **15**(3), 520–523.

Whitten, A. (1999). The evolution of deep social mind in humans. In M. C. Corballis & S. E. G. Lea, eds., *The Descent of Mind: Psychological Perspectives on Humanoid Evolution*. New York: Oxford University Press, pp. 173–193.

Zaki, J. (2014). Empathy: A motivated account. *Psychological Bulletin*, **140**, 1608–1647.

Zimbardo, P. (2007). *The Lucifer Effect: How Good People Turn Evil*. London: Rider.

35 Disordered Social Cognition
Alexithymia and Interoception

REBECCA BREWER AND GEOFFREY BIRD

35.1 INTRODUCTION

Alexithymia is a condition characterized by difficulties identifying and describing one's own emotional states (Nemiah, Freyberger, & Sifneos, 1976). Individuals with alexithymia are often aware that they are experiencing an emotion, but struggle to determine whether it is fear, excitement, or anger, for example. Alexithymia is therefore associated with difficulties describing how one would feel in particular emotional scenarios (Lane et al., 1990), as well as with difficulties regulating one's emotions (Stasiewicz et al., 2012; Venta, Hart, & Sharp, 2013). This chapter details the behavioral and neurological characteristics of alexithymia, its etiology (including whether it may be evolutionarily adaptive), and its role in emotional impairment across clinical populations. The relationship between alexithymia and interoception (the ability to perceive and recognize the internal state of one's body) is also discussed, alongside evidence that alexithymia may represent a general deficit of interoception. Finally, the theory is proposed that poor interoception (and therefore alexithymia) is a candidate for the "P" factor of susceptibility to psychopathology.

35.2 BEHAVIORAL AND NEUROLOGICAL CHARACTERISTICS OF ALEXITHYMIA

Although many see alexithymia as primarily reflecting cognitive impairment in the representation of emotions (e.g., Luminet et al., 2006; Suslow & Junghanns, 2002), others have argued for the central role of affective difficulties, characterized by decreased ability to experience emotions (e.g., Bermond, 1997; Vorst & Bermond, 2001). A number of subtypes of alexithymia have been proposed, defined in terms of affective and cognitive abilities. The term "Type I" alexithymia has been used to describe individuals who are impaired in both the affective and cognitive domains, whereas "Type II" defines those who suffer impairment only in the cognitive domain (Bermond, 1997). More recently, Moorman and colleagues also identified "Type III" alexithymics, with affective but not cognitive impairment – "lexithymics," with high performance in

the affective and cognitive domains, and "Modals," with average performance in both domains (Moormann et al., 2008). Impairment in the affective and cognitive domains has unsurprisingly been associated with different underlying neural atypicalities (Goerlich-Dobre et al., 2015). The majority of research focuses on the cognitive dimension and typically assesses alexithymia using the self-report Toronto Alexithymia Scale (TAS-20; Bagby, Parker, & Taylor, 1994), which has become the gold-standard tool for the assessment of alexithymia. Research into affective difficulties is becoming more common, however, with many researchers aiming to separate the cognitive and affective dimensions using the Bermond–Vorst Alexithymia Questionnaire (BVAQ; Vorst & Bermond, 2001; although see Bagby et al., 2009).

Interestingly, the impairments associated with alexithymia extend beyond difficulties recognizing and interpreting emotions in oneself. A plethora of evidence suggests that alexithymia is also associated with deficits in recognizing others' emotional states, such as from facial (Grynberg et al., 2012; Jessimer & Markham, 1997; McDonald & Prkachin, 1990; Parker, Taylor, & Bagby, 1993; Parker, Prkachin, & Prkachin, 2005; Prkachin, Casey, & Prkachin, 2009; Swart, Kortekaas, & Aleman, 2009), vocal (Delle-Vigne et al., 2014; Goerlich et al., 2011; Heaton et al., 2012), and body posture cues (Borhani et al., 2016). If one does not possess clear representations of individual emotions in the self and the situations and behaviors usually associated with these emotions, it follows that one would also have difficulty identifying how another would feel in the same situations or from relevant behavioral cues. Notably, emotion recognition difficulties appear to be general rather than specific; despite many researchers suggesting that recognition of negative emotions may be especially impaired, numerous studies have also found diminished recognition of positive emotions in alexithymia (Jessimer & Markham, 1997; Lane et al., 2000; Parker, Taylor, & Bagby, 1993).

Beyond emotion recognition difficulties, those with alexithymia demonstrate difficulties empathizing with others. Empathy has been defined in multiple ways, but is generally characterized by the sharing of another's emotional state, as well as attributing the emotion to the other and representing one's own emotion as being caused by that of the other (Bird & Viding, 2014; Singer & Lamm, 2009). Empathy therefore involves emotion contagion (affect sharing) and self–other distinction (representing the state of the self and the other separately). Self-reported empathy tends to decrease as alexithymia severity increases (Gleichgerrcht, Tomashitis, & Sinay, 2015; Grynberg et al., 2010; Guttman & Laporte, 2002; Neumann et al., 2014), although personal distress in response to others' negative states (likely reliant on the emotion contagion facet) increases (e.g., Guttman & Laporte, 2002; Moriguchi et al., 2007). Individuals with alexithymia also exhibit atypical responses in brain regions associated with empathy, such as the anterior insula (AI), when viewing others in pain (Moriguchi et al., 2007; Silani et al., 2008), although these findings are complicated by the fact that the pain stimuli used were likely to induce disgust in the observer, an emotion also associated with AI activity. Those with alexithymia also rate others' pain as less intense than do those with low alexithymia levels (Moriguchi et al., 2007). While the definition of empathy differs across studies, a consistent feature is the sharing of the other's affect. It is therefore unsurprising that alexithymia is associated with impaired empathy; if one cannot differentiate between one's own affective states, it is difficult both to associate those states with cues of affective states in others and to share the others' states (Bird & Viding, 2014). In line with difficulties empathizing and inferring emotions in oneself, alexithymia also seems to impact on one's ability to make typical moral judgments, a process shown to rely typically on emotion understanding. Individuals with alexithymia are more likely to make utilitarian moral decisions than those without alexithymia (Patil & Silani, 2014b; although see Gleichgerrcht et al., 2015) and are more likely to judge accidentally harming others to be acceptable (Patil & Silani, 2014a). Difficulties identifying emotions in oneself therefore appear to lead to further difficulties in the socio-affective domain.

At the neurological level, alexithymia has consistently been associated with the atypical structure and function of a number of brain regions. In particular, the anterior cingulate cortex (ACC) and AI, which play crucial roles in the processing of one's own emotions (e.g., Bush, Luu, & Posner, 2000; Etkin, Egner, & Kalisch, 2011; Kober et al., 2008; Lindquist et al., 2012; Phan et al., 2002), have been implicated. Much evidence suggests that alexithymia is associated with decreased gray matter volume of the ACC (Borsci et al., 2009; Grabe et al., 2014; Ihme et al., 2013; Koven et al., 2011; Paradiso et al., 2008; Sturm & Levenson, 2011; for a review, see van der Velde et al., 2013). Although one study found the opposite pattern of results (Gündel et al., 2004), the structural markers used in this study differed from others, and only individuals with low degrees of alexithymia were studied. ACC function

during affective tasks has also been consistently associated with alexithymia severity (Berthoz et al., 2002; Deng, Ma, & Tang, 2013; Frewen et al., 2008b; Heinzel et al., 2010; Jongen et al., 2014; Kano & Fukudo, 2013; Moriguchi & Komaki, 2013; Moriguchi et al., 2007), although the direction of the effect has varied across studies (Pouga et al., 2010; van der Velde et al., 2013). Relatedly, investigation of neurotransmitters in the ACC indicated atypical concentrations of GABA in individuals with alexithymia (Ernst et al., 2014).

Where the insula is concerned, studies have produced mixed findings concerning structural atypicalities; although the majority of studies observed decreased insula volume in individuals with alexithymia (Borsci et al., 2009; Grabe et al., 2014; Ihme et al., 2013), increased insula volume has also been reported (Goerlich-Dobre et al., 2014a; Zhang et al., 2011). Alexithymia also appears to be associated with atypical insula function during affective processing (Deng et al., 2013; Frewen et al., 2006; Heinzel et al., 2010; Kano et al., 2003, 2007; Moriguchi et al., 2007; Reker et al., 2010; Silani et al., 2008), although the direction of the effect has varied between studies (Craig, 2009). This atypical function may be explained in part by atypical glutamate neurotransmitter concentrations in the insula (Ernst et al., 2014). Besides the ACC and AI, alexithymia has been found to be associated with amygdala structure (Grabe et al., 2014; Ihme et al., 2013) and function (Goerlich-Dobre et al., 2014b; Kugel et al., 2008; Miyake et al., 2012; Reker et al., 2010; Zotev et al., 2011), as well as atypicalities in the orbitofrontal cortex (OFC), fusiform gyrus, dorsomedial prefrontal cortex, precuneus, and limbic, paralimbic, and premotor areas (for a review, see van der Velde et al., 2013).

35.3 ETIOLOGY AND DEVELOPMENT OF ALEXITHYMIA

Alexithymia is likely to be a neurodevelopmental condition, meaning that it occurs in the absence of brain damage (Lane et al., 2015), but can also develop following neurological trauma (Henry et al., 2006; Hogeveen et al., 2016; Neumann et al., 2014; Williams & Wood, 2010). The etiology of developmental alexithymia is not, thus far, well understood, but both genetic and environmental factors have been identified as being involved in its development. Evidence from twin studies suggests that alexithymia is heritable, with genetic factors contributing to approximately 33 percent of the variation in alexithymia severity (Heiberg & Heiberg, 1978; Jørgensen et al., 2007; Picardi et al., 2011; Valera & Berenbaum, 2001). Recent research has also identified candidate genes for the condition (Ham et al., 2005; Mezzavilla et al., 2015; Walter et al., 2011). Of particular interest are findings implicating dopaminergic function in cortical areas known to be atypical in alexithymia, with some genetic variants implicating decreased dopaminergic function in ACC and OFC activation (Ham et al., 2005; Walter et al., 2011). Beyond genetic influences, some research has investigated the environmental causes of alexithymia. Alexithymia has been associated, for example, with decreased perceived parental care and increased perceived overprotectiveness (Fukunishi et al., 1997; Kooiman et al., 2004), as well as child abuse (Berenbaum, 1996), rural upbringing, and large family size (Joukamaa et al., 2003). These findings are confounded, however, by the subjective nature of self-reported parenting styles (which may be affected by alexithymia), as well as the genetic relationship between parents and participants; parenting style (as well as decisions over family size and location of upbringing) may be associated with alexithymia traits in parents, which are inherited by participants. Little reliable evidence therefore exists concerning the precise role of environmental factors in alexithymia, but heritability estimates do suggest that environmental factors contribute to its development.

From an evolutionary perspective, it is clear that the ability to form emotional concepts and identify one's own and others' emotions is adaptive. It is necessary to identify whether another individual is feeling angry, for example, in order to predict and avoid aggressive behavior. Identifying fear, on the other hand, in oneself and others allows one to evade a threatening situation. The fact that alexithymia has a genetic component suggests that impairment in this ability may be adaptive in some situations, or at least that alleles associated with susceptibility to alexithymia may offer an evolutionary advantage. One way in which alexithymia may be adaptive is following trauma, or when one is experiencing a highly aversive emotion or period of negative emotion, such as during depression. A decreased tendency to recognize and attend to negative emotions may lead to less interference of one's emotional state with one's daily functioning. Indeed, it has been suggested that traumatic life events such as childhood maltreatment can lead to alexithymia tendencies (Berenbaum, 1996) and that alexithymia is highly associated with depression (Honkalampi et al., 2001). This is also in line with the proposal that, when environmental inputs are heterogeneous (as is the case for humans), conditional adaptation (developmental plasticity) occurs such that one's phenotype and behavioral strategies develop to reflect one's context, optimizing fitness in that particular environment (Del Giudice, 2014; Del Giudice, Ellis, & Shirtcliff, 2011). These authors propose early stress in particular, such as from childhood maltreatment, violence, and unpredictability, as a determinant of later adaptive behavioral strategies.

A second possibility is that a reduced ability to recognize negative emotions (e.g., signals of fear) in alexithymia allows one to take risks that others are not willing to, which may benefit the individual or the community as a whole. Indeed, alexithymia has been found to be associated with increased risk-taking (Barlow et al., 2015; Hahn, Simons, & Simons, 2016; Shishido, Gaher, & Simons, 2013), and while risky behavior increases the likelihood of harm to the individual, it can also lead to relatively large gains. It is possible that the survival advantages associated with alexithymia outweigh the disadvantages, as has been proposed for some psychological disorders, such as schizophrenia (for a review, see McClenon, 2011). It is also worth noting that alexithymia is associated with increased sexual risk-taking (including more sexual partners and intercourse without contraception), potentially leading to increased fertility in this population, causing the alexithymia phenotype to continue to be inherited.

Alternatively, the emergence of alexithymia may simply be due to polygenic mutation, as has been suggested to account for multiple psychological disorders (Keller & Miller, 2006). If this is the case, it is likely that the alexithymia phenotype simply emerges due to multiple combinations of mutations, rather than being specifically selected. Keller and Miller argue that, as over half of the genetic mutations carried by humans affect the brain, atypical neural functioning associated with psychological disorders may simply be attributable to a high number of mutations, which individually do not exert a large effect on behavior. Whether or not alexithymia is selected due to adaptive features therefore remains to be determined.

Alexithymia severity varies across the lifespan, with particularly high levels in young adolescence (Säkkinen et al., 2007). Alexithymia appears to be relatively stable in later adolescence (Karukivi et al., 2014), with severity tending to increase throughout adulthood (Mattila et al., 2006; for a review, see Murphy, Brewer, & Bird, 2017). Difficulties with the measurement of alexithymia across the lifespan, especially during childhood (Griffin, Lombardo, & Auyeung, 2016), however, can make it difficult to draw firm conclusions concerning the developmental trajectory of alexithymia. Future work should prioritize investigation into alexithymia across the lifespan and its relationship with associated difficulties; it is possible that alexithymia contributes to the development of varying forms of psychopathology at different stages of development (Murphy et al., 2017).

35.4 THE IMPACT OF ALEXITHYMIA ON SOCIAL AND EMOTIONAL ABILITIES ACROSS DISORDERS

Although alexithymia is an independent condition, it frequently co-occurs with multiple psychiatric disorders, such as autism spectrum disorder (ASD) (Hill, Berthoz, & Frith, 2004), feeding and eating disorders (EDs) (Cochrane et al., 1993), schizophrenia (Heshmati et al., 2010), depression (Honkalampi et al., 2001), posttraumatic stress disorder (Frewen et al., 2008a), and substance abuse (Mann et al., 1995), as well as neurological conditions such as Parkinson's disease (Assogna et al., 2012), frontotemporal dementia (Sturm & Levenson, 2011) and multiple sclerosis (Chahraoui et al., 2014; Prochnow et al., 2011) and physical illness such as diabetes (Abramson et al., 1991). Crucially, alexithymia is neither necessary nor sufficient for a diagnosis of these disorders and is not universal among individuals in these populations. Approximately 50 percent of individuals with ASD (Berthoz & Hill, 2005; Berthoz et al., 2013; Hill et al., 2004) and 60 percent of the ED population (Cochrane et al., 1993; for a review, see Nowakowski, McFarlane, & Cassin, 2013), for example, also have alexithymia, compared to approximately 10 percent of the typical population (Honkalampi et al., 2000; Kokkonen et al., 2001; Salminen, Saarija, & Rela, 1999). Furthermore, despite the association between alexithymia and depression (Honkalampi et al., 2000), evidence suggests that these are distinct constructs (Marchesi, Brusamonti, & Maggini, 2000; Parker, Bagby, & Taylor, 1991)

While it has often been suggested that alexithymia is a risk factor for the development of psychiatric disorders (e.g., De Beradis et al., 2014; Parker et al., 2005; van der Velde et al., 2015), the nature of the relationship is not well understood at present (Grabe, Spitzer, & Freyberger, 2004; Honkalampi et al., 2010; Nowakowski et al., 2013). Interestingly, a number of the behavioral difficulties associated with alexithymia, such as emotion recognition, are also commonly observed in clinical populations, such as ASD, ED, and schizophrenia, although the literature has been relatively inconsistent (Edwards, Jackson, & Pattison, 2002; Harms, Martin, & Wallace, 2010; Oldershaw et al., 2011). Similarly, the core neurological regions implicated in alexithymia, such as the ACC and insula, have often been reported to be structurally and functionally atypical in clinical groups, though findings are again variable (Anagnostou & Taylor, 2011; Bauman & Kemper, 2005; Franklin et al., 2002; Kaye, 2008; Mundy, 2003; Nagai, Kishi, & Kato, 2007; Rogers et al., 2009).

Recent evidence suggests that alexithymia may be responsible for a number of the emotional impairments often thought to be associated with clinical disorders. In those with ASD and EDs, for example, co-occurring alexithymia (rather than ASD or EDs per se) appears to explain difficulties recognizing others' facial expressions (Brewer et al., 2015b; Cook et al., 2013) and atypical empathy (Bird et al., 2010; Brewer et al., 2019). These findings suggest that individuals with ASD or an ED are only impaired in the emotional domain if they also have co-occurring alexithymia and that inconsistent previous findings may be attributable to varying levels of alexithymia in clinical samples. Alexithymia also appears to influence behaviors that are not explicitly emotional but rely on emotional processes, such as judgments of personality traits from novel faces (Brewer et al., 2015a). As alexithymia co-occurs with multiple conditions, it is likely that a range of emotional difficulties, including emotion recognition and empathy, are explained by alexithymia, rather than disorder diagnosis or symptom severity, across a range of clinical populations (Brewer et al., 2015c). Further research is required, however, into the impact of alexithymia and disorder presence and severity on more complex emotional processes, such as moral reasoning. Evidence suggests, for example, that those with alexithymia in the typical population tend to make atypical moral judgments,

while alexithymia is not associated with atypical moral judgments in individuals with ASD (Brewer et al., 2015d). Moral decision-making can involve both emotional and rational processes (Cushman, Young, & Greene, 2010; Ugazio, Lamm, & Singer, 2012), meaning that, as well as emotional abilities themselves, differences in the extent to which one relies on these two routes are likely to influence outcomes. The relationship between alexithymia and social abilities that rely on a combination of emotional and nonemotional processes therefore deserves further attention.

35.5 ALEXITHYMIA AS A GENERAL DEFICIT OF INTEROCEPTION

While the AI and ACC are clearly involved in emotion processing and alexithymia, these brain regions are also referred to as the "interoceptive cortex" due to their well-documented involvement in interoception (perception of the internal states of the body; Craig, 2002, 2003a, 2009; Critchley & Harrison, 2013; Garfinkel & Critchley, 2013). While interoception was initially defined as the perception of the condition of the viscera only (Sherrington, 1900), current definitions include the identification of numerous bodily states, such as heart rate, respiration, pain, temperature, itch, hunger, satiety, and fatigue (Craig, 2002, 2003a, 2009). The interoceptive cortex is centrally involved in the perception of these states and contributes to the human ability to subjectively represent one's own interoceptive condition, and therefore the "feeling self," via a spinothalamocortical pathway (Craig, 2002, 2003a). Recently, interoception has been divided into three distinct (though likely related) processes (Garfinkel & Critchley, 2013). Under this model, interoceptive sensitivity refers to one's ability to accurately perceive objective interoceptive states. Interoceptive sensibility, on the other hand, describes the propensity to introspect on internal states and subjective beliefs about one's own interoceptive states. Finally, interoceptive awareness refers to the accuracy of one's interoceptive sensitivity (one's metacognitive ability relating to interoception).

Studies of patients with insula lesions have generally observed interoceptive difficulties (for a review, see Ibañez, Gleichgerrcht, & Manes, 2010), and stimulation of the insula in humans has been found to elicit unpleasant sensations (Krolak-Salmon et al., 2003), gastrointestinal motility, abdominal sensations, and nausea (Penfield & Faulk, 1955), while insula inhibition has been associated with reduced awareness of one's heart beat and respiratory effort (Pollatos et al., 2016; but see Coll et al., 2017). While much evidence therefore suggests that the insula is integral to consciously representing feeling states, it should be noted that typical experience of interoceptive states, such as itch, tickle, pain, and temperature, as well as emotions, has been observed following bilateral insula lesion (Damasio, Damasio, & Tranel, 2013). Damasio and colleagues therefore emphasize the role of the brain stem and the thalamic, hypothalamic, and somatosensory regions (alongside interoceptive cortices) in representing these states. Similarly, ACC and AI lesions do not appear to impair the detection of heartbeats (Khalsa et al., 2009a), although it is possible that exteroceptive signals (e.g., from the chest wall) may contribute to this ability (Khalsa et al., 2009a). Overall, it is undeniable that interoceptive cortical areas, such as the AI and ACC, are involved in interoceptive awareness, but subcortical and somatosensory regions also contribute to the representation of internal states.

As alexithymia is associated with AI and ACC atypicalities, and as these brain regions are central to interoception, it is possible that alexithymia is associated with deficits of non-affective as well as affective interoception. Indeed, alexithymia is often described as involving difficulties differentiating one's emotions from other bodily sensations (Nemiah et al., 1976; Parker et al., 2003). It is therefore possible that alexithymia is better characterized as a general failure of interoception, rather than of affective interoception only (Brewer et al., 2015c; Brewer, Cook, & Bird, 2016). Some evidence does indeed suggest a negative relationship between alexithymia and identification of interoceptive states. Alexithymia appears to be associated, for example, with less accurate perception of one's heart rate (Herbert, Herbert, & Pollatos, 2011; Shah et al., 2016), as well as delayed responses to acute myocardial infarction (Carta et al., 2013). Erratic consumption of alcohol (Lyvers et al., 2012) and other substances (de Haan et al., 2014; Taylor, Parker, & Bagby, 1990) is also common in alexithymia, likely due to decreased awareness of the effects of these substances on the state of the body. Individuals with alexithymia also struggle to estimate their objective levels of arousal (Gaigg, Maurice, & Bird, 2018). Recent evidence suggests that increased alexithymia is associated with self-reported difficulties distinguishing between a range of nonemotional internal states, as well as differentiating these states from emotions (Brewer et al., 2016; Longarzo et al., 2015). While further investigation is needed into whether interoception is a unitary construct (Garfinkel et al., 2016a), it appears that alexithymia may be associated with numerous interoceptive difficulties beyond the emotional domain and may therefore be conceptualized as a general deficit of interoception, rather than one specific to emotion (Brewer et al., 2015c, 2016). Indeed, novel tests assessing the use of interoceptive information when judging one's speed of respiration and interoceptive accuracy in the domains of muscular effort and taste also reveal an association with levels of alexithymia (Murphy, Catmur, & Bird, 2018).

35.6 ALEXITHYMIA (POOR INTEROCEPTION) AS THE "P" FACTOR OF PSYCHOLOGICAL DISORDERS

As previous research suggests that alexithymia explains emotional impairments across clinical populations, alexithymia may also account for interoceptive impairments across the multiple disorder populations with which it co-occurs. Where ASD is concerned, for example, sensory atypicalities, such as hypersensitivity to touch, are often exhibited (Leekam et al., 2007; Tomchek & Dunn, 2007). As itch and slow, affective touch can be considered to be interoceptive states, alexithymia (and impaired interoception) may account for atypical sensory profiles in those with ASD. Gastrointestinal difficulties are also common in those with ASD (Torrente et al., 2002; Wakefield et al., 2000, 2005; White, 2003; Williams et al., 2011) and may be attributable to alexithymia due to the well-documented role of interoception in gastric sensitivity (Stephan et al., 2003; Vandenbergh et al., 2005) and atypical visceral sensitivity in alexithymia (Kano et al., 2007). Evidence also suggests atypical neural responses (implicating the ACC and insula in particular) during reward processing in those with ASD (Bellebaum, Brodmann, & Thoma, 2014; Cascio et al., 2012; Dichter et al., 2012; Kohls et al., 2013; Larson et al., 2011; Schmitz et al., 2008). Thus, difficulties in processing reward may also rely on interoceptive abilities (and therefore be affected by alexithymia), rather than ASD itself. Finally, although evidence concerning imitation in those with ASD has varied depending on task demands (Cook & Bird, 2012; Leighton et al., 2010; Press, Richardson, & Bird, 2010), it is possible that alexithymia can account for any differences between those with and without ASD due to the role of interoception in the ability to represent the self (Craig, 2009; Critchley et al., 2004; Quattrocki & Friston, 2014; Seth, Suzuki, & Critchley, 2011), and therefore likely the ability to distinguish between the self and other. Indeed, both alexithymia and interoceptive sensitivity to heartbeats have been found to predict control of the tendency to automatically imitate others' actions (Ainley, Brass, & Tsakiris, 2014; Sowden et al., 2016). It should be noted that the impact of interoceptive difficulties on the self–other distinction is also particularly relevant for schizophrenia, which is characterized by difficulties distinguishing the self from others (Ebisch et al., 2013; Jardri et al., 2011; Sass & Parnas, 2003).

Interoception and the interoceptive cortex, such as the AI and ACC, are also thought to play a role in processes such as the perception of time (Bushara, Grafman, & Hallett, 2001; Livesey, Wall, & Smith, 2007; Magnani et al., 2014), reward processing (Furl & Averbeck, 2011; Tanaka et al., 2004; Wittmann et al., 2010), cognitive control (Brass & Haggard, 2007; Cai et al., 2014; Cole & Schneider, 2007; Dosenbach et al., 2007; Eichele et al., 2008; Ghahremani, Rastogi, & Lam, 2015), metacognition (Fleming & Dolan, 2012), attention (Chen et al., 2015; Mason et al., 2007; Weissman et al., 2006), decision-making (Botvinick, 2007; Coricelli et al., 2005; Critchley, Mathias, & Dolan, 2001; Dunn et al., 2010), and cravings (relating to addiction; Gray & Critchley, 2007; Naqvi & Bechara, 2009, 2010; Naqvi et al., 2007). These abilities relate to Damasio's somatic marker hypothesis, which posits that interoceptive signals of arousal (including emotional responses) can influence cognition and behavior, particularly decision-making, through learned associations between behaviors and outcomes (Damasio, Tranel, & Damasio, 1991). It is likely that, if alexithymia is characterized by poor interoception, those with alexithymia may exhibit impairments in all of these domains.

Beyond ASD, a similar interoceptive explanation may also hold for impairments observed in other disorders. Difficulties with reward processing and decision-making, for example, are characteristic of those with depression (Davidson et al., 2002; Forbes & Dahl, 2012), substance abuse and other addiction disorders (Bechara & Damasio, 2002; Bechara, Dolan, & Hindes, 2002; Koob & Le Moal, 2001; Noble, 2000; Reuter et al., 2005; Schoenbaum, Roesch, & Stalnaker, 2006), EDs (Broft et al., 2011; Cowdrey et al., 2011; Keating et al., 2012; Wagner et al., 2007), and schizophrenia (Abi-Dargham et al., 2000; Heerey, Bell-Warren, & Gold, 2008; Simon et al., 2010), and may all be attributable to alexithymia (and therefore interoceptive impairment) rather than to disorder presence or severity per se. Indeed, alexithymia has been associated with poor decisions on gambling tasks in those with gambling disorder (Aïte et al., 2014), in line with the widely accepted role of interoception in addiction (Naqvi & Bechara, 2010). Similarly, atypical decision-making has been associated with alexithymia in the typical population (Shah, Catmur, & Bird, 2016). Alexithymia may also explain the interoceptive deficits characteristic of EDs, such as difficulties detecting signals of hunger and satiety (Fassino et al., 2004; Lilenfeld et al., 2006), as well as perceiving one's own heartbeat (Pollatos et al., 2008). Alexithymia may therefore predict these impairments across the broad range of disorders it co-occurs with and possibly contribute to the development of these disorders in some individuals. Determining whether the role of alexithymia in interoceptive impairment is comparable across populations should be a priority for future work.

In line with evidence for interoceptive difficulties across a number of disorders, we have recently proposed that alexithymia (and therefore poor interoception) encompasses the "P" factor of psychopathology (Brewer et al., 2016). The "P" factor has been described as a unitary factor that can account for general susceptibility to psychopathology, with symptom severity being associated with the degree of neural atypicality (Caspi et al., 2015; Lahey et al., 2012). The P factor model currently includes 11 disorder symptoms, comprising depression, anxiety, addiction, conduct disorder, obsessive–compulsive disorder, mania, and schizophrenia, the majority of which have been associated with interoceptive impairment (for a review, see Khalsa & Lapidus, 2016). Atypical interoception has been associated, for example, with depression (Harshaw, 2015), addiction (Naqvi & Bechara, 2010; Verdejo-Garcia, Clark, & Dunn, 2012), schizophrenia (Ardizzi et al., 2016), OCD (Lazarov et al., 2010; Stern, 2014), and anxiety (associated with atypically high rather than low interoceptive abilities; Domschke et al., 2010;

Ehlers & Breuer, 1992; Paulus & Stein, 2006), as well as with disorders that are not currently included, such as ASD (Garfinkel et al., 2016b) and EDs (Pollatos et al., 2008). Interoceptive impairment appears, therefore, to be a likely candidate for increased disorder susceptibility. In fact, as interoceptive awareness is required for physical as well as mental health, an extension of this hypothesis may suggest that alexithymia accounts for susceptibility to health issues in general. Indeed, elevated levels of alexithymia have been observed in diabetes, which is characterized by atypicalities in the implicit interoception required for homeostatic maintenance of blood glucose levels. Clearly, this hypothesis requires substantial testing, but existing evidence points to a role for interoception (and alexithymia) in at least a broad range of symptoms that are observed in multiple conditions (for a full description of this theory, see Murphy et al., 2017). As it is possible to improve interoceptive abilities through training (Schandry & Weitkunat, 1990), investigation of this theory may provide opportunities for therapeutic interventions that may be applied across psychological and medical disorders.

35.7 CONCLUSION

Alexithymia is a condition that has traditionally been defined in terms of difficulties understanding one's own emotions, and much research has focused on its relationship with affective abilities, such as emotion regulation, recognition of others' emotions, empathy, and moral decision-making. The neurological regions implicated in alexithymia (e.g., the ACC and AI) are also centrally involved, however, in more general interoceptive processes. This chapter reviewed evidence that alexithymia is responsible for the emotional difficulties experienced by many individuals with psychological disorders, such as ASD and EDs, and theorized that alexithymia may also account for the interoceptive difficulties often assumed to be core features of clinical disorders. We suggest that alexithymia may be characterized by a general deficit of interoception rather than an impairment of affective interoception alone, and that interoceptive deficits may be a reasonable candidate for the "P" factor, a unitary factor representing susceptibility to psychopathology. We therefore suggest that alexithymia, and by inference interoception, may explain both the symptom commonalities between psychiatric conditions and the symptom heterogeneity within conditions. If it is the case that alexithymia, or difficulties with interoception, predisposes one to developing psychological disorders, the question remains as to why this condition continues to be inherited. Due to its relatively high prevalence, it is possible that there are adaptive aspects of alexithymia. Determining whether this is the case should be a priority for future research.

REFERENCES

Abi-Dargham, A., Rodenhiser, J., Printz, D., et al. (2000). Increased baseline occupancy of D2 receptors by dopamine in schizophrenia. *Proceedings of the National Academy of Sciences*, **97**(14), 8104–8109.

Abramson, L., McClelland, D. C., Brown, D., & Kelner, S. (1991). Alexithymic characteristics and metabolic control in diabetic and healthy adults. *Journal of Nervous and Mental Disease*, **179** (8), 490–494.

Ainley, V., Brass, M., & Tsakiris, M. (2014). Heartfelt imitation: High interoceptive awareness is linked to greater automatic imitation. *Neuropsychologia*, **60**, 21–28.

Aïte, A., Barrault, S., Cassotti, M., et al. (2014). The impact of alexithymia on pathological gamblers' decision making: A preliminary study of gamblers recruited in "sportsbook" casinos. *Cognitive and Behavioral Neurology*, **27**(2), 59–67.

Anagnostou, E., & Taylor, M. J. (2011). Review of neuroimaging in autism spectrum disorders: What have we learned and where we go from here. *Molecular Autism*, **2**(1), 4.

Ardizzi, M., Ambrosecchia, M., Buratta, L., et al. (2016). Interoception and positive symptoms in schizophrenia. *Frontiers in Human Neuroscience*, **10**, 379.

Assogna, F., Palmer, K., Pontieri, F. E., et al. (2012). Alexithymia is a non-motor symptom of Parkinson disease. *American Journal of Geriatric Psychiatry*, **20**(2), 133–141.

Bagby, R. M., Parker, J. D. A., & Taylor, G. J. (1994). The twenty-item Toronto Alexithymia Scale – I. Item selection and cross-validation of the factor structure. *Journal of Psychosomatic Research*, **38**(1), 23–32.

Bagby, R. M., Quilty, L. C., Taylor, G. J., et al. (2009). Are there subtypes of alexithymia? *Personality and Individual Differences*, **47**(5), 413–418.

Barlow, M., Woodman, T., Chapman, C., et al. (2015). Who takes risks in high-risk sport?: The role of alexithymia. *Journal of Sport & Exercise Psychology*, **37**(1), 83–96.

Bauman, M. L., & Kemper, T. L. (2005). Neuroanatomic observations of the brain in autism: A review and future directions. *International Journal of Developmental Neuroscience*, **23**(2–3), 183–187.

Bechara, A., & Damasio, H. (2002). Decision-making and addiction (part I): Impaired activation of somatic states in substance dependent individuals when pondering decisions with negative future consequences. *Neuropsychologia*, **40**(10), 1675–1689.

Bechara, A., Dolan, S., & Hindes, A. (2002). Decision-making and addiction (part II): Myopia for the future or hypersensitivity to reward? *Neuropsychologia*, **40**(10), 1690–1705.

Bellebaum, C., Brodmann, K., & Thoma, P. (2014). Active and observational reward learning in adults with autism spectrum disorder: Relationship with empathy in an atypical sample. *Cognitive Neuropsychiatry*, **19**(3), 205–225.

Berenbaum, H. (1996). Childhood abuse, alexithymia and personality disorder. *Journal of Psychosomatic Research*, **41**(6), 585–595.

Bermond, B. (1997). Brain and alexithymia. In A. Vingerhoets, F. Bussel, & J. Boelhouwer, eds., *The (Non)expression of Emotions in Health and Disease*. Tilburg: Tilburg University Press, pp. 115–130.

Berthoz, S., & Hill, E. L. (2005). The validity of using self-reports to assess emotion regulation abilities in adults with autism spectrum disorder. *European Psychiatry*, **20**(3), 291–298.

Berthoz, S., Artiges, E., Van de Moortele, P. F., et al. (2002). Effect of impaired recognition and expression of emotions on fronto-cingulate cortices: An fMRI study of men with alexithymia. *American Journal of Psychiatry*, **159**(6), 961–967.

Berthoz, S., Lalanne, C., Crane, L., & Hill, E. L. (2013). Investigating emotional impairments in adults with autism spectrum disorders and the broader autism phenotype. *Psychiatry Research*, **208**(3), 257–264.

Bird, G., & Viding, E. (2014). The self to other model of empathy: Providing a new framework for understanding empathy impairments in psychopathy, autism, and alexithymia. *Neuroscience & Biobehavioral Reviews*, **47**, 520–532.

Bird, G., Silani, G., Brindley, R., et al. (2010). Empathic brain responses in insula are modulated by levels of alexithymia but not autism. *Brain*, **133**(5), 1515–1525.

Borhani, K., Borgomaneri, S., Làdavas, E., & Bertini, C. (2016). The effect of alexithymia on early visual processing of emotional body postures. *Biological Psychology*, **115**, 1–8.

Borsci, G., Boccardi, M., Rossi, R., et al. (2009). Alexithymia in healthy women: A brain morphology study. *Journal of Affective Disorders*, **114**(1–3), 208–215.

Botvinick, M. M. (2007). Conflict monitoring and decision making: Reconciling two perspectives on anterior cingulate function. *Cognitive, Affective & Behavioral Neuroscience*, **7**(4), 356–366.

Brass, M., & Haggard, P. (2007). To do or not to do: The neural signature of self-control. *Journal of Neuroscience*, **27**(34), 9141–9145.

Brewer, R., Collins, F., Cook, R., & Bird, G. (2015a). Atypical trait inferences from facial cues in alexithymia. *Emotion*, **15**(5), 637–643.

Brewer, R., Cook, R., Cardi, V., Treasure, J., & Bird, G. (2015b). Emotion recognition deficits in eating disorders are explained by co-occurring alexithymia. *Royal Society Open Science*, **2**(1), 140382.

Brewer, R., Happé, F., Cook, R., & Bird, G. (2015c). Commentary on "Autism, oxytocin and interoception": Alexithymia, not autism spectrum disorders, is the consequence of interoceptive failure. *Neuroscience & Biobehavioral Reviews*, **56**, 348–353.

Brewer, R., Marsh, A. A., Catmur, C., et al. (2015d). The impact of autism spectrum disorder and alexithymia on judgments of moral acceptability. *Journal of Abnormal Psychology*, **124**(3), 589–595.

Brewer, R., Cook, R., & Bird, G. (2016). Alexithymia: A general deficit of interoception. *Royal Society Open Science*, **3**(10), 150664.

Brewer, R., Cook, R., Cardi, V., et al. (2019). Alexithymia (and predictive coding) explains personal distress in individuals with eating disorders. *Journal of Affective Disorders*, **72**(7), 1827–1836.

Broft, A. I., Berner, L. A., Martinez, D., & Walsh, B. T. (2011). Bulimia nervosa and evidence for striatal dopamine dysregulation: A conceptual review. *Physiology and Behavior*, **104**(1), 122–127.

Bush, G., Luu, P., & Posner, M. I. (2000). Cognitive and emotional influences in anterior cingulate cortex. *Trends in Cognitive Sciences*, **4**(6), 215–222.

Bushara, K. O., Grafman, J., & Hallett, M. (2001). Neural correlates of auditory–visual stimulus onset asynchrony detection. *Journal of Neuroscience*, **21**(1), 300–304.

Cai, W., Ryali, S., Chen, T., Li, C.-S. R., & Menon, V. (2014). Dissociable roles of right inferior frontal cortex and anterior insula in inhibitory control: Evidence from intrinsic and task-related functional parcellation, connectivity, and response profile analyses across multiple datasets. *Journal of Neuroscience*, **34**(44), 14652–14667.

Carta, M. G., Sancassiani, F., Pippia, V., et al. (2013). Alexithymia is associated with delayed treatment seeking in acute myocardial infarction. *Psychotherapy and Psychosomatics*, **82**(3), 190–192.

Cascio, C. J., Foss-Feig, J. H., Heacock, J. L., et al. (2012). Response of neural reward regions to food cues in autism spectrum disorders. *Journal of Neurodevelopmental Disorders*, **4**(1), 9.

Caspi, A., Houts, R. M., Belsky, D. W., & Goldman-Mellor, S. J. (2015). The p factor: One general psychopathology factor in the structure of psychiatric disorders? *Clinical Psychological Science*, **2**(2), 119–137.

Chahraoui, K., Duchene, C., Rollot, F., Bonin, B., & Moreau, T. (2014). Longitudinal study of alexithymia and multiple sclerosis. *Brain and Behavior*, **4**(1), 75–82.

Chen, T., Michels, L., Supekar, K., et al. (2015). Role of the anterior insular cortex in integrative causal signaling during multisensory auditory–visual attention. *European Journal of Neuroscience*, **41**(2), 264–274.

Cochrane, C. E., Brewerton, T. D., Wilson, D. B., & Hodges, E. L. (1993). Alexithymia in the eating disorders. *International Journal of Eating Disorders*, **14**(2), 219–222.

Cole, M. W., & Schneider, W. (2007). The cognitive control network: Integrated cortical regions with dissociable functions. *NeuroImage*, **37**(1), 343–360.

Coll, M., Penton, T., Hobson, H., & Hobson, H. (2017). Important methodological issues regarding the use of transcranial magnetic stimulation to investigate interoceptive processing: A comment on Pollatos et al. (2016). *Philosophical Transactions of the Royal Society B: Biological Sciences*, **372**, 20160506.

Cook, J. L., & Bird, G. (2012). Atypical social modulation of imitation in autism spectrum conditions. *Journal of Autism and Developmental Disorders*, **42**(6), 1045–1051.

Cook, R., Brewer, R., Shah, P., & Bird, G. (2013). Alexithymia, not autism, predicts poor recognition of emotional facial expressions. *Psychological Science*, **24**(5), 723–732.

Coricelli, G., Critchley, H. D., Joffily, M., et al. (2005). Regret and its avoidance: A neuroimaging study of choice behavior. *Nature Neuroscience*, **8**(9), 1255–1262.

Cowdrey, F. A., Park, R. J., Harmer, C. J., & McCabe, C. (2011). Increased neural processing of rewarding and aversive food stimuli in recovered anorexia nervosa. *Biological Psychiatry*, **70**(8), 736–743.

Craig, A. D. (2002). How do you feel? Interoception: The sense of the physiological condition of the body. *Nature Reviews Neuroscience*, **3**(8), 655–666.

Craig, A. D. (2003a). Interoception: The sense of the physiological condition of the body. *Current Opinion in Neurobiology*, **13**(4), 500–505.

Craig, A. D. (2003b). Pain mechanisms: Labeled lines versus convergence in central processing. *Annual Review of Neuroscience*, **26**, 1–30.

Craig, A. D. (2009). How do you feel – Now? The anterior insula and human awareness. *Nature Reviews Neuroscience*, **10**(1), 59–70.

Critchley, H. D., & Harrison, N. A. (2013). Visceral influences on brain and behavior. *Neuron*, **77**(4), 624–638.

Critchley, H. D., Mathias, C. J., & Dolan, R. J. (2001). Neural activity in the human brain relating to uncertainty and arousal during anticipation. *Neuron*, **29**(2), 537–545.

Critchley, H. D., Wiens, S., Rotshtein, P., Ohman, A., & Dolan, R. J. (2004). Neural systems supporting interoceptive awareness. *Nature Neuroscience*, **7**(2), 189–195.

Cushman, F., Young, L., & Greene, J. (2010). Our multi-system moral psychology: Towards a consensus view. In J. Doris, G. Harman, S. Nichols, et al., eds., *The Oxford Handbook of Moral Psychology*. Oxford: Oxford University Press, pp. 47–72.

Damasio, A. R., Tranel, D., & Damasio, H. (1991). Somatic markers and the guidance of behaviour: Theory and preliminary testing. In H. S. Levin, H. M. Eisenberg, & A. L. Benton, eds., *Frontal Lobe Function and Dysfunction*. New York: Oxford University Press, pp. 217–229.

Damasio, A., Damasio, H., & Tranel, D. (2013). Persistence of feelings and sentience after bilateral damage of the insula. *Cerebral Cortex*, **23**(4), 833–846.

Davidson, R. J., Pizzagalli, D., Nitschke, J. B., & Putnam, K. (2002). Depression: Perspectives from affective neuroscience. *Annual Review of Psychology*, **53**, 545–574.

De Beradis, D., Conti, C., Iasevoli, F., et al. (2014). Alexithymia and its relationships with acute phase proteins and cytokine release: An updated review. *Journal of Biological Regulators & Homeostatic Agents*, **28**(4), 13–17.

de Haan, H. A., van der Palen, J., Wijdeveld, T. G. M., Buitelaar, J. K., & De Jong, C. A. J. (2014). Alexithymia in patients with substance use disorders: State or trait? *Psychiatry Research*, **216**(1), 137–145.

Del Giudice, M. (2014). An evolutionary life history framework for psychopathology. *Psychological Inquiry*, **25**(3–4), 261–300.

Del Giudice, M., Ellis, B. J., & Shirtcliff, E. A. (2011). The Adaptive Calibration Model of stress responsivity. *Neuroscience and Biobehavioral Reviews*, **35**(7), 1562–1592.

Delle-Vigne, D., Kornreich, C., Verbanck, P., & Campanella, S. (2014). Subclinical alexithymia modulates early audio-visual perceptive and attentional event-related potentials. *Frontiers in Human Neuroscience*, **8**, 106.

Deng, Y., Ma, X., & Tang, Q. (2013). Brain response during visual emotional processing: An fMRI study of alexithymia. *Psychiatry Research*, **213**(3), 225–229.

Dichter, G. S., Richey, J. A., Rittenberg, A. M., Sabatino, A., & Bodfish, J. W. (2012). Reward circuitry function in autism during face anticipation and outcomes. *Journal of Autism and Developmental Disorders*, **42**(2), 147–160.

Domschke, K., Stevens, S., Pfleiderer, B., & Gerlach, A. L. (2010). Interoceptive sensitivity in anxiety and anxiety disorders: An overview and integration of neurobiological findings. *Clinical Psychology Review*, **30**(1), 1–11.

Dosenbach, N. U. F., Fair, D. A., Miezin, F. M., et al. (2007). Distinct brain networks for adaptive and stable task control in humans. *Proceedings of the National Academy of Sciences*, **104**(26), 11073–11078.

Dunn, B. D., Galton, H. C., Morgan, R., et al. (2010). Listening to your heart. How interoception shapes emotion experience and intuitive decision making. *Psychological Science*, **21**(12), 1835–1844.

Ebisch, S. J. H., Salone, A., Ferri, F., et al. (2013). Out of touch with reality? Social perception in first-episode schizophrenia. *Social Cognitive and Affective Neuroscience*, **8**(4), 394–403.

Edwards, J., Jackson, H. J., & Pattison, P. E. (2002). Emotion recognition via facial expression and affective prosody in schizophrenia: A methodological review. *Clinical Psychology Review*, **22**, 789–832.

Ehlers, A., & Breuer, P. (1992). Increased cardiac awareness in panic disorder. *Journal of Abnormal Psychology*, **101**(3), 371–382.

Eichele, T., Debener, S., Calhoun, V. D., Spe et al. (2008). Prediction of human errors by maladaptive changes in event-related brain networks. *Proceedings of the National Academy of Sciences*, **105**(16), 6173–6178.

Ernst, J., Böker, H., Hättenschwiler, J., et al. (2014). The association of interoceptive awareness and alexithymia with neurotransmitter concentrations in insula and anterior cingulate. *Social Cognitive and Affective Neuroscience*, **9**(6), 857–863.

Etkin, A., Egner, T., & Kalisch, R. (2011). Emotional processing in anterior cingulate and medial prefrontal cortex. *Trends in Cognitive Sciences*, **15**(2), 85–93.

Fassino, S., Pierò, A., Gramaglia, C., & Abbate-Daga, G. (2004). Clinical, psychopathological and personality correlates of interoceptive awareness in anorexia nervosa, bulimia nervosa and obesity. *Psychopathology*, **37**(4), 168–174.

Fleming, S. M., & Dolan, R. J. (2012). The neural basis of metacognitive ability. *Philosophical Transactions of the Royal Society of London B: Biological Sciences*, **367**(1594), 1338–1349.

Forbes, E. E., & Dahl, R. E. (2012). Altered reward function in adolescent depression: What, when, and how? *Journal of Child Psychology and Psychiatry*, **53**(1), 3–15.

Franklin, T. R., Acton, P. D., Maldjian, J. A., et al. (2002). Decreased gray matter concentration in the insular, orbitofrontal, cingulate, and temporal cortices of cocaine patients. *Biological Psychiatry*, **51**(2), 134–142.

Frewen, P. A., Pain, C., Dozois, D. J. A., & Lanius, R. A. (2006). Alexithymia in PTSD: Psychometric and fMRI studies. *Annals of the New York Academy of Sciences*, **1071**, 397–400.

Frewen, P. A., Dozois, D. J. A., Neufeld, R. W. J., & Lanius, R. A. (2008a). Meta-analysis of alexithymia in posttraumatic stress disorder. *Journal of Traumatic Stress*, **21**(2), 243–246.

Frewen, P. A., Lanius, R. A., Dozois, D. J. A., et al. (2008b). Clinical and neural correlates of alexithymia in posttraumatic stress disorder. *Journal of Abnormal Psychology*, **117**(1), 171–181.

Fukunishi, I., Kawamura, N., Ishikawa, T., et al. (1997). Mothers' low care in the development of alexithymia: A preliminary study in Japanese college students. *Psychological Reports*, **80**(1), 143–146.

Furl, N., & Averbeck, B. B. (2011). Parietal cortex and insula relate to evidence seeking relevant to reward-related decisions. *Journal of Neuroscience*, **31**(48), 17572–17582.

Gaigg, S. B., Maurice, A. S. F., & Bird, G. (2018). The psychophysiological mechanisms of alexithymia in autism spectrum disorder. *Autism*, **22**(2), 227–231.

Garfinkel, S. N., & Critchley, H. D. (2013). Interoception, emotion and brain: New insights link internal physiology to social behaviour. Commentary on: "Anterior insular cortex mediates bodily sensibility and social anxiety" by Terasawa et al. (2012). *Social Cognitive and Affective Neuroscience*, **8**(3), 231–234.

Garfinkel, S. N., Manassei, M. F., Hamilton-Fletcher, G., et al. (2016a). Interoceptive dimensions across cardiac and respiratory axes. *Philosophical Transactions of the Royal Society B: Biological Sciences*, **371**, 20160014.

Garfinkel, S. N., Tiley, C., O'Keeffe, S., et al. (2016b). Discrepancies between dimensions of interoception in autism: Implications for emotion and anxiety. *Biological Psychology*, **114**, 117–126.

Ghahremani, A., Rastogi, A., & Lam, S. (2015). The role of right anterior insula and salience processing in inhibitory control. *Journal of Neuroscience*, **35**(8), 3291–3292.

Gleichgerrcht, E., Tomashitis, B., & Sinay, V. (2015). The relationship between alexithymia, empathy and moral judgment in patients with multiple sclerosis. *European Journal of Neurology*, **22**(9), 1295–1303.

Goerlich, K. S., Witteman, J., Aleman, A., & Martens, S. (2011). Hearing feelings: Affective categorization of music and speech in alexithymia, an ERP study. *PLoS ONE*, **6**(5), e19501.

Goerlich-Dobre, K. S., Bruce, L., Martens, S., Aleman, A., & Hooker, C. I. (2014a). Distinct associations of insula and cingulate volume with the cognitive and affective dimensions of alexithymia. *Neuropsychologia*, **53**, 284–292.

Goerlich-Dobre, K. S., Witteman, J., Schiller, N. O., et al. (2014b). Blunted feelings: Alexithymia is associated with a diminished neural response to speech prosody. *Social Cognitive and Affective Neuroscience*, **9**(8), 1108–1117.

Goerlich-Dobre, K. S., Votinov, M., Habel, U., Pripfl, J., & Lamm, C. (2015). Neuroanatomical profiles of alexithymia dimensions and subtypes. *Human Brain Mapping*, **36**(10), 3805–3818.

Grabe, H. J., Spitzer, C., & Freyberger, H. J. (2004). Alexithymia and personality in relation to dimensions of psychopathology. *American Journal of Psychiatry*, **161**(7), 1299–1301.

Grabe, H. J., Wittfeld, K., Hegenscheid, K., et al. (2014). Alexithymia and brain gray matter volumes in a general population sample. *Human Brain Mapping*, **35**(12), 5932–5945.

Gray, M. A., & Critchley, H. D. (2007). Interoceptive basis to craving. *Neuron*, **54**(2), 183–186.

Griffin, C., Lombardo, M. V., & Auyeung, B. (2016). Alexithymia in children with and without autism spectrum disorders. *Autism Research*, **9**(7), 773–780.

Grynberg, D., Luminet, O., Corneille, O., Grèzes, J., & Berthoz, S. (2010). Alexithymia in the interpersonal domain: A general deficit of empathy? *Personality and Individual Differences*, **49**(8), 845–850.

Grynberg, D., Chang, B., Corneille, O., et al. (2012). Alexithymia and the processing of emotional facial expressions (EFEs): Systematic review, unanswered questions and further perspectives. *PLoS ONE*, **7**(8), e42429.

Gündel, H., López-Sala, A., Ceballos-Baumann, A., et al. (2004). Alexithymia correlates with the size of the right anterior cingulate. *Journal of Psychosomatic Research*, **56**(6), 609–610.

Guttman, H., & Laporte, L. (2002). Alexithymia, empathy, and psychological symptoms in a family context. *Comprehensive Psychiatry*, **43**(6), 448–455.

Hahn, A. M., Simons, R. M., & Simons, J. S. (2016). Childhood maltreatment and sexual risk taking: The mediating role of alexithymia. *Archives of Sexual Behavior*, **45**(1), 53–62.

Ham, B. J., Lee, M. S., Lee, Y. M., et al. (2005). Association between the catechol O-methyltransferase Val108/158 Met polymorphism and alexithymia. *Neuropsychobiology*, **52**(3), 151–154.

Harms, M. B., Martin, A., & Wallace, G. L. (2010). Facial emotion recognition in autism spectrum disorders: A review of behavioral and neuroimaging studies. *Neuropsychology Review*, **20**(3), 290–322.

Harshaw, C. (2015). Interoceptive dysfunction: Toward an integrated framework for understanding somatic and affective disturbance in depression. *Psychological Bulletin*, **141**(2), 311–363.

Heaton, P., Reichenbacher, L., Sauter, D., et al. (2012). Measuring the effects of alexithymia on perception of emotional vocalizations in autistic spectrum disorder and typical development. *Psychological Medicine*, **42**(11), 2453–2459.

Heerey, E. A., Bell-Warren, K. R., & Gold, J. M. (2008). Decision-making impairments in the context of intact reward sensitivity in schizophrenia. *Biological Psychiatry*, **64**(1), 62–69.

Heiberg, A. N., & Heiberg, A. (1978). A possible genetic contribution to the alexithymia trait. *Psychotherapy and Psychosomatics*, **30**(3–4), 205–210.

Heinzel, A., Schäfer, R., Müller, H. W., et al. (2010). Increased activation of the supragenual anterior cingulate cortex during visual emotional processing in male subjects with high degrees of alexithymia: An event-related fMRI study. *Psychotherapy and Psychosomatics*, **79**(6), 363–370.

Henry, J. D., Phillips, L. H., Crawford, J. R., Theodorou, G., & Summers, F. (2006). Cognitive and psychosocial correlates of alexithymia following traumatic brain injury. *Neuropsychologia*, **44**(1), 62–72.

Herbert, B. M., Herbert, C., & Pollatos, O. (2011). On the relationship between interoceptive awareness and alexithymia: Is interoceptive awareness related to emotional awareness? *Journal of Personality*, **79**(5), 1149–1175.

Heshmati, R., Jafari, E., Hoseinifar, J., & Ahmadi, M. (2010). Comparative study of alexithymia in patients with schizophrenia spectrum disorders, non-psychotic disorders and normal people. *Procedia: Social and Behavioral Sciences*, **5**, 1084–1089.

Hill, E., Berthoz, S., & Frith, U. (2004). Brief report: Cognitive processing of own emotions in individuals with autistic spectrum disorder and in their relatives. *Journal of Autism and Developmental Disorders*, **34**(2), 229–235.

Hogeveen, J., Bird, G., Chau, A., Krueger, F., & Grafman, J. (2016). Acquired alexithymia following damage to the anterior insula. *Neuropsychologia*, **82**, 142–148.

Honkalampi, K., Hintikka, J., Tanskanen, A., Lehtonen, J., & Viinamäki, H. (2000). Depression is strongly associated with alexithymia in the general population. *Journal of Psychosomatic Research*, **48**(1), 99–104.

Honkalampi, K., Hintikka, J., Laukkanen, E., & Viinamäki, J. L. H. (2001). Alexithymia and depression: A prospective study of patients with major depressive disorder. *Psychosomatics*, **42**(3), 229–234.

Honkalampi, K., Koivumaa-Honkanen, H., Lehto, S. M., et al. (2010). Is alexithymia a risk factor for major depression, personality disorder, or alcohol use disorders? A prospective population-based study. *Journal of Psychosomatic Research*, **68**(3), 269–273.

Ibañez, A., Gleichgerrcht, E., & Manes, F. (2010). Clinical effects of insular damage in humans. *Brain Structure and Function*, **214**(5–6), 397–410.

Ihme, K., Dannlowski, U., Lichev, V., et al. (2013). Alexithymia is related to differences in gray matter volume: A voxel-based morphometry study. *Brain Research*, **1491**, 60–67.

Jardri, R., Pins, D., Lafargue, G., et al. (2011). Increased overlap between the brain areas involved in self-other distinction in schizophrenia. *PLoS ONE*, **6**(3), e17500.

Jessimer, M., & Markham, R. (1997). Alexithymia: A right hemisphere dysfunction specific to recognition of certain facial expressions? *Brain and Cognition*, **34**(2), 246–258.

Jongen, S., Axmacher, N., Kremers, N. A., et al. (2014). An investigation of facial emotion recognition impairments in alexithymia and its neural correlates. *Behavioural Brain Research*, **271**, 129–139.

Jørgensen, M. M., Zachariae, R., Skytthe, A., & Kyvik, K. (2007). Genetic and environmental factors in alexithymia: A population-based study of 8,785 Danish twin pairs. *Psychotherapy and Psychosomatics*, **76**(6), 369–375.

Joukamaa, M., Kokkonen, P., Veijola, J., et al. (2003). Social situation of expectant mothers and alexithymia 31 years later in their offspring: A prospective study. *Psychosomatic Medicine*, **65**(2), 307–312.

Kano, M., & Fukudo, S. (2013). The alexithymic brain: The neural pathways linking alexithymia to physical disorders. *BioPsychoSocial Medicine*, **7**(1), 1.

Kano, M., Fukado, S., Jiro, G., et al. (2003). Specific brain processing of facial expressions in people with alexithymia: An $H_2^{15}O$-PET study. *Brain*, **126**(6), 1474–1484.

Kano, M., Hamaguchi, T., Itoh, M., Yanai, K., & Fukudo, S. (2007). Correlation between alexithymia and hypersensitivity to visceral stimulation in human. *Pain*, **132**(3), 252–263.

Karukivi, M., Pölönen, T., Vahlberg, T., Saikkonen, S., & Saarijärvi, S. (2014). Stability of alexithymia in late adolescence: Results of a 4-year follow-up study. *Psychiatry Research*, **219**(2), 386–390.

Kaye, W. (2008). Neurobiology of anorexia and bulimia nervosa. *Physiology & Behavior*, **94**(1), 121–135.

Keating, C., Tilbrook, A. J., Rossell, S. L., Enticott, P. G., & Fitzgerald, P. B. (2012). Reward processing in anorexia nervosa. *Neuropsychologia*, **50**(5), 567–575.

Keller, M. C., & Miller, G. (2006). Resolving the paradox of common, harmful, heritable mental disorders: Which evolutionary genetic models work best? *Behavioral and Brain Sciences*, **29**(4), 385–452.

Khalsa, S. S., & Lapidus, R. C. (2016). Can interoception improve the pragmatic search for biomarkers in psychiatry? *Frontiers in Psychiatry*, **7**, 121.

Khalsa, S. S., Rudrauf, D., Feinstein, J. S., & Tranel, D. (2009a). The pathways of interoceptive awareness. *Nature Neuroscience*, **12**(12), 1494–1496.

Khalsa, S. S., Rudrauf, D., Sandesara, C., Olshansky, B., & Tranel, D. (2009b). Bolus isoproterenol infusions provide a reliable method for assessing interoceptive awareness. *International Journal of Psychophysiology*, **72**(1), 34–45.

Kober, H., Barrett, L. F., Joseph, J., et al. (2008). Functional grouping and cortical-subcortical interactions in emotion: A meta-analysis of neuroimaging studies. *NeuroImage*, **42**(2), 998–1031.

Kohls, G., Schulte-Ruther, M., Nehrkorn, B., et al. (2013). Reward system dysfunction in autism spectrum disorders. *Social Cognitive and Affective Neuroscience*, **8**(5), 565–572.

Kokkonen, P., Karvonen, J. T., Veijola, J., et al. (2001). Prevalence and sociodemographic correlates of alexithymia in a population sample of young adults. *Comprehensive Psychiatry*, **42**(6), 471–476.

Koob, G. F., & Le Moal, M. (2001). Drug addiction, dysregulation of reward, and allostasis. *Neuropsychopharmacology*, **24**(2), 97–129.

Kooiman, C. G., van Rees Vellinga, S., Spinhoven, P., et al. (2004). Childhood adversities as risk factors for alexithymia and other aspects of affect dysregulation in adulthood. *Psychotherapy and Psychosomatics*, **73**(2), 107–116.

Koven, N. S., Roth, R. M., Garlinghouse, M. A., Flashman, L. A., & Saykin, A. J. (2011). Regional gray matter correlates of perceived emotional intelligence. *Social Cognitive and Affective Neuroscience*, **6**(5), 582–590.

Krolak-Salmon, P., Hénaff, M. A., Isnard, J., et al. (2003). An attention modulated response to disgust in human ventral anterior insula. *Annals of Neurology*, **53**(4), 446–453.

Kugel, H., Eichmann, M., Dannlowski, U., et al. (2008). Alexithymic features and automatic amygdala reactivity to facial emotion. *Neuroscience Letters*, **435**(1), 40–44.

Lahey, B. B., Applegate, B., Hakes, J. K., et al. (2012). Is there a general factor of prevalent psychopathology during adulthood? *Journal of Abnormal Psychology*, **121**(4), 971–977.

Lane, R. D., Quinlan, D. M., Schwartz, G. E., Walker, P. A., & Zeitlin, S. B. (1990). The Levels of Emotional Awareness Scale: A cognitive–developmental measure of emotion. *Journal of Personality Assessment*, **55**(1–2), 124–134.

Lane, R. D., Sechrest, L., Riedel, R., Shapiro, D. E., & Kaszniak, A. W. (2000). Pervasive emotion recognition deficit common to alexithymia and the repressive coping style. *Psychosomatic Medicine*, **62**(4), 492–501.

Lane, R. D., Weihs, K. L., Herring, A., Hishaw, A., & Smith, R. (2015). Affective agnosia: Expansion of the alexithymia construct and a new opportunity to integrate and extend Freud's legacy. *Neuroscience & Biobehavioral Reviews*, **55**, 594–611.

Larson, M. J., South, M., Krauskopf, E., Clawson, A., & Crowley, M. J. (2011). Feedback and reward processing in high-functioning autism. *Psychiatry Research*, **187**(1–2), 198–203.

Lazarov, A., Dar, R., Oded, Y., & Liberman, N. (2010). Are obsessive–compulsive tendencies related to reliance on external proxies for internal states? Evidence from biofeedback-aided relaxation studies. *Behaviour Research and Therapy*, **48**(6), 516–523.

Leekam, S. R., Nieto, C., Libby, S. J., Wing, L., & Gould, J. (2007). Describing the sensory abnormalities of children and adults with autism. *Journal of Autism and Developmental Disorders*, **37**(5), 894–910.

Leighton, J., Bird, G., Orsini, C., & Heyes, C. (2010). Social attitudes modulate automatic imitation. *Journal of Experimental Social Psychology*, **46**(6), 905–910.

Lilenfeld, L. R. R., Wonderlich, S., Riso, L. P., Crosby, R., & Mitchell, J. (2006). Eating disorders and personality: A methodological and empirical review. *Clinical Psychology Review*, **26**(3), 299–320.

Lindquist, K. A., Wager, T. D., Kober, H., Bliss-Moreau, E., & Barrett, L. F. (2012). The brain basis of emotion: A meta-analytic review. *Behavioral and Brain Sciences*, **35**(3), 121–143.

Livesey, A. C., Wall, M. B., & Smith, A. T. (2007). Time perception: Manipulation of task difficulty dissociates clock functions from other cognitive demands. *Neuropsychologia*, **45**(2), 321–331.

Longarzo, M., D'Olimpio, F., Chiavazzo, A., et al. (2015). The relationships between interoception and alexithymic trait. The Self-Awareness Questionnaire in healthy subjects. *Frontiers in Psychology*, **6**, 1149.

Luminet, O., Vermeulen, N., Demaret, C., Taylor, G. J., & Bagby, R. M. (2006). Alexithymia and levels of processing: Evidence for an overall deficit in remembering emotion words. *Journal of Research in Personality*, **40**(5), 713–733.

Lyvers, M., Hasking, P., Albrecht, B., & Thorberg, F. A. (2012). Alexithymia and alcohol: The roles of punishment sensitivity and drinking motives. *Addiction Research & Theory*, **20**(4), 348–357.

Lyvers, M., Duric, N., & Thorberg, F. A. (2014). Caffeine use and alexithymia in university students. *Journal of Psychoactive Drugs*, **46**(4), 340–346.

Magnani, B., Frassinetti, F., Ditye, T., et al. (2014). Left insular cortex and left SFG underlie prismatic adaptation effects on time perception: Evidence from fMRI. *NeuroImage*, **92**, 340–348.

Mann, L., Wise, T., Trinidad, A., & Kohanski, R. (1995). Alexithymia, affect recognition, and five factors of personality in substance abusers. *Perceptual and Motor Skills*, **81**(1), 35–40.

Marchesi, C., Brusamonti, E., & Maggini, C. (2000). Are alexithymia, depression, and anxiety distinct constructs in affective disorders? *Journal of Psychosomatic Research*, **49**(1), 43–49.

Mason, M. F., Norton, M. I., Van Horn, J. D., et al. (2007). Wandering minds: The default network and stimulus-independent thought. *Science*, **315**(5810), 393–395.

Mattila, A. K., Salminen, J. K., Nummi, T., & Joukamaa, M. (2006). Age is strongly associated with alexithymia in the general population. *Journal of Psychosomatic Research*, **61**(5), 629–635.

McClenon, J. (2011). Evolutionary theories of schizophrenia: An experience-centered review. *Journal of Mind and Behavior*, **32**(2), 135–150.

McDonald, P. W., & Prkachin, K. M. (1990). The expression and perception of facial emotion in alexithymia: A pilot study. *Psychosomatic Medicine*, **52**(2), 199–210.

Mezzavilla, M., Ulivi, S., Bianca, M. L., et al. (2015). Analysis of functional variants reveals new candidate genes associated with alexithymia. *Psychiatry Research*, **227**(2–3), 363–365.

Miyake, Y., Okamoto, Y., Onoda, K., et al. (2012). Brain activation during the perception of stressful word stimuli concerning interpersonal relationships in anorexia nervosa patients with high degrees of alexithymia in an fMRI paradigm. *Psychiatry Research – Neuroimaging*, **201**(2), 113–119.

Moormann, P. P., Bermond, B., Vorst, H. C. M., et al. (2008). New avenues in alexithymia research: The creation of alexithymia types. In A. Vingerhoets, I. Nyklicek, & J. Denollet, eds., *Emotion Regulation*. New York: Springer, pp. 27–42.

Moriguchi, Y., & Komaki, G. (2013). Neuroimaging studies of alexithymia: Physical, affective, and social perspectives. *BioPsychoSocial Medicine*, **7**(1), 8.

Moriguchi, Y., Decety, J., Ohnishi, T., et al. (2007). Empathy and judging other's pain: An fMRI study of alexithymia. *Cerebral Cortex*, **17**(9), 2223–2234.

Mundy, P. (2003). Annotation: The neural basis of social impairments in autism: The role of the dorsal medial–frontal cortex and anterior cingulate system. *Journal of Child Psychology and Psychiatry*, **44**(6), 793–809.

Murphy, J., Brewer, R., & Bird, G. (2017). Interoception and psychopathology: A developmental neuroscience perspective. *Developmental Cognitive Neuroscience*, **23**, 45–56.

Murphy, J., Catmur, C., & Bird, G. (2018). Alexithymia is associated with a multi-domain, multi-dimensional failure of interoception: Evidence from novel tests. *Journal of Experimental Psychology: General*, **147**, 398–408.

Nagai, M., Kishi, K., & Kato, S. (2007). Insular cortex and neuropsychiatric disorders: A review of recent literature. *European Psychiatry*, **22**(6), 387–394.

Naqvi, N. H., & Bechara, A. (2009). The hidden island of addiction: The insula. *Trends in Neurosciences*, **32**(1), 56–67.

Naqvi, N. H., & Bechara, A. (2010). The insula and drug addiction: An interoceptive view of pleasure, urges and decision-making. *Brain Structure and Function*, **214**(5–6), 435–450.

Naqvi, N. H., Rudrauf, D., Damasio, H., & Bechara, A. (2007). Damage to the insula disrupts addiction to cigarette smoking. *Science*, **315**(5811), 531–534.

Nemiah, J. C., Freyberger, H. J., & Sifneos, P. E. (1976). Alexithymia: A view of the psychosomatic process. In O. W. Hill, ed., *Modern Trends in Psychosomatic Medicine*. London: Butterworths, pp. 430–439.

Neumann, D., Zupan, B., Malec, J. F., & Hammond, F. (2014). Relationships between alexithymia, affect recognition, and empathy after traumatic brain injury. *Journal of Head Trauma Rehabilitation*, **29**(1), E18–E27.

Noble, E. P. (2000). Addiction and its reward process through polymorphisms of the D2 dopamine receptor gene: A review. *European Psychiatry*, **15**(2), 79–89.

Nowakowski, M. E., McFarlane, T., & Cassin, S. (2013). Alexithymia and eating disorders: A critical review of the literature. *Journal of Eating Disorders*, **1**(1), 21.

Oldershaw, A., Hambrook, D., Stahl, D., et al. (2011). The socio-emotional processing stream in anorexia nervosa. *Neuroscience and Biobehavioral Reviews*, **35**(3), 970–988.

Paradiso, S., Vaidaya, J. G., McCormick, L. M., Jones, A., & Robinson, R. G. (2008). Aging and alexithymia association with reduced right rostral cingulate volume. *American Journal of Geriatric Psychiatry*, **16**(9), 760–769.

Parker, J. D. A., Bagby, R. M., & Taylor, G. J. (1991). Alexithymia and depression: Distinct or overlapping constructs? *Comprehensive Psychiatry*, **32**(5), 387–394.

Parker, J. D. A., Taylor, G. J., & Bagby, R. (1993). Alexithymia and the recognition of facial expressions of emotion. *Psychotherapy and Psychosomatics*, **59**(3–4), 197–202.

Parker, J. D. A., Taylor, G. J., & Bagby, R. M. (2003). The 20-Item Toronto Alexithymia Scale III. Reliability and factorial validity in a community population. *Journal of Psychosomatic Research*, **55**(3), 269–275.

Parker, J. D. A., Wood, L. M., Bond, B. J., & Shaughnessy, P. (2005). Alexithymia in young adulthood: A risk factor for pathological gambling. *Psychotherapy and Psychosomatics*, **74**(1), 51–55.

Parker, P. D., Prkachin, K. M., & Prkachin, G. C. (2005). Processing of facial expressions of negative emotion in alexithymia: The influence of temporal constraint. *Journal of Personality*, **73**(4), 1087–1107.

Patil, I., & Silani, G. (2014a). Alexithymia increases moral acceptability of accidental harms. *Journal of Cognitive Psychology*, **26**(5), 1–18.

Patil, I., & Silani, G. (2014b). Reduced empathic concern leads to utilitarian moral judgments in trait alexithymia. *Frontiers in Psychology*, **5**, 501.

Paulus, M. P., & Stein, M. B. (2006). An insular view of anxiety. *Biological Psychiatry*, **60**(4), 383–387.

Penfield, W., & Faulk, M. E. (1955). The insula: Further observations on its function. *Brain*, **78**(4), 445–470.

Phan, K. L., Wager, T., Taylor, S. F., & Liberzon, I. (2002). Functional neuroanatomy of emotion: A meta-analysis of emotion activation studies in PET and fMRI. *NeuroImage*, **16**(2), 331–348.

Picardi, A., Fagnani, C., Gigantesco, A., et al. (2011). Genetic influences on alexithymia and their relationship with depressive symptoms. *Journal of Psychosomatic Research*, **71**(4), 256–263.

Pollatos, O., Kurz, A. L., Albrecht, J., et al. (2008). Reduced perception of bodily signals in anorexia nervosa. *Eating Behaviors*, **9**(4), 381–388.

Pollatos, O., Herbert, B. M., Mai, S., & Kammer, T. (2016). Changes in interoceptive processes following brain stimulation. *Philosophical Transactions of the Royal Society of London B: Biological Sciences*, **371**, 20160016.

Pouga, L., Berthoz, S., De Gelder, B., & Grèzes, J. (2010). Individual differences in socioaffective skills influence the neural bases of fear processing: The case of alexithymia. *Human Brain Mapping*, **31**(10), 1469–1481.

Press, C., Richardson, D., & Bird, G. (2010). Intact imitation of emotional facial actions in autism spectrum conditions. *Neuropsychologia*, **48**(11), 3291–3297.

Prkachin, G. C., Casey, C., & Prkachin, K. M. (2009). Alexithymia and perception of facial expressions of emotion. *Personality and Individual Differences*, **46**(4), 412–417.

Prochnow, D., Donell, J., Schäfer, R., et al. (2011). Alexithymia and impaired facial affect recognition in multiple sclerosis. *Journal of Neurology*, **258**(9), 1683–1688.

Quattrocki, E., & Friston, K. (2014). Autism, oxytocin and interoception. *Neuroscience and Biobehavioral Reviews*, **47**, 410–430.

Reker, M., Ohrmann, P., Rauch, A. V., et al. (2010). Individual differences in alexithymia and brain response to masked emotion faces. *Cortex*, **46**(5), 658–667.

Reuter, J., Raedler, T., Rose, M., et al. (2005). Pathological gambling is linked to reduced activation of the mesolimbic reward system. *Nature Neuroscience*, **8**(2), 147–148.

Rogers, M. A., Yamasue, H., Abe, O., et al. (2009). Smaller amygdala volume and reduced anterior cingulate gray matter density associated with history of post-traumatic stress disorder. *Psychiatry Research: Neuroimaging*, **174**(3), 210–216.

Säkkinen, P., Kaltiala-Heino, R., Ranta, K., Haataja, R., & Joukamaa, M. (2007). Psychometric properties of the 20-item Toronto Alexithymia Scale and prevalence of alexithymia in a Finnish adolescent population. *Psychosomatics*, **48**(2), 154–161.

Salminen, J. K., Saarija, S., & Rela, E. A. A. (1999). Prevalence of alexithymia and its association with sociodemographic variables in the general population of Finland. *Journal of Psychosomatic Research*, **46**(1), 75–82.

Sass, L. A., & Parnas, J. (2003). Schizophrenia, consciousness, and the self. *Schizophrenia Bulletin*, **29**(3), 427–444.

Schandry, R., & Weitkunat, R. (1990). Enhancement of heartbeat-related brain potentials through cardiac awareness training. *International Journal of Neuroscience*, **53**(2–4), 243–253.

Schmitz, N., Rubia, K., Van Amelsvoort, T., et al. (2008). Neural correlates of reward in autism. *British Journal of Psychiatry*, **192**(1), 19–24.

Schoenbaum, G., Roesch, M. R., & Stalnaker, T. A. (2006). Orbitofrontal cortex, decision-making and drug addiction. *Trends in Neurosciences*, **29**(2), 116–124.

Seth, A. K., Suzuki, K., & Critchley, H. D. (2011). An interoceptive predictive coding model of conscious presence. *Frontiers in Psychology*, **2**, 395.

Shah, P., Catmur, C., & Bird, G. (2016). Emotional decision-making in autism spectrum disorder: the roles of interoception and alexithymia. *Molecular Autism*, **7**(1), 43.

Shah, P., Hall, R., Catmur, C., & Bird, G. (2016). Alexithymia, not autism, is associated with impaired interoception. *Cortex*, **81**, 215–220.

Sherrington, C. S. (1900). Cutaneous sensations. In E. A. Schäfer, ed., *Text-Book of Physiology*. Edinburgh: Pentland, pp. 920–1001.

Shishido, H., Gaher, R. M., & Simons, J. S. (2013). I don't know how I feel, therefore I act: Alexithymia, urgency, and alcohol problems. *Addictive Behaviors*, **38**(4), 2014–2017.

Silani, G., Bird, G., Brindley, R., et al. (2008). Levels of emotional awareness and autism: An fMRI study. *Social Neuroscience*, **3**(2), 97–112.

Simon, J. J., Biller, A., Walther, S., et al. (2010). Neural correlates of reward processing in schizophrenia – Relationship to apathy and depression. *Schizophrenia Research*, **118**(1–3), 154–161.

Singer, T., & Lamm, C. (2009). The social neuroscience of empathy. *Annals of the New York Academy of Sciences*, **1156**, 81–96.

Sowden, S., Brewer, R., Catmur, C., & Bird, G. (2016). The specificity of the link between alexithymia, interoception and imitation. *Journal of Experimental Psychology: Human Perception and Performance*, **42**(11), 1687–1692.

Stasiewicz, P. R., Bradizza, C. M., Gudleski, G. D., et al. (2012). The relationship of alexithymia to emotional dysregulation within an alcohol dependent treatment sample. *Addictive Behaviors*, **37**(4), 469–476.

Stephan, E., Pardo, J. V., Faris, P. L., et al. (2003). Functional neuroimaging of gastric distention. *Journal of Gastrointestinal Surgery*, **7**(6), 740–749.

Stern, E. R. (2014). Neural circuitry of interoception: New insights into anxiety and obsessive–compulsive disorders. *Current Treatment Options in Psychiatry*, **1**, 235–247.

Sturm, V. E., & Levenson, R. W. (2011). Alexithymia in neurodegenerative disease. *Neurocase*, **17**(3), 242–250.

Suslow, T., & Junghanns, K. (2002). Impairments of emotion situation priming in alexithymia. *Personality and Individual Differences*, **32**(3), 541–550.

Swart, M., Kortekaas, R., & Aleman, A. (2009). Dealing with feelings: Characterization of trait alexithymia on emotion regulation strategies and cognitive-emotional processing. *PLoS ONE*, **4**(6), e5751.

Tanaka, S. C., Doya, K., Okada, G., et al. (2004). Prediction of immediate and future rewards differentially recruits cortico-basal ganglia loops. *Nature Neuroscience*, **7**(8), 887–893.

Taylor, G. J., Parker, J. D. A., & Bagby, R. M. (1990). A preliminary investigation of alexithymia in men with psychoactive substance dependence. *American Journal of Psychiatry*, **147**(9), 1228–1230.

Tomchek, S. D., & Dunn, W. (2007). Sensory processing in children with and without autism: A comparative study using the short sensory profile. *American Journal of Occupational Therapy*, **61**(2), 190–200.

Torrente, F., Ashwood, P., Day, R., et al. (2002). Small intestinal enteropathy with epithelial IgG and complement deposition in children with regressive autism. *Molecular Psychiatry*, **7**(4), 375–382.

Ugazio, G., Lamm, C., & Singer, T. (2012). The role of emotions for moral judgments depends on the type of emotion and moral scenario. *Emotion*, **12**(3), 579–590.

Valera, E. M., & Berenbaum, H. (2001). A twin study of alexithymia. *Psychotherapy and Psychosomatics*, **70**(5), 239–246.

Vandenbergh, J., Dupont, P., Fischler, B., et al. (2005). Regional brain activation during proximal stomach distention in humans: A positron emission tomography study. *Gastroenterology*, **128**(3), 564–573.

van der Velde, J., Servaas, M. N., Goerlich, K. S., et al. (2013). Neural correlates of alexithymia: A meta-analysis of emotion processing studies. *Neuroscience and Biobehavioral Reviews*, **37**(8), 1774–1785.

van der Velde, J., Swart, M., van Rijn, S., et al. (2015). Cognitive alexithymia is associated with the degree of risk for psychosis. *PLoS ONE*, **10**(6), e0124803.

Venta, A., Hart, J., & Sharp, C. (2013). The relation between experiential avoidance, alexithymia and emotion regulation in inpatient adolescents. *Clinical Child Psychology and Psychiatry*, **18**(3), 398–410.

Verdejo-Garcia, A., Clark, L., & Dunn, B. D. (2012). The role of interoception in addiction: A critical review. *Neuroscience and Biobehavioral Reviews*, **36**(8), 1857–1869.

Vorst, H. C. M., & Bermond, B. (2001). Validity and reliability of the Bermond–Vorst Alexithymia Questionnaire. *Personality and Individual Differences*, **30**(3), 413–434.

Wagner, A., Aizenstein, H., Venkatraman, V. K., et al. (2007). Altered reward processing in women recovered from anorexia nervosa. *American Journal of Psychiatry*, **164**(12), 1842–1849.

Wakefield, A. J., Anthony, A., Murch, S. H., et al. (2000). Enterocolitis in children with developmental disorders. *American Journal of Gastroenterology*, **95**(9), 2285–2295.

Wakefield, A. J., Ashwood, P., Limb, K., & Andrew, A. (2005). The significance of ileo-colonic lymphoid nodular hyperplasia in children with autistic spectrum disorder. *European Journal of Gastroenterology and Hepatology*, **17**, 827–836.

Walter, N. T., Montag, C., Markett, S. A., & Reuter, M. (2011). Interaction effect of functional variants of the BDNF and DRD2/ANKK1 gene is associated with alexithymia in healthy human subjects. *Psychosomatic Medicine*, **73**(1), 23–28.

Weissman, D. H., Roberts, K. C., Visscher, K. M., & Woldorff, M. G. (2006). The neural bases of momentary lapses in attention. *Nature Neuroscience*, **9**(7), 971–978.

Werner, N. S., Jung, K., Duschek, S., & Schandry, R. (2009). Enhanced cardiac perception is associated with benefits in decision-making. *Psychophysiology*, **46**(6), 1123–1129.

White, J. F. (2003). Intestinal pathophysiology in autism. *Experimental Biology and Medicine*, **228**(6), 639–649.

Williams, B. L., Hornig, M., Buie, T., et al. (2011). Impaired carbohydrate digestion and transport and mucosal dysbiosis in the intestines of children with autism and gastrointestinal disturbances. *PLoS ONE*, **6**(9), e24585.

Williams, C., & Wood, R. L. (2010). Alexithymia and emotional empathy following traumatic brain injury. *Journal of Clinical and Experimental Neuropsychology*, **32**(3), 259–267.

Wittmann, M., Lovero, K. L., Lane, S. D., & Paulus, M. P. (2010). Now or later? Striatum and insula activation to immediate versus delayed rewards. *Journal of Neuroscience, Psychology, and Economics*, **3**(1), 15–26.

Zhang, X., Salmeron, B. J., Ross, T. J., et al. (2011). Factors underlying prefrontal and insula structural alterations in smokers. *Neuroimage*, **54**(1), 42–48.

Zotev, V., Krueger, F., Phillips, R., et al. (2011). Self-regulation of amygdala activation using real-time fMRI neurofeedback. *PLoS ONE*, **6**(9), e24522.

PART IX

APPLYING EVOLUTIONARY PRINCIPLES

As we have seen in previous chapters, many consider that the current Darwinian revolution has led to a big advancement in our understanding of the ultimate bases of human behavior and internal states. Through the work of evolutionists of various flavors, we now have a greater understanding of sexual strategies (and conflict), jealousy, the function of social anxiety, kin recognition, the functions of guilt, shame and deception, and how and why we make social judgments, to name but a few. Arguably of greater importance is that evolutionists have provided a metatheory for psychology, social–cultural anthropology, and behavioral biology that allows us to unite previously disparate fields of research and potentially end debates in which experts talk past each other. So can we "evolutionists" now relax and pat ourselves on the back? Perhaps not. Although the amount of research published in evolutionary psychology and human behavioral ecology has grown exponentially over the last 30 years, a relatively small proportion of this output can be described as "applied." This raises the question: In addition to understanding the roots of the human condition, might we not also use this understanding to tackle contemporary issues in human society and perhaps inform practice? In recent years, a growing number of researchers have begun to develop an applied evolutionary psychology (Barclay, 2012; Buunk & Van Vugt, 2013; Roberts, 2012). In Part IX, we examine a number of varied topics that have been brought together under the broad umbrella term of "applying evolutionary principles."

We begin with Abraham P. Buunk's consideration of whether or not evolutionary psychology can be used to provide solutions for social problems. Buunk, who is an internationally recognized expert on the relationship between evolution and social behavior, suggests that, while there are problematic theoretical debates among evolutionists, such approaches might still prove helpful in three ways: first, they provide insight into the evolutionary background of a social problem; second, they suggest which proximate variables contribute to specific problematic behaviors (which may then be considered when planning interventions); and third, evolutionary approaches can determine which behaviors are likely to be most difficult to change given their

adaptive significance. Despite these potentially positive factors, Buunk suggests evolutionary psychology may be most effective when linked with existing well-developed approaches, such as those used by social psychologists.

In our second contribution, Adrian Furnham and Satoshi Kanazawa consider the development of our understanding of personality factors within an evolutionary context. Furnham is an applied psychologist and is the most published living psychologist. Kanazawa coined the term "savanna principle," and this principle suggests that many of the societal difficulties human currently face arise because we are adapted to a savanna environment rather than the ones we encounter today. In this chapter, Furnham and Kanazawa examine the Five Factor model of personality in the light of developments in evolutionary psychology and behavior genetics. Whereas personality psychologists normally classify and describe individual differences, Furnham and Kanazawa suggest that applying evolutionary and behavioral genetics approaches may help us to explain both how and why such systematic individual differences arise.

Sandie Taylor has published two well-received books on criminological and forensic psychology and Lance Workman is well known for his coauthored books (with Will Reader) on evolutionary psychology. In their contribution, Taylor and Workman consider how evolutionary principles can be used to help explain criminality. Whereas social scientists explain criminal behavior on the bases of reactions to unfavorable societal factors, Taylor and Workman suggest that we can make use of well-developed evolutionary principles in order to generate a better understanding of why some people respond to these social factors with criminality. Such evolutionary principles include kin selection theory, reciprocation, sexual selection, and asymmetrical parental investment theory.

A modern understanding of Darwinism suggests that individuals have evolved to be inclusive fitness maximizers. Yet members of our own species regularly act as substitute parents, rearing the offspring of others. This universally observed practice appears to run counter to notions of promoting inclusive fitness. In their contribution to Part

IX, Martin Daly and Gretchen Perry examine this phenomenon. Daly, who is an influential founding figure in evolutionary psychology (Daly & Wilson, 1983, 1998), and Perry, who is an expert in adoption and caregiving, divide substitute parenting up into three categories: first, we see the substitute parenting of genetic relatives; second, we see stepparenting; and third, we see the adoption of nonrelatives. The first of these is perfectly conducive with kin section theory, and the second consists of part of the package that we take on when forming a reconstituted family, but it is the third that presents the greatest challenge to simple conceptions of evolutionary psychology. Daly and Perry suggest that this conundrum may be resolved in part by considering this custom as layered on top of previously exploitative practices that are now less likely given modern-day legislative screening of applicants.

Practicing historians are apt to treat human behavior as subject to constant change, with observable dates when such changes occur. Gregory Hanlon is a different kind of historian. As an expert on evolutionary approaches, he factors into his flavor of history the notion of an evolved human nature. To Hanlon, as a result of human adaptations, many responses change slowly or not at all. This notion of biologically based human universals helps to explain the extraordinary similarity of motivations and situations we observe over historical time. Because it embraces all facets of human activity, Hanlon suggests that evolutionary thinking can provide a powerful tool to examine a large range of historical problems. In his contribution, he considers four areas (reproduction, interpersonal conflict and violence, war, and aesthetics), each in some detail.

We end Part IX with a contribution from one of our editors, Jerome H. Barkow, as he considers the psychology of extraterrestrials. Barkow, who is a founder of evolutionary psychology (Barkow, 1989; Barkow, Cosmides, & Tooby, 1992), has developed a special interest in the search for extraterrestrial life. While this contribution is, by its nature, speculative, it is informed by what we know about the evolution of intelligent life on Earth. Barkow suggests that, if intelligent life-forms do exist elsewhere in the universe (and most experts consider this likely), then we will eventually encounter them. Prior to that occurrence, he argues, evolutionary experts will be best placed to at the very least ask the appropriate questions about such life-forms. In a wide-ranging chapter, Barkow makes a number of thought-provoking points. For example, we are apt to visualize alien species as being a psychologically uniform groups – all pretty much wearing, say, a "good guy" or "bad guy" hat. Yet we know from evolutionarily informed studies that there are good adaptive reasons why we should expect to observe a range of personality types in any intelligent species. The more we understand about how and why our own carbon-based form of intelligent life evolved, the better we will be able to predict the range of potential extraterrestrial intelligences.

REFERENCES

Barclay, P. (2012). The evolution of charitable behaviour and the power of reputation. In S. C. Roberts, ed., *Applied Evolutionary Psychology*. Oxford: Oxford University Press, pp. 149–172.

Barkow, J. H. (1989). *Darwin, Sex, and Status: Biological Approaches to Mind and Culture*. Toronto: University of Toronto Press.

Barkow, J. H., Cosmides, L. & Tooby, J., eds. (1992). *The Adapted Mind: Evolutionary Psychology and the Generation of Culture*. Oxford/New York: Oxford University Press.

Buunk, A. P., & Van Vugt, M. (2013). *Applying Social Psychology. From Problem to Solution*. London: Sage.

Daly, M. & Wilson, M. (1983). *Sex, Evolution and Behaviour*, 2nd ed. Belmont, CA: Wadsworth.

Daly, M., & Wilson, M. (1998). *The Truth about Cinderella: A Darwinian View of Parental Love*. London: Weidenfeld & Nicolson.

Roberts, S. C., ed. (2012). *Applied Evolutionary Psychology*. Oxford: Oxford University Press.

36 A Bridge Too Far?

Evolutionary Psychology and the Solution of Social Problems

ABRAHAM P. BUUNK

36.1 CRITIQUE ON THE APPLICATION OF THE TRADITIONAL SOCIAL AND BEHAVIORAL SCIENCES

The theme for this chapter as formulated by the editors was: "Can evolutionary psychology be used to solve social problems?" For the sake of simplicity, I use here a broad definition of evolutionary psychology as the evolutionary approach to human behavior, including approaches that have often been set apart, such as human sociobiology and human behavioral ecology (Laland & Brown, 2011). Although the question of the applicability of evolutionary thinking has occupied me for years and although I have applied such thinking to a variety of social problems, including occupational burnout, jealousy, and depression, I found it a real challenge to write a chapter on this issue. In fact, as I will argue, my answer to this question is not unequivocally positive. Instead, I will argue that evolutionary psychology does not provide clear solutions to all kinds of social problems. However, it may contribute to insights into the ultimate causes of many social problems, it may have a heuristic value in suggesting possible proximate causes, and in particular it may be useful in identifying those variables that will be difficult to change, given their adaptive significance.

My argument may not be generally accepted, as many evolutionary psychologists tend to think that the rise of their discipline represents a new Enlightenment in the social and behavioral sciences that will foster effective solutions to many social problems. According to Tooby and Cosmides (2015), in the past decades, ignoring, exceptionalizing, or explaining away evolutionary thinking by social and behavioral scientists has been severe. They suggested that "we may all have been complicit in the perpetuation of vast tides of human suffering – suffering that might have been prevented if the scientific community had not chosen to postpone or forgo a more veridical social and behavioral science" (p. 7), where the latter refers to a science based on evolutionary psychology. The obvious assumption here is that evolutionary psychology could help to reduce human suffering to a greater degree than traditional approaches in the social and behavioral sciences, and, moreover, that the latter – at least in part – bear the responsibility for the continuation of human suffering.

This assumption is related to the well-known critique of what Tooby and Cosmides (1992, 2015) refer to as the Standard Social Science Model, which they consider as completely wrong and misconceived. This model portrays the human mind as a blank slate on which culture and learning can write anything imaginable. It assumes that humans are infinitely malleable and shaped entirely by culture through a small number of general-purpose learning mechanisms. According to Tooby and Cosmides, this perspective has been strongly moralized within the scholarly world and, in the past decades, has been characterized by a strong resistance to acknowledging the new insights from evolutionary psychology. However, as noted by Barrett (2017), "Whether anyone, anywhere, ever subscribed to this grossly simplistic blank slate view is open to debate, but it served its rhetorical purpose" (p. 7). In fact, there have always been a large variety of perspectives within the social and behavioral sciences. The critique by Tooby and Cosmides seems in particular to concern anthropology that distanced itself in the beginning of the twentieth century from a biological approach. According to Franz Boas, the founder of cultural anthropology, this discipline had to focus on the study of culture, minimalizing the role of biological factors (even his most famous work was on the cultural malleability of skull size). The now controversial and clearly misinformed work by Boas' student Margaret Mead, especially on the assumed malleability of sex roles and on the sexual liberty among Samoan youngsters, could be found in many textbooks in the social and behavioral sciences (Buskes, 2006), and in this sense the Standard Social Science Model did indeed exist and was wrong in many ways, and this has resulted in all kinds of applications overestimating the effects of culture and upbringing. In part, this was due to strong resistance against the way in which the Nazis classified whole populations as biologically inferior.

Indeed, to some extent, Tooby and Cosmides (2005/2016) make an important point when arguing that the classic social sciences have contributed to the perseverance of human suffering by assuming the mind as a completely blank slate. As an extreme example, one might argue – with reason – that communist regimes were based on the assumption that the human mind was completely malleable and that society could strongly influence, if not completely change human nature. Indeed, many social scientists have been guilty of actively supporting this notion and the dictatorial regimes based on it. For example, the Dutch sociologist Willem Frederik Wertheim was an adamant admirer and supporter of Stalin, and later of Mao, and he virtually ignored the millions of deaths that occurred due to Mao's policies, first as a consequence of the forced collectivization of agriculture and later due to Mao's attempt to regain his power through the "Cultural Revolution." According to Wertheim, Mao's policies were an attempt to create a new type of human that was more altruistic and social (e.g., Wertheim, 1973). This may be an extreme example, but many students were influenced by Wertheim's ideas, and especially in sociology, radical Marxist ideas and communist sympathies were not at all uncommon in the past. Numerous humanistically oriented scientists have also strongly opposed the notion of an evolved human nature and expressed very optimistic views on the possibility of changing human behavior; for example, consider the Seville Statement on Violence, which states, among other things, that it is scientifically incorrect to say that we have inherited a tendency to make war from our animal ancestors. Ironically, this statement was made in a country that a couple of decades before had escaped from an extremely bloody civil war, and we can now quite certainly state that *that* statement was scientifically incorrect (e.g., Livingstone Smith, 2007; Wrangham & Peterson, 1996). By denying the potential evolutionary roots of intergroup violence, social and behavioral scientists have indeed often been naively optimistic about the possibility of eliminating or preventing such violence, may have overlooked the violent potential in intergroup conflicts, and may not have come up with the most appropriate social policies with respect to this issue.

36.2 DOES EVOLUTIONARY PSYCHOLOGY DO A BETTER JOB?

Despite the many misguided applications by "standard" social scientists, the history of applying Darwinian thinking to social problems is at least as problematic. At the end of the nineteenth and the beginning of the twentieth centuries, Darwinism *was* applied to social problems in the form of social Darwinism and eugenics (e.g., Plotkin, 2004). Like the work by Wertheim, these applications illustrate the role of political values in applying social and behavioral science. For Herbert Spencer, the founder of social Darwinism and coiner of the phrase "survival of the fittest," the greatest happiness for all was top priority. This might seem like a laudable humanist ideal, but according to Spencer,

progress could only be achieved by competition, and the role of the government should be limited to the protection of individual freedom and should not include help and support for the poor and the weak. Only the strongest would survive. Not surprisingly, Spencer's ideas evoked strong resistance among progressive and socialist societal reformers. A little later, in the first half of the twentieth century, eugenics was promoted as another application of Darwinism. According to Francis Galton, the human species could only progress if solely people with good characteristics were allowed to reproduce. Eugenics was originally viewed as a progressive idea, and it garnered much political and scientific support (among others from the Carnegie Foundation and scientists of Harvard), not least because many began to view the white race as vulnerable (Buskes, 2006; Plotkin, 2004). Particularly in the USA and Scandinavia, many genetically "inferior" individuals were sterilized. And, for example, the state of Alberta in Canada only repealed laws regarding eugenics and sterilization in 1972. Nazi Germany went a dramatic step further by killing around 200,000 mentally and physically handicapped individuals. In part as a consequence of these horrific acts, until long after World War II, genetic and biological approaches to human behavior were associated with Nazism.

In addition, recent applications of evolutionary psychology are not without their problems either. Beckstrom (1993), one of the first to explicitly address the application of recent Darwinistic thinking (in particular sociobiology) to social problems, suggested many solutions that may look quite awkward to our eyes today. For example, to prevent child abuse (which is committed relatively more often by stepfathers than by biological fathers), he suggested that single parents should marry a genetic relative, such as uncles or aunts and grandparents: "[T]he single parent's in-laws related to the child in those categories should typically be the best stepparent candidates … it should be better for children to have even fairly remote biological relatives as stepparents rather than strangers" (p. 27). While this may be considered as a well-intended advice to single parents, as a social policy it is completely unrealistic, if not unethical, and resistance to in-law marriage is very strong in Western societies (although it is accepted in many cultures; Buunk & Hoben, 2013). As another example, Beckstrom (1993) suggested to combat rape by increasing the "likelihood and/or severity of punishment for rape" (p. 54), referring to classical deterrence theory. This may be a reasonable proposal, as many men still get away with rape and women are often blamed for provoking it. Nevertheless, criminologists have noted for a long time that severe punishment often does not have a deterrent effect, and this would apply especially to sexual behavior with various partners, even when it may involve force, which seems to have evolved as a strong motive among males.

Politically and ethically, even more questionable policy recommendations have been made by well-known evolutionary psychologist Geoffrey Miller (2009) in his book about consumerism. While his ideas on the evolutionary background of conspicuous consumption are well grounded in research on sexual selection, his practical recommendations are rather bizarre. For example, Miller suggests that people should be obliged to tattoo their personality characteristics on their face to prevent them from using conspicuous consumption to show what type of person they are. Companies should be obliged to sell certain products only to consumers who score high or low on certain personality characteristics, or who show the "right" behavior. Certain condoms might only be sold to men who have regularly tested negatively for sexually transmitted diseases (one might ask: Why not to men who tested positively?). And we should not aim for ethnically or religiously mixed neighborhoods; people should live with like-minded people in the same neighborhood because only in this way can they correct each other's tendencies to consumerism. This looks like using evolutionary psychology to justify *apartheid*. Moreover, on the basis of social comparison theory, individuals would feel a stronger, not a weaker need to find ways to distinguish themselves from others when surrounded by similar people (Buunk & Gibbons, 2007). Miller also claims – without any evidence – that capitalism makes people unhappy. However, capitalism comes in many forms, and it is remarkable that, worldwide, the highest levels of happiness are found in liberal, capitalist countries like the Netherlands. Moreover, on the basis of this same research and theorizing, one might equally well conclude that conspicuous consumption is a part of human nature that we simply must accept, or that the wealth of nations increases when conspicuous consumption is encouraged. Moreover, Miller's recommendations seem to be based on a faulty interpretation of sexual selection theory. People do not use conspicuous consumption to display their personality, but to display their quality through the purchase of extremely expensive goods that are not necessarily useful and to indicate the group to which they belong.

36.3 THEORETICAL DISAGREEMENT WITHIN EVOLUTIONARY PSYCHOLOGY

The previous examples illustrate how confused evolutionary psychologists may become when they enter the realm of policy-making, and not without reason: there are many problems in applying evolutionary psychology to solving social problems. The first main problem is that evolutionary psychology is, though rapidly expanding, a new field, and many important issues are still subject to heavy theoretical debates and empirical controversies (e.g., Laland & Brown, 2011). Therefore, many topics lack a solid empirical basis for policy recommendations or for developing interventions. As noted by Barclay (2011), "Even among evolutionists, many bitter arguments have resulted from nothing more than differences in terminology or levels of analysis" (p. 150). For example, one relevant issue is the controversy in evolutionary psychology over individual

differences. According to Tooby and Cosmides (2005/2016), we all share a universal human nature, and individual differences are therefore just noise: "[T]raits caused by genetic variance are predominantly evolutionary noise, with little adaptive significance" (p. 39). This statement seems to ignore the vast literature on behavioral genetics, showing consistent, genetically determined individual difference. Moreover, in the animal kingdom, personality differences are widespread, especially in social species, and such differences may reflect distinct adaptive strategies (Buss, 1999; Figueredo et al., 2005). For example, there is considerable evidence for the existence of large individual differences in intrasexual competitiveness, which, among males, is independently predicted by extraversion and neuroticism – precisely the two characteristics that, according to Figueredo et al. (2005), were the first personality differences to emerge in freely moving species (Buunk & Fisher, 2009).

Additional debates concern, among other topics: whether people in modern society aim to enhance their reproductive success; whether reproductive success should be the "gold standard" by which to assess the functionality of behaviors; whether something like group selection can or does exist; and whether individuals do indeed have specialized mental mechanisms to solve different problems of survival and reproduction. With respect to the latter issue, Buunk and Park (2008) have argued that the human mind is characterized not by massive but by messy modularity: "Rather than being equipped with numerous specialized modules unique to the human species, many adaptive mechanisms were built on ancient mechanisms … instead of 'massively modular,' the human mind is better conceptualized as a mixed bag of mutual linked mechanisms and traits that reveal compromise and imperfection" (pp. 25–26). Of course, it seems that there are separate modules for particular domains, such as language and vision. Moreover, various specific motivational mechanisms may operate independently. Buunk and Dijkstra (2001), for example, found that sexual preference and rival assessment may operate independently: while gay and lesbian individuals have a different sexual preference from heterosexual men and women, the sex difference in the perception of threatening rivals was found to be similar, as the physical attractiveness of the rival was more threatening for lesbian women and the social dominance of the rival for gay men. This type of specific example does not, however, contradict the notion that, as noted by Laland and Brown (2011), domain-general cognitive mechanisms may be favored when they "do a good enough job at low cost, and domain-general processes are no less compatible with evolutionary theory than domain-specific processes" (p. 128).

A related – and particularly relevant – controversy is the debate over whether humans are individuals with a Pleistocene brain thrown into a modern world that presents very different adaptive challenges. Hunting and gathering is usually considered as the prototypical way of life in which the typical human characteristics developed, accounting for at least 100,000 years of our time on Earth. However, hunter–gatherers have lived – and still live – under extremely different conditions, and, as noted by Nicholson (2012), around 40,000 years ago a dramatic change occurred, often called "the Great Leap Forward" (Diamond, 1999), when humans emigrated to all parts of the planet and began to produce a large variety of decorative and artistic artifacts. According to various authors, it was only at this time that a sense of self emerged, and with it the ability to reflect upon one's goals, to plan ahead, and to manage one's moods. Nevertheless, even this may be overstated, given all of the emerging evidence on animal cognition that suggests, for example, that many animals can and do plan ahead, engage in self-control, and deceive themselves and others (de Waal, 2016). In fact, humans before the Great Leap Forward may have resembled other species in many ways. While according to Nicholson (2012) emotions such as "regret, self-doubt, intrapsychic conflict, and chronic anxiety are rare in tribal people" (p. 22), such emotions may actually be observed in other species (de Waal, 2016). Moreover, given the fact that evolution among humans has not stopped and, in fact, may have occurred especially rapidly over recent centuries, with many genes being subject to recent selection in distinct ways in different populations (Nielsen et al., 2007), along with recent work on epigenetics (Carey, 2012) suggesting that certain acquired characteristics may be inherited, our psychology may differ considerably in various ways from that of people who lived in the Pleistocene. Although the jury is still out on epigenetics, considerable caution is called for with the radical application of the notion that humans still possess a hunter–gatherer brain.

A second main problem with applying evolutionary psychology is that the discipline is concerned with the ultimate causes of human behavior, while interventions usually tend to target the proximate causes of behavior. Roberts (2012) suggested that having a fuller understanding of the evolutionary history or the adaptive value of particular behaviors might help us to identify which range of possible interventions is likely to be the most successful. However, Park and Buunk (2010) argued that simply applying an evolutionary perspective to psychological questions is both theoretically and methodologically complicated, as the process of evolution occurs over long stretches of time, and most psychologists are interested in cognitive and emotional processes operating within individuals at much shorter time intervals and in how such processes are influenced by specific situational and individual-level variables. A functional and phylogenetic view cannot necessarily identify the precise nature of proximate mechanisms. Viewed from this perspective, it is clear that the evolutionary approach does not – indeed, cannot – replace existing social scientific psychological approaches to human behavior. Rather, the evolutionary approach may primarily offer clues as to the ultimate origins of motives and needs (*why* humans have the motives they

have), which, as I will argue, enhances integration with existing disciplines such as social psychology.

Even when there is agreement about a specific mechanism, it is often not obvious what type of policy recommendation should be developed on the basis of it. For men more than for women, intergroup victories enhance their mating opportunities, such as access to mates and prestige gains (e.g., Buss, 1999). When confronted with aggressive rival groups or intergroup conflict, male group members cooperate in order to successfully defeat the rival group. Indeed, research on traditional societies shows that tribal warfare is almost exclusively the domain of men and that male warriors have more sexual partners and greater status within their communities than other men do (Chagnon, 1988). In a similar way, a US study on male street gangs revealed that gang members have above average mating opportunities (Palmer & Tilley, 1995). The phenomenon that men more than women increase intragroup cooperative behaviors during intergroup competition has been called the *male-warrior hypothesis* (Van Vugt, De Creemer, & Janssen, 2007). Thus, whereas male group members may compete with each other, when rival groups appear, intrasexual competition within the group decreases and men start to work together in an attempt to defeat the rival group. Does this imply that we should discourage women from joining the army, or that we should recommend strong measures to overcome women's – probably evolved – reluctance to participate in violent intergroup fights? I doubt it.

36.4 THE MANY CONTRIBUTIONS TO SOLVING SOCIAL PROBLEMS FROM THE STANDARD SOCIAL AND BEHAVIORAL SCIENCES

Critiques of the existing social and behavioral sciences tend to ignore the many successful contributions that the traditional social sciences have made to resolving social problems, and it is often not obvious which unique contribution evolutionary psychology can make to developing interventions. For example, it has been noted by Curtis and Aunger (2012) that from an evolutionary perspective most public health problems stem from a mismatch between the psychological makeup of humans and their environments that have changed drastically over the past 150 years with products such as refined sugars, edible oils, and salt, labor-saving devices, and fast cars. In fact, virtually all risk factors for the most prevalent current diseases can be ascribed to this mismatch, like tobacco, high blood pressure, alcohol, cholesterol, overweight, and physical inactivity. This is a valuable perspective for understanding the *origins* of most current diseases after the occurrence of most infectious diseases in Western countries has been considerably reduced. However, it is not clear if this perspective does directly lead to specific interventions other than those proposed by proximate approaches. Without using evolutionary perspectives, the example of the role of social inequality in making it difficult for people on low incomes to obtain fresh and healthy foods has been emphasized by many sociologists, and hundreds of behavioral interventions have been developed, often based on theories from social psychology that were developed in laboratory experiments, including the cognitive theories, self-regulation theories, social influence theories, and reinforcement theories so often derided and despised by evolutionary psychologists (cf. Barrett, 2012). Even more so, such interventions have been quite successful with respect to such divergent issues as improving blood pressure treatment, promoting safe driving, commercial and noncommercial marketing, promoting the integration of immigrants, preventing cervical cancer, promoting pro-environmental behavior, fostering women's careers, or promoting political participation (e.g., Bartholomew Eldredge et al., 2016; Steg et al., 2016). One may also question if we really need an evolutionary approach as a basis of the important work by Curtis and Aunger (2012) themselves, which focuses on promoting handwashing with soap by using peer pressure and inducing disgust to reduce the rates of diarrheal disease, the second biggest killer of children in developing countries.

Furthermore, in clinical psychology, cognitive–behavioral therapy – in part based on behaviorism – has been very successful in reducing human suffering in many domains, including phobias, depression, and anxiety, without much attention being paid to the ultimate causes of these behaviors (e.g., Dobson, 2009). In a chapter on the clinical applications of Darwinian psychiatry, Troisi (2012) noted that many psychiatrists agree that the evolutionary approach may and does deepen our insights into the origins of psychiatric disorders, but, at the same time, it is barely relevant to clinical practice. Even Randolph Nesse (2005), one of the founders of Darwinian medicine, has noted that there are no evolutionarily based treatments for mental disorders and that the main contribution of evolutionary thinking is that it offers an integrative theoretical perspective. In addition, for example, Troisi (2012) has proposed that one application of evolutionary thinking is the use of observation of nonverbal behavior in clinical interviews for diagnostic purposes. However, Troisi bases his argument on ethology and not on evolutionary theory, and long before the emergence of evolutionary psychology there was already a vivid observational tradition in clinical settings based on ethology, such as examining the nonverbal behavior of marital couples seeking therapy (e.g., Noller & Fitzpatrick, 1988).

36.5 THE RELEVANCE OF AN EVOLUTIONARY PERSPECTIVE FOR UNDERSTANDING SOCIAL PROBLEMS

36.5.1 Illuminating the Ultimate Background of Social Problems

Despite the many obstacles to applying evolutionary psychology in order to solve social problems, there are a number of ways in which evolutionary psychology *can*

contribute to solving such problems. First, it provides a better understanding of social problems by offering an ultimate perspective. In line with the already mentioned analysis by Curtis and Aunger (2012), and as suggested by Roberts (2012), evolutionary thinking may provide particular insights into applied issues that are missed by standard sociological or social psychological approaches. There are, of course, numerous examples of the contributions of an evolutionary approach in this respect, many of which will have been highlighted elsewhere in this volume, including explanations for: the higher risk of stepchildren being abused; earlier menarche under stressful conditions; the consistently higher prevalence of males in violent crimes; the omnipresent obsessive preoccupation of males with the extra-dyadic sex of their wives or girlfriends; the relatively high involvement of grandmothers with the daughters of their daughters compared to other grandchildren; the consistent occurrence of rape in the context of war; or the backgrounds of prejudice. To give just two examples in more detail, first, in an excellent chapter on the evolutionary and neural and hormonal basis of serial monogamy and clandestine adultery, Helen Fisher (2012) suggested that such knowledge may help professionals to understand the underlying evolutionary predispositions that lead to unstable partnerships, including the cross-culturally "high incidence of sexual jealousy, partner stalking, spousal abuse, love homicide, and clinical depression" (p. 104). As a second example, and as we have outlined elsewhere (Buunk & Dijkstra, 2012), applying evolutionary thinking on intrasexual competition to organizations may help us to understand why people behave the way they do, even if these behaviors seem counterproductive or irrational, as is the case when people feel envious in response to a coworker's success or when an organization faces financial problems due to the high costs of conspicuous consumption (i.e., luxury items that display high status). Intrasexual competition may also cause workers to bully each other and to steal from the company, induce recruiters to reject and negatively evaluate candidates because they are perceived as rivals, and eventually lead to burnout among employees when they feel that they have lost battles over status. In a similar vein, evolutionary psychological theorizing can be fruitfully applied to a host of different problems in society, such as bad leadership, intergroup prejudice, organ donation, and environmental pollution (e.g., Schaller & Duncan, 2007; Van Vugt & Hardy, 2010; Van Vugt, De Cremer, & Janssen, 2007).

36.5.2 Integrating Evolutionary and Social Psychology

The focus in the remainder of this chapter is on social psychology, as this discipline – more than most within the social and behavioral sciences – is characterized by a combination of laboratory experiments, applied research, and the development of practical interventions. It is definitely true that social psychology often assumes that all kinds of phenomena – like sex differences, pro-social behavior, or inter-partner violence – are simply the results of "sex roles," "socialization," and "culture," without further specifying how these socialization processes occur and without empirically testing these assumptions. It is also true that social psychology has rarely considered the possible adaptive significance of the phenomena studied and that social psychologists often resist evolutionary explanations as either biologically deterministic or superfluous. Nevertheless, I feel that a particular fruitful approach, which is becoming more and more common, is integrating evolutionary thinking with "classic" social psychology – an endeavor that started about 10 years ago with the pioneering volume by Schaller, Simpson, and Kenrick (2006).

As a nice example of such an approach, Kenrick et al. (2009) suggested that traditional psychological functions such as risk aversion, discounting of future benefits, and budget allocations to multiple goods vary depending on evolutionarily relevant motives. For instance, while humans are generally risk aversive, a mating motivation may lead men to display resources and status, which makes them more willing to take financial risks to achieve gains to impress desirable potential mates and possibly obtain more resources than would be possible by taking the safer route. Women, on the other hand, should not be expected to respond in the same way to a mating prime, knowing that men do not attach as much value to women's money and resources when considering them as potential mates. In general, when addressing social problems, evolutionary psychology cannot – and should not – ignore all of the knowledge in social psychology about, for example, cognitive illusions, overconfidence, status striving, emotional contagions, illusions of control, and especially social influence.

Another example comes from my own work. It has been widely suggested inside and outside of social psychology that *social comparison* – comparing one's accomplishments, attractiveness, possessions, or professional success with those of others – is a basic human feature that nevertheless may be detrimental to well-being and to the functioning of groups (Buunk & Gibbons, 2007). Why, then, would this be such a widespread phenomenon? We argued that there may be various evolutionary explanations for this (Buunk & Dijkstra, 2012). First, the tendency to engage in social comparisons may have evolved to monitor the fairness of the distribution of resources in the group and to assess if others engage in reciprocity. This information may help to determine who one can trust and who is to be punished. Indeed, processes of social comparison and assessing reciprocity are closely related (Buunk & Ybema, 1997). A second – and possibly major – explanation is that social comparison allows individuals to estimate their position in the status hierarchy and may induce them to refrain from challenging those who are clearly superior, but to challenge those who can be beaten (cf. Gilbert, 2006; Gilbert & Allan, 1998). Remarkably, the perception of

one's own status tends to be quite biased. As noted by Barkow (1989), human beings have evolved a strong desire to develop positive self-esteem by assigning themselves symbolic prestige and status in their reference group. Such a reference group need not actually be present and may even be cognitively constructed: "With human self-esteem others not only need not be physically present, they need not have physical existence" (p. 191). Over the course of evolution, humans have developed the potential for self-deception and cognitive distortions in building a positive self-concept. We have tried to demonstrate that many features characteristic of the social comparison processes of depressed people, including the tendency of depressed people not to view themselves as superior in the way that nondepressed people do, may be interpreted evolutionarily as the idea that depression stems from a perception of having a low status (Buunk & Brenninkmeijer, 2000).

36.5.3 From a Deductive to an Inductive Approach

Despite the value of providing insights into the ultimate causes of social problems, such insights do not lead to well-developed analyses of social problems, let alone adequate interventions. These approaches are primarily *deductive*: from a relevant theory some potential insights for interventions are derived. Roberts (2012) suggested that applied evolutionary psychology could explore, highlight, and analyze the ways in which evolutionary insights are already – or could potentially – be used in practical and beneficial ways within applied settings. At its best, most studies in applied evolutionary psychology are what De Ruiter et al. (2013) refer to as *problem-clarifying research* (i.e., providing a better understanding of a social problem). However, as noted by De Ruiter et al., *problem-solving research* requires a strongly *inductive* approach, starting with the description of the specific problem that needs to be resolved. The fact that there exist many controversies within evolutionary psychology and that, for many relevant issues, scientific knowledge is still developing does not prevent us from using evolutionary psychological theories to develop adequate interventions as long as we view many theories as having primarily a heuristic value. Typically, in a problem-oriented approach, various theories will be used to explain the problem and eventually develop interventions for it. Even though an adequate intervention requires insight into the proximate causes of behavior first and foremost, evolutionary psychology focuses mainly on ultimate causes and may offer innovative ideas about such causes. A nice example of an inductive approach that uses evolutionary and social psychological thinking to explain a specific problem comes from Nicholson (2012). He explained the 2008 financial crisis using a number of themes that had all been established within social and cognitive psychology and had received an ultimate explanation in evolutionary psychology, including the drive for status and reputation, loss aversion, the fear of being left behind, illusions of control, overestimation of oneself, coalitions acting in consort, diffusing responsibility, and herding (i.e., relying in informal communication on the wisdom of the crowds, which has been studied for a long time in social psychology as conformity).

In our book *Applying Social Psychology: From Problems to Solutions* (Buunk & Van Vugt, 2013), we – both of us evolutionary and social psychologists – presented a model for an inductive approach that provides a detailed method that takes us from global problem definitions to concrete interventions. With this method, one first needs to describe precisely *what* the problem is (e.g., "How can we prevent teenage pregnancies?", "How can we promote that people should take better care of their teeth?", "How can we reduce prejudice against ethnic minorities?", "How can we prevent traffic accidents?"). It is important to specify, among other factors, the target group one wants to address and to formulate precisely the outcome variable in terms of behaviors or attitudes that need to be changed, rather than refer to the problem in general terms like "promote altruistic behavior" or "combat crime." These more general problems may be the subject of theoretical analyses, but such analyses often do not result in effective interventions. After the problem definition, potential theoretical explanations need to be generated through brainstorming and through consulting not only the scientific literature that is directly relevant, but also general theories that may provide relevant insights. While in many cases the variables suggested by evolutionary psychology may also be derived from classic social psychology, evolutionary psychology may open the eyes to variables that are not obvious to classic social psychologists. As noted by Buunk and Van Vugt (2013), when using evolutionary psychology to generate possible causes of a problem, a social psychologist should ask him or herself what the function of a specific behavior may be in the context of human survival and reproduction. For example, when one wants to reduce car accidents among young male drivers, evolutionary theorizing suggests that young men are generally more likely to take physical and monetary risks, especially in the presence of females (Iredale, Van Vugt, & Dunbar, 2008; Wilson & Daly, 2004). In a similar vein, when looking for variables explaining the occurrence of professional burnout (which is usually attributed to the rather broad and vague variable "stress"), evolutionary theorizing on the importance of a high rank in the group and the painful consequences of a loss of social rank may be considered as particularly relevant predictors (for a review, see Gilbert, 2006). Indeed, Schaufeli and Buunk (2003) found that burned-out individuals tend to feel helpless, hopeless, and powerless and experience feelings of insufficiency, incompetence, and poor job-related self-esteem – all experiences that suggest a subjectively low status.

There are many more examples of social problems for which evolutionary psychology may offer innovative insights. For instance, when trying to explain and

eventually promote charitable behavior, one can exploit unconscious cues to kinship on the basis of inclusive fitness theory, such as having the same family name or facial resemblance (Barclay, 2012). Or take the more mundane example of taking care of one's teeth. Classic social psychology would consider good oral self-care as determined by one's attitude toward this behavior, the perceived norms in the environment, and one's sense of self-efficacy (Buunk-Werkhoven, Dijkstra, & Van der Schans, 2011). However, an evolutionary psychologist might think of the function of healthy teeth as affecting physical attractiveness and as an important sign of the health of a potential mate (e.g., Sugiyama, 2005) and would therefore be sensitive to sex differences in this respect. Indeed, women tend to find bad breath in a mate more repellent than men do, and this might eventually call for different types of interventions for men and women (Buunk-Werkhoven & Buunk, 2016). To give another example, when trying to explain vandalism in a specific neighborhood, rational choice theory explains vandalism by referring to the belief that it is easy to avoid punishment without explaining *why* someone endorses this belief or what they (unconsciously) gain from it. As noted by Buunk and Van Vugt (2013), an evolutionarily minded social psychologist may "dig" a little deeper and find out why the problem of vandalism is so persistent, pointing to youngsters feeling that they can impress their mates and potential sexual partners by engaging in such activities (Iredale et al., 2008).

A particular interesting example of the potential contribution of evolutionary perspectives to the explanation of social problems comes from recent evolutionary work on the background of prejudice. Social psychologists have long noted that humans have a basic tendency to categorize others in terms of in-groups and out-groups. However, as noted by Park (2012), "[S]ocial contexts that resemble tribal intergroup contexts should be characterized especially by competitiveness and hostility" (p. 192; see also Van Vugt & Park, 2009). While social psychologists from the USA, where prejudice against African-Americans is still quite prevalent, have argued that race is (in addition to gender and age) one of the three "primitive categories" that people pay attention to, there is little evidence for race as an evolutionarily evolved cue. Rather, features like language, accent, and ethnic background may be cues of coalitional alliance, or a lack thereof, and race may simply serve as a weak cue to such an alliance (Kurzban, Tooby, & Cosmides, 2001). In addition, as the work by Schaller and his colleagues on the behavioral immune system suggests (e.g., Schaller & Duncan, 2007), out-groups may be particularly stigmatized and avoided in the case of a threat of disease from previously segregated out-groups.

Through a careful selection of a limited number of potentially relevant variables, the next step in the approach of Buunk and Van Vugt (2013) is to develop a *model* integrating these variables and to test this in an applied empirical study. In this process, the potentially added value of a variable suggested by evolutionary theorizing can be tested. For example, in a study among Spanish teachers, Buunk et al. (2007) found that variables suggested by evolutionary theorizing (e.g., Barkow, 1989; Gilbert, 2006), such as a low subjective status, a loss of status, and a sense of defeat, were much more important predictors of burnout than classic occupational stress predictors such as role problems, relationships with others, and organizational climate. Finally, and most importantly, an *intervention* needs to be developed addressing the most powerful and modifiable predictors in the model. In developing the model, evolutionary thinking may already have suggested relevant variables to address in interventions, although these may often already have been used in practice without any influence of evolutionary theorizing. For example, Barclay (2011) suggested that one obvious way to promote charitable behavior is by inducing reciprocation, a major social influence tactic identified by Cialdini (1988) before evolutionary psychology came into existence and long used in practice. The same is true for competitive altruism (i.e., the notion developed in evolutionary psychology that people want to give more than others; Hardy & Van Vugt, 2006). This was already often being used in practice by practitioners who knew that people tend to give more when they know how much others had given.

Given all of the existing practical and theoretical knowledge on changing human behavior, evolutionary psychology may not always offer very many innovative insights into how to change behavior and may simply confirm existing notions. Rather, one of the most important contributions that evolutionary psychology can make in the development of interventions may be emphasizing the variables that may be very difficult to change, given their adaptive significance. An evolutionary perspective may explain why certain policy measures may not be effective, or may even backfire. A good illustration of this comes from research on the influence of sex ratios. For example, one may want to promote the idea that men in an African-Caribbean culture like Curaçao, where about 40 percent of children grow up without support from their father, should take more responsibility in providing care for their offspring (Van Brummen-Girigori & Buunk, 2016). However, in this case (and in many similar cases), this lack of paternal investment is in part an adaptive response to an unbalanced sex ratio, with women by far outnumbering men. When a female-based sex ratio exists, men have more opportunities to pursue women, tend to be less committed to their partners, and tend to invest less in their children, whereas women tend to be less intrusive toward their partners, tend to give their partners greater freedom, and tend to invest more in their own careers. Under such conditions, it may be very difficult to develop policies to increase the investment of men in their offspring (e.g., Griskevicius et al., 2012; Guttentag & Secord, 1983).

In addition, against the assumptions of classic social psychologists, the strategy of reducing prejudice by *decategorization* (i.e., by psychologically integrating the in-group and out-group within a single group) may work only or particularly with groups to whom no danger-connoting stereotypes are ascribed, such as women, the elderly, gay people, or obese people (Park, 2012). As rightfully noted by Park, prejudice against what he refers to as "tribal" out-groups (e.g., people from other ethnic backgrounds) differs in important ways from other kinds of prejudice, such as sexism or ageism, and consequently different types of prejudice cannot be reduced with the same strategies. As another example, on the basis of life history theory, Figueredo, Gladden, and Hohman (2012) suggested a number of co-concurring traits that could easily be integrated into a model of proximate factors affecting criminal behavior (i.e., desire for casual sex, high mating effort, low emotional attachment to romantic partners, tendencies toward risk-taking or impulsive behavior, decreased social and moral rule following, and decreased law-abidingness). However, given their rootedness in life history, these may not be easy to change.

Thus, policy measures that demand people resist their evolutionarily based motivations may have a low chance of success. A policy that is based on an understanding of the adaptive functions of many behaviors rather than on changing the basic motives of humans will be more effective. For example, although positive discrimination of women or ethnic minorities may be desirable from a societal or rational point of view, it may clash with workers' feelings of fairness and so result in envy or bullying. As noted by Buunk and Dijkstra (2012), when one wants to promote the entrance of women into the highest levels of an organization, one needs to realize that, for their reproductive success, women may be overall more concerned with the timing of and care for their offspring than men. Nevertheless, given that men more often drop out of school and are, on average, becoming less educated than women in Western societies, women may in the future become – and are already becoming – dominant in all kinds of professional jobs. In addition, the likelihood that an increasing number of men may, because of their lower educational levels, end up as homemakers may also foster the participation of women in organizations. However, though it seems quite likely that, in the future, women will be more successful than men on average, it may be the case that they will still be less likely to be found in top positions, as well as in the lowest positions. This would suggest that policies might be effectively directed toward attracting women to an organization itself, but less so to the highly demanding top positions. It may be that, given women's different psychological makeup from men, it is unrealistic to assume that complete equality in the work place is a reachable goal.

36.6 CONCLUSION

Although evolutionary psychology has developed into an active field of research and has provided many new and important insights into the ultimate bases of human behavior, applying this discipline in order to solve social problems is fraught with many pitfalls. One of the reasons for this is that there are still many basic theoretical disagreements in this field. In general, evolutionary psychology is not a panacea for solving such problems, as some scholars seem to assume, and it still has to prove its added value to the "standard" social and behavioral sciences. In these sciences, many effective interventions have been developed that aim at, for example, changing health behaviors, reducing criminal behavior, and improving mental health. While in the past "standard" social science has been guilty of promoting ideologically driven, even dictatorial interventions based on the assumption of the human mind as a blank slate, some recent applications of evolutionary psychology, such as those of Miller (2009), are equally bizarre and extreme and do not serve to promote evolutionary psychology as a valuable discipline for solving social problems. Nevertheless, evolutionary psychology is loaded with new and relevant insights, and it can contribute in a number of ways to solving social problems: first, by providing insights into the evolutionary background of a social problem; second, by suggesting proximate variables that contribute to specific behaviors related to the problem or that may be included in interventions; and third, by pointing to behaviors that will be very difficult to change, given their adaptive significance. In doing so, evolutionary psychology may be most effective if it joins up with existing approaches, particularly in social psychology.

36.7 ACKNOWLEDGMENTS

I thank Louise Barrett and Douglas Kenrick for their very helpful comments on this chapter.

REFERENCES

Barclay, P. (2011). The evolution of charitable behaviour and the power of reputation. In C. Roberts, ed., *Applied Evolutionary Psychology*. Oxford: Oxford University Press, pp. 149–172.

Barkow, J. H. (1989). *Darwin, Sex, and Status: Biological Approaches to Mind and Culture*. Toronto: University of Toronto Press.

Barrett, L. (2012). Why behaviorism isn't Satanism. In J. Vonk & T. K. Shackelford, eds., *The Oxford Handbook of Comparative Evolutionary Psychology*. New York: Oxford University Press, pp. 17–38.

Barrett, L. (2017). What is human nature (if it is anything at all)? In R. Joyce, ed., *The Routledge Handbook of Evolution and Philosophy*. London: Routledge, pp. 194–209.

Bartholomew Eldredge, L. K., Markham, C. M., Ruiter, R. A. C., et al. (2016). *Planning Health Promotion Programs. An Intervention Mapping Approach*. San Francisco, CA: Jossey-Bass.

Beckstrom, J. (1993). *Darwinism Applied: Evolutionary Paths to Social Goals*. Westport, CT: Praeger.

Buskes, C. (2006). *Evolutionair Denken: De Invloed van Darwin op Ons Wereldbeeld*. Amsterdam: Uitgeverij Nieuwezijds.

Buss, D. M. (1999). *Evolutionary Psychology*. London: Allyn & Bacon.

Buunk, A. P., & Brenninkmeijer, V. (2000). Social comparison processes among depressed individuals: Evidence for the evolutionary perspective on involuntary subordinate strategies? In L. Sloman & P. Gilbert, eds., *Subordination and Defeat: A Evolutionary Approach to Mood Disorders and Their Therapy*. Mahwah, NJ: Erlbaum, pp. 147–164.

Buunk, A. P., & Dijkstra, P. (2001). Evidence from a homosexual sample for a sex-specific rival-oriented mechanism: Jealousy as a function of a rival's physical attractiveness PRIVATE and dominance. *Personal Relationships*, **8**, 391–406.

Buunk, A. P., & Dijkstra, P. (2012). The social animal within organizations. In S. C. Roberts, ed., *Applied Evolutionary Psychology*. Oxford: Oxford University Press, pp. 36–51.

Buunk, A. P., & Fisher, M. (2009). Individual differences in intra-sexual competition. *Journal of Evolutionary Psychology*, **7**, 37–48.

Buunk, A. P., & Gibbons, F. X. (2007). Social comparison: The end of a theory and the emergence of a field. *Organizational Behavior and Human Decision Process*, **102**, 3–21.

Buunk, A. P., & Hoben, A. D. (2013). A slow life history is related to a negative attitude towards cousin marriages: A study in three ethnic groups in Mexico. *Evolutionary Psychology*, **11**, 442–458.

Buunk, A. P., & Park, J. H. (2008). Not massive, but messy modularity. *Psychological Inquiry*, **19**, 23–26.

Buunk, A. P., & Van Vugt, M. (2013). *Applying Social Psychology: From Problem to Solution*. London: Sage.

Buunk, A. P., & Ybema, J. F. (1997). Social comparisons and occupational stress: The identification-contrast model. In A. P. Buunk & F. X. Gibbons, eds., *Health, Coping and Well-Being: Perspectives from Social Comparison Theory*. Hillsdale, NJ: Erlbaum, pp. 359–388.

Buunk, A. P., Peíró, J. M., Rodríguez, I., & Bravo, J. M. (2007). A loss of status and a sense of defeat: An evolutionary perspective on professional burnout. *European Journal of Personality*, **21**(4), 471–485.

Buunk, A. P., Pollet, T. V., Dijkstra, P., & Massar, K. (2011). Intrasexual competition within organizations. In G. Saad, ed., *Evolutionary Psychology in the Business Sciences*. New York: Springer, pp. 41–70.

Buunk, A. P., Dubbs, S. L., & Van Hooff, J. A. R. A. M. (2012). Social influence on reproductive behavior in humans and other species. In D. Kenrick, N. Goldstein, & S. Braver, eds., *Six Degrees of Social Influence: Science, Application and the Psychology of Robert Cialdini*. Oxford: Oxford University Press, pp. 98–108.

Buunk-Werkhoven, Y. A. B., & Buunk, A. P. (2015). Fear of social rejection and oral self-care in men versus women. *International Dental Journal*, **65**(Suppl. S1), 13.

Buunk-Werkhoven, Y. A. B., Dijkstra, A., & Van der Schans, C. P. (2011). Determinants of oral hygiene behavior: A study based on the theory of planned behavior. *Community Dentistry and Oral Epidemiology*, **39**, 250–259.

Carey, N. (2012). *The Epigenetics Revolution*. London: Icon Books.

Chagnon, N. A. (1988). Life histories, blood revenge, and warfare in a tribal population. *Science*, **239**, 985–992.

Cialdini, R. B. (1988). *Influences: Science and Practice*. Glenview, IL: Scott-Freeman.

Curtis, V., & Aunger, R. (2012). Motivational mismatch: Evolved motives as the source of – and solution to – global public health problems. In S. C. Roberts, ed., *Applied Evolutionary Psychology*. Oxford: Oxford University Press, pp. 259–275.

De Ruiter, R. A. C., Massar, K., Van Vugt, M., & Kok, G. (2013). Applying social psychology to understanding social problems. In A. Golec de Zavala & A. Cichocka, eds., *Social Psychology of Social Problems: The Intergroup Context*. New York: Palgrave Macmillan, pp. 337–362.

De Waal, F. (2016). *Are We Smart Enough to Know How Smart Animals Are?* New York: W. W. Norton.

Diamond, J. (1999). *Guns, Germs, and Steel: The Fates of Human Societies*. New York: W. W. Norton.

Dobson, K. S. (2009). *Handbook of Cognitive Behavioral Therapies*. New York: Guilford.

Figueredo, A. J., Sefcek, J. A., Vasquez, G., et al. (2005). Evolutionary personality psychology. In D. M. Buss, eds., *The Handbook of Evolutionary Psychology*. Hoboken, NJ: John Wiley & Sons, pp. 851–877.

Figueredo, A. J., Gladden, P. R., & Hohman, Z. (2012). The evolutionary psychology of criminal behavior. In S. C. Roberts, ed., *Applied Evolutionary Psychology*. Oxford: Oxford University Press, pp. 201–221.

Fisher, H. E. (2012). Serial monogamy and clandestine adultery: Evolution and consequences of the dual human reproductive strategy. In S. C. Roberts, ed., *Applied Evolutionary Psychology*. Oxford: Oxford University Press, pp. 93–114.

Gilbert, P. (2006). Evolution and depression: Issues and implications. *Psychological Medicine*, **36**, 287–297.

Gilbert, P., & Allan, S. (1998). The role of defeat and entrapment (arrested flight) in depression: An exploration of an evolutionary view. *Psychological Medicine*, **28**, 585–598.

Griskevicius, V., Simpson, J. A., Durante, K. M., Kim, J. S., & Cantu, S. M. (2012). Evolution, social influence, and sex ratio. In D. Kenrick, N. Goldstein, & S. Braver, eds., *Six Degrees of Social Influence: Science, Application and the Psychology of Robert Cialdini*. Oxford: Oxford University Press, pp. 79–89.

Guttentag, M., & Secord, P. F. (1983). *Too Many Women? The Sex Ratio Question*. Beverly Hills, CA: Sage.

Hardy, C., & Van Vugt, M. (2006). Nice guys finish first: The competitive altruism hypothesis. *Personality and Social Psychology Bulletin*, **32**, 1402–1413.

Iredale, W., Van Vugt, M., & Dunbar, R. I. M. (2008). Showing off in humans: Male generosity as mating signal. *Evolutionary Psychology*, **6**, 386–392.

Kenrick, D. T., Griskevicius, V., Sundie, J. M., et al. (2009). Deep rationality: The evolutionary economics of decision making. *Social Cognition*, **27**, 764–785.

Kurzban, R., Tooby, J., & Cosmides, L. (2001). Can race be erased? Coalitional computation and social categorization. *Proceedings of the National Academy of Sciences*, **98**, 15387–15392.

Laland, K. N., & Brown, G. R. (2011). *Sense and Nonsense. Evolutionary Perspectives on Human Behavior*. Oxford: Oxford University Press.

Livingstone Smith, D. (2007). *The Most Dangerous Animal*. New York: St. Martin's Press.

Miller, G. (2009). *Spent: Sex, Status, and Consumer Behavior*. New York: Viking.

Nesse, R. M. (2005). Evolutionary psychology and mental health. In D. M. Buss, ed., *The Handbook of Evolutionary Psychology*. Hoboken, NJ: John Wiley & Sons, pp. 903–930.

Nicholson, N. (2012). The evolution of business and management. In S. C. Roberts, ed., *Applied Evolutionary Psychology*. Oxford: Oxford University Press, pp. 16–35.

Nielsen, R., Hellmann, I., Hubisz, M., Bustamante, C., & Clark, A. G. (2007). Recent and ongoing selection in the human genome. *Nature Reviews. Genetics*, **8**, 857–868.

Noller, P., & Fitzpatrick, M. A., eds. (1988). *Perspectives on Marital Interaction*. Philadelphia, PA: Multilingual Matters.

Palmer, C. T., & Tilley, C. F. (1995). Sexual access to females as a motivation for joining gangs: An evolutionary approach. *Journal of Sex Research*, **32**, 213–217.

Park, J. H. (2012). Evolutionary perspectives on intergroup prejudice: Implications for promoting tolerance. In S. C. Roberts, ed., *Applied Evolutionary Psychology*. Oxford: Oxford University Press, pp. 186–200.

Park, J. H., & Buunk, A. P. (2010). Interpersonal threats and automatic motives. In D. Dunning, ed., *Frontiers in Social Motivation*. New York: Psychology Press, pp. 11–35.

Plotkin, H. (2004). *Evolutionary Thought in Psychology: A Brief History*. Malden, MA: Blackwell Publishing.

Roberts, S. C., ed. (2012). *Applied Evolutionary Psychology*. Oxford: Oxford University Press.

Schaller, M., & Duncan, L. A. (2007). The behavioral immune system: Its evolution and social psychological implications. In J. Forgas, M. Haselton, & W. Von Hippel, eds., *Evolution and the Social Mind: Evolutionary Psychology and Social Cognition*. New York: Psychology Press, pp. 293–307.

Schaller, M., Simpson, J. A., & Kenrick, D. T. (2006). *Evolution and Social Psychology*. New York: Psychology Press.

Schaufeli, W. B., & Buunk, A. P. (2003). Burnout: An overview of 25 years of research and theorizing. In M. J. Schabracq, J. A. M. Winnubst, & C. L. Cooper, eds., *The Handbook of Work and Health Psychology*. Chichester: Wiley, pp. 383–425.

Steg, L. M., Keizer, K., Buunk, A. P., & Rothengatter, T., eds. (2016). *Applied Social Psychology*. Cambridge, UK: Cambridge University Press.

Sugiyama, L. S. (2005). Physical attractiveness in adaptionist perspective. In D. M. Buss, ed., *The Handbook of Evolutionary Psychology*. Hoboken, NJ: John Wiley & Sons, pp. 292–343.

Tooby, J., & Cosmides, L. (1992). The psychological foundations of culture. In J. Barkow, L. Cosmides, & J. Tooby, eds., *The Adapted Mind: Evolutionary Psychology and the Generation of Culture*. New York: Oxford University Press, pp. 19–136.

Tooby, J., & Cosmides, L. (2005/2016). Conceptual foundations of evolutionary psychology. In D. M. Buss, ed., *The Handbook of Evolutionary Psychology*. Hoboken, NJ: John Wiley & Sons, pp. 5–67.

Tooby, J., & Cosmides, L. (2015). The theoretical foundations of evolutionary psychology. In D. M. Buss, ed., *The Handbook of Evolutionary Psychology*. Hoboken, NJ: John Wiley & Sons, pp. 3–87.

Troisi, A. (2012). Mental health and well-being: Clinical applications of Darwinian psychiatry. In S. C. Roberts, ed., *Applied Evolutionary Psychology*. Oxford: Oxford University Press, pp. 276–289.

Van Brummen-Girigori, O., & Buunk, A. P. (2016). Intrasexual competitiveness and non-verbal seduction strategies to attract males: A study among teenage girls from Curaçao. *Evolution and Human Behavior*, **37**, 134–141.

Van Vugt, M., & Hardy, C. L. (2010). Cooperation for reputation: Wasteful contributions as costly signals in public goods. *Group Processes & Intergroup Relations*, **13**, 101–111.

Van Vugt, M., & Park, J. H. (2009). Guns, germs and sex: How evolution shaped our intergroup psychology. *Social and Personality Compass*, **3**, 927–938.

Van Vugt, M., De Cremer, D., & Janssen, D. P. (2007). Gender differences in cooperation and competition: The male-warrior hypothesis. *Psychological Science*, **18**, 19–23.

Van Vugt, M., Hogan, R., & Kaiser, R. B. (2008). Leadership, followership, and evolution: Some lessons from the past. *American Psychologist*, **63**, 182–196.

Wertheim, W. F. (1973). *Dawning of an Asian dream: Selected Articles on Modernization and Emancipation*. Amsterdam: Antropological–Sociological Center of the University of Amsterdam.

Wilson, M., & Daly, M. (2004). Do pretty women inspire men to discount the future? *Proceedings of the Royal Society B*, **271**, 177–179.

Wrangham, R. W., & Peterson, D. (1996). *Demonic Males: Apes and the Origins of Human Violence*. Boston, MA: Houghton, Mifflin and Company.

37 The Evolution of Personality

ADRIAN FURNHAM AND SATOSHI KANAZAWA

37.1 INTRODUCTION

Two presidents of the American Psychological Association pointed out in their state-of-the-art addresses that there seemed, in psychology, a great division between *experimental psychology*, which sought to discover universal laws of human behavior, and *correlational psychology*, which sought to describe and explain individual differences.

Cronbach (1957) noted that experimental psychologists are so embarrassed and annoyed by individual differences that they often treat them as "error variance." Yet this variability is the very essence of correlational or differential psychology. Eysenck (1981) pointed out that a science of psychology cannot properly function without *both* branches, which are indispensable to a proper understanding of people. Individual differences interact in almost every case with experimental and situational factors to produce results that are profoundly different for individuals of different personalities, different capacities, and different motivations. Eysenck believed that if we are to explain the major factors of personality in scientific terms, we must make appeal to the concepts used in experimental *and* correlational/differential psychology.

Today, we still observe this basic tension between universal human nature and individual differences/personality in the differences between the two otherwise closely allied fields of social psychology and personality psychology, and, more recently, between evolutionary psychology and behavior genetics. Unlike behavior genetics, which seeks to explore the genetic foundations and heritabilities of individual differences in phenotypes, evolutionary psychologists are usually "universalists," being little interested in individual differences that can seem hard to explain. Both differential and evolutionary psychologists acknowledge biological and genetic factors in explaining behavior, but they differ in their interests in systematic variability.

Evolutionary psychology seems little interested in personality and preferences. Some writers, however, have made an effort to integrate the two fields (Buss, 2009; Figueredo et al. 2005, 2009; Kanazawa, 2011; Nettle, 2005). At the heart of the debate is the idea that adaptations to the environment (through natural selection) are universal. In short, the "theory" goes like this: adaptations that emerge from the process of natural selection are universal and species-typical. But individual phenotypes are often the product of both the genes and the environment. To the extent that the development of phenotypes responds to particular conditions in the local environment, individual differences in the phenotypic expressions of the universal and species-typical adaptations can occur.

Thus, we have to explain why there are replicable and consistent *differences* in mate choice; for example, why do some males prefer and select females who are the very opposite of the highly fecund model in terms of the body mass index and waist-to-hip ratio that the evolutionary literature suggests? Or how can there be any evolutionary advantage in being Neurotic?

Differential psychology is itself split into two definable groups: those who study *personality* traits and those who study *ability/intelligence*. Attempts have been made to integrate evolutionary psychology and intelligence research. For example, Kanazawa (2010) used the Savanna–IQ Interaction Hypothesis to explain how intelligence accounts for individual differences in preferences and values. The idea is that intelligence is essential to comprehending and responding to complex and evolutionarily novel situations, but not necessarily to evolutionarily familiar situations. Thus, it is possible to explain a whole range of individual differences from television watching to political attitudes in terms of the intelligence of individuals. More intelligent individuals are more likely to acquire and espouse evolutionarily novel preferences and values such as political liberalism, atheism, and nocturnal lifestyles.

Personality psychologists who study individual differences spent most of the twentieth century discussing and debating the structure of personality. They were in search of "the periodic table of personality," but it was more a search for the Holy Grail. Some tried the hypothetico-deductive methodology to discover through observations and thence multivariate statistics the underlying structure of personality. They aimed to be both parsimonious and comprehensive. Others created theoretical models based on their "clinical insights."

37.2 THE STRUCTURE OF PERSONALITY

Without doubt, the most sophisticated trait personality theory is that of Hans Eysenck, which has been likened to finding St Pancras railway terminus (i.e., an elaborate, Victorian structure) in the jungle of personality theories. The theory, which has undergone various changes over a 30-year period, argues for the psychophysiological basis of personality and locates three major factors that relate to social behavior: extraversion, neuroticism, and psychoticism. The theory has been applied to a wide range of activities, including criminality, sex, smoking, health, and learning (Furnham, 2008; Furnham & Arnold, 1999).

Eysenck suggests that there are three fundamental (higher-order) unrelated (orthogonal) traits: extraversion, neuroticism, and psychoticism. These traits, which can be measured and described on a continuum, are biologically based and have many behavioral implications. Two traits – extraversion and neuroticism – have, however, been most investigated at the biological, information-processing, and motivational levels.

Research on extraversion suggests that it is substantially genetically inherited (it is explained in terms of cortical arousal and reward sensitivity) and that extraverts succeed in high-pressure jobs that involve considerable interaction with strangers. They handle overload and stress well and have task-focused coping, feelings of self-efficacy, and a good sense of well-being. Neuroticism, which is also substantially biologically based and inherited, is associated with stress vulnerability, sensitivity to punishment, and threat avoidance. Neurotics have highly selective biases in cognitive processes with considerable awareness of danger, cautious decision-making, a generally negative self-concept, and often depressed mood and pessimistic outlooks.

A review of the dimension presents an impressive array of findings. Introverts are more sensitive to pain than are extraverts; they become fatigued and bored more easily than do extraverts; excitement interferes with their performance, whereas it enhances performance for extraverts; and they tend to be more careful but slower than extraverts. Introverts do better in school than extraverts, particularly in more advanced subjects. Students withdrawing from college for academic reasons tend to be extraverts, whereas those who withdraw for psychiatric reasons tend to be introverts.

Equally, extraverts prefer vocations involving interactions with other people, whereas introverts tend to prefer more solitary vocations. Extraverts seek diversion from job routine; introverts have less need for novelty. Extraverts enjoy explicit, sexual, and aggressive humor, whereas introverts prefer more intellectual forms of humor, such as puns and subtle jokes. Extraverts are more active sexually, in terms of frequency and number of different partners, than are introverts. Extraverts are more suggestible than introverts.

Introverts are more easily aroused by events and more easily learn social prohibitions than extraverts. As a result, introverts are more restrained and inhibited. There also is some evidence that introverts are more influenced by punishments in learning, whereas extraverts are more influenced by rewards. It is hypothesized that individual differences along this dimension have both hereditary and environmental origins. Indeed, several studies of identical and fraternal twins suggest that heredity plays a major part in accounting for the differences between individuals in their scores on this dimension.

Opposing Eysenck for nearly 50 years was the work of R. B. Cattell, who developed his famous 16-trait measure 16PF (Cattell, 1971), published initially over 40 years ago. For Cattell, the theory and test have several advantages: the theory is unusually *comprehensive* in its coverage of personality dimensions; it is based on the *functional* measurement of previously located natural personality structures; and the measurements are relatable to an *organized* and *integrated* body of practice and theoretical knowledge in clinical education and industrial psychology. The test measures 16 dimensions of personality (and 6–9 second-order factors) that are supposedly independent and identifiable, as well as reliably and validly measurable.

The direct inheritor of the Eysenckian and Cattellian traditions are the Americans Costa and McCrae (McCrae & Costa, 1987), whose work in the 1980s and 1990s has revived the world of personality theory and testing. Working within the psychometric trait tradition, they settled on three and then five dimensions of personality. Now called the "five-factor approach" or "five-factor model," there is now broad agreement on their proposed personality structure from many sources, including those who adopt the lexical approach, who look at natural language and the relationships among everyday terms for personality traits (Goldberg, 1992). Indeed, there is an active psycho-lexical tradition in personality theory that attempts to "recover" the basic dimensions of personality through analysis of natural language. Researchers have found impressive evidence, across various different languages, of the emergence of similar factors that are analogous to the Big Five personality dimensions. What they have not done, however, is to look at the association between personality traits and work outcomes. There are vigorous critiques of the five-factor model, but these have not reduced its popularity among personality researchers (Block, 1995; Eysenck, 1992).

It is probably true to say that never as much as now has there been broad agreement among differential psychologists around the five-factor model. It is suggested that individual differences can be described parsimoniously on these five dimensions/super-traits/domains, which can also be described at a secondary, more detailed or facet level. It is also generally agreed that all traits have a considerable genetic basis and that personality is surprisingly stable over time. Further, there is little evidence that these traits are substantially influenced by culture, and there is considerable evidence that these traits are found in all societies (Furnham, 2008; Furnham & Cheng, 2015).

37.3 THE EVOLUTION OF PERSONALITY

Could one derive and test hypotheses as to why some traits are seen to be more socially and physically attractive than others? Nettle (2006) attempted a full and parsimonious description of the benefits and costs of each of the Big Five personality dimensions.

Extraversion. Extraverts are more sociable and sensation-seeking and have more social support. They are usually more socially skilled and interpersonally confident. Their attitude to, interest in, and experience of sex means that they tend to have more sexual partners, "mating successes," and offspring. However, this also means that they are more prone to infidelity (Nettle, 2005), which suggests that their children are more likely to be later exposed to stepparents, which is an established risk factor in their development (Daly & Wilson, 1985).

Introverts are less sociable but safer; they run the risk of a lower likelihood of finding mates and social support networkers, but leading a more secure lifestyle, which is better for childrearing. Certainly, it seems that Extraverts are rated as more interpersonally attractive. Social confidence, fun-loving activity preferences, and optimism make them more attractive to many people and for short-term social relationships and mating.

Neuroticism. Neurotics are more likely to be anxious, depressed, guilt-ridden, phobic, and hypochondriacal. They are less likely to have good long-standing and satisfying personal relationships and jobs. They are, however, socially vigilant, wary, and risk averse. Neurotics are very aware of subtle (and possibly threatening) social changes, which can be a strong survival mechanism in certain environments. Neurotics are very interested in their own and others' emotions, which can make them highly sensitive "readers of social situations."

Those who are very low in Neuroticism – labeled stable or highly adjusted – may have some disadvantages. They may be too trusting and eager to avoid social and physical hazards; they may underperform and strive less hard because they are afraid of failure. They may also be socially insensitive to the anxieties and worries of those around them and therefore have a small social support network. Overall, however, it does seem as if strongly neurotic people would be considered unattractive, too demanding, and high-maintenance.

Openness. Openness to experience is marked by creativity, cognitive complexity, imagination, and curiosity. Open individuals are attracted to the unusual and the unconventional. Openness is a good predictor of artistic and scientific achievement and innovation (Furnham, 2008). Open individuals are often thought of as creative. The flipside of novel thinking is delusions and occasionally supernatural and paranormal ideas. Creative individuals, when emotionally stable and associated with particular skills, especially in the arts, are highly attractive to others and therefore have many different mates and a wide relationship network.

However, those with "unusual" beliefs can easily be described as "mad" and rejected by society. Moderate to high levels of creativity are associated with attractiveness partly because creativity is highly valued in many settings. It also means that such individuals are thought of as very interesting.

Agreeableness. Agreeable people are empathic, trusting, kind, well-liked, respected, and valued as friends. They always seek and attempt to create harmony and concord. The trait is highly valued, and being sensitive ("emotionally intelligent") to others' moods is clearly advantageous. However, being too trusting, particularly of antisocial individuals, could be counterproductive. Being excessively attentive to the needs of others rather than oneself may also be less adaptive. Agreeable people may be easy to exploit and unable or unwilling to assert their rights. To take an extreme example, it is highly doubtful that the man who is known to have been most reproductively successful in recorded human history – Moulay Ismail the Bloodthirsty, the last Sharifian Emperor of Morocco, who had at least 1,042 children but was also reputed to have murdered 30,000 men by his own hands (Miller & Kanazawa, 2007, p. 35) – would be described as Agreeable.

Paradoxically, it appears that people who are called tough-minded, critical, and skeptical often do better in professions and business than those with high Agreeableness scores. Agreeableness is almost always rated as attractive in others. Disagreeable people are rated as egocentric, selfish, and unkind.

Conscientiousness. Conscientious individuals are hardworking, dutiful, and orderly. They show self-control and tend to be moral. They may be achievement oriented and highly diligent. It is no surprise, then, that this trait is one of the clearest makers of success in educational and occupational life. Conscientious people plan for the future and are happy to work constantly for desirable long-term payoffs. People like to work and study with conscientious people. They pitch up and pitch in.

The major downside of high conscientiousness is associated with perfectionism, rigidity, and social dogmatism. Conscientiousness may also be thought of as a reaction to low ability in competitive settings. For example, Conscientiousness is negatively associated with intelligence (Moutafi, Furnham, & Paltiel, 2004). Some students may learn to compensate or "make up" for their lack of ability. Thus, Conscientiousness is associated not with what in business circles is described as "doing the right thing," but with "doing the right thing right." Managers, teachers, and parents value Conscientiousness and attempt to instill it in those they know.

Nettle's (2006) ideas are summarized in Table 37.1, which is an adaptation and extension of his work. Nettle (2006) notes that his "trade-off" evolutionary account of traits may be useful partly because it is hypothesis-generating. Thus, for instance: Neuroticism may facilitate performance on particular perceptual motor tasks; highly Open people may be either culturally embraced or marginalized; Consientious people are slow to respond to affordances in the local environment; or Agreeable people are often regarded as "suckers" or victims of exploitative individuals. He suggests that the framework is not a "post hoc explanation of the past," but rather an engine for "predictors about the consequences of dispositional variation in the present" (Nettle, 2006, p. 629).

It may be possible that there is a reproductive niche for highly Introverted or Neurotic people. Equally, it may be that the highly Open, low Conscientious, creative persons serve a very useful evolutionary function for the group, even though they may have rather unhappy lives. However, it may be too early to judge. The marriage of dispositional and evolutionary psychology is young. It may well be that the former approach is able to generate interesting and important hypotheses, which the latter can test empirically.

37.4 HOW TO RECONCILE EVOLUTIONARY PSYCHOLOGY AND INDIVIDUAL DIFFERENCES

If a certain personality trait – say, Extraversion – is adaptive, why does its polar opposite – say, Introversion – exist simultaneously? Why does evolution by natural and sexual selection not eliminate variation and ensure that all individuals have the most adaptive trait? How can evolutionary psychology explain the existence of individual differences in personality traits? There are several ways to reconcile individual differences in personality and other traits – the main domains of personality psychology, behavior genetics, and differential psychology – with universal human nature, which is the main focus of social psychology and evolutionary psychology.

37.4.1 The Fluctuating Selection Model

As summarized in Section 37.3, Nettle (2005, 2006) is the primary proponent of the fluctuating selection model. The model focuses on *variations in selection pressure*, either over time or by geographical locations, and *trade-offs* as the causes of individual differences. Different ecological niches select for different personality types. For example, in safe and predictable environments that are relatively free of predators, Extraversion may be selected, because the

exploration of unknown environment may bring benefits in terms of additional resources and mating opportunities. However, the same tendency to explore the environment may harm Extraverts in dangerous and unpredictable environments, because Extraverts would then be more likely to come into contact with predators and other unknown dangers, so such environments may select for Introverts.

The same principle holds for temporal variations in the same geographical location. During times of plenty, the exploration of unknown environments may produce relatively little additional benefit and may have some cost in terms of predation, injury, and accidents, whereas, during times of scarcity, Extraverts may be more likely to locate new food sources, undiscovered and undiscoverable by Introverts.

The fluctuating selection model highlights the fact that few personality traits have unqualified and universal reproductive benefits under all circumstances; instead, their benefits depends on contexts and circumstances. Every personality trait has trade-offs in terms of costs and benefits. Different geographical locations and different historical times may select for different personality types. If, however, one aggregates across all ecological niches occupied and all historical times lived by humans, the average reproductive benefits of different personality types would be equal and the distribution of personality types would probably turn out to be normal, with sufficient variance to allow for individual differences in personality.

37.4.2 The Fitness Indicator Model

Miller (2000, 2009) is the primary proponent of the fitness indicator model of individual differences, principally in general intelligence. Miller argues that general intelligence and other fitness indicators such as height and physical attractiveness are indicators of underlying genetic and developmental health, which he calls the f factor. Those who have healthier and higher-quality genes are able to have higher general intelligence, greater height, and greater physical attractiveness than those who have less healthy and lower-quality genes. Since it is difficult for everyone to have perfectly healthy genes and since different individuals have different mutation loads, which simultaneously affect the quality of their genes and phenotypic fitness indicators (the greater the mutation load, the lower the general intelligence, for example), there must be natural variance in the levels of general intelligence and in the other fitness indicators. From this perspective, individual differences in intelligence and personality are the direct outcomes of individual differences in mutation loads.

Both men and women may prefer to mate with those who have better genes; so, for example, men prefer to mate with physically attractive women and women prefer to mate with tall and intelligent men. However, not everyone

Table 37.1 Examples of adjectives, Q-sort[a] items, and cost–benefits defining the five factors of personality.

| Factor | Factor definers | | Positive benefits | Negative costs |
	Adjectives	Q-sort items		
Extraversion	Active Assertive Energetic Enthusiastic Outgoing Talkative	Talkative Skilled in play, humor Rapid personal tempo Facially, gesturally expressive Behave assertively Gregarious	Big social networks Relationship and mating success Explorer of opportunities Happiness	Accidents and risk-taking Impulsivity and poor decision-making Relationship instability
Neuroticism	Anxious Self-pitying Tense Touchy Unstable Worrying	Thin-skinned Brittle ego defenses Self-defeating Basically anxious Concerned with adequacy Fluctuating moods	Hypervigilance Achievement striving Emotional sensitivity Competitiveness	Poor mental health Stress sensitivity Poor physical health
Openness	Artistic Curious Imaginative Insightful Original Wide interests	Wide range of interests Introspective Unusual through processes Values intellectual matters Judges in unconventional terms Aesthetically reactive	Social attractiveness Creativity Flexibility Change oriented	Mental illness Social exclusion Bizarre belief systems and lifestyles
Agreeableness	Appreciative Forgiving Generous Kind Sympathetic Trusting	Not critical, skeptical Behaves in a giving way Sympathetic, considerate Arouses liking Warm, compassionate Basically trustful	Psychological mindedness Social networks Strong relationships Valued group member	Vulnerable to exploitation Failure to maximize personal advantages Too conflict-avoidant Low assertiveness
Conscientiousness	Efficient Organized Planning ability Reliable Responsible Thorough	Dependable, responsible Productive Able to delay gratification Not self-indulgent Behaves ethically Has high aspirational level	Long-term planning Longer life expectancy Good citizenship Dependable and dutiful team member	Obsessionality and perfectionism Rigidity with poor flexibility Slow to respond

[a] Q-sorting is a way of sorting data, similar to an intuitive version of factor analysis. It is often a ranking of variables, usually presented as statements printed on small cards according to some specific instructions.

succeeds in mating with the ideal mates endowed with the perfect genes, and some individuals whose mate values are lower must necessarily mate with others with less ideal qualities. In addition, more mutations are introduced in every generation. Hence, there will always be natural variations and individual differences in all of the fitness indicators, such as general intelligence and personality traits.

37.4.3 The Random Quantitative Variations Model

Tooby and Cosmides (1990) and Kanazawa (2010) are the primary proponents of the random quantitative variations model. Some critics (Borsboom & Dolan, 2006) believe that individual difference traits cannot be evolved adaptations and evolved adaptations cannot exhibit individual differences. Adaptations are universal and constant features of a species shared by all of its members, the critics contend, so there cannot be any heritable individual differences in such universal features. These critics argue that evolved adaptations and individual differences are mutually exclusive.

These criticisms betray a profound misunderstanding of the nature of evolved adaptations. A trait could simultaneously be an evolved adaptation and an individual differences variable (Kanazawa, 2010, pp. 283–284; Sosis, 2009, pp. 326–327). In fact, *most evolved adaptations exhibit such individual differences*. Full-time bipedalism is a uniquely human adaptation, yet some individuals walk and run faster than others. The eye is a complex adaptation, yet some individuals have better vision than others. Language is an adaptation, yet some individuals learn to speak their native language at earlier ages and have greater linguistic facility than others.

Individual differences in evolved adaptations are what Tooby and Cosmides call "random quantitative variation on a monomorphic design." "Because the elaborate functional design of individuals is largely monomorphic [shared by all members of a species], our adaptations do not vary in their architecture from individual to individual (*except quantitatively*)" (Tooby & Cosmides, 1990, p. 37, emphasis added).

Intraspecific (interindividual) differences in such traits, which personality psychologists measure, study, and explain, pale in comparison to interspecific differences. It is therefore possible for traits to be both universal and species-typical evolved adaptations (exhibiting virtually no variations in the *architecture* of the evolved design in cross-species comparison) *and* to manifest vast individual differences in the quantitative performance among members of a single species (Kanazawa, 2010, pp. 283–284).

Tooby and Cosmides (1990, pp. 38–39) made this exact point using "a complex psychological mechanism regulating aggression" as an example. They contended that this mechanism is an adaptation, even though there are heritable individual differences in the mechanism's threshold of activation (whether one has a "short fuse" or not). Tooby and Cosmides suggested that a complex psychological mechanism regulating aggression "is (by hypothesis) universal and therefore has zero heritability" (p. 38), even though "the *variations* in the exact level at which the threshold of activation is set are probably not adaptations" (p. 39) and can thus vary between individuals.

The ability to run bipedally – faster than a sloth but slower than a cheetah – is a trait that is universally shared by all normally developing humans; it is a species-typical adaptation with zero heritability. But the exact speed at which a human can run is a heritable individual differences variable and is therefore not an adaptation. Similarly, Kanazawa (2004, 2010) proposes that general intelligence is an adaptation and has zero heritability (in the sense that all humans have the ability to think and reason and are more intelligent than sloths or cheetahs), even though the exact level of an individual's general intelligence ("IQ") is not an adaptation and is a highly heritable individual difference

variable. Tooby and Cosmides (1990) also contended that "nonadaptive, random fluctuations in the monomorphic design of a mental organ can give rise to heritable individual differences *in nearly every manifest feature of human psychology*" (p. 57, emphasis added). One would therefore expect some individual differences in all evolved traits, and the Big Five and other personality traits are no exceptions.

From this perspective, Extraversion and Introversion are not two polar opposites. Introversion instead is merely a lower level of Extraversion (if Extraversion is the evolved adaptation) or Extraversion is merely a lower level of Introversion (if Introversion is the evolved adaptation). The same is true of all other seemingly opposite personality traits. Explicitly recognizing that personality traits can simultaneously be an evolved, species-typical adaptation *and* an individual differences variable allows us to integrate evolutionary psychology – the study of universal human nature – and personality psychology – the study and measurement of heritable individual differences.

37.5 THE 50–0–50 RULE OF THE ORIGIN OF INDIVIDUAL DIFFERENCES IN PERSONALITY: PARENTS ARE IMPORTANT, BUT PARENTING IS NOT

"Do parents have any important long-term effects on the development of their child's personality? This article examines the evidence and concludes that the answer is no." Thus begins Judith Rich Harris's groundbreaking 1995 *Psychological Review* article "Where Is the Child's Environment? A Group Socialization Theory of Development."

Given that the main theoretical and empirical focus of evolutionary psychology is the species-typical universal human nature that all humans share (or male human nature that all men share and female human nature that all women share), the question of individual differences in personality and other traits often lies outside of its purview. This is where the sister field of behavior genetics takes over. Theory and research in behavior genetics has often demolished the common assumption that socialization and environment strongly influence how individuals come to be different from each other (Rowe, 1994).

For example, in her 1995 *Psychological Review* article and then in her 1998 book *The Nurture Assumption: Why Children Turn Out the Way They Do*, Harris methodically refutes the universally held assumption that how parents raise their children is a major determining factor in how they turn out. Harris instead argues that parental socialization has very little effect on children because they are mostly socialized and influenced by their peers. While Harris's conclusion was enormously controversial and widely condemned by politicians and the media alike, it is in fact corroborated by behavior genetics research.

Behavior geneticists decompose total variance in personality and other individual traits into three components: heritability (genes), shared environment (everything that happens within the family that makes siblings from one family similar to each other but different from those from other families), and non-shared environment (everything that happens within and outside the family that makes siblings from one family different from each other) (Plomin et al., 2012). Behavior geneticists contend that the rough rule of thumb when it comes to the determinants of adult personality and other traits is 50–0–50; that is, roughly 50 percent of the variance in personality, behavior, and other traits is heritable (influenced by genes), roughly 0 percent of the variance is due to the shared environment (what happens within the family that is experienced similarly by all siblings), and roughly 50 percent of the variance is due to the non-shared environment (what happens inside and outside the family that is not shared by siblings) (Kanazawa, 2012).

For example, partisan attachment in politics (how strongly individuals identify with their political party) is precisely 50–0–50 among both men and women; 50 percent of the variance in partisan attachment is heritable, 0 percent is due to the shared environment, and 50 percent is due to the non-shared environment (Hatemi et al., 2009). Another study estimates it as 46–0–54 (Settle, Dawes, & Fowler, 2009). Socio-sexual orientation – whether one is sexually restricted or unrestricted – follows 49–2–47 (Bailey et al., 2000). The remaining two percent of the variance in their model is attributable to age.

Largely as a result of the high heritability and low variance attributed to the shared environment in socio-sexuality, the risk of divorce follows the same pattern. One study estimates it to be 59–0–41 among women and 55–0–45 among men (Jockin, McGue, & Lykken, 1996). Another study estimates it to be 53–0–47 for women and 52–0–48 for men (McGue & Lykken, 1992). Circadian rhythm – whether one is a morning person or a night person – follows 45–0–55 (Hur, 2007).

Of course, the precise breakdown among the three components in the behavior genetic model varies slightly by the trait in question, and also by the population used to derive the estimates, but the general pattern across many traits in many populations appears to be 50–0–50. A major exception to the 50–0–50 rule is general intelligence, for which heritability is *higher*. There is wide consensus among intelligence researchers that the heritability of adult general intelligence is about 0.8 (Jensen, 1998).

Parental socialization normally (though not exclusively) falls under the shared environment; it therefore has virtually no influence on how children turn out. "The implication is that parents are often given too much credit for children who turn out well, and too much blame for children who turn out poorly. The source of causal influence is not in rearing variation, but in the genes and in unshared environmental variation" (Rowe, 1994, p. 92).

Harris's work highlights the importance of the non-shared environment (particularly peer socialization) on child development and individual differences in adult personality, and partly explains why siblings who share half of their genes and are raised by the same set of parents within the same family can often turn out to be very different – as different sometimes as children from different families. Of course, contrary to how the media portrayed (and viciously attacked) Harris's work, her conclusion decidedly does *not* mean that parents are not important to children's development. On the contrary, it means that parents are *enormously* important because children receive 100 percent of their genes from their parents. It simply means that, within broad limits, how parents raise and socialize their children may not be very important to adult personality. It also explains why adopted children often turn out to be very similar to their biological parents and not at all like their adoptive parents, even when the adopted children are raised entirely by their adoptive parents and may not even know their biological parents (or even that they were adopted) (Rowe, 1994, pp. 7–8). Children often do greatly resemble their biological parents in their personality, values, and behavior. Conservatives often beget conservatives; liberals often beget liberals. Football enthusiasts often beget football enthusiasts; opera aficionados often beget opera aficionados. But this is mostly because they share common genes, not because the parents raise their children in certain ways.

One recent and highly creative line of evidence for the strong effect of genes and negligible effect of shared environment comes from studies of doppelgängers (Segal, 2013; Segal, Graham, & Ettinger, 2013). Behavior genetics and twin research typically estimate the importance of genes for personality and thus the strength of heritability by demonstrating that monozygotic (MZ) twins, who share 100 percent of their genes, are much more similar to each other than are dizygotic (DZ) twins, who share only 50 percent of their genes. One persistent criticism of the twin methodology is that MZ twins are treated more alike by their parents, family members, and friends than are DZ twins because of their greater physical resemblance. Segal directly addresses and refutes this criticism of twin methodology in her study of doppelgängers (Segal, 2013; Segal, Graham, & Ettinger, 2013).

Doppelgängers are genetically unrelated individuals who, for unknown reasons, look identical to each other. If the critics of the twin methodology are correct and if MZ twins' greater similarity in their personality stems from their greater physical resemblance than DZ twins, then doppelgängers, who often look as similar to each other as MZ twins and much more similar than typical DZ twins, should develop similar personalities because of the similar treatments and reactions that they received from their respective families and friends. In large samples of doppelgängers from all over the world, Segal demonstrated that the correlation between doppelgängers on most personality dimensions averaged to be 0 (Segal, 2013; Segal, Graham, & Ettinger, 2013). Despite their identical appearance (like MZ twins), doppelgängers are not at all similar in their personalities (as MZ twins are). Segal's studies of doppelgängers strongly suggest that MZ twins' close similarities in personality stem from their shared genes, not from environmental factors.

37.6 CONCLUSION

Correlational or differential psychology has always emphasized how systematic individual differences interact with environmental factors to "produce" behavior. Personality and differential psychologists have made great advances in the last half-century to delineate the structure of personality, especially in establishing the Big Five personality dimensions (McCrae & Costa, 1987). More effort has gone into the classification and description of personality than into the explanation of the process by which traits lead to particular social behavior. The emphasis has been on describing stable universal individual differences in personality.

Evolutionary psychology primarily focuses on universal human nature and, as a result, individual differences in personality tend to be outside its purview. However, there are ways to reconcile evolutionary psychology and personality psychology. The primary ways to do so include the fluctuating selection model (Nettle, 2005, 2006), the fitness indicator model (Miller, 2000, 2009), and the random quantitative variations model (Kanazawa, 2010; Tooby & Cosmides, 1990). Behavior genetics and twin research, particularly the recent work on doppelgängers (Segal, 2013; Segal, Graham, & Ettinger, 2013), convincingly demonstrate the relative importance of genes and the relative unimportance of shared environment, including parental socialization, in determining individual differences in adult personality. Many personality traits appear to follow the 50–0–50 rule: roughly 50 percent of the variance in individual differences in personality is heritable (determined by genes), roughly 0 percent is attributable to the shared environment, and roughly 50 percent is due to the non-shared environment, including peer socialization (Harris, 1995, 1998).

As personality psychologists move away from classification and description onto the explanation of individual differences, they are likely to find the contributions of behavior genetics and evolutionary psychology to be much more important for understanding the fundamental question of how and why there are systematic differences in individuals.

REFERENCES

Bailey, J. M., Kirk, K. M., Zhu, G., Dunne, M. P., & Martin, N. G. (2000). Do individual differences in sociosexuality represent genetic or environmentally contingent strategies? Evidence from the Australian Twin Registry. *Journal of Personality and Social Psychology*, **78**, 537–545.

Block, J. (1995). A contrarian view of the five-factor approach to personality description. *Psychological Bulletin*, **117**(2), 187–215.

Borsboom, D., & Dolan, C. V. (2006). Why g is not an adaptation: A comment on Kanazawa (2004). *Psychological Review*, **111**, 433–437.

Buss, D. (2009). How can evolutionary psychology successfully explain personality and individual differences? *Perspectives on Psychological Science*, **4**, 359–366.

Cattell, R. B. (1971). *Abilities: Their Structure, Growth, and Action*. Boston, MA: Houghton Mifflin Harcourt.

Cronbach, L. (1957). The two disciplines of scientific psychology. *American Psychologist*, **12**, 671–684.

Daly, M., & Wilson, M. (1985). Child abuse and other risks of not living with both parents. *Ethology and Sociobiology*, **6**, 197–210.

Eysenck, H. J. (1981). Aim and scope. *Personality and Individual Differences*, **1**, 1–2.

Eysenck, H. J. (1992). Four ways five factors are not basic. *Personality and Individual Differences*, **13**(6), 667–673.

Figueredo, A. J., Sefcek, J. A., Vásquez, G., et al. (2005). Evolutionary personality psychology. In D. M. Buss, ed., *Handbook of Evolutionary Psychology*. Hoboken, NJ: Wiley, pp. 851–877.

Figueredo, A. J., Gladden, P. R., Vásquez, G., Wolf, P. S. A., & Jones, D. N. (2009). Evolutionary theories of personality. In P. J. Corr & G. Matthews, eds., *Cambridge Handbook of Personality Psychology: Part IV. Biological Perspectives*. Cambridge, UK: Cambridge University Press, pp. 265–274.

Furnham, A. (2008). *Personality and Intelligence at Work*. London: Routledge.

Furnham, A., & Cheng, H. (2015). The stability and change of malaise scores over 27 years: Findings from a nationally representative sample. *Personality and Individual Differences*, **79**, 30–34.

Furnham, A., & Heaven, P. (1999). *Personality and Social Behaviour*. London: Arnold.

Goldberg, L. R. (1992). The development of markers for the Big-Five factor structure. *Psychological Assessment*, **4**(1), 26–42.

Harris, J. R. (1995). Where is the child's environment? A group socialization theory of development. *Psychological Review*, **102**, 458–489.

Harris, J. R. (1998). *The Nurture Assumption: Why Children Turn Out the Way They Do*. New York: Free Press.

Hatemi, P. K., Alford, J. R., Hibbing, J. R., Martin, N. G., & Eaves, L. J. (2009). Is there a "party" in your genes? *Political Research Quarterly*, **62**, 584–600.

Hur, Y.-M. (2007). Stability of genetic influence on morningness–eveningness: A cross-sectional examination of South Korean twins from preadolescence to young adulthood. *Journal of Sleep Research*, **16**, 17–23.

Jensen, A. R. (1998). *The g Factor: The Science of Mental Ability*. Westport, CT: Praeger.

Jockin, V., McGue, M., & Lykken, D. T. (1996). Personality and divorce: A genetic analysis. *Journal of Personality and Social Psychology*, **71**, 288–299.

Kanazawa, S. (2004). General intelligence as a domain-specific adaptation. *Psychological Review*, **111**, 512–523.

Kanazawa, S. (2010). Evolutionary psychology and intelligence research. *American Psychologist*, **65**, 279–289.

Kanazawa, S. (2011). Evolutionary psychology and individual differences. T. Chamorro-Premuzic, S. von Stumm, & A. Furnham, eds., *The Handbook of Individual Differences*. Oxford: Blackwell-Wiley, pp. 353–376.

Kanazawa, S. (2012). *The Intelligence Paradox: Why the Intelligent Choice Isn't Always the Smart One*. New York: Wiley.

McCrae, R. R., & Costa, P. T., Jr. (1987). Validation of the Five-Factor Model of personality across instruments and observers. *Journal of Personality and Social Psychology*, **52**, 81–90.

McGue, M., & Lykken, D. T. (1992). Genetic influence on risk of divorce. *Psychological Science*, **3**, 368–373.

Miller, A. S., & Kanazawa, S. (2007). *Why Beautiful People Have More Daughters*. New York: Penguin.

Miller, G. F. (2000). *The Mating Mind: How Sexual Choice Shaped the Evolution of Human Nature*. New York: Doubleday.

Miller, G. F. (2009). *Spent: Sex, Evolution, and Consumer Behavior*. New York: Viking.

Moutafi, J., Furnham, A., & Paltiel, L. (2004). Why is conscientiousness negatively correlated with intelligence? *Personality and Individual Differences*, **37**, 1013–1022.

Nettle, D. (2005). An evolutionary approach to the extraversion continuum. *Evolution and Human Behaviour*, **26**, 363–373.

Nettle, D. (2006). The evolution of personality variation in humans and other animals. *American Psychologist*, **61**, 622–631

Plomin, R., DeFries, J. C., Knopik, V. S., & Neiderhiser, J. M. (2012). *Behavior Genetics*, 6th ed. New York: Worth.

Rowe, D. C. (1994). *The Limits of Family Influence: Genes, Experience, and Behavior*. New York: Guilford.

Segal, N. L. (2013). Personality similarity in unrelated look-alikes: Addressing a twin study challenge. *Personality and Individual Differences*, **54**, 23–28.

Segal, N. L., Graham, J. L., & Ettinger, U. (2013). Unrelated look-alikes: Replicated study of personality similarity and qualitative findings on social relatedness. *Personality and Individual Differences*, **55**, 169–174.

Settle, J. E., Dawes, C. T., & Fowler, J. H. (2009). The heritability of partisan attachment. *Political Science Quarterly*, **62**, 601–613.

Sosis, R. (2009). The adaptationist–byproduct debate on the evolution of religion: Five misunderstandings of the adaptationist program. *Journal of Cognition and Culture*, **9**, 315–352.

Tooby, J., & Cosmides, L. (1990). On the universality of human nature and the uniqueness of the individual: The role of genetics and adaptation. *Journal of Personality*, **58**, 17–67.

38 Applying Evolutionary Principles to Criminality

SANDIE TAYLOR AND LANCE WORKMAN

38.1 INTRODUCTION

In this chapter, we consider how developments in evolutionary theory might be applied to help us understand why some people gravitate toward criminal behavior. Sociologists generally explain the emergence of criminal behavior in terms of responses to unfavorable societal factors. In contrast, social and developmental psychologists explain such behavior as the outcome of the interaction between individual differences and a challenging rearing environment. Evolutionary psychologists, however, bring a whole new perspective to the problem of criminality (Taylor & Workman, 2019). They suggest that evolutionary principles such as kin selection theory, reciprocation, sexual selection, and parental investment theory can provide powerful tools to help explain criminality. Before considering how these principles might be applied, we need first to explore what is meant by the concept of "criminality."

Defining criminality and, in particular, what encapsulates criminal behavior is far from straightforward. For a definition of what constitutes criminal behavior and criminality as an underlying cognitive and personality dimension, it is appropriate to adopt a legal approach. Although legal approaches vary cross-culturally due to differences of cultural infrastructure, McGuire (2004) has highlighted the international congruency regarding the definition of serious crime and the similarity of responses from respective criminal justice systems. One concept that is found cross-culturally is that of *corpus deliciti*, which translated refers to "body of the crime"; in other words, the information and evidence that proves a criminal act has been committed. Two important elements of *corpus deliciti* relevant to our discussion here are *actus reus* (or a guilty act, such as murder) and *mens rea* (or a guilty mind, such as intending to do harm). *Webster's New World Law Dictionary* defines *actus reus* as a "voluntary and wrongful act or omission that constitutes the physical components of a crime" (Wild, 2010, p. 13). *Mens rea* is defined as a "defendant's guilty state of mind, as an element in proving the crime with which he or she is charged" (Wild, 2010, p. 178).

Actus reus and *mens rea* are two important concepts in determining whether a crime has been committed intentionally. This conceptual distinction is, however, not the only divide used to ascertain the nature of the crime committed. During the reign of Henry VII *mala in se* and *mala prohibita* were introduced and later legally formalized by Sir William Blackstone (1765–1769). This distinction offers a crude demarcation between serious and nonserious criminal behavior, respectively. Davis (2006) considers Blackstone's conception of *mala in se* crimes as contravening the human moral code that offends a higher supreme being (e.g., God in Christianity, Allah in Islam). These crimes are the most serious and harmful to potential victims and, as such, they include murder, sexual offenses, pedophilia, arson, aggravated burglary, and affray. For the perpetrators who commit these crimes, their behaviors and motivations lack a distinctive moral compass and appear to be devoid of empathy. Statutory laws, alternatively, vary cross-culturally but are there to prevent societal chaos and to promote rule acceptance through specific regulatory statutes prohibiting specified behaviors – break these and you are punished by set tariffs defined in *mala prohibita*.

If evolutionary principles are to be used to help explain the occurrence of criminal behavior, then it is necessary to consider these in the light of *mala in se* crimes. This is because such acts are considered to be criminal cross-culturally. Clearly, if evolutionary principles are to be utilized in this task, then it will be necessary to consider a biological/evolutionary foundation for such criminal behavior. Typically in the social and natural sciences, processes that could potentially drive an individual to commit serious crime can be considered as biologically or environmentally oriented. Important findings over the last 30 years by Karmiloff-Smith (1996, 1998) have begun to explore the interactive relationship between biological and environmental factors during development. According to Karmiloff-Smith, genetic or biologically oriented predispositions for violent behavior only develop to their full potential if the social environment is challenging. Here, Karmiloff-Smith is presenting an epigenetic model. This allows for environmentally contingent internal states and responses to develop that, while they may be socially inappropriate, might also be adaptive given a stressful social environment. This is a tenable explanation, and one that gains some support from research into scanning techniques and behavioral genetics (see Section 38.5). Hence, an understanding of the development of criminal behavior requires an understanding of the biological foundations, in addition to factors such as infant and peer attachment, social bonding, and socialization. We argue here that the underlying framework that can be used to connect these factors and explain why some individuals commit serious and often violent crimes is evolutionary psychology. Furthermore, we argue that the need to develop laws arose from a number of adaptive challenges that our ancestors faced. Selection pressures faced by our hominin ancestors initiated the need for individuals to live together in forager groups where cooperation generally promoted individual inclusive fitness. One of the ways to enable this, we consider, was the development of laws, however rudimentary, based on adaptive moral parameters (Krebs, 1998). Taylor (2016) highlights three plausible pathways for why simple rules/laws were created by our ancestral hominins:

1 Set rules and simple laws controlled behaviors and as a consequence antisocial behavior arose out of the definitions of unacceptable behavior.
2 Morality underlined the development of these rules and laws and helped guide individuals away from behaving antisocially.
3 Socialization was the underlying process encouraging individuals to behave morally and to follow the rules and laws set.

Points 1 and 2 are the most pertinent for our discussion. These will be considered in relation to evolutionary principles and might be thought of as part of our hominin legacy – discussed next.

38.2 LAWS: THE HOMININ LEGACY

The beginnings of an infrastructure for civilization stemmed from the point when our hominin ancestors, the australopithecines, first evolved in Equatorial Africa around four million years before present (MYBP). When the *Homo* line evolved from gracile australopithecines, fossils and the artifact record suggest that they dwelled in groups of between 20 and 200 individuals from around 12 up until 2 MYBP. Many of these were genetically related and others would often have formed friendship groups. Although these groups comprised relatively small numbers by our standards today, a sense of order and cooperation was required to organize successful foraging expeditions, to defend against raids by competing hominin bands, and to negotiate the complexities of social relationships in general. The successful continuation of cooperation among these early hominin dwellers would

be based on kinship and reciprocation (Barkow, 1989). Reciprocation (originally labeled "reciprocal altruism"), first considered by Trivers in 1972, was a way of creating a pro-social living environment where individuals benefitted from the delayed return of favors. Later, in 1985, Trivers outlined a number of important considerations regarding reciprocal altruism that are relevant to our understanding of criminality. First, a sharing and cooperative social environment is the norm, and therefore an individual not playing by the rules and who instead takes advantage of the system by failing to reciprocate (a free rider) needs to be easily detected. Second, by encouraging reciprocation of aid to others in the group, an emotional tie is developed. This emotional tie is entwined with empathy and morality, which underpin the principle of human reciprocal altruism. As we will see later in Sections 38.3 and 38.3.1, empathy and morality are important factors in the development of human pro-social behavior. It would appear that reciprocation underlies the development of many human laws. Some evolutionary psychologists, such as Trivers, argue that our moral codes developed as a consequence of the harsh open savanna environment early hominins experienced. Such moral codes led, over time, to the development of rules and laws that were based around what is considered to be appropriate social behavior in a given group. According to Krebs (1998), the development of many rules, laws, and moral codes arose to counteract the effects of free riders. Of course, individuals who fail to conform to rules and laws in modern societies are punished once caught. Free riders in modern societies, it is argued, might be retained in the population so long as their numbers are proportionally low (Mealy, 1995, 2005; Workman & Reader, 2014). It has been suggested that around three percent of the population are regular non-reciprocators due to having a "personality disorder" (possibly maintained at this ratio through frequency-dependent selection; see Section 38.4). We will return to the notion of personality disorders in relation to criminal behavior in Section 38.4.

38.3 MORAL BEHAVIOR: THE HOMININ LEGACY

Investigators have long sought to understand the roots of human moral behavior. One argument put across by Hobbes (1651) is that the natural human state is one of ruthless, selfish conflict. In Nayef Al-Rodhan's (2008) theory of "Emotional Amoral Egoism," he states that our behavior is directed by our emotional self-interest underpinned by "genetically coded survival instincts ... modified by the totality of our environment and expressed as neurochemically mediated emotions and actions" (p. 16). Like Hobbes, he believes that humans need rules and laws to create social order, and this is achieved through cooperation. Evolutionary psychologists have addressed this question by considering how humans evolved specific brain mechanisms to process the information used in our moral decision-making.

Ruse and Wilson (1985) have argued that our moral codes are shaped by our genes to increase our inclusive fitness. It would therefore make little sense to adopt a moral code or a reproductive strategy that would put the viability of offspring at risk. In order to explain this, Ruse and Wilson introduced the notion of epigenetic rules. Many of these gene-driven epigenetic rules influence how we respond to and interact with the social environment. An example of an epigenetic rule provided by Ruse and Wilson is the avoidance of incestuous relationships, as these increase the probabilities of having nonviable offspring. Interestingly, many cultures appear to conform to this epigenetic rule by labeling incest as an immoral act (generally sanctioned by criminal justice systems). This notion was expanded on by evolutionary psychologist Denis Krebs (1998). He highlighted three pathways leading to the evolution of morality: altruism via reciprocity, devotion, and deference to authority. Altruism via reciprocity was discussed in Section 38.2. In relation to devotion, Krebs (1998; see also Krebs, 2011, and Chapter 11) argues that ensuring offspring survival is best achieved through the development of strong attachment bonds to increase the longevity of the relationship. Strong attachments occur through devotion to one another. Regarding deference, Krebs argues that it is a superior survival strategy to show deference rather than to challenge a higher-status individual, which might result in injury or death. According to Krebs, the social hierarchical structure of early hominins contained dominant "alpha males." There were two possible responses to these males: show deference or defiance. Showing deference ensured survival, whereas defiance may have resulted in injury or death if the challenger was overwhelmed by the alpha male. Today, the concept of the alpha male has largely been replaced by societal infrastructure and, in particular, by the laws of the land. This means that whereas we once paid deference to the alpha male, we now do so to organizations such as the police force. Hence, following on from Krebs' model, many *mala in se* crimes can be traced back to transgressions of these three pathways. For example, abandoning a spouse or neglecting a child frequently leads to legal sanctions. Likewise, failure to defer to bodies such as the police force often impacts negatively on individuals.

38.3.1 The Neural Basis of Moral Behavior: The Use of Brain Scanning Technology

One approach to investigating the relationship between moral development and evolved brain mechanisms is to look at how the brain processes moral-oriented information. Neuroscience provides us with the software to observe which areas of the brain are involved when dealing with such information. Evolutionary psychologists have used moral dilemmas as tools for exploring areas of the brain that become activated during functional magnetic resonance imaging (fMRI) scanning. Studies using fMRI arguably provide evidence for the existence of evolved brain

mechanisms that facilitate moral decision-making. This, in turn, provides impetus to the notion of there being moral universals that, when transgressed, can be considered to be universal crimes. To explore this possibility, Hauser et al. (2007) presented four different moral scenarios to participants who had to decide whether an action is morally correct. These scenarios were based around the need to stop or divert a train in order to save lives. Under Scenario 1, "Denise" is traveling on a train on which the driver has fainted. She can steer the train to kill one person rather than the five people it is heading toward. Eighty-three percent of participants consider that it is right for Denise to kill the one rather than the five. In Scenario 2, "Frank" has to decide whether or not to push a hefty man in front of a train to stop it running into five people. In this case, only 12 percent of participants consider that pushing the hefty man is morally correct. In Scenario 3, "Ned" can pull a switch to divert a train onto a side track where it will kill one man instead of five. In this case, 56 percent of participants consider it morally justified to pull the switch. Finally, in Scenario 4, "Oscar" can also throw a switch setting a train off onto a side track to avoid killing five people, but in this case the train will be stopped by a heavy weight that has a man in front of it. This means that the weight will stop the train from killing five people but the man in front of it will be killed. In this final scenario, 72 percent of participants consider it morally correct for Oscar to throw the switch. Using these scenarios, Hauser, Young, and Cushman (2008) asked participants to decide how morally permissible it was to perform the actions suggested in the narrative. We can see that in each case one person is sacrificed to save five. It might seem strange, therefore, that the percentage of participants within each scenario considering the act to be appropriate varies so much. The big difference, however, between the "Frank scenario" and the other scenarios is that Frank physically shoves the man in front of the train. Under the other scenarios, no physical contact is made with the person who is sacrificed. In the case of the "Frank scenario," people perceive this as making a "moral-personal" decision. Participants are uncomfortable about this because the action involves direct contact. The actions of Denise and the others, alternatively, do not involve direct physical contact and are perceived as "moral-impersonal" acts. Hence, participants felt less uncomfortable about their actions. This difference between these scenarios really comes down to the degree of distance from the consequences of the decision made (and especially in the case of personal contact). This is interesting because research by Greene et al. (2001) using fMRI scans had previously shown that moral-personal and moral-impersonal decisions activate the brain differentially. They located specific areas of the brain (i.e., the medial frontal gyrus, posterior cingulate gyrus, and angular gyrus) that were more active during decisions of a moral-personal nature. Such findings may make a lot of sense when we think about how socially integrated early hominins must have been. It may well have paid them to have, in effect, safety mechanisms to avoid seriously harming the kith and kin that made up such groups except under the most extreme of conditions. Therefore, at some point in our ancestral history we are likely to have evolved a neural substrate that supported the two types of moral decision-making. Based on this and similar findings, evolutionists such as Hauser consider individuals who commit *mala in se* crimes to have a poorly developed neural substrate for moral decision-making.

While most of us avoid *mala in se* crimes, there are, however, some individuals who fail to develop the forms of empathy that are necessary to inhibit such behaviors. For example, psychopathy as an extreme subset of antisocial personality disorder (APD; see Sections 38.4 and 38.5) has been robustly associated with criminal behavior and an underdeveloped moral compass. Neurocognitive research has uncovered evidence demonstrating both structural and functional differences in the "wiring" of the brains of psychopaths. Psychopathy, although a personality dimension, contains traits that commonly act as risk factors for deviant behavior, hence crossing the boundary of society's consensus of morality and appropriate moral behavior. In Section 38.4, we consider the notion of a criminal personality type and how might it be maintained in a given population.

38.4 CAN EVOLUTIONARY PSYCHOLOGY EXPLAIN THE EXISTENCE OF PERSONALITY DISORDERS?

Evolutionary psychologists have considered personality disorders as having had an adaptive function in our ancestral past (Troisi & McGuire, 2000). Note that some of these adaptations may no longer be appropriate for modern-day environments (hence the mismatch hypothesis). In order to understand how evolutionary principles can account for personality disorders, we need to consider the different types of disorder and focus particularly on those connected with criminality. In the *Diagnostic and Statistical Manual of Mental Disorders*, 5th Edition (DSM-5; a manual describing different conditions and disorders used by psychiatrists and clinical psychologists), 10 different dimensions of personality disorder have been identified. These are divided into three clusters according to overlapping traits (American Psychiatric Association, 2013) (Table 38.1). A personality disorder is defined by the American Psychiatric Association as "an enduring pattern of inner experience and behavior that deviates markedly from the expectations of the individual's culture."

As Cluster B personality disorders are most commonly associated with criminal behavior (Table 38.2), these will be considered further. Most research into criminality has concentrated on personality disorders from Cluster B (although less so for histrionic personality disorder). It should be noted, however, that not all individuals with APD, for example, are criminal.

Research by Coid et al. (2006b) found the prevalence of personality disorders to be higher among males than

Table 38.1 Different personality disorders classified by cluster and likely criminal behavior.

Personality disorder clusters	Type of criminal behavior
Cluster A: paranoid, schizoid, and schizotypal	Mostly nonviolent but, due to suspicious and bizarre thoughts, they can commit serious violent behaviors
Cluster B: antisocial, borderline, histrionic, and narcissistic	Commonly associated with all types of criminal behavior; most likely of the three clusters to commit serious and violent offenses (*mala in se* crimes)
Cluster C: avoidant, dependent, and obsessive–compulsive	Least prone to criminal/violent behavior

Table 38.2 Breakdown of Cluster B personality disorders and criminal behavior.

Cluster B personality disorders	Type of criminal behavior
Antisocial: dishonest; destructive; callous; aggressive; irresponsible; manipulative; empathy, guilt, and shame deficit	Higher rates of institutional violence; aggression; violent offenses including homicide and sexual assault; burglary; larceny; theft; arson
Borderline: erratic emotions; impulsive; unpredictable; argumentative	Higher rates of institutional violence; aggression; violent offenses including homicide and sexual assault
Narcissistic: grandiose self-perception; egotistic; opinionated; empathy deficit	Violent offenses including homicide and sexual assault

females. Furthermore, Coid et al. (2006a) found that individuals diagnosed with Cluster B personality disorders were 10 times more likely than the general public to have a criminal conviction. This increased risk for criminality does not occur for those in Clusters A and C. Previously, Fazel and Danesh (2002) performed a meta-analysis of 62 studies across 12 countries for the prevalence of personality disorders among prisoners. Of the total of 22,790 prisoners studied, APD was most prevalent: 5,113 out of 10,797 in males (47 percent) and 631 out of 3,047 in females (21 percent). Clearly, the Cluster B personality disorders feature strongly among the prison population (and often for highly violent offenders). Psychopathy is considered to be a subset of individuals with APD. Descriptions of APD overlap with those of psychopathy, and there is debate concerning whether the two can be considered synonymously. Decuyper, De Fruyt, and

Buschman (2008) concluded that, although psychopathy is distinct from APD, behavior across these two types of personality disorder is consistent. For our purposes here, we will consider the two synonymously.

It has been argued that individuals with such personality disorders are behaving in a maladaptive way. McGuire and Troisi (1998), however, question whether they really are. The underlying genes involved in the development of such personality disorders might have offered an adaptive function when our ancestors encountered adverse environmental conditions. Being able to respond in a deceptive or antagonistic way may have been a successful strategy used to obtain resources when environmental conditions were challenging, such as during times of drought or famine. Under these conditions, the ability of psychopaths to free ride may well have been a boon for such individuals. We pointed out earlier how Trivers' conception of reciprocation also allowed individuals to be non-reciprocating free riders. McGuire and Troisi (1998) speculate that traits of the psychopathic personality, in particular their tactics to exploit others, might be an evolved strategy maintained within human populations in order to exploit this general reciprocation (see also Glenn, Kurzban, & Raine, 2011; Mealey, 1995, 2005). Some researchers have suggested that psychopathy may be maintained in the population by frequency-dependent selection; that is, an alternative strategy (non-reciprocation) is likely to be successful provided no more than a certain proportion of the population adopts it. In the case of psychopathy, it is suggested that a rate of three percent can be maintained in a given population of reciprocators (Mealy, 1995).

For this speculation to have substance, there needs to be empirical evidence for a genetic component in psychopathy. This is seen in familial research, including high concordance rate for psychopathy in monozygotic vs. dizygotic twins (Lyons et al., 1995), adopted children with biological vs. adoptive parents (Ge et al., 1996), and first-degree male relatives of psychopaths (Plomin et al., 2008). These and other studies suggest that psychopathy is heritable, with the genes accounting for around 50 percent of the variation in the disorder (Larsson, Andreshed, & Lichtenstein, 2006). According to McGuire and Troisi (1998), "In a society made up primarily of reciprocators, genes for cheaters can enter the population and remain, provided persons with such genes reproduce" (p. 191).

38.5 THE PSYCHOPATH: WHY MAINLY MALES?

There are sex differences in terms of the prevalence of psychopathy: more males exhibit psychopathic tendencies than females. This raises the question: Why does this sex difference exist? A number of reasons could explain this. First, in order to avoid detection as a free rider, moving from one group to another would make it difficult to be identified as a non-reciprocator. Moving from one group to another is easier for males given that females more often than not would be gestating, lactating, and/or caring

for offspring. Remaining in the same kinship group (where they may gain family support) might therefore have been more advantageous for females. Second, under challenging circumstances, specific psychopathic traits might have aided an individual to overcome competition, leading to successful mating (Buss, 2009; Glenn et al., 2011). Therefore, being devious and/or aggressive might have been a successful strategy for competing against other males. Third, the maintenance of psychopathy in the population suggests that it might have been sexually selected for in our evolutionary past (see Section 38.6).

If there is a genetic component to psychopathy, then it follows that there must also be a neurological substrate for such a disorder. In fact, neuroimaging techniques such as structural MRI and fMRI have helped forensic psychologists to understand the workings of the brains of murderers, many of whom were diagnosed as psychopathic. Using the Psychopathy Checklist – Revised (PCL-R; Hare & Neumann, 2006) to identify psychopaths, neuroscientists have uncovered differences in the workings of the brains of criminal psychopaths using both types of MRI scans. In such cases, reduced amygdala activity was discovered by Kiehl et al. (2001) and structural abnormalities of the amygdala were discovered by Yang et al. (2009). Violent criminals with high PCL-R scores have a reduced volume in the right amygdala (Gordon, Baird, & End, 2004). The connections between the amygdala and the orbitofrontal cortex, such as the uncinated fasciculus, were weakened or frayed in criminal psychopaths convicted for multiple rape, manslaughter, and attempted murder (Craig et al., 2009). This led Craig et al. to conclude that an abnormal amygdala–orbitofrontal pathway underpins the typical behaviors observed in criminal psychopaths. Dolan (2004) suggested that this might be a reason why UK imprisoned criminal psychopaths are responsible for committing 50 percent more *mala in se* crimes than their non-psychopathic counterparts. Shamay-Tsoory et al. (2010) argue that deficits in empathy observed in the social behavior of criminal psychopaths can be explained by the abnormal wiring of the amygdala–orbitofrontal pathway.

In the words of Brian Dugan, who was convicted of rape and murder and is now serving two life sentences, "I have empathy too – but it's like it just stops. I mean, I start to feel, but something just blocks it. I don't know what it is" (Hagerty, 2010). According to Hagerty (2010), the "emotional circuit" fails to stop psychopaths from killing, and she likens this to the brakes not working.

In addition to evidence from scanning techniques, recent developments in the field of neurogenetics have increased our understanding of how variant forms of specific genes are associated with disorders such as psychopathy. It has been found that a variant form of the gene labeled MAOA contributes to criminal behavior (Dutton, 2012; Raine, 2002). The variant form is responsible for low levels of the enzyme MAO-A, leading to high concentrations of spinal serotonin (5-HIAA). Interestingly, high concentrations of 5-HIAA are

Table 38.3 The cases of Waldroup and Fallon.

Cases	Description
Bradley Waldroup brutally murdered his wife's friend and attempted to murder his wife	Presence of the variant form of the MAOA gene; structural anomalies of the amygdala–orbitofrontal pathway
Professor James Fallon is a respected neuroscientist researching the brain scans of murderers. He has eight distant relatives who are allegedly murderers, but no personal history of criminal behavior	Presence of the variant form of the MAOA gene; structural anomalies of the amygdala–orbitofrontal pathway

associated with impulsive, aggressive behavior, as was found in the male lineage of a Dutch family (Brunner et al., 1993; McDermott et al., 2009). It has been suggested that the structural and functional anomalies of the criminal psychopath's brain coupled with the presence of the variant form of the MAOA gene predispose an individual to the typical traits observed in criminal psychopaths (Fallon, 2013, 2014). But is this the complete answer as to what creates a criminal psychopath? In Table 38.3, two individuals with the variant form of the MAOA gene and the same structural anomalies of the amygdala–orbitofrontal pathway are considered.

Considering Table 38.3, we see here two individuals with apparently the same genetic anomaly and brain structure, and yet their lifestyle choices are very different. If being a criminal psychopath is attributed only to our biology, then why is Professor Fallon a law-abiding individual? By all accounts, he too has a biological makeup predisposing him to the traits of a typical psychopath – yet he refrains from violence. Low levels of MAO-A as a singular explanation does not appear to be the answer. Studies that include the additional variable of the type of nurturance experienced, however, do provide some insight. Male children with low levels of MAO-A activity, for example, have been found to behave aggressively and delinquently if they had been maltreated or considered a failure at school (Buades-Rotger & Gallardo-Pujol, 2014; Caspi et al., 2002; Guo et al., 2008).

The difference between Waldroup and Fallon rests in the type of childhood they experienced – hence, nature interacting with nurture. Waldroup was brutally abused as a child, often experiencing extreme punishment leaving him covered in welts and bruises. Waldroup's genetic predisposition combined with an abusive upbringing culminated in his violent adult temperament. In contrast, Fallon's childhood was a happy experience that he described as loving and caring and full of fond memories. It is interesting, however, that despite his positive childhood experiences steering him away from a potentially destructive adulthood, his family – and he himself – have

noted personality traits suggesting indifference and impatience. Gao et al. (2010) emphasize the importance of a nature–nurture interaction in the development of psychopathy. Childhood abuse and neglecting parental nurturance were considered to be important factors in triggering a psychopathic phenotype. Gao et al. also argue that the contingent shift theory can account for the differences in behavior between Waldroup and Fallon. This means that the genes involved in the development of psychopathy will only lead an individual fully down the path of this disorder under seriously challenging environmental circumstances (for a full discussion of contingent shift theory and of frequency-dependent selection in relation to psychopathy, see Glenn et al., 2011). During our ancestral past, it is likely that the type of childhood experienced would be an indication of the nature of the anticipated adult environment. In the case of a socially aversive environment, it would arguably have paid some individuals to have followed the path of cold, calculating manipulation. In this way, the phenotypic psychopathic profile may have been adaptive for some individuals (and might continue to be so today). The contingent shift theory highlights how individuals adjust their interpersonal strategies according to their environmental circumstances. Having a good childhood protected Fallon from his "bad genes" and guided him toward a law-abiding interpersonal strategy. This was far from the case for Waldroup. Waldroup, however, is not an isolated case. In fact, in 2001, Frodi et al. (2001), in studying the personal histories of imprisoned psychopaths, found that such offenders were far more likely to have been physically abused as children than non-psychopathic offenders. Frodi et al. also found that a childhood of secure attachment was an extremely rare occurrence in psychopathic offenders.

In line with the contingent shift theory, Troisi (Chapter 33) argues that a dysfunctional style of insecure attachment should not be conceived of as a biological or psychosocial problem. It is the outcome of a behavioral system that has adjusted to specific nurturing experiences. Troisi asks the question, "Why do some psychosocial factors impact on the epigenome more than others?" (Chapter 33). He answers this question using a life history approach. Here, the early social interactions and experiences with the immediate environment provide the algorithms necessary to calibrate an individual's behavioral systems. It is this that prepares behavioral systems, such as attachment, to develop in an adaptive way to a probable future social environment. Likewise, Belsky (2016) highlights how a traumatic childhood can establish sequelae that promote survival through using adaptive coping mechanisms toward an unpredictable environment. In the case of Waldroup, he developed coping mechanisms that were adaptive to the type of social environment he experienced as a child given his genetic makeup. Morally, we can argue that his behavior as an adult is dysfunctional as it fails to conform to societal mores, values, and behavioral standards. In evolutionary terms, however, his attachment behavioral system was calibrated to help him cope with his circumstances and as a means of adapting to a potentially grim future social environment. Troisi (2015 and Chapter 33) argues that, in our evolutionary past, natural selection selected for genes that allowed adaptation to a multitude of diverse ecological niches. It is this diversity that influences our sensitivity to different aspects of our early lives. Note that Troisi directs his arguments toward an understanding of mental health, but it is clear that the same arguments also hold for personality disorders.

As we stated in Section 38.4, there are sex differences in the frequency of occurrence of personality disorders: more males have personality disorders than females (especially psychopathy). Putting aside personality disorders, it is also the case that males in general commit more *mala in se* crimes than females. Can we explain this difference using principles from evolutionary psychology?

38.6 SEXUAL SELECTION, ASYMMETRICAL PARENTAL INVESTMENT, AND SEX DIFFERENCES IN VIOLENT CRIMINAL BEHAVIOR

In 1877, sociologist Richard Dugdale documented a "family of deviants" – the Jukes – over a number of generations. He discovered that most of the crimes committed were by the male members of this family. Interestingly, Dugdale was one of the first to speculate that there may be both biological and environmental causes of such a sex difference. He did not, however, seek to apply Darwinian principles to explain this difference. In recent years, evolutionary psychologists have used the concepts of sexual selection and asymmetrical parental investment to explain why males, especially young males, are more aggressive than females. It is well established that, universally, men are more likely to engage in acts of physical aggression than women (Archer, 2004; Campbell, 1999; Van Vugt, 2009). Because women invest more time and effort in the production of offspring than men (i.e., like most mammals, our species has asymmetrical parental investment – think of gestation, parturition, and lactation), then, it is argued, injury sustained in violent encounters would have greater costs for them than for men. For this reason, evolutionists argue, women across cultures are more risk averse and less likely to engage in physical aggression (Campbell, 1999; see also Kanazawa, 2009, discussed later in this section). Conversely, for men, the costs (to their offspring) of engaging in physical aggression are lower, whereas the benefits, in terms of status and reputation (for them), are likely to have been relatively high. In fact, it has been argued that coalitional aggression had clear fitness benefits for our male hominin ancestors (Archer, 2004; Van Vugt, 2009). This argument has been labeled the "male warrior hypothesis" (Van Vugt, 2009). Interestingly, a recent study found that not only were men more likely than women to engage in violent robbery, but also the likelihood of injury to the victim of

the robbery is significantly higher when multiple male offenders are involved (Bourgeois & Fisher, 2018). This finding provides clear support for the male warrior hypothesis and its association with violent crime.

Although it is well established that men engage in a significantly higher proportion of violent crime than women (and that they tend to do so in groups), it is also clear that the likelihood of engaging in criminal aggression varies greatly as a function of age. Boyd (2000) considered the impact of age on male aggression in participants from the USA, Canada, and the UK. Violence was rarely seen in those aged 40 years and over, but was increasingly observed in males of 15 years and peaking before 20 years, then continuing to decline post-30 years of age. Boyd also claims that homicide co-occurs at an age when testosterone levels are at their highest. As testosterone plays an important role in male sexuality, it is at a peak during adolescence when sexual competition is high. And consequently, the potential for violence also coincides at this time. Coccaro et al. (2007) found a robust association between increased levels of testosterone in the male brain and impulsive and aggressive behavior.

Recently, Kanazawa (2009) has argued that the relationship between adolescence, testosterone, and aggression is an adaptive one. That is, at this age, males begin to strive for status and resources as an evolved means of attracting females. To Kanazawa, those males who were most able to gain status and resources at this age were also the ones most likely to boost their inclusive fitness. In this way, Kanazawa relates sexual selection theory directly to acts of criminal violence in young men. Prior to this, Kanazawa and Still (2000) had already suggested that the rapid fall in male violence beyond the age of 30 occurs because once men start to reproduce, the benefits of aggressive behavior begin to be outweighed by the costs. The reason for this, they argue, is because retribution or imprisonment is likely to impact on the welfare of the perpetrators' children. In this way, it may have paid some of our ancestors to engage in acts of criminal aggression in young adulthood and later to leave this strategy behind when offspring appear on the scene (Cartwright, 2016; Kanazawa & Still, 2000). For some, however, this shift toward a more paternal, caring mode fails to occur (as we have seen in Section 38.4, this might be due, in part, to having a personality disorder).

We tend to associate heinous crimes with psychopathy. There are, however, crimes that society abhors and yet are often committed by those who do not appear to have a personality disorder. Such crimes include infanticide. Is it possible that evolutionary theory can throw some light on infanticide?

38.6.1 The Cinderella Syndrome and Infanticide

Cross-culturally, acts of violence against children are perceived as some of the worst crimes a person can commit, and yet such acts are well documented in all cultures studied. We might ask: Can evolutionary principles be used to help explain such heinous crimes? Evolutionary psychologists suggest it can. Parental investment theory predicts that parents should be selective in their provisioning of others. The closer the kin, the more likely parents are to invest in them. It is therefore not surprising to find that, cross-culturally, adults provide the greatest degree of investment for their direct offspring and, when this is not an option, they tend to shift their investment to other close relatives, such as nephews, nieces, and younger cousins (Silk, 1980, 1990). Studies by social anthropologists suggest that, in forager societies, given their relatively short life expectancies, adoption of such younger kin is widespread (Dustin & Alcock, 2018; Gibson & Lawson, 2015; Silk, 1980, 1990). In the modern world, however, families in which stepparents and stepchildren share no genes by common descent are commonplace. In the vast majority of these "reconstituted" families, children of both parents are no doubt treated well. In some, however, this is not the case, and in a subsection of these, the stepchildren are treated heartlessly – a situation that is recognized cross-culturally and, following the fairy tale of a young girl's cruel treatment at the hands of her stepmother, has been given the label of "Cinderella syndrome." During the 1980s, two evolutionists, Martin Daly and Margo Wilson, decided to test whether the Cinderella syndrome stands up to scrutiny. Daly and Wilson reasoned that, following kin selection and parental investment theory, parental solicitude is likely to be discriminative with respect to the offspring's contribution to the inclusive fitness of said parents. What they uncovered surprised even these "hard-nosed" evolutionists. Daly and Wilson found that, while the majority of stepparents do not harm their stepchildren, the rate of recorded infanticide increased 120 times when a family contained a stepfather (Daly & Wilson, 1988). Moreover, a child of two years or younger is seven times more likely to be physically abused if there is one stepparent living in the home than in cases where there are two biological parents in the home (Daly & Wilson, 1998, 2005, 2007).

As you might imagine, such findings have been disputed. Biological philosopher (and general critic of evolutionary psychology) David Buller has suggested that Daly and Wilson's findings are illusionary and can be explained by abuse from biological parents being underreported (Buller, 2005). Putting aside the fact that, for Buller's argument to be correct, the underreporting rates would have to be 120 times lower for biological parents than for stepparents, his concerns are also weakened by a recent meta-review. In reviewing the literature, Archer found that in 9 out of 10 child infanticide studies, a higher level of stepparent violence was uncovered for stepparents than for biological ones (Archer, 2013). In fact, overall, Archer's meta-review provided strong support for the existence of the Cinderella syndrome as at least one explanation for the occurrence of the crimes of physical child abuse and infanticide. While most stepparents cope well with bringing up their unrelated children, evolutionary principles can help to provide an explanation for why this does not always work out.

38.7 CONCLUSIONS

It is perhaps natural to consider serious criminal behavior as deviant and even maladaptive. Using well-developed and tested evolutionary principles, including kin selection, sexual selection, reciprocation, and parental investment theory, we have attempted to demonstrate that much of criminality, while deviant, is not necessarily maladaptive (or might not have been during our ancestral past). The advantage of the evolutionary approach is that it reminds us that selective forces did not shape the human mind in order for us to live happily ever after, but rather to support decisions that are likely to have promoted inclusive fitness during our ancestral past. We need, then, to consider the adaptive significance of criminality if we wish to predict the conditions under which it is most likely to occur.

REFERENCES

Al-Rodhan, N. R. F. (2008). *Emotional Amoral Egoism: A Neurophilosophical Theory of Human Nature and Its Universal Security Implications*. Zurich: Lit Verlag.

American Psychiatric Association (2013). *Diagnostic and Statistical Manual of Mental Disorders*, 5th ed. Arlington, VA: American Psychiatric Publishing.

Archer, J. (2004). Sex differences in aggression in real-world settings: A meta-analytic review. *Review of General Psychology*, **8**, 291–322.

Archer, J. (2013). Can evolutionary principles explain patterns of family violence? *Psychological Bulletin*, **138**, 403–440.

Barkow, J. H. (1989). *Darwin, Sex and Status: Biological Approaches to mind and Culture*. Toronto: University of Toronto Press.

Belsky, J. (2016). The differential susceptibility hypothesis: Sensitivity to the environment for better and for worse. *JAMA Pediatrics*, **170**(4), 321–322.

Blackstone, W. (1765–1769). *Commentaries on the Laws of England: A Facsimile of the First Edition of 1765–1769*, Vol. 1. Chicago, IL: University of Chicago Press.

Bourgeois, C., & Fisher, M. (2018). More bro's, more woes? A look at the prevalence of coalitions in crimes of robbery. *Evolutionary Behavioral Science*, **12**(2), 126–131.

Boyd, N. (2000). *The Beast Within: Why Men Are Violent*. Vancouver: Greystone Books.

Brunner, H. G., Nelen, M., Breakefield, X. O., Ropers, H. H., & van Oost, B. A. (1993). Abnormal behaviour associated with a point mutation in the structural gene for monoamine oxidase A. *Science*, **262**(5133), 578–580.

Buades-Rotger, M., & Gallardo-Pujol, D. (2014). The role of the monoamine oxidase A gene in moderating the response to adversity and associated antisocial behavior: A review. *Psychology Research and Behavior Management*, **7**, 185–200.

Buller, D. J. (2005). *Adapting Minds: Evolutionary Psychology and the Persistent Quest for Human Nature*. Cambridge, MA: MIT Press.

Buss, D. M. (2009). How can evolutionary psychology successfully explain personality and individual differences? *Perspectives on Psychological Science*, **4**, 359–366.

Campbell, A. (1999). Staying alive: Evolution, culture, and women's intrasexual aggression. *Behavioral and Brain Sciences*, **22**, 203–214.

Cartwight, J. (2016). *Evolution and Human Behaviour: Darwinian Perspectives on the Human Condition*, 3rd ed. London: Palgrave Macmillan.

Caspi, A., McClay, J., Moffitt, T. E., et al. (2002). Role of genotype in the cycle of violence in maltreated children. *Science*, **297**, 851–854.

Coccaro, E. F., Beresford, B., Minar, P., Kaskow, J., & Geracioti, T. (2007). CSF testosterone: Relationship to aggression, impulsivity, and venturesomeness in adult males with personality disorder. *Journal of Psychiatric Research*, **41**(6), 488–492.

Coid, J., Yang, M., Roberts, A., et al. (2006a). Violence and psychiatric morbidity in a national household population – A report from the British Household Survey. *American Journal of Epidemiology*, **164**(12), 1199–1208.

Coid, J., Yang, M., Tyrer, P., Roberts, A., & Ullrich, S. (2006b). Prevalence and correlates of personality disorder in Great Britain. *British Journal of Psychiatry*, **188**, 423–431.

Craig, M. C., Catani, M., Deeley, Q., et al. (2009). Altered connections on the road to psychopathy. *Molecular Psychiatry*, **14**, 946–953.

Daly, M., & Wilson, M. (1988). *Homicide*. New York: Aldine de Gruyter.

Daly, M., & Wilson, M. (1998). *The truth about Cinderella: A Darwinian View of Parental Love*. New Haven, CT: Yale University Press.

Daly, M., & Wilson, M. (2005). The "Cinderella effect" is no fairy tale. *Trends in Cognitive Sciences*, **9**, 507–508.

Daly, M., & Wilson, M. (2007). Is the "Cinderella Effect" controversial? In C. Crawford & D. L. Krebs, eds., *Foundations of Evolutionary Psychology*. Mahwah, NJ: Erlbaum, pp. 383–400.

Davis, M. S. (2006). Criminal *mala in se*: An equity-based definition. *Criminal Justice Policy Review*, **17**(3), 270–289.

Decuyper, M., De Fruyt, F., & Buschman, J. (2008). A five-factor model perspective on psychopathy and comorbid Axis-II disorders in a forensic–psychiatric sample. *International Journal of Law and Psychiatry*, **31**(5), 394–406.

Dolan, M. (2004). Psychopathic personality in young people. *Advances in Psychiatric Treatment*, **10**, 466–473.

Dugdale, R. L. (1877). *The Jukes: A Study in Crime, Pauperism, and Heredity*. New York: Putnam.

Dustin, R. R., & Alcock, J. (2018). *Animal Behavior*, 11th ed. Sunderland, MA: Sinauer.

Dutton, K. (2012) *The Wisdom of Psychopaths: What Saints, Spies, and Serial Killers Can Teach Us about Success*. New York: Scientific American.

Fallon, J. (2013). *The Psychopath Inside: A Neuroscientist's Personal Journey into the Dark Side of the Brain*. Shelton, CT: Current Publishing.

Fallon, J. (2014). *The Psychopath Inside*. New York: Penguin Group.

Fazel, S., & Danesh, J. (2002). Serious mental disorder in 23000 prisoners: A systematic review of 62 surveys. *Lancet*, **359**(9306), 545–550.

Frodi, A., Dernevik, M., Sepa, A., Philipson, J., & Bragesjo, M. (2001). Current attachment representations of incarcerated offenders varying in degree of psychopathy. *Attachment and Human Development*, **3**, 269–283.

Gao, Y., Raine, A., Chan, F., Venables, P. H., & Mednick, S. A. (2010). Early maternal and paternal bonding, childhood physical abuse and adult psychopathic personality. *Psychological Medicine*, **40**(6), 1007–1016.

Ge, X., Conger, R. D., Cadoret, R. J., et al. (1996). The developmental interface between nature and nurture: A mutual influence model of child antisocial behaviour and parent behaviours. *Developmental Psychology*, **32**, 574–589.

Gibson, M. A., & Lawson, D. W. (2015). Applying evolutionary anthropology. *Evolutionary Anthropology*, **24**(1), 3–14.

Glenn, A. I., Kurzban, R., & Raine, A. (2011). Evolutionary theory and psychopathy. *Aggression and Violent Behavior*, **16**, 371–380.

Gordon, H. L., Baird, A. A., & End, A. (2004). Functional differences among those high and low on a trait measure of psychopathy. *Biological Psychiatry*, **56**, 516–521.

Greene, J. D., Sommerville, R. B., Nystrom, L. E., Darley, J. M., & Cohen, J. D. (2001). An fMRI investigation of emotional engagement in moral judgment. *Science*, **293**(5537), 2105–2108.

Guo, G., Ou, X.-M., Roettger, M., & Shih, J. C. (2008). The VNTR 2-repeat in *MAOA* and delinquent behaviour in adolescence and young adulthood: Associations and *MAOA* promoter activity. *European Journal of Human Genetics*, **16**(5), 626–634.

Hagerty, B. B. (2010). Inside a psychopath's brain: The sentencing debate. *NPR*. www.npr.org/templates/story/story.php?storyId=128116806.

Hare, R. D., & Neumann, C. S. (2006). The PCL-R assessment of psychopathy. In C. J. Patrick, ed., *Handbook of Psychopathy*. New York: Guilford Press, pp. 58–88.

Hauser, M. D., Cushman, F. A., Young, L., Kang-Xing, K., & Mikhail, J. (2007). A dissociation between moral judgements and justifications. *Mind and Language*, **22**(1), 1–21.

Hauser, M. D., Young, L., & Cushman, F. A. (2008). Reviving Rawl's linguistic analogy. In W. Sinnott-Armstrong, ed., *Moral Psychology*, Vol. 2. Cambridge, MA: MIT Press, pp. 107–143.

Hobbes, T. (1651). *Leviathan, or the Matter, Form, and Power of a Commonwealth, Ecclesiastical and Civil*, 1982 reprint. New York: Viking Press.

Kanazawa, S. (2009). Evolutionary psychology and crime. In A. Walsh & K. M. Beaver, eds., *Biosocial Criminology: New Directions in Theory and Research*. New York: Routledge, pp. 90–110.

Kanazawa, S., & Still, M. C. (2000). Why men commit crimes (and why they desist). *Sociological Theory*, **18**(3), 434–437.

Karmiloff-Smith, A. (1996). *Beyond Modularity: A Developmental Perspective on Cognitive Science*. Cambridge, MA: MIT Press.

Karmiloff-Smith, A. (1998). Development itself is the key to understanding developmental disorders. *Trends in Cognitive Sciences*, **2**(10), 389–398.

Kiehl, K. A., Smith, A. M., Hare, R. D., et al. (2001). Limbic abnormalities in affective processing by criminal psychopaths as revealed by functional magnetic resonance imaging. *Biological Psychiatry*, **50**, 677–684.

Krebs, D. L. (1998). The evolution of moral behaviour. In C. Crawford & D. L. Krebs, eds., *Handbook of Evolutionary Psychology: Ideas, Issues, and Applications*. Hillsdale, NJ: Erlbaum, pp. 337–368.

Krebs, D. L. (2011). *The Origins of Morality*. New York: Oxford University Press.

Larsson, H., Andershed, H., & Lichtenstein, P. (2006). A genetic factor explains most of the variation in the psychopathic personality. *Journal of Abnormal Psychology*, **115**(2), 221–230.

Lyons, M. J., True, W. R., Eisen, S. A., et al. (1995). Differential heritability of adult and juvenile antisocial traits. *Archives of General Psychiatry*, **52**, 906–915.

McDermott, R., Tingley, D., Cowden, J., Frazzetto, G., & Johnston, D. D. P. (2009). Monoamine oxidase A gene (MAOA) predicts behavioural aggression following provocation. *Proceedings of the National Academy of Sciences*, **106**, 2118–2123.

McGuire, J. (2004). *Understanding Psychology and Crime: Perspectives on Theory and Action*. Maidenhead: Open University Press.

McGuire, M., & Troisi, A. (1998). *Darwinian Psychiatry*. Oxford: Oxford University Press.

Mealey, L. (1995). Primary sociopathy (psychopathy) is a type, secondary is not. *Behavioural and Brain Sciences*, **19**, 579–599.

Mealey, L. (2005). Evolutionary psychopathology and abnormal development. In R. L. Burgess & K. MacDonald, eds., *Evolutionary Perspectives on Human Development*, 2nd ed. Thousand Oaks, CA: Sage, pp. 381–406.

Plomin, R., DeFries, J. C., McClearn, G. E., & McGuffin, P. (2008). *Behavioural Genetics*, 4th ed. New York: Worth Publishers.

Raine, A. (2002). The biological basis of crime. In J. Q. Wilson & J. Petersilia, eds., *Crime: Public Policies for Crime Control*. Oakland, CA: ICS Press, pp. 43–74.

Ruse, M., & Wilson, E. (1985). The evolution of morality. *New Scientist*, **1478**, 108–128.

Shamay-Tsoory, S. G., Harari, H., Aharon-Peretz, J., & Levkovitz, Y. (2010). The role of the orbitofrontal cortex in affective theory of mind deficits in criminal offenders with psychopathic tendencies. *Cortex*, **46**(5), 668–677.

Silk, J. B. (1980). Adoption and kinship in Oceania. *American Anthropologist*, **82**, 799–820.

Silk, J. B. (1990). Human adoption in evolutionary perspective. *Human Nature*, **1**, 25–52.

Taylor, S. (2016). *Crime and Criminality: A Multidisciplinary Approach*. Abington: Routledge.

Taylor, S., & Workman, L. (2019). The evolution of crime. In T. K. Shackelford & V. A. Weekes-Shackelford, eds., *Encyclopedia of Evolutionary Psychological Science*. https://doi.org/10.1007/978-3-319-16999-6_3209-1.

Trivers, R. L. (1972). Parental investment and sexual selection. In B. Campbell, ed., *Sexual Selection and the Descent of Man*. Chicago, IL: Aldine, pp. 139–179.

Trivers, R. L. (1985). *Social Evolution*. Menlo Park, CA: Benjamin/Cummings.

Troisi, A. (2015). The evolutionary diagnosis of mental disorder. *Wiley Interdisciplinary Reviews: Cognitive Science*, **6**(3), 323–331.

Troisi, A., & McGuire, M. T. (2000). Psychotherapy in the context of Darwin psychiatry. In P. Gilbert & K. G. Bailey, eds., *Genes on the Couch: Explorations in Evolutionary Psychotherapy*. London: Routledge, pp. 28–41.

Van Vugt, M. (2009). Sex differences in intergroup competition, aggression, and warfare: The male warrior hypothesis. *Annals of the New York Academy of Sciences*, **1167**, 124–134.

Wild, S. E., ed. (2010). *Webster's New World Law Dictionary*. Hoboken, NJ: Wiley Publishing, Inc.

Workman, L., & Reader, W. (2014). *Evolutionary Psychology: An Introduction*, 3rd ed. Cambridge, UK: Cambridge University Press.

Yang, Y., Raine, A., Narr, K. L., Colletti, P., & Toga, A. W. (2009). Localisation of deformations within the amygdala in individuals with psychopathy. *Archives of General Psychiatry*, **66**, 986–994.

39 Substitute Parenting

MARTIN DALY AND GRETCHEN PERRY

39.1 SUBSTITUTE PARENTING PRESENTS A PUZZLE FOR EVOLUTIONISTS

Evolutionary theory has straightforward relevance to parental behavior. The behavioral inclinations that natural selection favors are those that contribute to Darwinian fitness, that is, to one's expected genetic posterity (in the statistical, not the psychological, sense of "expected"). The primary avenue by which people and other creatures promote their fitness is by producing viable young who will eventually reproduce. Parental motives, emotions, and actions are therefore prime targets of selection.

"Parental investment" (Trivers, 1972) is a limited resource that parents have evolved to allocate in ways that can be expected to maximize the eventual reproductive success of one's total progeny (Clutton-Brock, 1991; Royle, Smiseth, & Kolliker, 2012). This means investing preferentially in young whose individual attributes predict that the investment will be most beneficial, but above all, it means investing preferentially in one's *own* young (Daly & Wilson, 1980). Why a Darwinian would predict that parents will avoid squandering their limited resources on unrelated young should be obvious: selection favors those genes and traits that enhance their carriers' fitness relative to the fitness of conspecific rivals.

As theory would lead us to expect, parents of many animal species indeed care discriminatively for their own young while spurning others, using a variety of complex psychophysiological adaptations to make the distinction (Daly & Wilson, 1988, 1995). And yet, despite abundant evidence that animal parents indeed care selectively for their own offspring, our own species is one in which nonparents often serve as children's primary caregivers, sometimes temporarily ("fosterage") and sometimes indefinitely or permanently ("adoption"). When and why this occurs are the focuses of this chapter.

The initially puzzling phenomenon of substitute parenting in *Homo sapiens* falls into three broad categories that require distinct treatments. One major subtype of substitute parenting entails genetic relatives, especially grandparents, stepping up to replace parents who cannot or will not care for their children, thereby promoting their own inclusive fitness. A second subtype is stepparenthood, which is most persuasively interpreted as a component of "mating effort." Both stepparenting and replacement care by genetic relatives are cross-culturally ubiquitous and almost certainly ancient, and the behavior of substitute parents in these contexts is therefore likely to exhibit evolutionary adaptation to the characteristic opportunities and pitfalls associated with these recurrent social dilemmas. The same cannot be said, however, for the third major subtype of substitute parenting, namely adoption by nonrelatives. Families sometimes adopt children to fill otherwise vacant social and familial roles or niches, and they foster or adopt children as a component of reciprocity and citizenship within close-knit communities. It is the modern practice of "adoption by stranger" that presents the greatest challenge to a simple conception of human beings as evolved fitness maximizers, necessitating that we ask why large numbers of people elect to treat unrelated children as if they were their own.

Each of these three broad categories of substitute parenting and their possible explanations will be discussed in a subsequent section of this chapter.

39.2 NEPOTISTIC ALLOPARENTING AND THE "GRANDMOTHER HYPOTHESIS"

In the non-state societies that provide our best models of the social circumstances in which humans evolved, babies are typically born into groups that consist largely of close relatives (Hrdy, 1999, 2009; Huber & Breedlove, 2007). The most assiduous parental helpers tend to be the children's grandparents, followed by aunts and older siblings (Hrdy, 2009; Kramer, 2005, 2010; Sear & Mace, 2008; Tanskanen & Danielsbacka, 2018).

In hunting and gathering societies, grandmothers are often more efficient food producers than their adult daughters and are committed, competent providers of direct childcare (Hawkes et al., 1997). Indeed, a case can be made that natural selection has "designed" human grandmothers to be specialized alloparents. Why, after all, should women cease to be potential reproducers when they can still function effectively in other domains? Women's reproductive capability comes to an end at about the same age as is the case in our nearest relatives, the great apes, but unlike female apes, women continue to be robust net economic producers for many years after their last child has been weaned. Furthermore, humans reproduce at shorter intervals than apes in spite of the burden imposed by our species' prolonged childhood dependency. How do women manage this feat? Hawkes (2003) has proposed that the contributions of grandmothers provide the answer, and that the inclusive fitness gains from grandmaternal investment explain the evolution of our species' exceptional postmenopausal lifespan.

Hawkes's "grandmother hypothesis" remains controversial for various reasons, the most important of which is that demographic data from natural-fertility populations, including some hunter–gatherers, indicate that grandmothering may not yield sufficient gains in inclusive fitness, on average, to offset the costs of ceasing to reproduce. However, age-specific mortality and grandmaternal impacts in past environments may have differed from what we see in any contemporary population. For the arguments and counterarguments, see the commentaries and reply following Hawkes et al. (1997). In any event, regardless of whether menopause itself or postmenopausal longevity is properly interpreted as an adaptation "for" alloparenting, there is no question that grandmothers indeed provide a lot of help to their adult daughters, help that often has substantial positive effects on child survival and functioning (Hrdy, 2009; Scelza, 2011; Sear & Mace, 2008, 2009).

Given their prominence among alloparental helpers, it is no surprise that grandmothers are also the relatives who are most likely to take over as primary caregivers when children cannot be cared for by their parents. One context in which grandmothers are prominent as primary caregivers is when official agencies in the developed world remove children from parents who have been abusive, neglectful, or dysfunctional. The preferred solution in such cases used to be placing the child with unrelated foster parents, but, for a combination of reasons, preferential placement with kin is now widely favored (Daly & Perry, 2011), and it turns out that "kin" mostly means grandmothers. For example, Perry, Daly, and Macfarlan (2014) analyzed kin placements at a Canadian child protection agency over a three-year period, and reported that in 318 placements with the focal child's genealogical relatives, a grandmother was the primary caregiver in 199 cases (63 percent). Such a predominance of grandmothers among nonparental caregivers is not exceptional (e.g., Coall & Hertwig, 2010; Hrdy, 2009; Zinn, 2010).

From an evolutionary perspective, this is unsurprising. The inclusive fitness returns from childcare depend on relatedness, and in an outbred population, grandparents are closer relatives of a focal child ($r = 0.25$) than anyone other than its full siblings and the parents themselves. Moreover, regardless of whether the trajectory of human female fertility is correctly interpreted as reflecting adaptation "for" grandmothering, that fertility trajectory has the effect that senior women typically lack options for promoting their fitness other than indirectly. Aunts and uncles who are full siblings of a focal child's parent are also relatives of degree 0.25, and they are in fact the next most common substitute caregivers in the child protection context after grandparents (e.g., Perry et al., 2014; Zinn, 2010). But the aunts and uncles of a child in need of care are usually of reproductive age themselves and are therefore likely to have more competing demands than is the case for the child's grandparents. And whereas the child's siblings are even closer kin, they are often too young to take over as primary caregivers, and if old enough, they are likely to have their own romantic and family lives to attend to. Even so, older siblings *are* extremely important alloparental helpers (Kramer, 2005, 2010), and in dire circumstances, such as in HIV/AIDS-decimated populations, even young children are likely to become the primary caregivers of their younger siblings (e.g., Mturi, 2012).

In the Perry et al. (2014) study, maternal grandmothers outnumbered paternal grandmothers as emergency caregivers of children removed from their parents by almost two to one (130 maternal vs. 69 paternal). This difference could derive from the specific circumstances of the child protection context, in which children's fathers may be unusually often absent or even unknown. However, the predominance of the maternal side was especially striking among grandmothers with major health problems and/or a lack of social support, and despite these challenges, placements with maternal kin were substantially more stable than those with paternal kin, with the latter being twice as likely to "break down" such that the child moved on to another temporary home. These facts suggest that maternal grandparents were simply more committed, on average, than their paternal counterparts, an interpretation that fits with considerable evidence that children's relationships with their mothers' mothers in the modern West are generally stronger and warmer than those with their fathers' mothers (Daly & Perry, 2017; Smith, 1991).

Why should this be so? Chapais (2008) has proposed that female solidarity within matrilines is an ancient primate adaptation that has been overlaid, rather than fundamentally revised, by the occasional advent of pair-bonds and paternal investment. Alexander (1974) was perhaps the first to explicitly argue that the uncertainty of paternity makes the progeny of one's daughters more reliable fitness vehicles than the progeny of one's sons. Following this line of thought, Smith (1981, 1988) proposed that maternal grandmothers should be the most solicitous grandparents, followed by maternal grandfathers and paternal grandmothers, each of whose putative genetic connection to the child has one uncertain paternal link, and then by paternal grandfathers, whose connection includes two uncertain links. The results of Smith's interview study indicated that retrospective recall of one's relationship with one's grandparents upheld the predicted ordering, as have several subsequent studies of differential closeness and investment (Chrastil et al., 2006; Danielsbacka et al., 2011; Euler & Weitzel, 1996). Whether uncertain paternity can really be implicated as a source of these rankings remains questionable, however, since the combination of strong mother–daughter ties and a sex difference in the inclination to nurture children could generate the same rank ordering even if paternity were as certain as maternity. Moreover, even if misattributed paternity were vanishingly rare, investing preferentially in a daughter's children might still be adaptive by virtue of helping maintain the daughter's capacity for further reproduction and nepotistic investment in her natal kin, in which grandmothers have a greater stake than in the corresponding capacities of their daughters-in-law (Perry & Daly, 2017).

In many human societies, newlyweds are expected to reside with the groom's family. Such "patrilocal" norms are typically associated with a cultural emphasis on kinship ties through fathers, while matrilineal links are downplayed. (Our modern Western society's normative use of patronyms is a vestige of this sort of patrilineal kinship system.) But even in patrilineal, patrilocal societies, women continue to play a role in the lives of their adult daughters' children. Rural Bangladesh provides an example. Patrilocality, *purdah* (the normative seclusion of women), poverty, and seasonal flooding all make it difficult for married women to maintain contact with their natal families, but virtually every young mother nevertheless visits her own mother regularly (Perry, 2017b). Young children usually co-reside in the same family compound as the paternal grandmother, often in the same household, and yet the maternal grandmother is more likely to take over as primary caregiver in the event of a divorce or the death of either parent (Perry, 2015). In intact families, the relatively accessible paternal grandmothers do provide more childcare assistance than maternal grandmothers, but the former help less than would be predicted on the basis of co-residence and proximity, and the latter help more (Perry, 2017a). Similarly, social bonds with and through the mother's mother remain surprisingly strong in some other patrilineal, patrilocal societies, too (reviewed by Daly & Perry, 2017).

According to one version of the "grandmother hypothesis," the psychology of grandmotherhood evolved to redirect older women's reproductive efforts toward support of their daughters' children, and the findings above suggest that senior women may feel a deeper emotional commitment to their daughters' children than to those of their sons. Could it be that it is only maternal grandmothers who are genuinely helpful? Some studies (e.g., Sheppard & Sear, 2016) suggest that the answer is yes. Two cross-cultural reviews (Huber & Breedlove, 2007; Strassmann & Garrard, 2011) have concluded that maternal grandparents have a beneficial effect on grandchild survival, whereas paternal grandparents have no demonstrable impact. Strassmann and Garrard's (2011) meta-analytic study focused on patrilineal and patrilocal societies, and its results suggest that even where paternal grandparents are more accessible than maternal grandparents, they are less beneficial to a grandchild's survival. Some results from European history (Voland & Beise, 2002) seem to have the same implication.

Fox et al. (2010) have proposed that our genetic sex-determination system may have had some surprising evolutionary effects on grandparenting. XX individuals develop as female and XY individuals as male; when a woman reproduces, she transmits either X chromosome, with equal likelihood, regardless of the child's sex, but a father necessarily transmits his only X to each daughter and his only Y to each son. The result is that the chances that a given grandmaternal X chromosome has a descendant copy in a grandchild vary: it has a 25 percent probability of appearing in any child of her daughter (just like a typical nuclear gene), but it has a 50 percent chance of being transmitted to a son's daughter, and it is never transmitted to a son's son. Suppose, then, that a mutation that affects how a woman responds to her grandchildren were to arise on the X. Such a mutation could be favored by selection if its effect were to make the grandmother invest in her son's daughters at the expense of their brothers. Indeed, such a mutation could, in principle, increase in prevalence all the way to universality, even if it yielded only a small gain in the fitness of one's granddaughters through sons at the expense of destroying their brothers! This specific sort of "selfish gene" effect is called "sexually antagonistic zygotic drive," and there are a number of phenomena that suggest that it really does operate in some nonhuman animals (Rice et al., 2008).

At first glance, the theory sounds preposterous. Wouldn't other interested parties keep paternal grandmothers away from their grandsons if their impacts were predictably harmful? And because nuclear genes have an equal fitness stake in all grandchildren and are vastly more numerous than X-chromosome genes, wouldn't the whole grandmaternal genome have evolved to suppress these renegade X chromosome effects? Those are indeed reasonable expectations, and yet there is some intriguing evidence that grandmothers "play favorites" in ways that match the theory. Fox et al. (2010) analyzed the association between grandmaternal presence and child survival in seven disparate data sets and found that the apparent impact of a grandmother was almost always positive *except* for the case of sons' sons, who survived less well in the grandmother's presence than in her absence in every society! These ostensible effects may yet be explained by unobserved variables – perhaps children who live near their paternal vs. maternal grandmothers differ systematically in other ways, for example – but in light of present knowledge, this initially far fetched theory certainly deserves further testing.

39.3 STEPPARENTHOOD

A lone parent is often in a difficult situation. Whether never married, widowed, divorced, or abandoned, mothers who find themselves without a supportive spouse commonly seek help from their natal families or even turn their children over to the care of their own mothers. The single mother who perseveres as her children's primary caregiver and also wishes to find a new partner will be disadvantaged in the mating market. From a suitor's perspective, the prospect of becoming a stepparent to a predecessor's child is treated as a cost, not a benefit, in remarriage negotiations, and stepchildren are sources of marital instability, conflict, and violence (Becker, Landes, & Michael, 1977; Campbell et al., 2003; Daly & Wilson, 1996; Daly, Singh, & Wilson, 1993; White & Booth, 1985;).

Stepparental investment has to be understood as a form of mating effort, not parental effort (Rohwer, Herron, & Daly, 1999). The obligations of co-parenting are undertaken as part of the give-and-take of establishing a sexual partnership, and it is the partnership – not the parental role – that the stepparent seeks. It follows that we should not expect the average stepparent to be as selfless and devoted as the average genetic parent, and indeed, although most stepparents provide adequate care and some go far beyond mere adequacy, there is abundant evidence that their contributions to children's well-being fall short, on average, of what genetic parents provide.

The most dramatic evidence of this is the much greater rates of abuse and death at the hands of stepparents than of genetic parents, hazards that apparently arise because some stepparents resent their obligations and are actively hostile to their stepchildren (Daly & Wilson, 1998, 2008). In the case of nonfatal child abuse, the overrepresentation of stepchildren as victims might, in principle, have been due to biases in detecting or recording abuse, rather than to real differences in incidence. If such biases were the whole story, however, they should be reduced or abolished in the most extreme and unequivocal cases such as fatal batterings, when in fact those are precisely the cases in which excess risk to stepchildren is maximal. The most thorough analyses are for Canada, where children under five years of age were beaten to death by stepfathers at a rate of 321.6 deaths per million child-years at risk (i.e., residing with stepfathers) in 1977–1990, compared to a death rate at the hands of birth fathers of 2.6 per million child-years at risk (Daly & Wilson, 2001). Data from Great Britain are similar in that they, too, indicate that the risk of fatal battering by a father figure is elevated more than 100-fold in stepfather households; Australian data indicate an even larger differential (Daly & Wilson, 2008).

Although elevated risk to stepchildren is the most extensively documented fact in the family violence literature, efforts to cast doubt on the phenomenon have been oddly persistent and vehement. Gelles and Harrop (1991) claimed to have debunked all prior evidence on the basis of a telephone survey in which interviewees were no more likely to admit assaulting their stepchildren in anger than their genetic children; this would hardly warrant mention were it not for the fact that the American Medical Association has notified clinicians that steprelationship is not, after all, a genuine risk factor for child maltreatment on the sole basis of this survey (Daly & Wilson, 1998)! Other writers have tried to explain away even the data on lethal abuse as reflecting nothing more than biased detection, a claim that is easily shown to be absurd: Child Fatality Review Panels have indeed uncovered large numbers of child-maltreatment deaths that were initially miscategorized, but stepparents are massively overrepresented as perpetrators in those cases, too, and even if *every* "accidental" infant death were really a successfully concealed paternally perpetrated murder, there are not enough such accidental deaths to raise the rate of fatal batterings by fathers to match the rate by stepfathers (Daly & Wilson, 2008).

The most concerted efforts to discredit the evidence that stepparents are more dangerous than birth parents have been those of a Swedish zoologist, Hans Temrin, and his collaborators. Temrin, Buchmayer, and Enquist (2000) initially claimed to have demonstrated that Swedish stepfathers are no more likely to kill children than birth fathers, but they had simply done the calculations wrong, and their own data in fact showed a substantial differential in the usual direction (Daly & Wilson, 2001). Grudgingly conceding the error, Temrin, Nordlund, and Sterner (2004) then presented new data indicating that although stepchildren incur excess risk when very young, this differential disappears, and furthermore that Swedish parents with both stepchildren and genetic offspring were actually slightly more likely to kill the latter. What this interesting result appears to reflect is the fact that hostile, assaultive child murders are extremely rare in Sweden and are overwhelmed numerically by a very different sort of case, namely murder–suicides by depressed parents who imagine themselves to be taking their loved ones with them (Somander & Rammer, 1991). This is certainly not the case elsewhere, and it raises the interesting question of whether stepparental antipathy and resentment are less severe in Sweden than in other developed countries, perhaps because the Swedish welfare state reduces the obligations of stepparents and hence their resentment. Finally, Temrin et al. (2011) purport to have demonstrated that the excess risk of death at the hands of stepparents can be entirely accounted for by the correlated attributes of becoming a stepparent. Hilton et al. (2015) have made the most direct attempt to see whether this interpretation can be upheld elsewhere, and they conclude that it cannot, at least in Canada. More generally, in all of the above work, Temrin and his collaborators have persisted in lumping together angry assaults and acts of suicidal depression, which are crucially different (Daly & Wilson, 1994; Harris et al., 2007; Weekes-Shackelford & Shackelford, 2004), and they have ignored the fact that, in every analysis except their own, parents who have both stepchildren and birth children in the same household have been found to be selectively violent toward the stepchildren, not the birth children (Daly & Wilson, 1985, 2008; Hilton et al., 2015).

It is important to note that although children are much more likely to be assaulted or killed by a stepparent than by a birth parent, such violence is nevertheless rare. Nonviolent manifestations of discrimination against stepchildren, by contrast, are not at all rare. Stepchildren routinely receive less financial assistance and other support, net of effects of the family's wealth, than children living with both birth parents (e.g., Anderson et al., 1999; Case & Paxson, 2001; Case, Lin, & McLanahan, 2000; Emmott & Mace, 2015; Sundström, 2013; Zvoch, 1999), suffer excess morbidity and mortality (e.g., Fergusson, Fleming, & O'Neill, 1972; Tooley et al., 2006; Wadsworth et al., 1983; but see Malvaso et al., 2015), and have poorer adult outcomes in many ways, not just in comparison to children living with two genetic parents, but even in comparison to those living with single mothers (e.g., Biblarz & Raftery, 1999; McLanahan & Sandefur, 1994). The examples cited above are from modern nation-states (including Sweden), but the available evidence (reviewed by Daly & Wilson, 2008) indicates that stepchildren often incurred even greater disadvantages in small-scale, non-state societies, as well as in the past. A telling indicator of the precarious nature of their experience is that stepchildren's levels of the stress hormone cortisol were chronically elevated relative to other children in similar material circumstances in a study of Dominican villagers (Flinn & England, 1995).

Hundreds of popular books offer advice on how to navigate the characteristic conflicts of stepfamily life, but their empirical content seldom if ever goes beyond anecdotes. Family counselors appear to be unanimous in cautioning against efforts to minimize or ignore the differences between stepfamily relationships and genetic family relationships, but there is no consensus on the essential nature of those differences. Most stepfamily research has been carried out in single countries, and explicit cross-national comparison of stepfamily functioning and stepchildren's disadvantages is needed. It seems, for example, that Swedish stepchildren suffer an elevated risk of violence to a lesser degree than is the case in several other countries, but no efforts have yet been made to compare the magnitude of this "Cinderella effect" cross-nationally or to seek its determinants. Such research may be especially likely to have policy implications.

39.4 ADOPTION

From an evolutionary perspective, cases in which substitute parents are not related to their wards are particularly puzzling. The idea that stepparenting is an investment in the mating relationship, not the child, provides a partial answer, but what about other cases of adoption? If natural selection favors investing one's efforts and resources in projects that are likely to result in the replication of one's own genes, not those of one's rivals, why does adoptive parenthood even exist?

Let us define "adoption" as the act of assuming parental responsibility for a dependent child who is not one's own, with both a presumption of permanence and some broader social recognition of the adopting party's status as the child's de facto and de jure parent. By this definition, adoption is certainly not peculiar to the modern world. Indeed, the proportion of children who are raised by adoptive parents rather than by birth parents is surprisingly high in many small-scale, traditional societies (e.g., Carroll, 1970; Damas, 1983; Decaluwe et al., 2015; Reghupathy et al., 2012; Silk, 1980). It must be noted, however, that in these face-to-face societies, in which adoptions are arranged privately without the involvement of governments or bureaucracies, adoptive parents are almost always the adoptee's close kin, especially grandmothers and aunts (Silk, 1987, 1990). Thus, despite some assertions to the contrary (e.g., Palacios & Brodzinsky, 2010), the modern practice of adoption by nonrelatives is neither cross-culturally universal nor truly ancient, and is arguably something for which our evolved psychology is not specifically prepared.

In societies like those in which we evolved, people who dwelt in close proximity were apt to be close kin, and even an indiscriminate inclination to nurture children within households, camps, and bands could have been functionally nepotistic. Might these considerations suffice to explain the human animal's willingness to adopt nonrelatives as a sort of "mistake": a by-product of a generalized beneficence toward children that was fitness promoting in ancestral social environments? There is surely something to this idea. A great many people find babies appealing regardless of whether they are related to them, so much so that Hrdy (2009) has proposed that babies constitute "sensory traps" for women and perhaps even for men. Preston (2013) has made the case that a generalized positive response to babies was adaptive in ancestral environments and was foundational to the evolution of the human animal's exceptional empathic and altruistic responsiveness. Nevertheless, parental-like solicitude is *not* indiscriminate. Women regularly report that after giving birth they experience a burgeoning sense that their babies are uniquely wonderful and worthy (Klaus & Kennell, 1976), and as for fathers, paternal affection can be shattered by a revelation of nonpaternity (Daly & Wilson, 1988). And of course, there is abundant evidence that stepparents systematically withhold investment relative to birth parents, as we have seen in Section 39.3.

So there is clearly more going on in parental responsiveness than mere reflex-like responses to the "sensory trap" of an appealing child, and yet the idea that people possess an evolved preparedness to alloparent may still be valid and of relevance to the puzzle with which we are concerned. Moreover, a strong urge to be a parent may be thwarted by infertility; 50 years ago, this was the prototypical context of nonrelative adoptions in the developed world, and adoption is still a frequent "second-best" recourse of infertile couples today (e.g., Bausch, 2006; Hollingsworth, 2000; Kirk, 1964; Park & Hill, 2014).

In the modern West, children who are adopted by nonrelatives tend to be well cared for, more or less as if they were their adoptive parents' genetic progeny (e.g., Judge & Hrdy, 1992). Outcome data may sometimes even indicate that unrelated adoptees receive more parental investment than genetic progeny, but adoptive parents are substantially more affluent, on average, than parents in general, and Hamilton, Cheng, and Powell (2007) have shown that controlling for parental means eliminates the adoptees' apparent advantage. Unfortunately, these authors portrayed their finding that adoptees are not discriminated against as "inconsistent" with "evolutionary science's kin selection theory," apparently supposing (as is all too common among social scientists) that inclusive fitness theory can be falsified by any demonstrated failure to choose the course of action that maximizes inclusive fitness. This is the same fallacy as supposing that voluntary childlessness falsifies Darwinism.

That said, any instance in which parental discrimination is lacking does indeed present a challenge for evolutionists (Daly & Wilson, 1988). In the case of human adoption, the puzzle is to some degree resolved when we recognize that it is only since the relatively recent introduction of legislated screening of applicants that adoption has ceased to be predominantly exploitative. For centuries, children were adopted to serve as cheap, controllable labor, and they were routinely prevented from marrying and obliged to repay their adoptive parents by providing eldercare (e.g., Boswell, 1988; Daly & Perry, 2011; Holt, 1994).

The modern practice of treating unrelated adoptees as if they were one's own children may be best understood as a novelty against which we have evolved no specific "defenses" because such adoptions never presented a recurrent threat to fitness in ancestral environments. Many interesting questions about the psychology of adoptive kinship that might speak to the adequacy of this interpretation remain unexplored. Are there systematic qualitative differences between adoptive and birth parent–child relationships with respect to sentiments and cognitions? Might it be the case that successful adoptive relationships are grounded less in a co-opting of evolved kinship psychology than in psychological processes appropriate to reciprocity and friendship? Are adoptive parents as eager to see their children reproduce and as smitten with the resultant grandchildren as genetic parents? These are sensitive issues that may be difficult to study without giving offense, but the answers could have real value.

Although exploitative adoption is largely a thing of the past, adoptive family relations continue to be fraught with difficulties, not all of which seem to be explicable as results of the challenges that the children confronted before they were adopted (e.g., Barth et al., 1988; Smith et al., 2006). As is the case with grandmothering and stepparenting, so, too, with modern adoption: we believe that there are opportunities for much more evolution-minded research on its psychological underpinnings and that such research could have applied utility.

REFERENCES

Alexander, R. D. (1974). The evolution of social behavior. *Annual Review of Ecology & Systematics*, **5**, 325–383.

Anderson, K. G., Kaplan, H., & Lancaster, J. B. (1999). Paternal care by genetic fathers and stepfathers. I: Reports from Albuquerque men. *Evolution & Human Behavior*, **20**, 405–431.

Barth, R. P., Berry, M., Yoshikami, R., Goodfield, R. K., & Carson, M. L. (1988). Predicting adoption disruption. *Social Work*, **33**, 227–233.

Bausch, R. S. (2006). Predicting willingness to adopt a child: A consideration of demographic and attitudinal factors. *Sociological Perspectives*, **49**, 47–65.

Becker, G. S., Landes, E. M., & Michael, R. T. (1977). An economic analysis of marital instability. *Journal of Political Economy*, **85**, 1141–1187.

Biblarz, T. J., & Raftery, A. E. (1999). Family structure, educational attainment, and socioeconomic success: Rethinking the "pathology of matriarchy." *American Journal of Sociology*, **105**, 321–365.

Boswell, J. (1988). *The Kindness of Strangers: The Abandonment of Children in Western Europe from Late Antiquity to the Renaissance*. Chicago, IL: University of Chicago Press.

Campbell, J. C., Webster, D., Koziol-McLain, J., et al. (2003). Risk factors for femicide in abusive relationships: Results from a multisite case control study. *American Journal of Public Health*, **93**, 1089–1097.

Carroll, V., ed. (1970). *Adoption in Eastern Oceania*. Honolulu, HI: University of Hawaii Press.

Case, A., & Paxson, C. (2001). Mothers and others: Who invests in children's health? *Journal of Health Economics*, **20**, 301–328.

Case, A., Lin, I.-F., & McLanahan, S. (2000). How hungry is the selfish gene? *Economic Journal*, **110**, 781–804.

Chapais, B. (2008). *Primeval Kinship: How Pair-Bonding Gave Birth to Human Society*. Cambridge, MA: Harvard University Press.

Chrastil, E. R., Getz, W. M., Euler, H. A., & Starks, P. T. (2006). Paternity uncertainty overrides sex chromosome selection for preferential grandparenting. *Evolution & Human Behavior*, **27**, 206–223.

Clutton-Brock, T. H. (1991). *The Evolution of Parental Care*. Princeton, NJ: Princeton University Press.

Coall, D. A., & Hertwig, R. (2010). Grandparental investment: Past, present and future. *Behavioral & Brain Sciences*, **33**, 1–19.

Daly, M., & Perry, G. (2011). Has the child welfare profession discovered nepotistic biases? *Human Nature*, **22**, 350–369.

Daly, M., & Perry, G. (2017). Matrilateral bias in human grandmothering. *Frontiers in Sociology*, **2**, 11.

Daly, M., & Wilson, M. I. (1980). Discriminative parental solicitude: A biological perspective. *Journal of Marriage & the Family*, **42**, 277–288.

Daly, M., & Wilson, M. I. (1985). Child abuse and other risks of not living with both parents. *Ethology & Sociobiology*, **6**, 197–210.

Daly, M., & Wilson, M. I. (1988). The Darwinian psychology of discriminative parental solicitude. *Nebraska Symposium on Motivation*, **35**, 91–144.

Daly, M., & Wilson, M. I. (1994). Some differential attributes of lethal assaults on small children by stepfathers versus genetic fathers. *Ethology & Sociobiology*, **15**, 207–217.

Daly, M., & Wilson, M. I. (1995). Discriminative parental solicitude and the relevance of evolutionary models to the analysis of motivational systems. In M. Gazzaniga, ed., *The Cognitive Neurosciences*. Cambridge, MA: MIT Press, pp. 1269–1286.

Daly, M., & Wilson, M. I. (1996). Evolutionary psychology and marital conflict: the relevance of stepchildren. In D. M. Buss & N. Malamuth, eds., *Sex, Power, Conflict: Feminist and Evolutionary Perspectives*. New York: Oxford University Press, pp. 9–28.

Daly, M., & Wilson, M. (1998). *The Truth about Cinderella: A Darwinian View of Parental Love*. London: Weidenfeld & Nicolson.

Daly, M., & Wilson, M. (2001). An assessment of some proposed exceptions to the phenomenon of nepotistic discrimination against stepchildren. *Annales Zoologici Fennici*, **38**, 287–296.

Daly, M., & Wilson, M. (2008). Is the "Cinderella effect" controversial? A case study of evolution-minded research and critiques thereof. In C. B. Crawford & D. Krebs, eds., *Foundations of Evolutionary Psychology*. Mahwah, NJ: Erlbaum, pp. 381–398.

Daly, M., Singh, L. S., & Wilson, M. I. (1993). Children fathered by previous partners: A risk factor for violence against women. *Canadian Journal of Public Health*, **84**, 209–210.

Damas, D. (1983). Demography and kinship as variables of adoption in the Carolines. *American Ethnologist*, **10**, 328–344.

Danielsbacka, M., Tanskanen, A. O., Jokela, M., & Rotkirch, A. (2011). Grandparental child care in Europe: Evidence for preferential investment in more certain kin. *Evolutionary Psychology*, **9**, 3–24.

Decaluwe, B., Jacobson, S. W., Poirier, M.-A., et al. (2015). Impact of Inuit customary adoption on behavioral problems in school-age Inuit children. *American Journal of Orthopsychiatry*, **85**, 250–258.

Emmott, E. H., & Mace, R. (2015). Direct investment by stepfathers can mitigate effects on educational outcomes but does not improve behavioural difficulties. *Evolution & Human Behavior*, **35**, 438–444.

Euler, H. A., & Weitzel, B. (1996). Discriminative grandparental solicitude as reproductive strategy. *Human Nature*, **7**, 39–59.

Fergusson, D. M., Fleming, J., & O'Neill, D. P. (1972). *Child Abuse in New Zealand*. Wellington: Government of New Zealand Printer.

Flinn, M. V., & England, B. G. (1995). Childhood stress and family environment. *Current Anthropology*, **36**, 854–866.

Fox, M., Sear, R., Beise, J., et al. (2010). Grandma plays favourites: X-chromosome relatedness and sex-specific childhood mortality. *Proceedings of the Royal Society B*, **277**, 567–573.

Gelles, R. J., & Harrop, J. W. (1991). The risk of abusive violence among children with nongenetic caretakers. *Family Relations*, **40**, 78–83.

Hamilton, L., Cheng, S., & Powell, B. (2007). Adoptive parents, adaptive parents: Evaluating the importance of biological ties for parental investment. *American Sociological Review*, **72**, 95–116.

Harris, G. T., Hilton, N. Z., Rice, M. E., & Eke, A. W. (2007). Children killed by genetic versus step-parents. *Evolution and Human Behavior*, **28**, 85–95.

Hawkes, K. (2003). Grandmothers and the evolution of human longevity. *American Journal of Human Biology*, **15**, 380–400.

Hawkes, K., O'Connell, J. F., & Blurton Jones, N. G. (1997). Hadza women's time allocation, offspring provisioning, and the evolution of long post-menopausal life spans. *Current Anthropology*, **38**, 551–577.

Hilton, N. Z., Harris, G. T., & Rice, M. E. (2015). The step-father effect in child abuse: Comparing discriminative parental solicitude and antisociality. *Psychology of Violence*, **5**, 8–15.

Hollingsworth, L. D. (2000). Who seeks to adopt a child? Findings from the National Survey of Family Growth (1995). *Adoption Quarterly*, **3**, 1–23.

Holt, M. I. (1994). *Orphan Trains: Placing Out in America*. Lincoln, NE: University of Nebraska Press.

Hrdy, S. B. (1999). *Mother Nature*. New York: Pantheon.

Hrdy, S. B. (2009). *Mothers and Others*. Cambridge, MA: Harvard University Press.

Huber, B. R., & Breedlove, W. L. (2007). Evolutionary theory, kinship, and childbirth in cross-cultural perspective. *Cross-Cultural Research*, **41**, 196–219.

Judge, D. S., & Hrdy, S. B. (1992). Allocation of accumulated resources among close kin: Inheritance in Sacramento, California, 1890–1984. *Ethology & Sociobiology*, **13**, 495–522.

Kirk, D. (1964). *Shared Fate*. New York: Free Press.

Klaus, M. H., & Kennell, J. H. (1976). *Maternal–Infant Bonding*. St. Louis, MO: C.V. Mosby.

Kramer, K. L. (2005). Children's help and the pace of reproduction: Cooperative breeding in humans. *Evolutionary Anthropology*, **14**, 224–237.

Kramer, K. L. (2010). Cooperative breeding and its significance to the demographic success of humans. *Annual Review of Anthropology*, **39**, 417–436.

Malvaso, C., Delfabbro, P., Proeve, M., & Nobes, G. (2015). Predictors of child injury in biological and stepfamilies. *Journal of Child & Adolescent Trauma*, **8**, 149–159.

McLanahan, S., & Sandefur, G. (1994). *Growing Up with a Single Parent*. Cambridge, MA: Harvard University Press.

Mturi, A. J. (2012). Child-headed households in South Africa: What we know and what we don't. *Development Southern Africa*, **29**, 506–516.

Palacios, J., & Brodzinsky, D. (2010). Adoption research: Trends, topics, outcomes. *International Journal of Behavioral Development*, **34**, 270–284.

Park, N. K., & Hill, P. W. (2014). Is adoption an option? The role of importance of motherhood and fertility help-seeking in considering adoption. *Journal of Family Issues*, **35**, 601–626.

Perry, G. (2015). Alloparental care in two societies: Who helps and in what circumstances? Unpublished PhD dissertation, University of Missouri.

Perry, G. (2017a). Alloparental care and assistance in a normatively patrilocal society. *Current Anthropology*, **58**, 114–123.

Perry, G. (2017b). Going home: How mothers maintain natal family ties in a patrilocal society. *Human Nature*, **28**, 219–230.

Perry, G., Daly, M., & Macfarlan, S. (2014). Maternal foster families provide more stable placements than paternal families. *Children & Youth Services Review*, **46**, 155–159.

Preston, S. (2013). The origins of altruism in offspring care. *Psychological Bulletin*, **139**, 1305–1341.

Reghupathy, N., Judge, D. S., Sanders, K. A., Amaral, P. C., & Schmitt, L. H. (2012). Child size and household characteristics in rural Timor-Leste. *American Journal of Human Biology*, **24**, 35–41.

Rice, W. R., Gavrilets, S., & Friberg, U. (2008). Sexually antagonistic "zygotic drive" of the sex chromosomes. *PLoS Genetics*, **4**(12), e1000313.

Rohwer, S., Herron, J. C., & Daly, M. (1999). Stepparental behavior as mating effort in birds and other animals. *Evolution & Human Behavior*, **20**, 367–390.

Royle, N., Smiseth, P. T., & Kolliker, M., eds. (2012). *The Evolution of Parental Care*. Oxford: Oxford University Press.

Scelza, B. (2011). The place of proximity: Social support in mother–adult daughter relationships. *Human Nature*, **22**, 108–127.

Sear, R., & Mace, R. (2008). Who keeps children alive? A review of the effects of kin on child survival. *Evolution & Human Behavior*, **29**, 1–18.

Sear, R., & Mace, R. (2009). Family matters: Kin, demography and child health in a rural Gambian population. In G. R. Bentley & R. Mace, eds., *Substitute Parents: Alloparenting in Human Societies*. Oxford: Berghahn Books, pp. 50–76.

Sheppard, P., & Sear, R. (2016). Do grandparents compete with or support their grandchildren? In Guatemala, paternal grandmothers may compete, and maternal grandmothers may cooperate. *Royal Society Open Science*, **3**, 160069.

Silk, J. B. (1980). Adoption and kinship in Oceania. *American Anthropologist*, **82**, 799–820.

Silk, J. B. (1987). Adoption and fosterage in human societies: Adaptation or enigma? *Cultural Anthropology*, **2**, 39–49.

Silk, J. B. (1990). Human adoption in evolutionary perspective. *Human Nature*, **1**, 25–52.

Smith, M. S. (1981). Kin investment in grandchildren. Unpublished PhD dissertation, York University, Toronto, Canada.

Smith, M. S. (1988). Research in developmental sociobiology: Parenting and family behavior. In K. B. MacDonald, ed., *Sociobiological Perspectives on Human Development*. New York: Springer, pp. 271–292.

Smith, P. K. (1991). Introduction. In P. K. Smith, ed., *The Psychology of Grandparenthood. An International Perspective*. London: Routledge, pp. 1–16.

Smith, S. L., Howard, J. A., Garnier, P. C., & Ryan, S. D. (2006). Where are we now? A post-ASFA examination of adoption disruption. *Adoption Quarterly*, **9**, 19–44.

Somander, L. K. H., & Rammer, L. M. (1991). Intra- and extra-familial child homicide in Sweden 1971–1980. *Child Abuse & Neglect*, **15**, 45–55.

Strassmann, B. L., & Garrard, W. M. (2011). Alternatives to the grandmother hypothesis: A meta-analysis of the association between grandparental and grandchild survival in patrilineal populations. *Human Nature*, **22**, 201–222.

Sundström, M. (2013). *Growing Up in a Blended Family or a Stepfamily: What Is the Impact on Education? Working Paper 2/2013*. Stockholm: Swedish Institute for Social Research (SOFI), Stockholm University.

Tanskanen, A. O., & Danielsbacka, M. (2018). *Intergenerational Relations: An Evolutionary Social Science Approach*. Abingdon: Routledge.

Temrin, H., Buchmayer, S., & Enquist, M. (2000). Step-parents and infanticide: New data contradict evolutionary predictions. *Proceedings of the Royal Society of London, Series B, Biological Sciences*, **267**, 943–945.

Temrin, H., Nordlund, J., & Sterner, H. (2004). Are stepchildren overrepresented as victims of lethal parental violence in Sweden? *Proceedings of the Royal Society of London, Series B, Biological Sciences (Suppl.)*, **271**, S120–S124.

Temrin, H., Nordlund, J., Rying, M., & Tullberg, B. S. (2011). Is the higher rate of parental child homicide in stepfamilies an effect of non-genetic relatedness? *Current Zoology*, **57**, 253–259.

Tooley, G. A., Karakis, M., Stokes, M., & Ozanne-Smith, J. (2006). Generalising the Cinderella effect to unintentional childhood fatalities. *Evolution & Human Behavior*, **27**, 224–230.

Trivers, R. L. (1972). Parental investment and sexual selection. In B. Campbell, ed., *Sexual Selection and the Descent of Man, 1871–1971*. Chicago, IL: Aldine, pp. 136–179.

Voland, E., & Beise, J. (2002). Opposite effects of maternal and paternal grandmothers on infant survival in historical Krummhörn. *Behavioral Ecology & Sociobiology*, **52**, 435–443.

Wadsworth, J., Burnell, I., Taylor, B., & Butler, N. (1983). Family type and accidents in preschool children. *Journal of Epidemiology & Community Health*, **37**, 100–104.

Weekes-Shackelford, V. A., & Shackelford, T. K. (2004). Methods of filicide: Stepparents and genetic parents kill differently. *Violence & Victims*, **19**, 75–87.

White, L. K., & Booth, A. (1985). The quality and stability of remarriages: The role of stepchildren. *American Sociological Review*, **50**, 689–698.

Zinn, A. (2010). A typology of kinship foster families: Latent class and exploratory analyses of kinship family structure and household composition. *Children & Youth Services Review*, **32**, 325–337.

Zvoch, K. (1999). Family type and investment in education: A comparison of genetic and stepparent families. *Evolution & Human Behavior*, **20**, 453–464.

40 Historians and the Evolutionary Approach to Human Behavior

GREGORY HANLON

40.1 INTRODUCTION

Darwinian approaches to human behavior are inherently historical; recounting the origin and development of *Homo sapiens* and the increasing complexity of human societies over many millennia is second nature to evolutionary psychologists and behavioral ecologists. Admitting the existence of biologically based human universals entails some considerable adjustment for historians, who are forced to retool by learning more about various social and biological sciences. Most historians have specialized interests by region, period, and topic; it is rare enough for practitioners to study the same problem across longer periods or in a different locale.

This contribution will try to speak to both communities of scholars, on the off chance that practicing historians will wish to rethink their problems afresh from the perspective of *human nature*. Historians often assume that all of our practices are subject to constant change, where we can apply a date to observable shifts in behavior. Human nature posits that some things change slowly or not at all, and this helps explain the uncanny similarity of the situations and motivations one finds across time and space. This section will pass in review some of the main acquisitions of a half-century of evolutionary inquiry for the benefit of historians, without pretending that scholars follow identical approaches or reach the same conclusions. I have no need here to take sides in the debates between evolutionary psychologists, human behavioral ecologists, or ethologists, or to choose between individual or group selection. Evolutionary thinking can inform a vast array of historical problems; indeed, it embraces every facet of human activity, but in this brief contribution, I will discuss only four areas at more length: reproduction, interpersonal conflict and violence, war, and aesthetics.

40.2 UNIVERSAL PEOPLE

The generation of psychologists after World War II, keen to break with Nazi racial science, repudiated the notion that humans shared any innate qualities at all. Humans were held to be capable of anything, given enough brainwashing,

guidance, and conditioning. For many, this stance was compatible with humanity's continual perfectibility and ultimate redemption. This behavioralist dogma was challenged tentatively during the 1950s with the work of Noam Chomsky, who postulated the existence of innate language-learning abilities in humans everywhere. Simultaneously, the first work on molecular biology (François Jacob, Francis Crick) and population genetics (Luca Cavalli-Sforza), which rested on wide-ranging research on the building blocks of life, prepared intellectuals for the notion that humans were part of the natural world. The breakthrough came in 1964, when William Hamilton, an evolutionary biologist, demonstrated the existence of cooperation and altruism in related insects. All of these developments occurred under the radar of historians. Only a few of them understood the implications raised when another entomologist, Edward O. Wilson, published his book, *Sociobiology: The New Synthesis*, in 1975, and its sequel devoted to humans, *On Human Nature*, in 1978.

If historians were unaware of the implications of this work for their discipline, anthropologists working out of several distinct traditions began to focus on the similarities of motivations and behaviors in societies that had no contact with each other. Napoleon Chagnon, Robin Fox, and Lionel Tiger figured as pioneers of evolutionary anthropology by focusing on human universals. Another field leading to the same realization was ethology, applied by Konrad Lorenz and his European followers first on a wide variety of bird and mammal species. These subjects did not possess language and could tell us no stories; we were forced to understand their behavior from their actions alone. The Dutch primatologist Frans de Waal, who studies higher primates in captivity, was able to describe and explain the rules that underpinned the daily interactions of chimpanzees, each of whom had their own personality and a specific place in the group. Primates have order without law and continuous communication without language. Reading de Waal, the connections between higher primates and humans are inescapable. It is impossible to think of human politics and human law as the results of rational discourse, for they are rooted in

489

a deep sense of fairness and cooperation that is much older than humanity.

Some kind of culture may be intrinsic to all animal life. Ethologists show us that animals are subjects too, be they birds or marine mammals. Animal communities display a variety of traits within the same species, so that we should imagine a population to be like an ethnic group for which we must undertake a proper ethnography (Lestel, 2001, pp. 9–11). The behaviors in many species suggest that the individuals have testable mental images of objects, events, and relationships, and therefore have self-knowledge and intentionality (p. 50). Many species function in rational ways (p. 208), able to choose from a range of possible actions given the context of the moment. If insect actions are collective and rigid, those of chimpanzees are individual and intelligent; humans are collective and intelligent.

By the 1980s, human ethology emerged as a separate specialization, particularly in the UK (Desmond Morris, Niko Tinbergen) and in the German-speaking world. This discipline came to prominence with the Austrian ethologist Irenaus Eibl-Eibesfeldt, whose textbook, *Human Ethology*, appeared in German in 1985, then in English four years later. Human ethology is the biology of human behavior, built upon close description and observation. Eibl-Eibesfeldt postulates that group selection plays an important role among humans; the war ethos, group sharing, and loyalty to authority were unlikely to depend on individual selection alone. All cultures identify the same kinds of emotions. Ethologists are justifiably wary of verbal explanations of behavior by their subjects (this is a common problem in anthropology), and they frequently film their subjects' interactions in order to draw a maximum of nuance, just as de Waal films his chimps over years in order to maximize detail.

Donald Brown tied much of this research together in a groundbreaking book, *Human Universals* (1991), which, discussing the fundamental similarity of behavior and social process around the world, brings it back to Darwin's discussion of the similarity of emotions and facial expressions in humans. Brown summarized these common traits in an important chapter entitled "Universal People," which begins to list behavioral traits that humans have in common everywhere and could only be the expression of innate capacities.

The likelihood of the existence of universal innate predispositions or instincts in humans across time and space constituted an invitation to psychologists to examine our mental structures more closely. Steven Pinker, a Canadian psychologist working from Chomsky's research on language acquisition, developed these themes in a pair of hugely influential books that pushed the evolutionary approach into the limelight. Pinker, who deploys considerable erudition, possesses a gift for vivid analogy and the entertaining turn of phrase. The mind is full of specialized modules dealing with one specific area of the outside world; the brain is equipped with specialized applications,

like a Swiss army knife. "The mind is a neural computer, fitted by natural selection with combinatorial algorithms for causal and probabilistic reasoning about plants, animals, objects and people. It is driven by goal states that served biological fitness in ancestral environments, such as food, sex, safety, parenthood, friendship, status and knowledge" (Pinker, 1997, pp. 524–525). This does not reduce thought to mechanical reactions to stimuli, for humans juggle multiple goals simultaneously, making continual trade-offs and seizing opportunities. Most of all, we are constrained by other peoples' goals, for humans are social animals who cannot survive outside the group. This predicament is common to all social animals, but the enlarged prefrontal cortex where social reasoning takes place renders humans more sophisticated than any other animal. Much of this activity is unconscious, whereby we act before we have processed our knowledge and come to a conscious decision. Pinker then sought to refute the prevailing social science view that the mind was a tabula rasa at birth. *The Blank Slate* is a treasure chest of accessible information available to the general public, engagingly written. The science writer Robert Wright, in *The Moral Animal* (1994), similarly brought together a great deal of research on how our behavior is goal directed to transmit our genes into the next generation. Citing the Cambridge biologist Richard Dawkins, Wright claims that humans are largely the puppets of the interests of their genes. The evolutionary psychologists (as sociobiologists were rebaptized) develop theories of how our genetic default settings have influenced human cultural evolution. John Tooby and Leda Cosmides claim that that these structures evolved to solve the problems of foragers in anatomically modern humans 50,000 years ago. "The past explains the present," they write. Following Lorenz, they claim that instincts and social environments fashioned in prehistoric times are possibly no longer adequate for our modern urban context.

The late twentieth century saw rapid advances in cognitive psychology, popularized by Michael Gazzaniga, Jean-Pierre Changeux, and Antonio Damasio, among others, who explain how chemical transformations or the coding of pathways and connections by electricity constitute our thought processes. They emphasize that there is no homunculus or conscious operator in the brain, but rather a number of specific knowledge systems that do not always require experience to function. There is no hierarchy of goals, which allows choice and exploration and permits a multiplicity of possible outcomes. But much of what we do is unconscious or semiconscious. We also possess a processing system in the left hemisphere that interprets information to make a convincing story. This is the way we were born.

Pinker's seductive synthesis did not win the allegiance of all psychologists, who were unable to locate what they called "massive modularity" (Panksepp & Panksepp, 2000). Many of these researchers place more emphasis on what they call "plasticity"; that is, the variability of behavior according to context. They criticize an excessive reliance on computer science analogies and prefer to focus more on hugely complex biological processes in the brain. These psychologists also wished to show the interaction between the enlarged human prefrontal cortex that solves social problems with the limbic system and subcortical systems we share with other animals, where emotions rule. We have a strong neurochemistry of attachment, for example (Quartz & Sejnowski, 2002). We are born with a sense of fairness and some other moral intuitions and emotional instincts that can be detected in very young infants. We are also wired for reciprocity, but only in our social group. Emotion and reason are tightly entwined, to the point where it is wrong to discuss them as separate entities autonomous from each other (Damasio, 1994). Daniel Kahneman and Amos Tversky built up a repertoire of ways in which the brain and human thinking display universal defects based on these emotions, which color our reasoning. We possess an intuitive system that we share with animals that jumps to conclusions, and a calculating system capable of more refined reflection, but that is lazy, typically proceeding with the least effort. This leads us to jump to conclusions from small samples and to exaggerate the consistency and coherence of what we see. These default settings of the brain are hugely important for historians, who must explain human motivations and actions. But one thing remains constant: to date, there is no evidence that the workings of the human mind or the emotions underlying them vary from one place to another, although there is considerable variation between individuals – and the same degree of variation – around the world. And what is constant around the world is likely to have been constant over time as well. There are also measurable differences between the sexes, with boys more prone to dominance and fearlessness, and girls twice as likely to show an anxiety disorder (Kagan, 1994).

The problem with a strong emphasis on human plasticity is that it cannot explain the similarity in human motivations and the many enduring behaviors of human societies everywhere throughout recorded history. The critique of "massive modularity" is perhaps justified, but it is doubtful, as Raymond Tallis claims, that we have been drifting away from our biological origins.

40.3 SOCIAL SCIENCES DISCOVER EVOLUTION

Evolutionary theory posits that there exist a large number of human universals and a set of "default settings" in our judgments that operate in all times and places. One need not fall back on contingent "culture" to explain the behavior one encounters in historical sources. Given the amount of literature that posits "culture" as an autonomous capacity that liberates us from our animal origins, there is little consensus on exactly what it entails (Kuper, 1999). Culture is a very fuzzy concept to pin down. It is often invoked generically to explain phenomena without going into detail as to how it operates. Culture has come to

be synonymous with relativism, the basic incommensurability of human groups atomized by their "otherness." Radical differences separate peoples, the partisans of social construction write, and only insiders can understand the meaning of behavior. Evolutionary anthropologists, on the other hand, suspect that the "other" is a fiction invented by philosophers. People can, and do, attach themselves to groups, nations, and linguistic communities not their own by birth, while others study exotic peoples closely. The French sociologist Raymond Boudon, reacting to the corrosive relativism of our time, similarly argues that social scientists should be able to explain observed phenomena in the light of individual goals that anyone can understand, given enough information on the specific context of their subjects. From the instant we can visualize the rationale behind any behavior, in any place or time, the concept of culture becomes largely redundant (Boudon, 2001, p. 22).

A few historians (Degler, 1981) saw the implications of E. O. Wilson's conclusions almost immediately, but the first attempts to combine history and the evolutionary studies came from sociobiologists with backgrounds in anthropology. An early attempt was Laura Betzig's *Despotism and Differential Reproduction: A Darwinian View of History* (1986). Despotism, or "the exercised right of heads of societies to murder their subjects arbitrarily and with impunity," should coincide with greatly enhanced access to nubile females by powerful men. Big men exploit their power to the end of reproduction and compete with others to achieve this goal. Working from a compendium of cross-cultural surveys of 186 societies compiled by Murdock (1949) and Murdock and White (1969), she concludes that all of Darwin's predictions hold true for humans. This compendium of research summaries drawn from a half-century of anthropology should be better known by historians. They will not always find it satisfying, for it treats each study equally, ignoring the length of time the anthropologist stayed in the community, their degree of familiarity with local languages, or whether or not they had access to all of the social groups in their subject population. Another drawback of anthropology is that simple conversation with an interlocutor contains its fair share of distortions. The host group is keenly aware of the presence of the anthropologist, who is not always able to catch them off guard.

Another noteworthy attempt to confirm Darwinian theory through historical narrative is Robert Wright's *Nonzero: The Logic of Human Destiny* (2000), which sums up in a single volume thousands of years of what one might call cultural evolution; that is, the development of specific societies by socially transmitted innovations that are analogous to biological evolution, but much quicker. The brain's plasticity enables people to engage with others in cooperative win–win exchanges (called reciprocal altruism) that are better than zero-sum (whatever I win, you lose in equal measure) or negative exchanges (where both are worse off than before). History progresses through people playing non-zero-sum "games" in ever greater numbers, while societies become more complex as a cumulative result of these benefits. We are social beings engaged in contact and cooperation (if only to procure mates outside our group), but we have an innate sense of fairness that triggers strong emotions of indignation toward non-reciprocators. Notwithstanding the innate basis of this understanding, peoples have moved along the "arrow" of history at different speeds, depending upon the abundance of resources (population size) and the ease of communication (contact with other groups). There are echoes here of Jared Diamond's famous book *Guns, Germs and Steel* (1997), who likewise explains the dramatic variations in human societies through the differential access to communication over the very long duration.

In Wright's convincing and entertaining telling, social groups surrender autonomy, often unwillingly (war is one conduit of influence), but the fear of conquest pushes groups into alliances that can shift according to short-term contingencies. In ancient times, literacy, money, and taxation were created independently around the globe. Societies resemble large, thick brains, "their neurons spreading incremental innovation rapidly and reliably, further spreading innovation" (Wright, 2000, p. 146). From the sixteenth to the eighteenth century, Europe was becoming a very large yet fast brain, catalyzed by competition among its numerous bellicose entities. This process led, by the twentieth century, to world-reaching governance and the voluntary submission of individual players to international and global authority. In this ongoing story, humans by no means cease to be evolved animals acting egotistically. Greed, hatred, and tribalism will not vanish, but core impulses can be tempered and redirected (p. 232). Global history is easier to advocate in theory than to practice convincingly. World history on this scale of millennia is not practiced by many, for it requires a degree of erudition few can acquire, especially if they can read only English. Moreover, it rarely permits a solid demonstration, usually being content to spur ideas and consolidate notions that others must confirm by more empirical means, each in their own corner specialization.

Historians have leveled another complaint against interlopers who wish to confirm evolutionary theory by scanning a panorama of secondary literature. Most, if they use primary sources at all, refer to translated literary documents of various kinds; Nordic sagas, chronicles, the Bible, each culled for examples that reinforce the argument. Even when the sagas can be quantified, this is only an illusion of precision. Chroniclers wrote about events that they felt were memorable, leaving out whatever they considered unremarkable or uncontroversial. Literary sources have few equals when it comes to evocative power, and many naive historians write narratives that string together anecdotes drawn from published texts in an impressionistic manner. Literary sources alone demonstrate nothing: they are at best merely illustrative of situations and processes that must be founded on more compelling evidence and proper empirical demonstrations. Even worse is recourse to fiction in the past to illustrate historical situations: Shakespeare is often trotted out to make a point. But this ignores that fiction requires dramatic pathos, artifice, and vivid contrasts. Non-historians and evolutionary psychologists in particular discard or neglect information that does not fit the theory. Most historians, more attuned to the best practices of the discipline, would prefer to subordinate the theory to the evidence, which is often messy. But this is not particular to history. Evidence is often ambiguous or difficult to interpret in all of the sciences, which is what makes them continually subject to revision.

The Darwinian anthropologist James Boone took a major step forward by attempting to test the Trivers–Willard theory whereby males will outnumber females if the mothers are in good condition and do less well if they are in poor condition. Human societies are everywhere hierarchical, even where private property is minimal. Boone analyzed the genealogies of Portuguese aristocrats to determine whether there existed family strategies to promote males or females differentially. The data drawn from seventeenth-century documents showed that daughters of the lower nobility were more likely to marry than their brothers. At the higher level, however, daughters disappeared into convents, whose dowry demands were much lower than those a husband of high rank could expect. Elite families were loath to marry their daughters with lower-born men and so dilute their pedigree, for aristocrats were nothing if not competitive. This small project began with a testable problem, resorted to a period source, and concluded with a demonstration. The professional historian is likely to demand the use not of genealogies, which were secondhand sources compiled by the users to buttress claims of social preeminence, but rather of parish registers of baptism, marriage, and burial, which the families did not control. Historians demand to know – with good reason – where the information comes from and why it was committed to writing.

Historians in the late twentieth century were often influenced by a generation of French historians who focused from the 1960s onward on what was called "*mentalité.*" The term originated in the writings of the anthropologist Lucien Levy-Bruhl in the 1920s, who sought to describe "*la pensée sauvage,*" or the prescientific modes of thinking of primitive societies. The critique of this notion by his peers led to his abandonment of much of the theory by the late 1940s, but historians could not fail to see the connections between extant prehistoric societies and the prescientific thought of antiquity, the medieval, and the early modern period. By the early 1970s, social historians began to speak of *mentalités*, or *structures mentales* specific to each time and place. The psychological literature underlying the psychological theory was almost completely nonexistent, however. Before long, the British classicist Geoffrey Lloyd exploded the theory as tautological, and the use of the term receded in France. Young historians then largely veered away from behavioral history.

This *"histoire des comportements"* that has existed in France since the 1960s nevertheless spurred historians to explore hitherto abundant, but neglected, sources from antiquity onward. While short on psychology, their analysis was frequently grounded in cultural anthropology (usually secondhand). The French approach (quickly dubbed the *"Ecole des Annales"* after the pioneering journal in Paris) convinced historians elsewhere to focus on local societies and the everyday behavior of anonymous people, quantifying the archival data whenever possible; in the UK, this was "history from below"; in Germany, *"all-tagsgeschichte"*; in Italy, single interesting events were illuminated by in-depth research on their contexts, a technique called "microhistory." It is conceptually very easy to pass from this kind of history to one inspired by Darwinian theory of the psychic unity of humankind, for we are obliged to find plausible explanations for the actions of our subjects.

The first – and therefore the bravest – academic historians to go down this path were a young American, Abel Alves, and an eminent Swiss classicist, Walter Burkert, studying completely different problems. Each produced an important book in 1996. Alves (1996) began by noting that all of human history reflects a constant dialectic of aggression and mutual aid, of both brutality and benevolence, springing from the same rules underpinning our existence in groups. Elites impose their power and maintain it by force, but they also create and manage systems of conflict resolution and charity in order to forestall rebellion. Subordinates also subject their leaders to an ongoing cost–benefit assessment. The Aztecs were able to impose a powerful state in Central Mexico during the fifteenth century, strong in the center but restive on the periphery. When a small group of Spanish adventurers landed seeking treasure, they attracted enough support from the latter to provoke the collapse of the empire. But the aliens provided a system of government that was in fundamental ways similar to what the Mexicans had experienced. What are fixed and enduring are human tendencies toward hierarchy, reciprocity, aggression, xenophobia, territoriality, display, altruism, and even culture creation, which Alves studies in the context of sixteenth-century Mexico. Walter Burkert's *Creation of the Sacred: Tracks of Biology in Early Religions* (1996) similarly posits the permanence of religious sentiment throughout the ages in such a way as to render it in many ways beyond culture. There exist phenomena common to all human civilizations and are universal in anthropology, and religion is one of them. It is important that Burkert rooted his understanding of the human condition in the ethology of Konrad Lorenz, not entirely compatible with the "inclusive fitness" of Anglo-American neo-Darwinians.

These books did not seem to hobble the careers or besmirch the reputations of their authors. In the years after 2000, the number of studies drawing inspiration from evolutionary thinking in psychology and anthropology began to multiply in Europe and North America, while the basic tenets of neo-Darwinian thinking moved into the mainstream. Two medievalists, the American Daniel Smail and the German Jörg Wettlaufer, focused on the power of universal emotions to shape behavior in the thirteenth and fourteenth centuries. Smail (2003) noted how people in face-to-face communities used talk to shape the reputations of everyone around them. No one who is familiar with the work of primatologist Frans de Waal can ignore the power of such interactions, often done through grooming. Humans use language to establish – and harm – the reputations of their peers, their underlings, and also their social superiors. My own work in this vein may have been the first to use evolutionary theory to explain a broad array of behavior applied to a specific group of villagers in seventeenth-century Tuscany, *Human Nature in Rural Tuscany* (2007). An exceptional wealth of sources of every description made it possible to identify by name practically every person living in a rural community over the space of two generations, with a considerable amount of information on a large number of individuals. Five hundred criminal trials portrayed the urgent predicaments of the principals and hundreds of eyewitnesses of both sexes, whose judges wrote down the testimony almost verbatim. Human universals saturated the institutions managing community affairs (participation and hierarchy), underpinned the cooperation of daily sociability and economic exchange, fed the competition pitting people of various social strata against each other as individuals or members of families, determined the villagers' mating and childrearing practices, and finally shaped the ways in which individuals, families, and communities coped with their time and the events they could not control.

The roundtable on history and biology hosted by *The American Historical Review* in 2014 may have given the neo-Darwinian approach more legitimacy by giving some of the leading figures in the American academy a conspicuous platform on which to convince the reader that practicing historians could no longer ignore the developments outside their field. Most of the contributors submitted theory pieces that stressed the feedback loop between the brain, the body, and the external environment; historians finally engaged with cognitive psychology. The classicist Walter Scheidel emphasized the ubiquity of the incest taboo and consequently determined that it was unlikely to be a cultural construction. Many claims by evolutionary psychology of universal human behavioral traits needed to be tested against the archival record, he argued. Michael D. Gordin reminded us that few problems have the full consensus of specialists, in any field. Science is in ongoing flux too, and some of its creations will soon be discarded, or simply laid aside until they are picked up again in the future. Historians should be wary of uncritically using biology, but not so wary that they should avoid it completely (Gordin, 2014).

As historians, our task is not just to explain differences over time. In fact, we must explain the behavior of specific communities in past time. Their context was different from ours, and we sometimes encounter phenomena that strike us as strange. But there is much in the motivation and the execution of behavior that is similar to our own. A recent attempt by W. Garry Runciman to explain the evolution of English society since the early eighteenth century underscores this very continuity: it was very different, but much the same (Runciman, 2015). Human universals explain human behavior in past times, everywhere, in a great many realms. Given the limited space here, I will briefly illustrate this point by discussing four fields with which I have some familiarity: reproduction, interpersonal violence, war, and display. Evolutionary theory suggests some fruitful avenues of analysis, but we need to propose testable questions, and select and deploy the right kind of sources to answer them.

40.4 REPRODUCTION

The ethologist Eibl-Eibesfeldt first teased out the universal sequence of human amorous behavior in the 1960s, with the aid of a hidden camera. Both females and males exhibit flirting gestures, she with coyness, he with expanding torso, leading to eye contact and smiles. Men signal that they are important, but that there is no risk in approaching them. Women seek to attract the man's gaze and selectively display signals of availability. As the acquaintance is struck and the affair proceeds, it universally includes music, and gifts to the women. Romantic love is not a Western invention, but an objectively measurable trait, having a chemical basis in the limbic system, phenylethylamine. It is not a permanent state, however, lasting between 18 months and 3 years, often then giving way to a calmer attachment (Fisher, 1992, p. 59). That is enough time to pass from mere mating to reproduction and infant care, wherein the newborn enjoys the advantages of two parents. Marriage, which is another human universal, is intended to provide children with two sets of blood kin and a public identity. There is therefore a biological basis of marriage, and its chief function is to raise children. This is true in polygamous societies (men with more than one wife simultaneously, exclusive of concubines or mistresses) and in the rare polyandrous ones too, in which the husbands are usually brothers.

Monogamy is the most common married state, but everywhere there is the risk of adultery. David Barash and Elizabeth Lipton (2001) have noted how widespread straying is in mammals and birds previously thought to be monogamous, and in humans also. Today, routine DNA testing of newborns in hospitals is teaching us lessons on human sexuality not revealed by questionnaires. Both males and females established in socially sanctioned couples imagine that they can do better. Some, even many, will eventually act upon a chafing sense of dissatisfaction. Straying is particularly dangerous in females, for it creates offspring whose very survival is an insult to the husband. Jealousy is yet another human universal, a spirited defense of claims on a mate, a highly possessive behavior toward

the partner that is most violent in males. Fear of adultery can lead to male sequestering behavior and physical abuse of the weaker female. At the base of sexual reproduction in humans are some timeless equations: women must attract – and keep – the attention of males in order to obtain resources and protection for themselves and their offspring. Each pregnancy ties them down for years at a time. Having multiple children exacerbates this dependency, but given historically high levels of infant and juvenile mortality, she would need to produce four or five offspring simply to maintain the population. In the event that she should live to an advanced age, her children would provide for her. So, faced with the necessity of reproducing, and given the high stakes involved, females will be choosy.

Males and females will have different kinds and degrees of attractiveness: youth and beauty for women, wealth and status for men. In societies where wealth can be amassed and inherited, men readily take brides below their status. (The inverse is less frequently the case.) Powerful males can take spouses and attract other mates in the form of concubines and mistresses, while still others with extra resources can pay willing females for sex. There is little obstacle and no cost in multiplying their copulation with whichever women will consent to it. Wives object to the deviation of resources that rightfully should go to them and their offspring. Neither males nor females can function in complete autonomy, for the egotistical desires of each are set against the myriad social connections with others. When a spouse strays – or even gives suspicion of it – people will talk. If a husband mistreats his wife, his reputation will suffer and her male kin may well retaliate. Marriage unites in alliance whole groups of kin, who have their social and reproductive interests to defend. Scolding may repel the husband, who makes himself scarce. He will be tempted by complete abandonment, which is easier for men than for women. The absence of children and sexual straying (especially of wives) are still the leading causes of divorce. Since the tensions inside couples are themselves timeless and universal, divorce exists in practically every society, even where it is legally forbidden.

Historians of the family would do well to read the work of anthropologist Sarah Blaffer Hrdy, who compares maternal instincts in humans and other primates. Real-life mothers in the wild are strategic thinkers, manipulators, and allies as much as nurturers, more interested in the maintenance of interpersonal relationships than power. Females lactate, and they have an innate threshold for responding more quickly to infant signals than men (Hrdy, 1999, p. 212). Males are not certain that the offspring are their own, so they do not respond as automatically as mothers, who continually make trade-offs between subsistence and reproduction in the hope of producing good-quality offspring. Females are not immune to status seeking, but will not challenge males physically in pursuit of it. Mothers in most social species avail themselves of helpers, kin, and friends, who assist her in many ways. For

Hrdy (2009), pro-social tendencies at the heart of child-rearing are probably the model for human cooperation generally, since aid to others is likely to be reciprocated and rewarded. Help comes quickly from kin, but the desire to cooperate seems to be innate, and nonkin will readily offer assistance as well. Females are more affiliative than males in most mammalian species. Slow-maturing infants like those of humans require the assistance of others – fathers, of course, but also older siblings, grandmothers, sisters, and other kin on the female side – who enable mothers to breed again sooner. Nurturing responses in human fathers are facultative; that is, situation dependent and expressed under certain conditions. A stepfather is least helpful to the child's survival, and often enough constitutes a threat.

Male humans care for young in every culture studied to date by offering social protection and material resources (Balshine, 2012). Parents continue to proffer support for their offspring, even once these are adults. Where adults live to a more advanced age, grandparents figure among the nurturers and protectors of their descendants, but not indiscriminately. Research on grandmothers in historical Germany, Japan, and New France (Beise & Voland, 2002) has noted and measured the beneficial impact of maternal grandmothers on their descendants, especially where they reside with their daughters. Paternal grandmothers influenced infant survival much less, for she was not certain that her son was the infant's father. The mother's death greatly imperiled the infant, much more than the father's. By the second or third year, the risk fell, especially if the maternal grandmother was present.

Social stratification influences parental investment in measurable ways. As we saw in Section 40.3 in the article by James Boone, parents in some groups have good reason to prefer boys, particularly in the upper classes. The males frequently took brides in the social strata just below their own, such that their sisters often remained unmarried. Lower-class parents who wished to have grandchildren were consequently better served by having daughters, even if their dowry capacities were only hypothetical. Voland et al. (1997) emphasized how economic rational-choice models operating within families do not contradict our innate predispositions. Where good population records are available, it becomes possible to study not only marriage and hypergamy, but also the differential incidence of juvenile mortality between males and females and the marriage outcomes of the offspring who reached adulthood. We have known since the 1970s that parents applied discernable strategies of social advancement that an array of good records can confirm. Alain Collomp, whose medical training imparted a notable rigor to his research, aligned in chronological order all of the available sources pertaining to scores of families living in seventeenth- and eighteenth-century Provence to show how parents oriented the destiny of their children depending on their number, their sex, and their abilities. Of course, the children, who had personalities and interests of their own often disappointed their parents (Collomp 1983)! Their association with same-sex peers (another universal) and the rigid observance of the rules of that group, or the desire to strike out on their own path in the face of parental opposition, are universal traits that generate considerable documentation.

Human mothers do not accept their newborn infants automatically, depending upon their quality and their prospects, nor do they treat all of their children equally. It is lactation, with its attendant nurturing and cooing, that establishes the bond between mother and infant, who quickly responds (Hrdy, 1999). Devereux (1955), using the Human Relations Resource Files (compiled during the first half of the last century), found abortion to be practiced in several hundred societies: absence of reports of it cannot be construed to demonstrate its absence, for anthropologists had different levels of access to the populations they studied. Abortion and infanticide appear to be part of the universal repertoire of human behavior and are much more widespread than we thought. The mothers appear everywhere to be the authors of them, and midwives are active or complicit in the phenomena with various degrees of concealment. Where laws exist against abortion or infanticide, they are often only fitfully applied. Trivers and Willard (1973) first formulated the theory that mammals might cull their newborn infants by sex, depending upon the condition of the mother: fit mothers would prefer males, while less fit mothers would benefit more from raising females. It is difficult to ignore the theory when encountering skewed sex ratios across different social classes, but the human situations vary quite a bit according to inheritance customs and economic possibilities. Routine infanticide is best known for East and South Asian societies, where demographers count "missing" females in the millions. Today, this is done by sex-selective abortion, but in previous times, deliberate neonatal infanticide went largely unchecked, even where it was illegal, as in China. Right across the social spectrum there, and among the high-caste families in Northern India, it was overwhelmingly practiced on girls (Campbell & Lee, 2010; Lee & Wang Feng, 1999). Elsewhere, parents killed boys as well as girls, but in numbers and proportions that are difficult to establish. In northeastern Japan, Drixler calculates that in difficult years, neonaticide might have been the fate of as many as 40 percent of live births, at a rate of 3 girls for 2 boys. He calculates millions of infanticides to 1870, then large numbers of fictitious stillbirths thereafter, until modern family planning in the early twentieth century made the practice obsolete (Drixler, 2013).

The long-standing opposition of Christianity to infanticide is held to represent the dominance of culture over nature. The historical study of infanticide in the West is pretty much restricted to single mothers, which is the social group overwhelmingly represented in the criminal archives where the practice is encountered. It is often assumed that Asian-style infanticide was unknown (Lynch, 2011). The anthropologists Marvin Harris and Eric Ross ventured that since European birth rates before the Demographic Transition of the nineteenth century were so low, mothers *must have* practiced some form of "death control," carried out perhaps not consciously, but rather "accidentally on purpose." Industrialization in the eighteenth century, along with wage labor, then gave more room to children of both sexes (Harris & Ross, 1987). This theory was proffered in the complete absence of empirical support. The lack of reliable positive evidence of neonaticide hampered the demonstrations of medievalists who suspected the practice of infanticide had continued unchecked since antiquity. Richard Trexler then uncovered scores of cases of "overlaying" in Renaissance Florence, where mothers were accused of accidentally killing their infants by sleeping with them. These mothers required absolution from the bishop of Fiesole and underwent a cursory investigation before the diocesan tribunal. In the absence of solid proof of murder, bishops and their magistrates did not desire to find the mothers guilty. Excessive harshness by Church authorities against the mothers would have placed their toddlers in great danger. But the occult phenomenon spurred Church authorities to establish foundling hospitals for bastards and unwanted infants (Trexler, 1973).

American sociologist Brigitte Bechtold (2001, 2006) discovered that the male sex ratio in nineteenth-century France, above 106 per 100 females in 1800, dropped to below 105 a century later. After 1880, when it became possible to study the phenomenon by *département*, agricultural districts had rates over 108 males per 100 females, while industrial districts displayed much lower rates, even below the natural ratio of about 105 males per 100 females. Parents could claim – with complete impunity for most of the century – that a victim of neonatal infanticide was born stillborn (Dansette, 1987). Even in nineteenth-century America, there were periodic waves of infanticides whenever parents or guardians could not divest themselves of children who were seen as an encumbrance. Randolph Roth's research was restricted to judicial archives, but American baptismal records have not yet been studied to establish sex ratios (Roth, 2001). In fact, routine infanticide may well have existed throughout the West, measurable in the discrepancy of males to females at baptism, with spikes in favor of males in difficult years (Hanlon, 2003). Neonaticide was not always aimed at girls: in seventeenth-century Parma and its district, the stable discrepancy between peasants (who preferred boys) and workers in textile neighborhoods (who preferred girls) ran as high as 30 percent (Hynes, 2011). Parents had good reasons for preferring males or females, but it seems to have been a measure applied mostly in dire circumstances. This kind of research, applied to England (Reynolds, 1979), Italy, France, and Acadia, fits very well into the theoretical framework of Harris and Ross (1987). Most of this is neonaticide, but Daly and Wilson (1995) and Voland and Beise (2002) have noted the differential mortality of juveniles who became an encumbrance, especially after the death of the father. A widow with toddlers had good reasons to let them die before remarrying. There is a lot of work showing that mercenary wet-nursing and abandonment of infants in foundling homes acted as forms of delayed infanticide. Wet-nurses made it possible for affluent mothers to reproduce more quickly, but mothers situated lower on the social scale took fewer precautions with their offspring. The hospices were in theory reserved for bastards, but in a crisis married parents might leave their children there as well, hoping to reclaim them when conditions improved. Historians place much emphasis on the high rates of mortality (as high as 80–90 percent, compared to 50 percent for infants kept at home), but in fact, very few bastards were likely to survive to adulthood if their mothers did not marry the father (Voland & Beise, 2002). Mothers can and do kill their children, but more than anyone else, they also keep their children alive. This research on routine infanticide in the West is still in its early stages, but there appears to be no sharp discrepancy between Christian countries (both Catholic and Protestant) and the great agricultural civilizations of Asia (Hanlon, 2016b).

40.5 INTERPERSONAL CONFLICT

No historian interested in interpersonal conflict and violence is likely to be unmoved by the work of Frans de Waal, who scrutinizes closely the social lives of apes in captivity. Without language and written law, social animals live by rules, whereby each member has expectations of behavior from all of the others. Free from the necessity to forage, captive animals spend much of their time socializing with their peers, under the watchful gaze of ethologists who study their every move. The individual behavior of primates reveals a great deal about what they are thinking and feeling, and apes use social manipulation in ways that betray thought and reason (de Waal, 1982, p. 49). These are knowing, wanting, and calculating beings, very close to humans. Our closest relatives, the chimpanzees live in bands in which dominance and submission are keys to stability. The alpha males have disproportionate access to females, but they cannot operate without the assistance of other males and strategic females. Rank orders bind individuals together in pacts of loyalty (de Waal, 1989, p. 46). Males desire status, and older males, operating rationally, intervene in disputes with a policy directed toward increasing their own power (de Waal, 1982, p. 196). Reciprocity among primates occurs in both the positive and negative sense, in which social reactions are remembered by the protagonists and the bystanders as well. Adult males wishing to rise in rank must challenge the dominant male repeatedly and get the better of him, but in so doing, they must solicit the ongoing aid of allies in the troop. These animals are extremely strong and could fight to the death, but they are also capable of restraining their violence and stopping before it goes too far. Some of this restraint is innate, while some of it springs from the active intervention of others – males and females – who intercede to calm things down (de Waal, 1982, p. 95). Social animals almost never fight to the death, for they need each other and value good relationships (de Waal, 1996, p. 29). Major fights are then followed by an upswing of grooming and other friendly contacts. These friendships, however, are situation based, making alliances unstable. The violence usually serves some purpose: dominant males must use their power to protect females and their young from the aggressive moves from other males. The alpha male can also use his power to allow access to females by his supporters and to close off such access to his adversaries, who then must take their copulations furtively. A stable hierarchy is a guarantee of peace and harmony in the group and reduces violence many times over if it exists, but this situation is constantly tested (de Waal, 1982, p. 118).

Aggressive behavior is a fundamental characteristic of all animal and human life, an innate capacity that requires some learning in both chimps and humans. De Waal deplores the blanket condemnation of aggression in research on conflict resolution, for it is not by itself antisocial. It is the basis of good social relationships, and it is not something that can or should be eradicated. The threat of resistance or retaliation keeps dominance within bounds and forces group members to make peace. Peace overtures may not be sincere: chimps, like humans, can compose the conflict in the short term only in order to unleash it anew when the chances of success are better. The belligerents making peace will remember what occurred and nurse grievances that express themselves as punitive or moralistic aggression against cheaters. Their indignation is real (de Waal, 1996, p. 159).

De Waal's research is rich in lessons for anthropologists and historians, for he shows us that our early ancestors, like these primates, were similarly guided by gratitude, obligation, indignation, and retribution long before they acquired the gift of language. Françoise Héritier (2004) notes that no human society exists in which there is an unlimited possibility of harming or killing any member of the group, or of fornicating with any other member, or of taking whatever they wish from others without some accountability. Wherever these events transpire, they give rise to a burning desire for justice – that is to say, the punishment of the offender. This is the matrix of universal ethics. The criminologist Maurice Cusson, a disciple of Raymond Boudon, explores what the objective sense of the gravity of an offense implies for the level of crime in any modern society (Cusson, 1990). Criminologists, like historians, work from statistics, which are only as good as the bureaucracies that produce them. People do not denounce all of the transgressions that happen to them, and the police do not act on all of the cases brought to their attention. Authorities tend to filter the denunciations and devote their energies to those that appear to be objectively the most serious. Their sensibility is likely to mirror that of the public as well. Some categories of crimes fluctuate in importance over time, but others (like homicide, to which we will return) are considered serious offenses everywhere.

The sociologist specialized in crime will immediately encounter a series of human universals. First, the principal actors in crime and other transgressive behaviors are overwhelmingly male, aged between 14 and 30. Second, the criminal action is usually committed with some prior planning, wherein the offender weighs the advantages of their actions against the drawbacks (Cusson, 1990, p. 43). Young offenders have a lot in common: they are usually in conflictual relations with their parents, or the latter are conspicuous by their lack of control. The youths also self-select their peer groups and run with others of their ilk. If they have access to social resources that aid and abet their actions, such as the likelihood of impunity, the crime rate will rise. The more crime is committed, the less likely it will appear to be serious or exceptional, and a fatalism and routine set in that demoralize the general population (Cusson, 1990, p. 133). Heavier controls and more surveillance by neighbors have the opposite effect: Cusson points to Switzerland and Japan as places where low crime rates reflect widespread social surveillance and neighborhood accountability (Cusson, 1990, p. 66). If society erects enough obstacles to crime (for the culprits are not perseverant and give up easily) or punishes the offenders regularly, the incidence will decline and the general public will be reassured. Bystanders deplore the impunity of cheaters, and they ardently desire that they should be punished. It is not so much the severity of the punishment that produces the positive effect, as the certitude of it. As the offender ages, he learns that the illicit short-term advantages are not worth it in the long run.

Cusson's work deals with universal features of crime and punishment, but it remains outside the biological framework of ultimate causes. The important work by Martin Daly and Margo Wilson on homicide constitutes a turning point. For these early sociobiologists (Daly & Wilson, 1988), homicide reflects deep universal issues, and there is less reporting bias than for other crimes. They proceeded like good historians, developing their theories not just from the global literature on homicide rates, but also by sifting through the case files in the archives of the Detroit police department when that city was the murder capital of North America. Both men and women kill, but their victims reflect their universal differences. Women killed their children, first and foremost, for the reasons outlined in Section 40.4. Daly and Wilson noted how males had good reason to be fearful of their women replacing them with rivals, which provoked in them a violent jealousy, particularly if the woman intended to leave. The situation might lead to simple sequestering, but irate men also beat and occasionally murder the women and the couple's children too, out of an intensely felt proprietary hold over both. Wherever the homicide rate is high, these domestic killings are not very numerous, but as the state represses violence between unrelated males, the proportion (if not the actual incidence) of domestic killing rises. Cusson claims that the incidence of family homicide today is still comparable to that of the Middle Ages (Cusson,

2000, p. 181). When the victims and killers are unrelated to each other, the perpetrators and their victims are everywhere overwhelmingly male. Where homicide rates are high, the act emerges out of arguments, insults, and rivalries between young males who feel that their reputations are at stake and must be defended. Men who engage in predatory violence have different time horizons than law-abiding men and wish to gratify short-term goals (Daly & Wilson, 1988, p. 170). Escalated "show-off" disputes are overwhelmingly male affairs, as is lethal retaliation for previous physical or verbal abuse. Whenever the monopoly of state violence is relaxed – and this is true everywhere in the world – predatory and deterrent killing by males soars. Men also display a universal predominance in crimes of plunder, particularly enacted by brazen confrontation. The motivation of blood revenge is perhaps the most powerful impulse of all. In societies without an objectively neutral police and courts, turning the other cheek to another's provocation is stupid or contemptible, or else an admission of impotence. One must stand one's ground or strike back at the offender to deter others from trying the same thing. One cannot do this alone, however; one must stand with the help of one's blood kin and other allies, who are bound to help. Men will often kill with the active support of their male blood kin or with the auxiliary assistance of their womenfolk (p. 129).

Unchecked violence entails high costs both for the participants and for bystanders. Therefore, both are keen to seize an opening to compose the conflict and to restore an equilibrium. One common measure was to unite in marriage offspring from each camp, who would produce children related by blood to both groups. But composing conflict was also the task of kings, who could punish recalcitrant belligerents with force of their own (p. 236). Punishment of offenders, which is central to our understanding of justice, satisfies a basic human desire for revenge. Moral sensibility – that is, righteous indignation, pangs of conscience, and a sense of duty – is a universal aspect of human nature (p. 255).

Historians influenced by evolutionary theory have been working on these issues since 2000. The Cambridge criminologist Manuel Eisner followed up the first findings that homicidal violence has declined in the West since the late Middle Ages. This "civilization process" was first described by the German sociologist Norbert Elias, who charted the growing phenomenon of "self-control" promoted in the writings of humanists like Erasmus circa 1500. German historians speak of a process of "social discipline" that followed on the growing strength of governments, the intensification of trade (which required trust), and the harangues of priests and pastors (Eisner, 2001). There is a second model following Durkheim, who studied the progressive detachment of people from the tight bonds of obligation and solidarity to kin from the Middle Ages onward, the onset of our modern "anomie." A certain inwardness was discernable in social elites by the seventeenth century that worked against automatic solidarity in

the conflicts of others and inhibited the desire to avenge oneself. The growing reach of the king's law and judicial distaste for people taking matters into their own hands also served to curtail violence (Nassiet, 2011).

An important collection of articles – more theoretical than empirical – appearing under the direction of Eisner in 2011 invited criminologists and historians of crime to take evolutionary lessons to heart. In addition, Oscar Di Simplicio revised an earlier book in order to incorporate insights from evolutionary theory to a single case, a depraved priest who was the alpha male for his rural parish in seventeenth-century Tuscany. This microhistory depicts the variety of social resources available to a dominant male in order to further his own power and access to females. Eager to impress his parishioners with his access to magic and occult powers, the delinquent priest attracted the curiosity of the powerful Holy Office of the Inquisition instead, which effectively clipped his wings (Di Simplicio, 2011).

Still in 2011, the psychologist Steven Pinker published a heavy tome chronicling the decline of violence in the West since the Middle Ages. It contains two parts: a commentary on the historical decline and a discussion of the wellsprings of human interpersonal aggression. Norbert Elias claims that Europeans toned up a mental faculty (self-control) that had always been an integral part of human nature (Pinker, 2011, p. 73). Why did this occur around 1500? For Pinker, the crucial turning points are the invention of printing, the Age of Reason, and the Enlightenment. North American intellectuals frequently offer idealist explanations that magnify the importance of Great Books and intellectual movements, something we might call the "humanist fallacy." It is not easy to improvise oneself as an historian, for convincing demonstrations require a familiarity with the extant sources on crime and an understanding of how they came into existence. Pinker relied instead on large "data sets" compiled by people – Rudolph Rummel and Matthew White – who were not themselves conversant with criminal archives. They relied on books, in English exclusively, filled with lurid descriptions of unpleasantness on a truly massive scale. Many of the numbers Pinker bandied about with some hyperbole were fantastically exaggerated, which had the effect of magnifying the decline in violence over the centuries. Another weakness was to place the low homicide rate of the eighteenth century on a different level than today's rate, forgetting that without the advent of ambulances, cell phones, and advanced surgical teams working in operating rooms, today's rates would be comparable to those of recent centuries past. The chronological decline across Western Europe from England to Italy is indisputable, but falls markedly everywhere from the seventeenth to the early eighteenth centuries. This "Civilizing Offensive" is unlikely to be the Age of Reason, and still less the Enlightenment of the *philosophes*; this was the Age of Absolutism, employing thousands of judges who staffed the administrative monarchies of their time.

These agents of power, with their police and militia forces in hand, enjoyed the support of social elites and clergies in order to submit powerful males to the law (Hanlon, 2002, 2004). The incidence of capital punishment declined precipitously as well, and the heightened security was followed by some measure of popular disarmament. This tends to confirm Randolph Roth's (and Frans de Waal's) claim that trust in government and stable hierarchy are the keys to containing conflict and curbing violence.

Pinker's entertaining book is more convincing when he discusses the several systems in the brain that are "wired" for violent action; but aggression is not a single entity, he shows. Violence has several distinct roots. There are systems for predation, for rage, for fear, and for seeking. Panksepp's "intermale aggression" system is studded with receptors for testosterone, which rises in the presence of attractive females and in anticipation of competition with other men. As is the case for primates, our readiness for violence is acutely situational (Pinker, 2011, p. 510). But then violence is but the flip side of communal solidarity, the desire to cooperate with other members of our group. "Most of the harm that people visit upon one another comes from motives found in every moral person." In fact, we might be too moral, inclined to mete out punishment to others, given that we see ourselves as more virtuous than we really are (p. 569).

40.6 WAR

Wars erupt naturally everywhere humans are present (Ghiglieri, 2000, p. 160). But it is not confined to humans, since chimpanzees patrol their boundaries like disciplined warriors and raid rival groups (Wrangham & Peterson, 1996, pp. 49–82). Some anthropologists and paleontologists speculate that war in humans had a more recent origin, around 10,000 years ago at the beginning of the Neolithic period, when unprecedented high population densities pushed humans into competition for precious resources. Jonathan Haas and Matthew Piscitelli take issue with the claim that "warfare is and always has been a prevalent [sic] part of human existence." They note, quite rightly, that anthropologists presume deep biological roots for war as judged from the ethnographic record of peoples living in recent centuries, which is a methodological error (Haas & Piscitelli, 2013, pp. 168–190). They claim that for the first 95 percent of human existence, low population densities rendered conflict over resources unlikely. The authors conducted an extensive survey of the literature (much of it in French) analyzing the skeletal remains of almost 3,000 individuals living over 10,000 years ago, to conclude that warfare was the "rare exception" (sic) prior to the Neolithic period. The objection is sensible, but war need not be "prevalent" in every era to remain part of the human behavioral repertoire. It could spring from biological roots that trigger a warlike response when it seems like a lesser evil. Some wars are necessary for at least one of the belligerents; the trick is to know, without the benefit of hindsight, which ones they are!

War and violence are not irresistible drives, but situational responses designed to defend the group from external threat (Gat, 2010, p. 170). Ghiglieri (2000) claims that most men are resistant and reluctant to kill, but will do so if they are convinced that their adversaries deserve it (p. 180). War is closely linked to the control of resources. The only solution to avoid depredation is deterrence, to dissuade others that encroachment on your rights, your resources, and your territory will be met with riposte, which is usually calibrated, not blind. Even weak groups will strike back against more powerful offenders at high cost to themselves, since there is emotional satisfaction in righteous retaliation (Ghiglieri, 2000, p.192). In foragers, coalitional violence can be meted out in the group in order to keep ambitious or homicidal alpha males in their place. This coalitional violence, under conditions of strong competition, hostility, and desire for retaliation, can be directed next against outsiders (Boehm, 2013). It makes good sense to stand one's ground to defend one's resources, but if you can broadcast your readiness to fight, then a potentially hostile group will think twice before trespassing (Gat, 2010).

Not everyone is suited for war, however, and it operates as a form of selection for leaders and participants. Strong, dominant males will stand to gain from its existence. All primate communities form status hierarchies, and individuals with a strong desire for dominance generally rise to the top. Roughhouse play is a male universal to establish dominance in boys (Rosen, 2005, pp. 23, 77). War is overwhelmingly a male activity: women may occasionally fight in defense of hearth, and occasionally individual women join the men in disguise, but the dominance of males is everywhere overwhelming (Tiger, 1969, p. 180). Anthropologist Lionel Tiger claimed that the young male has a biologically given need to prove himself as a physical individual, just as he needs to validate his place and his reliability to important peers and superiors, who are judging his actions (Tiger, 1969, pp. 182–183). Both raiding parties and large armies rely on male bonding, which strikes deep emotional chords in boys and men, many of whom have a vivid imagination for such things (Ghiglieri, 2000, p. 183). Even among men, there are considerable personality differences and robust variations in testosterone levels. Psychologists since Hans Eysenck have spotted a trait they call "sensation seeking," research that has been advanced by Marvin Zuckerman (2007). High sensation seekers live in the present and cannot abide boredom. They seek not mere novelty, but intensity of experience, danger for its own sake. This boldness is also related to dominance and aggressiveness, the expression of high levels of testosterone. There is no doubt that these traits are biological, for castration reduces dominance seeking in both humans and other animals (Zuckerman, 2007, p. 21). Men everywhere, and young men especially, score more highly in risk seeking than women. Then, however, the trait gradually recedes as the subject ages.

Individual reasons for making war do not prevent it from being a social activity, reliant on group cohesion. John Tooby and Leda Cosmides sketched the theoretical underpinnings of warlike behavior in an important article. The warrior coalition will not be stable unless the participants are rewarded or punished in proportion to the risks they have run, so there must be some way to identify free riders. There is a risk contract containing several clauses: there must be a random distribution of death, an equitable allocation of spoils, and the men must have a reasonable assurance of victory and a sense that they will not themselves face certain death. Confidence in success is crucial in encouraging coalitions to initiate war (Tooby & Cosmides, 1988, pp. 5–10). Most societies exercise strong control over intergroup aggression in order to avoid that hotheads should determine policy (Leblanc, 2007). Leaders, however, should be able to reach higher rank and reap more resources – both physical and sexual – than their underlings. They also accrue prestige among their peers and, not infrequently among their enemies, open admiration (Gat, 2010).

Anthropologists, often in the thrall of the "noble savage" theory, long tended to avoid studying warfare. The first anthropologists to look squarely at war in prehistoric societies were struck by the extremely high rates of violent death. They soon learned that war was a universal capacity with a very long past. The increase of population led to an increase in the scale of warfare in the Neolithic period around the globe (Leblanc, 2007). Napoleon Chagnon spent a lifetime among fierce South American tribes, where he encountered a Hobbesian world of primitive, endemic conflict. The wars themselves often started over trivial issues like women or small-scale fights between protagonists of a local group. Cumulative grievances of many kinds pit men against each other, and these escalated into major confrontations through chain reactions of vengeance killings (Chagnon, 2013, p. 222). The larger the group, the more easily it ensured security from potential rivals, but once the group reached 400 or 500 individuals, the internal rivalries and tensions led to a scission and a potentially inimical new group. Chagnon had the good sense to count, quantifying the events occurring in his population as often as he could. Not all of the men were killers. The most successful warriors, however, had the most children, the most relatives, and the most power (p. 93). In tribal society where no one possesses much property, the men vary a great deal in their ability to inspire confidence and issue commands. Yanomamo headmen or leaders had no use for money or land. Their power was expressed in social and political networks, which were bound up with blood ties. Patrilineal societies were also designed to keep the menfolk together as allies and born cooperators (p. 315). Meekness among them was not a virtue. The accusation of cowardice is social suicide for men, for it invites rivals to snatch away their women and resources. Defeated groups are pushed away from the best lands toward infertile margins.

The reasons for making war and the coalitional techniques in pursuing victory have not changed much since prehistory. Revenge is a major cause of fighting in pre-state societies, which is a rational motive where there are no valid arbiters composing antagonistic bands. Endemic competition and suspicion eventually lead to a public challenge, but one group may attempt to overcome their adversary with a preemptive strike (Gat, 2010, p. 170). Confidence in one's success is usually also decisive in obtaining general consensus to start a war. Historical societies differ in scale from prehistoric peoples, not in their underlying motivations. Readers will remember the strong case Western countries made to overturn the Iraqi invasion of Kuwait and to impose on Saddam Hussein a return to the status quo ante. A decade later, leaders in the Anglo-American world, with surprising ease, convinced the democratic press and population to embark on "regime change" in Iraq. Confident of success and expecting rich rewards, the coalition launched a preemptive strike – and vilified those countries warning them of the imponderables. Territorial integrity and tribal affirmation underpinned both the Argentinian invasion of the Falklands/Malvinas and their ejection by the British. Victory and conquest then rewarded the warriors of both nations with public acclaim. Status issues are central to war and peace, and we enjoy the process of victory. This holds true for polities as well as individuals (Rosen, 2005, p. 71).

A noteworthy interdisciplinary contribution to the understanding of war throughout time is Marco Costa's *Psicologia Militare*, which begins with interpersonal violence and the extension of its properties to a mass level. Costa illustrates the durable pertinence of psychological motivations over the centuries by drawing lessons from the leading military theorist of the seventeenth century, Raimondo Montecuccoli (Costa, 2003, pp. 115–123). Modern mass-market writers on war, like Gwyn Dyer and John Keegan, are increasingly open to evolutionary explanations of the phenomenon, and it is surprising that their insights have not been relayed by historians in more specific contexts. Dyer establishes three propositions: (1) that humans can kill cospecifics; (2) that their societies tend to push the environmental limits; and (3) that humans are no better at conserving their environment than any other animal (2004, p. 86). Almost every state that ever existed was destroyed by war or integrated into a larger unit in order to increase its security (p. 137). Larger states can raise larger armies, such that smaller states must enter into alliances in order to survive. In modern societies, warriors are only a small fraction of the whole, but this can change with the breakdown of public security (p. 87). There is no reason to expect that war-making will always be a frequent response in global affairs, but long periods of peace do not imply that humans have outgrown their ability to resort to organized force. Endemic war in Japan was followed by over two centuries of peace under the Tokugawa shogunate, but scant decades after the end of that regime in 1869, the empire flexed its muscles afresh.

Military history was long anathema in university departments; the study of recruitment and of army administration dates only from the 1960s (André Corvisier), that of logistics from the 1970s (Geoffrey Parker). Battle history was not judged worthy of a serious historian until John Keegan published his *Face of Battle* in 1976. Keegan's earliest book is cognizant of human and animal ethology, which readers gradually understood to be important. Battle history is the oldest historical genre, for participants at the time understood the event to be important and worthy of public remembrance. Keegan was the first to focus primarily upon the psychology of the combatants themselves and their motivations for acquitting themselves well (Keegan, 1976, pp. 50–51). He perceived that much of battle entails warlike posturing, which is not in itself bloody. Soldiers clearly had some natural inhibitions to kill, with some not firing and others aiming to miss (Keegan, 1976, pp. 167, 172). Fear could quickly transform the formation into an unruly crowd, and the collapse often came from the rear, since those in front were unable to move. In a pertinent article on combat motivation, Keegan reduced the ordeal to three things: inducement, coercion, and narcosis. Men must be forced to fight, but there were material and emotional rewards, like self-esteem and recognition by one's peers and grateful superiors. The regiment is a tribe, and belonging to it enhances its corporate spirit, while ethnic homogeneity helps the men fight better. But even mercenaries will put up a good fight if they have professional pride. Alcohol and drugs dampen the awareness of the risks involved. Once in combat, soldiers modulate their fears with forebrain processing of situational awareness. They take notice of the exertions of individuals who flourish in this environment and often enough imitate them. Once their frenzy is aroused, however, men can kill anything that moves (Keegan, 1997, p. 10).

Keegan's work drew heavily from a contemporary of Charles Darwin, the French colonel Charles Ardant du Picq, whose *Etudes sur le Combat*, written between 1865 and 1869, was published after his death in battle in 1870. One would like to imagine the colonel approving of the daring thesis of the English naturalist, had he but known of it. Starting with an analysis of the combat experience of armies in antiquity, Ardant du Picq was struck by how little changed throughout the centuries. The prolonged corps-à-corps with the enemy is extremely rare, and collapse is due more to panic than being overcome with arms. Melees of infantry or cavalry are the fruits of the poetic imagination, he wrote, on the basis of his own observations and on the results of questionnaires he passed out to his fellow officers. Real troops wished above all to maintain cohesion, to remain in contact with their friends. Attackers and defenders must do the utmost to induce the other side to refuse combat. The latter's firm stand would force an advance to slow down and stop as increasing numbers of attackers fell behind. Attackers who advanced resolutely with few stragglers were likely to induce their adversaries into retreating. "There are never,

ever, ever, two equal resolutions face to face" (Ardant du Picq, 2004). Long firefights produced astonishingly few casualties relative to the number of shots fired, for the men clearly aimed to miss. Fear is a central element in combat, and discipline serves to keep it at bay as long as possible. In the past, armor gave boldness to cavalry troopers, but those without it would not behave similarly. Readers today will be struck by the modernity and freshness of the prose and the clarity of the argument. It inspired generations of social science research that wished to understand the impact of combat on the eternal instincts of the participants (Hanlon, 2016a).

One American military sociologist inspired by the French colonel, S. L. Marshall, conducted interviews with soldiers in the firing line in the aftermath of combat to determine the small-group dynamics in the heat of battle. The object was to determine what measures might increase the confidence and participation of the men, most of whom could not be induced to shoot. Battles hinged on the active participation of a few fighters on both sides. Team weapons firing from farther away were different, however, and these killed and maimed increasingly large proportions of men. More recent research claims that Marshall just made up the fire ratios he cites, and that his book *Men against Fire*, published decades after the war, did not provide any interview statistics (Marshall, 1967). Marines psychologist Dave Grossman carried this research forward in two short books, *On Killing* and *On Combat*, both of which begin with the premise that soldiers have evolved instincts that inhibit killing their foes face-to-face. Only about two percent of soldiers, if pushed or given a good reason, will kill without regret or remorse (Grossman, 1995, p. 180). In conflict, the two options of fight or flight are doubled by two more: posturing and submission. Noise and intimidation are often sufficient to convince the enemy to withdraw. Soldiers may expend considerable ammunition in an exchange with the enemy, but hit rates have always been low, for face-to-face killing is almost unbearable. On the other hand, soldiers lose their inhibitions to kill if the enemy appears to be fleeing. This Grossman holds to be true throughout history. With the proper conditioning and proper circumstances, almost anyone can and will kill, but this is recent knowledge. Grossman has developed a very lucrative calling in instructing armies and police forces on how to induce their members to kill, a science he labels "killology."

Soldiers in the throes of combat undergo a roller coaster of strong emotions. An easy success will trigger a "parasympathetic backlash," wherein the elated men lose all capacity for resistance and are chased off by a rapid counterattack. Only fresh troops sent to help will be willing to put up a fight. The body goes through several phases of alertness and mobilization in the course of combat. Soldiers tune out all of the senses except the ones needed for survival, as a natural cognitive response to danger. Combat can shut down forebrain processing, wherein the soldiers go into autopilot, to the level of their training (Grossman, 2004, pp. 39–57, 77). The men who try to control their confusion and reduce unpredictability are the officers. The good ones are able to inspire a high degree of confidence in the men under their command. Officers must lead, however, which results in very high casualty rates for those who perform well. Men will also be influenced by the sight of large numbers of friends or enemy, like mobbing or schools of fish. The actual battle among men at close quarters is a process of posturing until one side turns and runs. "It is in the subsequent pursuit of a broken or defeated enemy that the vast majority of the killing happens. Dogs understand this too, and you will be pursued if you run from them" (Grossman, 2004). Grossman discusses the historical neglect of posturing as "a conspiracy of silence by historians, and cultural conspiracy of forgetfulness, distortions and lies" (Grossman, 1995, p. 36). These same points were made by Raimondo Montecuccoli in his writings derived from personal experience in the Thirty Years' War, and similar lessons can be teased out of battle descriptions in just about any era (Hanlon, 2016a).

Finally, prolonged exposure to combat is a debilitating experience. In battles past, men would use the immediate aftermath to recover their calm through talking (debriefing), once they were safe. Medical personnel have for centuries observed the psychological burnout of many soldiers exposed to danger over and over. Since the 1950s, we speak of stress, which in battle incites the men to action. Situations of extreme peril can induce panic and flight, or else inertia and immobility, where men do not react as enemies cut them down. Once the contagion of fear and flight infects a unit, there is no reasoning with the men, and they will flee until completely exhausted. Under normal stress, these sentiments dissipate after several hours or a few days at most (Crocq, 1999, pp. 75–87), but they may not be able to face the enemy again.

40.7 AESTHETICS AND DISPLAY

This is not an area where ethologists have concentrated their research, but we can sketch out some basic traits of the phenomenon that will interest historians. People everywhere sing, dance, decorate surfaces, and tell and act out stories. Dutton calls this an "art instinct," although it not based on a single genetically driven impulse. He describes it as an ensemble of sub-instincts that are absent in other animals, and even in *Homo sapiens* they originated only about 100,000 years ago (Dutton, 2009, p. 6). Paintings, jewelry, sculpture, and musical instruments date back 35,000 years in Europe, and probably longer elsewhere. All human cultures have expressive activities that European traditions would describe as artistic, but not every form exists everywhere (Dutton, 2009, p. 29). Art provides pleasurable experiences that are simultaneously perceptual, cognitive, emotional, and operational (Dissanayake, 1995b, p. 40). We are attracted by the emotional content of a work of art and its insight into the human condition, since these tap into the timeless tragedies of our biological predicament: our mortality, finite knowledge and wisdom, the differences between us, and our conflicts of interest with others (Pinker, 2002, p. 418). Music, singing and dancing, or reciting and listening to poetry and stories unites participants with one another, and this common experience enhances social solidarity (Dissanayake, 1995b, p. 24).

Philosopher Dennis Dutton enumerates 12 properties of uncontroversial works of art: (1) they should be valued for the pleasure they give in themselves; (2) there should be some measure of skill involved in their creation; (3) objects in all art forms are made in recognizable styles, in which artists create elements of novelty or surprise; (4) unpredictability and newness offset the rules and routine; (5) there is always some kind of critical language of judgment and appreciation; (6) art objects represent or imitate real and imaginary experiences of the world; (7) there are several ways in which mundane activities or objects are enhanced with a virtue of "specialness" to demand extra attention; (8) individuality can be a focus of attention and evaluation; (9) the experience of works of art is saturated with emotion or may be pervaded by a distinct emotional tone; (10) the best works of art tend to utilize both perceptual and intellectual capacities; (11) art objects must be judged by their place and significance in the history and traditions of their art; and (12) above all, art objects provide an imaginative experience for both producers and audiences. These 12 points are what he calls "cluster criteria," which are valid for the vast majority of artistic experiences (Dutton, 2009, pp. 50–56).

Nomadic people who own few material possessions usually decorate what they have. Red ochre for body decoration is one of the oldest embellishing cosmetics. People everywhere enhance themselves through ornament and decorate their space and their objects as well, but not indiscriminately so. Dissanayake calls it the act of "making special," a pleasurable capacity to place an object, person, moment, or place on a higher plane of experience. Humans often adopt non-natural shapes in this context, abstractions springing from the human mind, a preference for order. (Rhythm is the equivalent of order in sound, which, like decoration, is subject to variation, just as ornamentation and rhythm enhance speaking; Dissanayake, 1995a, pp. 56, 80–83.) These works of art are intended to delight, dazzle, or frighten the beholder. This urge to decorate is not specific to exotic cultures, since countless mothers and daughters embroidered, knitted, and embellished special objects in the West (Dutton, 2009, p. 65). Even art objects produced by traditional cultures elicit praise and appreciation by those of other cultures for the emotions they provoke (Pinker, 2002, p. 408).

These fundamentally social instincts might have derived from play, which has deep animal origins. The arts are commonly linked with ceremonial contexts and rituals, which use effective means to arouse, capture, and hold attention: they are deliberately non-ordinary and formalized (Dissanayake, 1995b, p. 46). The group that performs ritual ceremonies together stays together, working harmoniously at least in this (Dissanayake, 1995b, p. 130). All societies have rites of passage, performed to produce results, and these provide focal points for the arts. The ceremonies themselves are culturally learned, whose very evolved forms sometimes require special education to comprehend all of the elements, but beneath them there is an innate desire to "make special" (Dissanayake, 1995b, pp. 66–68). Dressing up is another universal expression of making the person and the occasion seem something out of the ordinary.

The arts squander brain power, resources, and time, which seems anathema to natural selection, but the real biological origins of art lie elsewhere. Darwin understood these to be features of sexual selection, an outgrowth of the instinct and desire to make oneself attractive to members of the opposite sex (Dutton, 2009, p. 138). Here, there are enduring and universal differences of priorities for males and females – different perceptual biases – with men seeking youth and beauty and women seeking resources and dominance. Irrespective of the cultural setting, women place much greater emphasis on their physical appearance. Men are more interested in money, power, and hierarchical ascendancy (Saad, 2007, p. 72). Ostentation in humans and other creatures is the whole point of a fitness display, which is designed to attract and seduce members of the opposite sex. People innately wish to be more attractive to them by signaling their best attributes. This makes it important for the intended target to distinguish authentic signals from deceptive ones (Dutton, 2009, pp. 145–153). This relation is aided and abetted by adornment, dramatic gestures, gift-giving (especially by the males), song and bodily movement, display of finery, etc. None of this generic behavior is specific to a place or a period, and poor people engage in it as well, up to and often beyond the limits of their resources (Miller, 2000).

Steven Pinker emphasizes another facet of display that is surely true: the hunger for status. Art is conspicuous consumption writ large, an instrument by which one can impress other people (Pinker, 2002, p. 406). Those who are able to design beautiful artifacts acquire special standing in their own communities, a recognized ability that sets them apart from and above the others (Dutton, 2009, p. 233). The objects themselves are frequently made of rare or expensive materials, and the person making them may have needed years to reach that level of skill (Dutton, 2009, p. 157). Activities enhanced by aesthetic objects and fine ceremonies were not conducted for aesthetic purposes alone, although this does not diminish the artistic qualities of the objects themselves. Extravagant architecture occurs as a means for rulers to send honest signals to their competitors and their subjects regarding their power and resources. In many societies, resources were channeled into religious ceremonies, buildings, and communities, where status seeking also prevails (Saad, 2007, pp. 211–213). Eminent individuals make gifts to the community, but such gifting is usually public. Imitation of the dress and material attributions of highly placed people by their social inferiors makes it necessary for the rich to invent new forms of distinction. Many social occasions involve such social signaling, like weddings and funerals, and lower-class people are no less immune to it. Some people treat particular material possessions as an extension of the self, items that symbolize belonging to the high-status group. People on the make are liable to deceptive signaling (renting or borrowing, rather than owning an object), which becomes possible once communities are too large for the social standing of each to be accurately registered. When pushed to extremes, this can lead to bankruptcy, but it is difficult to legislate against this (Saad, 2007, pp. 90–112).

In Baroque Italy, all of these cross-cultural traits seem to have been amplified, particularly the use of beauty to express power and status, not only of the governing elites, but also of anyone aiming to join their ranks. People availed themselves of a growing art market that supplied them with objects designed to please their taste (Feigenbaum, 2014). This was no less true of the members of the Catholic Church, a religion that emphasized that God's greatness should be proclaimed through the splendor of holy places and the lavishness of vestments and liturgical paraphernalia. All of these processes were codified by art criticism that established the parameters of what was beautiful. Even the most threadbare friars who ostentatiously rejected worldly glory devoted some of their meager resources to fashioning artistic liturgical objects in humble materials and to maintaining a minimal standard of decorum they called "decency." These canons of beauty and lavishness were then exported to almost every continent by missionaries and conquistadores.

There is little in this behavior that is specifically Western, and the impulses underpinning it are as old as modern humanity, so historians will immediately understand the relevance of the above for the periods and places they study. There is no lack of documentation for most historical periods, beginning with the surviving works themselves. Historians today mine the vast notarial archive holdings containing postmortem inventories and dowry lists, while testaments often contain clauses bequeathing specific objects to specific people as tokens of affection. Very little work has been done by art historians on the vast production of decoration or the kinds of practical and instrumental products rendered special by decoration and embellishment, like silver, textile hangings, and furniture, whose utilitarian functions were frequently overshadowed by their ornateness.

40.8 CONCLUSION

This brief exposition cannot discuss other important avenues of research that flow out of an evolutionary approach to the past, such as the evolutionary underpinnings and the universal themes developed in fiction by Joseph Carroll and Lynn Hunt. The psychologist Robin Dunbar and the anthropologist Christopher Boehm have illuminated in different ways the processes of governance that reveal some timeless patterns. As historian Carl Degler wrote in 1981, historians should be interested in the nature of human beings and the origins of many forms of behavior, which is the realm of evolutionary psychology. But even more than origins, the documents describing human behavior in past time – that is, owing more to the ethological approach – will not fail to reveal patterns recurrent across space and time. Humans are less exotic than partisans of cultural primacy and radical "otherness" have claimed. But we are still on the cusp of this discovery.

REFERENCES

Alves, A. A. (1996). *Brutality and Benevolence: Human Ethology, Culture and the Birth of Mexico*. Westport, CT: Greenwood Press.

Ardant du Picq, C. (2004). *Etudes sur le Combat: Combat Antique et Combat Moderne*, Paris: Economica.

Balshine, S. (2012). Patterns of parental care in vertebrates. In N. J. Royle, P. T. Smiseth, & M. Kolliker, eds. *The Evolution of Parental Care*. Oxford: Oxford University Press, pp. 62–80.

Barash, D., & Lipton, J. (2001). *The Myth of Monogamy: Fidelity and Infidelity in Animals and People*. New York: Holt Books.

Barkow, J. (1979). Human ethology: Empirical wealth, theoretical dearth. *Behavioral and Brain Sciences*, **2**, 1–57.

Barrett, L., Dunbar, R., & Lycett, J. (2002). *Human Evolutionary Psychology*. Princeton, NJ/Oxford: Princeton University Press.

Bechtold, B. H. (2001). Infanticide in 19th-century France: A quantitative interpretation. *Review of Radical Political Economics*, **33**, 165–187.

Bechtold, B. H. (2006). The changing value of female offspring in 19th-century France: Evidence from secondary sex ratios. In *Killing Infants: Studies in the Worldwide Practice of Infanticide*. Lewiston, NY: Edwin Mellen Press, pp. 315–335.

Beise, J. (2005). The helping and the helpful grandmother: The role of maternal and paternal grandmothers in child mortality in the seventeenth and eighteenth-century population of French settlers in Quebec, Canada. In *Grandmotherhood: The Evolutionary Significance of the Second Half of Female Life*. New Brunswick, NJ/London: Rutgers University Press, pp. 215–238.

Betzig, L. (1986). *Despotism and Differential Reproduction: A Darwinian View of History*. New York: Aldine.

Betzig, L. (1991). History. In M. Maxwell, ed., *The Sociobiological Imagination*. Albany, NY: SUNY Press, pp. 131–140.

Boehm, C. (2013). The biocultural evolution of conflict resolution between groups, In D. P. Fry, ed., *War, Peace and Human Nature: The Convergence of Evolutionary and Cultural Views*. Oxford/New York: Oxford University Press, pp. 315–340.

Boone, J. L., III (1988). Parental investment, social subordination and population processes among the 15th and 16th century Portuguese nobility. In *Human Reproductive Behavior: A Darwinian Perspective*. Cambridge, UK/New York: Cambridge University Press, pp. 201–219.

Boudon, R. (2001). *The Origin of Values: Sociology and Philosophy of Beliefs*. New Brunswick, NJ/London: Transaction Publishers.

Brown, D. E. (1991). *Human Universals*. Philadelphia, PA: Temple University Press.

Burkert, W. (1996). *Creation of the Sacred: Tracks of Biology in Early Religions*. Cambridge, MA: Harvard University Press.

Buss, D. M., & Shackelford, T. K. (1997). From vigilance to violence: Mate retention tactics in married couples. *Journal of Personality and Social Psychology*, **72**, 346–61.

Caldwell, J. C., & Caldwell, B. K. (2005). Family size control by infanticide in the great agrarian societies of Asia. *Journal of Comparative Family Studies*, **36**, 205–226.

Campbell, C. D., & Lee, J. Z. (2010). Fertility control in historical China revisited: New methods for an old debate. *History of the Family*, **15**, 370–385.

Chagnon, N. A. (2013). *Noble Savages: My Life among Two Dangerous Tribes – The Yanomamo and the Anthropologists*. New York/London: Simon and Schuster.

Collomp, A. (1983). *La Maison du Père: Famille et Village en Haute-Provence aux XVIIe et XVIIIe Siècles*. Paris: Presses Universitaires de France.

Costa, M. (2003). *Psicologia Militare: Elementi di Psicologia per gli Appartenenti alle Forze Armate*. Milan: Franco Angeli.

Crawford, S. (2010). Infanticide, abandonment and abortion in the Graeco-Roman and early medieval world: Archaeological perspectives. In L. Brockliss & H. Montgomery, eds., *Childhood and Violence in the Western Tradition*. Oxford: Oxbow Books, pp. 59–66.

Crocq, L. (1999). *Les Traumatismes Psychiques de Guerre*. Paris: Odile Jacob.

Cusson, M. (1990). *Croissance et Décroissance du Crime*. Paris: Presses Universitaires de France.

Cusson, M. (2000). Les homicides d'hier et d'aujourd'hui. In *L'Acteur et ses Raisons: Mélanges en l'Honneur de Raymond Boudon*. Paris: Presses Universitaires de France, pp. 43-58.

Daly, M., & Wilson, M. (1988). *Homicide*. New York: Aldine de Gruyter.

Daly, M., & Wilson, M. (1995). Discriminative parental solicitude and the relevance of evolutionary models to the analysis of motivational systems. In M. Gazzaniga, ed., *The Cognitive Neurosciences*. Cambridge, MA: MIT Press, pp. 1269–1286.

Damasio, A. R. (1994). *Descartes' Error: Emotion, Reason and the Human Brain*. New York: Putnam Press.

Dansette, M.-D. (1987). La mortinatalité et l'infanticide dissimulée dans le canton de Milly, 1780–1872. *Bulletin de la Société Historique et Archéologique de Corbeil, de l'Essonne et du Hurepoix*, **93**, 9–53.

Degler, C. N. (1981). Can a historian or social scientist learn anything from sociobiology? An attempt at an answer. *Historical Methods*, **14**, 173–179.

Devereux, G. (1955). *A Study of Abortion in Primitive Societies: A Typological, Distributional and Dynamic Analysis of the Prevention of Birth in 400 Preindustrial Societies*. New York: Julian Press.

de Waal, F. (1982). *Chimpanzee Politics: Power and Sex among Apes*. New York: Harper and Row.

de Waal, F. (1989). *Peacemaking among Primates*. Cambridge, MA: Harvard University Press.

de Waal, F. (1992). The Chimpanzee's sense of social regularity and its relation to the human sense of justice. In R. D. Masters & M. Gruter, eds., *The Sense of Justice: Biological Foundations of Law*. London: Sage Publications, pp. 241–255.

de Waal, F. (1996). *Good Natured: The Origin of Right and Wrong in Humans and Other Animals*. Cambridge, MA: Harvard University Press.

de Waal, F. (2010). Morality and its relation to primate social instincts. In H. Hogh-Olesen, ed., *Human Morality and Sociality: Evolutionary and Comparative Perspectives*. Basingstoke/New York: Palgrave.

Diamond, J. (1997). *Guns, Germs and Steel*. London: Vintage.

Di Simplicio, O. (2011). *Luxuria: Eros e Violenza nel Seicento*. Rome: Salerno Editrice.

Dissanayake, E. (1995a). Chimera, spandrel or adaptation: Conceptualizing art in human evolution. *Human Nature*, **6**, 99–117.

Dissanayake, E. (1995b). *Homo Aestheticus: Where Art Comes From and Why*. Seattle, WA/London: University of Washington Press.

Drixler, F. (2013). *Mabiki: Infanticide and Population Growth in Eastern Japan, 1660–1950*. Berkeley, CA/Los Angeles, CA/London: University of California Press.

Drixler, F., & Kok, J. (2016). A lost family-planning regime in 18th-century Ceylon. *Population Studies*, **70**, 93–114.

Dunbar, R. (1988). Darwinizing man: A commentary. In L. Betzig, N. Borgerhoff Mulder, & P. Turke, eds., *Human Reproductive Behaviour: A Darwinian Perspective*. Cambridge, UK/New York: Cambridge University Press, pp. 161–169.

Dutton, D. (2009). *The Art Instinct: Beauty, Pleasure and Human Evolution*. New York/London/Berlin: Bloomsbury Press.

Dyer, G. (2004). *War*, 2nd ed. Toronto: Random House.

Eibl-Eibesfeldt, I. (1989). *Human Ethology*. New York: Aldine de Gruyter.

Eisner, M. (2001). Modernization, self-control and lethal violence: The long-term dynamics of European homicide rates in theoretical perspective. *British Journal of Criminology*, **41**, 618–638.

Eisner, M. (2011). Introduction: Human evolution, history and violence. *British Journal of Criminology*, **51**, 473–478.

Engen, R. (2009). Tuer pour son pays: Nouveau regard sur l'homicidologie. *Revue Militaire Canadienne*, **9**, 120–128.

Feigenbaum, G., ed. (2014) *Display of Art in the Roman Palace, 1550–1750*. Los Angeles, CA: Getty Research Institute.

Fisher, H. (1992). *The Anatomy of Love: A Natural History of Mating, Marriage and Why We Stray*. New York: Batus.

Fox, R. (2011). *The Tribal Imagination: Civilization and the Savage Mind*. Cambridge, MA/London: Harvard University Press.

Gat, A. (2010). The causes of war in natural and historical evolution. In H. Hogh-Olesen, ed., *Human Morality and Sociality: Evolutionary and Comparative Perspectives*. Basingstoke/New York: Palgrave, pp. 160–190.

Gazzaniga, M. S. (2005). *The Ethical Brain*. New York/Washington, DC: Dana Press.

Gazzaniga, M. S. (2011). *Who's in Charge? Free Will and the Science of the Brain*. New York: HarperCollins.

Ghiglieri, M. P. (2000). *The Dark Side of Man: Tracing the Origins of Male Violence*. Cambridge, MA: Perseus Books.

Gordin, M. D. (2014). Evidence and the instability of biology. *American Historical Review*, **119**(5), 1621–1629.

Grossman, D. (1995). *On Killing: The Psychological Cost of Learning to Kill in War and Society*. Boston, MA/London: Little, Brown & Co.

Grossman, D. (2004). *On Combat: The Psychology and Physiology of Deadly Conflict in War and Peace* (with L. W. Christensen). Mascoutah, IL: PPCT Research Publications.

Haas, J., & Piscitelli, M. (2013). The prehistory of warfare: Misled by ethnography? In D. P. Fry, ed., *War, Peace and Human Nature: The Convergence of Evolutionary and Cultural Views*. Oxford/New York: Oxford University Press, pp. 168–190.

Hanlon, G. (2002). Violence and its control in the late Renaissance: An Italian model. In G. Ruggiero, ed., *A Companion to the Worlds of the Renaissance*. Oxford: Blackwell, pp. 139–155.

Hanlon, G. (2003). L'infanticidio dei coppie sposati nella Toscana della Contro-Riforma. *Quaderni Storici*, **38**, 453–498. (English version on the author's Academia.edu webpage.)

Hanlon, G. (2004). Justice in the Age of Lordship: A feudal tribunal in 17th-century Tuscany. *Sixteenth Century Journal*, **35**, 1005–1033.

Hanlon, G. (2007). *Human Nature in Rural Tuscany: An Early Modern Story*. Basingstoke/New York: Palgrave Macmillan.

Hanlon, G. (2009). The facts of life in rural Counter-Reformation Tuscany. *Journal of Interdisciplinary History*, **40**, 1–31.

Hanlon, G. (2013). The decline of violence in the West: From cultural to post-cultural history: Review article. *English Historical Review*, **128**, 367–400.

Hanlon, G. (2016a). *Italy 1636: Cemetery of Armies*. Oxford/New York: Oxford University Press.

Hanlon, G. (2016b). Routine infanticide in the West 1500–1800. *History Compass*, **14**, 535–548.

Harris, M., & Ross, E. B. (1987). *Death, Sex and Fertility: Population Regulation in Preindustrial and Developing Societies*. New York: Columbia University Press.

Héritier, F. (2004). Les fondements de la violence. Analyse anthropologique. In A. Touati, ed., *Violences: De la Réflexion à l'Intervention*. Antibes: Cultures en Mouvement. pp. 23–39.

Holmes, R. (2003). *Acts of War*, 2nd ed. London: Free Press.

Hrdy, S. B. (1999). *Mother Nature: A History of Mothers, Infants and Natural Selection*. New York: Pantheon Books.

Hrdy, S. B. (2009). *Mothers and Others: The Evolutionary Origins of Mutual Understanding*. Cambridge, MA/London: Belknap Press.

Hynes, L. (2011). Routine infanticide by married couples? An assessment of baptismal records from seventeenth-century Parma. *Journal of Early Modern History*, **15**, 507–530.

Jacob, F. (1998). *Of Flies, Mice and Men*. Cambridge, MA/London: Harvard University Press.

Kagan, J. (1994). *Galen's Prophecy: Temperament in Human Nature*. New York: Basic Books.

Kagan, J. (1998). *Three Seductive Ideas*. Cambridge, MA/London: Harvard University Press.

Kahneman, D. (2011). *Thinking, Fast and Slow*. Toronto: Anchor Canada.

Keegan, J. (1976). *The Face of Battle: A Study of Agincourt, Waterloo and the Somme*. Harmondsworth/New York: Penguin.

Keegan, J. (1993). *A History of Warfare*. New York: Alfred A. Knopf.

Keegan, J. (1997). Towards a theory of combat motivation. In P. Addison & A. Calder, eds., *Time to Kill: The Soldier's Experience of War in the West 1939–1945*. London: Pimlico, pp. 3–11.

Kuper, A. (1994). *The Chosen Primate: Human Nature and Cultural Diversity*. Cambridge, MA/London: Harvard University Press.

Kuper, A. (1999). *Culture: The Anthropologist's Account*. Cambridge, MA/London: Harvard University Press.

Lazarus, J. (2002). Human sex ratios: Adaptations and mechanisms, problems and prospects. In I. C. W. Hardy, ed., *Sex Ratios: Concepts and Research Methods*. Cambridge, UK: Cambridge University Press, pp. 287–313.

Leblanc, S. A. (2007). Why warfare? Lessons from the past. *Daedalus*, **136**, 13–21.

Lee, J. Z., & Wang, F. (1999). *One Quarter of Humanity: Malthusian Mythology and Chinese Realities, 1700–2000*. Cambridge, MA: Harvard University Press.

Lestel, D. (2001). *Les Origines Animales de la Culture*. Paris: Flammarion.

Lloyd, G. E. R. (1990). *Demystifying Mentalities*. Cambridge, UK: Cambridge University Press.

Lynch, K. A. (2011). Why weren't (many) European women "missing"? *History of the Family*, **16**, 250–266.

Marshall, S. L. A. (1967). *Men against Fire*. New York: William Morrow & Co.

Miller, G. (2000). *The Mating Mind: How Sexual Choice Shaped the Evolution of Human Nature*. New York: Random House.

Montecuccoli, R. (1752). *Memoires de Montecuculi, Généralissime des Troupes de l'Empereur, Divisé en Trois Livres*. Amsterdam: Wetstein.

Murdock, G. P. (1949). *Social Structure*. New York: Macmillan.

Murdock, G. P., & White, D. (1969). Standard cross-cultural sample. *Ethnology*, **8**, 329–369.

Nassiet, M. (2011). *La Violence, une Histoire Sociale: France, XVIe–XVIIIe Siècles*. Seyssel: Champvallon.

Panksepp, J., & Panksepp, J. B. (2000). The seven sins of evolutionary psychology. *Evolution and Cognition*, **6**, 108–127.

Pinker, S. (1997). *How the Mind Works*. Harmondsworth/New York: Penguin.

Pinker, S. (2002). *The Blank Slate: The Modern Denial of Human Nature*. Harmondsworth/New York: Penguin.

Pinker, S. (2011). *The Better Angels of Our Nature: The Decline of Violence and Its Causes*. London/New York: Allen Lane.

Quartz, S. R., & Sejnowski, T. J. (2002). *Liars, Lovers and Heroes: What the New Brain Science Reveals about How We Become What We Are*. New York: William Morrow.

Reynolds, G. (1979). Infant mortality and sex ratios at baptism as shown by reconstruction of Willingham, a parish at the edge of the Fens in Cambridgeshire. *Local Population Studies*, **22**, 31–37.

Rosen, S. P. (2005). *War and Human Nature*. Princeton, NJ/Oxford: Princeton University Press.

Roth, R. (2001). Child murder in New England. *Social Science History*, **25**, 101–147.

Roth, R. (2009). *American Homicide*. Cambridge, MA: Belknap Press.

Runciman, W. G. (2015) *Very Different, but Much the Same: The Evolution of English Society since 1714*. Oxford/New York: Oxford University Press.

Saad, G. (2007). *The Evolutionary Bases of Consumption*. Mahwah, NJ/London: Lawrence Erlbaum.

Scheidel, W. (2010). Greco-Roman sex ratios and femicide in comparative perspective. *Princeton/Stanford Working Papers in Classics*. https://papers.ssrn.com/sol3/papers.cfm?abstract_id=1505793.

Scheidel, W. (2014). Evolutionary psychology and the historian. *American Historical Review*, **119**, 1563–1575.

Sear, R., & Mace, R. (2008). Who keeps children alive? A review of the effects of kin on child survival. *Evolution and Human Behavior*, **29**, 1–18.

Smail, D. L. (2003). *The Consumption of Justice: Emotions, Publicity and Legal Culture in Marseille, 1264–1423*. Ithaca, NY/London: Cornell University Press.

Tallis, R. (2011). *Aping Mankind: Neuromania, Darwinitis and the Misrepresentation of Humanity*. Durham: Acumen.

Tiger, L. (1969). *Men in Groups*. New York: Random House.

Tooby, J., & Cosmides, L. (1988). The evolution of war and its cognitive foundations. *Institute for Evolutionary Studies Technical Report 88-1*. https://pdfs.semanticscholar.org/7f95/d9d117721df9e69b929b004d9d85ea6c560d.pdf.

Tooby, J., & Cosmides, L. (1989). Evolutionary psychology and the generation of culture. Part 1. *Ethology and Sociobiology*, **10**, 29–49.

Tooby, J., & Cosmides, L. (1990). The past explains the present: Emotional adaptations and the structure of ancestral environments. *Ethology and Sociobiology*, **11**, 375–424.

Trexler, R. C. (1973). Infanticide in Florence: New sources and first results. *History of Childhood Quarterly*, **1**, 98–116.

Trivers, R. L., & Willard, D. E. (1973). Natural selection of parental ability to vary the sex ratio of offspring. *Science*, **179**, 90–92.

Voland, E., & Beise, J. (2002). Differential infant mortality viewed from an evolutionary biological perspective. *History of the Family*, **7**, 515–526.

Voland, E., & Beise, J. (2005). "The husband's mother is the devil in the house": Data on the impact of the mother-in-law on stillbirth mortality in historical Krummhorn (1750–1874) and some thoughts on the evolution of postgenerative female life. In *Grandmotherhood: The Evolutionary Significance of the Second Half of Female Life*. New Brunswick, NJ: Rutgers University Press, pp. 239–255.

Voland, E., & Stephan, P. (2000). "The hate that love generated": Sexually-selected neglect of one's own offspring in humans. In C. P. van Schaik & C. H. Janson, eds., *Infanticide by Males and Its Implications*. Cambridge, UK: Cambridge University Press, pp. 447–465.

Voland, E., Dunbar, R. M., Engel, C., & Stephan, P. (1997). Population increase and sex-biased parental investment in humans: Evidence from 18th and 19th century Germany. *Current Anthropology*, **38**, 129–135.

Wells, R. (1990). *Human Sex Determination: An Historical Review and Synthesis*. Riverlea: Tharwa Australia.

Wilson, E. O. (2012). *The Social Conquest of Earth*. New York/London: Liveright Publishing.

Wrangham, R., & Peterson, D. (1996). *Demonic Males: Apes and the Origins of Human Violence*. Boston, MA/New York: Houghton Mifflin Company.

Wright, R. (1994). *The Moral Animal*. New York: Abacus.

Wright, R. (2000). *Nonzero: The Logic of Human Destiny*. New York: Vintage Books.

Zuckerman, M. (2007). *Sensation Seeking and Risky Behavior*. Washington, DC: American Psychological Association.

41 The Psychology of Extraterrestrials

The New Frontier?

JEROME H. BARKOW

41.1 THE WANING OF UNIQUENESS

We begin with very broad strokes: if we take the Ancient Greeks and Romans as founders of Western civilization and ignore the cosmologies of the many other peoples of the world, then from Ptolemy and the first century CE until sixteenth-century Europe, the Earth was the center of the universe and all else rotated around it. With the work of Copernicus, the Earth was demoted and the sun was given the honor of centrality. But by the early twentieth century, the very notion of centrality was abolished and our sun took its place as one among a vast number of stars in an expanding universe. In similar fashion, the centrality of *Homo sapiens* has steadily declined. In Judeo-Christian-Islamic mythology, intelligent humans were created and given dominion over the Earth by an all-powerful deity; the only other intelligences were that deity itself and its supernatural creations: angels and demons and devils and djinns. With the European Enlightenment of the eighteenth century came deism and the spread of the idea that, while there was some kind of creator, it was now noninterventionist: naturalism – the belief that natural laws governed the universe – came to dominate. By the twentieth century, theorists such as J. B. S. Haldane and Alexander Oparin argued that life itself came about through natural processes (Dick, 1999), and soon anthropologists and others (e.g., Barkow, 1989; Dobzhansky, 1963; Geertz, 1962; Wallace, 1961) theorized about how Darwinian evolution could eventually have produced intelligent, culture-bearing life on Earth.

If Earth was no longer unique, then the natural processes believed to have produced life, including complex organisms, could not be either (Levin et al., 2019): the early decades of the twenty-first century have seen a search for life (past or present) elsewhere in the solar system, including Mars and the apparently deep and ice-encrusted oceans of Jupiter's moon Europa and Saturn's moon Enceladus. At the same time, the discovery of thousands of "exoplanets" – planets that revolve around stars other than our own – has led to the conclusion that planets are the norm rather than the exception in the universe, making it seem likely that life, too, is widespread. It therefore stands to reason that extraterrestrial

intelligences (ETIs) should be out there on exoplanets circling other stars. Even as early as 1959, Cocconi and Morrison (1959) had proposed listening for radio communications from exoplanets. The astronomer Frank Drake (1961) soon proposed what became known as the "Drake Equation," which presented the variables whose values, were they known, would permit the calculation of the number of exoplanets with ETIs currently emitting detectable signals. In 1984, the Search for Extraterrestrial Intelligence (SETI) Institute was established in part to detect such signals, while in 2016 came the establishment of the Breakthrough Listen project (based at the University of California's Berkeley SETI Research Center), with a similar goal and $100 million in funding. An organization with a different but related goal was founded in 2015, Messaging Extraterrestrial Intelligence (METI) International, aiming to send out signals to possible ETIs rather than passively seeking only to detect them. A number of other organizations in various parts of the world are also involved in detecting signals from and transmitting signals to theoretically present extraterrestrials.

41.2 THE TECHNOLOGY FILTER

The waning of our species-centrism has permitted researchers to appreciate that intelligence is not rare, even on our own planet; see, for example, Garland and Rendell's chapter on cetaceans in this volume (Chapter 3), Bates' chapter on the cognitive capacity of elephants (Chapter 2), and Pepperberg's chapter comparing the cognitive abilities of our own species with those of birds (Chapter 1). Arguably, what distinguishes our own species from the other species with which we share Earth is not so much our intelligence as our reliance on *culture* to adapt to changing circumstances and the incredibly effective technology that our cultural capacity has made possible. Intelligence may be ubiquitous throughout the universe, but unless or until we develop some kind of interstellar transport (whether for ourselves or for automated equipment and artificial intelligences [AIs]), the only kinds of ETIs we will be in contact with are those that, like ourselves, have a technology sufficiently advanced to send and receive signals. Applying this technology filter to ETIs leads to some relatively straightforward predictions.

Any high-technology ETI will need some kind of manipulative appendage(s) and/or the ability to control another species that does have prehensile body parts. For example, our ETI may move objects with controlled puffs of air or perhaps with tentacles or a trunk. The species will need a distal sense, perhaps radar, sonar, or vision. It will likely not be aquatic because of the difficulty of developing a knowledge of chemistry without an easily manipulated source of heat. The technology filter makes no other predictions with regard to an ETI's Bauplan, but it would be a remarkable coincidence if it were to be physically similar to ourselves. For example, there is no evolutionary advantage to putting a brain in a part of the body used both for eating and for fighting, rendering it vulnerable. Our poorly designed skeletal support system for the upper body, the spine, requires several curves and is easily injured because we evolved from quadrupeds whose organs depended from it; otherwise evolution, like a good engineer, might have placed our support centrally. ETIs may be more fortunate in their ancestors. Gravity will be crucial in determining morphology, and a heavy-planet ETI will presumably be quite low to the ground (as with the ETIs created by science fiction writer Hal Clement, 1954), and if it has legs they will either be numerous or very thick (or perhaps both).

A high-technology species necessarily has some form of culture. By "culture" I mean a vast body of information conveyed socially both laterally and vertically (that is, acquired both from one's own generational cohort and from older generations) and constantly being revised, with some informational items deleted and others added in a permanent process. Technological knowledge is a subcategory of cultural information. Advanced technologies require massive amounts of specialized knowledge because the equipment to communicate with (or visit!) intelligences light-years distant (presumably) requires a huge number of components, each of which in turn has its own parts made from various feedstocks involving mining or extraction of chemicals from plants and animals (or their equivalents). Great though the body of technological knowledge may be, it would be useless without individuals cooperating on a massive scale. How that cooperation is achieved cannot be predicted; for example, it could involve fear, pheromones, socialization, something akin to religion, bureaucracy, or the haplodiploid genetic arrangement of terrestrial ants and bees (assuming analogs of genes).

A static culture would only be adaptive in a totally unchanging environment. Since environments do alter, new enemies may emerge and new opportunities arise: a static culture would soon be maladaptive (Barkow, 1989). Therefore, any culture-bearing species would have been selected for ways to edit and add to the pool of cultural information. One of the ways in which our species does this is a tendency in adolescence to "rebel" and to test received wisdom. Another way is through preferential attention to those who are high in status. Learning from the high rather than the low in prestige increases the likelihood of acquiring information and habits that may aid in raising one's own relative standing: ignoring or "unlearning" habits typical of lower-ranking individuals may similarly increase one's chances of success (Barkow, 2014a; Barkow, O'Gorman, & Rendell, 2012). While we cannot know how ETIs would edit their culture, the implication is that their culture would continuously change, as does our own. This means, given that communication over light-years would be very slow, a dialog could prove challenging: they, like us, could lose the thread. It also means that ETI species may, like ourselves, have multiple cultures (especially if they have colonized multiple planets).

There is the slight possibility of the genetic assimilation of culturally acquired behavior (Barkow, 1977, 1980a). For example, an ancient society with rulers who, for many, many generations, prevented change and did not permit nonconformists to breed could eventually genetically assimilate their culture. Individuals would have been artificially selected to readily acquire their assigned, inherited role. Eventually, they would appear to be acting culturally, but in fact would be incapable of behaving idiosyncratically or even learning something new. We would find that their interactions with us were always rigid and stereotyped.

41.3 EVOLUTIONARY PSYCHOLOGY AND THE PLANET OF HATS

Let us assume that our ETI has genes or something analogous to them, and that these are transmitted vertically and not laterally (though the latter possibility will be discussed shortly). If these two assumptions are valid, then we are forced to dismiss some of our favorite science fiction aliens because they dwell on planets of hats (All The Tropes Wiki, 2017).

The "planet of hats" trope harks back to the early westerns in which the good guy invariably wore a white hat and the bad guy a black one. In science fiction circles, it refers to entire ETI species characterized by a single distinctive behavioral trait. Thus, we may find a warrior species such as the Klingons, or one of greedy capitalists such as the Ferengi. We find little complexity or diversity and much psychological uniformity. Hat planet species often make for good entertainment, but they could not evolve anywhere evolutionary theory has validity. This is because a single dominant trait species would not be an evolutionarily stable strategy; that is, a strategy that, if adopted by most of a population, "cannot be bettered by an alternative strategy" (Dawkins, 1976, p. 74). A less greedy Ferengi would have a distinct reproductive advantage, and soon there would be many Ferengi with more complex psychologies, permitting them to not be greedy some portion of the time, or else less greedy all of the time. A Klingon who chose his or her battles with caution would likely have more offspring than one who never stopped being fierce and combative, again leading to an evolved psychology featuring some complexity.[1] We are as unlikely to meet hat planet extraterrestrials as we are to find a life-bearing planet with no variation in geography or climate (another science fiction standard). ETIs will have at least some measure of psychological complexity and as individuals will likely differ from one another to at least a certain extent.

41.4 WILL EXTRATERRESTRIALS BE DANGEROUS?

Could an extraterrestrial be an obligate xenophobe and therefore hostile and dangerous to our entire species? Could they be searching their skies for enemies?[2] There is indeed a slim chance that such an ETI could have evolved. Evolution can often be usefully thought of in terms of culling, with those culled obviously leaving fewer offspring than those who remain. One issue is: Who does the culling? For our own species, the culling would have been done in part by members of the culled's own respective bands and also by members of rival bands (Barkow, 1989). What traits would have led to culling? Presumably, the most likely to be culled would have been the slow to learn from others, the slow to recognize enemies, and the slow to cooperate. (Sexual selection, discussed in Section 41.5, would also have served to reduce the number of offspring such individuals would have produced.) The important point is that evolutionary culling would have been performed by conspecifics – members of our own species – rather than by members of other species. Other intelligent species did indeed exist in our evolutionary history (the Neanderthals and Denisovans, in particular), but these were Eurasian species; we apparently evolved in Africa and were fully modern humans long before we entered Eurasia and met with them (with at least some possible geographic movement back and forth and with some gene flow; (Boyd & Silk, 2017; Langdon, 2016).

That our psychology is a product of aggression among bands of conspecifics is compatible with a major facet of our evolved psychology: the ethnocentrism syndrome. Faced with external threat, we tend to set aside within-group conflicts and experience enhanced in-group solidarity, rallying around a leader; we readily learn to distrust and even hate, dehumanize, and murder members of other bands; we readily adopt band markers to identify ourselves as members of a specific band, the markers ranging from language and accent to dress, religion, hairstyle, scarification and tattoos, political and other beliefs, food, music, and no doubt many other differences in behavior or appearance (Barkow, 1989; LeVine & Campbell, 1972;

Reynolds, Falger, & Vine, 1987; Thienpont & Cliquet, 1999; Van den Berghe, 1981; Yuki & Yokota, 2009).

Let us suppose that our extraterrestrials achieved their intelligence and cultural capacity in a quite similar way, except that the culling was done by one or more rival species instead of by conspecifics (Barkow, 2014b). Rather than ethnocentrism, could they have been selected for xenophobia? Would they react to an alien intelligent species as if they were the ancient enemy and seek only to slaughter them? Might they not eventually construct self-replicating interstellar weapons designed to identify planets with high-technology alien species and destroy them – destroy *us*? Should we silence our electronic emissions and search not for asteroids in dangerously near-Earth orbits, but for implacable extraterrestrials and their weapons? While not totally impossible, this scenario is more a science fiction plot than it is a realistic danger because (in my opinion) it is quite unlikely.[3] To take ourselves as an example, we are not obligate ethnocentrics and are capable of alliances and even integration with other groups, accepting immigrants and forming multicultural societies. Xenophobic extraterrestrials, should they exist, may be equally likely to at least have learned to tolerate other species. The probability that even a xenophobic ETI would seek to eliminate extremely distant "enemies" seems exceedingly low. We humans are in far more danger from ourselves, with our wars and our reverse terraforming of the planet, than from ETIs.

41.5 SEX AND SEXUAL SELECTION

Sexual selection obviously requires that a species has two sexes, as is common on our own planet.[4] There are a number of often mutually compatible theoretical explanations for why two sexes evolve, one of the most powerful being that having two sexes speeds up the evolution of large species subject to much smaller parasites whose rapid generation time gives them the advantage of relatively rapid adaptive change (for a discussion, see Barkow, 2000; see also ScienceDaily, 2009). Having two sexes levels the playing field somewhat. This result could also be produced by larger species having very brief generation times. However, if we are talking about ETIs with culture, it is difficult to see how a very brief lifespan is compatible with the transmission and editing of huge amounts of information. Thus, one could make a reasonable argument that ETIs are likely to resemble us in having two sexes subject to sexual selection.

Sexual selection was apparently an important component of our development of the capacity for culture. Individuals who preferred mates more capable than others of succeeding in a troop or band with even a "protoculture" would have been more likely to have offspring and grandchildren who were similarly successful than were those

---

[1] This discussion is summarized from Barkow (2020).

[2] This topic is treated at considerably greater length in Barkow (2014b).

[3] Disclosure: I am on the Board of Directors of METI International. The name of the organization, "Messaging Extraterrestrial Intelligence," makes my position on this issue clear.

[4] This discussion in part summarizes arguments first presented in Barkow (2000).

who chose mates who were less successful. What traits would have made for success in early band environments? Probably the same ones that make for success today, presumably including (in no particular order): the capacity to control one's own temper and the ability to mediate disputes among others, thus minimizing the violence self and kin are exposed to; cooperating well in hunting and in other groups, increasing both the success of these groups and the likelihood that one is included in them; sharing whatever was gathered or hunted, increasing the probability of benefiting from reciprocal altruism; readily learning and even developing the techniques to make stone and wooden tools such as digging sticks, stakes, and spears, as well as skin containers and garments; the ability to communicate and to understand the communications of others; and, above all, the gift of acquiring knowledge from elders and age-mates alike and subsequently conveying it to offspring and other kin and eventually band members in general. This list is exemplary rather than exhaustive, but it does give an idea of which early individuals were most likely to become our ancestors and whose genes and phenotypic characteristics thereby live on in us (cf. Buss, 1989, 1992, 2003; Buss et al., 1990).

Mate preference is only part of sexual selection – there is also direct and indirect competition with others, generating (among social mammals, including primates) social hierarchies (Barkow, 1975, 1989, 2014a). Social rank can vary over the lifespan, but in general, higher rank tends to lead to greater reproductive success and (very likely) inclusive fitness. There is also competition for the ability to signal to potential mates that one has "good genes"; that is, genes favoring health, success in competition with others, and (in circular fashion) being attractive to the other sex. Typically, as Darwin and others (Cronin, Curry, & Singer, 1999; Darwin, 1859, 1874; Miller, 2000) have taught us, sexual selection results in conspicuous hypertrophies, the better to signal the likelihood of genetic superiority – one thinks of the elaborate courtship rituals of many species and bright plumage, including the peacock's feathers; often sheer physical size is a signal, as is superior coordination; and in humans, perhaps in males especially, the capacity to take risks and survive them (heroes, "bad boys"). Because we are a cultural species, the criteria in terms of which we compete and then allocate prestige – relative standing – can vary enormously (Barkow, 1975, 1980b, 1989, 2014a). What is universal is not the specific kind of competition, but competition itself, competition in terms of a standard of excellence. Even in Western society, domains of competition can range from displays of religious piety to success in athletics to skill in painting to the accumulation of wealth. Sexual selection quite likely shaped our species as much as did natural selection, so that our capacity for language, technology, and culture in general appears to be far greater than what unadorned natural selection would have produced: hunting and gathering is not rocket science; for that, we needed sex.

If we encounter extraterrestrials who, having had two sexes, were also in part products of sexual selection, we should not be surprised to find that we have much in common with them. For example, if we find their communications and/or artifacts unnecessarily complex, this is a likely indication that these ETIs are products of sexual selection and they are proudly showing us works of art. If we find nothing but the functional, the possibility must arise that we are dealing with a species with only a single sex, perhaps one that transmits genes (or gene equivalents) laterally rather than horizontally (as do bacteria and archaea) and is psychologically very different from ourselves. Human technology is in large measure a product of competition among individuals, coalitions, and large collectivities (such as nations): Could an extraterrestrial species develop such competition in the absence of sexual selection? Could evolution produce competition via a process other than the sexual? Can there be technological development without competition? Or could it be that some ETIs are vastly different from ourselves?

41.6 IS THE HUMAN EXPERIENCE IRRELEVANT TO UNDERSTANDING ETIS?

So far, this chapter has assumed that the laws of evolution as formulated on Earth are universals and that any high-technology extraterrestrials we contact or visit (or are visited by) will be sufficiently similar to ourselves for mutual understanding. Suppose this assumption is wrong. Here are some conjectures that postulate ETIs unimaginably different from at least the current version of *Homo sapiens*. In all of the cases below there are no real-world models in existence – they represent only imagination, at least so far.

(1) Perhaps we will encounter a single ETI with an intelligence so vast that it has created high technology entirely on its own. (2) Perhaps the ETI will be more ant- or termite-like than human, their "intelligence" being the property of a collectivity with individuals having little independent cognition. (3) Perhaps the ETI will be so heavily cyborged and/or so ancient that its cognitions are opaque to our young and unassisted brains. (4) Perhaps the ETI will be an AI with motivations incomprehensible to us. (5) Perhaps the answer to Fermi's Paradox (If the universe is so full of intelligence, why have we not been contacted yet?) is that the ETIs out there are just waiting for us to achieve a Kurzweilian "singularity" (Kurzweil, 2005) to invite us to join them in an existence beyond our current ken, but will be made possible by advances in nanotechnology, cybernetics, biotechnology, etc. (similar to [Case3], but, as we will see, akin to theology in its promise of imminent total change). Many more examples of speculation about ETIs could be harvested by a perusal of the science fiction print and film catalog, but these examples should be sufficient. Let us respond to each.

(1) The first conjecture is easily dismissed (Barkow, 2020): Why would an eternally solitary creature have a need to contact others? A nonsocial creature is as likely to crave company as an herbivore prey. How would such a being even conceive of the existence of others? What would motivate it to develop the smelting of metals and the creation of chemical feedstocks so that it could manufacture the supply chains needed for the hundreds of thousands of distinct parts needed to build even a jet airliner? The infrastructure permitting the bringing together of parts is huge, the shear organization and scheduling vast. Where would the motivation to do all of this come from?

(2) Termites on our own planet have sophisticated technology in their designed-for-thermoregulation mounds (Korb, 2003) and their organization of labor by caste. But if one thinks of a termite colony as a superorganism, then this "technology" pales in comparison to the physiological "technology" in the bodies of both termites and humans when it comes to providing cells with oxygen and energy, a nervous system to coordinate behavior, growth, and development, and so forth. In short, we may be overestimating the evolutionary accomplishments of our terrestrial ants, termites, and bees. Biological evolution is a process that accumulates information, so perhaps we should be asking whether biological evolution on exoplanets could produce a species as capable of reaching space and maneuvering there as an Earth bird is of flying in the atmosphere. The answer is "perhaps," but this possibility begs the question: How would it evolve the intelligence to construct the technology to contact us? Even if a kind of termite mound-building species evolved to build spaceflight-capable mounds in the ecology of an exoplanetary system, it still would require some other set of selection pressures to evolve the capacity and motivation to develop a SETI or METI.

(3) A cyborged entity seems less fanciful. Life on a planet other than the one on which it evolved, or simply living in space, would reasonably require cyborging (or else major changes in physiology and perhaps morphology). Our own technology is already moving in this direction. It is not clear whether, even after one million years, the results would be entirely alien. Thus far, we use new technology in the service of our Pleistocene-evolved psychology: the Internet is filled with pornography, with efforts to find and to impress potential mates, with efforts to present ourselves as so successful that users of Facebook may be prone to depression (Blease, 2015), and perhaps with successful efforts to recreate a level of mutual knowledge of the minute-by-minute activities of other members of one's circles previously only possible in band-level societies. These Internet activities and many more obviously reflect our evolved psychology: Why would cyborging create new psychologies unrelated to the old?

(4) Could ETIs be AIs? "Artificial" means that some biological organism originally constructed them. After all, a pure AI would presumably be inert, the motivations of organic lifeforms would have their ultimate origins in evolution, the human brain is superbly adapted to solving the problems of our ancestors, and the builders of any AI ETIs would presumably be similarly motivated in ways that would have solved the problems of their own ancestors. Thus, even if their originators have long been extinct, we would be well advised to treat AI ETIs as if they were very long-lived organic beings. Suppose, however, that the evolutionary process that produced our AIs is digital in nature, with no real connection to biological creators. As Lehman et al. (2018, p. 2) repeatedly illustrate, "digital evolution experiments often produce strange, surprising, and creative results." Therefore, the nature of any digitally evolved ETIs is, at this point, unknowable. That being the case, there seems little profit in discussing its possibility, beyond pointing out that there may be a plethora of intelligences in the universe and while we may only contact one, we might also contact dozens, some biological to varying extents and others not, each very different not only from ourselves but from one another. Some might be products of digital evolution.

(5) The idea of a Kurzweilian "singularity" is an updated millenarian movement, a cargo cult for those with a faith in transcendent technology rather than theology. Human societies have frequently believed that the end of the world as we know it is nigh, but the context is generally supernatural rather than scientific. There is no doubt that accumulating technological achievement will continue to change the way we live and relate to one another, but the process is an ongoing thing – if it is an explosion, it is a slow explosion. Thus, we may have to worry about the effects of humanoid robots raising children, some of us may soon be or already are interacting with similar robots easing the loneliness of our final years or meeting our sexual needs, and there is no doubt that many people have been and will continue to lose their means of livelihood because their work has been automated and computerized away. None of this makes for a sudden singularity, as opposed to phenomena calling for the efforts of psychologists and sociologists to understand what is happening and for wise leaders to teach us how to cope.

We are left with two positions: the first is that convergent evolution means that any ETI is likely to share a number of psychological attributes with us, at least opening the possibility of slow but real communication; and the second is that there are so many pathways to ETI that we are being narrow-minded in restricting ourselves to roads similar to the one our own ancestors happened to take. The first position is based on n = 1, and no conclusion based on a single case can be convincing. The only less persuasive argument is when n = 0, as it unfortunately is for the second position.

41.7 CONCLUSIONS

Should evolutionary and other psychologists and social scientists stick to their last and continue to focus on our own species? This is not a suggestion to abandon thinking about space communication and travel. Bases on the moon and Mars are already being actively planned as of this writing, while entrepreneurs are exploring space tourism and asteroid mining. Regardless of whether we eventually communicate with extraterrestrials, our own technology and ambitions are reaching the point where space settlements and colony ships are becoming feasible. We have nearly arrived at the space-faring species stage: What do evolutionists have to contribute? The answer is: an understanding of *Homo sapiens*. Freudian psychoanalysis once promised a universal theory of human psychology, from individual development to the primal circumstances that produced our psychological universals. Evolutionists are in fact producing an actual, data-based theory that does all that Freud promised, from an understanding of human evolution and cultural capacity to how Facebook can lead to depression (Blease, 2015). What does evolutionary psychology tell us about space settlements and colony ships, for example?

We human beings want more for our own children than we want for the children of others, an aspect of our psychology that many of us, being liberal-minded and concerned about the well-being of those less fortunate than ourselves, prefer not to discuss. Unfortunately, this is a deep part of our evolved psychology. Equally unfortunately, both space settlements and ships are extremely fragile environments: error or even simple sloppiness are likely to be deadly when a simple hole means one's atmosphere has vanished or the wrong chemical in the hydroponics section means hunger. Current work in space assumes the military model, a wise choice for a limited-duration mission because responsibility and authority are always clear and the obeying of orders has been deeply instilled. Now, change that mission to one that lasts generations. How many of those generations will it take before being an officer is hereditary rather than merit based? How long will it be before growing social inequality leads to revolution? Do we gamble that there will never be sufficient violence to destroy the settlement or ship, given the fragility of the environment and given human psychology and history? These are challenges that evolutionists can certainly help meet. Another challenge for evolutionists is that METI requires multigenerational organizations that maintain their goals regardless of external circumstances and changes in personnel. But human beings evolved to challenge the status quo, especially in our youth, when

doing so appears to open a pathway toward enhanced social rank. So the challenge is, once again, to design a culture or social organization that somehow works around our evolved psychology (a task likely impossible for those who do not understand that we have such a psychology).

Still, there remains an excellent chance that, one way or another, our species will eventually be in contact with an *other*, even if it is light-years away. There is an even better chance that space tourism and bases will lead to permanent settlements and colony vessels. Who is better qualified to understand the challenges of these endeavors than evolutionary experts on our species? We at least are likely to know what questions to ask. So we have the dual challenge of continuing to develop our understanding of our own evolved psychology while thinking deeply about that of ETIs.

REFERENCES

All The Tropes Wiki (2017). Planet of Hats. http://allthetropes .wikia.com/wiki/Planet_of_Hats.

Barkow, J. H. (1975). Prestige and culture: A biosocial approach. *Current Anthropology*, **16**(4), 553–572.

Barkow, J. H. (1977). Conformity to ethos and reproductive success in two Hausa communities: An empirical evaluation. *Ethos*, **5**(4), 409–425.

Barkow, J. H. (1980a). Biological evolution of culturally patterned behavior. In J. Lockard, ed., *The Evolution of Human Social Behavior*. New York: Elsevier, pp. 227–296.

Barkow, J. H. (1980b). Prestige and self-esteem: A biosocial interpretation. In D. R. Omark, F. F. Strayer, & D. G. Freedman, eds., *Dominance Relations: An Ethological View of Human Conflict and Social Interaction*. New York: Garland, pp. 319–332.

Barkow, J. H. (1989). *Darwin, Sex, and Status: Biological Approaches to Mind and Culture*. Toronto: University of Toronto Press.

Barkow, J. H. (2000). Do extraterrestrials have sex (and intelligence)? *Annals of the New York Academy of Sciences*, **907**, 164–181.

Barkow, J. H. (2014a). Prestige and the ongoing process of culture revision. In J. T. Cheng, J. L. Tracy, & C. Anderson, eds., *The Psychology of Social Status*. New York: Springer Science +Business Media, pp. 29–46.

Barkow, J. H. (2014b). Eliciting altruism while avoiding xenophobia: A thought experiment. In D. A. Vakoch, ed., *Extraterrestrial Altruism: Evolution and Ethics in the Cosmos*. New York: Springer, pp. 17–48.

Barkow, J. H. (2020). The evolutionary psychology of extraterrestrials. In D. A. Vakoch, ed., *Extraterrestrial Intelligence: Cognition and Communication in the Universe*. New York: Oxford University Press.

Barkow, J. H., O'Gorman, R., & Rendell, L. (2012). Are the new mass media subverting cultural transmission? *Review of General Psychology*, **16**(2), 121–133.

Blease, C. R. (2015). Too many "friends," too few "likes"? Evolutionary psychology and "Facebook depression." *Review of General Psychology*, **19**(1), 1–13.

Boyd, R., & Silk, J. B. (2017). *How Humans Evolved*, 8th ed. New York: W. W. Norton & Company.

Buss, D. M. (1989). Sex differences in human mate preferences: Evolutionary hypotheses tested in 37 cultures. *Behavioral and Brain Sciences*, **12**(1), 1–14.

Buss, D. M. (1992). Mate preference mechanisms: Consequences for partner choice and intrasexual competition. In J. H. Barkow, L. Cosmides, & J. Tooby, eds., *The Adapted Mind: Evolutionary Psychology and the Generation of Culture*. New York: Oxford University Press, pp. 249–266.

Buss, D. M. (2003). *The Evolution of Desire: Strategies of Human Mating*, revised ed. New York: Basic Books.

Buss, D., Abbott, M., Angleitner, A., Asherian, A., & Biaggio, A. (1990). International preferences in selecting mates. *Journal of Cross-Cultural Psychology*, **21**, 5–47.

Clement, H. (1954). *Mission of Gravity*. Garden City, NY: Doubleday.

Cocconi, G., & Morrison, P. (1959). Searching for interstellar communications. *Nature*, **184**(4690), 844–866.

Cronin, H., Curry, O., & Singer, P. (1999). *Darwinism Today: Natural Politics: A Darwinian Agenda for the Left*. London: Weidenfeld & Nicolson General.

Darwin, C. (1859). *On the Origin of Species by Means of Natural Selection, or the Preservation of Favoured Races in the Struggle for Life*, 1st ed. London: John Murray.

Darwin, C. (1874). *The Descent of Man, and Selection in Relation to Sex*, 2nd ed. London: Charles Murray.

Dawkins, R. (1976). *The Selfish Gene*. Oxford: Oxford University Press.

Dick, S. J. (1999). *The Biological Universe: The Twentieth Century Extraterrestrial Life Debate and the Limits of Science*. Cambridge, UK: Cambridge University Press.

Dobzhansky, T. (1963). Cultural direction of human evolution. *Human Biology*, **35**, 311–316.

Drake, F. (1961). The Drake Equation. The SETI Institute. www.seti.org/drakeequation.

Geertz, C. (1962). The growth of culture and the evolution of mind. In J. M. Scher, ed., *Theories of Mind*. Glencoe: Free Press, pp. 713–740.

Korb, J. (2003). Thermoregulation and ventilation of termite mounds. *Naturwissenschaften*, **90**(5), 212–219.

Kurzweil, R. (2005). *The Singularity Is Near: When Humans Transcend Biology*. New York: Viking Press.

Langdon, J. H. (2016). *The Science of Human Evolution*. New York: Springer International Publishing.

Lehman, J., Clune, J., Misevic, D., et al. (2018). The surprising creativity of digital evolution: A collection of anecdotes from the evolutionary computation and artificial life research communities. https://hal.inria.fr/hal-01735473.

Levin, S. R., Scott, T. W., Cooper, H. S., & West, S. A. (2019). Darwin's aliens. *International Journal of Astrobiology*, **18**(1), 1–9.

LeVine, R. A., & Campbell, D. T. (1972). *Ethnocentrism: Theories of Conflict, Ethnic Attitudes and Group Behavior*. New York: John Wiley and Sons.

Miller, G. (2000). *The Mating Mind: How Sexual Choice Shaped Human Nature*. New York: Doubleday.

Reynolds, V., Falger, V., & Vine, I., eds. (1987). *The Sociobiology of Ethnocentrism. Evolutionary Dimensions of Xenophobia, Discrimination, Racism and Nationalism*. London/Sydney: Croom Helm.

ScienceDaily (2009). Parasites may have had role in evolution of sex. www.sciencedaily.com/releases/2009/07/090706171542.htm.

Thienpont, K., & Cliquet, R. (1999). *In-Group/Out-Group Behaviour in Modern Societies: An Evolutionary Perspective*, Brussels: Vlaamse Gemeenschap.

Van den Berghe, P. (1981). *The Ethnic Phenomenon*. New York: Elsevier.

Wallace, A. F. C. (1961). The psychic unity of human groups. In B. Kaplan, ed., *Studying Personality Cross-Culturally*. New York: Harper and Row, pp. 129–163.

Yuki, M., & Yokota, K. (2009). The primal warrior: Outgroup threat priming enhances intergroup discrimination in men but not women. *Journal of Experimental Social Psychology*, **45**(1), 271–274.

PART X

EVOLUTION AND THE MEDIA

In the environment in which we evolved there were no automobiles, and yet most of us are drivers. With training, general abilities that evolved for other purposes – visual data processing, reflexive responses, habit formation – can be mobilized and integrated to create an evolutionarily novel ability. The automobile was a transformative technology, with consequences at the individual, societal, and ecological levels. Electronic media are today's transformative technology, with similarly vast consequences and similarly giving us new abilities by mobilizing capacities that evolved for other purposes. The contributors to Part X seek to understand both the evolutionary underpinnings of our capacity to use electronic media and the latter's consequences for current psychology and society.

In the environments in which we evolved, some kinds of information were more relevant to social competition, survival, and the production of strong offspring than were others, and we apparently were selected to find these topics fascinating even when we feel we should disdain them. These *attention attractors* crop up in diverse contexts, such as newspaper headlines, which tend to be similar across nations and history (e.g., Davis & McLeod, 2003). They are also the subjects we gossip about (Barkow, 1992; Davis et al., 2018; McAndrew & Milenkovic, 2002). In Part X, we begin with Schwender's finding that these attention-attracting subjects are the recurring topics of German talk shows. The shows in effect make us members of a community that shares its gossip with us.

Barrett builds on Tinbergen's concept of *supernormal stimuli* to explain how our attention can be not merely attracted but bound for long periods by technology designed to take advantage of our evolved psychology. The producers of television and cinematic productions and the makers of videos and commercials are expert in gaining and keeping our attention. Stimuli associated with sex and romance, socializing with friends, action, adventure, and even ethnocentrism (patriotism) are expertly simulated so that television may successfully compete with reality. For example, most of us have access to only a limited number of sexual and romantic partners, and

these are often not as alluring as the professional entertainers presented by the screen (readers of this volume and their partners represent exceptions to this generalization). Needless to say, films and television programs are replete with what we have been referring to as "attention attractors" (e.g., danger).

Barrett deliberately only touches on horror media because that subject is the focus of Clasen and Kjeldgaard-Christiansen's (engrossing rather than horrifying) chapter. These authors discuss a wide range of research findings and theory, including how we do not view horror films or read horror stories because they give us pleasure, but because they are attention attractors and traps. Learning about dangers from personal experience is expensive, but learning from the experiences of others is far less risky, so we were selected to pay strong attention to such information. This evolved mechanism makes possible the cultural practices not just of recounting adventures, but of making them up and ultimately creating horrifying stories, films, and computer games; we view them not in spite of but *because* they create a sense of danger and fear, which makes it difficult for us not to attend to them. The fear evoked by horror films, computer games, and narratives may prepare us for coping with actual danger in real life. Horror films are, of course, full of Barrett's "supernormal stimuli," while horror computer games promote vigilance.

Salmon, Fisher, and Burch finish Part X with a discussion of Internet pornography. They organize their chapter around the following questions: "Why do we spend time looking at online pornography? What sex differences are there in the consumption of online pornography? What can we say about individual differences between men and between women?" While Internet pornography is obviously evolutionarily novel, as Barkow (1992) has pointed out and as the authors document, with respect to electronic pornography, "beneath new culture is old psychology." Short-term mating strategies are more in a male's fitness interests than in a female's (until the invention of reliable birth control, she had much more of her total reproductive potential at risk in any given mating than did he). As a result, men today view online pornography more often than do women. While men's pornography focuses on actual sex acts, women, having been selected to focus on longer-term relationships and the male's willingness and ability to invest in any resulting offspring, apparently tend to prefer electronic and literary narratives that place sex in the context of a romantic or erotic relationship. The authors note that a relatively small proportion of research on pornography makes use of evolutionary theory, but the research findings nevertheless "strongly support an evolutionary perspective."

REFERENCES

Barkow, J. H. (1992). Beneath new culture is old psychology. In J. H. Barkow, L. Cosmides, & J. Tooby, eds., *The Adapted Mind. Evolutionary Psychology and the Generation of Culture.* New York: Oxford University Press, pp. 626–637.

Davis, A., Dufort, C., Desrochers, J., Vaillancourt, T., & Arnocky, S. (2018). Gossip as an intrasexual competition strategy: Sex differences in gossip frequency, content, and attitudes. *Evolutionary Psychological Science*, **4**, 141–153.

Davis, H., & McLeod, S. L. (2003). Why humans value sensational news: An evolutionary perspective. *Evolution & Human Behavior*, **24**, 208–216.

McAndrew, F. T., & Milenkovic, M. A. (2002). Of tabloids and family secrets: The evolutionary psychology of gossip. *Journal of Applied Social Psychology*, **32**(5), 1064–1082.

42 Daily Talk Shows as Virtual Gossip Communities

CLEMENS SCHWENDER

42.1 THE MEDIA DECOY

Behavioral biologists often rely on the use of physical decoys to test for stimuli that might elicit reactive behaviors. The range of acceptance of the stimulus is looked at, as well as its properties. The successful deployment of such decoys, be it by the scientist in the lab or the by hunter in the duck blind, demonstrates that it is possible to reduce the complexity of an animal's environment down to a set of singular response-trigger features that serve as perceptual indicators for the occurrence of more complex events. The advantages are clear: the simple algorithm suffices to trigger the appropriate response in an environment of evolutionary adaptedness.

Biologists work with fish, birds, and mammals. What about humans? Is it possible to fashion a decoy consisting of visual and auditory stimuli such that it will elicit an appropriate emotional response even when such stimuli are being conveyed electronically? Clearly, we are moved and affected by and laugh and cry over fictional persons and events as conveyed through their depictions in the media. Although we recognize the difference between real life and the impressions of life as conveyed to us by electronic media, we nonetheless react in a directly emotional way, as if such depictions were real. We even accept such illogical illusions as singing mice and cursing ducks and find ourselves emotionally moved by their conduct. Electronic media, however, are not windows to the world, but rather offer insights into our perceptions and ideas. They appeal to us visually and aurally in ways that induce us to accept their content as depictions of reality (in the case of nonfiction) or as models of reality (in the case of fiction). Ensuring that we fully accept such media events as perceptions of reality (or of a conceivable reality) requires that they be carefully constructed. In order to deceive the eye and mind and function as a representation of both the social and nonsocial environment, such media decoys need to be fashioned to share certain features in common with us, their intended subject.

The media decoy must feed the mind with the images and sound that will give rise to the necessary emotional and cognitive processing of the impressions being created; the content must address our psychological and communicative facilities and provide intellectual and emotional engagement and stimulation; and it should further reflect motives that are relevant for our social behavior. With not too much effort, it is possible to view the numerous genres and endless choice of electronic communications and mass media offerings today as a widely distributed experiment in response behavior to this media decoy effect. The recipients (e.g., television viewers) are offered an array of topics from which to choose those subject areas and topics they find the most appealing. One possible measure of the effect of such stimuli is the "viewer rating." Television viewer ratings, having been in use for some time and widely accepted, are capable of quantitatively measuring the attraction of such media events.

42.2 PREREQUISITES FOR VISUAL PERCEPTION

42.2.1 Theory of Mind and Perspective-Taking

Object permanence is not a result of perception, but rather is a construct of the brain. It was Jean Piaget whose work established that an infant's cognitive capacity for spatial perception develops only gradually. This is because perception itself is not a passive occurrence, but similarly is a construct of the brain. In a series of experiments especially relevant to visual perception, Piaget and Inhelder (1956, pp. 209–246) found that the newborn infant is initially fully egocentric with respect to its perception. The first developmental stage consists in the recognition by an infant that an object still exists even when it is no longer visible. If an object is removed from the line of sight of an infant aged four to five months, it is indeed "gone" – to see and to exist are synonymous. The next stage of development then arises with the child's recognition that its own perspective is not necessarily the same as that of another person (Piaget & Inhelder, 1956, p. 209).

Film and television programs, consisting of sequences of scenes broken down into various camera angles (points of view), rely on this perceptual ability. The master shot comes first: a wide shot showing an overview of the entire scene, allowing the viewer to establish the overall relationships of the various elements. The action within the scene is then broken down into the various points of view of the participants or actors. Despite the sequential nature and multiple viewpoints, we as viewers are able to visually reconstruct and maintain the entire scene in our mind, a skill that can only function properly with a neurally operational visual object constancy.

The next step in child cognitive development, according to Piaget and Inhelder, is self-awareness of one's own independent viewpoint: "A perspective system entails his relating the object to his own viewpoint, as one of which he is fully conscious. Here, as elsewhere, to become conscious of one's own viewpoint involves distinguishing it from the viewpoints of others and, by the same token, coordinating it with those of others" (Piaget & Inhelder, 1956, pp. 209–210). The difficulty lies in the abstract consideration of one's own perception and the imagined perspective of a possible other observer. The experimental attempt to test the mental faculties of children involved the use of three papier-mâché models – "mountains" – of differing sizes, shapes, and colors placed on a table and shown to the subject, who is allowed to view the array of mountains from a single point of view. A child's doll is placed into the scene at various locations and the child is asked to describe what the doll sees. The ability to recognize the various external viewpoints and reproduce them is called "perspective-taking." A similar exercise occurs in the visual media as each new camera position introduces a different viewpoint.

The ability to imagine a point of view other than our own is thus due to the social and cooperative nature of our coexistence, which allows us to better predict the intentions and actions of others. In communities that respond not only as a result of preprogrammed patterns of behavior, but also on the basis of subjective perceptions and decisions, standardized patterns of behavior are not enough.

Piaget's concept of perspective-taking has much in common with the concept of theory of mind – the ability to attribute thoughts, intentions, and beliefs to others, particularly in an effort to predict or explain their actions. Both deal with the awareness and then understanding of the viewpoints and perceptions of others. This ability begins at around six to eight months in human development, as young infants learn to follow the sight of other people and then direct their own gaze at things that others are looking at – a rudimentary form of visual perspective-taking. At around 14 months, babies start to recognize the significance of gestures and learn to understand the meanings they convey. Games of pretend play and "what if" begin at around 18–24 months, along with the comprehension of imaginary stories. Joint games of "let's pretend" can be seen as an example of intentional communication (Leslie, 1994, p. 141f.); when the mother says: "Telephone, it's for you," and hands her young child a banana – a seeming incongruity – the child knows not to take the statement literally; it is aware of the mother's intention: "Let's pretend that this is a telephone." By four to five years of age, this development is complete, and children have acquired a fully functional theory of mind.

Theory of mind is a shorthand way of referring to our ability to form a notion about someone else's mental state. In communicative situations involving any sort of negotiation or exchange, the ability to sense the other party's intentions can be very important. To do this, we have learned to rely on various forms of evidence: emotional expressions involving mimicry and gestures, as well as the interpretation of wording and tonal expression. We put ourselves in the proverbial other person's shoes and thus strive to develop our own interpretation of the other person's intentions. Theory of mind also plays a role in keeping our own intentions hidden. Lies and deception are only possible if we can form an image of the other party's mind-set. The ability to assume someone else's viewpoint, beliefs, and imagination is a fundamental requirement in differentiating between fictional and nonfictional accounts.

Imagination lies very close to perception. An idea can very much resemble a perception, and the transition between the two can be fluid. A 1910 experiment by the American psychologist Cheves West Perky demonstrates the similarity between these two concepts. In what became known as the Perky experiment, subjects were told that they would be participating in an experiment about imagining colors. They were instructed to focus on a point on a screen and to imagine a colorful object, "for example a tomato." Unbeknownst to the participant, a soft-edged, barely perceptual image of a shape roughly corresponding to the object being visualized was then projected onto the screen from the rear. The experimental subjects firmly believed that the apparent visualization was a product of their imagination, even though the image could also be seen by the attendants that were present. At the end of the experiment, the subjects were asked whether they were sure that they had truly imagined the object, and the mere act of asking triggered some uncertainty. The experiment demonstrates the soft-edged transition between our senses of imagination and perception. The direction of the image transfer is of note – the visually perceived images were interpreted as originating in the mind. In this case, "visual" represents not just the visible in this sense, but also the "imaginable." This is a central premise: visual cues and their contents can act as emotional triggers – they can serve as the basis for a "media decoy."

42.2.2 Thinking as a Form of Mental Rehearsal

The opportunity to contemplate and consider various options and choices in responding to a particular situation can be a disadvantage if a decision is not made quickly enough. Humans evolved in an environment in which decisions often had to be made instantly – an environment in which time for contemplation would often have been fatal. Thus, we learned to anticipate situations in advance, plan for them, and run through them first in our minds.

The ability to make a decision can also be a burden in and of itself. If we were to stop to think before each and every action undertaken, we would not accomplish much of anything. "That is the significance of the derived behavioral system: it allows the cognitive processes to be more easily and effectively executed than does action, which first requires us to stop and mentally review the pending deed or translate it into language, and then reason, telling ourselves what we should or want to do" (Aebli, 1981, p. 311).

The difference between acting and thinking in this regard is admittedly minor. The action objects are simply represented differently – namely through thought. Thus, it is just as easy to "master certain actions or behaviors by observing the behavioral models of others as it is to try to discover such things on your own: the advantage of having a clear model for action and the chance to mentally apply appropriate thought processes quickly offset the advantages of trying out such things oneself" (Aebli, 1993, p. 213). As a rule, such mental rehearsals only make sense for those situations in which the possible courses of action and strategy can actually be tested. Not every form of learning is governed by practice – only those where the pros and cons of various scenarios can actually be tested and considered. Strict stimulus–response schemes are impractical for complex social situations.

A lengthy analysis makes little sense in spontaneously arising situations of immediate danger; instead, emotions have since emerged as more adequate instruments of response (Cosmides & Tooby, 2000; Schwab & Schwender, 2010). In situations that allow for some planning, however, it is possible to play out or consider various behaviors and courses of action. According to Aebli (1981), thinking expresses itself as a cognitive activity in the "planning, doing, and reviewing of actions and the social processes and situations that lead to action" (p. 310). Aebli argues the evolutionary aspect when he references prehistoric living conditions:

Human survival certainly will ultimately be determined by behavior. Not only are we known by our deeds, they determine our fate. . . . And looking at the history of mankind, we see that practical activities consumed the greatest share of our expended effort. Granted, the practice of art was present early on and certainly religious beliefs and practices already were in existence, but these, too, were originally very closely associated with practical activities – sowing and harvest, reward and punishment, shelter and the procurement of provisions. (Aebli, 1981, p. 309f.)

Since Aebli, who was influenced by Piaget's ideas, is interested primarily in learning and the development of the human mind, he describes the tasks very generally. From an evolutionary psychological perspective, it is expected that mental rehearsals mainly concern social issues. And so it is as well with respect to the function of the mass media in general and talk shows in particular – they function in this sense as the visualization of such mental rehearsals. And that is an evolutionary victory: the ability to build an internal model of the social world in which one acts, plans, and mentally rehearses potential responses or courses of action in order to be ready for significant situations. Since ideas can be verbally triggered by the spoken and written language as well as by static and moving images, art and literature should be considered as special forms of social behavior or sociality. One can consider their communicative function as being much like such mental rehearsals. In this respect, verbal and fictional accounts have nothing to do with reality, but serve as references to imagined (rehearsed) actions or behavior.

In our mind, we are able to reflect on and consider many variations of an idea. Past events are not just remembered, but also replayed in order to learn from mistakes and optimize the results of future events. We rehearse these mentally, think through the consequences, and weigh potential results in order to prepare ourselves for any number of different potentialities, thus allowing us to take immediate action when required without undue consideration.

42.3 GOSSIP

Robin Dunbar (1996) was the first to systematically derive the human speech functions from evolutionary theory. One of the basic conditions for the evolution of the social-cognitive functions of the brain is the establishment of social groups, which provide the individual with security. Initially, the content of the language interaction is not important – what matters is not *what* is said, but rather *that* it is said. Moving beyond this initial phase, those interactions containing content that further advances social relationships should, within an evolutionary context, quickly find success; Robin Dunbar (2004, p. 104f.) sees four functions that they serve:

- Seeking advice or discussing hypothetical situations.
- Provision of a policing function to control those who fail to abide by the formal and informal agreements that underpin the society.
- Advertising ourselves.
- Deceiving and telling others what we think it would help us for them to know.

An important purpose of language, therefore, is that it makes possible the exchange of information about other people, thereby shortening the laborious process of ascertaining the significance of their behavior.

Wert and Salovey (2004) see value in gossip as a source of social information. They emphasize that not only is information created and disseminated, but it also serves as a form of social comparison. Gossip contains information about morality, as it not only reports but also can comment on observed behavior, thus answering the question of how moral behavior is organized in a preverbal society. The transmission of such gossip contributes to the formation and maintenance of in-groups and out-groups; members of the same in-group share the same code of conduct.

Baumeister, Zhang, and Vohs (2004) see a similar use of gossip in social learning. Participants implicitly agree on the standards of the social network; one thus learns how to behave in a social context. Much like humor (Schwender, 2006), gossip is an indirect form of aggressive behavior and can serve as a rebuke. But even though everyone participates in it, gossip nonetheless carries a negative connotation. This can be explained by the fact that it cannot be controlled; no one wants to become the focal point of negative publicity. One way to avoid this is the adoption of appropriate behavior standards. Gossip conveys such guidelines in an easily understood narrative form. Situations are described in ways that can be generalized as: How was someone shamed? How could the subjects have helped themselves? What should we be aware of if we find ourselves in a similar situation? What are the norms for proper behavior in these various situations? Gossip is thus a form of learning; it involves the appropriation of concepts that were not previously found in a person's range of behavior. The confirmation of positive or negative conduct is important for the refinement of our own behavioral models. The information we acquire through gossip allows us to examine the success or failure of an action in an observational model. Such observations furthermore provide guidance on how to optimize already learned behaviors.

Gossip is also a particularly powerful mechanism of control for the problem of free riders – those who enjoy social benefits without paying the costs (Dunbar, 2004, p. 100). Gossip conveys essential information regarding who will help us, who should expect help from us, who is to be believed, and who is lying. Such information clearly offers long-term advantages with respect to survival and success (Pinker, 1997, p. 540); this is especially true of shared information that is exclusive or confidential. Becoming the subject of gossip can be damaging; nonetheless, it is essential to social survival: only by being talked about are we able to promote ourselves to others. The gossip community is thus a conspiratorial circle, one that promotes alliances and cooperation through the reciprocal exchange of information. Those excluded from this circle will have difficulty building alliances and finding sexual partners within the in-group, as someone excluded from the group's care and protection is not especially attractive as a possible partner.

42.4 CONVERSATION AS A TELEVISION GENRE

If there is one single television genre that is best suited to testing evolutionary psychology's explanations of the functions of human language, then it is the daytime talk show; for example, *Oprah* (USA, 1986–2011), *Donahue* (USA, 1970–1996), etc. In order to define the "daytime talk show" as an object of study, certain formal requirements need to be addressed and discerned: the show, by virtue of the genre, is broadcast weekdays, Monday through Friday. As often as not, it carries the name of its host, thus already flagging the key role of the host, whose personality, style, and demeanor shape all aspects of the talk show, even the choice of topics. Each "episode" is generally monothematic (i.e., dedicated to a common theme or topic agreed upon in advance to which guests appropriate to the topic are then invited). Most important to their success are the personalities, emotions, firsthand experiences, and expressions of feelings, less so any factual aspects. The topics are personalized, and the focus is usually on the fate of a single person. The individual takes precedence over the universal, which serves to lend the presentation authenticity: true stories about real people.

Without the need for elaborately scripted fiction, the focus is for the most part on the direct exchange of gossip via face-to-face conversation – the statement that "a talk show is a conversation between two, three, or more people" does not really capture the special nature of this type of program. After all, we carry on such face-to-face conversations several times daily. The television talk show, being a staged production bound by the prerequisites of drama and the confines of a tightly scheduled format, follows its own rules. Manfred Sack offers a description of the genre that exactly captures the artificiality and staginess of what social linguists refer to as a speech-act form of entertainment

A talk show is not a discussion – and should one arise or somehow come about it lapses into the chaotic and will then be reined in by the insistent expectations of the host – nor an interview. In a talk show there are no conversations of a clarifying or instructive nature. ... A talk show consists of nothing more than a series of chats between a person hired by the producers to ask questions and a guest who answers them. (Sack, 1985, p. 32)

Despite the great artificiality of the situation, one must add that the character of the ordinary and everyday is very much preserved. The hosts see themselves serving a double function: "On the one side they play the role of the questioning journalist; on the other, they emphasize their function as 'host'" (Mast, 1978, p. 24). This seemingly contradictory combination nonetheless yields a stylish and enthusiastically received approach – though not necessarily for the guests. After all, what host at a party suddenly confronts his guests during small talk with their criminal records or youthful errors? The talk show appears, then, to offer an intimate view into the lives of the guests, their aesthetic and moral preferences, and attitudes.

As far as understanding viewer attachment goes, parasocial interaction theory (Giles, 2002) provides the most useful model: when people become lonely, they tend to build emotional attachments to television personalities; the greater their loneliness, the more willing they are to identify with these television personalities and to view them as intimate friends. The cheerful and friendly talk show host then is an ideal partner. He or she comes into your home daily and is courteous, receptive, and always prepared with a suitable response. With respect to our thesis on the media decoy effect of such programs, Hippel (1993, p. 129f.) notes four fundamental, interrelated aspects that are significant:

1 The performer addresses the camera as if in a face-to-face situation.
2 The viewer receives information in the same manner as in a face-to-face conversation.
3 The performer responds according to the *presumed* responses and reactions of the viewer.
4 The viewer, however, perceives the performer as responding to his or her reactions.

42.4.1 Who Are They Talking About?

As noted in the previous section, when we watch one of these talk shows, we are essentially watching people gossip; interestingly, it does not matter that they are not family or even distant acquaintances of ours. One cannot even speak of a proxy here; those we see on television are fully unrelated to our own acquaintances. Barkow (1992, p. 630) presumes that this is due to the medium and its surrogate nature – and the nature of our imagination and mental rehearsal. Barkow's expectations can be operationalized and measured by looking at the program announcements and content summaries for upcoming episodes as published in the television program guides. These listings, which provide information about upcoming topics and themes, usually include a title that attempts to summarize the content of the episode from the producers' point of view.

For our analysis of these program announcements, 438 talk show episodes on German television have been reviewed in a quantitative content analysis drawn from the four daytime talk shows with the highest viewer ratings: *Meiser* (2.52 million viewers on average), *Fliege* (1.46 million), *Arabella* (0.80 million), and *Vera am Mittag* (0.79 million). All four are considered mainstream programs; the use of abusive language or personal attacks by the guests is generally avoided. If we want to draw parallels with American talk show programs, particularly in terms of the host, *Meiser* could be considered a "Phil Donahue." *Fliege* skews toward an older audience and *Arabella* a younger one.

The episode announcements were categorized on the basis of the details drawn from the program guides, and the following categories, as derived from Barkow, were identified, each with examples noted:

- *Sexual partners and rivals*: "I love my husband – and my lover," "Sweetheart, I'm gay," "I buy my love." Rivals are included here in the same category as sexual partners,

as it became evident that both are frequently pitted against each other.
- *Partners in a social context*: "My neighbor is a slut," "Reporters ruined my life," "Schoolchildren in danger! Where are the good teachers?" As in the previous category, social groups are more likely to be involved when there is a need to iron out potential conflicts; the titles reflect this.
- *Children/parents/siblings*: "My parents don't like my boyfriend," "My mother is the greatest," "In my brother's shadow." The family affiliation classification could be reduced to immediate family – more distant relations were scarcely mentioned and will find their place in another category.
- *(High) status*: "Child stars," "I'm in contact with spirits," "I saw God in my living room." Status refers to all who enjoy certain renown. One also has to include in this category non-real personalities, as they, too, often achieve a certain status and occasionally even a certain level of influence.
- *Indiscernible*: "Too late – missed opportunities," "Revenge is sweet," "Arabella's surprise show," "Arabella's 500th show." No persons are mentioned in these episode titles; therefore, a classification – even one that might disprove the thesis – is not possible.
- *Other*: Any remaining uncategorized segment titles were initially placed in this residual category; it is possible that they hold clues or references that might be used to disprove the hypothesis.

Looking at these titles, one becomes aware of their systematic similarities; further examination led to the establishment of two additional categories, allowing us to reassign all remaining titles:

- *Animals*: "An animal saved my life," "Like cats and dogs – what animals say about our personality," "Animals are my best friends."
- *Me*: "No one loves me like I do," "Breast cancer – why me?" "My life in prison."

Thus, two categories were added to Barkow's hypothesis (Table 42.1). These, however, are not to be seen as a refutation, but rather as an addition.

This treatment of animals as a category equivalent to "people" requires some further explanation. Animals have variously been treated as infants, living partners, and social partners. If we consider the infant aspect, specific behaviors can be observed: they are petted, fed, and spoken to – though what they evidently understand is very minimal. Communication with animals is reduced to expressions of emotion by means of mime and gesture and tactile and vocal interaction, and less so verbal expressions. Their particular advantage as communication partners to us is that when we speak to them they seem to understand our language, yet rarely argue back. There is also the innate "babyness" aspect of many animals, including dogs and cats: they are small in size; we bend down to

Table 42.1 Who are they talking about?

Subject	Percentage of cases
The individual	43.2
Sexual partners and rivals	20.8
Partners in social context	15.5
Parents/children/siblings	13.7
(High) status	5.3
Animals	2.5
Other relatives	0.7
Not discernible	26.9
n = 438	

the host usually does not offer a personal opinion. The host is much more likely to offer a generous measure of understanding and sympathy, no matter how vague or contradictory the proffered viewpoints are.

In summary, we see that each of the categories for subject persons anticipated by Barkow is found in the program guide listings. With the exception of titles in which no one whatsoever is mentioned, there were no titles that could be judged to refute Barkow's hypothesis; in the context of an analysis of daytime talk shows, the hypothesis is corroborated.

42.4.2 What Are They Talking About?

If one looks at these various subjects of gossip from an evolutionary psychology viewpoint, one would expect to see fitness-relevant topics, as formulated by Jerome Barkow (1992, p. 628): "The short answer to 'what kinds of information' is relative standing and anything likely to affect it, control over resources, sexual activities, births and deaths, current alliances/friendships and political involvements, health, and reputation about reliability as a partner in social exchange."

As in the case of the classification by subject, Barkow's hypothesized categories of topics can easily be operationalized and tested using the program listings for the talk shows. A preliminary analysis showed that only a single modification was necessary: "family matters" is a category that he did not postulate, but frequently occurs in the daytime talk shows. Since the category "relatives" was anticipated and confirmed while reviewing the program guide listings for categorization by subject person, it was only logical that "family matters" should become a corresponding topic and immediately be considered as an initial addition to the hypothesis dealing with the conversational content.

Each of the following categories, listed here together with typical episode titles as taken from the program announcements, was identified:

- *Sexual activities*: "Everyone wants my wife," "I really go for fat women," "I can have any man I want."
- *Health, beauty, and body*: "Is my child a victim of mad cow disease?", "I am too beautiful – no one dares to get close to me," "I hate my body."
- *Political activities*: "War-game in Germany," "Bring back the wall," "Neo-Nazis – terror from the right." Included are topics on religion, since they also deal with involvements in larger groups.
- *Partnership*: "I'm fighting for my friend – I want to save my relationship," "My wife should be like my mother," "Don't marry him!"
- *Family affairs*: "Mommy, why did you throw me out?", "Puzzled parents: Where are their children?", "I don't understand my child anymore."
- *Birth and death*: "Adoption – mothers without children," "A murder has many victims," "I should be dead – living with HIV."

look into their eyes. Relative to their bodies, their limbs are short, their eyes large, and their faces round. In addition to the characteristics of vocal communication, tactile attention is also important (Keller et al., 1988). This likewise serves the conveyance and reception of emotional states, which in the case of "man's best friend" we easily learn to recognize: submissive and cowering behavior, wet-tongued greetings, and joyful tail-wagging.

People transfer human traits to animals. Concepts such as friendship, affection, communication, and understanding are projected onto the beloved family pet; we see in them a true companion, a member of the family, and a helper. Given this transference and the strong bonds of attachment that arise, it is no wonder that our house pets become a part of our social communications with others. We talk about the social idiosyncrasies, experiences, and behaviors of our house pets just as we do about those of our children, our partners, and our friends.

It is surprising that while Barkow was adding up protagonists and antagonists he overlooked the role of the self-directed speaker ("me"). Not infrequently, when we inquire about another person, we are doing it to express our own opinion or attitude. The subject of someone else's behavior is then used as a keyword with which to convey a comment and make an evaluation according to our own measure; our intent as speaker is often to establish our own stance within the network of moral and aesthetic beliefs and values (Barkow, 1989). This is where Dunbar's hypothesis comes into play: one of the functions of gossip is "[to advertise] one's own advantages as a friend, ally, or mate" (Dunbar, 2004, p. 105). This "me" function can be seen at work in the daytime talk show when a guest is given the chance or is even encouraged to express a personal opinion and to take a stance together with or against others. There is the question of power: Who is permitted to set the values that might serve as the moral standards? Approval and displeasure alike are spontaneously signaled by the live audience in the studio: clapping, cheers, boos, and even laughter provide comment and reaction to the subjective variations of behavior. Interestingly – and this needs to be thought about further –

- *Reputation and dependability*: "I risk my life for law and order – policemen," "Why should I be honest, if no one else is?", "Workplace bullying – war in the office."
- *Hierarchy and resources*: "Everyone knows my friend, but no one knows me," "Not beautiful, but rich," "Quality has its price – luxury."
- *Other*: "Very funny – the new comedy," "Crinoline petticoats and kidney-shaped tables – the wild fifties."
- *Not discernable*: "This day changed my life," "Arabella's 500th show," "Arabella's surprise show," "Welcome to my program."

After an initial review of the data, two additional categories were added: "aesthetics" and "danger." The topic of aesthetics plays a significant role. It involves the assessment of taste preferences, which are used for the purpose of group assignment and differentiation. Since aesthetics play a prominent role in the assessment of perceptions (Orians & Heerwagen, 1992; Voland & Grammer, 2003), it is to be expected that intensive debates are conducted in the context of gossip; after all, matters of aesthetics involve negotiations of behavior.

Danger plays a central role in evolutionary psychology theory. The necessity for cheater detection (Cosmides & Tooby, 1992) requires that we exchange information about possible cheaters. Examples of episode titles include "Drunk driving," "Suddenly everything's on fire!," and "A murder has many victims."

The episode titles refer to talk show episodes broadcast in Germany (the examples in this article are provided in translation), but an intercultural generalization should not be difficult. One need only compare them with program listings for the corresponding daytime talk shows.

The entries in Table 42.2 sound like items out of the tabloid news. In fact, Davis and McLeod (2003) were able to empirically verify the presence of these types of news items in newspapers. The categories demonstrated consistency across cultures and times, an indication that we are dealing here with evolutionarily determined topics. Stories about ordinary people with respect to social context can be categorized as follows: "murder/physical assault; robbery/vandalism; accidental or natural injury/death; altruism/heroism; suicide/self-inflicted injury; abandoned/destitute family; harm to a child; sexual assault/rape; taking a stand/fighting back; reputation; marital courtship anomalies" (Davis & McLeod, 2003, p. 213). It makes sense that one would want to be able to recognize and be prepared for perils affecting the individual as well as the group as a whole the moment that they occur. The correspondence with the list based on Barkow's hypotheses, which also underlies the study of talk show topics, is striking. Further studies confirm the results.

Since the viewing figures were available for the German talk shows being analyzed, a quantitative ranking was possible (Table 42.3). The question that naturally arises is whether specific topics attract greater viewing numbers than others.

Looking at the numbers of viewers arranged by category of topics, only two significant differences can be seen. Both and can be explained by viewer preferences for specific hosts. The average number of viewers (the numbers come from the appendix to Schlosser, 1998) varies between roughly 1.1 and 1.6 million.

All of the predicted categories are found; however, the two residual categories bear closer scrutiny. The category "not discernible" includes titles such as: "Arabella's surprise show," "Arabella's 500th show," "This day changed my life," "When fate knocked at my door – hotel stories," "Off the shore of Madagascar – cruise ships and love boats," "My fateful flight – stories between heaven and earth," "Off on holiday," and "My place at four"; that the viewing figures for these episodes do not vary from the average can be seen

Table 42.2 What are they talking about?

Topics	Percentage of cases
Reputation	28.1
Partnership	21.7
Sexual activities	21.0
Health, beauty, and body	17.1
Family affaires	13.2
Hierarchy and resources	13.2
Aesthetics	13.0
Birth and death	5.5
Politics and religion	5.0
Danger	3.7
Not discernible	4.8
$n = 438$	

Table 42.3 Topics by mean number of viewers.

Topics	Number of viewers (millions)
Family affaires	1.6
Reputation	1.5
Hierarchy and resources	1.5
Partnership	1.4
Health, beauty, and body	1.4
Birth and death	1.3
Politics and religion	1.2
Danger	1.2
Aesthetics	1.2[*]
Sexual activities	1.1[**]
Not discernible	1.4
Mean	1.4
$(n = 438)$	

[*] $p < 0.05$;
[**] $p < 0.01$.

as a first indication that other criteria are decisive for the audience size of a broadcast. A further indication is the relatively even distribution of the viewer ratings in relation to the individual topic listings. No specific topic stands out. The specific content seems to make little difference as long as the discussion is about topics of a predisposed relevance.

The catchall category "other" includes titles such as "Crinoline petticoats and kidney-shaped tables – the wild fifties," and "Talking about flowers – the secret powers of flowers." The question arises once more as to whether this implies a falsification of the hypothesis or the need for additional thematic categories. What the titles in the examples here share in common is an aesthetic aspect. The first episode, it would appear, deals with a nostalgia for older styles; the second with the effects of the appearances and fragrances of flowers (assuming that this is not about the preparation of tea, in which case it could be assigned to the category "health"). These are not the only episodes that appear to address topics of an aesthetic nature (e.g., "My husband has no taste," "I am too beautiful – nobody dares to get close to me," "Quality has its price – luxury." Aesthetics, then, must be considered a potential additional category.

Within the context of this discussion, the aspect of aesthetics and its significance from the perspective of evolutionary psychology need to be looked at more closely (Thornhill, 1998, p. 544). The basis of Darwinian aesthetics is the perception of beauty and the subconscious discernment of paths to optimal fitness. Beauty is the promise of successful functioning in the area in which it arises. Ugliness is the prospect of reproductive failure. Given the premise that our brain is an information-processing mechanism, ever ready to provide solutions to problems of survival and increased reproduction – and this is a premise of evolutionary psychology – aesthetic judgments about our surroundings will also play a role.

It would be useful if the outcome or success of a decision whose consequences will only be known in the future were somehow predictable. The investments that are involved are sometimes significant; an indicator for success would be an enormous advantage. This is also the case with the selection of a partner. Although we cannot predict what abilities and talents our as yet unborn offspring may enjoy, we can make choices with respect to the abilities and talents of their mother or father. Indicators for fitness, health, boldness, strength, and self-assertion could lead to decisions that might be advantageous for our future descendants. The chances that these traits are inherited, after all, are quite good. Such indicators are nothing other than aesthetic criteria (Buss, 1992, 1998).

42.4.3 The Talk Show Host

With the daytime talk shows, the person talking about a topic appears to be more important than the topic itself. In the context of these virtual gossip communities, it is therefore necessary to look more closely at the role of

Table 42.4 Talk show hosts and their mean numbers of viewers.

Talk show host	Number of viewers (millions)
Meiser	2.6
Fliege	1.5
Arabella	0.8
Vera	0.8
Mean	1.4
$n = 438$	

discourse in their creation. The correct choice of host is essential for successfully building an audience; this is reflected in the program viewing figures. Looking at the average number of viewers independent of host, no significant differences (ascertained by means of the simple factorial analysis of variance model) are found. The distribution of viewing figures by host, on the other hand, reveals a clear picture (Table 42.4).

The host, being at the center of the virtual gossip community, is accordingly a much better predictor of the viewing interests of the public than the choice of topics. With the exceptions of "Arabella" and "Vera," which show only a small difference ($p = 0.074$) as they attract similar-sized audiences, almost all comparisons by host show significant differences in numbers of viewers. Further verification of a possible hypothesis that the host is the more important factor in the viewer's decision to tune in is found when looking at the numbers of viewers of individual hosts. We single out Hans Meiser ($n = 118$), who has the largest viewing audience. Most differences in the number of viewers by topic do not vary significantly and cannot be explained systematically; deviations fall within the range of chance. The exception is "partnership," with a mean of 2.8 million viewers ($p = 0.013$).

Looking more closely at the role of the most successful hosts, we see that one of their most striking qualities is the ability to universally demonstrate empathy and understanding; in almost all situations, they are able to identify with and appear receptive to each and every guest. As hosts, however, they largely avoid taking their own stand; instead, adversaries are brought into the discussion. The resulting drama is developed in such a way that viewers can find support for their own positions and thereby feel that their own viewpoints have been confirmed. Viewers are able to identify with their own moral and aesthetic positions without feeling that they might be positioning themselves in opposition to the leader of the discussion. The host thus stands at the center of a community that can establish norms and condemn deviant behavior; participation in such a community therefore seems to be important. The host enjoys the trust and confidence of each and every member of the community and keeps them informed about other community members. Barkow et al. (2012) discuss Michael Chance's thesis that we preferentially attend to and learn from those who are high in status.

In the context of the daytime talk show, if we look at how conversations are initiated and conducted, a certain pattern becomes apparent. By means of questions and statements, the host is able to stir up conflict, sharpen contradictions, and, near the end of the program, offer appeasement and reconciliation. In the course of this, we as viewers are confronted with many of the elements that make gossip especially appealing to us: conflict between family members, partners, allies, and others from whom we presumably expect support but instead are shown anger or even hatred is one sort; another is the communal condemnation of antisocial behavior and the rallying around the collective good and pro-social ideals.

42.4.4 The Television Court Show

A special case of the talk show is the more or less staged judicial hearing; for example, *Judge Judy* (first aired 1996) on American television and *Richterin Barbara Salesch* (1999–2012) in Germany. Here, language is used as a pre-aggression problem solver. Viewers follow the clamor of argument and counterargument, claim and counterclaim, and reciprocal accusations of guilt and insults. Perhaps the most important rule that this format follows is that the authority of the host as arbitrator is unquestionable. The cases that these "television judges" generally deal with are usually quarrels of small consequence that require neither complex legal findings nor difficult decisions; but in contrast to the gossip and talk shows, where conflicts are only introduced and then discussed and possible solutions are generally external to the program, the cases in these staged trials are typically resolved (by prior consent of the participants) with legally binding decisions. While the host of a talk show, concerned with uniting a larger audience, may rarely profess an opinion, the television judge must clearly state an opinion and make an unequivocal finding of right and wrong.

From an evolutionary psychology perspective, differences between the two subgenres of the daytime talk show and the television courtroom are scarcely discernible; many topics could easily be handled in both. Much like the talk show format, the television courtroom often involves an assessment of deviant social behavior. Moral faculties of judgment are tested and sharpened; the viewer is called upon to separate right from wrong and proper from improper motives. In the case of the mock courtroom, the decision-making is supported and directed by a person in a judge's robe rather than a host or moderator, with the viewer participating in the casting of moral judgment and rebuke. This can lead to social comparisons. The viewer experiences a sense of superiority that arises when a participant on the show is pulled down a peg (Wills, 1981).

Court television also succeeds in fulfilling one of the tasks that Robin Dunbar postulates in his theory of gossip:

policing free riders and cheaters. In the debates that are presented, the viewer is provided with examples of deviant behavior and shown how they are wrong and how they can be recognized. The court television judge has the task of establishing guidelines and standards, something that the talk show host tries to avoid.

42.5 CONCLUSIONS

Two fundamental conditions make it possible for us to accept media presentations as effective representations of imagination or reality. (1) As humans, we are capable of taking on both the visual and the mental perspectives of others; we understand what others see, know, and feel, and we can sense their intentions. (2) Imagination and visual perception do not always provide a clear distinction; mental images and concepts can be easily triggered by visual (or here media) decoys.

Evolutionary psychology theory would suggest that the contents of such mental preoccupations are neither random nor arbitrary, but rather address issues that are of consequence for our own welfare and well-being. Hypotheses to this end can be derived from Barkow's discussion of the purpose of gossip; these are confirmed by a look at the issues and subject persons found in successful television talk shows. Furthermore, it can also be shown that argument and controversy serve valid functions in debates about common values within a community, as they aid in the negotiation of members' expectations with respect to both thought and deed.

The success of mass media formats such as the daytime talk show derives from their presentation of conflicting viewpoints – opinion and counter-opinion, argument and rebuttal – as advocates and opponents face off. Such debates serve to position the participants and their views within aesthetic, political, and moral spectra. They support the formation of groups through acceptance or disapproval of various actions and behaviors. Each and every viewer is almost certain to find his or her viewpoint addressed within the course of a talk show episode.

Further aspects of individual- and group-relevant behavior are also found in these media depictions. All of the anticipated topics and categories of persons are present; additional categories were even found. What becomes clear is that in television, too, how people speak and what they say is not as important as who is doing the talking: people gather around a central figure.

Meanwhile, we as viewers are sitting at home in the safety and comfort of our own easy chairs and watching these events with our eyes and in our minds. The opportunity that this presents for mental rehearsal and trial thinking has two advantages. First, viewers need not take the risk of themselves becoming the subject of gossip. They can safely observe the various points of view and subsequent responses that are put forth and then compare these with their own positions and viewpoints. Second, viewers need not expend great effort on acquiring such readily available social information themselves.

The mass media ultimately do not show us the way things are, but rather that which is important. The significance of this can be derived from and verified by evolutionary psychology expectations.

REFERENCES

Aebli, H. (1981). *Denken: Das Ordnen des Tuns, Band II: Denkprozesse*. Stuttgart: Klett-Cotta.

Aebli, H. (1993). *Denken: Das Ordnen des Tuns, Band I: Kognitive Aspekte der Handlungstheorie*. Stuttgart: Klett-Cotta.

Barkow, J. (1989). *Darwin, Sex, and Status: Biological Approaches to Mind and Culture*. Toronto: University of Toronto Press.

Barkow, J. (1992). Beneath new culture is old psychology: Gossip and social stratification. In J. Barkow, L. Cosmides, & J. Tooby, eds., *The Adapted Mind. Evolutionary Psychology and the Evolution of Culture*. New York/Oxford: Oxford University Press, pp. 627–637.

Barkow, J., O'Gorman, R., & Rendell, L. (2012). Are the new mass media subverting cultural transmission? *Review of General Psychology*, **16**(2), 121–133.

Baumeister, R. F., Zhang, L., & Vohs, K. D. (2004). Gossip as cultural learning. *Review of General Psychology*, **8**(2), 111–121.

Buss, D. M. (1992). Mate-preference mechanisms: Consequences for partner choice and intersexual competition. In J. Barkow, L. Cosmides, & J. Tooby, eds., *The Adapted Mind. Evolutionary Psychology and the Evolution of Culture*. New York/Oxford: Oxford University Press, pp. 249–266.

Buss, D. M. (1998). The psychology of human mate selection: Exploring the complexity of the strategic repertoire. In C. Crawford & D. L. Krebs, eds., *Handbook of Evolutionary Psychology: Ideas, Issues, and Applications*. Mahwah, NJ/London: Lawrence Erlbaum Associates, pp. 405–429.

Cosmides, L., & Tooby, J. (1992). Cognitive adaptations for social exchange. In J. Barkow, L. Cosmides, & J. Tooby, eds., *The Adapted Mind. Evolutionary Psychology and the Evolution of Culture*. New York/Oxford: Oxford University Press, pp. 163–228.

Cosmides, L., & Tooby, J. (2000). Evolutionary psychology and the emotions. In M. Lewis & J. Haviland-Jones, eds., *Handbook of Emotions*, 2nd ed. New York: Guilford, pp. 91–115.

Davis, H., & McLeod, S. L. (2003). Why humans value sensational news – An evolutionary perspective. *Evolution and Human Behavior*, **24**, 208–216.

Dunbar, R. I. M. (1996). *Grooming, Gossip, and the Evolution of Language*. London: Faber & Faber.

Dunbar, R. I. M. (2004). Gossip in evolutionary perspective. *Review of General Psychology*, **8**(2), 100–110.

Giles, D. C. (2002). Parasocial interaction: A review of the literature and a model for future research. *Media Psychology*, 4(3), 279–305.

Hippel, K. (1993). Parasoziale Interaktion als Spiel. Bemerkungen zu einer interaktionistischen Fernsehtheorie. *Montage/AV*, **2**(2), 127–145.

Keller, K., Scholmerich, A., & Eibl-Eibelsfeldt, I. (1988). Communication patterns in adult–infant interactions in Western and non-Western cultures. *Journal of Cross-Cultural Psychology*, **19**(4), 427–445.

Leslie, A. M. (1994). ToMM, ToBy, and agency: Core architecture and domain specification. In L. A. Hirschfeld & S. A. Gelman, eds., *Mapping the Mind*. Cambridge, UK: Cambridge University Press, pp. 119–148.

Mast, C. (1978). *Politische Öffentlichkeit. Untersuchung einer Parteisendung im Zweiten Deutschen Fernsehen*. Osnabrück: A. Fromm Verlag.

Orians, G. H., & Heerwagen, J. H. (1992). Evolved Responses to Landscapes. In J. Barkow, L. Cosmides, & J. Tooby, eds., *The Adapted Mind. Evolutionary Psychology and the Evolution of Culture*. New York/Oxford: Oxford University Press, pp. 555–579.

Perky, C. W. (1910). An experimental study of imagination. *American Journal of Psychology*, **21**, 422–452.

Piaget, J., & Inhelder, B. (1956). *The Child's Conception of Space*. New York: The Norton Library.

Pinker, S. (1997). *How the Mind Works*. New York/London: W. W. Norton & Company.

Sack, M. (1985). Stammtisch des Fernsehens. Die Talkshow – Diskussionsrunde, Gesprächsversammlung, Gruppeninterview oder "gesitteter Klamauk"? *Die Zeit*, **40**, 32.

Schlosser, A. (1998). TV-Genre Talk Show: Eine medienwissenschaftliche Studie. Unpublished thesis, TU Berlin.

Schwab, F., & Schwender, C. (2010). The descent of emotions in media: Darwinian perspectives. In K. Döveling, C. von Scheve, & E. Konijn, eds., *The Routledge Handbook of Emotions and Mass Media*. New York: Routledge, pp. 15–36.

Schwender, C. (2006). *Medien und Emotionen*. Wiesbaden: DUV.

Thornhill, R. (1998). Darwinian aesthtetics. In C. Crawford & D. L. Krebs, eds., *Handbook of Evolutionary Psychology: Ideas, Issues, and Applications*. Mahwah, NJ/London: Lawrence Erlbaum Associates, pp. 543–572.

Voland, E., & Grammer, K., eds. (2003). *Evolutionary Aesthetics*. Berlin/New York: Springer Verlag.

Wert, S. R., & Salovey, P. (2004). Introduction to the special issue of gossip. *Review of General Psychology*, **8**(2), 76–77.

Wills, T. A. (1981). Downward comparison principles in social psychology. *Psychological Bulletin*, **90**(2), 245–271.

43 Supernormal Stimuli in the Media

DEIRDRE BARRETT

What if you turned on Seinfeld, only to see Jerry and the gang locked in their separate apartments, watching television. Would that be a good show? White Dot challenged viewers in 1996, "Think about it: that's how you are living now. . . . You are alone in the dark, staring at a plastic box. This is like a science fiction horror story. . . . Jerry and Elaine, Kramer, none of them know you. They don't care whether you live or die. Why don't you get yourself some real friends?

(Whitedot.org, 1996)

This is a great example of what William James termed "making the natural seem strange" (James, 1890), which he suggested was the only way we could notice and reflect on human instinctual behavior. Entertainment media feel good; people do not question it. But why do we choose to sit in front of a plastic box? Evolutionary psychology has brought many insights from Darwin and ethological researchers to bear on such behavior. There is one concept that has been overlooked, however, which I believe has the very most to contribute to explaining why entertainment and news media enthrall: the supernormal stimulus. In this chapter, I will summarize the concept and then discuss how it can inform our thinking about modern human behavior in general, and especially how media operate.

Niko Tinbergen won the 1973 Nobel Prize in Biology for his research on instinctive behavior in animals. One major cluster of these studies used dummies to elicit nurturing, mating, and fighting responses. Tinbergen constructed fake eggs that birds sat on, artificial female butterflies that male butterflies courted, and models of male fish that other males attacked. Some of the impostors were ingeniously unrealistic to elucidate exactly what characteristics triggered the behaviors. Territorial male stickleback fish would not attack a fish-shaped model if its belly was not red, but they violently pursued crude rectangular forms when the underside was red (Tinbergen, 1951).

The most intriguing of Tinbergen's discoveries was that some dummies surpassed the power of any natural stimuli. A male silver-washed fritillary butterfly was more sexually aroused by a butterfly-sized rotating cylinder with horizontal brown stripes than it is by a real, live female if the cylinder's stripes were more pronounced than the butterfly's. Mother birds preferred feeding a fake baby bird beak on a stick if the dummy beak was wider and redder than a chick's (Tinbergen, 1953).

Tinbergen coined the phrase "supernormal stimuli" for these dummies that elicit stronger responses than those that occur naturally. The supernormal stimuli that he studied in the most detail were decorated plaster eggs. Tinbergen found that most birds preferred eggs that were larger than their own and those with exaggerated colors or markings. The oystercatcher, which lays small brown speckled eggs, will ignore them in favor of a giant brown plaster egg the size of the bird itself. Songbirds abandon their pale blue eggs dappled with gray to hop onto a black polka-dotted Day-Glo blue dummy so large that the valiant incubators repeatedly slide off and have to climb back on. Ground-nesting birds will retrieve eggs dislodged from the nest, rolling them back into place. The greylag goose will ignore its own egg as it struggles to retrieve a volleyball.

Animals encounter supernormal stimuli mostly when experimenters build them. We humans can produce our own: candy sweeter than any fruit, stuffed animals with eyes wider than any baby, pornography, propaganda about menacing enemies. In my book *Supernormal Stimuli* (Barrett, 2010), I argued that this concept helps explain phenomena from junk food to modern warfare. In this chapter, I will focus on how the media co-opt key instincts to manufacture more and more intense supernormal stimuli. I will start with television and film as prototypes for modern media and then address how video games, phone "apps," social websites, and other emerging media resemble these older forms and how they differ from them.

Before examining the content of television and films, there are ways in which supernormal stimuli are relevant to the medium itself. Humans have a basic instinct to pay attention to any sudden or novel stimulus, such as a movement or sound. Neurologist Ivan Pavlov (1927) named this the "orienting response." Shared with other animals, the orienting response is part of human evolutionary heritage. It evolved to help people spot and assess potential predators, prey, enemies, and mates. The orienting person or animal turns their eyes and ears in the direction of the stimulus and then freezes while parts of the brain associated with new learning become more active. Blood vessels to the brain dilate, those to muscles constrict, the heart slows, and alpha waves are blocked for a few seconds.

By the age of six months, babies orient when a television is turned on. Adults continue to do so. The visual techniques of television – cuts, zooms, pans, switches from one camera angle to another within the same visual scene, and sudden noises – all activate the orienting response (Reeves & Thorson, 1986). The effect persists for four to six seconds after each stimulus. Producers of educational television for children have found that judicious orchestration of these formal features can increase learning – presumably by keeping children focused on the screen. After a certain level of intensity, however, the orienting response is overworked and effects on learning and attention begin to reverse. This is what we see with advertisements, action sequences, and music videos, where formal features provoke orienting at the rapid-fire rate of one per second. Following prolonged bombardment with these stimuli, the viewer develops a mix of physiological signs of both high and low attention rarely seen in natural settings. Eyes remained focused, the body is still and directed toward the set, but learning and memory drop to lower levels than when not orienting (Rothschild et al., 1986).

Measurements of metabolism, including calorie burning, average 14.5 percent lower when watching television than when simply lying in bed. Electroencephalogram studies similarly find less mental stimulation, as measured by alpha brain waves, during television viewing than during reading or other quiet activities. This odd hybrid of attention and fatigue in response to supernormal orienting stimulation makes for a state in which viewers are disinclined to turn off their sets. Movie theaters, which make their money from tickets and concessions upon entering, have a different agenda – they terminate once you have plausibly gotten your money's worth and welcome the next audience. Television, operating off advertising revenue and having no real-estate expenses, never wants anyone to leave their network.

43.1 SCHEHERAZADE'S STRATEGY

Beyond these attentional characteristics, supernormal stimuli play an equally prominent role in what we usually notice about television and film: their content. Enthusiasm for a dramatic story – more eventful than daily life, with a hero/heroine involved in a romantic or survival challenge, climax, and resolution – may, in ancestral settings, have been adaptive for vicarious learning. Traditional tales are more efficient than each individual encountering survival challenges or social dilemmas anew. Classic fiction and drama raised questions about how an intriguing character would deal with an important situation, and then modeled an answer.

This can easily be co-opted with the strategy of Scheherazade – the crafty princess who survived "1001 Nights" not by completing a tale, making the point, imparting a lesson, but by constantly setting up new, unfulfilled questions and anticipations. A broadcast or cable network uses orienting to keep viewers watching right now. A given show needs to persuade them to tune in next week. Soap operas parallel Scheherazade most obviously, but other shows close with "scenes from next week," raising rather than answering questions. Films are the most covert as they must work as a whole, but increasingly blockbuster wannabes leave enough plotlines hanging so that if box office receipts meet expectations, "Blockbuster: Parts II and III" are givens.

The plots – and even names – of popular television shows tell us which instincts they are tugging at. *Friends* introduced the person in front of the plastic box to a group of lively roommates whose smiles, quips, and laughter caught them up in their camaraderie without our having to exercise any social effort. *Sex and the City* gave fans more vicarious romantic adventures than they would encounter in a lifetime. *Girls* invites viewers to hang out with a delightful group of young women.

Innumerable hits from the 1950s (e.g., *Leave It to Beaver*) and through to the 1970s (e.g., *The Brady Bunch*) provided idealized wise parents and adorable, agreeable youngsters. Currently, *Modern Family* varies race and sexual orientation in directions that would have been scandalous in Beaver's day, but otherwise sticks to the formula. The list of basic desires goes on: *Good Times, Mad About You, Homeland, How I Met Your Mother*, etc.

Sexuality is probably the single most catered-to drive, as actors are selected for physical attractiveness and most shows feature at least subplots of sex and romance that pull even when the main plot revolves around adventure or crime. Many shows are sustained entirely by dating dramas and sexual tension. For depiction of what that attraction leads to, however, there is the separate world of pornography. Catherine Salmon, Maryanne L. Fisher, and Rebecca L. Burch deal with porn in much more detail in Chapter 45, so I wish only to underline the medium's supernormal characteristics – hardly a big stretch from Tinbergen's animal research.

While sexual stimuli in mainstream film and television overlap with general social ones and tend to appeal to both genders, hard-core porn is designed largely for males. The same characteristics of general female prettiness hold: large eyes, small jaws, and narrow waists. But artificially enhanced breasts are even more common and are larger yet than in mainstream media. Behavior is not just the friendliness desired in general social drama; rather, the women are specifically enthusiastic about sex – always available, ready for anything or anyone. Romantic "chick flicks," meanwhile, feature a similarly supernormal version of characters swept away by romantic longings, with males desperate to protect, marry, and never be parted from the heroine (Salmon & Symons, 2001).

In Kurt Vonnegut's novel *God Bless You, Mr. Rosewater*, Fred Rosewater, a mild-mannered accountant, longs for approval from Harry, whom he views as a real man. So Fred shows him a scantily clad centerfold:

> "Like that, Harry?" he asked.
> "Like what?"
> "The girl there."
> "That's not a girl. That's a piece of paper."
> "Looks like a girl to me." Fred Rosewater leered.
> "Then you're easily fooled," said Harry.
> *(Vonnegut, 1965)*

This is Jamesian "making the ordinary seem strange" again. It does "look like a girl," and an extraordinarily desirable one at that – in the same way that Tinbergen's striped cardboard cylinders looked like quite a butterfly.

While Fred Rosewater's simple print media still churns out erotic images, pornography has always been at the forefront of every new medium – film, video, the Web – and now there is hard-core sexual anime, termed "hentai," and realistic computer-simulated sex. Both can achieve waist-to-hip ratios that plastic surgeons can only dream of. Both are becoming available in versions for 3D goggles. A recent article on the phenomena titled "Will 3D Hentai Kill the Human Porn Star?" concluded in the affirmative (Lhooq, 2013).

An interest in pornography seems normal to modern humans – indeed it is if "norm" means average behavior, as only 11 percent of American men deny ever looking at porn (Hite, 1981). But in terms of what the instinct developed for, if there is anything stranger than staring at the little box of wires as if one had friends in there, it is ejaculating to it.

43.1.1 Attack of the Supernormal Stimuli

The other instinct television and films plays to besides social ones is adventure. For a plot to succeed without sex, romance, buddies, or family ties, there must be exotic travel, wild animals, car chases, or mountain climbing. It is no coincidence that "cliff-hanger" is a colloquialism for a scene that captures audience attention.

The viewer of adventure films vicariously explores exotic locales, creates empires, and struggles in confrontations with nature, man, and beast. The films are often set in romantic periods of the past and may feature historic figures or literary heroes – Robin Hood, pirates, or kings. They involve expeditions for lost continents, jungle and desert epics, treasure hunts, and medieval quests. They may have futuristic or interplanetary settings – all supernormally stimulating to our curiosity and instinct to explore.

A related genre, the "action" film, puts more emphasis on continuous extreme activity – physical stunts, chase scenes, rescues, fights, and escapes – often in adversarial settings such as spy/counterspy or western cowboy/Indian. Knowing that physical competition appeals more to men, films like *The Wild Bunch* and *Deliverance* follow all-male groups in grueling adventures. Two variations on the action film are war films and disaster films.

War films stimulate many of the same instincts that leaders use to whip up real wars: vivid images of supernormal threats and portraying the enemy as a different pseudo-species. Pseudo-species is another neglected concept arising from ethology. An unlikely player, psychoanalyst Erik Erikson, became enamored with ethology after Freudian theory seemed lacking in ways to explain World War II. Erikson attended an ethology conference, talked with Konrad Lorenz, and came up with the thesis that all species have different sets of instinctive behaviors toward other species vs. their own, but humans are unique in their tendency to construct abstract differences that trigger out-species behavior toward a pseudo-species.

Any national, ethnic, or religious identity inevitably involves some sense of being the superior or chosen ones, Erikson observed. The dark side of this is projection of negative, inferior, or evil traits onto other groups. After inculcation with such an identity, man applies his instinctive rules about how he should behave toward his own species only to his own pseudo-species and "possessed by this combination of lethal weaponry, moral hypocrisy, and identity panic is apt to … turn on another subgroup with a ferocity generally alien to the 'social' animal world," Erikson (1968, p. 298) wrote. Obviously, this has implications for war itself, and I have written of that elsewhere (Barrett, 2010), but in this chapter, I am concerned with war films. Viewers can thrill to the hero committing supernormal amounts of bloodshed only when the blood is that of another pseudo-species (Figure 43.1).

Disaster films afford a unique window onto our coded instinctive fears. These films are never about a flu outbreak that kills 1 in 100 people (as the worst epidemic did), nor about a series of fatalities due to car crashes, nor the planetary damage that will happen from global warming over half a century. To grab our attention, a threat must be immediate and dramatic, and it is generally one that existed in the Stone Age – or at least one that can be easily extrapolated from something we feared then.

The list of disaster film titles reads as an inventory of our deeply instinctive fears: *Arachnophobia, Ticks, Empire of the Ants, Mosquito, The Swarm, The Savage Bees, Anaconda, Snakes on a Plane, Jaws, Rabid, Infested, The Rats, Killer Rats, Plague, Outbreak, Virus, The Andromeda Strain, Avalanche, Tremors, Backdraft, Tornado, Twister*, and *The Perfect Storm*. Some titles need images to connect them to a primal fear: our Stone Age ancestors might never have feared a *Meteor*, but the film's poster makes clear it is simply the largest rock that has ever come hurtling toward you.

Whether one has seen these films or just their posters, or even their titles, they stick in people's memory. The mere mention of these concepts or the image of looming teeth or swarming vermin grabs human attention. The effect may be most obvious for badly acted clunkers that no one would watch if they did not activate our alarm instincts. But consider great suspense films: Hitchcock's tales of murder, *Star Wars'* battles, or the engrossing world of *Avatar*. Accolades of "classic" or "archetypal" basically mean that these films use acting, cinematography, and

 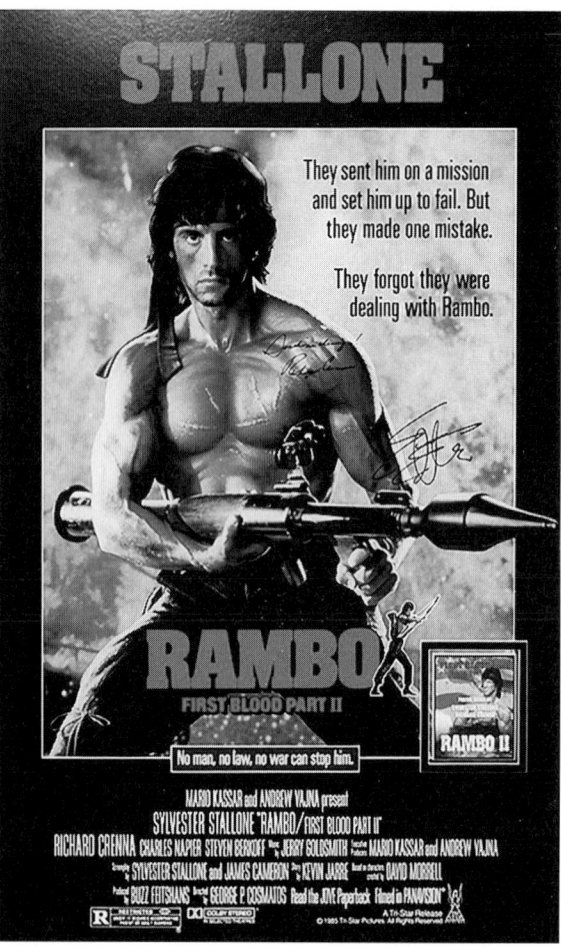

Figure 43.1 *The Seven Samurai* and *Rambo* both tell stories in which the enemy is viewed as what Erikson termed a pseudo-species.

plot twists to push our instinctive buttons more strongly than any others (Figure 43.2).

Throughout our evolutionary past, people needed to pay attention when a rabid animal was on the loose or a fire raged out of control. Any mention of such a threat triggers an impulse to "go find out more about this." Those genetically disposed to think, "Tornado ... man-eating tiger ... people dying of a mysterious disease? I don't care; I've got hunting and gathering to do," would not have survived to leave many offspring. Instincts will not let humans ignore celluloid killer bees or giant rats any more than Tinbergen's sticklebacks could disregard the red-bellied dummy intruder (Figure 43.3).

Fear lends itself at least as well as other instincts to supernormal stimulation. The shark of *Jaws* is already exaggerated in size and viciousness for a great white – and the poster gives us a more gigantic beast. But after the string of authorized *Jaws* sequels had exhausted itself, a spate of yet more supernormal sharks appeared in film. Scared of one great big mouth? How about *Attack of the Two Headed Shark*? *Monster Shark*? *Super Shark*? *Sand Shark*? *Snow Shark*? *Swamp Shark*? *Dinoshark*? *Sharktopus*? The low-budget made-for-television film *Sharknado* pulled 1.37 million viewers on its first airing. It trended on Twitter over the next few days and garnered an even larger audience later in the week (Payne, 2013). This propelled it into regular theaters and immediately kicked into production *Sharknado 2*. Meteorologists fielded calls about whether monster storms could actually carry angry great whites into cities. Recognizing that the supernormal storm was getting more attention than real ones, however, both the Red Cross and the National Weather Service (NWS) took advantage of the educational opportunity: "As with any waterspout or tornado," stated the NWS, "the best advice is to be in an interior part of the lowest floor of a sturdy building – and not outside, whether sharks are raining down or not" (Suebsaeng, 2013).

Perhaps scientists should have let the film's premise stand. The "Sharknado" was revealed to be caused by global warming. In the brilliantly titled essay "If Only Gay Sex Caused Global Warming," Daniel Gilbert (2006) pointed out that this environmental threat has not been stirring up anxiety proportional to its actual impact because we are hardwired to fear mainly events that: (1) are the products of human intention; (2) violate our moral sensibilities; (3) represent an immediate problem; and (4) appear suddenly or grow rapidly. Most horror film premises exploit multiples of these characteristics.

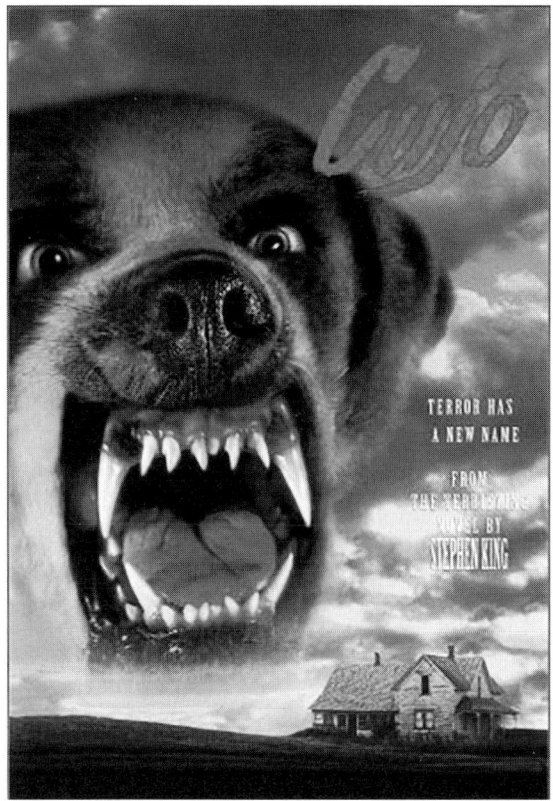

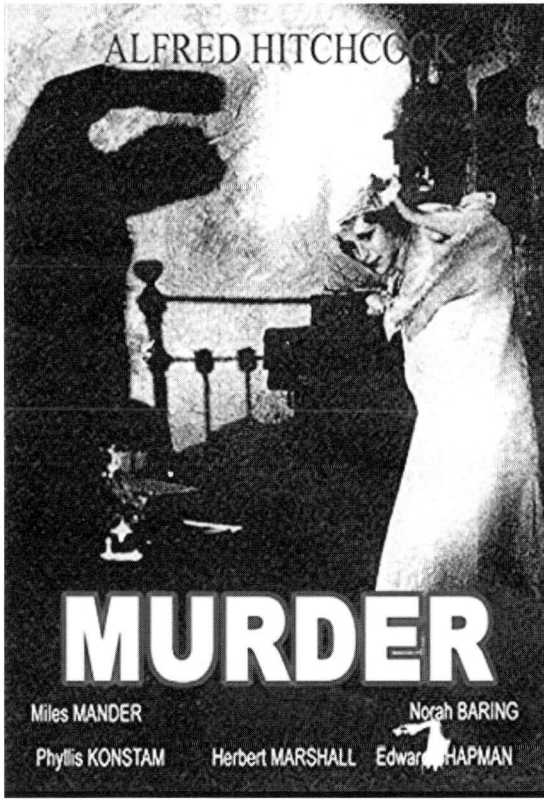

Figure 43.2 We have instinctive fears of rabid or other dangerous animals, human assailants, and deadly insects, which films play off.

Figure 43.3 When 30-foot gorillas or airborne sharks menace, we should know we are in the thrall of supernormal stimuli.

What else are King Kong, Godzilla, Frankenstein, vampires, werewolves, aliens, radioactive mutants, zombies, or the Devil if not supernormal stimuli – exaggerated versions of a menacing human or wild animal? The pull of horror films is often explained as the thrill of an adrenaline rush – indeed, getting scared sometimes has this payoff, as arousal of all types shares some common pathways. Roller coasters exploit this effect. But these films also end with the same clear victory as war films – Godzilla is defeated, the last radioactive ant is killed, and megalodon sharks are once again extinct. There is a ritual return to normality at the end.

It may not be safe to assume that all horror film audiences like the films in a simple sense, however. Some viewers experience negative aftereffects like nightmares, fear of the dark, and trouble sleeping. The supernormal fear stimulus does not necessarily make us happy – it makes us pay attention (cf. Chapter 44). As with ordinary wild animals, the mention of a genetically engineered *Tyrannosaurus rex* evokes the "go find out more" response.

43.1.2 Nonfiction Media: How Supernormal Stimuli Drive Broadcast Sports and News

In hunter–gatherer societies, most games and sports were mildly stylized competitions utilizing similar skills to those employed daily for survival, and there was a fluid boundary between audience and players (Gray, 2009). Once agriculture took hold, not everyone physically obtained their own food, built their own shelter, or made their own tools – and those that did often performed limited, repetitive movements in doing so. Most of the reward for speed and agility began to come from sports and games – purely for fun or as jostling for dominance within a group or between groups. Adults began to engage in activities that are mostly practice for children in hunter–gatherer societies: tag and races showcasing running speed and endurance (Barrett, 2007, ch. 4). Hide-and-seek provided practice for stalking prey and concealing oneself from enemies. Balls were early supernormal stimuli – capable of being thrown farther and more predictably than other projectiles and rolling more easily than the smoothest log – but games with these still developed strength, reaction times, and precision.

As population densities grew, sports quickly became spectator events – the impetus was channeled into vicarious experiences for many. Until recently, however, games often had between a 1:1 and 1:3 ratio of players to spectators. The viewers often moved around the perimeter to get a better view of the action – and they certainly walked to the arena. The next version of these activities subverted the "gaming instinct" yet further from exercise; crowds drove to huge stadiums serving beer and hot dogs. Now in the age of modern media, we sit in our home recliners using the remote to switch between two games while gobbling chips and dip.

As with fictional film and television, the vicarious experience of watching "real" sports gives viewers a false sense that they are engaging in activity and meeting challenges. A good part of the population, especially boys and men, spend hours of their week experiencing a vicarious sense of exercising while actually reclining motionlessly.

The full impact of supernormal stimuli lies not just in larger-than-life goals – what is most attention-grabbing, sexiest, cutest – it also affects an instinct to "rest when you don't need to be using energy," or what George Zipf (1949) termed the "Principle of Least Effort."

Zipf describes this in very precise mathematical terms, but he basically quantified what evolutionary psychologists have always observed: that we weigh effort primarily against instinct-driven short-term goals. Across a wide variety of endeavors, people do what is easiest rather than what we might achieve if we were not weighing conservation of effort as our ancestors did. Zipf observed that most people most of the time are turned back by modest hurdles that they know could be overcome – with effort. "To be habitual, an action must be relatively effortless or carry a particularly large psychic reward," he observed (Zipf, 1949, p. 8).

The principal of least effort has implications for all forms of media: why people may skip calling up a friend and turn on *Friends*, pass up a chance to meet a woman in order to surf porn, pass up a chance to meet a man to read a romance novel, postpone planning a real vacation while flipping on the Discovery Channel. When it comes to sports, however, least effort becomes the defining principal. There is no way the average male could have the experience of running as fast, throwing as far, or hearing the cheers that he does with televised sport's tracking shots and soundtracks. And he can do this while indulging the "rest when you don't need to move" impulse in a padded lounge chair and consuming supernormal stimuli for hunger or enjoying endogenous feel-good chemicals.

"News" is usually contrasted with "entertainment," but it is hardly a sharp distinction. Just as entertainment sometimes co-opts drives to get information, news often provides vicarious social experience. Almost everyone realizes that "news" of Amanda Bynes' drug and mental health struggles is information that they do not really need, despite instincts pushing them to learn more. Even the little boy who has fallen down a well in another state is not something the viewer is going to do anything about. Combing the planet for the day's most dramatic disasters has some of the same effect as presenting fictional ones and can be viewed as a type of supernormal stimulus itself. The intention of the media is to grab attention and boost ratings, but the indirect effect is to heighten fears about unlikely threats and to divert attention from less engaging

but more likely ones. "Flu epidemic hitting your area – flu shots have arrived" is a very, very small part of what we hear in the media.

43.2 *ANGRY BIRDS*, HAPPY COWS: THE INTERNET, GAMING CONSOLES, AND SMARTPHONE APPS

In the "Plan B" episode of *30 Rock*, as the cancelation of the fictional-television-show-within-the-real-television-show was under discussion, one character observed to another, "Our craft is dying while people are playing *Angry Birds* and poking each other on Facebook." Most media pundits also predicted that television viewing would drop as the Internet and video games expanded. However, this has not happened so far, unless one defines "television" as live broadcasts of shows viewed on old-fashioned television sets – and then it is true only for younger viewers. The 2012 Nielsen poll (*New Media*, 2012) reported that the average American over age 2 spends more than 34 hours a week watching live television, plus another 3–6 hours watching recorded programs. Their results show that average weekly viewing time has not changed much between 2008 and 2012, even though the Internet exploded during that period. Americans aged 12–34 spend somewhat less time in front of television sets. Children aged 2–11 watch an average of 24 hours of television a week, or 3.5 hours a day. That number dips to 22 hours a week for teens aged 12–17, then goes back up to 25 hours a week for 18–24-year-olds. Those aged 35 and older spent more time in front of a set in 2012 than 2008. Rates rise steadily until people over 65 averaged 48 hours of viewing television a week in 2012, or nearly 7 hours a day.

Over the same period, Internet use increased to being online for 2 hours and 35 minutes, and mobile phone use rose to 50 minutes per day (Stross, 2009). People often cannot bear to choose one screen over another – 31 percent of Internet use occurs in front of a television set. Viewers of popular sitcoms "tweet" about the show as well as texting other friends about unrelated matters as they watch. For the sports viewer in his lounge chair, there is Sports Yapper, an app where fans can post "yaps" (the equivalent of "tweets") or holler their approval in caps on a dozen other sports community sites or general ones such as Facebook.

If the Internet were replacing television rather than adding screen hours, it might be viewed as a slightly less artificial stimulus than television. While the Net has its share of canned, passive content, some of it is more interactive – and a subset of that interaction occurs with another human being, albeit mediated by technology (e.g., MMORPGs or "massively multiplayer online role-playing games," or online poker, discussed below). Some computer gaming programs have begun to require significant exercise, such as Nintendo's *Wii Sports*, with programs asking the player to swing a control standing in for a tennis racket or bat as simulated landscapes and balls respond. *Wii Fit* connects the computer to a balance board on which one performs aerobics, weight lifting, or yoga. *Dance Dance Revolution* comes with music and videos demanding a challenging sequence of steps on a footpad. These games can be played alone or as multi-player competitions.

But most popular games involve tremendous supernormal simulation of activity, with very little physical effort. "First-person shooter" games became a major industry from the 1990s onward as screens reached a level of detail allowing great realism for the player who sees through their character's eyes as they race around shooting villains. Thus, in *Wolfenstein*, the player fought Nazis (the game was banned in Germany). In *Doom*, a multiplayer level was introduced for the first time as players battled demons and the undead. In *Half-Life*, the player enacted the role of a nuclear physicist who had inadvertently opened a portal to another dimension, allowing alien monsters to flood into New Mexico. In *Halo*, players are menaced by "The Covenant" – a group of space aliens who worship an evil religion. One could read some nasty political metaphors into fighting aliens in New Mexico or doing battle with the wrong religion, but critics generally just focus on the extreme violence (Yenigun, 2013). All first-person shooter games feature the same CGI special effects of action films, but with the viewer participating. They all have a single-player mode – easier to be able to play when one wanted, especially before Wi-Fi was ubiquitous, but most since *Doom* have featured a choice between a single-player mode and a version where multiple players inhabit avatars in the same game, usually total strangers who happen to log on to the game at the same time. The intense, realistic fight scenes hook a mostly male audience, but a huge range of ages. Some of the games make use of computers' microphones, allowing players to hear each other. Grown men can find it disconcerting to shoot at a muscular hitman or science fiction monster and hear an eight-year-old crying or giggling in response – or hear someone's mother sternly summoning them away from the game. Some turn the real sound function off for this reason, preferring the synthesized soundtrack that is as supernormal as the visuals.

Smartphone games – or "entertainment apps" in the parlance of that medium – employ a different strategy. They must display on a small screen that can go anywhere. Instead of realistic, immersive detail, they derive their power from paring games down to just a few essential supernormal details. If first-person shooter games are the equivalent of a very realistic male stickleback dummy with an unusually red belly, then phone apps are the equivalent of the crudely shaped blob dummy with the fire truck red underside that nevertheless gets the maximum response. Smartphone apps' pared down cartoonish looks lend themselves to the pull of supernormal cuteness without sacrificing chasing baddies and flinging projectiles.

30 Rock singles out *Angry Birds* with good reason. It is as simple but extreme a set of stimuli as you can find. The "birds" consist of colorful ball-shaped faces with large eyes, short beaks, and rudimentary hints of bodies – protagonists designed to appeal to cuteness/nurturing instincts. The plot engages another set of instincts: the cute little birds frown; bad green pigs have stolen their eggs. The player uses a slingshot to launch the birds one by one at structures housing the thieves, toppling these to kill as many pigs as possible.

Primarily a smartphone app with spin-offs for PCs and gaming consoles, *Angry Birds* has been downloaded 1.7 billion times across all platforms. The total number of hours of *Angry Birds* playing worldwide is roughly 200 million minutes a day, which translates into 1.2 billion hours a year. For context, all person-hours spent creating and updating Wikipedia total about 100 million hours over the entire lifespan of that site (Gaudiosi, 2012). A visual comparison made in *PC Magazine*'s commentary on the game emphasizes the supernormal aspect with more gut-level impact (Yin, 2011). It shows a flock of real birds next to a larger flock of Angry Birds to illustrate that living birds on Earth total about 100 billion, while players of *Angry Birds* have already hurled more than 100 billion of the little cartoon birds. The *MIT Entrepreneurship Review* (2011) deemed *Angry Birds* "the largest mobile app success the world has seen so far."

As further testament to the birds' supernormal appeal, in January 2014, when media reported that Edward Snowden's documents revealed the National Security Agency (NSA) to be spying on people through their phone apps, none illustrated the article with images of Google Maps, which was what they actually described as giving away locations and navigation plans of social media sites that were leaking classifications of "gay" or "swinger." Instead, they all illustrated it with the cutest way NSA is spying on us: those little frowny feathered faces appeared next to headlines such as, "A Little (Angry) Bird Told NSA What You're Up To." The only app game beginning to overtake *Angry Birds* is *Candy Crush Saga*, which has garnered a limited-use trademark on the word "candy" (Kosner, 2014). *Candy Crush Saga* stimulates cuteness and food-foraging instincts at the same time; so far, it is the only mobile game to spawn an inpatient addiction treatment center dedicated solely to weaning patients from it (Van Grove, 2013).

Facebook, the other competitor for human attention that *30 Rock* decries, is interactive – with friend requests, messaging, and posting of updates and photos that users at least hope reach human eyes. Facebook is a medium of communication only in the sense that our old-fashioned dumb-phones were, not in terms of what is usually meant by "the media." I will not address those aspects in this chapter, as the role of supernormal stimuli may be less central. However, Facebook has also become a gateway into commercial entertainment media. The site itself posts advertising galore; users post links to YouTube music videos and stream episodes of their favorite television shows. Games are available for installation on Facebook home pages. These were originally lightly adapted from games in other venues, but more and more they are specifically designed for Facebook. Fifty-three percent of Facebook users have played games on the site, and 19 percent say they are "addicted." Unlike with gaming consoles, 69 percent of Facebook gamers are female. Fifty percent of Facebook log-ins are specifically to play games (All Facebook, 2010). Sixty-five percent of users play for 30 minutes or less per average social gaming session, while 27 percent of users play for 30–60 minutes, and 8 percent of users play for more than an hour (Quora, 2010).

Farmville is to Facebook what *Angry Birds* is to the smartphone. *Farmville* holds the record for the longest run as Facebook's #1 game, with 56 million people playing it at its height in 2010 (*BBC News*, 2009). It lost ground gradually, but appears reinvigorated by the recent introduction of *Farmville 2*. Unlike games in male-dominated venues, *Farmville* features no violence. Competition is still a strong theme, while the game also features cuteness and other aesthetic elements. There is a broad consensus among evolutionists that modern humans evolved on the African savanna. In my book *Supernormal Stimuli* (Barrett, 2010), I discussed how people nurture alien strains of lawn grass to keep yards from the Middle East to Scandinavia resembling savanna, install pools with bottoms painted blue to mimic especially clean water, and even buy the occasional long-legged plastic lawn bird. *Farmville* outdoes this with an intense supernormal version of our ancestral environment as a cartoon world on a little flat screen. Set in the agricultural era, it nevertheless bursts with green grass, lush fruits and vegetables to gather, and abundant animals. In the words of its maker, the game "allows you to become immersed in a vibrant 3D countryside where everything comes alive and reacts to every touch. Create, personalize and run your own farm ... beautiful trees, bountiful crops, and adorable animals!"

Farmville is a "freemium" game – the formula that more and more apps are employing whereby the basic version is a free download but the player is immediately offered upgrades for purchase. In the case of *Farmville*, upgrades include custom settings from "English Countryside" to "Hawaiian Paradise" (I would recommend "African Savannah") and decorations from topiaries to Christmas lights for the farmhouse. There are strict time limits in the freemium version in which one must harvest crops – and receive little icons showing one's abundant larder – or one will return to find brown, withered stalks in the cartoon field. Additional payment can buy players out of these tasks and avoid delays on the longer waits. There is a social component. One is largely competing for the success of one's farm, but friends on Facebook also can help you harvest crops, and you can give them gifts for their farms. The game encourages people to publicize every time you play on Facebook news feeds ("I'm harvesting my wheat") and to pester nonplaying friends to download *Farmville*. *Farmville* was named one of "The 50 Worst Inventions" in recent decades by *Time Magazine*, observing

that it consists of a "series of mindless chores on a digital farm" that somehow render it "the most addictive of Facebook games" (Fletcher, 2010).

Ian Bogost, a professor of digital media and designer of realistic, educational, and modest-selling video games (fortunately for him, he is a full-time member of the computer science faculty at Georgia Tech), was appalled by the success of *Farmville*. When Bogost was invited to speak at a New York University symposium entitled "Social Games on Trial," instead of a simple PowerPoint presentation listing the abuses in popular games, he decided to design a spoof illustrating all of the most offensive features of *Farmville*. In three days, he threw together *Cow Clickers*, which debuted at his talk. Listeners could download his game to their Facebook page and play it on the spot.

Cow Clickers was inanely simple. The game presented a square on one's Facebook page called "the pasture." In it was a static image of a cartoon cow. Players could place their cursor over the cow and "click" on it. One then heard a friendly "moo" and the player was awarded one point. Clicks in the free game were allowed only once every six hours. Players could invite up to eight friends into their "pasture"; whenever anyone within the pasture clicked on their cow, they all received a point. A leaderboard tracked the game's most prodigious clickers. The only prizes for huge numbers of clicks were goofy images of gold cowbells. There was an in-game currency, called "mooney," which players could buy with real cash and then use to buy more cows or more frequent clicks. Every single time a player clicked on a cow, an announcement – "I'm clicking a cow" – appeared on their Facebook news feed.

Bogost believed he had boiled *Farmville* down to the key components that would make its drawbacks obvious: reinforcement of action combined with partial frustration of action in the absence of anteing up real money, pestering friends to participate, and patently silly rewards. "I didn't set out to make it fun," Bogost said. "Players were supposed to recognize that clicking a cow is a ridiculous thing to want to do" (Tanz, 2011).

The audience at his talk found it hilarious and forwarded the *Cow Clickers* link to techie friends. The parody was covered by online site TechCrunch and discussed on techie chat boards. Social game critics and indie game fans delighted in generating an "I'm clicking a cow" news feed update as an ironic protest. Readers of this book can probably guess what happened next.

Even within the in-group banter, a few players mentioned that the game was actually "kinda fun." Friends outside tech critic circles saw those "I'm clicking a cow" updates and were curious. Others found *Cow Clicker* while browsing Facebook apps. Downloads shot up to 5,000, then 10,000. The new players did not treat it as a joke, but rhapsodized about how delightful the game was. They began to e-mail suggestions to Bogost: different kinds of cows would be nice. They were willing to pay for them. Bogost offered premium paisley cows and plaid cows. People bought them; player numbers climbed to 20,000. They requested vampire cows and werewolf cows. That generated more money, and player numbers rose to 56,000.

For a while, Bogost amused himself inventing parallels to mainstream games. There was a BP cow drenched in oil, a green tentacled Cowthulu, a $1,000 golden cow. He offered a $110 Cowclicktivist cow (emaciated, its ribs showing) whose purchase generated a donation to Oxfam America; this raised some money for the charity. Players still just clicked. "One thing that interested me," said Bogost, "was how many features could I add to the game without adding any gameplay?" (*Edge Magazine*, 2011). Gradually, however, he grew bored and annoyed that the bovine burlesque was taking up his time, taking up players' time, and outselling any of his serious games.

A clock appeared over each pasture with the notice that it was counting down to the "cowpocalypse." The term was not defined, but players did not like the sound of it. There was one last capitalist twist – one could pay to postpone the cowpocalypse. One dollar bought an extra hour for everyone's cows; an additional month cost $400. Repeatedly, players shelled out money to save the cows, but eventually, the clock ticked up to midnight without a ransom. All of the cows vanished from their pastures. The square box on players' Facebook pages remained, but if one clicked on it, there was no familiar "moo."

The original, ironic adopters had a good laugh. But its other players were not amused. Some did not understand that the game was over. "Hi Ian," wrote one. "I've noticed that the Cowpocalypse has happened and users have to pay to see their cow. Do you have a goal or timeframe of when this will be set back to normal?"

"There's no way to pay to see your cow," replied Bogost. "The cows got raptured."

The user threatened he would no longer play, as *Cow Clicker* was "not a very fun game" now.

"It wasn't very fun before :)," replied its creator (Alexander, 2011).

But of course it was fun. If one really wanted to design an un-fun game, you would stay away from supernormal stimuli. Bogost's cow was cute – which, in evolutionary terms, is merely the possession of traits of immaturity that trigger affection and nurturance. The little cartoon creature had almost all of the hallmarks of cuteness that Konrad Lorenz (1978/1981) had outlined: small body, large head, chubby, clumsy limbs, short snout, upturned nostrils, and a friendly sound. It also had one trait of cuteness not present in human children, but that we nevertheless sense as neotenous – floppy ears. Darwin noted in *The Origin of Species* that "not a single domestic animal can be named which has not in some country drooping ears" (Darwin, 1859, p. 11), a feature not found in any adult wild animal except the elephant, but that is extremely common among young mammals. The single significant hallmark of cuteness that the Bogost's cow does not possess is large eyes relative to its head size. It had squinty points for eyes, albeit ones that do achieve a bewildered, helpless expression. Perhaps a wide-eyed version would

have advanced the game's popularity right up there with *Farmville* and been spared the cowpocalypse.

Tinbergen's formula for identifying the key stimuli for an instinct and producing dummies exaggerating these beyond anything seen in nature will continue to determine the most successful media. It has been simply intuitive until now – producers of media are equipped with the same instincts as the rest of us. However, as the entertainment and advertising industries take a stronger interest in evolutionary psychology, the strategy will become more and more deliberate. The difficult aspect to predict is the relative appeal of the increasingly portable devices that enable the constant presence of a supernormal stimulus even while pursuing other mundane tasks vs. immersive technologies that maximize supernormal attention to themselves alone. Both are being explored aggressively. The current leader in the former category, the "smartphone," is being challenged by Google Glass – a light, wearable device with which everything on computer screens becomes transparently visible in one's upper-right visual field and controllable with voice commands. The leader in the latter category, the super-high-resolution screen, is about to be challenged by three-dimensional virtual reality goggles. Existing for two decades only as quarter-million-dollar prototypes, goggles that allow the viewer a changing view of a simulated world as they move their heads are suddenly affordable. After breaking Kickstarter records, the Oculus Rift goggles sold in late 2013 for $300 to game developers, and they are expected to be available to consumers for half that price within a year. Both early products will no doubt be challenged by numerous rivals. But whether immersive strategies, portable strategies, or some hybrid dominates the media of the next decade, content will continue to be supernormal stimuli that condense further and further all of the instinctive drives by which media capture and hold our attention.

REFERENCES

Alexander, L. (2011). The Life-Changing $20 Rightward-Facing Cow. *Kotaku*. http://kotaku.com/5846080/the-life+changing-20-rightward+facing-cow (accessed July 12, 2013).

All Facebook (2010). Graph compiled from surveys by Neilsen, InsideGames.com, and InsideFacebook.com. http://allfacebook.com/facebook-games-statistics_b19240 (accessed June 28, 2013).

Barrett, D. (2007). *Waistland*. New York: Norton.

Barrett, D. (2010). *Supernormal Stimuli*. New York: Norton.

BBC News (2009). Facebook farmers want India flag. http://news.bbc.co.uk/2/hi/south_asia/8298840.stm (accessed July 12, 2013).

Chang, J. (2014). A Little (Angry) Bird Told NSA What You're Up To. *ABC News*. http://abcnews.go.com/Technology/angry-bird-told-nsa-youre/story?id=22251583 (accepted January 27, 2014).

Darwin, C. (1859). *On The Origin of Species by Means of Natural Selection, or the Preservation of Favoured Races in the Struggle for Life*. London: John Murray.

Edge Magazine (2011). Cow Poke. www.edge-online.com/features/poking-cow-clicker (accessed June 28, 2013).

Erikson, E. (1968). *Identity: Youth and Crisis*. New York: W. W. Norton & Company.

Fletcher, D. (2010). Worst Inventions: *Farmville*. *Time Magazine*. http://content.time.com/time/specials/packages/article/0,28804,1991915_1991909_1991768,00.html (accessed July 12, 2013).

Gaudiosi, J. (2012). Rovio Execs Explain What Angry Birds Toons Channel Opens Up To Its 1.7 Billion Gamers. *Forbes.com*. www.forbes.com/sites/johngaudiosi/2013/03/11/rovio-execs-explain-what-angry-birds-toons-channel-opens-up-to-its-1-7-billion-gamers (accessed June 28, 2013).

Gilbert, D. (2006). If Only Gay Sex Caused Global Warming *Los Angeles Times*. http://articles.latimes.com/2006/jul/02/opinion/op-gilbert2 (accessed January 25, 2014).

Gray, P. (2009). Play as a foundation for hunter–gatherer social existence. *American Journal of Play*, 1(4), 476–522.

Hite, S. (1981). *The Hite Report on Male Sexuality*. New York: Alfred A. Knopf.

James, W. (1890). *Principles of Psychology*. New York: Henry Holt.

Kosner, A. (2014). Candy Crush Saga Has Trademarked "Candy" and Apple's App Store Is Helping Enforce It. *Forbes*. www.forbes.com/sites/anthonykosner/2014/01/20/candy-crush-saga-has-trademarked-candy-and-apples-app-store-is-helping-enforce-it (accessed January 25, 2014).

Lhooq, M. (2013). Will 3D Hentai Kill the Human Porn Star. *Vice*. www.vice.com/read/3d-hentai-killed-the-human-porn-star (accessed September 6, 2013).

Lorenz, K. (1978/1981). *The Foundations of Ethology*, English ed., transl. from the German ed. New York: Springer-Verlag.

MIT Entrepreneurship Review (2011). *Angry Birds* Will Be Bigger than Mickey Mouse and Mario. Is There a Success Formula for Apps? http://miter.mit.edu/article/angry-birds-will-be-bigger-mickey-mouse-and-mario-there-success-formula-apps (accessed June 28, 2013).

New Media (2012). Demographics. www.newmediatrendwatch.com/markets-by-country/17-usa/123-demographics (accessed July 12, 2013).

Pavlov, I. P. (1927). *Conditioned Reflexes: An Investigation of the Physiological Activity of the Cerebral Cortex*. Translated and edited by G. V. Anrep. Oxford: Oxford University Press.

Payne, E. (2013). Oh No, It's "*Sharknado*" and It's Ravaging Twitter. *CNN*. https://edition.cnn.com/2013/07/12/showbiz/sharknado-twitter/index.html (accessed July 12, 2013).

Quora (2010). What Percent of Facebook Users Play Games and What Percent of Time Spent on Facebook is Gaming? www.quora.com/Games/What-percent-of-Facebook-users-play-games-and-what-percent-of-time-spent-on-Facebook-is-gaming (accessed June 28, 2013).

Reeves, B., & Thorson, E. (1986). Watching television: Experiments on the viewing process. *Communication Research*, **13**, 343–361.

Rothschild, M. L., Thorson, E., Reeves, B. B., Hirsch, J. E., & Goldstein, R. (1986). EEG activity and the processing of television commercials. *Communication Research*, **13**(2), 182–220.

Salmon, C., & Symons, D. (2001). *Warrior Lovers*. London: Weidenfeild & Nicholson.

Stross, R. (2009). Why Television Still Shines in a World of Screens. *New York Times*. www.nytimes.com/2009/02/08/business/media/08digi.html (accessed August 28, 2019).

Suebsaeng, A. (2013). Can a "*Sharknado*" Really Happen? www.motherjones.com/mixed-media/2013/07/sharknado-sharknado-sharknado-sharknado-sharknado (accessed July 12, 2013).

Tanz, J. (2011). The Curse of Cow Clicker: How a Cheeky Satire Became a Videogame Hit *WIRED*. www.wired.com/magazine/2011/12/ff_cowclicker/all (accessed July 12, 2013).

Tinbergen, N. (1951). *The Study of Instinct*. Oxford: Clarendon Press.

Tinbergen, N. (1953). *The Herring Gull's World*. London: Collins.

Van Grove, J. (2013). Candy Crush: You Play, You're Hooked. Now What? *CNET*. http://news.cnet.com/8301-1023_3-57600850-93/candy-crush-you-play-youre-hooked-now-what (accessed September 6, 2013).

Vonnegut, K. (1965). *God Bless You, Mr. Rosewater*. New York: Delacorte Press.

Whitedot.org (1996). Plastic Box. http://whitedot.org/plas/1996/01/20 (accessed January 20, 1996).

Yenigun, S. (2013). Video Game Violence: Why Do We Like It, and What's It Doing to Us? *National Public Radio*. www.npr.org/2013/02/11/171698919/video-game-violence-why-do-we-like-it-and-whats-it-doing-to-us (accessed July 21, 2013).

Yin, S. (2011). Infographic: Why Is *Angry Birds* So Damn Popular? *PC Magazine*. www.pcmag.com/article2/0,2817,2392794,00.asp (accessed June 28, 2013).

Zipf, G. K. (1949). *Human Behavior and the Principle of Least Effort*. Boston, MA: Addison-Wesley.

44 An Evolutionary Approach to Horror Media

MATHIAS CLASEN AND JENS KJELDGAARD-CHRISTIANSEN

People generally try to avoid negative emotions such as fear, anxiety, and terror – and unsurprisingly so. These negative emotions evolved to motivate defensive avoidance behavior and are elicited by threatening stimuli. Common sense tells us that nobody desires to dwell on such stimuli, and yet a good chunk of the entertainment industry is devoted to manufacturing products designed to instill negative emotions in consumers. Horror cinema, horror literature, and horror video games are all thriving industries. Every season sees a new crop of novels, films, and video games depicting rotting monsters, moaning ghosts, and terrifying situations. Why do such media products draw large audiences? This chapter proposes to answer that question by invoking an evolutionary explanatory paradigm that builds on the evolutionary social sciences and encompasses sociological and historicist approaches to horror (Clasen, 2012, 2017).

Traditional sociological and cultural approaches to horror media, which are widespread in academic horror study, have failed to take into account an evolved and adapted human nature. Such approaches typically view works of horror purely as symptoms of sociocultural anxieties. For example, Skal (2001) writes about the giant mutants and other monsters populating Cold War science fiction–horror films: "Fifties monsters personified the Bomb as well as the Cold War itself" (p. 248). Symptomatic readings like Skal's contain some truth; most successful horror works resonate with culturally specific anxieties, and their monsters ooze symbolic significance. But, as we will argue here, even apparently idiosyncratic "fifties monsters" like Godzilla and giant ants target evolved psychological mechanisms. Thus, an integrative approach to horror combines attention to cultural context with attention to such evolved psychological mechanisms, specifically hazard-precaution mechanisms. Horror stories are populated by dangerous monsters and maniacs and feature highly unlikely but highly dangerous scenarios. These features do not reflect empirical reality, but rather the design specifications of human precautionary cognition.

In this chapter, we delineate the evolutionary approach to negative emotion and horror, discuss the psychological functions of horror fiction, and offer hypotheses about the adaptive functions of horror fiction. We then analyze examples of horror entertainment from video games, film, and literature to demonstrate the explanatory value of the evolutionary paradigm.

44.1 AN EVOLUTIONARY APPROACH TO HORROR

We use the phrase *horror fiction* to denote those fictional and narrative media presentations that are deliberately designed to evoke the negative emotions of fear, anxiety, and/or disgust in their audience, and we are particularly interested in supernatural horror fiction. Naturalistic horror fiction – the type of horror that features protagonists threatened by psychotic killers and other real-world dangers – is paradoxical in its own right, but supernatural horror fiction presents an additional paradox: Why would people who do not otherwise believe in supernatural agents react with genuine negative emotion to representations of supernatural monsters? Our focus in this chapter is on supernatural horror literature, film, and video games, but the theoretical paradigm we delineate could be applied to other media presentations as well, such as horror in visual art, horror theater, horror comics, and horror radio drama.

Horror fiction is affectively defined (according to audience reaction), and the target affect of horror fiction is negative emotion. In an innovative study, Davis and Javor (2004) empirically demonstrated that the most successful horror films (as measured by user ratings on the Internet Movie Database) are those that most effectively target domain-specific adaptations to danger. This result suggests that we need to understand the mechanics of negative emotion and the nature of domain-specific adaptations to danger in order to understand how and why horror fiction works. The "scariness" of a horror film, a horror story, or a horror video game is frequently used as a parameter of quality by consumers and as an explicit selling point by marketers. The tagline of David Cronenberg's 1986 film *The Fly* pithily illustrates the point: "Be afraid. Be very afraid." Horror consumers expect to be scared, just as extreme sports participants expect to feel fear when they engage in extreme sports activities. Such fear and arousal are central to the attraction of extreme sports (Brymer & Schweitzer, 2013), as they are central to horror fiction. In the following, we delineate the mechanics and evolutionary history of fear and related negative emotions by drawing on research on evolved human cognition and emotion.

44.1.1 The Nature of Fear and Anxiety

Fear is the prototypical negative emotion, and it is a human universal (Ekman, 1999; Panksepp & Biven, 2012). It is an adaptive response to imminent threat, and it is closely related to anxiety, the adaptive response to a distant or abstract threat. As Öhman (2008) argues, fear is usually associated with a fight-or-flight response, whereas anxiety is associated with probing or, alternatively, non-coping behavior. Both emotions entail calibrations of physiology (resources are directed away from irrelevant processes such as digestion and shunted toward large muscles in preparation for evasion or conflict) and cognition (attention is focused on the potential danger, and irrelevant concerns such as foraging and mating effort are ignored), and both emotions originate in "evolved mammalian defense systems" (Öhman, 2008, p. 709).

While fear may be an unwelcome distraction to many people going about their day-to-day lives, its role in evolutionary history can hardly be exaggerated. Humans, like other organisms, are adapted to the dangers common to the environments in which we evolved. Most modern people in industrialized nations have little reason to fear predation, in contrast to our evolutionary ancestors. They faced potentially lethal danger from carnivorous predators, invisible pathogens and toxins, hostile conspecifics, social exclusion, recurrent topological features such as cliffs and deep water, etc. In present-day hunter–gatherer communities, people very rarely die from sheer old age. A study looking at causes of death in seven groups of hunter–gatherers and forager–horticulturalists, comprising more than 3,000 individuals, found that only 9.5 percent of deaths were attributed to senescence; 72.4 percent died from disease, 11.0 percent from violence (including predation), and 5.2 percent from accidents (Gurven, 2012, p. 297). To the extent that present-day hunter–gatherer communities are valid proxies for ancestral human communities, we can conclude that our evolutionary ancestors led dangerous lives. Indeed, archaeological evidence shows that humans and our hominin ancestors have historically faced a very real danger of predation (Barrett, 2005; Hart & Sussman, 2009; Kruuk, 2002). Existence in such a dangerous world has exerted strong selection pressures on human danger-management adaptations.

In recent decades, scientists have adduced much evidence for evolved cognitive danger-management architecture (Barrett, 2005; Duntley, 2005; Öhman & Mineka, 2001).

This cognitive architecture, labeled the "Hazard-Precaution System" by Boyer and Liénard, is a "motivational system geared to the detection of and reaction to particular potential threats to fitness" (Boyer & Liénard, 2006, p. 1). The system is swift, automatic, biased to err on the side of caution, and designed for ontogenetic calibration. Biologically speaking, these design characteristics – hypersensitivity and swift, automatic processing – are adaptive. When the stakes are high and time is short (e.g., if one hears a strange sound in the dead of night), a false negative is vastly more costly than a false positive (Marks & Nesse, 1994). That strange sound may turn out to be the footfall of an approaching predator, in which case the fearful are much better off than those who ignore it. Over evolutionary time, natural selection favored anxious over fearless individuals, and humans today retain the tendency to react strongly to ambiguous cues of danger.

The evolutionary pedigree of the human precautionary system is also evident in the distribution of typical fear objects. People quickly acquire fears of evolutionarily recurrent, ancestrally dangerous objects and situations. The 2012 ChildFund Alliance report "Small Voices, Big Dreams," which quantified children's fears and dreams based on the responses of 5,100 individuals from 44 countries, found that the most common fear among children across developing and developed countries is the fear of "dangerous animals and insects" (ChildFund Alliance, 2012, p. 10). Even children growing up in industrialized, urban environments free of nonhuman predators easily acquire fears of dangerous animals. For example, we much more easily acquire fears of snakes and spiders than we do of cars (LoBue & DeLoache, 2008), although the latter figure much more prominently in present-day mortality statistics. According to the US National Safety Council, the lifetime odds of dying from a motor vehicle accident for a person born in 2007 were 1 in 88. In contrast, the odds of dying from contact with venomous spiders were 1 in 483,457, and the odds of dying from contact with venomous snakes or lizards were 1 in 552,522 (National Safety Council, 2011, pp. 35–36). Clearly, cars (and other evolutionarily novel hazards) are much more dangerous to contemporary citizens of the USA than are spiders and snakes, yet spider and snake fears are much more common. The danger posed by cars simply has not been around long enough to exert selection pressure on the human genotype.

Humans can acquire fears of objects by classical conditioning (learning from personal experience to associate a stimulus with an aversive outcome), by vicarious experience (by observing other humans react with fear to a stimulus), and by acquiring negative conceptual information about an object (being informed by other humans that a stimulus is dangerous) (Boyer & Bergstrom, 2011), but common to all forms of fear learning is that such learning takes place within a biologically constrained possibility space. This applies to other primates as well. For instance, Cook and Mineka (1990) demonstrated that rhesus monkeys learn to fear snakes by observing conspecifics reacting fearfully to snakes, whereas they do not acquire a fear of flowers by observing conspecifics reacting with fear to flowers in manipulated video presentations. As Marks and Nesse (1994) argued, our species' phylogeny gives rise to a "nonrandom distribution of fears ... stimuli that come to be feared are mostly ancient threats: snakes, spiders, heights, storms, thunder, lightning, darkness, blood, strangers, social scrutiny, separation, and leaving the home range. Most phobias are exaggerations of these natural fears" (p. 255). Experimental psychologists have demonstrated that such evolutionarily conserved, prepotent fear objects preferentially engage human attention. Humans are faster at detecting a snake embedded in an array of distractor stimuli, such as mushrooms and flowers, than vice versa, and humans easily acquire fears of snakes and spiders, but not of flowers and mushrooms (LoBue, 2010; LoBue & DeLoache, 2008; Öhman & Mineka, 2001; Soares et al., 2014). And proceeding from the hypothesis that ancestrally dangerous stimuli such as snakes have particular salience to humans, Penkunas and Coss (2013) used a visual search task to measure search time in detecting ancestrally dangerous animals embedded in a matrix of distractor images. Research participants spotted a snake embedded in a matrix of lizards faster than they spotted a lizard embedded in a matrix of snakes, and they spotted a lion embedded in a matrix of similarly colored antelopes faster than they spotted the antelope in a matrix of lion images. The human fear system, moreover, is wired for calibration, an adaptive feature given our species' historically nomadic lifestyle combined with the differential distribution of threats in different ecologies. Over evolutionary time, humans and our hominin ancestors have traveled the planet and faced a variety of local hazards. Given the unpredictability of risk in novel environments, a flexible fear system designed to be sensitive to input about local dangers is adaptive. In certain ecologies, children must learn to be wary of polar bears; in others, alligators. Common to such animal fears is that they originate in a predator-defense system with deep roots in mammalian evolution (Öhman, 2008). Moreover, a child from rural Romania may be scared of *strigoi* – the evil undead of Romanian folk belief – whereas an urban Baltimore child out after dark may be more afraid of gang violence. These children both fear assault, but the objects of their fears are variations on a theme; that is, the fear of conspecific violence.

Learning about danger from direct experience can be risky. As Öhman and Mineka (2001) pointed out, "if effortful trial-and-error learning was the only learning mechanism available, most animals would be dead before they knew which predators and circumstances to avoid" (p. 487). Hence, vicarious or social learning is an adaptive strategy in the domain of fear acquisition and precautionary behavior. Studies of children's nighttime fears show that such fears are overwhelmingly determined by negative information, rather than by conditioning or

observation (Muris et al., 2001). Direct exposure to a threat is not prerequisite to acquiring a fear of that threat – a scary movie or a creepy anecdote will do the trick. Adding to this picture, an emerging literature on adaptive memory shows that people more easily process and remember survival-relevant information (Nairne & Pandeirada, 2016; Nairne et al., 2012). Barrett and Broesch (2012), for example, provided evidence for the existence of an evolved social learning mechanism that causes children "to preferentially attend to and remember culturally transmitted information about danger" (p. 499). Children cross-culturally are better at remembering danger-related facts about animals than they are at remembering what these animals eat and what they are called. In addition, children's enthusiastic engagement in imaginative simulations of predator–prey relations – chase play – allows them to practice efficient evasion strategies and fear management in safe contexts (Steen & Owens, 2001). Such play behavior is a way of acquiring vital experience with danger and with one's own reactions to danger, but without substantial risk (Špinka, Newberry, & Bekoff, 2001). Historically, folktales and myths about dangerous agents probably had the function of teaching people, especially children, about dangerous animals and risky social interactions (Scalise Sugiyama & Sugiyama, 2011). Commenting on the profusion of deadly carnivores in human imaginative culture, Kruuk (2002) has argued that "there may be a survival value in this aspect of our culture. We are teaching others what is lethal in the environment, how its deadly forces work, and one might call it a cultural alarm system" (p. 179).

Horror films, video games, and literature exploit the human danger-management adaptations recounted in this section. The genre is designed to elicit negative emotions in its audience, and it does so prototypically by featuring human characters in dangerous situations involving horrible monsters. Such monsters reflect evolved fears, notably the fear of predators. Effective horror fiction introduces characters whose perspective the audience is encouraged to take, pits these characters against dangerous agents, and uses narrative and dramatic techniques to heighten suspense and facilitate the elicitation of negative emotion.

44.1.2 The Psychological Functions of Horror Fiction

Fictional narratives can serve the function of transmitting adaptively relevant information in dramatically compelling form (Scalise Sugiyama, 2001). It is not difficult to imagine how information about dangerous agents and situations could prove adaptively relevant, but the supernatural monsters of horror fiction are not the reptilian and mammalian monsters of the real world. Might our imaginative responses to unrealistic monsters still serve an adaptive function?

The desire to be scared in a safe context has outlets other than horror fiction. Practitioners of extreme sports, for example, expect intense arousal from their activities, yet do not expect to be killed in the process. Reporting on interviews with extreme sports enthusiasts, Brymer and Schweitzer (2013) note that for such practitioners, "facing fear in extreme sports and learning to participate despite the intensity of the fear facilitates the management of fears in other aspects of life," allowing them to "transcend [their] own limitations and invite new possibilities into [their] lives" (pp. 484, 485). It may be that horror consumers likewise "transcend their own limitations" and effectively test their own moxie by enduring fear- and anxiety-inducing media fare, in the process perhaps learning to manage the sometimes debilitating effect of fear on their own psyche. Whether horror fiction can have such empowering effects is an empirical question that has so far attracted no research. We do know, however, that the consumption of horror films may induce trauma – particularly in young audiences (Cantor, 2004). Research has documented spillover effects and shown that exposure to scary entertainment can adversely affect people's lives, disrupting sleep, increasing anxiety, and promoting hypervigilance (Cantor, 2004). From a phenomenological perspective, feeling compelled to check one's sleeping quarters for hidden monsters every night is undesirable, but from an evolutionary perspective, such behavior is adaptive if it improves one's chances of survival.

Fiction provides a framework for guided simulation, a virtual world in which we can get vicarious experience (Tooby & Cosmides, 2001), see the world from other people's perspectives (Mar & Oatley, 2008), bond with an imaginary community (Boyd, 2009; Gabriel & Young, 2011), and create imaginative and motivational order (Carroll, 2011; Gottschall, 2012). Stories have real effects in the real world; they shape people's worldviews and guide their behaviors (Correa et al., 2015; Gottschall, 2012; Laer et al., 2014). As Carroll (2006) has argued: "[T]he arts are a chief means through which humans organize their complex motivational dispositions and thus channel their own evolved motive dispositions into a functional program of behavior … The distinguishing characteristic of literature is that it creates an imaginative order in which simulated experience can take place" (pp. 43–44). Building on Carroll's hypothesis that the imaginative arts fulfill the adaptive function of providing a stimulus to psychological orientation and organization, we argue that horror stories can provide us with formalized worst-case scenarios and furnish our imagination of danger with concrete, emotionally saturated images. When we cognitively model the experience of being in great danger, most of us probably draw on imagery and emotions provided to us by horror stories. If asked what it might feel like to be lost and preyed upon in a hostile natural environment, many of us would probably recollect the film The Blair Witch Project (1999), which provided an engaging simulation of that particular scenario. As Woody and Szechtman (2011) suggested, "rare, high-consequence events [are] inherently unpredictable because they lie outside the realms of usual observation … even close observation of what happens in normal circumstances may provide very little information about what happens rarely, as outliers." Such "rare, high-consequence events" are thus "very difficult to map cognitively" (p. 1019), but fiction can map them vicariously. If we have read Richard Matheson's I Am Legend (1954), we know, subjectively and emotionally, what it feels like to be assaulted on all sides by hostile creatures and to be all alone in a dangerous world (Clasen, 2010). All effective horror exploits evolved danger-management adaptations and elicits physiological and cognitive arousal, which by itself is pleasurable to some individuals. While most commercial horror stories and games probably function as forgettable and adaptively useless stimulation technologies, some few works, including Matheson's genre-defining vampire story, allow for imaginative and emotional simulation and thus give us vicarious experience with the terrifying. Such vicarious experience may be adaptive. It may allow for a kind of implicit learning – a nonpropositional "knowing how" (Ryle, 1946) – that lets consumers of horror fiction make better choices under conditions of threat and uncertainty (Clasen, 2012; Nairne & Pandeirada, 2016). As mentioned, the extent to which horror media will prove adaptive in this way is an unexplored empirical question. We think it merits experimental investigation, which an evolutionary perspective could frame and direct.

Horror stories and horror video games need not be realistic to engage and sustain people's interest or to exert shaping force on their imaginative universes. Psychologists studying negative emotions have found clips from horror films to be efficient and ecologically valid elicitors of fear (Gross & Levenson, 1995). The neural hardware activated by horror fiction overlaps with the neural hardware responsible for handling actual threats in the real world, and the resulting fear is anything but fake. In Section 44.2, we provide examples of the way horror across media targets evolved hazard-precaution systems and analyze the function of horror in films, video games, and literature.

44.2 THREE CASES: FILM, VIDEO GAME, AND LITERATURE

44.2.1 Horror Film: *Halloween*

Horror is one of the most persistent film genres (Prince, 2004), never going out of fashion and consistently luring audiences with morbid spectacles and characters in great peril. One of the most well-known modern horror films, John Carpenter's 1978 picture *Halloween*, tells the surpassingly bleak and simple story of Michael Myers (Will Sandin, age six), a psychotic child who, at the age of six, kills his teenage sister with a kitchen knife. After 15 years of incarceration, Myers (Tony Moran and Nick Castle, age 21) escapes and returns on Halloween night to his hometown, where he lurks around before starting to kill off a small group of teenaged female babysitters. Of the three babysitters, only Laurie Strode (Jamie Lee Curtis) manages to survive. The film was shot in 21 days and on a very low budget (Rockoff, 2002), but quickly became one of the most financially successful independent productions of the 1970s (Newman, 2011, p. 198). The film has so far spawned seven sequels and two remakes, and it generated a wave of derivative slasher films featuring psychotic male killers stalking innocent victims, usually female teenagers (Rockoff, 2002). Even people who have not seen any of the *Halloween* films recognize the name Michael Myers and the characteristic costume consisting of a blue jumpsuit and a white, expressionless mask. The film and its antagonist have become cultural attractors; something about the film connected with moviegoers at the time of its release and continues to resonate with audiences today.

When adult Michael Myers arrives in his hometown of Haddonfield, Illinois, most characters are unaware of the danger he poses. The flow of story information is organized so that the audience frequently has access to more information than the characters, yet not the whole range of possible knowledge. This is a technique commonly used in thrillers and horror films to build suspense. For example, in one scene we see Annie (Nancy Loomis) talking on the phone with her back to a glass door. We notice the sudden appearance of Michael Myers behind the door, but Annie cannot see him and keeps chatting in a carefree manner into the phone. We in the audience know that Myers is lurking in the background and we know that Annie does not realize the imminent danger, so we feel anxiety about the fate of this character. Laurie Strode is the first major character to notice Michael Myers in Haddonfield (and the only survivor of his babysitter killing spree). She sees him standing on a pavement observing her, and then slipping behind a bush. Laurie's friend, Annie, fails to see Myers. She dismisses Laurie's anxiety, saying, "You're wacko, now you're seeing men behind bushes." Awareness of danger is a primary theme in this film. Pennington (2009) notes, "Usually in slasher movies, ignorance of danger diminishes a character's chances for survival" (p. 59). He rightly identifies this pattern as a genre convention, but the convention springs from actual ancestral patterns of predator–prey relations. Ignorance of danger reliably diminishes organisms' chances for survival, and *Halloween* is structured around this basic fact (Clasen, 2017).

The predator in *Halloween*, Michael Myers, is recognizably human, yet simultaneously sub- and super-human. Myers is ascribed no intelligible motivation, no remorse, and no reason. He is a human predator with whom one cannot reason or bargain. Moreover, he appears impossible to kill. During the course of the film, Myers is stabbed with a sewing needle, a steel coat hanger, and a kitchen knife, and he is shot six times at close range, falling wounded from a second-story balcony, yet he still keeps going. The film gives us no insight into Myers' mind and offers no clues to his motivation. This is characteristic of stories about evil monsters: the audience is denied access to the monster's perspective (Clasen, 2014; Kjeldgaard-Christiansen, 2016), which allows for a framing of the monster as an agent of unmotivated destruction. The signature mask worn by Myers further inhibits sympathy and understanding. The result is deeply unsettling and contributes to the characterization of Myers as a subhuman, unstoppable, invincible force of nature.

Fictional representations of antagonistic, homicidal humans such as those found in slasher movies exploit an adaptive fear of hostile strangers. As Joshua Duntley (2005) has argued, over the course of human evolutionary history other humans have been "one of the most pervasive hostile forces of nature" (p. 224). Duntley provides evidence for the existence of anti-homicide adaptations, such as cognitive biases that "lead people to overinfer homicidal intent in others" so that they "systematically overestimate the likelihood that they will be killed" (p. 241), particularly in situations of great uncertainty. This is another instance of the paradigmatic rule of survival in an uncertain world: a false positive is vastly less costly than a false negative. The selection pressure behind this logic has led to the design of a perceptual mechanism that makes people faster at detecting schematic representations of angry-looking faces than they are at detecting happy-looking, sad-looking, or neutral faces in arrays of distractor images (Öhman, Lundqvist, & Esteves, 2001), presumably because rapid detection of hostile conspecifics can mean the difference between life and death. Slasher films exploit such adaptive machinery by providing dramatic depictions of ancestrally dangerous situations involving predatory humans.

Halloween is designed to engage adaptive hazard-precaution mechanisms. The film features a dangerous, relentless antagonist and dwells on terrified protagonists in reaction shots that facilitate emotional contagion and perspective-taking. We feel their fear and fear for them when they do not realize the danger they are in. The film uses formal techniques to sustain emotional engagement and generate apprehension and suspense. For example, several shots are framed in such a way that the audience sees Myers standing in the foreground of a shot, watching a character who is unaware of his presence in the background. We know that Myers sees a potential victim and we know that the victim is unaware that she is being watched, and this prompts us to feel suspense and apprehension. The film ends with a montage, showing static images of places that have been haunted by Michael Myers and overlaid with the sound of Myers' breathing, suggesting that not only is he still alive, but he could be anywhere. The montage has the function of emphasizing the film's thematic focus on the intrusion of a homicidal monstrosity into pleasant, well-ordered, and recognizable suburbia (Murphy, 2009, pp. 142–146). It is thus in line with the overall aim of the film, which is to scare and unsettle viewers. In film critic Kim Newman's (2011) interpretation, the film's "only message is 'boo!'" (p. 201). Feminist and psychoanalytical critics, in contrast, have interpreted Myers as a moral agent of conservative gender politics and/or a castrating phallus-wielding symbol (Jones, 2002, pp. 114–120). Yet John Carpenter, the film's director, sees another significance in *Halloween*, one that accords with an evolutionary understanding of horror film as threat-simulation devices: "If there's any point to be made in the film, it's that you *can* survive the night … being aware of the possibility of evil is an important thing in life … the world can be bad and dark and danger-ous, but with a little luck and awareness you can survive" (Carpenter, 2003).

44.2.2 Horror Video Game: *Slender: The Eight Pages*

Horror video games have existed for almost as long as video games have existed and have their origins in early text-only games of the 1980s (Perron, 2009). Horror video games adopt many of the formal elements of horror films, but they allow for a more immersive experience, an experi-ence over which the audience has greater control (Krzywinska, 2002; Lynch & Martins, 2015). The type of horror video game known as "survival horror" (Perron, 2009) is exceedingly popular. Such games situate players in virtual environments teeming with threats and motivate them to negotiate these threats. As Rouse III (2009) observed, in such games, "death would no longer be some-thing happening to someone else [as in the movies], but instead to you, the player" (p. 15). And death, in survival horror games, has more immediate consequences, as it forces the player "to replay a section of the game in order to progress, giving death real stakes, unlike a movie, where the plot will keep going no matter who dies" (p. 20). Horror video games position the player in an intensely hostile virtual environment with a clear objective (basically, to survive in the game, but often also to acquire knowledge and/or some kind of in-game currency). Many horror video games use a first-person perspective to facilitate emotional engagement with the game avatar (the virtual character controlled by the player). Recurring elements include dark or otherwise obscured surroundings, an ominous backs-tory, and grotesque, hostile monsters (Lynch & Martins, 2015). One popular game is titled *Alone in the Dark* (Eden Games/Hydravision Entertainment, 2008). For an ultra-social species that depends on conspecifics and social alli-ances for survival (Dunbar, 2004) and that moreover is vulnerable in the dark because of poor night vision, the experience alluded to in this game's title is likely to gen-erate anxiety.

Horror video games generally promote and reward *vig-ilance* – the player is rewarded for detecting subtle cues of danger and for reacting swiftly and appropriately to these cues. They also generally promote and reward *persistence* in the face of fear- and anxiety-inducing circumstances – the player is not only rewarded for killing enemies and

completing the game, but is also challenged to solve puz-zles, explore danger-filled locales, and overcome obstacles along the way. In the beginning of the action horror game *Resident Evil 4* (Capcom Production/Studio 4, 2005), for example, the player traverses lonely woods and encounters desolate countryside in which rabid, zombie-like villagers (infected with a parasitic organism) attack with primitive weapons. The player must kill these villagers, after which he or she is free to walk around the village, looking for caches of treasure that will increase the odds of survival. In many horror games, the player starts out with a knowledge deficit and is driven by a narrative desire to reconstruct the backstory. For example, the independent hit game *Amnesia: The Dark Descent* (Frictional Games, 2010) pits the player's avatar against horrible, insanity-inducing monsters in a decrepit, dark castle in a quest to figure out who the avatar is and what has happened to put him in the castle. Thus, horror video games trigger evolved hazard-precaution mechanisms in the players, but are structured to override avoidance tendencies by rewarding the player for persistence and vigilance (Clasen & Kjeldgaard-Christiansen, 2016). Horror video games let players exer-cise their survival skills, which is pleasurable for the same reason that hide-and-seek is pleasurable to children and play fighting is pleasurable to kittens: natural selection tends to make adaptive behavior rewarding, and play behavior provides low-cost and low-risk experience with critical situations (Špinka et al., 2001; Steen & Owens, 2001).

The 2012 video game *Slender: The Eight Pages* (Parsec Productions) exemplifies salient features of horror video games. *Slender: The Eight Pages* exploits the urban folklore of "Slender Man," a monstrous humanoid invented in 2009 by Internet user Victor Surge in response to a challenge to create "paranormal images" with the aid of image manipulation software such as Photoshop (Chess, 2011). Surge initially produced two images featuring Slender Man. In the first image – a photograph of children playing happily in a playground – an abnormally tall and slender person lurks in the background, partly obscured by shadows. The person appears to have tentacles for arms and no distinct facial features. In the second image, a group of anxious-looking children seems to be fleeing from something. In the background, indistinctly, lurks the tall, faceless Slender Man. These images were quickly and widely circulated and inspired a profusion of user-generated "fakelore" surrounding Slender Man, as well as additional manipulated images featuring the character, and eventually video games (Yarish, 2013).

Slender: The Eight Pages features simple gameplay. Experiencing the game world from a first-person perspec-tive, the player is situated in a wood at night, equipped only with a flashlight with limited battery life. The player's objective is to collect eight pieces of paper that are scat-tered around the woods. Somewhere in the woods is Slender Man, a hostile entity who is stalking the player. Looking at Slender Man for more than a brief moment will cause the player's avatar to go insane, at which point the game ends. The player is forced to periodically switch off the flashlight to conserve battery life, and at any rate the flashlight provides limited illumination. As the player progresses in the game, the likelihood of encountering Slender Man increases. Slender Man's proximity is signaled by static-like white noise on the screen, which is a cue that the player must react quickly to avoid fatal contact.

Slender: The Eight Pages enacts a primal scene of great danger. The player's avatar is alone, at night, in an environment that is difficult to navigate, hunted by a highly dangerous agent. The game heightens the tension by loading on acousmatic sounds that keep the player guessing. At one point, faint, ominous, low-pitched booming sounds come at regular intervals, but their source is withheld. The player's general uncertainty is intensified by the suggestive drawings on the pages the player is trying to collect. On one page, we see the figure of Slender Man standing among trees that look like simple 2D renditions of those that feature in 3D in the player's virtual environment. Slender Man is almost fully camouflaged in the drawing – his long, slim body looks like another tree silhouette. The message is clear: Slender Man is out there, lurking in the dark, and you should pay close attention to your surroundings if you want to make it out alive. The combination of visual cues to danger and ambiguous sounds used in *Slender: The Eight Pages* induces anxiety and vigilance (Garner & Grimshaw 2011), and it potentiates the startle that inevitably results when Slender Man appears.

The gameplay of *Slender: The Eight Pages* hardly mirrors most people's ordinary experience, and Slender Man himself is not realistic. Yet the game engages evolved mechanisms for survival in dangerous environments. It is designed to reward vigilance, as the player cannot complete the game without paying close attention to the environment, and to reward persistence by setting up the subgoal of collecting the eight pages. The chief function of the game is to scare players, and players are attracted to the game precisely to experience fear. Survival horror games like *Slender: The Eight Pages* exemplify how horror entertainment provides intense negative emotional arousal and that this emotional experience is a primary attraction of the genre (Lynch & Martins, 2015). Such games offer little by way of narrative content, but they do allow players to simulate, with gut-wrenching conviction, the primal experience of being hunted by predators.

44.2.3 Horror Literature: "The Mist"

Just as horror video games have been around about since video games were popularized and as horror films are about as old as the medium itself, people have probably told each other made-up horror stories ever since they acquired the ability to give verbal shape to their worst fears and most terrible imaginings. With the rise of the novel in the eighteenth century, so-called Gothic romances – melodramatic horror stories – became viable as an art form and commercially successful (Botting, 1996). The popularity of horror literature has fluctuated ever since, never entirely going out of fashion. In modern times, the American author Stephen King has established himself as a master of the form. King entered the market in 1974 with the novel *Carrie* and is now one of the best-selling authors alive.

Written in 1976 and published in 1980, King's novella "The Mist" is one of his most well-known stories. "The Mist" is told from the first-person perspective of a young man named David Drayton, who lives in rural western Maine with his wife and five-year-old son. A freak summer storm hits the little town in which they live, cutting off electricity and landline telephones. The storm is followed by the encroachment of an unnaturally dense, white mist. Drayton and his son drive to a nearby supermarket to shop for supplies. Soon the impenetrable mist descends on the supermarket, trapping the shoppers inside. It turns out that horrible, predatory monsters cavort in the mist. Several characters are killed by the monsters in their attempt to escape from the supermarket. The people trapped inside begin to form antagonistic groups. One group decides to try to escape the supermarket in a car. Drayton and his son join this group. They drive carefully through the mist, heading south, meeting no people and seeing a few gigantic monsters moving through the mist on their way. The story ends on a bleak note, with Drayton explaining that the refugees have sought shelter in a hotel for the night. The reader is not told what becomes of the protagonists.

King places his characters in a situation fraught with danger and uncertainty and describes the horror of such a situation through the eyes of his narrator. The unnatural mist itself is anxiety-provoking to the narrator (King, 1980, p. 538), and King vividly evokes the horror of finding oneself in a visually impenetrable environment with hostile creatures lurking about. As the narrator puts it, "when I cracked the [car] window I could hear them in the woods, crashing and blundering about ... Overhead the mist darkened momentarily as some nightmarish and half-seen living kite overflew us" (p. 546). The monsters in the mist are characterized as predatory and dangerous. Several of these monsters evoke ancestral threats. The first visual description of one such squid-like monster subtly evokes a serpent, an ancient threat to ourselves and our mammalian ancestors (Isbell, 2006): "A tentacle came over the far lip of the concrete loading platform and grabbed Norm around the calf ... The tentacle tapered from a thickness of a foot – the size of a grass snake ... to a thickness of maybe four or five feet where it disappeared into the mist" (King, 1980, p. 466). Less noxious – but still disgusting – are the giant fly-like creatures swarming around in the mist; they are "two feet long, segmented, the pinkish color of burned flesh that has healed over" (p. 502). There are also flying reptilian monsters with "red eyes" and "heavy, hooked beak[s]" and a taste for human flesh (p. 504), and there are highly aggressive spider-like creatures – "so like the death-black spiders . . . in the shadows of our boathouse," except these spiders are "the size of a big dog" (p. 530). There are even bigger – and presumably more terrifying – monsters hiding in the mist, ones that go undescribed. The narrator describes the sound these monsters emit: "It was the sound of a big animal . . . low and tearing and savage" (p. 498). The fictional monsters in "The Mist," like so many horror monsters, are exaggerated versions of known predators. They function as supernormal stimuli, intensifications of evolutionarily relevant threats (Barrett, 2010; Clasen, 2012). Zoologically implausible yet psychologically compelling, they inspire real fear because they were designed to meet the input specifications of our evolved danger-management system.

The cause of the strange mist and the uncanny monsters is never explained in the story, but King is not so much interested in the cataclysm as in its social and psychological consequences. As King has stated on several occasions, "I'm . . . interested by ordinary people in extraordinary situations" (King, 2010, p. 365). King sets up an anxiety-provoking situation in his story and introduces dangerous monsters preying on protagonists to exploit our propensity to focalize and fear predators. He does so not because he wants to teach his readers about the morphology and eating habits of made-up monsters, but because he wants to tell an emotionally engaging story about human relations and human psychology. The story is about disaster bringing out the best and the worst in people, and by inviting us to share the perspective of his protagonists, King lets us experience their attempts to navigate dangerous physical and social environments. That form of emotionally saturated vicarious experience is a primary attraction of horror fiction, and the evolutionary approach explains why people are motivated to seek out such experiences.

"The Mist" is a horror story featuring outlandish monsters and implausible events, but the meaning of the story goes beyond mere B-movie depictions of novelty monsters. Readers take away a sense of what it would feel like to be trapped in confined quarters with both helpful and antagonistic people inside and dangerous predators outside; a sense of what, in a dangerous situation, would be good and bad, of how to handle such situations psychologically and behaviorally. As Joseph Carroll (2006) has argued, one of the chief functions of literature is to provide "paradigmatic and emotionally saturated images of the world and of human experience . . . it is through these images that people come to understand the emotional quality of the motives available to them" (pp. 42–43). The human appetite for made-up stories about made-up people and events is built into our nature precisely because such stories can help us navigate the world. Fiction does not need to be faithfully realistic to be a useful source of the emotionally saturated images that Carroll talks about. Supernatural horror fiction does use patently unrealistic or implausible props to engage readers, but the outlandish quality of such props should not blind us to the fact that serious horror fiction is about people and their perceptions of the world, about what makes people tick, and about the quality of human relationships and the qualities of extreme experience in an indifferent or even hostile world.

44.3 CONCLUSION

Fear and anxiety are crucial parts of our basic biological construction, and avoiding danger is a basic human motivation. At the same time, humans are drawn to situations that give them experience with danger in a safe context. Organisms need to learn effective coping strategies to survive in dangerous environments. Horror in different media can serve as venues for such vicarious learning, allowing consumers to immerse themselves in fictional universes teeming with danger and either test various behavioral strategies (in an interactive entertainment) or observe while fictional agents test behavioral strategies for them. The functions of horror range from the most primitive arousal activation produced by simple jump scares to providing emotionally saturated images of phenomenal experience in a dangerous world and thus shaping the way we see the world. Horror fiction, then, can teach us to cope. It does this partly by eliciting our own fear toward represented events and partly by our witnessing fictional agents' coping behaviors and their effectiveness. A film like *Halloween*, for instance, teaches viewers that awareness is paramount in a dangerous environment. A story like "The Mist" immerses readers in a hostile physical and social environment and invites them to consider strategies for navigating that environment. A video game like *Slender: The Eight Pages* teaches players to focus and persist in the face of danger. Horror across media generally facilitates perspective-taking with one or more characters in great peril and allows for mentally simulating dangerous events. The payoff is that we get to experience terrible situations with no real risk. Horror in video games, cinema, and literature works because we humans are constructed the way we are – and understanding why we are constructed that way requires an evolutionary perspective. That is why an evolutionary approach is indispensable in accounting for the forms and functions of horror.

REFERENCES

Barrett, D. (2010). *Supernormal Stimuli: How Primal Urges Overran Their Evolutionary Purpose*, 1st ed. New York: W. W. Norton & Co.

Barrett, H. C. (2005). Adaptations to predators and prey. In D. M. Buss, ed., *The Handbook of Evolutionary Psychology*. New York: Wiley, pp. 200–223.

Barrett, H. C., & Broesch, J. (2012). Prepared social learning about dangerous animals in children. *Evolution and Human Behavior*, **33**(5), 499–508.

Botting, F. (1996). *Gothic*. London: Routledge.

Boyd, B. (2009). *On the Origin of Stories: Evolution, Cognition, and Fiction*. Cambridge, MA: Belknap Press.

Boyer, P., & Bergstrom, B. (2011). Threat-detection in child development: An evolutionary perspective. *Neuroscience and Biobehavioral Reviews*, **35**(4), 1034–1041.

Boyer, P., & Liénard, P. (2006). Why ritualized behavior? Precaution systems and action parsing in developmental, pathological and cultural rituals. *Behavioral and Brain Sciences*, **29**(6), 595–613; discussion 613–650.

Brymer, E., & Schweitzer, R. (2013). Extreme sports are good for your health: A phenomenological understanding of fear and anxiety in extreme sport. *Journal of Health Psychology*, **18**(4), 477–487.

Cantor, J. (2004). "I'll never have a clown in my house" – Why movie horror lives on. *Poetics Today*, **25**(2), 283–304.

Carpenter, J., writer (2003). Audio commentary with writer/director John Carpenter [DVD], *Halloween*. Beverley Hills, CA: Anchor Bay Entertainment.

Carroll, J. (2006). The human revolution and the adaptive function of literature. *Philosophy and Literature*, **30**(1), 33–49.

Carroll, J. (2011). *Reading Human Nature: Literary Darwinism in Theory and Practice*. Albany, NY: State University of New York Press.

Chess, S. (2011). Open-sourcing horror. *Information, Communication & Society*, **15**(3), 374–393.

ChildFund Alliance (2012). *Small Voices, Big Dreams 2012: A Global Survey of Children's Hopes, Aspirations and Fears*. Richmond, VA: ChildFund Alliance.

Clasen, M. (2010). Vampire apocalypse: A biocultural critique of Richard Matheson's *I Am Legend*. *Philosophy and Literature*, **34**(2), 313–328.

Clasen, M. (2012). Monsters evolve: A biocultural approach to horror stories. *Review of General Psychology*, **16**(2), 222–229.

Clasen, M. (2014). Evil monsters in horror fiction: An evolutionary perspective on form and function. In J. Pennington & S. M. Packer, eds., *A History of Evil in Popular Culture: What Hannibal Lecter, Stephen King, and Vampires Reveal about America*. Santa Barbara, CA: ABC-CLIO/Praeger, pp. 39–47.

Clasen, M. (2017). *Why Horror Seduces*. New York: Oxford University Press.

Clasen, M., & Kjeldgaard-Christiansen, J. (2016). A consilient approach to horror video games: Challenges and opportunities. *Academic Quarter*, **13**, 128–143.

Cook, M., & Mineka, S. (1990). Selective associations in the observational conditioning of fear in rhesus monkeys. *Journal of Experimental Psychology: Animal Behavior Processes*, **16**(4), 372–389.

Correa, K. A., Stone, B. T., Stikic, M., Johnson, R. R., & Berka, C. (2015). Characterizing donation behavior from psychophysiological indices of narrative experience. *Frontiers in Neuroscience*, **9**, 301.

Davis, H., & Javor, A. (2004). Religion, death and horror movies: Some striking evolutionary parallels. *Evolution and Cognition*, **10**(1), 11–18.

Dunbar, R. I. M. (2004). *The Human Story: A New History of Mankind's Evolution*. London: Faber and Faber.

Duntley, J. D. (2005). Adaptations to dangers from other humans. In D. M. Buss, ed., *The Handbook of Evolutionary Psychology*. New York: Wiley, pp. 224–249.

Ekman, P. (1999). Basic emotions. In T. Dalgleish & M. Power, eds., *Handbook of Cognition and Emotion*. Chichester: John Wiley & Sons, pp. 45–60.

Gabriel, S., & Young, A. F. (2011). Becoming a vampire without being bitten: The narrative collective-assimilation hypothesis. *Psychological Science*, **22**(8), 990–994.

Garner, T. A., & Grimshaw, M. N. (2011). A climate of fear: Considerations for designing a virtual acoustic ecology of fear. In *AM '11: Proceedings of the 6th Audio Mostly Conference: A Conference on Interaction with Sound*. New York: Association for Computing Machinery, pp. 31–38.

Gottschall, J. (2012). *The Storytelling Animal: How Stories Make Us Human*. Boston, MA: Houghton Mifflin Harcourt.

Gross, J. J., & Levenson, R. W. (1995). Emotion elicitation using films. *Cognition & Emotion*, **9**(1), 87–108.

Gurven, M. (2012). Human survival and life history in evolutionary perspective. In J. C. Mitani, J. Call, P. M. Kappeler, R. A. Palombit, & J. B. Silk, eds., *The Evolution of Primate Societies*. Chicago, IL: University of Chicago Press, pp. 293–314.

Hart, D., & Sussman, R. W. (2009). *Man the Hunted: Primates, Predators, and Human Evolution*, expanded ed. Boulder, CO: Westview Press.

Isbell, L. A. (2006). Snakes as agents of evolutionary change in primate brains. *Journal of Human Evolution*, **51**(1), 1–35.

Jones, D. (2002). *Horror: A Thematic History in Fiction and Film*. London: Arnold.

King, S. (1980). The Mist. In K. McCauley, ed., *Dark Forces*. London: Futura, pp. 419–551.

King, S. (2010). *Full Dark, No Stars*, 1st Scribner hardcover ed. New York: Scribner.

Kjeldgaard-Christiansen, J. (2016). Evil origins: A Darwinian genealogy of the popcultural villain. *Evolutionary Behavioral Sciences*, **10**(2), 109–122.

Kruuk, H. (2002). *Hunter and Hunted: Relationships between Carnivores and People*. Cambridge, UK: Cambridge University Press.

Krzywinska, T. (2002). Hands-on horror. In G. King & T. Krzywinska, eds., *ScreenPlay: Cinema/Videogames/Interfaces*. London: Wallflower, pp. 206–223.

Laer, T. V., Ruyter, K. D., Visconti, L. M., & Wetzels, M. (2014). The extended transportation-imagery model: A meta-analysis of the antecedents and consequences of consumers' narrative transportation. *Journal of Consumer Research*, **40**(5), 797–817.

LoBue, V. (2010). And along came a spider: An attentional bias for the detection of spiders in young children and adults. *Journal of Experimental Child Psychology*, **107**(1), 59–66.

LoBue, V., & DeLoache, J. S. (2008). Detecting the snake in the grass: Attention to fear-relevant stimuli by adults and young children. *Psychological Science*, **19**(3), 284–289.

Lynch, T., & Martin, N. (2015). Nothing to fear? An analysis of college students' fear experiences with video games. *Journal of Broadcasting and Electronic Media*, **59**(2), 298–317.

Mar, R. A., & Oatley, K. (2008). The function of fiction is the abstraction and simulation of social experience. *Perspectives on Psychological Science*, **3**(3), 173–192.

Marks, I. M., & Nesse, R. M. (1994). Fear and fitness: An evolutionary analysis of anxiety disorders. *Ethology and Sociobiology*, **15**(5–6), 247–261.

Matheson, R. (1954). *I Am Legend*. New York: Fawcett Publications.

Muris, P., Merckelbach, H., Ollendick, T. H., King, N. J., & Bogie, N. (2001). Children's nighttime fears: Parent-child ratings of frequency, content, origins, coping behaviors and severity. *Behaviour Research and Therapy*, **39**(1), 13–28.

Murphy, B. M. (2009). *The Suburban Gothic in American Popular Culture*. New York: Palgrave Macmillan.

Nairne, J. S., & Pandeirada, J. N. (2016). Adaptive memory: The evolutionary significance of survival processing. *Perspectives on Psychological Science*, **11**(4), 496–511.

Nairne, J. S., Vanarsdall, J. E., Pandeirada, J. N., & Blunt, J. R. (2012). Adaptive memory: Enhanced location memory after survival processing. *Journal of Experimental Psychology: Learning, Memory, and Cognition*, **38**(2), 495–501.

National Safety Council (2011). *Injury Facts*, 2011 ed. Itasca, IL: National Safety Council.

Newman, K. (2011). *Nightmare Movies: Horror on Screen since the 1960s*, revised and updated ed. London: Bloomsbury Publishing.

Öhman, A. (2008). Fear and anxiety: Overlaps and dissociations. In M. Lewis, J. M. Haviland-Jones, & L. F. Barrett, eds., *Handbook of Emotions*, 3rd ed. New York: Guilford Press, pp. 709–729.

Öhman, A., & Mineka, S. (2001). Fears, phobias, and preparedness: Toward an evolved module of fear and fear learning. *Psychological Review*, **108**(3), 483–522.

Öhman, A., Lundqvist, D., & Esteves, F. (2001). The face in the crowd revisited: A threat advantage with schematic stimuli. *Journal of Personality and Social Psychology*, **80**(3), 381–396.

Panksepp, J., & Biven, L. (2012). *The Archaeology of Mind: Neuroevolutionary Origins of Human Emotions*, 1st ed. New York: W. W. Norton & Co.

Penkunas, M. J., & Coss, R. G. (2013). Rapid detection of visually provocative animals by preschool children and adults. *Journal of Experimental Child Psychology*, **114**(4), 522–536.

Pennington, J. (2009). The good, the bad, and Halloween: A sociocultural analysis of John Carpenter's slasher. *P.O.V.*, **28**, 54–63.

Perron, B. (2009). *Horror Video Games: Essays on the Fusion of Fear and Play*. Jefferson, NC: McFarland & Co.

Prince, S. (2004). Introduction: The dark genre and its paradoxes. In S. Prince, ed., *The Horror Film*. New Brunswick, NJ: Rutgers University Press, pp. 1–11.

Rockoff, A. (2002). *Going to Pieces: The Rise and Fall of the Slasher Film, 1978–1986*. Jefferson, NC: McFarland & Co.

Rouse III, R. (2009). Match made in hell: The inevitable success of the horror genre in video games. In B. Perron, ed., *Horror Video Games: Essays on the Fusion of Fear and Play*. Jefferson, NC: McFarland & Co., pp. 15–25.

Ryle, G. (1946). Knowing how and knowing that: The presidential address. *Proceedings of the Aristotelian Society*, **46**(1), 1–16.

Scalise Sugiyama, M. (2001). Food, foragers, and folklore: The role of narrative in human subsistence. *Evolution and Human Behavior*, **22**(4), 221–240.

Scalise Sugiyama, M., & Sugiyama, L. S. (2011). "Once a child is lost, he dies": Monster stories vis-a-vis the problem of errant children. In E. Slingerland & M. Collard, eds., *Creating Consilience: Integrating the Sciences and the Humanities*. New York: Oxford University Press, pp. 351–371.

Skal, D. J. (2001). *The Monster Show: A Cultural History of Horror*, rev. ed. New York: Faber and Faber.

Soares, S. C., Lindström, B., Esteves, F., & Öhman, A. (2014). The hidden snake in the grass: Superior detection of snakes in challenging attentional conditions. *PLoS ONE*, **9**(12), e114724.

Špinka, M., Newberry, R. C., & Bekoff, M. (2001). Mammalian play: Training for the unexpected. *Quarterly Review of Biology*, **76**(2), 141–168.

Steen, F. F., & Owens, S. A. (2001). Evolution's pedagogy: An adaptationist model of pretense and entertainment. *Journal of Cognition and Culture*, **1**(4), 289–321.

Tooby, J., & Cosmides, L. (2001). Does beauty build adapted minds? Toward an evolutionary theory of aesthetics, fiction, and the Arts. *SubStance*, **30**(1–2), 6–27.

Woody, E. Z., & Szechtman, H. (2011). Adaptation to potential threat: The evolution, neurobiology, and psychopathology of the security motivation system. *Neuroscience and Biobehavioral Reviews*, **35**(4), 1019–1033.

Yarish, B. (2013). Building a legend: The "skinny" on the Slender Man. *The University of Winnipeg Graduate Students Research Colloquium 2013*. http://winnspace.uwinnipeg.ca/handle/10680/433.

45

The Internet Is for Porn
Evolutionary Perspectives on Online Pornography

CATHERINE SALMON, MARYANNE L. FISHER, AND REBECCA L. BURCH

45.1 INTRODUCTION

Though it is probably surprising to many young people today, there was a time before the rise of the Internet when pornography was mostly found in magazines and videos. These were purchased via subscription or could be found in stores, either hidden away at the back of the magazine racks or located in seedy-looking "adult" stores with the windows blocked out. In the 1990s, there were fewer than 100 adult magazines available (including famous examples such as *Playboy, Hustler, Penthouse*, and *Jugs*), and most shops only carried a few of the most popular and best-selling magazines. In stark contrast, today there are millions of websites devoted to pornographic images, and four percent of the millions of most frequently visited sites in 2010 were sex related (Ogas & Gaddam, 2011). This shift has made pornography more accessible than it has ever been before. In 2012, in the USA, close to 75 percent of households had Internet access (gained via cellular phones, gaming systems, tablets, laptops, etc.), providing very high levels of easy access to sexually explicit materials (US Census Bureau, 2014). The number of naked bodies and sexual acts one can view in an hour today is likely far higher than people a couple of generations ago might have seen in their entire lifetimes. While much of the research attention on Internet pornography has been focused on men as consumers or its availability to children, recent evidence shows women are also consuming pornography online, though in ways that may differ from those of men (e.g., Ogas & Gaddam, 2011).

Since the early 2000s, the consumption of Internet pornography has increased substantially (Short et al., 2012). This growth is likely due in part to the greater privacy and feeling of anonymity viewers have in their own homes (i.e., not needing to go to a public place to acquire access). In addition, there is a large variety of online pornography that is free to access. According to TopTenReviews.com (2016), every 39 minutes a pornography video is created in the USA, while every second, US$3,075.64 is spent on porn, 28,258 people are viewing porn, and 372 people are typing adult search terms. TopTenReviews.com also

claimed that there are 4.2 million websites devoted to pornography (12 percent of all websites) and that there are 62 million daily pornography search engine requests (25 percent of all requests). The popular pornography website PornHub reported 28.5 billion visits in 2017, with 81 million visits each day (on average). This rate of consumption averaged out to 50,000 searches per minute, or 800 per second (www.pornhub.com/insights/2017-year-in-review).

In this chapter, we will focus on the following questions: Why do we spend time looking at online pornography? What sex differences are there in the consumption of online pornography? What can we say about individual differences between men and between women? We will approach these questions with an evolutionary or adaptationist perspective that we argue is well suited to examining sexual behavior in general. To support this claim, we begin with a short background on this approach and what we currently know about human sexuality.

45.2 ADAPTATIONS AND SEXUAL DESIRES

The nature of sexual reproduction over the course of human evolution guarantees that ancestral men and women faced different reproductive opportunities and constraints. Put another way, men and women faced different problems, and therefore evolution via selection can be expected to have produced distinct psychological solutions. These solutions are considered to be "adaptations," referring to the process whereby any change in the structure or function of a mechanism leads successive generations to be better suited to their environment (Daintith & Martin, 2010). Choosing and then acquiring "good" mates – ones high in mate value – was an adaptive problem faced by men and women. Potential mates vary in "mate value," which is composed of traits such as genetic quality, fertility, and the ability to provide resources or protection that are associated with greater reproductive success.

What constitutes high vs. low mate value is different for men and women. For example, physically strong and skilled men could contribute more resources to their families – and have subsequently higher mate value – than those that trip over their own feet and cannot catch dinner to save their own lives. Likewise, an 18-year-old woman, on average, has greater mate value than a 50-year-old woman, because the younger woman has many child-producing years ahead of her, while the older woman has few. As a result, sexual selection has shaped psychological adaptations that are specialized to process information about mate quality. Further, those adaptations are somewhat different for males and females because the specific information that is useful for mate choice differs between the sexes. Men evaluating women's attractiveness are largely influenced by cues of fertility and reproductive value (Confer, Perilloux, & Buss, 2010; Symons, 1995), such as the shape of her body, including both her waist-to-hip ratio (WHR) and her body mass index (Platek & Singh, 2010; Singh, 1993), as well as the quality of her hair and skin (Hinsz, Matz, & Patience, 2001; Sugiyama, 2005), which signal youth and fertility. In the context of short-term, uncommitted mating, men place higher priority on cues of high fertility and sexual accessibility.

When men are focused on long-term mating other traits also come into play, such as indicators of sexual fidelity (Buss & Schmitt, 1993), so that the men will not be faced with the problems of paternity uncertainty and raising another man's children as their own. Fertility and reproductive value are still important, but paternity certainty indicators increase in importance when high levels of paternal investment are likely, in contrast to the avoidance of any paternal investment inherent in short-term mating. Pornography, whether online or off, is focused on short-term mating, meaning that cues related to fertility and indicators of sexual promiscuity are attended to by males. Women signaling that they are promiscuous is important because it suggests that men do not have to invest a great deal in order to obtain a variety of mates. Men have a generally greater desire than women for short-term mating (Symons, 1979; Trivers, 1972) because of the possible reproductive payoff, and for this to be a successful strategy, there must be a variety of young, willing women available.

For ancestral women, the fertility of their mates was not a primary concern. Rather, the greater minimal investment required by women (Trivers, 1972) in their children means that they must be highly selective in their choice of long-term mates. In particular, women who secured mates who were willing and able to provide resources were more likely to have healthy, surviving children than those who did not find such mates. As a result, women's mating adaptations are sensitive to cues that men are capable of providing economic resources or have good prospects for doing so in the future (Barkow, 1989; Buss et al., 2001; Kenrick et al., 1990). Women also exhibit preferences for men of high status (Li, 2007), which is typically associated with access to resources, as well as men who are dependable (Buss et al., 1990) and demonstrate a willingness to invest in children (Roney et al., 2006). Physical height and

strength are also seen as desirable (Courtiol et al., 2010; Hughes & Gallup, 2003), as well as indicators of genetic quality such as facial symmetry and masculinity (Gangestad & Thornhill, 1998; Thornhill & Gangestad, 2006).

While women may have evolved a general preference for long-term mating strategies, they have sometimes followed short-term mating strategies as well (Buss & Schmitt, 1993), including exchanging short-term sex for resources or protection (Smuts, 1985; Symons, 1979) and strategies related to infidelity, engaging in mixed short- and long-term sexual relationships at the same time (Gangestad & Thornhill, 2008). Studies of heterosexual, naturally cycling college women's self-reported mate preferences show shifts around ovulation. Women's fertility spikes during this phase of their menstrual cycle, and this is the time in which women are most likely to consider a short-term liaison, placing emphasis on potential mates possessing cues for high genetic quality, high levels of morphological symmetry, and high masculinity (Gangestad, Thornhill, & Garver-Apgar, 2005; Pillsworth & Haselton, 2006). In other words, the short-term preferences of women show more value being placed on physical attractiveness than do their preferences for long-term mates.

45.3 WHO IS SHOPPING ONLINE FOR PORNOGRAPHY?

Before we can link evolutionary psychological theory with pornography consumption, it is useful to describe who a typical consumer may be, according to various reports. In general, studies examining factors that influence Internet access have reported greater use among those with higher education levels, higher incomes, and lower ages, with sex of the user having little effect on usage (Bimber, 2000). However, this pattern does not fully hold for all Internet use, as sex of the user does play a key role in online pornography consumption. Indeed, as one may expect due to the sex-specific mating strategies surrounding preferences for ease of access to short-term mates and arousal based on visual sexual stimuli, differences between men's and women's Internet use with regard to pornography are frequently reported, with men consuming it at a higher rate than women. One US study reported that 44 percent of college students (both sexes) sometimes accessed sexually explicit material online, while 3 percent reported being frequent users (Goodson, McCormick, & Evans, 2001). A similar study of Canadian college students reported that 72 percent of men and only 24 percent of women viewed sexually explicit material online (Boies, 2002).

In a study of 216 Portuguese women, 57 percent reported using pornography websites and 7 percent stated they spend more than 6 hours per week doing so, with the purposes of entertainment or curiosity, or due to feeling sexually aroused (Gaspar & Carvalheira, 2012). These latter statistics might seem surprising: many people seem to believe men are the only significant consumers of sexually related material online, but that is clearly not the case. Other studies have reported similar sex differences in frequency of traffic, with 75 percent of men reporting viewing pictures of or watching pornography, while only 41 percent of women downloaded or watched pornography online (Ferree, 2003). A study comparing online sexual activity between American and Peruvian college students documented significantly greater consumption of online sexual material by men cross-nationally (Velezmoro, Negy, & Livia, 2012). While the percentages vary, this brief review clearly indicates that the direction of the sex difference is inarguably showing men are always the larger proportion of online pornography consumers. We note that past researchers, such as Regnerus, Gordon, and Price (2015), have explored how various methodological approaches influence rates of documented pornography consumption, which is particularly important given that individuals may have difficulty in recalling their behavior, or they may answer inaccurately because they feel they are being judged. Their findings reveal that, in their opinion, the most accurate rate is that 46 percent of men and 16 percent of women between 18 and 39 years of age intentionally viewed pornography within a given week. This difference in access rates is not surprising from an evolutionary perspective, in that men's sexual psychology is primed to find short-term mating opportunities (and visual cues of such sexual opportunities) arousing (Symons, 1979).

As previously mentioned, it is not just that there are sex differences in frequencies in accessing pornography online, but there are also differences in what men and women are actually doing while accessing pornography. Data indicate that men are typically consuming pornographic videos and explicit photographs, while women are often engaging in cyber-chat and reading erotic fiction (Cooper, 2000; Ferree, 2003). This difference is not unexpected considering that men are more responsive to visual sexually arousing stimuli than women, while women's sexual psychology is more likely to be activated by the mating contexts described in written stories of sexual relationships (Hamann et al., 2004; Salmon & Symons, 2003). There are also corresponding sex differences in terms of using online material as a masturbatory aid. Boies (2002) reported that 72 percent of Canadian college men masturbated while online, while only 22 percent of women did so.

45.4 EVOLUTIONARY APPROACHES TO ONLINE SEXUAL MATERIAL

Previous work by Ogas and Gaddam (2011) has used evolutionary psychology to explain the data obtained from search engines, as well as other sources of Internet content such as personal advertisements, websites, and electronic erotic fiction. For example, they collected 400 million unique queries from the Dogpile search engine from July 2009 to July 2010, reporting that 55 million of these queries were ones seeking sexual content, and of these, the most common term was "youth" (13.5 percent). They propose that these queries represent the desires of about two million people, and the focus on terms related to youth clearly reflects men's evolved interest in young, fertile women. Additionally, they report that about 20 different categories account for 80 percent of all sexual content searches, which shows a surprising lack of diversity. From an evolutionary perspective, this homogeneity is reasonable, given that the mating-relevant problems (and resulting interests) individuals have faced tend to be rather similar, at least within each sex.

The ways in which we engage with the Internet and seek certain types of content reflect solutions to issues our ancestors faced during human evolution (as was discussed in Section 45.2). The premise of the field of Darwinian literary studies (e.g., Carroll, 1995) is that one may interpret texts (as well as other cultural artifacts; Fisher & Salmon, 2012) such that they reveal human's evolved emotions, motivations, and cognition. We create texts (and other artifacts) using our evolved brains, then interact with them using those same evolved brains.

Internet pornography can be viewed as a modern artifact that is created by humans with evolved brains and consumed by humans with these same evolved brains. Although the technological environment has changed rapidly over the last century, or even just over the last few decades, generations of evolutionary pressures leading to humans' cognitions, motivations, and emotions remain intact. Thus, men's and women's general behaviors and preferences reflect their evolutionary histories, but in the novel digital environment.

Of course, such desires may differ from the actual mating choices people make, as human behavior is often far more constrained than human desire. For example, a majority of men may desire a woman who resembles a young Angelina Jolie, but only a small minority of men will have high enough mate value to attract such an attractive woman. Our actual mate choices are constrained by our own mate value in a way that our desires are not.

If the adaptive problems that men and women faced in the sexual domain differed somewhat, we would expect them to show some differences in their online sexual interests. One way to look at this possibility is to examine the most popular websites visited by men and women. The top five sites for men (7.4–16.0 million visits/month) are all adult video sites (including PornHub and YouPorn), while the to five sites for women include four fiction sites, and the fifth is a video site (including FanFiction.net, eHarlequin.com, and AdultFanFiction.net) (Ogas & Gaddam, 2011). The takeaway here is that men are visiting sites specializing in visual content, while women are focusing on stories (comparable to the contrast between adult DVD and *Playboy* subscriptions vs. romance novels and so-called chick flicks). The fanfiction sites that are in the top five sites visited by women contain stories, often of a romantic and sexual nature, between fictional characters (from television, movies, novels, and so on) and include both heterosexual (think House and Cuddy from *House M.D.*) and homosexual (think House and Wilson from *House M. D.*) fiction (Salmon & Symons, 2003). Sites frequented by women that are not fanfiction focused tend to nevertheless provide romantic or erotic stories. Both the fanfiction and romance/erotica sites focus on the same aspects as do traditional romance novels: the adaptive problems women have faced (finding love, obtaining a mate willing to commit, finding a mate with good genes and who is a good provider), just as Internet pornography for men taps into the evolved psychology men developed to solve their problems of assessing fertile females.

Most online pornography falls into stable, predictable categories. Take, for example, PornHub. They categorize pornography in a variety of ways and report that the most frequently accessed categories are teen, lesbian, MILF, amateur, and ebony (Giannotta, 2013). PornHub does offer a long list of available categories beyond those most frequently used that can appeal to a wide assortment of tastes. Their categories include a number of individuals involved (e.g., orgies), type of activity (e.g., anal sex), or that vary by the characteristics of those participating (e.g., elderly, various ethnicities). The fact that these are not the categories primarily used by consumers supports the idea that, although pornography of some form presumably exists for almost everything one could imagine, less common forms can mostly be considered creative expressions of sexuality or individual differences. This is similar to the distribution of human phobias; while people may have irrational fears of anything from spiders to dill pickles, the most common phobias (heights, snakes, spiders, etc.) reflect dangers that have been real selection pressures for millennia (Öhman & Mineka, 2001). This "biological preparedness" also predicts that certain sexual preferences (youth, body shapes, etc.) will be far more popular than less evolutionarily relevant stimuli.

45.5 MALE PREFERENCES

Piazza and Bering (2009) suggest that evolutionary psychology allows one to frame Internet behavior in terms of input into domain-specific psychological mechanisms that have been shaped over human history. Given that men are the primary consumers of online pornography, it is to be expected that their sexual interests, which reflect their evolved history, are clearly evident in the content. For example, why is it that men (for the most part) look for young but not elderly women when searching online pornography? The most straightforward answer is that youth is a characteristic that indicates fertility and high reproductive potential. It must be noted that these preferences usually exist without conscious awareness of the underlying rationale. The preference for youth is inarguable; in addition to the aforementioned findings from PornHub, Ogas and Gaddam (2011) reported that the most frequent

search term was "youth" (followed by "gay," "MILFs," "breasts," and "cheating wives"). PornHub's globally most searched terms of 2017 included "teen" and "cheerleader," both indicators of youth, as well as those universally popular female body parts "big tits" and "big ass" (www.pornhub.com/insights/2017-year-in-review). No one is likely to be surprised that men spend time looking at breasts; research shows that breasts are a salient sexual signal linked with women's fecundity across a range of cultures, along with other gynoid fat deposits (e.g., Lassek & Gaulin, 2008; Singh & Young, 1995).

The popularity of MILFs and cheating wives may be a bit more surprising. However, Millford's (2013) study of adult film star statistics indicates that the average age of female stars in MILF films is 33 years, all of whom exhibit fertile-looking body shapes. In addition, some researchers have documented that women in their 30s have more sexual fantasies, are more likely to engage in casual sex, and exhibit a greater willingness to have one-night stands (Easton et al., 2010). These are all characteristics that would be highly desirable in a short-term mate, and younger men often exhibit an interest in women with sexual experience who will improve their own sexual skills, particularly when they are not looking for a long-term sexual partner (Kenrick et al., 1996).

The appeal of cheating wives may be found in studies that examine the features surrounding sperm competition. A study of sperm competition cues (see Chapter 23 for a discussion of sperm competition) in pornography by Pound (2002) demonstrated that men have a preference for viewing sexually explicit content that involves a female and multiple males (32–36 percent as opposed to 12–13 percent preferring one female and one male or 16–18 percent for multiple females and one male). The argument is that as pornography is often used as an aid to masturbation, stimuli showing multiple males with one female are stronger sources of arousal, presumably because they trigger a cue of male competition for sexual access that ancestrally might have translated into increased sexual behavior, which would have increased the chance of fertilization somewhat for all participating males (as when multiple males mate with females in estrus in some primates). In the modern world, men may be using such images to increase their arousal, as caused by cues of sperm competition. However, more work clearly needs to be done to explore the popularity of multi-male with one female scenes, as well as the draw of cheating wives as characters in online pornography.

This topic takes on another dimension when one questions how men view their relationships to the actors in the pornographic material. Vörös (2014) reports how each of his interviewees "transposed" pornographic depictions into their own memories and preferences. Indeed, one interviewee stated, "I like to transpose and to imagine I'm in his place when I watch the videos" (p. 9). Klein (2006) reports it this way: "When men watch porn, they're able to imagine 'yes, that's how I would be without the obstacles

I normally face'" (p. 251). One of Hite's (1981) subjects commented, "It does me good to think she wants me" (p. 783). Janssen et al. (2003) found that, for both men and women, imagining that they are an active participant in the sexual act was significantly correlated with sexual arousal. It is not surprising given this research that "point-of-view" pornography is a trending genre and that three-dimensional and virtual reality pornography technologies are on the rise (www.pornhub.com). This can also explain why male porn stars usually have larger-than-average penises, though interestingly there are more data on the sizes of gay male porn star penises (Brennan, 2018); the viewer can now imagine himself as having a larger penis and being free of issues surrounding erectile dysfunction (which varies according to definition; Schouten et al., 2005).

Klein (2006) also mentions how, within pornography, the typical evolutionary constraints of male desirability (e.g., wealth, status, etc.) are irrelevant; the female characters always find the men desirable. The reasons underlying men's preferences become obvious if they are imagining themselves in the lead roles in pornographic scenarios. Men search for plots involving a lead male who is so irresistible and with such high mate value that a married woman will cheat on her husband in order to be with him. "MILFs" are by definition mothers, sexy women with children (MILF stands for "Mom I would Like to Fuck") who appear willing to engage in short-term mating at low cost. Of the top US PornHub searches of 2017, terms 3–6 were "MILFs," "stepmom," "stepsister," and "mom." PornHub reported the top searches in 20 different countries, and the term "mom" was reported in 15 countries, "MILF" in 14, "stepmom" in 14, and "stepsister" in 9. Overall (worldwide), the top two search terms were "MILF" and "stepmom." When PornHub separates searches for men vs. women, "MILF" remains in the top position, followed by "stepmom" for men. All of these data indicate a cultural universal in men's pornographic preferences. The scenarios display a man with such high mate value that women will perform risky behaviors to engage in short-term sexual acts with him, a fantasy that certainly plays to the evolutionary interests of men. Other frequently searched terms in 2017 were "3D," "VR," and "POV," providing the viewer with the means of putting himself into the sexual act (www.pornhub.com/insights/2017-year-in-review).

This fantasy of extremely high mate value is illustrated in other popular plots; for example, a lead male who has such high mate value that women – many women – are willing not only to have sex with him, but also to share him sexually. The term "lesbian" was the top global search term in 2017 (as well as in 2016 and 2015, and reaching second position in 2014). Other terms such as "threesome" and "lesbians scissoring" involve multiple women. This finding brings us to the question: Why is lesbian pornography so often consumed by men? The fantasy of extremely high mate value would certainly reach its peak here; the fantasy

is that they are so irresistible that women behave outside of their sexual orientation, or become so aroused that they forgo their sexual orientation in order to engage in sexual acts. Although there exists little empirical work on this issue, a recent survey-based study indicates that men generally prefer their opposite-sex partners to have some same-sex sexual interest or experience (Apostolou & Christoforou, 2018).

Using the aforementioned PornHub "Year in Review," we can not only examine the content of searches, but also identify the most sought-after pornographic actors and actresses. PornHub released the names of the most searched actresses in 2017, both globally (top 20 searches) and across 20 separate countries (i.e., top 5 searches for each country). These 20 countries (Argentina, Australia, Belgium, Brazil, Canada, France, Germany, India, Italy, Japan, Mexico, the Netherlands, the Philippines, Poland, Russia, South Africa, Spain, Sweden, the UK, and the USA) account for 80 percent of the daily traffic on the PornHub website (www.pornhub.com/insights/2017-year-in-review). In total, there were 108 top-searched names, made up of 50 actresses (some names appeared multiple times across countries and some were actors). This overlap shows that while there is some variability, there are a number of women who are searched repeatedly regardless of country, indicating similarity in preferences across the world. The body measurements of these women can also be obtained through biographical websites. Over the 108 most popular searches, the average bra cup size was a D, the average bust size was near 36 inches, the average waist size was approximately 25 inches, and the average hip measurement was over 36 inches (Table 45.1).

As already mentioned, men evaluate female body shape closely, particularly WHR. Singh (1993) found that in both *Playboy* centerfolds and Miss America winners, WHR hovered around 0.7 (i.e., the waist size being 70 percent of the hips). Higher WHRs (e.g., 0.9 or 1.0) were rated as significantly less attractive. When examining the 108 porn actress searches (some women were searched repeatedly), WHRs ranged from 0.59 to 0.83, but averaged just 0.6881 (Table 45.1). When the 50 women were viewed separately (the duplicate searches were removed), the measurements for chest, waist, and hips decreased slightly and the WHR increased to 0.6994. This indicates that the women with larger chests and hips (and smaller WHRs) were searched more often. When the top 20 searched women in the world were examined, the WHR dropped to 0.6875, and when the top 5 women who appeared the most in rankings across the 20 listed countries were measured, their WHR dropped to 0.6691. This finding falls in line with the body of literature on men's WHR preferences across the globe (Singh, 2002; Singh & Luis, 1995; Sorokowski et al., 2014).

We examined chest measurements given that "breasts" was a frequently searched term. Although the most searched porn actresses ranged in breast size from an A to a D cup, the average cup size across all searches was

Table 45.1 Bodily measurements of 2017's most searched actresses on PornHub.

Category	Cup size	Chest	Waist	Hip	WHR
108 top searches	D	35.87 (SD 2.79)	24.96 (SD 1.86)	36.27 (SD 2.77)	0.6881 (SD 0.048)
50 most searched women	D	34.42 (SD 2.16)	24.57 (SD 1.56)	35.25 (SD 2.87)	0.6994 (SD 0.051)
Top 20 global searches	C	34.40 (SD 2.47)	24.50 (SD 1.70)	35.75 (SD 3.04)	0.6875 (SD 0.053)
Top 5 frequency in 20 countries	D	36.80 (SD 3.03)	24.60 (SD 2.70)	36.60 (SD 2.61)	0.6691 (SD 0.053)

a D cup. Chest measurements correlated with popularity of the actress; the larger the measurement, the higher the actress ranked in the global top 20 searches ($r = 0.481$, $p < 0.05$) and the number of top 5 rankings in the 20 listed countries ($r = 0.461$, $p < 0.05$). Interestingly, cup size and WHR did not correlate with any rankings, perhaps because the range of these variables was so low (i.e., few actresses had a bust size of a B cup or smaller, and the average cup size was a D; WHRs ranged from 0.59 to 0.83 and averaged 0.69 [SD of 0.048]).

In all of these searches, there were no significant differences across the listed countries for cup size, chest, waist, or hip measurements, or WHR. We believe this illustrates the similarities in male preferences for female bodies across the world.

45.6 FEMALE PREFERENCES

There has been far less attention, relatively speaking, paid to women's preferences for pornography as compared to the research that is available concerning their preferences for erotica and plot-focused fiction. If one purpose of fiction is to arouse our thinking about ancestral problems, the popularity of female-focused long-term mate choice stories, such as the romance, should not be surprising (Salmon, 2012). As reviewed in Section 45.2, women's greater investment in their children (Trivers, 1972) demands that they be selective about their long-term mates. Thus, women tend to seek men who are resourceful (Buss et al., 2001; Kenrick et al., 1990), possess status (Li, 2007), are dependable (Buss et al., 1990), and have great physical height and strength (Courtiol et al., 2010; Hughes & Gallup, 2003), as well as having indicators of high genetic quality such as facial symmetry and masculinity (Rhodes, 2006; Thornhill & Gangestad, 2008).

In general, romance is focused on the domain of long-term mating. The importance of high mate quality in heroes has been revealed through analysis of the titles of romance novels (Cox & Fisher, 2009), as well as the character development of heroes and heroines (Fisher & Cox, 2010). An examination of the 20 most frequently mentioned occupations in the titles of popular romances reveals 17 professions for the hero, falling into two

categories: resource based (e.g., doctor, CEO, royalty) and athletic or protector (e.g., cowboys, lawmen, soldiers). The majority of heroes are well equipped to provide stable resources and to have sufficient genetic quality and physical prowess to serve as a protector (not only of the female and any children, but also of their own resources). The hero falling in love with the heroine guarantees his long-term commitment to her, with commitment and reproduction being two other themes reflected in the titles. Even in the more esoteric romance genre of slash fiction, we see the same focus on highly masculine-appearing men, powerful heroes who are either physically powerful or high in status (or both) who can provide and protect their partners and who are intensely in love with them, illustrating the strength of their commitments (Salmon, 2012; Salmon & Fisher, 2018; Salmon & Symons, 2003).

This being said, however, in recent years women have been searching for, discussing, and writing about pornography, and pornographic companies have taken notice. PornHub reports that worldwide approximately 26 percent of its viewership consists of women. This ranges across the 20 most popular countries from 19 percent in Japan to 36 percent in the Philippines. The number one new trend in search terms on PornHub for 2017 was "Porn for Women." The term was reported to have increased from 2016 by 1400 percent (www.pornhub.com/insights/2017-year-in-review). Rhyne (2007) reports that the recognition of female consumption began almost 50 years ago: "Since the 1972 crossover success of *Deep Throat* the porn industry has recognized women as an increasingly lucrative audience demographic, and producers, particularly female producers, have cultivated an alternative aesthetic designed to appeal to this specialty market" (p. 42). This acknowledgment led the pornographic industry to focus on women as potential consumers, while also using them as the primary sexual objects within it (Williams, 1999).

So, what do we know about women's pornography preferences? As of this writing, little research has been done to determine how many women view pornography and to identify their actual preferences. Berkowitz (2006) and Hefley (2007) both acknowledged this omission and set out to measure women's visits to pornography/sex shops.

Hefley found that 40 percent of the customers were women, in varying groups (e.g., in mixed-sex pairs or all-female groups), while Berkowitz (using a far less structured methodology) found equal percentages of men and women to be customers. If women constitute such a high percentage of customers, what are they purchasing and how do these preferences map onto evolutionary theory? Who are they purchasing for: themselves, female partners, or male partners? The research on these issues at this point is sparse and vague. For example, Laumann and colleagues (1994) stated that women were twice as likely to prefer "other sex toys" over pornography, but did not specify what is meant by "other sex toys." Other researchers have either taken the stance that pornography is harmful to women or specifically studied women who shared that perspective (Attwood, 2005). Another theme in the existing literature is that while some women enjoyed some pornography, they did not view it as being for them; they felt that they were watching something created for men (Attwood, 2005).

Now that there is a pornography industry for women and women are searching specifically for "porn for women," what does this content entail? Schauer (2005) analyzed "porn for women" sites, but did not collect data from women on their preferences. Schauer's analysis concluded that "porn for women" material was often similar to gay pornography for men in some contexts and heterosexual pornography for men in others. Schauer viewed this outcome as a failure to accurately provide for female preferences. Woodard et al. (2008) investigated what types of erotic film clips women actually enjoyed using participant self-ratings of mental appeal and physical arousal. These ratings correlated with each other, and both were higher when heterosexual vaginal intercourse was depicted. Ratings were lowest during depictions of male homosexual behavior, fellatio, and anal intercourse (notice that they did not show depictions of female homosexual behavior).

If heterosexual intercourse is viewed as most arousing, what do women find most arousing within that genre? Heiman (1977) found that women were more aroused by audiotapes that described female-initiated (e.g., the female was more active than the male in initiating and performing sexual activities) and female-centered (e.g., description focused on the female's responses, enjoyment, and genitals) interactions. Heiman noted that films made by men evoked substantial genital vasocongestion but negative affect and low subjective sexual excitement. Laan et al. (1994) had European women view erotic clips made by women showing female-centered content with women initiating, as well as clips that were made by men with male-centered content and males initiating the sexual activities. They found that women had equal physiological arousal to both. However, women's subjective experiences were very different; self-rated sexual arousal was significantly higher

during the female-centered clips, while shame, guilt, and aversion were higher after being shown the male-centered clips. Finally, Mosher and McIan (1994) found the same pattern, with women being mildly aroused but experiencing more interest and enjoyment and less shame and distress when watching films made by women. So while physiological arousal remains the same, women psychologically prefer female-centered erotica. It is possible that this "female-centered" content is actually an indication of the male partner's interest, commitment, and desire to provide for the woman, which does follow evolutionary predictions. However, other forms of commitment in pornography (e.g., giving resources, romantic commitment, professing loyalty) have not been examined, and studies of female arousal in non-Western cultures should be pursued where possible.

Lykins et al. (2008) found that when watching erotic content, women paid more attention to the female characters. This supports the assumption that pornography viewers like to imagine themselves as active participants in the pornography and corroborates the PornHub list of popular searches: pornography featuring lesbians is by definition female focused, and the search terms "MILF" and "stepmom" do feature at least some background on the female lead (www.pornhub.com/insights/2017-year-in-review). It is also important to acknowledge female sexual fluidity and how this can affect preferences for lesbian and bisexual erotic content (Diamond, 2008). However, the surge in searches for "porn for women" shows that the female preference for heterosexual pornography remains (www.pornhub.com/insights/2017-year-in-review).

What other information regarding female preferences can we glean from PornHub? "Popular with women" was the 16th most searched category for 2017. The popularity of the number one search term "lesbian" is attributed not only to males, but also to increasing female viewership. "Lesbian" was the most searched term for women and the seventh most searched term for men. This was followed by "lesbians scissoring" and "threesome" (these are the same rankings as in 2016). The most searched categories were "lesbian," "threesome," "big dick," and "popular with women" (www.pornhub.com/insights/2017-year-in-review). In 2016, the most searched categories were "lesbian," "threesome," "big dick," and "female friendly" (www.pornhub.com/insights/2016-year-in-review). Women were far more likely to view videos that featured cunnilingus and videos labeled as "popular with women" and "lesbian" than men. These were the same results as in 2016, with "female friendly" instead of "popular with women." This is not to imply that women are only viewing female-only sexual encounters; in fact, they are also more likely to view content involving "gangbang," "double penetration," "hardcore," "rough sex," "threesome," "bondage," and "big dick" than men (www.pornhub.com/insights/2017-year-in-review). Although a great deal more detail is needed, this list supports the assertion from the academic literature that women enjoy depictions of heterosexual sex but prefer female-centered content.

Unfortunately, there is very little research on the content and themes in female-oriented pornography, making the mapping of preferences onto evolutionary theory difficult. We would posit that, due to the lack of female-focused pornography available, women are currently merely seeking out representation in the pornography they view. As more female-centric content is created, it may be possible to investigate this further. What we do know is what women prefer in other types of erotica, which does correspond to evolutionary theory. Perhaps the incorporation of these elements into pornography will create a more popular form of "porn for women." Such a form would presumably need to incorporate the committed relationship-focused storytelling aspect of female mate choice, pornography with a romance plot, as well as explicit sex between hero and heroine.

Even with these universal themes of what is preferred by men and women, there is no doubt that there is a wide range of pornographic material available online; one can easily satisfy almost any preference or curiosity with a few search terms. For example, along with these universal search terms, we also find terms reflecting trends such as "hentai," "Overwatch," "fidget spinner," "twerking," and "Rick and Morty." PornHub has content sections based on celebrities, fictional characters, and other popular culture references (www.pornhub.com/insights/2017-year-in-review). Indeed, we are left with no option but to agree with the popular Internet meme known as Rule 34, which states, "If it exists, there is porn of it." There is a definite need for more research into the factors that shape the variety of sexual material available online.

45.7 CONCLUSIONS

Many features of our modern environment are best understood using an evolutionary perspective; here, we have examined online pornography, but other readily available examples of other features of modern life exist. Take, for example, Fisher and Salmon (2012), who reviewed an evolutionary perspective of the obesity epidemic in the USA. In our evolutionary past, storage of fat was highly important for survival, as it helped stave off death due to starvation by providing a ready energy store. Food was more challenging to acquire during that time, whether due to problems in finding and obtaining food or processing it; rarely would there be a constant supply of easily consumed foods. Rare tastes such as fatty, sweet, and salty foods would have also been preferred, given that these tastes have ties to caloric intake and health (e.g., blood pressure). However, in today's environment, these tastes are immediately available via fast food. We are then faced with a mismatch: we prefer these tastes due to our evolved preferences, but we did not evolve the ability to stop overeating them as they were scarce in our past. Our fondness

for online pornography may be a similar mismatch made possible by modern technology.

The ways in which we can apply evolutionary theory to understanding modern situations is significant and noteworthy. Evolutionary theory is vastly underutilized in pornography studies; one simply needs to scan the articles in the journal *Porn Studies* or review titles of recent books in the field to see that this area continues along standard social science lines.

Nevertheless, though much of the research that has been done on pornography has not been evolutionarily informed, the results strongly support an evolutionary perspective. An adaptationist perspective on human sexual psychology makes sense not only of the content seen in sexual material that is available online, but also of how it is tailored to male and female audiences. Further, online pornography not only illustrates the differences between male and female sexual psychology, it also draws attention to the variability within the sexes, which may then provide evidence of the individual differences in viewing interests that reflect specific developmental experiences.

However, there are a number of questions relating to online and offline pornography that are unanswered, many of which focus on individual or cultural differences. For example, while many studies document greater online pornography consumption by men, there are women consumers in those studies. Are these women consuming more or less than men, and is what they are consuming, as we have suggested, essentially different from what men consume? Are there interesting individual differences between women that shape pornography consumption? Are the women who are consuming pornography online more likely to be following a short-term mating strategy that could be measured by something like the SOI (Sociosexual Orientation Inventory)? These questions will only be answered with future evolutionarily informed studies.

REFERENCES

Apostolou, M., & Christoforou, C. (2018). Same-sex attraction and contact in an opposite-sex partner: Exploring sex, religiosity, porn consumption and participation effects. *Personality and Individual Differences*, **131**(1), 26–30.

Attwood, F. (2005). What do people do with porn? Qualitative research into the consumption, use, and experience of pornography and other sexually explicit media. *Sexuality and Culture*, **9**(2), 65–86.

Barkow, J. H. (1989). *Darwin, Sex, and Status: Biological Approaches to Mind and Culture*. Toronto: University of Toronto Press.

Berkowitz, D. (2006). Consuming eroticism: Gender performances and presentations in pornographic establishments. *Journal of Contemporary Ethnography*, **35**(5), 583–606.

Bimber, B. (2000). Measuring the gender gap on the Internet. *Social Science Quarterly*, **81**, 868–876.

Boies, S. C. (2002). University students' uses of and reactions to online sexual information and entertainment: Links to online and offline sexual behavior. *Canadian Journal of Human Sexuality*, **11**, 77–89.

Brennan, J. (2018). Size matters: Penis size and sexual position in gay porn profiles. *Journal of Homosexuality*, **65**, 912–933.

Buss, D. M., & Schmitt, D. P. (1993). Sexual strategies theory: An evolutionary perspective on human mating. *Psychological Review*, **100**, 204–232.

Buss, D. M., Abbott, M., Angleitner, A., et al. (1990). International preferences in selecting mates: A study of 37 cultures. *Journal of Cross-Cultural Psychology*, **21**, 5–47.

Buss, D. M., Shackelford, T. K., Kirkpatrick, L. A., & Larsen, R. J. (2001). A half century of American mate preferences. *Journal of Marriage and the Family*, **63**, 491–503.

Carroll, J. (1995). *Evolution and Literary Theory*. St Louis, MO: University of Missouri Press.

Confer, J. C., Perilloux, C., & Buss, D. M. (2010). More than just a pretty face: Men's priority shifts toward bodily attractiveness in short-term mating contexts. *Evolution and Human Behavior*, **31**, 349–353.

Cooper, A. (2000). *Cybersex: The Dark Side of the Force: A Special Issue of the Journal of Sexual Addiction and Compulsion*. New York: Brunner-Routledge.

Courtiol, A., Ramond, M., Godelle, B., & Ferdy, J. (2010). Mate choice and human stature: Homogamy as a unified framework for understanding mate preferences. *Evolution*, **64**, 2189–2203.

Cox, A., & Fisher, M. (2009). The Texas billionaire's pregnant bride: An evolutionary interpretation of romance fiction titles. *Journal of Social, Evolutionary, and Cultural Psychology*, **3**, 386–401.

Daintith, J., & Martin, E. A., eds. (2010). *A Dictionary of Science*. Oxford Paperback Reference, 6th ed. New York: Oxford University Press.

Diamond, L. M. (2008). *Sexual Fluidity*. Boston, MA: Harvard University Press.

Easton, J. A., Confer, J. C., Goetz, C. D., & Buss, D. M. (2010). Reproduction expediting: Sexual motivations, fantasies, and the ticking biological clock. *Personality and Individual Differences*, **49**, 516–520.

Ferree, M. C. (2003). Women and the web: Cybersex activity and implications. *Sexual and Relationship Therapy*, **18**, 385–393.

Fisher, M., & Cox, A. (2010). Man change thyself: Hero versus heroine development in Harlequin romance novels. *Journal of Social, Evolutionary, and Cultural Psychology*, **4**, 305–316.

Fisher, M., & Salmon, C. (2012). Introduction: Human nature and pop culture. *Review of General Psychology*, **16**(2), 104–108.

Gangestad, S. W., & Thornhill, R. (1998). Menstrual cycle variation in women's preferences for the scent of symmetrical men. *Proceedings of the Royal Society of London B*, **265**, 727–733.

Gangestad, S. W., & Thornhill, R. (2008). Human oestrus. *Proceedings of the Royal Society of London B*, **275**, 991–1000.

Gangestad, S. W., Thornhill, R., & Garver-Apgar, C. E. (2005). Adaptations to ovulation implications for sexual and social behavior. *Current Directions in Psychological Science*, **14**, 312–316.

Gaspar, M. J., & Carvalheira, A. (2012). The consumption of pornography on the Internet in a sample of Portuguese women. *Psychology, Community and Health*, **1**(2), 163–171.

Giannotta, M. (2013). PornHub 2012 adult entertainment statistics. http://thecelebritycafe.com/feature/2013/01/pornhubcoms-2012-adult-entertainment-statistics (accessed August 29, 2013).

Goodson, P., McCormick, D., & Evans, A. (2001). Searching for sexually explicit materials on the Internet: An exploratory study of college students' behavior and attitudes. *Archives of Sexual Behavior*, **30**(2), 101–118.

Hamann, S., Herman, R. A., Nolan, C. L., & Wallen, K. (2004). Men and women differ in amygdala response to visual sexual stimuli. *Nature Neuroscience*, **7**, 411–416.

Hefley, K. (2007). Stigma management of male and female customers to a non-urban adult novelty store. *Deviant Behavior*, **28**(1), 79–109.

Heiman, J. R. (1977). A psychophysiological exploration of sexual arousal patterns in females and males. *Psychophysiology*, **14**(3), 266–274.

Hinsz, V. B., Matz, D. C., & Patience, R. A. (2001). Does women's hair signal reproductive potential? *Journal of Experimental Social Psychology*, **37**, 166–172.

Hite, S. (1981). *The Hite Report on Male Sexuality*. New York: Alfred A. Knopf.

Hughes, S. M., & Gallup, G. G. (2003). Sex differences in morphological predictors of sexual behavior: Shoulder to hip and waist to hip ratios. *Evolution and Human Behavior*, **24**, 173–178.

Janssen, E., Carpenter, D., & Graham, C. A. (2003). Selecting films for sex research: Gender differences in erotic film preference. *Archives of Sexual Behavior*, **32**(3), 243–251.

Kenrick, D. T., Sadalla, E. K., Groth, G., & Trost, M. R. (1990). Evolution, traits, and the stages of human courtship: Qualifying the parental investment model. *Journal of Personality*, **58**, 97–116.

Kenrick, D. T., Keefe, R. C., Gabrielidis, C., & Cornelius, J. S. (1996). Adolescents' age preferences for dating partners: Support for an evolutionary model of life-history strategies. *Child Development*, **67**, 1499–1511.

Klein, M. (2006). Pornography: What men see when they watch. In P. Lehman, ed., *Pornography: Film and Culture*. New Brunswick, NJ: Rutgers University Press, pp. 244–257.

Laan, E., Everaerd, W., van Bellen, G., & Hanewald, G. (1994). Women's sexual and emotional responses to male-and female-produced erotica. *Archives of Sexual Behavior*, **23**(2), 153–169.

Lassek, W. D., & Gaulin, S. J. C. (2008) Waist–hip ratio and cognitive ability: Is gluteofemoral fat a privileged store of neurodevelopmental resources? *Evolution and Human Behavior*, **29**, 26–34.

Laumann, E. O., Gagnon, J. H., Michael, R. T., & Michaels, S. (1994). *The Social Organization of Sexuality: Sexual Practices in the United States*. Chicago, IL: University of Chicago Press.

Li, N. P. (2007). Mate preference necessities in long- and short-term mating: People prioritize in themselves what their mates prioritize in them. *Acta Psychologica Sinica*, **39**, 528–535.

Lykins, A. D., Meana, M., & Strauss, G. P. (2008). Sex differences in visual attention to erotic and non-erotic stimuli. *Archives of Sexual Behavior*, **37**(2), 219–228.

Millford, J. (2013). Deep inside: A study of 10,000 porn stars and their careers. http://jonmillward.com/blog/studies/deep-inside-a-study-of-10000-porn-stars.

Mosher, D. I., & MacIan, P. (1994). College men and women respond to X-rated videos intended or male or female audiences: Gender and sexual scripts. *Journal of Sex Research*, **31**, 99–113.

Ogas, O., & Gaddam, S. (2011). *A Billion Wicked Thoughts*. New York: Dutton.

Öhman, A., & Mineka, S. (2001). Fears, phobias, and preparedness: Toward an evolved module of fear and fear learning. *Psychological Review*, **108**(3), 483–522.

Piazza, J., & Bering, J. M. (2009). Evolutionary cyber-psychology: Applying an evolutionary framework to Internet behavior. *Computers in Human Behavior*, **25**, 1258–1269.

Pillsworth, E. G., & Haselton, M. G. (2006). Women's sexual strategies: The evolution of long-term bonds and extrapair sex. *Annual Review of Sex Research*, **17**, 59–100.

Platek, S. M., & Singh, D. (2010). Optimal waist-to-hip ratios in women active neural reward centers in men. *PLoS ONE*, **5**, e9042.

Pound, N. (2002). Male interest in visual cues of sperm competition risk. *Evolution and Human Behavior*, **23**, 443–466.

Regnerus, M., Gordon, D., & Price, J. (2015). Documenting pornography use in America: A comparative analysis of methodological approaches. *Journal of Sex Research*, **53**(7), 873–881.

Rhodes, G. (2006). The evolutionary psychology of facial beauty. *Annual Review of Psychology*, **57**, 199–226.

Rhyne, R. (2007). Hard-core shopping: Educating consumption in SIR video production's lesbian porn. *Velvet Light Trap*, **59**(1), 42–50.

Roney, J. R., Hanson, K. N., Durante, K. M., & Maestripieri, D. (2006). Reading men's faces: Women's mate attractiveness judgments track men's testosterone and interest in infants. *Proceedings of the Royal Society of London B*, **273**, 2169–2175.

Salmon, C. (2012). The pop culture of sex: An evolutionary window on the worlds of pornography and romance. *Review of General Psychology*, **16**, 152–160.

Salmon, C., & Fisher, M. L. (2018). Putting the "sex" into "sexuality": Understanding online pornography using an evolutionary framework. *EvoS Journal: The Journal of the Evolutionary Studies Consortium*, **9**(2), 1–15.

Salmon, C., & Symons, D. (2003). *Warrior Lovers: Erotic Fiction, Evolution, and Female Sexuality*. New Haven, CT: Yale University Press.

Schauer, T. (2005). Women's porno: The heterosexual female gaze in porn sites "for women." *Sexuality and Culture*, **9**(2), 42–64.

Schouten, B., Bosch, J., Bernsen, R., et al. (2005). Incidence rates of erectile dysfunction in the Dutch general population: Effects of definition, clinical relevance and duration of follow-up in the Krimpen Study. *International Journal of Impotence Research*, **17**, 58–62.

Short, M. B., Black, L., Smith, A. H., Wetterneck, C. T., & Wells, D. E. (2012). A review of Internet pornography use research: Methodology and content from the past 10 years. *Cyberpsychology, Behavior, and Social Networking*, **15**(1), 13–23.

Singh, D. (1993). Adaptive significance of waist-to-hip ratio and female physical attractiveness. *Journal of Personality and Social Psychology*, **65**, 293–307.

Singh, D. (2002). Female mate value at a glance: Relationship of waist-to-hip ratio to health, fecundity and attractiveness. *Neuroendocrinology Letters*, **23**(Suppl. 4), 81–91.

Singh, D., & Luis, S. (1995). Ethnic and gender consensus for the effect of waist-to-hip ratio on judgment of women's attractiveness. *Human Nature*, **6**(1), 51–65.

Singh, D., & Young, R. K. (1995). Body weight, waist-to-hip ratio, breasts, and hips: Role in judgments of female attractiveness and desirability for relationships. *Evolution and Human Behavior*, **16**(6), 483–507.

Smuts, B. B. (1985). *Sex and Friendship in Baboons*. New York: Aldine de Gruyter.

Sorokowski, P., Kościński, K., Sorokowska, A., & Huanca, T. (2014). Preference for women's body mass and waist-to-hip ratio in Tsimane'men of the Bolivian Amazon: Biological and cultural determinants. *PLoS ONE*, **9**(8), e105468.

Sugiyama, L. (2005). Physical attractiveness in adaptationist perspective. In D.M. Buss, ed., *The Handbook of Evolutionary Psychology*. New York: Wiley, pp. 292–342.

Symons, D. (1979). *The Evolution of Human Sexuality*. New York: Oxford University Press.

Symons, D. (1995). Beauty is in the adaptations of the beholder: The evolutionary psychology of human female sexual attractiveness. In P. R. Abramson & S. D. Pinkerton, eds., *Sexual Nature: Sexual Culture*. Chicago, IL: University of Chicago Press, pp. 80–118.

Thornhill, R., & Gangestad, S. W. (2006). Facial sexual dimorphism, developmental stability, and susceptibility to disease in men and women. *Evolution and Human Behavior*, **27**, 131–144.

Thornhill, R., & Gangestad, S. W. (2008). *The Evolutionary Biology of Human Female Sexuality*. New York: Oxford University Press.

TopTenReviews.com (2016). Top Ten Statistics. Internet pornography statistics. http://internet-filter-review.toptenreviews.com/internet-pornography-statistics.html (accessed April 11, 2016).

Trivers, R. L. (1972). Parental investment and sexual selection. In B. Campbell, ed., *Sexual Selection and the Descent of Man: 1871–1971*. Chicago, IL: Aldine, pp. 136–179.

US Census Bureau (2014). Computer and Internet Access in the United States: 2012. www.census.gov/hhes/computer (accessed May 20, 2014).

Velezmoro, R., Negy, C., & Livia, J. (2012). Online sexual activity: Cross-national comparison between United States and Peruvian college students. *Archives of Sexual Behavior*, **41**, 1015–1025.

Vörös, F. (2014). Raw fantasies. An interpretative sociology of what bareback porn does and means to French gay male audiences. *Queering Paradigms*, **IV**, 321–344.

Williams, L. (1999). *Hard Core: Power, Pleasure, and the "Frenzy of the Visible."* Berkeley, CA: University of California Press.

Woodard, T. L., Collins, K., Perez, M., et al. (2008). What kind of erotic film clips should we use in female sex research? An exploratory study. *Journal of Sexual Medicine*, **5**(1), 146–154.

Index